KIRK-OTHMER

ENCYCLOPEDIA OF CHEMICAL TECHNOLOGY

Third Edition

VOLUME 15

**Matches
to
N-Nitrosamines**

KIRK-OTHMER

ENCYCLOPEDIA OF CHEMICAL TECHNOLOGY

THIRD EDITION

VOLUME 15

MATCHES
TO
N-NITROSAMINES

A WILEY-INTERSCIENCE PUBLICATION

John Wiley & Sons

NEW YORK • CHICHESTER • BRISBANE • TORONTO

Library of Congress Cataloging in Publication Data:

Main entry under title:
    Encyclopedia of chemical technology.

    At head of title: Kirk-Othmer.
    "A Wiley-Interscience publication."
    Includes bibliographies.
    1. Chemistry, Technical—Dictionaries. I. Kirk, Raymond
Eller, 1890–1957. II. Othmer, Donald Frederick, 1904--
    III. Grayson, Martin. IV. Eckroth, David. V. Title:
Kirk-Othmer encyclopedia of chemical technology.

TP9.E685   1978   660'.03   77-15820
ISBN 0-471-02068-0

Printed in the United States of America

CONTENTS

EDITORIAL STAFF
FOR VOLUME 15

Executive Editor: **Martin Grayson**
Associate Editor: **David Eckroth**
Production Supervisor: **Michalina Bickford**
Editors: **Galen J. Bushey**   **Caroline I. Eastman**   **Anna Klingsberg**
        Leonard Spiro   **Mimi Wainwright**

CONTRIBUTORS
TO VOLUME 15

Lester A. Alban, *Fairfield Manufacturing Company,Inc., Lafayette, Indiana,* Case hardening under Metal surface treatments

Lyle F. Albright, *Purdue University, West Lafayette, Indiana,* Nitration

R. A. Anderson, *Union Carbide Corporation, Tarrytown, New York,* Molecular sieves

D. H. Antonsen, *The International Nickel Company, Inc., New York, New York,* Nickel compounds

Philip J. Baker, Jr., *International Minerals and Chemical Corporation, Terre Haute, Indiana,* Nitroparaffins

Robert Q. Barr, *Climax Molybdenum Company, Greenwich, Connecticut,* Molybdenum and molybdenum alloys

H. F. Barry, *Climax Molybdenum of Michigan, Ann Arbor, Michigan,* Molybdenum compounds

W. E. Bastian, *CPS Chemical Company, Old Bridge, New Jersey,* Naphthenic acids

† **F. Benesovsky,** *Metallwerk Plansee, Reutte, Austria,* Nitrides

Jasjit S. Bindra, *Pfizer, Inc., Groton, Connecticut,* Memory-enhancing agents and antiaging drugs

Allen F. Bollmeier, Jr., *International Minerals and Chemical Corporation, Terre Haute, Indiana,* Nitro alcohols; Nitroparaffins

† **D. W. Breck,** *Union Carbide Corporation, Tarrytown, New York,* Molecular sieves

Robert H. Dewey, *International Minerals & Chemical Corporation, Terre Haute, Indiana,* Nitro alcohols

Harold J. Drake, *U.S. Bureau of Mines, Washington, D.C.,* Mercury

Hans Dressler, *Koppers Co., Inc., Monroeville, Pennsylvania,* Naphthalene derivatives

K. L. Dunlap, *Mobay Chemical Corporation, New Martinsville, West Virginia,* Nitrobenzene and nitrotoluenes

Herbert Ellern, *Consultant, Oakland, California,* Matches

Mary G. Enig, *University of Maryland, College Park, Maryland,* Mineral nutrients

P. Ettmayer, *Technical University, Vienna, Vienna, Austria,* Nitrides

R. M. Gaydos, *Koppers Company, Inc., Monroeville, Pennsylvania,* Naphthalene

R. B. Gengelbach, *Celanese Chemical Company, Inc., Dallas, Texas,* Methanol

Carl W. Hall, *Washington State University, Pullman, Washington,* Milk and milk products

W. L. Hallbauer, *Celanese Chemical Company, Houston, Texas,* Methanol

H. Stuart Holden, *Diamond Shamrock Technologies, S.A., Geneva, Switzerland,* Metal anodes

Alan S. Horn, *University of Groningen, Groningen, The Netherlands,* Neuroregulators

Timothy E. Howson, *Columbia University, New York, New York,* Nickel and nickel alloys

Larry A. Jackman, *Special Metals Corporation, New Hartford, New York,* Metal treatments

R. Kieffer, *Technical University, Vienna, Vienna, Austria,* Nitrides

Benjamin B. Kine, *Rohm and Haas Company, Philadelphia, Pennsylvania,* Methacrylic polymers

Lawrence S. Kirch, *Rohm and Haas Company, Philadelphia, Pennsylvania,* Methacrylic acid and derivatives

James M. Kolb, *Diamond Shamrock Corporation, Dallas, Texas,* Metal anodes

R. C. Krutenat, *Exxon Research and Engineering Company, Linden, New Jersey,* Survey under Metallic coatings

Leonard Lamberson, *Wayne State University, Detroit, Michigan,* Materials reliability

R. F. Mawson, *Colorado State University, Fort Collins, Colorado,* Meat products

G. Morel, *Université de Paris-Nord, Villetaneuse, France,* Membrane technology

Joseph W. Nemec, *Rohm and Haas Company, Philadelphia, Pennsylvania,* Methacrylic acid and derivatives

Daniel J. Newman, *Barnard and Burk, Inc., Mountainside, New Jersey,* Nitric acid

William E. Newton, *Charles F. Kettering Research Laboratory, Yellow Springs, Ohio,* Nitrogen fixation

R. W. Novak, *Rohm and Haas Company, Philadelphia, Pennsylvania,* Methacrylic polymers

Milton Nowak, *Troy Chemical Corporation, Newark, New Jersey,* Mercury compounds

James Y. Oldshue, *Mixing Equipment Co., Inc., Rochester, New York,* Mixing and blending

John M. Osepchuk, *Raytheon Company, Waltham, Massachusetts,* Microwave technology

Donald R. Paul, *University of Texas, Austin, Texas,* Membrane technology

Patrick H. Payton, *Teledyne Wah Chang Albany, Albany, Oregon,* Niobium and niobium compounds

Andrew Pocalyko, *E. I. du Pont de Nemours & Co., Inc., Coatesville, Pennsylvania,* Explosively clad metals under Metallic coatings

John A. Roberts, *Arco Ventures Co., Troy, Michigan,* Metal fibers

Carl L. Rollinson, *University of Maryland, College Park, Maryland,* Mineral nutrients

Glenn R. Schmidt, *Colorado State University, Fort Collins, Colorado,* Meat products

USP	*United States Pharmacopeia*
uv	ultraviolet
V	volt (emf)
var	variable
vic-	vicinal
vol	volume (not volatile)
vs	versus
v sol	very soluble
W	watt

Wb	Weber
Wh	watt hour
WHO	World Health Organization (United Nations)
wk	week
yr	year
(*Z*)-	zusammen; together; atomic number

Non-SI (Unacceptable and Obsolete) Units		*Use*
Å	angstrom	nm
at	atmosphere, technical	Pa
atm	atmosphere, standard	Pa
b	barn	cm^2
bar[†]	bar	Pa
bbl	barrel	m^3
bhp	brake horsepower	W
Btu	British thermal unit	J
bu	bushel	m^3; L
cal	calorie	J
cfm	cubic foot per minute	m^3/s
Ci	curie	Bq
cSt	centistokes	mm^2/s
c/s	cycle per second	Hz
cu	cubic	exponential form
D	debye	C·m
den	denier	tex
dr	dram	kg
dyn	dyne	N
dyn/cm	dyne per centimeter	mN/m
erg	erg	J
eu	entropy unit	J/K
°F	degree Fahrenheit	°C; K
fc	footcandle	lx
fl	footlambert	lx
fl oz	fluid ounce	m^3; L
ft	foot	m
ft·lbf	foot pound-force	J
gf den	gram-force per denier	N/tex
G	gauss	T
Gal	gal	m/s^2
gal	gallon	m^3; L
Gb	gilbert	A
gpm	gallon per minute	(m^3/s); (m^3/h)
gr	grain	kg
hp	horsepower	W
ihp	indicated horsepower	W
in.	inch	m
in. Hg	inch of mercury	Pa
in. H_2O	inch of water	Pa
in.-lbf	inch pound-force	J
kcal	kilogram-calorie	J

† Do not use bar (10^5Pa) or millibar (10^2Pa) because they are not SI units, and are accepted internationally only for a limited time in special fields because of existing usage.

Non-SI (Unacceptable and Obsolete) Units   *Use*

kgf	kilogram-force	N
kilo	for kilogram	kg
L	lambert	lx
lb	pound	kg
lbf	pound-force	N
mho	mho	S
mi	mile	m
MM	million	M
mm Hg	millimeter of mercury	Pa
mμ	millimicron	nm
mph	miles per hour	km/h
μ	micron	μm
Oe	oersted	A/m
oz	ounce	kg
ozf	ounce-force	N
η	poise	Pa·s
P	poise	Pa·s
ph	phot	lx
psi	pounds-force per square inch	Pa
psia	pounds-force per square inch absolute	Pa
psig	pounds-force per square inch gauge	Pa
qt	quart	m^3; L
°R	degree Rankine	K
rd	rad	Gy
sb	stilb	lx
SCF	standard cubic foot	m^3
sq	square	exponential form
thm	therm	J
yd	yard	m

BIBLIOGRAPHY

1. The International Bureau of Weights and Measures, BIPM (Parc de Saint-Cloud, France) is described on page 22 of Ref. 4. This bureau operates under the exclusive supervision of the International Committee of Weights and Measures (CIPM).
2. *Metric Editorial Guide (ANMC-78-1)* 3rd ed., American National Metric Council, 1625 Massachusetts Ave. N.W., Washington, D.C. 20036, 1978.
3. *SI Units and Recommendations for the Use of Their Multiples and of Certain Other Units (ISO 1000-1981)*, American National Standards Institute, 1430 Broadway, New York, N. Y. 10018, 1981.
4. Based on *ASTM E 380-79 (Standard for Metric Practice)*, American Society for Testing and Materials, 1916 Race Street, Philadelphia, Pa. 19103, 1979.
5. *Fed. Regist.*, Dec. 10, 1976 (41 FR 36414).
6. For ANSI address, see Ref. 3.

R. P. LUKENS
American Society for Testing and Materials

M *continued*

MATCHES

The word match is of uncertain origin. In common parlance, a match is a short, slender, elongated piece of wood or cardboard, suitably impregnated and tipped to permit, through pyrochemical action between dry solids with a binder, the creation of a small transient flame. The word match also is used for fuse lines which after ignition on one end serve as fire-transfer agents in fireworks and for explosives (qv). Such items belong in the field of pyrotechnics (qv).

The development of the ordinary match followed thousands of years of fire-making by laborious means. It has been perfected in the course of the last fifty years and its formulations have remained basically unchanged. Progress has been achieved in selection of modifying components, mainly in the control of the drying process of the freshly dipped matches, also in the mixing procedures. The mechanical equipment for cutting out the match stems and assembling the match books has become more and more efficient as to precision and speed of production.

The history of the modern match, well presented in detail in ref. 1, can be condensed here into the essential lines of development as follows: White (also called yellow) phosphorus, discovered by Hennig Brand (1669) and described as easily ignited on slight warming or rubbing, was first applied by Robert Boyle (1680) to the ignition of sulfur-tipped wood splints (see Phosphorus and phosphides). Between 1780 and 1830, numerous matchlike contrivances used this ignition principle. After the discovery of potassium chlorate by Berthollet (ca 1786), its combination with white phosphorus and modifying ingredients led to the manufacture of the friction ("strike-anywhere") match which became the most popular means of ignition in the United States until July 1, 1913 when an Internal Revenue tax of two cents per one hundred matches terminated its production. This act followed numerous prohibitions and similar punitive taxation in other countries because of the hazard that the phosphorus constituted to the health of the workers during the days of primitive hand-dipping methods. Vapors of the white phosphorus entering the body, mainly through defective teeth, caused a permanent necrotic destruction of the bones (phossy jaw). Another evil was that these matches were sometimes used in suicide attempts or caused the death of children.

A direct descendant of these matches is the nontoxic modern double-tipped SAW (strike-anywhere), the large "kitchen match" version, or the smaller "penny box" variety. It is based on the invention of two Frenchmen, Henri Sévène and Émile David Cahen, who used the nonpoisonous compound tetraphosphorus trisulfide (P_4S_3) as a phosphorus substitute and acquired a U.S. patent in 1898.

Other early matchlike devices were based on the property of various combustible substances mixed with potassium chlorate to ignite when moistened with strong acid (Chancel's instantaneous light box or *briquet oxygéné*, 1805, and Samuel Jones' promethean match, 1828) (see Chlorine oxygen acids and salts). More important was the property of chlorates to form mixtures with combustibles of low ignition point which were ignited by friction (John Walker, 1827). However, such matches containing essentially potassium chlorate, antimony sulfide and, later, sulfur ("lucifers"), rubbed within a fold of glass-powder-coated paper, were hard to initiate and not too reliable.

The modern safety match owes its qualities to the discovery by Schrötter (1844) of the red, nonpoisonous but easily ignitable variety of phosphorus called red phosphorus. Pasch in Sweden and Böttger in Germany (1845) prepared striking surfaces containing the new material, thus separating the two major fire-producing components, the chlorate in the matchhead as the oxidizer and the most sensitive fuel-type material in the striker. This type of match was much improved and made an article of commerce by J. E. Lundstrom in Jonköping, Sweden (1855). However, the United States was quite slow to accept this safety match. The "one-hand" phosphorus match and its successor, the double-tipped SAW match, were easier to handle and more reliable, whereas the early safety matches were often sputtering, hard-striking, and explosive.

The final step in the development of the modern match was the invention of the safety-type cardboard match ascribed to Joshua Pusey (1892), now called the book match. It now dominates the American match industry and is gaining in popularity in other countries although it was rather slow in gaining acceptance because it was somewhat more difficult to ignite than the wood-splint match.

Mechanism of Fire Production

The essential chemical reaction takes place on contact of potassium chlorate and red phosphorus which by itself is one of the most unpredictably hazardous dry reactions in pyrochemistry. It has been the cause of serious injury to chemistry students who mix the two materials without permission, only vaguely aware of their explosive potential. In the match head, and separately in the striker, each of two materials is embedded in a matrix of glue so that, on striking under mild friction, a few particles of both materials come harmlessly in contact and react with formation of well-contained sparks. The modifying materials in the match head function as sensitizers (sulfur or rosin), burning-rate modifiers (potassium dichromate or lead thiosulfate), and ash-formers (diatomaceous earth, powdered glass, etc); the latter serve to hold the glowing residue safely together by a sintering process. The glue, starch, and the paraffin in the stem below the head act as flame-forming fuels and the neutralizers account for the practically indefinite storage stability of well-made matches. In the striker, the glass powder controls proper bite and sensitivity. The binder is insolubilized to prevent staining of clothing caused by rain or perspiration.

The SAW match is similar to the safety match except that it is richer in fuel, and gives a billowing somewhat wind-resistant flame. The phosphorus sulfide in the tip provides the ignitability on any solid surface, and a little of the same material in the base bulb adds to wind resistance but, otherwise, the base is underbalanced in active materials to prevent self-ignition from rubbing during transportation.

Manufacture

The low price of book matches is mainly the result of high speed, mechanized production methods. Book matches are punched from 1-mm thick, lined chipboard in strips of one hundred splints of ca 3.2 mm width each. In an eight-hour shift, a single machine can produce about 20×10^6 match splints and deliver them half an hour later as completed, strikable matches, ready for cutting and stapling into books. In this half hour, the tips of the punched-out splints are first immersed in molten paraffin wax, without which no persistent flame and fire transfer is possible (see Waxes). Immediately following wax application, the tip composition is affixed by dipping the ends of the strips into a thick but smoothly fluid suspension carried on a cylinder rotating in a relatively small tank at the same speed as the match strips move over it in the clamps of an endless chain. As soon as an evenly rounded match tip has been formed, the matches enter a dryer where the main object is not so much the speedy removal of the water in the match composition as the prior congealing of the match head, which takes place at a temperature of about 24°C and a relative humidity of 45 to 55%. This drying technique is one of the most important parts of the manufacture, and it is responsible for the amazingly uniform quality and ease of ignition of the modern match (see Drying).

The match-cover board is an approximately 0.4-mm thick, coated or lined, chipboard, or sometimes a fancy grade of a variety of decorative and more expensive cardboards, on which a striking strip is printed by a roller-coating process from a thin slurry of a composition described below. The cover is more or less elaborately printed with an advertising message. Although the technique of printing varies, it is possible on some presses to apply all the colors, as well as a one-color underprint (inside of cover), on one printing press in one successive operation.

The most common size is a book of 20 or 28 (30) matches, and the 40-match size offers additional advertising area. "Ten strike" matches are included in military food packages and are sometimes used for advertising purposes.

Wooden matches can be made by a veneering method whereby aspen wood is peeled from a section of a log and cut into splints which have a square cross section of 0.25–0.4 cm thickness, depending on the length of the match. The alternative method consists in cutting round splints from selected blocks of white pine by means of rows of cutting dies each resembling a large darning needle of which the eye is the cutter. The splints in both types of operation are forced into holes in cast-iron plates and are thus transported through the various dipping operations.

A third type of commercial matches popular in South America is the wax vestas with a center of cotton threads or of a rolled and compressed thin and tough paper surrounded by and impregnated with wax; each match is a miniature candle of long (ca one minute) burning time.

Two processes precede the affixing of the heads for wooden matches. The first one is glow-proofing of the splint by impregnation with ammonium phosphate or a mixture of it with boric acid. (In paper matches, the impregnation is conveniently done during the fabrication of the paper.) This suppresses continuation of glowing of the carbonized splint after discard and also prevents the burned part with the still-hot tip from falling off and singeing clothing (see also Flame retardants). The second impregnation is the soaking up of paraffin wax into the stem for a certain length to assure flame forming and fire transfer to the wood. Head formation is similar to the process described for book matches except that for SAW matches a smaller second tip is affixed to the larger bulb. The rollers in the dipping tank over which the splints travel are grooved, the first roller deeply for the base tip, the second roller shallow-grooved for the SAW tip. The same equipment, simply leaving out the second dipping, can be used for wooden safety matches. A formaldehyde bath (sometimes also employed with paper matches) aids in congealing and subsequent proper drying of the match head.

The wooden match industry, still prevalent in Europe, is described with a wealth of detail in ref. 2.

Nonstandard and Military Matches.   Because match manufacture is a series of high speed and highly mechanized operations, any variation that involves dimensional or incisive procedural changes is a major undertaking which is only warranted if continual high production is to result. Hence, specialties that occasionally appear on the market are actually fireworks items, made laboriously and at relatively high cost by hand-dipping with very limited mechanization. Such matches produce a colored flame, give off perfume or fumigating vapors, or furnish a persistent glow or flame for the purpose of burning in a strong draft. In order to do these things effectively, an enlarged elongated bulb is necessary.

An interesting variation of the regular match is the pull match. It is a paper match, considerably thinner and narrower than a regular book match, since it needs very little stiffness when being used. The tip part of the match is enclosed in a strip of corrugated paper glued to a flat cardboard (such as a box of cigarettes) and the inside of the corrugated board is covered with striking material. On pulling the match fast enough out of the corrugation, the tip passes and engages the striker and becomes lit.

A curious item is the repeatedly ignitable match. It resembles a tiny pencil, the center part being a safety match composition which is surrounded by a cool-burning chemical mixture whose essential ingredient is nearly always metaldehyde. Notwithstanding the exaggerated claims for its performance, this match, although ignitable a few times in succession, and thus being an interesting curiosity, is economically not competitive with the book match and technically definitely inferior.

The principle of the safety match is also used in the pull-wire fuse lighter used to start a fuse train for the ignition of fireworks items or more frequently for blasting work. This is a reversed pull match whereby the striker material is coated on a pull wire, whereas the match-head material is within a small metal cup in a cardboard tube. Pulling the coated wire vigorously out of the device ignites the match mixture in the tube for fire transfer to the tubular fuse train.

Match buttons and strikers are built-in components of certain flares such as the well-known red-burning railroad fusee (3) and of some fire-starting devices invented during World War II to help marooned military personnel to light a fire with a minimum of effort.

During World War II, the Quartermaster Corps of the United States requested development of a SAW match that would withstand at least six hours submersion in water. Although no match will be strikable after prolonged exposure to extremely high humidity, it is possible to prevent temporarily infiltration of moisture, and especially attack by liquid water, by coating the match head and part of the stem with nitrocellulose lacquer. This is impractical for the safety match because it makes striking progressively more difficult the thicker the coating becomes. It is, however, possible to protect SAW matches so as to withstand 6 to 10 h of submersion in water and still be readily ignitable. Large numbers of matches protected in this manner were made during World War II by at least two large wood-match manufacturers in the United States. The most recent Federal specification for matches includes a Type III, Class 2, water-resistant strike-anywhere wooden match for performance after two hours submersion in water (4). For prolonged exposure to high humidity, the only safe protection is heat-sealing in plastic pockets and then canning (see also Waterproofing and water repellancy).

Since the largest user of matches is nowadays the person who smokes cigarettes, efforts have not failed to combine the safety match with cigarettes—an idea going back to 1835 when loco foco or self-lighting "segars" (5) were sold. More recently, a fairly effective cigarette of this type based on British patents (6) has appeared in commercial use.

Formulations. Since match manufacture has become a combination of high speed machinery design and operation, commercial art work and engraving, as well as of pyrotechnical processing, the latter is only a small though highly important part of the match business. Formulations are by no means a closely guarded secret, mainly since they are only a starting point on the way to producing a satisfactory match and striker adapted to the specific conditions of manufacture. The following formulations (7) with some adjustments permit the preparation of workable but not necessarily salable matches (Tables 1–3).

European matches, mostly of brown or black tips, are basically identical with U.S. matches in their formulations, except that they contain in addition red iron oxide or manganese dioxide of pigment grade in the match heads (2).

Table 1. Commercial Safety Match

	%
animal (hide) glue	9–11
starch	2–3
sulfur	3–5
potassium chlorate	45–55
neutralizer (ZnO, CaCO$_3$)	3
diatomaceous earth	5–6
other siliceous fillers	15–32
(powdered glass; "fine" silica)	
burning-rate catalyst	to suit
(K$_2$Cr$_2$O$_7$ or PbS$_2$O$_3$)	
water-soluble dye	to suit

Table 2. Striker of Commercial Safety Match

	%
animal glue or casein[a]	16
red phosphorus	50
calcium carbonate	5
powdered glass	25
carbon black	4

[a] Suitably insolubilized according to ref. 8, or lightly brushed with diluted formaldehyde after application.

Table 3. SAW (Strike-Anywhere) Match

	Tip composition, %	Base composition, %
animal glue	11	12
extender (starch)	4	5
paraffin		2
potassium chlorate	32	37
phosphorus sesquisulfide (P_4S_3)	10	3
sulfur		6
rosin	4	6
dammar gum		3
infusorial earth (diatomite)		3
powdered glass and other fillers	33	21.5
potassium dichromate		0.5
zinc oxide	6	1

A wealth of details on match materials, testing methods, and related matters is contained in ref. 9.

Economic Aspects

Production of matches in the United States in 10^6 was 105 in 1977, 96 in 1976, 82 in 1972, 66 in 1967, and 65 in 1963 (10). For 1976, the Consumer Product Safety Commission (11) estimated that the total of all flame producers—book matches (made from cardboard), individual (wooden) stick matches, and lighters (see below)—amounted to 645×10^9 of such fire resources or "lights." Paper matches accounted for about 65% or 420×10^9, lighters for 25% or 160×10^9, and wooden matches the remaining 10% or 65×10^9.

The rising popularity of mechanical lighters since the availability of the disposable butane lighter has made significant inroads in the use of book matches. By now, 98% of all these light sources are used for tobacco products, and notwithstanding the promoted curbs on smoking, their total use increases by about 3% per year.

Prior to the inflationary spiral, book matches retailed at as little as one cent for 100 lights or at one-fifth of a cent per book of twenty matches. The cost and selling price increase considerably with higher quality of the cover paper, the elaborateness of the printed messages on and inside the cover (and sometimes even on the splint), and the size of the order. In any case, the customer receives exactly the same high quality matches and striking strip.

Book matches are an important medium of advertising since they represent a truly utilitarian item which is more often given away than sold. Manufacturers of nationally sold products pay for the message on many millions of books without entering otherwise in the sale of the matches (resale match). Individual establishments, such as associations, banks, trucking companies, hotels, etc, buy the books with their personal message for their own distribution (special reproduction). A cross between both transactions is the distribution of a design advertising, eg, a brand of gasoline with the name of the dealer and other information about a filling station imprinted on the basic design (dealer imprint). On the smallest scale, a generalized so-called stock design permits someone who needs only a few thousand books a year to buy and give away matches carrying a message without a larger outlay for art work and printing plates.

Toxicity and Other Safety Aspects

Since small children often suck on matches, the question of toxicity is often raised and the lingering, vague, though unwarranted idea of phosphorus poisoning may cause concern to laymen and even to physicians. Potassium chlorate is the only active material that can be extracted in more than traces from a match head and only nine milligrams are contained in one head. This, even multiplied by the content of a whole book, is far below any toxic amount (12) for even a small child. No poisonous properties whatsoever can be imputed to the striking strip. SAW matches are similarly harmless but, because of their easy flammability, they should be entirely kept out of a household with smaller children. The same warning may apply to all wooden matches.

Safety-match strips arranged for decorative purposes in the form of flower pots are especially undesirable items, since an accidental ignition causes a dangerous flaring of many matches at one time.

Safety matches can be ignited by friction alone only when, with some deftness, they are rubbed on cardboard or glass. Accidental ignition is nearly always due to careless or absent-minded handling. In packaged condition, large numbers of safety matchbooks, if ignited in one spot, flare momentarily and harmlessly, often not igniting the adjoining matchbooks.

Exploding matches (other than for practical jokes) have virtually disappeared with modern manufacturing methods, ie, drying of the heads under controlled conditions. The occasional match that ignites and ejects sparking portions is generally a result of excessive pressure during lighting or of an accidentally cracked match head.

Sometimes a match ignites promptly but only a weak and unsatisfactory flame follows. This is the result of prolonged exposure of the matches to a temperature above 54°C in storage. The defect is caused by gradual dissipation of the paraffin wax throughout the splint and is evidenced by the disappearance of the line of demarcation, which is clearly visible in book matches. Otherwise, all safety matches tolerate exposure to elevated temperatures until about 177°C is reached. A new U.S. Government specification requires that safety matches withstand exposure for two hours at 90°C (4). Earlier specifications stipulated a minimum acceptable nonignition temperature of 170°C. SAW matches are much more heat sensitive but still tolerate heating below a self-ignition temperature of 120 to 150°C.

BIBLIOGRAPHY

"Matches" in *ECT* 1st ed., Vol. 8, pp. 819–824, by C. K. Wolfert, The Diamond Match Company; "Matches" in *ECT* 2nd ed., Vol. 13, pp. 160–166, by Herbert Ellern, UMC Industries, Inc.

1. M. F. Crass, Jr., *J. Chem. Ed.* 18(3,6,8–9), (1941).
2. H. Hartig, *Zündwaren,* 2nd revised and extended ed., VEB Fachbuchverlag, Leipzig, 1971.
3. *Specifications for Red Railroad Fusees or Red Highway Fusees,* Bureau of Explosives, revised May 1, 1959.
4. *Fed. Spec. EE-M-101 J,* 1978.
5. M. M. Mathews, ed., *Dictionary of Americanisms on Historical Principles,* University of Chicago, 1951.
6. Brit. Pats. 746,435 (1953), 758,858 (1953), 800,596 (1955), F. D. Capitani.
7. H. Ellern, *Military and Civilian Pyrotechnics,* Chemical Publishing Co., Inc., New York, 1968.
8. U.S. Pat. 2,722,484 (Nov. 1, 1955), I. Kowarsky (to UMC Corp.).
9. I. Kowarsky, "Matches" in F. D. Snell and L. S. Ettre, eds., *Encyclopedia of Industrial Chemical Analysis,* Vol. 15, John Wiley & Sons, Inc., New York, 1972.
10. *1977 Census of Manufacturers,* SIC 3999, U.S. Dept. of Commerce, Washington, D.C., 1978.
11. *Fed. Regist.* 43(223), (1978).
12. A. Osol and G. E. Farrar, *The Dispensatory of the United States,* J. B. Lippincott Company, Philadelphia, Pa., 1950.

HERBERT ELLERN
Consultant

MATERIALS RELIABILITY

Reliability is a parameter of design like a system's performance or load ratings and is concerned with the length of failure-free operation. It is difficult to conceptualize reliability as part of the usual design calculations. Further complications are the complexity of organizations needed to produce the large systems of today and the usual time and financial constraints on research and development. Reliability as it relates to products or equipment can be measured in various ways. Since it is a design parameter, it has to be addressed early in the design cycle.

Terminology

Reliability. The reliability of a system is defined as the probability that the system will perform its intended function satisfactorily for a specified interval of time when operating under stated environmental conditions. It has to be realized that supposedly identical products fail at different times. Thus, reliability can be quantified only as a probability. For any product there is some underlying function that describes this success pattern. Typical reliability functions are shown in Figure 1 for two different products. These products can be compared at the same reliability level R_1 or the reliability levels can be compared for any selected time period, t_2.

In applying the definition of reliability, the concept of adequate performance must be established clearly. Products usually do not fail suddenly, but degrade over time. Gasket leaks on equipment, for example, may start as a slow seep and increase in volume over time. The point at which this undesirable occurrence is called a failure must be clear before reliability can be measured objectively. Changing the failure definition for a product changes its reliability level, although the product itself has not changed.

The reliability level of a product also depends upon the operating or environmental conditions, which may produce a variety of failure modes. Reliability can only

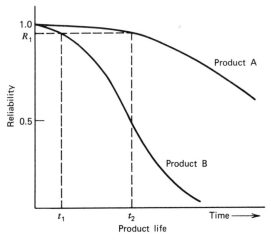

Figure 1.   Product reliability functions.

be assessed relative to a defined environment. Unless the above points are established clearly, confusion surrounds any quoted reliability number for a product.

Because of the interrelationship of the system measures, reliability should not be considered by itself since, if taken alone, it does not express the totality of attributes that contribute to system effectiveness. However, in practice, reliability has gained the most acceptance and uniformity of definition. The other concepts described below are not always defined uniformly from group to group and are sometimes used interchangeably. Further discussion of these concepts is found in refs. 1 and 2.

System Effectiveness. A system is designed to perform some intended function in a prescribed fashion. This overall capability is termed system effectiveness. Figure 2 illustrates the design tradeoffs that constitute the components of system effectiveness.

From the standpoint of a military product, system effectiveness is the probability that the system meets successfully an operational demand within a given time when operating under specified conditions. From the standpoint of commercial products, system effectiveness is harder to define, but basically means customer satisfaction. There are several system parameters that are important to the customer. Some of these parameters are defined below.

Maintainability. Maintainability is defined as the probability that a failed system is restored to a satisfactory operating condition in a specified interval of downtime. The ease of fault detection, isolation, and repair are all influenced by system design and are the main factors contributing to maintainability. Also contributing is the supply of spare parts, the supporting repair organization, and the preventive maintenance practices. Good maintainability may somewhat offset low reliability. Hence, if the desired reliability cannot be met because of performance constraints, improvements in the system maintainability should be considered. From the customer's viewpoint, low reliability may be acceptable if the system is repaired easily, inexpensively, and quickly, once it has failed (see also Maintenance).

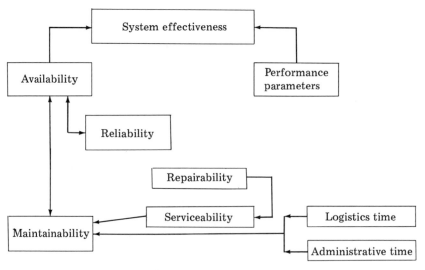

Figure 2. Components of system effectiveness.

Serviceability.   Serviceability is defined as the degree of ease (or difficulty) with which a system can be repaired. This measure specifically considers fault detection, isolation, and repair. Repairability considers only the actual repair time, and is defined as the probability that a failed system is restored to operation in a specified interval of active repair time. Access covers, plug-in modules, or other features to allow easy removal and replacement of failed components improve the repairability and serviceability (see also Electrical connectors).

Logistics Time.   Logistics time is the portion of downtime during which repair is delayed because of the unavailability of spare parts. It is dependent upon the maintenance support organization and the manufacturer's ability to predict and supply spare parts.

Availability.   The system attributes of maintainability and reliability must both be considered. The trade-offs are rather complex and difficult to capture with any one measure. However, the term availability has been used to quantify these attributes simultaneously. The availability is sometimes related by inherent availability:

$$A = \frac{MTBF}{MTBF + MTTR} \tag{1}$$

The mean time between failures $MTBF$ is used as a measure of system reliability, whereas the mean time to repair $MTTR$ is taken as a measure for maintainability. For example, a system with a $MTBF$ of 1200 h and a $MTTR$ of 25 h would have an availability of 0.98. Furthermore, if only a $MTBF$ of 800 h could be achieved, the same availability would be realized if the maintainability could be improved to the point where the $MTTR$ was 16 h. Such trade-offs are illustrated in Figure 3, where each curve is at a constant availability.

Availability, as defined above, does not take into account idle time, waiting time, and preventive-maintenance time. However, it is still a useful and easily interpreted parameter for equipment design. A more encompassing measure is operational availability given by

$$A_0 = \frac{MTBM + RT}{MTBM + RT + MDT} \tag{2}$$

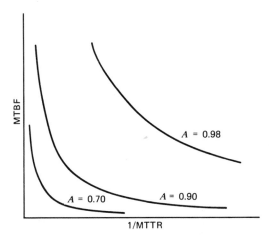

Figure 3.   System availability trade-off curves. MTBF = Mean time between failures; MTTR = Mean time to repair.

where $MTBM$ = mean interval between corrective and preventive-maintenance actions; RT = system average ready time; and MTD = (mean active corrective and preventive time) + (mean waiting time) + (mean logistic time) + (mean administrative time). Such a measure assumes a complete operational cycle given by $MTBM + MDT + RT$. In military hardware, this cycle is defined by the mission; however, in commercial hardware it may be more difficult to define a mission precisely because the equipment usually has multiple applications.

Administrative Time. Administrative time is a catchall category concerned with paper work, permits to perform work, waiting for official authorization to proceed with repairs, etc.

Design Reliability

The total picture of product reliability from the customer's viewpoint is rather complex. Since reliability and the related measures are essentially design parameters, improvements are most easily and economically accomplished early in the design cycle. Useful techniques for design reliability analysis are described below.

Design Review. A design review is a formalized, documented, and systematic audit of a design by senior company personnel. It addresses the complex design tradeoffs and assures early design maturity. It should be multiphased and performed at various stages of the product development cycle. The parameters contributing to product availability must be a recognized input to this process.

Definite and known procedures for follow-up must be provided for, with the design group assessing the value of each idea and suggestion presented by the review committee. The actions taken are known to the committee and subject to further review. With such organization, the trade-offs can be acted upon at the appropriate level.

Failure-Mode Analysis. The product design activity usually emphasizes the attainment of performance objectives in a timely and cost-efficient fashion. Failure-mode analysis (FMA) determines how the product might fail. The terms design FMA (DFMA) and failure-mode effects and criticality analysis (FMECA) also are used. This technique identifies and eliminates failure modes early in the design cycle and its success is well documented (3–4).

Failure-mode analysis begins with the selection of a subsystem or component and then documents all potential failure modes. Their effect is traced up to the system level. A documented worksheet is used on which the following elements are recorded:

Function.   A concise definition of the functions that the component must perform.

Failure Mode.   A particular way in which the component can fail to perform its function.

Failure Mechanism.   A physical process or hardware deficiency causing the failure mode.

Failure Cause.   The agent activating the failure mechanism; eg, saltwater seepage owing to an inadequate seal might cause corrosion as a failure mechanism.

Identification of Effects on Higher Level Systems.   This determines if the failure mode is localized or does it cause higher level damage or create an unsafe condition?

Criticality Rating.   A measure of severity and probability of failure occurrence used to assign priority design actions.

These procedures ensure early design maturity. Performing an FMA on purchased equipment may eliminate maintenance problems and provide a plan for spare-parts inventories.

Life-Cycle Cost.   The total cost of ownership of a system during its operational life can be accounted for. The cost of ownership not only includes the initial design and acquisition cost but also cost of personnel training, spare-parts inventories, repair, and operations, etc. A complete projection of system costs might point out the wisdom of investing more initially in order to forego high maintenance costs owing to poor reliability and serviceability, as illustrated in Figure 4.

System Reliability Models

Static reliability models are used in preliminary analyses to determine necessary reliability levels for subsystems and components. A subsystem is a particular low level grouping of components. Some trial and error is usually necessary to obtain reasonable groupings for any particular system. Early identification of potential system weaknesses facilitates corrective action.

A reliability block diagram can be developed for the system from the definition of adequate performance. The block diagram represents the effect of subsystem or component failure on system performance. In this preliminary analysis, each subsystem is assumed to be either a success or failure. A reliability value is assigned to each subsystem where the application and a specified time period are given. The reliability values for each subsystem and the functional block diagram are the basis for the analysis.

Series Systems.   The series configuration is the most commonly encountered in practice. In a series system, all subsystems must operate successfully for the system to be successful. The reliability block diagram is given in Figure 5. The system reliability is

$$R_s = \prod_{i=1}^{n} R_i \tag{3}$$

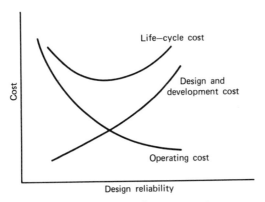

Figure 4.   Life-cycle cost concept.

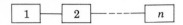

Figure 5.   Series block diagram.

where R_i = the reliability for the ith subsystem; and R_s = system reliability. It can be seen that

$$R_s \leq \min_i \{R_i\} \tag{4}$$

or the reliability of the system is never greater than the least reliable subsystem. In this analysis it is assumed that subsystems fail independently.

In a series system, if each subsystem had an exponential time to failure given by

$$f(t) = \lambda_i e^{-t\lambda_i}, \quad t \geq 0 \tag{5}$$

where λ_i is the failure rate for the ith subsystem. The system failure rate is

$$\lambda_s = \sum_{i=1}^{n} \lambda_i \tag{6}$$

Or if $MTBF$s are used, then

$$\frac{1}{\theta_s} = \sum_{i=1}^{n} (1/\theta_i) \tag{7}$$

where $\theta_s = 1/\lambda$. Failure rates are sometimes more convenient to use in high reliability systems and are simply apportioned by equation 6.

Example 1.   A gear pump is to be designed for use as an emergency backup system. The pump is driven by a small gasoline engine. Electronic sensing and starting circuitry are provided to automatically start the system during a power failure. Figure 6 gives a possible reliability block diagram for the system. For this application the reliability values are as follows: $R_1 = 0.9999$; $R_2 = 0.95$; $R_3 = 0.90$; $R_4 = 0.999$. This would give an overall system reliability of $R_s = 0.9999 \times 0.95 \times 0.90 \times 0.999 = 0.8541$.

If the information is insufficient to select the R_i values for this application, failure rates can be obtained from available sources such as refs. 5–6. The failure rates obtained might be as follows: $\lambda_1 = 2.67 \times 10^{-6}$/h, $\lambda_2 = 591 \times 10^{-6}$/h, $\lambda_3 = 9.03 \times 10^{-6}$/h, $\lambda_4 = 4.45 \times 10^{-6}$/h. Then:

$$\lambda_s = 607.15 \times 10^{-6}/\text{h}$$

λ_s = sum of λ_1, λ_2, λ_3, and λ_4. And for an operating period of 12 h the reliability as calculated from equation 12 is

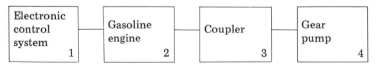

Figure 6.   Parallel block diagram.

$$R(12 \text{ h}) = \exp[-12 \times 607.15 \times 10^{-6}] = 0.9927$$

In using these failure rates an exponential distribution for time to failure was assumed. Such an assumption should be made with caution.

Parallel Systems. A parallel (or redundant) system is not considered to be in a failed state unless all subsystems have failed. The system reliability is calculated as

$$R_s = 1 - \prod_{i=1}^{n} (1 - R_i) \tag{8}$$

System reliability is improved by providing alternative means for performing the same task. For example, automobiles were equipped with hand cranks even though they had electric starters. This back-up equipment was provided because at that time starters were not too reliable. In contemporary system design, factors such as added cost, weight, and space may prohibit the use of redundant systems.

Systems can have both parallel and series subsystems. Reliability is calculated by successively reducing the system using the basic series or parallel formulas. This is illustrated in Example 2.

Example 2. Figure 7 shows a system block diagram indicating subsystem reliabilities. Applying equation 8 to part A of Figure 7 gives

$$R_a = 1 - (0.20)(0.25)(0.30) = 0.985$$

And for part B:

$$R_b = 1 - (0.40)(0.15) = 0.94$$

Then the series equation is applied to give the system reliability

$$R_s = 0.999 \times 0.985 \times 0.99 \times 0.94 = 0.916$$

Some systems cannot be represented by a simple combination of series and parallel subsystems. The systems are more complex in nature and the concept of coherent systems must be used in a more general and powerful treatment (7).

Reliability Measures

The reliability function $R(t)$ is defined as

$$R(t) = P(\mathbf{t} > t) = 1 - F(t) \tag{9}$$

where $\mathbf{t}$ is the time to failure random variable and $F(t)$ is the cumulative distribution. In terms of the probability density function $f(t)$, the reliability function is given by

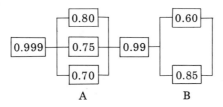

Figure 7. Parallel and series combinations.

$$R(t) = \int_t^\infty f(u)du \tag{10}$$

For example, if the time to failure is given as an exponential distribution, then

$$f(t) = \lambda e^{-\lambda t}, \quad t \geq 0, \quad \lambda > 0 \tag{11}$$

And the reliability function is found as follows:

$$R(t) = \int_t^\infty e^{-\lambda u}du = e^{-\lambda t}, \quad t \geq 0 \tag{12}$$

Life Expectancy of Devices. The expected or average life of devices is defined as

$$E(\mathbf{t}) = \int_0^\infty uf(u)du \tag{13}$$

where $f(t)$ is the probability density function (*pdf*) for the time-to-failure random variable **t**. The expected life also can be found from

$$E(\mathbf{t}) = \int_0^\infty R(t)dt, \quad t \geq 0 \tag{14}$$

The expected life is sometimes used as an indicator of system reliability; however, it can be a false indication and should be used with caution. In most test situations the chance of surviving the expected life is not 50% and depends upon the underlying failure pattern. For example, considering the exponential as used in equation 11, the expected life would be

$$E(\mathbf{t}) = \int_0^\infty te^{-\lambda t}dt = 1/\lambda \tag{15}$$

And the chance of surviving this time can be found from the reliability function

$$R(t = 1/\lambda) = e^{-1} = 0.368 \tag{16}$$

That is, in this case there is only a 36.8% chance of surviving the mean life. If the distribution were other than exponential, the chance of survival would change. Since the mean life is not associated with constant reliability, the expected life should not be the only indicator of reliability, particularly when comparing products.

Failure Rate and Hazard Function. The failure rate is defined as the rate at which failures occur in a given time interval. Considering the time interval $[t_1, t_2]$, the failure rate is given by

$$\frac{R(t_1) - R(t_2)}{(t_2 - t_1)R(t_1)} \tag{17}$$

And this is the rate of failure for those surviving at the beginning of the interval. This formula can be used to calculate failure rate from empirical life-test data.

The hazard function is defined as the limit of the failure rate as the interval of time approaches zero. The resulting hazard function $h(t)$ is defined by

$$h(t) = \frac{f(t)}{R(t)} \tag{18}$$

The hazard function can be interpreted as the instantaneous failure rate. The quantity

$h(t)\Delta t$ for small Δt represents the probability of failure in the interval Δt, given that the device was surviving at the beginning of the interval.

The failure rate changes over the lifetime of a population of devices. An example of a failure-rate vs product-life curve is shown in Figure 8 where only three basic causes of failure are present. The quality-, stress-, and wear-out-related failure rates sum to produce the overall failure rate over product life. The initial decreasing failure rate is termed infant mortality and is due to the early failure of substandard products. Latent material defects, poor assembly methods, and poor quality control can contribute to an initial high failure rate. A short period of in-plant product testing, termed burn-in, is used by manufacturers to eliminate these early failures from the consumer market.

The flat, middle portion of the failure-rate curve represents the design failure rate for the specific product as used by the consumer market. During the useful-life portion, the failure rate is relatively constant. It might be decreased by redesign or restricting usage. Finally, as products age they reach a wear-out phase characterized by an increasing failure rate.

In real-life applications, many other failure mechanisms are present and this type of curve is not necessarily obtained. For example, in a multicomponents system the quality-related failures do not necessarily all drop out early but might be phased out over a longer period of time.

Hazard function, pdf, and reliability function are related for any theoretical failure distribution. The relationships are

$$f(t) = h(t) \exp\left[-\int_0^t h(u)du\right] \tag{19}$$

and

$$R(t) = \exp\left[-\int_0^t h(u)du\right] \tag{20}$$

Conditional Failure Probability. The concept of conditional probability of failure is useful to predict the chances of survival for a device that has been in operation for a period of time and is not in a failed state. Such information is helpful for maintenance planning.

If a device has a reliability function $R(t)$ and has been successfully operating for a period of time T, the conditional reliability function is given by

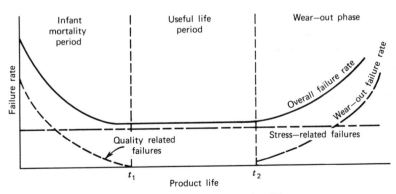

Figure 8. Failure rate vs product life.

$$R(t \,|\, \mathbf{t} > T) = \frac{R(t)}{R(T)}, \quad t > T \tag{21}$$

The use of this concept is illustrated in the following example.

    Example 3. A centrifugal pump moving a corrosive liquid is known to have a time-to-failure that is well-approximated by a normal distribution with a mean of 1400 h and a standard deviation of 120 h. A particular pump has been in operation for 1080 h. In order to plan maintenance activities the chances of the pump surviving the next 48 h must be determined.

    Applying equation 21 gives

$$R(1128 \text{ h} \,|\, \mathbf{t} > 1080 \text{ h}) = \frac{R(1128 \text{ h})}{R(1080 \text{ h})}$$

To determine $R(t)$ for the normal distribution, a standard normal variate must be calculated by the following formula:

$$z = \frac{t - \mu}{\sigma} \tag{22}$$

where μ = the mean time to failure; and σ = the standard deviation. Applying this formula for $t = 1080$ h gives

$$z = (1080 - 1400)/120 = -2.67$$

Then this value of z is used with any readily available normal table to find

$$R(1080 \text{ h}) = 0.99621$$

Similarly

$$R(1128 \text{ h}) = 0.98840$$

which is the unconditional probability of surviving 1128 h. The conditional probability of survival is then

$$R(1128 \text{ h} \,|\, t > 1080 \text{ h}) = 0.98840/0.99621 = 0.99216$$

In this application, based on the consequences, management has a rule to plan a replacement when the chances of failure over the next 48 h period drops below 0.99. In this case they would forego scheduling the replacement.

    Example 3 illustrated the use of the normal distribution as a model for time-to-failure. The normal distribution has an increasing hazard function which means that the product is experiencing wear-out. In applying the normal to a specific situation, the fact must be considered that this model allows values of the random variable that are less than zero whereas obviously a life less than zero is not possible. This problem does not arise from a practical standpoint as long as $\mu/\sigma \geqslant 4.0$.

Exponential Distribution

    The exponential distribution has proved to be a reasonable failure model for electronic equipment (8–13). Since the field of reliability emerged, owing to problems encountered with military electronics during World War II, exponential distribution has had considerable attention and application. However, like any failure model, it has limitations which should be well understood.

Basic Statistical Properties. The *pdf* for an exponentially distributed random variable **t** is given by

$$f(t,\lambda) = \lambda e^{-\lambda t}, \quad t \geq 0 \tag{23}$$

where λ is the failure-rate parameter. The quantity $\theta = 1/\lambda$ is the mean or expected life, also expressed as *MTBF*. The *pdf* is shown in Figure 9.

The reliability function is given by

$$R(t) = e^{-\lambda t}, \quad t \geq 0 \tag{24}$$

or

$$R(t) = e^{-t/\theta}, \quad t \geq 0 \tag{25}$$

whereas the hazard function is

$$h(t) = \lambda = \frac{1}{\theta} \tag{26}$$

The hazard function is a constant which means that this model would be applicable during the midlife of the product when the failure rate is relatively stable. It would not be applicable during the wear-out phase or during the infant mortality (early failure) period.

On complex systems, which are repaired as they fail and placed back in service, the time between system failures can be reasonably well modeled by the exponential distribution (14–15).

Point Estimation. The estimator for the mean life parameter θ is given by:

$$\hat{\theta} = \frac{T}{r} \tag{27}$$

where T = total accumulated test time considering both failed and unfailed (or suspended) items; and r = total number of failures. The reliability function is then estimated by:

$$\hat{R}(t) = e^{-t/\hat{\theta}}, \quad t \geq 0 \tag{28}$$

Example 4. A particular microprocessor (MPU) is assigned for a fuel-injection system. The failure rate must be estimated, and 100 MPUs are tested. The test is terminated when the fifth failure occurs. Failed items are not replaced. This type of testing, where n = the number placed on test and r = the number of failures are specified, is termed a Type-II censored life test.

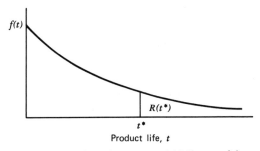

Figure 9. *Pdf* for the exponential failure model.

Assuming that the above test produces the following data (failure time in hours): 84.1; 240.1; 251.9; 272.2; 291.9—the *MTBF* is estimated by using equation 27 and is:

$$\hat{\theta} = \frac{84.1 + 240.1 + 251.9 + 272.2 + 291.9 + 95(291.9)}{5}$$

$$= 5774 \text{ h}$$

From equation 9 it was shown that the chance of surviving the mean life was 36.8% for the exponential distribution. However, this fact must be used with some degree of rationality in applications. For example, in the above situation the longest surviving MPU that was observed survived for 291.9 hours. The failure rate beyond this time is not known. What was observed was only a failure rate of $\hat{\lambda} = 1.732 \times 10^{-4}$ failures per hour over approximately 292 hours of operation. In order to make predictions beyond this time, it must be assumed that the failure rate does not increase because of wear-out and aging.

The reliability function in this example could be estimated as:

$$\hat{R}(t) = e^{-t \times 1.732 \times 10^{-4}/\text{h}}$$

Since these MPUs are used to control fuel-injection systems, it might be interesting to know the 24,000-km reliability (the warranty period). Assuming an average speed of 80 km/h, 300 h of use are obtained. The reliability would be estimated as:

$$\hat{R}(300 \text{ h}) = 0.949$$

Or about 5.0% failures can be expected over the warranty period.

Example 5. There are six dynamometers available for engine testing. The test duration is set at 200 h which is assumed to be equivalent to 20,000 km of customer use. Failed engines are removed from testing for analysis and replaced. The objective of the test is to analyze the emission-control system. Failure is defined as the time at which certain emission levels are exceeded.

The testing situation where the duration is specified (ie, time-truncated) is termed Type-I censored life testing.

Assuming that this test produces five failures, the *MTBF* would be estimated as

$$\hat{\theta} = \frac{120,000 \text{ km}}{5 \text{ failures}} = 24,000 \text{ km}$$

Or the failure rate is:

$$\hat{\lambda} = 4.17 \times 10^{-5} \text{ failures/km}$$

Again, these estimates must be used with caution. The system is obviously a mixture of electrical and mechanical components, and it can be assumed that wear-out starts well beyond the 20,000 km period. If this is a reasonable assumption based on experience, then reliability predictions can be made over the 20,000-km period. For example, the 6000-km reliability might be estimated as

$$R(6000 \text{ km}) = 0.79$$

However, a 50,000-km reliability estimate might not be reasonable based on this testing scheme.

Confidence-Interval Estimates.   Confidence-interval estimates for the expected life or reliability can be obtained easily in the case of the exponential. Here only the limits for failure-censored (Type II) and time-censored (Type I) life testing are given. It is possible to specify a test as either time- or failure-truncated, whichever occurs first. The theory for such tests is explained in refs. 16–17.

Time-Censored Life Tests.   In this case the total test time T is specified. From the test, r failures are observed. The $100(1 - \alpha)\%$ two-sided confidence interval for the expected life is

$$\frac{2\,T}{\chi^2_{\alpha/2,2(r+1)}} \leq \theta \leq \frac{2\,T}{\chi^2_{1-\alpha/2,2r}} \tag{29}$$

The quantities $\chi^2_{\beta,\nu}$ are the $(1 - \beta)$ percentiles of a chi square distribution with ν degrees of freedom and are found readily in chi square tables.

Frequently, only a one-sided lower confidence limit is desired. In this case the limit is

$$\frac{2\,T}{\chi^2_{\alpha,2(r+1)}} \leq \theta \tag{30}$$

This is a $100(1 - \alpha)\%$ lower confidence limit.

If the above limits on the expected life are designated by L and U for the lower and upper, respectively, then the $100(1 - \alpha)\%$ confidence interval on the reliability is

$$e^{-t/L} \leq R(t) \leq e^{-t/U} \tag{31}$$

Failure-Censored Life Tests.   In this testing situation, the number of failures r is specified with n items initially placed on test $(r \leq n)$. The test produces failure times $t_1, t_2 \ldots t_r$. The $100(1 - \alpha)\%$ confidence interval for the expected life is calculated by

$$\frac{2\,T}{\chi^2_{\alpha/2,2r}} \leq \theta \leq \frac{2\,T}{\chi^2_{1-\alpha/2,2r}} \tag{32}$$

Here again the quantity $\chi^2_{\beta,\nu}$ is the $(1 - \beta)$ percentile of a chi square distribution with ν degrees of freedom.

If only a $100(1 - \alpha)\%$ lower confidence limit is desired, it can be calculated from

$$\frac{2\,T}{\chi^2_{\alpha,2r}} \leq \theta \tag{33}$$

The confidence limits for the reliability function can be found from equation 31.

The Nonzero Minimum-Life Case.   In many situations, no failures are observed during an initial period of time. For example, when testing engine bearings for fatigue life no failures are expected for a long initial period. Some corrosion processes also have this characteristic. In the following it is assumed that the failure pattern can be reasonably well approximated by an exponential distribution.

The *pdf* for the two-parameter exponential distribution is given by

$$f(t,\theta,\delta) = \frac{1}{\theta}\,e^{-(t-\delta)/\theta}, \quad t \geq \delta \geq 0, \quad \theta > 0 \tag{34}$$

The reliability function is

$$R(t) = e^{-(t-\delta)/\theta}, \quad t \geq \delta \geq 0 \tag{35}$$

The expected life is $(\delta + \theta)$. The quantity δ is referred to as the minimum life parameter.

Point Estimation. This is a Type-II censored life-testing situation where n items are placed on test and the test is terminated at the time of the rth failure. The life test produces the ordered failure times $t_1, t_2 \ldots t_r$. The estimator for θ is

$$\hat{\theta} = \frac{\sum_{i=2}^{r} (t_i - t_1) + (n - r)(t_r - t_1)}{(r - 1)} \tag{36}$$

And the estimator for δ, the minimum life, is:

$$\hat{\delta} = t_1 - \frac{\hat{\theta}}{n} \tag{37}$$

The reliability function is then estimated as

$$\hat{R}(t) = e^{-(t-\hat{\delta})/\hat{\theta}}, \quad t \geq \hat{\delta} \tag{38}$$

Confidence Limits. The $100(1 - \alpha)\%$ confidence interval for the parameter θ is:

$$\frac{2(r - 1)\hat{\theta}}{\chi^2_{\alpha/2, 2(r-1)}} \leq \theta \leq \frac{2(r - 1)\hat{\theta}}{\chi^2_{1-\alpha/2, 2(r-1)}} \tag{39}$$

And the $100(1 - \beta)\%$ confidence interval for the minimum life δ is

$$t_1 - \frac{\hat{\theta}}{n} F_{\beta, 2, 2(r-1)} \leq \delta \leq t_1 \tag{40}$$

The quantity F_{β, ν_1, ν_2} is the $(1 - \beta)$ percentile of an F-distributed random variable with ν_1, ν_2 degrees of freedom and is readily obtainable from F-tables.

Example 6. A return spring used on a butterfly-valve mechanism must have a high reliability. In order to determine the spring reliability, fifty springs are randomly selected and placed on life test. The test is terminated when the tenth spring fails. The data are given in the left column of Table 1. For the right column, equation 36 is applied.

The estimate of θ is

Table 1. Cycles to Failure

$t_i \times 10^3$	$(t_i - t_1) \times 10^3$
61.0	0
64.1	3.1
64.6	3.6
66.2	5.2
73.9	12.9
75.0	14.0
77.4	16.4
79.8	18.8
80.5	19.5
83.6	22.6
	$\Sigma = 116.1$

$$\hat{\theta} = \frac{116.1 + (40)(22.6)}{9} = 113.3 \times 10^3 \text{ cycles}$$

And the minimum life is estimated from equation 37 as:

$$\hat{\delta} = 61.0 - \frac{113.3}{50} = 58.7 \times 10^3 \text{ cycles}$$

Since the minimum life is critical in this application, a confidence limit estimate would be more appropriate, which can be calculated with the help of equation 40. For a 90% confidence limit, the required value of F is

$$F_{0.10,2,18} = 2.62$$

And substituting into the confidence interval equation gives

$$\left[61.0 - \frac{113.3}{50}(2.62) \right] \times 10^3 \text{ cycles} \leq \delta \leq 61.0 \times 10^3 \text{ cycles}$$

Or:

$$55.1 \times 10^3 \text{ cycles} \leq \delta \leq 61.0 \times 10^3 \text{ cycles}$$

In order to ensure virtually failure-free operation, a policy of changing this spring at 50,000 cycles of operation might be adopted.

In test planning, the number to be placed on test n and the number of failures r must be determined. The operating characteristic curves in ref. 18 can be used to specify the test, and to control the errors.

The Weibull Distribution

The Weibull distribution is a more versatile failure model than the exponential one. It is a popular model and widely used to estimate product reliability because it can be analyzed graphically with Weibull probability paper. Although the graphical form of analysis is presented here, other procedures are available (19–21).

Basic Statistical Properties. The reliability function for the three-parameter Weibull distribution is given by

$$R(t) = \exp\left[-\left(\frac{t - \delta}{\theta - \delta} \right)^\beta \right], \quad t \geq \delta \geq 0, \quad \beta > 0, \quad \theta > \delta \tag{41}$$

where δ = minimum life; θ = characteristic life; and β = Weibull slope.

The two-parameter Weibull has a minimum life of zero and the reliability function is:

$$R(t) = e^{-(t/\theta)^\beta}, \quad t \geq 0 \tag{42}$$

The hazard function for the two-parameter Weibull is:

$$h(t) = \frac{\beta}{\theta^\beta} t^{\beta-1}, \quad t \geq 0 \tag{43}$$

This hazard function decreases with $\beta < 1$, increases with $\beta > 1$, and remains constant for $\beta = 1$. The value of β can give some indication of wear-out or infant mortality.

The expected life for the two-parameter Weibull distribution is:

$$\mu = \theta \, \Gamma(1 + 1/\beta) \tag{44}$$

where $\Gamma(\cdot)$ is a gamma function and can be found in gamma tables. The variance for the Weibull is

$$\sigma^2 = \theta^2 \left[\Gamma\left(1 + \frac{2}{\beta}\right) - \Gamma^2\left(1 + \frac{1}{\beta}\right) \right] \tag{45}$$

The characteristic life parameter θ has a constant reliability associated with it. Evaluating the reliability function at $t = \theta$ gives

$$R(\theta) = e^{-1} = 0.368$$

And this is the same for any parameter value. Thus, it is a constant for any Weibull distribution.

Parameter Estimation. Weibull parameters can be estimated using the usual statistical procedures; however, a computer is needed to solve readily the equations. A computer program based on the maximum likelihood method is presented in ref. 22. Graphical estimation can be made on Weibull paper without the aid of a computer; however, the results cannot be expected to be as accurate and consistent.

The two-parameter cumulative Weibull distribution is

$$F(t) = 1 - e^{-(t/\theta)^\beta} \tag{46}$$

which, after rearranging and taking logarithms twice becomes

$$\ln\left(\ln\frac{1}{1 - F(t)}\right) = \beta \ln t - \beta \ln \theta \tag{47}$$

This would give a straight line plot on rectangular graph paper. Weibull graph paper plots $[F(t),t]$ as a straight line. Figure 10 illustrates a typical Weibull paper.

In using Weibull graph paper, a plotting position $p_j = F(t_j)$ for the jth-ordered observation has to be decided. The mean or median are the principal contenders. The median can be conveniently approximated (23) by

$$p_j = \frac{j - 0.3}{n + 0.4} \tag{48}$$

And the mean is given by

$$p_j = \frac{j}{n + 1} \tag{49}$$

The failure points (t_j,p_j) are plotted and a straight line is fitted to estimate the Weibull population.

Example 7. In order to illustrate graphical parameter estimation, five failure times are considered: 24,000 km, 39,000 km, 52,000 km, 64,000 km, and 82,000 km. These times-to-failure were obtained by placing five items on test and allowing them to go to failure.

The median-rank plotting positions are obtained from equation 48. Tables such as found in ref. 24 can be used also. The data ready for plotting are given in Table 2 and are plotted on the Weibull paper in Figure 10. The Weibull slope parameter is estimated as $\hat{\beta} = 2.0$, and this implies an increasing failure rate. The characteristic life is estimated using the 63% point on the cumulative scale which gives $\hat{\theta} = 61,000$ km. Confidence limits can be also placed about this line; however, special tables are

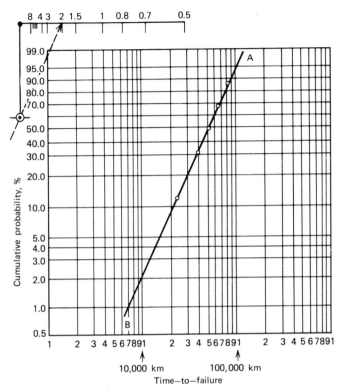

Figure 10.   Weibull probability paper: A, estimate of population; B, estimate of β (line drawn parallel to population line).

Table 2.   Weibull Paper Plotting Data

Order no., j	Failure times, t_j, km	Cumulative frequency, p_j, %
1	24,000	13
2	39,000	31
3	52,000	50
4	64,000	69
5	82,000	87

needed (24). The population line can be used to estimate either percent failure at a given time or the time at which a given percentage will fail.

In plotting on Weibull paper, a downward concave plot implies a nonzero minimum life. Values for $\hat{\delta} < t_1$ can be selected by trial and error. When they are subtracted from each t_i, a relatively straight line is produced. This essentially translates the three-parameter Weibull distribution back to a two-parameter distribution.

As can be seen from the above, the graphical method does provide a good visual means for analyzing life data and is easily understood and explained. If used with discretion, graphical analysis can provide a useful means for data analysis.

Binomial Distribution

To determine in the laboratory if a component survives in use, a test bogey is frequently established based on past experience. The test bogey is correlated with the particular test used to duplicate (or simulate) field conditions. The bogey can be stated in cycles, hours, revolutions, stress reversals, etc. A number of components are placed on test and each component either survives or fails. The reliability for this situation is estimated.

The failure model is the binominal distribution given by

$$p(y) = \binom{n}{y}R^y(1 - R)^{n-y}, \quad y = 0,1,2 \ldots n \tag{50}$$

where R = the product reliability; n = the total number of products placed on test; and y = the number of products surviving the test. Furthermore

$$\binom{n}{y} = \frac{n!}{y!(n - y)!}$$

The quantity $p(y)$ is the probability that exactly y out of n components survive the test where the component reliability is R.

Reliability Estimation.   Both a point estimate and a confidence interval estimate of product reliability can be obtained.

Point Estimate.   The point estimate of the component reliability is given by

$$\hat{R} = \frac{y}{n} \tag{51}$$

Confidence Limit Estimate.   An exact $100(1 - \alpha)\%$ lower confidence limit on the reliability is given by

$$R_L = \frac{y}{y + (n - y + 1)F_{\alpha,2(n-y+1),2y}} \tag{52}$$

where $F_{\alpha,2(n-y+1),2y}$ is easily obtained from tables for values of F.

A convenient approximate limit based on the normal distribution given by:

$$R_L = \frac{(y - 1)}{n + z_\alpha \sqrt{\dfrac{n(n - y + 1)}{(y - 2)}}} \tag{53}$$

where z_α is the upper $(1 - \alpha)$ percentile of the standard normal distribution as is readily obtained from normal tables.

Example 8.   There are 40 components placed on an accelerated 80-h life test. A 75% lower confidence limit on the reliability is desired.

To use equation 52, a value of F must be looked up. In this case, $n = 40$ and $y = 37$, and the required value is

$$F_{0.25,8,74} = 1.31$$

The lower confidence limit is calculated by

$$R_L = \frac{37}{37 + (4 \times 1.31)} = 0.876$$

or the 75% lower confidence limit on the reliability is

$$0.876 \le R$$

If the approximate limit given by equation 53 is used, the value for $z_{0.25}$ is 0.67. The limit would be calculated as

$$R_L = \frac{36}{40 + 0.67 \sqrt{\dfrac{40(4)}{35}}} = 0.869$$

As can be seen, the approximation is reasonably close. This approximation is better with large degrees of freedom for the value of F.

Success Testing. Acceptance life tests are sometimes planned with no failures allowed. This gives the smallest sample size necessary to demonstrate a reliability at a given confidence level. The reliability is demonstrated relative to the test employed and the testing period.

For the special case where no failures are allowed ($y = 0$) the $100(1 - \alpha)\%$ lower confidence limit on reliability is given by

$$R_L = \alpha^{1/n} \tag{54}$$

where α = the level of significance, and n = the sample size. If $C = 1 - \alpha$ is taken as the desired confidence level, then the required sample size to demonstrate a minimum reliability of R is

$$n = \frac{\ln(1 - C)}{\ln R} \tag{55}$$

For example, if a reliability level of $R = 0.85$ is to be demonstrated at 90% confidence, the required sample size is

$$n = \frac{\ln(0.10)}{\ln(0.85)} = 15$$

where no failures are allowed.

Probabilistic Engineering Design

The probabilistic approach to design is an attempt to quantify the design variables from a reliability standpoint (25–28).

The basic approach for probabilistic engineering design is to realize that a component has a certain strength which, if exceeded, results in failure. Broadly speaking, strength indicates the ability to resist failure, whereas stress indicates the agents that tend to induce failure. The factors determining strength or stress on a component are random variables. Conceptually both strength and stress distribution exist, as illustrated in Figure 11 where the shaded area indicates the potential of failure.

In order to illustrate this approach, it is assumed that the strength random variable **y** has a normal distribution with mean μ_y and standard deviation σ_y, whereas the stress random variable **x** has a normal distribution with mean μ_x and standard deviation σ_x. The component reliability is given by

$$R = P(\mathbf{y} > \mathbf{x}) \tag{56}$$

Or if one defines

$$w = y - x$$

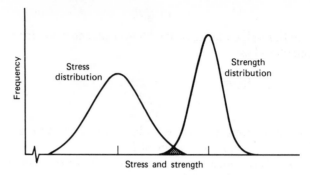

Figure 11.   Stress and strength distribution for a component.

Then

$$R = P(\mathbf{w} > 0) \tag{57}$$

A standard variate can be calculated as

$$z = \frac{\mu_x - \mu_y}{\sqrt{\sigma_x^2 + \sigma_y^2}} \tag{58}$$

And standard normal tables can be used to look up the reliability.

Example 9.   In a situation where $\mu_y = 12$ MPa (1740 psi), $\sigma_y = 1.2$ MPa (174 psi), $\mu_x = 8$ MPa (1160 psi), and $\sigma_x = 0.8$ MPa (116 psi), and the reliability is to be calculated under the assumption of normality, applying equation 58 gives

$$z = -2.77$$

And using normal tables

$$R = 0.977$$

The factor of safety *FS* might be defined as

$$FS = \frac{\mu_y}{\mu_x} \tag{59}$$

Hence, the factor of safety in the above example is

$$FS = 1.5$$

The reliability associated with a factor of safety depends on both the mean and standard deviations of the stress and strength distributions.

The following hypothetical example illustrates the application of this concept to a simple design problem. Furthermore, complex functions of random variables are handled by using a convenient approximation.

Example 10.   A simple design problem is illustrated in Figure 12. This is a highly idealized dynamic system where the shaft rotates at a constant speed of 1200 rpm. The rod retains the ball which travels on a frictionless surface. The downward force of the ball is 2 N (0.45 lbf) and the radius of travel is 10 cm. The rod diameter must be specified in such a manner that the stress in it does not exceed 12 MPa (1740 psi). The weight of the rod is neglected in the following analysis.

If variability is ignored, the problem might be solved as follows:

$$\omega = 1200(2\,\pi/60) = 40\,\pi \text{ rad/s}$$

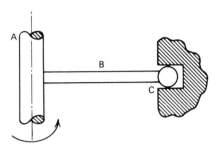

Figure 12. Example of design by reliability: A, shaft; B, rod; C, ball.

The force on the rod would be

$$F = ma = \frac{w}{g}\, \omega^2 R = \frac{2}{9.8}\, (40\,\pi)^2\, \frac{10}{100} = 322.27 \text{ N } (72.45 \text{ lbf})$$

Then the required area is

$$A = \frac{F}{\text{allowable stress}} = \frac{322.27}{1{,}200} = 0.269 \text{ cm}^2$$

Since $A = \pi d^2/4$, the required diameter would be

$$d = \left(\frac{4}{\pi} \times 0.269\right)^{1/2} = 0.585 \text{ cm}$$

Assuming that the manufacturing process introduces variability in producing this diameter, the standard deviation of this manufacturing process is 0.04 cm and the required reliability 0.98; ie, it is required that no more than 2% of the rods have a stress that exceeds the allowable.

The stress in the rod is given by

$$x = F/A = 4\,F/\pi d^2$$

d is now considered to be a random variable with standard deviation $\sigma_d = 0.04$ cm and the mean must be specified to obtain the required reliability.

A convenient Taylor's series (TS) approximation can be employed to handle nonlinear relationships frequently arising in engineering problems. When considering a random variable $\mathbf{x}$ with mean μ_x and standard deviation σ_x, the problem is to find the mean and standard deviation of y where

$$y = g(x) \tag{60}$$

The random variable y, when a first-order TS approximation is used, has mean

$$\mu_y \doteq g(\mu_x) \tag{61}$$

And variance

$$\sigma_y^2 \doteq [g'(\mu_x)]^2\, \sigma_x^2 \tag{62}$$

Applying this to the previous problem gives

$$\mu_x = 4\,F/\pi \mu_d^2$$

and

$$\sigma_x = (8 \, F / \pi \mu_d^3) \sigma_d$$

In order to satisfy the reliability requirement

$$R = 0.98 = P(x < 12 \text{ MPa or } 118 \text{ atm})$$

Using a simple normal approximation:

$$z = \frac{1{,}200 - \mu_x}{\sigma_x} = \left(1{,}200 - \frac{410.3}{\mu_d^2}\right) \Big/ \left(\frac{32.8}{\mu_d^3}\right)$$

Using normal tables with $R = 0.98$ gives a standard normal value of $z = 2.05$. Solving for the average diameter:

$$\mu_d = 0.654 \text{ cm}$$

The diameter increased because of manufacturing variability and the reliability requirement. Further complications would be introduced if the angular velocity and/or the material property was allowed to vary. The approach would be similar but more random variables would have to be handled (29).

BIBLIOGRAPHY

1. W. H. Von Alven, ed., *Reliability Engineering*, Prentice-Hall, Inc., Englewood Cliffs, N.J., 1964.
2. *AMCP 706-133, Engineering Design Handbook, Maintainability Engineering Theory and Practice*, U.S. Army Materiel Command, Washington, D.C., 1976.
3. *MIL-STD-1629 (SHIPS), Procedures for Performing a Failure Mode and Effects Analysis for Shipboard Equipment*, Department of the Navy, Naval Ship Engineering Center, Hyattsville, Md., 1974.
4. *ARP-926, Design Analysis Procedure for Failure Mode, Effects and Critically Analysis (FMECA)*, Society of Automotive Engineers, Inc., New York, 1967.
5. Reliability Analysis Center, *NPRD-1, Nonelectronic Parts Reliability Data*, Rome Air Development Center, Griffiss AFB, N.Y., 1978.
6. *MIL-STD-217B, Reliability Prediction of Electronic Equipment*, U.S. Superintendent of Documents, Washington, D.C., 1978.
7. R. E. Barlow and F. Proschan, *Statistical Theory of Reliability and Life Testing Probability Models*, Holt, Rinehart and Winston, Inc., New York, 1975.
8. B. Epstein, *Ann. Math. Stat.* **25**, 555 (1954).
9. B. Epstein, *Technometrics* **2**(4), 447 (1960).
10. Ref. 9, p. 435.
11. Ref. 9, p. 403.
12. B. Epstein and M. Sobel, *J. Am. Stat. Assoc.* **48**, 486 (1953).
13. B. Epstein and M. Sobel, *Ann. Math. Stat.* **25**, 373 (1954).
14. R. F. Drenick, *J. Soc. Ind. Appl. Math.* **8**(21), 125 (1960).
15. *Ibid.*, **8**(4), 680 (1960).
16. D. J. Bartholomew, *Technometrics* **5**, 361 (1963).
17. G. Yang and M. Sirvanci, *J. Am. Stat. Assoc.* **72**, 444 (1977).
18. K. H. Schmitz, L. R. Lamberson, and K. C. Kapur, *Technometrics* **21**(4), 539 (1979).
19. L. J. Bain, *Statistical Analysis of Reliability and Life-Testing Models*, Marcel Dekker, Inc., New York, 1978.
20. D. I. Gibbons and L. C. Vance, *A Simulation Study of Estimators for the Parameters and Percentiles in the Two-Parameter Weibull Distribution*, General Motors Research Publication No. GMR-3041, General Motors, Detroit, Mich., 1979.
21. N. Mann, R. Schafer, and N. D. Singpurwalla, *Methods for Statistical Analysis of Reliability and Life Tests*, John Wiley & Sons, Inc., New York, 1974.
22. D. R. Wingo, *IEEE Trans. Reliab.* **R-22**(2), (1973).
23. A. Benard and E. C. Bos-Levenbach, *Statistica* **7**, (1953).

24. K. C. Kapur and L. R. Lamberson, *Reliability in Engineering Design*, John Wiley & Sons, Inc., New York, 1977.
25. E. B. Haugen, *Probabilistic Approaches to Design*, John Wiley & Sons, Inc., New York, 1968.
26. D. Kececioglu and D. Cormier, *Proc. Third Ann. Aerospace Reliab. Maintainab. Conf.*, 546 (1964).
27. D. Kececioglu and E. B. Haugen, *Ann. Assurance Sci.—Seventh Reliab. Maintainab. Conf.*, 520 (1968).
28. C. Mischke, *J. Eng. Ind.* 537 (Aug. 1970).
29. A. H. Bowker and G. J. Lieberman, *Engineering Statistics*, 2nd ed., Prentice-Hall, Inc., Englewood Cliffs, N.J., 1972.

General References

R. E. Barlow and F. Proschan, *Mathematical Theory of Reliability*, John Wiley & Sons, Inc., New York, 1965.
I. Bazovsky, *Reliability Theory and Practice*, Prentice-Hall, Inc., Englewood Cliffs, N.J., 1961.
R. Billinton, *Power System Reliability Evaluation*, Gordon and Breach Science Publishers, New York, 1970.
R. Billinton, R. J. Ringlee, and A. J. Wood, *Power-System Reliability Calculations*, The MIT Press, Cambridge, Mass., 1973.
J. H. Bompas-Smith in R. H. W. Brook, ed., *Mechanical Survival: The Use of Reliability Data*, McGraw-Hill, New York, 1973.
DARCOM-P-702-4, *Reliability Growth Management*, U.S. Army Materiel Development and Readiness Command, Alexandria, Va., 1976.
D. K. Lloyd and M. Lipow, *Reliability: Management Methods and Mathematics*, Prentice-Hall, Inc., Englewood Cliffs, N.J., 1962.
M. L. Shooman, *Probabilistic Reliability: An Engineering Approach*, McGraw-Hill, New York, 1968.
U.S. Army Materiel Command, *Engineering Design Handbooks-Development Guide for Reliability*, *Part 2: Design for Reliability (AMCP 706-196); Part 3: Reliability Prediction (AMCP 706-197); Part 4: Reliability Measurement (AMCP 706-298)*; National Technical Information Service, Springfield, Va., 1976.

LEONARD LAMBERSON
Wayne State University

MATERIALS STANDARDS AND SPECIFICATIONS

Standards have been a part of technology since building began, both at a scale that exceeded the capabilities of an individual, and for a market other than the immediate family. Standardization minimizes disadvantageous diversity, assures acceptability of products, and facilitates technical communication. There are many attributes of materials that are subject to standardization, eg, composition, physical properties, dimensions, finish, and processing. Implicit to the realization of standards is the availability of test methods and appropriate calibration techniques. Apart from physical or artifactual standards, written or paper standards also must be examined, ie, their generation, promulgation, and interrelationships.

A standard is a document, definition, or reference artifact intended for general use by as large a body as possible; whereas a specification, although involving similar technical content and similar format, usually is limited in its intended applicability and its users. The International Organization for Standardization (ISO) defines a standard as the result of the standardization process, "the process of formulating and applying rules for an orderly approach to a specific activity for the benefit and with the cooperation of all concerned and in particular for the promotion of optimum overall economy taking due account of functional conditions and safety requirements" (1). Standardization involves concepts of units of measurement, terminology and symbolic representation, and attributes of the physical artifact, ie, quality, variety, and interchangeability. A specification, however, is defined as "a document intended primarily for use in procurement which clearly and accurately describes the essential technical requirements for items, materials, or services including the procedures by which it will be determined that the requirements have been met" (2). The corresponding ISO definition is: "A specification is a concise statement of a set of requirements to be satisfied by a product, a material or a process indicating, whenever appropriate, the procedure by means of which it may be determined whether the requirements given are satisfied. Notes—(1) A specification may be a standard, a part of a standard, or independent of a standard. (2) As far as practicable, it is desirable that the requirements are expressed numerically in terms of appropriate units, together with their limits." A specification is the technical aspects of the legal contract between the purchaser of the material, product, or service and the vendor of the same and defines what each may expect of the other.

There are psychological barriers to making intelligent selections and applications of standards. For some, a standard has attributes of superiority; it is considered an unattainable ideal which only can be approached but never met. For others it represents the opposite, ie, a lowest common denominator to which all must be reduced. For most, however, the terms impart a sense of a condition from which all creativity and individuality have been removed.

Standards

Objectives and Types. The objectives of standardization are economy of production by way of economies of scale in output, optimization of varieties in input material, and improved managerial control; assurance of quality; improvement of interchangeability; facilitation of technical communication; enhancement of innovation and technological progress; and promotion of the safety of persons, goods, and the environment. The likely consequences of choosing a material that is not standard, other than in very exceptional circumstances, are that the selected special would be unusually costly; require an elaborate new specification; be available from few sources; be lacking documentation for many ancillary properties other than that for which it was chosen; be unfamiliar to others, eg, purchasers, vendors, production workers, maintenance personnel, etc; and contribute to the proliferation of stocked varieties and, thus, exacerbate problems of recycling, mistaken identity, increased purchasing costs, etc.

Physical or artifactual standards are used for comparison, calibration, etc, eg, the national standards of mass, length, and time maintained by the NBS or the Standard Reference Materials collected and distributed by NBS. Choice of the standard is determined by the property it is supposed to define, its ease of measurement, its stability with time, and other factors (see also Fine chemicals).

Paper or documentary standards are written articulations of the goals, quality levels, dimensions, or other parameter levels that the standards-setting body seeks to establish.

Value standards are a subset of paper standards and usually relate directly to society and include social, legal, political, and to a lesser extent, economic and technical factors. Such standards usually result from federal, state, or local legislation.

Regulatory standards most frequently derive from value standards but also may arise on an *ad hoc* or consensus basis. They include industry regulations or codes that are self-imposed; consensus regulatory standards that are produced by voluntary organizations in response to an expressed governmental need, especially where well-defined engineering practices or highly technical issues are involved; and mandatory regulatory standards that are developed entirely by government agencies. Examples of regulatory standards from the materials field include safety regulations, eg, those of the OSHA; clean air and water laws of the EPA; or rulings related to exposure to radioactive substances. Regulatory standards deliberately may be set in advance of the state of the art in the relevant technology, in contrast to the other types of standards (see Regulatory agencies).

Voluntary standards are especially prevalent in the United States and are generated by various consortia of government and industry, producers and consumers, technical societies and trade associations, general interest groups, academia, and individuals. These standards are voluntary in their manner of generation and in that they are intended for voluntary use. Nonetheless, some standards of voluntary origin have been adopted by governmental bodies and are mandatory in certain contexts. Voluntary standards include those which are recommended but which may be subject to some interpretation and those conventions as to units, definitions, etc, that are established by custom.

Product standards may stipulate performance characteristics, dimensions, quality factors, methods of measurement, and tolerances; and safety, health, and environmental protection specifications. They are introduced principally to provide for interchangeability and reduction of variety. The latter procedure is referred to as rationalization of the product offering, ie, designation of sizes, ratings, etc, for the attribute range covered and the steps within the range. The designated steps may follow a modular format or a preferred number sequence.

Public and private standards also may be distinguished. Public standards include those produced by government bodies and those published by other organizations but promoted for general use, eg, the ASTM standards. Private standards are issued by a private company for its own interests and, generally, are not available to parties other than its vendors, customers, and subcontractors.

Consensus standards are the key to the voluntary standards system since acceptance and use of such standards follow directly from the need for them and from the involvement in their development of all those who share that need. Consensus standards must be produced by a body selected, organized and conducted in accordance with "due process" procedures. All parties or stakeholders are involved in the development of the standard and substantial agreement is reached according to the judgment of a properly constituted review board. Other aspects of "due process" involve proper issuance of notices, record keeping, balloting, and attention to minority opinion.

Generation, Administration, and Implementation.   The development of a good standard is a lengthy and reiteratively involved process, whether it be for a private organization, a nation, or a international body. The generic aspects of the development of a standard are shown in Figure 1. Once it has been determined that a need for a standard exists, information relevant to the subject must be gathered from many sources, eg, libraries and specialists' knowledge, field surveys, and laboratory results. Multidisciplinary and multifunctional teams must digest this information, array and analyze options, achieve an effective compromise in the balanced best interests of all concerned, and participate in the resolution of issues and criticisms arising from the reviews and appeals process. Among its functions, the administrative arm of the pertinent standards organization sets policy, allocates resources, establishes priorities, supervises reviews and appeals procedures, and interacts with organizations external to itself. The affected entities usually comprise a very large, diverse, and overlapping group of interests, ie, economic sectors: industry, government, business, construction, chemicals, energy, etc; functions: planning, development, design, production, maintenance, etc; and organizations and groups of individuals: manufacturers, consumers, unions, investors, distributors, etc. No standard can be fully effective in meeting its objectives unless attention is paid to the implementation function which includes promulgation, education, enforcement of compliance, and technical assistance. Usually in the choice of a standard for a given purpose, the more encompassing the population to which the standard applies, ie, from the private level to the trade association or professional society to the national level or to the international level, the more effective and the less costly the application of the standard is. Finally it must be recognized that standardization is a highly dynamic process. It cannot function without the continued feedback from all affected parties and it must provide for constant review and adaptation to changing circumstances, improved knowledge and control.

Standard Reference Materials.   An important development in the United States, relative to standardization in the chemical field, is the establishment by the NBS of standard reference materials (SRMs), originally called standard samples (4). The objective of this program is to provide materials that may be used to calibrate measurement systems and to provide a central basis for uniformity and accuracy of measurement. SRMs are well-characterized, homogeneous, stable materials or simple artifacts with specific properties that have been measured and certified by NBS. Their use with standardized, well-characterized test methods enables the transfer, accuracy, and establishment of measurement traceability throughout large, multilaboratory measurement networks. More than 1000 materials are now included, eg, metals and

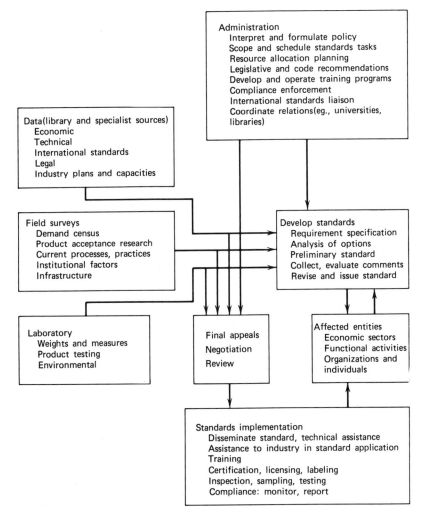

Figure 1.   Flow chart of the standardization process (3).

alloys, ores, cements, phosphors, organics, biological materials, glasses, liquids, gases, radioactive substances, and specialty materials. The standards are classified as standards of certified chemical composition, standards of certified physical properties, and engineering-type standards. Although most of these are provided with certified numerical characterizations of the compositions or physical properties for which they were established, some others are included even where provision of numerical data is not feasible or certification is not useful. These latter materials do, however, provide assurance of identity among all samples of the designation and permit standardization of test procedures and referral of physical or chemical data on unknown materials to a known or common basis.

Major shifts in the nature of the materials included in the SRM inventory are expected. In the compositional SRMs, increased attention is anticipated to trace-organic analysis for environmentally, clinically, or nutritionally important substances;

to trace-element analysis in new high technology materials, eg, alloys, plastics, and semiconductors; and for bulk analyses in the field of recycled, nuclear and fibrous materials (see also Trace and residue analysis). It is expected that there will be concern not just for certification of the total concentration of individual elements but for the levels of various chemical states of those species. With regard to the development of physical property SRMs, density standards, dimensional standards at the micrometer and submicrometer level, and materials relative to standardization of optical properties will be among the more active areas. Major developments in SRMs for engineering properties will include materials suitable for nondestructive testing (qv), evaluation of durability, standardizing computer and electronic components, and workplace hazard monitoring.

Standard reference materials provide a necessary but insufficient means for achieving accuracy and measurement compatibility on a national or international scale. Good test methods, good laboratory practices, well-qualified personnel, and proper intralaboratory and interlaboratory quality assurance procedures are equally important. A systems approach to measurement compatibility is illustrated in Figure 2. The function of each level, I–VI, is to transfer accuracy to the level below and to help provide traceability to the level above. Thus, traversing the hierarchy from bottom to top increases accuracy at the expense of measurement efficiency.

Analytical standards imply the existence of a reference material and a recommended test method. This subject with reference to analytical, electronic-grade, and reagent chemicals has been discussed (see Fine chemicals). Analytical standards other than for fine chemicals and for the NBS series of SRMs have been reviewed (5). Another sphere of activity in analytical standards is the geochemical reference standards maintained by the U.S. Geological Survey and by analogous groups in France, Canada, Japan, South Africa, and the GDR (6).

Chronological standards are needed for an extremely diverse range of fields, eg, astrophysics, anthropology, archaeology, geology, oceanography, and art. The techniques employed for dating materials include dendrochronology, thermoluminescence, obsidian hydration, varve deposition, paleo-magnetic reversal, fission tracks, racemization of amino acids, and a variety of techniques related to the presence or decay of radioactive species, eg, ^{14}C, ^{10}Be, ^{18}O, and various daughter products of the U and Th series. Since the time periods of interest range from decades to millions of years and the available materials may be limited, no one technique presents a general solution. Although some progress has been made on age standardization and calibration through the efforts of the Sub-Commission on Geochronology of the International Union of Geological Sciences, much remains to be done (7). A great benefit could be extended to a broad range of scientific and cultural communities by the establishment of a physical bank of chronological standards that are analogous to those standards set by the NBS and the U.S. Geological Survey for compositional and physical properties.

Standard Reference Data. In addition to standard reference materials, the materials scientist or engineer frequently requires access to standard reference data. Such information helps to identify an unknown material, describe a structure, calibrate an apparatus, test a theory, or draft a new standard specification. Data are defined as that subset of scientific or technical information that can be represented by numbers, graphs, models, or symbols. The term, standard reference data, implies a data set or collection that has passed some screening and evaluation by a competent body and

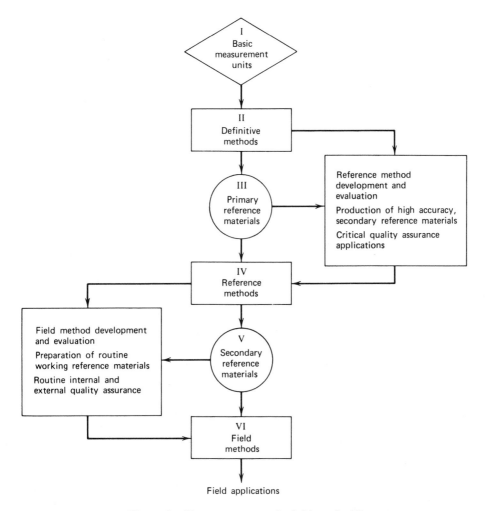

Figure 2.   Measurements standards hierarchy (4).

warrants their imprimatur and promotion. Such a data set may be generated expressly for this purpose by especially careful measurements made on a standard reference material or other well-characterized material, eg, the series of standard x-ray diffraction patterns generated by the NBS. In some cases, a reference data set may not represent any specific set of real experimental observations but a recommended, consistent set of stated reliability that is synthesized from limited, fragmentary, and conflicting literature data by review, analysis, adjustment, and interpolation (8). A convenient reference summarizing and classifying data sources relevant to materials has been published (9).

Standards for Nondestructive Evaluation (NDE). Nondestructive evaluation standards are important in materials engineering in evaluating the structure, properties, and integrity of materials and fabricated products. Such standards apply to test methods, artifactual standards for test calibration, and comparative graphical or pictorial references. These standards may be used as inspection guides, to define terms describing defects, to describe and recommend test methods, for qualification and certification of individuals and laboratories working in the NDE field, and to specify materials and apparatus used in NDE testing. NDE standards have been reviewed with regard to what standards are available, what are satisfactory, what are lacking, and what need improvement (10). Other references include useful compilations of standards and specifications in NDE (11–12) (see Nondestructive testing).

Traceability. Measurements are traceable to designated standards if scientifically rigorous evidence is produced on a continuing basis to show that the measurement process is producing data for which the total measurement uncertainty is quantified relative to national or other designated standards through an unbroken chain of comparisons (13). The intent of traceability is to assure an accuracy level sufficient for the need of the product or service. Although calibration is an important factor, measurement traceability also requires consideration of measurement uncertainty that arises from random error, ie, variability within the laboratory, and from systematic error of that laboratory relative to the reference standard. Although the ultimate metrological standards are those of mass, length and time, maintained at the NBS and related to those defined by international standardizing bodies (see below), there are literally thousands of derived units, for only a fraction of which primary reference standards are maintained.

Traceability also is used by materials engineers for the identification of the origin of a material. This attribution often is necessary where knowledge of composition, structure, or processing history is inadequate to assure the properties required in service. Thus, critical components of industrial equipment may have to be related to the particular heat of the steel that is used in the equipment apart from having to meet the specification. The geographical derivation of certain ores and minerals often is specified where analytical measurements are inadequate, and much recycled materials can be used only if traceability to the original form can be established.

Basic Standards for Chemical Technology. There are many numerical values that are standards in chemical technology; a brief review of a few basic and general ones is given below. Numerical data and definitions quoted in this section are taken from refs. 14–16 (see also Units) and are expressed in the International System of Units (SI).

Atomic Weight. The present definition of atomic weights (1961) is based on ^{12}C, which is the most abundant isotope of carbon and whose atomic weight is defined as exactly 12 (see Isotopes).

Temperature. Temperature is the measurement of the average kinetic energy, resulting from heat agitation, of the molecules of a body. The most widely used scale, ie, Celsius, uses the freezing and boiling points of water as defining points. The ice point is the temperature at which macroscopic ice crystals are in equilibrium with pure liquid water under air that is saturated with moisture at standard atmospheric pressure (101.325 kPa). One degree on the Celsius scale is 1.0% of the range between the melting and boiling points of water. The unit of thermodynamic temperature is the Kelvin, defined as 1/273.16 of the thermodynamic temperature of the triple point of water. The relation of the Kelvin and Celsius scales is defined by the International Practical Temperature Scale of 1968 (17) which is related to the temperature standard reference points published by ASTM (16) and others. The international temperature scale between −190 and 660°C is based on a standard platinum resistance thermometer and the following formula for the resistance R as a function of temperature t

$$\text{below } 0°C \quad R_t = R_o \left[1 + At + Bt^2 + C(t - 100)t^3\right]$$

$$\text{above } 0°C \quad R_t = R_o \left[1 + At + Bt^2\right]$$

where A, B, and C are arbitrary constants (see Temperature measurement).

Pressure. Standard atmospheric pressure is defined to be the force exerted by a column of mercury 760 mm high at 0°C. This corresponds to 0.101325 MPa or 14.695 psi. Reference or fixed points for pressure calibration exist and are analogous to the temperature standards cited above (18). These are based on phase changes or resistance jumps in selected materials. For the highest pressures, the most reliable technique is the correlation of the wavelength shift $\Delta\lambda$ with pressure of the ruby R_1 fluorescence line and is determined by simultaneous specific-volume measurements on cubic metals correlated with isothermal equations of state which are derived from shock-wave measurements (19). This calibration extends from 6–100 GPa (0.06–1 Mbar) and may be represented

$$P(\text{Mbar}) = \frac{(19.04)}{5} \left\{\left[\frac{\lambda_o + \Delta\lambda}{\lambda_o}\right]^5 - 1\right\}$$

where λ_o is the wavelength measured at 100 kPa (1 bar) (see Pressure measurement).

Length. One meter is defined as exactly 1,650,763.73 wavelengths of the radiation in vacuum corresponding to the unperturbed transition between the levels $2p_{10}$ and $5d_5$ of krypton-86 (the orange-red line).

Mass. The unit of mass is the kilogram and is the mass of a particular cylinder of Pt–Ir alloy which is preserved in France by the International Bureau of Weights and Measures.

Time. The unit of time in the International System of units is the second: "The second is the duration of 9,197,631,770 periods of the radiation corresponding to the transition between the two hyperfine levels of the fundamental state of the atom of cesium-133" (20). This definition is experimentally indistinguishable from the ephemeris-second which is based on the earth's motion.

Standard Cell Potential. A very large class of chemical reactions are characterized by the transfer of protons or electrons. Substances losing electrons in a reaction are said to be oxidized, those gaining electrons are said to be reduced. Many such reactions can be carried out in a galvanic cell which forms a natural basis for the concept of the half-cell, ie, the overall cell is conceptually the sum of two half-cells, one corresponding to each electrode. The half-cell potential measures the tendency of one reaction, eg, oxidation, to proceed at its electrode; the other half-cell of the pair measures the corresponding tendency for reduction to proceed at the other electrode. Measurable cell potentials are the sum of the two half-cell potentials. Standard cell potentials refer to the tendency of reactants in their standard state to form products in their standard states. The standard conditions are 1 M concentration for solutions, 101.325 kPa (1 atm) for gases, and for solids, their most stable form at 25°C. Since half-cell potentials cannot be measured directly, numerical values are obtained by assigning the hydrogen gas–hydrogen ion half reaction the half-cell potential of zero V. Thus, by a series of comparisons referred directly or indirectly to the standard hydrogen electrode, values for the strength of a number of oxidants or reductants can be obtained (21), and standard reduction potentials can be calculated from established values (see also Electrochemical processing; Batteries).

Standard cell potentials are meaningful only when they are calibrated against an emf scale. To achieve an absolute value of emf, electrical quantities must be referred to the basic metric system of mechanical units. If the current unit A and the resistance unit Ω can be defined, then the volt may be defined by Ohm's law as the voltage drop across a resistor of one standard ohm (Ω) when passing one standard ampere (A) of current. In the ohm measurement, a resistance is compared to the reactance of an inductor or capacitor at a known frequency. This reactance is calculated from the measured dimensions and can be expressed in terms of the meter and second. The ampere determination measures the force between two interacting coils while they carry the test current. The force between the coils is opposed by the force of gravity acting on a known mass; hence, the ampere can be defined in terms of the meter, kilogram, and second. Such a means of establishing a reference voltage is inconvenient for frequent use and reference is made to a previously calibrated standard cell.

Ideally, a standard cell is constructed simply and is characterized by a high constancy of emf, a low temperature coefficient of emf, and an emf close to one volt. The Weston cell, which uses a standard cadmium sulfate electrolyte and electrodes of cadmium amalgam and a paste of mercury and mercurous sulfate, essentially meets these conditions. The voltage of the cell is 1.0183 V at 20°C. The a-c Josephson effect, which relates the frequency of a superconducting oscillator to the potential difference between two superconducting components, is used by the NBS to maintain the unit of emf, but the definition of the volt remains the Ω/A derivation described above (see Superconducting materials).

Concentration. The basic unit of concentration in chemistry is the mole which is the amount of substance that contains as many entities, eg, atoms, molecules, ions, electrons, protons, etc, as there are atoms in 12 g of ^{12}C, ie, Avogadro's number $N_A = 6.022045 \times 10^{23}$. Solution concentrations are expressed on either a weight or volume basis. Molality is the concentration of a solution in terms of the number of moles of solute per kilogram of solvent. Molarity is the concentration of a solution in terms of the number of moles of solute per liter of solution.

A particular concentration measure of acidity of aqueous solutions is pH which, usually, is regarded as the common logarithm of the reciprocal of the hydrogen-ion concentration (qv). More precisely, the potential difference of the hydrogen electrode in normal acid and in normal alkali solution (-0.828 V at 25°C) is divided into 14 equal parts or pH units; each pH unit is 0.0591 V. Operationally, pH is defined by pH = pH (soln) + E/K, where E is the emf of the cell:

$$H_2 | \text{solution of unknown pH} \| \text{saturated KCl} \| \text{solution of known pH} | H_2$$

and $K = 2.303 \, RT/F$, where R is the gas constant, 8.314 J/(mol/K) (1.987 cal/(mol·K)), T is the absolute temperature, and F is the value of the Faraday, 9.64845×10^4 C/mol. pH usually is equated to the negative logarithm of the hydrogen-ion activity, although there are differences between these two quantities outside the pH range 4.0–9.2:

$$-\log q_{H^+} m_{H^+} = \text{pH} +0.014 \, (\text{pH} - 9.2) \text{ for pH} > 9.2$$

$$-\log q_{H^+} m_{H^+} = \text{pH} +0.009 \, (4.0 - \text{pH}) \text{ for pH} < 4.0$$

Energy. The SI unit of energy is the joule which is the work done when the point of application of a force of one newton is displaced a distance of one meter in the direction of the force. The newton is that force which, when applied to a body having a mass of one kilogram, accelerates a body one meter per second squared. The calorie is the quantity of heat absorption of water per gram per degree Celsius at 15°C and it is equal to 4.184 J (see Units).

Specifications

Objectives and Types. A specification establishes assurance of the fitness of a material, product, process, or service for use. Such fitness usually encompasses safety and efficiency in use as well as technical performance. Material specifications may be classified as to whether they are applied to the material, the process by which it is made, or the performance or use that is expected of it. Product or design specifications are not relevant to materials. Within a company, the specification is the means by which the engineering function conveys to the purchasing function what requirements it has for the material to be supplied to manufacturing. It has its greatest utility prior to and at the time of purchase. Yet a properly written and dated specification with accompanying certificates of test, heat or lot numbers, vendor identification, and other details pertinent to the actual material procurement constitutes an important archival document. Material specification records provide information regarding a proven successful material that can be used in a new product. Such records also are useful in the rebuilding of components and as defense evidence in a liability suit.

Content. Although formats of materials specifications may vary according to the need, the principal elements are title, statement of scope, requirements, quality assurance provisions, applicable reference documents, preparations for delivery, notes, and definitions. The scope statement comprises a brief description of the material, possibly its intended area of application, and categorization of the material by type, subclass, and quality grade. Requirements may include chemical composition, physical properties, processing history, dimensions and tolerances, and/or finish. Quality assurance factors are test methods and equipment including their precision; accuracy and repeatability; sampling procedures; inspection procedures, ie, acceptance and rejection criteria; and test certification. Reference documents may include citation of well-established specifications, codes and standards, definitions and abbreviations, drawings, tolerance tables, and test methods. Preparations for delivery are the instructions for packing, marking, shipping mode, and unit quantity of material in the shipment. The notes section is intended for explanations, safety precautions, and other details not covered elsewhere. Definitions are specifications of terms in the document which differ from common usage.

Strategy and Implementation. Great reliance used to be placed on compositional specifications for materials, and improvements in materials control were sought by increasing the number of elements specified and by decreasing the allowable latitude, eg, maximum, minimum, or range, in their concentration. However, the approach is fallible: the purchaser assumes enough knowledge about materials behavior to completely and unerringly associate the needed properties with composition; analyses must be made by purchaser, vendor, or both for each element specified; and the purchaser bears responsibility for materials failures when compositional requirements have been met. Property requirements alone or in combination with a less exacting compositional specification usually is a more effective approach. For example, the engineer may specify a certain class of low alloy steel and call for a particular hardenability but leave the vendor considerable latitude in determining composition to achieve the desired result.

The most effective specification is that which accomplishes the desired result with the fewest requirements. Properties and performance should be emphasized rather than how the objectives are to be achieved. Excessive demonstration of erudition on the part of the writer or failure to recognize the usually considerable processing expertise held by the vendor results in a lengthy and overly detailed document that generally is counterproductive. Redundancy may lead to technical inconsistency. A requirement that cannot be assessed by a prescribed test method or quantitative inspection technique never should be included in the specifications. Wherever possible, tests should be easy to perform and highly correlatable with service performance. Tests that indicate service life are especially useful. Standard test references, eg, ASTM methods, are the most desirable, and those that are needed should be selected carefully and the numbers of such references should be minimized. To eliminate unnecessary review activity by the would-be complier, the description of a standard test should not be paraphrased or condensed unless the original test is referenced.

Effective specification control often can be established other than through requirements placed on the end-use material, ie, the specification may bear upon the raw materials, the process used to produce the material, or ancillary materials used in its processing. Related but supplementary techniques are approved vendor lists, accredited testing laboratories, and preproduction acceptance tests.

Economic Aspects

The costs and benefits associated with standardization are determined by direct and indirect effects. A proper assessment depends on having suitable base-line data with which to make a comparison. Several surveys have shown typical dollar returns for the investment in standardization in the range of 5:1–8:1 with occasional claims made for a ratio as high as 50:1.

Savings include reduced costs of materials and parts procurement; savings in production and drafting practice; reduction in engineering time, eg, design, testing, quality control, and documentation; and reduction in maintenance, field service, and in-warranty repairs. It is curious fact that in most companies a very small number of individuals are authorized to write checks on the company's funds, but a large number of people are permitted to specify materials, parts, processes, services, etc, which just as definitely commit company resources. Furthermore, actions involving specification and standard setting frequently lack adequate control and may not be monitored regularly. Thus, awareness, appreciation, and involvement of top management in any industrial standardization program are essential to its success.

The DOD estimates conservatively that materials and process specifications represent almost 1% of the total hardware-acquisition costs. The operation of a single ASTM committee dealing with engine coolants has been estimated at over \$150,000/yr (2). Costs of generation of a single company specification range from a few hundred to several thousand dollars. The total U.S. cost for material and process specifications is greater than 3×10^8. Since these costs can be so great it is imperative to ensure that monies are not spent unwisely in the specification and standardization field. Although there are justifiable instances for "specials" or documents intended to fill the needs of an individual company or other institution, savings will usually be realized by adopting a standard already established by an organization at a higher hierarchical level—trade association, national, international.

The ideal specification regards only those properties required to assure satisfactory performance in the intended application and properties that are quantitative and measurable in a defined test. Excessively stringent requirements not only involve direct costs for compliance and test verification but constitute indirect costs by restricting the sources of the material. Reducing the margin between the specification and the production target increases the risk that an acceptable product is unjustly rejected because the test procedure gives results that vary from laboratory to laboratory. A particularly effective approach is to recognize within a specification or related set of specifications the different levels of quality or reliability required in different applications. Thus, *Military Handbook V* recognizes class A, class B, and class S design allowables where, on the A basis ≥99% of values are above the designated level with a 95% confidence; on the B basis, ≥90% are above with a 95% confidence; and, on the S basis, a value is expected which exceeds the specified minimum (22).

From the customer's point of view, there is an optimal level of standardization. Increased standardization lowers costs but restricts choice. Furthermore, if a single minimal-performance product standard is rigorously invoked in an industry, competition in a free market ultimately may lead the manufacturer of a superior product to save costs by lowering his product quality to the level of the standard, thus denying other values to the customer. Again, excessive standardization, especially as applied to design or how the product performance is to be achieved, effectively can limit technological innovation.

Legal Aspects

The increasing incidence of class action suits over faulty performance, the trend toward personal accountability and liability, and the increasing role of consumerism have affected standardization. Improvement in the technical quality of standards, the involvement of all of the possible stakeholders in their creation, and their endorsement by larger standardizing bodies help to minimize the legal exposure of the individual engineer or of his company. A particular embodiment of these attitudes is the certification label, ie, a symbol or mark on the product indicating that it has been produced according to the standards of a particular organization. The Underwriters Laboratories seal on electrical equipment is a familiar indication that the safety features of the product in question have met the exacting standards of that group. Similarly, the symbol of the International Wool Secretariat on a fabric attests to the fiber content and quality of that material, and the API monogram on piping, fittings, chain, motor-oil cans, and other products carries analogous significance.

Antitrust laws sometimes have been invoked in opposition to the collaborative activities of individual companies or private associations, eg, ASTM, in the development of specifications and standards. Although such activities should not constitute restraint of trade, they must be conducted so that the charge can be refuted. Therefore, all features of due process proceedings must be observed. Actions aimed at strengthening the voluntary standards system have begun (23). A recommended national standards policy has been generated by an advisory committee that was initiated by, but is independent of, the ANSI (24). The Federal Office of Management and Budget issued a circular establishing a uniform policy for federal participation and the use of voluntary standards (25). In general, the circular calls for federal agency participation in the development, production, and coordination of voluntary standards and encourages the use, whenever possible, of applicable voluntary standards in Federal procurement.

Education

Since few universities or engineering schools incorporate much explicit treatment of the subject of specifications and standards in their curricula, resort must be made to seminars, workshops, and short courses sponsored by professional societies and trade associations. Far too little is being done in this area. The needed training encompasses sources of information in the field, how to prepare specifications and standards, how to tailor requirements for cost effectiveness, and the cross-referencing and correlation of specifications and standards.

Outlook

International trade is increasing rapidly in volume and in its significance to individual national economies and, thus, will require more extensive adoption of international standards and, in the immediate future, more extensive cross-referencing of equivalent national specifications. International trade also is increasing in complexity as well as volume. No longer is international trade comprised principally of raw materials sold by undeveloped countries to industrialized countries in exchange for manufactured products. Consider for example, the U.S.-designed car that is equipped with a German engine and French tires, built in part from Japanese steel and Dutch plastics, and the needs that the composite implies for materials standardization. Further, there is always the hazard of the intentional or unintentional use of standards as technical barriers to trade. The recent international treaty, the General Agreement on Tariffs and Trade (GATT Code) is intended primarily to prevent just such a possibility. The implications of this Code for the predominantly voluntary standards programs used in the United States are reviewed in ref. 26.

New technology, eg, nuclear energy, introduces demands for new and better standards and specifications. Extension of temperature and pressure capabilities in the laboratory and the factory demand new accepted standards of calibration. Pressure equipment in the GPa range and tokamak nuclear fusion apparatus with 10^6 °C operating temperatures are being built but the state of the art of standards in these fields is far behind such values (see Fusion energy). Microminiaturization of the active components of electronic equipment requires updated standards for smoothness and dimensional and compositional measurement and control within tens of nanometers. Standards and specifications work may soon affect the biomaterials field, eg, regarding laboratory-created microorganisms (see Genetic engineering). The current and future impact of the computer on standardization is realized through enormously increased automation with computer-aided design and computer-aided manufacturing (CAD/CAM) and the integration of testing with automated manufacturing.

New environments, a selection of which is presented in Table 1, and in which all the usual engineering functions must be performed, also pose problems and opportunities for material standards. Sensors must measure the attributes of the new environment, construction materials must withstand new exposure regimes, and new performance criteria must be specified.

As the SI system is adopted increasingly worldwide, the elimination of confusion from contemporary but distinctly different French, German, Italian, etc, metric units will be realized. Only Brunei, Burma, Yemen, and the United States have not formally

Table 1. Characteristics of Some New Environments[a]

Space	Ocean	Human body	Nuclear reactor	Laboratory
extreme vacuum	high pressure	moist	high temperature	high temperature
radiation	nearly constant	complex and	neutron flux	high pressure
nonpenetrating	temperature	diverse electro-	reactive coolants	high magnetic fields
penetrating	saline water	chemistry of	radioactive	plasmas
temperature	silt and colloidal	various body	sources	
ascent	suspensions	fluids	high thermal flux	
reentry	marine life	complex flexural	inaccessibility	
ambient	mechanical	behavior		
lack of normal	instability	multicomponent		
gravitational	waves	composite,		
field	tides	highly damped		
micrometeorites	currents	in the		
		mechanical		
long-term missions	opaque to EM[b]	and electrical		
inaccessibility	radiation	sense		
	inaccessibility	multielement		
		constitution of		
		body fluids		
		gases		
		wastes		
		nutrients		
		antibodies		
		hormones		
		enzymes		
		reactive to foreign		
		materials		
		inaccessibility		

[a] Ref. 27.
[b] EM = electromagnetic.

adopted the SI as the predominant national system of metrology. The participation of the United States in the metrication movement has been favored by the passage of the Metric Acts of 1866 and 1975 and the subsequent establishment of the American National Metric Council (private) and the U.S. Metric Board (public) to plan, coordinate, monitor, and encourage the conversion process (16) (see Units and conversion factors).

Changing environments of civilization introduces the increased requirement for quality and reliability in all products and especially in those of high dollar value and in components of highly integrated technological systems. Increasing concern over the environment and safety issues has led to new standards for exposure of organisms to certain materials or to noise or electromagnetic radiation (see Industrial hygiene and toxicology; Noise pollution). The decreasing availability of natural resources forces industry to make use of leaner ores and more frugal applications of materials in short supply; both will result in new analytical standards and compositional specifications. The use of specifications in coping with problems of residual and additive elements in both virgin and recycled materials has been reviewed (28) (see Recycling).

Sources

There are many hundreds of standards-making bodies in the United States. These comprise branches of state and Federal government, trade associations, professional and technical societies, consumer groups, and institutions in the safety and insurance fields. The products of their efforts are heterogeneous, reflecting parochial concerns and different ways of standards development. However, by evolution, blending, and accreditation by higher level bodies, many standards originally developed for private purposes eventually become *de facto*, if not official, national standards. Individuals seeking access to standards and specifications are referred to the directories listed in refs. 29–30. Selected references, principally from these two sources, that are especially relevant to chemistry and chemical technology, are listed below (see also Information retrieval).

Equipment and Instrumentation Standards

Instrument Society of America
400 Stanwix Street
Pittsburgh, Pa. 15222
Standards and Practices for Instrumentation 5th ed., 1977. Instrumentation standards and recommended practices abstracted from those of 19 societies, the U.S. Government, the Canadian Standards Association, and the British Standards Institute. Covers control instruments, including rotameters, annunciators, transducers, thermocouples, flowmeters, and pneumatic systems.

American Institute of Chemical Engineers
345 East 47th Street
New York, N.Y. 10017
Standard testing procedures for plate distillation columns, evaporators, solids mixing equipment, mixing equipment, centrifugal pumps, dryers, absorbers, heat exchangers, etc.

Scientific Apparatus Makers Association
1140 Connecticut Ave., NW
Washington, D.C. 20036
Standards for analytical instruments, laboratory apparatus, measurement and test instruments, nuclear instruments, optical instruments, process measurement and control, and scientific laboratory furniture and equipment.

General Sources

American National Standards Institute (ANSI)
1430 Broadway
New York, N.Y. 10018
ANSI, previously the American Standards Association and the United States of America Standards Institute, is the coordinator of the U.S. federated national standards system and acts by: assisting participants in the voluntary system to reach agreement on standards needs and priorities, arranging for competent organizations to undertake standards development work, providing fair and effective procedures

for standards development, and resolving conflicts and preventing duplication of effort.

Most of the standards-writing organizations in the United States are members of ANSI and submit the standards that they develop to the Institute for verification of evidence of consensus and approval as American National Standards. There are ca 11,000 ANSI-approved standards, and they cover all types of materials from abrasives to zirconium as well as virtually every other field and discipline. Presently, ANSI adopts the standard number of the developing organization, eg, ASTM, unless they (ANSI) have sponsored the standards coordinating activity. ANSI also manages and coordinates participation of the U.S. voluntary-standards community in the work of nongovernmental international standards organizations and serves as a clearinghouse and information center for American National Standards and international standards (31).

MTS Systems Corp.
Box 24012
Minneapolis, Minn. 55424
Standards Cross-Reference List, 2nd ed., 1977 includes standards issued by an agency but adopted and renumbered or redesignated by another one; compiled and cross-referenced to aid in their location and identification.

The American Society for Testing and Materials (ASTM)
1916 Race Street
Philadelphia, Pa. 19103
The ASTM *1980 Annual Book of ASTM Standards* comprises over 48,000 pages and contains all currently formally approved (ca 6000) ASTM standard specifications, test methods, classifications, definitions, practices, and related materials, eg, proposals. These are arranged in 48 parts as follows:
Part 1—Steel piping, tubing, and fittings.
Part 2—Ferrous castings, ferroalloys.
Part 3—Steel: plate, sheet, strip, and wire, metallic coated products, fences.
Part 4—Steel, structural, reinforcing, pressure vessel, railway, fasteners.
Part 5—Steel-bars, forgings, bearings, chain, springs.
Part 6—Copper and copper alloys.
Part 7—Die-cast metals: aluminum and magnesium alloys.
Part 8—Nonferrous metals: nickel, lead, tin alloys; precious, primary, and reactive metals.
Part 9—Electrodeposited coatings, metal powders, sintered P/M structural parts.
Part 10—Metals: physical, mechanical, corrosion testing.
Part 11—Metallography, nondestructive testing.
Part 12—Chemical analysis of metals and metal-bearing ores.
Part 13—Cement, lime, ceilings and walls, manual of cement testing.
Part 14—Concrete and mineral aggregates, manual of concrete testing.
Part 15—Road: paving bituminous materials, traveled surface characteristics.
Part 16—Chemical-resistant materials: vitrified clay and concrete, asbestos, cement products, mortars, masonry.

Part 17—Refractories, glass, ceramic materials, carbon and graphite products.

Part 18—Thermal insulation, building seals and sealants, fire tests, building constructions, environmental acoustics.

Part 19—Soil and rock: building stones, peats.

Part 20—Paper: packaging, business-copy products.

Part 21—Cellulose, leather, flexible barrier materials.

Part 22—Wood, adhesives.

Part 23—Petroleum products and lubricants (I) D 56–D 1660.

Part 24—Petroleum products and lubricants (II) D 1661–D 2896.

Part 25—Petroleum products and lubricants (III) D 2891—latest; aerospace materials.

Part 26—Gaseous fuels, coal and coke, atmospheric analysis.

Part 27—Paint: tests for formulated products and applied coatings.

Part 28—Paint: pigments, resins, and polymers.

Part 29—Paint: fatty oils and acids, solvents, miscellaneous, aromatic hydrocarbons, naval stores.

Part 30—Soap, coolants, polishes, halogenated organic solvents, activated carbon, industrial chemicals.

Part 31—Water.

Part 32—Textiles: yarns, fabrics, and general test methods.

Part 33—Textiles: fibers, zippers.

Part 34—Plastic pipe and building products.

Part 35—Plastics: general test methods, nomenclature.

Part 36—Plastics: materials, film, reinforced and cellular plastics, high modulus fibers and composites.

Part 37—Rubber: natural and synthetic, general test methods, carbon black.

Part 38—Rubber products: industrial specifications and related test methods, gaskets, tires.

Part 39—Electrical insulation: test methods, solids, and solidifying fluids.

Part 40—Electrical insulation: specifications, test methods for solids, liquids, and gases, protective equipment for liquids and gases.

Part 41—General test methods, nonmetal: laboratory apparatus, statistical methods, space simulation, durability of nonmetallic materials.

Part 42—Analytical methods: spectroscopy, chromatography, computerized systems.

Part 43—Electronics.

Part 44—Magnetic properties; metallic materials for thermostats, electrical resistance heating, and contacts; temperature measurement; illuminating standards.

Part 45—Nuclear standards.

Part 46—End-use and consumer products.

Part 47—Test methods for rating motor, diesel, and aviation fuels.

Part 48—Index: subject index, numeric list.

General Services Administration
Federal Supply Service
18th and F Streets
Washington, D.C. 20406
Publishes *Index of Federal Specifications and Standards*, *41CFR 101-29.1.*

Defense Supply Agency

Publishes *Department of Defense Index of Specifications and Standards*, a monthly with annual accumulations; available from Superintendent of Documents, GPO, Washington, D.C. 20402.

Global Engineering Documentation Services, Inc.
3301 W. MacArthur Boulevard
Santa Ana, Calif. 92704
An information broker, not an issuer of standards. The world's largest library of government, industry, and technical society specifications and standards, including obsolete documents dating from 1946. Publishes an annual *Directory of Engineering Documentation Sources*.

National Bureau of Standards (NBS)
Standards Information & Analysis Section
Standards Information Service (SIS)
Bldg. 225, Room B162
Washington, D.C. 20234
Maintains a reference collection on standardization, engineering standards, specifications, test methods, recommended practices, and codes obtained from U.S., foreign, and international standards organizations. Publishes various indexes and directories, including: *(1) An Index of U.S. Voluntary Engineering Standards*, NBS Special Publication 329, 1971, Supplement 1, 1972, and Supplement 2, 1975; *(2) World Index of Plastics Standards*, NBS Special Publication 352, 1971; *(3) Tabulation of Voluntary Standards and Certification Programs for Consumer Products*, NBS Technical Note 948, 1977; *(4) An Index of State Specifications and Standards*, NBS Special Publication 375, 1973; *(5) Directory of United States Standardization Activities*, NBS Special Publication 417, 1975; *(6) Index of International Standards*, NBS Special Publication 390, 1974; *(7) Index of U.S. Nuclear Standards*, 1977; and *(8) Directory of Standards Laboratories*, 1971.

National Standards Association, Inc.
5161 River Road
Washington, D.C. 20016

Visual Search Microfilm Files (VSMF)
Information Handling Services
15 Inverness Way East
Engelwood, Col. 80150
VSMF carries government specifications, ASTM, AMS, and many other specifications and standards. Copies of these may be obtained on an individual basis or broad categories of this service may be obtained on a subscription basis.

Journals

ANSI Reporter and Standards Action
American National Standards Institute
The biweekly *ANSI Reporter* provides news of policy-level actions on standardization taken by ANSI, the international organizations to which it belongs, and

the government. *Standards Action*, also biweekly, lists for public review and comment standards proposed for ANSI approval. It also reports on final approval actions on standards, newly published American National Standards, and proposed actions on national and international technical work. These two publications replace *The Magazine of Standards* which ANSI, formerly The American Standards Association, discontinued in 1971.

ASTM Standardization News (formerly *Materials Research and Standards* and, earlier, *ASTM Bulletin*)
American Society for Testing and Materials
A monthly bulletin which covers ASTM projects, national and international activities affecting ASTM, reports of new relevant technology, and ASTM letter ballots on proposed standards.

Journal of the American Society of Safety Engineers
American Society of Safety Engineers
850 Busse Hwy.
Park Ridge, Ill. 60068
A monthly that reviews safety standards.

Journal of Research of National Bureau of Standards
The journal is published in four parts: (*1*) physics and chemistry, (*2*) mathematics and mathematical physics, (*3*) engineering and instrumentation, and (*4*) radio science.

Journal of Physical and Chemical Reference Data
American Chemical Society
1155 16th St., NW
Washington, D.C. 20036: quarterly

Journal of Testing and Evaluation
American Society for Testing and Materials
A bimonthly in which data derived from the testing and evaluation of materials, products, systems, and services of interest to the practicing engineer are presented. New techniques, new information on existing methods, and new data are emphasized. It aims to provide the basis for new and improved standard methods and to stimulate new ideas in testing.

Metrologia
International Committee of Weights and Measures (CIPM)
Pavillon de Breteuil
Parc de St. Cloud, France
Includes articles on scientific metrology worldwide, improvements in measuring techniques and standards, definitions of units, and the activities of various bodies created by the International Metric Convention.

Standards Engineering
Standards Engineering Society
6700 Penn Avenue South
Minneapolis, Minn. 55423

A bimonthly in which general news and technical articles dealing with all aspects of standards and U.S. and foreign articles on standard materials and calibration and measurement standards are presented.

Materials

Biochemical Compounds
National Research Council
Committee on Biological Chemistry
National Academy of Science
Washington, D.C. 20418
Specifications and Criteria for Biochemical Compounds, 2nd ed., 1967.

Carbides
Cemented Carbide Producers Association
712 Lakewood Center North
Cleveland, Ohio 44107
Standards developed by Cemented Carbide Producers Association, ie, standard shapes, sizes, grades, and designation and defect classification.

Castings
Investment Casting Institute
8521 Clover Meadow
Dallas, Texas 75243

American Die Casting Institute
2340 Des Plaines Ave.
Des Plaines, Ill. 60060

Steel Founders Society of America
20611 Center Ridge Rd.
Cast Metals Federation Bldg.
Rocky River, Ohio 44116

Cement
Cement Statistical and Technical Association
Malmo, Sweden
Review of the Portland Cement Standards of the World (1961). Describes standards in the 42 countries that have issued their own national specifications. Chemical, physical, and strength requirements of each country are reviewed.

Chemicals
Chemical Manufacturers' Association
1825 Connecticut Ave., NW
Washington, D.C. 20009
Manual of Standard and Recommended Practice for chemicals, containers, tank car unloading and related procedures.

Chemical Specialties Manufacturers Association
1001 Connecticut Ave., NW
Washington, D.C. 20036
Standard Reference Testing Materials for insecticides, cleaning products, sanitizers, brake fluids, corrosion inhibitors, antifreezes, polishes, and floor waxes.

Color Association of the United States
24 East 39th Street
New York, N.Y. 10016
Color standards for fabrics, paints, wallpaper, plastics, floor coverings, automotive and aeronautical materials, china, chemicals, dyestuffs, cosmetics, etc.

Friction Materials
Friction Materials Standards Institute
E210, Rte. 4
Paramus, N.J. 07652

Leather
Tanners' Council of America
411 Fifth Avenue
New York, N.Y. 10016
Standards for color, hide trim, leather weight or thickness.

American Leather Chemists' Association
c/o University of Cincinnati
Cincinnati, Ohio 45221
Chemical and physical test methods for leather.

Metals and Alloys
Aluminum Association
818 Connecticut Ave., NW
Washington, D.C. 20006
Standards for wrought and cast aluminum and aluminum alloy products, including composition, temper designation, dimensional tolerance, etc.

Society of Automotive Engineers (SAE)
400 Commonwealth Drive
Warrendale, Pa. 15096
SAE Handbook. An annual compilation of more than 500 SAE standards, recommended practices, and information reports on ferrous and nonferrous metals, nonmetallic materials, threads, fasteners, common parts, electrical equipment and lighting for motor vehicles and farm equipment, power-plant components and accessories, passenger cars, trucks, buses, tractor and earth-moving equipment, and marine equipment.

AMS Index. A listing of more than 1000 SAE Aerospace Material Specifications (AMS) on tolerances; quality control and process; nonmetallics; aluminum, magnesium, copper, titanium, and miscellaneous nonferrous alloys; wrought carbon steels; special purpose ferrous alloys; wrought low alloy steels; corrosion- and heat-resistant steels and alloys; cast iron and low alloy steels; accessories, fabricated parts, and assemblies; special property materials; refractory and reactive materials.

Copper Development Association
405 Lexington Avenue
New York, N.Y. 10017
Standards for wrought and cast copper and copper alloy products; a standards handbook is published with tolerances, alloy data, terminology, engineering data, processing characteristics, sources and specifications cross-index for 6 coppers and 87 copper-based alloys that are recognized as standards.

Tin Research Institute
483 West 6th Avenue
Columbus, Ohio 43201

Zinc Institute
292 Madison Avenue
New York, N.Y. 10017

Lead Industries Association
292 Madison Avenue
New York, N.Y. 10017

American Iron & Steel Institute
1000 16th Street, NW
Washington, D.C. 20036
Standards for steel compositions, steel products manufacturing tolerances, inspection methods, etc.

Ferroalloys Association
1612 K Street, NW
Washington, D.C. 20006

Metal Powder Industries Federation
PO Box 2054
Princeton, N.J. 08540

Silver Institute
1001 Connecticut Ave., NW
Washington, D.C. 20036

International Magnesium Association
c/o Bell Publicom
1406 Third National Bldg.
Dayton, Ohio 45402

Paper
Technical Association of the Pulp and Paper Industry
One Dunwoody Park
Atlanta, Ga. 30338

Tappi Standards and *Yearbook* cover all aspects of pulp and paper testing and associated standards.

American Paper Institute
260 Madison Avenue
New York, N.Y. 10016
Physical standards, sizes, gauges, definitions of paper and paperboard.

Petroleum Products
Institute of Petroleum
61 New Cavendish Street
London, W1, England
Annually publishes *Institute of Petroleum Standards for Petroleum and its Products*.

American Petroleum Institute
1801 K Street, NW
Washington, D.C. 20006
Fosters development of standards, codes, and safe practices in petroleum industries and publishes the same in its journals and reference publications.

Steam
International Association for Properties of Steam
c/o Dr. H. White
National Bureau of Standards
Washington, D.C. 20234

Treating and Finishing
Metal Treating Materials
1300 Executive Center, Suite 115
Tallahassee, Fla. 32301

National Association of Metal Finishers
111 E. Wacker Drive
Chicago, Ill. 60601

Welding
American Welding Society
2501 N.W. Seventh Street
Miami, Fla. 33125
Codes, Standards, Specifications. A complete set of codes, standards, and specifications is published by the Society and is continuously updated. Covers fundamentals, training, inspection and control, and process and industrial applications.

Wood
American Lumber Standards Committee
20010 Century Blvd.
Germantown, Md. 20767

American Wood Preservers Bureau
2740 S. Randolph Street
Arlington, Va. 22206

National Hardwood Lumber Association
332 S. Michigan Ave.
Chicago, Ill. 60604

National Standards, Worldwide. Most countries have a national standards organization that both leads the standardization activities in that country and acts within its own country as sales agent and information center for the other national standardizing bodies. In the United States, the ANSI performs that function. The organizations for the leading industrial countries of the world are as follows:

Australia: SAA, AS, Standards Association of Australia, 80-86 Arthur Street, North Sidney NSW 2060.

Austria: ON, ONORM, Oesterreichlisches Normungsinstitut, Leopoldsgasse 4, A-1021 Wien 2.

Belgium: IBN, Institut Belge de Normalisation, 29 Avenue de la Brabanconne B-1040 Bruxelles 4.

Brazil: ABNT; NB, EB, Associacao Brasileira de Normas Tecnicas, Caixa Postal 1680, Rio de Janeiro.

Canada: CSA, Standards Council of Canada, 2000 Argentia Road, Suite 2-401, Mississauga, Ontario.

China: China Association for Standardization, PO Box 820, Beijing, People's Republic of China.

Czechoslovakia: Urad pro normalizaci a mereni, Vaclavske namesti 19, 113 47 Praha 1, Czechoslovakia.

Denmark: DS, Dansk Standardiseringsraad, Aurehjvej 12, DK-29000, Hellerup.

Finland: SFS, Suomen Standardisoimisliitto, Box 205 SF-00121 Helsinki 12.

France: AFNOR, NF, Assoc. Francaise de Normalisation, Tour Europe, Cedex 7, 92080 Paris-La Defense.

FRG: DNA, DIN, Deutsches Institut fur Normung, 4-10 Burggrafenstrasse, D-1000 Berlin 30.

India: ISI, IS, Indian Standards Institution, Manak Bhavan, 9 Bahadur Shah Zafar Marg, New Delhi 110002.

Iran: Institute of Standards and Industrial Research of Iran, Ministry of Industries and Mines, PO Box 2937, Tehran.

Ireland: IIRS, I.S., Institute for Industrial Research and Standards, Blasnevin House, Ballymun Road, Dublin-9.

Israel: SII, Standards Institution of Israel, 42 University St., Tel Aviv 69977.

Italy: UNI, Ente Nazionale Italiano de Unificazione, Piazza Armando Diaz 2, 120123 Milano.

Japan: JISC, JIS, Japanese Industrial Standards Committee, Agency of Industrial Science and Technology, Ministry of International Trade and Industry, 1-3-1 Kusumigaseki Chiyoda-Ku, Tokyo 100.

Mexico: DGN, Diraccion General de Normas, Tuxpan No. 2, Mexico 7, D.F.

Netherlands: NNI, Nederlands Normalisatie-instituut, Polakweg, 5 Rijswijk (ZH)-2280.

Poland: Polski Komitet Normalizacji, Miar i Jakosci, Ul. Elektoraina 2, 00-139 Warszawa.

Romania: Institutul Roman de Standardizare, Casuta Postala 63-87, Bucarest 1.

Spain: Instituto Nacional de Racionalizacion y Normalizacion, Aurbano 46, Madrid 10.

Sweden: SIS, Standardiseringskommission i Sverige, Tegnergatan 11, Box 3295, Stockholm S 10366.

United Kingdom: BSI, BS, British Standards Institution, 2 Park Street, London W1 A 2BS, England.

USSR: GOST, Gosudarstvennyj Komitet Standartov, Soveta Ministrov S.S.S.R., Leninsky Prospekt 9b, Moskva 11 7049.

Nuclear Standards

American Nuclear Society (ANS)
555 N. Kensington Avenue
La Grange Park, Ill. 60525
Information center on nuclear standards.

American National Standards Institute (1974)
1430 Broadway
New York, N.Y. 10018
Catalog of Nuclear Industry Standards

National Bureau of Standards
Index of U.S. Nuclear Standards
W. I. Slattery, ed.
National Bureau of Standards, Special Pub. 483 (1977)
Washington, D.C.

Safety Standards

The American Society of Mechanical Engineers (ASME)
United Engineering Center
345 East 47th Street
New York, N.Y. 10017

The ASME Boiler and Pressure Vessel Code, under the cognizance of the ASME Policy Board, Codes and Standards, considers the interdependence of design procedures, material selection, fabrication procedures, inspection, and test methods that affect the safety of boilers, pressure vessels, and nuclear-plant components, whose failures could endanger the operators or the public (see Nuclear reactors). It does not cover other aspects of these topics that affect operation, maintenance, or nonhazardous deterioration.

American Insurance Association (AIA)
85 John Street
New York, N.Y. 10038

Handbook of Industrial Safety Standards, Association of Casualty and Surety Companies, New York, 1962. Compilation of industrial safety requirements based on codes and recommendations of the ANSI, the National Fire Protection Association (now part of AIA), the ASME, and several government agencies.

National Fire Protection Association
470 Atlantic Avenue
Boston, Mass. 02210
National Fire Codes.   1980 ed., issued in 16 volumes: Volume 13 is devoted exclusively to hazardous chemicals, but most other volumes have some coverage of material hazards, use of materials in fire prevention or extinguishing, hazards in chemical processing, etc. More than 200 standards are described.

American Public Health Association
1015 18th St., NW
Washington, D.C. 20036
Standard Methods for Examination of Water and Wastewater, 14th ed., 1975; *Methods of Air Sampling and Analysis*, 2nd ed., 1977.

National Safety Council
444 N. Michigan Avenue
Chicago, Ill. 60611
Industrial safety data sheets on materials and materials handling and safe operation of equipment and processes.

Underwriters Laboratories
207 East Ohio Street
Chicago, Ill. 60611
Standards for Safety is a list of more than 200 standards that provide specifications and requirements for construction and performance under test and in actual use of a broad range of electrical apparatus and equipment, including household appliances; fire-extinguishing and fire protection devices and equipment; and many other nongenerally classifiable items, eg, ladders, sweeping compounds, waste cans, and roof jacks for trailer coaches.

Safety Standards
U.S. Department of Labor
GPO
Washington, D.C. 20402
Industrial safety hazards.

American Conference of Governmental Industrial Hygienists
P.O. Box 1937
Cincinnati, Ohio 45201
Practices, analytical methods, guides to codes and/or regulations, threshold limit values.

Factory Mutual Engineering Corporation
1151 Boston-Providence Turnpike
Norwood, Mass. 02062

Standards for safety equipment, safeguards for flammable liquids, gases, dusts, industrial ovens, dryers, and for protection of buildings from wind and other natural hazards.

Code of Federal Regulations
Title 49, Transportation, Parts 100 to 199
Superintendent of Documents
GPO
Washington, D.C. 20402
Safety regulations related to transportation of hazardous materials and pipeline safety.

Code of Federal Regulations
Title 29, Occupational Safety and Health
Superintendent of Documents
GPO
Washington, D.C. 20402
Safety regulations and standards issued by OSHA.

Code of Federal Regulations
Title 40, Environmental Protection Administration
Superintendent of Documents
GPO
Washington, D.C. 20402
Safety regulations and standards issued by the EPA.

Code of Federal Regulations
Title 21, Radiological Health
Superintendent of Documents
GPO
Washington, D.C. 20402

Weights and Measures

National Conference on Weights and Measures
c/o National Bureau of Standards
Washington, D.C. 20234

National Conference on Standards Laboratories
c/o National Bureau of Standards
Boulder, Col. 80303

U.S. Metric Association
Sugarloaf Star Rte.
Boulder, Col. 80302

U.S. Metric Board
1815 N. Lynn Street
Arlington, Va. 22209

American National Metric Council
1625 Massachusetts Ave.
Washington, D.C. 20036

Metrology and Fundamental Constants
A. F. Milone and P. Giacomo, eds.
North Holland Publishing Co.
Amsterdam, 1980
Proceedings of an international course that was organized to review comprehensively metrology and to illustrate links between metrology and the fundamental constants. Status of research is presented and future work and priorities are outlined.

International Bureau of Weights and Measures
Pavillion de Breteuil
F-92310
Sevres, France

International Standardization.   International standardization began formally in 1904 with the formation of the International Electrotechnical Commission (IEC) and involves the national committees of more than 40 member nations who represent their countries' interests in electrical engineering, electronics, and nuclear energy. In 1947, the International Organization for Standardization (ISO) was formed to review standardization activities in fields other than electrical; it is comprised of more than 80 member countries. Both organizations are autonomous but they maintain a coordinating committee to answer jurisdictional questions and both occupy the same building in Geneva. With the recent rapid growth in world trade, the activities of the IEC and ISO have increased many fold. The United States is represented in ISO by ANSI and in IEC by a U.S. National Committee that is a part of ANSI. Although the two organizations are dominant in drafting documentary standards, the influence and activities of other international organizations are substantial. Among these are IUPAC; NATO; the European Economic Community (EEC); COPANT, the Pan American Standards Commission; and CODATA, a subsidiary of the International Council of Scientific Unions. In contrast to the other groups, CODATA concentrates its attention on the evaluation of data and the methodology of compilation, presentation, manipulation, and dissemination of data in all fields of science and technology. Much of its work consists of appraisal of standard data and standards for presentation of data (32). The role of the U.S. government in international standardization activities has been examined by a special ASTM task force (33). The addresses of these organizations or their subsidiary standards groups are as follows:

International Bureau of Weights and Measures
Pavillon de Brateuil
F-92310, Sevres, France

International Electrotechnical Commission (IEC)
1, rue de Varembe, 1211 Geneve 20 Switzerland/Suisse

International Organization for Standardization (ISO)
1, rue de Varembe, CH 1211 Geneve 20 Switzerland/Suisse

North Atlantic Treaty Organization (NATO)
Military Committee, Conference of National Armament Directors
1110 Brussels, Belgium

European Economic Community (EEC)
200 rue de la Loi
1049 Brussels, Belgium

COPANT
Av. Pte, Roque Soenz Pena 501
7 Piso
OF 716, Buenos Aires, Argentina

CODATA
51, Boulevard de Montmorency
75016 Paris, France

International Union for Pure and Applied Chemistry (IUPAC)
Bank Court Way, Cowley Centre
Oxford OX4 3YF, England
Among its publications in the standards field are *Manual of Symbols and Terminology for Physico-chemical Quantities and Units*, Buttersworths, London, 1970; and *Nomenclature of Inorganic Chemistry*, Buttersworths, London, 1970.

BIBLIOGRAPHY

1. *ISO Standardization Vocabulary*, Geneva, 1977 (available from ANSI).
2. N. E. Promisel and co-workers, *Materials and Process Specifications and Standards*, NMAB Report-33, Washington, D.C., 1977.
3. D. Lebel and K. Schultz, *Technos*, 4 (Apr.–June 1975).
4. G. A. Uriano, *ASTM Standardization News* **7,** 8 (Sept. 1979).
5. G. W. Latimer, Jr. in C. T. Lynch, ed., *Handbook of Materials Science*, Vol. I, CRC Press, 1974, p. 667.
6. F. J. Flanagan, *Geochim. Cosmochim. Acta* **37,** 1189 (1973).
7. R. H. Steiger and E. Jaeger, *Planet. Sci. Lett.* **36,** 359 (1977).
8. C.-Y. Ho and Y. S. Touloukian, *Proceedings 5th Biennial International CODATA Conference*, Boulder, Col., 1977, pp. 615–627.
9. J. H. Westbrook and J. D. Desai, *Ann. Rev. Mater. Sci.* **8,** 359 (1978).
10. H. Berger, "Nondestructive Testing Standards—A Review," *Symposium Report No. STP624*, ASTM, Philadelphia, Pa., 1977.
11. *Handbook for Standardization of Nondestructive Testing Methods*, Vols. I and II, MIL HDBK-33, 1974.
12. R. E. Englehardt, "Bibliography of Standards, Specifications and Recommended Practices" in *Nondestructive Testing Information Analysis Center Handbook*, Mar. 1979.
13. B. C. Belanger, *ASTM Standardization News*, 8 (Sept. 1979).
14. E. R. Cohen and co-workers, *CODATA Bull. 11*, (Dec. 1973).
15. "Quantities, Units, Symbols, Conversion Factors, and Conversion Tables" *ISO Reference 31*, 15 sections, Geneva, 1973–1979.
16. "Standard for Metric Practice," *ASTM E 380-79*, Philadelphia, Pa., 1979.

17. *Metrologia* **12,** 7 (1976).
18. F. P. Bundy and co-workers in B. D. Timmerhaus and M. S. Barber, eds., *High Pressure Science and Technology*, Vol. 1, Plenum Press, New York, 1979, pp. 773, 805.
19. H. K. Mao, P. M. Bell, J. W. Shaver, and D. J. Steinberg, *J. Appl. Phys.* **49,** 3276 (1978).
20. *Proceedings 1967 General Conference of Weights and Measures*, International Bureau of Weights and Measures, BIPM, Parc de Saint-Cloud, France.
21. J. F. Hunsburger, "Electrochemical Series" in R. C. Weast, ed., *Handbook of Chemistry and Physics*, CRC Press, Boca Raton, Fla., 1980–1981.
22. "Metallic Materials and Elements for Aerospace Vehicle Structures," *Military Handbook V*, Department of Defense, Dec. 15, 1976.
23. *The Voluntary Standards System of the United States of America—An Appraisal by the American Society for Testing and Materials*, ASTM, Philadelphia, Pa., 1975.
24. *ASTM Standardization News* **16,** 8 (May 1978); *Fed. Regist.*, 7 (Dec. 1978).
25. *ASTM Standardization News* **8,** 21 (Mar. 1980); *ANSI Rep.* **14**(2), (1980).
26. D. L. Peyton, *Implementing the GATT Standards Code*, ANSI, New York, 1979.
27. J. H. Westbrook in A. B. Bronwell, ed., *Science and Technology in the World of the Future*, John Wiley & Sons, Inc., New York, 1970, p. 329.
28. J. H. Westbrook, *Phil. Trans. Roy. Soc. London* **A295,** 25 (1980).
29. S. J. Chumas, "Directory of United States Standardization Activities," NBS SP 417, Washington, D.C., 1975.
30. E. J. Struglia, *Standards and Specifications—Information Sources*, Gale Research, Detroit, Mich., 1965.
31. *ANSI Progress Report*, PR 35, New York, 1980.
32. S. A. Rossmassler and D. G. Watson, eds., *Data Handling for Science and Technology*, North Holland Publ. Co., Amsterdam, 1980.
33. *ASTM Standardization News* **8,** 16 (Apr. 1980).

J. H. WESTBROOK
General Electric Company

MEAT PRODUCTS

Meat, an excellent source of protein, iron, and B vitamins (1), was processed as early as prehistoric times, probably by drying in the sun and later by smoking and drying over wood fires. Homer, in 850 BC, recorded procedures for smoking and salting of meat. The purpose of meat processing was to prepare products that could be stored for considerable time periods at ambient temperatures. The high salt concentration that was essential for meat preservation before the widespread use of refrigeration is no longer needed or desired. Today, meat is processed with salt, color-fixing ingredients, and seasonings in order to impart desired palatability traits to intact and comminuted meat products. Intact meat products include bacon, corned beef, ham, smoked butt, and pork hocks. Comminuted meat products include all types of sausage items. Products intermediate to these categories are sectioned, or chunked and formed meats (see Food processing).

A USDA survey of meat production is given in Table 1, per capita consumption in Table 2. The amount of meat products processed under Federal inspection in 1978 is given in Table 3 and the amount of meat products canned in the same year in Table 4.

United States per capita consumption of livestock products is shown in Figure 1 and of fats and oils in Figure 2. In the United States, fresh meat is classified by a complex grading system (3).

Curing

Meat-curing agents include sodium chloride, nitrite, ascorbate or erythorbate, and possibly sodium phosphate, sucrose, dextrose, or corn syrup, and seasonings (4). The salt content of processed meats varies 1–12%, according to the type of product. Salt is used for flavor, preservation, and extraction of myofibrillar protein, whereas nitrite promotes color development, flavor, and preservation by inhibiting the growth of microorganisms and fat oxidation. Erythorbate acts as a color stabilizer, reduces fat oxidation, and inhibits undesirable nitrite reactions. Phosphates facilitate myofibrillar protein extraction, inhibit fat oxidation, and improve color development (5). Sugars and seasonings are used principally for flavor (see Food additives).

Color Changes. The most obvious characteristic of cured meats is the development of the characteristic pink color on heat processing compared to the brown color of uncured cooked meat. The color results from a reaction between the heme protein myoglobin and nitrite. Other heme proteins, such as hemoglobin and cytochromes, react similarly but are present in much smaller amounts. The reaction sequence is given in Figure 3.

Only 20–30 ppm of nitrite are required for color development but higher concentrations are needed to maintain color.

Current regulations permit the use of sodium or potassium nitrates and nitrites in meat and poultry products. However, the status of these chemicals as food additives is in doubt because of tests indicating the possibility of carcinogenic properties. Nitrates are generally no longer used except in some dry cured products and are not permitted in pumped bacon. Nitrites are limited to 120 ppm ingoing in pumped bacon and must be accompanied by sodium ascorbate or erythorbate at no less than 550 ppm. Neither nitrites nor nitrates are permitted in baby foods (see N-Nitrosamines).

Table 1. Worldwide Meat Production[a], Thousand Metric Tons

Region and country	1960	1970	1975
North America	15,747.5	19,837.2	20,131.1
Canada	1,100.5	1,482.6	1,595.5
Mexico	585.3	983.9	1,326.6
United States	13,850.6	17,020.3	16,800.3
Central America	211.1	350.4	408.7
South America	5,279.3	7,423.7	7,802.9
Argentina	2,275.4	3,094.4	2,845.7
Brazil	1,766.4	2,538.4	3,015.0
Colombia	359.8	524.4	577.3
Uruguay	345.2	484.7	434.0
Venezuela	147.2	246.9	322.8
other countries	385.3	534.9	608.1
Western Europe	12,218.4	16,438.4	18,521.6
EEC countries	10,191.9	13,518.8	14,995.1
Belgium–Luxembourg	468.2	725.3	928.1
Denmark	759.9	911.5	970.5
France	2,644.0	3,008.5	3,400.8
Federal Republic of Germany	2,656.0	3,585.9	3,777.0
Ireland	248.1	402.9	567.7
Italy	1,023.0	1,750.0	1,798.7
Netherlands	673.6	1,014.1	1,235.3
United Kingdom	1,719.1	2,120.6	2,317.0
Austria	375.3	434.0	503.0
Finland	132.6	216.9	242.4
Greece	121.7	228.9	348.0
Spain	554.4	953.2	1,208.3
Sweden	333.0	404.3	429.0
Switzerland	240.5	338.7	382.1
other countries	269.0	343.6	413.7
Eastern Europe	3,491.8	4,479.1	6,590.5
Bulgaria	195.7	294.6	402.1
Czechoslovakia	536.3	689.7	854.6
German Democratic Republic	676.1	893.8	1,189.5
Hungary	376.5	437.0	649.7
Poland	1,155.3	1,482.8	2,609.4
Yugoslavia	551.9	681.2	885.2
USSR	6,295.4	9,161.1	10,987.7
Africa	619.5	835.9	875.2
Asia	1,985.9	3,378.6	4,114.9
Republic of China (Taiwan)	163.5	359.9	358.7
Japan	317.5	1,023.9	1,397.6
Philippines	286.6	451.4	591.9
Turkey	412.4	544.3	613.5
other countries	805.9	999.1	1,153.2
Oceania	2,087.6	2,941.7	3,449.4
Australia	1,359.5	1,948.8	2,416.1
New Zealand	728.1	992.9	1,033.3
Total	47,725.4	64,495.7	72,473.3

[a] Includes beef and veal, pork, lamb, mutton and goat meat, and horse meat on carcass weight basis.

Table 2. Per Capita Meat Consumption[a], kg/Person

Region and country	1965	1970	1975
North America			
Canada	66.3	70.9	73.2
Mexico	16.8	18.2	21.8
United States	81.0	88.1	83.1
South America			
Argentina	82.2	96.5	99.1
Brazil	24.4	25.8	27.2
Chile	28.6	31.9	28.4
Colombia	23.4	24.1	23.2
Peru	15.8	15.5	11.6
Uruguay	109.6	109.6	111.9
Venezuela	21.6	23.0	25.9
Western Europe			
EEC countries			
Belgium–Luxembourg	53.3	61.7	74.9
Denmark	57.7	61.3	53.6
France	59.6	62.7	66.2
Federal Republic of Germany	55.4	63.8	67.9
Ireland	54.6	60.8	66.5
Italy	28.3	39.7	42.2
Netherlands	45.5	48.9	55.5
United Kingdom	63.9	63.3	57.8
average	51.8	57.4	59.1
Austria	55.6	60.4	67.8
Finland	37.0	42.8	52.6
Greece	30.4	39.0	44.1
Norway	36.5	37.4	45.4
Portugal	19.9	24.2	30.2
Spain	21.8	31.3	35.9
Sweden	45.4	48.4	51.8
Switzerland	54.3	62.2	64.4
Eastern Europe			
Bulgaria	34.9	34.1	40.1
Czechoslovakia	47.7	54.7	57.5
German Democratic Republic	50.0	54.3	69.3
Hungary	39.6	42.4	50.5
Poland	37.2	41.5	72.4
Yugoslavia	24.3	28.0	37.3
USSR	33.6	37.9	44.7
South Africa	32.2	32.5	30.2
Asia			
Republic of China (Taiwan)	17.6	24.4	22.6
Iran	7.7	9.3	11.7
Israel	17.8	21.8	17.8
Japan	8.0	13.1	17.1
Philippines	12.0	12.3	13.9
Turkey	14.1	15.3	15.0
Oceania			
Australia	93.6	93.2	108.3
New Zealand	116.8	109.0	103.0

[a] Carcass weight basis, includes horse meat.

Table 3. Meat Products Processed Under Federal Inspection, 1978

Product	Thousand metric tons	Product	Thousand metric tons
cured		sausage (cont.)	
beef briskets	69	uncooked cured	5
other beef	76	*Total fresh*	*449*
pork	1914	dried	102
other meats	10	semidried	48
smoked-dried or cooked hams		*Total dried/semidried*	*150*
bone-in	57	franks and weiners	
bone-in, water added	159	regular, retail	433
semiboneless	9	regular, bulk	117
semiboneless, water added	53	with extenders, retail	38
boneless	69	with extenders, bulk	34
boneless, water added	195	with variety meats, retail	11
sectioned and formed	49	with variety meats, bulk	4
sectioned and formed, water added	93	with extenders and variety meats, retail	29
hams, dry-cured	41	with extenders and variety meats, bulk	17
Total hams	*725*	*Total franks and weiners*	*684*
pork, regular	79	bologna	
pork, water added	164	regular	265
Total other pork	*243*	with extenders	29
bacon	709	with variety meats	41
Total pork	*1676*	with extenders and variety meats	29
beef, cooked	123	*Total bologna*	*365*
beef, dried	14	liver sausage	60
other meats	129	cured meat loaves	43
fresh-frozen products		other cooked items	338
beef cuts	4767	*Total liver, loaves and other*	*441*
beef boning	3144	nonspecific loaves	85
pork cuts	5126	other formulated products	80
pork boning	1302	*Total other sausage*	*165*
other cuts	237	*Total sausage*	*2254*
other boning	126	sliced products	
steaks, chops, roasts	1475	bacon, retail	463
steaks, chops (chopped-formed)	135	bacon, bulk	180
meat patties	189	*Total bacon*	*643*
hamburger/ground beef	1299	ham	122
other fresh-frozen	323	sausage, <0.34 kg	148
convenience foods		sausage, >0.34 kg	300
pizza	220	other products	154
pies	81	*Total other sliced*	*725*
dinners	119	fats and oils	
entrees	164	lard rendered	439
other	126	lard refined	294
sausage		edible tallow	679
fresh beef	11	compound with animal fat	447
fresh pork	332		
fresh other	101		

Table 4.   Meat Products Canned Under Federal Inspection, 1978 (U.S.)

Product	Thousand metric tons
canned hams	
under 1.36 kg	5
1.36–2.72 kg	76
over 2.72 kg	42
Total	*123*
luncheon meat	127
chili con carne	152
meat stew	71
hash products	39
pasta–meat products	193
pork picnic and loins	7
vienna sausage	53
franks and weiners	1
miscellaneous sausage products	13
deviled ham	7
potted products and spreads	21
tamales	14
sliced dried beef	2
chopped beef–hamburger	4
vinegar–pickled products	9
by-products, not pickled	2
corned beef	1
all other, +20% meat	80
−20% meat	143
Total canned meats[a]	*1060*

[a] Figures exclude soup but include all government purchases.

Pickle Curing. Pickle-cured hams are generally pumped before being placed in the curing vat, that is, pickling solution (pickle) is injected into the ham in order to hasten diffusion of the curing ingredients throughout the ham. For short cures, hams and pork shoulders are generally artery- or stitch-pumped. In artery pumping, a special needle is inserted into the end of the exposed artery and pickling solution (about 15 wt % of the ham depending on product and process) is pumped into the vascular system. Pump pressures are 340–380 kPa (3.4–3.8 atm) (6).

In stitch or spray pumping, the brine enters the meat through numerous perforations in the walls of hollow needles. Most hams and bacon cured in the United States are injected with automatic, multineedle injectors through a number of fine hollow-stemmed needles arranged in a bank which automatically moves the needles into the product as it moves on conveyor belts. The amount of water added to cured meat products is determined by government regulations. Cured and smoked hams and bacon shall not contain more than uncured hams. Canned cured pork products may not contain more than 8% added water, whereas noncanned cured products may contain added water up to 10 wt % of the uncured product. To accelerate the distribution of the pickling solution within the product and improve cure uniformity, boneless hams may be tumbled or massaged after injection (4).

Dry Curing. This is an older method in which curing agents are rubbed in dry form over the surface of the cut of meat. The cuts are then stored and allowed to cure. For large cuts, the cure must be applied several times. Dry curing is now used only on specialty items such as country-cured hams and bacons.

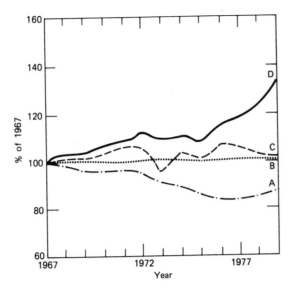

Figure 1.　United States per capita consumption of livestock products: A, eggs; B, dairy (excludes butter); C, meat; D, poultry (2).

Comminution

Comminuted meat may be cured or a fresh product. The degree of comminution varies considerably from one product to another. Sectioned or chunked and formed products may be composed of particles that weigh more than 450 g each, whereas finely comminuted meats are chopped to a pastelike texture of very small particles. Comminution equipment includes grinders, silent cutters, emulsion mills, and flaking machines. In addition to comminution, the meat is blended with other ingredients. Blenders, mixers, tumblers, and massagers are used to subject the meat protein to mechanical action in the presence of salt. This causes the salt to extract the principal myofibrillar protein, myosin, from the muscle. The extracted myosin gels when the comminuted meat is heated to form a matrix which entraps water and fat and binds the meat particles to each other (7).

Comminution reduces the raw meat material to small meat pieces, chunks, chips, or slices. Sausages are comminuted, seasoned meat products that may also be cured, smoked, molded, and heat-processed (8). They are classified according to the processing methods used for their manufacture (see Table 5).

Large particles or chunks of meat can be massaged or tumbled in the presence of salt and phosphate (usually sodium tripolyphosphate or hexametaphosphate) to extract salt-soluble proteins that form a tacky exudate which acts as a heat-set glue to bind the chunks of meat together after cooking. This method is used to prepare chunked and formed hams, roasts, and steaks. This is a growing area of production because it permits the manufacture of products that have the composition, shape, and size preferred by the consumer (10).

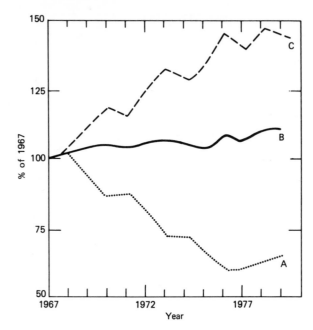

Figure 2. United States per capita consumption of fats and oils: A, animal fats (includes butter); B, total; C, vegetable oils (2).

Table 5. Classification of Sausages[a]

Classification	Products
fresh	fresh pork sausage
uncooked, smoked	smoked pork sausage, metwurst, Italian pork sausage
cooked, smoked	frankfurter, bologna, knackwurst, mortadella, berliner
cooked	liver sausage, beer salami
dry or fermented	summer sausage, cervelat, dry salamis, pepperoni
cooked meat specialties	luncheon meats, sandwich spreads

[a] Ref. 9.

Smoking

Many intact and comminuted, cured meat products are smoked to impart a desirable smoked flavor and color. The smoking process may also include a drying or cooking cycle, depending on the product.

The smoking process imparts a characteristic flavor to products. In addition, some phenolic compounds present in smoke provide protection from fat oxidation. Further protection is provided by the bacteriostatic effect of smoke components along with the drying effect that inhibits bacterial growth on the dried surface.

Modern processes use forced-air smoking chambers with close control of time,

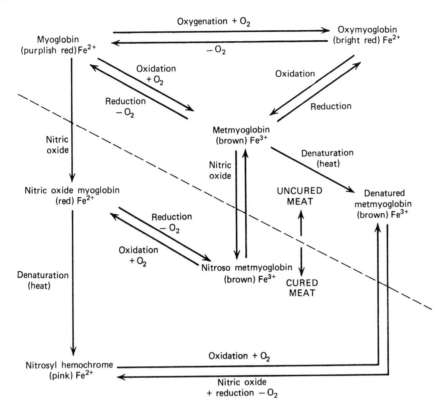

Figure 3.   Color changes taking place in uncured and cured meat products.

temperature, and humidity. The processing cycle may include predrying, smoking, cooking, drying, and cooling. Smoke is generated by electrical smoke-heated generators which offer close control over temperatures and hence smoke composition or liquid smoke may be used as an atomized spray or regenerated smoke. Smoking chambers are designed for batch or continuous processes (11).

Oil- or water-based liquid smoke can be added directly to the products as a flavoring in lieu of the smoking process. Oil-based liquid smokes are used when the product is sensitive to the low pH of water-based liquid smokes and to ensure penetration of the smoke components into the fat phase.

Canning

Canned meats may be processed to be commercially sterile or semipreserved. The objective of commercial sterilization is to destroy all harmful bacteria, or bacteria that may cause spoilage of the product under normal unrefrigerated storage. However, the process does not kill the spores of all heat-resistant bacteria. Therefore, it is essential to cool the cans rapidly after processing and avoid storage above 35°C. The most persistent type of bacteria are sporeforming organisms. The common vegetative and nonsporeforming pathogenic bacteria are killed with adequate processing and are of little or no importance in spoilage.

The amount of heat, time, and temperature required for a given degree of sterility depends upon the nature of the product, the pH, curing salts, shape and size of the can, and the type of heat-processing retort used. Some products are packed into the can hot and others cold. Hot filling eliminates the need to apply a vacuum during can closure. The vacuum is necessary to avoid excessive strain on the can and its seams during processing and to minimize oxidative spoilage of the product.

Semipreserved or pasteurized products depend on curing and chilled storage for their preservation.

Freezing

Frozen meat can be kept at low temperatures for many months. Freezing and subsequent thawing produce changes in the structure of meat that affect its physical properties. If meat is frozen very rapidly at low temperatures, the ice crystals are small and form within the fibers. The drip loss upon thawing is generally greater in slow-frozen than in quick-frozen meat. The rate of freezing is determined largely by temperature, freezing system, and size and shape of the cut. Holding a quick-frozen product at high and fluctuating temperatures encourages the growth of ice crystals and protein denaturation and the advantage of low drip loss from rapid freezing is lost. The amount of drip in meat is further affected by temperature and length of storage, meat surface, thawing rate, and the physiological condition of the muscle at slaughter and time of freezing.

Packaged frozen meat cuts are sold in the retail markets to a limited extent and stored by the consumer in home freezers. Beef retains good quality for a year or more and pork for six months if properly packaged and stored at -18 to $-23°C$. Freezer burn or dehydration and discoloration can be prevented with packaging systems that cling closely to the product surface and restrict movement of moisture from the product surface and the diffusion of oxygen into the product (12).

Dehydration

Freeze-Drying. Freeze-drying of meat results in a product with a spongelike appearance, practically devoid of moisture but resembling the original product. The composition and type of meat influence the acceptability and stability of freeze-dried meat.

Freeze-drying meat extends shelf life and reduces weight. The meat is readily defrosted by immersing in water before cooking. Under optimum processing and storage conditions, reconstituted meats have acceptable flavor, color, texture, and nutrient retention.

Freeze-drying does not entirely eliminate changes in meat products during processing and storage. Enzymic changes are reduced but not completely eliminated. Deteriorative changes include some loss of vitamins, protein denaturation, browning, and fat rancidity. To avoid deterioration, it is essential that the product is packaged under an inert gas, in moisture-, gas-, and light-impermeable packaging.

Air-Drying. Precooked comminuted lean meat is dried in a forced-air rotary or tunnel dryer to less than 10% moisture for use as an ingredient in dried soups and stews. The dried product is often compressed to reduce its volume to about one third of the fresh meat volume for shipment.

By-Products

By-products in the meat-packing industry represent a substantial part of the sales value of the production derived from the slaughter of animals. By-products include variety meats, edible and inedible fats, and hides and other inedibles. The value of meat or by-products depends upon the species and age of the animal, degree of finish, and price.

Approved portions of the carcasses of clean, sound meat animals yield edible fats including lard (from swine), edible tallow (from cattle and sheep), and related products such as oleo stock. These may be consumed in the form of shortening for baking and similar food applications, or as frying fats.

Inedible fats are used widely in animal feeds and for other industrial uses. The principal products of this type are inedible tallow and grease (see Fats and fatty oils).

Among the important nonfat by-products are hides, skins, hair, wool, dried blood, bone, glue, gelatin, casings, sutures, tennis-racket strings, pharmaceuticals, enzymes, and hormones.

The efficient utilization of by-product is essential for the economic operation of a packing plant.

Fat Extraction. *Wet or Steam Rendering.* In this procedure, the raw materials are cooked in a closed vertical tank under pressure by direct steam injection, typically at 380–500 kPa (3.8–4.9 atm) for 3–6 h. The tallow is drawn off from the top of the vessel after it has been allowed to settle for several hours. The process was used mostly for hard raw materials such as bones but is being replaced by modern methods.

Dry Rendering. Both batch and continuous processes are available for dry rendering. The material is heated in a horizontal steam-jacketed cooking vessel equipped with rotating arms to agitate the material. The moisture driven off is usually vented through a condenser to recover heat and control atmospheric pollution. At the completion of the rendering operation, the fat is separated from the cooked material (tankage) by basket centrifuges or expellers in batch processes, and by continuous centrifugation in continuous processes (see Centrifugal separation). Bones are coarsely crushed before rendering. Viscera should be washed of their contents and should be rendered with minimum delay to maximize tallow quality in terms of color and free fatty acid content. Inedible tallow is sold on the basis of color and free fatty acid content.

Low Temperature Rendering. High-quality fats are produced by semibatch or continuous processes and the defatted residues may be used in some processed meat products. The raw material is first comminuted, and then heated to about 45–50°C. The fat is then separated by centrifugation (12).

Fat Processing. After rendering, the fat is usually further processed outside the meat-processing plant. Depending on the desired product fat (13), the fats may be hydrogenated, bleached, deodorized, plasticized, interesterified or fractionated.

Unsaturation in the fatty acids is eliminated by hydrogenation in the presence of a catalyst under carefully controlled temperature, pressure, and mixing conditions. Hydrogenation raises the melting point of the fat and modifies the plastic range.

Bleaching can be achieved by liquid extraction, hydrogenation or treatment with absorbents.

Odiferous substances are removed by steam-stripping at 150–250°C under high vacuum. Free fatty acids also are removed by this process which yields a bland fat with a high smoke point.

Plasticizing imparts desirable textural properties to fats. The process often involves chilling the fat through its melting range to promote the formation of crystal nuclei, followed by mechanical working to inhibit the growth of large fat crystals. The fat is then tempered by holding it for a certain time at an appropriate temperature. Tempering inhibits subsequent crystal growth because of fluctuating temperatures below the fat's melting point.

Interesterification rearranges the distribution of fatty acids on the glycerol moiety of the fat molecule. It is used particularly in shortening manufacture where it imparts desirable baking properties to the fat.

Fractionation involves separating animal fat by fractional crystallization into oils and fats which have specific applications.

Meat Meal. The defatted tankage from dry rendering is milled to form meat and bone meal which is utilized as animal feed and sold on the basis of protein content. A minimum concentration of 50% protein defines meat and bone meal. Tankage from wet rendering must be dried by a process similar to dry rendering and then milled. The protein level and quality is generally low and it is often sold as bone meal. Unground tankage may be sold on the basis of its nitrogen content. Meat meals have to be certified with regard to specified minimum heat treatment.

Blood meal is prepared by cooking and drying blood by a variety of batch and continuous processes. Batch processes frequently use wet and dry rendering equipment. In continuous processes, the blood is heat-coagulated by steam injection as it is pumped down a pipe. The coagulated blood is centrifuged and dried, often by a ring drier. Blood meal is sold as an animal feed on the basis of nitrogen or protein content (see Pet and other livestock feeds).

Soluble dried blood, used in glue manufacture, is prepared by spray or fluidized-bed driers. Ultrafiltration may be used as a preconcentration step and the blood must be treated to prevent clotting either by adding an anticoagulant at the collection point or by homogenizing (see Glue).

Blood plasma is used to a limited extent in some sausage products. It is prepared from blood collected from clean beef or pork animals by centrifugation and may be preserved by freezing or spray-drying (14).

Organs and Glands. The importance of these materials as sources for pharmaceuticals is, perhaps, declining with the development of synthetic drugs and more convenient raw material sources. Table 6 gives materials that are still in demand and the substances extracted from them.

Organs and glands which are processed for food purposes (15), often as ethnic food specialties, include brains, hearts, kidneys, liver, spleen, sweetbreads (thymus glands, calves, and lambs), tongue, oxtails, pig's tails, pig's head, pig's ears, cheek and head trimmings, sheep's eyes, bull testes, sheep and beef stomachs (tripe), pig's stomach (sausage container), small and large intestines (sausage casings), pig's large intestine (chitterlings), blood, bones, and skin trimmings (gelatin extraction) (see Gelatin).

Hides and Skins. Cattle hides are cured for tanning by dry-salting or brine pickling (see Leather). The processes involves trimming, machine fleshing to remove any adhering fat, and washing to remove manure and dirt. In dry-salting, the hides are stacked interlayered with rock salt, about 1 kg salt per kg hide. If the cured hides are to be

Table 6.   Meat By-Products as Sources of Pharmaceuticals

Source	Product
pituitary glands	adrenocorticotrophic and other hormones
adrenal glands	epinephrine
pancreas	insulin, trypsin, chymotrypsin
ovaries	estrogens
testes	hyaluronidase
thyroid glands	L-thyroxine and calcitonin hormones
parathyroid glands	parathyroid hormones
blood	albumin, serum, and amino acids
bones	calcium and phosphorus dietary supplements
intestines	surgical sutures and heparin
stomachs of suckling calves	rennet
pigs stomachs	mucin, pepsin
bile	taurocholic acid[a]

[a] Intermediate in steroid hormone synthesis.

stored, a preservative may be added to the salt to inhibit halophilic spoilage organisms. In the pickling process, which is gaining in popularity, the pelt is tumbled in a drum with a salt and chlorinated lime (or other preservative) solution. Tumbling proceeds for about 2 h, then the hides are put aside to drain before stacking for shipment.

Sheep skins are trimmed and preserved either by curing with salt, if they are to be used as woolly skins, or treated with a sodium sulfide-and-lime depilatory paint, the wool removed, and the pelt treated again with depilatory, then washed and baited. Baiting involves treatment with ammonium salts and pancreatic enzymes. The pelts are rewashed and preserved with a salt-and-sulfuric acid pickling solution. The pulled wool is scoured and dried. The pelt is processed either in open dollies stirred by wooden paddles with the pelts being transferred between the dollies for each stage of the process or in a tumbling drum with the solutions being changed at each stage (3).

BIBLIOGRAPHY

"Meat and Meat Products" in *ECT* 1st ed., Vol. 8, pp. 825–839, by H. R. Kraybill, American Meat Institute Foundation; "Meat and Meat Products" in *ECT* 2nd ed., Vol. 13, pp. 167–184, by W. J. Aunan and O. E. Kolari, American Meat Institute Foundation.

1. B. K. Watt and A. L. Merrill, *Composition of Foods–Raw, Processed, Prepared, Agriculture Handbook No. 8*, United States Department of Agriculture, Washington, D.C., revised 1963, reprinted 1975.
2. *Handbook of Agricultural Charts*, USDA, U.S. Government Printing Office, Washington, D.C., 1979.
3. J. R. Romans and P. T. Ziegler, *The Meat We Eat*, The Interstate Printers and Publishers, Inc., Danville, Ill., 1977.
4. J. C. de Holl, *Encyclopedia of Labeling Meat and Poultry Products*, Meat Plant Magazine, St. Louis, Mo., 1978.
5. R. Hamm, "Interactions Between Phosphates and Meat Proteins," in J. M. Deman and P. Melnycnyn, eds., *Symposium: Phosphates in Food Processing*, AVI Publishing Co., Westport, Conn., 1970, Chapter 5, p. 65.
6. W. E. Kramlich, A. M. Pearson, and F. W. Tauber, *Processed Meats*, The AVI Publishing Co., Westport, Conn., 1973.

7. J. Schut, "Meat Emulsions" in S. Friberg, ed., *Food Emulsions*, Marcel Dekker, Inc., New York, 1976, pp. 385–459.
8. E. Karmas, *Sausage Products Technology*, Noyes Data Corporation, Park Ridge, N.J., 1977.
9. S. L. Komarik, D. K. Tressler, and L. Long, *Food Products Formulary*, Vol. 1, The AVI Publishing Co., Westport, Conn., 1974.
10. G. R. Schmidt, "Sectioned and Formed Meat," in *Proc. 31st Ann. Recip. Meat Conf.*, American Meat Science Association and National Live Stock and Meat Board, 1978, pp. 18–24.
11. E. Karmas, *Processed Meat Technology*, Noyes Data Corporation, Park Ridge, N.J., 1976.
12. J. E. Price and B. S. Schweigert, *The Science of Meat and Meat Products*, W. H. Freeman and Co., San Francisco, Calif., 1971.
13. A. E. Bailey in D. E. Swern, ed., *Industrial Oil and Fat Products*, Wiley-Interscience, New York, 1964.
14. J. Wismer-Pedersen, *Food Technol.* **33**(8), 76 (1979).
15. J. C. Forrest, E. D. Aberle, H. B. Hedrick, M. D. Judge, and R. A. Merkel, *Principles of Meat Science*, W. H. Freeman and Company, San Francisco, Calif., 1975.

GLENN R. SCHMIDT
R. F. MAWSON
Colorado State University

MECHANICAL TESTING.   See Materials reliability.

MEDICAL DIAGNOSTIC REAGENTS

In the absence of disease processes, interrelationships of subcellular parts and intra- and extracellular chemical environments are maintained. Escape of vital constituents, eg, enzymes and proteins, into the extracellular compartment where they normally are present in different concentrations or activities is controlled exquisitely. Disease processes frequently cause changes in the internal and surrounding cellular environmental chemical composition.

The need for expeditious and reliable testing has been increasing in the field of laboratory medicine and medical diagnostic reagents have been provided increasingly by commercial sources. Striking advances also have been made to provide diagnostic reagent systems in the fields of microbiology, serology, hematology, and parasitology (see also Biomedical automated instrumentation).

Generally, a medical diagnostic reagent is a product used to measure an analyte (material undergoing analysis), either in concentration or by activity in a biological matrix, and thereby to help assess health or disease in humans or animals.

Chemical reagent systems have increasing use because of the convenience and uniform standards of a product produced under rigorous control. Despite the attention to detail and the care of reagent-system manufacturers, the user should be educated in the selection criteria for a reagent system. Detailed discussions of these criteria are the subjects of other technical publications (1–4).

The requirements of the intended method, ie, its practicality, cost, and characteristics, must be defined and information about the test must be supplied by the manufacturer. Consideration of the scientific literature as a source of the methodology of choice for measurement of a particular analyte should be given. Prices, amounts, and stability of the reagents are as varied as the types of analytes to be measured. Kits cost from $25.00 per hundred tests to several hundred dollars for less than 50 tests. The economic consideration that the user must make is the cost of technician labor versus the convenience of the kit in terms of a calculation of net cost per test. Once the method has been selected, evaluation of the reagent system is completed by consideration of its precision, accuracy, specificity, and sensitivity as related to the medical usefulness of the test (5). A list of abbreviations is included at the end of the article.

Tests

Measurements of Clinical Body-Fluid Enzymes. One of the important constituents of life is the protein in particular enzymes (qv). Enzymes are biochemical catalysts without which all life would cease. Under conditions of health and absence of pathology, a quantity of cellular enzyme leaks into the body-fluid compartments, eg, the blood, urine, and cerebrospinal fluid (6–7). Reference values for these enzymatic activities have been established by method. In many disease and other abnormal states, eg, myocardial infarction, hepatitis, pancreatitis, pregnancy, and prostatic carcinoma, rises in particular enzyme activities are noted and measured in serum and other body fluids (8–13).

Cardiac Enzymes. The heart is heavily dependent on oxidative metabolism and, as a result, is rich in vascularities and mitochondria containing large quantities of cytochromes (14–26). Ultimately, the blood vessels of the heart separate into the very small branches of the coronaries. Occlusion of these blood vessels deprives the heart of oxygen and, in its early phases, this deprivation is manifested by angina. Later phases of anoxia result in cell death and coronary or myocardial infarction (see Cardiovascular agents). When the cells of the heart die, enzymes that are within them enter the blood (27–28). Among the more frequently measured enzymes that are measured by medical diagnostic reagents are serum oxaloacetic transaminase (SGOT, aspartate transferase or AST); creatine phosphokinase isozymes; lactic dehydrogenase (LDH) and its isozymes; and α-hydroxybutyric acid dehydrogenase (α-HBD). An isoenzyme is an enzyme that exists in more than one form with differing pH optima, temperature optima, and substrate specificities (37–41).

There are three basic ways of measuring enzyme activity, eg, kinetic, static, and immunological coupled with kinetic. Another method which soon will be available is the measurement of enzyme concentration rather than activity (29–32). Kinetic enzyme methods involve rate measurements of enzyme activity, static enzyme methods are end-point analyses, and the immunological-coupled-to-kinetic enzyme measurement is a special example of immunological methods (see Immunological Measurements of Hormones and Drugs). The latter involve immunoprecipitation of one or more isozymes by specific antibodies and the measurement of activity by kinetic methods of the remaining isozymes (33–36).

Serum Oxaloacetic Transaminase (SGOT). A method for measuring SGOT activity in patients with acute myocardial infarction by coupling the production of oxaloacetate to the oxidation of nicotinamide adenine dinucleotide (NADH) [53-84-9] with excess malic dehydrogenase (MDH) [9001-64-3] is reported in refs. 37–38. An increase in measured transaminase activity can be obtained by increasing the concentrations of L-aspartate [56-84-8] and NADH (39). Both of these methods require incubation of the serum with all reagents except α-ketoglutaric acid [328-50-7] for 10–30 min in order for the LDH in the serum to destroy endogenous pyruvate; this is followed by initiation of the transaminase with L-ketoglutarate. A kit is available which contains 200 IU/L of LDH which destroys pyruvate within 1 min, thereby eliminating the need for a separate solution of L-ketoglutarate (40).

Glutamic oxaloacetic transaminase [9000-97-9] catalyzes the transamination of asparate and L-ketoglutarate to produce oxaloacetate and L-glutamate:

L-aspartate + α-ketoglutarate $\xrightarrow{\text{SGOT}}$ oxaloacetate + glutamate

MDH and NADH are used to measure the oxaloacetate produced by SGOT from aspartate and α-ketoglutarate:

oxaloacetate + NADH + H⁺ $\xrightarrow{\text{MDH}}$ malate + NAD⁺

The rate of disappearance of NADH and the resulting decrease in absorbance at 340 nm is directly proportional to the amount of oxaloacetate used by the system.

Pyruvate, which is in the serum sample and which would otherwise interfere with the measurement of SGOT, is destroyed by lactic dehydrogenase in the reagent:

pyruvate + NADH + H⁺ $\xrightarrow{\text{LDH}}$ lactate + NAD⁺

Creatinine Phosphokinase (CPK). Elevated levels of specific serum enzymes, particularly creatine phosphokinase (CPK), are valuable in diagnosing acute myocardial infarction and other diseases (41). The importance of serum CPK as a diagnostic indicator, particularly for acute myocardial infarction, has been advanced by the development of techniques to separate CPK into its three isoenzymes, eg, MM, MB, and BB (42).

MB isoenzyme is cardiac specific and is most highly concentrated in the myocardium; elevated MB isoenzyme concentrations provide a more definitive indication of myocardial cell damage or death then total CPK alone. The clinical usefulness of determining MB has been described in a number of published reports (43–45). In one report, determinations of CPK isoenzymes in 376 patients admitted to a coronary care unit with diagnoses of possible myocardial infarction and the time course of total CPK and MB levels after the onset of symptoms were described. The appearance of MB isoenzyme in the plasma shortly after the onset of symptoms permitted early diagnosis of acute myocardial infarction (46).

Until recently, separation of the serum CPK isoenzyme usually has been by conventional electrophoretic techniques, eg, separations on cellulose acetate, agarose, and polyacrylamide gel. Routine use of electrophoretic and fluorescent staining techniques for separation and quantitative measurement has been limited because of technical and reagent sensitivity difficulties (47–48).

The separation of CPK isoenzymes in human serum can be performed by a simple and rapid anion-exchange column chromatographic technique (49). Serum that is layered on a column of DEAE-Sephadex is eluted with TRIS-buffered NaCl solution. MM isoenzyme is retained slightly on the anion-exchange gel; the MB isoenzyme is retained until a specified increase in chloride concentration decreases its ionic interaction with the gel.

Quantitative assay of the CPK activity of the MB isoenzyme effluent is performed (9,50). The assay is based on the following reactions:

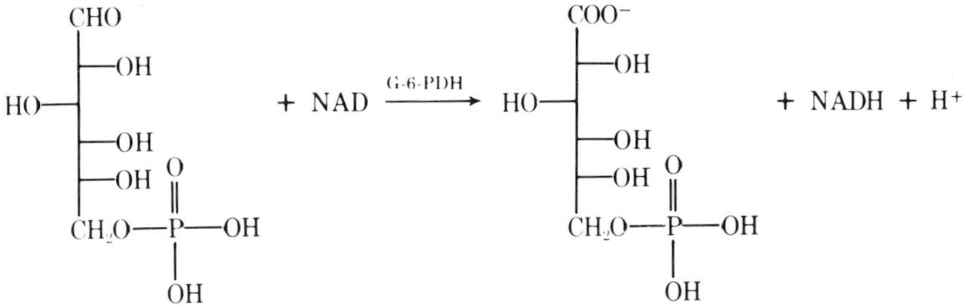

adenosine diphosphate (ADP) + creatine phosphate $\overset{\text{CPK}}{\rightleftharpoons}$ creatine + adenosine triphosphate (ATP)

ATP + glucose $\overset{\text{hexokinase}}{\rightleftharpoons}$ glucose-6-phosphate + ADP

The level of reduced NADH is measured at 340 nm.

Since total CPK and MB isoenzyme values may be elevated by noncardiac conditions, it has been suggested that certain criteria be applied when establishing normal and abnormal values (51). The following guidelines should be applied when using the modified CPK–CS cardiac-specific CPK isoenzyme system: if total CPK is 100 IU/L or greater, any result in which the MB isoenzyme level is greater than 3% of the total can be considered abnormal, ie, MB isoenzyme activity is caused primarily by myocardial involvement; if total CPK is less than 100 IU/L, an absolute MB isoenzyme value of ≥ 3 IU/L is considered abnormal (15,53).

Lactic Dehydrogenase (LDH). The measurement of LDH activity may be performed as either a forward or reverse reaction kinetic method. The forward reaction involves the conversion of lactate to pyruvate by LDH with concomitant reduction of NAD to NADH (1). The rate of NADH formation thus is proportional to the LDH activity of the sample. With lactate as the substrate, enzyme inhibition attributed to high concentrations of pyruvate does not occur. Because of the kinetic methodology, a blank is not required, and uncertainties resulting from interfering substances which may affect some other procedures are not encountered. The level of LDH in serum has clinical significance in the study of myocardial infarction, liver disease, renal disease, carcinoma, and pernicious anemia (52–53).

In the sample, LDH catalyzes the conversion of lactate to pyruvate with the concomitant reduction of NAD to NADH. The presence of NADH is characterized by marked absorption of ultraviolet light at 340 nm in contrast to NAD which shows no absorption at this wavelength. Thus, the rate of NADH formation, as measured by the rate of absorbance increase at 340 nm, is directly proportional to LDH activity of the sample.

The analytical reaction is:

$$\text{L-lactate} + \text{NAD}^+ \xrightarrow{\text{LDH}} \text{pyruvate} + \text{NADH} + \text{H}^+$$

A sample of this reagent system for measuring LDH is available (40).

α-Hydroxybutyric Dehydrogenase (α-HBD). The ratio of LDH to α-hydroxybutyric dehydrogenase between sera, and electrophoretically separated functions of individual sera are of different ratios (54). α-HBDH was measured at 25°C and pH 7.4 using 3.3 mmol α-ketobutyrate [600-18-0] as substrate. The utility of this test in diagnosing myocardial infarction is that LDH isoenzyme, which is released into the serum following infarction, shows greater activity with α-ketobutyrate as a substrate than does the LDH isoenzyme released in liver disease (55–57).

Optimum assay temperature is 37°C and the method optimized at 37°C is comparable in diagnostic accuracy but superior in precision to the original method (6,58–59).

In the procedure, α-HBD catalyzes the reduction of α-ketobutyrate to α-hydroxybutyrate by NADH. Thus, the rate of decrease in absorbance at 340 nm is directly proportional to the amount of α-HBD in the serum:

α-ketobutyrate + NADH + H⁺ → (α-HBD) α-hydroxybutyrate + NAD⁺

A commercial version of this test system is available (40).

Liver Enzymes. Serum enzyme elevations indicate liver damage (60). The three major enzymes that are measured are alkaline phosphatase, alanine transferase (SGPT), and γ-glutamyl transpeptidase.

Alkaline Phosphatase. Alkaline phosphatase catalyzes the following hydrolytic reaction, which is activated by magnesium [7439-95-4]:

p-nitrophenylphosphate + H₂O → (Mg²⁺) p-nitrophenolate + H₂PO₄⁻

The product of the reaction, p-nitrophenolate, has a much higher molar absorptivity at 415 nm than the substrate, p-nitrophenylphosphate. Therefore, the reaction may be followed by measuring the increase in absorbance at this wavelength. The rate of hydrolysis is directly related to the activity of the alkaline phosphatase (61).

Alanine Transferase (SGPT). Glutamic pyruvic transaminase [9000-86-6] catalyzes the transamination of alanine and α-ketoglutarate to produce pyruvate and glutamate:

$$\text{L-alanine} + \alpha\text{-ketoglutarate} \xrightarrow{\text{SGPT}} \text{pyruvate} + \text{L-glutamate}$$

Lactic dehydrogenase and NADH are used to measure the pyruvate produced by SGPT from alanine:

$$\text{pyruvate} + \text{NADH} + \text{H}^+ \xrightarrow{\text{LDH}} \text{lactate} + \text{NAD}^+$$

The rate of disappearance of NADH and the resulting decrease in absorbance at 340 nm is directly proportional to the amount of pyruvate produced in the system (62).

γ-Glutamyl Transpeptidase (GGTP). The substance γ-glutamyl-p-nitroanilide [47018-63-3] is split by γ-glutamyl transpeptidase into glutamate and p-nitroaniline. The rate of formation of p-nitroaniline is determined kinetically at 405 or 415 nm and is proportional to GGTP activity.

The increase in absorption at 415 nm is proportional to enzyme activity (40,63).

Pancreatic Enzymes. Acute pancreatitis is manifested by a rise in serum and urine pancreatic enzymes. The major enzyme activities that are measured are amylase and lipase (64–65).

Amylase. A polysaccharide dye is hydrolyzed to a soluble blue dye fragment, and the amount of solubilized dye is proportional to the enzyme activity. A kit is available commercially for carrying out this procedure (66).

$$\text{polysaccharide dye (insoluble)} \xrightarrow{\text{amylase}} \text{dye fragment (soluble)}$$

High levels of amylase may indicate perforated peptic ulcer, perforated gallbladder, acute nonperforated cholecystitis, choledocholithiasis, biliary dyskinesia, pneumonia, bronchogenic carcinoma, dissecting aneurysm of the aorta, intestinal obstruction, uremia, mumps, ruptured ectopic pregnancy and afferent loop obstruction following gastrectomy. Injections of morphine and codeine phosphate also produce marked elevation of serum amylase.

Low amylase levels may be found in extensive marked destruction of the pancreas and in severe liver damage (67–68).

Lipase. An elevated serum lipase is indicative of acute pancreatitis or obstruction of the pancreatic duct (69).

One turbidimetric method permits measurement of serum lipase according to a decrease in absorbance of an olive oil emulsion after incubation at 37°C for 20 min. The olive oil substrate is prepared in TRIS buffer (pH 8.8) which has sodium deoxycholate [302-95-4] as the emulsifying agent. The lipase activity is reported as Cherry-Crandall units.

$$
\begin{array}{c}
\underset{\text{triglyceride}}{
\begin{array}{l}
CH_2OCR \\
CHOCR' \\
CH_2OCR''
\end{array}}
\xrightarrow[OH^-, H_2O]{\text{lipase}}
\underset{\substack{\text{diglyceride} \\ + \\ \text{fatty acid I}}}{
\begin{array}{l}
CH_2OH \\
CHOCR' \\
CH_2OCR''
\end{array}}
\xrightarrow[OH^-, H_2O]{\text{lipase}}
\underset{\substack{\text{monoglyceride} \\ + \\ \text{fatty acid III}}}{
\begin{array}{l}
CH_2OH \\
CHOCR' \\
CH_2OH
\end{array}}
\xrightarrow{\text{isomerization}}
\underset{\text{monoglyceride}}{
\begin{array}{l}
CH_2OH \\
CHOH \\
CH_2OCR'
\end{array}}
\xrightarrow{\text{lipase}}
\underset{\substack{\text{glycerol} \\ + \\ \text{fatty acid II}}}{
\begin{array}{l}
CH_2OH \\
CHOH \\
CH_2OH
\end{array}}
\end{array}
$$

Other methods utilizing fluorimetric and colorimetric substrates have been employed (70–74) (see also Glycerol).

Prostatic Acid Phosphatase. Acid phosphatase is produced primarily in the red blood cells and the prostate. Elevation of the acid phosphatase level in females is rare and is of red-cell origin. The activity is high in newborns, declines during the first 2 weeks of life, remains at a medium level to age 13 then declines until ca age 17. A steady increase occurs with age to 60 yr. Acid phosphatase is elevated in men with carcinoma of the prostate and with bone metastases. Increased acid phosphatase levels also are observed in patients with bone-invading carcinoma from the breast, thyroid, or colon. Occasionally, acid phosphatase is elevated in association with nephritis, hepatitis, Paget's disease, and hyperparathyroidism. Prostatic acid phosphatase may be inhibited by L-tartrate [87-69-4], and the latter is used to determine phosphatase activity (75–76).

The difference between the total amount of inorganic phosphorus, after incubation of the serum with a sodium glycerophosphate [1555-56-2] substrate, and the amount of serum inorganic phosphorus without such incubation is used as a measure of phosphatase activity. The substrate is buffered at pH 5.0 for acid phosphatase. A Bodansky unit of phosphatase activity is that amount of activity that liberates 1 mg of inorganic phosphorus per 100 mL serum from the sodium glycerophosphate substrate at a defined pH during 1 h of incubation at 37°C. Various other materials have been used as the substrate for acid phosphatase, eg, disodium phenylphosphate [3729-54-7] and disodium p-nitrophenyl phosphate [4624-83-9]. A commercial kit is available (77–78).

Measurements of Abnormal Levels of Chemical Constituents. Many medical diagnostic reagent systems are designed for measurement of levels of chemical constituents as well as enzymatic activity. Abnormalities of these constituents have been associated with pathology. Among the substances measured by reagents in kit form are blood urea nitrogen (BUN), albumin, total protein, cholesterol, triglycerides, bilirubin, calcium, glucose, creatinine, and phosphorus.

Blood Urea Nitrogen (BUN). In 1965, an enzymatic method for measuring urea nitrogen in serum or protein-free filtrates of blood was described (79). Urease [9002-13-5] splits urea into ammonia and carbon dioxide. The ammonia combines quantitatively with α-ketoglutarate in the presence of glutamic dehydrogenase (GLDH) and reduced NADH to yield glutamate and NAD. The amount of absorbance decrease at 340 nm, resulting from the conversion of NADH to NAD, is quantitatively related to the amount of ammonia formed which, in turn, is quantitatively related to the amount of urea that was present initially:

α-ketoglutarate   glutamate

Elevations in the level of BUN has been associated with kidney and liver disease (80).

Albumin. Abnormalities in albumin levels in the blood and urine generally are used as indexes of renal or hepatic damage. Damage to the kidney tubula results in high urine levels of albumin and low serum albumin values; damage to the liver also can impair synthesis of this protein (81).

bromcresol green (BCG)

When albumin is bound to BCG [76-60-8], color shift from green to blue results and is measured at 550 nm; the color is proportional to the amount of albumin. A kit containing this reagent is available and can be used with an ABA 100 or a manual spectrophotometer (82).

Total Protein. Total serum determination is used for detection of hypoproteinemia, associated with edema, bleeding, neoplastic disease, and malnutrition. Detection of hyperproteinemia in those suffering from dehydration and hemoconcentration, and of hyperglobulinemia of multiply myeloma, lymphogranuloma, liver disease, and parasitic infections are other uses for this parameter. Protein determinations are important in postoperative patients, and burn victims. In many instances, however, exact concentrations of protein fractions must be determined in addition to total protein concentration (83).

Polypeptides containing at least two peptide bonds, eg,

react with the biuret reagent, eg, copper tartrate [52327-55-6], lithium hydroxide [1310-65-2], and sodium tartrate [868-18-6]. The blue color results from the formation of a coordination complex between cupric ion or protein nitrogen in an alkaline medium. Albumins and globulins react almost equally with the reagent (84–85).

Cholesterol. Cholesterol is a lipid that is present ubiquitously in the body (see Steroids). Elevated levels of cholesterol in serum or plasma are used as an index of hyperlipidemia, which has been implicated in vascular pathology including coronary disease.

Cholesterol esters in serum are hydrolyzed to free cholesterol by cholesterol ester hydrolase [9026-00-0]. The free cholesterol that is produced is oxidized by cholesterol oxidase [9028-76-6] to cholest-4-en-3-one with simultaneous production of hydrogen peroxide which oxidatively couples with 4-aminoantipyrine [83-07-8] and phenol [108-95-2] in the presence of peroxidase [9001-05-2] to yield a quinoneimine dye with an absorption maximum at 500 nm. The amount of color produced is directly proportional to the total cholesterol content of the sample (86). The reaction sequence is as follows:

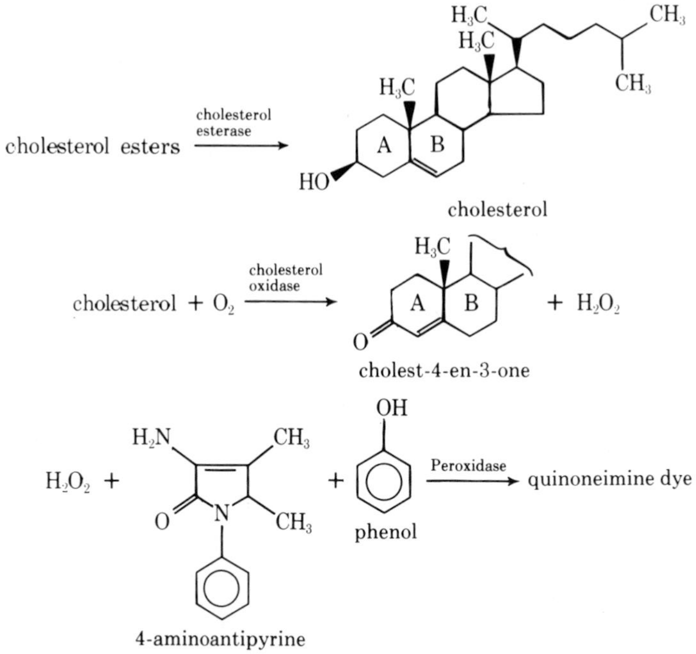

Triglycerides. Serum or plasma triglycerides also are used to assess hyperlipidemic conditions. Their measurement in conjunction with serum or plasma cholesterol is used to define phenotypes and to assess cardiovascular pathology risk (87).

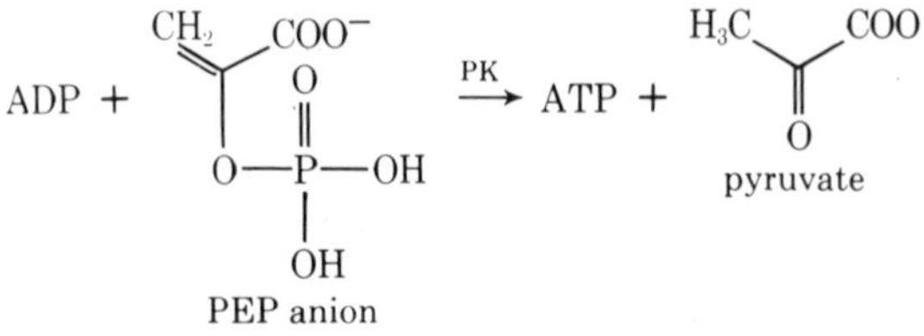

$$\text{triglycerides} \xrightarrow{\text{lipase}} \text{glycerol} + \text{free fatty acids (FFA)}$$

$$\text{glycerol} + \text{ATP} \xrightarrow{\text{GK}} \text{GP} + \text{ADP}$$

$$\text{pyruvate} + \text{NADH} \xrightarrow{\text{LDH}} \text{lactate} + \text{NAD}$$

Triglycerides in the sample are enzymatically hydrolyzed by lipase [9001-62-1] to glycerol and free fatty acids (FFA). The glycerol is then phosphorylated, a molar equivalent of ADP is produced and reacts as shown above; simultaneously, a molar equivalent of reduced NAD is oxidized. The change in absorbance at 340 nm is prportional to the concentration of glycerol that is liberated from triglycerides and of free glycerol in the sample.

Bilirubin. Bilirubin is a breakdown product of hemoglobin. Its metabolism is directly affected by the integrity of the liver and the integrity of the red cells within the blood. Pathology of the liver or the erythrocyte cause elevation of bilirubin in serum or plasma (88).

Bilirubin couples with the diazonium salt of 2,4-dichloroaniline, ie, in the presence of sulfamic acid, and 50% methanol, to form "azobilirubin"; the intensity of the latter is measured at 540 nm. The absorbance is used to determine the corresponding bilirubin concentration from a standard curve.

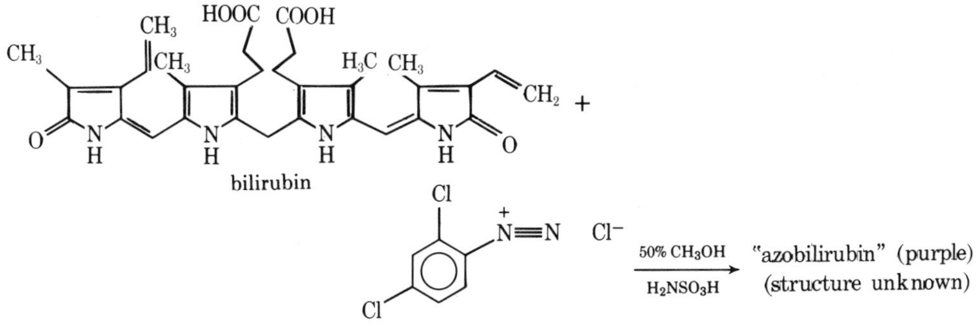

bilirubin

2,4-dichlorophenyldiazonium
chloride
[13617-98-6]

$$\xrightarrow[\text{H}_2\text{NSO}_3\text{H}]{50\% \text{ CH}_3\text{OH}}$$ "azobilirubin" (purple) (structure unknown)

This reagent system is available in a kit which can be used manually or as an automated test (40,89).

Calcium. Serum calcium measurements are used to assess parathyroid gland function and other aspects of calcium metabolism. Normal values for adults, children, and infants are 9–11 mg/100 mL, 10–11.5 mg/100 mL, and 10.5–12.0 mg/100 mL, respectively. Calcium levels increase with elevated serum proteins and decrease in patients with hypoproteinemia. Phosphorus levels are reciprocal to calcium values. Diseases causing elevated calcium values are polycythemia, hyperparathyroidism, multiple myeloma, and metastatic carcinoma of the bone. Decreases in calcium levels are seen in hypoparathyroidism, renal rickets, tetany, sprue, nephritis, pneumonia, and pregnancy.

Serum calcium forms a colored chromophore with o-cresolphthalein [596-27-0] in alkaline solution. The presence of 8-quinolinol sulfate [2149-36-2] in the reagent prevents interference by magnesium. This method permits determination of serum calcium without prior treatment of the serum sample (40,90–92).

Inorganic Phosphorus. Normal adult serum levels of inorganic phosphorus are 0.24–0.47 mg/L; children's values are slightly higher. Elevations occur in patients with nephritis, hypoparathyroidism, and healing bone fractures. Depressions occur from hyperparathyroidism, rickets, and ether and chloroform anesthesia (93). The most commonly used methods are based on the reaction of ammonium molybdate [12027-67-7] with a phosphate to form compounds, eg, ammonium molybdiphosphate, which are reduced to molybdate blue (94). Various reducing agents have been utilized, eg, ferrous sulfate [7720-78-7], stannous chloride [7772-99-8]. ascorbic acid [50-81-7], Elon (p-methylaminophenol) [150-75-4], Semidine (N-phenyl-p-phenylenediamine) [101-54-2], and 1-amino-2-naphthol-4-sulfonic acid [116-63-2]. These reducing agents differ with respect to the intensity and stability of the color formed.

An automated, single-step procedure has been developed for centrifugal analysis and permits quantification of the unreduced phosphomolybdate at 340 nm. This heteropolyacid complex has maximum absorption in the uv region. The use of a surfactant with this method obviates the need for the preparation of a protein-free filtrate (95).

Glucose. Perhaps among the most frequently analyzed parameters in clinical laboratory medicine is glucose (see Sugar). Changes in levels of glucose both as elevations or depressions from normal are diagnostically important (40,96). Elevations of glucose or hyperglycemic changes may be indicative of diabetes. Such increases in glucose are confirmed, among other tests, with a glucose tolerance test which depends on the ingestion of a load of glucose, usually 100 g, and the study of the decay curve of sugar level with time. Diabetics have a slow decay curve and usually a high peak level of glucose. Hypoglycemia is characterized by lower than normal glucose blood levels which frequently can lead to episodes of mental aberration, coma, and death.

Glucose in the serum sample is phosphorylated by hexokinase (HK) and excess ATP in the presence of magnesium ions to glucose-6-phosphate:

$$\text{glucose} + \text{ATP} \xrightarrow[\text{Mg}^{2+}]{\text{HK}} \text{glucose-6-phosphate} + \text{ADP}$$

Glucose-6-phosphate is oxidized in the presence of NAD by glucose-6-phosphate dehydrogenase (G-6-PDH) to 6-phosphogluconate:

$$\text{glucose-6-phosphate} + \text{NAD}^+ + \text{H}_2\text{O} \xrightarrow{\text{G-6-PDH}} \text{6-phosphogluconate} + \text{NADH} + \text{H}^+$$

The degree of reduction of NAD to NADH, which is measured at 340 nm, provides a quantitative analysis of the amount of glucose present.

Measurements of Hormones. Although many of the hormones of clinical importance are measured by other techniques, eg, glc, hplc, and RIA, many laboratories still use colorimetric analytical techniques (see Hormones, survey). These techniques can give clinically useful results but, in general, they are less specific than the techniques employing chromatography or immunoisotopic principles. Hormones, eg, 17-hydroxycorticosteroids (see Hormones, adrenocortical), 17-ketosteroids, vanilmandelic acid, and hydroxyindolacetic acid, can be indexed using colorimetric analysis. In most instances, urine is the body fluid used for analytical purposes (97–100). The analysis of 17-ketosteroids is illustrative of the test principles used in the measurement of all the hormones mentioned above.

17-Ketosteroids. The glucuronide conjugates of the 17-ketosteroids are cleaved by mild acid hydrolysis to yield free 17-ketosteroids. The free 17-ketosteroids are extracted into ethyl ether and the extract is washed with alkali to remove interfering compounds. The extract is evaporated to dryness and the 17-ketosteroid react with m-dinitrobenzene [99-65-0] in alkaline solution to form a reddish-purple complex.

Immunological Measurements of Hormones and Drugs. With development of radioimmunoassay (RIA) as an analytical technique, the field of analytical endocrinology and immunopharmacology has burgeoned (101) (see Radioactive tracers). Radioimmunoassay permits measurement of picogram quantities of hormones or drugs. The three prominent permutations of this technology are RIA, EMIT (enzyme-multiplied immunotechnique), and ELISA (enzyme-linked immunosorbent analysis). Competition occurs between antigen without a label, ie, the analyte, and a labeled analyte for a fixed amount of a specific antibody for that antigen:

$$Ab + Ag \rightarrow Ab\text{–}Ag$$

$$Ab\text{–}Ag + Ag^* \rightarrow Ab\text{–}Ag^* + Ag$$

where Ab = antibody, Ag = unlabeled antigen (free unlabeled), Ag^* = labeled antigen (free labeled), $Ab\text{–}Ag$ = antibody–antigen complex unlabeled (bound unlabeled, zeolite type), and $Ab\text{–}Ag^*$ = antibody–antigen complex labeled (bound labeled, zeolite type).

The $Ab\text{–}Ag^*$ is bound and can be separated from the free Ag^* by charcoal, polyethylene glycol, Z-gel, or sephadex. The bound or free fraction of the labeled material is proportional to the amount of unlabeled analyte that is present in the sample. The difference between immunoassay methods depends on the type of label and the method of separation.

Illustrative of labels other than isotopes are fluorescence probes, viruses, and enzymes (66,102). As indicated above, techniques are required to separate free from bound isotopic species in order to allow analysis; however, enzymatic probes obviate this need:

$$Ab + Ag^*E \rightarrow Ab\text{–}Ag^*E \text{ (no enzyme activity)}$$

$$Ab\text{–}Ag^*E + Ag \rightarrow Ab\text{–}Ag + Ag^*E \text{ (enzyme activity)}$$

where Ag^*E = enzyme coupled to labeled antigen, Ab = antibody, and Ag = unlabeled antigen.

From the second equation, it can be seen that a reagent system to measure the activity of E gives results proportional to Ag; this is the basis of all immunoassay techniques. Examples of the types of analytes measured by these techniques include: hormones, drugs, antibodies, receptors, specific proteins, and enzymes (66,102–103).

Many factors affect selection of drug dosage, eg, body size, renal function, electrolyte status, thyroid function, malabsorption, and concurrent medication. Monitoring the drug levels in urine, blood, or other tissues provides the critical feedback to ensure that an adequate dose is delivered with minimum risk of toxicity. The drugs most commonly measured by RIA are digoxin and gentamicin. Enzyme-labeled immunoassays are more widely used than RIA techniques for measuring levels of anticonvulsant drugs (104).

Gentamicin. Gentamicin is an aminoglycoside antibiotic that is effective against certain gram-negative and gram-positive bacteria. Since gentamicin is poorly absorbed from the gastrointestinal tract, it is given by intramuscular or intravenous routes. Peak blood levels are reached ca 0.5 h after the completion of an intravenous dose and approximately 1.5 h after an intramuscular dose. The drug is eliminated by glomerular filtration. In persons with normal renal function, the decline in serum levels following the peak is exponential, with a half-time of ca 2–3 h.

Peak serum levels of 8–10 μg/mL and occasionally higher are necessary for adequate bactericidal effect, but prolonged exposure to serum levels exceeding 15 μg/mL is associated with serious ototoxicity and potentially life-threatening nephrotoxicity.

Several techniques are available for measuring serum gentamicin, eg, RIA with ^{125}I [*14158-31-7*] and ^{3}H [*10028-17-8*] labels, enzyme immunoassay, radioenzymatic assay, and bioassay. An immunoassay or radioenzymatic assay is preferred because of speed, technical simplicity, and lack of interference from concurrently administered antimicrobials.

Digoxin. Digoxin is a cardiac glycoside that is prescribed for the control of congestive heart failure and certain abnormalities in cardiac rhythm (105–106). The therapeutic index for digoxin is very low with only a very narrow difference between therapeutic and toxic dosages. There are no specific cardiac arrhythmias owing to digoxin toxicity that cannot result from underlying heart disease. Therefore, the physician often is presented with the therapeutic choice of either increasing his patient's digoxin dosage to improve cardiac function with the hope of abolishing cardiac arrhythmias or withholding digoxin. Digoxin levels may be difficult to predict because of variation in absorption of oral doses (107) and variation in nonrenal excretion (108). Patients with disturbed renal function also may require more careful monitoring of serum digoxin levels (109). Monitoring serum digoxin levels is a valuable aid in decreasing toxicity risk and in detecting underdigitalization.

Digoxin is excreted unchanged in the urine at a rate that is proportional to glomerular filtration. Tissue and plasma levels decrease exponentially with a half-time of ca 36 h. Endogenous creatinine [60-27-5] is the best renal function measurement adjusting digoxin dosage. Use of the creatinine level alone is hazardous, owing to marked variations resulting from sex, age, and body size. The relationship between serum digoxin levels at steady-state and toxicity in terms of frequency of arrhythmias for a euthyroid, normokalemic patient (4.5 mg/L) in sinus rhythm is

Serum digoxin levels, mg/mL	Approximate incidence of toxicity, %
1.1	5
1.4	7
2.0	19
2.7	38
4.4	71

Digoxin also can be measured by ELISA which is the basis of Organon's OREIA-DIG kit (110).

In the measurement of digoxin levels by immunoassay techniques, digoxin in a serum sample reacts with an antibody that is specific for digoxin and a fixed amount of labeled digoxin. During an incubation period, digoxin and labeled digoxin compete for a restricted number of binding sites on the digoxin antibody; attachment occurs in proportion to the relative concentrations of the digoxin and labeled digoxin in the reaction mixture. Separation of the unbound digoxin from the antibody-bound digoxin and subsequent measurement of the labeled fraction of the bound phase completes the test. By comparing results of the unknown sample with those obtained from a series of digoxin calibrators, an accurate measurment of the concentration of digoxin in the sample may be obtained. All of the nonisotopic immunoassays can be used for many analytes.

In radioimmunoassay systems, the digoxin generally is labeled with radioactive iodine, ^{125}I, and the amount of labeled digoxin that is bound to the antibody is measured with a γ-ray counting instrument in RIA systems; enzymes are used instead of radioactive isotopes, and the readout is a spectrophotometeric measurement of the enzyme activity in the bound fraction. In the ELISA test, digoxin is conjugated to the enzyme horseradish peroxidase [9003-99-0]. A serum sample is mixed with labeled digoxin, ie, digoxin-enzyme conjugate, and the digoxin–antibody complex which is covalently bound to latex particles. An incubation period follows during which digoxin and labeled digoxin compete for the limited binding sites on the digoxin antibody complex. Following the incubation period, the antibody-bound digoxin is separated from the unbound digoxin by centrifugation and decantation of the supernatant. A solution containing a peroxidase substrate, eg, urea peroxide [35220-04-3], and a colorless chromogenic reducing agent, eg, o-phenylenediamine [95-54-5] is mixed with the antibody-bound digoxin. The enzyme-labeled digoxin which is attached to the digoxin antibody converts the substrate–chromogen mixture to a colored product. The intensity of the color is proportional to the amount of enzyme present and, therefore, is related inversely to the amount of unlabeled digoxin present in the sample. By reference to a series of digoxin calibrators that have been processed similarly, the concentration of digoxin in the unknown sample can be measured.

A second type of nonisotopic analytical procedure is EMIT. The EMIT assay is a homogeneous enzyme immunoassay technique developed for the microanalysis of specific compounds in biological fluids. A drug is labeled with an enzyme and, when the enzyme-labeled drug is bound to an antibody against the drug, the activity of the enzyme is reduced. The unbound drug in a sample competes with the enzyme-labeled drug for antibody binding sites and, thereby, decreases the antibody-induced inactivation of the enzyme. Enzyme activity correlates with the concentration of the free drug and is measured by an absorbance change resulting from the enzyme's catalytic action on a substrate.

In the performance of an EMIT digoxin assay, serum or plasma is mixed with a sodium hydroxide [1310-73-2] solution (the serum pretreatment reagent) to destroy interfering proteins. Reagent A, which contains antibodies to digoxin and the substrates for the enzyme glucose-6-phosphate dehydrogenase (G-6-PDH), is added to the treated serum and binding occurs to any drug in the serum or plasma that is recognized by the antibody. The digoxin labeled with the enzyme G-6-PDH then is added. The labeled digoxin combines with any remaining unfilled antibody binding sites and, thereby, the enzyme activity is proportionately reduced. The residual enzymatic activity is directly related to the concentration of the drug present in the serum or plasma (111–114). The active enzyme catalyzes the reduction of NAD to NADH, resulting in an absorbance change that is measured spectrophotometrically. The use of bacterial G-6-PDH (*Leuconostoc mesenteroides*) prevents interference from serum G-6-PDH, an enzyme which does not function efficiently with the NAD coenzyme.

The use of radioimmunoassays for therapeutic drug and toxic drug level monitoring remains popular. Foremost among these technologies are the Abuscreen kits. In principle, these tests are illustrated by the barbiturate test. The Abuscreen radioimmunoassay for barbiturates is based on the competitive binding to an antibody of radiolabeled antigen and unlabeled antigen in proportion to their concentration in the solution. Unlabeled antigen displaces radioactive antigen from the antibody present. An unknown specimen is added to a test tube containing known amounts of barbiturate antibodies and radiolabeled antigen. After precipitation and centrifugation, the supernatant fluid, which contains the free antigen, is transferred to test tubes for counting in a gamma scintillation counter. A positive specimen is identified qualitatively when the radioactivity is equal to or greater than that of the positive control and quantitatively by comparision to a standard curve.

The Abuscreen technology (trademark of Roche Diagnostics, Nutley, N.J.) has been applied to morphine, cocaine, amphetamine, methadone hydrochloride, and phencylidine measurement.

Chromatographic Kits for Drug Measurement. Thin-layer chromatography (tlc) is used to screen urine specimens for drugs of abuse. Many prepackaged kits are available for this purpose; they contain plates, sprays, applicators, and reference tables for determining color reactions and R_f (distance ratio) values for the compounds of interest (see Analytical methods). The most common drugs screened with this type of procedure are salicylates, barbiturates, opiates, pentachlorophenol (PCP), benzodiazepines, and amphetamines. Many other compounds can be detected. Positive findings obtained using tlc should be verified with another methodology (115–116).

Urine is extracted with an organic solvent and a concentrated aliquot is applied in a small spot to a tlc plate which is coated with cellulose or another coating material. The plate with the dried spots is immersed in a tank containing a mobile phase, the mobile phase moves up the tlc plate by capillary action, and the constituents of the mixture contained in the spot are carried across the plate with the mobile phase. In transit, they partition between the solvent and the particles of the coating. This partition retards some constituents and allows others to move more rapidly in comparision with the solvent front. The plate is dried after the solvent has moved a sufficient distance up the plate. A series of sprays is applied and characteristic colors and mobilities are observed for compounds of interest (117).

Future Trends

Immunological techniques relating to the determination of antibodies and to the concentration rather than activity of enzymes will be developed in the 1980s. The development of these techniques for detecting receptor proteins and other types of specific proteins of sufficient specificity so as to be clinically relevant in malignancy and other disease states is probable. The development of new and better labeling probes that allow sensitive analyses without the use of isotopic technology in many cases is expected. The worldwide sales of cancer-detection kits probably will be 2×10^9 by 1990 (78). The combination of chemistry and immunology to produce better antigen antibody, separation, and detection techniques is the vanguard of the progress of kits for medical diagnostics.

Nomenclature

ABA 100	= Abbott Bichromatic Analyzer 100
ADP	= adenosine diphosphate [58-64-0]
ALT	= alanine transferase [9000-86-6]
AST	= aspartateamino transferase [9000-97-9]
ATP	= adenosine triphosphate [56-65-5]
BB	= CPK isoenzyme [9001-15-4] (related to cardiac damage)
BCG	= bromocresol green [76-60-8]
BUN	= blood urea nitrogen
CPK	= creatine phosphokinase [9001-15-4]
DEAE	= diethylaminoethyl
ELISA	= enzyme-linked immunosorbent analysis
EMIT	= enzyme-multiplied immunotechnique
G-6-PDH	= glucose-6-phosphate dehydrogenase [9001-40-5]
GGTP	= γ-glutamyl transpeptidase [9046-27-9]
GK	= glycerolkinase [9030-66-4]
glc	= gas–liquid chromatography
GLDH	= glutamic dehydrogenase [9001-46-1]
GP	= glycerol phosphate [12040-65-2]
α-HBD	= α-hydroxybutyric acid dehydrogenase [9028-38-0]
HK	= hexokinase [9001-51-8]
hplc	= high pressure liquid chromatography
IU	= international unit
LDH	= lactic dehydrogenase [9001-60-9]
MB	= CPK isozyme [9001-15-4] (related to cardiac damage)
MDH	= malic dehydrogenase [9001-64-3]
MM	= CPK isoenzyme [9001-15-4] (related to cardiac damage)
NAD	= nicotinamide–adenine dinucleotide [53-84-9]
NADH	= reduced nicotine amide–adenine dinucleotide [58-68-4]
PEP	= phosphoenolpyruvic acid [138-08-9]
PK	= pyruvatekinase [9001-59-6]
RIA	= radioimmunoassay
SGOT	= serum oxaloacetic transaminase (see also AST) [9000-97-9]
SGPT	= alanine transferase (see also ALT)
tlc	= thin layer chromatography
TRIS	= tris(hydroxymethyl)methylamine [77-86-1]

BIBLIOGRAPHY

1. J. E. Logan and D. R. Toada, *Clin. Biochem.* **10**, 133 (1977).
2. *Approved Standard ASL-1*, National Committee for Clinical Laboratory Standards, NCCLS, 1975.
3. J. O. Westgard, *CAP Aspen Conference 1976: Analytical Goals in Clinical Chemistry*, College of American Pathologists, Skokie, Ill., pp. 105–114.
4. *Can. Gaz.* (*Part II*) **109**(18), 2491 (1975).
5. J. E. Logan, "Criteria for Kit Selection in Clinical Chemistry" in *Clinical Biochemistry*, Academic Press, New York, in press.
6. M. Asada and J. T. Galambos, *Gastroenterology* **45**, 578 (1963).
7. T. R. C. Boyde, *Enzymol. Biol. Clin.* **9**, 385 (1968).
8. A. S. Carlson, A. M. Siegelman, and T. Robertson, *Am. J. Clin. Path.* **38**, 260 (1962).
9. R. J. Clermont and T. C. Chalmers, *Medicine* **46**, 197 (1967).
10. F. DeRitis, M. Coltori, and G. Giusti, *Clin. Chem. Acta.* **2**, 70 (1957).
11. S. Posen, *Clin. Chem.* **16**, 71 (1970).
12. J. H. Wilkinson, *Clin. Chem.* **16**, 882 (1970).
13. W. Bloom and D. W. Fawcett, *A Textbook of Histology*, W. B. Saunders, Philadelphia, Pa., 1962.

14. J. Brachet, *Biochemical Cytology*, Academic Press, Inc., New York, 1957.
15. N. S. Cohn, *Elements of Cytology*, Harcourt, Brace and World, New York, 1969.
16. J. K. Grant, *Biochem. Symp.*, 23 (1963).
17. E. M. Crook, *Bio. Soc. Symp.*, 16 (1959).
18. A. Lehninger, *The Mitochondrion*, Benjamin, New York, 1964.
19. A. Allfrey, "The Isolation of Subcellular Components" in J. Brachet and A. E. Brachet, eds., *The Cell*, Vol. 1, Academic Press, New York, 1959, p. 193.
20. A. Claude, *Harvey Lectures* **48**, 121 (1948).
21. C. De Duve, C. R. Wattiaux, and P. Baudhuin, *Advan. Enzymol.* **24**, 291 (1962).
22. C. De Duve, *J. Theoret. Biol.* **6**, 33 (1964).
23. C. De Duve, *Lectures* **59**, 49 (1965).
24. A. E. Mirsky and S. Osauda, "The Interphase Nucleus" in J. Brachet and A. E. Brachet, eds., *The Cell*, Vol. II, Academic Press, New York, 1961, p. 677.
25. G. E. Palade, *Proc. Natl. Acad. Sci.*, 613 (1964).
26. W. C. Schneider, *Advan. Enzymol.* **21**, 1 (1959).
27. J. A. Halsted, ed., *The Laboratory in Clinical Medicine*, W. B. Saunders, Philadelphia, Pa., 1976, pp. 280–281.
28. A. M. Weissler, ed., *Noninvasive Cardiology*, Greene and Stratton, New York, 1974.
29. M. K. Schwartz, G. Kessler, and O. Bodansky, *J. Biol. Chem.* **236**, 1207 (1961).
30. P. E. Strandford and K. Clayson, *Clin. Chem.* **10**, 635 (1964).
31. E. Amador, L. E. Dorfman, and E. C. Wacker, *Clin Chem.* **12**, 406 (1966).
32. S. Morgenstern, R. Flor, G. Kessler, and B. Klein, *Anal. Biol. Chem.* **13**, 149 (1965).
33. J. F. Burd, Usategui, and M. Gomez, *Biochem. Biophys. Acta* **310**, 238 (1973).
34. J. F. Burd, Usategui, M. Gomez, A. Fernandez des Castro, and F. M. Yeager, *Clin. Chem. Acta* **46**, 205 (1973).
35. Usategui, M. Gomez, R. W. Wicks, and M. Warshaw, *Clin. Chem.* **25**, 729 (1979).
36. F. Lui, Usategui, M. Gomez, and G. Reynoso, "Immunochemical Determination of LDH," *paper presented at the Scientific Assembly*, ASCP-CAP, St. Louis, Mo., Sept. 14–22, 1978.
37. A. Karmen, F. Worblenski, and J. S. LaDue, *J. Clin. Invest.* **34**, 126 (1955).
38. A. Karmen, *J. Clin. Invest.* **34**, 131 (1955).
39. R. J. Henry, N. Chiamori, O. J. Golub, and S. Berkman, *Am. J. Clin. Path.* **34**, 381 (1960).
40. Abbott Diagnostics, South Pasadena, Calif.
41. D. M. Goldberg and D. A. Winfield, *Br. Heart J.* **34**, 597 (1972).
42. K. J. Van der Veen and A. F. Willebrand, *Clin. Chem. Acta* **13**, 312.
43. R. S. Galen and co-workers, *J. Am. Med. Assoc.* **232**, 145 (1975).
44. D. W. Mercer, *Clin. Chem.* **20**, 36 (1974).
45. A. Konttinen and H. Somer, *Am. J. Cardiol.* **29**, 817 (1972).
46. G. S. Wagner and co-workers, *Circul.* **47**, 263 (1973).
47. M. A. Varat and D. W. Mercer, *Circul.* **51**, 855 (1975).
48. R. Roberts and B. E. Sobel, *Arch. Intern. Med.* **136**, 421 (1976).
49. *Item 43149*, Roche Diagnostics, Nutley, N.J.
50. S. B. Rosalki, *J. Lab. Clin. Med.* **69**, 696 (1967).
51. S. M. Sax and co-workers, *Clin. Chem.* **22**(1), 87 (1976).
52. W. E. C. Wacker, D. D. Ulmer, and B. L. Vallie, *New Eng. J. Med.* **255**, 449 (1956).
53. G. J. Race, *Laboratory Medicine*, Vol. 1, Harper & Row, New York, 1977, Chapt. 6, p. 33.
54. S. B. Rosalki and J. H. Wilkinson, *Nature* **188**, 110 (1960).
55. B. A. Elliott and J. H. Wilkinson, *Lancet i*, 698 (1961).
56. B. A. Elliott, E. M. Jepson, and J. H. Wilkinson, *Clin. Sci.* **23**, 305 (1962).
57. S. B. Rosalki, *Br. Heart J.* **25**, 795 (1963).
58. G. Ellis and D. M. Goldberg, *Am. J. Clin. Path.* **56**, 627 (1971).
59. R. J. Spooner, and D. M. Goldberg, *Clin. Chem.* **19**, 1387 (1973).
60. J. H. Wilkinson, *An Introduction to Diagnostic Enzymology*, Ed. Arnold Ltd., London, 1962.
61. O. A. Bessey, O. H. Lowry, and M. J. Brock, *J.B.C.* **164**, 321 (1946).
62. F. Wroblewski and J. S. La Due, *Proc. Soc. Exptl. Biol. Med.* **91**, 569 (1956).
63. G. N. Bowers, Jr. and R. B. McComb, *Clin. Chem.* **12**, 70 (1966).
64. B. Klein, J. A. Foreman, and R. L. Searcy, *Clin. Chem.* **16**, 32 (1970).
65. S. Take, J. E. Berk, and L. Fridhandler, *Clin. Chem. Acta* **26**, 533 (1969).
66. Roche Diagnostics, Nutley, N.J.

67. J. B. Flege, Jr., *Arch. Surg.* **92,** 397 (1966).
68. J. B. Gross and co-workers, *Mayo Clin. Proc.* **26,** 81 (1951).
69. J. Wallach, *Interpretation of Diagnostic Tests*, Little, Brown and Co., Boston, Mass., 1970, p. 37.
70. E. M. Gindler, *Clin. Chem.* **17,** 633 (1971).
71. M. Fleisher and M. K. Schwartz, *Clin. Chem.* **17,** 417 (1971).
72. N. W. Tietz and E. A. Fiereck, "Measurement of Lipase Activity in Serum" in G. R. Cooper, ed., *Standard Methods of Clinical Chemistry*, Vol. 7, Academic Press, New York, 1972, p. 19.
73. C. G. Massion and D. Seligson, *Am. J. Clin. Path.* **48,** 307 (1967).
74. J. S. Yang and H. G. Biggs, *Clin. Chem.* **17,** 512 (1971).
75. R. H. Goodale, *Clinical Interpretation of Laboratory Tests*, 4th ed., F. A. Davis and Co., Philadelphia, Pa., 1959, p. 107.
76. R. J. Henry, *Clinical Chemistry: Principles and Techniques*, Harper and Row Publishers, Inc., 1962, p. 482.
77. G. Y. Shinowara, L. M. Jones, and H. L. Reinhart, *J. Biol. Chem.* **142,** 921 (1942).
78. *Chem. Week*, 56 (Nov. 26, 1980).
79. H. Talke and G. E. Schubert, *Klin. Wochenskr.* **43,** 174 (1965).
80. W. R. Faulkner and J. W. King in N. Teitz, ed., *Fundamentals of Clinical Chemistry*, W. B. Saunders Publisher, Philadelphia, Pa., 1976, p. 991.
81. T. J. Peters, "Serum Albumin" in O. Bodansky and C. P. Stewart, eds., *Advances in Clinical Chemistry*, Vol. 13, Academic Press, New York, 1970, p. 37.
82. *Spec. Tru BCG Albumin*, Pierce Chemical Company, Rockford, Ill.
83. J. E. Young and J. Roberts, *Laboratory Medicine*, Vol. 1, Harper and Row, New York, 1977, p. 23.
84. A. G. Gornall, C. J. Bardawill, and M. N. David, *J. Biol. Chem.* **177,** 751 (1949).
85. J. Booij, *Clin. Chem. Acta* **38,** 355 (1972).
86. C. C. Allain, L. Poon, S. G. Chan, W. Richmond, and P. Fu, *Clin. Chem.* **20,** 470 (1974).
87. *Item #LO 3048*, Calbiochem., La Jolla, Calif.
88. J. I. Routh in N. Teitz, ed., *Fundamentals of Clinical Chemistry*, W. B. Saunders Co., Philadelphia, Pa., 1976, p. 1026.
89. R. N. Rand and di Pasqua, *Clin. Chem.* **8,** 570 (1962).
90. H. V. Connerty and A. R. Briggs, *Am. J. Clin. Path.* **45,** 290 (1966).
91. L. G. Morin, *Am. J. Clin. Path.* **61,** 114 (1974).
92. W. R. Moorehead and H. G. Biggs, *Clin. Chem.* **20,** 1458 (1974).
93. W. L. White and S. Frankel, *Seiverds Chemistry for Medical Technologists*, 2nd ed., C. V. Mosby Company, 1965, p. 224.
94. M. Reiner, ed., *Standard Methods of Clinical Chemistry*, Vol. 1, Academic Press, New York, 1953, p. 84.
95. J. A. Daly and G. Ertingshaesen, *Clin. Chem.* **18,** 263 (1972).
96. R. Richterlick and H. Dauwalder, *Schweiz. Med. Wochenschr.* **101,** 860 (1971).
97. J. K. Norymberski, *Nature* **170,** 1974 (1952).
98. W. Zimmerman, Z. *Physiol. Chem.* **233,** 257 (1935).
99. R. D. Eastham, *Biochemical Values*, 4th ed., Williams and Wilkins, Baltimore, Md., 1971.
100. *Cuvette Cortiset*, Houston, Tex.
101. S. A. Berson and R. S. Yalow, *Clin. Chem. Acta* **22,** 51 (1968).
102. *AED Dilantin*, Syva Diagnostics, Palo Alto, Calif.
103. *Code #100F*, Organon Diagnostics, West Orange, N.J.
104. D. K. Oxley in G. J. Race, ed., *Laboratory Medicine*, Harper and Row, New York, 1977, pp. 29–31.
105. C. K. Moe and A. E. Farah in L. S. Goodman and A. Gilman, eds., *The Pharmacological Basis of Therapeutics*, MacMillan Co., New York, 1970.
106. R. H. Goldman, *J. Am. Med. Assoc.* **229,** 331 (1974).
107. J. E. Doherty, *Am. J. Med. Sci.* **225,** 382 (1968).
108. P. M. Bloom and W. F. Nelp, *Am. J. Med. Sci.* **251,** 133 (1966).
109. T. W. Smith, V. P. Butler, Jr., and E. Haber, *New Eng. J. Med.* **284,** 989 (1971).
110. *OREIA–DIG*, Organon Diagnostics, A Division of Organon, Inc., West Orange, N.J.
111. K. E. Rubenstein, R. S. Schneider, and E. F. Ullman, *Biochem. Biophys. Res. Commun.* **47,** 846 (1972).
112. R. J. Bastiani, R. C. Phillips, R. S. Schneider, and E. F. Ullman, *Am. J. Med. Tech.* **39,** 211 (1973).
113. R. S. Schneider, E. Lindquist, K. E. Rubenstein, and E. F. Ullman, *Clin. Chem.* **19,** 821 (1973).
114. *EMIT–CAD Digoxin*, Syva Diagnostics, Palo Alto, Calif., 1977.

115. B. Davidow, N. L. Petri, and Quame, *Tech. Bull. Reg. Med. Tech.* **38,** 714 (1968).
116. *LQD*, Quantum Industries, Fairfield, N.J.
117. D. K. Oxley in G. J. Race, ed., *Laboratory Medicine*, Vol. 1, Harper and Row Publishers, New York, 1977, p. 13.

HERBERT E. SPIEGEL
Hoffmann-La Roche Inc.

MELTING AND FREEZING TEMPERATURES. See Temperature measurement; Calorimetry.

MEMBRANE TECHNOLOGY

Membranes rapidly are gaining an important place in chemical technology. Specific membrane processes are covered as separate topics in the Encyclopedia; it is the purpose of this article to provide an overview of membrane technology to assist with background for specific topics such as Dialysis, Hollow-fiber membranes, Reverse osmosis, Ultrafiltration, Water supply and desalination, Ion exchange, Microencapsulation, Pharmaceuticals controlled-release, Barrier polymers, Contact lenses, and Diffusion separation methods.

This article deals with the mechanisms of transport processes in membranes, membrane structure and fabrication, how membranes can be formed into functional modules, and important applications of this technology. Because separation processes represent a large use of membranes, many of the current and potential concepts for separation and purification via membrane technology are presented. Membrane technology also is used in a variety of other contexts; a sampling of these is presented to suggest the possibilities.

Mechanisms of Transport

Transport of fluids or solutes through membranes can occur by any of several different mechanisms, depending on the structure and nature of the membrane. In all cases, transport of any species through the membrane is driven by a difference in free energy or chemical potential of that species across the membrane. These driving forces may result from differences in pressure, concentration, electrical potential, or combinations of these factors between the fluid phases on the upstream and downstream sides of the membrane. Species fluxes can be related to these driving forces through generalized abstract relationships from irreversible thermodynamics; however, various specific relationships applicable to individual cases are introduced here.

Most often, the transport of a permeant through a membrane can be specified in terms of a permeability coefficient P defined as

$$P = \frac{Jl}{\Delta\phi} \tag{1}$$

where J = permeant flux in appropriate units, l = membrane thickness, and $\Delta\phi$ is the difference in hydrostatic pressure, partial pressure, concentration, or other potential between the upstream and downstream fluid phases.

Solution-Diffusion Model. A very important and fundamental means by which a species can be transported through a membrane involves dissolving of the permeate molecules into the membrane at its upstream surface followed by molecular diffusion down its concentration gradient to the downstream face of the membrane. There it is evaporated or dissolved into the adjacent fluid phase. This solution-diffusion mechanism is applicable when the membrane does not contain pores and may be regarded for thermodynamic purposes as a fluid phase (1). This mechanism is operative only in dense membranes or in the dense skin of asymmetric membranes. In most cases, thermodynamic equilibrium partitioning of the penetrant species between the membrane surface and the adjacent fluid phase takes place. Thus, the permeability defined by equation 1 is the product of a thermodynamic partition coefficient and a molecular diffusivity for the permeate through the membrane.

Fick's first law, a specialized result of irreversible thermodynamics, can be used to describe the molecular diffusion within the membrane mentioned above. A general form for the flux of species 1 in mass units, relative to stationary coordinates, can be written as follows for a binary system (2):

$$n_1 = -\rho D_{12}\nabla w_1 + w_1(n_1 + n_2) \tag{2}$$

where w_1 = mass fraction of species 1, ρ = density of the mixture, and D_{12} = the binary mutual diffusion coefficient. The second term on the right is a frame of reference term (3) that is important in some instances. When component 2 is the membrane, it is stationary and $n_2 = 0$. Then equation 2 becomes

$$n_1 = -\frac{\rho D_{12}}{1 - w_1}\frac{dw_1}{dx} \tag{3}$$

for uniaxial transport in the x-direction. When the mass fraction of 1 is small compared to unity, equation 3 can be reduced to its more familiar form

$$n_1 = -\rho D_{12}\frac{dw_1}{dx} = -D_{12}\frac{d\rho_1}{dx} \tag{4}$$

where ρ_1 = mass of species 1 per unit volume. Other units and notations for flux and concentration in equation 4 are often used.

Gases and Vapors. Gas sorption and transport in rubbery amorphous polymers represents one of the simplest examples of the solution-diffusion mechanism. The equilibrium partitioning of gas molecules between the membrane and the external gas phase follows Henry's law (see left side of Fig. 1)

$$C = Sp \tag{5}$$

where C = gas concentration in the polymer, p = partial pressure of the particular gas in the gas phase, and S = a solubility coefficient. Transport follows the simple form of Fick's law expressed by equation 4 with a diffusion coefficient that is independent of gas concentration D. At steady state the resulting equation can be integrated and combined with equation 5 to obtain

$$J = D\frac{C_o - C_l}{l} = DS\frac{p_o - p_l}{l} \tag{6}$$

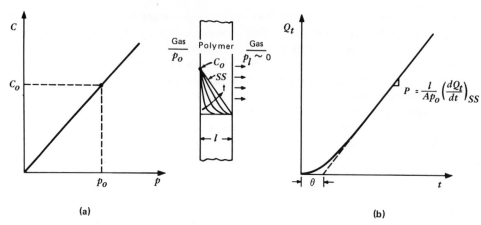

Figure 1.   Gas sorption (**a**) and permeation (**b**) in polymer membranes. Q_t = amount of gas that has permeated through the membrane in time t.

where subscripts l and o refer to the downstream and upstream sides of the membrane, respectively. Comparison of equations 1 and 6 reveals that the permeability, in this case, is given by

$$P = DS \tag{7}$$

and, thus, depends on both mobility and solubility of the gas in the membrane. These factors can be determined by the transient permeation experiment shown schematically in Figure 1. The steady-state slope gives the permeability; the time lag θ defined there can be used to calculate the diffusivity from (4)

$$D = \frac{l^2}{6\,\theta} \tag{8}$$

There are extensive tabulations of the permeability coefficient P for various gas–polymer systems (4–7), but information on D and S separately are available less readily (see Barrier polymers). For a given polymer, S increases as the condensibility of the gas (boiling points or critical temperatures provide a convenient measure) increases (4), and D decreases as the molecular diameter of the gas increases. For a given gas the main factor is the segmental mobility of the polymer. Both P and D increase rapidly with temperature, whereas S is less temperature sensitive and may increase or decrease depending on the system.

For polymers below their glass transition temperature, T_g, an additional sorption mechanism, which follows Langmuir's isotherm, is operative (8–9), making it necessary to replace equation 5 by

$$C = k_D p + \frac{C'_H bp}{1 + bp} \tag{9}$$

where k_D, C'_H, and b are parameters describing the nonlinear sorption isotherm. Transport is described best by a dual mobility model (9), which replaces equations 7 and 8 by more complex relations. The origin of these additional mechanisms of sorption and transport lie in microvoids that result from the nonequilibrium nature of the glassy state.

Vapors are much more soluble in polymers than are simple gases which cause plasticizing. The sorption isotherm is described best by the Flory-Huggins theory (10). Plots of concentration vs vapor activity (p/p^*, where $p^* =$ vapor pressure) show upward curvature and are not very temperature dependent. In general, the diffusion coefficient depends upon vapor concentration because of plasticization, and simple results like equation 6 cannot be used directly.

Hydraulic Permeation of Liquids. Hydraulically driven permeation of liquids through dense membranes can occur by a solution-diffusion mechanism (11–13). The rate of this process is strongly dependent on the extent that the permeating liquid swells the polymer membrane. The upper part of Figure 2 shows a membrane adjacent to a porous support plate which freely transports liquid component 1. Application of a pressure differential, $p_o - p_l$, causes a flux of liquid 1 through the membrane that is related by the definition of the hydraulic permeability

$$J_1 = K(p_o - p_l)/l \tag{10}$$

Mechanical arguments can be used to show that there is no pressure gradient within the membrane provided that it is nonporous but that the pressure is uniformly p_o as shown in Figure 1. However, the pressure differential does induce a concentration gradient of component 1 in the membrane. A thermodynamic analysis shows that the volume fraction of 1 in the membrane at the upstream face, v_{1o}, remains at its equilibrium value v_1^*; the volume fraction of 1 at the downstream face is reduced to v_{1l}. This

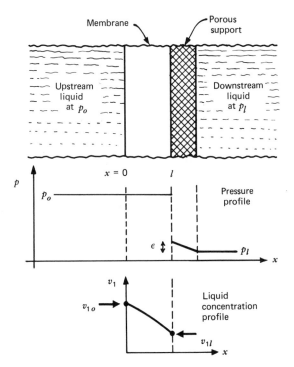

Figure 2. Schematic representation of hydraulic permeation through membranes. When the transport is by a solution-diffusion mechanism, the middle part shows the pressure profile and the lower part shows the liquid or solvent concentration profile in the membrane (12). Courtesy of Marcel Dekker, Inc.

reduction occurs because the pressure difference between the membrane and fluid phases at $x = l$ results in a reduced activity in the membrane phase for component 1, a_{1l}^m

$$a_{1l}^m = a_{1l}^L \exp\left[-(p_o - p_l)V_1/RT\right] \tag{11}$$

compared to its activity in the fluid phase a_{1l}^L. Combining knowledge of the sorption isotherm, ie, $v_1 = f(a_1)$, with equation 11 permits calculation of v_{1l}.

A form of equation 3 in terms of volumetric fluxes and volume fractions:

$$J_1 = -\frac{D}{(1 - v_1)}\frac{dv_1}{dx} \tag{12}$$

can be simply integrated when D is a constant to relate the flux to the concentration differential $(v_{1o} - v_{1l})$. The result is made somewhat complex by the $(1 - v_1)$ term in the denominator, but if the differential is not large, the result can be written as

$$J_1 = \frac{D}{(1 - v_1^*)}\frac{(v_1^* - v_{1l})}{l} \tag{13}$$

Such small differentials would mean that K in equation 10 does not depend on $(p_o - p_l)$, and in this limit, the hydraulic permeability can be written as

$$K = \frac{DV_1 v_1^*}{RT(1 - v_1^*)\left(\dfrac{\partial \ln a_1}{\partial \ln v_1}\right)} \tag{14}$$

where D = the mutual diffusion coefficient, V_1 = the molar volume of liquid 1, and the derivative in the denominator is a thermodynamic factor that can be computed from the sorption isotherm. In many cases, it is permissible to define a thermodynamic diffusion coefficient D_1 as

$$D = D_1\left(\frac{\partial \ln a_1}{\partial \ln v_1}\right) \tag{15}$$

and rewrite equation 14 as

$$K = \frac{D_1 V_1 v_1^*}{RT(1 - v_1^*)} \tag{16}$$

The diffusion coefficient D_1 is strongly dependent on the degree to which the polymer membrane is swollen by the permeating liquid as shown by the generalized plot (12,14) in Figure 3.

Note that equation 16 contains a product of a diffusivity and a thermodynamic solubility factor just as in the similar equation 7. If liquid 1 did not swell the polymer at all ($v_1^* = 0$), then it would not exhibit any hydraulic transport through the membrane.

The analysis of hydraulic permeation can be extended to mixtures of liquids (15); however, the more common multicomponent situation is a single-liquid solvent in which one or more solutes are dissolved (see Reverse osmosis).

Solutes. Frequently in membrane technology, the liquid phases on either side of the membrane contain one or more solutes that can be transported across the membrane by a solution-diffusion mechanism (16–17). These solutes are often solids in the pure state and may be inorganic salts, dyes, drugs, etc. The solute concentration

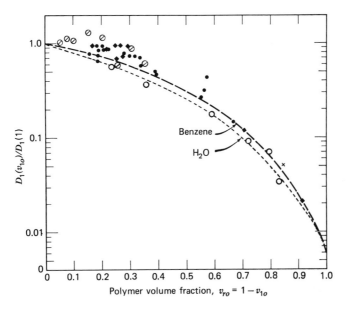

Figure 3. Effect of polymer volume fraction (or degree of swelling) on the thermodynamic diffusion coefficient of liquid, species 1, in a polymer membrane. Points were computed from hydraulic permeabilities for a variety of different membrane-liquid systems. Solid lines represent tracer measurements. Note that $v_{1o} = v_1^*$ in this case and D_1 (1) is the self-diffusion coefficient of the pure liquid (12). Courtesy of Marcel Dekker, Inc.

in the liquid phases is usually small, and in many cases there is a finite solubility limit of the solute in the solvent. In the simplest cases, the equilibrium partition of the solute between the membrane and liquid phases follows a linear distribution relation:

$$C_S^m = K_S C_S^l \tag{17}$$

where C_S^m is the concentration of solute in the membrane phase that is in equilibrium with a liquid phase containing solute at concentration C_S^l.

Fick's law for solute diffusion can be written for most cases as

$$J_S = D_S(C_{So}^m - C_{Sl}^m)/l \tag{18}$$

where the subscripts o and l denote conditions at the upstream and downstream surfaces of the membrane, respectively. Combining equations 17 and 18 yields

$$J_S = D_S K_S(C_{So}^L - C_{Sl}^L)/l \tag{19}$$

where the product $D_S K_S$ represents the solute permeability.

Both D_S and K_S depend on the natures of the solute and the membrane, and an important factor is the degree to which the solvent swells the membrane. Generally, the more the solvent swells the membrane, the larger are both D_S and K_S (18). As a first approximation, the partition coefficient K_S for inorganic salts is directly proportional to the water content in the membrane (19).

When the solute has a solubility limit in the solvent $C_{S, \text{sat}}^L$, there is an upper limit on the solute flux through the membrane that can be achieved:

$$J_{S, \text{max}} = \frac{D_S K_S C_{S, \text{sat}}^L}{l} \tag{20}$$

This case has important consequences in applications of membrane technology to controlled release of drugs and other agents.

Pore Transport Models. If the membrane has interconnecting static pores larger in diameter than molecular segment dimensions, then transport may no longer involve any molecular participation by the membrane material itself. Rather, the local environment in the pores is similar to the external fluid phase, and transport can occur in this medium in response to imposed concentration, pressure, or electrical potential gradients.

The nature of these transport processes depends on the size of the pores in relationship to the mean free path of the molecules in the fluid within the pores. If the pores are large compared to the mean free path, then continuum models can be used to describe transport within the pores. However, the complexity of the pore structure of most membranes often precludes development of *a priori* predictive equations. Figure 4 schematically illustrates the irregularity of pore structures typical of most membranes. A molecule must follow a very tortuous path that is longer than the membrane thickness by a tortuosity factor τ. Furthermore, this molecule is constrained to be within the fraction of the membrane volume ϵ (porosity) that the pores comprise (20). In the absence of a pressure gradient, simple unrestricted molecular diffusion of a solute may occur in this region if the pores are large compared to the molecular dimensions of the diffusing species. In this case, equations 17 and 18 may be replaced by

$$C_S^m = \epsilon C_S^l \tag{21}$$

$$J_S = D_S^l \, (C_{So}^m - C_{Sl}^m)/\tau l \tag{22}$$

and, thus, the permeability is given by

$$P_S = D_S^l \, \epsilon/\tau \tag{23}$$

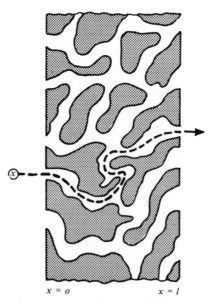

$$x = o \qquad\qquad x = l$$

Figure 4.   Illustration of tortuous diffusion path through a porous membrane.

where D_S^l is the solute diffusion coefficient in the fluid phase. Thus, selective transport in the membrane would be based only on the relative rates at which two solutes would diffuse in the fluid phase without the membrane. However, as the membrane pores become closer in size to the dimensions of the solute, the membrane imposes some restrictions on the transport (21–25); the result of this effect is, in general, to increase the membrane selectivity. Certain macromolecules and particles may be larger than the membrane pores and be excluded completely from the pores (26–28). This principle is used extensively in development of dialysis and filtration membranes (see Filtration).

A pressure differential across this type of membrane causes a hydrodynamic flow of the fluid in the pores if they are of large enough diameter. For a membrane with n uniform, cylindrical pores per unit area, the mass flux of fluid can be expressed in terms of the Hagen-Poiseuille relationship (2,29) provided the Reynold's number is less than 2100:

$$J = n\rho \frac{\pi r^4}{8\,\mu l}\,(p_o - p_l) \tag{24}$$

where r = pore radius, μ = fluid viscosity, and ρ = fluid density (evaluated at the mean pressure $p_o + p_l/2$ for gases). The hydraulic permeability based on volumetric flux is

$$K = n\pi r^4/8\,\mu = \epsilon r^2/8\,\mu \tag{25}$$

Even for complex pore geometries, Darcy's law (30):

$$J = \frac{K}{\mu}\frac{(p_o - p_l)}{l} \tag{26}$$

has been found applicable. The coefficient in Darcy's law may be interpreted in terms of a capillary model using average pore dimensions and tortuosities. Flow of this type results in no separation of components unless one of the components is physically excluded from the pores because of size or electrical charge.

When the fluid is a gas, the mean free path may be larger than the pore diameter. As a result, equations 23 or 26 are not applicable. Rather than colliding with each other, the gas molecules collide more often with the pore walls, and transport is of the Knudsen type (31) rather than Poiseuille flow or simple molecular diffusion. From kinetic theory it has been shown that in the Knudsen regime, the molar rate of gas flow through a long cylindrical capillary is given by

$$\text{molar rate} = \frac{8\,\pi r^3}{3(2\,\pi RMT)^{1/2}}\frac{(p_o - p_l)}{l} \tag{27}$$

where M is the molecular weight of the gas. Analogous relationships can be developed for other pore geometries. For microporous membranes, the result can be expressed generally as

$$\frac{\text{molar rate of gas flow}}{\text{cross-sectional area of pores}} = \frac{G}{(2\,\pi RMT)^{1/2}}\left(\frac{p_o - p_l}{l}\right) \tag{28}$$

where G is a factor related to pore geometry and M is the molecular weight of the gas. These relations can be recast in the form of diffusion equations, which upon integration yield the following permeability expressions for gas component i:

$$J_i = D_{Ki}\frac{\Delta C_i}{l} = \frac{D_{Ki}}{RT}\frac{\Delta p_i}{l} \tag{29}$$

The Knudsen diffusion coefficient for component i in a microporous membrane can be written in terms of model equations that involve the average pore radius, tortuosity, and porosity:

$$D_{Ki} = k\frac{\bar{r}\epsilon}{\tau}\sqrt{\frac{T}{M_i}} \tag{30}$$

For a multicomponent gas, each component diffuses independently since the collisions are mainly with the pore walls and not with other gas molecules. Thus, this mechanism can lead to separation of components owing to the difference in molecular weights. Ideally, the separation factor based on mol fractions

$$\alpha_{12} = \frac{(y_{1l}/y_{1o})}{(y_{2l}/y_{2o})} \tag{31}$$

is given by (29)

$$\alpha_{12} = \sqrt{\frac{M_2}{M_1}} \tag{32}$$

This principle is used to accomplish gas separations.

There is a transition region in which neither the Knudsen nor Poiseuille regime is strictly applicable; however, this will not be dealt with here.

Electrochemical Phenomena.   When the solute or membrane is ionized, electrical potential can play a role in transport just as do concentration and pressure (29,32). An extension of Fick's law for an ionic species i in an electrical field $d\phi/dx$ gives

$$J_i = D_i\left[\frac{dC_i}{dx} + \frac{Z_i C_i F}{RT}\frac{d\phi}{dx}\right] \tag{33}$$

where Z_i = ionic valence (including sign) and F = the Faraday constant. This is the Nernst-Planck equation in a form that assumes thermodynamic ideality of the solution. The system must maintain electrical neutrality, hence

$$\sum_i Z_i C_i = 0 \tag{34}$$

Application of an electrical field on such a system may induce a transport flux of ions which, in general, is opposed by molecular diffusion from the concentration gradient induced. Conversely, molecular diffusion by an imposed concentration gradient of ion i induces an electrical field, or streaming potential, that opposes this transport and affects all other ions in the system.

A membrane whose structure includes ionizable groups can have a net concentration of fixed charges X of valence Z. This may have important consequences for equilibrium and transport of ionizable species. Such ion-exchange membranes may contain negative ions (eg, carboxyl or sulfonic acid groups) or positive ions (eg, quaternary ammonium groups) and are referred to as cation- or anion-exchange membranes, respectively (see Ion exchange). The condition of net electrical neutrality within the membrane

$$\sum Z_i C_i^m + ZX = 0 \tag{35}$$

tends to exclude sorption of ions of the same charge as the fixed ions attached to the membrane, ie, co-ions. A complete thermodynamic analysis that equates chemical potentials of each component in the membrane to those in the external fluid phase leads to a Donnan potential and distribution relations for each free component. At low concentrations, anion-exchange membranes allow transport of anions but do not pass cations. Similarly, cation-exchange membranes allow transport of cations but not anions. The extent of exclusion of the co-ions depends on the relative concentrations of the co-ion and the fixed charge.

Details of electrical phenomena such as the Donnan potential, electroosmosis, streaming potential, etc, may be found in many of the references given in the General Bibliography. However, one important but simple example of the Donnan effect is as follows: if electrolyte solutions are placed on the right, R, and left, L, sides of a membrane permeable to cations but not to anions or water, then the anion concentrations of the two solutions will not change. However, the cations will redistribute between the right and left solutions until an equilibrium is established. If the cations are a mixture Na^+ and Ca^{2+}, then at equilibrium

$$\frac{[Ca^{2+}]_R}{[Ca^{2+}]_L} = \left(\frac{[Na^+]_R}{[Na^+]_L}\right)^2 \tag{36}$$

Thus, by placing a dilute Ca^{2+} solution on one side of the membrane and a concentrated Na^+ solution on the other, one can develop a more concentrated Ca^{2+} solution by what has been referred to as Donnan dialysis (33).

Other Mechanisms. Membrane phenomena such as osmosis, reverse osmosis, and electrodialysis can be understood and analyzed in terms of mechanisms and models presented earlier. Temperature gradients may lead to species transport that can be explained in terms of irreversible thermodynamics. Mechanistic models of thermal diffusion can be developed in certain instances, eg, in the Knudsen regime (34).

When diffusional phenomena are coupled with chemical reactions, the resultant transport rates may be greatly affected. For example, in facilitated transport, carrier molecules present in a membrane react or complex with the permeating species to create an additional gradient that contributes to transport (35). A well-known example is hemoglobin in blood which has been shown to increase oxygen transport rates manyfold compared to the rates possible in its absence (36). Such carriers may be looked upon as a means of increasing the solubility of the penetrant in the membrane. Facilitated transport does not affect the ultimate equilibrium state but simply acts as a catalyst which increases the rate at which equilibrium is approached.

In living systems, there are other phenomena that may allow species to be transported against a gradient of composition or electrical potential or in the direction opposite to simple thermodynamic equilibrium. Such processes do not violate thermodynamic principles but simply involve a source of chemical energy that drives the transport; this is usually called active transport in contrast to the passive transport involved in all previous discussions here. Accumulation of potassium ions in living cells is a popular example of this process. References in the General Bibliography provide further discussion of this subject.

Physical Structure of Membranes

The most important synthetic membranes are formed from organic polymers. They perform functions that also could be performed by metals, carbon, inorganic glasses, and other materials, but because of their predominant importance in current membrane technology, the focus here is on organic polymers. Early artificial membranes were based on natural polymers such as cellulose (qv), and these still are used (37). Because of the demand for more versatile and highly tailored membranes, membrane technology currently employs a wide range of other polymeric materials, some synthesized especially for this purpose (38). The chemical structures of these polymers range from simple hydrocarbons (like polyethylene or polypropylene) to polar structures (like polyamides) or ionic structures in which cations or anions are attached to the backbone (see Olefin polymers; Polyamides). Performance, therefore, may depend upon physiochemical interactions between the permeating species with the membrane material including strong ionic interactions, weaker dipolar interactions, and quite weak van der Waals forces (39). However, in all cases, both the physical microstructure and the macro-form of the membrane are important considerations.

Basically there are two microstructural forms—porous and nonporous. In the extreme cases, there is no difficulty in classifying membranes as one or the other of these, although there is a middle ground that is difficult to categorize. Membranes may have an asymmetry of physical structure characterized by a gradation of pore sizes through the membrane or, in the extreme, a dense layer on top of a porous substructure. Composite membranes may be formed in which two materials are arranged in series or other combinations. Such composite membranes potentially can be highly tailored to meet specific demands, including both transport and mechanical properties (20).

The polymer alternatively may be wholly amorphous or partially crystalline. In either case, the amorphous material may be above or below its glass-transition temperature. For porous membranes, transport performance generally does not depend on these issues; however, mechanical performance and chemical resistance do. For dense membranes, transport behavior is affected greatly by all of these issues.

The macroscopic form of the membrane may be that of a flat film, a thin-walled but large-diameter tube, or fine hollow fibers in which the walls constitute a significant fraction of the diameter. Each of these macroforms may employ any of the microstructural forms of the polymer described above.

Dense Films.   In its usage here, the term dense implies that there are no pores of microscopic dimensions and that all unoccupied volume is simply free space between the segments of the macromolecular chains. In an amorphous polymer above its glass-transition temperature, the chain segments undergo motion similar to the thermal motions of molecules in liquids. In fact, the local structure may be regarded as identical to that of a simple liquid. Therefore, transport through any membrane of this type must be by a solution-diffusion mechanism. The penetrating molecules may interact strongly or weakly with the polymer segments depending upon the structure of each; however, the rate of transport is strongly affected by the cooperative molecular motions of the polymer segments. Temperature has a large affect on these motions, and transport parameters frequently follow an Arrhenius relationship. As a result, this is frequently referred to as activated diffusion (40).

As the temperature of an amorphous polymer is reduced, part of its volume may crystallize. The crystals are characterized by denser and more ordered packing which apparently excludes penetrating molecules. Hence, virtually no transport occurs in crystalline regions of polymers. When polymers crystallize, there remains an amorphous phase that in practical cases comprises of the order of half the volume. Transport may continue to occur in this amorphous phase below the melting point; however, the crystalline regions act as obstructions that decrease the area for diffusion and increase the diffusional path length. Thus, the morphological arrangement of these crystalline and noncrystalline phases plays an important role in the macroscopic transport behavior of polymer membranes. Clearly, this two-phase picture is only an approximation to reality, and the boundaries between amorphous and crystalline regions are not clear. Abnormal packing defects can be envisioned in the amorphous regions of semicrystalline polymers.

Upon further cooling, whether the polymer crystallizes or not, the temperature will fall below the glass transition of the polymer, at which temperature rotational motion of the chain segments cannot occur on the time-scale of conventional observations. The polymer becomes a rigid glass whose local segmental packing still resembles that of a liquid but which is no longer an equilibrium. Thus, spaces between chains no longer have the same dynamic character as in liquids; they more nearly may resemble pores although their dimensions are still molecular (9). Transport by solution-diffusion mechanisms still occurs in the glassy state; however, these processes are somewhat more complex as described above.

An important consideration in dense membranes is the extent to which the adjacent fluid phase is imbibed into it or acts as a solvent for the polymer. As described above, solubility in the membrane is crucial to the transport mechanism (41–42); however, in the extreme case, the membrane may be completely dissolved by the solvent. The extent of sorption is determined by the interaction of the solvent with the polymer segments—a well-known problem in considerations of polymer solubility and chemical resistance. Sorption of solvent may reduce the melting point of the polymer and its glass-transition temperature. Sorption of significant quantities of solvents, vapors, or gases plasticizes the polymer by increasing the mobility of its segments (43) and its diffusion coefficients.

There is no detailed or universal theory to predict transport rates of various penetrants in dense polymers. However, there are many limited relationships useful for selecting membrane materials. For separation purposes, selectivity among penetrants is a vital issue as is productivity. Generally, materials with high productivity intrinsically have very poor selectivity, and vice versa.

Dense membranes can be formed by a wide variety of techniques including solution techniques, melt processing, or direct polymerization (20). Solution methods generally involve casting a film and then completely evaporating the solvent; care must be exercised not to introduce microscopic voids or pores. In some cases, the type of solvent and method of its removal significantly affect the membrane properties (44), especially when the polymer is crystallizable or forms a glass whose structure is dependent on past history (45). Melt-extrusion techniques follow the same strategy as in plastics processing. However, many important polymers used in membrane formation cannot survive such processing without chemical degradation and, therefore, require solution-processing methods (see Film and sheeting materials; Plastics processing).

If the membrane must be cross-linked, then simple solution or melt processing is often not applicable. Rather, the membrane is formed directly by polymerization with a cross-linking agent. In some cases the membrane can be made by melt or solution methods followed by cross-linking as a secondary reaction (20).

Chemical modifications of the polymer after the membrane has been formed offers a way to tailor the membrane uniquely to its intended function. Such reactions might include the creation of ionic groups on the chain or the grafting of another monomer onto the existing backbone (46).

Because transport rates through dense films are inherently very slow, they are not attractive for large-scale application in separation processes where productivity is of paramount concern. On the other hand, such films are ideal barriers for use in packaging foods and other items where certain components are to be kept in and others kept out. There is a large industry for melt processing polyethylene, polystyrene, polypropylene, poly(vinyl chloride), poly(vinylidene chloride), etc, for these purposes. Dense films can also be used to regulate the rate of passage of components as in controlled-release applications described below. Attempts have been made to create ultrathin dense films with thickness of 5–5000 nm; however, such materials are exceedingly difficult to handle.

Porous Membranes. Membranes containing voids that are large in comparison with molecular dimensions are considered porous. In these membranes, the pores are interconnected, and the polymer may comprise only a few percent of the total volume. Transport, whether driven by pressure, concentration, or electrical potential, occurs within these pores. The essential transport characteristics are determined by the pore structure with selectivity being governed primarily by the relative size of solute molecules or particles compared to the membrane pores (47). The mechanical properties and chemical resistance of the membrane are greatly affected by the nature of the polymer.

The simplest pore geometry to envision is that of parallel capillaries of uniform dimension extending throughout the thickness of the membrane. Such structures can be approximated by bombarding a polymer film with certain fragments from radioactive decay which cause tracks of damaged material that can be etched chemically to form a uniform pore (48) (see Particle-track etching). Ionotropic gels with highly regular cylindrical pores also have been described (49). However, most other processes for creating porous membranes do not approximate this geometry, and produce structures similar to a sponge. One of the most versatile techniques is to effect phase separation within a polymer solution by adding a nonsolvent (see example in Fig. 5) or by changing the temperature (20). If a continuous porous membrane is to result, the two phases must be interpenetrating. Other techniques for forming porous membranes include controlled hot and cold stretching of polymer films, extraction of a soluble phase, or sintering of fine particles. As an example of the latter, porous polytetrafluoroethylene films are obtained from an emulsion containing some volatile or heat-degradable compounds easily removed while sintering the solid particles (see Fluorine compounds, organic). Table 1 summarizes various methods of forming microporous membranes and gives some examples of their commercial uses.

The pore structure of the typical microporous membrane has only statistical significance. Information about pore sizes and distributions can be obtained by a variety of techniques including electron microscopy. Other techniques include measurement of porosity, of surface area by BET sorption, and of average pore radius by hydraulic permeation, all of which give some information about average pore geometry. However, it frequently is useful to know the largest effective pore dimensions. Measurement of the bubble pressure commonly is used for this purpose (ANSI/ASTM E 128-61 and F 316-70). In this method, a gas is used to displace a liquid from the pores and the pressure required to do this is related to pore radius and surface tension as follows:

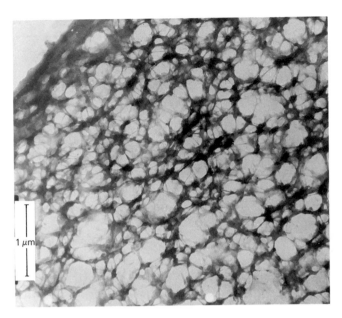

Figure 5. Transmission electron photomicrograph of a thin section of a microporous membrane made by coagulation.

Table 1. Examples of Microporous Membranes

Polymer	Trade name (company)	Process	Refs.
cellulose esters, poly(vinyl chloride), high temperature aromatic polymers, etc	(Millipore) (Gelman)	coagulation of polymer solution by a nonsolvent	20, 50–51
polytetrafluoroethylene	Gore-Tex (W. L. Gore)	sintering fine polymer particles	52
polypropylene	Celguard (Celanese)	controlled stretching of polymer films	53
polycarbonate	Nuclepore (Nuclepore)	chemical etching of tracks formed by radiation damage	20
polypropylene, polystyrene, polyethylene, nylon, etc	Accurel (Armak division of Akzona)	liquid-phase separation of polymer solution by cooling	54

$$r = \frac{2\,\gamma}{p} \qquad (37)$$

Commercial microporous membranes have pore dimensions of ca 0.005–20 μm. They are made from a wide variety of polymers in order to provide a range of chemical and solvent resistances. Some are fiber or fabric reinforced to obtain the required mechanical rigidity and strength. The operational characteristics of the membrane are defined sometimes in terms of particles or molecules that will pass through the membrane pore structure. Figure 6 shows the general relationship between percent retention and molecular size of a series of macromolecular compounds. It is common to speak in terms of a cut-off molecular weight which can be varied by changing the dimensions of the membrane pores. The shape of the retention curve is affected by

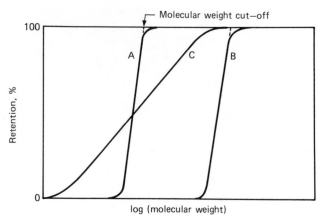

Figure 6.   Typical solute retention characteristics. Membranes A and B exhibit sharp cut-offs; C shows a more diffuse cut-off.

pore size distribution. Such diagrams should be used with considerable caution since most macromolecular compounds in solution are flexible rather than of fixed geometry and molecular size is not determined solely by molecular weight.

Microporous membranes are often used as filters. Those with relatively large pores are used in separating coarse disperse, suspended substances such as particulate contamination in refined sugar, oil pumps, hydraulic oils, etc. Membranes with smaller pores are used for sterile filtration of gases, separation of aerosols, and sterile filtration of pharmaceutical, biological, and heat-sensitive solutions (see Sterile techniques; Ultrafiltration). The very finest membranes may be used to separate or purify soluble macromolecular species. The term microfiltration describes such processes in which the particles are typically 0.1–10 μm; ultrafiltration refers to processes involving small particles or macromolecular compounds.

Porous membranes also are used in dialysis applications such as removing waste from human blood (hemodialysis), for separation of biopolymers with molecular weights ranging from 10,000 to 100,000, and for analytical measurement of polymer molecular weights. Microporous membranes also may be used as supports for very thin, dense skins or as containers for liquid membranes.

Asymmetric and Composite Membranes.   The ability of dense membranes to transport species selectively makes possible molecular separation processes such as desalination of water or gas purification, but with normal thicknesses these rates are extremely slow (see Water, supply and desalination). In principle, the membranes could be made thin enough that the rates would be attractive, but such thin membranes would be very difficult to form and to handle, and they would have difficulty supporting the stresses imposed by the application. Conversely, microporous membranes have high transport rates but very poor selectivity for small molecules. One of the greatest advances in the field of membrane technology was the resolution of this dilemma through use of asymmetric membranes in which a very thin, dense membrane is placed in series with a porous substructure.

The first step towards this concept was the preparation of a cellulose diacetate membrane with an asymmetric structure by an empirically developed casting procedure (55). This membrane recipe was shown to result in a thin dense layer about 0.2

μm thick which rests on a much thicker porous supporting layer. The preparation begins with a concentrated solution of cellulose diacetate, ca 25 wt %, in a mixture of acetone and formamide. The solution is spread on a glass plate to the desired thickness, and the solvent is allowed to evaporate for a brief period. This evaporation creates a gradient of polymer concentration that is responsible for the subsequent asymmetric structure. After the evaporation period, the plate is immersed in ice water to precipitate the polymer. Water is soluble in the acetone–formamide mixture but not in the polymer. As water progressively diffuses into the swollen polymer, nucleation of a polymer phase occurs. This coagulation process leads to a porous polymer structure with the porosity and pore size being related to the polymer concentration prior to coagulation (56–57). Thus, because of the concentration gradient created during the evaporation step, there is a gradation of porosity and pore size across the membrane thickness. The subsequent treatment of the membrane culminates in heat treatment in water typically at 80–85°C. This annealing causes some collapse of the smaller pores and a reordering of the polymer, and thus, the final structure of the membrane depends on the time and temperature of this procedure. The result may be a composite in which there is an effectively dense skin over a porous substrate. The dense skin is primarily responsible for the transport properties of the membrane—high selectivity because the skin is dense but high rates because the skin is thin. The porous sublayer has little affect on the transport characteristics of the membrane but gives the membrane mechanical integrity and strength. Asymmetric membranes of this type were used first for reverse-osmosis desalination but they are now used in many other areas of membrane technology.

Cellulose acetate asymmetric membranes still have a wide range of uses, in spite of such drawbacks as poor bacteriological resistance and inferior mechanical properties (see Cellulose derivatives). Other polymers have been developed that can form the Loeb-Sourirajan-type asymmetric structure through the proper choice of solvents, nonsolvents, and casting conditions (58–63).

Composite membranes of two different materials form a related structure. Here, a porous membrane is coated with another polymer or monomer that covers the surface and penetrates into the pores. One approach uses a microporous polysulfone membrane as the substrate with various reactive treatments including polyethylenimine cross-linked by toluene diisocyanate to create a highly cross-linked, salt-rejecting, interfacial layer useful for reverse osmosis (63) (see Imines, cyclic; Polymers containing sulfur).

When asymmetric membranes are operated with high applied pressure, a gradual decline in flux generally occurs. This is probably the result of collapse or compaction of part of the porous substructure, effectively making the dense layer thicker. This phenomenon is sometimes referred to as permeability creep. It can be combatted by proper selection of substrate materials and optimization of its pore structure by manipulation of the physiochemical variables of its formation.

Hollow Fibers. The economics of using membranes for separation processes dictates the development of a high membrane surface area per unit of volume of container. An ideal geometry for this purpose is fine hollow fibers. Moreover, hollow fibers may be self-supporting and thus eliminate the need for expensive support hardware. Hollow fibers with inside diameters as small as 10 μm can be formed using spinning technology adapted from the synthetic fiber industry. Basically, the polymer is extruded through an annular hole, and an appropriate fluid is injected into the bore to prevent collapse. Early work on hollow fibers was directed toward reverse-osmosis applications but now these geometries are used in a variety of other membrane applications (64).

Hollow fibers with totally dense walls can be readily prepared by melt spinning of the selected polymer. However, they are relatively uninteresting because the thick wall results in very slow transport rates. Adaptation of solution spinning technology has produced an asymmetric Loeb-type structure in hollow-fiber form. Initial efforts were devoted to cellulose acetate materials but now a wide variety of polymers can be converted to asymmetric hollow-fiber membranes. The approach involves extruding a solution of the polymer in an appropriate solvent. Coagulation is accomplished by passing the filament through a bath of a suitable nonsolvent. The evaporation step can be achieved by providing an air space between the spinnerette face and coagulation bath in a process called dry jet-wet spinning which combines features of both wet and dry spinning. With this approach, the pore structure of the fiber can be varied over a wide range to create membranes for specific purposes. In addition, composite membranes can be prepared by coating with other materials (see Hollow-fiber membranes).

Membrane Module Configurations

The first requirement in any membrane process is a membrane capable of the function needed, but successful implementation requires packaging the membrane in a module whose configuration is engineered for the specific application. Membranes may be formed as flat sheets, tubes of relatively large diameter, or fine hollow fibers. Modules have been developed to accommodate each of these. Important economic considerations in their design and operation include the cost of the supporting and containing vessels (which is largely determined by the ratio of membrane area per module volume that can be achieved), power consumption in fluid pumping, and how much of the module hardware can be reused when the membrane is replaced.

The module configuration has a significant affect on the fluid dynamics and, thus, the spatial concentration pattern in the field phases (29). Two separate issues affect the performance of membrane processes. The first of these is concentration polarization, local concentration gradients in the upstream fluid phase normal to the membrane surface caused by selective passage of components (65). This phenomenon affects the driving forces across the membrane and results in poorer selectivity and productivity in processes like reverse osmosis. It can be combated by introducing fluid motions that help erase these gradients by convective transport or by reducing the system dimensions so that molecular diffusion can eliminate them more effectively. The second factor in performance is formation of macroscopic fluid-phase concentration gradients along the module length in the absence of complete back-mixing. For most purposes, it is desirable to achieve plug-flow behavior and to avoid back-mixing. For plug flow, the two streams may be countercurrent or cocurrent to each other. The former is more efficient for separations.

Flat Sheet.   The initial development of a membrane invariably involves small flat sheets because of the simplicity of their preparation and subsequent testing. This is especially important for screening polymers available only in small quantities. In some cases, flat-sheet modules are scaled up and translated into commercial practice. Figure 7 shows simple configurations that can be operated in batch or continuous-flow modes.

If any pressure is to be applied, the membrane must be supported. Such a design has an inherently low ratio of membrane area to module volume. However, in practice, flat-sheet modules are much more complex than that shown in Figure 7. Some employ multiplate cartridges (66); others resemble plate-and-frame filter presses. All practical designs take considerable care to control fluid motions above the membrane in order to combat concentration polarization and back-mixing.

Tubular. Membranes can be prepared in tubular form by joining flat sheets with an appropriate seam or by direct casting onto a cylindrical form. One approach to the latter is to introduce the casting solution inside a tube or pipe that is spun about its axis so that centrifugal forces generate a uniform film on the inside wall. Generally, a tubular module design employs a porous tube of 0.5–5 cm diameter, which serves both as the membrane support and as a pressure vessel. In principle, the membrane could be placed either on the inside or outside of the porous support tube. In the schematic design of Figure 8, the membrane is inside the porous tube where the feed passes under pressure. Solutes or purified liquid permeate the tubular membrane along its entire length, and rejected fluid is discharged at the exit of the tube. An appropriate number of membrane-tube units are connected in parallel inside a modular container; manifolds provide connections for feed, reject, and permeate. One or more container modules may be connected in series or parallel as a treatment plant.

Tubular supports are most commonly made of rigid, porous fiberglass. Perforated metal tubes or nonporous tubes containing a grooved liner also may be used. The membrane is reinforced with an appropriate backing material to protect it from damage against the pores or grooves in the support. Because of the large diameter, the feed pressure drop down the tube is small. Thus, long tubes can be used, and viscous or concentrated feeds can be handled with relative ease. The tubes can be cleaned *in*

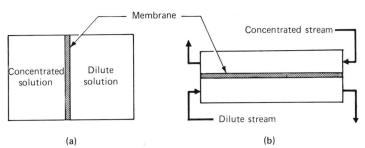

Figure 7. Schematic illustration of a flat membrane configuration as might be used for batch (**a**) or continuous (**b**) dialysis.

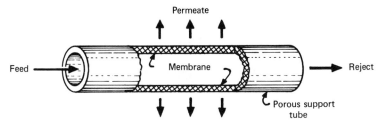

Figure 8. Tubular membrane configuration.

situ. Given the proper support, high pressure can be employed. The velocity of the feed in the tube affects the extent of concentration polarization.

Spiral Wound.   The spiral-wound concept (Fig. 9) permits use of flat sheet membranes in a configuration that avoids some of the limitations of flat membranes (67). In this form, a laminated structure is wound around a central tube prior to installation in an ordinary pipe which serves as the containing vessel. The laminate consists of two membranes separated by porous feed and permeate spacers through which these fluids flow. The laminate is sealed on three edges by an adhesive; a fourth edge is open and connected to a central, perforated permeate-collection tube on which the laminate is wrapped. The feed fluid flows axially along the module in the channels between laminates created by the porous feed spacers. Permeate spirals inward along the channel between membranes created by the porous permeate spacer until it reaches the central tube where it flows axially from the module.

Hollow Fibers.   Hollow-fiber membranes are used in modules that have some similarity to shell-and-tube heat exchangers. Fiber ends are embedded in a tube sheet frequently of an epoxy formulation. Often, each module has only one tube sheet with the fibers either looped (Fig. 10) or plugged at one end. In some cases, there may be two tube sheets as would be necessary if the feed were to flow down the fiber bore rather than on the shell side. Internal pressurization of hollow fibers requires a sturdier fiber wall than does shell-side pressurization since the former involves tensile stresses and the latter only compressive stresses. Because of the small dimensions of the fibers and the distance between them, molecular diffusion within the shell-side fluid helps reduce

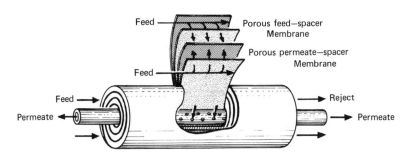

Figure 9.   Spiral wound membrane configuration.

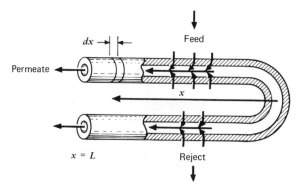

Figure 10.   Looped-end hollow fiber.

the problem of concentration polarization. Hollow-fiber modules generally offer the highest membrane area and productivity per module volume of all the configurations considered here (see Table 2). However, other configurations have advantages that suit them better to some applications.

Because of the small diameter of the fiber bore, there may be a significant pressure drop along its length in order to force the permeate from the bore. For pressure-driven processes, this leads to a loss in driving force across the fiber wall and consequent loss in productivity and separation selectivity. The relationship between the flow rate of fluid and pressure gradient in the bore at position x (Fig. 10), is approximated by the Hagen-Poiseuille equation:

$$Q_x = - \frac{\pi r^4}{8 \mu} \frac{dp_x}{dx} \tag{38}$$

The next increment of length dx adds an additional amount of fluid to the bore dQ_x by permeation through the wall:

$$dQ_x = K(P - p_x)2 \, \pi r dx \tag{39}$$

where P is shell-side pressure (assumed constant) and K is an appropriately defined permeability. These two equations can be combined into a differential equation and solved with appropriate boundary conditions (69). From the solution, one can calculate the pressure in the bore at the looped end p_o compared to that at the exit, p_L:

$$p_o - p_L = (P - p_L) \left[1 - \frac{1}{\cosh aL} \right] \tag{40}$$

and the flow from one end of the fiber bore:

$$Q_L = \frac{\pi r^{5/2}}{2} \sqrt{\frac{K}{\mu}} (P - p_L) \tanh aL \tag{41}$$

where $a \equiv 4\sqrt{\mu K/r^3}$. An efficiency factor that accounts for the permeation reduction by bore-side pressure gradients can be defined as follows:

$$\text{efficiency} = \frac{Q_L}{2 \, \pi r L K(P - p_L)} = \frac{\tanh aL}{aL} \tag{42}$$

This phenomenon limits the length of fibers that can be used practically in modules for separation processes, and the above analysis can be used along with other considerations to make such design decisions.

Table 2. Comparison of Various Membrane-Module Configurations[a]

Module type	Packing density, m²/m³	Flux density m³/(m³·d)
plate and frame	330	270[b]
tubular	220	180[b]
spiral wound	660	540[b]
hollow fiber	9200	700[c]

[a] Ref. 68.
[b] Assumes flux of 0.813 m³/(m²·d) at 4.1 MPa (600 psi).
[c] Assumes flux of 0.073 m³/(m²·d) at 2.7 MPa (400 psi).

Separation Processes

Separation or purification processes consume energy, and any reduction in this consumption is very attractive in view of the rapidly rising costs for energy. Even a thermodynamically reversible process must expend an amount of energy equivalent to the free energy change on mixing to drive the spontaneous mixing process backwards:

$$\Delta G_{mix} = \Delta H_{mix} - T\Delta S_{mix} \tag{43}$$

Figure 11 illustrates this point by showing the minimum energy expenditure to recover fresh water from the sea (70) as a function of the fraction of the water recovered. Real processes operate with finite driving forces and are, therefore, irreversible. As a result, they consume more energy than this minimum (see also Separation systems synthesis).

Membrane processes offer the potential for energy-efficient separations. As a result, most membrane applications are in this area, particularly water purification. This section illustrates a selected variety of membrane uses in separation processes. Some of these already have significant commercial importance, others appear to be on their way, and some have had no economic impact at all.

Reverse Osmosis (Hyperfiltration). When solutions of different solute contents are placed adjacent to each other, solute diffuses between the two until a uniform concentration results. However, if these two fluid phases are separated by a semipermeable membrane, this route to equilibrium is blocked—semipermeable membranes in this context are rather impermeable to solutes (salts) but transport solvent

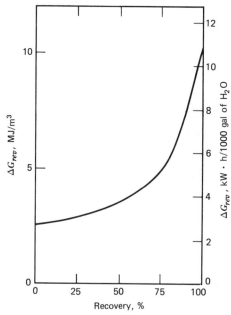

Figure 11. Reversible free-energy requirement to produce fresh water from sea water (3.5% NaCl).

(water) more readily. Instead, solvent is transferred in an attempt to dilute the solute content of the more concentrated solution; this process is called osmosis. Another equilibrium can be reached by establishing a pressure on the solution phase, called the osmotic pressure π, which exactly balances the reduction in the chemical potential of the solvent caused by adding solute. This leads to the following thermodynamic definition of osmotic pressure for a solution:

$$\pi = -\frac{RT}{V_1} \ln a_1 \tag{44}$$

where V_1 = solvent molar volume and a_1 = activity of solvent in the solution. For dilute, ideal solutions this exact result may be replaced by the van't Hoff relation:

$$\pi = \frac{RTC}{M_2} \tag{45}$$

where M_2 = solute molecular weight. For NaCl–H$_2$O mixtures, a rule of thumb is that each unit increase in NaCl weight percent raises π by 700 kPa (ca 7 atm or 100 psi) (70). For macromolecular solutions, osmotic pressures are much lower (see equation 44); for a 10% solution of a polymer with $M = 10^4$, $\pi \approx 70$ kPa (10 psi).

Reverse osmosis (RO) is simply the application of pressure on a solution in excess of the osmotic pressure to create a driving force that reverses the direction of osmotic solvent transfer. If the membrane is adequately semipermeable, this results in purified solvent and is, thus, an attractive way to desalt water. Although osmotic phenomena have been known for two centuries, it was not until the early 1960s that serious thought was given to using reverse osmosis for desalination. By the 1970s the reverse-osmosis industry was estimated to be valued at more than 10^8/yr (71). All of the membranes in current use are asymmetric ones configured into tubular, spiral-wound, or hollow-fiber modules.

The transport behavior in reverse osmosis can be analyzed elegantly in terms of the general theories of irreversible thermodynamics (72); however, a simplified solution-diffusion model proves more instructive here since this seems to reflect quite well the actual mechanism involved in most RO systems. Most successful RO membranes sorb ca 5–15% water at equilibrium. Thus, it is a good approximation to assume thermodynamic ideality, ie, $(\partial \ln a_1/\partial \ln v_1) = 1$, and $(1 - v_1^*) \approx 1$. This assumption considerably simplifies the hydraulic permeability expression of equation 14. Since the driving force for hydraulic permeation is the pressure differential in excess of the osmotic pressure differential, the water flux can be approximated using the simplified version of equation 14 as follows:

$$J_w = \frac{D_w V_w v_w^*}{RTl} (\Delta p - \Delta \pi) \tag{46}$$

where D_w is the molecular diffusion coefficient of water in the membrane. The solute or salt flux will be given by equation 19 provided the pressure differential does not contribute significantly to the driving force for its transport—the usual case.

It is common to state the selectivity characteristics of an RO membrane system in terms of a rejection coefficient R defined as

$$R \equiv 1 - \frac{C_{Sl}^L}{C_{So}^L} \tag{47}$$

where C_{So}^L and C_{Sl}^L are the solute concentrations in the upstream and downstream liquid phases, respectively. The salt content on the downstream side of the membrane is determined by the ratio of salt and water fluxes through the membrane. Combining results from the solution-diffusion model with the definition of the rejection coefficient gives

$$R = \left(1 + \frac{D_S K_S R T}{D_w V_w v_w^* \left(\Delta p - \Delta \pi\right)}\right)^{-1} \tag{48}$$

Note that the rejection is independent of the membrane thickness, but the flux increases as the membrane becomes thinner. This is precisely the advantage of asymmetric membranes; ie, they have the high rejections intrinsic to the dense membrane with simultaneous high fluxes owing to the very thin, dense active skin. Equation 48 predicts that rejection increases with applied pressure as is observed experimentally. However, the perfect rejection that equation 48 predicts at very large pressures is not observed in practical membranes, apparently owing to imperfections.

Current commercial membranes are made of cellulose acetate, various aromatic polyamides, and certain composites (73). Rejections are of the order of 99% for NaCl, depending upon the relative transport properties of the membrane to salt and water. Fluxes as high as 1.0 m³/(m²·d) can be obtained through optimized asymmetric structures. Rejection of other ions such as calcium and magnesium are much higher than those for sodium. Most units are sold for brackish water purification but some for seawater purification have been installed. The largest plant installed to date produces 12,000 m³/d (3.2×10^6 gal/d) of fresh water (74). Other water purification applications are not yet as significant as desalination.

Of course, reverse osmosis competes with other technologies for desalination, and economics is the major factor in determining which is used. On the basis of energy consumption, reverse osmosis compares very well (see Table 3). Operating costs, amortization of investment, and membrane replacement are more difficult factors that best are dealt with elsewhere (75).

Since the membrane rejects salt, concentration polarization tends to occur at the upstream membrane surface as mentioned above. This decreases water flux (larger $\Delta \pi$) and increases salt flux (larger ΔC_S^L). In the extreme cases of solutes of limited solubility, such as calcium salts, concentration polarization may result in formation of a precipitate and membrane fouling. Concentration polarization effects depend mainly on the fluid-phase hydrodynamics and module design and must be minimized by proper engineering (see Reverse osmosis; Water, supply and desalination).

Table 3.   Energy Consumption for Desalting Brackish Water (10,000 ppm) By Various Technologies[a]

	Energy consumption[b], MJ/m³ H₂O (kW·h/1000 gal H₂O)	Water quality, ppm	Recovery, %
vapor-compression distillation	86 (90)	5	57
reverse osmosis	19 (20)	150	50
flash distillation	133 (140)	5	25
electrodialysis	38 (40)	500	33

[a] Ref. 75.
[b] See Figure 11 for thermodynamic minimum energy requirements.

Ultrafiltration and Microfiltration.   Hydraulically driven membrane-separation processes resemble ordinary filtration, differing mainly in the size of the substances not passed by the filter. There are no precise boundaries between the various terms used to describe filtration processes but the following provide rough guides. Microfiltration applies to the removal of very fine particles from about 100 nm to larger than micrometer size which do not develop any osmotic pressure. Driving pressures are low, of the order of 101 kPa (1 atm), and fluxes are quite high. Ultrafiltration (qv) applies primarily to dissolved or suspended macromolecular species which often do generate a small osmotic pressure. Driving pressures are usually several hundred kilopascals (several atmospheres) and resultant fluxes are an order of magnitude less than in microfiltration. Reverse osmosis can be thought of in the same context, and the term hyperfiltration is often used. Here, the filtration is on a micromolecular scale, less than 1 nm, and osmotic pressures are quite high. As a result, driving pressure must be 1–10 MPa (10–100 atm), and specific fluxes are very low.

In ultrafiltration and microfiltration, the membrane is always microporous, and solvent transport is by a viscous-flow mechanism through the membrane pores. In the simplest case, solute or particle rejection occurs because the pores are too small to allow them to pass. However, in practice these processes are not that simple. Instead, factors outside the membrane may control both solvent flux and rejection. One model envisions the build-up of a cake or gel layer at the upstream membrane surface created by rejection or concentration polarization of the solute or particulate material (65). This layer then plays a large role in performance of the filtration process. Salts and similar solutes are passed readily through the membranes used in these processes.

Membranes for ultra- and microfiltration may be combined into any of the module types described above. Hydrodynamics plays a very important role in the performance owing to the dominating influence that resistances external to the membrane may have. High cross-flow fluid velocities are an important feature of some filtration processes of this kind (76).

Both ultrafiltration and microfiltration have become commercially significant. Some of the main applications of ultrafiltration include recovery of paint from electrocoat paint rinse tanks (77); concentration of the whey generated as a by-product of cheese making (78); concentration and purification of enzymes, proteins, and albumin (79–80); and the recovery of lignosulfonate from sulfite liquors in the pulp and paper industry (81). Key uses of microfiltration involve sterile filtration. Bacteria can be removed from drinking water, heat-sensitive foods, or solutions for intravenous uses (82) (see Sterile techniques).

Dialysis.   Dialysis (qv) involves selective transport of solutes through a membrane as a result of concentration differences between the two fluid phases; there is usually little difference in pressure between the two phases. The purpose is usually separation or purification of a solution in some solutes that are transferred more rapidly than others as a result of the relative permeabilities of these species through the membrane (83). The dialysis operation may be carried out in batch or continuous modes (84). The rate of dialysis of each component is, of course, influenced by the membrane properties and geometry but it may also be affected by boundary-layer resistances in the external fluid phases and the extent of backmixing in continuous-flow configurations. For batch dialysis the concentration differential of a given species i across the membrane changes in time as follows:

$$\ln \frac{(\Delta C_i)_t}{(\Delta C_i)_o} = -\frac{P_i}{l} A \left(\frac{1}{V_o} + \frac{1}{V_l}\right) t \tag{49}$$

where A = membrane area, V_o = upstream fluid-phase volume, V_l = downstream fluid-phase volume, and P_i = the specific permeability of the membrane to species i provided boundary-layer resistances have been made small by adequate stirring in each compartment. In flow systems, steady-state conditions can be obtained.

One of the most important commercial uses of this technique is hemodialysis, in which human blood is cleansed of metabolic wastes, such as urea, creatinine, and uric acid, while retaining essential higher molecular weight constituents and blood cells (85). This technique is widely used as a life-saving measure for sufferers of partial or complete loss of normal kidney functions. Numerous artificial-kidney membrane systems are commercially available. The trend has been to reduce costs and to develop disposable membrane modules, some of which now are hollow-fiber units. The membranes are of the microporous type tailored to clear the waste products without excessively depleting the blood of other required constituents and to have a low hydraulic permeability to minimize fluid exchange by any pressure differentials required to pump the fluids through the device. A new technique that employs the peritoneal membrane in the abdomen for this purpose is called continuous ambulatory peritoneal dialysis, CAPD (86).

Pervaporation. When a mixture of liquids is evaporated, the composition of the vapors is in thermodynamic equilibrium with the liquid mixture. It is this equilibrium relationship that determines what degree of separation can be obtained in this simple process or in multistage distillation. The separation is often inadequate or, in the case of azeotropes, nonexistent. In some cases, a more complete separation could be obtained on a kinetic basis where the vapor composition is determined instead by the relative rates at which the different species permeate a membrane. This combination has been called pervaporation and is schematically illustrated in Figure 12 (87). The driving force for transport of each species is its concentration gradient in the membrane. For each species, equilibrium partitioning takes place between the membrane and the liquid-phase upstream and the vapor-phase downstream. Selective transport of species then depends on both the sorption equilibrium relationship between the components and the membrane and on the relative diffusion coefficients for each species in the membrane. The left side of Figure 13 shows a simple solution-diffusion model formulation for species 1; analogous relations apply for each component.

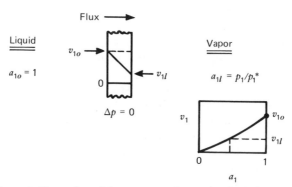

Figure 12. Schematic illustration of the pervaporation mode of membrane operation for a single component. The concentrations at the membrane surfaces are determined by the activities in the fluid phases and the sorption equilibrium relationship for that component as shown (12). Courtesy of Marcel Dekker, Inc.

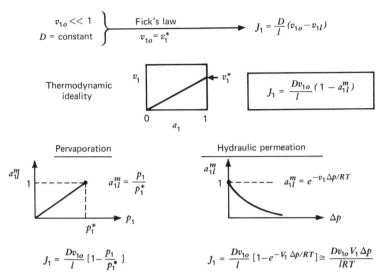

Figure 13. Summary of results for the solution-diffusion model for the idealized conditions defined at the top of the figure (12). Courtesy of Marcel Dekker, Inc.

No commercial process uses this concept, primarily because of the low productivity rates inherent to relatively thick, dense membranes. However, the advent of high flux asymmetric membranes may lead to future applications. Much research has been done in tailoring membranes for achieving high degrees of selectivity in this mode of operation. Both selectivity and permeability are related to specific and preferential chemical interactions between the components of the mixture to be separated and the polymer (88–89). For this reason, highly swollen polymer membranes such as rubber for separations of hydrocarbons (90) or hydrophilic grafted membranes for organic-mixture separations (91–92) have been developed. Physical factors such as density and cross-linking of the polymer constituting the membrane also affect selectivity (93).

There is an important relationship between the processes of hydraulic permeation and pervaporation illustrated in Figure 13 for a single fluid component. Application of a pressure upstream reduces the concentration of this species at the downstream membrane surface as described earlier. A similar reduction occurs when the downstream activity is reduced by having the component in the vapor state at less than its saturation vapor pressure. When the downstream concentration is reduced to zero, the maximum possible gradient in the membrane and the largest possible flux of this component are obtained. In pervaporation, this condition is reached when the partial pressure of the downstream vapor p is reduced to zero. In hydraulic permeation, this condition would be reached in principle by applying an infinite pressure upstream. Thus, there exists a ceiling flux in hydraulic permeation which is the same flux observed when the system is operated in the pervaporation mode with a perfect vacuum downstream (12). This relationship applies only when transport is by a solution-diffusion mechanism. The general hydraulic permeation equations shown in Figure 13 reduce to those used in the simple model for reverse osmosis (eq. 46) when $V_1 \Delta p/RT$ is small (94).

Gas Separations. Using membranes to separate gases is a very old idea that has stimulated much research on gas permeation through polymers (95–96). Adapting equation 6 for gas mixtures gives the following ratio of fluxes of gas 1 to that of gas 2 through the membrane:

$$\frac{J_1}{J_2} = \frac{P_1\,(p_o - p_l)_1}{P_2\,(p_o - p_l)_2} = \frac{P_1\,y_{1o}p_o - y_{1l}p_l}{P_2\,y_{2o}p_o - y_{2l}p_l} \tag{50}$$

From the definition of the separation factor given in equation 31 and the fact that $J_1/J_2 = y_{1l}/y_{2l}$, equation 50 can be recast as

$$\alpha_{12} = \alpha_{12}^{*}\frac{1 - \left(\dfrac{p_l y_{1l}}{p_o y_{1o}}\right)}{1 - \left(\dfrac{p_l y_{2l}}{p_o y_{2o}}\right)} \tag{51}$$

where $\alpha_{12}^{*} = P_1/P_2$. When the downstream pressure is much lower than the upstream pressure, ie, $p_o \gg p_l$, then the separation factor achieved is the ideal one given by the ratio of permeabilities of the two gases through the membrane, α_{12}^{*}. However, at finite downstream pressure $\alpha_{12} < \alpha_{12}^{*}$.

Table 4 lists the permeabilities of selected gases through several polymers. Some polymers offer very attractive separation factors for certain gas pairs (see Barrier polymers). Gas separation by polymer membranes has received much attention. As for pervaporation, the low productivity inherent to transport through relatively thick, dense membranes has limited commercial use of this process. Some portable units for producing oxygen-enriched air for medical uses have been placed on the market (99) and pilot processes for other uses have been announced (100). However, Monsanto has recently developed a line of hollow-fiber modules, called Prism separators, based on asymmetric coated polysulfone fibers which offer both high separation factors and economical productivities (101). These units are expected to have an important impact on hydrogen recovery and similar applications.

The primary economic advantage of membrane separation of gases is their much lower energy consumption compared to technologies such as cryogenic processes (see Cryogenics).

Table 4.   Gas Permeabilities in Various Polymers [a]

Polymer	Temperature, °C	Permeability, $P \times 10^{12}$ (cm³(STP)·cm)/(cm²·s·Pa) [b]					
		He	H₂	CO₂	CH₄	O₂	N₂
silicone rubber	25	22.5	41.3	20.3	60.1	37.6	18.8
natural rubber	25	2.33	3.68	9.85	2.25	1.8	0.61
butyl rubber	25	0.63	0.54	0.39	0.058	0.097	0.025
low density polyethylene	25	0.37		0.95	0.218	0.218	0.073
polycarbonate	30		0.90	0.6		0.105	0.022
poly(phenylene oxide)	30	5.87	8.5	5.7		1.18	0.28
nylon-6,6	25	0.039	0.075	0.013		0.0025	0.0006
polystyrene	25	1.4	1.75	0.79		0.197	0.059
cellulose acetate	30	1.02 [c]	0.263 [c]	1.7		0.06	0.02

[a] Refs. 6, 97–98.
[b] To convert (cm³·cm)/(cm²·s·Pa) to (cm³·cm)/(cm²·s·mm Hg), divide by 7.5×10^{-3}.
[c] At 20°C.

Reactive Dialysis. In ordinary dialysis, solutes are transferred from a concentrated to a more dilute solution; hence, this process cannot be used to concentrate a process stream as would be useful for recovery of a valuable chemical or to avoid environmental pollution. This restriction can be removed if it is possible to couple an appropriate chemical reaction with the dialysis process. In this extension of dialysis, the solute reacts with another chemical in the downstream fluid. The reaction creates a sink condition for the solute on the downstream side of the membrane so that the necessary concentration gradient of solute across the membrane is present. Thus, the downstream fluid may be more concentrated in the reaction product than the upstream fluid in terms of the unreacted solute. Of course, the membrane should be relatively impermeable to the reacted form of the solute and the reaction step must be reversed in a separate operation if the solute is to be recovered in its original chemical state.

This concept might be applied to weak acids, such as phenol, that can be converted to a salt by reaction with a strong base. For example, a dilute phenol stream might be dialyzed against a concentrated solution of NaOH to produce a concentrated solution of sodium phenolate. There seem to be no instances of large-scale commercial use of this potentially powerful membrane technique for solving difficult separation problems.

Liquid Membranes. Most membranes are solids but, in some cases, liquids can form useful membranes. Although diffusion through liquids is usually much more rapid than through solids, the diffusion process in liquids is not usually as selective as in solids. Thus, selectivity in liquid membranes generally stems from solubility considerations. Two quite different approaches to the formation and use of liquids as membranes have been developed. In the first, the liquid membrane is dynamically formed at the interface between two fluid phases. Mass transfer of components occurs between these two phases at rates governed by the membrane. In the usual case, the membrane is a surfactant-containing mixture which naturally locates at the interface when one fluid phase is mechanically dispersed in another. This approach has been developed for a number of applications (102–103). Commercial utilization of this technology is being sought.

In the second approach, a liquid is imbibed into the pores of a microporous membrane to form a system that resembles conventional membrane processes. Frequently, the liquid is doped with additives that chemically facilitate transport of certain solutes (104–108). Examples include crown ethers which complex with potassium ions but not with lithium ions (107), concentrated aqueous mixtures of HCO_3^-/CO_3^{2-} to facilitate CO_2 transport (109), and K_2CO_3 to facilitate H_2S transport (110). So-called liquid ion-exchangers, eg, oximes or amines, which exchange metal and hydrogen ions depending on pH, also have been used (see Chelating agents). Such systems can be used to concentrate metal ions by dialysis against a concentrated acid solution (97,106). Care must be exercised that the liquid membrane is not leached, evaporated, or forced out by pressure from its microporous support. Immobilized liquid membranes of this type show considerable promise, and some commerical uses for them are expected.

Electrochemical Membrane Processes. An entire class of membrane processes may be built around the use of ion-exchange membranes in which the driving force for transport is an electrical potential (29), eg, electrodialysis (qv). A scheme for water desalination requires the alternation of anion membranes (membranes that pass anions but not cations) and cation membranes (those that pass cations but not anions). The impressed potential provides a driving force for cations to migrate towards the cathode and for anions to migrate in the opposite direction. As a result of the anion-selective characteristics of the membranes, electrolyte is removed from the feed stream and concentrated in the stripping stream. Electrodialysis is used commercially on a limited scale (111). Serious problems include the formation of stable, high flux ion-exchange membranes and the control or minimization of electrochemical polarization which limits productivity and increases energy consumption.

An important advance in the use of electrochemical membrane processes is occurring in the production of chlorine and caustic soda (112,113) (see Alkali and chlorine products; Metal anodes). The United States market value for these products is in excess of 10^9 dollars. These materials have been traditionally produced by electrolytic processes employing either mercury or diaphragm cells, which are competitive on an economic basis. However, in recent years, DuPont has developed a line of cation-exchange membranes based on a poly(tetrafluoroethylene) backbone with sulfonic acid groups attached at the end of short side chains based on the perfluoropropylene ether unit (114):

$$-(CF_2CF_2)_x\!-\!(CF_2CF)_y\!-$$
$$|$$
$$OCF_2CFCF_3$$
$$|$$
$$OCF_2CF_2SO_3H$$

Membranes formed from this polymer are called Nafion, and a process for producing chlorine gas and caustic soda has been developed which essentially replaces the diaphragm with a Nafion membrane (112–113).

In diaphragm-cell technology, the brine solution is fed to the anode compartment where chloride ions are discharged to form chlorine gas. A delicate balance of hydrostatic pressure is used to force the partially depleted brine through the asbestos diaphragm into the cathode compartment where water is electrolyzed to hydrogen gas and hydroxyl ions (to form an NaOH solution). The diaphragm acts as a barrier to prevent (1) mixing of hydrogen and chlorine gas that would pose fire or explosion hazards and (2) hydroxyl ions from migrating into the anode compartment and producing loss of electrolysis efficiency by generation of oxygen. The diaphragm does transport dissolved chlorine gas, and the blockage of hydroxyl transport is achieved as a result of the convective liquid flow through the diaphragm. The ion-exchange character of the Nafion membrane allows passage of sodium ions but essentially prevents transport of all anions. The membrane constitutes an impermeable barrier to prevent mixing of hydrogen and chlorine for improved process safety and operational simplicity compared to the diaphragm cell. Since chloride ions are not transported, the caustic formed in the cathode compartment is essentially salt free, obviating the purification steps required in the diaphragm cell processes. This results in a potentially substantial reduction in investment and energy cost. Commercialization of this membrane technology is underway (112–113). Other general uses of electrically driven membrane processes have been described (29).

Piezodialysis. Reverse osmosis suffers the conceptual disadvantage that it is the main component, water, that is forced through the membrane rather than the minor component, salt. Thus, more material has to be transported than if salt were the species forced through the membrane. A process for achieving desalination by salt transport would reduce greatly the problems associated with membrane rupture since this event would result only in the loss of some feed water rather than contamination of product water. For these reasons, piezodialysis (115) has attracted interest although no practical membrane has been developed for utilization of this concept.

Piezodialysis requires a charge-mosaic membrane composed of cation-exchange regions adjacent to anion-exchange regions (Fig. 14). In principle, the positive and negative ions in the feed are transported through the membrane via the cation and anion-exchange regions of the membrane. This spatial division of charged species sets up circulating currents as shown; and as shown by the theory for this process, the size of the zones must be small (115). In addition, the membrane must give a high flux of salt and be relatively impermeable to water. These are quite stringent demands for the membrane, and to date none has approached what is required for a practical process although many synthesis concepts have been explored (115).

Analytical Uses of Membranes

Membrane processes can be used for a variety of analytical purposes. One of the earliest and most widely used examples of such use is the measurement of osmotic pressure of dilute polymer solutions to determine the molecular weight of polymers (116). The osmotic pressure of a solution is related to the activity of the solute by the rigorous thermodynamic relation given in equation 43. Measurement of the osmotic pressure requires a membrane that is permeable to the solvent but not to the solute. It is not difficult to find membranes that closely approximate this ideal when the solute is a high molecular weight polymer. To obtain the molecular weight from osmotic pressure determination requires relating solvent activity to this quantity by a thermodynamic relationship. For dilute polymer solutions, this is done by the virial expression (116):

$$-\ln a_1 = V_1 \left(\frac{C}{M_2} + A_2 C^2 + \ldots \right) \tag{52}$$

where the first term on the right contains the solute molecular weight; A_2, the second virial coefficient, expresses the thermodynamic nonideality of the solution; and C is the polymer concentration of the solution in mass per unit volume. It is necessary to measure the osmotic pressure at finite polymer concentrations, and then extrapolate these observations to zero concentration to eliminate the effects of solution nonideality. As can be seen from equations 43 and 52, the relationship in this limit is

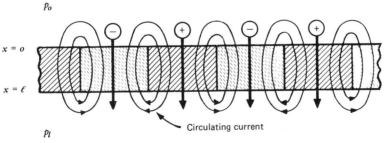

Figure 14. Illustration of charge-mosaic membrane for piezodialysis. ▨, cation-exchange region; ▧, anion-exchange region.

$$\left(\frac{\pi}{RTC}\right)_{C\to0} = \frac{1}{M_2} \qquad (53)$$

which is the same as the ideal relation given by equation 44. The extrapolation can be done graphically by plotting (π/C) vs C. When the polymer is polydisperse, osmometry yields the number-average molecular weight, $\overline{M}_n$.

Since osmotic pressure is inversely proportional to molecular weight, the technique is not accurate at very high molecular weights, and at very low molecular weights the technique may be compromised because the membrane permits passage of these small polymer molecules. Modern commercial osmometers largely solve these problems of sensitivity and leakiness so that the technique can be applied to a range of polymer molecular weights. These instruments are relatively fast and provide means of diluting the polymer solution automatically to give the data necessary for extrapolation to zero concentration. Of course, osmometry can be applied only to polymers soluble in solvents that do not attack the membrane.

A class of analytical devices known as ion-selective electrodes (qv) has developed wide use in recent years. The principles of electrochemistry permit the design of electrodes whose electrical potentials are related to the activity of the ionic solution to which they are exposed. If such an electrode is encased in a membrane that passes only specific ions (29), then the potential developed is a measure of the activity of this ion alone. This concept was originally developed for pH measurement but the scope of application has now been extended to include ions such as Na^+, K^+, Li^+, NH_4^+, and Ag^+. The membrane may be made of certain types of glass, metals, crystals, or liquids. The principle of ion selectivity may be based on ion-exchange concepts, carriers, or specific vacancy-diffusion mechanisms. Calibration is required to relate concentration to the measured potential and to correct for a variety of effects. This concept may be applied to other kinds of analyses. Enzyme-substrate electrodes (29,117) and specific gas probes (29) for oxygen, carbon dioxide, and sulfur dioxide, have been described, but none is widely available on a commercial basis. The ion-selective electrodes mentioned above are available commercially.

Monitoring of stack gases is an important step in air-pollution abatement. Such analysis may be tedious, time-consuming, and the presence of vapors, mists, or other chemical species may interfere with accurate determination of pollutant content. An experimental in-stack polymer membrane permits continuous monitoring and avoids some of the above-mentioned problems (118–119). The membrane is an FEP Teflon tube (see Fluorine compounds, organic). A clean carrier gas that passes continuously through the bore of the tube picks up SO_2 which has permeated the tube wall. The carrier-gas stream flows through a continuous analyzer which determines the SO_2 content. The membrane screens particles and condensible vapors that would interfere with this measurement. A scheme for calibration and temperature correction is incorporated in the device (see Air pollution control methods).

A sophisticated dual-membrane device that has recently been described (120) offers considerable promise for analysis of gas mixtures. The technique makes use of the pressure change caused by unequal permeation rates of gases through membranes. The device uses a matched pair of membranes to form three chambers. A reference gas and the test gas are passed through the two extreme chambers while a pressure transducer monitors the pressure in the center chamber. Analysis of binary combinations of N_2, O_2, CO_2, NO_2, NH_3, and SO_2 have been made with an accuracy of ±1%. The advantages and disadvantages of this approach have been described (120) (see also Analytical methods).

Controlled-Release Technology

Conventionally, chemicals such as drugs, pesticides, fertilizers, and herbicides, are administered in a periodic fashion, causing temporal concentration variations. The concentration extremes may range from dangerously high to ineffectively low and achieve the optimum level for only short periods of time. This cyclic application inefficiently utilizes these chemicals and results in added expense and possibly in certain side effects. During the last two decades, the concept of controlled delivery or release has emerged as a means of solving many of these problems. Often this technology involves the use of membranes (121) (see Microencapsulation; Pharmaceuticals, controlled-release).

The simplest of the methods developed for timed release of chemicals are the so-called matrix devices in which the solute to be delivered is compounded in a matrix, usually a polymer, which may be in the form of a slab, rod, sphere, or other shape. This agent then may be released from the matrix by diffusion. In every case, the rate of release decreases with time. When the solute is completely soluble in the matrix, ie, the solute concentration A is less than the saturation limit $C_{S,\,sat}^m$, then the release rate is given by

$$\text{release rate} = AS\sqrt{\frac{D}{\pi t}} \qquad (54)$$

until about 60% of the solute has been lost. Then the rate falls more rapidly. The terms in this equation have the following meanings: S = surface area of the device, and D = diffusion coefficient of solute through the matrix. The solute concentration in the matrix A may exceed the solubility limit, $C_{S,\,sat}^m$; in this case the excess solute above that which is dissolved at equilibrium is dispersed in the matrix as small particles. The details of the extraction may be complicated but for the usual case in which diffusion is rate-limiting a simple model suffices. Figure 15 shows the physical picture in which there is a core containing undissolved solute surrounded by regions depleted of undissolved solute (122–123). It is diffusion through the latter regions which controls the rate of release. A simple pseudosteady-state analysis yields the following relationship:

$$\text{release rate} = S\sqrt{\frac{DC_{S,\,sat}^m\,(A - \tfrac{1}{2}\,C_{S,\,sat}^m)}{2\,t}} \qquad (55)$$

which describes the behavior of slab geometries quite well as long as there is a core of undissolved solute (124). The release in this situation follows the same square root of time pattern as when $A < C_{S,\,sat}^m$; however, the dependence on solute loading A is not the same. This model has been extended to other geometrical shapes (123,125) and to include boundary-layer resistances (124,126). Various devices of this type have been marketed (121).

The release rate for the monolithic matrix devices described above is not constant in time, or zero order, which may be desirable in many situations. This constancy of release rate can be achieved by the so-called reservoir or depot devices, in which a rate-controlling membrane completely encloses a cavity that contains an active agent appropriately dispersed. This device is a most useful situation when the reservoir contains a suspension of the solute in a fluid since this reservoir has constant solute activity until the solute in excess of the saturation limit has been removed. Diffusion

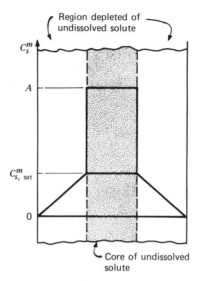

Figure 15. Concentration profile in a matrix-type system in which the solute loading exceeds solubility limit according to Higuchi model.

across the membrane gives a constant steady-state release where the solute flux is given by equation 20.

Reservoir devices are commercially available for the delivery of drugs (121,127) and fertilizers (121) (qv). The former are macroscopic; the latter are often made by microencapsulation technology. The membranes may be dense or porous. Release rates are determined by appropriate selection of materials and design specifications.

Osmotic pumps are another class of devices that can affect controlled delivery rates. As shown in Figure 16 one version of this concept employs a flexible reservoir bag containing a liquid or solution to be pumped at constant rate. Outside this bag is a salt suspension contained in a rigid, semipermeable membrane that permits passage of water but not salt. When this unit is introduced into an aqueous environment, water is imbibed at a rate controlled by the membrane characteristics and by the driving force for transport created by the osmotic agent. As a consequence of the water uptake, the volume of the salt compartment increases and reduces the volume of the flexible reservoir bag. Hence, the contained liquid is forced through the delivery portal at a rate equal to the water imbibition. Under ideal circumstances, zero-order release can be achieved for long periods. Devices of this general type are commercially available (127).

Miscellaneous Uses of Membranes

Energy-storage batteries require some sort of permeable barrier or separator that allows an adequate rate of ion transport and prevents penetration by electrode reactants, which if deposited on the electrodes, would cause an undesirable drain on available energy. The separator must be an electronic insulator to avoid a slight electronic conduction leak. In certain cases, the separator must permit convective passage of gases, ions, or liquids that might otherwise accumulate at one electrode.

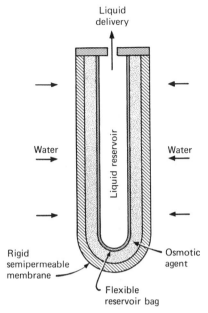

Figure 16.　Osmotic pump device.

Battery separators must be made of chemically inert materials and have certain membrane transport characteristics (128). Some are made from woven or nonwoven fibers; others are microporous sheets similar in structure to the porous membranes described above. Nylon and cotton fabrics are used in NiOOH/KOH/Cd and HgO/KOH/Zn cells, respectively. $PbO_2/H_2SO_4/Pb$ batteries require acid-resistant materials such as poly(vinyl chloride), polyolefins, or polytetrafluoroethylene in the form of porous sheets (see Batteries).

A novel approach to power generation using a process called pressure-retarded osmosis recently has been reported (129–130). This process recovers some of the free energy of mixing concentrated and dilute saline water streams by using the same type of membrane systems developed for reverse osmosis. A low pressure, dilute salt water stream is fed to one side of the membrane, and a more concentrated salt solution at a pressure that is higher but still less than the osmotic pressure is on the opposite side of the membrane. Osmosis, retarded by the pressure differential, causes a flow of water to the high pressure side. Thus, the net effect is pressurization of the transported water. This stream is then depressurized through a hydroturbine to produce power, a clever way of converting the free energy of mixing into mechanical energy when the appropriate salt water streams are available. However, this concept has no practical application at the present time.

There are many fabrics which have been developed to perform semipermeable membrane functions. For example, the ideal raincoat would not permit passage of liquid water but would allow transport of water vapor for the comfort of the wearer. Gore-Tex film (see Table 1) made of microporous polytetrafluoroethylene has these characteristics. Water vapor can diffuse through its pores but because of the hydrophobic nature of the polymer, liquid water will not wet the material and, thus, does not enter the pores. Laminates of this material with conventional fabrics are sold for this purpose in certain types of sportswear. Surgical gowns ought to similarly transport water vapor for comfort reasons but not allow passage of bacteria from the medical personnel to the patient (a major source of post-operative infections). The latter is prevented if water does not wet the garment or if the garment's membrane pores are smaller than bacteria (minimum size, ca $0.2~\mu$m).

Materials for contact lenses (qv) must meet certain basic criteria: correct optical properties, physiological inertness, and the mechanical characteristics needed to assure stability during the complex motions of the eye and eyelid. Since the cornea is a vascular tissue with metabolic activity, it must receive a certain supply of oxygen (131) which can reach the cornea surface either by permeation through the lens or by pumping oxygenated tears under the lens with each blink of the eyelid. Thus, permeability characteristics are an important factor in materials selection and lens design. It has been estimated that the ideal lens material should have an oxygen permeability of at least 2.25×10^{-12} (cm^3(STP)·cm)/(cm^2·s·Pa) [300×10^{-12} (cm^3·cm)/(cm^2·s·mm Hg)] (132). Poly(methyl methacrylate), the most commonly used material for hard contact lenses, has an oxygen permeability of 0.015×10^{-12}. Thus, lack of transport through the lens must be compensated by making the lens diameter small and relying on the pumping mechanism mentioned above. Cellulose acetate butyrate, which has a permeability of 0.28×10^{-12} (cm^3(STP)·cm)/(cm^2·s·Pa) [37.3×10^{-12} (cm^3·cm)/(cm^2·s·mm Hg)] currently is undergoing clinical tests as an alternative material for lens construction (133). The soft contact lenses now in wide use are cross-linked hydrogels made primarily from 2-hydroxyethyl methacrylate (134). Several commercial versions are available which have oxygen permeabilities ranging from 0.376×10^{-12} to 6×10^{-12} (cm^3(STP)·cm)/(cm^2·s·Pa) [50–800×10^{-12} (cm^3·cm)/(cm^2·s·mm Hg)] depending on the extent they are swollen by water (135). The soft lenses are made with larger diameters than the hard ones and are more comfortable for the wearer. They also adhere more closely to the cornea surface, reducing the oxygen supply by pumping; hence, their high oxygen permeabilities are essential to their use.

Nomenclature

A	= solute concentration in matrix devices
A_2	= second virial coefficient
a_i	= activity of component i
a_{il}^L	= activity in the liquid phase for component i at $x = l$ (downstream)
a_{il}^m	= activity in the membrane phase for component i at $x = l$ (downstream)
b	= affinity constant
C_i^m	= concentration of i in membrane
C_l, C_o	= concentration in the membrane at the downstream and upstream sides, respectively
C_H'	= Langmuir capacity constant
C_S^m, C_S^L	= concentration of solute in membrane phase and liquid phase, respectively
D	= diffusion coefficient
D_{Ki}	= Knudsen diffusion coefficient of component i
D_S^L	= solute diffusion coefficient in the liquid phase
F	= Faraday constant
G	= pore geometry coefficient
ΔG_{mix}	= free energy of mixing
ΔH_{mix}	= enthalpy of mixing
J	= flux (units vary)
K	= hydraulic permeability coefficient
K_S	= partitioning coefficient of solute

k	= proportionality constant
k_D	= Henry's law coefficient
L	= liquid phase
l	= membrane thickness
M	= molecular weight
$\overline{M}_n$	= number-average molecular weight
m	= membrane phase
n	= number of pores per unit area
n_i	= flux of i in mass units
P	= permeability
p	= pressure
p_o, p_l	= pressure in the upstream and downstream fluid phases, respectively
Q_t	= amount of gas which has permeated in time t
Q_x	= bore flow rate at x
R	= rejection coefficient
r	= pore radius
$\bar{r}$	= average pore radius
S	= solubility coefficient
S	= surface area (controlled release equations)
T	= temperature
t	= time
V_i	= molar volume of liquid i
V_o, V_l	= upstream and downstream fluid-phase volumes, respectively
v_i	= volume fraction of i
v_i^*	= volume fraction of i at equilibrium
v_{io}, v_{il}	= volume fraction of i in the membrane at the upstream face and downstream face, respectively
w_i	= mass fraction of i
X	= concentration of fixed charges
x	= position on the x axis, Cartesian coordinates
y_{io}, y_{il}	= mole fraction of i in the membrane at the upstream phase and downstream phase, respectively
Z	= ionic valence
α_{12}	= separation factor
α_{12}^*	= ideal separation factor
γ	= surface tension
ϵ	= porosity
θ	= time lag
μ	= fluid viscosity
π	= osmotic pressure
ρ	= mass density
ρ_i	= mass concentration of i
τ	= tortuosity factor
$\Delta\phi$	= general potential difference
$\dfrac{d\phi}{dx}$	= electrical field

BIBLIOGRAPHY

1. K. J. Laidler and K. E. Schuler, *J. Chem. Phys.* **17,** 851 (1949).
2. R. B. Bird, W. E. Stewart, and E. N. Lightfoot, *Transport Phenomena*, John Wiley & Sons, Inc., New York, 1960.
3. D. R. Paul in H. B. Hopfenberg, ed., *Permeability of Plastic Films and Coatings*, Plenum Press, New York, 1974, p. 35.
4. G. J. van Amerongen, *Rubber Chem. Technol.* **37,** 1065 (1964).
5. S. M. Allen, M. Fujii, V. Stannett, H. B. Hopfenberg, and J. C. Williams, *J. Membr. Sci.* **2,** 153 (1977).
6. H. Yasuda and V. Stannett in J. Brandrup and E. H. Immergut, eds., *Polymer Handbook*, 2nd ed., John Wiley & Sons, Inc., New York, 1975, p. 111.

7. H. J. Bixler and O. J. Sweeting in O. J. Sweeting, ed., *The Science and Technology of Polymer Films*, Vol. II, John Wiley & Sons, Inc., New York, 1971, p. 85.

8. W. R. Vieth, J. M. Howell, and J. H. Hsieh, *J. Membr. Sci.* **1,** 177 (1976).

9. D. R. Paul, *Ber. Bunsenges. Phys. Chem.* **83,** 294 (1979).

10. C. E. Rogers in E. Baer, ed., *Engineering Data for Plastics*, van Nostrand-Reinhold, New York, 1964, Chapt. 9.

11. U. Merten in U. Merten, ed., *Desalination by Reverse Osmosis*, M.I.T. Press, Cambridge, Mass., 1966, Chapt. 2, p. 15.

12. D. R. Paul, *Sep. Purif. Methods* **5,** 33 (1976).

13. D. R. Paul, *J. Polym. Sci. Polym. Phys. Ed.* **11,** 289 (1973).

14. D. R. Paul and O. M. Ebra-Lima, *J. Appl. Polym. Sci.* **19,** 2759 (1975).

15. D. R. Paul and J. D. Paciotti, and O. M. Ebra-Lima, *J. Appl. Polym. Sci.* **19,** 1837 (1975).

16. M. E. Heyde and J. E. Anderson, *J. Phys. Chem.* **79,** 1659 (1975).

17. H. K. Lonsdale, U. Merten, and R. L. Riley, *J. Appl. Polym. Sci.* **9,** 1341 (1965).

18. D. R. Paul, M. Garcin, and W. E. Garmon, *J. Appl. Polym. Sci.* **20,** 609 (1976).

19. H. G. Burghoff and W. Pusch, *Polym. Eng. Sci.* **20,** 305 (1980).

20. R. E. Kesting, *Synthetic Polymeric Membranes*, McGraw-Hill, New York, 1971.

21. E. M. Renkin, *J. Gen. Physiol.* **38,** 225 (1954).

22. R. E. Beck and J. S. Schultz, *Science* **170,** 1302 (1970).

23. J. A. Quinn, J. L. Anderson, W. S. Ho, and W. J. Petzny, *Biophys. J.* **12,** 990 (1972).

24. J. L. Anderson and J. A. Quinn, *Biophys. J.* **14,** 130 (1974).

25. J. L. Anderson and J. A. Quinn, *J. Chem. Soc. Faraday Trans. I* **68,** 608 (1972).

26. E. F. Casassa, *Macromolecules* **9,** 182 (1976).

27. C. K. Colton, C. N. Satterfield, and C. J. Lai, *AIChE J.* **21,** 289 (1975).

28. C. N. Satterfield, C. K. Colton, B. DeTurckheim, and T. M. Copeland, *AIChE J.* **24,** 937 (1978).

29. S. T. Hwang and K. Kammermeyer, *Membranes in Separations*, Vol. VII in *Techniques of Chemistry*, A. Weissberger, ed., John Wiley & Sons, Inc., New York, 1975.

30. A. E. Scheidegger, *The Physics of Flow in Porous Media*, University of Toronto Press, Toronto, Can., 1957.

31. M. Knudsen, *Ann. Phys.* **28,** 75 (1909).

32. F. Helfferich, *Ion Exchange*, McGraw-Hill, New York, 1962.

33. R. M. Wallace, *Ind. Eng. Chem. Proc. Des. Dev.* **6,** 423 (1967).

34. A. Katchalsky and P. E. Curran, *Nonequilibrium Thermodynamics in Biophysics*, Harvard University Press, Cambridge, Mass., 1967.

35. J. S. Schultz, J. D. Goddard, and S. R. Suchded, *AIChE J.* **20,** 417, 625, 831 (1974).

36. F. P. Scholander, *Science* **131,** 585 (1960).

37. A. F. Turbak, ed., "Membranes from Cellulose and Cellulose Derivatives," *Applied Polymer Symposia*, Vol. 13, John Wiley & Sons, Inc., New York, 1970.

38. P. R. Keller, *Membrane Technology and Industrial Separation Techniques*, Noyes Data Corp., London, Eng., 1976.

39. P. Meares in P. Meares, ed., *Membranes Separation Processes*, Elsevier Scientific Publishing Co., New York, 1976, Chapt. 1.

40. C. A. Kumins and K. K. Kwei in J. Crank and G. S. Park, eds., *Diffusion in Polymers*, Academic Press, Inc., New York, 1968, Chapt. 4.

41. I. Cabasso, *Am. Chem. Soc. Div. Org. Coat. Plast. Chem. Prepr.* **37,** 110 (1977).

42. P. Aptel, J. Cuny, J. Josefonvicz, G. Morel, J. Neel, and B. Chaufer, *Eur. Polym. J.* **14,** 595 (1978).

43. R. H. Huang and V. J. C. Lin, *J. Appl. Polym. Sci.* **12,** 2615 (1968).

44. A. Keller, *Phil. Mag.* **2,** 1171 (1952).

45. A. H. Chan and D. R. Paul, *J. Appl. Polym. Sci.* **24,** 1539 (1979).

46. G. Bex, A. Chapiro, M. B. Huglin, A. M. Jendrychowska-Bonamour, and T. O'Neil, *J. Polym. Sci.* **22C,** 493 (1968).

47. R. L. Goldsmith, *Ind. Eng. Chem. Fundam.* **10,** 113 (1971).

48. R. Fleischer, P. B. Price, and R. Walker, *Science* **143,** 249 (1964).

49. H. Thiele and K. Hallich, *Kolloid Z.* **163,** 115 (1959).

50. W. Elford, *Proc. R. Soc. London Ser. B.* **106,** 214 (1930).

51. M. Vaughn, *Nature* **183,** 43 (1959).

52. Brit. Pat. 1,475,316 (March 19, 1973), (to Sumitomo Electric Industries, Ltd.).

53. H. S. Bierenbaum, R. B. Isaacson, M. L. Druim, and S. G. Plovan, *Ind. Eng. Chem. Prod. Res. Dev.* **13,** 2 (1974).
54. *Chem. Eng. News* **56,** 23 (Dec. 11, 1978).
55. S. Loeb and S. Sourirajan, *Adv. Chem. Ser.* **38,** 117 (1962).
56. M. A. Frommer, I. Feiner, O. Kedem, and R. Block, *Desalination* **7,** 393 (1970).
57. J. P. Craig, J. P. Knudsen, and V. F. Holland, *Text. Res. J.* **32,** 435 (1962).
58. R. McKinney in H. K. Lonsdale and H. E. Podall, eds., *Reverse Osmosis Membrane Research*, Plenum Press, New York, 1972, p. 253.
59. U.S. Pat. 3,615,924 (Oct. 1971), A. S. Michaels (to Amicon Co.).
60. A. S. Michaels, H. J. Bixler, R. W. Housslein, and S. M. Fleming, *Polyelectrolytes Complexes as Reverse Osmosis and Ion Selective Membranes*, U.S. Dept. Interior, Office of Saline Water, Research and Development Progress Report 149, U.S. Government Printing Office, Washington, D.C., 1965.
61. I. S. Model and L. A. Lee in H. K. Lonsdale and H. E. Podall, eds., *Reverse Osmosis Membrane Research*, Plenum Press, New York, 1972, p. 285.
62. H. S. Strathmann, K. Koch, P. Amar, and R. W. Baker, *Desalination* **16**(2), 179 (1975).
63. L. T. Rozelle, J. E. Cadotte, K. E. Cobian, and C. V. Kopp in S. Sourirajan, ed., *Reverse Osmosis and Synthetic Membrane*, National Research Council, Ottawa, Can., 1977, p. 249.
64. B. Baum, W. Holley, Jr., and R. A. White in ref. 39, Chapt. 15, p. 187.
65. A. S. Michaels, *Chem. Eng. Prog.* **64**(12), 31 (1968).
66. P. R. Klinkowski, *Chem. Eng.*, 164 (May 8, 1978).
67. S. S. Kremen in S. Sourirajan, ed., *Reverse Osmosis and Synthetic Membranes*, National Research Council of Canada, Ottawa, 1977.
68. H. E. Podall in N. N. Li, ed., *Recent Developments in Separation Science*, Vol. II, C.R.C. Press, Cleveland, Ohio, 1972, p. 185.
69. M. R. Doshi, W. N. Gill, and V. N. Kabadi, *AIChE J.* **23,** 765 (1977).
70. R. W. Stoughton and M. H. Lietzke, *J. Chem. Eng. Data* **10,** 254 (1965).
71. H. K. Lonsdale, *J. Membr. Sci.* **5,** 263 (1979).
72. W. Pusch in *Polymer Science and Technology*, Vol. 6, Plenum Press, New York, 1974, p. 233.
73. H. K. Lonsdale, *Desalination* **13,** 317 (1973).
74. A. Al-Gholaikah, N. El-Ramly, I. Jamjoom, and R. Seaton, *paper presented at the Sixth International Symposium on Fresh Water from the Sea, Las Palmas, Gran Canaria, Spain, Sept. 1978; World Water*, March 1979, p. 38.
75. G. F. Leitner, *paper presented at National Water Supply Improvement Association, 7th Annual Conference, New Orleans, La., Sept. 17, 1979.*
76. L. Svarovsky, *Chem. Eng.*, 69 (July 30, 1979).
77. F. Forbes, *Prod. Finish.* **23,** 24 (1970).
78. F. E. McDonough and W. A. Mattingly, *Food Technol.* **24,** 88 (1972).
79. R. Schmitthausler, *Process Biochem.* **12**(8), 13 (1977).
80. M. C. Porter, *Biotechnol. Bioeng. Symp.* **3,** 115 (1972).
81. R. E. Lacey and S. Loeb in R. E. Lacey and S. Loeb, eds., *Industrial Processing with Membranes*, Wiley-Interscience, New York, 1972, pp. 71, 223.
82. M. C. Porter, "Membrane Filtration" in P. A. Schweitzer, ed., *Handbook of Separation Techniques for Chemical Engineers*, McGraw-Hill Book Co., New York, 1979, Section 2.
83. J. A. Lane and J. W. Riggle, *Chem. Eng. Prog. Symp. Ser.* **55**(24), 127 (1959).
84. S. B. Tuwiner in *Diffusion and Membrane Technology*, Reinhold, New York, 1962, Chapt. 14.
85. C. K. Colton, *A Review of the Development and Performance of Hemodialysers*, Artificial Kidney–Chronic Uremia Program, National Institute of Arthritis and Metabolic Diseases, National Institutes of Health, Bethesda, Md., 1967.
86. R. P. Popovich, J. W. Moncrief, K. D. Nolph, A. J. Ghods, Z. J. Twardowski, and W. K. Pyle, *Ann. Intern. Med.* **88,** 449 (1978).
87. R. C. Binning and F. E. James, *Pet. Refiner* **37,** 214 (1958).
88. R. F. Sweeny and A. Rose, *Ind. Eng. Chem. Prod. Res. Dev.* **4,** 248 (1965).
89. P. Aptel, J. Cuny, J. Josefonvicz, G. Morel, and J. Neel, *J. Appl. Polym. Sci.* **16,** 1061 (1972).
90. C. Larchet, J. P. Brun, and M. Guillou, *C. R. Acad. Sci. Ser. C*, **287,** 31 (1978).
91. A. Chipiro, *Desalination* **24,** 83 (1978).
92. G. Morel and J. Josefonvicz, *J. Appl. Sci.* **24,** 771 (1979).
93. G. Morel, J. Josefonvicz, and P. Aptel, *J. Appl. Polym. Sci.* **23,** 2397 (1979).
94. D. R. Paul and J. D. Paciotti, *J. Polym. Sci. Polym. Phys. Ed.* **13,** 1201 (1975).

95. K. Kammermeyer, *Chem. Eng. Prog. Symp. Ser.* **55**(24), 115 (1959).
96. V. Stannett, *J. Membr. Sci.* **3,** 97 (1978).
97. V. T. Stannett, W. J. Koros, D. R. Paul, H. K. Lonsdale, and R. W. Baker, *Adv. Polym. Sci.* **32,** 69 (1979).
98. V. Stannett, *Polym. Eng. Sci.* **18**(15), 1129 (1978).
99. W. Browall and S. Kimura, *Permselective Membranes for Oxygen Enrichment*, brochure, General Electric Co., Fairfield, Conn., May 1976.
100. *Chem. Eng. News*, 13 (May 21, 1979).
101. J. M. S. Henis and M. K. Tripodi, *paper presented at Symp. on Separation Science and Technology for Energy Applications, Gatlinberg, Tenn., Oct. 31, 1979.*
102. N. N. Li, *AIChE J.* **17,** 459 (1971).
103. U.S. Pat. 3,410,794 (Nov. 12, 1968), N. N. Li (to Esso Research and Engineering Co.).
104. W. J. Ward and C. K. Neulander, *Immobilized Liquid Membranes for Sulfur Dioxide Separation, P.B. 191769*, U.S. Dept. of Commerce, Washington, D.C., 1970.
105. J. W. Frankenfeld and N. N. Li in N. N. Li, ed., *Recent Developments in Separation Science*, Vol. III. B, C.R.C. Press, Cleveland, Ohio, 1977, p. 285.
106. R. W. Baker, M. E. Tuttle, D. J. Kelly, and H. K. Lonsdale, *J. Membr. Sci.* **2,** 213 (1977).
107. F. Caracciolo, E. L. Cussler, and D. F. Evans, *AIChE J.* **21,** 160 (1975).
108. T. Largman and S. Sifniades, *Hydrometallurgy* **3,** 153 (1978).
109. W. J. Ward in ref. 105, Vol. 1, 1972, p. 153.
110. S. L. Matson, C. S. Herrick, and W. J. Ward, *Ind. Eng. Chem. Proc. Des. Dev.* **16,** 370 (1977).
111. W. A. McRae and F. B. Leitz in ref. 105, Vol. II, 1972, p. 157.
112. D. J. Vaughan, *DuPont Innovation*, **4**(3), 10 (1973).
113. P. R. Savage, *Chem. Eng.*, 63 (Nov. 5, 1979).
114. T. Takamatsu, M. Hashigama, and A. Eisenberg, *J. Appl. Polym. Sci.* **24,** 2199 (1979).
115. F. B. Leitz, "Piezodialysis" in ref. 39, p. 261.
116. F. W. Billmeyer, *Textbook of Polymer Science*, 2nd ed., Wiley-Interscience, New York, 1971, p. 64.
117. D. Thomas and S. R. Caplan, "Enzyme Membranes" in ref. 39, p. 351.
118. L. C. Treece, R. M. Felder, and J. K. Ferrell, *Environ. Sci. Technol.* **10,** 457 (1976).
119. R. M. Felder, G. W. Miller, and J. K. Ferrel, *AIChE Symp. Ser.* **75** (188), 215 (1979).
120. B. M. Kim, D. J. Graves, and J. A. Quinn, *J. Membr. Sci.* **6,** 247 (1980).
121. D. R. Paul, "Polymers in Controlled Release Technology" in D. R. Paul and F. W. Harris, eds., *Controlled Release Polymeric Formulations*, ACS Symposium Series 33, American Chemical Society, Washington, D.C., 1976, p. 1.
122. T. Higuchi, *J. Pharm. Sci.* **50,** 874 (1961).
123. *Ibid.*, **52,** 1145 (1963).
124. D. R. Paul and S. K. McSpadden, *J. Membr. Sci.* **1,** 33 (1976).
125. T. J. Roseman and W. I. Higuchi, *J. Pharm. Sci.* **59,** 353 (1970).
126. Y. W. Chien, H. J. Lambert, and D. E. Grant, *J. Pharm. Sci.* **63,** 365, 515 (1974).
127. K. Heilmann, *Therapeutic Systems*, Georg Thieme Publishers, Stuttgart, FRG, 1978.
128. J. A. Lee, W. C. Maskell, and F. L. Tye, "Separators and Membranes in Electrochemical Power Sources" in ref. 39, p. 399.
129. S. Loeb, *J. Membr. Sci.* **1,** 49 (1976).
130. S. Loeb, F. Van Hessen, and D. Shahaf, *J. Membr. Sci.* **1,** 249 (1976).
131. K. A. Polse and M. R. Mandell, *Arch. Ophtal.* **84,** 505 (1970).
132. W. Timmer, *Chemtech*, 175 (March 1979).
133. M. F. Refojo, F. T. Holly, and F. L. Leong, *Contact Intraocular Lens Med. J.* **3**(4), 27 (1977).
134. O. Wichterle and D. Lim, *Nature* **185,** 117 (1960).
135. M. F. Refojo and F.-L. Leong, *J. Membr. Sci.* **4,** 415 (1979).

General Bibliography

R. M. Barrer, *Diffusion In and Through Solids*, 2nd ed., Cambridge University Press, Cambridge, Mass., 1951.
M. Bier, ed., *Membrane Processes in Industry and Biomedicine*, Plenum Press, New York, 1971.
J. Crank, *The Mathematics of Diffusion*, 2nd ed., Oxford University Press, London, Eng., 1974.
J. Crank and G. S. Park, eds., *Diffusion in Polymers*, Academic Press, Ltd., London, Eng., 1968.

J. F. Danielli, M. D. Rosenberg, and D. A. Cadenhead, eds., *Progress in Surface and Membrane Science*, Vol. 6, Academic Press, Inc., New York, 1973.

J. E. Flinn, ed., *Membrane Science and Technology*, Plenum Press, New York, 1970.

F. Helfferich, *Ion Exchange*, McGraw-Hill, New York, 1962.

H. B. Hopfenberg, ed., "Permeability of Plastic Films and Coatings," *Polymer Science and Technology*, Vol. 6, Plenum Press, New York, 1974.

S. T. Hwang and K. Kammermeyer, *Membranes in Separations*, Vol. VII in A. Weissberger, ed., *Techniques of Chemistry*, John Wiley & Sons, Inc., New York, 1975.

H. K. Lonsdale, ed., *Journal of Membrane Science*, published since 1976 by Elsevier Scientific Publishing Co., Amsterdam, The Netherlands.

A. Katchalsky and P. F. Curran, *Nonequilibrium Thermodynamics in Biophysics*, Harvard University Press, Cambridge, Mass., 1967.

P. R. Keller, *Membrane Technology and Industrial Separation Techniques*, Noyes Data Corporation, London, Eng., 1976.

R. E. Kesting, *Synthetic Polymer Membranes*, McGraw-Hill, New York, 1971.

R. E. Lacey and S. Loeb, eds., *Industrial Processing with Membranes*, Wiley-Interscience, New York, 1972.

N. Lakshminarayanaiah, *Transport Phenomena in Membranes*, Academic Press, Inc., New York, 1969.

N. N. Li, ed., *Recent Developments in Separation Science*, Vols. I–II, Chemical Rubber Company, Cleveland, Ohio, 1972.

H. K. Lonsdale and H. E. Podall, eds., *Reverse Osmosis Membrane Research*, Plenum Press, New York, 1972.

P. Meares, *Polymers: Structure and Bulk Properties*, Van Nostrand Company, Ltd., London, Eng., 1965.

P. Meares, *Membrane Separation Processes*, Elsevier Scientific Publishing Company, New York, 1976.

Membrane Processes for Industry—Proceedings of the Symposium, South Research Institute, Birmingham, Alabama, May 19–20, 1966.

U. Merten, ed., *Desalination by Reverse Osmosis*, M.I.T. Press, Cambridge, Mass., 1966.

R. McGregor, *Diffusion and Sorption in Fibers and Films*, Vol. 1, Academic Press, Inc., New York, 1974.

D. R. Paul and F. W. Harris, eds., *Controlled Release Polymeric Formulations*, ACS Symposium Series 33, Washington, D.C., 1976.

E. S. Perry, ed., *Progress in Separation and Purification*, Vol. 1, Wiley-Interscience, New York, 1968.

R. N. Rickles, *Membrane Technology and Economics*, Noyes Development Corp., Park Ridge, N.J., 1967.

C. E. Rogers, ed., *Permselective Membranes*, Marcel Dekker, Inc., New York, 1971.

S. Sourirajan, *Reverse Osmosis*, Logos Press, London, Eng., 1970.

S. Sourirajan, ed., *Reverse Osmosis and Synthetic Membranes*, National Research Council, Ottawa, Can., 1977.

V. T. Stannett, W. J. Koros, D. R. Paul, H. K. Lonsdale, and R. W. Baker, "Recent Advances in Membrane Science and Technology," *Adv. Polym. Sci.* **32,** 69–121 (1979).

A. F. Turbak, ed., "Membranes from Cellulose and Cellulose Derivatives," *Appl. Polym. Symp.* **13,** (1970).

S. G. Tuwiner, *Diffusion and Membrane Technology*, Reinhold, New York, 1962.

DONALD R. PAUL
University of Texas

G. MOREL
Université de Paris-Nord

MEMORY-ENHANCING AGENTS AND ANTIAGING DRUGS

Memory-Enhancing Agents

The understanding of learning and memory is rather limited, but there is general agreement that drugs can impair or facilitate the acquisition (registration), consolidation (storage), and retrieval of a learned task. Recent research concerning drug influences on learning and memory has been summarized (1–7).

The processes underlying recent or short-term memory probably are different from those underlying long-term memory. There is considerable evidence that, although the short-term processes alter the synaptic conductance within the brain, protein and macromolecular synthesis is related intimately to the longer-term processes of memory retention, particularly consolidation (1–4,8–10). There are several ways in which macromolecules might be involved in memory storage, including the synthesis of new molecules or a change in the composition or conformation of preexisting macromolecules. Neural coding also has been related to memory storage as a result of the chemical identification in animals of a behavior-inducing peptide, scotophobin [33579-45-2] (11).

Many types of tasks and training procedures are used in studies of learning and memory. Mazes, discrimination tasks, and active and passive (inhibitory) avoidance tasks are used most commonly and may be based on different types of appetitive and aversive motivation. Drugs can be given at different times, either before or after training, to study the drug's effects on acquisition, memory storage, and retention (4).

Neuropeptides. A number of peptide hormones act directly on the brain to affect motivation, learning, and memory processes (12–14) (see Hormones, brain oligopeptides). The pituitary hormones adrenocorticotrophic hormone [9002-60-2] (ACTH) and melanocyte-stimulating hormone [9002-79-3] (MSH) appear to function as modulators of neuronal activity and facilitate several types of conditioned behaviors in animals (see Hormones, anterior-pituitary). Vasopressin [11000-17-2], which is synthesized in the hypothalamus and is stored and released by the pituitary, appears to have a long-term effect on learned behavior (see Hormones, posterior-pituitary). The actions of peptides related to ACTH and of those related to vasopressin differ primarily with respect to the duration of their effects: the ACTH-related peptides have a short-term behavioral effect, whereas those related to vasopressin have a long-term one. Thus, in animals receiving ACTH, which affects motivation processes, a learned avoidance response lasts only for a few hours, whereas animals receiving vasopressin, which is physiologically involved in long-term memory consolidation processes and retrieval, retains the avoidance response for several days.

ACTH and Related Peptides. The behavioral effects of ACTH and MSH are very similar, and both peptides share a common core of a sequence of seven amino acids (15–17). Structure–activity studies (16) have confirmed that this heptapeptide sequence, ACTH$^{4-10}$ (H–Met–Glu–His–Phe–Arg–Tyr–Gly–OH), is the CNS-active component. Systematic modifications of ACTH$^{4-10}$ indicate that the tetrapeptide ACTH$^{4-7}$ (H–Met–Glu–His–Phe–OH) bears the essential elements required for the behavioral activity of ACTH and MSH. A modification of ACTH$^{4-10}$, which includes deletion of the glycine residue and simultaneous replacement of arginine with D-lysine, tryptophan with phenylalanine, coupled with oxidation of the methionine to the corresponding sulfoxide, furnishes an orally active ACTH[4–9] analogue with a 1000-fold increase in behavioral potency over ACTH (16–17). This increase in potency partially results from increased resistance against enzymatic degradation (17). The corresponding sulfone, H–Met(O_2)–Glu–His–Phe–D–Lys–Phe–OH (Org 2766), is as active as the sulfoxide (18). It is noteworthy that the behavioral effects of these analogues and hormone fragments are independent of the well-known endocrine effects of MSH and ACTH. Thus, the fragment ACTH[4–10], which is as effective in the behavioral assays as the intact molecule, does not elicit formation of adrenal corticoids *in vivo*.

There is much data supporting the early suggestion that ACTH, MSH, and their fragments influence memory. Treatment of rats with ACTH and related peptides delays extinction of active- and passive-avoidance responses. They also facilitate memory retrieval in animals suffering from retrograde amnesia that is induced by CO_2 and electroconvulsive shock (18–22). However, more recent evidence indicates that these neuropeptides facilitate attention, ie, acquisition, and not memory, ie, consolidation (23–24). Accordingly, in several studies with human subjects, ACTH[4–10] was shown to reduce performance deficits, increase general arousal, and to enhance attention and short-term visual memory (23–26). However, in a number of studies, little or no effect of ATCH[4–10] was found in geriatric volunteers or in children with learning difficulties (27–28). Thus, not all the results are impressive. The orally active ACTH analogues may induce more reliable and pronounced effects on motivation and attention.

The mechanism by which ACTH and MSH exert their behavioral effects is not clear. There is some evidence that they increase brain RNA and protein metabolism (29), elevate brain cyclic AMP (adenosine monophosphate) (30), and/or effect noradrenaline turnover (31) (see Neuroregulators).

Vasopressin and Related Peptides. Vasopressin protects against puromycin-induced amnesia and facilitates acquisition of passive avoidance and sexually motivated learning behavior, and increases resistance to extinction of newly acquired experience (32–33). Brattleboro rats, which lack the ability to synthesize vasopressin, exhibit impaired memory. The memory function in these rats is improved by vasopressin and its analogues. These experiments have established that it is the consolidation of acquired response that is selectively impaired in the absence of vasopressin, rather than acquisition. Vasopressin also appears to facilitate retrieval processes (32–33).

The brain functions that are involved in the consolidation of acquired avoidance responses are modulated in opposite ways by vasopressin and oxytocin (see Hormones, posterior-pituitary); thus, oxytocin may be regarded as an amnesic peptide. The effect of these neuropeptides on memory appears to result from modulation of catecholamine turnover in specific brain areas (34).

Lysine[8]-vasopressin [*50-57-7*] (LVP) and its analogue desglycinamide[9]-lysine[8]-vasopressin [*37552-33-3*] (DG-LVP) delay extinction of active avoidance behavior (35–36). Arginine[8]-vasopressin [*11000-17-2*] (AVP) is the most potent peptide with respect to delaying extinction of active avoidance behavior (37). Removal of the C-terminal part of AVP yields a peptide with a decreased activity. Desglycinamide[9]-arginine[8]-vasopressin [*37552-33-3*] (DG-AVP) has ca 50% and pressinamide [*27559-18-8*] (PA) has 10% of the behavioral activity of AVP. The C-terminal part of vasopressin (PAG) only slightly mimics the action of vasopressin but it counteracts amnesia (38). Although the ring structure of vasopressin is important in avoidance behavior, the C-terminal straight chain part is more important in retrieval mechanisms. The LVP and AVP fragments are practically devoid of classical endocrine and metabolic effects, eg, on blood pressure, water retention, and pituitary adrenal activity. The amino sequence of vasopressin and related peptide fragments is illustrated in Figure 1.

Clinical studies confirm the beneficial effect of vasopressin on memory function. A marked improvement in several attention and memory tests was observed in a double-blind study in elderly patients given vasopressin nasal spray (39). Beneficial effects also have been reported in patients suffering from posttraumatic amnesia (40). Further studies are warranted with fragments of AVP and LVP which may become the drugs of choice for the treatment of mental disorders of cognitive origin.

Cholinergic Mechanisms.   Central cholinergic mechanisms have long been implicated in learning and memory consolidation (41–42). Changes in free acetylcholine [60-31-1] (ACh) after stress or learning have been observed in different brain regions of mice (43) (see Choline; Cholinesterase inhibitors). Differences in ACh metabolism in the temporal lobes of good- and poor-performing mice have been reported (44). Newly synthesized ACh in the hippocampus accompanies the acquisition of passive avoidance (45).

Agents that increase the availability of ACh at or promote transmission of impulses across cholinergic synapses are expected to facilitate learning. Deanol [108-01-0] (2-dimethylaminoethanol), an ACh precursor capable of penetrating the blood–brain barrier more effectively than choline, was reported to improve performance of trained rats in a maze but only when administered during the tasks (46). However, there is some controversy as to whether Deanol can raise levels of ACh in the brain (47).

Cholinesterase inhibitors, eg, physostigmine [57-47-6] and diisopropyl fluorophosphate [55-91-4] (DFP) which prevent the destruction of ACh at the synapse, facilitate maze learning and enhance weak or nearly forgotten memory (48–49), but they have a blocking effect upon well-established memory (49). Human subjects receiving physostigmine (50), arecholine [63-75-2] or choline [62-49-7] (51) show improvement of long-term memory processes and serial learning. Thus, therapies to boost functioning of the cholinergic system may be valuable in reversing memory deficits resulting from aging (52).

Evidence that cholinergic neurons may be selectively damaged in Alzheimer-type dementia, ways by which the availability of choline and lecithin [8002-43-5] influence acetylcholine synthesis, and preliminary attempts to treat Alzheimer's disease with these compounds have been reviewed (53).

Arginine-vasopressin (AVP)   H–Cys–Tyr–Phe–Gln–Asn–Cys–Pro–Arg–Gly–NH$_2$
                                     1   2   3   4   5   6   7   8   9

Desglycinamide-arginine-   H–Cys–Tyr–Phe–Gln–Asn–Cys–Pro–Arg–OH
  vasopressin (DG-AVP)

Pressinamide (PA)   H–Cys–Tyr–Phe–Gln–Asn–Cys–NH$_2$

Prolyl-argyl-glycinamide (PAG)   H–Pro–Arg–Gly–NH$_2$

Figure 1.   Amino acid sequence of vasopressin and related peptide fragments.

Catecholaminergic Mechanisms. The beneficial role of catecholamines in memory consolidation has been suggested (54). Drugs that deplete the biogenic amines of the central nervous system have been found to impair the consolidation of memory (55), suggesting that normal levels of catecholamines during a critical period of memory storage may be essential for retention (56). Post-training administration of exogenous norepinephrine (NE) (57), or of amphetamine [300-62-9] (58), which releases brain catecholamine (CA), improves memory in several learning tests. Available data are consistent with the hypothesis of an involvement of one of the catecholamines either in memory processes or in events parallel and related to memory processes (56). The catecholamine precursor, L-Dopa [59-92-7], administered to depressed patients improved long but not short-term verbal memory (59). Performance decrements in the elderly may be due to a decline in brain NE, possibly due to age-related increases in monoamine oxidase (60). Unlike the catecholamines, 5-hydroxytryptamine [50-67-9] (5-HT) is believed to exert an inhibitory effect on memory processes (61). There is some evidence that the effects of 5-HT upon memory consolidation may be related to protein-synthesis inhibition (62).

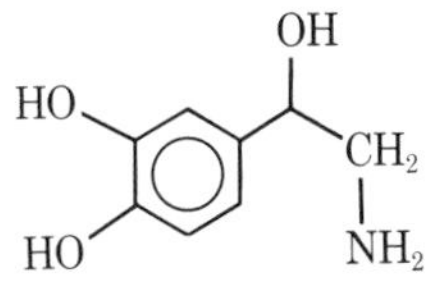

norepinephrine

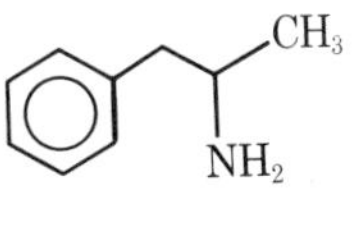

amphetamine

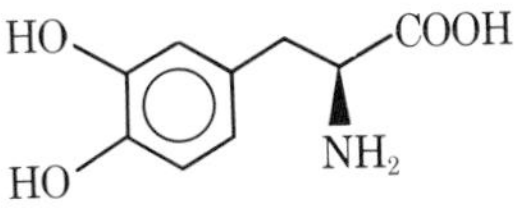

L-Dopa

CNS Stimulants and Analeptics. Numerous studies using subconvulsive doses of CNS stimulants on learning facilitation have been reported (4). Experiments using strychnine, as well as other stimulants such as picrotoxin, bemegride, pentylenetetrazole and xanthines (see Alkaloids), have been interpreted as indicating that these agents affect learning by increasing attention and acquisition or enhancing storage processes (63). Most convulsants impair memory when administered in convulsive doses.

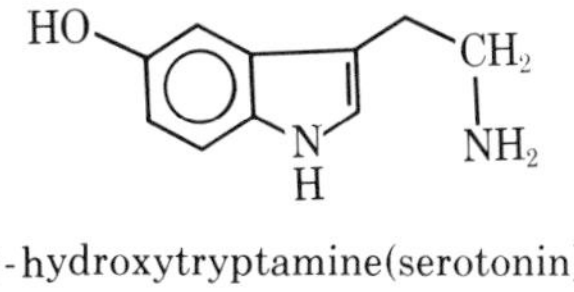

5-hydroxytryptamine(serotonin)

Drugs Affecting Protein Synthesis. RNA and protein synthesis undoubtedly play an important role in memory formation, particularly the consolidation phase. However, attempts to facilitate learning or memory processes by treatment with exogenous RNA have been subject to several criticisms on theoretical as well as methodological grounds (2). Studies therefore have been focused on the disruptive effects of inhibitors of protein synthesis, particulary puromycin, cycloheximide, and anisomycin (64–66) on memory consolidation. These drugs produce amnesia for a specific learning task when given shortly before or after training. Although the results are controversial, it is generally believed that this action is due to inhibition of cerebral protein synthesis. Of course, complete inhibition of protein synthesis *in vivo* can hardly be obtained without toxic doses, and the inhibitors also seem to influence other enzymatic processes (68–69). More encouraging is the beneficial influence on consolidation and retention of different learned behaviors by RNA precursors such as orotic acid [65-86-1] (5,70). It appears that the supply of pyrimidine nucleotides in the central nervous system probably controls macromolecular syntheses involved in the consolidation of a memory trace.

Miscellaneous Agents.   Pemoline (2-amino-5-phenyl-4-oxazolidinone) is a mild central nervous system stimulant approved in the United States for use as adjunctive therapy for management of hyperkinetic behavior in children (72). Although its precise mechanism of action is not known, pemoline is believed to act through stimulation of dopamine turnover. Its reported effects on memory- and performance-enhancing properties in animals probably result from its amphetaminelike psychostimulant properties. Clinical trials for the alleged antifatigue-, antidepressant-, and performance-enhancing effects of pemoline have not yet succeeded in demonstrating its efficacy for these purposes (72).

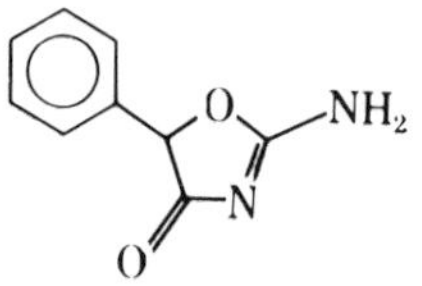

pemoline [2152-34-3]

Piracetam (nootropyl) is a nonstimulant, nonadrenergic agent reported to facilitate learning in animals (73,74). It protects rats against hypoxia-induced amnesia (75) but does not protect them against electroconvulsive shock-induced amnesia (76). In clinical trials, with some exceptions (77), piracetam increased alertness, counteracted fatigue, and increased verbal interactions in geriatric patients (78–80).

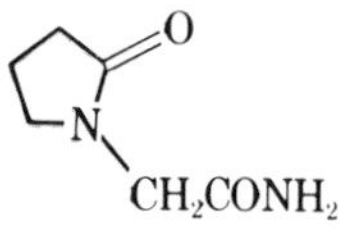

piracetam [7491-74-9]

Antiaging Drugs

Scientists from many disciplines have begun to focus on aging as a disease process (81). Several books highlighting various aspects and individual theories have been published, and many of the theories regarding the cause of aging have been comprehensively reviewed (82–92). Several of the theories on aging simply attempt to explain obvious deleterious changes associated with senescence and many are based on extremely limited evidence or on pure speculation. One of the problems of gerontology is its numerous hypotheses with very little supporting data.

Increasingly, aging processes are thought of as originating at the cellular or subcellular level, even though there is no general agreement as to the nature of these processes. An attractive general hypothesis of aging is that senescence is primarily a cellular information-loss phenomenon originating at the molecular level. Interference with the flow of cellular information could occur anywhere in the sequence from DNA to RNA to protein synthesis to Orgel's cytoplasmic error catastrophe which results from impaired protein synthesis (93). Alternatively, impaired cellular function caused by breakdown of essential cellular components, rampant lysosomal enzyme activity, autoimmune reactions, increased cross-linking, and accumulation of lipofuscin pigments and waste products also could result in cellular information loss, which ultimately manifests itself as biological aging.

Agents Protecting Against Free-Radical Damage and Cell-Debris Accumulation. Free radicals normally are present in all biological systems and can arise from a number of sources in the living system (94). Free-radical-mediated damage of unsaturated fatty acids and other cellular components can result in cumulative degeneration of cellular function which expresses itself as aging (95). One consequence of free-radical attack on cell components is lipofuscin production. Lipofuscin granules, referred to as age pigments, are fluorescent lipoprotein debris from peroxidation of polyunsaturated lipids and their cross-linking with proteins in subcellular membranes (96). However, there is as yet no direct evidence that lipofuscin interferes with normal cell processes (see Hydrocarbon oxidation; Initiators).

Antioxidants and Free-Radical Scavengers. Vitamin E and several chemically

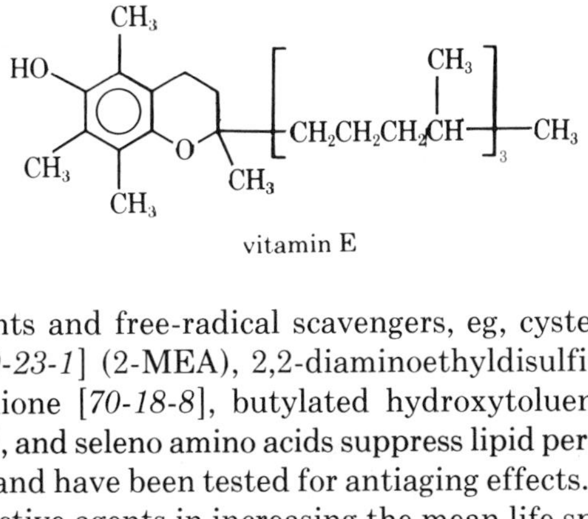

vitamin E

unrelated antioxidants and free-radical scavengers, eg, cysteine [52-90-4], 2-mercaptoethylamine [60-23-1] (2-MEA), 2,2-diaminoethyldisulfide [51-85-4], vitamin C [50-81-7], glutathione [70-18-8], butylated hydroxytoluene [128-37-0] (BHT), ethoxyquin [91-53-2], and seleno amino acids suppress lipid peroxidation of biological membranes *in vitro* and have been tested for antiaging effects. BHT and 2-MEA are among the most effective agents in increasing the mean life span in animal colonies but have no significant effect on maximum life span (95,97). Ethoxyquin produces a 20% increase in life span of C3H mice (98). Several antioxidants that were experimentally ineffective in increasing survival may have achieved inadequate tissue dis-

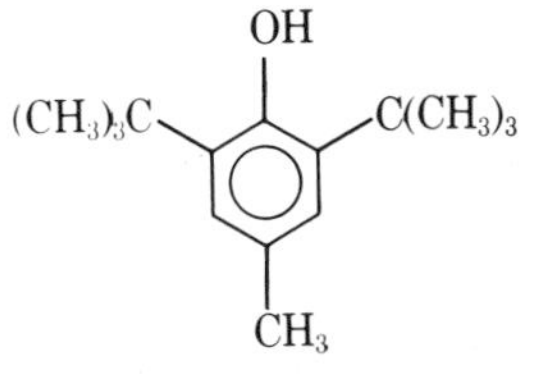

BHT

$HSCH_2CH_2NH_2$

2-MEA

tribution or concentration. Various combinations of vitamins C and E (see Vitamins), cysteine, glutathione, and selenium compounds are being investigated for antioxidant therapy in humans (99) (see Antioxidants).

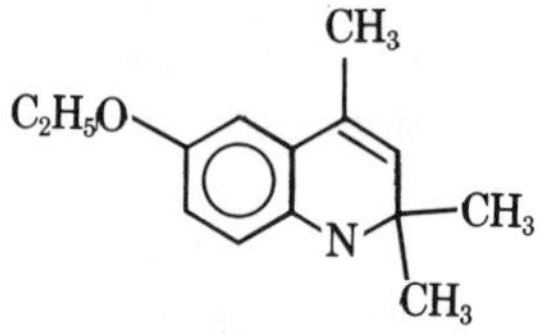

ethoxyquin

Agents Protecting Against Senility. *Lipofuscin Inhibitors.* Meclofenoxate [51-68-3] (centrophenoxine), the p-chlorophenoxyacetyl ester of dimethylaminoethanol, is a geriatric therapeutic agent useful for treatment of patients suffering from confusional states, Parkinsonism, and other senile mental disorders. Meclofenoxate reverses the accumulation of lipofuscin pigments in neurons of senile guinea pigs, and it has been suggested that reduction of lipofuscin may be one of the ways by which the drug exerts its beneficial effects on the CNS (100–101).

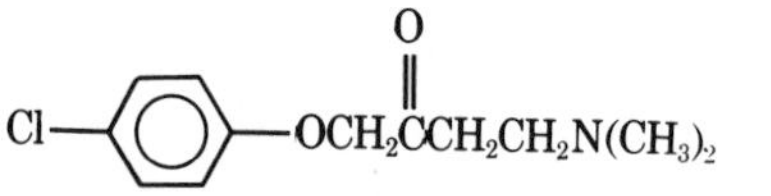

meclofenoxate

Agents that improve blood flow and anticoagulants that inhibit blood clogging have some success in alleviating senile symptoms (see Blood, coagulants and anticoagulants). FDA approval is being sought for nylidrin hydrochloride [849-55-8] (102), a vasodilator, as a specific treatment for senility, a disorder that afflicts 10^6 people over the age of 65 in the United States (103).

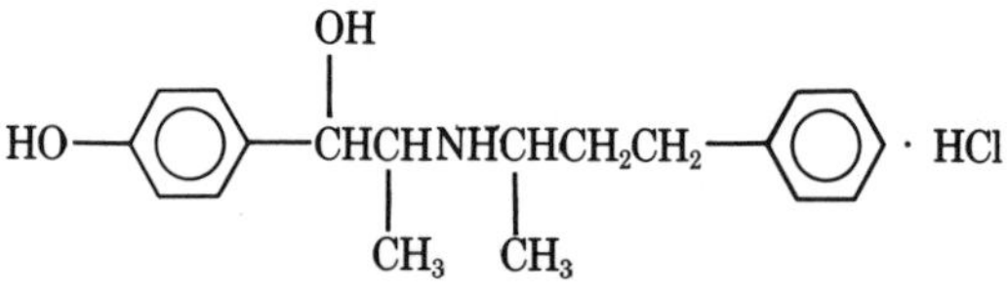

nylidrin hydrochloride

Agents Protecting Against Membrane Damage. Labilization of the lysosomal membrane and rampant lysosomal action have been suggested as causes of cellular death during aging (104). Leakage of lysosomal enzymes into the cytoplasm could, in principle, be responsible for several processes suggested as primary aging mechanisms, eg, reduced fidelity of protein synthesis (105). Agents, eg, dimethylaminoethanol [108-01-0] (DMAE), which stabilize lysosomal membranes, produce significant increases in the life span of senile mice (106). FDA-approved clinical studies with DMAE are believed to be underway (107).

Agents Affecting Cross-Linking of Connective Tissue. *Lathyrogens.* Agents that interfere with cross-linking of peptide chains in fibrous connective tissue macromolecules and collagen are expected to retard biological aging (108). Therefore, lathyrogenic compounds, eg, β-aminopropionitrile [151-18-8], penicillamine [52-67-5], and semicarbazide hydrochloride [563-41-7], which prevent maturation of collagen, have been suggested as antiaging drugs. Some evidence indicates that these drugs extend the average life span but not the maximum life span of rats if given in doses sufficient to retard cross-linking but not enough to cause undesirable side effects (26). The use of penicillamine is accompanied by an inhibition of wound healing and by increased skin fragility (109).

Agents Affecting Immunity.   The idea that aging is primarily an autoimmune phenomenon has been proposed and reviewed (110–114). There is an age-dependent decline of immunocompetence which is manifested as a progressive failure of the organism's cells to distinguish self from foreign cells with damaging immunological reaction against self cells or the development of damaging autoimmune cell clones arising from somatic mutation of immunocytes and a decreased responsiveness to extrinsic antigens. Decreased immunological surveillance may originate in the thymus-dependent segment of the immune system (113) (see Immunotherapeutic agents).

Immunosuppressants.   Immunosuppressants, eg, azathioprine [446-86-6] (Imuran) and cyclophosphamide [50-18-0], have been tested on aging mice, but do not affect maximum life span. However, mean survival times that were longer than those of controls were observed (115). Most likely, immunosuppressant therapy may be applied too late to be effective and, moreover, suppression of the immune system may accelerate death from other causes.

azathioprine

cyclophosphamide

Agents Affecting Protein Synthesis.   The error theories are based on the assumption that aging is a consequence of progressive accumulation of errors in protein synthesis. The errors may occur either at the DNA level by mutation, single-strand breaks, or other structural changes or at subsequent stages during transcription and translation (116–117). Introduction of errors during translation of the DNA message into RNA and protein sequences has been suggested as a more likely basis for aging than changes in the DNA molecule. The mechanism by which errors in the synthesis of RNA and information-handling enzymes can accumulate progressively, decrease the fidelity of protein synthesis, and eventually jeopardize the entire protein-synthesizing machinery (error catastrophe) have been reviewed (93,118). The interrelationship between cell function and protein biosynthesis in aging also has been reviewed (119).

Several agents that enhance learning and memory have been examined as antiaging drugs because they act on brain RNA and protein synthesis. However, no satisfactory evidence for their antiaging effects has been obtained. Procainamide [614-39-1] has been tested for learning enhancement activity in experimental animals of different ages and has been claimed to improve behavior and learning deficits in aging rats. Procainamide was reported to increase brain RNA content and to redistribute intracellular amino acids from the free pool to an RNA-bound form (120).

$$H_2N—\bigcirc—\overset{\overset{O}{\|}}{C}NHCH_2CH_2N(C_2H_5)_2$$

procainamide

Procaine Hydrochloride (Gerovital H₃). The highly controversial use of procaine hydrochloride [51-05-8] for retarding the aging process was first disclosed in the 1950s (121). Repeated injections of procaine hydrochloride were claimed to improve skin texture, recall, psychomotor activity, and muscle strength in senescent individuals. However, the studies are characterized by a lack of scientific control and evaluation, and several controlled human trials by others have not substantiated the rejuvenation claims (122–123). The subsequent popular use of Gerovital H₃, a formulation of procaine hydrochloride containing benzoic acid and potassium metabisulfate, as a rejuvenating agent appears to be equally dubious (121). Gerovital H₃ has weak monoamine oxidase inhibitor properties, and this probably accounts for its reported benefits in treating depressive disorders in the elderly (124).

$$H_2N—\bigcirc—\overset{\overset{O}{\|}}{C}CH_2CH_2N(C_2H_5)_2 \cdot HCl$$

procaine hydrochloride

Conclusion. On the basis of experimental evidence, it is unreasonable to advance a chemotherapeutic approach to aging, and the available rejuvenating drugs are of dubious utility. A clearer understanding of the fundamental aging process is expected to provide at least a rational chemical approach to retarding human aging during the next two decades.

BIBLIOGRAPHY

1. E. Glassman, *Ann. Rev. Biochem.* **38,** 605 (1969).
2. W. B. Essman, *Adv. Pharmacol. Chemother.* **9,** 241 (1971).
3. R. G. Rahwan, *Agents Actions* 2/3, 87 (1971).
4. J. L. McGaugh, *Ann. Rev. Pharmacol.* **13,** 229 (1973).
5. H. Matthies, *Life Sci.* **15,** 2017 (1974).
6. J. A. Gylys and H. A. Tilson, *Ann. Rep. Med. Chem.* **10,** 21 (1975).
7. P. E. Gold, *Ann. Rep. Med. Chem.* **12,** 30 (1977).
8. G. Ungar, *Agents Actions*, 1/4, 155 (1970).
9. G. Ungar, *Biochem. Pharmacol.* **23,** 1553 (1974).
10. G. Ungar, *Internat. Rev. Neurobiol.* **17,** 37 (1975).
11. G. Ungar, *Naturwissenschaften* **60,** 307 (1973).
12. D. de Wied, *Life Sci.* **20,** 195 (1977).
13. L. H. Miller, C. A. Sandman, and A. J. Kastin, eds., *Neuropeptide Influences on the Brain and Behaviour*, Raven Press, New York, 1977.
14. C. A. Sandman, L. H. Miller, and A. J. Kastin, *Pharmacol. Biochem. Behav.* 5(Suppl. 1), (1976).
15. D. de Wied, *Hosp. Pract.*, 123 (1976).
16. D. de Wied, A. Witter, and H. M. Greven, *Biochem. Pharmacol.* **24,** 1463 (1975).
17. A. Witter, H. M. Greven, and D. de Wied, *J. Pharmacol. Exp. Ther.* **193,** 853 (1975).
18. H. Rigter, R. Janssens-Elbertse, and H. van Riezen, *Pharmacol. Biochem. Behav.* 5(Suppl. 1), 53 (1976).
19. D. de Weid, *Proc. Soc. Exp. Biol.* **122,** 28 (1966).

20. C. A. Sandman, A. J. Kastin, and A. V. Schally, *Physiol. Behav.* **6,** 45 (1971).
21. H. Rigter and H. van Riezen, *Physiol. Behav.* **14,** 563 (1975).
22. H. Rigter, H. van Riezen, and D. de Wied, *Physiol. Behav.* **13,** 381 (1974).
23. C. A. Sandman, J. M. George, J. D. Nolan, H. van Riezen, and A. J. Kastin, *Physiol. Behav.* **15,** 427 (1975).
24. M. M. Ward, C. A. Sandman, J. M. George, and H. Shulman, *Physiol. Behav.* **22,** 669 (1979).
25. L. H. Miller, A. F. Kastin, C. A. Sandman, M. Fink, and W. van Veen, *Pharmacol. Biochem. Behav.* **2,** 663 (1974).
26. C. A. Sandman, J. George, T. R. McCanne, J. D. Nolan, J. Kaswan, and A. F. Kastin, *J. Clin. Endocr. Metab.* **44,** 884 (1977).
27. L. H. Miller, S. A. Fisher, G. A. Groves, M. E. Rudrauff, and A. F. Kastin, *Pharmacol. Biochem. Behav.* **7,** 417 (1977).
28. J. G. Small, I. F. Small, V. Milstein, and D. A. Dian, *Acta. Psychol. Scand.* **55,** 241 (1977).
29. A. J. Dunn and W. H. Gispen, *Behav. Rev.* **1,** 15 (1977).
30. B. E. Leonard, *Arch. Intl. Pharmacol. Therap.* **207,** 242 (1974).
31. V. M. Wiegant, A. J. Dunn, P. Schotman, and W. H. Gispen, *Brain Res.* **168,** 565 (1979).
32. J. M. van Ree, B. Bohus, D. H. G. Versteeg, and D. de Wied, *Biochem. Pharmacol.* **27,** 1793 (1978).
33. D. de Wied and D. H. G. Versteeg, *Fed. Proc.* **38,** 2348 (1979).
34. G. L. Kovács, B. Bohus, D. H. F. Versteeg, E. R. de Kloet, and D. de Wied, *Brain Res.* **175,** 203 (1979).
35. D. de Wied, *Nature (London)* **232,** 58 (1971).
36. B. Bohus, R. Ader, and D. de Weid, *Horm. Behav.* **3,** 191 (1972).
37. D. de Wied, B. Bohus, I. Urban, T. B. van Wimersma Greidanus, and W. H. Gispen in R. Walter and J. Meienhofer, eds., *Peptides Chemistry Structure and Biology*, Ann Arbor Sci. Publ., Ann Arbor, Mich., 1975, p. 635.
38. R. Walter, P. L. Hoffman, J. B. Flexner, and L. B. Flexner, *Proc. Nat. Acad. Sci.* **72,** 4180 (1975).
39. J. J. Legros and co-workers, *Lancet i,* 41 (1978).
40. J. C. Oliveros and co-workers, *Lancet i,* 42 (1978).
41. S. N. Pradhan and S. N. Dutta, *Int. Rev. Neurobiol.* **14,** 173 (1971).
42. J. A. Deutsch in J. A. Deutsch, ed., *The Physiological Basis of Memory*, Plenum Press, New York, 1973, p. 59.
43. A. G. Karczmar, *Adv. Neuropharmacol.*, 455 (1971).
44. P. Mandel, G. Ayad, J. Hermetet, and A. Edel, *Brain Res.* **72,** 65 (1974).
45. S. Glick, T. Mittag, and J. Green, *Neuropharmacol.* **12,** 291 (1973).
46. A. J. Karoly and L. J. Hunt, *Fed. Proc.* **24,** 296 (1965).
47. G. Pepeu, D. X. Freedman, and N. J. Giarman, *J. Pharmacol. Exptl. Therap.* **129,** 291 (1960).
48. J. A. Izquierdo, *Prog. Drug Res.* **16,** 334, 1972.
49. J. A. Deutsch, *Science* **174,** 788 (1971).
50. K. L. Davis, R. C. Mohs, and J. R. Tinklenberg, *Science* **201,** 272 (1978).
51. N. Sitaram, H. Weingartner, and J. G. Christian, *Science* **201,** 274 (1978).
52. S. Kent, *Geriatrics* (7), 77 (1979).
53. R. D. Terry and P. Davies, *Ann. Rev. Neurosci.* **3,** 77 (1980); J. H. Growdon and S. Corkin in J. O. Cole and J. Barrett, eds., *Psychopathology in the Aged*, Raven Press, 1980.
54. S. F. Zornetzer in M. A. Lipton, A. DiMascio, K. F. Killam, eds., *Psychopharmacology A Generation of Progress*, Raven Press, New York, 1978, p. 637.
55. R. K. Dismukos and A. V. Rake, *Psychopharmacologia* **23,** 17 (1972).
56. P. Kurtz and T. Palfai, *Biobehav. Rev.* **1,** 25 (1977).
57. L. Stien, J. D. Belluzzi, and C. D. Wise, *Brain Res.* **84,** 329 (1975).
58. I. Izquierdo, D. G. Beanish, and H. Anisman, *Psychopharmacol.* **63,** 173 (1979).
59. D. Murphy, G. Henry and H. Weingarter, *Psychopharmacologia* **27,** 319 (1972).
60. D. Robinson, *Fed. Proc.* **34,** 103 (1975).
61. W. B. Essman, *Adv. Biochem. Psychopharmacol.* **11,** 265 (1974).
62. W. B. Essman, *Pharmacol. Biochem. Behav.* **1,** 7 (1973).
63. M. J. Brennan and W. C. Gordon, *Pharmacol. Biochem. Behav.* **7,** 451 (1977).
64. D. Daniels, *Nature* **231,** 395 (1971).
65. R. G. Scrota, R. B. Roberts, and L. B. Flexner, *Proc. Nat. Acad. Sci. U.S.* **69,** 340 (1972).
66. L. R. Squire and S. H. Barondes, *Proc. Nat. Acad. Sci. U.S.* **69,** 1416 (1972).
67. L. B. Flexner and R. H. Goodman, *Proc. Nat. Acad. Sci. U.S.* **72,** 4660 (1975).

68. L. B. Flexner, R. G. Scrota, and R. H. Goodman, *Proc. Nat. Acad. Sci. U.S.* **70,** 354 (1973).
69. P. Lundgren and L. A. Carr, *Pharmacol. Biochem. Behav.* **9,** 559 (1978).
70. T. Ott and H. Matthies, *Psychopharmacologia* **20,** 16 (1971).
71. A. T. Dren and R. S. Janicki in M. E. Goldberg, ed., *Pharmacological and Biochemical Properties of Drug Substances*, American Pharmaceutical Assoc., Washington, D.C., 1977, p. 33.
72. A. T. Dren and R. S. Janicki in M. E. Goldberg, ed., *Pharmacological and Biochemical Properties of Drug Substances*, Amer. Pharm. Assoc., Washington, D.C., 1977, p. 33.
73. C. Giurgea, D. Lefevre, C. Lescrenier, and M. David-Remacle, *Psychopharmacologia* **20,** 160 (1971).
74. O. L. Wolthuis, *Eur. J. Pharmacol.* **16,** 283 (1971).
75. S. J. Sara and D. Lefevre, *Psychopharmacologia* **25,** 32 (1972).
76. S. J. Sara and M. David-Remacle, *Psychopharmacologia* **36,** 59 (1974).
77. F. S. Abuzzahab, G. Merwin, and M. Sherma, *Pharmacologist* **15,** 237 (1973).
78. A. J. Stegink, *Arzneim. Forsch.* **22,** 975 (1972).
79. A. Voelkel, *Arzneim. Forsch.* **24,** 1127 (1974).
80. S. J. Dimond and E. Y. M. Brouwers, *Psychopharmacologia* **49,** 307 (1976).
81. J. S. Bindra, *Ann. Rep. Med. Chem.* **9,** 214 (1974).
82. A. Comfort, *Aging: The Biology of Senescence*, Holt, Rinehart and Winston, New York, 1964; *Nature* **217,** 320 (1968).
83. C. G. Kormendy and A. D. Bender, *J. Pharm. Sci.* **60,** 167 (1971).
84. *Chem. Eng. News*, 13 (July 24, 1971); *Chem. Eng. News*, 15 (Mar. 28, 1974).
85. L. Hayflick, *Fed. Proc.* **34,** 9 (1975); *New Eng. J. Med.* **295,** 1302 (1976).
86. S. Goldstein, *New Eng. J. Med.* **285,** 1121 (1971).
87. F. S. La Bella in A. A. Rubin, ed., *Search for New Drugs*, Marcel Dekker, New York, 1972, p. 347.
88. M. Rockstein, ed., *Theoretical Aspects of Aging*, Academic Press, Inc., New York, 1974.
89. C. E. Finch and L. Hayflick, eds., *Handbook of the Biology of Aging*, Van Nostrand Reinhold, New York, 1977.
90. A. Rosenfeld, *Prolongevity*, A. A. Knopf, New York, 1976.
91. G. J. Thorbecke, *Fed. Proc.* **34,** 4 (1975).
92. S. Kent, *Geriatrics* **31**(2), 135 (1976).
93. L. E. Orgel, *Nature* **243,** 441 (1973).
94. W. A. Pryor, *Sci. Am.* **223,** 70 (1970).
95. D. Harman, *Agents Action* **1,** 3 (1969).
96. P. Gordon in Ref. 88, p. 61.
97. D. Harman, *Am. J. Clin. Nutr.* **25,** 839 (1972).
98. A. Comfort, I. Youhotsky-Gore, and K. Pathmanathan, *Nature* **229,** 254 (1971).
99. R. A. Passwater and P. A. Welker, *Am. Lab.* **3,** 21 (1971).
100. K. Nandy and G. H. Bourne, *Nature* **210,** 313 (1966).
101. K. Nandy, *J. Gerontol.* **23,** 82 (1968).
102. *The United States Pharmacopeia XX* (*USP XX–NF XV*), The United States Pharmacopeial Convention, Inc., Rockville, Md., 1980, p. 562.
103. *Chem. Week*, 39 (Oct. 22, 1980).
104. A. Comfort, *Lancet ii*, 1325 (1966).
105. R. Hochschild, *Exp. Gerontol.* **6,** 153 (1971).
106. *Ibid.*, **8,** 177, 185 (1973).
107. *Med. World News*, 13 (Jan. 11, 1974).
108. F. S. LaBella, *Gerontologist* **8,** 13 (1968).
109. M. E. Nimmi and L. A. Baretta, *Science* **150,** 905 (1965).
110. R. L. Walford, *Lancet* **ii,** 1226 (1970).
111. R. L. Walford, *Immunologic Theory of Aging*, Munksgaard, Copenhagen, 1969.
112. F. M. Burnet, *Immunological Surveillance*, Pergamon Press, New York, 1970.
113. F. M. Burnet, *Lancet ii*, 358 (1970).
114. R. L. Walford, *Fed. Proc.* **33,** 2020 (1974).
115. R. L. Walford, *Symp. Soc. Exp. Biol.* **21,** 351 (1967).
116. K. L. Yielding, *Perspect. Biol. Med.* 201 (1974).
117. H. J. Curtis, *Biological Mechanisms of Aging*, C. C. Thomas, Springfield, Mass., 1966.
118. L. E. Orgel, *Nature* **67,** 1476 (1970).
119. V. V. Frolkis, *Gerontologia* **19,** 189 (1973).

120. P. Gordon, *Recent Adv. Biol. Psychiatry* **10,** 121 (1968).
121. A. Aslan in Ref. 70, p. 145.
122. G. C. Chiu, *J. Am. Med. Assoc.* **175,** 502 (1961).
123. A. D. Bender, C. G. Kormendy, and R. Powell, *Exp. Gerontol.* **5,** 97 (1970).
124. T. M. Yau in Ref. 88, p. 157.

JASJIT S. BINDRA
Pfizer, Inc.

MENDELEVIUM. See Actinides and transactinides.

MERCURY

Mercury [*7439-97-6*], Hg, atomic number 80, also called quicksilver, is a heavy, silvery-white metal that is liquid at room temperature. Below its melting point, mercury is a white solid and above its boiling point a colorless vapor. The symbol Hg is taken from the Latin word *hydrargyrum*, meaning liquid silver. In nature, mercury occurs mainly in combination with sulfur to form more than a dozen different minerals. Commercially, the most important one is the red sulfide, cinnabar, HgS (86.2% mercury and 13.8% sulfur). Mercury metal produced from mining operations is called prime virgin mercury and is usually more than 99.9% pure. It has a clean, bright appearance, and contains less than 1 ppm of any base metals. Prime virgin mercury is acceptable for most industrial uses. Higher-purity mercury necessary for some applications is obtained by multiple distillation or electrolytic refining. Mercury is used in numerous and varied applications dependent mainly on its physicochemical properties such as uniform volume expansion, electrical conductivity, toxicity, and ability to alloy with other metals.

Mercury and cinnabar have been known and used since antiquity. The first recorded mention of mercury was by Aristotle in the fourth century BC, when it was used in religious ceremonies. Earlier, cinnabar (vermillion) is known to have been used as a pigment for cave and body decoration (see Pigments, inorganic). The ancient Egyptians, Greeks, and Romans used mercury for cosmetic and medical preparations and for amalgamation.

The development of the Patio process in the 16th century for the recovery of silver by amalgamation greatly increased the consumption of mercury (1).

In 1643, Torricelli invented the barometer using mercury to measure the pressure of the atmosphere, and in 1720, Fahrenheit invented the mercury thermometer. Other scientific applications followed, such as sealing off water-soluble gases in gas analysis. Mercury fulminate, $Hg(ONC)_2$, a detonator for explosives, was first prepared in 1799 (see Explosives).

Continued research on mercury's physical and chemical properties resulted in rapidly expanding industrial use after 1900, particularly in electrical applications, which offset the sharp decline in its use in amalgamation. The invention of the mercury battery in 1944 immediately caused a sharp and continuous rise in mercury consumption (see Batteries).

Occurrence

Mercury is estimated to occur in continental rocks in a range of 10–1000 ppb in contrast to a typical ore grade of 5,000,000 ppb or 5.1 kg of mercury per metric ton of ore (0.5%) (2). In petroleum it occurs in a range of 2000–20,000 ppb. Although native or metallic mercury is found in very small quantities in some ore deposits, mercury usually occurs as a sulfide, and occasionally as a chloride or an oxide, generally in conjunction with base and precious metals. In addition to cinnabar, which is by far the predominate mercury mineral in ore deposits, corderoite ($Hg_3S_2Cl_2$), livingstonite ($HgSb_4S_7$), montroydite (HgO), terlinguaite (Hg_2OCl), calomel ($HgCl$), and meta-cinnabar, a black form of cinnabar, are commonly found in mercury deposits (3). The numerous other mercury minerals are rare and not commercially significant.

Mercury ore deposits occur in faulted and fractured rocks, such as limestone, calcareous shales, sandstones, serpentine, chert, andesite, basalt, and rhyolite. They are mostly epithermal in character, ie, they have been deposited by rising warm solutions at comparatively shallow depths ranging from a meter to about a thousand meters (4).

Properties

Mercury has a uniform volume expansion over its entire liquid range which, in conjunction with its high surface tension and therefore an inability to wet and cling to glass, makes it extremely useful for barometers, manometers, and thermometers, as well as many other measuring devices. This ability is enhanced by the liquidity of mercury at room temperature. Mercury also has a propensity to form alloys (amalgams) with almost all other metals except iron, and at high temperatures even with iron. Because of its low electrical resistivity, mercury is rated as one of the best electrical conductors among the metals. Mercury has a high thermal-neutron-capture cross section (360×10^{-28} m^2 or 360 barns), enabling it to absorb neutrons and act as a shield for atomic devices; its high thermal conductivity also permits it to act as a coolant. The physical and chemical properties of mercury are given in Table 1, and the distribution of the stable isotopes in Table 2. A number of artificial isotopes also are known (see Radioisotopes).

The volume expansion of mercury, one of its important properties, is calculated over its entire liquid range by the formula (t, °C)

$$V_t = V_o(1 + 0.18182 \times 10^{-3}\, t + 0.0078 \times 10^{-6}\, t^2)$$

The specific heat varies with the temperature; in solid mercury it increases but in liquid mercury it drops; however, not at a uniform rate (see Table 3). The specific heat at 210°C is the same as that at −75°C. Up to 50°C, it is given closely by the formula $0.00339 - 0.0001038(t + 36.7)$; for temperatures from 50 to 150°C, the formula should be modified by a corrective factor of $-0.000006(t - 50)$; and between 150 and 250°C a further correction is needed, $-0.000003(t - 150)$. Values derived from these formulas

Table 1. Physical and Chemical Properties of Mercury[a]

Property	Value
atomic weight	200.59
accommodation coefficient, at -30 to $60°C$, $(T_3 - T_1)/(T_2 - T_1)^b$	1.00
angle of contact of glass at $18°C$	128
atomic distance, nm	3.0
melting point, $°C$	-38.87
boiling point, $°C$	356.9
triple point, $°C$	$-38.84168°$
bp rise with pressure, $°C/kPa^c$	0.5595
compressibility (volume), at $20°C$, per MPa^d	39.5×10^{-6} ^{e}
condensation on glass, $°C$	-130 to -140
conductivity, thermal, $W/(cm^2 \cdot K)$	0.092
critical density, g/cm^3	3.56
critical pressure, MPa^d	74.2
critical temperature, $°C$	1677
crystal system	rhombohedral
density, g/cm^3, at $20°C$	13.546
at melting point	14.43
at $-38.8°C$ (solid)	14.193
at $0°C$	13.595
electrode reduction potentials, normal, V	
$Hg^{2+} + 2\,e \rightleftharpoons Hg$	0.851
$Hg_2^{2+} + 2\,e \rightleftharpoons 2\,Hg$	0.7961
$2\,Hg^{2+} + 2\,e \rightleftharpoons Hg_2^{2+}$	0.905
emf, relative to Pt-cold junction, $0°C$; hot, $100°C$, mV	-0.60
expansion coefficient (volume) of liquid, at $20°C$, per $°C$	182×10^{-6}
freezing temperature, $°C$	-38.87
fusion, latent heat of, J/g^f	11.80
hydrogen overvoltage, V	1.06
ionization potentials, V	
1st electron	10.43
2nd electron	18.75
3rd electron	34.20
4th electron	(72)
5th electron	(82)
magnetic moment, Hg^{199}, J/T (μ_B)	4.63×10^{-24} (0.4993)
magnetic volume susceptibility at $18°C$, cm^3/g (cgs)	1.885×10^{-6} (-0.15×10^{-6})
potential, V,	
contact Hg/Sb	-0.26
contact Hg/Zn	$+0.17$
pressure, internal, MPa^d	1.317
reflectivity, degree at 550 nm	71.2
refractive index at $20°C$	1.6–1.9
temperature coefficient of resistance at $20°C$	0.9×10^{-3}
resistivity, $\Omega \cdot cm$, at $20°C$	95.8×10^{-6}
sol in water, $\mu g/L$	20–30
surface tension temp coefficient, $mN/(m \cdot °C)$ (= $dyn/(cm \cdot °C)$)	-0.19
viscosity, $mPa \cdot s$ (= cP), at $20°C$	1.55

[a] Ref. 5.

[b] T_1 is the temperature of a gas molecule striking a surface which is at temperature T_2, and T_3 is the temperature of a gas molecule leaving the surface.

[c] To convert kPa to mm Hg, multiply by 7.5.

[d] To convert MPa to atm, divide by 0.101.

Table 2.   Distribution of Stable Isotopes[a]

Isotope	CAS Registry Number	Abundance, %	Isotope	CAS Registry Number	Abundance, %
Hg^{196}	[14917-67-0]	0.146	Hg^{201}	[15185-19-0]	13.22
Hg^{198}	[13891-21-0]	10.02	Hg^{202}	[14191-86-7]	29.80
Hg^{199}	[14191-87-8]	16.84	Hg^{204}	[15756-14-6]	6.85
Hg^{200}	[15756-10-2]	23.13			

[a] Ref. 5.

above 150°C are not as accurate as those for lower temperatures, but suffice for most work when a table of specific heats is not available. Thermodynamic properties are given in Table 3.

The vapor pressure of mercury also behaves irregularly. For accurate values, a standard table may be obtained from the following formulas: from 0 to 150°C, $\log P = (-3212.5/T) + 7.150$; from 150° to 400°C, $\log P = (-3141.33/T) + 7.879 - 0.00019 \, t$, where P is vapor pressure in kPa (101.3 kPa = 760 mm Hg), T is the absolute temperature in K, and t is the temperature in °C. The accuracy of these formulas is believed to be within one percent.

Another valuable property of mercury is its relatively high surface tension, 480.3 mN/m (= dyn/cm) at 0°C, compared with 75.6 mN/m for water. Because of its high surface tension, mercury does not wet glass and exhibits a reverse miniscus in a capillary tube.

Table 3.   Thermodynamic Properties of Mercury[a]

Property	Value
entropy (S_{298}), J/mol[b]	76.107
heat of fusion, J/atom[b]	2,297
heat of vaporization (ΔH_v), J/atom[b]	59,149
liquid mercury, 25–357°C	
heat capacity (C_P), J/mol[b]	27.66
$H_T - H_{298}$ K	$-1971 + 6.61 \, T$
$F_T - H_{298}$ K	$-1971 + 6.61 \, T \ln T + 26.08 \, T$
gaseous mercury, 357–2727°C	
heat capacity, J/mol[b]	20.79
$H_T - H_{298}$ K	$13{,}055 + 4.969 \, T$
$F_T - H_{298}$ K	$13{,}055 - 4.969 \, T \ln T - 8.21 \, T$
vaporization	
heat of, J/g[b] atom, at 25°C	61.38
latent heat of, J/g[b]	271.96
specific heat, J/g[b]	
solid	
−75.6°C	1.1335
−40°C	0.141
−263.3°C	0.0231
liquid	
−36.7°C	0.1418
210°C	1.1335

[a] Ref. 6.
[b] To convert J to cal, divide by 4.184.

At ordinary temperatures, mercury is stable and does not react with air, ammonia, carbon dioxide, nitrous oxide, or oxygen. It combines readily with the halogens and sulfur, but is little affected by hydrochloric acid, and is attacked only by concentrated sulfuric acid. Either dilute or concentrated nitric acid dissolves mercury, forming mercurous salts when the mercury is in excess or no heat is used, and mercuric salts when excess acid or heat is used. Mercury reacts with hydrogen sulfide in the air and always should be covered.

For the diffusion of some metals in mercury, see Table 4.

The only metals having good or excellent resistance to corrosion by amalgamation with mercury are vanadium, iron, niobium, molybdenum, cesium, tantalum, and tungsten (6).

Table 4. Diffusion of Some Metals in Mercury[a]

Metal	Diffusion rate, $cm^2/s \times 10^{-5}$	Metal	Diffusion rate, $cm^2/s \times 10^{-5}$
lithium	0.9	silver	1.1
sodium	0.9	gold	0.7
potassium	0.7	zinc	1.57
rubidium	0.5	cadmium	2.07
cesium	0.6	mercury	0.007
calcium	0.6	thallium	1.03
strontium	0.5	tin	1.68
barium	0.6	lead	1.16
copper	1.06	bismuth	0.99

[a] Ref. 5.

Production and Shipment

Primary. Mercury ore is mined by both surface and underground methods; the latter furnish about 90% of the world's production. Mercury is recovered also as a by-product in the mining and processing of precious and base metals. In past years, small quantities of mercury have been produced by processing ground under and adjacent to the sites of less efficient ore-burning furnaces used in early-day mercury recovery operations. Mercury is also produced by working mine dumps and tailing piles, particularly those accumulated during turn-of-the-century mining operations. The average grade of mercury ore mined from large mines throughout the world in recent years has ranged from 4 to 16–20 kg/t with total recovery approaching 95 percent. The average grade of ore has generally declined over the years partly as a result of the practice of mining the richest parts of ore bodies to realize a higher profit and partly because prices generally have increased over the years, thus allowing lower-grade ores to be exploited.

Secondary. Smaller quantities of mercury are produced each year from industrial scrap and waste materials, such as discarded dental amalgams, batteries, lamps, switches, measuring devices, control instruments, and wastes and sludges generated in laboratories and electrolytic refining plants (7). Mercury in these materials may be recovered *in situ* or by firms specializing in secondary recovery (see Recycling). Secondary production depends on the price of mercury; it is more economical to purchase primary mercury on the market when the price is low and more economical to recycle when the price is high.

Shipping. Prime virgin-grade metal is packaged in wrought iron or steel flasks containing 34.5 kg of mercury. Mercury of greater purity, produced by multiple distillation or other means, may be marketed in flasks but is packaged usually in small glass or plastic containers.

Processing

Primary. Mercury metal is produced from its ores by standard methods throughout the world. The ore is heated in retorts or furnaces to liberate the metal as vapor which is cooled in a condensing system to form mercury metal (8). Retorts are inexpensive installations for batch-treating concentrates and soot and require only simple firing and condensing equipment. For larger operations, either continuous rotary kilns or multiple-hearth furnaces with mechanical feeding and discharging devices are preferred. With careful control at properly designed plants, 95 percent or more of the mercury in the ore can be recovered as commercial grade, 99.9 percent purity, mercury.

Other recovery methods (9) include leaching the ores and concentrates with sodium sulfide and sodium hydroxide, and subsequent precipitation with aluminum or by electrolysis (10). In another process, the mercury in the ore is dissolved by a sodium hypochlorite solution, the mercury-laden solution is then passed through activated carbon to absorb the mercury, and the activated carbon heated to produce mercury metal. These methods, however, are not being used at present. Mercury can be extracted from cinnabar by electrooxidation (11–12).

Secondary. Scrap material and industrial and municipal wastes and sludges containing mercury are treated in much the same manner as ores to recover mercury. Scrap products are first broken down to liberate metallic mercury or its compounds. Heating in retorts vaporizes the mercury which upon cooling condenses to high purity mercury metal. Industrial and municipal sludges and wastes may be treated chemically before roasting.

Economic Aspects

Annual world production averaged 3078 metric tons from 1800 to 1900 (13), and 5590 t from 1900 to 1959. Table 5 gives a breakdown by country after 1959. After a high in 1971, production declined as price, grade of ore, and consumption dropped, recycling increased, and mining operations were curtailed by the application of stringent environmental protection regulations. World producers further reduced mine production and sold mercury from excessively high inventories that depressed the price in recent years.

Before 1850, mercury was mined mostly in Spain, Yugoslavia, and Peru. In the second half of the 19th century, mines in Italy and California became prominent, whereas more recently, mines in the USSR, Algeria, and the People's Republic of China have produced significant quantities.

Mercury production in the United States began about 1850, and peaked in 1877 with 2755 t. California contributed 85% of production.

Temporary reversals of the declining trend, which had set in at the outset of World War I, occurred during wartime and during the U.S. government's purchase program during 1954–1958. In 1975, a new deposit of cinnabar and corderoite averaging 4.5 kg/t

Table 5. Mercury World Production 1959–1978 (Metric Tons)

Country	1959	1962	1965	1968	1971	1975	1978[a]	Percent of total
Algeria					246	965	1,034	
Canada			1	196	638	414		
People's Republic of China[b]	793	896	896	689	896	896	586	9
Czechoslovakia[b]	25	25	27	4	194	203	179	
Federal Republic of Germany[b]					70	110	84	
Italy	1,580	1,879	1,976	1,838	1,469	1,092	3	16
Japan	206	145	162	175	192			
Mexico	566	650	662	593	1,220	490	76	7
Peru	87	120	107	108	119	53		
Philippines	122	95	82	122	173	8		
Spain	1,782	1,820	2,574	1,963	1,752	1,517	1,070	22
Turkey	51	93	95	161	361	187	59	
USSR[b]	862	1,207	1,379	1,551	1,724	1,896	2,068	18
United States	1,078	906	675	995	616	254	883	8
Yugoslavia	460	561	566	510	572	584		6
other	76	49	32	47	122	30	143	
Total[b]	7,688	8,446	9,234	8,952	10,364	8,699	6,051	

[a] Preliminary.

[b] Estimate.

ore and containing 13,800 t mercury was discovered and developed by open-pit methods at McDermitt, Nevada (14).

In recent years, the mercury industry in the United States has been characterized by static consumption, price instability, inadequate mine production, oversupply, high import reliance, declining ore grade and reserves, and sharply reduced mining operations (see Tables 6 and 7). The number of producing mines has fallen from a high

Table 6. Mercury, U.S. Economic Statistics, 1959–1978

Year	1959	1962	1965	1968	1971	1975	1978
production, metric tons							
primary	1078	906	675	995	616	254	833
secondary	171	200	514	509	376	260	123
Total	1249	1106	1189	1504	992	514	956
government releases, t			1095	676	199	17	196
imports for consumption, t	1039	1088	560	801	981	1512	1681
exports and reexports, t	41	17	277	262	249	17	34
stocks, industry, t	468	514	703	790	581	881	1336
consumption, t	1892	2251	2536	2600	1801	1753	1681
import reliance[a]	43	51	52	41	49	62	64
price[b], dollars per flask[c]:							
New York	227	191	571	536	292	158	153
London	209	173	608	547	282	130	132
number of mines	71	56	149	87	56	13	1
grade of ore, kg/t	3.9	6.1	1.9	2.3	2.2	3.1	4.5

[a] Net import reliance as a percent of apparent consumption.

[b] Average annual price.

[c] A flask contains 34.5 kg.

Table 7.   U.S. Mercury Consumption by Uses (Metric Tons)

Uses	1959	1962	1965	1968	1971	1975	1978
agriculture	110	147	107	118	51	21	a
amalgamation	9	10	9	9		<0.5	a
catalysts	33	30	32	66	35	29	a
dental preparations	95	113	110	106	81	81	18
electrical applications	426	485	651	677	582	585	619
chlorine-caustic production	201	252	302	602	419	525	385
general laboratory	38	60	80	69	62	12	14
industrial and control instruments	351	306	356	275	168	159	120
paint:							
antifouling	34	4	9	14	14		
mildew proofing	87	157	283	351	282	239	309
paper and pulp	150	90	21	14	<0.5		
pharmaceuticals	59	116	14	15	24	15	a
other[b]	298	479	562	285	83	60	216
unknown uses					<0.5	28	
Total[c]	1892	2251	2536	2600	1801	1753	1681

[a] Withheld to avoid disclosing individual company proprietary data; included in other.

[b] Mostly inventory metal for use in installation and expansion of chlorine and caustic soda plants.

[c] Yearly data may not add to totals because of independent rounding.

of 149 in 1965 to one in 1978 as a result of falling prices, rising costs, and an inability to comply with environmental protection regulations. The United States has gone from a net exporter of mercury in the second half of the 19th century to a net importer. Import reliance for 1959–1978 ranged from a low of 25% in 1961 to 87% in 1974 and averaged 54% per year (see Table 6). Historically, Spain, Italy, and Yugoslavia have been the principal suppliers; Algeria, Canada, and Mexico were the principal sources in recent years.

World reserves and other resources were estimated at 160,000 t (see below) and 431,000 t, respectively, at the end of 1978 (see Table 8). The U.S. Bureau of Mines in collaboration with the U.S. Geological Survey began evaluating U.S. mercury reserves and resources in 1979; they are located primarily in Nevada and California (13). Mercury recoverable from mercury-cell plants producing chlorine and caustic soda constitutes an above-ground reserve estimated, at the end of 1978, at ca 5000 t.

Price.   The price history of mercury is marked by sharp, wide-ranging fluctuations, strong price competition between producing nations, long periods of price depression, and numerous attempts to support prices. Price instability, evidenced by the wide-ranging fluctuation, was usually caused by reduced demand, overproduction, competition, and lately, fears that governmental action to protect the environment would sharply reduce mercury consumption. Table 9 gives the history of U.S. mercury prices.

Governmental Policy.   The U.S. government, through various actions, has stockpiled mercury, controlled prices, imports, and exports, restricted uses, instituted a government purchase program, and explored domestic ore deposits. As of December 31, 1979, inventories in the national stockpile totaled 6,598 t (191,391 flasks) with a goal or need of 1,862 t (54,004 flasks) established on October 1, 1976. The surplus 4,736 t (137,387 flasks) was available for disposal which required Congressional approval,

Table 8. Mercury: World Resources[a] (Thousands of Metric Tons)

	Reserves	Other resources	Total	Percent of total
North America:				
United States	14	16	28	5
Canada	4	10	14	
Mexico	9	16	25	4
South America	1	9	10	
Europe:				
Italy	14	55	69	
Spain	52	138	190	33
USSR	17	86	103	17
Yugoslavia	17	52	69	12
other	1	1	2	
Africa	12	1	13	
Asia:				
People's Republic of China	17	34	51	9
Japan	1	6	7	
Philippines	1	2	3	
Turkey	2	5	7	
World total	*160*	*431*	*591*	

[a] In collaboration with the U.S. Geological Survey. Source: U.S. Bureau of Mines.

Table 9. Price Trend of Mercury in the United States

	1850–1900	1901–1920	1921–1930	1931–1940	1941–1950	1951–1960	1961–1970	1971–1978
$/flask[a], av	49	44	79	89	125	233	384	206

[a] A flask contains 34.5 kg.

which had not been given at year end 1979. The U.S. government offers financial assistance for mercury and other mining ventures through the Departments of Agriculture and Commerce and the Small Business Administration. Mercury miners are granted a depletion allowance of 22% on domestic production and 14% on foreign production.

A U.S. duty of $0.25/lb ($19 per flask) on imports of mercury, established by the Tariff Act of 1922, has been gradually reduced over the years to $0.119/lb ($9.04 per flask) on January 1, 1980, for those countries officially designated as Most Favored Nation (MFN). This duty will be reduced to $0.075/lb ($5.70 per flask) by January 1, 1987. Imports from those nations not designated MFN, the Communist countries, are dutiable at the statutory rate established in 1922. The duty on waste and scrap mercury has been suspended until June 30, 1981.

Grades, Specifications, and Quality Control

The commercial grade, marketed as prime virgin-grade mercury (99.9%), has a clean bright appearance and contains less than 1 ppm of dissolved base metals. Prime virgin of lower purity is brought up to specification by filtration, redistillation, or electrolytic processing.

Triple-distilled mercury is of highest-purity, commanding premium prices. It is produced from primary and secondary mercury by numerous methods including mechanical filtering, chemical and air oxidation of impurities, drying, electrolysis, and most commonly, multiple distillation.

The purity of mercury can be estimated by its appearance. Because mercury has such a high specific gravity, almost all impurities, including amalgams, are lighter and float on the surface causing the bright, mirrorlike surface to become dull and black. The coating of oxides of base metal is called film and scum and its presence or absence indicates the degree of purity. A clean, bright appearance usually signifies an impurity content of <5 ppm (15–16). Other tests measure the residue remaining after the mercury is evaporated (17–18).

Analytical Methods

Ore. The assay of mercury ores is not satisfactory because of difficulties encountered in obtaining representative ore samples. The brittle nature of crystalline cinnabar causes it to break loose from adjacent rock and fall into the sample being collected. This uncontrollable salting of the sample can give results as much as several hundred percent over the actual mercury content of the sample.

The two preferred methods employ distillation–amalgamation or distillation–titration techniques. In the former, the weighed sample is heated with a flux such as iron filings to volatize the mercury which is then amalgamated with silver or gold foil. The mercury content is calculated from the change in foil weight. In the titration method, the sample is heated and the mercury volatilized and collected as metal, which is then dissolved in hot nitric acid. Potassium permanganate is added to oxidize the mercury and a peroxide to destroy excess permanganate. After adding ferric sulfate, a nitrate indicator, the solution is titrated with standard potassium thiocyanate solution to a faint pink end point. When a 0.5-g sample is titrated with a 1/400-N solution, each milliliter of titrating solution is equivalent to 0.41 kg of mercury per metric ton.

Commercial Product. The efficiency of furnacing and condensing operations can be checked by spot-testing the condensed mercury for precious and base-metal content, the most common contaminants from mining operations. Gold is determined by dissolving a 10-g sample of mercury in equal parts of distilled water and 15 N nitric acid; as much as 0.1 mg of gold present is visible as a small particle. The residue is evaporated with aqua regia; if no precipitate results with sodium hydroxide and hydrogen peroxide, no gold is present. Silver content is determined by spot-testing with trivalent manganese reagent. Base metals are determined by dissolving the sample in concentrated hydrochloric acid, volatilizing the mercuric chloride, and analyzing the residue gravimetrically.

The purity of ultrapure mercury (triple distilled) is commonly tested by evaporation or spectrographic analysis. In the former, a composite sample is evaporated and the residue weighed. In spectrographic analysis, a sample is dissolved and evaporated, the residue mixed with graphite, and the emission spectrum determined with a spectrograph.

Trace Mercury. In recent years, the number and variety of quantitative or qualitative methods and instruments used to determine trace quantities of inorganic and organic mercury occurring in natural or synthetic substances have increased greatly (see also Trace and residue analysis) (19). This proliferation came about after a series of mercury poisonings alerted the world to the dangers of mercury from industrial discharges, mercury-containing products, and wastes to the environment. Trace element analysis literature describing numerous techniques and a myriad of mercury-containing substances is voluminous and only the most commonly used methods are described here (20).

Atomic Absorption Spectroscopy. Mercury is separated from a measured sample by various methods and passed as vapor into a closed system between an ultraviolet lamp and a photocell detector or into the light path of an atomic absorption spectrometer. Ground-state atoms in the vapor attenuate the light which decreases the current output of the photocell in an amount proportional to the concentration of the mercury. The light absorption can be measured at 253.7 nm and compared with previously established calibrated standards (21). A mercury concentration of 0.1 ppb can be measured by atomic absorption.

Neutron Activation Analysis. A measured sample is activated by neutron bombardment and emits gamma rays that are used to determine the mercury content by proton-spectrum scanning. Mercury concentrations as low as 0.05 ppb have been determined by this method.

X-ray Fluorescence Method. The sample containing mercury is exposed to a high intensity x-ray beam which causes the mercury and other elements in the sample to emit characteristic x rays, the intensity of which is directly proportional to the element concentration in the sample (22). Mercury content below 1 ppm can be detected by this method.

X-ray diffraction analysis is ordinarily used for the qualitative but not the quantitative determination of mercury.

Dithizone (Diphenylthiocarbazone) Method. A finely powdered sample is treated with sulfuric acid, hydrobromic acid, and bromine to give a solution adjusted to pH 4 that is treated with dithizone in *n*-hexane to form mercuric dithizonate (23). It is isolated as an amber-colored solution whose color intensity can be compared to standard solutions to determine the mercury concentration of the sample. Concentrations below 0.02 ppm have been measured by this method.

Other Methods. Numerous other techniques similar to those mentioned above exist but are not widely used. Methods include chromatography, micrometry, radiometry, spectrography, and titrimetry.

Health and Safety Factors

Mercury metal, its vapors, and most of its organic and inorganic compounds are protoplasmic poisons that can be fatal to humans, animals, and plants (24). The most toxic are the organic mercury compounds, such as the alkyl types. Factors that determine the effect of mercury poisoning on humans are the amount and rate of absorption, the physicochemical properties of the compounds, and individual susceptibility. Mercury poisoning may occur in mining and recovery, in the manufacture and use of mercury products, from industrial waste discharges, and by concentration in the human food chain (25). Mercury may enter the body through the skin, gastrointestinal tract, and respiratory tract. Chronic poisoning may develop gradually without conspicuous warning signs.

The immediate causes of industrial mercury poisoning are usually the absorption and retention of small quantities of mercury metal, vapor, or compounds over a long period of time. Recommended safety measures include the use of efficient respirators, adequate ventilation and air-exhaust systems, employee warning signs and messages, training in accident emergency procedures, immediate, thorough, and safe cleanup of spills, air-tight storage of mercury, wastes, and soiled clothing, and frequent monitoring of mercury levels in the work area (26). Additional precautions include coverall-type work clothes, keeping floors, work surfaces, and equipment free of cracks, crevices, and indentations that might hold spilled mercury, adequate washing facilities and showering at the end of the work period. Medical and dental examination at frequent intervals are also recommended.

The U.S. government has declared that mercury and its compounds are hazardous substances and has proposed and issued regulations controlling their emission into the environment. Emission from mercury ore-processing facilities and chlorine-caustic soda plants are limited to 2.3 kg per day per plant. Most biocide uses for mercury were cancelled and registrations for alkyl and nonalkyl fungicides for rice seed and laundry products and as marine antifouling paint were suspended. Cosmetic use of mercury was prohibited except in special cases, as was dumping wastes containing more than trace amounts of mercury. Emission of mercury from incinerating or drying wastewater sludges was limited to 3.2 kg per day and a limit of 0.1 mg mercury vapor per cubic meter air was accepted as the standard for occupational exposure. The maximum amount of mercury in public drinking water was recommended to be 2 μg per liter.

Uses

Mercury consumed in the United States has been utilized historically for a number of general purposes (see Table 7), each with numerous distinct applications. Over the years, some uses have declined sharply or ceased altogether because of federally imposed bans.

Agriculture.   Mercury compounds are used in agriculture and related applications as fungicides, pesticides, bactericides, and disinfectants (see Mercury compounds). Most of the mercury-based pesticides and fungicides have been banned as being hazardous substances.

Catalysts.   Mercury is used as a catalyst for the production of vinyl chloride monomers, urethane foams, anthraquinone derivatives, and other products. Urethane foams and vinyl chloride monomers are the principal consumers of mercury catalysts (see Urethane polymers).

Electrical Applications.   The main use for mercury is in electrical applications which include batteries, electrical lamps, and wiring and switching devices.

Batteries, the largest specific use, include the primary-type zinc–carbon cell, alkaline–manganese dioxide cell, mercury cell, carbon–zinc air cell and the storage-type zinc–silver oxide cell. Mercury batteries are used in many products, including hearing aids, cameras, toys, portable radios, calculators, measuring devices, smoke alarms, self-winding watches, radio microphones, and in guided missiles and space craft.

Electrical lamps, or mercury lamps, employ an electric discharge tube usually made of fused silica that contains varying volumes of mercury vapor. Mercury lamps are more efficient and produce more lumens-per-watt than some types of outdoor lighting and as a result are used widely for this purpose. The two main types of mercury lamps are the low pressure, or fluorescent, lamp as it is commonly called, and the high pressure mercury lamp used in industrial plants and workshops, in large, high ceilinged buildings such as aircraft hangers, and in street lighting and floodlighting. Other mercury vapor lamps are used in motion-picture projection, photography, heat therapy, and other minor applications.

Wiring and switching devices using mercury include high and low voltage mercury-arc rectifiers, oscillators, power control switches for motors, phanotrons, thyratrons, ignitrons, reed switches, silent switches in homes and offices, thermostats, and cathode tubes used for radios, radar, and telecommunications equipment.

Electrolytic Preparation of Chlorine and Caustic Soda. The preparation of chlorine and caustic soda is an important use for mercury metal. Mercury is used as a flowing cathode in an electrolytic cell into which a sodium chloride solution (brine) is introduced which is then subjected to an electric current. Chlorine gas evolves and is collected at the anode and an alkali metal amalgam is formed with the mercury cathode. The amalgam is subsequently decomposed with water to form caustic soda and hydrogen and relatively pure mercury metal which is recycled to the cell (see Alkali and chlorine products).

Industrial and Control Instruments. Mercury is used in many industrial and medical instruments to measure or control reactions and equipment functions, including thermometers, manometers (flow meters), barometers and other pressure-sensing devices, gauges, valves, seals, and navigational devices (see Instrumentation and control).

Paint. Mercury is used in paint (qv) as a fungicide to prevent mildew after the paint has been applied and as a bactericide or preservative to prevent bacterial attack while the paint is in storage. Most commonly used are phenylmercuric acetate and phenylmercuric oleate, both of which are highly toxic (see Fungicides; Industrial antimicrobial agents).

Laboratory Uses. Mercury and its compounds have been used widely in laboratories for experimental and research work but such usage has declined because of a growing awareness of the toxicity of mercury. Standard laboratory uses include reagents, indicators, calibration, sealing, radioactive diagnosis, and tissue fixative.

Pharmaceuticals. Mercury compounds are employed as diuretics, antiseptics, skin preparations, and preservatives. Included in this category are mercuric oxide, mercuric chloride, mercuric cyanide, mercuric amide chloride, mercuric iodide, mercurous chloride, and others. However, pharmaceutical use has declined sharply (see also Disinfectants and antiseptics; Diuretics).

Other Uses. Small quantities of mercury are used in a number of applications, including heat transfer, pigments, refining lubricating oils, and certain kinds of research.

BIBLIOGRAPHY

"Mercury" in *ECT* 1st ed., Vol. 8, pp. 808–882, by G. A. Roush, *Mineral Industry*; "Mercury" in *ECT* 2nd ed., Vol. 13, pp. 218–235, by George T. Engel, U.S. Department of the Interior Bureau of Mines.

1. C. N. Schuette, *U.S. Bur. Mines Bull.* **335,** 4 (1931).
2. E. H. Bailey, A. L. Clark, and R. M. Smith, *U.S. Geol. Surv. Pap.* **820,** 401 (1973).
3. V. A. Cammarota, *U.S. Bur. Mines Bull.* **667,** 669 (1975).
4. E. H. Bailey, A. L. Clark, and R. M. Smith, *U.S. Geol. Surv. Prof. Pap.* **820,** 407 (1973).
5. C. L. Gordon and E. Wichers, *Ann. N.Y. Acad. Sci.* **65,** 382 (1957).
6. C. E. Wicks and R. E. Block, *U.S. Bur. Mines Bull.* **605,** (1962).

7. W. E. Clark and W. Fulkerson, *Survey of the Mercury Reprocessing Industry*, Oak Ridge National Laboratory, ORNL NSF-EP-22, 1972.
8. J. W. Pennington, *U.S. Bur. Mines Inform. Circ.* **7941,** 29 (1959).
9. "Mercury," in *U.S. Bureau of Mines Minerals Yearbook*, Vol. 1, Washington, D.C., 1964–1977.
10. J. N. Butler, *Studies in the Hydrometallurgy of Mercury Sulfide*, Nevada Bureau of Mines, Rept. No. 5.
11. B. J. Scheiner, D. L. Pool, and R. E. Lindstrom, *U.S. Bur. Mines Rept. Invest.* **7660,** 1972.
12. E. S. Shedd, B. J. Scheiner, and R. E. Lindstrom, *U.S. Bur. Mines Rept. Invest.* **8083,** (1975).
13. "Mercury Potential of the United States," *U.S. Bur. Mines Inform. Circ.* **8252,** 22 (1965).
14. V. A. Cammarota, *U.S. Bur. Mines Bull.* **667,** 671 (1975).
15. E. Wichers, *Chem. Eng. News* **20,** 1111 (1942).
16. *Reagent Chemicals—American Chemical Society Specifications*, 6th ed., American Chemical Society, Washington, D.C., 1980, p. 368.
17. The United States Pharmacopeia XX (*USPXX–NFXV*), The United States Pharmacopeial Convention, Inc., Rockville, Md., 1980, p. 1073.
18. *Am. Dental Assoc. J.*, 409 (Mar. 1932).
19. R. G. Smith, "Methods of Analysis for Mercury and its Compounds: A Review," in R. Hartung and B. D. Dinman, eds., *Environmental Mercury Contamination*, Ann Arbor Science Publishers, Inc., Ann Arbor, Mich., 1972, pp. 97–136.
20. F. N. Ward, *U.S. Geol. Surv. Prof. Pap.* **713,** 46 (1970).
21. I. R. Jonasson, J. J. Lynch, and L. J. Trip, *Geol. Surv. Can. Pap.* **73-21,** 2 (1973).
22. H. H. Heady and K. G. Broadhead, *U.S. Bur. Mines Inf. Circ.* **8714r,** 4 (1977).
23. F. N. Ward, *U.S. Geol. Surv. Prof. Pap.* **713,** 47 (1970).
24. Q. R. Stahl, *Preliminary Air Pollution Survey of Mercury and its Compounds*, U.S. Department of Health, Education, and Welfare, 1969.
25. *Compilation of Air Pollution Factors*, U.S. EPA, AP-42, 1972.
26. *Occupational Exposure to Inorganic Mercury*, U.S. Department of Health, Education, and Welfare, 1973, pp. 7–13.

General References

M. J. Ebner, "A Selected Bibliography on Quicksilver 1811–1953," *U.S. Geol. Surv. Bull.* **1019-A,** (1954).

G. M. Caton, D. P. Oliveira, C. J. Oen, and G. U. Ulrikson, *Mercury in the Environment and Annotated Bibliography*, Oak Ridge National Laboratory, ORNL-EIS-71-8, 1972.

C. N. Schuette, "Quicksilver," *U.S. Bur. Mines Bull.* **335,** (1931).

"Mercury Potential of the United States," *U.S. Bur. Mines Inf. Circ.* **8252,** (1965), contains extensive list of mines by name and location.

Materials Balance and Technology Assessment of Mercury and Its Compounds on National and Regional Bases, Fin. Rept. EPA-560/3-75-007, U.S. EPA, 1975.

E. H. Bailey, A. L. Clark, and R. M. Smith, "Mercury," *United States Mineral Resources*, U.S. Geol. Surv. Prof. Paper **820,** (1973).

F. N. Ward, "Analytical Methods for Determination of Mercury in Rocks and Soils," *Mercury in the Environment*, U.S. Geol. Surv. Prof. Pap. **820** (1973).

R. G. Weast ed., *Handbook of Chemistry and Physics*, 61st ed., The Chemical Rubber Co., Boca Raton, Fla., 1980–1981.

V. A. Cammarota, "Mercury," *Mineral Facts and Problems*, U.S. Bur. of Mines Bull. **667,** 669 (1975).

J. W. Pennington, "Mercury A Materials Survey," *U.S. Bur. Mines Inf. Circ.* **7941,** (1959).

Mercury (Quicksilver) Rept. Invest. No. 32, U.S. International Trade Commission, 1958.

C. L. Gordon and E. Wichers, "Purification of Mercury and Its Compounds," *Ann. N.Y. Acad. Sci.* 65, 382 1957.

R. Hartung and B. D. Dinman, *Environmental Mercury Contamination*, Ann Arbor Mich., 1972.

Mercury, 3rd ed., Roskill Information Services Ltd., London, 1974.

"Mercury," in *Minerals Yearbook*, U.S. Bureau of Mines, annual.

HAROLD J. DRAKE
U.S. Bureau of Mines

MERCURY COMPOUNDS

Mercury salts exist in two oxidation states: mercurous (valence +1) and mercuric (valence +2). Mercurous compounds exist as double salts; eg, mercuric chloride is Hg_2Cl_2 in both solution and the solid states, as shown by conductance studies and x-ray analysis. The standard oxidation electrode potentials are

$$2\,Hg \rightleftharpoons Hg_2^{2+} + 2\,e \quad -0.7961 \text{ V}$$

$$(Hg_2)^{2+} \rightleftharpoons 2\,Hg^{2+} + 2\,e \quad -0.905 \text{ V}$$

$$Hg \rightleftharpoons Hg^{2+} + 2\,e \quad -0.851 \text{ V}$$

Many mercury compounds are volatile, and often may be purified by sublimation. They are labile and easily decomposed by light, heat, and reducing agents. Organic compounds of weak reducing activity, such as amines, aldehydes, and ketones, often break them down to compounds of lower oxidation state and mercury metal. This lack of stability makes it relatively easy to recover the mercury values from the various wastes that accumulate with the production of compounds of economic and commercial importance (see Recycling).

The toxic nature of mercury and its compounds has caused government agencies to be concerned with environmental pollution and to impose severe restriction on their release to waterways and the air. In recent years, methods of precipitation and agglomeration of mercurial wastes from process water have been developed. These methods generally depend on the formation of relatively insoluble compounds such as mercury sulfides, oxides, thiocarbamates, etc, together with metallic mercury that invariably is formed as a by-product. The use of co-precipitants is frequently helpful, in that adsorption of mercury on their surface facilitates its removal. As a final step, activated carbon or ion-exchange principles may be used to eliminate final traces before discharge.

Mercury from these accumulated wastes generally is recovered best by total degradation in stills, where metallic mercury is condensed and collected. The recovery costs are compensated amply by the value of the metal recovered while eliminating, or at least severely diminishing disposal problems.

Although mercuric oxide may be red or yellow, and mercuric sulfide is black, most other nonbasic compounds are colorless. Solutions not stabilized with an excess of acid tend to hydrolyze to form yellow to orange basic hydrates. Frequently, they are adsorbed onto the walls of containers, and may interfere with analytical results when low levels (ppm) of mercury are determined.

Concurrent with the requirements for low levels of mercurials in discharge water is the problem of their determination. The older methods of wet chemistry are inadequate, and total reliance is now placed on physical methods. Of these, the most popular is atomic absorption spectrophotometry, most recently relying on the absorption of light at 253.7 nm by mercury vapor, produced by a method described in ref. 1. Detailed procedures can be obtained from instrument suppliers (see Analytical methods; Trace and residue analysis).

The covalent character of mercury compounds and their ability to complex with various organic compounds explains their unusually wide solubility characteristics, including alcohols, ethyl ether, benzene, etc. Small amounts of chemicals such as amines, ammonia, ammonium acetate, can have a profound solubilizing effect (see also Coordination compounds).

Because of recent legal restrictions, the uses of mercury compounds have contracted. Agricultural applications now are confined largely to use on turf against specific fungal diseases (see Fungicides). Seed treatment is no longer permitted in the United States, although bis(2-methoxyethyl mercuric) silicate is still used in parts of Europe. Mercuric oxide may not be used as a component of antifouling paints in the United States. The pulp and paper industry now uses a variety of nonmercury-containing slimicides (see Industrial antimicrobial agents).

Some pharmaceutical uses remain, eg, in ophthalmic preparations, antiseptics (see Disinfectants), diuretics (qv), etc.

The most important areas that remain are as preservatives and fungicides for latex paints, as catalysts, and as intermediates in the formation of other compounds. All mercury compounds should be stored in amber bottles or otherwise protected from light. An alphabetical list of mercury compounds mentioned in the text with their CAS Registry Numbers is given at the end of this article.

Mercury Salts

Mercuric Acetate. Mercuric acetate, $Hg(C_2H_3O_2)_2$, is a white, water-soluble, crystalline powder, also soluble in many organic solvents. It is prepared by dissolving mercuric oxide in warm 20% acetic acid. A slight excess of acetic acid is helpful in reducing hydrolysis. Glass-lined equipment is preferred, although stainless steel may be used. Stainless steel may cause some discoloration at high temperatures if concentrated acetic acid is used.

Another method of preparation is the oxidation of mercury metal with peracetic acid dissolved in acetic acid. Careful control of the temperature is extremely important since the reaction is quite exothermic. A preferred procedure is the addition of approximately half to two thirds of the peracetic acid solution required to a dispersion of mercury metal in acetic acid to obtain the mercurous salt, followed by the remainder of the peracetic acid to form the mercuric salt. The exothermic reaction is carried to completion by heating slowly and cautiously to reflux which also serves to decompose excess peracid. It is possible to use 50% hydrogen peroxide instead of peracetic acid but the reaction does not go quite as smoothly; however, it may be more economical.

The primary use of mercuric acetate is as a starting material for the manufacture of organic mercury compounds.

Mercuric Carbonate (Basic). Mercuric carbonate, $HgCO_3 \cdot 3HgO$, may be prepared by the addition of sodium carbonate to a solution of mercuric chloride. The brown precipitate lacks usefulness, and generally it is not isolated. Rather, the slurry is refluxed, whereupon the carbonate decomposes to red mercuric oxide.

Mercuric Cyanide. Mercuric cyanide, $Hg(CN)_2$, is a white tetragonal crystalline compound, little used today, except to a small degree as an antiseptic. It is prepared by reaction of an aqueous slurry of yellow mercuric oxide (red is less reactive) with excess hydrogen cyanide. The mixture is heated to 95°C, filtered, crystallized, isolated, and dried. Its solubility in water is 10% at RT.

Mercuric Oxycyanide. Mercuric oxycyanide, $Hg(CN)_2 \cdot HgO$, or basic mercuric cyanide is prepared in the same manner as the normal cyanide, except that the mercuric oxide is present in excess. The compound is white and crystalline but only $\frac{1}{10}$ as soluble as the normal cyanide. Since the compound is explosive, it normally is supplied as a 1:2 mixture of oxycyanide to cyanide.

Mercuric Fulminate. Mercuric fulminate, $Hg(ONC)_2$, is used as a catalyst in the oxynitration of benzene to nitrophenol. Its most common use is as a detonator for explosives (qv).

Mercury Fluorides. See Fluorine compounds, inorganic.

Mercurous Chloride. Mercurous chloride, Hg_2Cl_2 (calomel), is a white powder insoluble in water. It sublimes when heated in an open container, but this probably occurs at least in part as the result of the dissociation reaction:

$$Hg_2Cl_2 \rightarrow Hg + HgCl_2$$

Its relatively low toxicity is probably owing to its very low solubility (0.002 g/L) and its lack of reactivity with acidic (HCl) digestive liquids.

The compound generally is prepared in one of two ways. The purer grade is made by the direct oxidation of mercury by a quantity of chlorine gas insufficient to produce mercuric chloride, which is always a side product. The chlorine gas is run into a heated silica retort whose mouth enters a large chamber made of chlorine-resistant material. Lead frequently is used but great care must be exercised to exclude moisture, otherwise lead chloride could be formed as well. The chamber should be large, approximately 5.1 m³ (180 ft³), since cooling is by convection and dissipation of heat by conduction through the walls.

The mercury burns with a green flame, and the product settles to the floor of the chamber, from which it is removed. The correct balance of mercury to chlorine yields about 70–80% mercurous chloride, and the remainder is mercuric chloride. There is no unreacted mercury, with proper operation, which would give a grayish tinge to the product and which cannot be washed out in the following step.

The material from the chamber is slurried with water and washed several times by decantation. It is filtered and washed on the filter until free of soluble chloride.

The second method of preparation is the precipitation of mercurous chloride from a cold acidic solution of mercurous nitrate by sodium chloride solution. It is isolated after washing in a manner similar to the chamber method described above. This product generally contains small amounts of occluded sodium nitrate, and difficulty may be encountered in having it pass NF or reagent-grade specifications (see Fine chemicals). As a technical-grade material, it is satisfactory.

For the preparation of mixtures of mercurous and mercuric chlorides used for turf-fungus diseases, the precipitated product of the second method may be mixed with the required amount of mercuric chloride; alternatively, the chamber material, if the ratios are correct, may be used directly.

The mercury contained in the mother liquor and washings is recovered by treatment with sodium hydroxide solution. Yellow mercuric oxide is precipitated and filtered. The filtrate is treated further to remove last traces of mercury before it is discarded.

Mercuric Chloride. Mercuric chloride, $HgCl_2$, also is known as the corrosive sublimate of mercury or mercury bichloride. It is extremely poisonous, and is particularly dangerous because of its water solubility, 71.5 g/L at 25°C, and high vapor pressure. It sublimes without decomposition at 300°C, and has a vapor pressure of 13 Pa (0.1 mm Hg) at 100°C and 400 Pa (3 mm Hg) at 150°C. The vapor density is high (9.8 g/cm³) and, therefore, mercuric chloride vapor dissipates slowly (2).

In addition to its high solubility in water, 7% at 30°C, and 38% at 100°C, it is very soluble in methyl alcohol, 53% at 36°C, in ethyl alcohol, 34% at 31°C; and about 10%

in amyl alcohol at 30°C. It also is soluble in acetone, formic acid, acetic acid, the lower acetate esters, etc.

The preparation of mercuric chloride is identical to the chamber method for mercurous chloride except that an excess of chlorine is used to ensure complete reaction to the higher oxidation state. Excess chlorine is absorbed by sodium hydroxide in a tower. Very pure product results from this method.

Mercuric chloride is used widely as a catalyst for the preparation of red and yellow mercuric oxide, ammoniated mercury USP, mercuric oxide, mercuric iodide, and as an intermediate in organic synthesis. It is a component of agricultural fungicides. It is used in conjunction with sodium chloride in photography, in batteries (qv), and it has some medicinal uses as an antiseptic.

Mercurous Bromide. Mercurous bromide, Hg_2Br_2, is a white, tetragonal crystalline powder, very similar to the chloride, and prepared in much the same way: by the direct oxidation of mercury by bromine or by precipitation from mercurous nitrate by sodium bromide. It is sensitive to light, less stable than the chloride, and is not of appreciable commercial importance.

Mercuric Bromide. Mercuric bromide, $HgBr_2$, is a white, crystalline powder, considerably less stable than the chloride, and also much less soluble in water, 0.6% at 25°C. Therefore, it is prepared easily by precipitation, using mercuric nitrate and sodium bromide solution. Drying of the washed compound is carried out below 75°C. It has a few medicinal uses.

Mercurous Iodide. Mercurous iodide, Hg_2I_2, is a bright yellow, amorphous powder, extremely insoluble in water, and very sensitive to light. It has no commercial importance but may be prepared by precipitation, using mercurous nitrate and potassium iodide. Care must be taken to exclude mercuric nitrate, which may cause the formation of the water-insoluble mercuric iodide.

Mercuric Iodide. Mercuric iodide, HgI_2, is a bright red, tetragonal powder, only slightly soluble in water. It dissolves in alkalies to form complex salts: Na_2HgI_4 or K_2HgI_4. It is made by precipitation from a solution of mercuric chloride by potassium iodide. The compound is used in the treatment of skin diseases and as an analytical reagent.

Complex Halides. Mercuric halides (except the fluoride) form neutral complex salts with metallic halides. Those made with alkali metal salts frequently are more soluble in water than the mercuric halide itself (see Mercuric Iodide), and take the form of $M(HgX_3)$ and $M_2(HgX_4)$.

Potassium Iodomercurate Dihydrate. Potassium iodomercurate dihydrate, $K_2(HgI_4).2H_2O$, is a yellow, water- and alcohol-soluble compound prepared by dissolving one mole of mercuric iodide in a solution of two moles of potassium iodide in distilled water. After filtering any insolubles, the filtrate is evaporated to dryness to isolate the compound. The filtrate, known as Mayers reagent, has uses as an antiseptic and as a precipitant for alkaloids. In strongly alkaline solution, called Nessler's reagent, it is used for the detection and determination of low levels of ammonia.

Cuprous Iodomercurate. Cuprous iodomercurate, $Cu_2(HgI_4)$, is a bright red, water-insoluble compound prepared by precipitation from a solution of $K_2(HgI_4)$ with cuprous chloride. It is used in temperature-indicating paints since it reversibly changes color to brown at 70°C (see Chromogenic materials).

Silver Iodomercurate. Silver iodomercurate, $Ag_2(HgI_4)$, a bright yellow compound, is prepared similarly to the cuprous salt with silver nitrate as the precipitant. It is used for the same application since it darkens reversibly at 50°C.

Mercurous Nitrate. Mercurous nitrate, $Hg_2(NO_3)_2$, is a white, monoclinic crystalline compound. It is not very soluble in water but it hydrolyzes to form a basic, yellow hydrate. It is, however, soluble in cold, dilute nitric acid, and this solution is used as starting material for other water-insoluble mercurous salts. It is difficult to obtain in the pure state directly since some mercuric nitrate formation is almost unavoidable. Mercury is dissolved in hot dilute nitric acid, and the technical product crystallizes on cooling. The use of excess mercury is helpful in reducing mercuric content but then an additional separation step is necessary. More concentrated nitric acid solutions should be avoided since they oxidize the mercurous to mercuric salt.

Reagent-grade material is obtained by recrystallization from dilute nitric acid in the presence of excess mercury.

Mercuric Nitrate. Mercuric nitrate, $Hg(NO_3)_2$, is a colorless, deliquescent crystalline compound prepared by the exothermic dissolution of mercury in hot, concentrated nitric acid. The reaction is complete when a cloud of mercurous chloride is not formed when treated with sodium chloride solution. The product crystallizes upon cooling.

Mercuric nitrate is used as the starting material for a great many other mercuric products and in organic syntheses.

Mercuric Oxide. Mercuric oxide, HgO, is a red or yellow, water-insoluble powder, rhombic in shape when viewed microscopically. The color and shade depends on the particle size, with the finer particles (under 5 μm) appearing yellow and the coarser particles (over 8 μm) redder. The product is soluble in most acids, organic and inorganic, but the yellow form, with its greater surface area, is more reactive and dissolves more readily. It decomposes at 332°C, and has a high specific gravity, 11.1.

Yellow mercuric oxide may be obtained by precipitation from solutions of practically any water-soluble mercuric salt through the addition of alkali. The most economical are mercuric chloride or nitrate.

Although the compound has some medicinal value in ointments and other such preparations, its primary use is as a starting raw material for other mercury compounds, eg, Millon's base, Hg_2NOH, is formed by the reaction of aqueous ammonia and yellow mercuric oxide.

Red mercuric oxide generally is prepared in two ways: by the heat-induced decomposition of mercurous nitrate and by hot precipitation. Both methods require careful control of reaction conditions.

In the calcination method, mercury and a deficiency of hot, concentrated nitric acid react to form mercurous nitrate:

$$6\,Hg + 8\,HNO_3 \rightarrow 3\,Hg_2(NO_3)_2 + 2\,NO + 4\,H_2O$$

After the water and nitrogen oxide are driven off, continued heating drives off vapors of nitric acid, additional water, NO_2, and some mercury-metal vapor.

$$Hg(NO_3)_2 \rightarrow 2\,HgO + 2\,NO_2$$

This secondary reaction starts at about 180°C but the mass must be heated to 350–400°C to bring it to completion and produce a nitrate-free product. The off-gases are extremely corrosive and poisonous, and considerable attention and expense is required for equipment maintenance and caustic-wash absorption towers. Treatment of the alkaline wash liquor for removal of mercury is required for economic reasons and to comply with governmental regulations for mercury in plant effluents.

In the hot precipitation method, sodium carbonate solution is added slowly to a refluxing solution of mercuric chloride, followed by an additional reflux period of 1–2 h. The washed precipitate is then dried. A variation allows the substitution of mercuric nitrate for the chloride if substantial quantities of sodium chloride are used. Sodium hydroxide, in the presence of sodium carbonate, is the precipitant.

Red mercuric oxide is identical chemically to the yellow form but is somewhat less reactive and more expensive to produce. An important use is in the Ruben-Mallory dry cell, where it is mixed with graphite to act as a depolarizer (see Batteries). The overall cell reaction is

$$Zn + HgO \rightarrow ZnO + Hg$$

Yellow mercuric oxide is considerably less suitable because an advantage of these batteries is their small size. The yellow oxide is less dense and would not permit adequate packing in the cell casing.

Mercurous Sulfate. Mercurous sulfate, Hg_2SO_4, is a colorless to slightly yellowish compound, sensitive to light, and slightly soluble in water, 0.05 g/100 g H_2O. It is more soluble in dilute acids. The compound is prepared by precipitation from acidified mercurous nitrate solution with dilute sulfuric acid. The precipitate is washed with dilute sulfuric acid until nitrate free. Its most important use is as a component of Clark and Weston types of standard cells.

Mercuric Sulfate. Mercuric sulfate, $HgSO_4$, is a colorless compound soluble in acidic solutions, but decomposed by water to form the yellow, water-insoluble, basic sulfate, $HgSO_4 \cdot 2HgO$. It is prepared by reaction of a freshly prepared and washed, wet filter cake of yellow mercuric oxide with sulfuric acid in glass or glass-lined vessels. The product is used as a catalyst and with sodium chloride as an extractant of gold and silver from roasted pyrites.

Mercuric Sulfide. Mercuric sulfide, HgS, exists in two stable forms, the black cubic tetrahedral form obtained when soluble mercuric salts and sulfides are mixed, and the red hexagonal form found in nature as cinnabar (vermilion) pigment (3). Both forms are very insoluble in water (see Pigments, inorganic).

Red mercuric sulfide is made by heating the black sulfide in a concentrated solution of alkali polysulfide. The exact shade of the pigment varies with concentration, temperature, and time of reaction.

Organomercury Compounds

Phenylmercuric Acetate. Phenylmercuric acetate,

$$C_6H_5HgO\overset{\displaystyle O}{\overset{\|}{C}}CH_3$$

PMA, melts at 149°C, is slightly soluble in water but much more soluble in solutions of ammonium acetate in aqueous ammonia. Such solutions are articles of commerce and contain 30% PMA (PMA-30) or the equivalent of 18% mercury as metal (PMA-18). Phenylmercuric acetate is also soluble in various organic solvents.

The compound is prepared by refluxing a mixture of mercuric acetate and acetic acid in a large excess of benzene in what is generally referred to as a mercuration reaction. The large excess of benzene is necessary because more than one hydrogen on the benzene ring can be replaced:

$$\mathrm{C_6H_6} \ + \ n\left(\underset{\displaystyle \mathrm{HgOCCH_3}}{\overset{\displaystyle \overset{O}{\|}}{}}\right)_2 \ \longrightarrow \ \mathrm{C_6H_{6-n}}\left(\underset{\displaystyle \mathrm{HgOCCH_3}}{\overset{\displaystyle \overset{O}{\|}}{}}\right)_n \ + \ n\,\underset{\displaystyle \mathrm{HOCCH_3}}{\overset{\displaystyle \overset{O}{\|}}{}}$$

n may be 1–4 but it is generally desirable to limit the amount of polymercurated benzene formed. The technical grade of PMA contains about 85% pure compound and the remaining 15% is di- and trimercurated product. These polymercurated products are less soluble than mono-PMA and are removed by recrystallization. Solvents, such as water, acetone, benzene, ethylene glycol monoethyl ether, etc, may be used.

The reaction is complete in ca 15 h, as indicated by the formation of a white precipitate of phenylmercuric sulfide when sodium sulfide is added to an ammoniacal solution. The product is isolated after distillation of excess benzene and acetic acid.

The ammoniacal solution of phenylmercuric acetate referred to above contains polymercurates, which serve the purpose of stabilizing these solutions. Moreover, far from being undesirable, their lack of solubility is an advantage in exterior coatings, eg, where activity is unimpaired and they are slower to leach.

The primary use for phenylmercuric acetate is in latex paint; it is used at low levels as a preservative to prevent putrefaction of the liquid paint, or at higher levels to protect the dry film from fungal attack or mildew. It can be used for these same purposes in similar aqueous systems, such as inks, adhesives, tape muds, caulking compounds, etc.

Phenylmercuric acetate is used as the starting material in the preparation of many other phenylmercury compounds which are generally prepared by double-decomposition reactions with the sodium salts of the desired acid groups in aqueous solution. However, the lower alkylates, such as the propionate and butyrate, are prepared directly in the same manner as the acetate. Another double-decomposition method is use of phenylmercuric hydroxide, which is prepared by reaction of PMA with hot dilute sodium hydroxide. Other phenylmercury compounds in use are the oleate, dodecenyl succinate, propionate, nitrate, and dimethyldithiocarbamate. Although all are toxic, the dimethyldithiocarbamate is the least soluble and exhibits the highest tolerated level in humans. The phenylmercury compounds are less toxic than soluble inorganic mercury, perhaps because they form the highly insoluble phenylmercuric chloride in the stomach. In addition to their use as biocides, phenylmercury compounds serve as catalysts for the manufacture of certain polyurethanes (see Urethane polymers).

3-Chloro-2-methoxypropylmercuric Acetate. 3-Chloro-2-methoxypropylmercuric acetate,

$$\mathrm{ClCH_2\underset{\displaystyle \overset{|}{OCH_3}}{CH}CH_2HgO\overset{\displaystyle \overset{O}{\|}}{C}CH_3}$$

is difficult to isolate and generally is sold as an ammoniacal solution containing 10% mercury as metal in much the same way as phenylmercuric acetate solution. It is prepared by the reaction of allyl chloride, methanol, and mercuric acetate in acetic acid, followed by the addition of ammonia and water. It has many of the same applications as PMA-30 (or PMA-18), and is a preservative or bactericide for aqueous systems. Because of its superior solubility and compatibility, the compound is not precipitated by anionic dispersants, eg, as is phenylmercuric acetate with its cationic character, and, therefore, lower levels of mercury often may be used to achieve the same protective effect.

Alkyl Mercuric Compounds. Alkyl mercuric compounds, RHgX, are no longer manufactured in most of the world because of the long-lasting toxic hazards they present, and their destructive effect on the brain and central nervous system, where they tend to accumulate. They were, until recent years, widely used as seed disinfectants. They have some utility in organic synthesis and in the preparation of other organometallics (qv).

In general, they are white, stable solids of appreciable volatility, and are often prepared by a Grignard reaction in ethyl ether:

$$RMgX + HgX_2' \rightarrow RHgX' + MgXX'$$

Miscellaneous Mercury Compounds of Pharmaceutical Interest

Antiseptics. *Ammoniated Mercury.* Ammoniated mercury, $Hg(NH_2)Cl$, is a white, odorless powder, sp gr 5.38. It is formed by the reaction of aqueous ammonia and mercuric chloride.

o-Chloromercuriphenol. o-Chloromercuriphenol (1), mercarbolide, is prepared in the same way as PMA, but from phenol instead of benzene. The hydroxyl group is highly activating and the reaction proceeds very quickly. The product is precipitated by sodium chloride and purified from hot water as a white, leafy crystalline compound.

Merbromin. Merbromin (2), disodium 2,7-dibromo-4-hydroxymercurifluorescein (mercurochrome), is prepared by refluxing dibromofluorescein with mercuric acetate in acetic acid. The precipitate is dissolved in water containing the stoichiometric amount of sodium hydroxide and evaporated.

Merthiolate. Merthiolate (3), sodium ethylmercurithiosalicylate, is prepared from a 1:1 molar ratio of ethylmercuric chloride and disodium thiosalicylate in ethanol. After removal of the sodium chloride by filtration, the free acid is precipitated by acidification with dilute sulfuric acid. Purification is achieved by recrystallization from 95% ethanol, and the product, merthiolate, is obtained by neutralization with a stoichiometric amount of sodium hydroxide.

Nitromersol and Mercurophen. Nitromersol (4) and mercurophen (5) are prepared by the same mercuration reaction as phenylmercuric acetate from 4-nitro-o-cresol or o-nitrophenol instead of benzene. The second step is reaction with sodium hydroxide to form the anhydride or sodium salt, respectively.

Miscellaneous. Other organic mercurials used as antiseptics include mercocresol, acetomeroctol, acetoxymercuri-2-ethylhexylphenolsulfonate, and sodium 2,4-dihydroxy-3,5-dihydroxymercuribenzophenone-2-sulfonate (see Disinfectants and antiseptics).

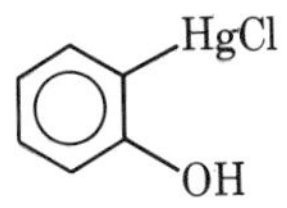

(1)

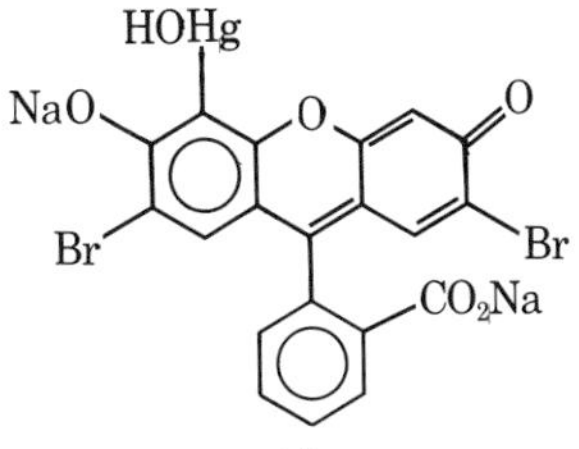

(2)

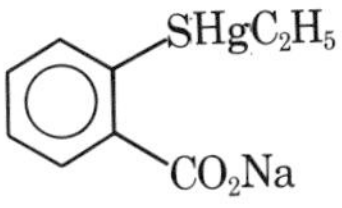

(3)

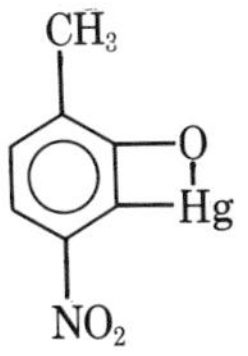

(4)

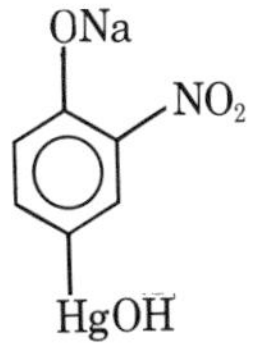

(5)

Antisyphilitics. Mercuric salicylate (**6**) and mercuric succinimide (**7**) are simple salts prepared by the reaction of mercuric oxide and either salicylic acid or succinimide, respectively, in water.

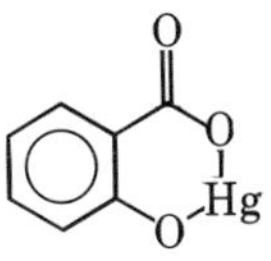

(6)

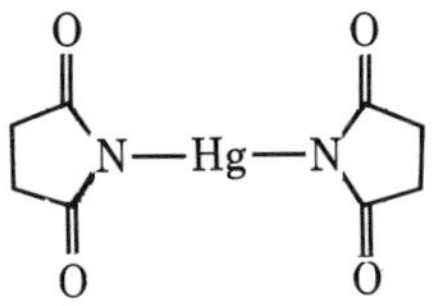

(7)

Diuretics. Chlormerodrin (**8**), methoxy(urea)propylmercuric chloride, is prepared in the same sort of reaction used for chlormethoxypropylmercuric acetate. Allyl urea is used instead of allyl chloride, together with methanol and mercuric acetate. The product, after dilution with water and neutralization, is precipitated with sodium chloride:

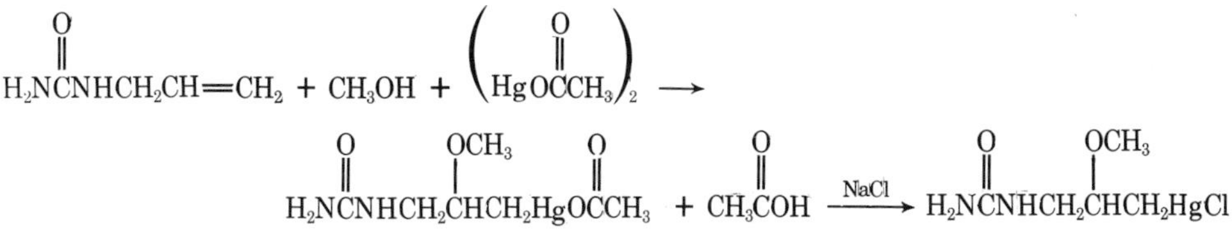

Other organic mercurials similar in chemical structure to chlormerodrin are meralluride, mercaptomerin, and mersalyl (see Diuretics).

Health and Safety Factors

With the discovery of the biomethylation of mercury in 1968 (4–5), the entire complex involving the manufacture, use and application of mercury and mercury compounds underwent an extensive change. The rapid development of powerful analytical tools capable of detection, identification and analysis of compounds and elements in the ranges of fractional parts per billion (ppb) and the coincident realization of the role of certain chemical moieties, not only as toxic sources, but also as potential carcinogens and mutagens has had a significant affect on the entire chemical industry. When it was realized that mercury compounds were being accumulated in the environment because of their resistance to biodegradation, the EPA was charged with the development and enforcement of rules and regulations concerning the safe preparation, use, and disposal of chemicals with particular reference to their environmental effects.

The toxic effects of mercury and mercury compounds as well as their medicinal properties have been known for many centuries. In the first century AD a Roman writer, Pliny, indicated the use of mercuric sulfide (cinnabar or vermilion) in medicine and in cosmetics. This compound probably was known to the Greeks in the time of Aristotle (6).

Galen, who died about 200 AD, believed that mercurials were toxic, and did not use any mercury compound therapeutically. Galen's prestige as a physician was such that his views were respected in the medical world for more than a thousand years. However, as a result of Arabian influence, the therapeutic uses of mercury were slowly recognized by Western Europe. In the 13th century mercury ointments were prescribed for treating chronic diseases of the skin.

Mercury and its compounds, such as mercurous chloride, mercuric oxide, mercuric chloride, and mercuric sulfide were used widely from the 15th to 19th centuries, and to some extent in the 20th century. During the first half of the 20th century, the primary therapeutic uses of mercury included bactericidal preparations, such as mercuric chloride, mercuric oxycyanide, and mercuric oxide, and diuretics, such as arylHgX (Novasural) and mercurated allyl derivatives (7).

Alkyl mercury compounds were used widely as seed disinfectants until their use in the United States was prohibited by the EPA in 1970. Subsequently, in 1972 the EPA prohibited the use of all mercury compounds in agriculture (8). At present only mercuric chloride and mercurous chloride are permitted for use on turf to control specific fungi.

Inorganic mercury compounds, aryl mercury compounds, and alkoxy mercurials are generally considered to be quite similar in their toxicity. Alkyl mercury compounds are considered to be substantially more toxic and hazardous.

Mercury and its compounds can be absorbed by ingestion, absorption through the skin, or by inhalation of the vapor.

After inorganic mercuric salts are absorbed and dissociated into the body fluids and in the blood, they are distributed between the plasma and erythrocytes. Aryl mercuric compounds and alkoxy mercuric compounds are decomposed to mercuric ions which behave similarly.

Alkyl mercury compounds in the blood stream are found mainly in the blood cells, and only to a small extent in the plasma. This is probably the result of the greater stability of the alkyl mercuric compounds, as well as their peculiar solubility characteristics. Alkyl mercury compounds affect the central nervous system and accumulate in the brain.

The elimination of alkyl mercury compounds from the body is somewhat slower than the inorganic mercury compounds, including the aryl and alkoxy mercurials. Methylmercury is eliminated from humans at a rate indicating a half-life of 50–60 d (3). Inorganic mercurials can be demonstrated to leave the body according to a half-life pattern of 30–60 d (9). Elimination rates are dependent not only on the nature of the compound but also on the dosage, method of intake, and rate of intake.

In the context of present-day governmental regulations, the control, recovery, and disposal of mercury-bearing waste products are as important as the manufacturing process. The difficulties involved in removing mercury from waste-product streams and the problems of recovery or disposal have resulted in a substantial reduction in the number of manufacturers of mercury compounds as well as the variety of mercury compounds being manufactured.

The manufacturing process used for preparing a mercury compound may not be necessarily the most efficient or economical from the viewpoint of raw material cost, total labor, etc. The choice may depend on the nature of the by-products, the toxic hazard of the process, and the ease of recovery of the mercury from the waste-product stream.

The MAC (maximum atmospheric concentration) for mercury—in all forms except alkyl compounds—is 0.1 mg Hg/m^3 air (10). The ACGIH proposed a TLV (threshold limit value) of 0.05 mg Hg/m^3 air. For alkyl mercury compounds the TLV is set at 0.01 mg Hg/m^3 air.

Suitable ventilating equipment, consisting mainly of carbon absorbers which effectively absorb mercury vapor from recirculated air, must be employed to maintain standards below the low value permitted in the atmosphere. When the possibility of higher exposures exist, small disposable masks utilizing a mercury-vapor absorbent may be employed.

Most inorganic mercury compounds have very low vapor pressures, and generally do not contribute to high mercury-vapor readings. Metallic mercury is the most potent and troublesome in this respect; organic mercurials also contribute to mercury-vapor readings, possibly by virtue of the presence of extremely small amounts of metallic mercury present as an impurity.

To safeguard the health of persons working in plants producing mercurials, the following precautions should be observed: adequate ventilation; use of disposable uniforms, so that a contaminated uniform is not a source of absorption through the skin; use of disposable mercury-vapor-absorbing masks; careful attention to good housekeeping, eg, avoidance of spills, and prompt and proper cleaning if a spill occurs; all containers of mercury and its compounds are kept tightly closed; floors are washed on a regular basis with dilute calcium sulfide solution or other suitable reactant; floors should be nonporous; all workers directly involved in the plant operation should shower thoroughly each day before leaving; and periodic medical exams including analysis of blood and urine for amount of mercury present for all workers directly involved in production of mercurials, or otherwise exposed to contact with mercury compounds or mercury vapor.

Mercury spills should be cleaned up immediately by use of a special vacuum cleaner. Then the area should be washed with a dilute calcium sulfide solution. Small quantities of mercury can be picked up by mixing with copper metal granules or powder, or with zinc granules or powder. To avoid or minimize spills, some plants use steel trays as pallets so that a spill, whether of mercury or a mercury compound, is contained on the steel tray.

Mercury vapor discharge from vents of reactors or storage tanks at normal atmospheric pressure are controlled readily by means of activated carbon. Standard units [208-L (55-gal) drums] of activated carbon equipped with proper inlet and outlet nozzles can be attached to each vent. To minimize the load on the carbon-absorbing device, a small water-cooled condenser is placed between the vent and the absorber.

The control of mercury in the effluent derived from the manufacturing processes used in the preparation of inorganic and organic mercurials is mandated by law in the United States. The concentrations and the total amounts vary with the industry and the location but generally it is required that the effluent contain not more than 0.5 ppm with a total Hg output of 230 g/d. However, individual states and individual publicly owned sewage-treatment plants have set up their own standards.

Removal of Mercury from Gases.   Removal of mercury from ambient air in the workplace and prevention of discharge of mercury vapor from reactions involving this element are of the greatest importance in the protection of the health of the worker.

Mercury vapor (as metallic mercury or in the form of its various compounds) can be absorbed readily by a number of different media (see Air pollution control methods). Activated charcoal adsorbs mercury but is not particularly efficient. Specially treated charcoal (containing sulfur compounds) is fairly effective but cannot be regenerated after it is saturated (see Adsorption separation, gases). Because of the problems involved in disposing of mercury-containing wastes such as unregenerated adsorbents, it is preferable to use other systems.

Because mercury is amalgamated readily with gold and silver, systems have been developed to use these metals distributed on various carriers as a means of removing mercury vapor from an air stream. When the system is saturated, the mercury can be removed easily and recovered by heating the unit and condensing the mercury. Other metals such as copper and zinc also can be used.

When the mercury present in the atmosphere is primarily in the form of an organic mercury compound, it may be preferable to utilize an aqueous scrubber. This method is particularly useful for control of emissions from reactors and from dryers. For efficient and economical operation, an aqueous solution of caustic soda, sodium hypochlorite, or sodium sulfide, etc, is recirculated through the scrubber until the solution is saturated with the mercury compound.

Removal of Mercury from Liquids.   The chlor–alkali producers who employ mercury-cathode electrolytic cells for the production of chlorine and caustic soda face the greatest problem in removal of mercury from aqueous effluent streams, and most of the patent literature is concerned with the processes for treatment of mercury-containing brine so produced (see Alkali and chlorine products).

One procedure involves the use of a bed of activated carbon impregnated with silver (11). The alkali–brine solution is passed through a bed of this material which is combined with supporting material, such as nickel turnings, polyethylene shreds, etc.

Another process relates to a method whereby the brine solution is allowed to contact a strong anion-exchange organic resin of the quaternary ammonium cross-linked type (12) (see Ion exchange).

Table 1. **Alphabetical List of Mercury Compounds Referred to in the Text**

Name	CAS Registry No.
acetomeroctol	[584-18-9]
acetoxymercuri-2-ethylhexylphenolsulfonate	[1301-13-9]
ammoniated mercury	[10124-48-8]
bis(2-methoxyethyl mercuric) silicate	[000-000]
chlormerodrin	[62-37-3]
3-chloro-2-methoxypropylmercuric acetate	
o-chloromercuriphenol (mercarbolide)	[90-03-9]
cuprous iodomercurate	[13876-85-2]
ethylmercuric chloride	[107-27-7]
meralluride	[104-20-5]
merbromin (mercurochrome)	[129-16-8]
mercaptomerin	[20223-84-1]
mercocresol	[8063-33-0]
mercuric acetate	[1600-27-7]
mercuric bromide	[7789-47-1]
mercuric carbonate, basic	[76963-38-7]
mercuric chloride	[7487-94-7]
mercuric cyanide	[592-04-1]
mercuric fulminate	[20820-45-5]
mercuric iodide	[7774-29-0]
mercuric nitrate	[10045-94-0]
mercuric oxide	[21908-53-2]
mercuric oxycyanide	[1335-31-5]
mercuric salicylate	[5970-32-1]
mercuric succinimide	[584-43-0]
mercuric sulfate	[7783-35-9]
mercuric sulfate, basic	[7783-35-9]
mercuric sulfide	[1344-48-5]
mercurophen	[52486-78-9]
mercurous bromide	[15385-58-7]
mercurous chloride	[10112-91-1]
mercurous iodide	[7783-30-4]
mercurous nitrate	[10415-75-5]
mercurous sulfate	[7783-36-0]
mersalyl	[486-67-9]
merthiolate (thimerosal)	[54-64-8]
methylmercury	[16056-34-1]
Millon's base	[12529-66-7]
nitromersol	[133-58-4]
phenylmercuric acetate	[62-38-4]
phenylmercuric butyrate	[2440-29-1]
phenylmercuric chloride	[100-56-1]
phenylmercuric dodecenyl succinate	
phenylmercuric hydroxide	[100-57-2]
phenylmercuric nitrate	[55-68-5]
phenylmercuric oleate	[104-60-9]
phenylmercuric propionate	[103-27-5]
phenylmercuric sulfide	[20333-30-6]
phenylmercuric N,N-dimethyldithiocarbamate	[32407-99-1]
potassium iodomercurate	[7783-33-7]
silver iodomercurate	[36011-71-9]
sodium 2,4-dihydroxy-3,5-dihydroxymercuribenzophenone-2-sulfonate	[6060-47-5]
sodium iodomercurate	[7784-03-4]

Soluble sulfides such as sodium sulfide, potassium sulfide and calcium polysulfides have been used to precipitate mercury salts from alkaline solutions. When using this procedure, caution is required to maintain the pH in the alkaline range to prevent evolution of H_2S. Because the solubility of mercuric sulfide in water is 12.5 μg/L at 18°C or 10.7 ppb of mercury, use of this method of removal of mercury would be considered adequate for most purposes. However, the presence of excess alkali, such as sodium hydroxide or sodium sulfide, increases the solubility of mercuric sulfide:

Na_2S, g/100 g soln	HgS, g/100 g soln
0.95	0.21
1.50	0.57
2.31	1.45
3.58	2.91
4.37	4.12
6.07	7.27
9.64	15.59

Thus, at a concentration of 0.95 g Na_2S/100 g solution, the solubility of mercuric sulfide has increased to 2100 ppm. It is, therefore, customary to use no greater than a 20% excess of the alkali sulfide.

Because the particle size of the precipitated mercuric sulfide is so small, it is helpful to add a ferric compound such as ferric chloride or ferric sulfate to effect flocculation. Sometimes other flocculating agents (qv) may also be added, such as starch or gum arabic.

Another method of removing mercury compounds from aqueous solution is to treat them with water-soluble reducing agents, thus liberating metallic mercury (13). The use of formaldehyde at a pH of 10–12 also is recommended.

Problems of removal of mercury from aqueous effluents are more complicated in those plants that manufacture a variety of inorganic and organic mercury compounds; it is generally best to separate the effluent streams of inorganic and organic mercurials. When phenylmercuric acetate is precipitated from its solution in acetic acid by addition of water, the filtrate is collected and re-used for the next precipitation. This type of recycling is necessary not only for economic reasons but also to minimize recovery operations.

When an aqueous effluent stream containing organo-mercurials cannot be recycled, it may be treated with chlorine to convert the organo-mercury to inorganic mercury. The inorganic compounds thus formed are reduced to metallic mercury with sodium borohydride. The mercury metal is drained from the reactor, and the aqueous solution discarded. The process utilizing sodium borohydride is known as the Ventron process (14).

Table 1 is an alphabetical list of mercury compounds referred to in the text.

BIBLIOGRAPHY

1. Hatch and Ott, *Anal. Chem.* **40,** 2085 (1968).
2. J. W. Mellor, *Comprehensive Treatise on Inorganic and Theoretical Chemistry*, Vol. 4, p. 820.
3. G. F. Nordberg, M. H. Berlin, and C. A. Grant, *Proc. 16th Int. Congr. Occup. Health, Tokyo, Japan, Sept. 22–27, 1969*, p. 234.
4. A. Jermelov, *Fisheries Research Board of Canada Translation Series No. 1352; Vatten* **24**(4), 360 (1968).

5. S. Jensen and A. Jernelov, *Nature* **223,** 753 (Aug. 16, 1969).
6. L. Clendening, *Source Book of Medical History*, Dover Publications, New York, 1960; W. Singer and E. A. Underwood, *A Short History of Medicine*, 2nd ed., Oxford University Press, New York, 1962.
7. H. L. Friedman, *Ann. N.Y. Acad. Sci.* **65,** 461 (1957).
8. *Fed. Reg.* **37**(61), 9373 (Mar. 29, 1972).
9. T. Rahola, T. Hattula, A. Kosclainen, and J. K. Miettinen, *Scand. J. Clin. Lab. Forest.* **Abstr. 27**(Suppl. 116), 77 (1971).
10. *Hygienic Guide Series*, American Industrial Hygiene Association, Westmont, N.J.
11. U.S. Pat. 3,502,434 (Mar. 24, 1970), J. B. MacMillan (to Canadian Industries Ltd., Canada).
12. U.S. Pat. 3,213,006 (Oct. 19, 1965), G. E. Grain and R. H. Judice (to Diamond Alkali Company).
13. U.S. Pat. 2,885,282 (May 5, 1959), M. P. Neipert and C. D. Bon (to The Dow Chemical Co.).
14. *Chem. Eng.*, 71 (Feb. 27, 1971).

WILLIAM SINGER
MILTON NOWAK
Troy Chemical Corporation

MERCURY, RECOVERY BY ELECTROOXIDATION.   See Mercury.

MESITYLENE, 1,3,5-$(CH_3)_3C_6H_3$.   See (Polymethyl)benzenes.

MESITYL OXIDE, $CH_3COCH = C(CH_3)_2$.   See Ketones.

METAL ALKYLS.   See Organometallics.

METAL ANODES

In any electrolytic process, the anode is the positive terminal through which electrons pass from the electrolyte. The selection of the materials of construction depends on reaction requirements and the specifics of the process. In practice, the choice of available materials is limited and they tend to be expensive. Soluble or insoluble metal anodes are used in a variety of processes, such as electroplating, electrorefining, cathodic protection, etc, whereas in other processes, such as batteries, fuel cells, etc, the dimensional stability of the anode is retained (see Batteries, primary; Electroplating).

The two largest electrochemical processes, the production of aluminum and of chlorine and caustic, use graphite as anode material (see Aluminum; Alkali and chlorine products). In both processes, the graphite anode is consumed, and, in the case of aluminum production, considerable research efforts to date have yielded no satisfactory substitute material.

For the chlor–alkali industry, however, alternatives to graphite have been developed and are now in commercial production. Today, coated titanium anodes are used for most of the world's chlorine and caustic production. They also are applied to a wide variety of new electrochemical processes.

An excellent review of many types of electrodes and their applications in industrial processes is contained in ref. 1 (see Electrochemical processing).

COATINGS FOR TITANIUM

The value of platinum metal as an anode was identified in the early days of electrochemistry and was employed in the original development of systems to produce chlorine and caustic. In spite of its high cost, platinum was the anode material used in the first commercial production of chlorine from diaphragm cells and in some of the first designs of chlor–alkali cells using mercury cathodes.

The introduction of synthetic graphite—using the Acheson process—established graphite as the universally used anode material for commercial chlorine production after 1900. Although graphite gradually is eroded and lost through reactions with oxygen and other agents, its low cost and availability made it preferrable to anodes based on noble metals. Developments in graphite and improved production methods provided anodes to meet industry's requirements, although considerable research effort continued in an effort to find a practical metal anode that would retain dimensional stability while generating chlorine.

Cladding platinum to copper and other substrate materials proved uneconomical because of the thickness of platinum required to give a complete and reliable cover. More recent advances in roll bonding of metals and edge-sealing techniques have resulted in metal structures of tantalum, niobium, and titanium with continuous platinum coatings of 1 μm (2). With these methods uniform platinum cladding and bonding to the substrate material is achieved.

In 1913, electrodes formed by applying noble-metal coatings to tungsten and tantalum were patented (3–4). The oxide-forming characteristics of these substrates eliminated the need for the coating to be pinhole free. Numerous methods were developed for applying the noble metals including evaporation techniques, thermal decomposition of compounds in organic bases, and a variety of electroplating systems using baths of many compositions. The most common systems today use platinum or its alloys applied under controlled conditions to give specific alloy characteristics.

In response to the needs of the aerospace industry, an important technological breakthrough in the development of metal anodes took place in the 1950s when titanium became commercially available in large quantities. The excellent corrosion resistance of titanium in a variety of solutions and its self-oxidizing, valve-metal characteristic quickly were recognized to be of value for electrochemical systems. Titanium as an anode does not pass current satisfactorily because of the buildup of noncorrodible oxide coatings on the surface, but with the addition of a noncorrodible metal coating, a useful anode can be produced. Extensive research work culminated in the filing of patents in 1957 in the Netherlands and the U.K. (5–6) in 1958, which led to a group of patents (7–9) where oxides of noble metals are used in the coating of titanium, in particular, ruthenium oxide in combination with other metals and oxides. These precious metal oxide coatings have received worldwide acceptance in the chlor–alkali industry and have resulted in considerable power savings in the production of chlorine. By optimizing the characteristics of these anodes, new cell designs and technology for the production of chlorine have been developed.

A second group of patents (10–12) covers platinum and mixtures of platinum–iridium deposited thermally or electrolytically on the titanium substrate. Such anodes have their main application in cathodic protection and the production of sodium chlorate.

These composite anodes, with titanium as the base metal, have been described variously as precious metal anodes (PMA), noble-metal-coated titanium (NMT), dimensionally stable anodes (DSA), and platinized titanium anodes (PTA).

Spurred by the success of these metal-anode systems, research efforts have been directed toward development of an alternative titanium coating for use in the chlor–alkali industry. The Dow Chemical Co. has developed a nonprecious metal coating based on cobalt oxide with spinel structure (13–14). A series of platinum metal oxides have been utilized by the C. Conradty Co. of the Federal Republic of Germany (15). The catalysts are referred to as platinates; eg, lithium platinate has the composition $LiPt_6O_8$.

Other metal anodes have limited use. The electrochemical properties of lead dioxide formed on a lead anode have resulted in the use of lead and lead alloy anodes for the electrowinning of metals such as copper, nickel, zinc, etc. Lead dioxide also may be deposited on graphite.

More recently, lead dioxide on titanium or precoated titanium has been developed (16). Several methods of applying manganese dioxide to titanium have been reported (17–18) but appear to have, as yet, very limited application.

Performance Characteristics

The operating behavior of metal anodes, indeed of any electrode, depends to the greatest extent on the electrolytic conditions under which they operate. Factors such as cell design, electrolyte flow, concentration of electrolyte and its pH, operating temperature, current density, and the presence or absence of impurities affect the voltage and operating lifetime of the anode. Only a laboratory comparison with standard testing conditions can evaluate fairly the differences among various anodes, and even then there is no guarantee that equivalent differences will occur under the conditions of commercial electrolytic processes. However, with over a decade of commercial use, several generalized features of each type of metal anode have been identified.

Platinum–Iridium. With respect to chloride electrolysis, the principal characteristic of the standard 70/30 wt % platinum–iridium coating is a low overpotential for chlorine evolution, together with a relatively high overpotential for oxygen evolution (19). The result is an anode coating that exhibits high current efficiency for production of chlorine. Changes occur in the coating, and thus in its electrochemical characteristics, over a period of time (19). The so-called active phase of operation is characterized by a gradual, preferential dissolution of the platinum metal from the coating. In this phase, there is no significant change of electrode potential with time. At high current densities, and at coating weights below a critical loading of 4–5 g/m^2 platinum–iridium, the coating exhibits passive characteristics, attributed to passivation of the platinum. In this phase, the electrode potential gradually increases with time.

In the electrolytic production of sodium chlorate, a process in which platinum–iridium coatings have their widest application, current densities range up to 6 kA/m^2. To prevent passivation, minimum coating weights of 7–10 g/m^2 are used, and at current densities of 6 kA/m^2 data accumulated over a period of several years indicate the wear rate to be less than 45 mg per metric ton of chlorate. Indications are that platinum selectively dissolves relative to the iridium dioxide (19) (see Chlorine oxygen acids and salts, chloric acid and chlorates).

The limited use of platinum–iridium coatings in commercial production of chlorine and caustic undoubtedly is due to their high cost. In addition, selective dissolution of platinum metal occurs. In diaphragm cells, however, the initial coating loss is relatively high, most likely because of mechanical loss of relatively loose coating. Platinum–iridium anodes display the small inherent current inefficiency of diaphragm cells as a high level of chlorate in the cell liquor, rather than as oxygen in the chlorine gas. The high oxygen overpotential, referred to previously, results in a higher steady-state concentration of hydroxyl ions available for chemical reaction with chlorine to form chlorates.

In the production of sodium hypochlorite by the electrolysis of a weak brine solution (30 g/L NaCl), platinum–iridium anodes exhibit wear rates of 0.3 $\mu g/(A \cdot h)$ (19). Although these wear rates are much higher than wear rates in chlorine or chlorate cells, they are low compared to platinum-metal coatings previously used extensively in this application. They also have the advantage of higher current efficiency for hypochlorite production.

Ruthenium–Titanium Oxides. Mixed oxide coatings of the DSA type exhibit stable operation for long periods of time at low operating voltages. In the production of chlorine, this results in a reduction in labor required to maintain cell operation, a reduction in downtime for cell cleaning and adjusting, and energy savings of 20% over graphite, the previous anode material of choice.

Anode lifetime depends on the current density at which the cell operates. At relatively low current densities of about 1.5 kA/m^2, typical of diaphragm cells, DSAs have operated well in excess of ten years. At 10 kA/m^2, common in mercury cells, these anodes operate for about two years before the structure must be removed from the cell and a new coating applied.

The mixed oxide coatings do not dissolve in mercury, and the coating surface is not easily wetted by mercury. These unique features permit operation of mercury cells with extremely small gaps between the anode and cathode, commonly as low as 3 mm. In diaphragm cells, anode–cathode gaps of 12–15 mm are common. With the use of newly developed polymer-modified asbestos diaphragms, this gap can be reduced to 3–5 mm.

The mechanism of coating failure appears to depend on the type of cell in which the anode is operated. The unavoidable occurrence of minor short circuits, through contact with the mercury cathode, causes gradual physical wear of the coating in mercury cells, which limits the lifetime to one to two years. In the absence of physical wear, such as in membrane or diaphragm cells, the limiting factor appears to be passivation of the anode, preceded by a very gradual dissolution of the precious metal oxide. In this case, the voltage begins to rise because of the buildup of nonconducting oxides in the interfacial layer between the titanium base metal and the coating. This effect is related to the percentage of oxygen evolved at the anode, which in turn, is determined by brine strength and electrolyte pH. In cells producing hypochlorite, where the brine strength is 30 g/L NaCl or below, oxygen production at 1.5 kA/m^2 averages 8–10%, and the anode lifetime is on the order of 1.5–2 yr. This compares to oxygen values of less than 0.5% and a lifetime in excess of ten years in chlorine-producing diaphragm cells operating at the same current density.

The DSA coating used in chlorine production is not useful at very low brine strengths, or in nonchloride electrolytes in which oxygen evolution is the sole anode reaction. Alternatively, mixed oxide–DSA coatings containing different valve metal–precious metal combinations have been developed which have an acceptable lifetime in the electrowinning of metals (20) (see Extractive metallurgy).

Other Coatings. The spinel cobalt oxide coatings have initial overvoltages for chlorine evolution of 30 mV in 300 g/L NaCl at 70°C and 0.8 kA/m^2. Continuous operation at 1.5 kA/m^2 over a period of several years results in an increase in overvoltage of about 18 mV/yr. Based on laboratory tests, Dow projects lifetimes of 6–8 yr in commercial use in diaphragm cells (21).

Anode coatings of the metal platinate types are reported to be extremely stable in mercury cell operation, and to have lifetimes of 3–5 yr in diaphragm chlorine cells and 2–4 yr in commercial sodium chlorate cells (22).

Coating Structure and Morphology

The crystal structure of metal-anode coatings has been investigated with x-ray diffraction studies, x-ray fluorescence analysis, and microprobe studies in conjunction with scanning electron micrographs. However, the role of the titanium metal substrate and its oxides, formed when the substrate is anodized or heated in air, remains unclear for coatings that contain titanium in the coating solutions. Most coatings, even those not containing titanium in the coating solution, exhibit traces of titanium dioxide, present either in rutile or anatase form.

Platinum–Iridium. The composition of standard 70/30 wt % platinum–iridium coatings has been reported to be platinum metal containing some iridium, partly in solid solution and partly as iridium oxide, with one or both of the forms of titanium dioxide, depending on the baking temperature (19). The titanium dioxide is present in amounts of a few weight percent. The iridium is present mostly as iridium dioxide with a small amount of the metal present in solid solution with platinum metal. Depending on the coating composition, the lattice parameters of iridium dioxide can be slightly distorted, most likely because of chloride-ion inclusion in the lattice of coatings made from chloride-based solutions. The analysis of these coatings suggests that the platinum metal acts as a binder for a uniform layer of iridium dioxide, which acts as the electrocatalyst for chlorine discharge (19). X-ray data obtained under different heat treatments indicates the presence of iridium dioxide as a film of constant thickness on a platinum metal core whose thickness changes with heat treatment. Crystal sizes of the platinum and iridium dioxide are 5–15 μm; platinum is larger and more sensitive to crystal growth with temperature.

Ruthenium–Titanium Oxides. The x-ray diffraction studies of these coatings show that the coating components are present as the metal dioxides, each in the rutile form and in solid solution with each other (23). The development of the crystal structure begins to occur at a bake temperature of about 400°C. By following the d_{110} diffraction line for the rutile structure, an increase in crystallinity can be seen up to 600–700°C. Above these temperatures, the d_{110} peak begins to separate into two separate peaks, indicative of phase separation into individual rutile oxides, one rich in ruthenium and one rich in titanium.

It appears that the titanium metal substrate on which the coating is deposited plays an important role in the structure and morphology of the coating. The surface layer of rutile titanium dioxide normally found on oxidized titanium metal apparently acts as a seed to initiate growth of the rutile form of the oxide, rather than the anatase form. Interfacial layers of titanium suboxides, known to be electrically conductive, also act to effect a gradual transition from pure metal to pure rutile oxides. Without this interfacial layer, the large stresses in the titanium crystal structure in going from metal to oxide would occur over a sharp boundary and reduce the adhesion of the coating to the metal. Thus, careful control of the bake temperature, and its rate of change (temperature profile), are necessary to ensure optimum adhesion of the oxides to the base metal.

Additional x-ray studies indicate some degree of lattice distortion in coatings prepared from chloride-containing coating solutions. This correlates with an analysis of 3–5% chlorine in the coating, which is reduced to near zero if the coating is heated to 800°C.

Coatings produced at bake temperatures of 400–600°C are incompletely crystallized solid solutions of rutile titanium dioxide and ruthenium dioxide, with lattice defects caused by the presence of chlorine. The high degree of crystalline disorder contributes greatly to the electrical conductivity of the coating, the presence of active catalytic sites, and the high surface area of the coating.

Scanning electron micrographs of ruthenium–titanium oxide coatings show a characteristic microcracked surface (23). This cracking occurs early in the coating preparation, as solvent evaporates from the surface to form a gel of unreacted ruthenium and titanium compounds. As the coating is baked at higher temperatures, the cracks increase in size because of volume contraction of the gel. A fully baked anode coating has the appearance shown in Figure 1 and a surface-area factor of 180–230 times the geometrical area, as measured by BET (Brunauer-Emmett-Teller) nitrogen adsorption. This large surface area contributes to the low chlorine discharge potential of these types of coatings, providing a large number of catalytic sites for gas evolution while minimizing concentration polarization.

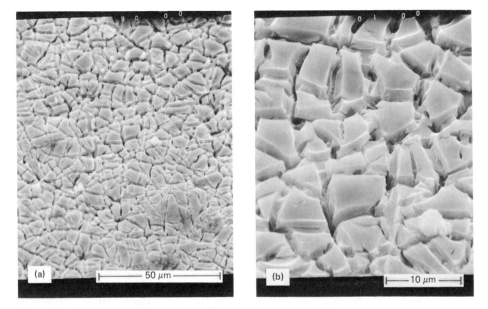

Figure 1.   Scanning electron microscope photographs of DSA oxide coating showing typical cracked surface.

Spinel Cobalt Oxides.   The oxide Co_3O_4 has the spinel structure, containing Co(II) and Co(III) ions. The x-ray analysis shows that divalent zinc ions substitute for Co(II) in the tetrahedral sites of the lattice causing an expansion of the lattice and formation of a normal $Zn_x Co_{3-x} O_4$ spinel structure (21). In coatings containing zirconium, a separate, partially crystalline phase of zirconium dioxide occurs.

The coating has large microscopic pores providing a high surface area. This appears to be related to the presence of zirconium, since spinel oxides applied without the presence of zirconium are dense and closely packed coatings. Thermal studies indicate that decomposition of zirconyl nitrate, the source of zirconium in the coating, produces NO_2 which is released into the cobalt–zinc oxide as it is forming. This presumably produces the open-pore, spongy morphology of the coating. Surface-area measurements confirm this model. The presence of zirconium increases the surface-area factor from approximately 250 to 1880, determined by BET measurements (21).

Manufacture

Generally, the details of preparation of commercial metal-anode coatings are considered proprietary by the manufacturer. Nevertheless, a review of the patent literature gives a fair idea of the general technique.

Platinum–Iridium.   The high chlorine overpotential of electroplated platinum metal on titanium was considered uneconomical for commercial chlorine–caustic production, which led to attempts to prepare such coatings by thermal decomposition of precious metal salts dissolved in suitable solvents. This technique allowed the preparation of a wide variety of coatings, and a wide range of mixed-metal and metal oxide-coating compositions. The combination 70% platinum–30% iridium was chosen to offset the high cost of iridium with a lower cost material such as platinum. The presence of 30% iridium seems to be considered an economic optimum from the standpoint of voltage and lifetime in commercial chlorine-caustic cells.

In the late 1920s, organometallic paints were used for platinum mirrors (24). Today, platinum–iridium coatings are prepared from paints composed of platinum and iridium compounds, generally halide salts, dissolved in organic solvents, with various natural oils added. The solution is oily in consistency and dark brown in color (19).

Coatings are prepared by applying the solution to a treated titanium surface, followed by a heat treatment. Chemically etching the titanium surface prior to application improves the adhesion of the coating. Methods of application include roller coating, dip coating, brushing, and spraying (see Coating processes). The coating usually is applied in many thin layers, each followed individually by heat treatment. A final heat treatment, or annealing step, sometimes is applied after the final coating thickness is obtained (10–12).

Ruthenium–Titanium Oxides.   Precious metal oxide coatings used in chlorine–caustic production are based primarily on ruthenium oxide. Initially, the ruthenium salts and an organotitanium compound were dissolved in alcohol (7). Subsequently, aqueous coating solutions were developed using titanium halides (8–9). The coating solutions in both cases are made acidic to maintain solubility of the components, and are applied to the prepared titanium substrate by brushing, rolling, or spraying. Identical oxide coatings are obtained from either the alcohol-based solution or the aqueous system.

In the usual procedure, an etched titanium sheet is covered with the coating solution, the solvent is evaporated, and the substrate baked at elevated temperatures (max 400–500°C) in air. Successive coatings are applied until the desired thickness is obtained. In some cases, posttreatment heating further stabilizes the coating.

Other Coatings.   Aqueous coating solutions, prepared from the nitrates of cobalt, zinc, and zirconium are used to produce the spinel cobalt oxide coatings (13–14). The mole ratio of cobalt to zinc ranges from 30:1 to 2:1. After each coating, the titanium substrate is heated in air at 250–475°C for about 10 min to form the oxides. After the final coat has been applied, the coated titanium is heated at the same temperature for about one hour. Few details of the commercial metal platinate coatings are available, but a patent (15) describes an electrocatalytic material prepared separately from the titanium metal substrate, using standard high temperature solid-state preparation methods. The catalyst material is ground and pulverized to a fine powder, and then applied to the prepared titanium substrate using low melting glass materials as binders.

Commercial Aspects.   Detailed production statistics for metal anodes are not published. It is estimated that about one half of the world's production of chlorine and caustic, and 55% of all sodium chlorate, were produced on metal anodes by the end of the 1970s.

Platinum–iridium and DSA metal anodes generally are supplied to the chlorine–caustic industry through a lease arrangement. Anodes are designed and fabricated specifically for the customers' cell, installed in the cell, and recoated as needed during the term of the lease. For other applications, anodes may be leased or sold.

Platinum–iridium coatings are supplied by IMI Marston, Ltd., in England. Patents covering the DSA coatings are held by Diamond Shamrock Technologies, S.A., Geneva, Switzerland, and have been licensed worldwide.

Health and Safety Factors

Metal anodes do not have adverse health or safety effects. In fact, in the chlorine–caustic and related industries, their use has eliminated several hazardous plant operations associated with the use of graphite, such as handling of molten lead and hot asphalt for mounting and sealing the anodes in the cell base.

With respect to metal-anode manufacturing plants, nasal ulcerations in a worker exposed to ruthenium and platinum salts have been reported (25). The problems were apparently minor and cleared up after vapor and dust-control measures were initiated.

Applications

Chlorine–Caustic. The widest application and most rapid acceptance of metal anodes has been in the chlorine–caustic industry, where ruthenium–titanium oxide DSA coatings are used. In the mid 1960s, chlorine producers were shifting worldwide to mercury cells to take advantage of the high current densities attainable. For this reason, and because of more favorable economics, the DSA initially was first operated commercially in mercury cells. Anode structures were designed to replace graphite anodes and conversion could be completed without modification of the cells (see Alkali and chlorine products).

Figure 2 is a schematic drawing of a typical mercury-cell anode with an expanded titanium mesh as active surface. Other types of mercury-cell anodes use small diameter titanium rods or thin titanium blades as active surface. Electrical connection in the cell is made through the boss using solid copper rods protected with a riser tube of titanium (Fig. 3). A typical arrangement of anodes in a mercury cell is shown in Figure 4.

The conversion of existing diaphragm cells from graphite to metal anodes was spurred by rising power costs in the early 1970s. Although several commercial instal-

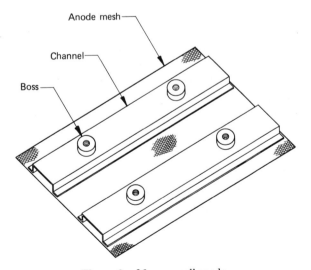

Figure 2.   Mercury-cell anode.

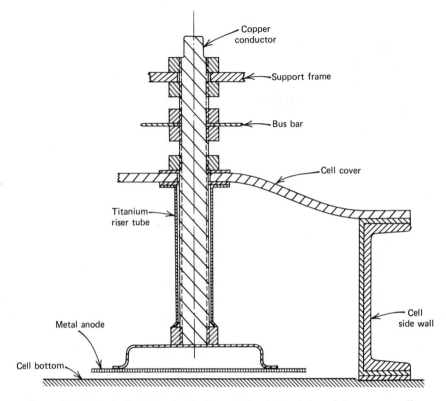

Figure 3. Method of making electrical connection to a metal anode in mercury cells.

lations have used other coatings, the DSA has been preferred by most chlorine producers. Anodes were designed to enable direct replacement of graphite (Fig. 5). Expanded titanium mesh is used for the active, coated surface, and current is carried through a copper-cored titanium conductor bar.

An expandable anode (Fig. 6) has been developed by Diamond Shamrock Corporation for use in its MDC series of cells, where the asbestos diaphragm is stabilized with polymer modifiers. The cell is assembled with the anode in the retracted position. After assembly, the active-anode mesh surfaces expand toward the diaphragm-covered cathode, resulting in a minimum gap between anode and cathode.

The development of ion-exchange membranes capable of operating efficiently in chlorine–caustic production provided an opportunity to design cells based on the unique properties of coated titanium metal anodes. Both monopolar and bipolar membrane cells have been developed in which the anode–cathode gap is minimized, and chlorine and high strength caustic are produced at high efficiencies (see Ion exchange; Membrane technology).

Chlorate Cells. In contrast to the experience in the chlor–alkali industry, the conversion of existing chlorate cells to titanium-based metal anodes has been slow. Direct replacement of graphite in cells of bipolar construction was not possible because of the limitations of hydride buildup on titanium when operated as a cathode. Furthermore, the operating and electrical conditions associated with the cells did not permit the full utilization of the capabilities of coated titanium anodes. In the late 1970s

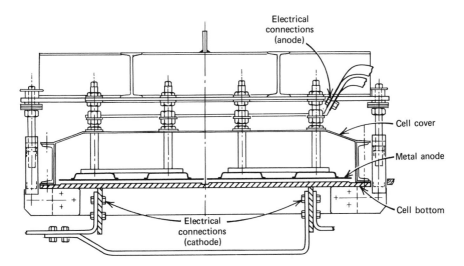

Figure 4.   End view of a mercury cell showing typical arrangement of metal anodes.

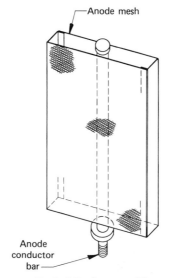

Figure 5.   Typical diaphragm-cell box anode.

considerable expansion of the world chlorate capacity took place to meet the ever-increasing application of chlorine dioxide technology by pulp and paper producers. New chlorate cell designs incorporate coated titanium anodes capable of operating at higher temperatures and current densities and with lower power consumption. In some cases, composite anode–cathode structures in bipolar form offer satisfactory solutions to the problem of dissimilar-metal junctions. Improved titanium manufacturing techniques permit a wide range of anode designs with close tolerances in the cell and anode-to-cathode gaps of 3 mm and less. Increasing electrical power costs have provided some of the incentive to replace magnetite and graphite anodes. Modern

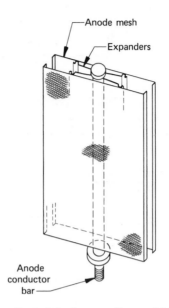

Figure 6.   Typical diaphragm-cell expandable anode.

titanium-anode-based cells, operating at current densities around 2.5 kA/m^2, produce chlorate at power consumptions 15% below that of the best graphite-anode cells where current density was limited to 0.8 kA/m^2 (26).

Excessive wear of graphite at operating temperatures above 40°C necessitated the use of large quantities of cooling water. With titanium anodes, acceptable cell temperatures have been raised to the 60–95°C range, increasing the reaction rate, reducing the required reactor volume and cooling water quantity, improving the water balance in the plant, and permitting the cells to operate at lower voltage.

Several types of titanium-anode coatings are used commercially, including platinum–iridium coatings (19). Platinate coatings of the nonstoichiometric form are used in new cells to convert existing plants (22). Metal oxide coatings are used in monopolar and bipolar form with anode coating life in excess of four years.

Hypochlorite Cells.   The ever-increasing use of low strength hypochlorite solutions to sterilize drinking water and prevent fouling in cooling systems has resulted in the development of specialized electrolytic cells capable of producing the hypochlorite at point of use from low strength brine solutions or from seawater. Such cells are also used in odor-control installations, secondary oil recovery, and for the destruction of cyanides.

As the salt concentration in the brine used is in the range of 20–40 g/L and the electrolyte contains relatively high levels of calcium and magnesium, the anode and its design must have specific special capabilities. In general, cell designs create relatively high velocity of electrolyte across the electrode surface to minimize the buildup of deposits on the cathodes. In one such arrangement, the anodes are rotated by independent drive.

Comparative data on the various types of anode coatings used by different anode manufacturers are not available. Commercial installations apparently use platinum-coated titanium, anodes with titanium to which platinum–iridium alloy has been applied, various mixed oxide coatings specifically developed for low brine strengths, lead dioxide coatings on metals and graphite, and complex solutions of platinates. The operating temperature of such hypochlorite cells is of critical importance. Very high losses of precious metal have been reported with platinum or platinum-coated titanium at operating temperatures below 10°C (27) (see Chlorine oxygen acids and salts, chlorine monoxide, hypochlorous acid and hypochlorites).

BIBLIOGRAPHY

1. A. T. Kuhn and P. M. Wright, "Electrodes for Industrial Processes" in A. T. Kuhn, ed., *Industrial Electrochemical Processes*, Elsevier Publishing Co., Amsterdam, The Netherlands, 1971, pp. 525–574.
2. R. Baboian, "Clad Metal Anodes," *paper presented at the Chlorine Bicenten. Symp.*, San Francisco, Calif., May, 1974.
3. U.S. Pat. 1,077,894 (Nov. 4, 1913), R. H. Stevens (to U.S. Smelting, Refining & Mining Co.).
4. U.S. Pat. 1,077,920 (Nov. 4, 1913), R. H. Stevens (to U.S. Smelting, Refining & Mining Co.).
5. Brit. Pat. 855,107 (April 3, 1958), H. B. Beer (to N. V. Curacaosche Exploitatie Maatschappij Uto).
6. Brit. Pat. 877,901 (Feb. 14, 1958), J. B. Cotton, E. C. Williams, and A. H. Barber (to Imperial Chemical Industries Ltd.).
7. S. Afr. Pat. 66/2667 (May 9, 1966), H. B. Beer (to Chemnor Aktiengesellschaft).
8. U.S. Pat. 3,632,498 (Jan. 4, 1972), H. B. Beer (to Chemnor Aktiengesellschaft).
9. U.S. Pat. 3,711,385 (Jan. 16, 1973), H. B. Beer (to Chemnor Corp.).
10. Brit. Pat. 885,819 (Dec. 28, 1961), C. H. Angell and M. G. Deriaz (to Imperial Chemical Industries Ltd.).
11. Brit. Pat. 984,973 (Mar. 3, 1965), C. H. Angell and M. G. Deriaz (to Imperial Chemical Industries Ltd.).
12. U.S. Pat. 3,177,131 (Apr. 6, 1965), C. H. Angell, S. Coldfield, and M. G. Deriaz (to Imperial Chemical Industries).
13. U.S. Pat. 3,977,958 (Aug. 31, 1976), D. L. Caldwell and R. J. Fuchs, Jr. (to The Dow Chemical Co.).
14. U.S. Pat. 4,061,549 (Dec. 6, 1977), M. J. Hazelrigg, Jr., and D. L. Caldwell (to The Dow Chemical Co.).
15. U.S. Pat. 4,042,484 (Aug. 16, 1977), G. Thiele, D. Zollner, and K. Koziol.
16. U.S. Pat. 4,040,939 (Aug. 9, 1977), B. A. Schenker and C. R. Franks (to Diamond Shamrock Corporation).
17. U.S. Pat. 4,028,215 (June 7, 1977), D. L. Lewis, C. R. Franks, and B. A. Schenker (to Diamond Shamrock Corporation).
18. U.S. Pat. 3,855,084 (Dec. 17, 1974), N. G. Feige, Jr.
19. P. C. S. Hayfield and W. R. Jacob, "Platinum–Iridium-Coated Titanium Anodes in Brine Electrolysis," *paper presented at Advances in Chlor–Alkali Technology, London, Eng., 1979.*
20. M. R. Tighe, Electrode Corporation, private communication, April, 1980.
21. D. L. Caldwell and M. J. Hazelrigg, "Cobalt Spinel-Based Chlorine Anodes," *paper presented at Advances in Chlor–Alkali Technology, London, Eng., 1979.*
22. *Conradty Metal Anodes*, C. Conradty Co., G.m.b.H., Nuremburg, FRG.
23. K. J. O'Leary and T. J. Navin, "Morphology of Dimensionally Stable Anodes," *paper presented at the Chlorine Bicenten. Symp., San Francisco, Calif., May, 1974.*
24. G. F. Taylor, *J. Opt. Soc. Am.* **18**, 138 (1929).
25. S. Harris, *J. Soc. Occup. Med.* **25**, 133 (1975).
26. T. J. Navin, Electrode Corporation, private communication, May, 1980.
27. C. Marshall and J. P. Millington, *J. Appl. Chem.* **19**, 298 (1969).

H. STUART HOLDEN
Diamond Shamrock Technologies, S.A.

JAMES M. KOLB
Diamond Shamrock Corporation

METAL-CONTAINING POLYMERS

In the past 20 years a number of new polymers containing metals as an integral part of the repeating structural unit have been prepared and their properties studied. Several monographs have been published and three ACS symposia have been devoted to these materials (see General references). Much of the research described is still in the early stages and many of the materials have not been characterized fully (see also Inorganic high polymers; Epoxy resins; Silicon compounds; Siloxanes; Polymers, conductive).

Transition-Metal Polymers

Cyclopentadienyl and arene metal π-complexes act as electron-rich aromatic systems that undergo many reactions typical of benzene and other aromatic compounds (see Organometallic compounds). At the same time they possess unusual properties introduced by the metal atom such as the possibility of variable oxidation states, ligand exchange on the metal atom, enhanced absorption of ultraviolet and visible radiation, electrical conductivity, and the ability to liberate finely divided particles of metal or metal oxide upon pyrolysis. Therefore, many attempts have been made to incorporate them into polymers.

Vinylic Polymers.   A recent comprehensive review (1) summarizes the extensive research on vinylic metal π-complexes. The compounds studied are shown in Table 1 (1–2). Typical polymers formed from these monomers are shown in Table 2, and typical copolymers in Table 3.

Properties.   Vinyl metallocenes such as compounds (1–3,7,9–10) undergo either cationic or radical polymerization but not anionic polymerization (22) to form soluble polymers with molecular weights of 2,000–20,000. The high ratios of weight-average molecular weight $\overline{M}_w$, to number-average molecular weight, $\overline{M}_n$, for polymers of (9) and (10) indicate highly branched structures, probably the result of extensive chain transfer. A few monomers, such as (11) and (14), do not homopolymerize. Polymers of compounds (1–10) are soluble in aromatic and chlorinated solvents but decompose in chlorinated solvents upon standing, particularly in the presence of light. Solutions of the polymers, when evaporated, produce transparent, highly brittle films. Thermomechanical properties of the polymers have been studied by differential scanning calorimetry (dsc) and torsional braid analysis (tba). Poly(vinylferrocene), the polymer of structure (1), is the most extensively studied and shows transitions at 170–190°C, 225–247°C, and softening at 280–290°C by dsc (23) but no transition below 260°C by tba (24). Detailed degradation studies of metallocene polymers have been conducted by tga and mass spectroscopy (25). The polymers of compounds (1–10) showed little weight loss below 200°C but loss of ligands such as CO or NO at 200–300°C and of cyclopentadiene above 350°C, and complete decomposition at 500–600°C to metals or metal oxides.

More flexible polymers with higher molecular weight may be obtained if the organometallic group is moved away from the growing chain. For instance, polymers of (13) have been prepared (26) with mol wt up to 7×10^5. The glass-transition temperature, T_g, is 140–145°C over the range $\overline{M}_n = 1.3 \times 10^4$ to 1.5×10^5 but can be lowered to 90°C by addition of a plasticizer (24). Thermomechanical properties are improved by incorporating comonomers (Table 3) but at the expense of metal content. The properties of monomers (1–26) in copolymerization reactions are discussed below.

Electrical Conductivity. The electrical conductivity of ferrocene-containing polymers has been investigated extensively (27). Poly(vinylferrocene) is an insulator $[10^{-15}/(\Omega\cdot\text{cm})]$ but can be oxidized by mild oxidizing agents, such as silver, benzoquinone, or dichlorodicyanobenzoquinone (DDQ), to a mixed-valence ferrocene–ferrocenium polymer (eq. 1) in which electrons probably move from one ferrocenyl (Fc) group to the next. The conductivity rises to a maximum of $10^{-8}/(\Omega\cdot\text{cm})$, with 20–70% oxidation.

$$\left[\text{CH}_2\text{CHCH}_2\text{CH}\right]_n \xrightarrow{\text{DDQ}} \left[\text{CH}_2\text{CHCH}_2\text{CH}\right]_n \tag{1}$$
$$\underset{\text{Fc} \quad \text{Fc}}{} \qquad \underset{\text{Fc}^+ \quad \text{Fc}}{}$$
$$\text{DDQ}^{\cdot}$$

Polymers of compound (3), which have a conjugated chain linking the ferrocenyl groups, can be oxidized to form charged polymers with conductivity up to $10^{-5}/(\Omega\cdot\text{cm})$ (8).

$$n\ \text{Fc—C}\equiv\text{CH} \xrightarrow{\text{AIBN}} \left[\text{CH}=\text{CCH}=\text{C}\right]_n \xrightarrow{\text{DDQ}} \left[\text{CH}=\text{CCH}=\text{C}\right]_n \tag{2}$$
$$\quad (3) \qquad\qquad\qquad \underset{\text{Fc} \quad \text{Fc}}{} \qquad\qquad \underset{\text{Fc}^+ \quad \text{Fc}}{}$$
$$\text{DDQ}^{\cdot}$$

Increasing the distance between ferrocenyl groups, as in polymers of compounds (18–21), or incorporation of a comonomer markedly reduces the electrical conductivity. Ferrocene is oxidized readily and reversibly and appears to be unique among metallocenes in this respect. Attempts to prepare conducting polymers from other monomers in Table 1 have failed.

Preparation. Vinylic polymers may be prepared from monomers (1–8) by cationic or radical polymerization (1). Cationic polymerization may be initiated by Lewis Acids such as $\text{BF}_3\cdot\text{O}(\text{C}_2\text{H}_5)_2$, $(\text{C}_2\text{H}_5)_2\text{AlCl}/\text{M}(\text{CH}_3\text{COCHCOCH}_3)_2$ (M = Ni, Cu, VO) or H_2SO_4. Low molecular weight, highly branched polymers are obtained along with varying amounts of insoluble cross-linked material. Because of the great ability of transition metal π-complexes to stabilize carbonium ions in the α-position (28), attack occurs preferentially upon the vinyl group. Attack of the carbonium ion upon the metallocene rings would lead to branching and cross-linking (eq. 4). Internal hydride transfer also could lead to branching (eq. 5).

Metallocenyl radicals are less stable than the corresponding aryl radicals because the high electron density does not permit easy delocalization of the unpaired electron but these radicals are sufficiently stable to permit radical-initiated homopolymerization. The most successful initiator has been 2,2′-azobisisobutyronitrile (AIBN). Peroxide initiators frequently cause oxidation of the metal atom and extensive decomposition. Homopolymerization is conducted best with 0.5–1% AIBN in the molten monomers (1–8) at 70–80°C in vacuum for 24 h. Better yields and higher molecular

Table 1. Transition Metal Vinyl Monomers

Structure	Name	CAS Registry No.	Formula	Refs.
Vinyl metallocenes				
(1)	vinylferrocene	[1271-51-8]	$C_5H_5FeC_5H_4CH=CH_2$	1
(2)	vinylruthenocene	[76082-22-9]	$C_5H_5RuC_5H_4CH=CH_2$	3
(3)	ethynylferrocene	[1271-47-2]	$C_5H_5FeC_5H_4C\equiv CH$	1, 4
(4)	1,1'-divinylferrocene	[1291-66-3]	$Fe(C_5H_4CH=CH_2)_2$	1, 5
(5)	diisopropenylferrocene	[12289-78-0]	$Fe[C_5H_4C(CH_3)=CH_2]_2$	
(6)	1,3-butadienylferrocene	[12126-86-2]	$C_5H_5FeC_5H_4CH=CHCH=CH_2$	1, 6
(7)	(1-methyleneallyl)ferrocene	[12504-78-8]	$C_5H_5FeC_5H_4CCH=CH_2$ $\|\ CH_2$	7
(8)	poly(3-vinyl-1,1'':1',1'''-bisferrocene)	[53775-36-3]		8
(9)	tricarbonyl[(1,2,3,4,5-η)-1-vinyl-2,4-cyclopentadiene-1-yl]-manganese	[12116-27-7]	$(CO)_3MnC_5H_4CH=CH_2$	1–2
(10)	dicarbonyl[(1,2,3,4,5-η)-1-vinyl-2,4-cyclopentadien-1-yl]-nitrosylchromium	[64539-45-3]	$(CO)_2NOCrC_5H_4CH=CH_2$	9
(11)	tricarbonyl[(1,2,3,4,5,6-η)-vinylbenzene]chromium	[31870-79-8]	$(CO)_3CrC_6H_5CH=CH_2$	10–11
(12)	dicarbonyl[(1,2,3,4,5-η)-1-vinyl-2,4-cyclopentadien-1-yl]cobalt	[73231-00-2]	$(CO)_2CoC_5H_4CH=CH_2$	12
(13)	tricarbonylmethyl[(1,2,3,4,5-η)-1-vinyl-2,4-cyclopentadien-1-yl]		$(CO)_3CH_3WC_5H_4CH=CH_2$	12
(14)	tricarbonyl[(1,2,3,4-η)-1,3,5-hexatriene]-iron, stereoisomer	[50277-87-7]	$Fe(CO)_3$	13

No.	Name	CAS	Structure	Ref.
(15)	chloro(4-vinylphenyl)bis(tributylphosphine)palladium	[54407-93-1]	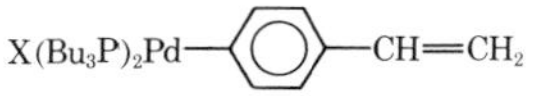 $X(Bu_3P)_2Pd$—[C₆H₄]—$CH{=}CH_2$; X = Cl, Br, CN, C_6H_5	14
(16)	chloro(4-vinylphenyl)bis(triphenylphosphine)platinum	[76082-13-8]	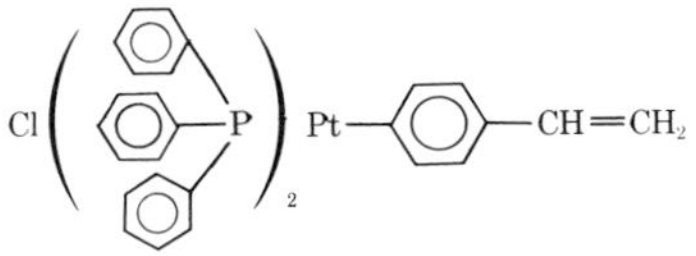 $Cl(Ph_3P)_2Pt$—[C₆H₄]—$CH{=}CH_2$	14
(17)	chloro(4-vinylphenyl)bis(tributylphosphine)platinum	[76095-57-3]	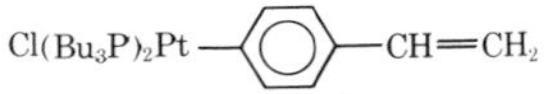 $Cl(Bu_3P)_2Pt$—[C₆H₄]—$CH{=}CH_2$	14

Acrylates and methacrylates

No.	Name	CAS	Structure	Ref.
(18)	[[acryloyl]methyl]ferrocene	[31566-60-6]	$C_5H_5FeC_5H_4CH_2OOCCH{=}CH_2$	2, 15
(19)	[[methacryoyl]methyl]ferrocene	[31566-61-7]	$C_5H_5FeC_5H_4CH_2OOCC(CH_3){=}CH_2$	2
(20)	[2-[acryloyl]ethyl]ferrocene	[34802-13-6]	$C_5H_5FeC_5H_4CH_2CH_2OOCCH{=}CH_2$	2
(21)	[2-[methacryoyl]ethyl]ferrocene	[34802-12-5]	$C_5H_5FeC_5H_4CH_2CH_2OOCC(CH_3){=}CH_2$	2
(22)	tricarbonyl[(η⁶-phenyl)methylacrylate]chromium	[35004-37-6]	$(CO)_3CrC_6H_5CH_2OOCCH{=}CH_2$	2
(23)	tricarbonyl[2-(η⁶-phenyl)ethylmethacrylate]chromium	[53761-68-5]	$(CO)_3CrC_6H_5CH_2CH_2OOCC(CH_3){=}CH_2$	2
(24)	tricarbonyl[[(2,3,4,5-η)-2,4-hexadienyl]acrylate]iron	[51745-70-1]	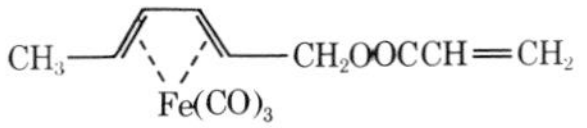 CH_3—...—$CH_2OOCCH{=}CH_2$, $Fe(CO)_3$	16
(25)	chloro-bis-[(1,2,3,4,5-η)-2,4-cyclopentadienyl][(methacryloyl)]-titanium		$(C_5H_5)_2Ti(Cl)OOCC(CH_3){=}CH_2$	17
(26)	chloro-bis-[(1,2,3,4,5-η)-2,4-cyclopentadien-1-yl][acryloyl]titanium		$(C_5H_5)_2Ti(Cl)OCH_2CH{=}CH_2$	18

Group IVA monomers

No.	Name	CAS	Structure	Ref.
(27)	3,3-dimethyl-1-butene	[558-37-2]	$(CH_3)_3CCH{=}CH_2$	19
(28)	trimethylvinylsilane	[754-05-2]	$(CH_3)_3SiCH{=}CH_2$	19
(29)	trimethylvinylgermane	[753-97-9]	$(CH_3)_3GeCH{=}CH_2$	19
(30)	vinyltrimethylstannane	[754-06-3]	$(CH_3)_3SnCH{=}CH_2$	19

Table 2. Homopolymerization of Vinylic Organometallic Monomers

Monomer structure	Name of homopolymer	CAS Registry No.	Solvent	Initiator[a]	Temperature, °C	Yield, %	M_n	M_w	$\eta = gf/(cm \cdot s)$ or Pa·s/10 (= P)	Ref
(1)	poly(vinylferrocene)	[34801-99-5]	neat	AIBN	70	69	11,400	119,000	6.80	2
(2)	poly(vinylruthenocene)	[76082-17-2]	neat	AIBN	70	93	19,660	118,950		3
(3)	poly(ethynylferrocene)	[33410-56-9]	neat	AIBN	80–150	82	2,700			4
(8)	poly(3-vinyl-1,1''':1,1'''-bisferrocene)	[53775-36-3]	benzene	AIBN	70	31	8,000			8
(9)	poly[tricarbonyl[(1,2,3,4,5-η)-1-vinyl-2,4-cyclopentadien-1-yl]manganese]	[35443-49-3]	benzene	AIBN	70	25	9,900	17,000		2
(10)	poly[dicarbonyl[(1,2,3,4,5-η)-1-vinyl-2,4-cyclopentadien-1-yl]nitrosylchromium]	[64552-93-8]	neat	AIBH	70	95	18,500	236,000		9
(18)	poly[[acryloyl]methyl][ferrocene]	[35443-50-6]	benzene	AIBN	70	57	20,100	80,000	12.76	2
(19)	poly[[methacryloylmethyl][ferrocene]	[35560-97-5]	benzene	AIBN	70	67	35,500	211,000	12.44	2
(20)	poly[[2-[acryloyl][ethyl][ferrocene]	[35560-98-6]	benzene	AIBN	70	86	36,000	81,000		2
(21)	poly[2-[methacryloyl][ethyl][ferrocene]	[35560-99-7]	benzene	AIBN	70	89	24,000	65,000		2
(22)	poly[tricarbonyl[(η^6-phenylmethyl)acrylate]-chromium]	[33338-60-2]	benzene	AIBN	70	90	60,000			20

[a] AIBN = azobisisobutyronitrile.

Table 3.   Copolymerization of Organometallic Vinyl Monomers with Styrene and Methyl Acrylate in Benzene at 60–70°C

	Monomer 1, g	Monomer 2, g[a]	AIBN, g	Yield, %	M_1, mol % in copolymer	CAS Registry No. of polymer	Refs.
(1)	3.63	STY, 1.03	0.014	22	33	[57604-89-4]	2
	1.82	MA, 2.00	0.011	66	27	[35139-57-2]	2
(2)	2.00	STY, 1.03	0.020	57	30		3
	1.00	MA, 1.00	0.020	76	20		3
(9)	2.28	STY, 1.88	0.027	44	14	[35443-58-4]	2
	2.49	MA, 1.46	0.020	20	61	[35443-59-5]	2
(10)	2.20	STY, 1.08	0.010	30	41		9
(11)	0.30	STY, 0.30	0.010	49	32	[35443-56-2]	2
	0.30	MA, 0.26	0.010	44	26	[35443-57-3]	2
(15)	1.67	STY, 0.17	0.023	21	46		14
(18)	4.32	STY, 2.50	0.006	60	24	[32612-76-3]	2, 16
	2.85	MA, 2.85	0.014	40	23	[32612-78-5]	2, 16
(19)	12.78	STY, 0.28	0.008	60	58	[32612-77-4]	2
	2.85	MA, 2.85	0.014	40	23	[32612-79-6]	2
(20)[b]		STY,				[35561-01-4]	2
		MA,				[35561-04-7]	2
(21)[b]		STY,				[35561-02-5]	2
		MA,				[35443-63-1]	2
(22)	2.50	STY, 0.37	0.004	68	60	[33338-61-3]	20
	1.25	MA, 0.36	0.009	56	40	[33338-62-4]	20
(23)	1.25	STY, 1.01	0.005	77	33	[53775-41-0]	21
	1.25	MA, 0.83	0.009	77	29	[53775-40-9]	21
(24)	1.23	STY, 0.81	0.020	30	22	[51745-73-4]	16
	1.02	MA, 0.56	0.016	63	33	[51796-23-7]	16

[a] STY = styrene, MA = methyl acrylate.
[b] Data not available.

189

$$R^+ + CH_2{=}CH \longrightarrow RCH_2CH^+ \longrightarrow R{-}[CH_2CH{-}]_n^+ \qquad (3)$$
$$\qquad\qquad\;\; | \qquad\qquad\quad | \qquad\qquad\quad\;\; |$$
$$\qquad\qquad\;\; Fc \qquad\qquad\quad Fc \qquad\qquad\quad Fc$$

$$R{-}[CH_2CH{-}]_n^+ + \left[\begin{array}{c} \text{Fe} \end{array}\right]_n \longrightarrow R{-}[CH_2CH{-}]_{n-1}CH_2CH{-}\left[\begin{array}{c}\text{Fe}\end{array}\right]_n \qquad (4)$$

branched polymer

R = alkyl

$$R{-}[CH_2CH{-}]_n^+ \xrightarrow{\text{hydride transfer}} R{-}[CH_2CH{-}]_m CH_2\overset{+}{C}{-}[CH_2CH{-}]_{n-m-1}H \longrightarrow \text{branched} \qquad (5)$$
$$\qquad | \qquad\qquad\qquad\qquad\quad | \qquad\quad | \qquad\quad | \qquad\qquad\qquad \text{polymers}$$
$$\qquad Fc \qquad\qquad\qquad\qquad\quad Fc \quad Fc \quad Fc$$

weights are obtained, particularly in the case of (3), if further amounts of initiator are added periodically throughout the reaction (4). Polymerization may also be conducted with AIBN in inert solvents such as benzene. Yields and molecular weights are generally somewhat lower (see Initiators).

The kinetics of polymerization of vinylferrocene (1) in dioxane and benzene has been studied extensively (28). In dioxane, typical reaction kinetics of first order in (1) and half order in AIBN were obtained. In benzene, the kinetic order was greater than first order in (1) and first order in AIBN. The polymer was found by uv, Mössbauer, and paramagnetic susceptibility measurements to contain one paramagnetic Fe^{3+} per polymer molecule that was not in the form of ferricinium ion. An unusual electron-transfer termination mechanism was proposed in which an electron was transferred from the ferrocenyl group to the radical site, forming a ferricinium ion and a carbanion (eq. 7). The carbanion could abstract a proton from the solvent to leave a ferricinium ion (eq. 8) or cleave the bond between ring and iron to release a substituted fulvene and a coordinatively unsaturated Fe^{3+} complex (eq. 9). Termination mechanisms of this type do not appear to be operating for (2, 5), or other monomers, possibly because of the greater difficulty in oxidizing the metal atom. Direct proximity of the metallocene group and the radical site is also necessary, for monomers (18–26) do not show this mechanism.

Extensive chain transfer occurs within the polymer chain (eq. 10) to produce highly branched polymers. This is particularly pronounced for monomers such as (3) and (9) and leads to high $\overline{M}_w/\overline{M}_n$ ratios. Chain transfer to monomer (eq. 11) during polymerization of (1) is extensive: $C_m = 8 \times 10^{-3}$ in benzene and 1.2×10^{-2} in dioxane at 60°C. These values correspond to maximum values of degree of polymerization, dp = 125 and 80, and $\overline{M}_n = 2.5 \times 10^4$ and 1.8×10^4 in benzene and dioxane, respectively. They are in good agreement with experimental results. Chain transfer to monomer is also observed for (9) and several other monomers (1). Thus, high polymers of (1) cannot be obtained unless a method to lower C_m is found.

$$R\cdot + CH_2{=}CH \longrightarrow RCH_2CH^{\bullet} \longrightarrow R{\dashv}CH_2CH{\vdash_n^{\bullet}} \tag{6}$$
(with Fc substituents)

$$R{\dashv}CH_2CH{\vdash_n^{\bullet}} \longrightarrow R{\dashv}CH_2CH{\vdash_{n-1}}CH_2CH^{-} \tag{7}$$

$$R{\dashv}CH_2CH{\vdash_{n-1}}CH_2CH^{-} \xrightarrow{H^{+}} R{\dashv}CH_2CH{\vdash_{n-1}}CH_2CH_2 \tag{8}$$

$$R{\dashv}CH_2CH{\vdash_{n-1}}CH_2CH^{-} \longrightarrow R{\dashv}CH_2CH{\vdash_{n-1}}CH_2CH{=}\bigcirc + Fe(C_5H_5)^{2+} \tag{9}$$

$$R{\dashv}CH_2CH{\vdash_n^{\bullet}} \xrightarrow{\text{chain transfer}} R{\dashv}CH_2CH{\vdash_m}CH_2\overset{\bullet}{C}{\dashv}CH_2CH{\vdash_{n-m-1}}H \tag{10}$$

$$R{\dashv}CH_2CH{\vdash_n^{\bullet}} + CH_2{=}CH \longrightarrow R{\dashv}CH_2CH{\vdash_n}H + CH_2{=}\overset{\bullet}{C} \tag{11}$$

Copolymerizations of monomers (1–26) with organic monomers, such as styrene, methyl acrylate (MA), methyl methacrylate (MMA), acrylonitrile, and N-vinylpyr-rolidinone, have been studied (1). Reactivity ratios r_1 and r_2 have been calculated and Q and e values obtained for the monomers. Q is a measure of the resonance stabilization of the free radical formed from the monomer and e is a measure of electron density in the vinyl group (a negative value means electron rich and a positive value electron poor relative to ethylene). Values of r_1 and r_2 are given in Table 4 and of e in Table 5. In general, monomers (1–11) show lower reactivity than commonly used organic monomers because of steric hindrance and resonance stabilization of the free radical. The reactivity ratio for homopolymerization r_1 is particularly low. Monomers (8) and (14) do not homopolymerize at all ($r_1 = 0$). Removal of the organometallic group from direct interaction with the radical site enhances reactivity and, therefore, (20–21, 23) show reactivity ratios typical of acrylate monomers. The values of e in Table 5 are average values and vary depending on the reaction employed (see Table 6). Monomers (1, 9–10) have some of the highest e values ever studied. The Pd- and Pt-containing monomers (15–17) have a lower vinyl group electron density than (1, 9–10) but still greater than dimethylaminostyrene. It is not surprising that the electron-rich monomers (1–17) encounter difficulty in homopolymerizing and couple more readily with more electron-deficient comonomers. The type of copolymers formed can be determined readily from the reactivity ratios. The low r_1 and high r_2 values of (1) with styrene show that (1) enters copolymers with styrene primarily as single units followed by short chains of polystyrene. This explains why the glass transition of copolymers of (1) with styrene rises abruptly as the molar percentage of (1) increases. Adjacent units of (1) are found only with high concentrations of (1). Values of r_1 and r_2 near unity for (1) with MA and MMA indicate formation of random copolymers. Low values of both r_1 and r_2 for (1) with acrylonitrile and maleic anhydride indicate a strong tendency toward alternation. Indeed, uv spectroscopy shows evidence of formation of charge complexes between (1) and electron-deficient monomers. Attack of the growing radical on the charge-transfer complex would lead to formation of an alternating polymer.

Table 4.   Reactivity Ratios for Copolymerization of Organometallic Vinyl Monomers with Organic Monomers[a]

Monomer	Comonomer	r_1	r_2
(1)	styrene	0.08	2.70
	N-vinylpyrrolidinone	0.67	0.33
	methyl acrylate	0.73	0.61
	methyl methacrylate	0.56	1.25
	acrylonitrile	0.11	0.17
	maleic anhydride	0.02	0.19
	(9)	0.49	0.44
(6)	styrene	0.7	2.3
(9)	styrene	0.10	2.5
	N-vinylpyrrolidinone	0.14	0.09
	vinyl acetate	2.35	0.06
	methyl acrylate	0.19	0.47
	acrylonitrile	0.19	0.22
	(1)	0.44	0.49
(10)	styrene	0.30	0.82
	N-vinylpyrrolidinone	5.34	0.079
(11)	styrene	0.00	1.35
	methyl acrylate	0.00	0.70
(18)	styrene	0.02	2.3
	methyl acrylate	0.14	4.46
	methyl methacrylate	0.08	2.9
	vinyl acetate	1.44	0.46
(19)	styrene	0.03	3.7
	methyl acrylate	0.08	0.82
	methyl methacrylate	0.12	3.37
	vinyl acetate	1.52	0.20
	acrylonitrile	0.30	0.11
(20)	styrene	0.41	1.06
	methyl acrylate	0.76	0.69
	vinyl acetate	3.4	0.07
(21)	styrene	0.08	0.58
	methyl methacrylate	0.20	0.65
	vinyl acetate	8.8	0.06
(22)	styrene	0.10	0.34
	methyl acrylate	0.56	0.62
(23)	styrene	0.04	1.35
	methyl methacrylate	0.90	1.19
	acrylonitrile	0.07	0.79
(24)	styrene	0.26	1.81
	methyl acrylate	0.30	0.74
	acrylonitrile	0.34	0.74
	vinyl acetate	2.0	0.05

[a] Ref. 1.

This behavior is also reflected in the marked change in e values calculated for (1) when electron-deficient monomers are used (see Table 6). This behavior is similar to the change in slope of a Hammett $\rho\sigma$ plot when a change in reaction mechanism occurs.

Anionic polymerization is not possible for monomers (1–5, 8–13) since metallocenes provide little stability for carbanions, but anionic initiation occurs readily with

Table 5. Values of e for Organometallic and Standard Monomers[a]

Monomer	e[b]	Monomer	e[b]
(1)	−2.1	$CH_2{=}CH_2$	0.00
(9)	−1.99	$CH_2{=}C(CH_3)CO_2CH_3$	0.40
(10)	−1.96	(29)	0.43
$(p\text{-}CH_3OC_6H_4)_2C{=}CH_2$	−1.96	$CH_2{=}CHCO_2CH_3$	0.58
(16)	−1.62	(30)	0.96
$(CH_3)_2NC_6H_4CH{=}CH_2$	−1.37	$CH_2{=}CHCN$	1.21
$C_6H_4CH{=}CH_2$	−0.80		
(27)	−0.63	 	2.25
(28)	−0.14		

[a] Ref. 1.
[b] e is the Alfrey-Price polarity function (29).

Table 6. Variation of e Calculated for Vinylferrocene in Copolymerizations with Organic Monomers[a]

Monomer	e for vinylferrocene[b]
N-vinylcarbazole	−2.4
p-N,N-dimethylaminostyrene	−2.2
1,3-butadiene	−2.1
styrene	−2.1
N-vinylpyrrolidinone	−2.1
methyl methacrylate	−0.20
methyl acrylate	−0.32
acrylonitrile	−0.81
maleic anhydride	−0.11

[a] Ref. 1.
[b] e is the Alfrey-Price polarity function (29).

the acrylates (18–24). Linear polymers with relatively uniform molecular weights up to 7×10^5 may be obtained by initiation with $LiAlH_4$ (26). Block copolymers may also be formed with styrene and with methyl methacrylate (30). Some of these polymers show two glass transitions for the two portions of the chain (24). In contrast to monomers (1–5, 8–13), monomer (6) undergoes butyllithium-initiated anionic polymerization to form polymers with $\overline{M}_n$ up to 4×10^4 (31).

Disubstituted ferrocenes such as (4–5) undergo either normal vinylic polymerization leading to highly cross-linked materials or cyclopolymerization, which leads to soluble materials, $\overline{M}_n = 2$–8×10^4, when either cationic or radical initiators are employed (1,5,7) (eq. 12). Variables such as the nature of the initiator, temperature, concentration of the monomer, and polarity of the solvent influence the ratio of cyclic polymerization to normal polymerization. The ratio of the rate constant for cyclic polymerization k_c and the rate constant for vinyl polymerization, k_v, k_c/k_v, is 0.67 in toluene, 0.97 in chloroform, and 1.49 in acetonitrile (1).

Ladder polymers may be formed by anionic polymerization of (3) (eq. 13) or cationic polymerization of 1,2-ferrocenylbutadiene [1291-68-5] (7) (eq. 14).

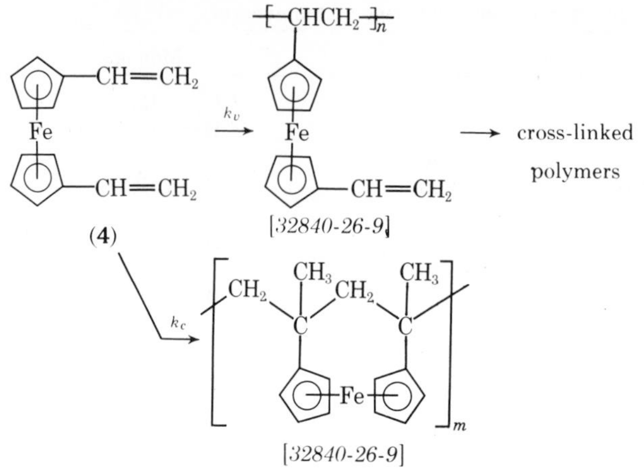

(12)

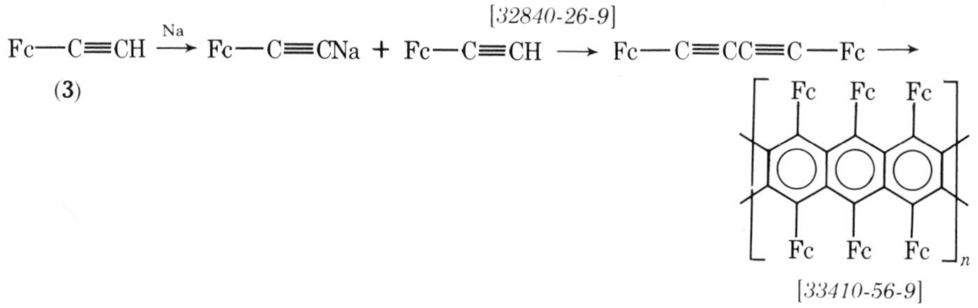

(13)

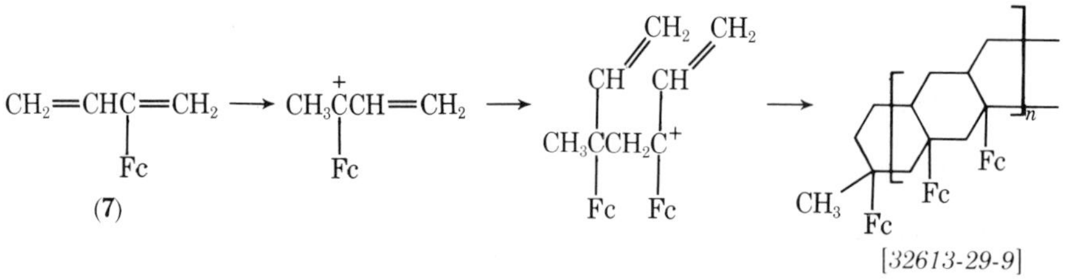

(14)

Applications. Applications of the vinylic polymers of compounds (1–14) have proved limited because of the poor thermomechanical properties of these polymers. Ferrocene and ruthenocene-containing polymers have been of interest because of their ability to absorb uv radiation without being degraded. Use as uv or radiation-resistant coatings has been proposed.

Monomers with pendant CO or NO ligands, such as compounds (9–14), are capable of losing these ligands to form coordinatively unsaturated metallocenes that can undergo reactions with olefins or arenes (see Catalysis). Species such as cyclopentadienyl dicarbonyl cobalt have been used extensively as catalysts for olefin trimerization (32) and for stereospecific synthesis of complex molecules (33–34). When anchored to a polymer chain, $C_5H_5Co(CO)_2$ shows enhanced performance as a Fischer-Tropsch catalyst (35). Recently reported polymers of (12) (12) are likely to have similar applications. Polymers of (11) also are good candidates for use as homogeneous catalysts. Replacement of one of the CO ligands would create a chiral center useful for catalyzing assymmetric synthesis (9). Photolysis of polymers containing chromium tricarbonyl moieties, such as (11, 22–23), produces finely divided particles of chromium oxide or chromium metal dispersed in a polymeric matrix (36–38). Similarly, thermal decomposition of polymers containing iron or manganese carbonyl units, such as (**9, 14, 24**), produces finely divided metal oxide in a polymer matrix (37). Since many of these residues are paramagnetic, applications as magnetic tape (qv) may be a possibility.

Monomers with Vinyl–Metal Bonds. Attempts to polymerize vinylic monomers, where the vinyl group is directly bonded to a transition metal have failed so far. The resulting polymers (eq. 15) rapidly eliminate metal hydrides:

$$L_nMCH{=}CH_2 \longrightarrow \left[CH_2CH{\underset{\underset{\displaystyle ML_n}{|}}{}}\right]_n \longrightarrow L_nMH + \left[CH{=}CH\right]_n \tag{15}$$

$$L_n = \text{ligand}$$

Polymerization of vinyl derivatives of main-group metals, particularly those of Group IVA, has also proven difficult (19). Monomers (**27–30**) could not be homopolymerized by γ irradiation, or cationic or radical initiation but (**29**) underwent anionic polymerization initiated by butyllithium to form oligomers, dp ca 8. Monomers (**27–30**) can be copolymerized with styrene and methyl acrylate. The e values in Table 5 were calculated from the reactivity ratios obtained in these studies. There is a continuous increase in electron deficiency of the vinyl group in passing from C ($e = -0.63$) to Sn ($e = 0.96$). It is, indeed, unusual that the vinyl group becomes more electron deficient as the metal becomes more electropositive. This effect is attributed (19) to increased resonance interaction between the metal d-orbitals and the vinyl group.

Metallocenemethylene Polymers. Metallocenemethylene polymers (**31**) analogous to the phenol–aldehyde resins (see Phenolic resins) have been prepared by a number of investigators:

(**31**)

M = Fe[*34801-99-5*], Ru[*76082-17-2*]

Properties. Polymers of this type are generally dark brown to black amorphous solids, partially soluble in aromatic or chlorinated solvents. The soluble fractions have molecular weights of 2000 to 20,000 and show varying degrees of branching, depending on the method of synthesis. The insoluble material is likely to be cross-linked. These polymers soften at 220–300°C and decompose above 300°C. Thermogravimetric analysis shows 75% weight retention at 425°C and 50% at 550°C. Ruthenocene polymers showed markedly greater stability than ferrocene polymers.

Preparation. The synthesis of metallocenemethylene polymers proceeds in two steps: generation of the α-metallocenyl carbonium ion (**32**) (eq. 16) and thermal polymerization (eq. 17) in the presence of an acidic catalyst such as BF_3, $ZnCl_2$, or H_2SO_4. Common precursors have been the metallocene and an aldehyde or ketone or compounds such as (33a–d).

$$M + RCHO \longrightarrow \quad \overset{R}{\underset{H}{C}OH} \longrightarrow \quad \overset{R}{\underset{H}{C^+}} \qquad (16)$$

M = Fe, Ru   R = alkyl   (32)

$$\qquad \longrightarrow \qquad \qquad \text{or} \qquad \qquad (17)$$

(33a) R = CH$_2$OH
(33b) R = CHOHCH$_3$
(33c) R = CH$_2$N$^+$(CH$_3$)$_3$
(33d) R = CH$_2^+$ BF$_4^-$

M = Fe, Ru

	M = Fe	M = Ru
(33a) R = CH$_2$OH	[1273-86-5]	[33270-45-0]
(33b) R = CHOHCH$_3$	[1277-49-2]	[12089-08-8]
(33c) R = CH$_2$N$^+$(CH$_3$)$_3$	[1271-86-9]	[33293-45-7]
(33d) R = CH$_2^+$ BF$_4^-$	[62341-75-7]	[76082-15-0]

Linkage in the polymer may be 1,2; 1,3; or 1,1′. When difunctional metallocene derivatives are used, exclusively cross-linked materials are obtained. The best yields and highest molecular weight soluble polymers have been obtained (39) when a salt of the carbonium ion (33d) is isolated in a pure form and allowed to polymerize at elevated temperatures.

The low molecular weight polymers formed as described may be mixed with cross-linking agents such as 1,1′-bis(hydroxymethyl)ferrocene and an acidic catalyst, laminated with fiberglass or carbon fibers and cured at 200–300°C and 2.76 MPa (400 psi) (40). The resulting laminates have tensile strength up to 207 MPa (3 × 10^4 psi) and moduli of 13.8–34.5 GPa ((2–5) × 10^6 psi). When exposed to high temperatures, they char rather than melt or soften.

Applications. Polymers of this type have been prepared at the Materials Research Center at Wright-Patterson Air Force Base for possible use as ablative materials (qv) for space-capsule heat shields. Ruthenocene derivatives were studied as radiation shields for communication satellites. However, before these materials were perfected, cheaper ceramic materials were found to perform satisfactorily. So far no large-scale applications have been found where the metallocene polymers are sufficiently superior to currently used materials to justify the higher cost.

Metallocenylene Polymers. Metallocenylene polymers can exist as 1,1′; 1,2; 1,3; or 1,1′,3,3′ isomers (34–37). These materials are of considerable interest because of their high rigidity and thermal stability and because partial oxidation could produce conjugated mixed-valence polymers such as (38) which would act either as a semiconductor or as a metallic conductor.

Properties. The synthesis of (34a–b) in high purity has been reported recently (41–42). The physical properties of (34) are summarized in Table 7.

The ir, uv, nmr, and Mössbauer spectra of samples of (34a) with $\overline{M}_n$ = 500, 1,500,

Table 7. Physical Properties of 1,1'-Poly(ferrocenylene) (34a)

Property	Value
color	yellowish orange
mp, °C	infusible below 350
solubility	soluble in aromatic and halocarbon solvents
tga[a], wt retention, %	
at 600°C	85
at 800°C	75
M_n	3,000 to 10,000
ir, peaks at cm^{-1}	3100, 1420, 1120, 1005, 1010, 825, 490
uv, λ_{max}, nm	
ϵ 17,800	228
ϵ 6,600	269
ϵ 5,300	308
ϵ 650	461
nmr	broad singlet 3.8–4.3 ppm relative to TMS
Mössbauer, ^{57}Co(Cu), mm/s	
δ^b	0.42
Δ Eq	2.34
Γ	0.26

[a] In argon.
[b] δ is relative to Fe foil.

3,000, and 10,000 are virtually identical. Thus, long-range conjugation between the units is absent.

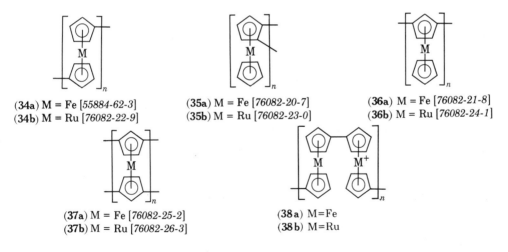

(**34a**) M = Fe [55884-62-3]
(**34b**) M = Ru [76082-22-9]

(**35a**) M = Fe [76082-20-7]
(**35b**) M = Ru [76082-23-0]

(**36a**) M = Fe [76082-21-8]
(**36b**) M = Ru [76082-24-1]

(**37a**) M = Fe [76082-25-2]
(**37b**) M = Ru [76082-26-3]

(**38a**) M=Fe
(**38b**) M=Ru

1,1'-Poly(ruthenocenylene) (**34b**) was prepared by procedures similar to those used for the Fe compound (42). It is a colorless material with high melting point and thermal stability. The highest $\overline{M}_n$ so far obtained is 2450. The ir and nmr spectra of (**34b**) are very similar to those of (**34a**) and there is little change in ir, uv, and nmr spectra with increasing molecular weight.

Efforts to prepare mixed polymers containing both Fe and Ru are currently under way.

Electrical Conductivity.   Pure (**34a**) and (**34b**) are diamagnetic and are insulators with conductivity of ca $10^{-14}/\Omega\cdot cm$. When oxidized with iodine or DDQ to the extent of 30–70%, (**34a**) forms a blackish mixed-valence polymer (**38a**) which acts as a semiconductor (conductivity $10^{-7}/\Omega\cdot cm$) (27,42). The properties of (**38a**) are summarized in Table 8. Since ruthenocene degrades readily upon oxidation, polymers of the type (**38b**) have not been studied extensively.

Mixed-valence polymers may be divided into Class I, essentially no interaction between valence centers; Class II, weak interaction between valence centers as indicated by valency transfer bands in the near-visible ir region of the spectrum; and Class III, strong interaction. Class III contains average-valency materials, which usually possess conductivity approaching that of metals. The presence of valence-transfer bonds and weak electrical conductivity place (**38a**) in Class II (43).

Preparation.   Synthesis of (**34**) is accomplished by linking (**39a**) or (**39b**) by the Ullman reaction, removing the mercury in (**39c**) in the presence of Ag, or coupling (**39d**) or (**39e**) in presence of $CoCl_2$. In the best procedure, polymers free of undesirable end groups or paramagnetic impurities are produced and (**39a**) is coupled with a slight excess of (**39f**) in THF or dimethyl ether (41–42). Excess lithium end groups are removed by hydrolysis.

(**39a**) X = I [12145-93-6]
(**39b**) X = Br [1293-65-8]
(**39c**) X = HgCl [37328-05-5]
(**39d**) X = Li [33272-09-2]
(**39e**) X = Cu [76082-12-7]
(**39f**) X = Li TMEDA complex [32677-72-3]
TMEDA = tetramethylethylene diamine

Poly(ferrocenylenes) are prepared by direct polymerization of ferrocene in the presence of a stoichiometric amount of di-*tert*-butyl peroxide. The procedure is much simpler and less expensive but produces a highly branched and sometimes cross-linked material containing a mixture of structures (**34–36**), containing methyl groups, *tert*-butoxy groups, and up to 30% of aliphatic linkages between the ferrocenyl units.

Table 8.   Mössbauer Parameters and Valence-Transfer Bands for Ferrocene, Ferrocenium, and Partially Oxidized Polyferrocenylene

Compound	$\overline{M}_n$	$(\delta\cdot s)/m^a$	$(\Delta\ Eq'\cdot s)/mm$	$(\Gamma\cdot s)/mm$	λ max, nm
ferrocene	186	0.68	2.37		
ferricinium	186	0.57	0.1		1900
ferrocenylferricinium	457	0.45	2.34	0.28	1830
		0.47	0.42	0.30	
polyferrocenylene					
50% oxidized	2500	0.46	2.47	0.34	1800
		0.48	0.43	0.46	
	5000	0.46	2.46	0.36	1795
		0.48	0.41	0.46	
25% oxidized	2300	0.44	2.28	0.35	1820
		0.45	0.42	0.43	
	4600	0.42	2.24	0.35	1820
		0.43	0.44	0.42	

^{a} δ is relative to Fe foil.

Applications of Poly(ferrocenylene). Linear poly(ferrocenylene) (**34a**) has the highest thermal stability and greatest structural rigidity of any of the known ferrocene-containing polymers. If (**34a**) with high molecular weight can be prepared, it may meet the need for a polymer with stability in the 400–600°C range. Applications of such materials have been described under ferrocenylenemethylene polymers. Preparation by direct polymerization of ferrocene is much simpler and cheaper and polymers have thermal stability comparable to that of the ferrocenemethylene polymers. They can be made into laminates comparable to those of ferrocenemethylene polymers when cured by a similar procedure (44).

Investigations of the conductivity and magnetic susceptibility of (**38a**) are still in their early stages. Since this compound can be prepared in high purity with known structure, and, in contrast to the other ferrocene-containing polymers, preparation of single crystals may be possible. By preparing mixed-metal polymers with varying degrees of oxidation, semiconductors with a wide range of properties could be produced. However, the low solubility of these materials in common solvents presents problems, and they have been described as "brick dust."

Polysilanyl and Polysiloxanyl Ferrocenes. Ferrocenylene–silylene polymers (**40**) and ferrocenyl–siloxanylene polymers (**41**) combine the advantages of the ferrocene group, such as high thermal stability and resistance to uv and γ radiation, with the flexibility over a wide temperature range provided by organosilicon polymers. The best materials of this type have greater thermal stability than the vinylic polymers and much better flexibility and film-forming characteristics.

Properties. The first materials of type (**41**) produced were low molecular weight oils (**42a–e**) with possible application as high temperature lubricating oils or hydraulic fluids (45). Physical properties of (**42a–e**) are summarized in Table 9.

Poly(ferrocenyl siloxanes) have been prepared (46) with sufficiently high mo-

$$\left[\!\!\left[R'\!-\!Fc\!-\!\underset{\overset{\displaystyle |}{\underset{\displaystyle R}{|}}}{\overset{\overset{\displaystyle R}{|}}{Si}}Z\right]\!\!\right]_n$$

(**40**)

R = alkyl, aryl
R′ = covalent bond, alkylene, arylene
Z = alkylene, arylene

$$\left[\!\!\left[R_2Si\!-\!Fc\!-\!SiR_2O(SiR'RO)_x\right]\!\!\right]_n$$

(**41**)

R = alkyl, aryl

Table 9. Physical Properties of Ferrocenyl-1,1′-bissiloxanes

Structure	CAS Registry No.	$bp_{Pa}{}^a$, °C	n_D^{25}	d_4^{25}	Mol wt
(**42a**)	[*12321-04-9*]	$200–205_4$	1.5473	1.1063	602
(**42b**)	[*12243-60-6*]	$220–223_{20}$	1.5162	1.0796	750
(**42c**)	[*12321-18-5*]	$245–255_{5.3}$	1.4850	1.0308	898
(**42d**)	[*38903-97-8*]	$107–110_{1.33}$	1.4940	1.2591	478
(**42e**)	[*12321-03-8*]	$263–265_{16}$	1.5675	1.0863	742

^{a} To convert Pa to mm Hg, multiply by 0.0075.

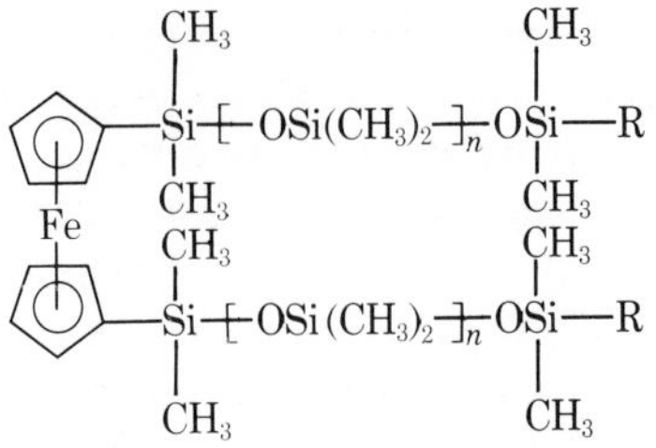

(**42a**) R = C_6H_5 $n = 0$ [*12321-04-9*]
(**42b**) R = C_6H_5 $n = 1$ [*12243-60-6*]
(**42c**) R = C_6H_5 $n = 2$ [*12321-18-5*]
(**42d**) R = CH_3 $n = 0$ [*38903-97-8*]

(**42e**) R = 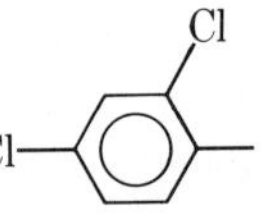 $n = 0$ [*12321-03-8*]

lecular weight to be cast as tough flexible films that exhibit good fiber-forming characteristics when drawn from melts. The physical properties of three such polymers
(**43d–f**) and their phenyl analogues (**43a–c**), prepared by the same procedure, are
shown in Table 10. The ferrocenyl polymers appear to be at least as stable as their
phenyl analogues. In air, (**43e**) showed only 4.5% weight loss by tga at 400°C, whereas
(**43a**) showed 8.7%. Neither polymer showed significant hydrolytic degradation after
refluxing for one hour in 50% aqueous THF.

Polymers with structures (**44a**) and (**44b**) can be prepared easily with $\overline{M}_n$ about
10^4 from 1,1′-bishydroxymethylferrocene and dichlorosilanes (47). These materials
showed marked instability toward hydrolysis. The great stability of the R—Fc—CH_2^+
carbonium ion permits ready cleavage by an S_N1 mechanism. Cleavage of this type
is not possible for polymers (**43d**) and (**43f**).

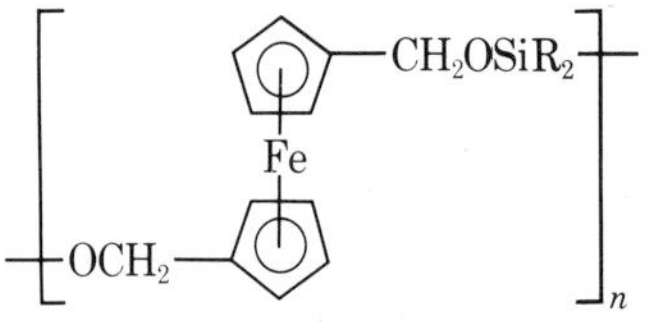

(**44a**) R = alkyl, aryl

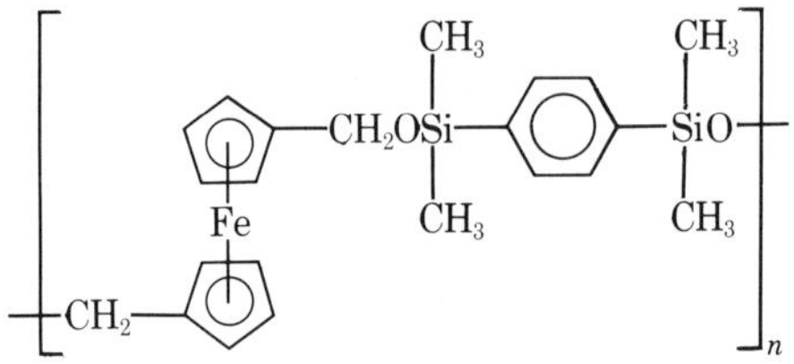

(**44b**) [*76082-30-9*]

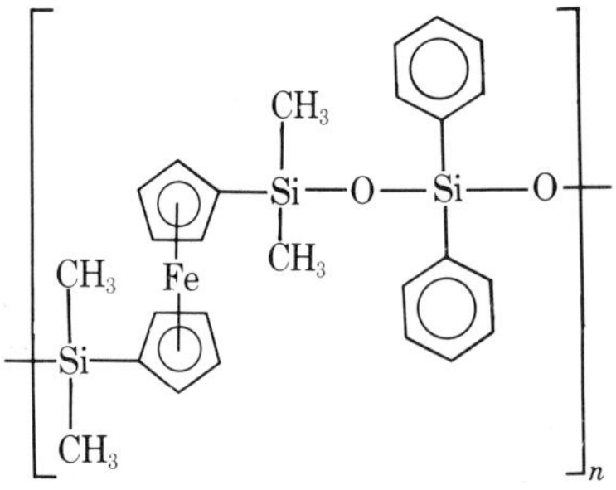

(**43d**) [*52436-24-5*]

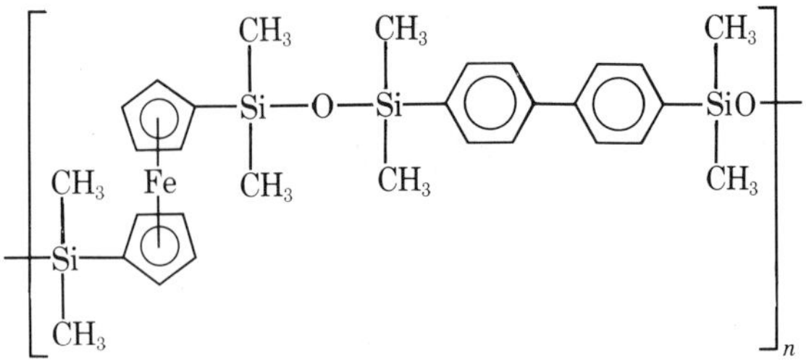

(**43f**) [*52436-25-6*]

Polymers with structures (**45–47**) have been prepared and their thermooxidative
stability measured (Table 11) (48). A wide variety of other low molecular weight sili-

Table 10. Properties of Polymers Prepared by Diol–Diaminosilane Polycondensation[a]

Structure[b]	CAS Registry No.	$\overline{M}_w$ × 10⁻⁴	$\overline{M}_n$ × 10⁻⁴	T_g, C°	T_m	Tga 10% wt loss 4° min⁻¹, N₂	Dsc exotherm N₂
(43a) $+SiR_2C_6H_4SiR_2O)_n$	[54811-84-6]	28.5	11.1	−25		475	
(43b) $+SiR_2C_6H_4SiR_2OSiR_2O)_n$	[41205-84-9]	25.5	10.9	−61		480	445
(43c) $+SiR_2C_6H_4SiR_2OSi(C_6H_5)_2O)_n$	[41205-85-0]	23.3	10.8	1		445	490
(43d) $+SiR_2—Fc—SiR_2OSi(C_6H_5)_2O)_n$	[52436-24-5]	1.84	0.92	37	59		
(43e) $+SiR_2—Fc—SiR_2OSiR_2C_6H_4SiR_2O)_n$	[52436-23-4]	4.68	1.81	−2	77	440	
(43f) $+SiR_2Fc—SiR_2OSiR_2C_6H_4—C_6H_4SiR_2O)_n$	[52436-25-6]	5.10	2.02	7	78		

[a] Ref. 46.

[b] R = CH₃; Fc = —⬡—Fe—⬡—

201

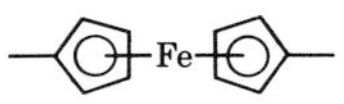

con-containing polymers have also been prepared (45); they are, however, not sufficiently characterized to be included here.

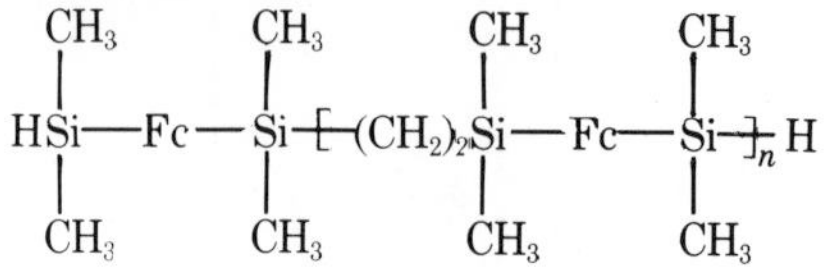

$$\underset{\underset{CH_3}{|}}{\overset{\overset{CH_3}{|}}{HSi}}-Fc-\underset{\underset{CH_3}{|}}{\overset{\overset{CH_3}{|}}{Si}}\underset{\underset{CH_3}{|}}{\overset{\overset{CH_3}{|}}{\underset{n}{+(CH_2)_2Si}}}-Fc-\underset{\underset{CH_3}{|}}{\overset{\overset{CH_3}{|}}{Si}}\overset{}{+_n}H$$

(45) *[76082-28-5]*

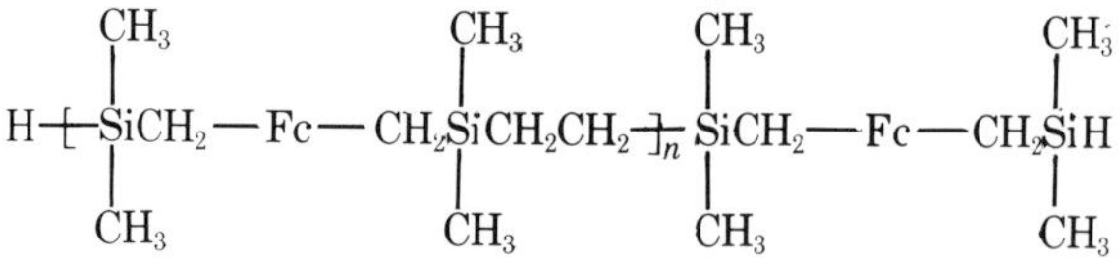

$$H+\underset{\underset{CH_3}{|}}{\overset{\overset{CH_3}{|}}{SiCH_2}}-Fc-CH_2\underset{\underset{CH_3}{|}}{\overset{\overset{CH_3}{|}}{SiCH_2CH_2}}+_n\underset{\underset{CH_3}{|}}{\overset{\overset{CH_3}{|}}{SiCH_2}}-Fc-\underset{\underset{CH_3}{|}}{\overset{\overset{CH_3}{|}}{CH_2SiH}}$$

(46) *[76082-29-6]*

$$H+\underset{\underset{CH_3}{|}}{\overset{\overset{CH_3}{|}}{SiCH_2}}-Fc-CH_2\underset{\underset{CH_3}{|}}{\overset{\overset{CH_3}{|}}{SiCH_2CH_2Si}}-O-\underset{\underset{CH_3}{|}}{\overset{\overset{CH_3}{|}}{SiCH_2CH_2}}+_n\underset{\underset{CH_3}{|}}{\overset{\overset{CH_3}{|}}{SiCH_2}}-Fc-\underset{\underset{CH_3}{|}}{\overset{\overset{CH_3}{|}}{CH_2SiH}}$$

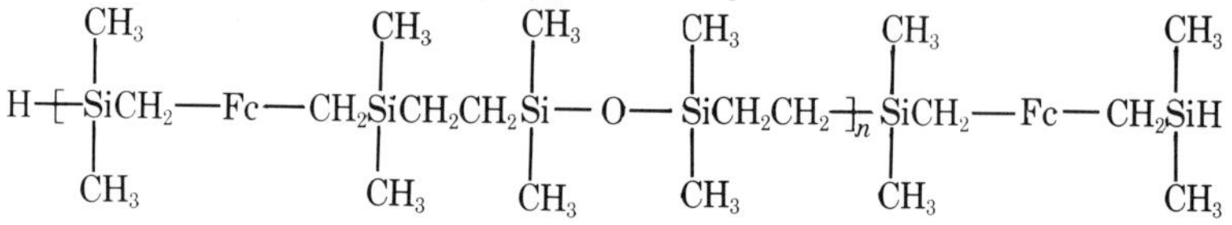

(47) *[76095-58-4]*

Preparation.   Precursors for **(40)** and **(41)** can be made by attaching the silanyl
or siloxanyl group to cyclopentadiene, followed by reaction with $FeCl_2$ to form the
ferrocenyl group (eqs. 18–19). The materials described in Table 11 were prepared by
this method.

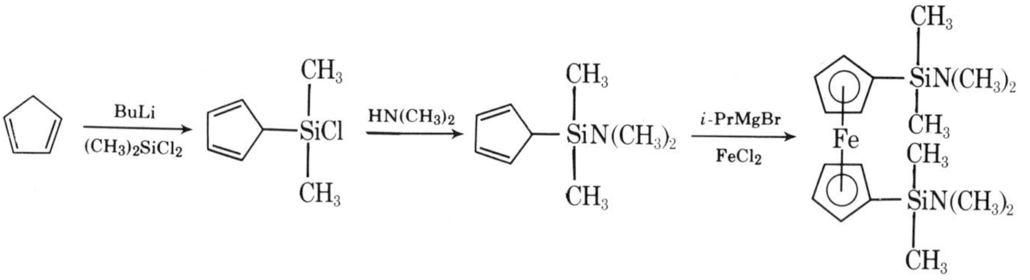

(18)

(48) *[35625-88-8]*

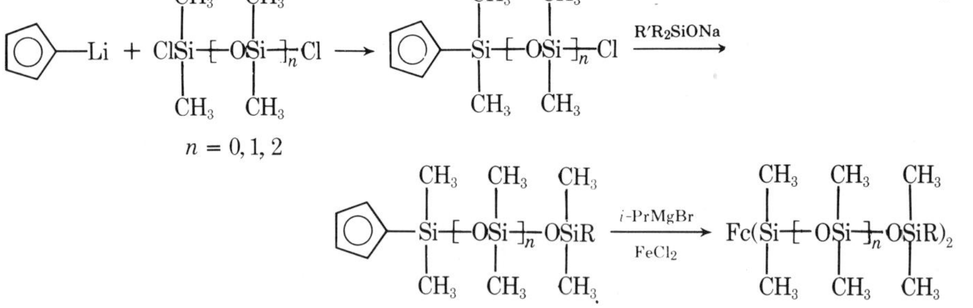

$n = 0, 1, 2$

(19)

R = alkyl or aryl

Table 11.   Thermooxidative Stability of Polysilanyl and Polysiloxanyl Ferrocenes

Structure	CAS Registry No.	Softening point, °C	tga wt loss, %	
			6 h at 300°C	2 more h at 350°C
(45)	*[76082-28-5]*	92–94	0.3	17.1
(46)	*[76082-29-6]*	90–96	5.1	13.0
(47)	*[76095-58-4]*	oil at 25	4.3	11.2

Better yields and a smoother process are obtained if the 1,1′-dilithiumferrocene–TMEDA complex (39f) is used (eqs. 20–21).

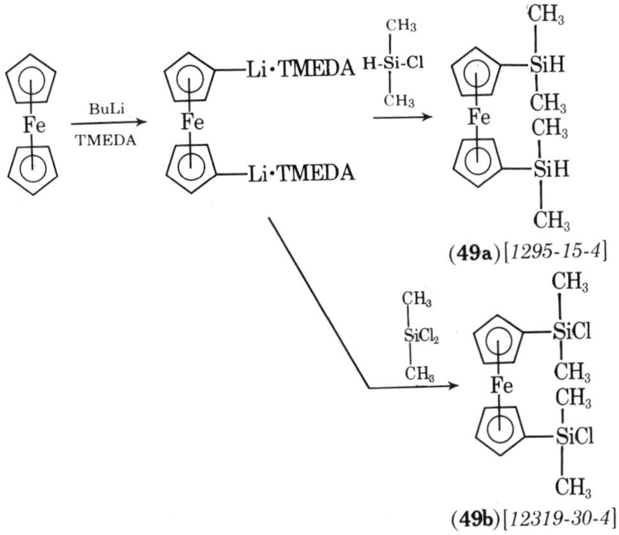

(49a) [1295-15-4]   (20)

(49b) [12319-30-4]

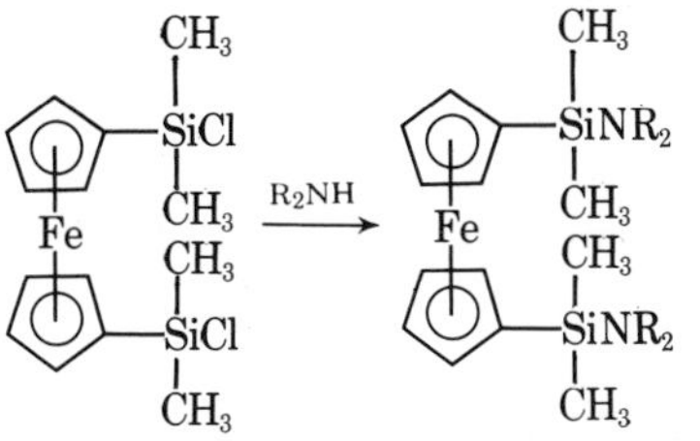

(21)

(49b)   (48) [35625-88-8] (R = CH₃)

R = alkyl or aryl

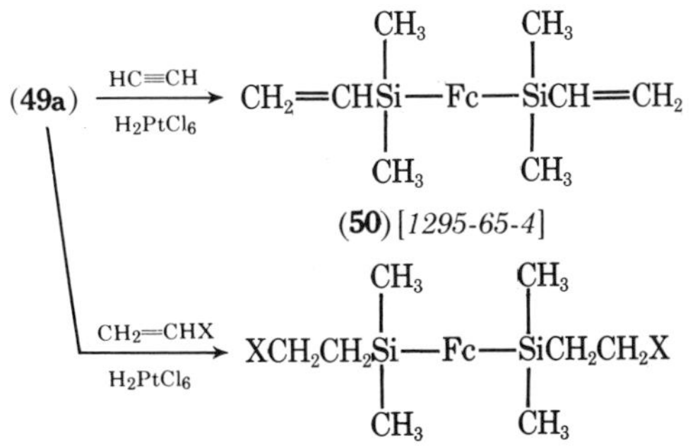

$$(49a) \xrightarrow[\text{H}_2\text{PtCl}_6]{\text{HC}\equiv\text{CH}} CH_2=CHSi-Fc-SiCH=CH_2$$

(50) [1295-65-4]   (22)

$$\xrightarrow[\text{H}_2\text{PtCl}_6]{\text{CH}_2=\text{CHX}} XCH_2CH_2Si-Fc-SiCH_2CH_2X$$

(51a) X = CO₂CH₃ [1295-84-7]
(51b) X = CH₂OH [1295-75-6]
(51c) X = CH₂NH₂ [12190-11-3]
(51d) X = CH₂OSi(CH₃)₃ [1295-94-9]

(51e) X = CH₂$\left(\text{OSi}\right)_3$SiCl

[1295-99-4]

(51f) X = CH₂OCH₂CH—CH₂
                         \O/

[1295-93-8]

Compounds of type (49a) have been used as precursors for a wide variety of monomers (50–51a–f) (48).

Ferrocenylene silylene polymers (52) can be prepared by copolymerizing (49a) and (50) with H_2PtCl_6 as initiator (48).

$$HSiR_2FcSiR_2[CH_2CH_2SiR_2FcSiR_2]_nH$$
$$(52)$$

Polymers (43d–f) (Table 11) were obtained by the reaction of 1,1′-bis(aminosilanyl)-ferrocenes (48) with disilanols (53a–c):

$$(48) + HOR'OH \longrightarrow (43d-f) \qquad (23)$$
$$(53)$$

R′
(53a) $Si(C_6H_5)_2$ [947-42-2]
(53b) $Si(CH_3)_2 C_6H_4Si(CH_3)_2$ [29036-28-0]
(53c) $Si(CH_3)_2 C_6H_4C_6H_4 Si(CH_3)_2$ [4852-15-7]

Preparation of polymers from monomers such as (54) proves unsatisfactory because intramolecular cyclization to form (55) takes place (eq. 24). If longer or more rigid functional groups are interposed, cyclization becomes negligible.

(54) [12320-47-7]   (55) [1272-10-2]   (24)

Polymers Containing Trialkyltin Esters

Polymers containing trialkyltin esters have been investigated extensively by several groups (50–52). The monomers listed in Table 12 (56–62) and cross-linking agents in Table 13 (64–70) were employed in preparing polymers with structures (71–75) (eqs. 25–29).

Table 12. Monomers Containing Trialkyltin Ester Groups

Structure	CAS Registry No.	Structure	CAS Registry No.
(56) $(Bu_3Sn)_2O$	[56-35-9]	(60) $Bu_3SnO_2CCH_2NH_2$	[45213-29-4]
(57) $Pr_3SnO_2CC(CH_3){=}CH_2$	[24154-35-2]	(61) $Bu_3SnO_2C(CH_2)_3NH_2$	[76082-06-9]
(58) $Bu_3SnO_2CCH{=}CH_2$	[13331-52-7]	(62) $Bu_3SnO_2C(CH_2)_5NH_2$	[76082-07-0]
(59) $Bu_3SnO_2CC(CH_3){=}CH_2$	[2155-70-6]	(63) $Bu_3SnO_2C(CH_2)_{10}NH_2$	[76082-08-1]

$$(\mathbf{57-59}) + CH_2{=}CRCO_2CH_3 \longrightarrow \left[\begin{array}{c} CHCH_2 \\ | \\ CO_2SnR_3 \end{array}\right]_m \left[\begin{array}{c} CHCH_2 \\ | \\ CO_2CH_3 \end{array}\right]_n \qquad (25)$$

$$(\mathbf{71})\, R = Pr\ [30444\text{-}62\text{-}3]$$

$$(\mathbf{57-59}) + (\mathbf{64}) \longrightarrow \left[\begin{array}{c} CHCH_2 \\ | \\ CO_2SnR_3 \end{array}\right]_m \left[\begin{array}{c} CHCH_2 \\ | \\ CO_2CH_2CH{-}CH_2 \\ \diagdown_O\diagup \end{array}\right]_n \qquad (26)$$

$$(\mathbf{72})\, R = Bu\ [65289\text{-}97\text{-}6]$$

$$(\mathbf{67\text{-}70}) + (\mathbf{72}) \longrightarrow \left[\begin{array}{c} CHCH_2 \\ | \\ CO_2SnR_3 \end{array}\right]_m \left[\begin{array}{c} CHCH_2 \\ | \\ CO_2CH_2CHCH_2 \\ | \quad | \\ HO \quad NH \\ | \\ polymer \\ chain \end{array}\right]_n \qquad (27)$$

$$(\mathbf{73})\, R = Bu$$

$$(\mathbf{60\text{-}63}) + (\mathbf{66}) \longrightarrow R_3SnO_2C(CH_2)_n NHCH_2CHCH_2O{-}\bigcirc{-}\underset{CH_3}{\overset{CH_3}{C}}{-}\bigcirc{-}OCH_2CH{-}CH_2 \qquad (28)$$
$$\underset{OH}{}$$

$$(\mathbf{74})\, R = Bu$$

$$(\mathbf{56}) + \left[\begin{array}{c} CHCH_2 \\ | \\ OCH_3 \end{array}\right]_m \left[\begin{array}{c} CH{-}CH \\ | \quad | \\ CO_2H\ CO_2H \end{array}\right]_n \longrightarrow \left[\begin{array}{c} CHCH_2 \\ | \\ OCH_3 \end{array}\right]_m \left[\begin{array}{c} CH{-}\quad CH \\ | \qquad | \\ CO_2SnBu_3\ CO_2SnBu_3 \end{array}\right]_n \qquad (29)$$

$$(\mathbf{75})$$

Table 13. Cross-linking Agents Employed in Preparing Polymers Containing Trialkyltin Esters

Structure	CAS Registry No.	Structure	CAS Registry No.
(64) CH$_2$—CHCH$_2$O$_2$CCH=CH$_2$ (epoxide)	[106-90-1]	(67) H$_2$NCH$_2$CH$_2$NHCH$_2$CH$_2$NH$_2$	[111-40-0]
(65) CH$_2$—CHCH$_2$O$_2$CC(CH$_3$)=CH$_2$ (epoxide)	[106-91-2]	(68) H$_2$NCH$_2$CH$_2$NHCH$_2$CH$_2$- NHCH$_2$CH$_2$NH$_2$	[112-24-3]
		(69) 1,3-(NH$_2$)$_2$C$_6$H$_4$	[108-45-2]
(66) $\left(CH_2{-}CHCH_2O{-}\bigcirc{-}\right)_2 C(CH_3)_2$ (diepoxide)	[1675-54-3]	(70) (4-H$_2$NC$_6$H$_4$)$_2$CH$_2$	[101-77-9]

Properties. Physical properties of polymers with structures (**71, 75**) are summarized in Table 14.

Improvement in the qualities of the films formed can be achieved by cross-linking (**72**). The properties of the films formed depend on the nature of the cross-linking agents and the molar ratio used (Table 15). Cross-linking was conducted at 60°C.

Table 14.   Physical Properties of Polymers (71, 75)

Struc-ture	Polymer	CAS Registry No.	$\overline{M}_n$	$\overline{M}_w$	$\overline{M}_w/\overline{M}_n$	Properties[a]
(71a)	1,1,1-copolymer of (57), (59), and methyl methacrylate		134,000	131,600	0.98	good solubility in organic solvents, good film-former[b]
(71b)	copolymer of (59) and methyl methacrylate		109,000	160,800	1.46	good film-former[b]
(75)			1,430,000			soluble only in ketone solvents, forms poor films, remains tacky upon drying (rejected)

[a] For ir, nmr, and Mössbauer spectra, see refs. 53–54.
[b] Suitable for incorporation into paint.

Oligomers formed from 1 mol of (60) to (63) and 4 mol of (66) typically have an equivalent weight of 340 g per epoxy equivalent and 2000 g per Sn equivalent (see Epoxy resins).

Preparation.   Polymers of type (71) can be formed by radical-initiated copolymerization of monomers (57–59) with methyl acrylate (MA) or methyl methacrylate (MMA). Polymers of type (72) are formed by copolymerizing (57–59) with (64) or (65). Reactivity ratios for these copolymerizations are shown in Table 16. Monomer (59) is more reactive than (58). Since the reactivity ratios differ, it is possible to create either blocks of Sn monomers separated by random copolymers; blocks of cross-linking agent separated by random copolymers; or purely random copolymers. Mixtures containing 46 mol % of (59) and 54 mol % of (65) showed azeotropic polymerization. Cross-linking studies in Table 15 were conducted at this composition. When aliphatic amines (67–68) were employed as cross-linking agents in benzene solution, curing took place in 24 h at 60°C. When aromatic amines (69) and (70) were used, the films were cured for 5 h at 150°C.

Trialkyltin amino esters were prepared by refluxing a trialkyltin oxide (56) with two moles of the appropriate amino acid. The diglycidylether of bisphenol-A (66) is added in a 4:1 molar ratio and the solvent evaporated *in vacuo* at 50°C. The resulting oligomers were mixed with the appropriate number of equivalents of (67) (3 equivalents of amine per mol) and allowed to cure at room temperature. Details for preparation of samples for thermomechanical testing are given in ref. 51.

Applications.   Polymers containing trialkyltin esters have been used primarily in antifouling paint to prevent growth of fungi and barnacles on ship bottoms and shore installations (55–59) (see Coatings, marine). The goal has been to produce a paint that would hydrolyze slowly to release minute amounts of trialkylstannanols, toxic to marine organisms, at a controlled rate sufficient to prevent fouling but not so great as to cause a hazard to other marine life. In general, a highly cross-linked polymer degrades more slowly. Polymers formed from (74) degrade more slowly if they are homopolymerized than if they are cross-linked with amines. After the tin compounds have been released, the remaining polymer is water soluble and slowly dissolves, exposing a fresh surface. Several polymers described here have prevented fouling for periods up to five years in static tests at Pearl Harbor, Hawaii. They are superior to

Table 15. Extractables and Tin Content of Cross-linked Azeotropic Copolymer of Tri-*n*-butyltin Methacrylate with Glycidyl Methacrylate

Cross-linking agent	Concentration of cross-linking agent[a] %	Extractables, %	Tin content after extraction, wt %	Tin content before extraction, wt %	Characteristics estimated by inspection (all are hard and nontacky)
(67)	75	18.84	16.59	21.01	tough
	85	19.13	16.26	20.81	tough
	100	20.75	14.89	20.62	very brittle
	110	22.61	14.48	20.43	very brittle
(68)	75	17.60	15.55	20.81	tough
	85	18.30	16.06	20.62	tough
	100	18.03	15.84	20.43	very brittle
	110	18.36	14.59	20.25	very brittle
(69)	75	12.67	19.55	20.62	tough
	85	11.83	18.96	20.43	very tough
	100	11.36	18.40	20.25	brittle, cracks on bending
	110	10.82	17.69	19.88	very brittle, cracks on bending
(70)	75	8.47	17.49	19.36	tough
	85	8.46	17.04	19.04	very tough
	100	7.63	16.86	18.71	brittle, cracks on bending
	110	7.59	16.99	18.40	very brittle, cracks on bending

[a] Percent of theoretical stoichiometric amount required.

Table 16. Reactivity Ratios for Copolymerization of (58) and (59)

Comonomer (M_2)		(58)	(59)
(64)	r_1	0.30	1.36
	r_2	1.41	0.37
(65)	r_1	0.34	0.75
	r_2	4.29	0.79
N-methylolacrylamide	r_1	0.98	4.23
	r_2	1.26	0.38

currently employed paints containing trialkyltin oxides since the rate of release of tin compounds and attendant marine toxicity can be controlled more carefully. Widespread commercial application of these materials is likely within the next five years.

Condensation Polymers

Interest in organometallic condensation polymers (60–61) arose from the observation that metal halide bonds possessing a high degree of covalent character, as in Group IVA, VA, and IVB metals, can behave in a manner similar to an organic acid chloride. Condensation polymers can, therefore, be prepared by copolymerizing a difunctional metal halide with a difunctional Lewis base that may contain a metallocene. The structures of typical polymers are shown below (76–83).

$$R_2MCl_2 + HO_2CZCO_2H \rightarrow \{MR_2O_2CZCO_2\}_n \tag{30}$$
$$\text{(76)}$$

$$R_2MCl_2 + RNHNH_2 \rightarrow \{MR_2NRNH\}_n \tag{31}$$
$$\text{(77)}$$

$$R_2MCl_2 + HSZSH \rightarrow \{MR_2SZS\}_n \tag{32}$$
$$\text{(78)}$$

$$R_2MCl_2 + H_2NZNH_2 \rightarrow \{MR_2NHZNH\}_n \tag{33}$$
$$\text{(79)}$$

$$R_2MCl_2 + HOZOH \rightarrow \{MR_2OZO\}_n \tag{34}$$
$$\text{(80)}$$

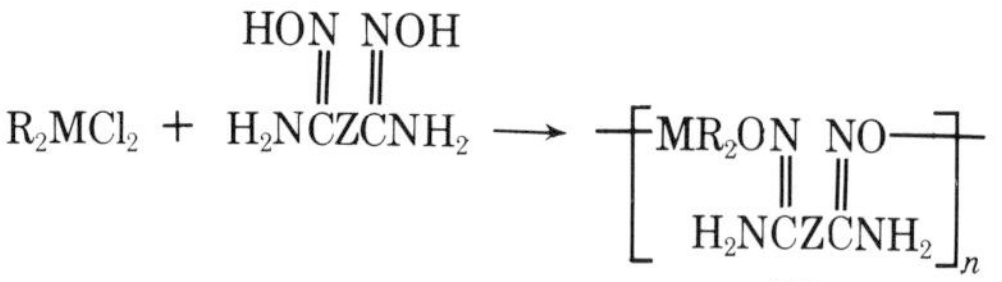

(81)

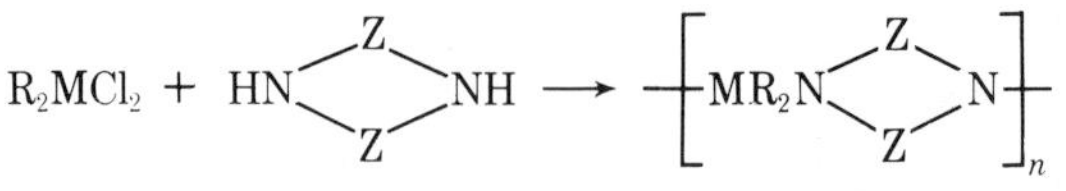

(82)

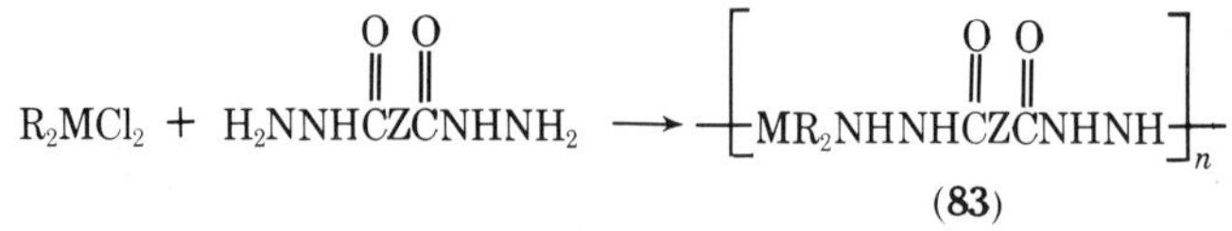

(83)

R = alkyl, aryl, or C_5H_5
Z = alkylene, arylene

Properties. Polymers of the Group IVA and VA metals are usually flaky, white powders with dp = 3–50. Infrared spectroscopy shows MOH groups as the predominant end groups. Solubility decreases in the order Si > Ge > Sn > Pb. Silicon polyesters (**76**) are generally soluble in polar solvents such as DMF, DMSO, and chloroform, whereas tin polyesters have only limited solubility in higher polar solvents such as trifluoroethanol, chloroethanol, and formic acid. Solubility may be enhanced by preparing mixed polymers containing two or more different metal-containing monomers or comonomers. Because of their limited solubility, the Group IVA and VA polymers do not form good films when cast from solution. Plasticizers, such as long-chain esters or alkyl phosphates, enhance the solubility of the polymers and greatly improve the quality of the films produced.

Thermal stability of the Group IVA and VA polymers is similar to that of the monomers. Thermogravimetric analysis in either nitrogen or air shows degradation commencing below 200°C and rising to 75–80% weight loss at 300°C (see Fig. 1). The polymers, like the monomers from which they are formed, are particularly susceptible to hydrolysis. Resistance to hydrolysis is greatest for polymers substituted with aryl groups, partly because they are more hydrophobic and partly because the aryl groups contribute electron density to the vacant d-orbitals of the metal atom which normally are the sites for attack by hydroxide ions. When wetting agents are added that reduce the hydrophobicity of the polymers, they hydrolyze rapidly to $R_2M(OH)_2$ and the comonomers.

Polymers of Group IVB metals have been prepared primarily from the metallocene dihalides. Titanocene polymers are yellowish orange, whereas zirconocene and hafnocene polymers are white or gray. In contrast to the halides of Group IVA metals, hydrolysis is slow and reversible. For $(C_5H_5)_2TiCl_2$ hydrolysis requires several days, whereas for $(C_5H_5)_2HfCl_2$ it occurs in 30 min. Polyethers (**80**) of titanocene form flexible films with molecular weights up to 10^5. Polyesters (**76**) also possess high molecular weights and a tendency to form fibers when films of the polymer are scratched.

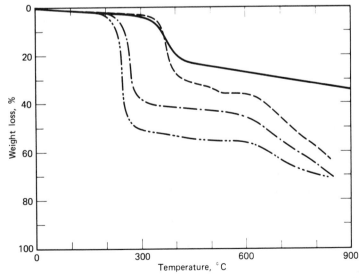

Figure 1. Thermogravimetric analysis of copolymers of $(C_5H_4CO_2)_2Co(III)PF_6$ with $(C_5H_5)_2TiCl_2$, ——; $(C_5H_5)_2ZrCl_2$, - - -; $(C_4H_9)_2SnCl_2$, ·····; and $(C_6H_5)_3SbCl_2$, ·—·— under N_2, heated at 10°C/min.

This fiber-forming tendency is indeed unusual and is not shared by zirconocene and hafnocene polymers.

Solubility of the polymers in organic solvents is extremely low. In an effort to enhance solubility, salts of cobalticinium-1,1'-dicarboxylate were copolymerized with titanocene dichloride. The resulting polymers had molecular weights up to 7×10^5 when PF_6^- was used as the counterion. Solubility remained low. When samples of the polymer were stirred with aqueous salt solutions, exchange of the PF_6^- ions occurred followed by rapid hydrolysis. When the bromide salt of cobalticinium-1,1'-dicarboxylate was used, polymers did not form. Thus the Group IVB polymers, like the IVA and VA polymers, remain stable toward hydrolysis only if they are hydrophobic.

Polymers of Group IVB show far greater thermal stability than those of Group IVA (see Fig. 1). The degradation of the titanocene polyesters has been studied by tga and mass spectroscopy (62). Decomposition begins at 350°C with loss of one or both cyclopentadienyl groups. Fragmentation occurs gradually (70% weight retention at 800°C and 60% at 1000°C) until complete degradation at 1300°C. Analysis of residues isolated after pyrolysis shows that substantial portions of the organic chain remain intact even at 800–1000°C. The thermal stability of the Ti polymers varies with the comonomer. It generally is higher with aryl diamines, dicarboxylates, and diols than with aliphatic compounds. The Zr and Hf polymers are less stable than the Ti polymers but still far more stable than the Group IVA and VA polymers.

Uranyl salts form polymers with dicarboxylic acids and dioximes (63–64). The carboxylate groups act as bidentate ligands in (84); whereas in (85) the oximes act as monodentate ligands.

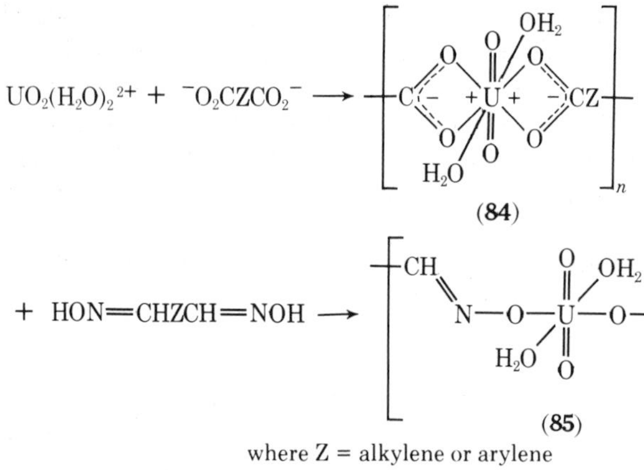

$$UO_2(H_2O)_2{}^{2+} + {}^-O_2CZCO_2{}^- \longrightarrow \quad (38)$$

(84)

$$UO_2(H_2O)_2{}^{2+} + HON{=}CHZCH{=}NOH \longrightarrow \quad (39)$$

(85)

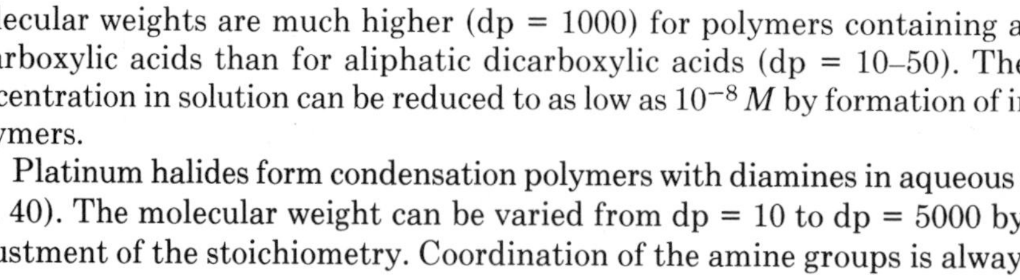

where Z = alkylene or arylene

Molecular weights are much higher (dp = 1000) for polymers containing aromatic dicarboxylic acids than for aliphatic dicarboxylic acids (dp = 10–50). The uranyl concentration in solution can be reduced to as low as $10^{-8}\,M$ by formation of insoluble polymers.

Platinum halides form condensation polymers with diamines in aqueous solution (eq. 40). The molecular weight can be varied from dp = 10 to dp = 5000 by careful adjustment of the stoichiometry. Coordination of the amine groups is always cis.

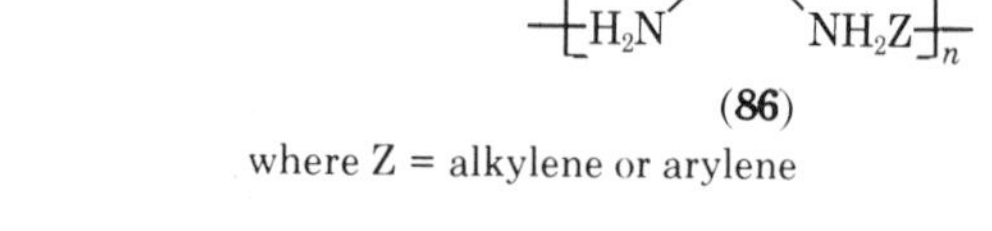

$$PtCl_4{}^{2-} + H_2NZNH_2 \longrightarrow \quad (40)$$

(86)

where Z = alkylene or arylene

The polymers (86) are insoluble in water but are soluble readily in DMSO and, once dissolved, can be diluted with water.

Preparation. Attempts to prepare condensation polymers in aqueous solution have led to cyclization or degradation of the starting materials and production of low molecular weight oligomers. Under equilibrium conditions, increase in entropy favors formation of small cyclic units rather than long chains. These problems may be avoided by interfacial polymerization, in which the polymer is formed under nonequilibrium conditions. In a typical procedure, the metal halide is dissolved in a nonpolar solvent such as benzene or hexane and the comonomer used as a pure liquid or dissolved in water, methanol, or a polar aprotic solvent (acetonitrile, nitrobenzene, or sulfolane). If the metal halide is particularly susceptible to hydrolysis, both phases must be an-hydrous. Stirring at $(1-5) \times 10^4$ rpm permits rapid mixing and formation of the polymer at the interface. The reaction is usually complete in 15–120 s; further stirring may lead to degradation.

Polymer yields are dependent on the initial concentrations of the reagents, the stirring rate, the reaction time and in aqueous solutions, the pH and the number of moles of excess base added (63). They are higher with aryl groups and bulky alkyl groups which shelter the sensitive bonds from cleavage. In general, maximum yields are obtained with monomer concentrations of 0.01–0.05 M; a twofold excess of base, either NaOH or triethylamine, which gives a solution with pH 9–11; reaction time of 30–120 s, and stirring rates of $(1-2.5) \times 10^4$ rpm.

Polymers of the Group IVB metallocene dihalides can be prepared by dissolving the metallocene dihalides in basic aqueous solutions at pH 9–11 and stirring with an organic phase containing an alcohol or amine, or by dissolving the metallocene dihalide in chloroform or nitrobenzene and stirring with an aqueous solution of a dicarboxylate salt. Polyamines and polyesters usually form in the organic phase, whereas polyoximes, polythiols, and polyethers form at the interface. When base is needed, the order of efficiency is $(C_2H_5)_3N > OH^- > (CH_3)_3N$. Aqueous buffers offer no advantage over a simple 2:1 base-to-metal ratio. Stirring rates and times are similar to those used for Group IVA and VA polymers. In all cases, the polymers are insoluble and can be col-lected by filtration, washed, and dried *in vacuo* at 25–50°C. So far, these polymers have been synthesized only on a 0.5–5 g scale. Large-scale processes will have to be developed for materials with promising applications.

Applications. Development of condensation polymers is still in an early stage. These materials represent some of the first attempts to incorporate elements such as Sb, Pb, Bi, or U into polymers. A wide range of applications has been proposed (61) but only a few have been tested. The following list gives some examples:

Bactericides and Fungicides. Polymers containing Sn are used as antifouling paint for ship bottoms. Polymers containing As and Sb may also have useful appli-cations in this area.

Catalysts. Titanocene and zirconocene are extremely useful Ziegler-Natta catalysts (qv), and polymers containing them may also prove useful.

Semiconductors. Measurements of the electrical conductivity, magnetic sus-ceptibility, and dielectric constants of these polymers are currently under way. Several have shown useful properties as semiconductors (qv).

Polymers with High Thermal Stability. If their mechanical properties can be improved, titanocene polyesters will be good candidates for applications where heat stability is required.

Recovery of Uranium from Wastewater. The effectiveness of dicarboxylate ions in precipitating uranyl ions from aqueous solution makes this a promising technique for recovery of uranium from wastewater from reactors or mines. The uranium can be recovered and the effluent rendered nontoxic.

Controlled-Release Chemotherapeutic Agents. The cis-platinumdiamine polymers have been shown to be effective as cancer chemotherapeutic agents in mice, with longer periods of effectiveness and lower toxicity than monomeric cis platinum complexes (64). Investigations of pharmacological properties of these materials are in progress (see Chemotherapeutics, antimitotic; Pharmaceuticals, controlled-release).

Polymeric Dyes. Dyes such as phenolphthalein that possess two basic functional groups may be copolymerized with metal halides to form a wide variety of polymers with high stability (65). The absorbance and fluorescence of these materials is being investigated with the hope that they may be utilized in lasers. Many of these polymeric dyes blend with polymers such as polyethylene to produce materials with better color stability than monomeric dyes. When formulated with latex resins, they are used as paint pigments. Incorporation of titanocene improves the stability toward uv radiation.

Coordination Polymers

Coordination of a metal atom and an organic ligand changes the properties of both in a variety of ways. Frequently, the organic ligand is rendered more stable toward hydrolysis, oxidation, chemical reagents, or high temperatures. In an effort to obtain polymers with both extremely high thermal stability and good thermomechanical properties, a number of investigators have studied polymeric coordination complexes (66–70). Considerations employed in designing coordination polymers have been summarized as follows (69):

Little plasticity can be expected from the metal ion or its immediate environment; thus plasticity must arise from the organic linkages.

The metal ion can stabilize only ligands in its immediate vicinity; thus organic links must be short and strong.

Thermal, oxidative, and hydrolytic stability are not related directly; the polymer must, therefore, be designed specifically for the properties desired.

Coordinate bonds have at least partial ionic character; hence, they rearrange more readily than ordinary covalent bonds. Partial charges on metal and ligand exert interchain attractive forces that tend to give rigidity. For plastic polymers, bonds must be almost purely covalent.

Ionic polymers are much less plastic than neutral polymers.

Polymeric cations with small anions tend to have lower thermal stability since the anions can attack the metal to displace the polymeric ligand. Anions such as NO_3^- or ClO_4^- can act as oxidizing agents at elevated temperatures.

Chelation enhances stability.

The coordination number and stereochemistry of the metal ion determine whether polymers are linear, planar, or three-dimensional.

Reactants must be pure and present in exact stoichiometric amounts.

If a solvent is used for the polymerization, it must not be capable of strong coordination to the metal ion.

Polymers from Bis Chelating Agents. The most common arrangement for preparing linear polymers are a bisbidentate ligand and a metal with coordination number four, such as Be, Cu, Ni, or Zn. Beryllium offers certain advantages since it is invariably tetracoordinate and the Be—O bonds are highly covalent. Ligands may be β-diketones, 1,2-dioximes, diamines, amino acids, hydroxyquinolines, hydroxy-Schiff bases, thiopicolinimides, and others. Structures of typical polymers are shown below (87–92):

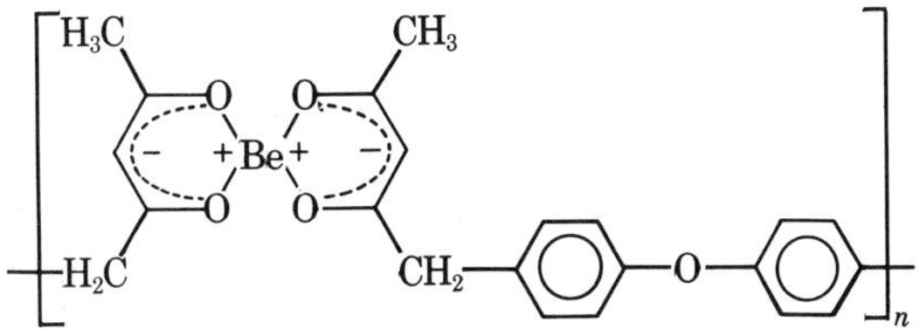

(**87**) [76082-19-4]

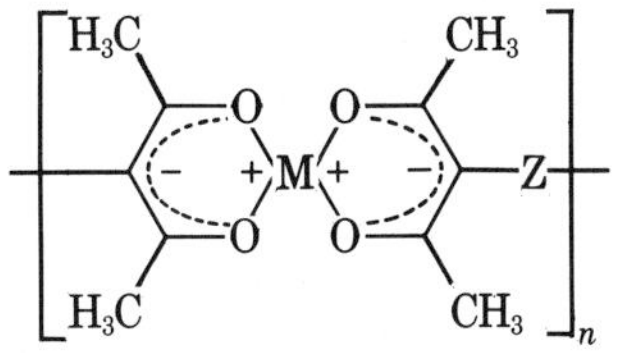

(**88**)

M = Mg, Ni, Co, Zn, Cd, Be

Z = $(CH_2)_{6-10}$

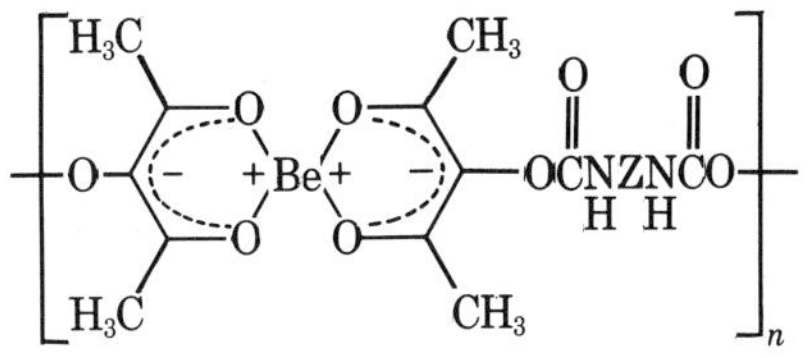

(**89**)

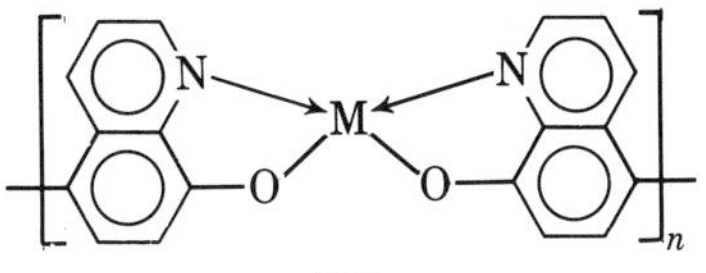

(**90**)

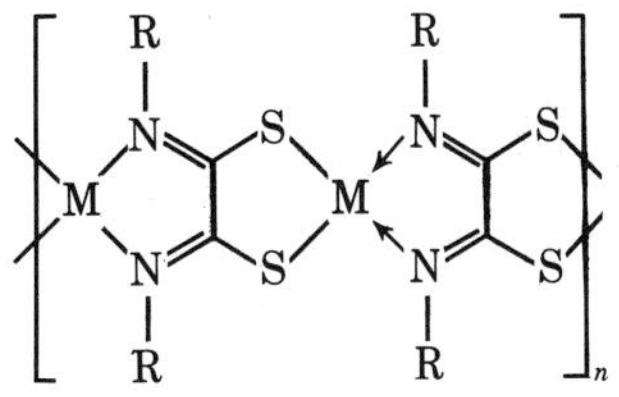

(**91a**) R = H
(**91b**) R = CH_3
(**91c**) R = $C_6H_5CH_2$
(**91d**) R = $n\text{-}C_{18}H_{37}$
(**91e**) R = CH_2CH_2OH
(**91f**) R = $(CH_2)_3N(CH_3)_2$

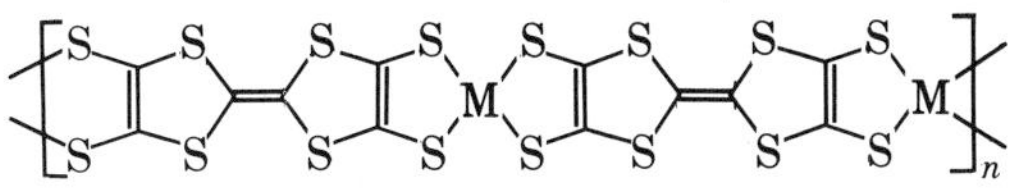

(**92**) M = Ni, Cu, Pd, Fe

Properties. Polymer (87) is soluble in organic solvents and can be prepared with molecular weights up to 1.26×10^5 by thermal polymerization of oligomers at 110–140°C. At slightly higher temperatures depolymerization occurs (71). Polymers with structure (88) are degradated by tga at 225–350°C with stability in the order Mg > Ni > Co > Cu > Zn > Cd, when Z = covalent bond (72). When M = Be and Z = $(CH_2)_{6–10}$ flexible polymers with good solubility in organic solvents, $T_g \approx 25°C$, tensile modulus (25°C) of 841 MPa (122,000 psi), tensile strength of 18.7 MPa (2710 psi) and melt index (190°C) of 1–20 g/min were obtained (73). However, these materials depolymerized above 200°C. Polymer (89) differs since it is formed by linking preformed complex units. Molecular weights and thermal stabilities comparable to the other Be polymers were obtained.

Hydroxyquinoline polymers (90) containing Ni, Cu, or other metals can be obtained by heating the metal acetylacetonates with the bis(hydroxyquinoline) ligands at 290°C (69). They are highly colored, insoluble, rigid materials that are thermally stable up to at least 500°C.

Rubeanate polymers (91) may be prepared from rubeanic acid and the appropriate metal ions, most commonly nickel, in aqueous solution (74–75). The polymers obtained are oligomeric, dp = 3–10, and highly insoluble. Increasing the size of R increases solubility. Compound (91d) dissolves readily in hot nitrobenzene and other high boiling organic solvents. The softening point drops from 400°C for (91a) to 200°C for (91d). Compound (91c) can be acetylated with acetyl chloride or cross-linked with bis(acid chlorides). Polymers (91a–f) have been used as color sources in duplicating processes, as molecular-oriented, dichroic stains in polarizing films, and as dyes for plastics.

Polymers with structure (92) currently are being investigated in the search for long-chain donor-acceptor complexes with appreciable electrical conductivity (76). Stacked polymers containing tetrathiafulvalene units alternating with electron acceptors such as tetracyanoquinodimethane (TCNQ) are well-known one-dimensional semiconductors. The materials produced so far, containing Ni, Cu, Pd, or Fe as the central metal, are low molecular weight oligomers insoluble in most organic solvents. Conductivity ranges from 30/(Ω·cm) for Ni to 1×10^{-5}/(Ω·cm) for Fe (see Polymers, conducting; Semiconductors, organic).

Polymeric Metal Phosphinates

Polymeric metal phosphinates have been studied extensively (77). Materials have been prepared with one, two, or three bridging phosphinate ligands containing metals such as Al, Be, Co, Cr, Ni, Ti, and Zn.

Properties. The metal phosphinate polymers are mostly powders, and soluble in organic solvents such as benzene and chloroform. Film-forming characteristics are marginal but can be improved greatly with plasticizers such as tricresyl phosphate or chlorinated biphenyls. Thermal stability is good, and many polymers are stable up to 450°C. Degradation commences with loss of alkyl or aryl groups as free radicals which may attack other phosphinate groups. There is little difference in the thermal stability in air and under nitrogen.

The thermomechanical properties of the Cr and Zn polymers can be improved greatly by copolymerizing low molecular weight oligomers with alkyl or aryl phosphinates. The resulting polymers have greater solubility and flexibility and still retain their high thermal stability (78).

Singly bridged:

$$(93)$$

Doubly bridged:

$$(94)$$

Triply bridged:

$$(95)$$

R = alkyl

$$Zn(C_2H_3O_2)_2 + 2(C_8H_{17})_2P(O)OH \xrightarrow{C_2H_5OH} 1/x\{Zn[OP(C_8H_{17})_2O]_2\}_x \qquad (41)$$

$$CoCl_2 + 2\ NaOP \xrightarrow{H_2O} 1/x[Co(OP)_2]_x + 2\ NaCl \qquad (42)$$

Preparation. Polymeric metal phosphinates may be prepared from metal salts and dialkyl- or diarylphosphinic acids either in solution or in the melt phase. Yields and molecular weights are nearly the same for both methods. Typical syntheses are shown in equations 41–42. The Cr(III) polymers are obtained by oxidizing Cr(II) polymers in the presence of water. The resulting $[Cr(H_2O)(OH)(OP(R)(R')O)_2]_n$ is thermally polymerized or copolymerized with alkyl or aryl phosphinic acids (78).

Applications. Metal phosphinate polymers have been used primarily as additives rather than pure materials. Chromium(III) phosphinates thicken silicones for form greases and improve the properties of the silicones at extremely high pressures. Both Cr and Ti phosphinates impart antistatic properties to plastics. Other applications are being investigated (see Inorganic high polymers; Phosphorus compounds).

Phthalocyanine Polymers

Phthalocyanine polymers (**97**) may be prepared from bisaryl-1,2-dinitriles (**96**) and metal salts, either in high boiling solvents such as nitrobenzene or in the molten state (eq. 43) (79). The resulting polymers possess a sheet structure and are highly colored, brittle, insoluble materials. They are thermally stable at 500–700°C either in air or in nitrogen. Many of these materials have potential uses as semiconductors, with conductivities of 10^{-9} to $10^{-5}/(\Omega \cdot cm)$.

Electrical conductivity of phthalocyanines has been enhanced greatly by preparing polymers containing Si, Ge, or Sn linked face to face with oxygen bridges (80). Partial

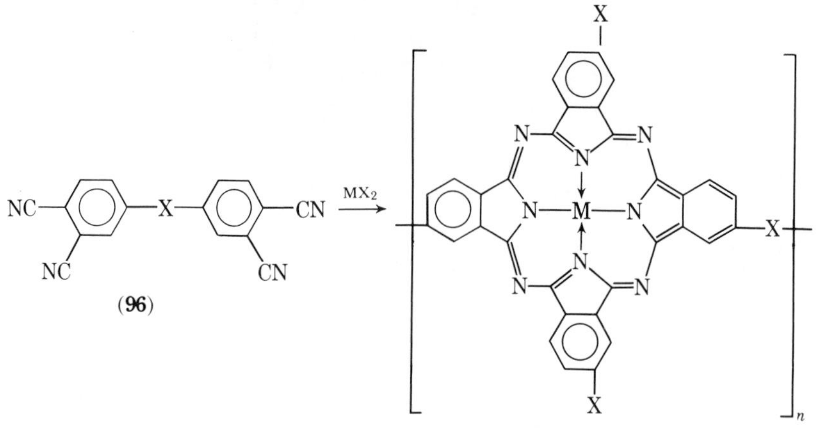

$$(43)$$

(96)   (97)

X = covalently bonded $(CH_2)_n$, CO, SO_2, O

oxidation with Br_2 or I_2 produces a mixed-valence material containing stacks of phthalocyanine units with I_3^- or Br_3^- ions intercalated in the channels between them. Conductivity of the Si and Ge polymers is in the range $10^{-2}-1/(\Omega\cdot cm)$ along the axis. Single crystals of these materials should have even greater conductivity along the axis but very little conductivity crosswise. Thus, they function as one-dimensional conductors analogous to $(SN)_x$, tetracyanoplatinates, and other inorganic polymers. The Sn polymer acts as a semiconductor (conductivity $2 \times 10^{-4}/(\Omega\cdot cm)$ because the distance between the phthalocyanine units is too great for good conductivity (see Phthalocyanine compounds).

Incorporation of Metals into Polymer Films

A wide variety of cross-linked or cross-linkable polymeric films have been prepared which contain coordinated metal atoms (80–82). The metal is usually incorporated during formation of the film:

Polymer (98) is a typical polymeric Schiff base formed by condensation of dialdehyde and a diamine (81). Polymerization in the presence of metal salts at elevated temperatures leads to a highly cross-linked, thermally stable film containing uniformly dispersed metal. Even greater thermal stability can be obtained with polyimide films

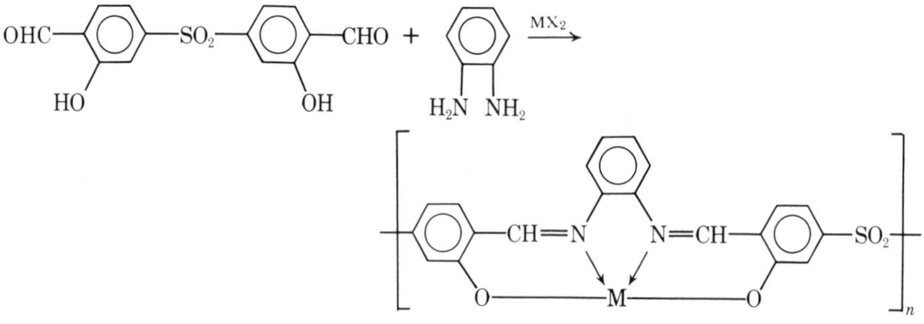

$$(44)$$

(98)

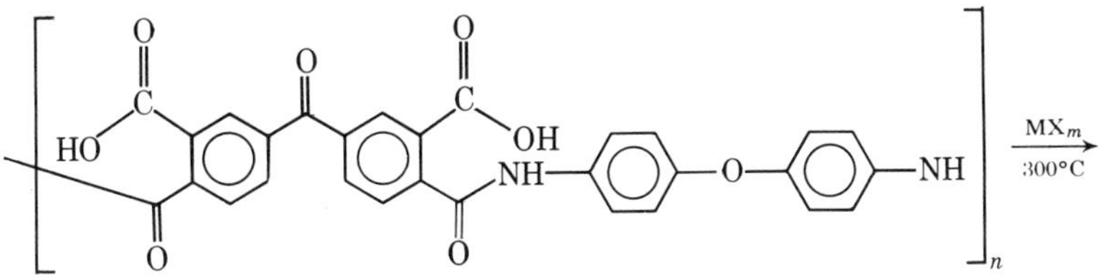

(**99**) polyamic acid

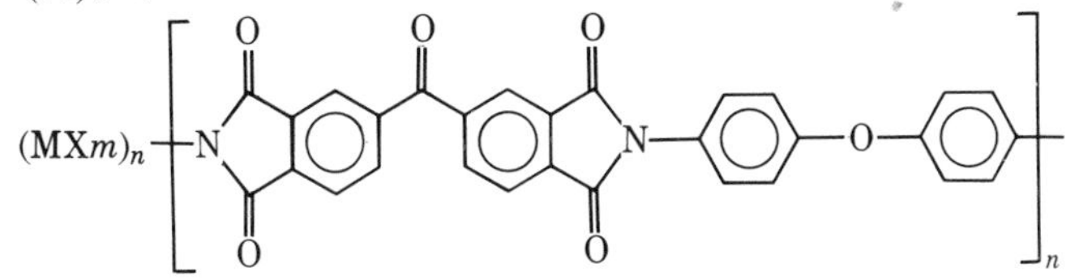

(**100**) polyimide film

$MXm = AuI_3, Ag, H\,AuCl_4, Pd, and\ PdCl_4$

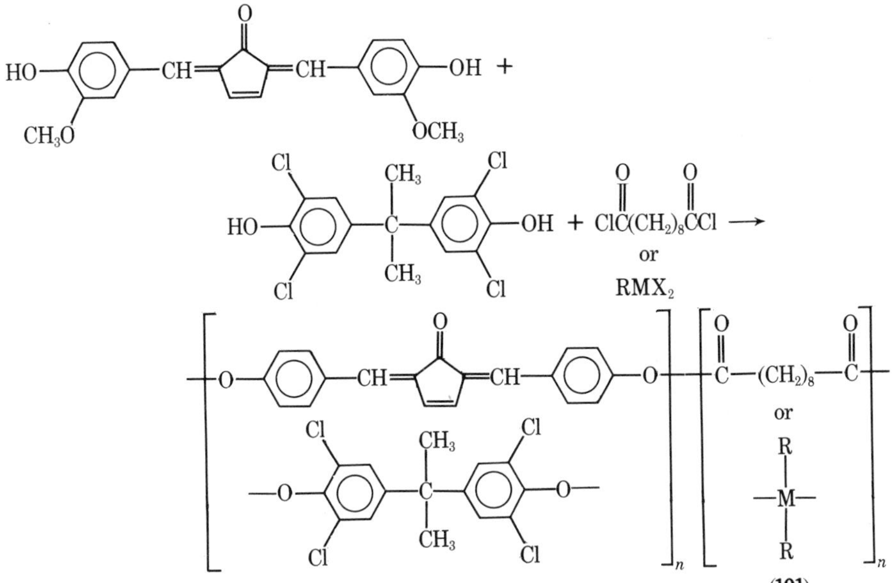

(45)

R = alkyl

(46)

(**101**)

(**100**) prepared by high temperature cyclization of polyamic acid (**99**) (**82**) (see Poly-imides). Polymer (**101**) can be formed at room temperature and is soluble in organic solvents (**83**). Exposure of films of this polymer to light produces a highly cross-linked insoluble coating. A total of 17 different metals have been incorporated. Many of the films have semiconductive properties. Printed circuits may be produced by exposing part of the film to light, then dissolving the unexposed polymer.

BIBLIOGRAPHY

1. C. U. Pittman, Jr., in E. Becker and M. Tsutsui, eds., *Organometallic Reactions*, Vol. 6, Plenum Press, New York, 1977, pp. 1–62.

2. C. U. Pittman, Jr., *J. Paint Technol.* **43**, 29 (1971).
3. J. E. Sheats and T. C. Willis, *Org. Coat. Plast. Chem.* **41**, 33 (1979).
4. C. U. Pittman, Jr., Y. Sasaki, and P. Grube, *J. Macromol. Sci. Chem.* **A8**, 923 (1974).
5. T. Kunitake, T. Nakashima, and C. Asao, *Macromol. Chem.* **146**, 79 (1971).
6. D. C. Van Landuyt, *J. Polym. Sci. B.* **10**, 125 (1972).
7. V. V. Korshak and S. L. Sosin in C. E. Carraher, Jr., J. E. Sheats, and C. U. Pittman, Jr., eds., *Organometallic Polymers*, Academic Press, Inc., New York, 1978, pp. 25–38.
8. C. U. Pittman, Jr., and B. Suryanarayanan, *J. Am. Chem. Soc.* **96**, 7916 (1974).
9. C. U. Pittman, Jr., T. D. Rounsefell, E. A. Lewis, J. E. Sheats, B. H. Edwards, M. D. Rausch, and E. A. Mintz, *Macromolecules* **11**, 560 (1978).
10. M. D. Rausch, G. A. Moser, E. J. Zaiko, and A. L. Lipman, Jr., *J. Organometal. Chem.* **23**, 185 (1970).
11. C. U. Pittman, Jr., P. L. Grube, O. E. Ayers, S. P. McManus, Jr., M. D. Rausch, and G. A. Moser, *J. Polym. Sci. A-1* **10**, 379 (1972).
12. W. P. Hart, D. W. Macomber, and M. D. Rausch, *J. Am. Chem. Soc.* **102**, 1196 (1980).
13. M. Anderson, A. D. H. Clague, L. P. Blaauw, and P. A. Couperus, *J. Organometal. Chem.* **56**, 307 (1973).
14. N. Fujita and A. Sonogashira, *J. Polym. Sci. Chem. Ed.* **12**, 2845 (1974).
15. C. U. Pittman, Jr., J. C. Lai, and D. P. Vanderpool, *Macromolecules* **3**, 105 (1970).
16. C. U. Pittman, Jr., O. E. Ayers, and S. P. McManus, *J. Macromol. Sci. Chem.* **A7**, 1563 (1973).
17. R. Ralea, C. Ungereanu, and I. Maxim, *Rev. Roum. Chem.* **12**, 523 (1967).
18. V. V. Korshak, A. M. Sladkov, L. K. Luneva, and A. S. Girshovich, *Vysokomol. Soedin.* **5**, 1284 (1963).
19. Y. Minoru and Y. Sakanaka, *J. Polym. Sci. A-1* **4**, 2757 (1966); **5**, 2927 (1967); **7**, 3287 (1969).
20. C. U. Pittman, Jr., and R. L. Voges, *Macromol. Synth.* **4**, 175 (1972).
21. C. U. Pittman, Jr., O. E. Ayers, and S. P. McManus, *Macromolecules* **7**, 737 (1974).
22. C. U. Pittman, Jr., and C. Lin, *J. Polym. Sci. Polym. Chem. Ed.* **17**, 271 (1979).
23. Y. Sasaki, L. L. Walker, E. L. Hurst, and C. U. Pittman, Jr., *J. Polym. Sci. Polym. Chem. Ed.* **11**, 1213 (1973).
24. Y. Ozari, J. E. Sheats, T. N. Williams, Jr., and C. U. Pittman, Jr., in ref. 7, pp. 53–66.
25. C. E. Carraher, H. M. Molloy, M. L. Taylor, T. O. Tiernan, R. O. Yelton, J. A. Schroeder, and M. R. Bogdan, *Org. Coat. Plast. Chem.* **41**(2), 197 (1979).
26. C. U. Pittman, Jr., and A. Hirao, *J. Polym. Sci. Polym. Chem. Ed.* **15**, 1677 (1977).
27. C. U. Pittman, Jr., and Y. Sasaki, *Chem. Lett.*, 383 (1975).
28. G. F. Hayes and M. H. George in ref. 7, pp. 13–24; *J. Polym. Sci. Polym. Chem. Ed.* **14**, 475 (1976).
29. F. W. Billmeyer, *Textbook of Polymer Science*, 2nd ed., Wiley-Interscience, New York, 1971, p. 347.
30. C. U. Pittman, Jr., and A. Hirao, *J. Polym. Sci. Polymer Chem. Ed.* **16**, 1197 (1978).
31. D. C. van Landuyt, *J. Polym. Sci. B.* **10**, 125 (1972).
32. K. P. C. Vollhardt, *Acc. Chem. Res.* **10**, 1 (1977).
33. R. L. Hilliard, III, and K. P. C. Vollhardt, *J. Am. Chem. Soc.* **99**, 4058 (1977).
34. R. L. Funk and K. P. C. Vollhardt, *J. Am. Chem. Soc.* **101**, 215 (1979).
35. P. Perkins and K. P. C. Vollhardt, *J. Am. Chem. Soc.* **101**, 3985 (1979).
36. C. U. Pittman, Jr., and G. V. Marlin, *J. Polym. Sci. Chem.* **11**, 2753 (1973).
37. C. U. Pittman, Jr., and P. L. Grube, *J. Polym. Sci. A-1* **9**, 3175 (1971).
38. C. U. Pittman, Jr., R. L. Voges, and J. Elder, *Macromolecules* **4**, 302 (1971).
39. A. Gal, M. Cais, and D. H. Kohn, *J. Polym. Sci. A-1* **9**, 1833 (1971).
40. U.S. Pats. 3,560,429 (Feb. 2, 1971); 3,640,963 (Feb. 8, 1972), N. Bilow and H. Rosenberg (to U.S. Dept. of the Air Force).
41. E. W. Neuse and L. Bednarik, *Org. Coat. Plast. Chem.* **41**(2), 158 (1979).
42. L. Bednarik, *The Synthesis of Polyferrocenylene and Derived Oxidation Products Exhibiting Mixed Valency Characteristics*, Ph.D. thesis, University of the Witwatersrand, Johannesburg, S. Afr., 1978.
43. G. M. Brown, T. J. Meyer, D. O. Cowan, C. LeVanda, F. Kaufman, P. V. Roling, and M. D. Rausch, *Inorg. Chem.* **14**, 506 (1975).
44. H. Rosenberg, Materials Research Center, Wright-Patterson AFB, Dayton, Ohio, private communication, 1976.

45. A. H. Gerber and E. F. McInerney, eds., *Survey of Inorganic Polymers, National Aviation and Space Administration, NASA CR-159563 HRI-396 Contract NAS 3-21369*, Horizon Research, Cleveland, Ohio, June 1979, 167–198.

46. C. U. Pittman, Jr., W. J. Patterson, and S. P. McManus, *J. Polym. Sci. Polym. Chem. Ed.* **12,** 837 (1974); **14,** 1715 (1976).

47. C. U. Pittman, Jr., W. J. Patterson, and S. P. McManus, *J. Polym. Sci. A-1* **9,** 3187 (1971).

48. G. Greber and M. L. Hallensleben, *Makromol. Chem.* **83,** 148 (1965); **92,** 137 (1966); **104,** 77 (1967).

49. R. L. Schaaf, P. T. Kan, C. T. Lenk, and E. P. Deck, *J. Org. Chem.* **25,** 1986 (1960).

50. R. V. Subramanian, B. K. Garg and J. Corredor in ref. 7, pp. 181–194.

51. R. V. Subramanian, R. S. Williams, and K. N. Somasekharan, *Org. Coat. Plast. Chem.* **41**(2), 38 (1979).

52. W. L. Yeager and V. J. Castelli in ref. 7, pp. 175–180.

53. J. F. Hoffman, K. C. Kappel, L. M. Frenzel, and M. L. Good in ref. 7, pp. 195–205.

54. E. J. O'Brien, C. P. Monaghan, and M. L. Good in ref. 7, pp. 207–218.

55. E. J. Dyckman, J. A. Montemarano, and E. C. Fischer, *Naval Eng. J.* **85**(6), 33 (1973); **86**(2), 59 (1974).

56. J. A. Montemarano and E. J. Dyckman, *J. Paint Technol.* **47,** 59 (1975).

57. J. A. Montemarano and E. J. Dyckman, *U.S.N.T.I.S., A.D. Rep. No. 903592/4GA* (1972); *Chem. Abstr.* **83,** 81342 (1975).

58. J. A. Montemarano, S. A. Cohen, *U.S.N.T.I.S., A.D. Rep. AD–A 020153* (1976); *Chem. Abstr.* **85,** 22463 (1976).

59. E. J. Dyckman and J. A. Montemarano, *Am. Paint J.* **58**(5), 66, 70 (1973).

60. C. E. Carraher, Jr., in F. Millich and C. E. Carraher, Jr., eds., *Interfacial Syntheses*, Marcel Dekker, Inc., New York, 1977, pp. 367–416.

61. C. E. Carraher, Jr., in ref. 7, pp. 79–86.

62. C. E. Carraher, Jr., H. M. Molloy, M. L. Taylor, T. O. Tiernan, R. O. Yelton, J. A. Schroeder, and M. R. Bogdan, *Org. Coat. Plast. Chem.* **41**(2), 197 (1979).

63. C. E. Carraher, Jr., G. F. Peterson, J. E. Sheats, and T. Kirsch, *J. Macromol. Sci. Chem.* **A8,** 1009 (1974).

64. C. E. Carraher, Jr., *Org. Coat. Plast. Chem.* **42**(1), 427 (1980).

65. C. E. Carraher, Jr., R. A. Schwartz, M. Schwartz, and J. A. Schroeder, *Org. Coat. Plast. Chem.* **42**(1), 23 (1980).

66. B. P. Block, "Coordination Polymers" in N. M. Bikales, ed., *Encyclopedia of Polymer Science and Technology*, Vol. 4, John Wiley & Sons, Inc., New York, 1966, pp. 150–165.

67. R. J. Cotter and M. Matzner, *Ring Forming Polymerizations—Part A: Carbocyclic and Metallorganic Rings*, Academic Press, Inc., New York, 1969.

68. J. C. Bailar, Jr., *Inorganic Polymers, Special Publication No. 15*, The Chemical Society, London, Eng., 1961, p. 51.

69. J. C. Bailar, Jr., in W. L. Jolly, ed., *Preparative Inorganic Reactions*, Vol. 1, Interscience Publishers, New York, 1964, pp. 1–27.

70. J. C. Bailar, Jr., in ref. 7, pp. 313–321.

71. V. V. Korshak and S. V. Vinogradova, *Dokl. Akad. Nauk SSSR* **138,** 1353 (1961).

72. R. G. Charles, *J. Polym. Sci. A-1*, **2,** 267 (1963); *J. Phys. Chem.* **64,** 1747 (1960).

73. R. Kluiber and J. Lewis, *J. Am. Chem. Soc.* **82,** 5777 (1960).

74. R. N. Hurd, G. De La Mater, G. D. McElheny, R. J. Turner, and V. H. Wallingford, *J. Org. Chem.* **26,** 3980 (1961).

75. R. N. Hurd, G. De La Mater, G. C. McElheny, and L. V. Peiffer, *J. Am. Chem. Soc.* **82,** 4455 (1960).

76. E. M. Engler, N. Martinez-Rivera, and R. R. Schumaker, *Org. Coat. Plast. Chem.* **41,** 52 (1979).

77. B. P. Block, *Inorg. Macromol. Chem. Rev.* **1**(2), 115 (1970).

78. J. K. Gillham, *J. Appl. Polym. Sci.* **16,** 917 (1972).

79. A. A. Berlin and A. I. Sherle, *Inorg. Macromol. Chem. Rev.* **1**(3), 235 (1971).

80. K. F. Schock, B. R. Kundalkar, and T. J. Marks, *Org. Coat. Plast. Chem.* **41,** 127 (1979).

81. C. S. Marvel and N. Tarkoy, *J. Am. Chem. Soc.* **79,** 6000 (1957).

82. V. C. Carver, T. A. Furtsch, and L. T. Taylor, *Org. Coat. Plast. Chem.* **41,** 150 (1979).

83. D. G. Borden in ref. 7, pp. 115–127.

General References

References 7 and 45 are General References.
Org. Coat. Plast. Chem. **33,** (1971); **39,** (1977); **41,** (1979).
E. W. Neuse and H. Rosenberg, *Metallocene Polymers*, Marcel Dekker, Inc., New York, 1970, 145 pp.

JOHN E. SHEATS
Rider College

METALDEHYDE, $(C_2H_4O)_n$.   See Acetaldehyde.

METAL FIBERS

A wide range of products, eg, textile products, paper, floor covering, insulation products, and many composites, depend on fiber manipulation technology for their economical manufacture and on the inherent fiber characteristics for their end-use properties. However, not many products have embodied the use of metal fibers. Until fairly recently, metals and alloys generally have been available at reasonable cost only in the form of wire of diameters significantly greater than natural fibers. The large diameter and the inherently high elastic modulus of metals and alloys has limited the use of many well-developed and economical fiber-manipulation processes; exceptions are woven wire products for which special looms and knitting equipment have been developed. Also, processes for producing metals and alloys with acceptable properties and dimensions have been developed slowly and are incomparable in scale and volume to those used in the synthetic organic fiber or the glass-fiber industry (see Fibers, chemical; Glass). The most significant reason for the slow development of economical processes is the difficulty of forming metal filament directly from the liquid phase. This arises from the unusually high ratio of surface tension to viscosity, which results in liquid-jet instability, and from the inability to attenuate significantly the stream outside the spinnerette as is the usual practice in the synthetic-fiber industry. Despite these shortcomings, a metal-fiber industry is emerging and will have a large impact on product and process design and performance.

Properties

Fiber. Fiber properties result from a combination of the material properties, the effect of processing the material into fiber form and, in some cases, the geometry of the final fiber. The mechanical, physical, and chemical characteristics of metals and alloys are high modulus and high strength, high density, high hardness, good electrical and thermal conductivity, can be magnetic, high temperature stability, good oxidation resistance, and corrosion resistance to varied chemical groups.

It may be that only one of the characteristics determines the choice of a particular metal or alloy for a specific application. For example, where a high specific modulus, ie, modulus-to-weight ratio, is desirable, beryllium with its very low density but high modulus may be the preferred material. In the generic material group, a metal or alloy usually can be formed and have properties that are functionally and economically acceptable for a particular application. However, for metal fibers, the choice of material is restricted since only a small number of metals and alloys are available in fiber form, particularly those with diameters <50 μm.

Most of the metal-fiber industry is directed to markets where one or more of the following properties predominate: strength–stiffness, corrosion resistance, high temperature oxidation resistance, and electrical conduction. These markets generally are satisfied by the following materials: carbon and low alloy steels, stainless steels, iron, nickel, and cobalt-based superalloys.

Property modifications can be made. For example, commercial stainless steels made according to standard melting practice contain small quantities of impurities which manifest themselves in the solidified ingot as small, nonmetallic inclusions. These inclusions are relatively unimportant in the majority of applications of the material because of their small size. However, in the case of stainless steel fiber, the diameter of the fiber can be of the same order as that of the inclusion and, consequently, the inclusion can significantly affect the strength of the fiber. The apparent strength of the fiber tends to decrease and the probability of finding an inclusion in the test sample increases with an increase in the length of the test fixture. The breaking strength of 12-μm fibers processed from five different heats of steel are plotted in Figure 1 as a function of the length of the fiber in the test machine. The gauge length effect is significantly more pronounced for heats G, H, and I as compared to heats E and F. The inclusion count for the five steels is presented in Table 1 and, as the data indicate, steels with low inclusion counts show the least sensitivity to fiber length.

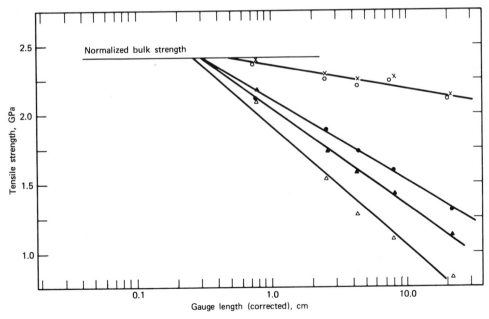

Figure 1. Gauge length dependence of the tensile strength of stainless steel filaments (diameter = 12 μm): X, heat E; O, heat F; ●, heat G; ▲, heat H; △, heat I. To convert GPa to psi, multiply by 145,000.

Table 1.   Inclusion-Count Data for Stainless Steel Heats

Mean inclusion diameters, μm	Number of inclusions for various heats				
	E	F	G	H	I
1	5	16	170	224	219
2		3	21	41	39
3		1	11	22	22
4			2	9	14
5			3	2	11
6				1	10
7					5
8				1	
9					
10					2

The normalized bulk strength value identified in Figure 1 corresponds to the intrinsic value for the specific stainless steel as measured in sufficient diameters, ie, >50 μm, to be uninfluenced by inclusion defects. The normalization is to a base-line chemistry since each heat shows small composition variations within the type specification. Even with the ultraclean, ie, low inclusion-count steels, breaking-load discrepancies within a fiber may exist resulting from small variations in fiber cross-section area. These cross-section variations, which are inherent to the bundle-drawing process, typically may result in a coefficient of variation in breaking load of about 6–8% for 12-μm diameter fiber, but only 3–4% for 25-μm fiber. Thus, fine-filament manufacturers using, eg, the bundle-drawing process must not only carefully control the process parameters but must establish strict specifications for the starting material in order to provide a quality product. Similar requirements are necessary for other fiber-forming techniques.

In a few cases, smallness improves material properties, eg, the mechanical properties of metal whiskers (1). The preparation of metal whiskers by vapor deposition or decomposition of a gaseous compound results in a slow buildup of material at high temperature. The resulting fiber is essentially free from dislocations and, consequently, exhibits mechanical properties which approach theoretical, ie, values related to the actual breaking of metal-to-metal bonds rather than the sliding of atoms or atom planes past each other as is the usual deformation mechanism.

Physical and chemical properties of metal fibers also may tend to be modified or exaggerated because of the small fiber diameter which results in high surface-to-volume ratios. For example, in electrical conduction, direct current is carried uniformly through a fiber cross section, whereas at high frequencies, electricity is carried close to the conductor (fiber) surface; the surface current density increasing with increasing frequency. Thus, the ratio of alternating-current to direct-current resistance of a small-diameter fiber is close to unity to significantly high frequency ranges; consequently, the fine fiber is an efficient high frequency conductor compared to large diameter wire.

In terms of chemical characteristics, the high surface-to-volume ratio is advantageous where the fiber serves as a catalyst but is disadvantageous where it is desired to minimize reaction of the fiber with the environment. The latter situation becomes important in many applications of fibers, eg, filtration (qv), seals, and acoustic treatment (see Insulation, acoustic). Fibers are chosen carefully to provide long life for the product.

Assembly. The properties of any assemblage of fibers often are determined by the particular arrangement of the fibers in that structure. For example, the mechanical tensile strength of a number of fibers or filaments in a yarn depends on the degree of twist imparted to the group of fibers during the spinning of the yarn. Too low a twist results in the fibers being loaded nonuniformly with application of tension and in progressive fiber failure which produces a low yarn strength. At optimum twist, the load applied to the yarn is equally distributed on all fibers and the yarn exhibits strength which is the sum of the breaking loads of all of the fibers.

The mechanical properties of sintered-fiber, randomly oriented structures, as exemplified by elastic modulus and tensile strength, have been reported by a number of investigators and there is considerable variation even for the same alloy fiber. A composite of tensile data obtained from a variety of samples of 304-type stainless steel fiber mat is shown in Figure 2. The samples include porous structures prepared by press and sinter techniques using 12-, 25-, and 125-μm fibers with aspect ratios of 25, 62, and 187, as well as sintered air-laid web of 12-μm fiber calendered to a density of about 20% of theoretical. The curve in Figure 2 signifies the probable maximum attainable strength for 304-type stainless steel sintered-fiber structures as a function of structure density for the particular fiber types employed, ie, chopped conventionally drawn wire and bundle-drawn fiber.

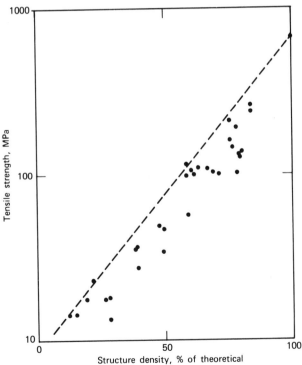

Figure 2.   Tensile strength of porous, stainless steel structures as a function of structure density. To convert MPa to psi, multiply by 145.

Measurement of the elastic modulus of sintered-fiber porous structures, again of 304-type stainless steel, have been reported over a density range of about 40–85% (2). The data was generated from samples composed of 50- and 100-μm diameter fiber, 4 mm in length. The data is reasonably consistent with the theoretical relation between elastic modulus and structure density that was established for porous materials (3):

$$E_m = E_0\,(1 - 1.9\,P + 0.9\,P^2)$$

where E_0 = modulus of elasticity (in tension) of 100% dense material, E_m = derived structure modulus (in tension) of the porous structure, and P = fractional porosity. The compressive modulus of elasticity is approximately one order of magnitude lower than the tension data. When compressed the fibers tend to buckle at low stress levels resulting in low structure modulus.

Measurement of the thermal conductivity of various types of porous materials has been reported (4). The samples range from sintered spherical powder of OFHC copper and 304L stainless steel in the 10–30% porosity range to fibrous stainless steel mats composed of 12-μm diameter filament with a porosity of 78%. The empirical relationship that best fits the data (5) is

$$\frac{\lambda}{\lambda_0} = \frac{1 - \epsilon}{1 + 11}\,\epsilon^2$$

where λ = sample thermal conductivity, λ_0 = solid material thermal conductivity, and ϵ = fractional porosity. The relationship is reliable for porosities up to about 80%. An attempt to derive thermal conductivity values from the measurement of electrical conductivity using a modified Wiedemann-Franz relationship is not successful in the case of stainless steels of >40% porosity (6).

The complications arising from changes in flow velocity; media geometry, including pore size and tortuosity; and fluid properties often require empirical approaches in order to satisfactorily describe observed results. All media present a finite resistance to this flow of fluids. In many cases, it is the magnitude of this resistance that is a determining economic factor in the choice of a particular porous material to perform a specific function, whether it be a filtration application or a lubricating device.

The resistance to flow is expressed in terms of the pressure drop across the medium per unit of length and the flow rate per unit area. The simplest relationship involving only viscous flow is Darcy's Law:

$$\frac{\Delta p}{L} = \frac{\mu}{K}\frac{Q}{A}$$

where Δp = pressure drop (101.3 kPa = 1 atm); L = media thickness, cm; μ = viscosity, mPa·s (= cP); Q = flow rate, cm^3/s; A = cross-section area, cm^2; K = permeability, darcys. The relationship defines the common permeability unit, the darcy (units of cm^2). When fluid flow involves inertial energy losses and viscous drag, the expression is modified to account for the increased energy loss:

$$\frac{\Delta p}{L} = \frac{\mu}{K}\cdot\frac{Q}{A} + \frac{\rho}{K^1}\cdot\frac{Q^2}{A^2}$$

where ρ = fluid density, and K^1 = permeability factor associated with kinetic energy loss. Compressible fluids require additional energy-loss terms to account for the work performed in fluid compression.

Thus for each porous structure and each specific fluid, there is a unique relationship defining the permeability of the medium under a particular set of conditions; however, in many cases the relationship cannot be derived from first principles. For the same medium type but with differing porosity values, the permeability is different. A relationship for laminar-flow conditions has been derived and satisfactorily predicts the change of permeability with change in porosity for the same fiber and structure geometry (fiber orientation):

$$K \propto \frac{\epsilon^3}{(1 - \epsilon)^2}$$

where ϵ = fractional porosity. The permeability of randomly oriented fiber structures can be very high when compared to other porous materials primarily because low density structures of good mechanical integrity can be fabricated. The use of fine fibers reduces viscous drag and, thus, compliments the special structure features.

The high permeability properties of randomly oriented fiber structures are particularly attractive in filtration applications. Comparative data for three nominal 20-μm rated filter media composed of sintered powder metal, woven wire cloth, and a randomly oriented fiber structure illustrate the advantage of the latter: the measured nitrogen permeability, normalized to the fiber structure data, are 0.12 cm, 0.42 cm, and 1.00 cm, respectively.

In addition to the high permeability values, the randomly oriented fiber structure exhibits another desirable property with regard to filtration, ie, the high dirt-holding capacity or on-stream life. This property can be quantified by the time taken to develop a particular cut-off Δp and can be as much as a factor of two compared to sintered powder media and a factor of 1.5 compared to the woven-wire structure.

Manufacture and Processing

Often there are two distinct elements in the fabrication of fiber products, ie, the formation of the fiber and the assembly of the fibers into a useful structure or form. The majority of commercial applications involve a large degree of secondary processing by various fiber-manipulation techniques.

Certain fiber-forming processes yield free fiber and others tend to produce a primitive fiber assembly, eg, a tow or mechanically interlocked bundles comparable to the bale of natural fiber. Thus, the commercially available form depends on the type of forming process employed. It also may depend on the business strategy of the manufacturer, who may limit the availability of the primitive form in order to attain the benefits of the value added by further in-house processing.

Fiber dimensions, as defined by the natural- and synthetic-fiber industry, tend to be restricted to diameters or equivalent diameters for noncircular cross sections of less than 250 μm. This would correspond to normal limitation for further processing by common textile-fiber manipulation techniques. This limitation also applies to metal fibers.

Fiber Forming. The principal methods that have been developed for metal-fiber forming relate to the basic starting material form. Mechanical processing incorporates processes that rely on plastic deformation to produce a fiber from a solid precursor.

In liquid-metal processing or casting, the fiber is formed directly from the liquid phase.

Mechanical Processing. The mechanical processes involve material attenuation by gross deformation or they involve the parting of material from a source, eg, a strip or rod. The first group encompasses wire-drawing techniques and solid-state extrusion, and the second group consists of cutting or scraping-type operations, ie, slitting, broaching, shaving, and grinding.

The various processes are identified in Figure 3 in relation to their source material; both current and potential commercial processes are given. For processes, eg, conventional wire drawing, the manufacturing cost is highly dependent on the diameter of the final product. Material attenuation must take place in a series of small, usually equal steps of ca 20% area reduction per step. Since drawing speed cannot be continually increased to compensate for the lower through-put per step, succeeding reducing steps involve lower and lower efficiency, ie, quantity produced per unit time. The modified wire-drawing process, ie, bundle drawing, circumvents this strong diameter dependency by drawing many wires or filaments simultaneously through a reduction die. Consequently, the mass flow (kg/h) of the wire-drawing machines is increased dramatically and results in a large reduction in manufacturing cost. Some of this ad-

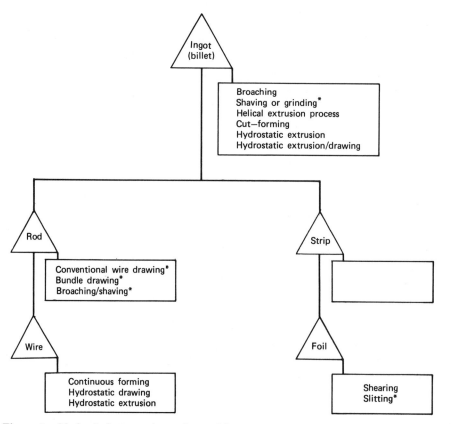

Figure 3. Mechanical attenuation and material separation processes for metal filament forming: Δ, stock material form; □, fiber-forming processes; *, commercially available product.

vantage, however, is offset by the equipment and processing that are necessary to form the initial bundle of wires and that are required to separate the fibers at the conclusion of the process. On balance, the overall economy improvement factor is highly favorable and can be as much as 40 or 50 to one for fibers of ca 12 μm in diameter. The bundle-drawing process (BDP) is an important source of quality fiber. A schematic process flow diagram comparing conventional wire drawing is shown in Figure 4.

The wire drawing, modified wire drawing, and extrusion processes produce continuous filaments of basically circular cross section and of unique properties that are partly a consequence of the processing history. Attenuation or constraining the material in the reducing die produces changes in the internal structure or morphology of the metal, eg, reduction in grain size, development of preferred orientation and,

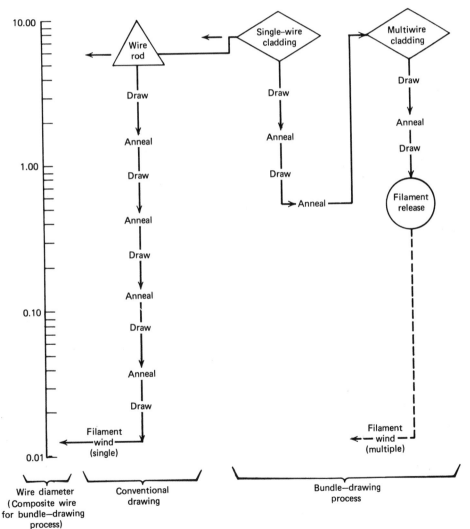

Figure 4.   Flow diagram comparing conventional and bundle-drawing processes: △, starting material; ◇, ○, special operation.

in some cases, induced phase transformation. In multiple-step reduction processes, the effect of such changes in material properties causes succeeding steps to become increasingly more difficult and usually it is necessary to eradicate the work-hardening effect by annealing the material. The drawing–annealing schedule can be arranged to provide efficient processing and desirable mechanical properties in the final fiber product.

The slitting, broaching, shaving, and grinding machines and devices designed to convert standard mill forms of metal to fiber have been described (7–9). The wire-shaving technique produces a fiber product at low cost and with wide application (10). Modern machines operate automatically on a multiplicity of wires and can produce metal wool with an equivalent diameter of <25 μm. The process involves the shaving of wire, which may be ca 12 mm in diameter, in a series of steps by serrated, chisel-type tools; the size and spacing of the serrations determining the cross-section dimensions of the individual fibers. The industry standards define fiber-wool grades as grade 3 (largest) through 0 to the very fine 0000 grade, with the extremes corresponding to a mean fiber width of 178–241 μm and 15–25 μm, respectively (11). The cross section of the shaved fiber generally is triangular with the apices defining sharp edges along the fiber length. The coarser grades contain very long fiber and, even in the finer grades, the fiber is of sufficient length to cling together without unraveling in the rolls or pads, which are the usual form for shipping. Material compositions are mainly ferrous alloys, either low carbon steel, or 400 series ferrite stainless steel.

Many variations of the wire-shaving process have been developed. A group of devices or machines have been designed for the shearing or shaving of stacks of thin metal foil, although no such commercial fiber product is on the market (8). The product produced by these processes usually is rectangular or square in cross section and the length depends on the process employed: for edge-shearing or slitting the fiber length can be equal to the foil length, whereas for end-shearing the width of the foil is the limiting factor. From the material standpoint, the versatility of mechanical processing is limited only by the required availability of the foil form. The slitting of metal foil in single thicknesses to provide continuous filament usually is unattractive because of the associated low production rate. For some specialized applications, however, eg, for decorative textile thread, the cost may be justified especially if the foil costs are reduced by substitution of metal coated plastics.

Other mechanical processing techniques identified in Figure 3 either are not used for fiber production or are emerging technology. Hydrostatic extrusion and extrusion-drawing techniques offer opportunities for fiber-forming materials which are difficult to work or deform (12). These processes also allow for comparatively high reduction ratios, ie, the ratio of the diameter of the metal wire or rod stock to the diameter of the final wire or filament. Helical extrusion and continuous forming offer potential for a low cost product but it is too early to predict the lower limit of product dimensions consistent with economic production (13–14). A 1 mm diameter copper wire can be produced directly from a 150 mm diameter and 1 m long billet by helical extrusion and a 1 mm diameter aluminum wire can be converted directly from 10 mm diameter wire stock by continuous forming; the cut-form process is a combination of the two processes (15). The first step is the formation of a metal chip from a billet; the second step involves feeding the chip into a forming die. The advantages of this process are the attractive economics of step one and the improved product quality associated with the second step. The process is in the early developmental stage.

The fiber characteristics and relative economics of the commercially significant fiber-forming techniques by mechanical processing are summarized in Table 2. Scanning electron micrographs of representative fiber products are presented in Figure 5.

Liquid-Metal or Casting Processes. Melt spinning of glass and certain polymers is an established technique for mass production of fine filaments or fibers. The liquid material is forced through a carefully designed orifice or spinerette and solidifies in a cooled environment, usually after considerable attenuation, before being wound on a spool. However, this process cannot be adapted easily to metals. The development of liquid-metal fiber-forming processes has revolved around overcoming the inherently low viscosity of molten metals. The low viscosity and the high surface tension of liquid metals make it extremely difficult to establish free-liquid jet stability over a length sufficient to allow freezing of the metal into a fiber before the jet separates into droplets. The problem has been solved with varying degrees of success by one of the following approaches: altering the surface of the liquid jet by chemical reaction; promoting jet stabilization by indirect physical means, eg, an electrostatic field; or accelerating the removal of heat from the jet to promote solidification before breakup occurs. An alternative approach which has not been commercially exploited is that of placing a glass envelope around the molten metal to control the formability and thus circumvent the problem of jet stabilization (16–17). However, the final application of the product may require the removal of the glass envelope, which would add considerably to the product cost since it may account for 75 vol % of the material produced. The various liquid-metal fiber-forming processes of historical and commercial interest are identified in Figure 6.

Melt spinning involving a free-liquid jet is used for a variety of low-melting-point metals or alloys including Pb, Sn, Zn, and Al. Various techniques are involved in the cooling and quenching process including co-current gas flow, mists, and liquid media. The processes permit production of fiber of 25–250 μm in diameter and of continuous length. Melt-spin processes applied to higher-melting-point metals, particularly the

Table 2. Mechanical Fiber-Forming Processes and Related Fiber Characteristics

Process	Typical fiber diameter, μm	Typical length	Materials	Cross-section shape	Economics
conventional wire drawing	≥12	continuous	all ductile metals and alloys	round (other sections possible)	304 stainless steel ca $3.00/kg at 250 μm; ca $3000/kg at 12 μm
bundle drawing	≥4; typically 8 or 12	continuous	ductile metals and alloys	rough surface	304 stainless steel ca $100/kg at 12 μm
broaching or shaving, eg, wire rod, and billet	≥8	short to continuous	most ductile metals and alloys	generally triangular	low carbon steel $2–4/kg depending on grade
slitting and shaving, eg, foil and sheet	ca 25 and greater	short (0.0004–4 cm) or continuous	most ductile metals and alloys	ductile square or rectangle	shaving fiber <$10/kg (not available currently)

Figure 5. Scanning electron micrographs of fibers produced by mechanical fiber-forming processes: (a) 25 μm stainless steel, conventional wire drawing; (b) 19 μm 304-type stainless steel, bundle-drawing process; (c) 0000-grade steel wool, shaving process; (d) transverse shaving marks of 0000-grade steel wool, shaving process; (e) felt metal (FM), 1100 series, 347 stainless steel; (f) sheared, low carbon, steel fiber edge.

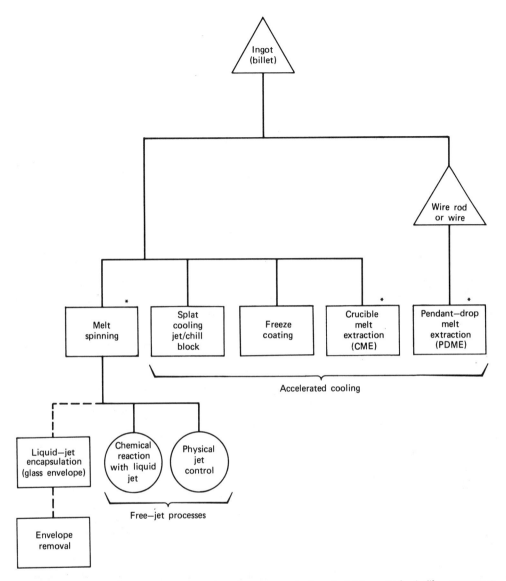

Figure 6. Liquid-metal, fiber-forming processes: △, stock material form; □, basic fiber process; ○, process modifications; *, commercially available fiber.

ferrous alloys, requires the removal of very large quantities of heat from the liquid-metal jet in a very short time. Generally, those techniques that are successful for the low melting alloys cannot be used for the high melting alloys. An alternative approach involves the reaction of the liquid jet and the cooling medium, or some addition to that medium which results in the formation of a case or envelope of sufficient strength to prevent breakup of the liquid-metal jet.

In some cases, the chemical composition of the melt must be adjusted prior to

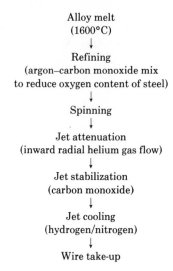

Alloy melt
(1600°C)
↓
Refining
(argon–carbon monoxide mix
to reduce oxygen content of steel)
↓
Spinning
↓
Jet attenuation
(inward radial helium gas flow)
↓
Jet stabilization
(carbon monoxide)
↓
Jet cooling
(hydrogen/nitrogen)
↓
Wire take-up

Figure 7.   Flow diagram for silicon steel and aluminum steel melt spinning with jet stabilization.

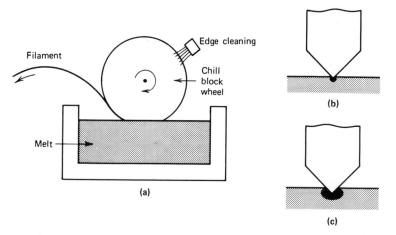

Figure 8.   (**a**) Crucible melt-extraction process (CME); (**b**) and (**c**) crescent shaped, larger filaments.

spinning in order to prevent dissolution of the stabilizing film in the molten jet. A typical process flow diagram for the melt spinning of steel wire is shown in Figure 7 (18). The process is designed for two steel compositions, an aluminum steel (1.00 wt % Al, 0.36 wt % C) and a silicon steel (2.40 wt % Si, 0.32 wt % C). Carbon monoxide is the stabilizing medium and surface films of aluminum oxide or silicon oxide are formed. A continuous filament of 75–200 μm in diameter is spun at 240–120 m/min with continuous operation to 100 h. A useful degree of attenuation of the molten-metal jet outside the orifice is achieved by using a radially directed helium gas flow. Consequently, the final filament diameter can be controlled independently of the spinning orifice or spinerette geometry.

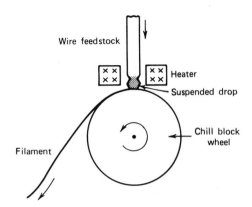

Figure 9.　Pendant-drop, melt-extraction process (PDME).

Table 3.　Liquid-Metal Fiber-Forming Processes and Related Fiber Characteristics

Process	Typical fiber diameter	Typical length	Materials	Cross-section shape	Economics
melt spin	$\geq 25\ \mu m$	continuous	Pb low melting point metals and alloys	circular	plumbers wool (lead) <$2/kg
melt spin with jet stabilization (chemical)	$\geq 75\ \mu m$	continuous	most metals and alloys (special compositions)	circular	
crucible melt extraction (CME)	$\geq 25\ \mu m$	continuous or controlled length	most metals and alloys	small diameter, circular; large diameter, crescent shaped	stainless steel $2–7/kg
pendant-drop melt extraction (PDME)	$\geq 25\ \mu m$	continuous or controlled length	most metals and alloys	small diameter, circular; large diameter, crescent shaped	

The alternative approach to fiber forming from the melt is by greatly accelerated heat transfer (19–20). The processes essentially are based on the quenching of the molten-metal jet on a chill plate positioned close to the orifice or at a point prior to the onset of jet instability and breakup. A number of different configurations for the chill plate have been proposed, ie, from rotating drums to concave disks. The quench rate of the molten jet on the chill plate is as high as 10^6 °C/s. Quench rates of this magnitude can produce nonequilibrium conditions in certain alloys and a characteristic grain structure associated with the impressed unidirectional heat flow and the kinetics of the nucleation–growth process. The fibers produced by these chill-plate processes generally are ribbonlike but can be continuous.

The crucible melt extraction (CME) process is an extension of the chill-plate processes and is illustrated in Figure 8a (21). There is no spin orifice and, thus, no need for forming a liquid stream. The chill surface, the shaped rim of a revolving disk, is dipped into the crucible so that the shaped edge just contacts the liquid-metal surface.

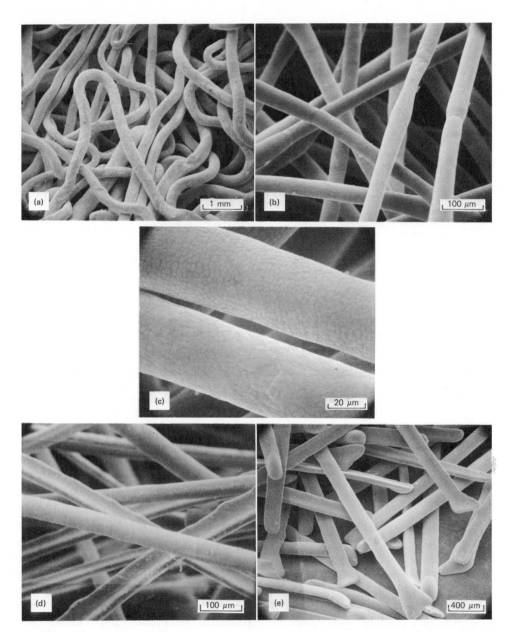

Figure 10. Scanning electron micrographs of fibers produced by liquid-metal or casting processes: (a) 250-μm lead filament, melt spinning; (b) 40-μm aluminum filament, melt spinning; (c) 35-μm melt spin aluminum filament showing surface structure; (d) 50 × 100-μm stainless steel filament, PDME; (e) 150-μm stainless steel filament (3 mm long), PDME.

Table 4.   Miscellaneous Metal-Fiber Forming Processes

Filament material source	Process	Product	Status
metal powder	slurry forming, extrusion with binder followed by sintering	potentially most metals and alloys; specifically W–Ni (250 μm in dia)	experimental
any metal form	vapor deposition on glass fiber or plastic strip which is subsequently slit[a]	aluminum-coated product for conductive or decorative applications	commercially available
	freeze coating on glass or other fiber	aluminum on glass	experimental
inorganic chemical or mineral	electroplating on a helical mandrel[b]	50×12-μm Ni continuous filament	experimental
	chemical decomposition in an organic fiber precursor[c]	Ni–Cr alloy fiber with dimensions equal to precursor	experimental
	halide reduction by hydrogen[d]	whiskers of many metals, eg, Cu, Ag, Fe, Ni, and Co; length, 3–20 mm	mostly experimental

[a] Ref. 22.
[b] Ref. 26.
[c] Ref. 25.
[d] Ref. 23–24.

The cool disk edge immediately causes solidification of a small volume of liquid metal, which is carried out of the crucible and is ejected from the wheel by centrifugal action. The shape of the fiber cross section is dependent on the wheel-edge geometry and the depth of immersion, but it can be made circular for small diameter filaments (25–75 μm); larger filaments tend to be crescent shaped, as indicated in Figures 8b–8c.

A further evolution in the chill-plate concept is the pendant-drop melt-extraction process (PDME) (21). In this process, the orifice for jet forming is eliminated and the crucible is replaced by a suspended drop held by surface tension to the feedstock, which is a wire or wire rod. The PDME process conveniently circumvents jet stability problems and solves the basic material compatibility problems associated with the molten-metal-orifice and the molten-metal-crucible interfaces. The basic elements of this process are schematically represented in Figure 9. The process can be applied to a wide range of metals and alloys and can produce fibers with equivalent diameters as small as ca 25 μm.

The PDME and CME processes have been varied to provide discrete fiber lengths as opposed to continuous filament. The introduction of discontinuities on the chill-wheel rim at fixed intervals results in the casting of short fibers with the discontinuity spacing defining the fiber length. The short fibers tend to have a dog-bone configuration which, for certain applications, is advantageous. A summary of the liquid-metal fiber processes in terms of fiber characteristics and comparative economics is presented in Table 3. Scanning electron micrographs of representative fibers produced by the various casting techniques are reproduced in Figure 10.

Table 5.   Special Metal-Fiber Characteristics and Processability

Fiber characteristics	Process parameter	Remedy
surface roughness high hardness	friction–wear wear	surface treatment: lubrication; change material of rubbing surfaces in equipment
irregular shape high density	fluid dynamics	modify process media conditions, eg, flow velocity, density, viscosity
high modulus	general processability	use fiber with smaller cross-section (lower section modulus)
high yield	resistance to take permanent set (spring back)	anneal to reduce yield strength

Table 6.   Fiber-Structure Fabrication Processes

Technology	Preferred fiber form	Process	Product structure
papermaking	free fiber; length >12 mm for fibers 12–250 μm	wet slurry casting on screen	green random-fiber mat
textile processing	continuous filament	twisting	continuous filament yarn
		breaking	sliver–roving
	sliver–roving	drafting–spinning	staple yarn (blends)
		carding	random web
		air laying	random web
	yarn (staple or continuous filament)	weaving	fabric sheet
		knitting	fabric sheet or tube
		braiding	ribbon or tube
		tufting	carpetlike structures
	random web	needle punching	feltlike products
miscellaneous	chopped fiber; length >4 mm	flocking	short-pile structure on substrate
	continuous filament yarn	filament winding	composite structures
	random web, woven or knitted structures	vacuum or gravity infiltration of matrix material	composite structures

Miscellaneous Processes.   Production of low cost metal filaments or products with unique properties has been reviewed (7,22–26). A sampling of significant technology other than mechanical and liquid-metal processes is presented in Table 4. Many of these processes could, with additional development and the use of recent advances in material science and related technology, lead to attractive processes.

Assembly.   The various assembly processes of metal fibers into structures are similar to those developed for natural and synthetic fibers. The main fiber characteristics and the process parameters most likely to be affected by those characteristics,

Table 7. Secondary or Special Processing Techniques

Process	Typical application or particular advantage
annealing	softens fiber, increases fiber ductility, assists in further processing
sintering	fuses fibers at contact points, provides increased mechanical strength in fibrous structures
fiber alignment	produces anisotropic characteristics in web structures or composite materials
plating	provides protection from corrosive environments
coatings	provides interfiber lubrication
calendering (rolling)	precise pore size and density control, particularly in nonwovens
pressing	precise pore size and density control, particularly in nonwovens
crimping	geometrical elasticity in fiber, yarn, or nonwovens

and suggested solutions or remedies to alleviate processing problems are given in Table 5. Applications of metal-fiber products may require a combination of fiber types, eg, metal fibers and organic fibers. Thus, processes must be adapted to accomodate blends of fibers of vastly differing properties and processing characteristics.

The basic fiber-processing techniques in relation to the preferred fiber form for each process and the resulting products are summarized in Table 6. In textile processing particularly, the product of one process often becomes the preferred fiber form for a second process; thus, sequential processing is common for the more advanced products. In addition to the principal fiber-structure processing techniques enumerated in Table 5, a variety of secondary or special processes are important in modifying or optimizing the fiber-structure properties; these are listed in Table 7 with their typical applications.

Uses

Applications for metal fibers are of two types. One is the substitution for other fiber types or for metal powder in the case of porous metals to improve performance or to provide a cost benefit, eg, high temperature oxidation resistance or improved permeability in porous structures. The other application is in the development of new products that are based on the unique fiber properties or property combinations of metals, eg, in high gradient magnetic separation (see Magnetic separations).

A representative listing of applications, with brief descriptions and main advantages associated with the use of metal fibers, is presented in Table 8. The fiber structure applications are exploitations of the structure and the inherent fiber properties. The composite applications benefit from both the fiber and fiber structure characteristics and their interaction with the matrix material (see Composite materials; Laminated and reinforced metals).

Table 8. Metal-Fiber Applications

Application	Description	Status[a]	Special advantages and principal fiber function
Textile products and porous structures			
filters			
surface	screen or wire-mesh products in disk or cartridge configuration used for low contamination level fluids, eg, hydraulic fluids	C	mechanical strength, nonmigrating, corrosion resistant
depth	nonwoven or random web structures in disk or cartridge configuration; general industrial applications	C	high permeability, high dirt-holding capacity
electrostatic	particle-capture augmentation of charged particles using low density web structures	R	improved filtration efficiency, electrical conduction
	filter-cake density control using designed electric-field configuration in bag-house filters	D	improved cake permeability
magnetic (HGMS)	magnetic attraction of ferromagnetic or paramagnetic particles through field distortion associated with ferromagnetic fibers in a uniform magnetic field; used for decontamination of kaolin	C	high field distortion associated with fine filaments
seals			
abradable	zero-clearance seal formed by turbine blade tips mating with deformable, porous-mat engine-housing lining	C	improved engine performance (efficiency)
gaskets	rope-type structures and nonwoven mats which conform to surface irregularities under compression; general industrial high temperature applications	C	high temperature, resilient structures
abrasion	metal-wool abrasive products for material removal and polishing	C	sharp cutting edges, shaved fibers
antistatic			
textiles	blended yarn (stainless steel–nylon fiber) up to 10 wt % steel woven into fabric for clothing applications	C	electrical conduction and flexibility
carpets	blended yarn introduced into tufting operation to provide <1% steel fiber in face yarns	C	electrical conduction and flexibility
filter bags	needle-punched blend of organic fiber and steel to provide static control in bag-house filtration applications	C	electrical conduction and flexibility
brush	steel-fiber brush for static control of paper in photocopying devices	C	electrical conduction and flexibility
insulation	low density web (<5%) with suitable attachments used for aerospace applications, eg, rocket-engine nozzle insulation	C	high temperature stability, nonfriable
acoustics	acoustic impedance control for tuned resonator-duct liners in jet engines, etc	C	fine fiberweb structures provide improved linearity in absorption characteristics
catalysts	worn wire-mesh stacks for nitric acid production (75 μm diameter, Pt–Rh wire)	C	optimum surface-to-volume ratio

238

Table 8 (*continued*)

Application	Description	Status[a]	Special advantages and principal fiber function
electromagnetic interference control and field production	Faraday suits of stainless steel/nylon fabric for high voltage transmission-line maintenance: heat shrinkable cable-termination shields embodying air-laid web	C	electrical conduction, flexibility
fluid flow	a family of products including, flow restrictors, snubbers, silencers, vents, demisters, homogenizer plugs, etc, using porous fiber structures	C	mechanical integrity, controlled permeability
heated fabrics	built-in flexible electrical conductors for heating, clothing, etc	C	electrical conduction, flexibility
electrodes	battery plaques, current collectors for high energy batteries, fuel cells; electrodes for electrolytic capacitors	D	high surface area, distributed conductor network
bearings	liquid lubricant or air bearings utilizing porous fiber structures	D	controlled porosity, mechanical integrity
wicks	capillary structure for liquid-phase transport in heat pipes	D	high temperature stability, controlled size, interconnected pores
shock mounts	friction damping in shock and vibration isolation mounts	C	high fiber-to-fiber friction, resilience
flame trap	protection of flammable fluids using sintered fiber-metal porous plugs	C	high thermal capacity and heat conductivity
transpirational cooling	surface cooling by controlled fluid flow through component structure, eg, turbine blades	D	structural integrity with designed permeability
fluidizer plate	support and diffuser plate for fluidized bed, eg, grain cars		porous-plate structure with load-bearing capabilities
ceramic attachment	compliant layer of fibers for ceramic attachment to metal surfaces	C	resilient fiber structure accomodating differential thermal expansion coefficients of substrate and ceramic
yarn and cable products	yarn and cable applications demanding flexibility with high strength, eg, in instrument control cables and medical sutures	C	good hand, high knot strength, biocompatibility, high flexibility
Composite applications tire cord	steel tire cord containing 0.15–0.25-mm, brass-plated filament in designed cord configurations	C	strength, dimensional stability, flexibility, heat conduction, bonding to rubber
tire tread	tread impregnation with chopped wire for off-the-road vehicle and aircraft tires	C	abrasion and cut resistance
timing belts	reinforced, nonstretching belts	?	dimensional stability, strength
brake lining	high friction composite material with dispersed short metal fiber for heat conduction	C	improved operating performance by temperature control using conductive fibers

239

Table 8 (*continued*)

Application	Description	Status[a]	Special advantages and principal fiber function
refractory bricks (linings)	furnace-lining reinforcement with castable refractories embodying chopped fiber or melt extraction process fiber to reduce friability	C	mechanical integrity, crack arrestor function
concrete	reinforcement with short fibers to increase load-bearing capability	C	crack arrestor
conductive plastics	plastic housings for electrical equipment to control electromagnetic interference (EMI)	C	electrical conduction
	fabric containing short metal fiber for shielding–reflecting microwave radiation	C	efficient coupling
superconductors	continuous filament, small-diameter superconductors embedded in high thermal conductivity matrix	C	more efficient use of superconductivity material and protection during conduction state transition

[a] C = commercial, D = development, and R = research.

BIBLIOGRAPHY

1. R. V. Coleman, *Metall. Rev.* **9**(35), 261 (1964).
2. P. Ducheyne, E. Aernoudt, and P. De Meester, *J. Mater. Sci.* **13**, 2650 (1978).
3. J. K. Mackenzie, *Proc. Phys. Soc.* (*London*) **B63**, 2 (1950).
4. R. P. Tye, *A.S.M.E. Publication No. 73-HT-47*, American Society of Mechanical Engineers, New York, 1973.
5. J. Y. C. Koh and A. Fortini, *NASA Publication No. CR-120854*, 1972.
6. R. W. Powell, *Iron and Steel Institute Special Report No. 43*, p. 315.
7. C. Z. Carroll-Porczynski, *Advanced Materials*, Chemical Publishing Co., Inc., New York, 1969.
8. U.S. Pat. 3,122,038 (Feb. 25, 1964), J. Juras.
9. W. M. Stocker, Jr., *Am. Mach.* (May 15, 1950).
10. L. E. Browne, *Steel* (Feb. 25, 1946).
11. *Federal Specification FF-S-740a*, U.S. Government Printing Office, Washington, D.C., Oct. 1965.
12. H. Ll. D. Pugh and A. H. Low, *J. Inst. Met.* **93**, 201 (1964–1965).
13. D. Green, *J. Inst. Met.* **99**, 76 (1971).
14. *Ibid.*, **100**, 295 (1972).
15. T. Hoshi and M. C. Shaw, *J. Eng. Ind. Trans. ASME*, 225 (Feb. 1977).
16. G. F. Taylor, *Phys. Rev.* **23**, 655 (1924).
17. U.S. Pat. 1,793,529 (Feb. 24, 1931), G. F. Taylor.
18. *AIChE Symp. Ser.* **74**(180), (1978).
19. U.S. Pat. 2,879,566 (1959), R. B. Pond.
20. U.S. Pat. 2,976,590 (1961), R. B. Pond.
21. R. E. Maringer and C. E. Mobley, *J. Vac. Sci. Technol.* **11**, 1067 (1974).
22. J. F. C. Morden, *Met. Ind.*, 495 (June 17, 1960).
23. S. S. Brenner, *Science*, **128**, 569 (1958).
24. S. S. Brenner, *Acta Met.* **4**, 62 (1956).
25. W. H. Dresher, *Technical Report AFML-TR-67-382*, WPAFB, Dec. 1967.
26. E. H. Newton and D. E. Johnson, *Technical Report AFML-TR-65-124*, WPAFB, April 1965.

JOHN A. ROBERTS
Arco Ventures Co.

METALLIC COATINGS

SURVEY

Metallic coatings provide a basis material with the surface properties of the metal being applied as coating. The functional composite so produced has an appearance or utility not achieved by either component singly and in fact becomes a new material (see also Composite materials). The base material almost always provides the load-bearing function and the coating metal serves as a corrosion- or wear-resistant protective layer (see Corrosion and corrosion inhibitors). The base material most often is another metal, but it can be a ceramic, paper, or a synthetic fiber (1). The bulk of all metallic coatings provide a protective function in one of five principal ways: they are anodic to iron and can protect it by cathodic protection, eg, Al [7429-90-5], Mg [7439-95-4], Zn [7440-66-6], and Cd [7440-67-7]; they form highly protective, passive films in aqueous media, eg, Cr [7440-47-3], Ni [7440-02-0], Ti [7440-32-6], Ta [7440-25-7], and Zr [7440-67-7]; their oxides are slow growing and adherent and, therefore, protect at high temperatures, eg, Al, Cr, and Si [7440-21-3]; they are noble metals and corrode little or not at all and function as barriers to corrosive agents, eg, Au [7440-57-5], Ag [7440-22-4], Cu [7440-50-8], Pt [7440-06-4], Rh [7440-16-6], Pd [7440-05-3], etc; and their compounds, which are formed by uniting with the basis metal or one of its constituents, are very hard and provide wear resistance, eg, B$_4$C [12069-32-8], SiC [409-21-2], TiC [56780-56-4], WC [12070-12-1], Cr$_2$O$_3$ [1308-38-9], etc. In certain coating processes, eg, chemical vapor deposition (CVD), the hard-phase compounds are grown directly on the substrate.

The protective function is primary. Decorative and reflective metallic coatings must be highly corrosion resistant, at least in ambient environment, to maintain a visually satisfying appearance and to provide satisfactory service over the anticipated useful life of the composite. The corrosion resistance of individual metallic elements and their principal alloys is discussed in refs. 2–5. Types are reviewed in refs. 6–9. Coating technology is discussed in refs. 10–14 and a discussion of high temperature oxidation- and sulfidation-resistant coatings primarily for aerospace applications is given in refs. 15–17 (see also Electroplating).

Metallic coatings have been used since ancient times. One of the earliest known applications involved the cementation of copper or bronze with arsenic to produce a silvery coating of Cu$_3$As [12005-75-3] on art objects (18). Between 1 and 600 AD, Andeans plated copper objects by galvanic displacement by which 0.5–2 μm thick films of gold or silver are deposited (19).

Metallic coatings provide functionality at low cost. Sudden changes in availability of materials have led to a critical shortage of and, hence, a large increase in the cost of cobalt. This increase in cost has strongly affected the technology of hard facing alloys and has promoted the development of several new substitutes. Strict government regulations on pollution control have delayed some UK plating firms' efforts to establish effective treatment measures and, in the United States, has led to the closing of some firms. However, the new regulations have had much less affect on some new coating technologies, which have grown rapidly. Strict regulations have caused a general stagnation in the metal-finishing industry for the last five years (1975–1980). However, continued growth in the electroplating of plastics and printed circuit boards is expected. Thus, it is the continued appearance of new applications for coatings coupled with emerging coating technologies that provide improved quality and cost effectiveness. Many of the new techniques will have enormous technological impact in the coming years because of growth in supporting industries, eg, electron-beam dissociation has been developed at IBM for depositing conducting metal coatings only 0.05-μm thick to form the elements of the Josephson-effect transistor (see Integrated circuits).

Diffusion

In the various processes for diffusion coating, the basis metal is contacted with the coating metal, which is in the liquid or solid state or is brought to the surface by vapor transport. The two materials are held at an elevated temperature for a sufficient time to allow lattice interdiffusion of the two materials. A solid-solution alloy forms and may be accompanied by the formation of one or more intermetallic compounds that provide a compositional transition zone within the coating. The growth of the coating often is limited by diffusion of one species through one of the intermetallic layers, resulting in a parabolic rate of coating thickness increase with time. Whether the coating metal is brought into contact as a liquid, or is carried by another solvent, or is transported by a vapor-phase mechanism, surface contamination and oxide films must be removed by a suitable fluxing process either prior to or during the contact period and oxidation-free conditions must be maintained during diffusion. Three basic techniques are used in diffusion coating: hot dipping, cementation, and use of liquid carriers. In hot dipping, an excess of liquid-coating metal or alloy usually is carried on the product which may be finished with steam or air (20) or it is wiped on rollers as it moves out of the bath to control the total thickness of the final coating. In many cases, the coating either is thicker on one side than the other or the coating is on one side only. In cementation and liquid-bath diffusion processes, the coating consists only of intermetallic compound and solid-solution regions, the excess coating metal having been removed by a postcoating process. Postcoating heat treatments often are used to equilibrate these intermetallics.

Hot-Dipped Coatings. *Aluminum.* Hot-dipped aluminum coatings on steel strips have been produced for 35 yr and, in 1976, 300,000–500,000 metric tons was produced in the United States (21). Aluminum forms a very protective oxide film with outstanding high temperature oxidation resistance and it provides the attractive appearance of pure aluminum. It also provides galvanic protection for steel (qv) but must be about twice the thickness of zinc if it is to impart acceptable corrosion resistance. The high melting point of aluminum, relative to zinc, causes recrystallization and, therefore, produces softening of the cold-worked steel during hot dipping. Nevertheless, hot dipping is the most widely used method of coating steel, primarily plain carbon, low alloy, and stainless steel, with aluminum.

During the process, the aluminum is maintained at 680–720°C and the steel strip is immersed for ca 5–15 s. Commercial practice is to utilize the Sendzimir line which first uniformly oxidizes the iron and then passes it to a reducing atmosphere in which a layer of pure iron quickly forms on the surface (22). On immersion, the iron layer reacts quickly with the aluminum. After dipping, the material is finished by jet finishing, rolling, and quenching. For high temperature applications, the bulk of the coating should consist of an outer layer of Fe–Al and Fe_3Al [12004-62-5] at the steel–coating interface with an average surface concentration of <50% aluminum and, ideally, 12%. Generally, this is accomplished by heat treating the strip at 820–930°C after dipping. For low temperature applications and to promote formability, the thickness of the intermetallic layer should be minimized. Silicon, although it detracts from the coating's corrosion resistance, impedes the growth of the intermetallic region. Aluminum hot-dip processing is difficult to control because the activation energy for the exponential growth of the intermetallic layer is 170–180 kJ/mol (41–43 kcal/mol). Also, simultaneous growth and spalling is possible if dipping times exceed 60 s (23). If the above activation energy values are used at a bath temperature of 680°C and an immersion time of 3.5 s, a 10-μm intermetallic layer develops. At 700°C, the immersion time must be less than 2 s. Cleanliness of the strip prior to immersion is essential to avoid barrier films of Al_2O_3. Commercial practice involves hydrogen treating, fluxing, and precoating treatments using a thin layer of copper or a film of ethylene glycol.

An aluminized material, Aluma-Ti (Inland Steel Co.), was designed to have heat-resistant properties equivalent to type 409 stainless steel for applications up to 815°C, eg, in motor vehicle exhaust systems. The material is basically a type 1 aluminized coating, eg, aluminum with a small addition of silicon, but the steel basis metal is an aluminum-killed steel with sufficient titanium to tie up residual carbon and nitrogen and to provide ca 0.3% titanium in solution. Aluma-Ti resists formation of a porous intermetallic layer and a subsurface iron oxide layer which tends to promote spalling of the intermetallic layer. Loss of the intermetallic layer and the presence of a continuous subsurface oxide stops the continued diffusion of aluminum into the basis steel, which is necessary to provide long term protection. Aluma-Ti in tests has shown superior spalling resistance to both type 1 aluminized and 409 stainless steel. Mechanical properties of Aluma-Ti compare well with those of type 409.

Armco Steel Corp. has produced a steel alloy which, after being aluminized, also is superior to 409 stainless. Its composition includes 2% Cr, 2% Al, 1% Si, and 0.5% Ti. In practice, the steel surface is sufficiently oxidized under controlled conditions in the oxidizing section of the Sendzimir line so that a region near the surface and including the iron is oxidized. When the oxidized strip passes to the reducing section, the iron oxide is reduced to iron which contains a fine dispersion of the stable oxides, ie, of Ti, Al, and Si. Subsequent aluminizing is satisfactory as long as sufficient oxidation of iron has taken place so that a continuous oxide film of the more stable oxides cannot form.

Aluminum hot-dipped steel sheet products are particularly useful in appliances for their heat-reflective qualities and in exhaust system components where temperatures are as high as 538°C. Because of their high surface quality, sheet products are used in buildings and other applications as a replacement for more expensive alloy steels.

Lead, Terne, and Tin. Hot dipping of pure lead [7439-92-1] is not used extensively despite its low cost which is less than tin and zinc. Lead does not alloy with steel and must be alloyed with other metals, eg, tin, antimony, zinc, or silver which improve fluidity and facilitate bonding with iron. Terne metal [39428-85-8, 54938-78-2] is more commonly used and contains 15–50% tin. It is applied where atmospheric corrosion resistance in the absence of abrasion is required at low cost. It is made predominantly in continuous electroplating lines up to 1067 mm wide with a coating weight of 150 g/m^2 and a nominal one-side thickness of 15 μm. British Steel Corporation produces 18,000 t/yr (24).

Hot-dipping of tin also has been superseded largely by electrolytic coating techniques. However, changes in effluent standards for electroplate wastes may revive hot dipping to some extent. Hot dipping of tin coatings usually is done in a two-compartment cell or pot that is partitioned in the upper region only, so that the plate enters through a flux layer, continues into the molten tin on one side of the partition, and leaves the tin bath emerging through a palm oil layer at the other end of the cell. Rollers are immersed in the hot oil layer and they control the thickness of the finished plate. Most of the oil is removed after the plated steel leaves the cell; the small amount that remains protects against in-storage discoloration and acts as a lubricant in subsequent forming operations. The coating is produced by the formation of a layer of $FeSn_2$ [12023-01-7]; its growth is limited by diffusion to practical thicknesses of ca 0.5 μm. The thickness of the outer coating of pure tin depends on the speed of the plate leaving the pot and the pressure of the rollers. Products, eg, wire, are tinned in a similar fashion.

A new terne-forming process is being implemented by Broderick Structures, Ltd. The material is a composition formed by cold-roll bonding at high pressure. Prior to the roll bonding, the sheet is treated resulting in the formation of a thin terne plate which, on contact with lead sheet during the rolling operation, becomes integrally and permanently bonded without deformation of the underlying steel. This process is the most economical solution to meet the need for structural lead sheet. The process should broaden the applications of lead, particularly for architectural uses.

Terne is used in the auto industry for gasoline tanks; in roofing eg, flashing; in plumbing, eg, for laboratories; and as a gasket material. More than 90% of tin plate is used as tin cans for food packaging. Tin-dipped wire resists corrosion by sulfur from rubber insulation layers.

Zinc. Over 40% of all zinc that is produced is used to protect steel products; the corrosion rate of iron is 25 times that of zinc in the atmosphere and in water. Zinc is anodic to its normal metal impurities and to steel, has a high hydrogen overvoltage, and forms insoluble basic salts. World production of galvanized steel is ca 1.4×10^7 metric tons per year and involves over 100 galvanizing lines of which ca 65 are Sendzimir lines (25). About 5×10^6 t/yr of galvanized sheet are produced in the United States.

The most important advance in hot-dip galvanizing is Sendzimir's process by which the surface is preoxidized at 650°C and then hydrogen-reduced at 850–950°C. The temperature is lowered to 400°C with the strip still protected in hydrogen until it enters the zinc bath. In this way, flux at the entrance to the bath is avoided and small amounts of aluminum are used to inhibit formation of zinc–iron intermetallic intermediate layers. The bath temperature is maintained at 450–460°C by the sensible heat of the incoming strip.

The important intermetallic region consists of three successive layers on the steel, namely, Fe_3Zn_{10} [12182-98-8], $FeZn_7$ [12023-07-3], and $FeZn_{13}$ [12140-55-5], followed by a thicker layer of pure zinc. The intermetallic region is one tenth of the total coating region. The most modern lines involve jet finishing rather than rolls because line speeds can be increased from 76–92 m/min to 185 m/min which increases the economic benefits of the process (20). As the strip rises vertically out of the zinc bath, it carries an entrained viscous layer of molten zinc. A row of horizontal jets of air are impinged perpendicularly to the strip with one on each side, as shown in Figure 1, and cause a return flow of liquid metal into the bath. Sensors above the row of air jets meter the thickness of the coating and adjust the velocity of air flow by electronic feedback circuits so that the desired thickness on each side can be maintained continuously throughout the run.

Armco Steel Corp. adds 0.01–0.10% magnesium in the galvanizing coating with 0.2–0.4% aluminum or 0.3% chromium which provide significant improvement in the atmospheric and marine corrosion resistance of galvanized steel (see also Coatings, marine). In a test involving 200 wet–dry cycles in seawater at pH 8.4, a threefold reduction in weight loss corrosion is observed for Zn–0.29 wt % Cr–0.4 wt % Mg which compares well with standard galvanizing (0.17 wt % Al–0.2% wt Pb).

The use of standard two-sided galvanized steel, eg, in automotive bodies, causes some difficulties in welding and the adherence and brightness of paint coatings is not as good as on bare steel. Consequently, many producers make a one-side, hot-dipped, galvanized product or electrolytically strip the thinner side of a differentially coated product. Paint adherence properties of a galvanized product have been improved by galvannealing which consists of heat treating the coating in-line to reduce the surface zinc to an iron–zinc intermetallic layer so that it accepts paint more readily than normal galvanized surfaces. Inland Steel produces sheet of which one side is galvannealed and the other side is hot-dipped zinc.

With regard to paint adherence on a standard galvanized product, research has shown that paint retention depends on the orientation of the zinc crystals in the spangle

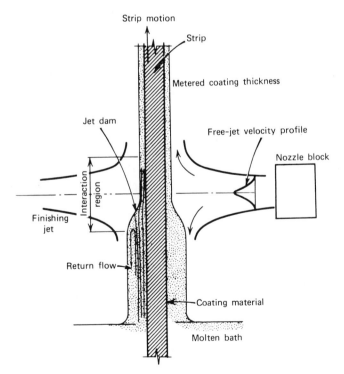

Figure 1. Schematic representation of jet-finishing process for hot-dip coating of strip.

(26). The crystal orientation affects the amount of carbonaceous residue which causes poor paint adherence. The sulfur-containing organic matter present on the surface is more readily removed by alcohol washing when the basal plane of the crystals is parallel to the steel substrate. Therefore, processing of a galvanized product to produce such a spangle orientation is preferred and undoubtedly will become accepted practice where the product use requires painting.

The characteristic spangle of galvanized sheet results from the rate of crystallization of the molten zinc, which depends on the condition of the starting steel and the presence of minor additions to the melt. The latter lower the melting point of the zinc and, thereby, lower the cooling rate of the molten layer. Large producers, eg, National Steel, offer regular spangle, minimized spangle, and flat bright; the latter is recommended for painted parts. Commercial grades of standard galvanized sheet, eg, G90, has 0.275 kg/m^2 of zinc amounting to a 0.19-μm thick layer on each side. The heaviest grade, G235, has ca 3 times as much zinc as the G90 material.

The various methods of applying zinc to steel surfaces have been compared with regard to capital costs and maintenance, and it was concluded that hot-dip galvanizing was the most advantageous method of application (25). A recent report included analyses of emissions from 17 hot-dip galvanizing lines; total particulate emissions from hot-dip galvanizing operations in the United States is estimated to be ca 1600 t/yr (27). Waste disposal from one galvanizing wire-coating operation involves two 36,000-L sulfuric acid pickling tanks and one dip-type rinse tank in an ammonium-fluxed, zinc-coating line. After being collected, neutralized, and aerated, the rinse water is discharged into the city sewer (28). Oxidized, solid sludge also is acceptable for discharge, eliminating the need for clarification equipment. Based on these limited examples, it can be tentatively concluded that no significant environmental hazards are associated with hot-dip zinc galvanizing.

Aluminum–Zinc. Within the past several years a hot-dip-processed coating has been introduced by Bethlehem Steel Corporation. The product is a 55% Al–Zn coating on steel; it is called Galvalume in the United States and Zincalume in Australia (29–30). The coating offers the resistance of aluminum with the galvanic protection of zinc to protect cut edges. The coating is applied by dipping at 593°C. The coating thickness on each side of the sheet normally is 20 μm and the thickness is controlled by automated jet finishing. The coating consists of a multiphase outer region containing 80 vol % α-Al, 22 wt % Al–Zn eutectoid, and a silicon-rich minor constituent. A layer of a quaternary intermetallic Fe–Al–Si–Zn (10% of the total thickness) is bonded to the steel at the steel interface. These regions are depicted in the micrograph of the coating cross section shown in Figure 2. Corrosion tests conducted over 14 yr show that the equivalent thickness of 55 wt % Al–Zn offers 2 to 6 times longer life than regular galvanized steel in a variety of atmospheres, including marine exposure (31). In moist condensate, 55 wt % Al–Zn outlasts galvanized samples after 4 yr. At temperatures up to 700°C, the 55 wt % Al–Zn is equivalent to type 1 Al-coated samples. Current production rates of 350,000 t/yr will increase to 750,000 t/yr in the early 1980s. The principal uses are for building roofing and siding and for high temperature parts used in appliances and automotive parts. In many products, it has replaced both aluminum-coated and galvanized sheet because of its superior performance and lower cost. Its salt-corrosion resistance makes it suitable for automotive body parts and its high temperature resistance allows its use in mufflers and exhaust-system components. Because it can be offered in high strength grades and has exceptional corrosion re-

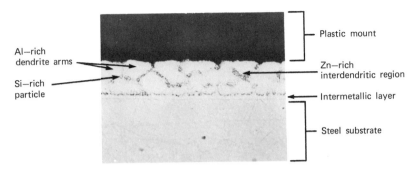

Figure 2. Random cross section of a 55 wt % Al–Zn coating (×500, Amyl-Nital etch). The steel substrate is bonded by a thin intermetallic layer which is overlaid with a solidified layer containing aluminum-rich dendrites, a zinc-rich interdendritic region, and silicon-rich particles.

sistance, it is expected that the number of uses will continue to grow as more information on its performance becomes available.

Cementation Coatings. The cementation process is conducted in a mixture of inert diluent particles, eg, alumina or sand; the coating metal in powder form; and a halide activator, which is poured or packed into a metal container with the part to be coated. The container usually is sealed against air entrainment but is arranged so that volatilization of the activator can drive the air from the pack mix. The pack is heated to 800–1100°C and held for ca 1–24 h, depending on the thickness of the desired coating. The coating metal is transferred to the basis material by the formation of a volatile metal halide which is transferred through the pack mix by volume diffusion. Decomposition of the halide at the part surface provides a coating metal which diffuses into the basis metal, thus forming compounds as dictated by the equilibrium phase diagram but limited by the activity of the coating metal. This activity often is controlled by using alloys or intermetallic compounds of the coating metal. The size of the part that can be coated in this manner is somewhat limited by the time required to heat a large pack vessel and the ability to heat the contents uniformly. Small parts, eg, turbine blades, screws, and nuts, are handled commercially. However, pipe and tubing as long as 14 m and up to 2 m in diameter that is intended for refinery and chemical-plant service have been coated by pack aluminizing (32). The pack cementation process has been applied primarily to Al, Cr, B [7440-42-8], and Zn (Sherardizing), although some work has been done with Si, Ti, and Mo [7439-98-7] and many other elements on an experimental basis. Details regarding many of the diffusion coating techniques and possible elements that can be utilized have been reported (33). Two new pack cementation processes have been described; one involves magnesium in the development of an adherent, diffused, sacrificial coating anodic to the substrate (34) and another involves manganese-rich layers which work-harden rapidly on running-in, to form a wear-resistant, adherent coating on carbon steel (35).

Aluminizing. Pack-aluminized coatings on superalloy turbine blades and vanes increase the parts' resistance to oxidation and sulfidation at high temperatures. This high temperature resistance to gaseous corrosion results from the formation of an adherent Al_2O_3 protective oxide film which grows very slowly at high temperatures and acts as a barrier to further metal loss.

There are two very different coating structures that can be obtained for diffusion aluminum coatings, depending on the activity of the aluminum in the pack mix: inward and outward-diffusing types. When the aluminum activity in the pack is low, nickel is the predominant diffusing constituent ($D_{Ni}/D_{Al} = 3$ where D is the diffusion coefficient. Units are cm^2/s.), whereas at the stoichiometric composition, this ratio changes abruptly and aluminum becomes the predominant diffusing species in high aluminum, NiAl [12003-78-0] ($D_{Ni}/D_{Al} = 0.1$) (36–38). Similar studies on Al–Fe interdiffusion have been carried out (39).

One of the problems with pack cementation processing is the cost of removal of the pack constituent from the coated parts, particularly those parts having intricate passages or reentrant angles. A pack process that has good throwing power for complex parts with internal passages can be obtained with parts that are held over the pack mixture rather than being immersed in it; this also alleviates the problem of pack-mix removal. A pack process has been developed which involves either $NaCl:AlCl_3$ at the stoichiometric weight ratio of 3:7 ($NaAlCl_4$) or Na_3AlF_6 as the activator (40). Gas-phase deposition of aluminum in the pack process increases with decreasing partial pressure of the transporting agent, and the rate of deposition is highest for the most stable aluminum alkaline-earth halide. This process undoubtedly will be commercially important for the coating of advanced turbine hardware with intricate air-cooling passages.

Diffusion aluminizing of steels for petrochemical applications, eg, the protection of ethylene pyrolysis tubes and downstream heat exchangers from coke formation and carburization is being considered by Alon Processing, Inc. Applications are expected to increase for coal-gasification and -liquefaction systems to provide resistance to sulfidation, carburization, and abrasion.

A new application of the pack aluminizing process has been developed by the Alloy Surfaces Co., which is using the process to prepare catalytic surfaces called BD catalysts. These materials are made by first subjecting a nickel metal or nickel-alloy substrate, eg, a gauze, sheet, screen, or other desirable configuration, to the pack aluminizing treatment to form an intermetallic compound, eg, NiAl. A leaching operation partially removes the aluminum to form a fragmented, high surface area coating. The catalyst, which is similar to Raney nickel, can be used in methanation; it imparts high durability and the opportunity for exceptional space velocity. Other basis metals and diffusing elements also have been used to form catalytic surfaces on metals, eg, platinum, silver, rhodium, iron, palladium, stainless steel, and titanium. Because aluminum diffuses readily in Ni, Cu, Fe, Ti, Pt, and Pd and readily forms intermetallic compounds with them, a significant emerging source of novel catalytic materials will be available which is based on the pack cementation process.

Chromizing.   Chromizing is a very popular and economical process by which the corrosion and wear resistance of low cost steels can be improved. The pack chromizing process, like aluminizing, is a relatively simple method of providing a diffused surface region on a steel part or on a nickel-base alloy. The chromizing process is carried out in a pack containing 30–60% chromium; a few percent of an activator, eg, NH_4Cl or, preferably, NH_4I; and inert diluent, eg, Al_2O_3. Heating the pack for 20–24 h at 950–1100°C produces a diffused coating layer ca 150–200 μm thick. During the initial period in the pack, the process is controlled by decomposition of the gas but, during the remainder of the treatment, the growth of the diffused layer is governed by the diffusion of chromium into the base metal. The chromizing of Armco iron has been studied (41). At the pack temperature, the initial diffusion process begins in the gamma phase (fcc) but, when the chromium content reaches ca 12%, a transformation to alpha occurs. This results in a moving front of alpha phase that has a higher diffusivity for chromium than for the gamma phase (at 1273 K, Cr diffuses 3.5 times faster in alpha than in gamma) so that a sharp discontinuity in chromium concentration occurs at the phase boundary. Chromium also diffuses more rapidly in the grain boundaries and tends to precipitate a $Cr_{23}C_6$ carbide phase and, at the same time and because of the affinity of chromium for carbon, the region below the coating is decarburized. These changes in composition affect the mechanical properties of the coated object, eg, the fatigue resistance; depending on the steel, the chromizing process does not provide any increase in fatigue life and may substantially deteriorate it. Nevertheless, sheet metal components of low carbon steel can undergo extensive forming operations after chromizing without cracking, spalling, or peeling even if bent 180°. If the steel to be chromized contains >0.3 wt % carbon, a chromium carbide forms on the surface of the steel as a result of the diffusion of carbon within the steel toward the chromium-rich surface; the resultant coating is not as corrosion resistant. If corrosion resistance is desired on these steels, they can be decarburized prior to chromizing, or it is possible to use steels with small percentages of the carbide stabilizing elements Nb, Ti, or Zr, eg, the high strength, low alloy steels, to achieve the required strength.

The hard chromium carbide surface coating may be desirable for wear resistance. The carbide grows outwardly in contrast to the inward diffusion of chromium in low carbon alloys, and the thickness of carbide depends on the rate of carbon diffusion through the carbide. The rate of diffusion of Cr in steel, like that of aluminum in NiAl, also depends on the stoichiometry since lowering the chromium activity in the pack increases the extent of diffusion. The type of carbide which forms depends on the chromium and carbon content in the steel. High Cr, low C steels form 23:6 carbide. High Cr, high C steels form a 7:3 carbide with a possible 23:6 outer layer. Low Cr, high C steels may form 23:6 and 7:3 carbides but will form a Fe_3C-based cementite between the 7:3 carbide and the substrate. Low Cr, low C steels may form either 23:6 or 7:3 carbides in equilibrium with the gamma phase.

A growing market for chromized steel is for automotive hardware, eg, mufflers. Pump components, which have been made from cast iron or stainless steel, can be made from chromized gray cast iron which increases component life by a factor of 10. Output of tube-drawing dies has been improved by chromized–carbide layers. Small components, eg, nuts, bolts, washers, etc, can be chromized by utilizing a rotary furnace. New and service-worn gas-turbine blades of both nickel and cobalt-base alloys are conditioned for service by chromizing.

Chrom-Aluminizing. Although dry corrosion resistance of either aluminized or chromized basis metals and alloys is excellent, further improvements are effected using a two-step process of chromizing followed by aluminizing. Above ca 900°C, chromized materials begin to rediffuse which results in a reduction in the surface chromium concentration. Aluminizing following chromizing can change the nature of the protective oxide from Cr_2O_3 to Al_2O_3; the latter is more stable and is less volatile at high temperatures. Thus, improved scaling resistance is possible above 1000°C and performance is superior to 18-8 stainless steels and Inconel (42). Two-step chrom-aluminizing treatments have been optimized for a number of nickel- and cobalt-based alloys used in high performance turbine engines. However, these duplex pack cementation coatings are more expensive and, therefore, are not widely used for common applications.

Boronizing. Boronizing is a diffusion process by which a boron-rich layer, which is similar to a carbided or nitrided surface layer, can be formed on steel or on another substrate. Extremely hard surface layers result, provided borides are formed. The pack cementation process is one of several techniques for boronizing and is carried out in closed retorts into which the parts to be boronized are packed with a mixture of 50% boron powder, 49% alumina and 1% $NH_4F(HF)$. Ammonium bifluoride acts as an activator to carry boron to the surface of the pack. After being sealed, the retort is heated at 800–900°C for 6–24 h. The steel is heat-treated for mechanical strength after boronizing and the nature of the coating is not affected because its hardness is a result of the formation of boride intermetallics. There are two iron borides, FeB [12006-84-7] and Fe_2B [12006-85-8], at 16.25 and 8.84 wt % boron, respectively; the former compound is the harder of the two but generally is not used because of its increased brittleness. The diffusional thickness of the boron compound follows a parabolic relation with time and is greatest for low alloy steels. It is difficult to form boronized layers exceeding 10 μm on alloy steels with Cr or refractory elements, eg, a high speed steel. On plain carbon steels, a 150-μm layer can be formed in ca 6 h. Boron has limited terminal solubility in the allotropic iron structures; thus, Fe_2B is the only constituent of the coating regardless of the application of postcoating treatments designed to modify it. If the activity of boron in the pack is high, FeB forms in the saturated Fe_2B. The integrity of the coating depends on the nature of the interface between the basis metal and Fe_2B. If there is a jagged saw-toothed interface, as occurs with plain carbon steels, the interfacial strain is accommodated more effectively and the coating is more adherent and less prone to microcracking. On the other hand, the smooth interface resulting from boronization of alloy steels leads to less desirable adhesion, and spallation and microcracking are likely, particularly at corners and edges. Thus, the corrosion resistance of boronized plain carbon and alloy steels differs widely.

The outstanding feature of the boronized coating is its hardness, relative to nitriding and carburizing layers, on steel. The Vickers microhardness of boronized steel typically is between 1600 and 2000, whereas a nitride layer is ca 600–900 and a carburized surface is ca 700–800. Heating to 1000°C affects the hardness of the latter coatings but not that of the boronized layer. The great affinity of boron for oxygen results in a thin oxide layer on the boride; the layer appears to provide an antiwelding surface which reduces the interaction of other metals with it. Therefore, boronizing is a particularly effective measure to improve the wear characteristics of steel surfaces (43). In recent tests of the relative cost effectiveness of various materials exposed to a number of abrasives, it has been demonstrated that boronized low alloy steel is as much as three times better than plasma Ni–Co–Cr carbide and significantly better than manual arc-coated WC (44). One company in the UK (Ronson Products Ltd.) uses boronized mild-steel jigs to carry parts through an abrasive polishing process.

Boronized parts, eg, Borofused wire dies, also show chemical corrosion resistance to HCl, HF, and/or H_2SO_4. However, there is no evidence of protection against corrosion by HNO_3 nor is there significant improvement in regard to conventional rusting of ferrous alloys, probably because of the tendency to microcracking which results from the brittle borides. Molten zinc is very corrosive to mild steel parts; one solution in galvanizing is the substitution of boronized mild-steel fixtures for the previously used more expensive titanium.

Siliconizing. In siliconizing from the pack, the source of silicon can be elementary silicon, ferrosilicon, or silicon carbide. The inert diluent material usually is Al_2O_3 and the activator generally is NH_4Cl at ca 2–5% of the weight of the pack mixture. Siliconizing of Armco iron or low carbon steel is not an efficient process; eg, 10 h of siliconizing at 1100–1200°C is required and there is considerable consumption of the pack source. Iron articles are limited in the silicon content of the coating layer to 5–12% silicon at the expense of a considerable increase in grain size. Although the silicon-rich surface improves the corrosion resistance of the material in weakly corrosive media, the high temperature corrosion resistance above ca 700°C of silicon-containing coatings that are applied in this way is not satisfactory, the improvement over untreated material being less than a factor of two. However, siliconizing followed by chromizing results in thicker diffusion layers, because silicon aids in stabilizing the alpha phase in which chromium diffuses more rapidly. Siliconizing is more effective as a deposition process for coating the refractory metals, eg, Ti, Nb, Ta, Cr, Mo, and W, where silicides can form. The pack silicide coating for Mo is one of the oldest and best-studied coating systems. The coating depends on the formation of $MoSi_2$ [1317-33-5] and the oxide responsible for protection is SiO_2 [7631-86-9]; the latter effectively protects molybdenum in air up to 1700°C. Small $MoSi_2$-coated molybdenum rocket engines were used very successfully in the Apollo program for attitude control of the command module and Lunar Excursion Module (LEM) (45). Where very high temperature oxidation resistance is required, the silicide-coated refractory metals offer the best performance.

Sherardizing. Sherardizing or dry galvanizing is the oldest and the most widely used pack cementation-type diffusion process (46). Parts to be coated are cleaned and packed in zinc dust in a metal container which, after sealing, is rotated slowly and heated to 350–375°C for 3–12 h. The diluent phase is comprised of zinc oxide and other impurities which are obtained by recovering used dust and by innoculating the new charge. Iron and iron oxide are undesirable contaminants and must be removed periodically from the charging inventory.

The coating consists of a zinc-rich intermetallic, $FeZn_7$ [12023-07-3], which can be accompanied by $FeZn_3$ [60383-43-9] if the process is carried out at >375°C or, if the part is subsequently heat treated. Because it is a diffused coating, the surface structure of the part is replicated in the coating and uniformly applied over the surface. As with other intermetallics formed by diffusion, the coating tends to be microcracked. However, the $FeZn_7$ is anodic to the steel and protects it sacrificially.

Because of its uniformity of coverage, the Sherardizing process is unexcelled in providing protection on steel hardware, eg, nuts, bolts, washers, and other close-fitting parts. It is likely that a steady market will persist for this coating. However, with the advent of mechanical plating and ion plating as techniques for fastener improvement, the range of materials coated by Sherardizing will depend on the specific application and the relative economies of these processes.

Liquid-Carrier Diffusion Coatings. Diffusion coatings also can be made by immersion of the basis metal in a liquid bath containing the dissolved coating metal. The bath can be composed of fused salt mixtures or liquid metals, eg, Ca or Pb, which can dissolve small amounts of the coating metal but in which the basis metal is not dissolved. The coating takes place because of the difference in activity of the coating metal in the bath and that of the basis metal. Generally, because of limitations in the amount of material that can be dissolved in such baths, the surface composition of the coating metal in the basis metal is lower than for other coating processes. Other limitations are related to the physical problems of containing and operating high temperature baths which may corrode containers and sometimes damage the basis metal.

Fused Salt. Transfer of metal can be accomplished in molten salt mixtures. One such bath for chromizing has a composition of 40 mol% NaCl, 40 mol % KCl, and 20 mol % $CrCl_2$ (47). The $CrCl_2$ is formed by adding $CrCl_3$ and sufficient Cr metal to reduce the trichloride *in situ* to $CrCl_2$. Carbon steels at 1000°C form chromium carbides in accordance with the Fe–Cr–C-phase diagram. In 30 h of exposure, 11 μm of $M_{23}C_6$-type carbide is formed at the outer surface over a 24-μm thick layer of M_7C_3. The carbides grow mainly by diffusion of carbon from the interior of the steel through the carbide layer to the outer surface where it meets chromium and forms the carbide. One difficulty with the operation of fused-salt baths is the necessity to scrupulously avoid contamination by oxygen. This is usually accomplished by purging with purified argon.

Toyota Central Research and Development Laboratories has announced a carbide-coating process which involves the immersion of parts into a molten-borax bath at 800–1200°C in ambient atmosphere (48). Metals are placed in the bath in the form of powders. Carbides, eg, Cr_7C_3, VC [12070-10-9], and NbC [12069-94-2], are formed on the surface of carbon-containing parts. Articles, eg, metal-working dies, knives, and machine components, have been coated with carbide layers that are characterized by excellent wear, abrasion, and corrosion resistance. Applications for this technique are expected to grow worldwide.

Boriding of low carbon steels can be done in fused mixtures of boric acid and potassium borate and diffused coatings containing titanium carbide and boron carbide can be formed on carbon steels from fused mixtures of boric acid, TiO_2, and potassium fluoride at 1020–1170°C (49). Diffusion layers 200–250 μm thick can be formed in 1 h or less.

Liquid Calcium. One liquid-metal diffusion-coating technique involves the use of molten calcium as the transfer agent (50). A solubility of only 0.1% for the coating element is required in the calcium bath for rapid transfer of the solute to the part. An argon-protected calcium bath with ca 10 wt % of the desired coating element is prepared, heated to 1100°C, and agitated, and the part to be coated is immersed for ca 1 h. In this manner, a 50-μm thick coating of Cr is obtained on iron in which the surface concentration of Cr is 45%. On removal from the bath, the part can be cooled in air or quenched in oil and, once it is cool, excess calcium may be removed in hot water or dilute HCl. Multiple diffusions can be made simultaneously by addition of the appropriate concentrations of the required elemental constituents. In this way, Ti, V, Cr, Mn, Co, and Ni have been simultaneously diffused into iron. However, with iron substrates, it usually is difficult to obtain coatings with less than 50% iron at the surface. Outward diffusion of carbon in iron also can occur and, if the bath contains chromium, carbides form in the bath until the chromium is saturated in carbon; then, chromium carbide ($Cr_{23}C_6$) forms on the surface of the iron. It is possible to select chromium or chromium carbide as the coating for substrates with high carbon or cast iron because calcium also is an effective decarburizing agent. Liquid metals are efficient media for transfer of metallic and nonmetallic elements between two metals and diffusion can be as rapid as when the two metals are in direct contact.

The surface of austenitic stainless steel can be transformed to a ferritic layer by removing the Ni into the bath and increasing the chromium level by treatment in a calcium–chromium bath. Such a ferritic layer can improve the chloride stress-cracking resistance of the austenitic steel. High carbon steel and cast iron can be decarburized at the surface and coated with a ductile layer of corrosion-resistant chromium. Oxidation-resistant coatings containing Al, Cr, Si, and Ni can be applied to steel from calcium baths. Refractory metals, eg, Mo, can be coated readily with Al, Si, or Cr. An example of a multielement coating composition on steel is 45 wt % Cr–52 wt % Fe–2 wt % Ni–1 wt % Al; this coating has excellent resistance to the CASS (copper accelerated salt spray) test (3). The practical exploitation of this process can be expected to be retarded by the problems of fire safety and control of alkali fumes associated with the finishing process. It is more likely that less hazardous liquid-metal baths, eg, lead, will provide a more economical approach.

Lead Bath. Materials Sciences Corp. has developed a process called Dilex 101 to chromize steel using molten lead baths in ca 1100°C for 4 h (51). The optimum chromium content of the bath is 0.85%, based on the weight of lead. The alloy bath may involve other diffusing elements, eg, Co, Ni, Y, Mo, Ti, Nb, V, Ta, W, Si, and Mn. Multiple-element, diffusion coatings, eg, Cr–Al coatings for heat-exchanger materials, have been applied successfully by this method.

It is not likely that this process can be used for parts that are required to operate at high temperatures because of the possibility of contaminating the part with lead which can deteriorate the creep-rupture properties of nickel-based alloys and high alloy steels. However, many other applications are being investigated, eg, coating powder metallurgy (qv) parts, wire, tubing, valves, fittings, etc.

Spraying

Sprayed coatings generally are applied to structures or parts which either are not conveniently coated by other means because of their size and shape or are susceptible to damage by the heating requirements of other coating techniques. Slurry coatings and electrostatic powder coatings require heating to the fusion temperature either by massive heating of the part or by localized heating, eg, by induction, electron-beam, or laser techniques and generally in a protective atmosphere.

Flame-spraying and arc-spraying techniques are used in a large variety of industrial applications, eg, in both shop and field situations because equipment usually is portable and can be taken to the work site. However, laser, electrostatic, and slurry coatings must be formed in the shop.

Flame Spraying. *Oxyacetylene.* Flame spraying is the simplest of the thermal spray techniques; it is used where heating the substrate above 315°C would cause undesirable tempering, recrystallization, oxidation, or warping. Flame spraying uses oxyacetylene or oxypropane flames with flame temperatures at ca 2750°C which is adequate to spray most ferrous and nonferrous coatings and oxides, eg, alumina and zirconia, but only to densities from 85–95%. Either wire or powder is fed into the flame. The heat of the flame melts the coating material and accelerates it toward the workpiece where particles fuse as interlocking laminates, each layer fused to the previous one. The flame oscillates over the part surface, giving a uniform coating. More than one pass can be made with several materials as required; thus, worn parts can be salvaged by building up material in the work area and machining back to the original dimension (52). One utility has found that rebuilding their transit-department vehicles has been consistently less expensive than the cost of replacement (53). Worn shafts, axles, packing sleeves, and journals are sprayed with a nickel–aluminum bond coat followed by aluminum bronze to a 3.2-mm thickness before remachining. Flame-sprayed materials, usually Zn or Al, also have been applied to large storage tanks and to at least one bridge for corrosion protection (54–55).

Detonation Gun. The detonation gun, invented by Union Carbide, overcomes many of the limitations of the flame-spray process. The detonation gun is a cannonlike device that detonates metered mixtures of oxygen and acetylene in a combustion chamber. Powder particles of the coating material, which are carried in a nitrogen stream, also are metered into the chamber prior to spark ignition. The shock wave leaving the barrel at ca 2770 m/s accelerates the powder particles to ca 770 m/s. Particles also are heated by the 3000°C combustion gases at which temperature most coating materials melt. The high temperature of the particles at this characteristic velocity produces a coating having exceptionally high bond strengths, ie, >98 MPa (>14,200 psi), and porosities <1%. The detonations are repeated 4–8 times each second and are accompanied by short nitrogen purges to clean the barrel.

The majority of coating materials applied by the detonation gun (D-gun) are oxides and carbide mixtures with various bond metals, eg, NiCr and Co. The coatings provide exceptional resistance to wear (56). A disadvantage of the process is the supersonic velocities that are produced and that require double-walled, soundproof cubicles and remote-control operation. The process also is line-of-sight, which limits the type of product that can be produced. The D-gun process has been estimated to be the most expensive of the spray techniques, depending on part size, fixturing, masking, and use (57). However, initial cost is compensated by the superior life of these coatings, which often outlast conventional metal spraying and weld surfacing by 8 to 1.

Arc Spraying. **Wire-Arc.** In wire-arc spraying, two wires are fed to a gun through two electrical conduits which bring the wires together at a 30° angle. On contact, an arc is struck and melts the wire ends. Compressed air drives the liquid metal forward to the work. As the arc is broken, the wires are advanced to repeat the process. The arc temperature (ca 3800°C) causes deposition of molten droplets ca 3–8 times faster and with more fluidity than oxyacetylene flame-spray units. At an arc current of 250 A, 7, 12, or 42 kg/h of Al, stainless steel, or zinc, respectively, can be sprayed. The bond quality is better than that for flame-sprayed material. The equipment is light and portable and is simple to use with any coating material which can be made into wire form. One arc-spraying facility includes an enclosure 7 × 13 × 3-m high which is purged and filled with purified argon which also is the gas used to drive the molten material from the arc-spray unit (58). Operators are provided with space suits equipped with independent breathing apparatus. Large components up to 2.75 m in diameter have been coated with 1-mm thick coatings of titanium. One experimental heat-exchanger tube sheet which was coated in this protective chamber has been running for more than 5000 h in a seawater desalination plant with no problems. The high quality of the coating product facilitates machining, welding, and forming operations. Moreover, freedom from internal porosity is obtained. These coatings have been produced without the substrate heating requirements of weld surfacing yet they are characterized by good bonding and high density.

A definite advantage of the arc-wire spray is the simplicity of operation and the absence of the degree of noise which is characteristic of plasma torches. Inability to deposit material that is not fabricated into wire is perhaps a disadvantage but more of the material reaching the substrate surface is molten leading to, in many cases, much better bonding and excellent high density.

Plasma. The plasma-spraying process utilizes the available energy in a controlled electric arc to heat gases to $\geq 8000°C$. The low voltage arc is ignited between a water-cooled tungsten cathode and a cylindrical water-cooled copper anode. Argon, nitrogen, or hydrogen or suitable mixtures of these gases are heated in the annulus and are expelled at high velocity and temperature into a characteristic flame. Powder material, either metallic or nonmetallic, is fed into the flame just downstream of the anode. Particles of the powder are melted and accelerated toward the work to be coated. Since it is hotter than the oxygen–gas flames or arc-wire systems, the plasma device can deposit W, Mo, tungsten carbide, and numerous ceramic materials. Usually the particle velocity is 125–300 m/s but, with plasma guns working in subatmospheric chambers, velocities of up to 460 m/s and extremely fine, dense, and wear-resistant coatings can be obtained. Metco, Inc. has supplied a significant portion of plasma-spray equipment, powdered material, and applications technology. The structure differences and mechanical behavior of plasma- and detonation-gun coatings have been reviewed (59). Plasma-sprayed metallic coatings have characteristic porosities of 5–15%. A great deal of effort has been made to increase the particle velocity and thereby reduce the porosity to improve the corrosion resistance of these coatings, particularly at high temperatures. An improved plasma process called Gator-gard is characterized by a particle velocity of 1230 m/s, which provides coating densities >99%. Deposit density and efficiency also is related to the uniformity and purity of the powder materials. A narrow distribution produces the best results because large particles may pass unmelted through the flame, whereas small particles are vaporized and lost. Many companies can provide spherical, uniform, high quality powder material by various techniques. If a metallic envelope can be provided over the sprayed material then hot isostatic pressing can be used to heal internal defects and, thereby, improve the finished density of plasma-sprayed coatings (60).

The plasma-spray coating technique has been used to deposit molybdenum and Cr on piston rings, cobalt alloys on jet-engine combustion chambers, tungsten carbide on blades of electric knives, wear coatings for computer parts, etc. The technique is used to increase component life and to reduce machinery down time (61–62) (see Plasma technology).

Laser Coating. Laser power is applied to produce sprays of powdered material for coating purposes (see Lasers). Powder particles can be accelerated in the laser beam and melted before striking the substrate material where rapid solidification takes place (63). Power from a 25-kW CO_2 laser is directed into an evacuated chamber and focused on the substrate surface. A carrier gas, eg, helium, is used to transport powdered material through a gold-plated water-cooled nozzle which projects powder into the laser beam. The accelerated molten particles impact on the substrate and, depending on the power density at the substrate, either coat the substrate with a quenched structure or are incorporated into a region of the substrate which has been melted by the beam. Solid particles, eg, carbides, in the latter process can be incorporated into the molten matrix with little dissolution, thereby producing a modified surface region that is up to 1 mm thick and impregnated with hard particles for wear resistance. A similar process, called laser alloying, uses high power laser energy to melt a thin layer of material at the surface of a part, to which alloying elements are added, to produce a chemically modified region on subsequent solidification (64). In one example, a paint that contains an alloy with 30% chromium is placed on the outer edge of an engine valve. A ring-shaped laser beam scans the painted area and produces a chromium-rich region in the valve. A cost analysis of laser alloying versus conventional hard-facing techniques indicates that an 80% savings can be achieved by the laser process (65). In laser cladding, a prepositioned coating material is melted so that it is bonded to the substrate surface but not alloyed with it, much as in braze surfacing, conventional hard facing by plasma techniques, or flame spraying (66).

For example, a Stellite Alloy No. 1 is prepared as a hard surface material by the oxyacetylene flame technique and has a structure with massive carbide particles with high (Rockwell) hardness, ie, HRC 70, imbedded in a soft matrix of HRC 45. The average hardness is HRC 51. However, the laser-clad Stellite Alloy No. 1 produces a uniquely homogeneous structure with extremely fine carbide particles, resulting in an average hardness of HRC 60. The cost/benefit of laser-clad coatings results not only from the unique microstructures produced but from the use of lower cost powder material, reduced amounts of necessary coating, rapid processing rate leading to high production rates, and lack of postcladding machining or clean-up operations. The projected in-production coating rate for the laser-cladding operation on a valve seat is 5 s compared with 15 s for other processes. The laser beam can be time-shared at many work stations to prorate the high cost of the laser equipment. In addition, laser beams can be directed into small holes that are inaccessible by other spray-coating techniques. However, the cost of the laser and optical system is a present limitation.

The laser will be used increasingly in the generation of amorphous coatings, eg, by the Laserglaze technique (67). Combinations of laser processing with other techniques may be useful in certain instances. Laser fusing of flame-sprayed Metco 15F (Ni–17 wt % Cr–4 wt % Fe–4 wt % Si–3.5 wt % B–1 wt % C) produces high quality coatings up to 0.25 mm thick on mild steel 10 times faster than by torch methods (68). Evaporation of high melting metals, eg, Pt, which, because of their high thermal conductivity, are difficult to evaporate by conventional electron-beam techniques can be accomplished with lasers. Some success with platinum laser-vapor deposition has been reported for the fabrication of beam-lead integrated circuits using a 20 W YAG laser (69).

Electrostatic Powder Coating. A relatively new application of an old technique is the electrostatic deposition of powders, eg, Al, Cr, Ni, and Cu. British Steel Corp. has been developing a line for an electrostatic aluminum-deposition process (Elphal) with an ultimate capacity of 75,000 t/yr of sheet steel (70). Cleaned strip or sheet is electrostatically coated with metal powder, cold-rolled to compact the coating, and sintered by heating to develop a bond to the steel. The principal advantage is the protection provided at 500°C in service, eg, heating appliances, automotive exhaust systems, and heat exchangers. USSR work on this type of process also is underway. Aluminum alloy powder that is sprayed onto aqueous alkali silicate-coated steel has been deposited to 150 g/m² at 7 kV/cm. The strip, which is dried at 350°C, is heat-treated at 500°C for 1 h to develop optimum corrosion resistance (71). The chief advantages of the electrostatic method are the wide variations in coating composition available through prealloying of the powder, the absence of any hot-metal baths, and the conservative use of coating material resulting from the self-leveling and low over-spray characteristics of the process. However, because of the relatively limited markets for specialty alloy coatings on rolled sheet, the role of this type of process in sheet and strip coating probably will not be as large as for hot-dipped or electrocoated material.

A related process involves electrophoretic deposition from a liquid mixture of powdered material (72). Usually, 0.04-mm (325-mesh) powders of the metallic coating material are added to a stirred solution of isopropanol and nitromethane at between 2 and 10%. Addition of ionic agents stabilizes the dispersion by electrostatic repulsion. A d-c potential of ca 100 V is applied between two stainless steel electrodes and the item to be coated. Deposition rates are high, eg, 0.005–0.05 mm/min. Particles, not ions, are plated out by electrophoretic migration. Uniformity of such coatings are excellent as a result of the self-leveling effect on the initial deposit which attaches at points of high field strength. The insulating effect of the coating reduces the rate of deposition in these regions in favor of points with lower field strength. Once applied, the coatings are densified by hydrostatic pressing followed by sintering or sintering alone if coatings are not to be handled in the green state. The sintering temperatures can be as high as 1100°C, depending on the coating and basis material. Among the advantages cited for this method are the ambient temperature of application, uniformity, versatility regarding the material used in the coating, ease of automation, and low cost which is one half that of pack cementation coatings and one tenth that of vapor deposited coatings that offer similar protection (see Powder coating).

Slurry Coatings. Metallic coatings can be applied simply by applying powders of the desired metal or alloy in a paint medium and by brushing, dipping, or spraying it onto the basis material. The coating is cured and then fired, during which time, the organic portion vaporizes and the metallic particles fuse to form a dense, metallurgically bonded coating. The coatings are versatile, both as to their composition and the shapes of the basis materials that can be coated. Slurry coatings were developed for commercial use in the 1960s for gas-turbine hardware and for coatings on niobium alloys. A Si–20 wt % Cr–20 wt % Fe, fused slurry coating is used commercially on niobium-alloy afterburner components for the F-100 jet engine and on niobium-alloy rocket nozzles for the space shuttle where the coatings are exposed to temperatures of 1370–1650°C. Similar coatings were developed for NASA by Lockheed for tantalum–tungsten alloy components (73). A family of nickel–chromium slurry coatings, Nicrocoat, was developed for high temperature, and corrosion and abrasion resistance by Wall Colmonoy Corp. for furnace fixtures, heater tubes, thermocouple protection tubes and automotive parts (74). More recently, Lockheed Palo Alto Laboratories has developed a slurry coating of 33 wt % Cr–63 wt % Al–4 wt % Hf for IN 800 alloy which is exceptionally resistant to sulfidation (75). Homogenization after coating by laser-fusion treatment has increased the corrosion and abrasion resistance of these coatings.

Combinations of slurry-fusion and pack cementation techniques have been developed for experimental evaluation by the Solar Division of International Harvester Company for advanced turbine-engine applications designed specifically for operation in marine environments where protection against sulfidation attack is critical. Slurries of Ni–Co–Cr have been applied to parts and allowed to react in cementation packs to deposit aluminum. The coatings that are produced protect against hot corrosion and oxidation as well as contemporary, vapor-deposited coatings. The low cost and versatility of combination methods of this kind is promising for other technologies where hot corrosion problems are recognized as limitations, eg, in high sulfur crude processing and coal-gasification units for synthetic fuel production (see Fuels, synthetic).

Mechanical and Liquid-Metal Cladding

Metallic coatings also can be applied by mechanical methods where the coating material is forced into intimate contact with the basis metal such that the forces at the interface disrupt and disperse the boundary oxide films existing on each of the constituent metals. Formation of a metallurgical bond may be augmented by mechanical attachment and thermal interdiffusion. Metallic coatings also can be melted into place by a weld-surfacing or casting operation. These techniques generally are applied to large, heavy basis metals in the form of plates or large forgings and usually produce quite thick coatings in comparison to other types of processes.

The selection criterion for a given method of producing a composite plate is the required thickness of the desired protective material and the thickness of the basis material to which it is applied. If 1 cm is adequate, the solid coating (cladded) alloy often is used alone. For a basis material greater than 1 cm and up to 6 cm thick, roll-bonding techniques are used. Explosion bonding is most often used when the basis metal thickness is ca 6–8 cm (see Metallic coatings, explosively clad metals). Beyond a 10-cm basis-metal thickness, only weld-overlay techniques or electroslag-casting techniques are practical.

Roll Bonding. The main sources of roll-bonded, clad plate are the Lukens Steel Company and the Phoenix Steel Corporation. In roll bonding heavy plate, a four-ply sandwich is made in which two basis-metal backer plates enclose two of the clad metal plates which are nickel plated on the sides exposed to the backer, and a parting agent is between the clad so that, when the rolling is complete, the sandwich can be opened to produce two clad plates (see Abherents). The nickel electroplate metallurgically bonds the clad to the backer. The sandwich is sealed around the edges by strips that are welded to the backers before hot rolling.

One of the most important applications of roll-bonded heavy sheet is in the fabrication of clad vessels. One of the largest and most experienced fabricators in the United States is the Nooter Corporation of St. Louis, Mo. Most of the vessels are used in chemical plants, eg, for acid-gas removal from process streams. A growing market is clad vessels for coal-gasification plants. Another new market is in shipbuilding. The 62-m ship Copper Mariner has been built with roll-bonded 90 wt % Cu–10 wt % Ni cupronickel [11114-42-4] on steel plate, with the clad occupying 10% of the composite thickness. Fuel savings and lower maintenance costs resulting from lack of fouling with marine organisms is between $5,000 and $10,000/yr. In terms of the compensated tonnage coefficient benefits, LNG carriers should benefit the most from clad construction. A minimum practical thickness for normal shipyard handling is 0.15 cm clad which is fabricated ideally by roll bonding. For certain specialty service in chemical-process equipment, tantalum-lined vessels are being produced with 0.25–0.38 mm thick elastomer-bonded tantalum [7440-25-7] sheet on steel plate.

Skive inlaying is another roll-bonding process which usually involves a precious metal bonded to a freshly skived or machined surface by pressure rolling, as shown in Figure 3 (76). The freshly skived surface is atomically bonded to the precious metal surface which contacts it for a short time in comparison to that required to develop an oxide film. The process provides savings of 50–90% in comparison to electroplating because the precious metal that is used is restricted to the contact area. Many inlays may be atomically bonded without heat, including solder, Cu–Ni–Si alloy, copper, silver, and gold. These inlays can be bonded into steel, stainless steel, aluminum, brass,

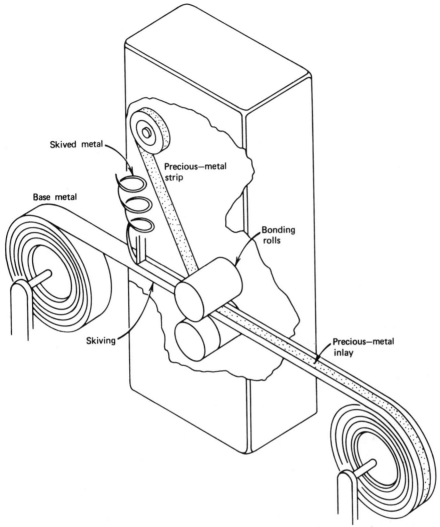

Figure 3. Pictorial view of skive-inlaying technique. A base-metal strip is skived by a blade to reveal atomically clean metal. A precious-metal strip is roll-bonded immediately into the skived groove to produce the inlay. No adhesives or other bonding is required.

bronze, and beryllium–copper. Coils up to 3050 m long and 1.6–102 mm wide have been made. Configurations to suit almost any design requirement can be produced, including edgelays, multiple stripe, and full-width cladding (76). Some of the applications for this new technology include contact springs for data-processing equipment and bonding pads for automobile voltage regulators.

Powder Rolling. Clad strip also can be produced from two kinds of powder by introducing two different powders on opposite sides of a baffle plate that extends longitudinally between two rollers. As the powders fall into the rolls, they are pressure-bonded into a strip with a green density that is dependent on roll speed and roll gap; thin sections have higher green density than thick sections. Sintering at 800–

1000°C follows. Copper and iron have been produced in a 0.1-mm composite strip and characterized by an ultimate strength of 275 MPa (40,000 psi) and an elongation of 25% when sintered to 900°C (77). Chromized steel strip, a development of Bethlehem Steel, has been produced by passing the steel strip, which is lightly coated with tridecyl alcohol, over a fluidized pumping bed of ferrochrome powder. The bed contains brushes which mechanically impinge loose powder up onto the sheet from the freeboard of the fluid bed. The powdered sheet is roll-bonded after being reversed so that the opposite side of the sheet can be coated over a second fluidized bed (78). The strip must be annealed before rolling so that it is soft enough to be penetrated by the powder on contact with the rolls. After rolling, the open-wound coils are diffused at 885°C in hydrogen for 28 h. The coil then is wound tightly to complete the process. The chromium content after heat treatment is 25% at the surface and tapers to 15% at a depth of 0.05 mm.

Roll bonding of Al–Zn alloys is accomplished in a similar way by passing steel sheet that is treated lightly with tridecyl alcohol through a powder bed of a mixture of 26-μm Al and Zn powders. About 325 g/m^2 attaches to the work which then is compacted by work rolls. The strip is heated to 399°C in an 18% H_2–N_2 atmosphere for 5 min to bond the coating. An ambient-temperature skin pass finishes the sheet by providing a smooth attractive surface. The coating is ductile, adherent, and metallurgically bonded (79). New powder compositions also include silicon additions which improve the interface alloying properties.

Mechanical Impingement Coatings. The mechanical plating process developed by 3M Company is the cold welding of a ductile metal onto the surface of a metal substrate by mechanical energy. The process is designed for small parts, eg, fasteners, and is a barrel process in which parts, water, glass shot, and proprietary additives are tumbled with fine metal powder. The glass shot and powder sizes are 0.2–5.5 mm and ca 4 μm, respectively. The plate thickness usually is 0.00245–0.0177 mm but can be as high as 0.076 mm. Alloys can be plated by mixing powders in the barrel. Generally, the coatings are limited to Cd, Zn, Sn, Pb, In, Ag, Cu, brass, and tin/lead solder (80–81). The main advantages of mechanical plating are the absence of hydrogen embrittlement; the coatings can be made with alloys; there is a simple, time-dependent control of coating thickness; the energy requirements are low; it is a room temperature, nonfuming process; and it requires little predisposal waste treatment of used solutions. However, it is limited to small parts and cannot produce a cosmetic coating with fine surface finish; nor is it possible to deposit nonductile powders, eg, Cr or Ni, or active metals, eg, Al.

Mechanical plating is used to coat ca 3 × 10^6 critical parts per month in one plant and provides corrosion protection without hydrogen embrittlement failures; costs are about the same as for electroplating with afterbake (82).

Mechanical vapor plating, which was developed by Alloying Surfaces Co., involves modulated pulsations which expose powder particles of material to as much as 4425°C at the particle–basis metal interface. Materials, eg, WC, TiC, Mo, Ni, Cr, and Ni borides, are deposited to a 0.45-μm finish at 6 cm^2/min (83). Applications for this process include wear-resistant surfaces and special tooling, eg, metal-forming dies and cutting tools.

Explosive Bonding. *Weld Surfacing.* Weld-deposited overlay coatings for small area requirements usually are achieved using manual stick electrodes or metallic inert-gas-shielded arc (MIG) or tungsten inert-gas-shielded arc (TIG) techniques (see

Welding). For larger areas, single or multiple submerged arc, strip cladding, and automated plasma-transferred arc spraying of powders or wire generally are used. The choice of cladding material and the method of application is determined largely by the intended purpose; the product or process to be contained or the general surface property requirement; metallurgical compatibility; relative coefficients of expansion, possible unfavorable circumstances required of the site for the cladding operation, eg, access to the surface; and the inspection requirements for the operation.

Plasma-Transferred Arc Cladding. Plasma and arc techniques can be used for clad overlay deposition. One technique uses a plasma torch in the transferred-arc mode to control basis-metal melting, whereas powders which are introduced in the normal plasma-spray method provide the overlay material. Greater efficiency and deposition rate over oxyacetylene and gas tungsten arc methods are obtained. Deposit rates are 4.5–5.5 kg/h. In gas–metal–plasma arc-weld cladding, the plasma arc is transferred to the work but by an independently controlled power supply (84). A second supply source powers an arc between the two advancing wires and the plasma to melt filler metal. Figure 4 is a schematic diagram of the plasma unit, the current path from power source no. 1 which controls basis-metal melting, the arc-wire circuit powered by source no. 2, and the mechanism for advancing the overlay metal in wire form. In this process,

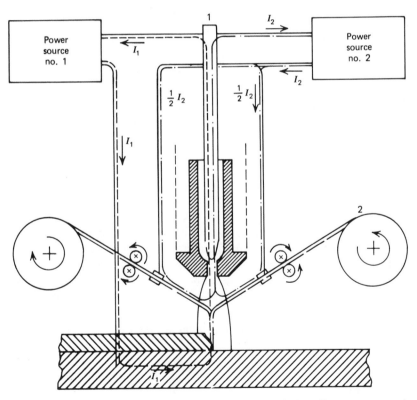

Figure 4. Sketch of gas–metal–plasma arc-weld surfacing technique. Power source no. 1 controls the heat from the plasma into the basis metal. Power source no. 2 controls the heat input from the plasma to the advancing filler wires, thereby independently providing control of melt rate and degree of interface mixing. 1, plasma torch; 2, wire feed. I_1 = current of power source no. 1. I_2 = current of power source no. 2.

which does not utilize a slag cover but depends on the plasma gas and an outer plenum gas as a pool shield, a mechanical oscillator moves the weld head perpendicular to the welding direction to produce a wide weld bead, which can be observed through a window in the protective gas-shield cap. The melting process and metal transfer to the weld pool are smooth and stable even at production deposition rates up to 80 kg/h. The unique control that is possible with independent supplies should make this process widely applicable to situations permitting wire-fed overlay material in a mechanized manner.

Electroslag Cladding. A technique similar to the submerged-arc process, but which works with more massive equipment and produces much thicker coatings, is the electroslag weld-overlay coating process. The basic process was developed in the USSR's Paton Electric Welding Institute in Kiev (85). The heat energy necessary to melt the basis metal and the filler alloy is generated by electrical current in a molten, electrically conducting slag which also purifies and protects the filler metal as it advances through the slag layer into the molten pool. The molten slag and metal pool are held against the basis metal by water-cooled copper retainers, which must conform to the contour of the object to be clad. One application for this process is in the refurbishing of mill rolls which can be performed on ESR (electroslag remelted) ingot-production units with only minor change (86). In operations, eg, cladding of steel arbors, deposition rates of 16 kg/h per electrode are possible with up to 45 electrodes used in unison and positioned appropriately around the circumference of the roll. Rolls up to 5.5 m long and 0.6 m in diameter have been produced.

A comparison of costs for submerged-arc and electroslag processes for similar cladding situations has been made (87). Electroslag is ca 25% more efficient than submerged arc, largely because of the lower labor requirements of the former process. An additional advantage of electroslag processing is cleanliness and reduced slag inclusions which contribute to a reduced coating spallation rate that often results in longer service life of the roll cladding.

Chemical Coatings

Chemical Vapor Deposition. Chemical vapor deposition (CVD) is the gas-phase analogue of electroless plating (qv): CVD is catalytic, occurs on surfaces, and involves a chemical reduction of a species to a metallic or compound material which forms the coating; the reactions are temperature dependent but occur at much higher temperatures (10–13,88–93). A CVD process involving a metal carrier compound, which is reduced by a gaseous reducing agent to deposit a metallic coating, is distinguished from the purely thermal decomposition of an unstable compound into its parts, one of which is a metal.

Chemical Deposition. The types of chemical reactions utilized in CVD are reduction reactions, eg,

$$WF_6 + 3\,H_2 \rightarrow W + 6\,HF$$

displacement reactions, eg,

$$SiCl_4 + CH_4 \rightarrow SiC + 4\,HCl$$

and disproportionation reactions, eg,

$$2\,GeI_2 \rightarrow Ge + GeI_4$$

These reactions generally require temperatures from 500 to 1200°C and often as high as 1500°C. Moreover, the structure of a given deposit may be different at different reaction temperatures. Much of the development work has centered on reactions that allow the deposition to proceed at lower temperatures. For example, a W_3C [12012-18-9] coating on steel can be produced at 300°C (94). Because few basis-metal substrates can tolerate high temperatures, the key to wider application of CVD lies in lowering the deposition temperature to make the process applicable to common structural materials.

Chemical vapor deposition coatings tend to be purer than non-CVD-produced coatings. Control of chemical composition is a matter of controlling the gaseous reactants entering the reactor; graded coatings and mixed (sequential) coatings are possible through selection of appropriate gases. Because the reactants are gaseous, the throwing power of the process is excellent. However, the kinetics of the deposition involves hydrodynamic flow and diffusion processes which, if not taken into account, can affect the quality of the deposit.

The largest bulk industrial application of CVD coatings is in wear-resistant overlays, eg, TiC [12070-08-5] and TiN [25583-20-4] on cemented-carbide cutting tools, but they must be applied at 800°C. A layer of TiC that is 4–8 μm thick can increase the tool life fivefold; α-alumina coatings deposited by CVD on cutting tools increases the wear life still further. However, the most technically sophisticated application of CVD techniques is in the production of electronic materials, where the extreme purity of CVD deposits and the variety of elements and compounds that can be deposited are a distinct advantage. The types of possible electronic materials produced by CVD include semiconductors (qv), insulators, conductors, magnetic materials (qv), and superconductors (95–96). However, the lack of relevant experimental data to characterize the various operational parameters is a major obstacle to the acquisition of a fundamental insight into CVD processes that must develop in order to broaden the utilization of this technique for electronic purposes. Notwithstanding, the applications of CVD to a variety of technologically important applications in the industrial sector continues. Tantalum coatings 20–30 μm thick which impart adequate acid corrosion resistance have been deposited on the inner surface of long carbon steel pipes by hydrogen reduction of tantalum pentachloride (97). A CVD silicon coating that is 50 μm thick has been developed for a nickel-based superalloy (Nimonic 105) by hydrogen reduction of $SiCl_4$ at 1090°C (98). The coating is ductile and corrosion resistant at high temperatures. The need for wear-resistant coatings in components for coal-gasification pilot plants has lead to the development of CVD techniques for deposition of TiN on low carbon steel at and below 1000°C. Ball seat valves up to 0.3 m in diameter have been coated (99). Chemical vapor deposition may be utilized in the production of solar absorber stacks because continuous multicomponent fabrication techniques, which can be achieved by changing the fractional composition of the reactant gas stream, can produce a deposit sequence, ie, silicon, silicon nitride, silicon oxynitride, silicon dioxide (100) (see Solar energy). Such graded refractive-index profiles provide antireflection properties over a large incidence angle and, thereby, provide superior performance in thermal-collector designs involving large optical acceptance angles. By substitution of electron kinetic energies for thermal energy, plasma discharges promote CVD at low temperatures (101). Lasers also have been used as a heat source in CVD because of the localized nature of the heat source and because of the avoidance of excessive substrate heating. Films of TiO_2 from mixtures of $TiCl_4$, H_2, and CO_2 have been prepared in this way (102) (see Film deposition techniques).

Thermal Decomposition. A CVD-related process is the thermally induced decomposition of a compound into a metal and a gaseous by-product. It differs from CVD in that no reducing agent is required; also the reagent need not be a gas. For example, when solid films of previously evaporated silver chloride are contacted in vacuum by an electron beam, they can be made to dissociate to form metallized silver. In this way circuit paths for integrated microelectronic devices that are 0.43 μm wide and 0.7 μm wide are made (103).

The decomposition technique is particularly attractive for *in situ* deposition on powder assemblages prior to pressure densification. Tungsten-coated Eu_2O_3 powders have been produced by decomposition of $W(CO)_6$. In a slightly different way, tantalum-coated powder is made by first blending the Eu_2O_3 powder with a slurry of $TiH_{0.5}$ in amyl acetate which acts as a binder. Thermal decomposition of the hydride produces tantalum-coated oxide particles which can be pressed and sintered to form a cermet (104) (see also Glassy metals). Vapor-formed nickel deposits on plastic injection-molding dies have been prepared commercially (105). The carbonyl is fed from storage containers to a vaporizer–mixer chamber where it is diluted with a carrier gas and fed into the coating chamber at atmospheric pressure. Nickel is deposited on the mold and is characterized by excellent throwing power, and brightness and its cost is one third that of producing comparable machined steel molding dies. A similar process involving the decomposition of nickel carbonyl has been used to produce 20-μm thick dendritic nickel coatings for selective photothermal energy absorbers at \$0.76–1.79/m^2 (1980) (106). Because of the low temperatures of these types of decomposition reactions in comparison with other CVD reactions, new applications for this coating method should continue.

Vacuum Coatings

In vacuum deposition, the desired coating metal is transferred to the vapor state by a thermal or ballistic process at low pressure. The vapor is expanded into the vacuum toward the surface of the precleaned basis metal. Diffusion-limited transport and gas-phase prenucleation of the coating material is avoided by processing entirely in a vacuum that is sufficiently low to ensure that most of the evaporated atoms arrive at the basis metal without significant collisions with background gas. This usually requires a background pressure of 0.665–66.5 mPa (0.005–0.5 μm Hg). At the basis metal, the arriving atoms of coating metal are condensed to a solid phase. The condensation process involves surface migration, nucleation of crystals, growth of crystals to impingement, and often renucleation. Thermal sources based on resistance (I^2R) heating, induction heating, electron-beam heating, and laser irradiation have been used to vaporize the coating material. These processes are physical vapor deposition techniques, ie, thermal energy raises the material to its melting point or above, whereupon it is vaporized by evaporation and adiabatic expansion. In high rate depositions, the reaction forces of atoms leaving the surface form a depression in the surface of the liquid. In these processes, the energy of the physically evaporated atom usually is fractions of an electron volt, depending on the physical properties of the coating material, ie, melting point. If the evaporated coating metal is made to intercept atoms or ions of a special background gas with which it may react to form a compound and then strikes the basis metal, the process is reactive evaporation.

The coating material also can be maintained in a solid form and then suitably bombarded by positive ions of rare gas generated by a glow discharge or other ion source. The coating material or target generally is negatively biased by several hundred to a few thousand volts. The high velocity ions that impinge on the coating material dislodge surface atoms by sputtering. Sputtered atoms are ejected from the coating metal surface with energies of between one and ten electron volts. In the sputtering process, the high energy ions impart a fraction of their energy in a collision cascade within the target that reaches back toward the coating metal surface. About five percent of the energy in the collision cascade reaches the surface and results in sputtered atoms; the rest of the energy is dissipated in the target as heat. Sputtering usually is carried out in an inert gas at 0.13–1.3 Pa (0.001–0.01 mm Hg) so that a glow discharge can be supported by electron impact which provides a source of ions to maintain a steady-state process. If the sputtered atoms are made to intercept a reactive gas species on their way to the substrate basis metal or other object so that a compound is formed which subsequently condenses on the substrate, the process is reactive sputtering. Alternatively, the coating material first can be evaporated or sputtered to vapor and then partially converted to the ionized state by electron bombardment. In ion plating, the ionized portion of the vapor cloud can be accelerated to high energies, eg, a few thousand electron volts, by maintaining a negative bias on the substrate so that ions sputter on arrival at the basis-metal substrate.

If the accelerating voltages are high, eg, 80–100 kV, the ions become permanently embedded as atoms in the near-surface region of the crystal lattice of the basis metal and the process is ion implantation (qv). At these energies, the sputtering process is minimal because the collision cascade does not impart sufficient energy at the surface to eject many surface atoms. Having been neutralized by conduction electrons, the ions become part of the surface of the target. If the ions are not of a rare gas but ions of metal atoms, they chemically alter the alloy's near-surface region. The properties of the near surface may be drastically changed by this internal alloying by the implantation process and, in certain circumstances, the near-surface region of the basis metal can be made amorphous in the affected region.

An enormous variety of elemental metallic films, compounds, semiconductors, insulators, and amorphous coatings have been made by these processes. Metallic coatings have been applied by vacuum-deposition techniques to a host of substrate materials, including paper, cloth, metal strip, metal parts of various sorts, plastic, glass, and semimetals. These techniques have been used most widely in the complex multistep coating processes involved in generating electronic microcircuits and memory elements (see Integrated circuits; Vacuum technology).

Physical Vapor Deposition. *Evaporation.* Electron-beam vapor sources are used extensively for metallic coating because they can be fed continuously with solid evaporant material either as wire or rod (107–108). Evaporation of alloys requires that a constant rate and inventory of liquid is maintained. When melting begins, the more volatile elements are gradually depleted in the inventory of liquid. With time, the liquid becomes enriched in the less volatile elements and the vapor composition becomes constant and identical to the composition of the feed (109). In this way, many alloy compositions have been successfully evaporated to form alloy coatings (110). The process is used to coat high temperature, corrosion-resistant coatings, eg, CoCrAlY [59299-14-8], for gas-turbine blades and vanes, and coating rates of up to 30 μm/min are achieved. Dense coatings with minimal defects are produced by glass-bead peening and heat treating to at least 950°C (111). However, two disadvantages occur with evaporated coatings: the atoms arrive essentially on line-of-sight from a virtual point source, necessitating that the objects to be coated are suitably rotated if uniform coatings are desired; and the temperature gradients in the source cause chemical and density gradients in the vapor cloud so that, depending on the position of the substrate, the composition of the deposit may vary. For microelectronic-device manufacturing where electron-beam sources are used, elaborate part-rotation operations involving planetary devices ensures uniform coating thickness and chemistry on the silicon chips.

A continuous electron-beam evaporation system with a total evaporation power of 150 kW has been developed to coat industrial 3 × 3.6-m plate glass for architectural applications (112). The cost of decorative evaporated coating, depending upon the part design, is as cost effective as electroplating and, in some circumstances, less expensive (113).

Vacuum metallizing is potentially effective on ABS (acrylonitrile–butadiene–styrene), nylon, polycarbonate, or Noryl that has been base-coated with polyester or urethane using a Cr–15 wt % Fe–5 wt % Ti metallization coating followed by an acrylic or urethane top coat (114). A cost of ca $0.25/m^2 of coated surface for such coatings or about one half that for electroplating can be achieved. Cost advantages of vacuum metallizing are chiefly in coating large parts (115).

Reactive Evaporation. In reactive evaporation, a compound deposit is produced by the reaction of the evaporant and a chemically active background gas (116). In activated reactive evaporation, an electrical discharge from a probe above the evaporation source, as shown in Figure 5, increases the reaction cross section, thereby increasing the probability of compound-producing collisions (117). Compounds, particularly carbides, eg, TiC, can be formed at high deposition rates with adjustable stoichiometry, depending on the preselected ratio of hydrocarbon gas and evaporated titanium atoms. A boron alloy evaporated in the presence of ammonia has produced a coating containing 20% of cubic boron nitride, a form of hard nitride previously formed only at very high pressures (118). The properties of coatings produced by reactive evaporation and related processes are given in refs. 12–14.

Ion Plating. Ion plating is conducted with either evaporation or sputtering to provide a vapor deposit of material on a basis metal or substrate which is maintained at a negative potential (119–121). The background gas (eg, argon) pressure in evaporative ion plating is increased to 0.1–0.15 Pa (0.75–1.15 μm Hg) in order to create a negative glow around the object to be coated. Argon ions bombard the substrate and remove undesirable oxides and other contamination from the surface. When evaporation or sputtering begins, most of the atoms from the vapor source enter the dark space, ie, the ion-accelerating region near the surface, around the part; and become partially (1%) ionized and, therefore, accelerated in the dark space toward the substrate or part surface. The energy over the thermal energy acquired by the coating atoms provides substrate heating and high surface mobility of the coating atoms. The energy is sufficient to imbed some of the atoms 0.1–0.2 nm beneath the surface. The material in the near surface region is highly disrupted and defects are introduced, which promotes coating–substrate interdiffusion. The result of this surface activity is greatly improved coating adherence, uniformity, and the ability to deposit a substantial fraction on the dark side of the vapor source, eg, on the surfaces not in line-of-sight with the source. The introduction of high rate, electron-beam evaporation sources and

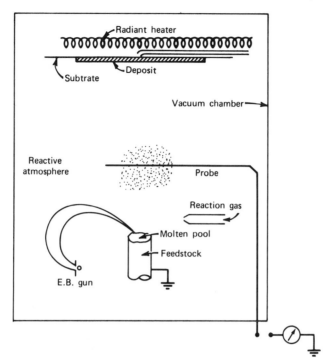

Figure 5.   Schematic illustration of activated reactive-evaporation process. An electron-beam vapor source produces an expanding vapor cloud. The reaction gas and metal vapor are partially ionized by glow discharge, which is maintained by current from the probe, to facilitate compound formation. The compound metal coating is condensed on the heated substrate.

induction-heated vapor sources have improved the technology (122–123). A major commercial application, called Ivadizer, is the production coating of aircraft fasteners with aluminum by ion plating (124). Cost of the process is competitive with many other conventional processes, and there are no toxic wastes or fumes associated with it. Also, no hydrogen embrittlement is possible as with electrochemical techniques.

Sputtering and Ion Implantation.   When the source material is maintained in the solid state and is bombarded with ions from a gas discharge, surface atoms are dislodged with high energy. If the discharge is created between the substrate and the coating metal or target, the process is diode sputtering. If a hot filament and anode circuit provide a separate low voltage electron-beam discharge, the process is triode sputtering. Alternatively, application of r-f power to the target results in r-f diode sputtering. Dual targets allow a-c sputtering. If the process is carried out in a strong transverse magnetic field it is magnetron sputtering. Many other variations and combinations exist, but the fundamental process of atom ejection remains the same. Addition of substrate biasing capability provides precoating etching and sputter–ion-plating possibilities. Background gas dopants that are added to the discharge gas provide reactive sputtering. All of these variants have been utilized to provide better deposit properties, ie, the ability to sputter dielectric materials; the achievement of a high deposition rate; and the production of compounds, highly pure materials, or epitaxially grown deposits. Glow-discharge sputtering has been reviewed recently (125).

Magnetron Sputtering. A magnetron discharge is used to achieve high rates of deposition by sputtering. Recently, an enormous growth in the applications of the process has taken place. The applications of magnetron sputtering have been reviewed for cylindrical magnetron sources (126) and for planar magnetron sources by Waits (127). Deposition power as high as 80 kW into sputtering targets as large as 0.35 × 1.8 m can be achieved (128). Because radiation damage by fast electrons and neutrals is so much less in magnetron sputtering than with previously used sputtering techniques, the former process is being considered for thermal-sensitive semiconductor device fabrication as a supplement to evaporated coatings. Magnetron sputtering also is used to produce experimental coatings on laser-fusion target microspheres that are 100–500 μm in diameter (see Fusion energy). Sputtering rate increases from 0.75–12.6 kA/min for platinum are expected to open up many new applications for magnetron sputtering of platinum-group metals. Selective solar-absorber coatings that are deposited by magnetron sputtering appear to be inexpensive (129). Automotive applications, chiefly for exterior trim, are expected to increase to $1 × 10^7$ in 1980 largely because of an 80% reduction in energy costs over electroplated coatings and the very low environmental effects of the technique.

The use of magnetron sputtering for coating gas-turbine components has been studied (130). A special advantage of sputtering for these components is the ability to coat very complex alloy systems that contain elemental constituents with both very high and very low vapor pressures; these alloys are difficult to apply by other techniques. The structural metallurgy associated with high rate sputtered deposits for coatings applications has been discussed (131). Substrate temperature, smoothness of the interface, and rate of deposition are the principal factors which influence coating structure. Future advantages for the process are the suppression of undesirable phase constituents and the development of dispersoid distributions by the incorporation of rapid solidification processes into the magnetron-sputtering technique.

Ion Implantation. When coating atoms are ionized and achieve sufficient energy, they can be driven into the surface of the basis metal to form a near-surface region where the chemistry has been altered by the addition of the ions, which are neutralized on entry into the host lattice (132). Because this process requires sophisticated ion-beam accelerators which function at 80–150 kV, the cost is very high, ie, $1–5/cm^2 (133). Ion-beam doping of GaAs, GaP, and other compound semiconductor materials is common. A 30-kV, high throughput ion-implantation system for B, P, As, and Sb doping of 7.6 cm thick wafers through a mask has been operating since 1974 and involves the use of beam currents of 5 mA; 450 wafers per hour are produced with doses up to 10^{15} ions per square centimeter (134). Application of ion implantation to the formation of metastable phases in metals has been suggested (135). Because the lattice temperature that is produced by an incoming energetic ion may be 500–1000 K, it has been calculated that a cooling rate of 10^{12}–10^{15} K is probable and is far in excess of the 10^6 K/s rate that is characteristic for liquid quenching of metals. Many atoms have been implanted in a copper host lattice as substitutional solid solutions, eg, Ag, Sb, I, Xe, W, Pt, Au, Hg, Tl, Pb, and Bi (136). If the dose rate of the ion is increased, for example Ta in copper, the surface structure becomes completely amorphous. In another example, phosphorus that is implanted into 316 stainless steel induces a surface amorphous state (137).

The applications of ion implantation to tribology and corrosion science have been reviewed (138). Wear rates of a nitrogen-implanted nitriding steel are an order of magnitude lower than the unimplanted material. Nitrogen that is implanted into cutting knives for paper and high speed steel taps for phenolic plastics increase the useful lives of the parts two- to fivefold. The fatigue life of AISI 1018 steel is increased twofold by nitrogen implantation. The oxidation of a chromium-rich stainless steel at high temperature is reduced by implantation of Y ions which cause a reduction in the rate of oxide spallation (139). Ion-implanted Cr in steel produces the same corrosion resistance as an Fe–Cr alloy (140). A commercial Fe–18 wt % Ni–8 wt % Co maraging steel can be implanted with Cr to improve corrosion resistance without the problems associated with prolonged heating of the alloy structure during chromium cementation coating (141). A number of other studies have been made to demonstrate the effect of ion implantation on the corrosion resistance of stainless steels (142), aluminum, and titanium. For example, titanium implanted with 10^{16} Pd atoms per square centimeter at 90 kV produces a 5-at %, Pd-enriched, subsurface region. The corrosion potential of the Pd-implanted sample in boiling, $1 M$ H_2SO_4 is ca 1000 mV more noble than the pure Ti and quite close to that of pure Pd; a dramatic reduction of the corrosion rate also is observed (143). A possible but untried combination for near-surface materials modification is the incorporation of laser annealing, laser fusion, and ion-implantation techniques (144). Because of its cost, ion implantation likely will remain largely a microelectronics fabrication method. It has already become an important part of magnetic bubble memory technology in that it is used to control bubble states, flux cap the bubbles, induce desirable increases in the local lattice parameter, and increase the etching rate (145). Because it is a clean, well-controlled process, ion implantation may be used eventually to mitigate corrosion, particularly where very costly implantable materials can be made to function as protective coatings (see also High temperature alloys).

Health and Safety Factors

All coating techniques are based on technologies that have inherent hazards, eg, high temperature, the use of liquid-metal or molten salt baths, high voltages, and often, the use of toxic chemicals. However, the coatings industry and the suppliers of coating equipment have provided carefully engineered systems, which if handled according to recommended practice, are relatively benign with respect to operator and user personnel hazards.

Abrasive cleaning can create oxide and metallic particulates which must be collected and carefully disposed of. Each abrasive process must be considered separately along with the basis-metal requirements for the established air-loadings limits for personnel exposure. Acid-etching baths and organic solvent washes must be constructed, ventilated, and controlled to avoid operator hazards. Information regarding established exposure levels to the types of materials discussed below are reviewed in ref. 146.

Diffusion coating can involve liquid-metal baths which must be isolated from operator contact but, more importantly, must be clear of water sources which could provide an explosion hazard. Corrosion of metal containers for metal baths also must be considered and accepted practice must be followed to avoid unexpected spills of liquid metal. In coating processes involving lead, vapors, excess coating material, and cleaning effluents must be properly controlled and disposed of. Tin, which has a low melting point and low vapor pressure and is nontoxic, is the most hazard-free of the hot-dipped coatings. Zinc, aluminum, and aluminum–zinc hot-dipped coatings require some care because of their higher melting points, greater reactivities and the nature of the fluxes required in their related processes.

Cementation coating is safe because of the use of closed retorts. Care must be exercised in some pack-dismantling operations regarding the use of fluoride activators. Proper inventory of pack materials, adequate ventilation, and dust-avoiding work stations are required and should be monitored to ensure compliance with published requirements. Similarly, care must be exercised with liquid-metal carrier baths and fused-salt baths to avoid moisture and to prevent undesirable release of either the bath materials, their fluxing agents, or their coating constituents.

Sprayed coatings are less hazardous than hot-dipped ones since the volume of liquid metal produced at any instant is quite low. However, sprayed coatings are largely hand operations involving torches. Aerosols are of concern mainly with plasma spraying and sometimes arc-wire spraying; water walls or properly vented enclosures are necessary. The concerns in this type of coating operation are not unlike those in welding operations and the same general rules for safe practice apply. NIOSH has concluded that dust and fumes from metallizing by flame spraying at Shell Oil Company's refinery at Wood River, Illinois, does not constitute a health hazard. Measures have been taken to avoid personnel contact with materials which, in some instances, had caused dermatitis (147). Arc and plasma methods have the additional hazards of radiation burns resulting from exposure to the uv radiation that is released during operation; welding practices for operator safety must be followed. Laser-coating operations are similar, in that direct contact with the radiation must be avoided and care is required to avoid difficult-to-detect stray reflections. Laser-beam processing in conjunction with established standards can be considered a safe operation. In those operations, eg, flame spraying, plasma spraying, electrostatic powder coating, D-gun and slurry coating, powder rolling, etc, which involve the use of submicrometer powders of the coating material, care is required in the handling, storage, and use of these powders to prevent dust and possible dust explosions and, in many cases, reactions with water or other reagents that might otherwise be nonreactive for bulk forms of these materials. Specialists in the coatings technology area should always be consulted to determine the precautions required for each powder material.

In CVD processes, the coating-metal carrier often is a volatile, reactive liquid or gas, often a fluoride or other halide which is easily hydrolyzed to acid halides. However, these reagents have established storage and delivery requirements and manufacturers can supply details regarding the proper utilization of these reagents and the measures required to safeguard against operator exposure.

Vacuum coatings, of necessity, are produced in well-controlled environmental chambers where air, moisture, and radiation are isolated by the walls of the device. Often high voltages are used as in electron-beam evaporation and sputtering and in the latter, r-f power at high voltage is common. Proper control of electron-beam evaporation and sputtering power sources and their distribution and entry points into equipment is necessary; properly engineered and maintained equipment is required and reputable manufacturers of such equipment should be consulted when necessary to ensure continued immunity against the possible operator hazards involved.

BIBLIOGRAPHY

"Metallic Coatings" in *ECT* 1st ed., Vol. 8, pp. 898–922, by W. W. Bradley, Bell Telephone Laboratories, Inc.; "Metallic Coatings" in *ECT* 2nd ed., Vol. 13, pp. 249–284, by William B. Harding, The Bendix Corporation.

1. A. Pinto, *Mod. Packag.* **52,** 25 (1979).
2. "Properties and Selection of Nonferrous Alloys and Pure Metals" in *Metals Handbook*, 9th ed., Vol. 2, American Society for Metals, Metals Park, Ohio, 1979.
3. H. H. Uhlig, *Corrosion Handbook*, John Wiley & Sons, Inc., New York, 1948.
4. F. L. LaQue and H. R. Copson, *Corrosion Resistance of Metals and Alloys*, 2nd ed., Reinhold Publishing Corp., New York, 1963.
5. "Metal/Environment Reactions" in L. L. Shrier, ed., *Corrosion*, 2nd ed., Vol. 1; "Corrosion Control" in L. L. Shrier, ed., *Corrosion*, 2nd ed., Vol. 2, Butterworths, London, Eng. available from American Society for Metals, Metals Park, Ohio; H. H. Uhlig, *Corrosion and Corrosion Control*, 2nd ed., John Wiley & Sons, Inc., New York, 1971.
6. R. D. Gabe, *Principles of Metal Surface Treatment and Protection*, 2nd ed., Pergamon Press, Inc., Elmsford, N.Y., 1978.
7. V. E. Carter, *Metallic Coatings for Corrosion Control (Prof. Conf.)*, Newnes-Butterworths, London, Eng., 1977.
8. *Materials and Coatings to Resist High Temperature Corrosion (Proc. Conf.)*, Verein Deutscher Eisenhuttenleute Dusseldorf, FRG, New York, May 1977, Applied Science Publishers, Ltd., 1978.
9. B. Chapman and J. C. Anderson, *Science and Technology of Surface Coatings*, Academic Press, New York, 1974.
10. R. C. Krutenat, ed., *Proc. of Conference on Structure/Property Relationships in Thick Films and Bulk Coatings*, San Francisco, Calif., Jan. 28–30, 1974, American Institute of Physics, 1974, LC 74-82950; *J. Vac. Sci. Technol.* **11,** 633 (1974).
11. R. E. Reed, ed., *Proceedings, 2nd Conference on Structure/Property Relationships in Thick Films and Bulk Coatings*, American Institute of Physics, New York, 1975 (LC 75-18563); *J. Vac. Sci. Technol.* **12,** 741 (1975).
12. R. F. Bunshah, ed., *Metallurgical Coatings 1976, Proceedings of the International Conference*, Elsevier Sequoia S. A. Lausanne, 1977; *Thin Solid Films* **39,** 1 (1976); *Thin Solid Films* **40,** 1 (1977).
13. R. F. Bunshah, ed., *Metallurgical Coatings, 1977, Proceedings of the International Conference*, Elsevier Sequoia S.A. Lausanne, 1977; *Thin Solid Films* **45,** 1 (1977).
14. R. F. Bunshah, ed., *Metallurgical Coatings, 1978, Proceedings of the International Conference*, Elsevier Sequoia S.A. Lausanne, 1978; *Thin Solid Films* **53,** 1 (1978); *Thin Solid Films* **54,** 1 (1978).
15. *High Temperature Oxidation-Resistant Coatings*, National Academy of Sciences, Washington, D.C., 1970 (LC 78-606278).
16. H. N. Hausner, ed., *Coatings of High-Temperature Materials*, Plenum Press, New York, 1966.
17. J. Huminik, Jr., ed., *High Temperature Inorganic Coatings*, Reinhold, New York, 1963.
18. C. S. Smith in W. J. Young, ed., *Application of Science in Examination of Works of Art*, Museum of Fine Arts, Boston, Mass., 1973, p. 96.
19. H. Lechtman, *J. Met.* **31**(12), 154 (1979).
20. J. A. Thornton and H. F. Graff, *Met. Trans. B* **7,** 607 (1976).
21. J. C. Zoccola and co-workers, "Atmospheric Corrosion Behavior of Al-Zn Alloy Coated Steel," *ASTM Symposium on Atmospheric Corrosion*, May 1976.
22. S. G. Denner and co-workers, *Iron Steel Int.* **48,** 241 (1974).
23. S. G. Denner and R. D. Jones, *Met. Tech.* **4**(3), 167 (1977).
24. K. Gale, *Steel Times* **206,** 896 (1978).
25. L. Pugazhenthy, *Corros. Maint.* **1**(2), 153 (1978).
26. D. Kim and H. Leidheiser, Jr., *Surface Technol.* **5,** 379 (1977).
27. P. J. Drivas, *Contract EPA-68-01-3156, Pacific Environmental Services, Inc.*, Santa Monica, Calif., Mar. 1976.
28. J. K. Fuller, *Wire J.* **10**(1), 60 (1977).
29. J. B. Horton, A. R. Borzillo, N. Kuhn and G. J. Harvey, *12th International Conference on Hot-Dip Galvanizing*, Paris, Fr., May 20–23, 1979.
30. G. J. Harvey, *Met. Australia* **8**(8), 176 (1976).
31. J. B. Horton, A. R. Borzillo, N. Kuhn, and G. J. Harvey, *12th International Conference on Hot-Dip Galvanizing*, Paris, Fr., May 20–23, 1979; R. D. Jones, *Iron Steel Inst.* **51**(3), 149 (1978); *Plat. Surf. Finish.* **65**(9), 30 (1978).
32. W. A. McGill and M. J. Weinbaum, *Met. Prog.* **116**(2), 26 (1979).
33. N. S. Gorbunov, *Diffuse Coatings on Iron and Steel*, Academy of Sciences, Moscow, USSR, 1958, translated by S. Friedman, A. Artman and Y. Halprin, OTS 60-21148.
34. U.S. Pat. 4,125,646 (Nov. 14, 1978), M. F. Dean and R. L. Blize (to Chromalloy American Corp. Off. Gaz.).
35. *Prod. Finish.* **30**(2), 10 (1977).

36. A. K. Sarkhel and L. L. Seigle, *Met. Trans.* **7A**, 899 (1976).
37. B. K. Gupta, A. K. Sarkhel, and L. L. Seigle, *Thin Solid Films* **39**, 313 (1976).
38. S. Shankar and L. L. Seigle, *Met. Trans.* **9A**, 1467 (1978).
39. R. W. Heckel, M. Yamada, C. Duchi, and A. J. Hickel, *Thin Solid Films* **45**, 367 (1977).
40. R. S. Parzuchowski, *Thin Solid Films* **45**, 349 (1977).
41. B. Weiss and M. R. Meyerson, *Trans. Met. Soc. AIME* **245**, 1633 (1969).
42. A. N. Mukherji and P. Prathakaram, *Anticorr. Methods Mater.* **25**(1), 5, 12 (1978).
43. T. S. Eyre, *Wear* **34**, 383 (1975).
44. R. H. Biddulph, *Thin Solid Films* **45**, 341 (1977).
45. Ref. 15, p. 126.
46. Brit. Pat. 5,647 (1900), S. Cowper-Coles; S. Trood, *Trans. Am. Inst. Met.* **9**, 161 (1919).
47. L. Zancheva and co-workers, *Met. Trans.* **9A**, 909 (1978).
48. *Met. Prog.* **117**(1), 89 (1980).
49. V. A. Parenov and F. K. Aleinikov, *Liet. TSR Mokslu Akad. Darb.* [B] **4**(107), 93 (1978).
50. G. F. Carter, *Met. Prog.* **93**(6), 124 (1968).
51. B. K. Granat, *Metalworking News*, 32 (Oct. 11, 1976).
52. *Weld. J. Miami Fl.* **55**, 675 (1976).
53. *Weld. J. Miami Fl.* **56**(8), 41 (1977).
54. "Corrosion Prevention with Thermal Sprayed Zn and Al Coatings," *Third International Congress on Marine Corrosion and Fouling*, Oct. 2–6, 1972, Sponsored by AWS Committee on Thermal Spraying, Special NBS Publication.
55. *Bull. Routier* **4**(3), (Nov.–Dec. 1978).
56. E. P. Cashon, *Tribology* **8**(3), 111 (1975).
57. R. R. Irving, B. D. Wakefield, and T. C. Dumond, *Iron Age* **211**(15), 65 (1973).
58. H. Kayser, *Thin Solid Films* **39**, 243 (1976); *Proc. International Conf. on Metallurgical Coatings*, Elsevier-Sequoia S.A. Lausanne, 1976.
59. R. C. Tucker, Jr., *J. Vac. Sci. Technol.* **11**, 725 (1974).
60. U.S. Pat. 4,182,223 (July 13, 1977), F. J. Wallace, N. S. Bornstein, and M. A. DeCrescente (to United Technologies Corporation).
61. G. J. Robson, *Australian Conf. on Manufacturing Engineering*, Aug. 17–19, 1977, pp. 211–216.
62. F. J. Wallace, *Plat. Surf. Finish.* **62**, 559 (1975).
63. R. J. Schaefer, J. D. Ayers, and T. R. Tucker, *NRL Report #3953*, Naval Research Laboratory, Washington, D.C., Mar. 30, 1979.
64. U.S. Pat. 4,015,100 (March 29, 1977), D. S. Gnanamuthu and E. V. Locke (to Avco Everett).
65. D. Belforte, *High Power Laser Surface Treatment*, SME Technical Paper IQ-77-373, Soc. Mfg. Eng., Dearborn, Mich.
66. U.S. Pat. 3,952,180 (April 20, 1976), D. S. Gnanamuthu (to Avco Everett).
67. D. A. Van Cleave, *Iron Age* **219**(5), 25, 30 (1977).
68. G. C. Irons, *Weld. J. Miami Fl.* **57**(12), 29 (1978).
69. M. S. Hess and J. F. Milkosky, *J. Appl. Phys.* **43**, 4680 (1972).
70. K. Gale, *Engineer* **234**(6064), 34 (1972).
71. V. A. Paramonov and co-workers, *Korroz. Zashch.* (8), 24 (1973).
72. *Iron Age*, 80 (May 7, 1964).
73. C. M. Packer and R. A. Perkins, *J. Less Common Met.* **37**, 361 (1974).
74. F. M. Miller and N. T. Bredzs, *Met. Prog.* **103**(3), 80, 82 (1973).
75. P. R. Clark, C. M. Packer and R. A. Perkins, *Quarterly Report, DOE Contract No. EF-77-C-012592*, Lockheed Palo Alto Research Laboratory, Palo Alto, Calif., Apr. 1–June 30, 1979, p. 9.
76. R. J. Russell, *Mater. Eng.* **81**(3), 48 (1975).
77. K. Tamura and N. Miyamoto, *J. Jpn. Soc. Powder Metall.* **20**(1), 10 (1973).
78. E. H. Mayer and R. M. Willison, *presentation*, *75th General Meeting American Iron and Steel Institute*, May 25, 1967, pp. 1–12.
79. V. Siran and co-workers, *Met. Finish.* **76**(9), 48 (1978).
80. E. A. Davis in *Coatings for Corrosion Prevention*, *1978 Symposium ASM Materials and Processing Congress*, Philadelphia, Pa., Nov. 9, 1978, American Society for Metals, Metals Park, Oh., 1979, p. 35, LC-79-17292.
81. *Prod. Finish.* (*London*) **31**(7), 22 (1978).
82. *Prod. Finish* (*Cincinnati*) **40**(1), 65 (1975).
83. R. A. Serlin, *Cutting Tool Eng.* **30**(7–8), 6 (1978).

84. E. Smars and G. Backstrom, *Proc. International Conf. on Exploiting Welding in Production Technology*, Welding Institute of Cambridge, Eng., 1975, pp. 179–187.

85. P. Blas'kovic, *Schweisstechnik* (9), (1975); J. S'Kcianiar, *Metals Technology Conference B*, Australian Institute of Metallurgy, 1976, paper 16-3-1.

86. C. Kubisch, P. Pressler, P. Machner, and O. Kleinhagauer in G. K. Bhat, ed., *Fifth International Symposium on Electroslag and Other Special Melting Technology, 1974*, Pittsburgh, Pa., Carnegie-Mellon Institute, 1975.

87. W. R. Foley and W. R. Huber, *Iron Steel Inst.* **51**(4), 72 (1971).

88. W. A. Bryant, *J. Mater. Sci.* **12**, 1285 (1977).

89. K. K. Yee, *Int. Met. Rev.* **23**(1), 19 (1978).

90. A. C. Schaffhauser, ed., *Proc. of Conf. on Chemical Vapor Deposition of Refractory Metals, Alloys and Compounds*, American Nuclear Society, Hinsdale, Ill., 1967.

91. J. M. Blocher, Jr., and J. C. Withers, eds., *Proc. of Second Int. Conf. on Chemical Vapor Deposition*, Electrochemical Society, New York, 1970.

92. F. A. Glaski, ed., *Proc. of Third Int. Conf. on Chemical Vapor Deposition*, American Nuclear Society, Hinsdale, Ill., 1972.

93. G. F. Wakefield and J. M. Blocher, Jr., eds., *Proc. of Fourth Int. Conf. on Chemical Vapor Deposition*, Electrochemical Society, Princeton, N.J., 1973.

94. N. J. Archer and K. K. Yee, *Wear* **48**, 237 (1978).

95. T. L. Chu and R. K. Smeltzer, *J. Vac. Sci. Technol.* **10**(1), 1 (1973).

96. J. J. Tietjen, *Ann. Rev. Mat. Sci.* **3**, 317 (1973).

97. C. Beguin, E. Horrath, and A. J. Perry, *Thin Solid Films* **46**, 209 (1977).

98. P. Felix and Erdos, *Werkst. Korr.* **23**, 627 (1972).

99. J. B. Stephenson, D. M. Soboroff, and H. O. McDonald, *Thin Solid Films* **40**, 73 (1977).

100. B. O. Seraphin, *J. Vac. Sci. Technol.* **16**(2), 193 (1979).

101. M. J. Rand, *J. Vac. Sci. Technol.* **16**, 410 (1979).

102. S. D. Allen and M. Bass, *J. Vac. Sci. Technol.* **16**(2), 431 (1979).

103. J. P. Ballantyne and W. C. Nixon, *J. Vac. Sci. Technol.* **10**, 1094 (1973).

104. C. S. Morgan, *Thin Solid Films* **39**, 305 (1976).

105. R. O. Betz, *Iron Age* **220**(1), 47 (1977).

106. D. P. Grimmer, K. C. Herr, and W. J. McCreary, *J. Vac. Sci. Technol.* **15**(1), 59 (1978).

107. T. Santala and C. M. Adams, Jr., *J. Vac. Sci. Technol.* **7**, 522 (1970).

108. R. F. Bunshah and R. S. Juntz, *Trans. Vac. Met. Conf.*, American Vacuum Society, New York, 1965, p. 200.

109. R. Nimmagadda, A. C. Raghuram, and R. F. Bunshah, *J. Vac. Sci. Technol.* **9**, 1406 (1972).

110. H. R. Smith, Jr. and C. D'A. Hunt, *Trans. Vac. Met. Conf.*, American Vacuum Society, 1964, p. 227.

111. S. Shen, D. Lee, and D. H. Boone, *Thin Solid Films* **53**, 233 (1978).

112. A. D. Grubb, *J. Vac. Sci. Technol.* **10**(1), 53 (1973).

113. C. C. Storms, *Proc. Soc. of Vac. Coaters, 19th Annual Conference*, 1976, pp. 1–3; A. Mock, *Mater. Eng.* **87**(4), 51 (1978).

114. D. M. Lindsey, *Prod. Finish. (Cincinnati)* **43**(10), 34 (1979).

115. S. Kut, *Finish. Ind.* **1**(7), 17 (1977).

116. G. K. Wehner, *Phys. Rev.* **102**, 690 (1956); *Phys. Rev.* **114**, 1270 (1959); *J. Appl. Phys.* **30**, 1762 (1959); G. S. Anderson and G. K. Wehner, *J. Appl. Phys.* **31**, 2305 (1960).

117. R. F. Bunshah and A. C. Raghuram, *J. Vac. Sci. Technol.* **9**(6), 1385 (1972).

118. *Battelle Today* (13), 1 (June 1979).

119. D. M. Mattox, *Electrochem. Tech.* **2**(9–10), 295 (1964).

120. D. M. Mattox, *J. Vac. Sci. Technol.* **10**(1), 47 (1973).

121. D. M. Mattox, *IPAT 79, Proc. Inter. Conf. Ion Plating and Allied Techniques*, London, July 1979, CEP Consultants, Ltd., Edinburgh, UK, p. 1.

122. C. T. Wan, D. L. Chambers, and D. C. Carmichael, *Proc. of Fourth Inter. Conf. on Vac. Met.*, The Iron and Steel Institute of Japan, Tokyo, 1974, p. 231.

123. G. W. White, *SAE paper 730546*, Detroit, Mich., May 14–18, 1973.

124. *Finish. Ind.* **1**(7), 19 (1977).

125. W. D. Westwood, *Prog. Surf. Sci.* **7**(2), 71 (1976).

126. J. A. Thorton, *J. Vac. Sci. Technol.* **15**(2), 171 (1978).

127. R. K. Waits, *J. Vac. Sci. Technol.* **15**(2), 179 (1978).

128. J. L. Hughes, *J. Vac. Sci. Technol.* **15**(4), 1572 (1978).

129. J. A. Thornton, *Proceedings, AES Coatings for Solar Collectors*, Atlanta, Ga., Nov. 9–10, 1976, American Electroplaters Society, Inc., Winter Park, Fla., 1976, pp. 63–77.

130. R. J. Hecht, *Air Force Materials Laboratory Contract No. F33615-78-C-5070*, Project No. 212-8, Wright Patterson AFB, Ohio, Pratt and Whitney Aircraft Group, West Palm Beach, Florida, Quarterly Progress Reports, 1978–1979, FR-11081, FR-11649, FR-11906, FR-12170 and FR-12557.

131. J. A. Thornton, *Ann. Rev. Mater. Sci.* **7**, 239 (1977).

132. W. L. Brown, ed., *Ion-Implantation—New Prospects for Materials Modification*, IBM Laboratories, Yorktown Heights, New York, sponsored by American Vacuum Society; *J. Vac. Sci. Technol.* **15**, 1629 (1978).

133. T. C. Wells, *Surf. J. (London)* **9**(4), (1978).

134. J. G. McCallum, G. I. Robertson, A. F. Rodde, B. Weissman, and N. Williams, *J. Vac. Sci. Technol.* **15**, 1067 (1978).

135. J. A. Borders, *Ann. Rev. Mater. Sci.* **9**, 313 (1979).

136. J. A. Borders and J. M. Poate, *Phys. Rev. B* **13**, 969 (1976).

137. W. A. Grant, *J. Vac. Sci. Technol.* **15**, 1644 (1978).

138. J. K. Hirronen, *J. Vac. Sci. Technol.* **15**, 1662 (1978).

139. J. E. Antill and co-workers in B. L. Crowder, ed., *Proc. Third Inter. Conf. on Ion Implantation in Semiconductors and Other Materials*, IBM, Yorktown Heights, 1972, Plenum Press, New York, 1973.

140. B. D. Sartwell, A. B. Campbell, and P. B. Needham in F. Chernow, J. A. Borders, and D. Brice, ed., *Proc. Fifth Inter. Conf. on Ion Implantation in Semiconductors and Other Materials*, Boulder, Co., Aug. 1976, Plenum Press, New York, 1977.

141. B. S. Corino, Jr., P. B. Needham, Jr., and G. R. Connor, *J. Electrochem. Soc.* **125**, 370 (1978).

142. S. B. Agarwal, Y. F. Wang, C. R. Clayton, and H. Herman, *Thin Solid Films* **63**, 19 (1979).

143. J. K. Hirronen, *J. Vac. Sci. Technol.* **15**, 1667 (1978).

144. H. S. Rupprecht, *J. Vac. Sci. Technol.* **15**, 1674 (1978).

145. J. C. North, R. Wolfe, and T. J. Nelson, *J. Vac. Sci. Technol.* **15**, 1675 (1978).

146. N. I. Sax, *Dangerous Properties of Industrial Materials*, 4th ed., Van Nostrand Reinhold Co., New York, 1975.

147. R. S. Kramkowski and E. Shmunes, *Final Report No. NIOSH-TR-058-74; HHE-72-87-58*, NIOSH, Cincinnati, Oh., June 1973.

R. C. KRUTENAT

Exxon Research and Engineering Company

EXPLOSIVELY CLAD METALS

Explosives (qv) were used increasingly in the 1950s in metal-working operations because the explosives provided an inexpensive source of energy and precluded need for expensive capital equipment (1–2). Research in explosively clad metals began during the same period (3–7).

Explosive cladding, or explosion bonding and explosion welding, is a method wherein the controlled energy of a detonating explosive is used to create a metallurgical bond between two or more similar or dissimilar metals. No intermediate filler metal, eg, a brazing compound or soldering alloy, is needed to promote bonding and no external heat is applied. Diffusion does not occur during bonding.

In 1962, the first method for welding metals in spots along a linear path by explosive detonation was patented (8); however, the method is not used industrially (see Welding). In 1963, a theory that explained how and why cladding occurs was published (9). Research efforts resulted in 27 U.S. process patents which standardized industrial explosion cladding. Several of the patents describe the use of variables involved in parallel cladding which is the most popular form of explosion cladding (10–13).

During the 1960s and early 1970s, research and development work on explosive cladding was conducted throughout the world. In the United States, much of the research occurred at Battelle Memorial Institute (14), Drexel University and Frankford Arsenal (15), DuPont (16–24), Stanford University (25–27), and the University of Denver (28). Research in other countries includes that conducted in Japan (29), Ireland (30–31), and the USSR (32–33). Several excellent reviews on metal cladding have been published (34–36). From 1964 to 1980, ca 200,000 metric tons of clad metals have been produced in noncommunist nations.

Advantages and Limitations

The explosive-cladding process provides the following advantages over other metal-bonding processes:

(1) A metallurgical, high quality bond can be formed between similar metals and between dissimilar metals that are incompatible for fusion or diffusion joining. Brittle, intermetallic compounds, which form in an undesirable continuous layer at the interface during bonding by conventional methods, are minimized, isolated, and surrounded by ductile metal in explosion cladding. Examples of these systems are titanium–steel, tantalum–steel, aluminum–steel, titanium–aluminum, and copper–aluminum. Immiscible metal combinations, eg, tantalum–copper, also can be clad.

(2) Explosive cladding can be achieved over areas that are limited only by the size of the available cladding plate and by the magnitude of the explosion that can be tolerated. Areas as small as 1.3 cm^2 (37) and as large as 27.9 m^2 (18) have been bonded.

(3) Metals with tenacious surface films that make roll bonding difficult, eg, stainless steel/Cr–Mo steels, can be explosion clad.

(4) Metals having widely differing melting points, eg, aluminum (660°C) and tantalum (2996°C), can be clad.

(5) Metals with widely different properties, eg, copper/maraging steel, can be bonded readily.

(6) Large clad-to-backer ratio limits can be achieved by explosion cladding. Stainless steel-clad components as thin as 0.025 mm and as thick as 3.2 cm have been explosion clad.

(7) The thickness of the stationary or backing plate in explosion cladding is essentially unlimited. Backers over >0.5 m thick and weighing 50 t have been clad commercially.

(8) High quality, wrought metals are clad without altering their chemical composition.

(9) Different types of backers can be clad; clads can be bonded to forged members, as well as to rolled plate.

(10) Clads can be bonded to rolled plate that is strand-cast, annealed, normalized, or quench-tempered.

(11) Multiple-layered composite sheets and plates can be bonded in a single explosion, and cladding of both sides of a backing metal can be achieved simultaneously. When two sides are clad, the two prime or clad metals need not be of the same thickness nor of the same metal or alloy.

(12) Nonplanar metal objects can be clad, eg, the inside of a cylindrical nozzle can be clad with a corrosion-resistant liner.

(13) The majority of explosion-clad metals are less expensive than the solid metals that could be used instead of the clad systems.

Limitations of the explosive bonding process are listed below.

(1) There are inherent hazards in storing and handling explosives and undesirable noise and blast effects from the explosion.

(2) Obtaining explosives with the proper energy, form, and detonation velocity is difficult.

(3) Metals to be explosively bonded must be somewhat ductile and resistant to impact. Alloys with as little as 5% tensile elongation in a 5.1-cm gauge length, and backing steels with as little as 13.6 J (10 ft·lbf) Charpy V-notch impact resistance can be bonded. Brittle metals and metal alloys fracture during bonding.

(4) For metal systems in which one or more of the metals to be explosively clad has a high initial yield strength or a high strain-hardening rate, a high quality bonded interface may be difficult to achieve. Metal alloys of high strength, ie, >690 MPa (10^5 psi) yield strength, are difficult to bond. This problem increases when there is a large density difference between the metals. Such combinations often are improved by using a thin interlayer between the metals.

(5) Geometries that are suited to explosive bonding promote straight-line egression of the high velocity jet emanating from between the metals during bonding, eg, for the bonding of flat and cylindrical surfaces.

(6) Thin backers must be supported, thus adding to manufacturing cost.

(7) The preparation and assembly of clads is not amenable to automated production techniques, and each assembly requires considerable manual labor.

Theory and Principles

To obtain a metallurgical bond between two metals, the atoms of each metal must be brought sufficiently close so that their normal forces of interatomic attraction produce a bond. The surfaces of metals and alloys must not be covered with films of oxides, nitrides, or adsorbed gases. When such films are present, metal surfaces do not bond satisfactorily.

In fusion welding, the surfaces of the two metals are melted by heating and the contaminated surface layers are brought to the surface of the melt pool. However, pressure-welding processes do not involve melting for removal of surface contaminants. Instead, the work pieces are plastically deformed, often after they have been heated to an appropriate temperature, which breaks the surface films and creates fresh uncontaminated areas where bonding can occur. Cold pressure welding, inertia welding, and ultrasonic welding are variations of this process.

Explosive bonding is a cold pressure-welding process in which the contaminant surface films are plastically jetted from the parent metals as a result of the high pressure collision of the two metals. A jet is formed between the metal plates, if the collision angle and the collision velocity are in the range required for bonding. The contaminant surface films that are detrimental to the establishment of a metallurgical bond are swept away in the jet. The metal plates, which are cleaned of any surface films by the jet action, are joined at an internal point by the high pressure that is obtained near the collision point.

Parallel and Angle Cladding. The arrangements shown in Figures 1 and 2 illustrate the operating principles of explosion cladding. Figure 1 illustrates angle cladding that is limited to cladding for relatively small pieces (38–39). Clad plates with large areas cannot be made using this arrangement because the collision of long plates at high stand-offs, ie, the distance between the plates, on long runs is so violent that metal cracking, spalling, and fracture occur. The arrangement shown in Figure 2 is by far the simplest and most widely used (10).

Jetting. A layer of explosive is placed in contact with one surface of the prime metal plate which is maintained at a constant distance from and parallel to the backer plate, as shown in Figure 2**a**. The explosive is detonated and, as the detonation front moves across the plate, the prime metal is deflected and accelerated to plate velocity V_P; thus, an angle is established between the two plates. The ensuing collision region progresses across the plate at a velocity equal to the detonation velocity D. When the

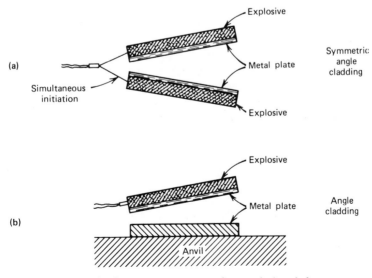

Figure 1.   Angle arrangements to produce explosion clads.

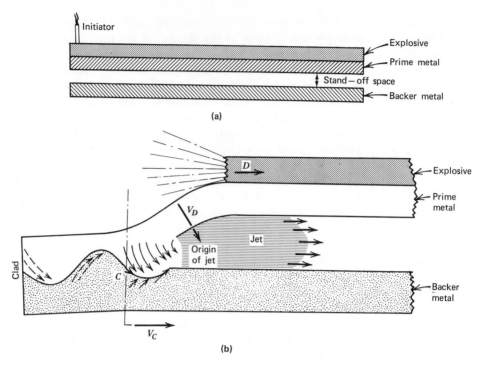

Figure 2.   Parallel arrangement for explosion cladding and subsequent collision between the prime and backer metals that leads to jetting and formation of wavy bond zone.

collision velocity V_c and the angle are controlled within certain limits, high pressure gradients ahead of the collision region in each plate cause the metal surfaces to flow hydrodynamically as a spray of metal from the apex of the angled collision. Jetting is the flow process and expulsion of the metal surface (9). Photographic evidence of jetting during an explosion-bonding experiment is given in Figures 3a, **b**, and **c** (19). The jet, which moves in the direction of detonation, is observed between the deflected prime metal and the backer metal.

Typically, jet formation is a function of plate collision angle, collision-point velocity, cladding-plate velocity, pressure at the collision point, and the physical and mechanical properties of the plates being bonded. For jetting and subsequent cladding to occur, the collision velocity has to be substantially below the sonic velocity of the cladding plates, usually ca 4000–5000 m/s (16). There also is a minimum collision angle below which no jetting occurs regardless of the collision velocity. In the parallel-plate arrangement shown in Figure 2, this angle is determined by the stand-off. In angle cladding, the preset angle determines the stand-off and the attendant collision angle (see Figure 1).

The amplitude and frequency of the bond-zone wave structure varies as a function of the explosive and the stand-off, as shown in Tables 1 and 2 (12).

Bond Nature.   It is preferable that commercial, explosively bonded metals exhibit a wavy, bond-zone interface. Bond-zone wave formation is analogous to fluid flowing around an obstacle (9). When the fluid velocity is low, the fluid flows smoothly around the obstacle but, above a certain fluid velocity, the flow pattern becomes turbulent,

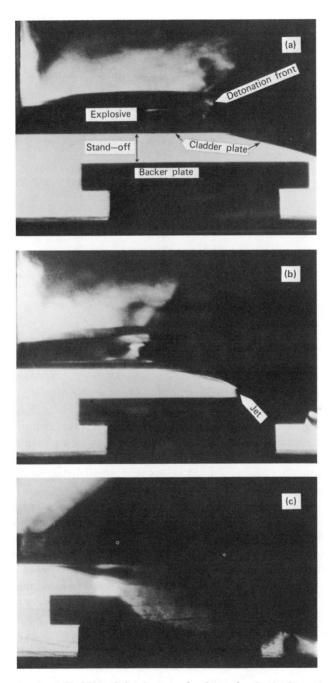

Figure 3. (a–c) Cladding of aluminum to aluminum showing jet formation (19).

279

Table 1. Measured Explosion-Cladding Parameters and Bond-Zone Characteristics for Cladding 3.2-mm Grade A Nickel to 12.7-mm AISI 1008 Carbon Steel[a]

| Collision velocity, m/s | Parallel stand-off, mm | Flyer-plate velocity, m/s | Flyer-plate angle, deg | Bond-zone characteristics | | |
				Type	Wavelength, μm	Amplitude, μm	Equivalent melt thickness, μm
1650	1.14	215	7.4	straight and wavy	112	10	<1
2000	1.14	250	7.0	wavy	103	11	<1
2500	1.14	270	6.6	wavy	236	39	1.6
3600	1.14	410	6.5	wavy	254	38	5.1
2000	2.16	310	8.7	wavy	318	41	<1
2500	2.16	337	8.2	wavy	425	76	3.9
3600	2.16	510	8.25	wavy	590	96	9.8
1650	3.96	325	11.2	wavy	520	52	<1
2000	3.96	372	10.5	wavy	567	88	<1
2500	3.96	407	9.7	wavy	671	121	6.0
3600	3.96	625	9.95	wavy	739	146	28.8
2000	6.35	420	11.8	wavy	790	132	<1
2500	6.35	462	10.8	wavy	895	171	9.0
3600	6.35	700	11.2	wavy	965	162	24.0
1650	10.54	425	14.8	wavy	1018	169	<1
2000	10.54	460	13.0	wavy	623	97	<1
3600	10.54	775	12.5	wavy	1333	284	59.2
1650	17.78	465	16.5	straight			not detectable

[a] Ref. 12.

as illustrated in Figure 4. In explosion bonding, the obstacle is the point of highest pressure in the collision region. Because the pressures in this region are many times higher than the dynamic yield strength of the metals, they flow plastically, as evidenced by the microstructure of the metals at the bond zone. Electron microprobe analysis across such plastically deformed areas shows that no diffusion occurs because there is extremely rapid self-quenching of the metals (16).

Under optimum conditions, the metal flow around the collision point is unstable and it oscillates, thereby generating a wavy interface. Typical explosion-bonded interfaces between nickel plates made at different collision velocities are illustrated in Figure 4 (24). A typical explosion-bonded interface between titanium and steel is shown in Figure 5. Small pockets of solidified melt form under the curl of the waves; some of the kinetic energy of the driven plate is locally converted into heat as the system comes to rest. These discrete regions are completely encapsulated by the ductile prime and base metals. The direct metal-to-metal bonding between the isolated pockets provides the ductility necessary to support stresses during routine fabrication.

The quality of bonding is related directly to the size and distribution of solidified melt pockets along the interface, especially for dissimilar metal systems that form intermetallic compounds. The pockets of solidified melt are brittle and contain localized defects which do not affect the composite properties. Explosion-bonding parameters for dissimilar metal systems normally are chosen to minimize the pockets of melt associated with the interface.

Table 2. Measured Explosion-Cladding Parameters and Bond-Zone Characteristics for Cladding 3.2-mm Grade 1 Titanium to 12.7-mm AISI 1008 Carbon Steel[a]

Collision velocity, m/s	Parallel stand-off, mm	Flyer-plate velocity, m/s	Flyer-plate impact angle, deg	Bond-zone characteristics				
				Type	Wavelength, μm	Amplitude, μm	Equivalent melt thickness, μm	Tensile elongation, %
2000	1.14	330	9.9	wavy	103	8	<1	32
2500	1.14	400	9.7	wavy	254	19	1.2	32
3600	1.14	580	9.3	b	250	23	14.0	
2000	2.16	420	12.2	b	215	17	<1	34
2500	2.16	465	11.0	b	482	47	3.0	29
3600	2.16	710	11.3	b	468	59	11.6	28
2000	3.96	495	14.2	b	373	31	<1	31
2500	3.96	520	12.1	b	768	89	3.1	29
3600	3.96	845	13.4	b	868	122	9.2	
2000	6.35	530	15.2	b	610	53	<1	32
2500	6.35	565	13.0	b	1009	130	8.2	29
3600	6.35	945	15.0	b	1228	189	18.5	
2000	10.54	560	15.5	b	1013	96	<1	32
2500	10.54	600	14.0	b	1300	167	3.6	27
3600	10.54	1040	16.5	b	1360	230	21.8	23

[a] Ref. 12.
[b] Melted layer and waves.

When cladding conditions are such that the metallic jet is trapped between the prime metal and the backer, the energy of the jet causes surface melting between the colliding plates. In this type of clad, alloying through melting is responsible for the metallurgical bond. As shown in Figure 6, solidification defects can occur and, for this reason, this type of bond is not desirable.

The industrially useful combinations of explosively clad metals that are available in commercial sizes are listed in Figure 7. The list does not include triclads or combinations that corrosion or materials engineers or equipment designers may yet envision. The combinations that explosion cladding can provide are virtually limitless (14).

Processing

Explosives. The pressure P generated by the detonating explosive that propels the prime plate is directly proportional to its density ρ and the square of the detonation velocity, V_D^2 (40):

$$P = \frac{1}{4}\,\rho\,V_D^2$$

The detonation velocity is controlled by adjusting the packing density or the amount of added inert material (41).

The types of explosives that have been used include the following (14,41) (see

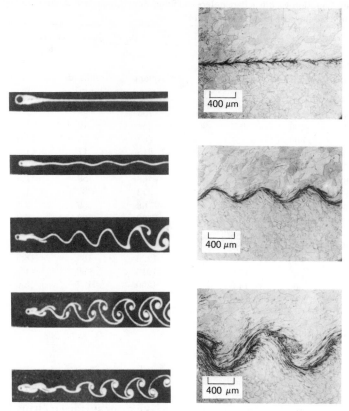

Figure 4. At left are photographs of fluid flow behind cylinders at increasing flow velocities top to bottom. At right are photomicrographs of nickel–nickel bond zones made at increasing collision velocities; top, about 1600 m/s; middle, about 1900 m/s; bottom, about 2500 m/s (24).

Explosives):

High velocity (4500–7600 m/s)	*Low–medium velocity (1500–4500 m/s)*
trinitrotoluene (TNT)	ammonium nitrate
cyclotrimethylenetrinitramine (RDX)	ammonium nitrate prills
pentaerythritol tetranitrate (PETN)	sensitized with fuel oil
composition B	ammonium perchlorate
composition C_4	amatol
plasticized PETN-	amatol and sodatol diluted
based-rolled sheet	with rock salt to 30–35%
and extruded cord	dynamites
primacord	nitroguanidine
	diluted PETN

Metal Preparation. Preparation of the metal surfaces to be bonded usually is required because most metals contain surface imperfections or contaminants that undesirably affect bond properties. The cladding faces usually are surface-ground using an abrasive machine and then are degreased with a solvent to ensure consistent bond

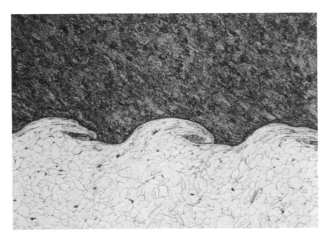

Figure 5. Photomicrograph of titanium, top, to carbon steel, bottom, explosion clad (100×).

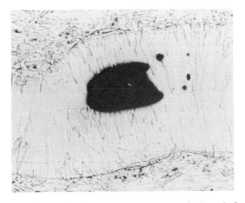

Figure 6. Solidification defects in the copper–copper explosion clad evidence the occurrence of melting at the interface (9) (100×).

strength (41). In general, a surface finish that is ≥ 3.8 μm is needed to produce consistent, high quality bonds.

Fabrication techniques must take into account the metallurgical properties of the metals to be joined and the possibility of undesirable diffusion at the interface during hot forming, heat treating, and welding. Compatible alloys, ie, those that do not form intermetallic compounds upon alloying, eg, nickel and nickel alloys, copper and copper alloys, and stainless steel alloys clad to steel, may be treated by the traditional techniques developed for clads produced by other processes. On the other hand, incompatible combinations, eg, titanium, zirconium, or aluminum to steel, require special techniques designed to limit the production at the interface of undesirable intermetallics which would jeopardize bond ductility.

Assembly, Stand-off. The air gap present in parallel explosion cladding can be maintained by metallic supports that are tack-welded to the prime and backer plates or by metallic inserts that are placed between the prime and backer (41–43). The inserts

	Zirconium	Magnesium	Stellite 6B	Platinum	Gold	Silver	Niobium	Tantalum	Hastelloy	Titanium	Nickel alloys	Copper alloys	Aluminum	Stainless steels	Alloy steels	Carbon steels
Carbon steels	•	•			•	•	•	•	•	•	•	•	•	•	•	•
Alloy steels	•	•	•					•	•	•	•	•	•	•	•	
Stainless steels			•		•	•	•	•		•	•	•	•	•		
Aluminum [7429-90-5]		•			•	•	•			•	•	•	•			
Copper alloys					•	•	•			•	•	•				
Nickel alloys		•	•	•				•	•	•	•					
Titanium [7440-30-6]	•	•				•	•	•		•						
Hastelloy										•						
Tantalum [7440-25-7]					•		•	•								
Niobium [7440-03-1]				•			•									
Silver [7440-22-4]						•										
Gold [7440-57-5]																
Platinum [7440-06-4]				•												
Stellite 6B																
Magnesium [7439-95-4]		•														
Zirconium [7440-67-7]	•															

Figure 7. Explosion-clad metal combinations that are commercially available.

usually are made of a metal that is compatible with one of the cladding metals. If the prime metal is so thin that it sags when supported by its edges, other materials, eg, rigid foam, can be placed between the edges to provide additional support; the rigid foam is consumed by the hot egressing jet during bonding (41–44) (see Foamed plastics). A moderating layer or buffer, eg, polyethylene sheet, water, rubber, paints, and pressure-sensitive tapes, may be placed between the explosive and prime metal surface to attenuate the explosive pressure or to protect the metal surface from explosion effects (14).

Facilities. The preset, assembled composite is placed on an anvil of appropriate thickness to minimize distortion of the clad product. For thick composites, a bed of sand usually is a satisfactory anvil. Thin composites may require a support made of steel, wood, or other appropriate materials. The problems of noise, air blast, and air pollution (qv) are inherent in explosion cladding, and clad-composite size is restricted by these problems (see Noise pollution). Thus, the cladding facilities should be in areas that are remote from population centers. Using barricades and burying the explosives and components under water or sand lessens the effects of noise and air pollution (14).

An attractive method for making small-area clads using light explosive loads employs a low vacuum, noiseless chamber (14). Underground missile silos and mines also have been used as cladding chambers (see also Insulation, acoustic).

Analytical and Test Methods

When the explosion-bonding process distorts the composite so that its flatness does not meet standard flatness specifications, it is reflattened on a press or roller leveller (see ASME-SA-20). However, press-flattened plates sometimes contain localized irregularities which do not exceed the specified limits but which, generally, do not occur in roll-flattened products.

Nondestructive. Nondestructive inspection of an explosion-welded composite is almost totally restricted to ultrasonic and visual inspection. Radiographic inspection is applicable only to special types of composites consisting of two metals having a significant mismatch in density and a large wave pattern in the bond interface (see Nondestructive testing).

Ultrasonic. The most widely used nondestructive test method for explosion-welded composites is ultrasonic inspection. Pulse-echo procedures (ASTM A 435) are applicable for inspection of explosion-welded composites used in pressure applications (see Ultrasonics).

The acceptable amount of nonbond depends upon the application. In clad plates for heat exchangers, >98% bond usually is required (see Heat-exchange technology). Other applications may require only 95% of the total area to be bonded. Configurations of a nonbond sometimes are specified, eg, in heat exchangers where a nonbond area may not be >19.4 cm^2 or 7.6 cm long. The number of areas of nonbond generally is specified. Ultrasonic testing can be used on seam welds, tubular transition joints, clad pipe and tubing, and in structural and special applications.

Radiographic. Radiography is an excellent nondestructive test (NDT) method for evaluating the bond of Al–steel electrical and Al–Al–steel structural transition joints. It provides the capability of precisely and accurately defining all nonbond and flat-bond areas of the Al–steel interface, regardless of their size or location.

The clad plate is x-rayed perpendicular from the steel side and the film contacts the aluminum. Radiography reveals the wavy interface of explosion-welded, aluminum-clad steel as uniformly spaced, light and dark lines with a frequency of one to three lines per centimeter. The waves characterize a strong and ductile transition joint and represent the acceptable condition. The clad is interpreted to be nonbonded when the x ray shows complete loss of the wavy interface (see X-ray technology).

Destructive. Destructive testing is used to determine the strength of the weld and the effect of the explosion-welding process on the parent metals. Standard testing techniques can be utilized on many composites; however, nonstandard or specially designed tests often are required to provide meaningful test data for specific applications.

Pressure-Vessel Standards. Explosion-clad plates for pressure vessels are tested according to the applicable ASME Boiler and Pressure Vessel Code Specifications. Unfired pressure vessels using clads are covered by ASTM A 263, A 264, and A 265; these include tensile, bend, and shear tests.

Tensile tests of a composite plate having a thickness of <3.8 cm require testing of the joined base metal and clad. Strengthening does occur during cladding and tensile strengths generally are greater than for the original materials. Some typical shear-strength values obtained for explosion-clad composites covered by ASTM A 263, A 264, A 265, which specify 138 MPa (20,000 psi) minimum, and B 432, which specifies 83 MPa (12,000 psi) minimum, are listed in Table 3 (see High pressure technology).

Chisel.   Chisel testing is a quick, qualitative technique that is widely used to determine the soundness of explosion-welded metal interfaces. A chisel is driven into and along the weld interface, and the ability of the interface to resist the separating force of the chisel provides an excellent qualitative measure of weld ductility and strength.

Ram Tensile.   A ram tensile test has been developed to evaluate the bond-zone tensile strength of explosion-bonded composites. As shown in Figure 8 the specimen is designed to subject the bonded interface to a pure tensile load. The cross-section area of the specimen is the area of the annulus between the OD and ID of the specimen.

Table 3.   Typical Shear Strengths of Explosively Clad Metals

Cladding metal on carbon-steel backers	Shear strength[a], MPa[b]
stainless steels	448
nickel and nickel alloys	379
Hastelloy alloys	391
zirconium	269
titanium[c]	241
cupronickel	251
copper	152
aluminum (1100-Ah4)	96

[a] See ASTM A 263, A 264, A 265, and B 432.
[b] To convert MPa to psi, multiply by 145.
[c] Stress-relief annealed at 621°C.

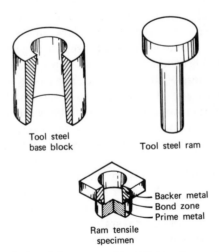

Tool steel
base block

Tool steel ram

Backer metal
Bond zone
Prime metal

Ram tensile
specimen

Figure 8.   Machined explosion-clad tests sample and fixture for ram tensile test of bond zone.

The specimen typically has a very short tensile gauge length and is constructed so as to cause failure at the bonded interface. The ultimate tensile strength and relative ductility of the explosion-bonded interface can be obtained by this technique.

Mechanical Fatigue. Some mechanical fatigue tests have been conducted on explosion-clad composites where the plane of maximum tensile stress is placed near the bond zone (20).

Thermal Fatigue and Stability. Explosion-welded plates have performed satisfactorally in several types of thermal tests (18). In thermal-fatigue tests, samples from bonded plate are alternately heated to 454–538°C at the surface and are quenched in cold water to less than 38°C. The three-minute cycles consist of 168 s of heating and 12 s of cooling. Weld-shear tests are performed on samples before and after thermal cycling. Stainless steel clads have survived 2000 such thermal cycles without significant loss in strength (18). Similarly welded and tested Grade 1 titanium–carbon steel samples performed in similar satisfactory fashion.

Metallographic. The interface is inspected on a plane parallel to the detonation front and normal to the surface. A well-formed wave pattern without porosity generally is indicative of a good bond. The amplitude of the wave pattern for a good weld can vary from small to large without a large influence on the strength, and small pockets of melt can exist without being detrimental to the quality of the bond. However, a continuous layer of melted material indicates that welding parameters were incorrect and should be adjusted. A line-type interface with few waves indicates that the collision velocity of the plate was not great enough and/or that the collision angle was too high for jetting to occur. A well-defined wave pattern in which the crest of the wave is bent over to form a large melt pocket with a void in the swirl is indicative of a poor bond. In this case, the plate velocity is too high as is the collision angle.

In some materials, eg, titanium and martensitic steels, shear bands are adjacent to the weld interface if the cladding variables are excessive. This is the result of thermal adiabatic shear developed from excessive overshooting energy, and a heat treatment is required to eliminate the hardened-band effect. When the cladding variables and the system energy are optimum, thermal shear bands are minimized or eliminated and heat treatment after cladding is not required. Several types of metal composites require heat treatment after cladding to relieve stress, but intermetallic compounds can form as a result of the treatment. A metallographic examination indicates if the heat treatment of the explosion-bonded composite has resulted in the formation of intermetallic compounds.

Hardness, Impact Strength. Microhardness profiles on sections from explosion-bonded materials show the effect of strain hardening on the metals in the composite (see Hardness). Figure 9 illustrates the effect of cladding a strain-hardening austenitic stainless steel to a carbon steel. The austenitic stainless steel is hardened adjacent to the weld interface by explosion welding, whereas the carbon steel is not hardened to a great extent. Similarly, aluminum does not strain harden significantly.

Impact strengths also can be reduced by the presence of the hardened zone at the interface. A low temperature stress-relief anneal decreases the hardness and restores impact strength (20). Alloys that are sensitive to low temperature heat treatments also show differences in hardness traverses that are related to the explosion-welding parameters, as illustrated in Figure 10 (36). Low welding-impact velocities do not develop as much adiabatic heating as higher impact velocities. The effect of the adiabatic heating is to anneal and further age the alloys. Hardness traverses in-

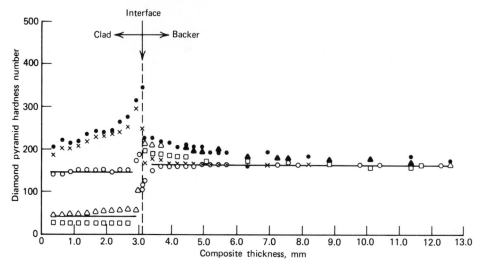

Figure 9. Microhardness profile across interfaces of two types of explosion clads that show widely divergent response resulting from their inherent cold-work hardening characteristics (17). 3.2 mm type 304L stainless/28.6 mm, A 516-70: — control (before cladding). ● = clad + flat. X = clad + stress-relief annealed at 621°C + flat. O = clad + normalize at 954°C. 3.2 mm 1100-H14 aluminum/25.4 mm, A 516-70: — control (before cladding). △ = clad + flat. □ = clad + flat + stress-relief annealed at 593°C (17).

dicate the degree of hardening during welding and what, if any, subsequent heat treatment is required after explosion bonding. Explosion-bonding parameters also can be adjusted to prevent softening at the interface, as shown in Figure 10.

Safety Aspects

All explosive materials should be handled and used following approved safety procedures either by or under the direction of competent, experienced persons and in accordance with all applicable federal, state, and local laws, regulations, and ordinances. The Bureau of Alcohol, Tobacco, and Firearms (BATF), the Hazardous Materials Regulation Board (HMRB) of the Department of Transportation (DOT), the Occupational Safety and Health Agency (OSHA), and the Environmental Protection Agency (EPA), Washington, D.C., have federal jurisdiction on the sale, transport, storage, and use of explosives. Many states and local counties have special explosive requirements. The Institute of Makers of Explosives (IME), New York, provides educational publications to promote the safe handling, storage, and use of explosives. The National Fire Protective Association (NFPA), Boston, Mass., similarly provides recommendations for safe explosives manufacture, storage, handling, and use.

Uses

Cladding and backing metals are purchased in the appropriately heat-treated condition because corrosion resistance is retained through bonding. It is customary to supply the composites in the as-bonded condition because hardening usually does not affect the engineering properties. Occasionally, a postbonding heat treatment is used to achieve properties required for specific combinations.

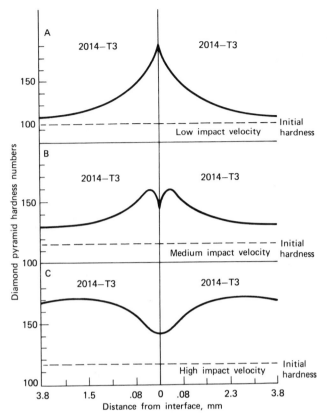

Figure 10. Microhardness profiles across interface of explosion-clad age-hardenable aluminum alloys (36).

Vessel heads can be made from explosion-bonded clads, either by conventional cold- or hot-forming techniques. The latter involves thermal exposure and is equivalent in effect to a heat treatment. The backing metal properties, bond continuity, and bond strength are guaranteed to the same specifications as the composite from which the head is formed.

Applications such as chemical-process vessels and transition joints represent ca 90% of the industrial use of explosion cladding.

Chemical-Process Vessels. Explosion-bonded products are used in the manufacture of process equipment for the chemical, petrochemical, and petroleum industries where the corrosion resistance of an expensive metal is combined with the strength and economy of another metal. Applications include explosion cladding of titanium tubesheet to Monel (Fig. 11), hot fabrication of an explosion clad to form an elbow for pipes in nuclear power plants, and explosion cladding titanium and steel for use in a vessel intended for terephthalic acid manufacture (Fig. 12).

Precautions must be taken when welding incompatible clad systems, eg, hot forming of titanium-clad steel plates must be conducted at 788°C or less. The preferred technique for butt welding involves a batten-strap technique using a silver, copper, or steel underlay (see Fig. 13). Precautions must be taken to avoid iron contamination

Figure 11.   Tubesheet of titanium explosively clad to Monel. Tubes were titanium and shell was Monel 400. Courtesy Nooter Corporation.

Figure 12.   Titanium–carbon steel vessel made from explosively clad plate for use in terephthalic acid manufacture. Courtesy of Nooter Corporation.

of the weld either from the backer steel or from outside sources. Stress relieving is achieved at normal steel stress relieving temperatures, and special welding techniques must be used in joining tantalum–copper–steel clads (45–46).

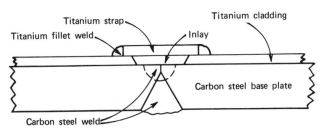

Figure 13. Double-v inlay, batten-strap technique for fusion welding of an explosion-clad plate containing titanium and zirconium.

Conversion-Rolling Billets. Much clad plate and strip have been made by hot and cold rolling of explosion-bonded slabs and billets. Explosion bonding is economically attractive for conversion rolling, because the capital investment for plating and welding equipment needed for conventional bonding methods is avoided. Highly alloyed stainless steels and some copper alloys, which are difficult to clad by roll bonding, are used for plates made by converting explosion-bonded slabs and billets. Conventional hot-rolling and heat-treatment practices are used when stainless steels, nickel, and copper alloys are converted. Hot rolling of explosion-bonded titanium, however, must be performed below ca 843°C to avoid diffusion and the attendant formation of undesirable intermetallic compounds at the bond interface. Hot-rolling titanium also requires a stiff rolling mill because of the large separation forces required for reduction.

Perhaps the most extensive application for conversion-rolled, explosion-bonded clads was for U.S. coinage in the 1960s (47). Over 15,900 metric tons of explosion-clad strip that was supplied to the U.S. Mint helped alleviate the national silver-coin shortage. The triclad composites consist of 70–30 cupronickel/Cu/70–30 cupronickel.

Transition Joints. Use of explosion-clad transition joints avoids the limitations involved in joining two incompatible materials by bolting or riveting. Many transition joints can be cut from a single large-area flat-plate clad and delivered to limit the temperature at the bond interface so as to avoid undesirable diffusion. Conventional welding practices may be used for both similar metal welds.

Electrical. Aluminum, copper, and steel are the most common metals used in high current–low voltage conductor systems. Use of these metals in dissimilar metal systems often maximizes the effects of the special properties of each material. However, junctions between these incompatible metals must be electrically efficient to minimize power losses. Mechanical connections involving aluminum offer high resistance because of the presence of the self-healing oxide skin on the aluminum member. Because this oxide layer is removed by the jet, the interface of an explosion clad essentially offers no resistance to the current. Thus, welded transition joints, which are cut from thick composite plates of aluminum–carbon steel, permit highly efficient electrical conduction between dissimilar metal conductors. Sections can be added by conventional welding. This concept is routinely employed by the primary aluminum-reduction industry in anode-rod fabrication. The connection is free of the aging effects that are characteristic of mechanical connections and requires no maintenance. The mechanical properties of the explosion weld, ie, shear, tensile, and impact strength, exceed those of the parent-type 1100 aluminum alloy.

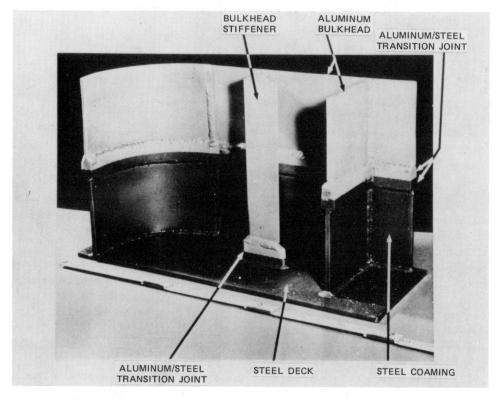

Figure 14.   Sample showing typical aluminum superstructure and deck connection made possible by use of explosion clad aluminum–carbon steel transition joint.

Usually, copper surfaces are mated when joints must be periodically disconnected because copper offers low resistance and good wear. Junctions between copper and aluminum bus bars are improved by using a copper–aluminum transition joint that is welded to the aluminum member. Deterioration of aluminum shunt connections by arcing is eliminated when a transition joint is welded to both the primary bar and the shunting bar.

The same intermetallic compounds that prevent conventional welding between aluminum and copper or steel can be developed in an explosion clad by heat treatment at elevated temperature. Diffusion can be avoided if the long-term service temperature is kept below 260°C for aluminum–steel and 177°C for copper–aluminum combinations. Under short-term conditions, as during welding, peak temperatures of 316°C and 232°C, respectively, are permissible. Bond ductility is maintained, although there is a reduction in bond strength as the aluminum is annealed. Bond strength, however, never falls below that of the parent aluminum; therefore, nominal handbook values for type 1100 alloy aluminum may be used in design considerations. The bond is unaffected by thermal cycling within the recommended temperature range.

Marine.   In the presence of an electrolyte, eg, seawater, aluminum and steel form a galvanic cell and corrosion takes place at the interface. Because the aluminum superstructure is bolted to the steel bulkhead in a lap joint, crevice corrosion is masked and may remain unnoticed until replacement is required. By using transition-joint

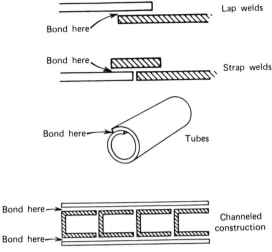

Figure 15. Explosion-clad welding applications (7).

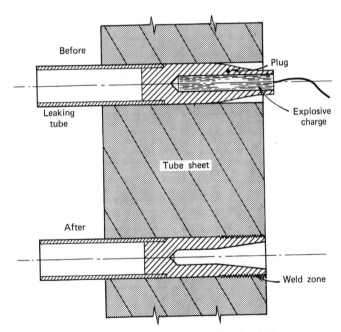

Figure 16. Tube-to-tubesheet plugging (48).

strips cut from explosion-welded clads, the corrosion problem can be eliminated. Because the transition is metallurgically bonded, there is no crevice in which the electrolyte can act and galvanic action cannot take place. Steel corrosion is confined to external surfaces where it can be detected easily and corrected by simple wire brushing and painting.

Explosion-welded construction has equivalent or better properties than the more complicated riveted systems. Peripheral benefits include weight savings and perfect electrical grounding. In addition to lower initial installation costs, the welded system requires little or no maintenance and, therefore, minimizes life-cycle costs. Applications of structural transition joints include aluminum superstructures that are welded to decks of naval vessels and commercial ships as illustrated in Figure 14.

Tubular. Explosion welding is a practical method for providing the means to join dissimilar metal pipes, eg, aluminum, titanium, or zirconium to steel or stainless steel, using standard welding equipment and techniques. The process provides a strong metallurgical bond which assures that the transition joints provide maintenance-free service throughout years of thermal and pressure/vacuum cycling. Explosion-welded tubular transition joints are being used in many diverse applications in aerospace, nuclear, and cryogenic industries. They operate reliably through the full range of temperatures, pressures, and stresses that normally are encountered in piping systems. Tubular transition joints in various configurations can be cut and machined from explosion-welded plate, or they can be made by joining tubes by overlap cladding. Standard welding practices are used to make the final joints.

Nonplanar Specialty Products. The inside walls of hollow forgings that are used for connections to heavy-walled pressure vessels have been metallurgically bonded with stainless steel. These bonded forgings, or nozzles, range from 50 to 610 mm ID and are up to 1 m long. Large-clad cylinders and internally clad, heavy-walled tubes have been extruded using conventional equipment. Other welding applications have been demonstrated, including those shown in Figure 15.

Tube Welding and Plugging. Explosion-bonding principles are used to bond tubes and tube plugs to tube sheets. The commercial process resembles the cladding of internal surfaces of thick-walled cylinders or prressure vessel nozzles, as shown in Figure 16; angle cladding is used (48). Countersink machining at the tube entrance provides the angled surface of 10–20° at a depth of 1.3–1.6 cm. The exploding detonator propels the tube or tube plug against the face of the tube- sheet to form the proper collision angle which, in turn, provides the required jetting and attendant metallurgical bond. Tubes may be welded individually or in groups. Metal combinations that are welded commercially include carbon steel–carbon steel, titanium–stainless steel, and 90–10 cupronickel/carbon steel.

BIBLIOGRAPHY

1. J. Pearson, *J. Met.* **12,** 673 (1960).
2. R. S. Rinehart and J. Pearson, *Explosive Working of Metals*, MacMillan, New York, 1963.
3. J. J. Douglass, *New England Regional Conference of AIME*, Boston, Mass., May 26, 1960.
4. *Ryan Reporter*, Vol. 21, No. 3, Ryan Aeronautical Company, San Diego, Calif., 1960, pp. 6–8.
5. C. P. Williams, *J. Met.*, 33 (1960).
6. "High Energy Rate Forming" in *Product Engineering and American Machinist/Metalworking Manufacturing*, McGraw Hill, New York, 1961 and 1962.
7. A. H. Holtzman and C. G. Rudershansen, *Sheet Met. Ind.* **39,** 401 (1961).
8. U.S. Pat. 3,024,526 (Mar. 13, 1962), V. Philipchuk and F. Le Roy Bois (to Atlantic Research Corp.).
9. G. R. Cowan and A. H. Holtzman, *J. Appl. Phys.* **34**(Pt. 1), 928 (1962).
10. U.S. Pat. 3,137,937 (Jun 23, 1964), G. R. Cowan, J. J. Douglass, and A. H. Holtzman (to E. I. du Pont de Nemours & Co., Inc.).
11. U.S. Pat. 3,233,312 (Feb. 8, 1966), G. R. Cowan and A. H. Holtzman (to E. I. du Pont de Nemours & Co., Inc.).
12. U.S. Pat. 3,397,444 (Aug. 20, 1968), O. R. Bergmann, G. R. Cowan, and A. H. Holtzman (to E. I. du Pont de Nemours & Co., Inc.).

13. U.S. Pat. 3,493,353 (Feb. 3, 1970), O. R. Bergmann, G. R. Cowan, and A. H. Holtzman (to E. I. du Pont de Nemours & Co., Inc.).
14. V. D. Linse, R. H. Wittman, and R. J. Carlson, *Defense Metals Information Center*, Memo 225, Columbus, Ohio, Sept. 1967.
15. J. F. Kowalick and D. R. Hay, *Met. Trans.* **2**, 1953 (1971).
16. A. H. Holtzman and G. R. Cowan, *Weld. Res. Counc. Bull. No. 104*, Engineering Foundation, New York, Apr. 1965.
17. A. Pocalyko and C. P. Williams, *Weld. J.* **43**, 854(1964).
18. A. Pocalyko, *Mater. Prot.* 4(6), 10 (1965).
19. O. R. Bergmann, G. R. Cowan, and A. H. Holtzman, *Trans. Met. Soc. AIME* **236**, 646 (1966).
20. J. L. DeMaris and A. Pocalyko, *Am. Soc. Tool and Mfg. Engrs. Paper AD66-113*, Dearborn, Mich., 1966.
21. O. R. Bergmann, *ASM Met. Eng. Quart.*, 60 (1966).
22. T. J. Enright, W. F. Sharp, and O. R. Bergmann, *Met. Prog.*, 107 (1970).
23. C. R. McKenney and J. G. Banker, *Mar. Technol.*, 285(1971).
24. G. R. Cowan, O. R. Bergmann, and A. H. Holtzman, *Met. Trans.* **2**, 3145 (1971).
25. D. E. Davenport and G. E. Duvall, *American Soc. Tool & Mfg. Engrs. Tech. Paper SP60-161*, 1960–1961.
26. G. R. Abrahamson, *J. Appl. Mech.* **28**, 519 (1961).
27. D. E. Davenport, *Am. Soc. Tool & Mfg. Engrs. Tech. Paper SP62-77*, Dearborn, Mich., 1961–62.
28. H. E. Otto and S. H. Carpenter, *Weld. J.* **51**, 467 (1972).
29. T. Onzawa and Y. Tshii, *Trans. Jpn. Weld. Soc.* **6**, 28 (1975).
30. A. S. Bahrani, T. J. Black, and B. Crossland, *Proc. Roy. Soc.* **296A**, 123 (1967).
31. B. Crossland and J. D. Williams, *Met. Rev.* **15**, 80 (1970).
32. A. A. Deribas, V. M. Kudinov, and F. I. Matveenkov, *Fiz. Goreniya Vzryva* **3**, 561 (1967).
33. S. K. Godunov, A. A. Deribas, A. V. Zabrodin, and N. S. Kozin, *J. Comp. Phys.* **5**, 517 (1970).
34. B. Crossland and A. S. Bahrani, *Proc. First Int. Conf. Center High Energy Forming*, University of Denver, Denver, Co., 1967.
35. A. A. Ezra, *Principles and Practices of Explosives Metal Working*, Industrial Newspapers, Ltd., London, 1973.
36. S. H. Carpenter and R. H. Wittman, *Ann. Rev. Mater. Sci.* **5**, 177 (1975).
37. J. L. Edwards, B. H. Cranston, and G. Krauss, *Met. Effects at High Strain Rates*, Plenum Press, New York, 1973.
38. U.S. Pat. 3,264,731 (Aug. 9, 1966), B. Chudzik (to E. I. du Pont de Nemours & Co., Inc.).
39. U.S. Pat. 3,263,324 (Aug. 2, 1966), A. A. Popoff (to E. I. du Pont de Nemours & Co., Inc.).
40. M. A. Cook, *The Science of High Explosives*, Reinhold Publishing Corp., New York, 1966, p. 274.
41. A. A. Popoff, *Mech. Eng.* **100**(5), 28 (1978).
42. U.S. Pat. 3,140,539 (July 14, 1964), A. H. Holtzman (to E. I. du Pont de Nemours & Co., Inc.).
43. U.S. Pat. 3,205,574 (Sept. 14, 1965), H. M. Brennecke (to E.I. du Pont de Nemours & Co., Inc.).
44. U.S. Pat. 3,360,848 (Jan. 2, 1968), J. J. Saia (to E. I. du Pont de Nemours & Co., Inc.).
45. U.S. Pat. 3,464,802 (Sept. 2, 1969), J. J. Meyer (to Nooter Corporation).
46. U.S. Pat. 4,073,427 (Feb. 14, 1978), H. G. Keifert and E. R. Jenstrom (to Fansteel, Inc.).
47. J. M. Stone, *Paper presented at a Select conference on Explosive Welding*, Hove, Eng., Sept. 1968, pp. 29–34.
48. R. Hardwick, *Weld. J.* **54**(4), 238 (1975).

General References

The Joining of Dissimilar Metals, DMIC Report S-16, Battelle Memorial Institute, Columbus, Ohio, Jan. 1968.
S. H. Carpenter, *Nat. Tech. Info. Ser. Rept. No. AMMRC CTR74-69*, Dec. 1978.
C. Birkhoff, D. P. MacDougall, E. M. Pugh, and G. Taylor, *J. Appl. Phys.* **19**, 563 (1948).
L. Zernow, I. Lieberman, and W. L. Kincheloe, *Am. Soc. Tool. & Mfg. Engr.*, *Tech. Paper SP60-141*, Dearborn, Mich., 1961.
R. H. Wittman, *Metallurgical Effects at High Strain Rates*, Plenum Press, New York, 1973.
R. H. Wittman, *Amer. Soc. Tool & Mfg. Engrs. Tech. Paper AD-67-177*, Dearborn, Mich.
G. Bechtold, I. Michael, and R. Prummer, *Gold Bull. Chamber Mines S. Afr.* **10**(2), 34 (1977).
B. H. Cranston, D. A. Machusak, and M. E. Skinkle, *West. Electr. Eng.*, 26 (Oct. 1978).

A. A. Popoff, *Amer. Soc. of Mfg. Engrs.*, *Tech. Paper AD77-236*, Dearborn, Mich., 1977.

T. Z. Blazynski, *International Conference on Welding and Fabrication of Non-Ferrous Metals*, Eastbourne, May 2 and May 3, 1972, Cambridge, Eng., The Welding Institute, 1972.

J. F. Kowalick and D. R. Hay, *Second International Conference of the Center For High Energy Forming*, Estes Park, Co., June 23–27, 1969.

J. Ramesam, S. R. Sahay, P. C. Angelo, and R. V. Tamhankar, *Weld Res. Suppl.*, 23s (1972).

L. F. Trueb, *Trans. Met. Soc. AIME* **2,** 147 (1971).

ANDREW POCALYKO
E. I. du Pont de Nemours & Co., Inc.

METALLIC SOAPS.   See Driers and metallic soaps.

METALLOCENES.   See Organometallics.

METAL PLATING.   See Electroplating; Metallic coatings.

METAL SURFACE TREATMENTS

Cleaning, pickling, and related processes,   296
Chemical and electrochemical conversion treatments,   304
Case hardening,   313

CLEANING, PICKLING, AND RELATED PROCESSES

Cleaning

Cleaning usually is necessary before painting, bonding, plating, or other surface treatments in the manufacture of metal products. It involves the removal of unwanted surface materials by a chemical or physical process or by a combination of these methods (1). The primary characteristics of important cleaning methods are given in Table 1. Additional information is given in refs. 2–4.

Solvent Cleaning. Solvent-cleaning techniques are used to remove fats, oils, waxes, and other organic materials from surfaces. They generally are ineffective against oxides, sulfides, and other inorganic materials. Usually cleaning solvents are flammable and/or toxic. They also leave a hydrophobic residue on the surface. This residue may provide a measure of corrosion resistance for a short time, but it also may interfere with subsequent surface treatment. Thus, an alkaline cleaning step may be required before additional processing.

Table 1.   Metal-Cleaning Methods

Metal	Common cleaning methods[a]	Avoid	Comments
Ferrous alloys			
low carbon steel	1, 3, 4, 5, 6	2, temperature over 77°C	use 4, 5, or 6 to clean greasy surfaces
cast iron	1, 3, 4, 5, 6	2, temperature over 49°C	remove scale and tarnish with 1
other alloys	1, 3, 4, 5, 6	2	
Stainless steel			
12% or less Cr	1, 3, 4, 5, 6	2, check temperature limits	watch for different stainless alloys used in the same assembly
400 series	1, 3, 4, 5, 6	2	
300 series	2, 3, 4, 5, 6	water containing chloride and moist atmospheres if chlorinated solvents are used	nonoxidizing acids except HCl may be used on 300 stainless
Copper alloys			
copper	1, 3, 4, 5, 6	2	
brass	1, 3, 4, 5, 6	2	
bronze	1, 3, 4, 5, 6	2	
aluminum	5, 6	1, except H_3PO_4 with used chromate	some formulations of 2, 3, and 4 may be used; check with supplier
ternplate	1, 3, 4, 5, 6	2	
zinc, galvanized steel	5, 6	1, 2, 3, 4; sometimes very mild alkalies may be used	zinc is a rather reactive chemical

[a] Cleaning-method code: 1 = inhibited, nonoxidizing acids, such as HCl, H_3PO_4, some H_2SO_4; 2 = HNO_3; 3 = organic acids; 4 = alkaline cleaners; 5 = chlorinated degreasing solvents; 6 = other organic solvents.

Wiping.   Moderately large metal objects, eg, door panels, metal stampings, and welded assemblies, often are cleaned by wiping with solvent-soaked rags. Solvents of low flammability and toxicity typically are used. Wiping is simple but has a high labor cost and presents quality control problems. Wiping usually is preferred for objects that are too large to be combined in baskets or for parts that require cleaning of only one surface.

Emulsifiable Solvents.   Emulsifiable solvents, eg, kerosene, contain an emulsifying agent (see Emulsions). They are applied by spray or immersion until the soil is dissolved and/or softened. Once the soil has been loosened, the solvent is rinsed away with water. The agent(s) that are present keep the solvent and removed organic materials suspended in the water, thus forming an emulsion. Excess water must be avoided in order to prevent breaking the emulsion by dilution and, thus, recontaminating the clean surface. Emulsifiable solvents generally have flash points of 43°C or less. Those containing concentrated emulsifiers must be diluted with a solvent prior to use.

Emulsion Cleaning.   In emulsion cleaning, a solvent is dispersed in water: the solvent concentration is typically 5–10%. The dispersion is applied to a metal surface by spray or immersion. Temperatures vary from ambient to 54°C. Spray application is a more effective technique because the physical force of impingement and agitation assists in removing and suspending the soil. Emulsion cleaners often contain a rust inhibitor which prevents corrosion during in-process storage (see Corrosion and corrosion inhibitors). The cleaners sometimes are combined with alkaline cleaners for improved effectiveness.

Diphasic Chlorinated Solvents.   Diphasic chlorinated solvents are composed of an aqueous and an organic layer. Depending on the formulation, the upper layer may be the aqueous or the organic phase. An aqueous upper phase is more common, since it retards evaporation of the organic solvent. These cleaners frequently include extremely powerful solvents, eg, methylene chloride, and often are effective against difficult-to-remove soils and paints. Flash points for these systems may be as high as 32°C, and the solvents typically leave a thin, corrosion-resistant, hydrocarbon film on the part surface (see Chlorocarbons).

Vapor Degreasing.   The object to be cleaned is placed in a zone of warm solvent vapor (5–7). The vapor condenses on the metal and subjects the surface to a solvent-flushing action as it flows downward. The liquid drops are collected in a reservoir and are revaporized, typically through the use of steam-heating coils. Thus, the surface is continually rinsed with fresh solvent. Virtually no oily residue remains once the part is removed from the vapor zone. A typical vapor degreaser is shown in Figure 1. Cooling coils are used around the freeboard of the tank in order to minimize vapor spill. Nonetheless, solvent replenishment and eventual replacement is necessary. Trichlo-

Figure 1.   Two-dip immersion degreaser.

roethylene traditionally has been used in vapor degreasers but, because of its toxicity, it is being supplanted by 1,1,1-trichloroethane and Freon compounds. Other chlorinated hydrocarbons sometimes are used. Corrosion inhibitors are added in order to counteract the small amounts of hydrochloric acid which inevitably are formed from these compounds. Some vapor-degreasing machines involve a liquid-immersion stage prior to the vapor action. Vapor degreasing is an ideal technique for use with small parts which are only moderately soiled and can be conveyed in baskets. The process requires little labor but, because of the cost of the solvent and the required ventilation, it is a moderately costly technique.

Ultrasonic. High energy sound waves, ie, ≥ 25 kHz, frequently are included in solvent-immersion cleaning in order to remove finely divided solid soils from surfaces, especially recesses. The sound waves produce a high degree of local agitation which helps to dislodge the soil. Equipment for ultrasonic cleaning is expensive and, therefore, is used primarily for small parts or in situations where the higher cost of a very thorough cleaning can be justified. Ultrasonic cleaning also can be used with aqueous cleaners (see Ultrasonics).

Alkaline. Years of experimentation and experience have indicated that alkaline cleaning solutions are effective, economical means of preparing metal surfaces for subsequent processing. Most proprietary formulations include several alkaline salts, soaps, synthetic surfactants, and various inhibitors (see Soap; Surfactants and detersive systems). Alkaline cleaners usually are used with an immersion or spray process.

Immersion. Immersion or soak cleaning is widely used for moderately soiled surfaces that are processed at low rates. The parts are allowed to soak in the cleaning solution which may be warmed to 38°C. Cleaner concentration commonly is 30–90 g/L. Immersion cleaning depends upon the ability of the alkaline solution to saponify vegetable oils and emulsify mineral hydrocarbons. Without agitation, this process is time consuming; thus, high concentrations and temperatures frequently are used.

Inhibited cleaners are used when corrosion or discoloration could be a problem; this is especially true with aluminum, zinc, and cuprous alloys. Thorough rinsing after immersion cleaning is extremely important to the success of subsequent surface treatments.

Spray. Spraying increases both the speed and effectiveness of alkaline cleaning. Despite the fact that pumps and filters are required and a potential foaming problem exists, spray cleaning is the method of choice for difficult-to-remove soils or machinery used for high production processes. Heat losses are higher with the spray process than with immersion. Cleaner concentrations typically are 4–32 g/L at up to 71°C, and 1–3 min usually is required at 100–210 kPa (15–30 psi).

The design of risers and nozzle locations is critical to successful spray cleaning. Solution-intake pipes should be located above the tank bottom to reduce the likelihood of circulating debris. As cleaning proceeds, soaps may be formed by saponification of the soil so that the surface tension of the cleaning solution is reduced. As a result, foaming may become a problem over a period of time, and defoamers (qv) may be needed. Spray cleaners must not be inadvertently admitted to the rinse chamber of the cleaning machine; curtains sometimes are used to reduce the occurrence of this problem. Corrosion of the part by the cleaner also must be considered. Properly inhibited products should be chosen on the basis of test runs with the actual parts and equipment to be used.

Ultrasonic Alkaline.   Ultrasonic techniques can be used with alkaline-cleaning solutions to remove oily residues.

Steam.   Steam (qv) cleaning depends upon the increase of vapor pressure of oily materials when they are heated by exposure to steam. There is also a mechanical removal effect as a result of the steam pressure. Usually super-heated water is sprayed with small concentrations of detergents at 350–2100 kPa (50–300 psi). Typical temperatures are 121–149°C with nozzle-to-part distances of up to 1 m. Detergent concentrations are 0.4–2.0 g/L, and contact time between the part and the steam-cleaner solution is ca ≤30 s. The cleaning solution is not reused and, thus, low concentrations are required to ensure economy. Plant steam or steam from portable or stationary steam generators is used. Detergents may be siphoned into the equipment or supplied directly by metered pumps. The detergent supply can be shut off at the spray head so that only water and steam are supplied for rinsing, thus assuring a residue-free surface. Steam cleaning is most commonly used with very large parts, eg, industrial and agricultural equipment, railroad cars, or fabricated structures, eg, bridges and storage tanks.

Electrocleaning.   In electrocleaning, steel parts form the anode and brass or zinc parts form the cathode of an electrical cell; dc is used in each case. Water is decomposed during the cleaning process with the formation of oxygen at the anode or hydrogen at the cathode. These gases, which are formed as minute bubbles beneath the soil on the metal surface, lift particles of soil from the metal surface. Strongly alkaline cleaners usually are used to obtain maximum conductivity. Metals that are sensitive to alkaline attack require inhibitive cleaners or lower concentrations. Alkaline solutions also generally are formulated to yield a surface layer of foam, which reduces the loss of caustic solution from the tank as a result of the vigorous rising of gas bubbles to the surface. Cleaner concentrations are 60–105 g/L and temperatures vary from 54 to 95°C. Depending on the metal, immersion times of up to 2 min at 220–1665 A/m^2 are typical.

Pickling

Pickling is the removal of oxides (8–9) by converting them to soluble compounds with acids or alkaline solutions or molten alkali salts.

Acid.   Traditional acid-pickling solutions are based upon inhibited sulfuric, hydrochloric or phosphoric acids (10). The process parameters are concentration, temperature, time, and agitation. An increase in any of these variables increases the rate of oxide removal. Higher temperatures, however, also result in the generation of corrosive fumes and, thus, frequently require a compromise between temperature and concentration.

Ferrous Metals Other than Stainless Steels.   Sulfuric acid, 5–10 vol % at 60–71°C, is used to remove heavy mill or fire scale produced during hot working of the metal. Hydrochloric acid diluted with 1–3 parts by volume of water and used at room temperature provides rapid removal of light scale or rust with minimal pickling smut, eg, ferric carbide, etc, and may be used subsequent to sulfuric acid pickling to remove such smut.

Phosphoric acid treatment usually is for light scale or light rust removal, particularly when rerusting is to be minimized or when the metal is to be prepared for painting. It may follow sulfuric or hydrochloric acid treatment. The minimum concentration should be ca 25 wt % (15 vol %) of the commercial 75% acid at room temperature. Heavier scale may be removed with unmodified phosphoric acid at temperatures up to 93°C.

Scale removal partially involves penetration of cracks that are formed in the brittle oxide during cooling and handling. Mechanical action to loosen or to crack heavy scale is effective in reducing pickling time.

Stainless Steels and High Nickel Alloys. Stainless steel and high nickel alloy scales are tightly adherent and difficult to remove with the acids used for plain steel, although either hot 10% sulfuric acid containing 1–2% sodium thiosulfate or hydrosulfite, or 2% hydrofluoric acid with 6–8% ferric chloride often is effective. Moderate or light scale is removed with 20% nitric acid containing 2–4% hydrofluoric acid. Nitric acid is a widely used pickling agent and does not affect the stainless character of the steel; in fact, these steels are passivated in nitric acid of greater than 20% concentration. Hydrochloric acid may be used as an activator or as a first treatment to remove scale, but it tends to pit the metal and is not recommended (see also Corrosion).

Some of the stainless steels and high nickel alloys are not pickled adequately in acidic solutions and molten alkali baths may be required (see Molten Salt Treatment).

Cuprous Metals (Copper, Brasses, and Bronzes). Hydrochloric acid or sulfuric acid is used for scale removal; the former at 0–3 parts dilution for ca 1 min or the latter at 5–10% for 20 min at 60–71°C. Sometimes brightening is used simultaneously with light scale removal though it often follows hydrochloric acid descaling (see Electropolishing and Brightening). Nitric acid baths can be used for brightening, especially when modified with chromic acid.

Aluminum and its Alloys. Hydrochloric acid is avoided because of its pitting tendencies. Acidic deoxidation often is accomplished with sulfuric acid plus sufficient chromic acid to prevent attack on the metal; this type of product may be conveniently available as a proprietary powder of sodium hydrogen sulfate and dichromate or chromic acid. This type of deoxidation normally is required prior to spot welding. Phosphoric acid may be used to remove oxide from aluminum. Much aluminum is pickled or etched in alkaline solutions (see Aluminum Etching).

Magnesium. Chromic acid or baths containing sulfuric or nitric and chromic acids are used most frequently. Scaled magnesium is pickled prior to application of a paint-bonding treatment or to assist in the removal of embedded graphite, which is used as a lubricant in metal forming at elevated temperatures. Hydrofluoric acid is added when silica is present in the alloy.

Zinc and Galvanized Steel. The voluminous oxide formed in weathering may be removed by sulfuric, hydrochloric, or phosphoric acid treatment. Conditions are mild because of the high reactivity of zinc and the solubility of zinc sulfate or zinc chloride. Galvanized steel often is treated with phosphoric acid to remove oxide and prepare the metal for painting. This treatment forms a layer of tightly held zinc phosphate crystals, which behaves as a corrosion-resistant, very adhesional paint base.

Alkaline Pickling. Light scale and light-to-medium rust can be removed by aqueous, strongly caustic solutions with sequestrants, eg, sodium gluconate or ethylenediaminetetraacetic acid, usually in proprietary formulation. The solutions are used at concentrations of ca 121–363 g/L, at 71–91°C, for 15 min to several hours. Action is considerably slower than in acidic pickling. Agitation is desirable and the use of electric current, either with the work as the anode or with the periodic reversal of polarity, is very effective. The same baths may be used for cleaning and paint stripping.

Aluminum is treated in fairly dilute caustic soda solutions to remove oxide and to impart an etched surface prior to anodizing, lacquering, or prepaint treatments.

Molten-Salt Treatment. Difficult oxides may be removed in molten caustic alkali baths containing sodium hydride or sodium nitrate or with application of anodic or cathodic electric current; these processes are proprietary. The drag-out of undiluted electrolyte is expensive and the high temperature operation requires suitable equipment and safety precautions. The metal is quenched in water and then may be acid pickled. These baths may be used for heat treatments and other purposes, eg, the residue may act as a lubricant in the drawing operation or the bath may be used to remove porcelain enamel or tough paints.

Related Processes

Aluminum Etching. Aluminum etching is a widely practiced commercial procedure utilizing modified aqueous, caustic soda baths, that often are proprietary. Concentrations are ca 32–60 g/L and processing temperatures and times are 60–82°C and 0.5–10 min, respectively. Conditions of 40 g/L at 71°C for 5 min usually is sufficient to remove about 0.025 mm from the surface, thereby reducing surface imperfections and providing a satin or matte finish. The type of finish is influenced primarily by the alloy but considerable modification can be obtained by varying conditions and additives.

The large volumes of hydrogen that are generated during alkaline aluminum etching entraps considerable caustic spray. Foaming agents are added to yield a blanket that holds the bubble of gas at the surface long enough for the caustic solution to drain back into the bath; excessive foam is detrimental to the process. Selection of the proper agent is difficult because many surfactants are defoamed as soil and/or sodium aluminate builds up; yet higher concentrations yield too much foam in a fresh bath.

The etching rate decreases as the sodium aluminate concentration increases. At concentrations of 111–146 g/L of sodium aluminate, the bath becomes viscous and prone to scaling and should be discarded. The higher viscosity interferes with rinsing; moreover, the drag-out can hydrolyze to yield insoluble, white aluminum hydroxide. Problems can sometimes be reduced by an alkaline, ie, pH 9–11, first rinse.

Etching Other Metals. Almost all metals other than aluminum are etched in acids and often in baths containing chlorides, eg, hydrochloric acid, ferric chloride, etc. Some aluminum is etched in phosphoric acid with chloride added for deep etches. Steel occasionally is etched in hydrochloric acid to improve adhesion of paint or other materials. Etching is reviewed in ref. 11.

Blasting. Blasting involves the mechanical hurling of hard materials, eg, sand, metal shot, abrasive grit, metal or ground nut shells against the surface to be cleaned (see Abrasives). The abrasive is propelled by air, suspended in water for spraying, or spun centrifugally from a high speed wheel. Cleaning depends upon the cutting action of the abrasive against the material being removed. The relative hardness of the abrasive, the oxide, and the base metal is such that oxide removal proceeds at a uniform rate which is more rapid than the loss of base metal. Blasting techniques often are used as an initial step prior to other cleaning processes. When blasting is the only surface-preparation technique, it is particularly important to guard against contamination of the abrasive media by previously removed soil and grease.

Electropolishing and Brightening. Electropolishing and brightening involve removal of the metal at the high spots of irregularities with little or no dissolution of metal in the low spots or valleys. Scratches resulting from abrasive polishing are avoided, but the base metal must be resonably smooth to produce a satisfactory finish. Stainless steel, steel, brass, aluminum, silver, nickel, copper, zinc, chromium, and gold are electropolished. Most of the baths are covered by patents, although some of these have expired. Highly concentrated solutions of sulfuric and/or phosphoric and/or chromic acids are used frequently for electropolishing. Brightening involves the use of more dilute solutions of oxidants, eg, chromic or nitric acid or hydrogen peroxide. Other acids also may be present. Brightening involves less loss of metal than electropolishing (see also Electroplating).

BIBLIOGRAPHY

"Solvent Cleaning, Alkali Cleaning, and Pickling and Etching" in *ECT* 1st ed., under "Metal Surface Treatment," Vol. 9, pp. 1–9, by August Mendizza, Bell Telephone Laboratories, Inc.; "Cleaning, Pickling, and Related Processes" in *ECT* 2nd ed., under "Metal Surface Treatments," Vol. 13, pp. 284–292, by S. Spring, Oxford Chemical Division, Consolidated Foods Corp.

1. W. L. Wageman, *FC-79*, Association of Finishing Processes of SME, 1979, p. 703.
2. "Heat Treating and Finishing" in *Metals Handbook*, Vol. II, 8th ed., American Society for Metals, Metals Park, Oh., 1964.
3. *Metal Finishing Guidebook and Directory 79*, Metals and Plastics Incorporated, Hackensack, N.J., 1979.
4. R. C. Snogren, *Handbook of Surface Preparation*, Palmerton Press, New York, 1974.
5. R. Monahan, *Met. Finish.* **75**(11), 26 (1977).
6. K. Suprenant, *Prod. Finish.* **43,** 66 (Mar. 1977).
7. R. N. Harvey, *Prod. Finish.* **42,** 78 (Dec. 1978).
8. J. Mazia, *Met.Finish.* **77**(9), 65 (1979).
9. J. Mazia, *Met. Finish.* **77**(10), 75 (1980).
10. N. Eldakar and K. Nobe, *Corros.* (National Association of Corrosion Engineers) **33,** 128 (1977).
11. D. C. Simpkins, *Inst. Met. Finish.* **57,** 11 (1979).

G. L. SCHNEBERGER
General Motors Institute

CHEMICAL AND ELECTROCHEMICAL CONVERSION TREATMENTS

Phosphating

Phosphating is the treatment of a metal surface to provide a coating of insoluble metal phosphate crystals which strongly adhere to the base material. Such coatings affect the appearance, surface hardness, and electrical properties of the metal. They may provide some corrosion resistance, but they are not sufficiently protective to be used by themselves in most corrosive atmospheres. The coatings, however, are somewhat absorptive and, thus, provide an excellent base for impregnation with oils, lacquers, and paint finishes. Phosphating is of major industrial importance in the production of iron and steel (qv) surfaces, eg, in the automotive and appliance industries. They have also been used to protect zinc-, aluminum-, cadmium-, and magnesium-based metals.

Modern phosphate technology began in England in the early twentieth century with the use of a phosphoric acid solution to produce a phosphate film on steel. A less violent reaction and a better coating results when a phosphoric acid solution of ferrous dihydrogen phosphate, $Fe(H_2PO_4)_2$, is used. Later developments included the uses of phosphoric acid solutions of zinc(II) dihydrogen phosphate, $Zn(H_2PO_4)_2$, and manganese(II) dihydrogen phosphate to deposit zinc and manganese phosphate crystals, respectively (1).

Contemporary phosphating solutions contain one or more of the phosphates of iron, manganese, or zinc in dilute phosphoric acid. They also may contain the phosphates of alkali and alkaline-earth metals and a variety of accelerators. Process times typically are from less than one to several minutes. Major development work in phosphating is carried out by companies specializing in the preparation of proprietary phosphating solutions (2–3).

Coating Formation. Commercial phosphating products are complex and proprietary and, thus, it often is difficult to classify precisely the reactions that take place during phosphate coating deposition. Simplified equations, however, may be used to illustrate the basic chemistry. When a ferrous surface is treated with a phosphating solution, it is attacked by the free phosphoric acid:

$$Fe + 2\,H_3PO_4 \rightarrow Fe(H_2PO_4)_2 + H_2$$

Hydrogen is liberated and iron is introduced into the solution as soluble, primary ferrous phosphate. The primary zinc or manganese iron phosphates, which are present as bath constituents, hydrolyze readily in aqueous solutions and produce the less soluble secondary and tertiary metal phosphates, according to the following equations:

$$M(H_2PO_4)_2 \rightleftharpoons M(HPO_4) + H_3PO_4$$

$$3\,M(H_2PO_4)_2 \rightleftharpoons M_3(PO_4)_2 + 4\,H_3PO_4$$

$$3\,M(HPO_4) \rightleftharpoons M_3(PO_4)_2 + H_3PO_4$$

where M = zinc, manganese, or iron.

The phosphoric acid that is produced in these reactions is consumed by dissolving iron from the treated part. The equilibrium shifts from left to right because of the precipitation of the sparingly soluble secondary and tertiary phosphates. The phosphoric acid is consumed primarily at the metal surface where the concomitant disturbance of the equilibria also occurs. The insoluble metal phosphates precipitate from solution and onto the surface of the iron to form a tightly adherent, highly interlocked crystalline layer.

Initially, a phosphating bath may contain only the primary phosphates of zinc or manganese. Consequently, the coating may consist only of the secondary or tertiary zinc or manganese phosphates. However, as the steel is processed, iron enters solution as primary ferrous phosphate. Inevitably, ferrous phosphate precipitates on the work with zinc or manganese phosphates. An equation representing the formation of a zinc phosphate coating on an iron surface may be expressed as follows:

$$3\,Zn(H_2PO_4)_2 + Fe + 4\,H_2O \rightarrow Zn_3(PO_4)_2.4H_2O + FeHPO_4 + 3\,H_3PO_4 + H_2$$
$$\text{(as coating)}$$

Iron, zinc, zinc–calcium, and manganese phosphates are the most commonly applied phosphate coatings and each has distinct characteristics and areas of application. Accelerators are included in the processing to produce coatings of superior and uniform quality; rapid processing times, ie, seconds; and with appropriate chemical replenishment, the prolongation of the effective life of the bath to several months.

Oxidizing agents are the most important accelerators which are routinely added to phosphating formulations. They are believed to exert a beneficial influence by depolarizing the hydrogen by the action of phosphoric acid and acid phosphate, and by converting the ferrous ion passing into solution to ferric ion. If oxidants are not included, ferrous ion concentration builds rapidly to levels which progressively deteriorate coating quality. If oxidants are included in the bath, the ferrous ion content builds more slowly, generally reaching equilibrium concentrations, and the ferric phosphate that forms precipitates because of its low solubility. Other techniques, eg, the use of dc, periodically reversed dc, and ac accelerate the phosphate coating and improve coating quality (4). Various adjuvants, eg, fluorides, complex fluorides, and chromates and heavy metals, eg, nickel and copper, etc, have been included in baths to improve coating on galvanized metal and on aluminum and its alloys.

Process Parameters. The complete phosphating process cycle, as conducted on a commercial scale, generally consists of the following steps: preparation of the surface to be processed, ie, cleaning, rinsing to remove cleaning agents, and special pretreatments; application of phosphate-coating process; rinsing of the coated surface, ie, water and posttreatment rinsing; and drying. The characteristics of the coating depends to a large extent on the conditioning of the surface prior to coating. The basic requirement is a clean surface which is free of harmful contaminants and receptive to subsequent treatment. Surface oxide scale or corrosion products must be removed by pickling or sand blasting. Ordinarily, the metal is alkali-cleaned. Sometimes special procedures, eg, steam cleaning or vapor degreasing are used. Generally, the work is rinsed with water, which often is deionized, after cleaning.

The pretreatment also influences phosphate crystal size, eg, alkaline surface residues produce increased crystal size and sparkle, and pickling and acid pretreatment also produce large crystals. However, brushing or mechanically wiping the metal tends to produce fine-crystal formation.

Generally, smaller crystals result in the wiping effect, ie, decreased porosity and improved corrosion resistance, and much effort has been expended to produce this effect by chemical means. Predips in oxalic acid solutions, sodium nitrite solutions, solvent mixtures, and solutions containing heavy metals, eg, copper and nickel, have been used with some success as has the application of Jernstedt salts (5). The pretreating solution consists essentially of a 1% solution of disodium phosphate containing ca 0.01% titanium as the anhydrous phosphate salt. The treatment consists of immersion or spray application which is followed by phosphating. The mechanism of action of the titanium disodium phosphate salts is not completely understood; however, the salt probably exists in colloidal form or as a complex phosphate and is thought to activate the surface by providing sites for crystal formation.

There are certain parameters in the operation of a phosphating bath that must be controlled within narrow limits in order to achieve a suitable coating and proper bath maintenance. The ratio of free to combined phosphoric acid and the total acid, metal-ion, and accelerator concentrations must be in proper balance so that the phosphate coating is readily formed at the metal–solution interface. Operating conditions must be controlled to minimize hydrolysis in the bath. If the free-acid concentration is too high, hydrolysis is repressed, pickling occurs, and only thin coatings are produced. If the total acid concentration is too high, the increase in pH as metal attack proceeds is slow, the coating process is inhibited, and only thin coatings are produced. If the proportion of free acidity is too low, acid attack of the steel is reduced and a poorly adherent coating may be formed. If the concentration of the solution is too low, the consumption of free acid resulting from attack of the metal causes the pH of the bath to rise rapidly so that precipitation is not confined entirely to the surface of the metal but also occurs in the body of the solution; consequently, a heavy sludge build-up occurs.

The make-up of a typical zinc-based phosphating bath might consist of a 2% solution of the following concentrate: 758 g zinc oxide, 1.89 L phosphoric acid (75%), and 1.89 L water. A solution of sodium nitrite is added until its bath concentration is ca 0.02%. Coating time for immersion processing may be 5–10 min at a bath operating temperature of 71–82°C. When spray coating is used, processing for ca 0.5 min at 49–60°C is sufficient to form a uniform conversion coating of insoluble phosphates.

After the work is treated, it must be given a water rinse and a final rinse in a passivating solution. The final rinse may be heated to facilitate drying and may consist of up to 0.25% chromic acid or up to 1% alkali metal chromates, chrome complexes, or combinations of phosphoric acid and chromic acid. The influence of this final rinse on enhancing corrosion resistance and promoting paint adhesion cannot be overemphasized. Concerns about the environmental effect of chrome which may be discharged by processing plants using a chromate final rinse have led to the development of nonchrome-containing passivating rinses. In general, these have been somewhat less effective than their chrome-containing counterparts, but they are suitable for some applications where corrosion is not a major concern.

Various accelerated test methods are commonly used to rate the performance of a coating system. These include the salt-spray test (ASTM B 117-49T) and the humidity (ASTM D 2247), water-immersion (ASTM D 870), and conical-mandrel impact tests (ASTM D 522), etc. A recent development has been the use of infrared absorption techniques to determine phosphate crystal weights (6).

Uses.   The largest use for phosphate coatings is as a base for bonding paint to metal. Phosphating of iron and steel and of zinc, cadmium, aluminum, and their alloys is common. Selection of a particular process depends on the specific use of the coating and the production facilities available. Iron, zinc, and microcrystalline zinc–calcium phosphates are used most commonly as underpaint coatings for steel. Iron phosphates generally are applied at 0.30–1.1 g/m² to stampings for cabinets and metal furniture. Zinc phosphates and microcrystalline zinc–calcium phosphates having coating weights of 1.1–6.5 g/m² usually are recommended for automobile bodies, refrigerators, and appliances. Heavy zinc phosphates and manganese phosphates ranging in weight from 11–43 g/m² are used on hardware, nuts, bolts, guns, cartridge clips, etc. This type of coating often is used in conjunction with protective oils and waxes.

Phosphate coatings have been employed widely as an aid in drawing and cold-forming metals. Manganese phosphates and zinc phosphates are used to a large extent for this purpose, although complex oxalates have been recommended for application to stainless and other high alloy steels. Phosphate coatings of ca 2.2–11 g/m² generally are applied. Soaps, molybdenum disulfide, and other lubricants often are used in conjunction with the coatings to reduce friction between the metal and the die, eliminate galling and seizure of the metal, and permit faster drawing (see Lubrication and lubricants).

Phosphate coatings have been used on bearings and moving parts to reduce scoring and friction. In the automotive industry, a variety of parts, eg, piston rings, cylinder lines, rocker arms, valve tappets, gears, cam shafts, etc, are coated with manganese phosphates. These coatings are applied at ca 11–43 g/m² and act as oil reservoirs and are beneficial during break-in periods.

Cadmium and zinc are phosphated largely for improvement of paint adhesion. Paint does not bond well to zinc surfaces which are painted directly without prior treatment. This is partially because of the reaction of zinc with the paint vehicle resulting in the formation of zinc soaps at the paint–metal interface, thereby weakening the bond. An insoluble phosphate coating formed on the surface improves bond strength by virtue of increased mechanical anchorage of the paint and because it imposes a barrier which retards interaction between the paint vehicle and the metal. A similar situation is thought to exist with cadmium.

Phosphating solutions that are used to coat aluminum generally contain fluoride, and some processes involve chromate as well as phosphate. The phosphate–chromate coatings are amorphous and range from nearly colorless to light blue-green to dark green. The weights of the greenish coatings are 2.2–2.7 g/m². They are electrically nonconductive and attractive and are used extensively in the unpainted form for decorative purposes. Typical product applications include aircraft components and assemblies, appliance parts, siding, roofing, chain-link fencing, and screening.

An alkaline phosphating process has been used for producing a protective coating on tin. The process is said to prevent sulfide blackening of the interior of tin cans and to retard rusting of the outside of the can when it is stored in a damp environment.

Chromate conversion coatings have been used widely in the metals industry to provide corrosion protection and to promote paint adhesion. Zinc, cadmium, magnesium, and aluminum are the most commonly treated metals. As in the case of phosphating processes, the chromate-treating processes are proprietary and are developed largely by specialists in the field.

Chromate-treating baths have two common basic ingredients, ie, hexavalent chromium and a mineral acid. Some contain organic or inorganic constituents as activators. Upon immersion, the metal reacts with the mineral acid, resulting in an increase in solution pH immediately adjacent to the surface. Concurrently, hexavalent chromium is reduced to the trivalent form. At a suitable pH, the trivalent chromium and some hexavalent chromium co-precipitate on the metal surface as a complex chromium chromate.

A wide variety of processes are available that produce protective films which range in color from iridescent yellow to brown, olive drab, and black. When in contact with water, the hexavalent chromium portion of the film is slowly released and provides a local supply of dissolved chromate which inhibits corrosion of the metal surface. The insoluble trivalent chromium portion of the film provides general protection by excluding, to a large extent, moisture from the metal surface.

Chromate-type conversion coatings are produced primarily by simple immersion processes. Freshly formed films obtained by the immersion process initially are soft and fragile and are susceptible to damage from abrasion and handling. A 12–24-h aging period generally is required before the films attain sufficient toughness to withstand normal shop handling. Chromate coatings may be dried or baked at 66°C and higher or dehydrated by desiccation in the presence of silica gel or calcium chloride. Extensive heating, however, at over 66°C reduces the protective value of the coating. Chromate conversion coatings are porous and absorbent when first formed and can be colored. The dyes generally are not fast and are used primarily for identification purposes.

There are a wide range of proprietary chromate treatments for zinc and cadmium; many are referred to as chemical bright dips. These treating solutions may contain chromic acid; one or two mineral acids, such as sulfuric and nitric; and often some activating compounds. The thickness and color of the film depends on solution composition, temperature, pH, and length of treatment. Generally, solutions prepared at 21–38°C and pH 1.5 or less, produce lustrous clear to light iridescent yellow coatings of moderate corrosion resistance. Solutions prepared at lower concentrations, at 16–32°C, and operating at pH 1.0–3.5 produce medium to heavy films ranging in color from iridescent yellow to bronze to black, and these show superior corrosion resistance.

Special passivating treatments are used in strip-line galvanizing operations to prevent the formation of white rust during storage and shipment of the sheet or coil. Treatment usually is applied immediately after the coil is rolled and water-quenched and may be applied by spray or roll coater. The treatment produces a clear finish that does not change the surface appearance. Processes producing chromate prepaint treatments are available for high speed galvanized strip. Heavier, colored coatings are produced which give a high degree of corrosion protection and good paint bonding.

Chromate conversion treatments for aluminum produce an amorphous coating ranging from a light iridescent gold to a light tan. Coatings are electrically conductive, provide improved corrosion resistance, and contribute to better paint adhesion. They may be applied by spray, dip, brush, or roll coater. Typical product applications include aircraft components and assemblies, appliances, windows and screening, and electronic components.

Passivation of magnesium, copper, silver, and their alloys also has been achieved by chromating. Because of its active nature, practically all magnesium is chemically treated and usually in a chromate solution. Generally this is done at the metal-production plant to ensure storage without deterioration. Chromate treatments also are used in chemical polishing. The treatment is used as a final finish and sometimes as a substitute for buffing prior to nickel or chromium plating. In the case of silver electrodeposits, chromate surface treatments aid in retarding sulfide tarnishing.

Anodizing

Anodizing involves the formation of an oxide surface on nonferrous metal by electrochemical means. These surface oxide films supplement the natural oxide which occurs in very thin layers on such metals and results in a significant increase in their corrosion resistance. Aluminum, in particular, forms a thin, tenaciously adhering, oxide film which provides an excellent barrier against corrosion.

Anodizing of aluminum involves electrochemical conversion of the surface to aluminum oxide; the aluminum serves as the anode and the oxygen is provided by the electrolytic dissociation of water. Chromic acid, sulfuric acid, and oxalic acid electrolytes have been widely used. Other electrolytes, eg, borates, citrates, carbonates, sulfamic acid, and phosphoric acid, have been used in specific applications. The structure of the coating normally is amorphous, although in certain electrolytes, eg, boric acid, a crystalline structure sometimes is observed. Treatment of the coating in boiling water causes partial hydration of the anodically formed oxide to a crystalline mono- or trihydrate. As a result, the porosity of the coating is progressively reduced, resulting in improved corrosion resistance. This sealing operation must be carefully controlled to obtain optimum results. Other sealing treatments involve immersion in a hot dilute dichromate or sodium silicate solution. The chromate is particularly effective in imparting a high degree of corrosion resistance to the coating.

The properties of the anodic coatings depend on the type of electrolyte, its concentration and temperature, the current and processing time, and the basic metal used. Generally, porous coatings that are produced at low temperatures in moderately concentrated electrolyte have high abrasion resistance, whereas coatings formed at elevated temperatures in more concentrated electrolytes are relatively soft, absorbent, and usually of low abrasion resistance. Anodized aluminum coatings may be colored by impregnation with dyes or by precipitating colored pigments within the pores. The coatings then are sealed in boiling water or in a hot dilute solution of nickel or cobalt acetates; the latter sealants improve the fastness of the dyes. Anodic coatings also may be saturated with oils, greases, and waxes to enhance corrosion resistance or to improve dielectric strength. An oxide coating 0.013 mm thick may exhibit breakdown values as high as 500 V.

Sulfuric acid, chromic acid, and oxalic acid are the most common electrolytes used in producing corrosion-resistant anodic coatings. Film color, porosity, flexibility, and other characteristics differ depending on the electrolyte used.

The sulfuric acid process, which is common, has the advantages of lower cost, rapid action, and comparatively low operating voltage. The latter is a consequence of the high conductivity of the electrolyte and its ability to penetrate the film to the underlying metal. The electrolyte has a solvent action upon the film which affects its character. A wide range of thickness, hardness, and porosity may be produced by adjustment of acid concentration, temperature, and time. A typical process involves a 15–20% solution of sulfuric acid and generally is operated at room temperature and at an anodizing current density of ca 130 A/m^2 at 15–22 V dc. Processing times vary from ca 15 to 60 minutes, depending on the alloy treated and film thicknesses desired. Anodized coatings on pure aluminum are relatively hard and transparent and exhibit a glazelike finish and thicknesses of ca 0.005–0.010 mm generally are obtained. A disadvantage of the process is the danger of entrapment of the corrosive electrolyte in tapped areas, blind holes, porous castings, etc. For such suspected situations, the chromic acid process is preferred.

Usually anodic coatings that are formed in chromic acids are thinner and some-what softer than those formed in sulfuric acid; the coatings have fewer pores but the pores are larger in diameter. Despite the thinness of the coatings, they are highly protective and electrically insulating. The thin film produced by chromic acid pro-cessing normally has a light gray color on pure aluminum, but may be dark gray to black on silicon- and copper-bearing alloys. Commercial anodizing of wrought alloys involves a 30–45 min treatment at 40 V dc in a bath containing 50–100 g/L CrO_3 maintained at 35°C. The bath composition is controlled by determination of pH and solution density. Chloride content as NaCl is kept below 2.8 g/L since chloride interferes with proper formation of the anodic film. A low starting voltage, eg, 5 V, minimizes initial current surge and possible burning. The voltage is raised gradually to 40 V over 5–10 min and is maintained at that level for at least 30 min. At 40 V, the current density usually is 11–53 A/m^2. Generally, thicknesses of ca 0.0025–0.005 mm are obtained. Anodizing in chromic acid is not suitable for aluminum alloys that contain more than 5% copper. The chromic acid process is a batch operation and is specified particularly for parts subject to stress and assembled parts containing recesses in which anodizing solution may be entrapped.

The oxalic acid process initially was designed to provide an insulating coating for aluminum; it has, however, equally important protective properties. The film is golden-yellow, often is transparent, and has excellent hardness and wear resistance. Like the sulfuric acid process, this method is continuous and has the added advantage that ac can be used. In some cases a combination of dc and ac is used and provides a coating of better hardness and corrosion resistance than when ac is used alone. A common process involves a bath of 2–6% oxalic acid, maintained at 16–29°C, and op-erating at 60–100 V; coatings are ca 0.025 mm thick.

Anodized aluminum coatings that are used in electrolytic condensers usually are prepared in a buffered mixture of borax and boric acid pH 6.8. The thickness of the coating depends largely on the applied voltage: at 100 V applied potential, thickness is ca 0.0001 mm. In anodizing aluminum in noncorrosive electrolytes, the current is initially high, but it decreases rapidly as the oxide coating builds, and the total cell resistance increases until growth is prevented. Electrolytic valve action is then complete and current passes through the coating in the reverse direction only. In practice, alu-minum is anodized to a predetermined voltage that depends on the ultimate use of the anodized foil in a particular condenser.

A number of anodic treatments have been applied to magnesium alloys to maxi-mize corrosion protection and to improve paint-adhesion properties. The anodic treatments produce a dense, hard, and abrasion-resistant coating of a high degree of electrical resistance. Commercial baths consist of modified acid fluoride or caustic electrolytes.

Electrolytic brightening or polishing is an anodic process wherein special elec-trolyte systems are used so that the anodic film is dissolved essentially as fast as it is formed. This tends to smooth out surface irregularities and produces a bright, reflective surface.

Metal Coloring

Metal surfaces can acquire a wide variety of attractive shades and colors by heat treatment or chemical immersion treatments. Although the primary purpose of coloring metal is for esthetic value, often the surface treatments impart other favorable properties, eg, improved corrosion resistance and better abrasion and wear characteristics. Coloring of steel, copper, aluminum, and their alloys is done on a fairly large commercial scale by chemical dip methods (see also Colorants; Dyes).

Hundreds of formulations are available for the coloring of metals and the majority are proprietary. Foremost among the oxide film-forming processes for steel are the alkaline blackening treatments. The treating bath consists of a strong alkaline solution containing oxidizing agents, eg, nitrites, nitrates, or chlorates. A typical solution composition may consist of sodium hydroxide, 6.82 kg; sodium nitrate, 140 g; and water up to 3.78 L. The processing temperature may range from 135 to 160°C and immersion time may vary from 5 to 30 min. Coatings formed on steel by these treatments are largely magnetic oxide and are ca 0.0008–0.0018 mm thick. The coating color and characteristics are largely a function of the alloy being treated, its surface characteristics, concentration of the bath, and temperature and time of immersion. Black oxide coatings are porous and not particularly corrosion resistant. However, they provide a base for greases, oils, waxes, and organic finishes. An oil-coated film may last 24–50 h in salt spray without appreciable rusting. The major attributes of the process, other than the coloring effect, are its speed and simplicity, reasonably good abrasion resistance and wear characteristics of the treated parts, and avoidance of hydrogen embrittlement. The coating is unsuitable for parts with recesses or folds because of difficulty in rinsing trapped alkali.

There are several other popular methods for tinting steel. Black, blue, or brown finishes may be produced in various shades. The processes involve heating the metals in air, steam, or oil, in baths of molten salts, or in low melting point alloys. The oxide films that are formed exhibit color resulting from interference effects. The color that develops is a function of temperature and duration of treatment. At 221°C, the film exhibits a pale straw color; at 260°C, the color is blue; and at 338°C, it is steel-gray. The films are not very protective; however, they absorb greases, oils, waxes, etc, and provide some temporary corrosion resistance.

Aluminum does not form many colored compounds, so direct chemical coloring is limited. A few proprietary chemical treatments are available by which black or bluish colors are possible; the coatings, however, are not very durable. Virtually the entire color spectrum is obtainable on copper and its alloys. The colors are easily obtained, but the quality is greatly influenced by the alloying elements and surface texture. Thus, a thin electrodeposit of copper frequently is applied prior to coloring. Coloring procedures for magnesium, tin, nickel, gold, lead, and other metals are used only to a limited extent.

Energy Considerations

Because of recent concern over the cost and the availability of fuel, industrial users are anxious to reduce the energy requirements of all of the preceding processes. Many companies and organizations are developing alternative or equivalent processes that would be less fuel demanding than has been traditional. As a general rule, many of the common chemical and electrochemical conversions processes can be carried out at lower temperatures provided that concentrations of reactants or treatment times are increased. Most chemical processes approximately double in rate for each 10°C temperature rise. This rule of thumb, however, admits many variations and exceptions and should not be incorporated in any industrial process without adequate experimentation and testing. Another important consideration is the fact that, at higher concentrations, competing side reactions may assume increased prominence with various effects on the process or the product. Almost invariably, higher chemical concentrations are accompanied by increased reaction times. The effect which this situation has on production rates should not be overlooked; ie, production lines must be slowed or additional space provided. Boilers and ovens are available that burn either gas or oil. Developments in these directions will, no doubt, continue and even accelerate as the availability and cost of fuel becomes more and more tenuous.

BIBLIOGRAPHY

"Chemical and Electrochemical Conversion Treatments" in *ECT* 1st ed., under "Metal Surface Treatment," Vol. 9, pp. 9–23, by August Mendizza, Bell Telephone Laboratories, Inc.; "Chemical and Electrochemical Conversion Treatments" in *ECT* 2nd ed., under "Metal Surface Treatments," Vol. 13, pp. 292–303, by Louis F. Schiffman, Amchem Products, Inc.

1. U.S. Pat. 1,069,903 (Aug. 13, 1913), R. G. Richards (to F. Richards, Coventry, England).
2. *Metal Finishing Guidebook and Directory—1979*, Metals and Plastics Publications Inc., Hackensack, N.J.
3. *Finishing Industry Yellow Pages*, Special Technical Publications Inc., Oxnard, Calif.
4. U.S. Pat. 2,132,439 (Oct. 11, 1938), G. Romig (to Amchem Products Inc., Ambler, Pa.).
5. G. Jernstedt, *Trans. Electrochem. Soc.* **83,** 361 (1943).
6. G. D. Cheever, *Am. Paint Coating J.* **62,** 14 (1977).

G. L. SCHNEBERGER
General Motors Institute

CASE HARDENING

Case hardening is a metal-treatment process that produces a hard surface (the case) on a metal (the core) which remains relatively soft. The product is a hard, wear-resistant case backed by a strong, ductile, and tough core. Because of increased energy scarcity and the need for cost control, case depths should be minimized without sacrificing operational quality. A well-documented approach for determining case depth requirements with an optimum carburizing cycle has been reported (1). Wear characteristics are achieved by a hard but not necessasily deep case. However, deeper cases with strong and resilient cores are needed when the metal must support extremely heavy loads.

The effective case depth is standard for a carburized case and is the depth from the surface of a carburized area to the point at which the case is characterized by a Rockwell hardness (HRC) of 50. For induction- or flame-hardened parts, the effective case is based upon a depth of hardness that is controlled by the percentage of carbon in the basic material.

Carbon content, %	Effective case-depth hardness, HRC
0.28–0.32	35
0.33–0.42	40
0.43–0.52	45
≥0.53	50

A microscope is used to measure the depth of the altered microstructure in microcased parts.

Processes

Carburizing. Carburizing, which is the most common of the case-hardening processes, adds and diffuses nascent carbon into a steel surface (see Steel). Subsequent hardening by quenching at >790°C produces a hard case in the high carbon areas. Carburizing generally is limited to low carbon steels, ie, below 0.30 wt % C. The low alloy steels do not harden to a significant depth and lower core hardnesses result. The depth of hardness increases with increasing alloy content and stronger and more uniform cores result. Because of their increased hardenability, alloy steels do not need to be quenched as severely and are less susceptible to distortion. Core tensile strengths as high as 1.55 GPa (2.25 × 10⁵ psi) are obtained with low carbon, alloy steels.

Carburizing is performed between 845 and 955°C. At these temperatures, nascent carbon is most soluble in steel because of the austenitic phase and the diffusion rate is sufficient for economical use. At ca 845°C, the diffusion rate is slow and, consequently, there is more time to increase the carbon content of shallow cases to 0.5 mm and to maintain control of case depth. The major benefit of using higher temperatures, ie, >845°C, for carburizing steel is to increase absorption and diffusion rates and, thereby, reduce processing times.

Case depth is controlled primarily by controlling carburizing temperature and time. The following formula represents the effect of time and temperature on case depth (2):

$$CD = \frac{803\,t}{10^{(3722/T)}} \text{ (SI units)}$$

where CD = case depth, mm; t = time, h; and T = absolute temperature, K. In the English system, where CD is in inches, t is in h and T is in Rankine (°F + 460):

$$CD = \frac{31.6\,t}{10^{(6700/T)}} \text{ (non-SI units)}$$

By substituting various carburizing temperatures, the formula can be graphed as in Figure 1.

Carbon is absorbed into the surface of the steel only when it is in the form of nascent carbon. The origin of the nascent carbon depends upon the particular carburizing process used, ie, gas, liquid, or pack (solid) carburizing.

Gas. Gas carburizing is the most common case-hardening process used. Nascent carbon is produced from a carburizing atmosphere which may be any of several types of carrier gases, composed principally of nitrogen, carbon monoxide, and hydrogen, to which a hydrocarbon gas or vaporized hydrocarbon liquid has been added. At carburizing temperatures, the hydrocarbon or enriching gas breaks down into carbon monoxide, carbon dioxide, methane, and water vapor; the carbon monoxide and methane are the sources of nascent carbon. The nitrogen and hydrogen dilute the atmosphere and the chemical balances between carbon monoxide and carbon dioxide and between hydrogen and water vapor control the carbon potential of the mixture.

Natural gas, manufactured gas, and propanes are the most commonly used gaseous sources of carbon (see Gas, natural; Hydrocarbons, C_1–C_6). Occasionally, butane is used, but it should be the normal commercial grade containing about 7% propane. Isobutane never should be used. Natural gas is predominantly methane. Liquid hydrocarbons usually are proprietary compounds and range in composition from pure hydrocarbons, eg, terpenes, dipentene, or ligroin, to oxygenated hydrocarbons, eg, alcohols, glycols, or ketones. These liquids are fed in droplet form to a target plate in the furnace where they volatilize instantly. The thermal dissociation produces carbon monoxide, carbon dioxide, methane, and water vapor.

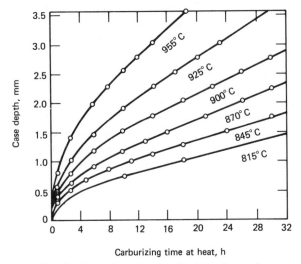

Figure 1. Case depth as a function of carburizing time and temperature.

The circulation resulting from gas flow alone is not sufficient to produce a uniform case; therefore, it is essential that each furnace has a high volume fan for forced circulation of the gases throughout all the parts of the work load. Gas-carburizing furnaces are of two types: batch and continuous. Both are built of the same materials but differ in size, shape, and method of operation (see Furnaces). In batch furnaces, the work is charged and discharged as a single unit or batch, and the furnace can be adapted to a variety of cycles and sizes of work pieces. With continuous furnaces, work is fed on a continuing basis at the charge end and is received at the discharge end for subsequent processing. In general, continuous furnaces are favored if production rates are high, case depth requirements are consistent, and material grades are compatible.

Liquid. A molten-salt bath containing a special grade of carbon is the source of nascent carbon in liquid carburizing. Carbon particles are dispersed in the molten salt by mechanical agitation, eg, small propeller stirrers. The chemistry involved is not fully understood but is thought to involve carbon monoxide adsorbed on carbon particles. The carbon monoxide is generated by the reaction between the carbon and carbonates of the molten salt. The absorbed carbon monoxide reacts with the steel surface in the same way as in gas or pack carburizing.

Operating temperatures for this type of bath are 900–955°C. Temperatures as low as 870°C are not recommended, as decarburization may take place. The most economical case depths are <1.3 mm. Temperatures >955°C produce more rapid carbon penetration and do not adversely affect noncyanide baths, as there is no cyanide to break down. Equipment deterioration is the chief factor that limits the use of higher temperatures.

Pack. Pack carburizing produces nascent carbon from carbon monoxide obtained by solid carburizing compounds that are packed around the parts in carburizing boxes. Hardwood charcoal and coke are the source of carbon for the process. A mixture of 10–20% energizers is bonded to the coarse mixture of charcoal and coke with oil, tar, or molasses. Generally, at least half of the energizer is composed of barium carbonate with the remainder made up with sodium or calcium carbonate.

The carburizing compound surrounds the work and is packed in steel, aluminized steel, or heat-resistant alloy boxes. The process produces essentially a sealed system within which carbon monoxide is produced from the carbon. When heated, the carbon monoxide produces nascent carbon by the same equilibrium reaction as in gas carburizing. Any CO_2 that is produced during the reaction immediately reacts with the charcoal and coke to produce more CO. The energizers enhance the reaction within the box by producing carbon dioxide which reacts to form carbon monoxide. The cycle continues as long as charcoal and coke are present to react with any carbon dioxide formed. As carburizing compounds are used, additions can be mixed in so as to maintain or regenerate the compound.

The temperature range for pack carburizing is 900–955°C. Because the work is surrounded by carburizing compound within a box, the heating rate is slow, processing times are long, and it is difficult to control the case depth to <0.25 mm. Although the carbon potential in the carburizing compound can be controlled to some extent, the potential is more easily controlled by varying the carburizing temperature, ie, the temperature is raised to achieve a higher carbon potential. Because pack carburizing does not require special furnaces or atmosphere generators, it is a relatively inexpensive process. Small lots and large parts can be processed economically and efficiently.

Carbonitriding. In carbonitriding, carbon and alloy steels are held at an elevated temperature in a gaseous atmosphere from which they absorb both carbon and nitrogen simultaneously. This process is used primarily to produce a hard, wear-resistant case, generally from 76 to 760 μm deep. A carbonitrided case has better hardenability than a carburized case; consequently, the former can be produced at less expense within the same case depth range.

Gas. Gas carbonitriding is a modified gas-carburizing process by which ammonia is fed into the carburizing gas atmosphere which then breaks down to form the nascent nitrogen which is diffused into the steel. The composition of the gas-carbonitrided case depends on temperature, time, and composition of the atmosphere. The higher the temperature, the less effective the ammonia is. Lower temperatures favor increased nitrogen concentrations near the steel surface but, as temperatures decrease, so does the nitrogen diffusion rate. The optimum temperature is 845°C. The two principal advantages of gas carbonitriding over liquid cyaniding are that any gas-carburizing furnace can be modified easily to accept ammonia, and there is no waste-disposal problem.

Cyaniding. Cyaniding is the liquid-bath form of carbonitriding. The most common production cyaniding-bath composition is 30% sodium cyanide, 40% sodium carbonate, and 30% sodium chloride. Baths containing 97%, 75%, and 45% of sodium cyanide also are used. Oxygen from the air oxidizes the sodium cyanide to sodium cyanate which, at high temperatures, decomposes to form carbon monoxide and nascent nitrogen:

$$2\,NaCN + O_2 \rightarrow 2\,NaCNO$$

$$4\,NaCNO \rightarrow Na_2CO_3 + 2\,NaCN + CO + 2\,N$$

The carbon monoxide that reacts to form nascent carbon produces carbon dioxide which reacts with the cyanide to produce more carbon monoxide:

$$2\,CO \rightarrow CO_2 + C$$

$$NaCN + CO_2 \rightarrow NaCNO + CO$$

The reactions leading to the decomposition of cyanate are

$$NaNCO + C \rightarrow NaCN + CO$$

and either

$$4\,NaNCO + 2\,CO_2 \rightarrow 2\,Na_2CO_3 + 2\,CO + 4\,N + 2\,C$$

or

$$4\,NaNCO + 4\,CO_2 \rightarrow 2\,Na_2CO_3 + 6\,CO + 4\,N$$

Both the nascent carbon and CO products of these reactions are utilized in the carburizing action. Nascent nitrogen is absorbed by the steel and increases the metal's surface hardness.

The average cyanide content of low temperature cyaniding baths generally is 20%. The carbon content of the case is controlled by varying the temperature and cyanide ranges. At low temperatures and cyanide content, the carbon content of the case decreases but the nitrogen content increases. All liquid carbonitriding baths should be checked frequently for cyanide. As drag-out and carbon depletion occur, special salt compositions are added to replenish and regenerate the bath.

Gas- or oil-fired, externally heated salt pots are the most common equipment used for cyaniding because of their versatility and low initial cost. The pots that contain the salt bath usually are made from pressed or welded low carbon steel or iron–nickel–chromium alloy. Use of aluminized low carbon steel pots is increasing because their life is near that of alloy pots but their cost is less.

High temperature baths from 900–955°C usually are used to achieve deep case depths with rapid carbon penetration. Low temperature baths from 650–730°C are used to produce light case depths with only nitrogen penetration. Temperatures for liquid carbonitriding are ca 760–870°C. A combination treatment also is used, ie, initiating a deep-carburized case with a high temperature bath and then transferring the work load to a low temperature bath to provide a high nitrogen-surface effect.

The principal advantages of liquid carbonitriding are rapid heating rate and flexibility. Work is brought to heat rapidly and uniformly through immersion in the bath and a thin layer is quickly carburized on the steel. This is especially important for processing low case depths where the time needed to reach the operation temperature may represent a large portion of the total cycle time. The flexibility of the process is illustrated by the many small batches requiring different cycle times that can be processed simultaneously. A major disadvantage is in the handling of cyanides; special precautions must be observed at all times because of the health hazard involved (see Health and Safety Factors).

Nitriding. Nitriding case-hardens steel by addition and diffusion of nascent nitrogen into the surface of the steel where it reacts to form nitrides. The process temperatures are 495–595°C, which are less than the temperature at which the transformation to austenite occurs. A case hardness is produced directly and, therefore, quenching is unnecessary. Nitriding is accomplished both by using a gas atmosphere or a liquid bath.

Although some growth occurs in nitriding, there generally is very little distortion. Before nitriding, machined parts should be stress-relieved at least 10°C over the nitriding temperature to minimize distortion.

Gas. In gas nitriding, nitrogen is introduced into the surface of steel when the steel is in contact with a nitrogenous gas, usually ammonia. Nascent nitrogen is produced during the dissociation of ammonia by the reaction

$$NH_3 \rightarrow N + 3 H$$

These products are unstable and react rapidly to form hydrogen and inert nitrogen,

$$2 N \rightarrow N_2 \quad 2 H \rightarrow H_2$$

Only nascent nitrogen can be absorbed into the steel, and only ammonia dissociating at the surface of the steel can supply the nascent nitrogen. The absorbed nascent nitrogen diffuses into the steel and reacts to form precipitates of the nitrides of iron and alloying elements. The precipitation creates compressive stresses which result in the case hardness.

Alloy steels containing aluminum, chromium, molybdenum, vanadium, or tungsten in solid solution are necessary for good nitriding results. Plain carbon steels produce a case that is extremely brittle and that spalls readily. Many stainless steels are successfully nitrided. A number of alloy steels have been developed especially for the nitriding process.

Since no quenching is required after nitriding, all hardenable steels must be heat-treated to produce the required core properties prior to nitriding. The common alloy steels used contain 0.25–0.5 wt % carbon and are quenched and tempered to the core hardness necessary for the strength required. Care must be exercised in the selection of the steel to be nitrided to ensure that the tempering temperature required for the core hardness is at least 30°C above the nitriding temperature to prevent tempering during nitriding. The case hardness of some alloy steels is partially dependent upon the core hardness, since a higher core hardness results in a higher case hardness.

The single-stage process occurs at 510°C and the anhydrous ammonia dissociation rate is from 15 to 30%. This process produces a nitrogen-rich layer, the white layer, which is somewhat brittle. A double-stage process, the Floe process (3–4), is used to reduce the thickness of the white nitride layer but is limited primarily to the aluminum-containing steels. After the first stage (which is the single-stage process) is complete at 510°C, the temperature is raised to 550°C with a dissociation rate of 80%. There is no advantage to the double-stage process unless the white nitride layer is intolerable.

Liquid. The same temperature range is used in liquid nitriding as in gas nitriding and, as in liquid carbonitriding, the medium is molten cyanide. A nitriding salt bath may be 60–70% sodium salts, mainly sodium cyanide, and 30–40% potassium salts, mainly potassium cyanide. The nitriding action is the result of 5% minimum content of cyanate (NaCNO or KCNO) which is produced by aging the salt at 565–595°C for at least 12 h prior to use. The ratio of cyanide to cyanate is critical but it varies depending on the salt bath and process used.

For best results in nitriding, special chromium–aluminum low alloy steels should be used. Because of the low temperatures that are involved, nitriding is not a rapid process. The time required for a case depth of 0.25 mm is from 15 to 30 h and the time required for a 0.38-mm depth is from 20 to 40 h.

Generally, the same equipment that is used for cyaniding can be used for nitriding; the equipment merely is cleaned and the salts changed. However, if only nitriding is to be used, high alloy pots are not necessary. All baths must be periodically relieved of oxidation products, which promote unfavorable temperature gradients; overheating, ie, above 595°C, should be avoided.

Microcasing. Microcasing processes are, in most cases, highly proprietary. They usually are used for surface or slightly subsurface treatments. The processes yield excellent wear resistance, gliding properties, ductility, and fatigue strength in a superficial layer.

Ionitriding. Ionitriding involves a current-carrying substance between an anode and cathode. The substance is a low pressure, nitrogen gas which, by means of high voltage, ie, 500–1000 V, is excited and ionized. Because of the glow associated with the excited gas state (plasma), the phenomenon is called glow discharge (see Ion implantation; Plasma technology).

The positive nitrogen ions produced inside the glow discharge are attracted toward the negatively connected workpieces. They collide with the surfaces, are occluded, and heat the pieces to the diffusion temperature. Thus, an ionitriding furnace is a container within which is a technical vacuum and which serves as the anode for the glow discharge; there are no classical heating devices. Regulation of the nitrogen–carbon concentrations in the atmospheres produces a number of different white layer structures, eg, from a thick white nitride layer to no layer (the latter occurs with pure diffusion). The iron–nitride compound that is formed can be controlled to the epsilon phase or to the gamma-prime phase. The layers can be as deep as 0.015 mm and are very uniform.

Workpieces are made from carbon steels, low alloy steels, nitriding steels, tool steels, heat-resistant steels, high speed steels, stainless steels, cast iron, and sintered iron. Products include extrusion cylinders, lead screws for machine tools, gears, spindles, punch dies, and cutting tools. In general, ionitrided surfaces are used where wear resistance and antifriction properties are important (see Bearing materials; Nitrides).

Siliconizing. Siliconizing produces a case with a unique combination of corrosion- and wear-resistance properties. The corrosion resistance of the case is exceptional in dilute nitric acid and normal atmospheric environments free of chlorides and sulfates. Although siliconized cases are not file hard, the surface is nongalling and resists frictional wear. Wrought or cast low carbon steels are necessary for best results. Siliconization occurs more slowly with increasing carbon content of the steel. Also, sulfur is deleterious to the corrosion properties of the case.

Siliconizing adds and diffuses silicon into the surface of the steel. Like nitriding, a case hardness is produced and does not require a subsequent hardening operation. The work is heated to ca 1010°C with silicon carbide in an atmosphere of silicon tetrachloride. The silicon tetrachloride decomposes on the surface of the steel and liberates chlorine,

$$SiCl_4 \rightarrow Si + 2\,Cl_2$$

The chlorine that is produced is removed from the atmosphere by reaction with the silicon carbide which produces more silicon:

$$2\,Cl_2 + SiC \rightarrow SiCl_4 + C$$

The silicon accumulates in the steel surface to ca 14%. Siliconizing is performed in a rotation retort; nitrogen is used as a protective atmosphere. At the reaction temperature, the nitrogen flow is shut off and silicon tetrachloride liquid is vaporized and is introduced into the retort. When the proper case depth is obtained, the heating unit is removed from around the rotating retort and the work is allowed to cool. Parts are then cleaned in slightly acidulated water.

Siliconizing at 1010°C produces a case depth of 0.38 mm in 1 h and 0.76 mm in 3 h.

Cases are not machined readily and generally are ground. Once ground, the case absorbs oil at 120–150°C and is similar to a self-lubricated bearing. The absorption of oil can be increased by etching in hydrofluoric acid.

Boronizing. In boronizing, a boride is added onto and into the surface of steel by impingement and diffusion; the basic material is proprietary and is available as a paste. The paste is enveloped around or over the areas to be treated and then is heat-treated between 705 and 1150°C. In one treatment, a low carbon steel part is heated at 895°C for 2 h and a 0.005-mm case is produced. A more popular method is a solid pack-cementation process using boron carbide (5). Boron is transferred to the metal surface by a gas-phase mechanism. The furnace cycles last from 1 to 6 h at 790–895°C and boronized case depths of 0.013–0.13 mm are produced.

The boride layer is resistant to corrosion by both alkalies and acids, with the exception of hot sulfuric acid and aqua regia. A marked advantage is improved wear with longer life which result from the case's low coefficient of friction. The paste medium permits processing in automated, continuous lines and a carburized and boronized smooth journal surface performs better than a smooth, hard alloy pad (6). The equivalent hardness of the bonded layer is HRC 62–67. The European automotive industry makes use of the boronizing process.

Tufftriding. Tufftride salt bath nitriding is used to upgrade wear and fatigue properties of ferrous materials. The process involves a short-time immersion in an aerated molten salt bath at $570 \pm 5°C$ and can be utilized either for batch or continuous operation. The bath contains primarily cyanide and cyanate compounds which liberate specific quantities of carbon and nitrogen in the presence of ferrous materials. Nitrogen, which is more soluble than carbon, diffuses into the steel, whereas the carbon forms iron carbide particles at or near the surface. These particles act as nuclei which precipitate some of the diffused nitrogen to form a tough compound zone of carbon-bearing, ϵ-iron nitride. The compound zone is no deeper than 0.013 mm and is formed in 90 min. It is highly resistant to wear, galling, and seizure and is not brittle.

Tufftriding increases the tensile strength of the surface fiber. It has been used successfully in crankshafts, gears, disk brakes, connecting rods and links, diecasting and extrusion dies, and powdered metal components.

Triniding. Triniding (Midland-Ross Corp.) is carried out at the standard nitro-carburizing temperature of 570°C for up to 4 h. It involves treating steel parts in an inert nitrogen-based, exothermic carrier gas with controlled, low percentages of ammonia. The compound layer is an ϵ-iron nitride phase but without the carbon additive that is found in the tufftride process. The trinide layer has good wear characteristics and fatigue resistance. Its application has been successful for the same type of parts that are subject to tufftriding.

Applied Energy. Applied-energy case hardening is a selective hardening process that produces a case by locally heating and quenching an area. The very rapid application of heat results in the surface being heated to the hardening temperature but very little heat being conducted inward. Since no carbon or nitrogen is added during the process, the carbon content of the ferrous metal determines the hardness response. A 0.30 wt % carbon steel that is properly case hardened should result in a case hardness of HRC 45–55; for 0.40 wt % carbon steels, the range should be HRC 52–58; and for 0.50 wt % carbon steels, the range is HRC 56–64. All ferrous materials that harden upon being quenched can be case hardened with selective hardening.

In some instances, surfaces that have been completely carburized can be selectively hardened in the areas requiring a case hardness. Thus, the remaining area remains relatively soft for subsequent machining.

Induction Hardening. Induction hardening is a selective hardening process in which the localized heat is produced by electromagnetic induction. A high frequency, alternating current is passed through the inductor or coil. The area to be heated is placed in or near the coil and becomes the secondary in what is fundamentally a transformer. The high frequency current in the coil produces a highly concentrated, rapidly fluctuating magnetic field which produces an electrical potential in the part. The resulting induced current is resisted by the metal, which causes heating by I^2R losses.

The rate of heating by the induction coils depends upon the strength of the magnetic field at the area being heated. The related induced currents and the material's resistance determines the heating rate. Heating is confined to the surface because of the skin effect resulting from the high frequency, and the higher the frequency, the more shallow the heating effect is. Shallow case depths up to 1.5 mm require frequencies from 10 to 2000 kHz and greater case depths require frequencies from 1 to 10 kHz. The depth of heating is determined by the duration of heating, the frequency used, and the power density. The power density is a function of the power source or generator, the coupling or distance separating the inductor and the work, and the design of the inductance coil.

The most common types of generators used for selective hardening are the solid-state inverters and the vacuum-tube units. Motor generator units and solid-state inverters operating at up to 10 kHz also are used for greater case depth ranges. Vacuum-tube units operate from 200 kHz to the megahertz range. Both the coil design and the quenching arrangement are important to the success of the process. Automatic timing is necessary for control because of the short cycles involved.

Flame Hardening. Flame hardening is rapid heating in an area through impingement of a high temperature flame or high velocity combustion-product gases. The high temperature ($\geq 2500°C$) flame is produced by the combustion of either oxygen and fuel or air and fuel, in burners or flame heads. High velocity connection-type burners utilize the same fuels but are essentially miniature furnaces that discharge hot gases at ca 1600°C.

Case depths from 0.79 to 6.4 mm or greater are obtainable. Shallow case depths, ie, <3.2 mm, require the use of an efficient burner with oxygen–fuel mixtures. Air–fuel mixtures are more desirable for heavy cases because there is less danger of overheating the surface. As with induction, the quenching arrangements are important to the success of the process. Many burners are designed with a water spray on one side for quenching as an area is hardened. Flame hardening is somewhat less expensive but much slower and less versatile than the induction processes.

Other. Generally, it is economically favorable to case harden wherever permitted by design considerations. In all other instances, special processing or protective coatings must be used to prevent casing in core areas. Protective coatings, or stop-off, protect the core areas by acting as a diffusion barrier.

Plating with another metal is one of the most common methods of producing a stop-off to protect core areas from carbon and/or nitrogen. A typical procedure for selective case hardening is to rough-machine the part, thereby allowing extra metal in the areas to be cased, plate the entire part, and grind the area to be cased to remove the plating and expose the surfaces to be cased. Other processes that produce localized plating are lacquering or waxing the areas to be cased. The lacquer or wax which prevents plating in these areas is removed after plating so as to expose the surfaces to be cased (see Electroplating).

For carburizing and carbonitriding, a minimum of 0.013 mm of copper is the most common stop-off. Generally, ca 0.025 mm of copper is required for nitriding but the same thickness of bronze is more effective. Copper-based stop-off should not be used in cyanide-based salt baths because the cyanide slowly dissolves the copper and, as the copper concentration increases in the bath, it replates onto the work. A special 10% sodium cyanide bath is used with copper plate for liquid carburizing.

Proprietary paints are used frequently, especially in heat-treatment operations. In carburizing, the paints consist of a copper base suspended in lacquer or water glass. A tin base is used for nitriding. The paint is applied to the core area with a brush. Ceramic coatings also are used as a stop-off and are applied as are the paints.

Excessive stock can be left on areas that are to remain as the core. In grinding the case-hardened part to required dimensions, the case is removed and the core areas are exposed. This procedure is relatively costly because the grinding must be controlled to prevent damage to the adjacent case-hardened areas. With carburizing, the procedure can be modified to reduce costs by adding an annealing step to the end of the carburizing cycle to assure machineability of the carburized surface. The excess stock in the core areas is easily machined off, and the case is removed before the part is hardened.

Where the stop-off is not required to be fully effective, mechanical means can be used, ie, stainless steel or copper plugs, capscrews, and nuts can be inserted to block off areas. A lubricant should be utilized to facilitate disassembly after casing. There should be no air pockets or the heat-induced pressure may force the plugs out of place.

Hardening.   For most of the processes, the work is quenched directly from the processing temperature. Frequently, products from gas and pack carburizing also are reheated for quenching. Only nitriding produces its case hardness without the necessity of a quench. However, nitrided work must be heat-treated prior to the nitriding process if any core properties are required.

Because of the cost involved in reheating prior to quenching, work is directly quenched wherever possible. In carburizing, direct quenching may not always be possible because of the equipment, the steel being processed, or the sequence of manufacturing operations. Some carburizing equipment, like pit batch furnaces, are not designed for direct quenching. Some high nickel grades of carburizing steel require reheating to the proper hardening temperature to control the amount of retained austenite in the case. When machining or special processing is required prior to hardening, reheating for hardening is necessary for proper operations sequence. Reheating also is required where fixturing or press quenching is performed to control dimensions.

Instead of quenching directly from the carburizing temperature of 925°C, the temperature usually is lowered and equalized to 815–845°C prior to quenching; this gives the best combination of case and core properties. The quenching medium depends upon the material and case process involved; the medium and the process also help to minimize distortion.

Water or brine is used where a severe quench is required to produce the desired properties; however, distortion tends to result. Oil quenching is less drastic and produces less distortion. The temperature of the oil-quench bath may vary from 25 to 205°C but the most common range is from 50 to 70°C. Quench baths at the higher temperatures reduce distortion but generally require special oils. Quench baths above 205°C usually are nitrate–nitrite molten salt baths.

High temperature quench baths usually are used with the martempering process, where the bath temperature is selected to be just above the temperature where transformation to martensite occurs. Upon being quenched, the parts cool to the bath temperatures and harden uniformly after being removed from the quench bath. Martempering minimizes distortion. Fixturing or press quenching is used to control distortion during quenching; however, these methods are expensive. Gas or atmosphere quenching is possible with thin parts having highly hardenable cases. All parts must be uniformly exposed to the quenching atmosphere.

Subzero treatments as low as −75°C are used in all carburizing and hardening operations, eg, to transform retained austenite into martensite. If a subzero cabinet is not available, the parts can be placed into a tank of trichloroethylene and cooled quickly with dry ice which sublimates at $-80°C$. Subsequent low temperature tempering, eg, at 120–190°C, is mandatory.

Tempering. Most parts are tempered to increase case toughness. Nitrided cases are an exception since the hardness is not produced by quenching. Low temperatures, ie, from 120 to 190°C, generally are used to minimize any loss in case hardness. Because nitrogen in steel resists tempering, carbonitrided cases can be tempered at slightly higher temperatures than those used for carburized cases.

Economic Aspects

In general, a case-hardening process is not established on the basis of cost, but upon the physical and metallurgical requirements of the surface for the functions it is to perform. However, the greatest cost of any process is bringing the furnace and the parts to be processed up to the operating temperature and maintaining that temperature. The different gases, compounds, or preparations that are used as the process medium are of minor cost relative to the total.

The bulk of all the heat-treating costs can be attributed to fuel, ie, electricity and natural gas, and labor. As illustrated in Figure 2, although the cost rate of natural gas has shown the greatest increase, it is still the least expensive energy source in most parts of this country. Direct labor costs have not risen drastically because productivity through better working conditions, newer equipment, and more experienced personnel has increased.

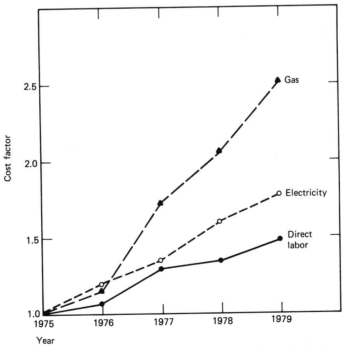

Figure 2. Relative fuel and labor costs associated with case hardening.

Health and Safety Factors

Heat-treatment equipment purchased since 1970 must comply with OSHA regulations (7). In general, applicable sections are those dealing with occupational noise exposure, ventilation, accident-prevention signs, and machine guards over moving or hot equipment.

Any chemical used in the processes must meet the control standards and the waste-disposal requirements of the Toxic Substance Control Act (TSCA) of 1976 (8).

BIBLIOGRAPHY

"Nonmetal Cementation Process" in *ECT* 1st ed., under "Metal Surface Treatments," Vol. 9, pp. 23–31, by Earl S. Greiner, Bell Telephone Laboratories, Inc.; "Case Hardening" in *ECT* 2nd ed., under "Metal Surface Treatments," Vol. 13, pp. 304–315, by Alex J. Schwarzkopf, National Aeronautics and Space Administration.

1. V. K. Sharma, G. H. Walter, and D. H. Breen, *J. Heat. Treat.* **1**(1), 48 (June 1979).
2. F. E. Harris, *Met. Prog.* **44**, 265 (1943).
3. U.S. Pat. 2,437,249 (Mar. 9, 1948), C. F. Floe (to Nitroalloy Corp.).
4. C. F. Floe, *Source Book on Nitriding*, American Society for Metals, Metals Park, Oh., 1977, pp. 144–171.
5. D. N. Guy, *Boronizing—A Surface Treatment for Critical Wear Surfaces*, Lindberg Technology Center, Melrose Park, Ill., p. 1.
6. *Met. Prog.* **117**, 82 (Jan. 1980).
7. *Occupational Safety and Health Act*, Public Law 91-596, Title 29, Chapt. XVII, Dec. 1970.
8. *Toxic Substance Control Act*, Public Law 94-469, Oct. 11, 1976.

General References

"Heat Treating, Cleaning and Finishing" in *Metals Handbook*, Vol. 2, American Society for Metals, Metals Park, Oh., 1964.
Carburizing and Carbonitriding, American Society for Metals, Metals Park, Oh., 1977.
Source Book on Nitriding, American Society for Metals, Metals Park, Oh., 1977.
G. M. Enos and W. E. Fontaine, *Elements of Heat Treating*, John Wiley & Sons, Inc., New York, 1953.
The Making, Shaping, and Treating of Steel, 9th ed., United States Steel, Pittsburgh, Pa., 1971.
"Heat Treatment of Metals," *Q. J. Wolfson Heat Treatment Centre* **6**, 3 (1979).

LESTER E. ALBAN
Fairfield Manufacturing Company, Inc.

METAL TREATMENTS

Operations performed on consolidated metals and alloys are referred to as metal treatments. Most of these treatments are mechanical and/or thermal. Mechanical treatments involve shape changes by forming or machining. Forming entails plastic deformation which changes the microstructure and, therefore, properties. In thermal treatments, heat is applied to alter structures and properties. Metal treatments such as joining and coating of metals are not discussed here (see Welding; Metallic coatings).

Mechanical Treatments

Forming processes and techniques available for a particular alloy depend upon its workability, which is the ability to be plastically deformed.

Workability Testing. Workability tests measure the amount of deformation that can be tolerated without fracture or the development of an instability. Instabilities may be either buckling or necking as illustrated in Figure 1. Buckling or wrinkling occurred at the flange of the cup in Figure 1 because it was too thin or was insufficiently supported. Local thinning or necking occurred in the walls of the cup. In most workability tests a specimen is deformed to failure at a constant load rate or strain rate in tension, compression, torsion, shear, or bending. The most common technique is a tensile test at a constant strain rate with load and elongation measured continuously. Stress and strain are calculated from load and elongation and presented in the form of an engineering stress–engineering strain diagram, as shown in Figure 2. Yielding represents the transition from elastic deformation, where atomic bonds are being stretched, to plastic or nonrecoverable deformation, where atomic slip is occurring. The yield stress specifies this transition. It is defined usually as the stress (load per area) that produces a small permanent strain, usually 0.002 (0.2% offset yield stress). Following yielding, the stress required for further strain increases but, unlike the elastic region, stress and strain are not linearly related. Increasing stress with increasing strain in the plastic region is termed strain hardening. The decrease in stress following the

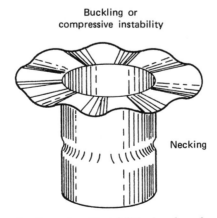

Buckling or
compressive instability

Necking

Figure 1.   Examples of instabilities in a deep-drawn cup.

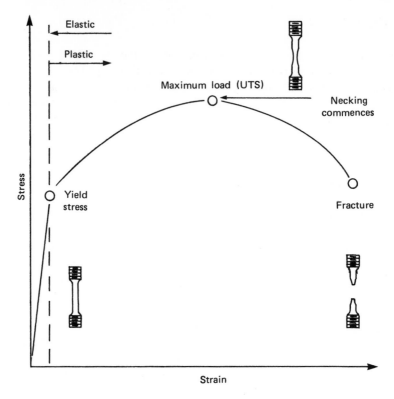

Figure 2.   Schematic stress–strain diagram.

maximum load is owing to necking or localized deformation; engineering stress is decreasing in Figure 2 only because the area in the necked region is rapidly decreasing. The true stress or actual load per unit area continues to increase with strain all the way to fracture. The onset of necking is defined by the ultimate tensile stress which is the maximum load divided by the initial area. Ductility is measured by % elongation (El) and % reduction in area (RA) which are defined below:

$$\% \text{ El} = \left(\frac{L - L_0}{L_0}\right) \times 100$$

where L_0 = initial length; L = final length.

$$\% \text{ RA} = \left(\frac{A_0 - A}{A_0}\right) \times 100$$

where A_0 = initial area; A = final area.

Both % El and % RA are frequently used as a measure of workability. Workability information also is obtained from parameters such as strain hardening, yield strength, ultimate tensile strength, area under the stress–strain diagram and strain-rate sensitivity.

Tensile testing frequently is used to assess mechanical properties other than workability. However, the strain rate is usually much faster when workability is being measured in order better to simulate forming processes. Standard testing is done at about 10^{-3} s^{-1} compared to strain rates up to 10^2 s^{-1} for workability testing. An indication of strain-rate sensitivity is given in Figure 3 for a commercial titanium alloy (1).

Stress–strain diagrams are not unique to tensile testing. They are also generated by other testing modes such as compression, torsion, shear, and bending.

Temperature strongly influences stress–strain behavior, as is also shown in Figure 3. Therefore, evaluating hot-workability entails testing over a range of temperatures. This can be appreciated from Figure 4, where hot-torsion data are presented for two nickel-base superalloys, Nimonic 90 and Nimonic 115, and for a 0.48% C steel (2). Strength is indicated by torque, and ductility is measured by the number of revolutions to failure. It can be seen that the hot-ductility of the nickel-base alloys, particularly N-115, is significantly less than that of steel. Furthermore, the required stresses are substantially greater.

When determining the temperature range for hot working, it is usually not sufficient to merely heat directly to various temperatures. Instead, it is also necessary to acquire cooling data on specimens that are tested after cooling from a temperature corresponding to the furnace temperature in a hot-working operation. Both heating and cooling tensile data are shown in Figure 5 for a Nimonic 115 ingot. The ductility on testing is lower after cooling from 1105°C and 1135°C compared to that determined on heating. These tensile tests were performed on a Gleeble machine, where the specimen is resistance-heated.

Plastic Deformation.   When plastic deformation occurs, crystallographic planes slip past each other. Slip is facilitated by the unique atomic structure of metals, which consists of an electron cloud surrounding positive nuclei. This structure permits

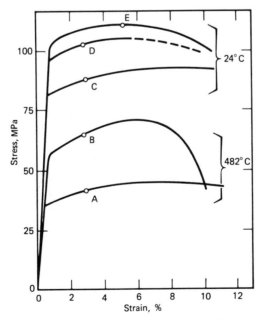

Figure 3.   Effect of temperature and strain rate on stress–strain diagram of Ti–5% Al–2.5% Sn (1). A, 5×10^{-1} s^{-1}; B, 1.6×10^{-4} s^{-1}; C, 5×10^{-1} s^{-1}; D, 5×10^{-6} s^{-1}; E, 1.6×10^{-4} s^{-1}. To convert MPa to psi, multiply by 145.

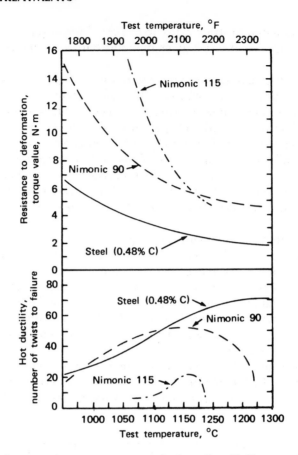

Figure 4.   Torsion properties versus temperature for three alloys (2). To convert N·m to ft·lbf, divide by 1.35.

shifting of atomic position without separation of atomic planes and resultant fracture. The stress required to slip an atomic plane past an adjacent plane is extremely high, if the entire plane moves at the same time. Therefore, the plane moves locally, which gives rise to line defects called dislocations. These dislocations explain strain hardening and many other phenomena.

Dislocation may be edge type or screw type (see Fig. 6). In slipping from (**a**)–(**d**) under the application of the shear stress shown, two distinctively different intermediate stages are possible. The stage shown in (**b**) represents an edge dislocation, whereas in (**c**) a screw dislocation is shown. An edge dislocation contains an extra half plane of atoms at the slip plane as shown in (**e**) which is a front view of (**b**). The term screw dislocation is derived from the fact that following this line defect from above and below the slip plane generates a screw pattern as shown in (**f**) where circles are above the slip plane and dots are atoms below this slip plane. A given dislocation line can be pure edge, pure screw, or any combination of edge and screw components.

An analogy to slip dislocation is the movement of a caterpillar where a hump started at one end moves toward the other end until the entire caterpillar moves forward. Another analogy is the displacement of a rug by forming a hump at one end and moving it toward the other end.

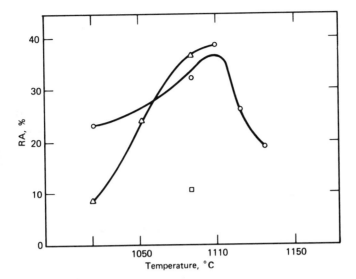

Figure 5. Hot-workability of cast Nimonic 115 as determined by tensile testing with a Gleeble machine. O, Heating; □, cooling 1135°C; △, cooling 1105°C. RA = reduction in area.

Strain hardening occurs because the dislocation density increases from about 10^7 dislocations per cm^2 to as high as $10^{13}/cm^2$. This makes dislocation motion more difficult because dislocations interact with each other and become entangled.

Slip tends to occur on more closely packed planes in close-packed directions.

Hot Working. Plastic deformation at temperatures sufficiently high that strain hardening does not result is termed hot working. The temperature range for successful hot working depends on composition and other factors such as grain size, previous cold working, reduction, and strain rate. Typical hot-working ranges are presented in Figure 7 for various alloys.

The lack of strain hardening is due to sufficient thermal energy for recrystallization, which refers to the formation of new grains. Because the new grains have relatively low dislocation densities, strength is not increased. The driving force for recrystallization is the large strain energy associated with the high dislocation density generated during deformation. Recrystallization is shown schematically in Figure 8 for hot rolling. The recrystallized grain size is small but growth occurs with time at temperature.

Hot working permits forming of relatively brittle materials that cannot readily be cold-worked. Other advantages are grain refinement, reduction of segregation, healing of defects, such as porosity, and dispersion of inclusions. Disadvantages are the formation of oxide surface scales and the requirement of heating facilities.

Cold Working. Cold working involves plastic deformation well below the recrystallization temperature. Required stresses for cold working are greater than for hot working and the amount of strain without heat treatment is limited. Advantages are close dimension control, good surface finish, and increased low temperature strength because of strain hardening. Grain refinement can be achieved by annealing, which entails heating after cold working to temperatures where recrystallization occurs. The effect of cold working on tensile properties and grain structure and subsequent annealing are shown schematically in Figure 9.

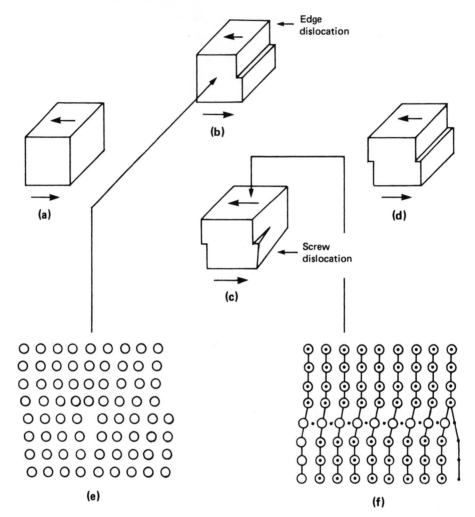

Figure 6.   Continuum (**a–d**) and atomic (**e–f**) representations of edge and screw dislocations. For the screw dislocations, open circles and dots are above and below the slip plane, respectively.

Primary Forming Processes.   Primary forming operations are usually hot-working operations directed toward converting cast ingots into wrought blooms, billets, bars, or slabs (see Fig. 10). In primary working operations the large grains typical of cast structures are refined, porosity is reduced, segregation is reduced, inclusions are more favorably distributed, and a shape desirable for subsequent operations is produced. Figure 11 illustrates the principal operations used for ingot breakdown, ie, forging, extruding, and rolling. Extrusion differs from forging and rolling in that more deformation occurs in one pass. Forging and rolling include many passes and some reheating. In addition, intermediate conditioning is sometimes necessary. This makes extrusion attractive but it is an expensive operation.

Secondary Forming Processes.   The objective of secondary forming processes— either cold or hot working—is to form a shape. Such processes include rolling, open- and closed-die forging, upset forging, extruding, roll forging, ring rolling, deep drawing, spinning, bending, stretching, stamping, drawing, and high velocity forming.

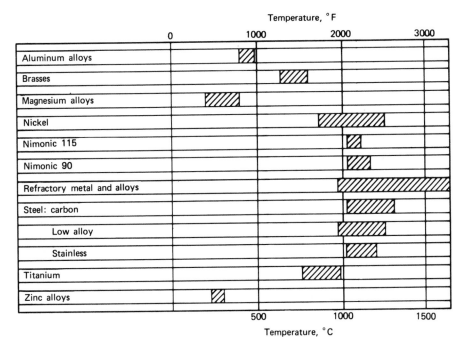

Figure 7. Hot-working ranges of various metals and alloys.

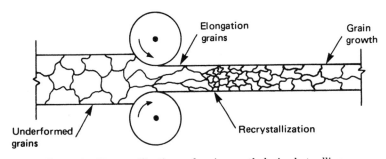

Figure 8. Recrystallization and grain growth during hot rolling.

Sheet and semifinished products such as round, rectangular, and shaped bars are produced by rolling.

Flat, V-shaped, and swaging dies are used for open-die forging. In closed-die forging, the metal is forced to flow into die cavities to form the impressions of dies attached to the anvil and ram. Forging is performed on both hammers and presses. Hammers have a strain rate of $1-100$ s^{-1} as compared to $0.05-5$ s^{-1} for presses. The energy or stress required for deformation is greater for hammers because of the faster strain rate. Die contact time and, therefore, die chilling are shorter for hammers. However, the faster strain rate can more readily cause an excessively increased temperature locally, resulting in localized grain growth and possibly incipient melting. The propensity for cracking caused by brittle second-phase particles is less for the slower strain rates of presses.

Typical steps used in the manufacture of a forged turbine blade are schematically

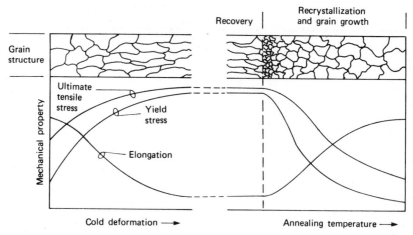

Figure 9. Variation of tensile properties and grain structure with cold working and annealing.

Figure 10. Distinctions between various intermediate wrought shapes.

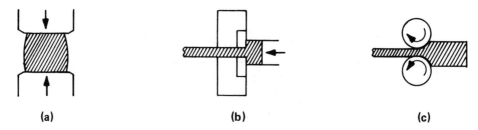

Figure 11. Primary hot-working processes for ingot breakdown: (**a**) forging; (**b**) extruding; (**c**) rolling.

shown in Figure 12. The first three operations involve gathering material for subsequent closed-die forging. Gathering has been accomplished by one extrusion and two upsetting operations. Initial forging results in a preform that is of the correct volume to produce the finished forging in the final operation. A lubricant must be used during closed-die forging to minimize sticking to the die and to promote metal flow. Forging operations must be designed to provide adequate metal flow to prevent critical grain growth.

Extrusion for gathering and producing shapes can entail significantly more than direct forward extrusion. An example of backward extrusion is given in Figure 13(**a**).

Roll forging differs from rolling, producing a short length of varying cross section,

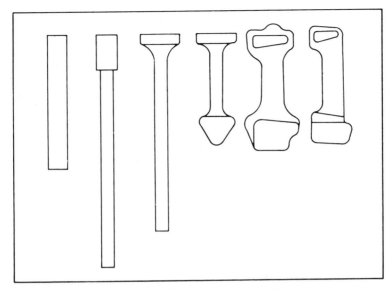

Figure 12.   Typical steps in forging a turbine blade.

as opposed to the long, uniform cross sections produced by rolling. In ring rolling, ring-shaped forgings are produced. A seamless ring is produced by reducing the cross section and increasing the circumference of a heated, doughnut-shaped blank between two rotating rolls.

Deep drawing, spinning, bending, stretching, and stamping are cold-working processes applicable to the forming of shapes from sheet and strip (see Fig. 13(**b**)–(**f**)). Deep drawing uses a shaped punch to force sheet metal into a die or through a die opening. The drawing metal must have excellent ductility and several draws may be required with annealing between draws. In spinning, a tool is pressed against one side of a sheet-metal die which is rotated at high speed. The tool gradually makes the disk conform to the shape of the forging instrument that it is forced against. Spinning is used in place of deep drawing if production does not justify the high cost of deep drawing punches and dies. Stamping is used for cutting blanks from sheet and strip, and usually precedes deep drawing and spinning. Embossing and coining are also stamping operations. In embossing, the impressions of punch and die match each other, whereas in coining they do not, as shown in Figure 13(**g**) and (**h**). Therefore, embossing results in more bending and less flow of metal than coining.

Drawing is a method of reducing the diameter of wire, rod, and tubing. It is similar to extrusion, except that the metal is pulled through the die instead of pushed through it as shown in Figure 13(**i**).

Tubes may be seamed or seamless. Seamed tubes are produced by bending plate, sheet, or strip into the appropriate shape and welding longitudinally. Seamless tubes are manufactured by opening the center of an ingot or billet and working the resulting shell or by working a cast hollow ingot. This may involve extrusion and subsequent drawing. Another method for producing a seamless tube is rotary piercing where metal is rolled over a mandrel, as shown in Figure 13(**j**).

High velocity forming has become successful in recent years for alloys with poor workability. Examples of such processes are explosive forming, electromagnetic

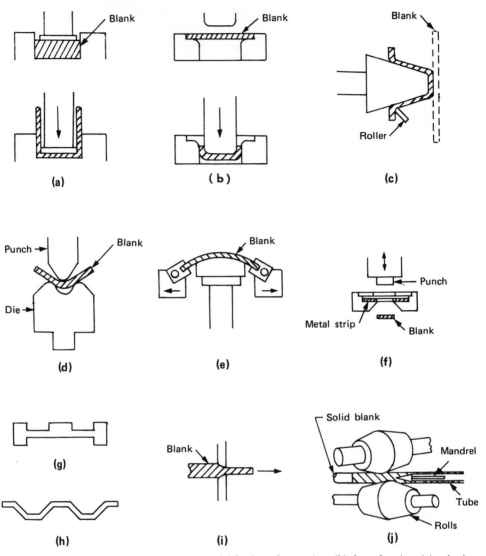

Figure 13. Various forming operations: (**a**) backward extrusion; (**b**) deep drawing; (**c**) spinning; (**d**) bending; (**e**) stretching; (**f**) stamping; (**g**) coining; (**h**) embossing; (**i**) drawing; (**j**) rotary piercing.

forming, and electrohydraulic forming. Explosive forming is shown schematically in Figure 14. In this process, an explosive charge is detonated in a water tank containing the workpiece and die. Shock waves from the explosion propagate throughout the liquid and impact the workpiece with sufficient energy to force it into the die (see Metallic coatings, explosively-clad metals).

Thermal Treatments

Annealing. In annealing, a cold-worked material is heated to soften it and improve its ductility. The three stages of annealing are recovery, recrystallization, and grain

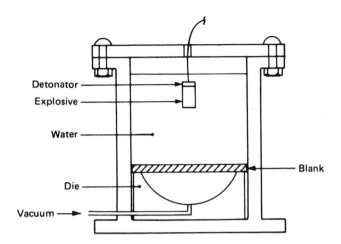

Figure 14. Explosive forming.

growth (see Fig. 9). Recovery occurs at relatively low temperature and may result in some softening caused mainly by the arrangement of dislocations into a more favorable distribution. Recrystallization is the formation of new grains with a relatively low dislocation density and little internal strain which replaces strained grains with high dislocation densities. At increasing temperature, the newly formed grains exhibit grain growth. Prolonged exposure at a given temperature also tends to promote grain growth.

Precipitation Hardening. In precipitation hardening, also called age hardening, fine particles are precipitated from a supersaturated solid solution. These particles impede the movement of dislocations, thereby making the alloy stronger and less ductile. In order for an alloy to exhibit precipitation hardening, it must exhibit partial solid solubility and decreasing solid solubility with decreasing temperatures.

An example of the many alloy systems satisfying these requirements is the aluminum–copper system. The diagram in Figure 15 shows a portion of the equilibrium phase diagram for the binary Al–Cu system, including the phases existing under equilibrium conditions at various temperatures as Cu is added to Al. At about 500–600°C, an alloy with 4.5% Cu consists only of alpha, a solid solution of Cu in Al. Below 500°C, the phase $CuAl_2$ exists in addition to alpha. The objective of precipitation hardening is to distribute the second phase ($CuAl_2$) as fine particles which are effective in blocking dislocation motion.

Precipitation hardening consists of solutioning, quenching, and aging. Solutioning entails heating above the solvus temperature in order to form a homogeneous solid solution.

Rapidly quenching to room temperature retains in solid solution a maximum amount of alloying element (Cu). The cooling rate required varies considerably with various alloys. For some alloys, air cooling is sufficiently rapid, whereas other alloys require water quenching. After cooling, the alloy is in a relatively soft metastable condition referred to as the solution-treated condition.

In aging, the alloy is heated below the solvus to permit precipitation of fine particles of a second phase ($CuAl_2$). The solvus represents the boundary on a phase di-

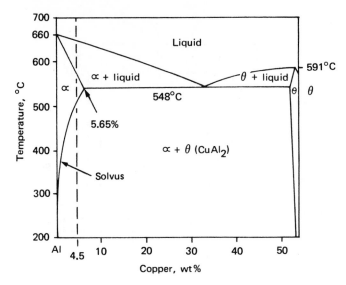

Figure 15.   Aluminum-rich portion of aluminum–copper-phase diagram.

agram between the solid-solution region and a region consisting of a second phase in addition to the solid solution.

The precipitation-hardening process and resulting structures are shown in Figure 16. Particles formed initially during aging tend to fit into the lattice of the matrix solid solution, which distorts the lattice at the particle–matrix interface. This accommodation to the matrix phase is termed coherency, and contributes significantly to dislocation blockage, and, therefore, strengthening.

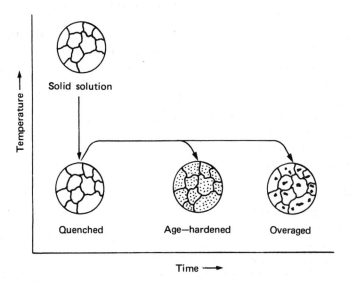

Figure 16.   Microstructures during precipitation hardening.

An optimum combination of temperatures and time generates particle size and spacing for the best combination of properties. If aging temperatures are too high, and times too long, particles coalesce and lose coherency resulting in decreased strength, as shown in Figure 17.

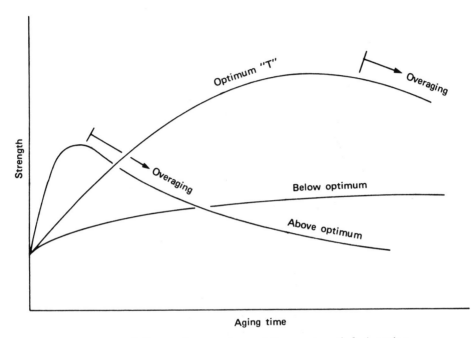

Figure 17. Influence of temperature and time on strength during aging.

Table 1 gives the effect of precipitation hardening on a 95.5% Al–4.5% Cu alloy. Heat treatment consisted of a solution treatment at 515°C with a water quench, followed by aging at 155°C for 10 h.

An example of a precipitation-hardened microstructure is presented in Figure 18 for Udimet 700, which is a precipitation-strengthened nickel-base superalloy. The precipitated phase is $Ni_3(Ti,Al)$, which is called gamma prime and is the primary strengthening phase in many commercial superalloys. Both coarse and fine gamma prime are present owing to high and low aging temperatures.

Table 1.   Effect of Heat Treatment on Tensile Properties of Al–4.5% Cu

Treatment	Yield strength, MPa[a]	Ultimate tensile strength, MPa[a]	Elongation, %
solution treated	103	241	30
age hardened	331	414	20
overaged	69	117	20

[a] To convert MPa to psi, multiply by 145.

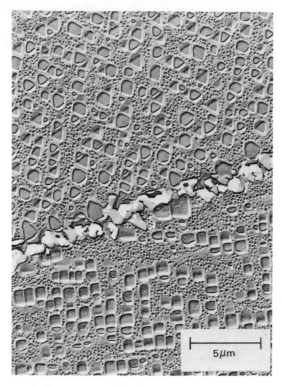

Figure 18. Structure of U-700 after precipitation hardening treatment of 1168°C/4 h + 1079°C/4 h + 843°C/24 h + 760°C/16 h with air cooling from each temperature. A grain boundary with precipitated carbides is passing through the center of the electron micrograph. Matrix precipitates are γ'-[$Ni_3(TiAl)$].

Heat Treatment of Steel. Steels are alloys with up to about 2% carbon in iron plus other alloying elements. The vast application of steels is mainly owing to their ability to be heat-treated to produce a wide spectrum of properties. This occurs because of a crystallographic or allotropic transformation which takes place upon quenching. This transformation and its role in heat treatment can be explained by the crystal structure of iron and by the appropriate phase diagram for steels (see Steel).

Iron exists in three allotropic modifications, each of which is stable over a certain range of temperatures. When pure iron freezes at 1538°C, the body-centered-cubic (bcc) δ modification forms, which is stable to 1394°C as shown in Figure 19. Between 1394 and 912°C, the face-centered-cubic (fcc) γ modification exists. At 912°C, bcc α-iron forms and prevails at all lower temperatures. These various allotropic forms of iron have different capacities for dissolving carbon. γ-Iron can contain up to 2% carbon, whereas α-iron can contain a maximum of only about 0.02% C. This difference in solubility of carbon in iron is responsible for the unique heat-treating capabilities of steel. The solid solutions of carbon and other elements in γ-iron and α-iron are called austenite and ferrite, respectively.

The appropriate phase diagram for the iron-rich side of the iron–carbon system is shown in Figure 20. Actually, this diagram does not truly represent equilibrium because cementite, Fe_3C, is a metastable phase. The stable phase for carbon is graphite

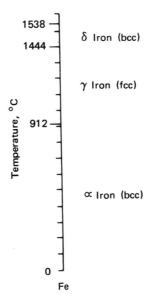

Figure 19. Crystal structure of iron as a function of temperature.

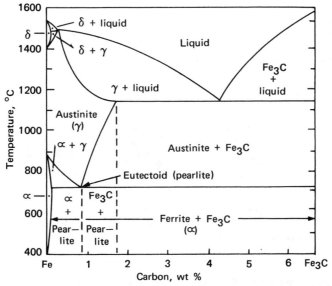

Figure 20. Iron–iron carbide phase diagram.

but the decomposition of the metastable Fe_3C to graphite and iron is so sluggish that Fe_3C must be treated as the stable phase for most practical purposes. At a composition of 0.77% C and a temperature of 727°C, a eutectoid reaction (Solid 1 → Solid 2 + Solid 3) occurs. The product of this reaction is pearlite, which has a lamellar structure consisting of alternate plates of ferrite and cementite. As the cooling rate increases, the interlamellar spacing decreases and the pearlite becomes finer, as indicated in

Figure 21. Pearlite is not a phase in the thermodynamic sense but rather a constituent consisting of two phases, ie, ferrite and cementite.

If the quenching rate is so rapid that pearlite does not form because diffusion cannot occur, another phase, termed martensite, forms, as shown in Figure 21. Martensite is a supersaturated solid solution of carbon in α-iron. It has a body-centered tetragonal crystal structure. Carbon retained in solution distorts the lattice in one edge direction. The strains generated by the carbon in solution impede the movement of dislocations, which results in tremendous strengthening. As shown in Table 2, the strength can increase by almost a factor of 3 by quenching rapidly to form martensite

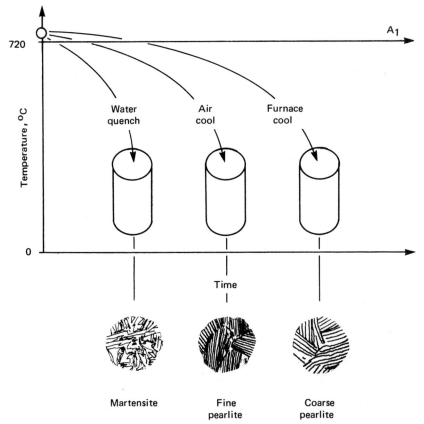

Figure 21.   Effect of cooling rate on structure of a eutectoid steel.

Table 2.   Properties of Steel Structures for a Eutectoid Steel

Structure	Yield strength, MPa[a]	Ultimate tensile strength, MPa[a]	Elongation, %
coarse pearlite	372	621	24
fine pearlite	524	1010	20
martensite		1724	low

[a] To convert MPa to psi, multiply by 145.

as compared to cooling slowly to form coarse pearlite. Since martensitic structures have such low ductilities, they usually are tempered following quenching. Tempering entails heating above 100°C to precipitate Fe_3C.

Other common heat-treatment processes are annealing and normalizing. Annealing is usually applied to produce softening and involves heating and cooling. Normalizing is a process in which a steel is heated into the austenite region and then air cooled. The objective is to obliterate the effects of any previous heat treatment or cold working and to ensure a homogeneous austenite on reheating for hardening or full annealing.

Structures that form as a function of temperature and time on cooling for a steel with a given composition are usually represented graphically by continuous-cooling and isothermal-transformation diagrams. Another constituent that sometimes forms at temperatures below that for pearlite is bainite. Like pearlite, it consists of ferrite and Fe_3C, but in a less well-defined arrangement. There is not sufficient temperature and time for carbon atoms to diffuse long distances and a rather poorly defined acicular or feathery structure results.

Homogenization. When alloys solidify, substantial segregation occurs. Therefore, ingots are sometimes given a high temperature heat treatment to generate a more homogeneous structure by diffusion in the solid state. Also, homogenization may eliminate undesirable phases that are present in the segregated cast structure.

Thermomechanical Processing. It is possible to develop desirable structures and, therefore, properties by uniquely combining thermal treatments with forming operations. An example of such thermomechanical processing is Minigrain processing (3) to produce fine grains in alloys such as Inconel 718 and Incoloy 901. These alloys are representative of iron–nickel base alloys that precipitate a second phase in addition to that primarily responsible for precipitation strengthening. Figure 22 shows an example of Minigrain processing for Incoloy 901. The primary strengthening phase for

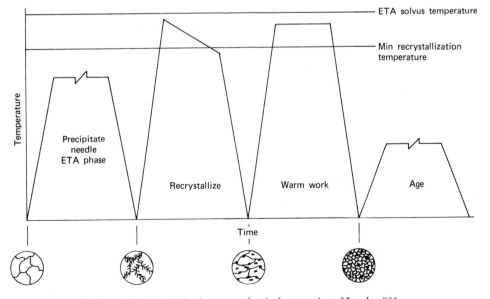

Figure 22.   Minigrain thermomechanical processing of Incoloy 901.

Incoloy 901 is gamma prime (fcc Ni_3Ti), whereas the phase used for grain size control is eta (hexagonal Ni_3Ti). A conditioning heat treatment is applied which precipitates eta in a needlelike registered pattern. This heat treatment is on the order of eight hours at 899°C. Working is carried out at about 954°C, which is below the eta solvus temperature, with final deformation occurring below the recrystallization temperature. A fine-grained structure is generated by subsequent recrystallization below the solvus. The needlelike eta phase has become spherical and restricts grain growth. Aging follows standard procedure to precipitate gamma prime. The resulting fine-grained structure has unique properties, such as excellent low-cycle fatigue properties.

Recent Developments and Outlook

Principal challenges facing the metals industry are the energy shortage; cost and availability of raw materials; and the need for new alloys and processing to meet the demands of new technologies. Most companies in the metals industry are now practicing some form of energy conservation. One approach has been recirculation and reuse of waste heat. Another approach has been the development of more energy-efficient processes. An example is powder-metallurgy strip rolling and strip casting where sheet and strip are produced directly from the melt. All hot-rolling steps are eliminated. Another example where operations are eliminated is continuous casting. About 17% of United States crude steel production now comes from continuous casting.

The raw materials situation is of great concern to all metallurgists, although it does not have the public awareness of the energy problem. The metals industry relies heavily on imports for many strategic elements such as chromium, nickel, manganese, cobalt, niobium, and tungsten. Many of these metals are mined in countries where political climates are unstable. For example, the primary source of chromium is southern Africa; no significant chromium deposits exist in the United States. The loss of chromium would be disastrous to the stainless steel, specialty steel, and superalloy industries. Efforts to alleviate raw material shortages include stockpiling, extensive use of recycling, improved melting and processing methods, alloy development and substitution, and use of cheap raw materials. The widespread use of duplex melting utilizing methods such as argon–oxygen decarburization (AOD) have helped greatly in these efforts.

Advanced materials and processes directed toward enhancing the performance and efficiency of aircraft turbines are an example of developments to meet new demands. These include new powder metallurgy techniques, directionally solidified eutectics and single crystals, and composites such as tungsten-reinforced superalloys.

Another related challenge faced particularly by the steel industry is foreign competition (4). During the past few years, imports have accounted for about 14% of domestic consumption in the United States. Competition from some European countries is from plants which are partially or fully government owned; competition from countries such as Japan and Korea is based on high efficiency achieved by modern facilities and low wage rates.

The application of powder metallurgy to superalloys will now be presented as an example of metal treatments which conserve energy and raw materials and can result in improved materials.

Powder Metallurgy of Superalloys. Although powder metallurgy has been an established manufacturing method in the metals industry for many years, its successful application to superalloys is relatively recent and is rapidly expanding. Higher yields and reduced energy requirements are among its attractive features. In addition, powder metallurgy allows higher alloying additions in wrought products because of reduced segregation, and, therefore, superplastic structures are readily achieved. This is a significant factor for superalloys which are heavily alloyed to provide good mechanical properties, and oxidation and corrosion resistance for rotating parts exposed to the high operating temperatures of aircraft turbine engines.

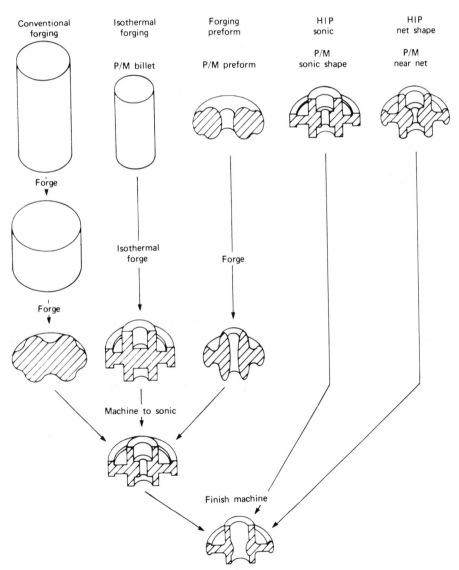

Figure 23. Powder metallurgy processing of superalloys indicating materials savings relative to conventional processing.

Superalloy powders are produced by argon atomization, the rotating electrode process, and soluble gas-vacuum atomization (5) (see Powder coating). Consolidation of the powder is accomplished by vacuum hot pressing, forging, extruding, or hot isostatic pressing. In vacuum hot pressing, loose powder is compacted in a cylindrical die with end plugs. Because of the long times involved and the small relative movement between particles, this process is seldom used. In forging, extruding, and hot isostatic pressing, the loose powder is put in a can, usually stainless steel, which is evacuated and sealed. Forging and extrusion are similar to conventional processes. Hot isostatic pressing (HIP) is becoming increasingly popular, as evidenced by the installation of new HIP units. In this process, the can is placed in a resistance furnace located inside a water-cooled pressure vessel. Isostatic pressure is applied by pumping argon gas into this sealed vessel. Pressures are usually around 103 MPa (15,000 psi) but may be as high as 200 MPa (29,000 psi) and temperatures are up to 1260°C. The unique advantage of this process is its ability to produce near net shapes for complex parts. Novel canning techniques to achieve these shapes are under intense investigation.

The various processing routes available by powder metallurgy are represented in Figure 23 for a superalloy disk. A sonic shape is significant because ultrasonic inspection usually is not possible on the final part since its complex shape lacks parallel and flat surfaces. Consequently, ultrasonic inspection is usually accomplished at a preceding stage (see Ultrasonics). Figure 23 schematically illustrates the substantial material savings achievable by powder metallurgy; eg, Pratt and Whitney Aircraft (UT Inc.) is producing JT8D disks that require about 35% less material than disks made by conventional forging. Where 195 kg of material has been required for a finish disk, the HIP process needs only 127 kg. Pratt and Whitney expect to reduce this to 130 kg, a saving of up to 40% in material (6).

The isothermal forging process indicated in Figure 23 utilizes superplasticity, which refers to the ability of some alloys to exhibit extensive ductility with elongations greater than 2000% being shown in tensile testing. The main prerequisite for superplasticity is an extremely fine and stable grain size, which is readily produced by powder metallurgy techniques. The dies are heated to the temperature of the workpiece. They are usually made of molybdenum in order to withstand elevated temperature and, therefore, must be protected from oxidation by forging under vacuum or inert gas atmosphere. Compared with conventional forging, strain rates are very low. Isothermal forging has the advantage over conventional forging in that it permits the formation of complex shapes in one operation.

A promising development in superalloy powder metallurgy is oxide dispersion-strengthened (ODS) material, which contains a finely dispersed oxide that is stable at elevated temperatures. The dispersed oxide provides strength at elevated temperatures where precipitated phases, such as gamma prime, are dissolved. Thoria is the dispersed phase in TD nickel and TD nichrome but current emphasis is on dispersion of yttria in alloys such as MA 754 (a Ni, Cr alloy). Oxide dispersion-strengthened materials are produced by mechanical mixing of the oxide with appropriate metal powders in an attritor. Consolidation of these alloys is challenging because it involves thermomechanical processing to achieve large recrystallized grains that are elongated and of a specific crystallographic texture.

Another interesting development is rapidly solidified powder. This is accomplished by rapid cooling of superalloy powders by rotary atomization to achieve solidification rates of 10^5–10^6°C/s. Resulting structures have excellent chemical homogeneity and promise superior properties.

BIBLIOGRAPHY

"Metal Treatments" in *ECT* 2nd ed., Vol. 13, pp. 315–331, by H. Herman, University of Pennsylvania, and J. G. Byrne, University of Utah.

1. D. K. Allen, *Metallurgy Theory and Practice*, American Technical Society, Chicago, Ill., 1969, p. 503.
2. W. Betteridge and J. Heslop, *The Nimonic Alloys and Other Nickel-Base High-Temperature Alloys*, Crane, Russak & Company, Inc., New York, 1974, p. 130.
3. E. E. Brown, R. C. Boettner, and D. L. Ruckle, *Superalloys—Processing, Proceedings of the Second International Conference*, Seven Springs Champion, Pa., Metals and Ceramics Information Center, 1972, pp. L-1–L-12.
4. J. Szekely, *Technol. Rev.* **81**(4), 23 (1979).
5. C. T. Sims and W. C. Hagel, *The Superalloys*, John Wiley & Sons, Inc., New York, 1972, pp. 427–450.
6. *Mater. Eng.* **91**(2), 9 (1980).

General References

J. Datski, *Material Properties and Manufacturing Processes*, John Wiley & Sons, Inc., New York, 1966.
L. E. Doyle, *Manufacturing Processes and Materials for Engineers*, Prentice-Hall, Inc., Englewood Cliffs, N.J., 1969.
M. M. Eisenstadt, *Introduction to Mechanical Properties of Materials*, The Macmillan Company, New York, 1971.
A. G. Guy, *Introduction to Materials Science*, McGraw-Hill, Inc., New York, 1972.
R. W. Hanks, *Materials Engineering Science*, Harcourt, Brace & World, Inc., New York, 1970.
C. A. Keyser, *Materials Science in Engineering*, Charles E. Merrill Co., Columbus, Ohio, 1968.
F. A. McClintock and A. S. Argon, *Mechanical Behavior of Materials*, Addison-Wesley Publishing Co., Inc., Reading, Penn., 1966.
W. J. Patton, *Materials in Industry*, Prentice-Hall, Inc., Englewood Cliffs, N.J., 1968.
R. E. Reed-Hill, *Physical Metallurgy Principles*, D. Van Nostrand Company, Inc., Princeton, N.J., 1964.
L. H. Van Vlack, *Materials Science for Engineers*, Addison-Wesley Publishing Co., Reading, Penn., 1970.

LARRY A. JACKMAN
Special Metals Corporation

METANILIC ACID, m-NH$_2$C$_6$H$_4$SO$_3$H.   See Amines, aromatic, aniline and its derivatives.

METHACROLEIN, CH$_2$=C(CH$_3$)CHO.   See Acrolein and derivatives.

METHACRYLIC ACID AND DERIVATIVES

Methacrylic acid (MAA), the simplest branched-chain unsaturated organic acid, first was prepared in 1865 from ethyl methacrylate which was obtained by dehydrating ethyl α-hydroxyisobutyrate (1). Commercial interest in the polymers of MAA began in the early 1930s (2–5) (see Methacrylic polymers). The publication of Otto Rohm's doctoral thesis in 1901 marks the beginning of the eventual commercialization of acrylics. Rohm investigated polymers of methyl methacrylate and described the preparation of colorless, clear, rubbery films which he felt could be of commercial interest. In 1909, he and Otto Haas formed a company to manufacture leather-treatment enzymes and in 1927 began to produce acrylates in Germany.

Commercial production of acrylates and methacrylates, which began in 1931 and 1933, respectively, from acetone cyanohydrin (ACN) involved conversion of acetone cyanohydrin to α-hydroxybutyrate ester followed by dehydration to the methacrylate with PCl_5 (6). In 1934, a process was patented for converting acetone cyanohydrin to methacrylamide sulfate [29194-31-8] which is hydrolyzed and esterified to the methacrylate ester (7); this process forms the basis for all contemporary commercial methacrylate production.

Physical Properties

The physical properties of the commercially available methacrylates are summarized in Table 1 (8–11). Thermodynamic properties and vapor pressures for these compounds are provided in Tables 2 and 3, respectively, and data on other methacrylic acid derivatives are listed in Table 4.

Azeotrope data and reciprocal solubilities of MAA with water are presented in Tables 5 and 6, respectively, and solubilities for other methacrylates in water are summarized in Table 7. Data on solubility in the ternary system methyl methacrylate–methanol–water and the quaternary system methyl methacrylate–methacrylic acid–methanol–water have been published (12–14). Methacrylic acid and its common esters are soluble in most organic solvents. Vapor–liquid equilibrium data for the methacrylic acid–water system also have been published (15). A minimum boiling azeotrope at 6 mol % methacrylic acid has been observed. Binary azeotropes do not form in the following systems: methyl methacrylate–methacrylic acid; methyl methacrylate–butyl methacrylate; and methyl methacrylate–2-ethoxyethyl methacrylate (9). Numerous other reports on vapor–liquid equilibria involving methacrylates have been published (16–20). A summary of spectral data on the methacrylates also is available (9). Physical properties of other methacrylic compounds are listed in Tables 8 and 9. Data on other aminoalkyl methacrylates have been published (21).

Chemical Properties

Reactions of Methacrylic Acid Esters. Methacrylic acid esters undergo reactions typical of their constituent functional groups. The sites of chemical reactivity are the terminal vinylic carbon, the double bond, the allylic methyl group, the ester moiety, and the functional groups in the nonmethacrylate moiety.

Table 1. Physical Properties of Commercially Available Methacrylates, $CH_2{=}C(CH_3)COOR$ [a]

Compound	CAS Registry No.	Mol wt	mp, °C	bp, °C	Refractive index, n_D^{25}	Density, d_5^{25} g/cm³	Flash point, °C COC[b]	TOC[c]	Typical inhibitor level[d], ppm
methacrylic acid	[79-41-4]	86.09	14	159–163[e]	1.4288	1.015	77		100 MEHQ
methyl methacrylate	[80-62-6]	100.11	−48	100–101[e]	1.4120	0.939		13	10 MEHQ
ethyl methacrylate	[97-63-2]	114.14		118–119[e]	1.4116	0.909	35	21	15 MEHQ
n-butyl methacrylate	[97-88-1]	142.19		163.5–170.5[e]	1.4220	0.889	66		10 MEHQ
isobutyl methacrylate	[97-86-9]	142.19		155[e]	1.4172	0.882		49	10 MEHQ
isodecyl methacrylate	[29964-84-9]	226		120[f]	1.4410	0.878	121		10 HQ + MEHQ
lauryl methacrylate[g]	[142-90-5]	262	−22	272–343[e]	1.444	0.868	132		100 HQ
stearyl methacrylate[h]	[32360-05-7]	332	15	310–370[e]	1.4502	0.864	>149		100 HQ
2-hydroxyethyl methacrylate	[868-77-9]	130.14	−12	95[i]	1.4505	1.064	108		1200 MEHQ
2-hydroxypropyl methacrylate	[923-26-2]	144.17	<−70	96[i]	1.4456	1.027	121		1200 MEHQ
2-dimethylaminoethyl methacrylate	[2867-47-2]	157.20	ca −30	68.5[i]	1.4376	0.933		74	200 MEHQ
2-t-butylaminoethyl methacrylate	[3775-90-4]	185.25	<−70	93[i]	1.4400	0.914	11		1000 MEHQ
glycidyl methacrylate	[106-91-2]	142.1		75[i]	1.4482	1.073		84	25 MEHQ
ethyl dimethacrylate[j]	[97-90-5]	198.2		96–98[l]	1.4520	1.048	113		60 MEHQ
1,3-butylene dimethacrylate[j]	[1189-08-8]	226		110[f]	1.4502	1.011	124		200 MEHQ
trimethylolpropane trimethacrylate[k]	[3290-92-4]	338	−14	155[m]	1.471	1.06	>149		90 HQ

[a] Refs. 8–11.
[b] Cleveland open cup.
[c] Tag open cup.
[d] MEHQ = monomethyl ether of hydroquinone; HQ = hydroquinone.
[e] At 101 kPa (1 atm).
[f] At 0.4 kPa (3 mm Hg).
[g] Prepared from a mixture of higher alcohols, predominantly lauryl alcohol.
[h] Prepared from a mixture of higher alcohols, predominantly stearyl alcohol.
[i] At 1.3 kPa (9.8 mm Hg).
[j] Diester.
[k] Triester.
[l] At 0.53 kPa (4.0 mm Hg).
[m] At 0.13 kPa (1.0 mm Hg).

347

Table 2.   Thermodynamic Properties of Methacrylates [a]

	Heat of vaporization, kJ/g[b]	Heat capacity, J/(g·K)[b]	Heat of polymerization, kJ/mol[b]	Dissociation constant
methacrylic acid		2.1–2.3	56.5	pK_a = 4.66
methyl methacrylate	0.36	1.9	57.7	
ethyl methacrylate	0.35	1.9	57.7	
n-butyl methacrylate		1.9	56.5	
2-hydroxyethyl methacrylate			49.8	
2-hydroxypropyl methacrylate			50.6	
2-dimethylaminoethyl methacrylate	0.31			pK_b = 5.6
t-butylaminoethyl methacrylate				pK_b = 4.6

[a] Ref. 9.
[b] To convert J to cal, divide by 4.184.

Table 3.   Vapor Pressure Data for Methacrylates [a]

| Pressure, kPa[a] | bp, °C | | |
	Methacrylic acid	Methyl methacrylate	n-Butyl methacrylate
101	161	100	162
53	142	80	
27	124		
25		60	
13.3	107		
12.0			98
10.8		40	
8.0	95		
6.8			84
5.3	86		
3.9		20	
2.7	73		
1.3	60		
1.1		0	
0.67	49		
0.27		−20	
0.13	26		

[a] To convert kPa to mm Hg, multiply by 7.50.

Terminal Vinylic Carbon.   Nucleophiles, eg, aliphatic amines, displace vinylic halogen groups to form enamines (22–23).

$$F_2C{=}C(CF_3)CO_2CH_3 + NH(CH_3)_2 \rightarrow [(CH_3)_2N]FC{=}C(CF_3)CO_2CH_3$$
[685-09-6]   [76392-08-0]

$$HBrC{=}C(CH_3)CO_2CH_3 + NH(C_2H_5)_2 \rightarrow [(C_2H_5)_2N]HC{=}C(CH_3)CO_2CH_3$$
[40053-01-8]   [61423-50-5]

Double Bond. A variety of nucleophiles add to the double bond to form α-methyl and β-substituted propionates (24–27).

$$CH_2\!\!=\!\!C(CH_3)CO_2CH_3 + HCN \rightarrow NCCH_2CH(CH_3)CO_2CH_3$$

$$CH_2\!\!=\!\!C(CH_3)CO_2CH_3 + CH_3OH \rightarrow CH_3OCH_2CH(CH_3)CO_2CH_3$$

$$2\ CH_2\!\!=\!\!C(CH_3)CO_2CH_3 + C_6H_5PH_2 \rightarrow C_6H_5P[CH_2CH(CH_3)CO_2CH_3]_2$$

$$CH_2\!\!=\!\!C(CH_3)CO_2CH_3 + CH_3CH(NO_2)CH_3 \rightarrow (CH_3)_2C(NO_2)CH_2CH(CH_3)CO_2CH_3$$

Electrophilic addition to the double bond occurs with halogens and dihalocarbenes (28–29).

$$\underset{\text{MMA}}{CH_2\!\!=\!\!C(CH_3)CO_2CH_3}\ +\ Br_2\ \text{or}\ NR_4^+Br_3^-\ \longrightarrow\ CH_2BrCBr(CH_3)CO_2CH_3$$

$$CH_2\!\!=\!\!C(CH_3)CO_2(CH_2)_2CH_3 \xrightarrow[\text{CHCl}_3]{\text{base}} \underset{\underset{CCl_2}{\diagdown\diagup}}{CH_2\!-\!C(CH_3)CO_2(CH_2)_2CH_3}$$

Diels-Alder reactions occur with 1,3-butadienes, cyclopentadiene, and *trans*-piperylene (30–35).

Table 4. Physical Properties of Methacrylic Acid Derivatives

Compound	CAS Registry No.	mp, °C	bp, °C	Refractive index, n_D	sp gr
methacrylic acid		14	159–163[a]	1.4288^{25}	1.015^{25}_4
methacrolein	[78-85-3]	−81	68[a]	1.4144^{20}	0.837^{20}_4
methacrylonitrile	[126-98-7]	−36	90[a]	1.3989^{25}	0.800^{20}_4
methacrylamide	[79-39-0]	110			
methacrylic anhydride	[760-93-0]		75[b]	1.4520^{25}	
methacrylic acetic anhydride	[76392-13-7]		63–73[c]	1.4183^{25}	
methacryloyl chloride	[920-46-7]		96–98[a]	1.4435^{25}	1.087^{25}_4
methacryloyl bromide	[6997-65-5]		60–63[d]		

[a] At 101 kPa (1 atm).
[b] At 0.67 kPa (5.0 mm Hg).
[c] At 2.7 kPa (20 mm Hg).
[d] At 13.3 kPa (100 mm Hg).

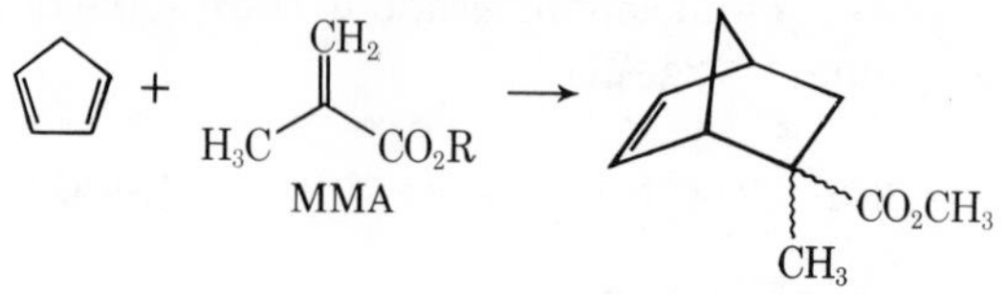

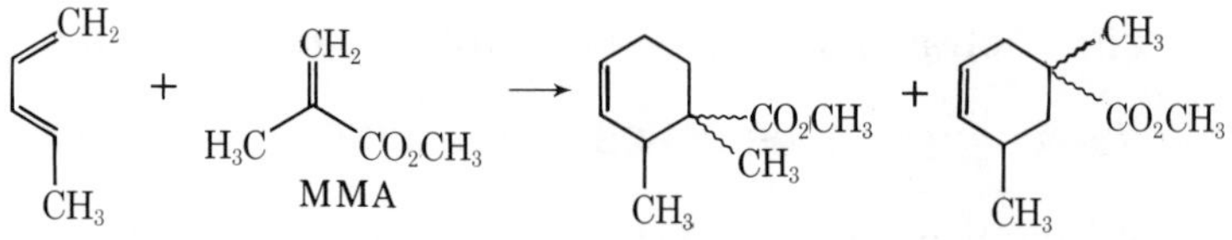

1,3-Dipolar cycloadditions are observed with organic azides (36).

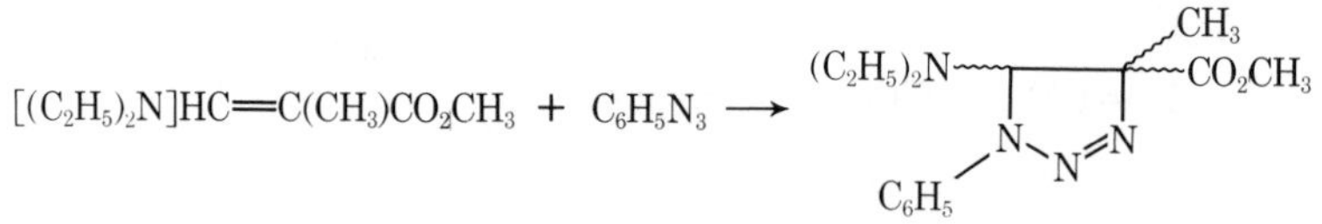

$$[(C_2H_5)_2N]HC{=}C(CH_3)CO_2CH_3 + C_6H_5N_3 \longrightarrow$$

Free-radical-induced cyclizations are observed with di- and trimethacrylates in the presence of azobisisobutyronitrile (AIBN) (37–39).

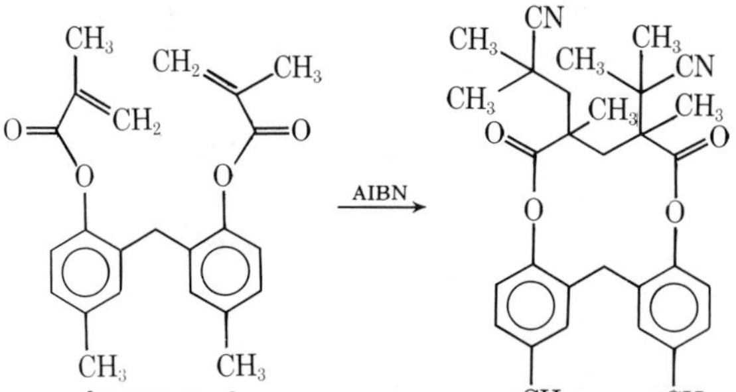

[18507-45-4]

The cyclic co-oligomerization of methyl methacrylate with 1,3-butadiene is catalyzed by nickel complexes in the presence of phosphites and triethylaluminum (40–41).

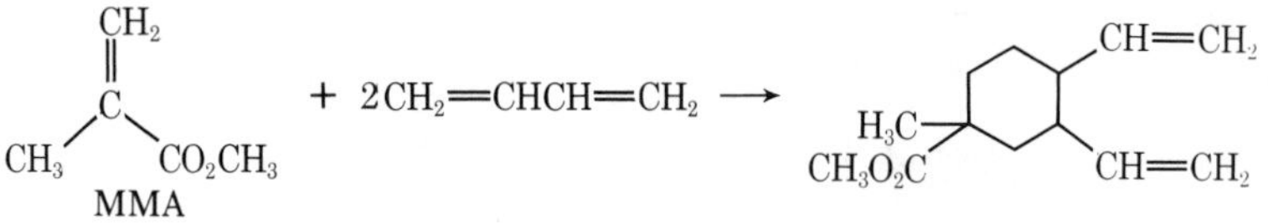

The catalyst system WCl_6–$Sn(CH_3)_4$ promotes the co-metathesis reaction of methyl methacrylate with symmetrical alkenes (42).

$$
\begin{array}{l}
CH_3CH_2CH{=}CHCH_2CH_3 \\
\qquad + \qquad\qquad \longrightarrow \quad CH_3CH_2CH{=}CH_2 + CH_3CH_2CH{=}C(CH_3)CO_2CH_3 \\
CH_2{=}C(CH_3)CO_2CH_3
\end{array}
$$

However, methyl methacrylate has a low reactivity towards head-to-head homometathesis.

Allylic Methyl. The reaction of methyl methacrylate with nitric acid containing various concentrations of dinitrogen trioxide results in the substitution of an allylic hydrogen with nitro or nitroso groups (43).

$$CH_2\!\!=\!\!C(CH_3)CO_2CH_3 + HNO_3(N_2O_3) \rightarrow CH_2\!\!=\!\!C(CH_2NO_2)CO_2CH_3 + CH_2\!\!=\!\!C(CH_2NO)CO_2CH_3$$
$$[51914\text{-}94\text{-}4] \qquad\qquad [51915\text{-}02\text{-}7]$$

Table 5. Azeotropic Mixtures with Methyl Methacrylate[a]

	Pressure, kPa[b]	bp, °C	Ester, %
water	101	83	86
	27	49	88.4
methanol	101	64.2	15.5
ethanol	27	34.5	18

[a] Ref. 9.
[b] To convert kPa to mm Hg, multiply by 7.50.

Table 6. Reciprocal Solubilities of Methyl Methacrylate and Water[a]

Temperature, °C	Methyl methacrylate in water, g/100 g solvent	Water in methyl methacrylate, g/100 g solvent
0	1.85	0.85
10	1.72	0.99
20	1.59	1.15
30	1.50	1.34
40	1.43	1.56
50	1.43	1.80
60	1.49	2.07
70	1.60	2.38
80	1.80	2.74

[a] Ref. 8.

Table 7. Solubility of Methacrylates in Water at 25°C[a]

Compound	Solubility, g/100 mL soln
methyl methacrylate	1.5
ethyl methacrylate	insol
n-butyl methacrylate	insol
isobutyl methacrylate	insol
lauryl methacrylate	insol
stearyl methacrylate	insol
2-hydroxyethyl methacrylate	infinitely sol
2-hydroxypropyl methacrylate	limited
2-dimethylaminoethyl methacrylate	very sol
t-butylaminoethyl methacrylate	1.8
methacrylic acid	infinitely sol

[a] Ref. 9.

Ester Moiety. The reactions of the ester group in methacrylates are similar to those of simple carboxylic acid esters. These include saponification, transesterification, and reduction with Grignard reagents (see Esters).

Functional Groups in the Nonmethacrylate Moiety. The functional groups in the alkoxy moiety of methacrylates undergo the usual reactions of their simpler analogues. The following reactions illustrate the reactivity of the monomers of β-hydroxyethyl methacrylate and its phosphonic acid ester; RCO_2 implies the methacrylate group (44–47).

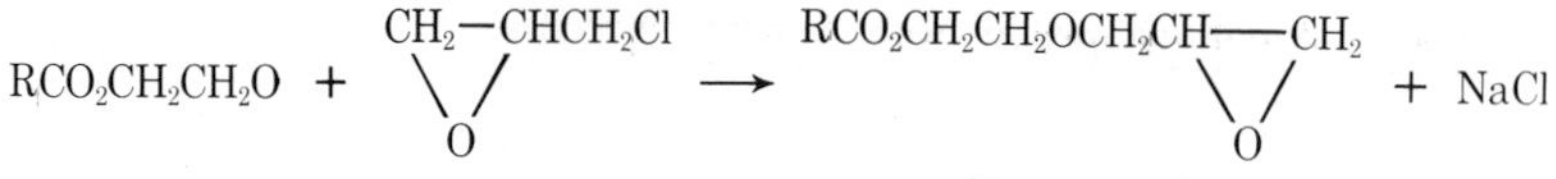

$$RCO_2CH_2CH_2O + \underset{O}{CH_2-CHCH_2Cl} \longrightarrow RCO_2CH_2CH_2OCH_2CH{-\!\!-\!\!}CH_2 + NaCl$$

[30491-79-3]

$$3\,RCO_2CH_2CH_2OH + 2\,H_3PO_4 \longrightarrow RCO_2CH_2CH_2OP\overset{OH}{\underset{O}{\diagdown}}_{OH} + (RCO_2CH_2CH_2O)_2POH + 3\,H_2O$$

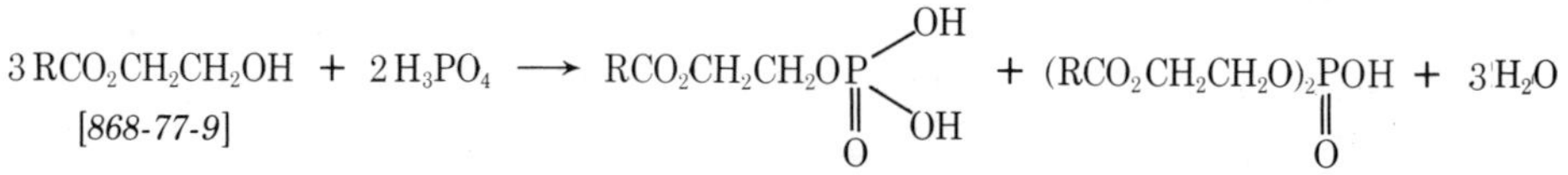

[868-77-9]   [24599-21-1]   [32435-46-4]

$$RCO_2CH_2CH_2OH + COCl_2 \longrightarrow RCO_2CH_2CH_2OCOCl + HCl$$

[13695-27-7]

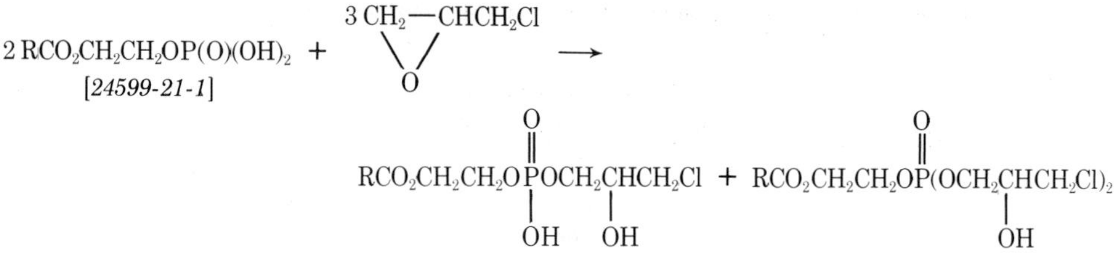

$$2\,RCO_2CH_2CH_2OP(O)(OH)_2 + 3\,\underset{O}{CH_2-CHCH_2Cl} \longrightarrow$$

[24599-21-1]

$$\underset{OH\ \ \ OH}{RCO_2CH_2CH_2OPOCH_2CHCH_2Cl} + \underset{OH}{RCO_2CH_2CH_2OP(OCH_2CHCH_2Cl)_2}$$

[53439-46-6]   [53502-14-0]

β-Dialkylaminoethyl methacrylates undergo quaternization with a variety of organic halides (48–50).

$$RCO_2CH_2CH_2N(CH_3)_2 + CH_3Cl \rightarrow RCO_2CH_2CH_2\overset{+}{N}(CH_3)_3Cl^-$$
[2867-47-2]   [5039-78-1]

$$RCO_2CH_2CH_2N(CH_3)_2 + ClCH_2CO_2H \rightarrow RCO_2CH_2CH_2\overset{+}{N}(CH_3)_2(CH_2CO_2H)Cl^-$$
[45156-26-1]

$$RCO_2CH_2CH_2N(CH_3)_2 + C_6H_5CH_2Cl \rightarrow RCO_2CH_2CH_2\overset{+}{N}(CH_3)_2(CH_2C_6H_5)Cl^-$$
[46917-07-1]

Glycidyl methacrylate,

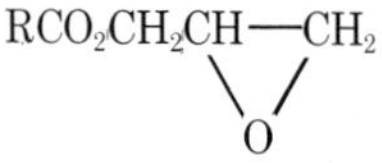

$$RCO_2CH_2CH{-\!\!-\!\!}CH_2 \atop O$$

undergoes the following reaction:

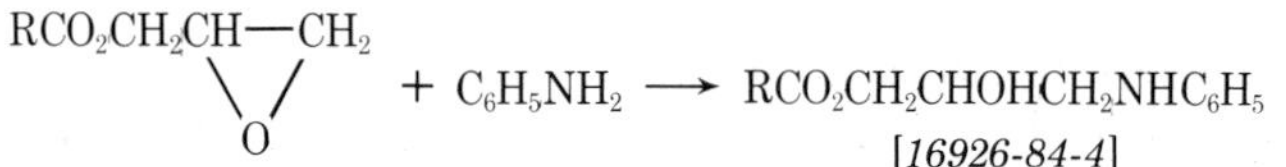

$$RCO_2CH_2CH{-\!\!-\!\!}CH_2 + C_6H_5NH_2 \longrightarrow RCO_2CH_2CHOHCH_2NHC_6H_5$$
O   [16926-84-4]
[106-91-2]

Table 8. Properties of Methacrylic Esters, $CH_2\!=\!C(CH_3)COOR$

Compound	CAS Registry No.	bp, °C (at kPa[a])	Refractive index, n_D	sp gr
Alkyl methacrylates, R				
methyl		100 (101)	1.4120^{25}	0.939^{25}_4
ethyl		119 (101)	1.4116^{25}	0.909^{25}_4
propyl	[2210-28-8]	141 (101)	1.4183^{20}	0.902^{20}_4
isopropyl	[4655-34-9]	125 (101)	1.4334^{25}	0.885^{25}_4
n-butyl		162–163 (101)	1.4220^{25}	0.889^{25}_4
isobutyl		155 (101)	1.4197^{20}	0.882^{25}_4
s-butyl	[2998-18-7]	72–73 (6.7)	1.4195^{25}	
t-butyl	[585-07-9]	52 (4.7)	1.4120^{25}	
n-hexyl	[142-09-6]	204–210 (101)	1.4310^{20}	0.894^{25}_4
n-octyl	[2157-01-9]	114 (1.9)	1.4373^{20}	
isooctyl	[28675-80-1]	68 (0.087)	1.4386^{20}	
2-ethylhexyl	[688-84-6]	47 (0.013)	1.4380^{20}	
n-decyl	[3179-47-3]	99–100 (0.17)	1.4425	
tetradecyl	[2549-53-3]	147–154 (0.093)	1.4480^{20}	
Unsaturated alkyl methacrylates, R				
vinyl	[4245-37-8]	63 (16.4)		
allyl	[96-05-9]	32 (1.3)	1.4328^{25}	
oleyl	[13533-08-9]	165–170 (0.0027)	1.4607^{20}	
2-propynyl	[13861-22-8]	47–49 (1.5)	1.4483^{20}	
Cycloalkyl methacrylates, R				
cyclohexyl	[101-43-9]	44 (0.040)	1.4583^{20}	
1-methylcyclohexyl	[76392-14-8]	94–98 (1.3)	1.4588^{25}	
3-vinylcyclohexyl	[76392-15-9]	63–70 (0.013)	1.4692^{25}	
3,3,5-trimethylcyclohexyl	[75673-26-6]	51–52 (0.013)	1.4548^{20}	
bornyl	[4647-84-1]	68–70 (0.040)	1.4739^{25}	
isobornyl	[7534-94-3]	112–117 (0.33)	1.4748^{25}	0.980^{25}_4
cyclopenta-2,4-dienyl	[76741-96-3]	115 (0.093)	1.4990^{25}	
Aryl methacrylates, R				
phenyl	[2177-70-0]	58–61 (0.13)	1.5184^{20}	
benzyl	[2495-37-6]	119–121 (0.1–0.2)	1.5095^{25}	
nonylphenyl	[76391-98-5]	120–127 (0.0040)	1.5020^{25}	
Hydroxyalkyl methacrylates, R				
2-hydroxyethyl		87 (0.67)	1.4505^{25}	1.064^{25}_4
2-hydroxypropyl	[923-26-2]	87 (0.67)	1.4456^{25}	
3-hydroxypropyl	[2761-09-3]	67–69 (0.013)	1.4496^{25}	
3,4-dihydroxybutyl	[62180-57-8]	110–111 (0.033)		
Methacrylates of ether alcohols, R				
methoxymethyl	[20363-82-0]	54 (2.0)	1.4233^{20}	
ethoxymethyl	[76392-16-0]	87–88 (7.3)	1.4216^{20}	
allyloxymethyl	[49978-33-8]	80–82 (2.7)	1.4422^{20}	
2-ethoxyethoxymethyl	[76392-17-1]	113–114 (2.3)	1.4302^{20}	
benzyloxymethyl	[76392-18-2]	136–137 (0.67)	1.5067^{20}	
cyclohexyloxymethyl	[76392-19-3]	91–92 (0.27)	1.4599^{20}	
1-ethoxyethyl	[51920-52-6]	64–65 (2.7)	1.4182^{25}	
2-ethoxyethyl	[2370-63-0]	85 (2.5)		
2-butoxyethyl	[13532-94-0]	104 (2.0)	1.4304^{25}	
1-methyl-(2-vinyloxy)ethyl	[76392-20-6]	50 (0.13)	1.4400^{20}	
methoxymethoxyethyl	[76392-21-7]	68–70 (0.27)	1.4310^{25}	
methoxyethoxyethyl	[45103-58-0]	67–75 (0.13)	1.4397^{20}	
vinyloxyethoxyethyl	[76392-22-8]	80–82 (0.13)	1.4515^{20}	
1-butoxypropyl	[76392-23-9]	51–53 (0.13)	1.4309^{20}	
1-ethoxybutyl	[76392-24-0]	85–88 (3.1)	1.4223^{20}	
tetrahydrofurfuryl	[2455-24-5]	59–62 (0.080)	1.4552^{20}	

353

Table 8 (*continued*)

Compound	CAS Registry No.	bp, °C (at kPa[a])	Refractive index, n_D	sp gr
furfuryl	[3454-28-2]	62–63 (0.40)	1.4770[25]	
Oxiranyl methacrylates, R				
glycidyl		75 (1.3)	1.4482[25]	1.073[25]
2,3-epoxybutyl	[68212-07-7]	45–50 (0.033)	1.4422[25]	
3,4-epoxybutyl	[55750-22-6]	55–56 (0.11)	1.4472[25]	1.038[25]
2,3-epoxycyclohexyl	[76392-25-1]	70–73 (0.040)	1.4671[25]	
10,11-epoxyundecyl	[23679-96-1]	115–119 (0.0027)	1.4553[25]	0.949[25]
Aminoalkyl methacrylates, R				
2-dimethylaminoethyl		97.5 (5.3)	1.4396[20]	0.933[25]
2-diethylaminoethyl	[105-16-8]	49 (0.040)	1.4442[20]	
2-*t*-butylaminoethyl		97 (1.6)	1.4400[25]	0.914[25]
2-*t*-octylaminoethyl	[14206-24-7]	138–139 (1.6–1.7)	1.4345[25]	0.933[25]
N,N-dibutylaminoethyl	[2397-75-3]	110 (0.13)	1.4474[20]	
3-diethylaminopropyl	[17577-32-1]	105 (0.20)	1.4770[20]	
7-amino-3,4-dimethyloctyl	[76392-26-2]	115–120 (0.10)	1.4570[25]	0.922[25]
N-methylformamidoethyl	[25264-39-5]	121–123 (0.16)	1.4693[25]	
2-ureidoethyl	[4206-97-7]	74–76 (0.40)		
Glycol dimethacrylates, R				
methylene	[4245-38-9]	54–57 (0.024)	1.4520[20]	
ethylene glycol	[97-90-5]	96–98 (0.53)	1.4520[25]	
1,2-propanediol	[7559-82-2]	68–72 (0.13)	1.4450	
1,3-butanediol	[1189-08-8]	78–79 (0.053)	1.4523[20]	
1,4-butanediol	[2082-81-7]	88 (0.027)	1.4872[20]	
2,5-dimethyl-1,6-hexanediol	[76392-00-2]	125–127 (0.13)	1.4567[20]	
1,10-decanediol	[6701-13-9]	170–178 (0.27)	1.4577[25]	
diethylene glycol	[2358-84-1]	120–125 (0.27)	1.4550[25]	
triethylene glycol	[109-16-0]	155 (0.13)	1.4604[20]	
Trimethacrylates				
trimethylolpropane trimethacrylate	[3290-92-4]	155 (0.13)	1.471	1.06[25]
Carbonyl-containing methacrylates				
R = carboxymethyl	[76391-99-6]	108–111 (0.033)		
R = 2-carboxyethyl	[13318-10-0]	104–106 (0.013)	1.4546[25]	
R = acetonyl	[44901-95-3]	40 (0.067)	1.4437[20]	
R = oxazolidinylethyl	[46235-93-2]	83–87 (0.067)	1.4688[25]	
N-(2-methacryloyloxyethyl)-2-pyrrolidinone	[946-25-8]	115–128 (0.067)	1.4872[20]	
N-(3-methacryloyloxypropyl)-2-pyrrolidinone	[76747-97-4]	127 (0.073)	1.4860[20]	
N-methacryloyl-2-pyrrolidinone	[23935-37-7]	98–103 (0.16–0.20)	1.4994[20]	
N-(methacryloyloxy)formamide	[76392-27-3]	127 (0.13)	1.4782[20]	
N-methacryloylmorpholine	[5117-13-5]	67–70 (0.040)	1.4933[20]	
tris(2-methacryloxyethyl)amine	[13884-43-0]	155–165 (0.0067)	1.4768[25]	
Other nitrogen-containing methacrylates				
2-methacryloyloxyethylmethyl-cyanamide	[76392-28-4]	126–132 (0.39)	1.4635[25]	
methacryoyloxyethyl-trimethylammonium chloride	[5039-78-1]			
N-(methacryloyloxyethyl)-diisobutylketimine		93–97 (0.027)	1.4543[25]	
cyanomethyl methacrylate	[7726-87-6]	79–80 (1.3)	1.4381[25]	
2-cyanoethyl methacrylate	[4513-53-5]	72 (0.067)	1.4459[20]	

354

Table 8 (*continued*)

Compound	CAS Registry No.	bp, °C (at kPa[a])	Refractive index, n_D	sp gr
Methacrylates of halogenated alcohols, R				
chloromethyl	[27550-73-8]	54–56 (3.1)	1.4434^{25}	
1,3-dichloro-2-propyl	[44978-88-3]	58–60 (0.027)	1.4670^{25}	
4-bromophenyl	[36889-09-5]	80–85 (0.013)		
2-bromoethyl	[4513-56-8]	65 (0.67)	1.4750^{20}	
2,3-dibromopropyl	[3066-70-4]	70–76 (0.0040)	1.5132^{25}	
2-iodoethyl	[35531-61-4]	112–119 (4.0)		
Sulfur-containing methacrylates				
methyl thiolmethacrylate	[52496-39-6]	57 (4.01)		
butyl thiolmethacrylate	[54667-21-9]	57 (0.27)	1.4828^{20}	
ethylsulfonylethyl methacrylate	[25289-10-5]	120–132 (0.020)		
ethylsulfinylethyl methacrylate	[3007-24-7]	116–119 (0.013)	1.4902^{25}	
thiocyanatomethyl methacrylate	[76392-29-5]	59 (0.020)	1.4899^{25}	
methylsulfinylmethyl methacrylate	[76392-30-8]	110–112 (0.04–0.06)	1.4963^{25}	
4-thiocyanatobutyl methacrylate	[76392-31-9]	102 (0.015)	1.4861^{25}	
bis(methacryloyloxyethyl) sulfide	[35411-32-6]	115–125 (0.067)	1.4894^{25}	
2-dodecylthioethyl methacrylate	[14216-26-3]	155–160 (0.020)	1.4731^{25}	
Phosphorus-boron- and silicon-containing methacrylates				
2-(ethylenephosphito)propyl methacrylate	[76392-32-0]	73–80 (0.0027)	1.4635^{25}	
diethyl methacryloylphosphonate	[76392-33-1]	125–165 (0.027)	1.4668^{25}	
dimethylphosphinomethyl methacrylate	[41392-09-0]	80–85 (0.0053)	1.4347^{25}	
dimethylphosphonoethyl methacrylate	[22432-83-3]	78–84 (0.0093)	1.4426^{25}	
dipropyl methacryloyl phosphate	[76392-34-2]	75–76 (1.6)	1.4411^{20}	1.017^{25}
diethyl methacryloyl phosphite	[3729-12-2]	64 (0.19)	1.4438^{25}	1.046^{25}
2-methacryloyloxyethyl diethyl phosphite	[817-44-7]	83 (0.13)	1.4483^{20}	1.063^{25}
diethylphosphatoethyl methacrylate	[814-35-7]	115–124 (0.067)	1.4340^{25}	1.14^{25}
2-(dimethylphosphato)propyl methacrylate	[76392-35-3]	85–90 (0.0027)	1.4359^{25}	
2-(dibutylphosphono)ethyl methacrylate	[3729-11-1]	130–140 (0.0027)	1.4390^{25}	
2,3-butylene methacryloyl-oxyethyl borate	[76392-36-4]	88–94 (0.0027)	1.4451^{25}	
methyldiethoxymethacryloyl-oxyethoxysilane	[76392-37-5]	75–82 (0.0027)	1.4216^{25}	

[a] To convert kPa to mm Hg, multiply by 7.50.

The product can react further with $C_6H_5N_2^+Cl^-$ to form a polymerizable azo dye (51–52) (see Azo dyes).

$RCO_2CH_2CH(OH)CH_2NH$—⟨○⟩ + Cl^- $\overset{+}{N_2}$—⟨○⟩ →

[16926-84-4]

$RCO_2CH_2CH(OH)CH_2NH$—⟨○⟩—$N{=}N$—⟨○⟩ + HCl

[16323-28-7]

Table 9. Properties of Amides and Nitriles of Methacrylic Acid

Compound	CAS Registry No.	bp, °C (at kPa[a])	mp, °C	Refractive index, n_D
N-methylmethacrylamide	[3887-02-3]	88 (0.47)		1.4740[20]
N-isopropylmethacrylamide	[13749-61-6]	112 (15.3)	90–91	
N-phenylmethacrylamide	[1611-83-2]		84–85	
N-(2-hydroxyethyl)methacrylamide	[5238-56-2]	147–157 (0.15)		1.5002[25]
1-methacryloylamido-2-methyl-2-propanol	[74987-95-4]	100–104 (0.013)	74–76	
4-methacryloylamido-4-methyl-2-pentanol	[23878-87-1]	114–119 (0.053)		1.4732[25]
N-(methoxymethyl)methacrylamide	[3644-12-0]	78–82 (0.03–0.05)		1.4707[25]
N-(dimethylaminoethyl)methacrylamide	[13081-44-2]	85–92 (0.1–0.4)		1.4744[20]
N-(3-dimethylaminopropyl)-methacrylamide	[5205-93-6]	92 (0.0053)		1.4789[20]
N-acetylmethacrylamide	[44810-87-9]	76 (0.16)		1.4835[25]
N-methacryloylmaleamic acid	[76392-01-3]		192–193	
methacryloylamidoacetonitrile	[65993-30-8]	114–118 (0.053)	32–34	
N-(2-cyanoethyl)methacrylamide	[24854-94-2]		46–48	
1-methacryloylurea	[20602-83-9]		138	
N-phenyl-N-phenylethylmethacrylamide	[76392-02-4]		63–64	
N-(3-dibutylaminopropyl)-methacrylamide	[76392-03-5]	125 (0.017)		1.4731[20]
N,N-dimethylmethacrylamide	[6976-91-6]	68 (0.80)		1.4600[20]
N,N-diethylmethacrylamide	[5441-99-6]	71–72 (0.33)		
N-(2-cyanoethyl)-N-methyl-methacrylamide	[76392-04-6]	113–116 (0.15)		1.4755[25]
N,N-bis(2-diethylaminoethyl)-methacrylamide	[76392-05-7]	122 (0.053)		1.4702[20]
N-methyl-N-phenylmethacrylamide	[2918-73-2]	88–96 (0.27)	50	
N,N'-methylenebismethacrylamide	[2359-15-1]		163–164	
N,N'-ethylenebismethacrylamide	[6117-25-5]		170 (dec)	
N-(diethylphosphono)-methacrylamide	[76392-06-8]	105–110 (0.0067)		1.4412[25]
N-t-butyl-N-(diethylphosphono)-methacrylamide	[76392-07-9]	80–85 (0.020)		1.4296[25]

[a] To convert kPa to mm Hg, multiply by 7.50.

$$RCO_2CH_2CH\!\!-\!\!CH_2 \;\;\underset{O}{\diagdown\!\diagup}\;\; + \;SO_2\; + \;NR'_3 \;\longrightarrow\; RCO_2CH_2CHCH_2\overset{+}{N}R'_3 \underset{|}{} OSO_3^-$$

Propargyl methacrylate $RCO_2CH_2C\!\equiv\!CH$ undergoes the Mannich reaction with formaldehyde and secondary amines to form the corresponding aminoacetylenic esters (53–54).

$$RCO_2CH_2C\!\equiv\!CH + CH_2O + NHR'_2 \rightarrow RCO_2CH_2C\!\equiv\!CCH_2NR'_2$$

[13861-22-8]

Manufacture and Processing

Methyl methacrylate is produced by Rohm and Haas; E. I. du Pont de Nemours & Co., Inc.; and CY/RO Industries in the United States. Foreign producers include ICI, Rohm G.m.b.H., Mitsubishi Rayon, and Asahi Chemical.

In about 1960, the Escambia Chemical Corporation in the United States and Nissan Chemical Industries in Japan developed syntheses based on the liquid-phase oxidation of isobutylene with nitrogen tetroxide and nitric acid to α-hydroxyisobutyric acid and the dehydration of the intermediate to methacrylic acid (55–56). The yields for these processes, based on isobutylene, were ca 65–75%. The Nissan process was not commercialized; Escambia operated a plant for a short period. However, a number of companies claim to have developed attractive alternative technology and several have announced plans to commercialize C_4-oxidation technology (57).

Proposed routes and the commercial method for the manufacture of methyl methacrylate and methacrylic acid are summarized in Figure 1, which illustrates the dependence of the method on specific hydrocarbon raw materials. Methanol is a common raw material for all processes for the methyl ester. Hydrogen cyanide is obtained as a by-product in the ammoxidation of propylene to acrylonitrile or is produced from the catalytic ammoxidation of natural gas (see Acrylonitrile). For a processing route to be commercially attractive, it must have costs that are superior to those of the existing acetone cyanohydrin (ACN) route. Process comparisons involve evaluations of factors, eg, raw material cost and utilization, operating costs with particular attention to energy-related charges, waste-disposal costs, environmental impact, and plant investment. Natural-gas costs are expected to increase steadily during the next decade and, since ethylene costs are dependent on the cost of crude oil, they will continue to increase. Propylene may be considered to be a by-product from the manufacture of ethylene and its cost also should follow that of crude oil but should remain below that of ethylene. Lead times of ca five years are required for the process development and construction of a plant based on novel technology.

C_4-oxidation technology is the most probable basis for future methyl methacrylate expansions. This technology requires isobutylene or t-butyl alcohol as the primary raw material. Isobutylene is a major component in C_4 streams from ethylene plants. After butadiene is extracted from the mixed C_4 stream, isobutylene is separated as such or as t-butyl alcohol from the linear butenes. t-Butyl alcohol also is available as a by-product from the Oxirane process for the manufacture of propylene oxide (qv) from the reaction of propylene with t-butyl hydroperoxide (58). The Oxirane Corp. intends to produce the ether of methanol and t-butyl alcohol to be blended at 5% in gasoline (59). However, this ready outlet for t-butyl alcohol may limit its availability as a feedstock for the C_4-oxidation process (see Gasoline).

Evaluations based on reasonable performance and economic assumptions suggest the potential for more attractive processes than the conventional acetone cyanohydrin process. Further process development and improvement may be required but it is likely that novel technology will replace the long-entrenched ACN process.

Acetone Cyanohydrin (ACN) Process. The commercial process for the manufacture of methyl methacrylate is based on well-established technology and on the available basic raw materials acetone, hydrogen cyanide, methanol, and sulfuric acid. However, the raw material costs for several of the alternative processes may be lower than those of the ACN process. Although the long-range outlook for acetone supply is optimistic,

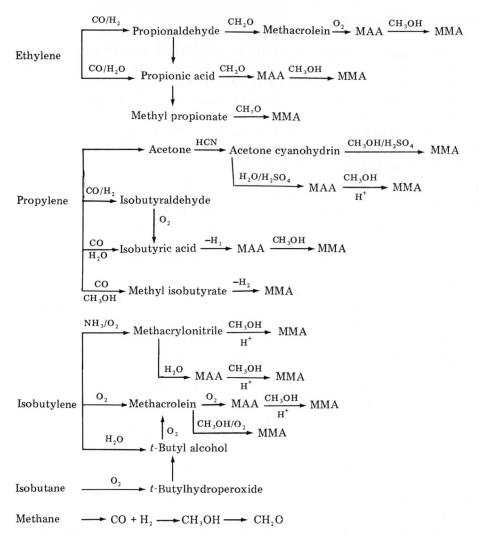

Figure 1. Routes to methyl methacrylate (MMA) and methacrylic acid (MAA).

the future supply of by-product hydrogen cyanide appears to be limited because increasingly selective catalysts which produce less HCN are used in the manufacture of acrylonitrile and there is a slower growth rate of acrylonitrile relative to methyl methacrylate usage. The ACN process also is characterized by a very large waste-acid stream.

The chemistry of the process involves the reaction of acetone cyanohydrin with excess concentrated sulfuric acid to form methacrylamide sulfate. The requirement for the excess sulfuric acid may be explained by the proposed α-sulfato imino intermediate that is formed from two moles of sulfuric acid and one mol of ACN. The intermediate is hydrolyzed by the water which is formed in the initial reaction of ACN with sulfuric acid; the resulting amide intermediate eliminates sulfuric acid to form methacrylamide sulfate.

$$
\underset{\underset{OH}{|}}{\overset{\overset{CH_3}{|}}{CH_3CCN}} + H_2SO_4 \longrightarrow \underset{\underset{OSO_3H}{|}}{\overset{\overset{CH_3}{|}}{CH_3CCN}} + H_2O \xrightarrow{H_2SO_4}
$$

$$
\left[\underset{\underset{OSO_3H}{|}}{\overset{\overset{CH_3}{|}}{CH_3CC}}\overset{+}{=}NH.HSO_4^-\right] \xrightarrow{H_2O} \underset{\underset{OSO_3H}{|}}{\overset{\overset{CH_3}{|}}{CH_3CCONH_2.H_2SO_4}}
$$

$$
\underset{\underset{OSO_3H}{|}}{\overset{\overset{CH_3}{|}}{CH_3CCONH_2.H_2SO_4}} \longrightarrow \overset{\overset{CH_3}{|}}{CH_2{=}CCONH_2.H_2SO_4} + H_2SO_4
$$
[3351-73-3]

It is important that both the acetone cyanohydrin and the sulfuric acid be anhydrous since the α-sulfato amide can be hydrolyzed to the α-hydroxy amide. The presence of the latter compound requires a thermal cracking step for the formation of methacrylamide sulfate. Hydroxy material that is not dehydrated can be converted to acetone or it remains in the crude MMA as methyl α-hydroxyisobutyrate. By-products from this step include acetone disulfonic acid and carbon monoxide (60).

$$
\underset{\underset{OSO_3H}{|}}{\overset{\overset{CH_3}{|}}{CH_3CCONH_2.H_2SO_4}} + H_2O \longrightarrow \underset{\underset{OH}{|}}{\overset{\overset{CH_3}{|}}{CH_3CCONH_2.H_2SO_4}} + H_2SO_4
$$

$$
\underset{\underset{OH}{|}}{\overset{\overset{CH_3}{|}}{CH_3CCONH_2.H_2SO_4}} \longrightarrow \overset{\overset{CH_3}{|}}{CH_2{=}CCONH_2.H_2SO_4} + H_2O
$$

The methacrylamide sulfate stream reacts with aqueous methanol to form methyl methacrylate by a combination hydrolysis–esterification reaction:

$$
\overset{\overset{CH_3}{|}}{CH_2{=}CCONH_2.H_2SO_4} + CH_3OH \longrightarrow \overset{\overset{CH_3}{|}}{CH_2{=}CCO_2CH_3} + NH_4HSO_4
$$
$$
\text{MMA}
$$

$$
\overset{\overset{CH_3}{|}}{CH_2{=}CCONH_2.H_2SO_4} + H_2O \longrightarrow \overset{\overset{CH_3}{|}}{CH_2{=}CCO_2H} + NH_4HSO_4
$$
$$
\text{MAA}
$$

$$
\overset{\overset{CH_3}{|}}{CH_2{=}CCO_2H} + CH_3OH \longrightarrow \overset{\overset{CH_3}{|}}{CH_2{=}CCO_2CH_3} + H_2O
$$
$$
\text{MMA}
$$

By-products from the reaction include dimethyl ether, methyl formate, acetone,

α-hydroxyisobutyric acid, methyl α-hydroxyisobutyrate, and methyl β-methoxyiso-butyrate.

The preparation of methacrylamide sulfate is performed continuously in a series of two stirred reactors at 80–110°C with a residence time of ca 1 h. The reaction mixture is circulated through external coolers to control the temperatures. The reactor effluent is heated from 125 to 145°C to complete the dehydration and then is cooled during transfer to the esterification reactor. The feed ratio to the first reactor is 1.5–1.9 mol of 99% sulfuric acid per mol of anhydrous ACN. Inhibitors are added at specific processing steps to prevent polymerization.

Modifications of the esterification step are variations in the procedures for the recovery of the crude ester and for the separation of methanol and methacrylic acid for recycling. In addition, process conditions vary in terms of the feed ratios of methanol and water to methacrylamide and the reactor temperatures and residence times.

In one version, the methacrylamide sulfate stream, excess aqueous methanol, and recycled streams react continuously in a series of steam-jacketed esterification reactors at 80–110°C with a 2- to 4-h residence time (61). The reactor effluent is distilled in an acid-stripping column to give crude methyl methacrylate, methanol, and water. The waste-acid ammonium sulfate bottoms either are treated with ammonia for conversion to fertilizer-grade ammonium sulfate or are burned in an acid recovery plant for reconversion to sulfuric acid for recycling (see Fertilizers). The sulfur dioxide from the combustion of the acid residue is oxidized to sulfur trioxide and then is converted to sulfuric acid using conventional technology. The water-washed crude ester is purified in a multicolumn distillation system, and the aqueous methanol is distilled to recover methanol for recycling. Based on ACN, the yield of methyl methacrylate is ca 90% and, based on methanol and depending on the process scheme and reaction conditions, it is 80–90%.

In another version of the esterification step, the reaction is performed under pressures of up to 790 kPa (100 psig) at 100–150°C with residence times of ≤ 1 h, depending on the reaction temperature. The product may be recovered as described above or the reactor effluent may be separated into organic and waste-acid phases. The lower acid layer is distilled in an acid stripper to recover organic compounds for recycling to the reactor. The bottoms from the acid stripper are sent to an acid regeneration plant to provide sulfuric acid for recycling to the first step of the process.

Light ends (low-boiling fractions) are removed from the organic layer in a flash column. The crude ester then is washed with water or with aqueous ammonia to remove methanol and some methacrylic acid; the aqueous raffinate is recycled to the esterification reactor. The washed crude ester is purified in a three-column distillation system. In the first-stage column, water and methanol are taken overhead and are recycled to the esterification reactor. The finished product is taken overhead in the product column. The bottoms from the product column are stripped to recover compounds for recycling and the residue is incinerated (see Fig. 2).

Methacrylic acid is manufactured by hydrolysis of the methacrylamide stream using facilities and conditions similar to those used in the esterification step. The reactor effluent is separated into two phases. The upper organic layer is distilled to provide high purity methacrylic acid. The lower waste-acid layer is steam-stripped to recover the dilute methacrylic acid overhead for recycling to the process and the

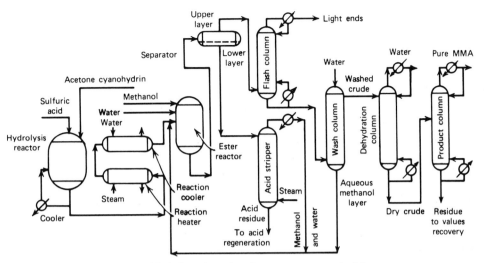

Figure 2. Methyl methacrylate from acetone cyanohydrin process.

bottoms which are sent to a sulfuric acid recovery plant. The aqueous recycle stream provides the water for the hydrolysis reaction.

C$_4$-Oxidation. The proposed C$_4$-oxidation process is based on the two-stage oxidation of isobutylene or t-butyl alcohol to methacrylic acid. It is similar to the highly successful process for the manufacture of acrylic acid from the catalytic oxidation of propylene (see Acrylic acid and derivatives). A basic hydrocarbon raw material is used to produce methacrylic acid and the only other required reactant is air. Isobutylene is oxidized in the first stage to methacrolein and, in the second stage, the methacrolein is oxidized to methacrylic acid. The methacrylic acid is separated from the by-products and is esterified with methanol to methyl methacrylate. For this route to be attractive, as compared with the ACN process, both catalysts must be selective and have a life

$$CH_2{=}\overset{\overset{\textstyle CH_3}{|}}{C}CH_3 + O_2 \longrightarrow CH_2{=}\overset{\overset{\textstyle CH_3}{|}}{C}CHO + H_2O \xrightarrow{\ 1/2\ O_2\ }$$

$$CH_2{=}\overset{\overset{\textstyle CH_3}{|}}{C}CO_2H \xrightarrow{\ CH_3OH\ } CH_2{=}\overset{\overset{\textstyle CH_3}{|}}{C}CO_2CH_3 + H_2O$$
$$\qquad\quad\text{MAA}\qquad\qquad\qquad\qquad\text{MMA}$$

of more than one year, and low cost, ie, relative to other organic feedstocks, isobutylene must be available.

Specific conditions, particularly temperature, depend on catalyst performance but, in a typical procedure, a mixture of 5% isobutylene, 30% steam, and 65% air are passed over the first-stage catalyst at about 350°C. Isobutylene conversion is ca 98% and selectivity to methacrolein plus methacrylic acid is 80–85%. In the preferred process configuration, the effluent from the first reactor, with or without additional air or diluent added to the interstage, is fed directly to the second-stage reactor. Second-stage operating conditions depend on catalyst properties and the temperature is ca 310°C. Where conversions are high, ie, >90%, selectivities are about 80%. Some catalysts give selectivities >80% but with conversions of 60–80%. In the latter case, unconverted methacrolein is recovered from the reactor effluent and is recycled to the second stage. The performance of other second-stage catalysts is best with a feed of methacrolein, air, and steam; methacrolein is isolated from the first-stage effluent and is fed with air and steam to the second stage. The products are isolated as an aqueous solution and the overall yield of methacrylic acid is 55–65%, based on isobutylene. By-products include acetic acid, acetaldehyde, acetone, acrylic acid, and waste gas (CO and CO_2) (see also Hydrocarbon oxidation).

t-Butyl alcohol may be used instead of isobutylene with the same results. *t*-Butyl alcohol is available as a by-product from the Oxirane process, and it is prepared readily in high yield by the hydration of isobutylene in mixed C_4 streams from ethylene plants. After the butadiene has been extracted, the resulting stream, which contains 40–50% isobutylene, is hydrated in acetic acid in the presence of a strong acid ion-exchange resin in high selectivity to *t*-butyl alcohol (62).

It has been suggested that the methyl *t*-butyl ether process be used as the procedure for the removal of isobutylene from a mixed C_4 stream (63). The ether is formed by the selective reaction of the isobutylene in the C_4 stream with methanol. The methyl *t*-butyl ether then is selectively cracked to high purity isobutylene which is the required feedstock for the C_4-oxidation process. The separation procedure is not corrosive nor does it create a waste-acid stream.

Catalysts.   The performance of the catalysts is principal to the economics of the C_4-oxidation process. The performance is measured by activity (conversion of reactant), selectivity (conversion to desired product), productivity (production of product per unit of reactor volume per unit of time), and catalyst life (time on stream before significant loss of activity or selectivity (see Catalysis). Catalyst performance depends on composition, method of preparation, physical properties, and use conditions. The catalysts are extruded or pelletized for use in the fixed-bed reactors.

The first-stage catalysts generally are complex oxides of molybdenum, bismuth, cobalt, iron, nickel, alkali metal, antimony, tellurium, phosphorus, and tungsten. The second-stage catalysts vary in composition but may be heteropolyacids of molybdenum or phosphomolybdates with oxides, eg, bismuth, antimony, thorium, chromium, copper, vanadium, or zirconium oxides. Patents claiming specific catalysts and processes for their use are given in refs. 64–73.

The commercial process is expected to be essentially the same as the propylene oxidation process for acrylic acid. The process flow sheets for the acrylic acid process are representative of those anticipated for the oxidation, separation, and esterification steps of the C_4-oxidation process for methacrylic acid (see Acrylic acid and derivatives).

The oxidation process is operated at the lowest temperature that is consistent with maximum conversion. The gaseous effluent from the second-stage reactor is fed into the absorber to remove the product from the gas stream. The absorber off-gas is transferred to an incinerator where all residual organic material is burned. The aqueous absorber effluent may be extracted with a solvent, eg, butyl acetate, toluene, or dibutyl ketone, which is chosen for high selectivity of methacrylic acid and low solubility of water. The methacrylic acid is recovered from the extract by fractional distillation. Solvent, light ends, and water are taken overhead in the first column. Crude acetic acid is removed overhead in the foreruns column; however, it may be purified in a separate column to obtain a salable product. The bottoms from the foreruns column are fed into the product column where the methacrylic acid is removed overhead. The bottoms from the product column are stripped to recover compounds for recycling and the residue is burned. Methacrylic acid does not react with itself nearly as readily as the self-reaction of acrylic acid and losses during recovery are low. Inhibitors are fed to the columns to prevent polymer formation. Product purity is ca 99% and recovery is >95%.

The process for the manufacture of methyl methacrylate essentially is that used for ethyl acrylate. The reaction of methacrylic acid and methanol is catalyzed by sulfuric acid, an organic sulfonic acid, or a strong acid ion-exchange resin. Product purity is >99% and yields of methanol and methacrylic acid are >95%.

A modification of the C_4-oxidation route involves a standard first-stage oxidation of isobutylene to methacrolein in ca 80% yield. The methacrolein is isolated from the reactor effluent and then is converted directly to methyl methacrylate by the liquid-phase, catalytic oxidation of a dilute solution in methanol (74). Process details have not been described but yields from batch examples are reported >90% (75). If this concept could be developed and equivalent results could be obtained in a large-scale continuous operation, the yield of methyl methacrylate based on isobutylene would be >70%.

Methacrylonitrile Process. The proposed methacrylonitrile route is related to the C_4-oxidation process in that it is based on a vapor-phase catalytic conversion of isobutylene; it also is related to the ACN process since methacrylonitrile may be used in the feed in place of acetone cyanohydrin. The methacrylonitrile process gives superior results to those of the ACN process.

The preparation of methacrylonitrile from the ammoxidation of isobutylene is similar to the commercial process for the manufacture of acrylonitrile (qv) from propylene. Reaction by-products are acetonitrile, hydrogen cyanide, waste gas, and a small amount of methacrolein. The yield of methacrylonitrile, based on isobutylene, is ca 70% (76).

$$\underset{\underset{\displaystyle CH_2=CCH_3}{|}}{CH_3} + NH_3 + 1.5\,O_2 \longrightarrow \underset{\underset{\displaystyle CH_2=CCN}{|}}{CH_3} + 3\,H_2O$$

The methacrylonitrile is converted to methyl methacrylate by a modified ACN process: 85% sulfuric acid is used instead of 99% acid. Essentially, no excess sulfuric acid is required in the methacrylonitrile process compared with more than 1.5 mol of sulfuric acid per mole of acetone cyanohydrin, and the yield of the intermediate methacrylamide is almost quantitative instead of being ca 95%. The esterification step is equivalent with both raw materials; however, the overall yield of methyl methacrylate is slightly higher based on methacrylonitrile because of the higher selectivity in the initial step (77). Also, disposal is less of a problem than with the ACN process because of the smaller quantity of sulfuric acid used.

The methacrylonitrile process is not competitive with the C_4-oxidation process where the yield of methacrylic acid is above 65% because of the higher raw material costs owing to the use of excess ammonia in the ammoxidation step and because of the higher costs associated with the large waste-acid stream.

A proposed process combines features of the methacrylonitrile and the ACN processes (78). The ammoxidation of isobutylene gives about 72 mol of methacrylonitrile, 35 mol of hydrogen cyanide, and 30 mol of acetonitrile per 100 mol of isobutylene that react. The product mixture is treated with acetone in the presence of an alkaline catalyst to convert the hydrogen cyanide to acetone cyanohydrin. The neutralized mixture is stripped to remove the acetonitrile, and the residual mixture of methacrylonitrile and acetone cyanohydrin is converted to methyl methacrylate using a modification of the conventional ACN process.

A version of the above concept involves treatment of the mixture of methacrylonitrile and acetone cyanohydrin with phosphoric anhydride to dehydrate the acetone cyanohydrin to methacrylonitrile (79).

$$\underset{\underset{OH}{|}}{\overset{\overset{CH_3}{|}}{3\,CH_3CCN}} + P_2O_5 \longrightarrow \overset{\overset{CH_3}{|}}{3\,CH_2{=}CCN} + 2\,H_3PO_4$$

Propionate–Formaldehyde Route.   The ethylene-based propionate–formaldehyde route involves the reaction of either propionic acid or methyl propionate with formaldehyde. The first step is preparation of propionic acid by the carbonylation of ethylene or by the hydroformylation of ethylene to propionaldehyde followed by oxidation of the aldehyde to propionic acid (see Oxo process). Commercial processes are available for the manufacture of propionic acid and methyl propionate. Also, a one-step synthesis is claimed for the preparation of methyl propionate by the rhodium-catalyzed reaction of ethylene and carbon monoxide in the presence of methanol (80).

$$
\begin{aligned}
CH_2{=}CH_2 + CO + H_2O \;\; Ni(CO)_4 & \\
CH_2{=}CH_2 + CO + H_2 \rightarrow CH_3CH_2CHO &
\end{aligned}
\;\;\nearrow\!\!\!\searrow CH_3CH_2CO_2H
$$

The second step involves the vapor-phase catalytic condensation of propionic acid or methyl propionate with formaldehyde to form methacrylic acid or methyl methacrylate. The catalyst must provide high selectivity at high conversions. Catalyst

$$CH_3CH_2CO_2H + CH_2O \longrightarrow \underset{MAA}{\overset{\overset{CH_3}{|}}{CH_2{=}CCO_2H}} + H_2O$$

$$CH_3CH_2CO_2CH_3 + CH_2O \longrightarrow \underset{MMA}{\overset{\overset{CH_3}{|}}{CH_2{=}CCO_2CH_3}} + H_2O\;\cdot$$

life should be at least six months and the ratio of reactants should be such as to minimize recycling requirements. Effective catalysts include alkali metal or alkaline-earth metal aluminosilicates, potassium hydroxide- or cesium hydroxide-treated pyrogenic silica, alumina, and lanthanum oxide (81–83). Results vary over a range of conversions and selectivities. Generally, best results are at conversions of about 50% where selectivities are >80%. However, the reported data were obtained in short runs and it appears that additional catalyst development is required for this concept to be attractive.

Improved results have occurred with the use of methylal (dimethoxymethane) in place of formaldehyde in the condensation reaction (84–85). High selectivities and conversions have been achieved based on methylal with excess methyl propionate.

$$CH_3CH_2CO_2CH_3 + CH_3OCH_2OCH_3 \longrightarrow \underset{MMA}{\overset{\overset{CH_3}{|}}{CH_2{=}CCO_2CH_3}} + 2\,CH_3OH$$

No data were given on the selectivity or recovery of the excess methyl propionate. Methylal is formed directly from methanol by a modified conventional formaldehyde process or from the reaction of formaldehyde with methanol (84).

Isobutyrate Dehydrogenation. Isobutyrate dehydrogenation is the most extensively explored of the several proposed methods of MMA manufacture based on propylene. The principal step is the catalytic dehydrogenation of isobutyric acid or methyl isobutyrate to MAA or MMA. There are several routes to the intermediate isobutyric derivative. One involves the liquid-phase oxidation of isobutyraldehyde to isobutyric acid. The acid-catalyzed addition of carbon monoxide to propylene has been investigated for the direct formation of the isobutyric intermediate. High yields of isobutyric acid have been obtained using an aqueous hydrogen fluoride catalyst (86–87). Systems containing boron trifluoride also are reported to be effective (88).

Catalyst performance in the vapor-phase dehydrogenation step is critical to the development of an attractive process. Improved catalyst performance has been obtained with the development of oxidative dehydrogenation technology. The catalysts are similar to those used in vapor-phase, hydrocarbon oxidation processes, eg, tungsten–molybdenum systems (89), phosphomolybdic acid catalysts with various additives (90), and molybdenum–vanadium systems (91). Single-pass yields are 65–75%.

$$\overset{\underset{\displaystyle |}{CH_3}}{CH_3CHCO_2H} + 1/2\,O_2 \longrightarrow \overset{\underset{\displaystyle |}{CH_3}}{CH_2{=}CCO_2H} + H_2O$$
$$\text{MAA}$$

Propylene Oxycarbonylation. Alkoxycarbonylation of propylene has not provided satisfactory selectivity to methacrylic acid or to methyl methacrylate. In addition to the desired methacrylate, both crotonate and saturated by-products are formed (92). However, propylene oxycarbonylation could be an effective route if a more selective catalyst system could be developed or if conditions could be defined for high selectivity to methacrylate and to a desirable co-product.

$$CH_3CH{=}CH_2 + CO + CH_3OH \longrightarrow$$

$$CH_3CH{=}CHCO_2CH_3 + CH_3CH_2CH_2CO_2CH_3 + \overset{\underset{\displaystyle |}{CH_3}}{CH_2{=}CCO_2CH_3} + \overset{\underset{\displaystyle |}{CH_3}}{CH_3CHCO_2CH_3}$$
$$\text{MMA}$$

Liquid-Phase Oxidation of Methacrolein. Liquid-phase oxidation of methacrolein with air or oxygen provides high yields of methacrylic acid (93). Both methacrolein and sodium hydroxide are added to the reactor; as the methacrylic acid is formed, it is converted to the sodium salt. The product is recovered from the acidified solution of the sodium salt. The economics of the proposed route are not attractive because of the consumption of a mol of caustic soda and a mol of inorganic acid for each mol of methacrylic acid, the problems associated with disposal of the salt, and the high catalyst cost.

$$2\,\overset{\underset{\displaystyle |}{CH_3}}{CH_2{=}CCHO} + 2\,NaOH + O_2 + H_2SO_4 \longrightarrow 2\,\overset{\underset{\displaystyle |}{CH_3}}{CH_2{=}CCO_2H} + Na_2SO_4 + 2\,H_2O$$
$$\text{MAA}$$

Liquid carbon dioxide has been suggested for recovery of methacrylic acid from the aqueous sodium methacrylate solution derived from the silver-catalyzed oxidation of methacrolein in aqueous sodium hydroxide (94).

The liquid-phase air oxidation of methacrolein to methacrylic acid in the absence of a base has been unsatisfactory because of polymerization problems and the formation of by-product peroxides of the unsaturated acid and aldehyde. Catalytic levels of thallium and other salts or fluorine-containing organic compounds in the reaction mixture provide fair yields of methacrylic acid (95–96).

Isobutane Route. A proposed process for the preparation of methacrylic acid involves the use of isobutane and oxygen as the only reactants (97). The reaction steps are oxidation of isobutane to form t-butyl hydroperoxide and t-butyl alcohol; reaction of the t-butyl hydroperoxide with methacrolein to give methacrylic acid and t-butyl alcohol; dehydration of t-butyl alcohol to isobutylene; and oxidation of isobutylene to methacrolein.

Although the stated yields for each step appear to be satisfactory, the overall yield is probably too low, because of the large number of reaction steps, for the process to be developed.

Carbonylation of Methylacetylene. The nickel carbonyl-catalyzed carbonylation of methylacetylene provides either methacrylic acid or methyl methacrylate, depending on the conditions (98). However, significant amounts of crotonic acid or methyl crotonate also are formed. In addition, methylacetylene is not readily available. Methylacetylene and allene are formed by the pyrolysis of propylene. Methylacetylene and allene are equivalent in the carbonylation reaction.

$$\underset{}{CH_3C\equiv CH} + CO + H_2O \longrightarrow \underset{MAA}{CH_2=\overset{\overset{\displaystyle CH_3}{|}}{C}CO_2H}$$

Direct Esterification. The methods described previously for the preparation of methyl methacrylate generally can be used for other lower alkyl esters. Direct esterification of methacrylic acid with simple alcohols using acid catalysts, eg, sulfuric acid, sulfonic acids, or strong acid resins, gives high yields of methacrylate esters. For the preparation of higher alkyl esters and functionalized alcohols, a solvent is used which permits operation at moderate temperatures and facilitates removal of water of reaction by azeotropic distillation. Methacrylate esters have been prepared by vapor-phase esterification of methacrylic acid with methanol and other lower alcohols with good selectivity and good but not complete conversion (99).

A reaction related to direct esterification involves the acid-catalyzed addition of olefins to methacrylic acid to form branched alkyl esters (100).

$$\underset{MAA}{CH_2=\overset{\overset{\displaystyle CH_3}{|}}{C}CO_2H} + CH_3CH=CH_2 \longrightarrow CH_2=\overset{\overset{\displaystyle CH_3}{|}}{C}CO_2\overset{\overset{\displaystyle CH_3}{|}}{C}HCH_3$$

Transesterification. A wide variety of methacrylate esters are prepared by transesterification of methyl methacrylate. The methanol is removed from the reaction mixture as the methyl methacrylate azeotrope.

$$\underset{MMA}{CH_2=\overset{\overset{\displaystyle CH_3}{|}}{C}CO_2CH_3} + ROH \longrightarrow CH_2=\overset{\overset{\displaystyle CH_3}{|}}{C}CO_2R + CH_3OH$$

This and the use of excess methyl methacrylate provides high conversion of the alcohol to the ester. The catalyst that is used depends on the nature of the alcohol and the kinetics of the reaction. Effective catalysts include dialkyl tin oxides (101), sodium phenoxide (102), zinc chloride (103), and magnesium methoxide (104).

Transesterification is an excellent route for batch processing of functionalized alkyl methacrylates. A mixture of excess methyl methacrylate and the alcohol, catalyst, and polymerization inhibitor are heated at the reflux temperature and the methanol–methyl methacrylate azeotrope is removed overhead as it forms. The reaction continues until the desired conversion is achieved. Excess methyl methacrylate is removed by distillation and the product is isolated by distillation or by another purification method. Yields generally are excellent (see Esterification).

Addition of Methacrylic Acid to Alkylene Oxides. Hydroxyalkyl methacrylates are prepared by the addition of methacrylic acid to the epoxide ring of alkylene oxides to form hydroxyalkyl methacrylates, ie, β-hydroxyethyl and hydroxypropyl methacrylate. Oxides other than ethylene oxide give isomeric products; commercial hydroxypropyl methacrylate has about a 2-to-1 ratio of 2-hydroxypropyl methacrylate to 1-methyl-2-hydroxyethyl methacrylate.

$$\underset{\text{MAA}}{CH_2\!\!=\!\!CCO_2H} + \underset{O}{\overset{CH_2\!-\!CH_2}{\diagdown\diagup}} \longrightarrow \underset{[4664\text{-}49\text{-}7]}{CH_2\!\!=\!\!CCO_2CH_2CH_2OH}$$

Effective catalysts include tertiary amines (104), anion-exchange resins (105), ferric chloride (106), and lithium methacrylate [13234-23-6] (107).

Monomer Recovery from the Polymer. Crude methyl methacrylate is obtained by thermal cracking of poly(methyl methacrylate). Scrap or off-grade polymer is heated to >300°C and the monomer yield is >80%. The crude ester is purified in a standard methyl methacrylate recovery system. The thermal conversion behavior of methacrylate polymers to monomers is unusual. Polyacrylates give a complex mixture when pyrolyzed (see also Recycling, plastics).

Economic Aspects

Table 10 shows the growth of domestic methyl methacrylate production from 1967 through 1977; the decreased production in 1975 resulted from a general slowdown in the economy in that year. Prices of methacrylates have risen in response to inflation, and the price histories of some selected methacrylates are given in Table 11.

Table 10. Domestic Production of Methyl Methacrylate, 1967–1977 [a]

Year	Production, 10^3 t
1967	137
1969	196
1971	210
1973	302
1975	247
1977	338

[a] Ref. 108.

Table 11. Maximum Bulk List Prices of Selected Methacrylates (1974–1979), $/kg

Year	Methacrylic acid	Methyl methacrylate	Ethyl methacrylate	n-Butyl methacrylate	Isobutyl methacrylate	n-Lauryl methacrylate
1974	0.95	0.73	1.41	1.06	1.06	
1975	1.04	0.79	1.43	1.08	1.08	
1976	1.10	0.84	1.65	1.15	1.12	2.03
1977	1.15	0.90	1.65	1.19	1.19	2.31
1978	1.21	0.95	1.72	1.26	1.25	2.43
1979	1.34	1.01	1.82	1.41	1.40	2.80

The U.S. producers of methyl methacrylate are

	Capacity, 10^3 metric tons
Cy/Ro Industries, Fortier, La.	36
DuPont, Belle, W.Va.	54
DuPont, Memphis, Tenn.	109
Rohm and Haas, Houston, Tex.	299

Analysis

Chromatographic techniques are the primary tools used for analysis of the methacrylate monomers, eg, gas chromatography (glc), high performance liquid chromatography (hplc), and gel-permeation chromatography (gpc) (9,109). Gas chromatography is used extensively for assaying monomer purity and for analysis of process streams in monomer manufacture (110–111). Methods have been developed for determination of methacrylates in mixtures with other monomers, in solvents including water, in biological fluids, and in polymers (112–119). High performance liquid chromatography (hplc) is replacing the colorimetric techniques employed for analysis of inhibitors. Concentrations of hydroquinone (HQ) and the methyl ether of hydroquinone (MEHQ) below 1 ppm can be detected in methacrylate monomers by hplc (120). Colorimetric methods are simpler and work well for systems containing a single inhibitor, but they fail when mixtures of inhibitors or other colored species are present (9,109). High performance liquid chromatography works well when specificity and selectivity are required, eg, in analyses involving two or more inhibitors or complex mixtures, eg, process streams. Other applications for hplc include measurement of residual monomer in polymer (121). The major drawback to the widespread application of hplc is the high cost of the instrumentation.

Polymer content in methacrylate monomers can be determined qualitatively by a haze test, which normally is sufficient for the pure materials that are available commercially (9,109). Gel-permeation chromatography is the preferred method for more quantitative analyses, eg, in the detection of oligomers and soluble polymer and in the analysis of process streams.

Thin-layer chromatography (tlc) (122–124), polarography (125–126), and spectrometry (127) are used for solution measurements. Methacrylates in air have been analyzed by tlc (128), polarography (129), and colorimetry (130). Polarography has been used for determination of any residual monomer in the polymer (131–133). A variety of spectroscopic techniques, eg, nmr (134–135), ir (136), and Raman spectroscopy (137) also have been used, particularly for analysis of surgical cements and dental restorative resins (see Analytical methods).

Table 12. Toxicity of Methacrylic Monomers

Monomer	Acute oral toxicity in rats, LD$_{50}$, g/kg	Inhalation toxicity in rats — LC$_{50}$, mg/L[a]	Inhalation toxicity in rats — ppm[a]	Time, h	Acute percutaneous toxicity in rabbits, LD$_{50}$, g/kg	Acute intraperitoneal toxicity in rats, LD$_{50}$, g/kg[b]	Remarks
methyl methacrylate	7.9[c]	29	7093	4[d]	>7.5[e]	1.1	dermal sensitizer[f]
ethyl methacrylate	13.5[c]	12.4–15.0	2700–3200[g]	8[c]	>9.1[c]	1.112	dermal sensitizer[f]
n-butyl methacrylate	20.3[h]	>5	>860	8[h,i]	10.2[h]	1.49	dermal sensitizer[f]
isodecyl methacrylate	>5.0[g]		saturated air	1[j]	>3.0	3.238	mild skin irritant; very mild eye irritant
lauryl methacrylate	>5.0		saturated air	1[j,k,l]	>3.0	21.6	slight skin irritant
isobornyl methacrylate	2.4		saturated air	1[j]	>3.0	3.64	slight skin irritant
1,3-butylene dimethylacrylate	14.3		saturated air	1[i]			no irritation of skin
trimethylolpropane trimethacrylate	>21.2		saturated air	8[i,m]	17.0[m]	2.89	moderate to severe skin irritant
glycidyl methacrylate	0.826		saturated air	i,l,m	0.483	0.528	no irritation of skin
2-hydroxyethyl methacrylate	4.23		saturated air	i,l	>3.0		no irritation of skin
hydroxypropyl methacrylate	11.2		saturated air		>3.0		slight skin irritation
t-butylaminoethyl methacrylate	1.517		saturated air	1[n]	>3.0	0.174	skin and eye irritant
dimethylaminoethyl methacrylate	1.55		saturated air	1[j,l,m,n]	>3.0	0.097	skin and eye irritant
methacrylic acid	2.20		100 / >1875 / 300	1[j,l,m,n]	<2.0[n]	0.048	corrosive to eye; strong skin irritant

[a] Concentration in air.
[b] Ref. 138.
[c] Ref. 139.
[d] Ref. 140.
[e] Ref. 141.
[f] Ref. 142.
[g] Lethal at top of range.
[h] Ref. 143.
[i] Not lethal.
[j] No toxic effects.
[k] Ref. 144.
[l] 16 or 20 6-h exposures.
[m] Ref. 145.
[n] Killed all animals.

369

Comprehensive reviews of the traditional analytical methods, involving assay by unsaturation or carboxylate functionality, and measurements of physical properties have been published (9,109).

Storage and Handling

Methacrylic acid and its esters can polymerize violently and, therefore, must be stabilized to be handled safely. Inhibitors are added to ensure safety and product quality and their minimum levels have been established (10). Once stabilized, the methacrylates can be handled as flammable liquids.

The monomethyl ether of hydroquinone frequently is used as inhibitor and low inhibitor grades of the most common methacrylate monomers are available commercially. MEHQ levels of 10–15 ppm generally are adequate for the lower alkyl methacrylates, eg, the methyl and ethyl esters, but substantially higher concentrations are needed for methacrylic acid, the higher alkyl esters, and the methacrylates containing additional functional groups. Hydroquinone (HQ) is another common inhibitor for these materials. Typical inhibitor levels for the commercially available methacrylates are included in Table 1.

The effectiveness of phenolic inhibitors, eg, MEHQ and HQ, depends on the presence of oxygen. Monomers inhibited with these materials must be stored in air rather than in an inert atmosphere. Contamination must be avoided; eg, moisture may cause rust-initiated polymerization.

Temperatures during storage must be kept low to minimize formation of peroxides and other oxidation products. In particular, the aminomethacrylates, eg, t-butyl-aminoethyl methacrylate and dimethylaminoethyl methacrylate, are more susceptible to polymerization than the alkyl methacrylates, and storage temperatures for these materials generally should not exceed 25°C in order to avoid the slow build-up of soluble oligomers and polymers. Storage temperatures below 30°C are recommended for the polyfunctional methacrylates.

Methacrylic acid has a high freezing point, ie, 14°C, and the inhibitor may not distribute uniformly between phases if frozen acid is partially thawed. If the liquid phase is inadequately inhibited, it could polymerize and initiate violent polymerization. Provisions should be made to maintain the acid in the liquid phase. If freezing should occur, melting should take place at room temperature, ie, 25°C; material should not be withdrawn until it is entirely thawed and well mixed.

In general, the methacrylate monomers should not be stored for longer than one year. Shorter storage times are recommended for the aminomethacrylates, ie, three months, and the polyfunctional methacrylates, ie, six months. Many of these compounds are sensitive to uv light and should, therefore, be stored in the dark. The methacrylic esters may be stored in mild steel, stainless steel, or aluminum. However, methacrylic acid is corrosive to many metals, including mild steel, and should be stored in glass, stainless steel, aluminum, or polyethylene-lined equipment.

For most applications, inhibitors do not have to be removed. The low inhibitor grades of the methacrylate monomers are suitable for the manufacture of polymers without pretreatment. Methyl methacrylate can be shipped uninhibited but only under carefully controlled conditions of temperature and shipping time. If removal of inhibitor is necessary, it should be done by adsorption with ion-exchange resins or other adsorbents. Phenolic inhibitors may be removed from esters with an alkaline brine

wash. Undried, washed monomers may be used in emulsion processes. Washed uninhibited monomers are less stable and should be used promptly.

Health and Safety Factors

The low flash points of the lower methacrylates create a fire hazard, and these
compounds can form explosive mixtures with air; explosive limits in air at 25°C and
101 kPa (1 atm) are 2.1–12.5% for methyl methacrylate and 1.8% to saturation for ethyl
methacrylate.

Of the methacrylate esters in commercial use, those listed in Table 12 have been
examined for toxicological activity. Methacrylates exhibit low to moderate acute
toxicity, produce slight to moderate skin and eye irritation, and generally are considered to be sensitizers with cross-sensitization potential (142). In addition, depending
upon volatility, these materials also may be sensory irritants.

Methacrylic acid is more acutely toxic and more severely irritating to the eyes
and skin than are the esters of alcohols with nonreactive groups. The toxicity of
methacrylates with reactive functional groups, eg, amino or glycidyl, on the alcohol
portion of the ester should be considered individually because the biological activity
of a particular methacrylate monomer may be determined by the reactive functional
group rather than by the methacryl portion of the molecule.

Methyl methacrylate is the only methacrylate for which a workplace exposure
limit, ie, 100 ppm TWA, has been established by ACGIH (146). Methyl methacrylate
at inhalation exposures of 400 ppm for two years with rats and 18 months with ham-
sters is noncarcinogenic and, at inhalation exposures of 1000 ppm in rats and 400 ppm
in mice, MMA is nonteratogenic (147–148). Additional studies have shown that MMA
is rapidly and extensively degraded, mainly to CO_2, and is expired from the lungs with
a small amount of methyl malonate (149).

Table 13.   Methacrylate Consumption, 1969–1978, Thousands of Metric Tons[a]

Year	Cast sheet	Other grades	Coatings	Molding/extrusion powder	Miscellaneous	*Total*
1969	77		29	38	15	*159*
1972	100		42	46	21	*209*
1975	49	23	44	57	20	*193*
1978	63	35	42	60	25	*225*

[a] Ref. 150.

Table 14.   Uses of Methacrylates (1969–1978), Thousands of Metric Tons[a]

Year	Appliances	Transportation	Building Glazing and skylights	Building Lighting fixtures	Building Panels and siding	Plumbing/bath fixtures
1969	1.8		8.2	19.5	4.1	3.2
1972	1.9	16.8	18.2	24.9	5.8	4.2
1975	1.9	15.6	29	13	3	12
1978	2.7	29.9	29.0	9.1	5.4	10.0

[a] Ref. 150.

Uses

The most important products prepared from methacrylate compounds are polymers. The physical properties of polymers of methacrylate esters can be modified for different uses by suitable changes in the alkyl portion of the molecule, eg, by varying carbon chain length or by introducing a functional group. The largest use of methacrylates is in plastics. Methacrylic sheet is used in glazing, signs and displays, and building materials. Molding powder is used largely for automobile parts. The consumption pattern of methacrylates is indicated in Table 13, and some of the major markets for methacrylates are listed in Table 14 (see Methacrylic polymers).

BIBLIOGRAPHY

"Methacrylic Compounds" in *ECT* 2nd ed., Vol. 13, pp. 331–363, by F. J. Glavis and J. F. Woodman, Rohm and Haas Company.

1. E. Frankland and B. F. Duppa, *Ann.* **136,** 1 (1865).
2. Fr. Englehorn, *Berichte* **13,** 433 (1880); L. Balbiano and A. Testa, *Berichte* **13,** 1984 (1880).
3. E. H. Riddle, *Monomeric Acrylic Esters*, Reinhold Publishing Corp., New York, 1954.
4. M. Kitahara, *Yuki Gosei Kagaku* **33,** 710 (1975).
5. M. Salkind, E. H. Riddle, and R. W. Keefer, *Ind. Eng. Chem.* **51,** 1232 (1959).
6. U.S. Pat. 2,100,933 (Nov. 30, 1977), H. A. Bruson (to Rohm and Haas Co.).
7. Brit. Pat. 405,699 (Feb. 12, 1934), J. W. C. Crawford (to Imperial Chemical Industries).
8. L. S. Luskin, "Acrylic Acid, Methacrylic Acid and the Related Esters" in E. C. Leonard, ed., *High Polymers, Vinyl and Diene Monomers*, Vol. 24, Part I, Wiley Interscience, New York, 1970.
9. L. S. Luskin, "Acrylic and Methacrylic Acids and Esters" in F. D. Snell and C. L. Hilton, eds., *Encyclopedia of Industrial Chemical Analysis*, Vol. 4, John Wiley & Sons, Inc., New York, 1967, p. 181.
10. *Storage and Handling of Acrylic and Methacrylic Esters and Acids*, Bulletin CM-17, Rohm and Haas Company, Philadelphia, Pa., 1975.
11. *Acrylic and Methacrylic Monomers, Typical Properties and Specifications*, Bulletin CM-16, Rohm and Haas Company, Philadelphia, Pa., 1978.
12. J. Kooi, *Rec. Trav. Chim.* **68,** 34 (1949).
13. A. F. Frolov, M. M. Yarovikova, B. F. Ustavshchikov, and N. S. Nikitina, *Isv. Vysshikh Uchebn Zavedenii Khim. Khim. Teknol.* **8,** 570 (1965).
14. A. F. Frolov, M. A. Loginova, and B. F. Ustavshchikov, *Zh. Obshch. Khim.* **36,** 180 (1966).
15. A. F. Frolov, M. A. Loginova, A. V. Saprykina, and A. B. Kondakova, *Zh. Fiz. Khim.* **36,** 2282 (1962).
16. E. A. Frolova, B. F. Ustavshchikov, and S. Yu Pavlov, *Zh. Prikl. Khim.* (*Leningrad*) **48,** 900 (1975).
17. G. A. Chubarov, S. M. Danov, and R. V. Efremov, *Zh. Prikl. Khim.* (*Leningrad*) **47,** 2130 (1974).
18. E. A. Frolova, B. F. Ustavshchikov, and S. Yu. Pavlov, *Zh. Fiz. Khim.* **48,** 1865 (1974).
19. G. A. Chubarov, S. M. Danov, and R. V. Efremov, *Zh. Fiz. Khim.* **48,** 1047 (1974).
20. S. Yu. Pavlov, A. B. Kirnos, S. P. Pavlova, and V. E. Lazaryants, *Zh. Prikl. Khim.* (*Leningrad*) **45,** 614 (1972).
21. L. S. Luskin, "Basic Monomers: Vinyl Pyridines and Aminoalkyl Acrylates and Methacrylates" in R. H. Yocum and E. B. Nyquist, eds., *Functional Monomers, Their Preparation, Polymerization and Application*, Vol. 2, Marcel Dekker, New York, 1974.
22. I. L. Knunyants, E. M. Rokhlin, U. Utebaev, and E. I. Mysov, *Izv. Akad. Nauk SSSR Ser. Khim.* **1,** 137 (1976).
23. F. Texier and J. Bourgois, *Bull. Soc. Chim. Fr.* **3–4**(Pt. 2), 487 (1976).
24. S. I. Mekhtiev and Yu D. Safarov, *Azerb. Khim. Zh.* **5–6,** 33 (1973).
25. S. I. Mekhtiev, Yu D. Safarov, and Ch. S. Aslanov, *Azerb. Khim. Zh.* **2,** 57 (1974).
26. I. N. Azerbaev, B. M. Butin, and Yu. G. Bosyakov, *Zh. Obshch. Khim.* **45,** 1730 (1975).
27. U.S. Pat. 3,642,843 (Feb. 15, 1972), J. W. Nemec (to Rohm and Haas Co.).
28. R. H. Stevens, *Diss. Abstr. Int. B* **33,** 3007 (1972).

29. M. Makosza and I. Gajos, *Bull. Acad. Pol. Sci. Ser. Sci. Chim.* **20**(1), 33 (1972).

30. E. G. Mamedov, I. M. Akhmedov, and M. M. Guseinov, *Dokl. Akad. Nauk Az. SSR* **33**(1), 34 (1977).

31. F. A. Martirosyan, R. M. Ispiryan, and V. O. Babayan, *Arm. Khim. Zh.* **24**, 884 (1971).

32. L. V. Khokhlova, M. A. Korshunov, V. A. Bukhareva, O. P. Yablonskii, and L. F. Lapuka, *Zh. Org. Khim.* **6**, 2404 (1970).

33. Ger. Offen. 2,217,623 (Oct. 26, 1972), A. Oshima, K. Tsuboshima, and N. Takahashi (to Kojin Ltd.).

34. A. I. Konovalov, G. I. Kamasheva, and M. P. Loskutov, *Zh. Org. Khim.* **9**, 2048 (1973).

35. E. G. Mamedov, I. M. Akhmedov, M. M. Guseinov, S. R. Piloyan, and E. I. Klabunovskii, *Izv. Akad. Nauk SSSR Ser. Khim.* **4**, 883 (1978).

36. B. A. Rudenko, S. Ya. Metlyaeva, and V. F. Kucherov, *Izv. Akad. Nauk SSSR, Ser. Khim.* **11**, 2500 (1972).

37. J. Bourgois, F. Tonnard, and F. Texier, *Bull. Soc. Chim. Fr.* **11–12**, 2025 (1976).

38. H. Kaemmerer and J. Pachta, *Makromol. Chem.* **178**, 1659 (1977).

39. H. Kaemmerer and V. Steiner, *Makromol. Chem. Suppl.* **1**, 133 (1975).

40. G. A. Tolstikov, O. S. Vostrikova, and U. M. Dzhemilev, *Izv. Akad. Nauk SSSR Ser. Khim.* **6**, 1459 (1975).

41. O. S. Vostrikova, *Khim. Vysokomol. Soedin. Nef-ekhim.*, 29 (1975).

42. E. Verkuijlen, R. J. Dirks, and C. Boehouwer, *Rec. Trav. Chim. Pays Bas* **96**(11), 86 (1977).

43. A. A. Stotskii, V. V. Kirichenko, and L. I. Bagal, *Zh. Org. Khim.* **9**(12), 2464 (1973).

44. USSR Pat. 361,169 (Dec. 7, 1972), F. Ibragimov, T. I. Usmanov, T. G. Gafurov, and D. Mukhamadaliev.

45. U.S. Pat. 3,855,364 (Dec. 17, 1974), R. Steckler (to Alcolac, Inc.).

46. Ger. Offen. 2,754,709 (June 15, 1978), M. D. A. Piteau and J. P. G. Senet (to Societe Nationale des Poudres et Explosifs).

47. Jpn. Kokai 74 49,922 (May 15, 1974), H. Yoshikawa and S. Nakahara (to Nippon Kayaku Co., Ltd.).

48. Jpn. Kokai 77 31,017 (Mar. 9, 1977), T. Nishikaji, K. Watenebe, S. Sawayama, and S. Matsumoto (to Mitsubishi Chemical Industries Co., Ltd.).

49. Jpn. Kokai 77 27,713 (Mar. 2, 1977), Y. Kametani (to Nitto Chemical Industry Co., Ltd.).

50. Jpn. Kokai 77 27,726 (Mar. 2, 1977), Y. Kametani and I. Yasou (to Nitto Chemical Industry Co., Ltd.).

51. Jpn. Kokai 73 26,370 (Aug. 9, 1973), Y. Iwakura, K. Uno, Y. Nakahara, and M. Makita (to Asahi Chemical Industry Co., Ltd.).

52. Jpn. Kokai 79 05,918 (Jan. 17, 1979), R. Mizuguchi, A. Takahashi, S. Ishikura, and S. Uenaka (to Nippon Paint Co., Ltd.).

53. A. G. Makhsumov, Sh. U. Abdullaev, and U. A. Abidov, *Khim. Atsetilena Tekhnol. Karbida Kal'tsiya*, 93 (1972).

54. USSR Pat. 303,868 (Dec. 25, 1975), A. G. Makhsumov, A. Safaev, Sh. U. Abdullaev, A. Abdurakhimov and P. Il'khamdzhanov (to Central-Asian Scientific-Research Institute of the Petroleum Refining Industry).

55. E. F. Schoenbrunn and J. H. Gardner, *J. Am. Chem. Soc.* **82**, 4905 (1960); U.S. Pat. 2,847,453 (Aug. 12, 1958), J. H. Gardner and T. T. Steadman (to Escambia Chemical Corporation); U.S. Pat. 2,847,454 (Aug. 12, 1958), W. R. Boehme and J. Nichols (to Ethicon, Inc.); U.S. Pat. 2,847,465 (Aug. 12, 1958), N. C. Robertson and T. R. Steadman (to Escambia Chemical Corporation); Brit. Pat. 855,880 (Dec. 7, 1960), J. H. Gardner and co-workers (to Escambia Chemical Corporation); Brit. Pat. 852,664 (Oct. 26, 1960), (to Escambia Chemical Corporation); *Chem. Week* **84**, 91 (June 20, 1959).

56. Belg. Pat. 619,895 (Jan. 7, 1963), (to Nissan Chemical Industries, Ltd.).

57. *Chem. Eng.* **85**, 25 (July 3, 1978).

58. *Hydrocarbon Process.* **57**, 108 (Nov. 1978).

59. *Chem. Eng. News*, 13 (Jan. 14, 1980).

60. W. G. Grot, *J. Org. Chem.* **30**, 515 (1965).

61. M. Salkind, E. H. Riddle, and R. W. Keefer, *Ind. Eng. Chem.* **51**, 1328 (1959).

62. Ger. Offen. 2,430,470 (Feb. 6, 1974), H. Matsuzawa and co-workers (to Mitsubishi Rayon Company, Ltd.).

63. A. Clementi and co-workers, *Hydrocarbon Process.* **58**, 109 (Dec. 1979).

64. Jpn. Kokai 76 63 112 (June 1, 1976), Y. Sikakura, F. Sakai, and H. Shimizu (to Nippon Kayaka Company, Ltd.); Belg. Pat. 850,425 (July 14, 1977), M. Ogawa and T. Kojima (to Nippon Kayaka Company, Ltd.).

65. U.S. Pat. 4,171,328 (Oct. 16, 1979), S. Umemura, K. Ohdan, K. Suzuki, and T. Hisayuki (to Ube Industries Ltd.); Jpn. Kokai 77 51,317 (Apr. 25, 1977), S. Umemura, K. Odan, K. Suzuki, F. Adachi, and T. Hogami (to Ube Industries, Ltd.).

66. Jpn. Kokai 74 117,425 (Nov. 9, 1974), Y. Nonaka, T. Arima, and K. Kihara (to Toyo Soda Mfg. Co. Ltd.); Jpn. Kokai 77 95,609 (Aug. 11, 1977), M. Sakamoto, T. Nagahama, S. Nakamura, and K. Kihara (to Toyo Soda Mfg. Co. Ltd.).

67. Jpn. Kokai 76 101,911 (Sept. 8, 1976), Y. Odam, K. Uchida, and T. Morimoto (to Asahi Glass Co., Ltd.); Jpn. Kokai 78 90,214 (Aug. 8, 1978), Y. Oda, K. Uchida, and T. Morimoto (to Asahi Glass Co., Ltd.).

68. Jpn. Kokai 79 3,008 (Jan. 11, 1979), N. Ando, M. Arawaka, and T. Harada (to Japan Synthetic Rubber Co., Ltd.).

69. Ger. Offen. 2,427,670 (Jan. 2, 1975), H. Ishii, H. Matsuzawa, M. Kobayashi, and K. Yamada (to Mitsubishi Rayon Co., Ltd.); Jpn. Kokai 71 115,414 (Oct. 12, 1976), H. Sonobe, M. Kato, H. Matsuzawam, H. Ishii, and M. Kobayashi (to Mitsubishi Rayon Co., Ltd.).

70. U.S. Pat. 3,928,462 (Dec. 23, 1975), T. Shiraishi, S. Kiwishiwada, S. Shimizu, S. Honmara, H. Ichihasi, and Y. Nagaoka (to Sumitomo Chemical Company, Ltd.); Jpn. Kokai 78 109,889 (Sept. 26, 1978), H. Ichihashi, Y. Kikuzono, Y. Nagaoka, and M. Otsuki (to Sumitomo Chemical Co., Ltd.).

71. U.S. Pat. 4,155,938 (May 22, 1979), H. Yamamoto, N. Yoneyama, and S. Akiyama (to Nippon Zeon Co., Ltd.); Ger. Offen. 2,454,587 (May 28, 1975), S. Akiyama and H. Yamamoto (to Nippon Zeon Co., Ltd.).

72. U.S. Pat. 4,162,234 (July 24, 1979), R. K. Grasselli, S. D. Suresh, and H. F. Hardman (to Standard Oil Company (Ohio)); U.S. Pat. 4,136,110 (Jan. 23, 1979), J. F. White and J. R. Rege (to Standard Oil Company (Ohio)).

73. U.S. Pat. Pending (filed Oct. 12, 1979), L. S. Kirch and W. J. Kennelly (to Rohm and Haas Company).

74. *Jpn. Chem. Daily*, (Nov. 20, 1979).

75. Ger. Offen. 2,848,369 (May 23, 1979), N. Tamura, Y. Fukuoka, S. Yamamatsu, Y. Suzuki, R. Mitsui, and T. Ibuki (to Asahi Chemical Industry Company, Ltd.).

76. Ger. Offen. 2,356,089 (May 15, 1975), S. I. Mekhtiev and co-workers; Fr. Demande 2,233,315 (Jan. 10, 1975), E. C. Milberger, S. R. Dolhyj, and J. F. White (to Standard Oil Company); Ger. Offen. 2,146,466 (Mar. 23, 1972), T. Shiraishi and co-workers (to Sumitomo Chemical Company, Ltd.); M. A. Dalin, S. I. Mekhtiev, and T. I. Rasulbekova, *Dokl. Akad. Nauk. SSSR* **154,** 854 (1964).

77. S. I. Mekhtiev and co-workers, *Dokl. Akad. Nauk SSSR* **190,** 180 (1970).

78. Fr. Pat. 2,040,793 (Jan. 22, 1971), S. I. Mekhtiev, M. A. Dalin, and A. Guseinova.

79. U.S. Pat. 3,878,238 (Apr. 15, 1975), S. I. Mekhtiev and co-workers.

80. Ger. Offen. 2,324,765 (Nov. 29, 1973), D. E. Morris (to Monsanto).

81. U.S. Pat. 3,247,248 (Apr. 19, 1966), V. A. Sims and J. F. Vitcha (to Cumberland Chemical Corporation).

82. Ger. Offen. 2,349,054 (May 9, 1974), F. W. Schlaefer (to Rohm and Haas Company).

83. U.S. Pat. 3,701,798 (Oct. 31, 1972), T. C. Snapp, Jr., A. E. Blood, and J. H. Hagemeyer, Jr. (to Eastman Kodak Company); Ger. Offen. 1,804,469 (May 14, 1969), J. H. Hagemeyer, Jr., A. E. Blood, and T. C. Snapp, Jr. (to Eastman Kodak Company).

84. Ger. Offen. 2,702,187 (July 27, 1978), W. Gaenzler, K. Kabs, and G. Schroeder (to Roehm G.m.b.H.).

85. Ger. Offen. 2,615,887 (Oct. 20, 1977), G. Fouquet and co-workers (to BASF A.-G.).

86. Y. Takezaki and co-workers, *Bull. Jpn. Pet. Inst.* **8,** 31 (1966).

87. U.S. Pat. 2,975,199 (Mar. 14, 1961), B. S. Friedman and S. M. Cotton (to Sinclair Refining Company); *J. Am. Chem. Soc.* **26,** 3751 (1961).

88. S. Pawlenko, *Chem. Ing. Tech.* **40,** 52 (1968).

89. Ger. Offen., (May 26, 1977), W. Gruber and G. Schroeder (to Roehm G.m.b.H.).

90. Jpn. Kokai 77 105,113 (Sept. 3, 1977), T. Onoda and M. Ohtake (to Mitsubishi Chemical Industries, Ltd.); Jpn. Kokai 73 76,812 (Oct. 16, 1973), T. Otaki and co-workers (to Mitsubishi Chemical Industries Company, Ltd.).

91. Jpn. Kokai 77 105,112 (Sept. 3, 1977), T. Onoda and M. Ohtake (to Mitsubishi Chemical Industries Company, Ltd.); Jpn. Kokai 77 108,918 (Sept. 12, 1977), T. Onoda and M. Ohtake (to Mitsubishi Chemical Industries Company, Ltd.).

92. J. Tsuji and co-workers, *J. Am. Chem. Soc.* **86,** 4350, 4851 (1964); P. J. Steinwand, *J. Org. Chem.* **37,** 2034 (1972); U.S. Pat. 3,621,054 (Nov. 16, 1971), K. L. Olivier.

93. U.S. Pat. 2,930,801 (Mar. 29, 1960), A. E. Montagna and L. V. McQuillen (to Union Carbide Corporation).

94. U.S. Pat. 4,180,681 (Dec. 25, 1979), J. J. Leonard and H. Shalit (to Atlantic Richfield Company).

95. U.S. Pat. 4,097,523 (June 27, 1978), J.-L. Kao and J. J. Leonard (to Atlantic Richfield Company).

96. U.S. Pat. 4,147,884 (Apr. 3, 1979), M. N. Sheng and J.-L. Kao (to Atlantic Richfield Company).

97. U.S. Pat. 3,470,239 (Sept. 30, 1969), J. L. Russel (to Halcon International, Inc.).

98. Jpn. Pat. 73 44,212 (June 26, 1973), (to National Lead).

99. Jpn. Kokai 74 43,934 (Apr. 25, 1974), S. Komura, K. Nakamura, and M. Takenaka (to Otsuka Chemical Company, Ltd.); Ger. Offen. 2,555,901 (June 24, 1976), T. Onoda and M. Ohtake (to Mitsubishi Chemical Industries Company, Ltd.).

100. U.S. Pat. 3,087,962 (Apr. 30, 1963), N. M. Bortnick (to Rohm and Haas Company).

101. Ger. Offen. 2,752,109 (June 1, 1978), Y. Kametani and Y. Ino (to Nitto Chemical Industry Company, Ltd.).

102. U.S. Pat. 4,059,617 (Nov. 22, 1977), T. Foster and T. S. Dawson (to American Cyanamid Company).

103. Ger. Offen. 2,311,007 (Sept. 13, 1977), S. Fukuchi, N. Shimizu, and T. Ohara (to Japan Catalytic Chemical Industry Company, Ltd.).

104. Jpn. Pat. 73 21,929 (July 2, 1973), G. Honma and S. Yoshinaka (to Mitsubishi Gas-Chemical Company, Inc.).

105. Brit. Pat. 1,120,301 (July 17, 1968), E. J. Percy and J. A. Wickings (to Distillers Company, Ltd.).

106. Ger. Offen. 2,027,444 (Dec. 17, 1970), M. Murayama and K. Abe (to Japan Gas-Chemical Company).

107. Jpn. Pat. 72 51,328 (Dec. 23, 1972), Y. Tanizaki and Y. Kubo (to Japan Oil and Fats Company, Ltd.).

108. *Synthesis Organic Chemicals, U.S. Production and Sales*, U.S. International Trade Commission.

109. *Analytical Methods for the Acrylic Monomers*, Bulletin CM-18, Rohm and Haas Company, Philadelphia, Pa., 1976.

110. A. Szocik and M. Linkiewicz, *Chem. Anal. (Warsaw)* **22,** 353 (1977).

111. I. Sociu, M. Tomescu, and V. Parausanu, *Rev. Chim. (Bucharest)* **24,** 641 (1973).

112. S. Z. Rosina, A. S. Turaev, R. A. Muminova, and Sh. Nadzhimutdinov, *Uzb. Khim. Zh.* (1), 13 (1978).

113. A. Bechtel, H.-G. Willert, and H.-A. Frech, *Chromatographia* **6,** 226 (1973).

114. A. M. Rijke, R. A. Johnson, and E. R. Oser, *J. Biomed. Mater. Res.* **11,** 211 (1977).

115. J. Ruhnke, A. Eggert, and H. Huland, *Chromatographia* **7,** 55 (1974).

116. S. L. Mel'nikova, V. T. Tishchenko, and V. V. Sazonenko, *Lakokras. Mater. Ikh. Primen.* (4), 56 (1977).

117. R. Holtmann and J. R. Souren, *Kuntstoffe,* **67,** 776 (1977).

118. S. Z. Rosina, A. S. Turaev, R. A. Muminova, and A. Kh. Usmanov, *Metody Anal. Kontrolya Kach. Prod. Khim. Prom-sti.* (2), 8 (1978).

119. M. S. Klescheva, V. I. Nesterova, and T. V. Smirnova, *Plast. Massy* (11), 67 (1972).

120. G. A. Pasteur, *Anal. Chem.* **49,** 363 (1977).

121. K. Aitzetmueller and W. R. Eckert, *J. Chromatogr.* **155,** 203 (1978).

122. U. A. T. Brinkman, T. A. M. Van Schaik, G. DeVries, and A. C. DeVisser, *ACS Symp. Ser.* **31,** 105 (1976).

123. A. I. Subbotina, L. V. Voronina, M. R. Leonov, and I. G. Sumin, *Tr. Khim. Khim. Tekhnol.* **3,** 150 (1973).

124. R. S. Ekhina, *Gig. Sanit.* **2,** 78 (1977).

125. Yu. P. Ponomarev, O. V. Meshkova, L. I. Ryzhova, and V. N. Dmitrieva, *Fiz.-Khim. Metody Ochistki Anal. Stochnykh Vod. Prom. Predpr.*, 99 (1974).

126. O. V. Meshkova and V. N. Dmitrieva, *Zavod. Lab.* **40,** 28 (1974).

127. Z. A. Krotova, *Zavod. Lab.* **40,** 263 (1974).

128. N. I. Kaznina, *Gig. Sanit.* **37,** 63 (1972).

129. V. N. Dmitrieva and L. A. Kotok, *Gig. Sanit.* (4), 59 (1976); V. N. Dmitrieva, L. A. Kotok, and N. S. Stepanova, *Gig. Sanit.* (12), 73 (1976).

130. E. Matyas and J. Burian, *Munkavedelem* **20,** 5 (1974).

131. A. I. Kalinin, V. N. Komleva, and L. N. Mol'kova, *Metody Anal. Kontrolya Kach. Prod. Khim Prom-sti.* **11,** 62 (1977).

132. L. A. Mirkind, N. V. Zaitseva, and V. S. Sporykhina, *Lakokrasoch Mater. Ikh. Primen.* (1), 46 (1972).

133. V. S. Sporykhina, A. I. Zavadskaya, and L. A. Mirkind, *Lakokrasoch Mater. Ikh. Primen.* (5), 61 (1977).

134. E. Asmussen, *Acta Odontol. Scand.* **33,** 129 (1975).

135. E. B. Sheinin, W. R. Benson, and W. L. Brannon, *J. Pharm. Sci.* **65,** 280 (1966).

136. I. E. Ruyter and S. A. Svendsen, *Acta Odontol. Scand.* **36,** 75 (1978).

137. D. N. Waters and E. D. Schmid, *Proceedings International Conference Raman Spectroscopy, 5th,* Hans Ferdinand Schulz Verlag, Freiburg, FRG, 1976, p. 500.

138. W. H. Lawrence, G. E. Bass, W. P. Purcell, and J. Autian, *J. Dent. Res.* **51,** 526.

139. W. Deichman, *J. Ind. Hyg. Toxicol.* **23,** 343 (1941).

140. M. F. Tansy, W. E. Landin, and F. M. Kendall, "LC$_{50}$ values for rats acutely exposed to methyl methacrylate monomer vapors," *Meeting abstract, International Association for Dental Research and American Association for Dental Research,* 1979.

141. W. H. Lawrence, M. Malik, and J. Autian, *J. Biomed. Mater. Res.* **8,** 11 (1974).

142. C. W. Chung and A. L. Giles, *J. Invest. Dermat.* **68,** 187 (1977).

143. H. F. Smyth, Jr., C. P. Carpenter, C. S. Weil, H. C. Pozzani, J. A. Striegel, and J. S. Nycum, *Am. Ind. Hyg. Assoc. J.* **30,** 470 (1969).

144. J. C. Gage, *Brit. J. Ind. Med.* **27,** 1 (1970).

145. C. P. Carpenter, C. S. Weil, H. F. Smyth, Jr., *Toxicol. Appl. Pharmacol.* **28,** 313 (1974).

146. *Threshold limits values for chemical substances in workroom air adopted by ACGIH for 1979,* American Conference of Governmental Industrial Hygienists, Cincinnati, Oh.

147. *BIBRA Bull.,* 124 (Apr. 1980).

148. J. M. Smith and co-workers, "Methyl Methacrylate: Subchronic, chronic and oncogenic inhalation safety evaluation studies," *Abstracts of the Eighteenth Annual Meeting of the Society of Toxicology,* New Orleans, La., 1979.

149. H. Bratt and D. E. Hathaway, *Br. J. Cancer* **36,** 114 (1977).

150. *Mod. Plast.* **47–56,** (1970–1979).

JOSEPH W. NEMEC
LAWRENCE S. KIRCH
Rohm and Haas Company

METHACRYLIC POLYMERS

Methacrylic ester monomers have the generic formula $CH_2{=}C(CH_3)COOR$, and it is the nature of the R group that generally determines the properties of the corresponding polymers. Methacrylates differ from acrylates in that the α hydrogen of the acrylate is replaced by a methyl group (see Acrylic ester polymers). It is the α methyl group of the polymethacrylate that imparts the stability, hardness, and stiffness of methacrylic polymers. The methacrylate monomers are unusually versatile building blocks since they are moderate-to-high boiling liquids that readily polymerize or copolymerize with a variety of other monomers. All of the methacrylates readily copolymerize with each other and with the acrylate series; thus, extreme ranges of properties can be built into the polymer.

Hard methacrylates easily copolymerize with soft acrylates to form polymers having a wide range of hardness; thus, polymers that are designed to fit specific application requirements can be tailored readily from these versatile monomers. The properties of the polymers can be varied to form extremely tacky adhesives (qv), rubbers, tough plastics, and hard powders. Although higher in cost than many other common monomers, the methacrylates' unique stability characteristics, ease of use, efficiency, and the associated high quality products more than compensate for their expense.

Development of the methacrylate esters has followed that of the acrylates. One of the first applications for acrylates was as the laminating resin for safety glass (1). The acrylates were copolymerized with ethyl methacrylate and the copolymer yielded optimum properties and led to the first commercial production of ethyl methacrylate in 1933 (2). In the development of safety-glass interlayering, methyl methacrylate was polymerized in a mold formed by two sheets of glass. Following polymerization, the two glass plates were separated from the polymer that had formed revealing a thin sheet of poly(methyl methacrylate) (see Laminated materials, glass). Methyl methacrylate soon became the most important member of the acrylate–methacrylate series. Research on cast sheets of poly(methyl methacrylate) was carried out in the 1930s in the United States by the Rohm and Haas Company and E. I. du Pont de Nemours & Company, in Germany by Rohm and Haas A.-G., and in England by Imperial Chemicals Industries Ltd. (1). By 1936, methyl methacrylate was used to produce an organic glass by a cast polymerization process (2). The demand from the plastics industry for a methacrylate molding powder to match the performance, clarity, and strength of the methacrylate sheet resulted in the introduction of molding powders for both compression and injection molding in 1938 (2). During the 1930s, the demand for methyl methacrylate increased because of its use in adjusting the hardness of acrylate copolymers produced by emulsion and solution polymerization.

The uniqueness of methyl methacrylate as a plastic component accounts for the large production volume of methyl methacrylate compared to the combined volumes of all of the other methacrylates. Methacrylate polymers are used in lubricating oil additives, surface coatings, impregnates, adhesives, binders, sealers, and floor polishes. The total production capacity of methyl methacrylate in the United States is estimated at 500,000 metric tons per year (3).

Physical Properties

The nature of the alcohol group of the methacrylate monomer unit within the polymer chain and the molecular weight of the polymer largely determine the physical and chemical properties of methacrylate ester polymers. Typically, the mechanical properties of methacrylate polymers improve with increasing molecular weight; however, beyond a critical molecular weight, eg, ca 100,000–200,000 for amorphous polymers, the increase in properties is slight and tends to level off asymptotically. The atactic (random) configuration is the result from polymerization unless there are special circumstances favoring stereospecific addition to form isotactic (cis) or syndiotactic (trans) chains (see Olefin polymers; Polymerization). Physical and chemical properties of the ordered or stereospecific polymer can differ significantly from the atactic type.

Glass-Transition Temperature. The glass-transition temperature T_g, or second-order transition temperature, is the temperature at which a polymer changes from a rubbery to a brittle material, ie, to a glassy state. The T_g is influenced primarily by the nature of the alcohol group and, to a lesser extent, by the stereoregularity of the backbone chain (see Tables 1–2).

The T_g indicates the mechanical properties that may be expected of polymers within an approximate temperature range. Methods used to determine the glass-transition temperature and the values for numerous polymers have been reported (5–6,15,17).

Low transition-temperature acrylates frequently are used to plasticize harder methacrylates. Examples of common commercial acrylates are given in Table 3 (see Acrylic ester polymers). Once polymerized, the acrylic monomer cannot be volatilized or extracted from the polymer. This is contrary to a fugitive plasticizer which is not incorporated into the polymer backbone. Fugitive plasticizers (qv) also lower the transition temperature of a polymer; however, they can be lost through volatilization, by diffusion into contacting surfaces, or by extraction (18).

Mechanical and Thermal. Substitution on the main chain of the methyl group in the methacrylates for the α hydrogen of the acrylates results in restricted freedom of rotation and motion of the polymer backbone yielding harder polymers of higher tensile strength and lower elongation than the acrylate counterparts (see Table 4).

At room temperature, the first member of the linear aliphatic series, poly(methyl methacrylate), is a hard, fairly rigid material which can be sawed, carved, or worked on a lathe. When heated above its T_g, poly(methyl methacrylate) is a tough, pliable, and extensible material that is easily bent or formed into complex shapes and that can be molded or extruded. Primarily, it is the length, flexibility, bulk, and degree of crystallinity of the side chain that determines the T_g of conventional polymethacrylates of a given molecular weight. With increasing length of the side chain, there is a decrease in the T_g of the polymer. By comparison, poly(n-hexyl methacrylate) ($T_g = -5°C$) is rubberlike and it is an order of magnitude softer and more extensible. With a further increase in the ester side-chain length, the T_g continues to decrease through poly(lauryl methacrylate) ($T_g = -65°C$). Further increase in side-chain length allows crystallization of the side chains; further lowering of the T_g of the homopolymer is masked by the onset of crystallinity in the side chain which restricts chain motion and a more brittle polymer results (20), eg, poly(octadecyl methacrylate) has a brittle temperature of 36°C (5).

Table 1. Physical Properties of Conventional Methacrylate Polymers

Homopolymer	CAS Registry No.	$T_g{}^a$, °C	Density[b] (at 20°C), g/cm³	Solubility parameter[c], $(J/cm^3)^{1/2}$	Refractive index, $n_D^{20}{}^d$
poly(methyl methacrylate)	[9011-14-7]	105	1.190	18.6	1.490
poly(ethyl methacrylate)	[9003-42-3]	65	1.119	18.3	1.485
poly(n-propyl methacrylate)	[25609-74-9]	35	1.085	18.0	1.484
poly(isopropyl methacrylate)	[26655-94-7]	81	1.033		1.552
poly(n-butyl methacrylate)	[9003-63-8]	20	1.055	17.8	1.483
poly(sec-butyl methacrylate)	[29356-88-5]	60	1.052		1.480
poly(isobutyl methacrylate)	[9011-15-8]	53	1.045	16.8	1.477
poly(t-butyl methacrylate)	[25213-39-2]	107	1.022	17.0	1.4638
poly(n-hexyl methacrylate)	[25087-17-6]	−5	1.007²⁵	17.6	1.4813
poly(2-ethylbutyl methacrylate)	[25087-19-8]	11	1.040		
poly(n-octyl methacrylate)	[25087-18-7]	−20	0.971²⁵	17.2	
poly(2-ethylhexyl methacrylate)	[25719-51-1]	−10			
poly(n-decyl methacrylate)	[29320-53-4]	−60			
poly(lauryl methacrylate)	[25719-52-2]	−65	0.929	16.8	1.474
poly(tetradecyl methacrylate)	[30525-99-6]	−72			1.47463
poly(hexadecyl methacrylate)	[25986-80-5]				1.47503
poly(octadecyl methacrylate)	[25639-21-8]	−100		16.0	
poly(stearyl methacrylate)	[9086-85-5]			16.0	
poly(cyclohexyl methacrylate)	[25768-50-7]	104	1.100		1.50645
poly(isobornyl methacrylate)	[28854-39-9]	170 (110)	1.06	16.6	1.5000
poly(phenyl methacrylate)	[25189-01-9]	110	1.21		1.571
poly(benzyl methacrylate)	[25085-83-01]	54	1.179	20.3	1.5680
poly(ethylthioethyl methacrylate)	[27273-87-0]	−20			1.5300
poly(3,3,5-trimethylcyclohexyl methacrylate)	[75673-26-6]	79			1.485

[a] Refs. 4–14.
[b] Refs. 5, 14–15.
[c] Refs. 5, 14–15. To convert $(J/cm^3)^{1/2}$ to $(cal/in.^3)^{1/2}$, multiply by 2.0.
[d] Refs. 10, 14–15.

Table 2. Glass-Transition Temperatures (T_g) of Atactic, Syndiotactic, and Isotactic Polymethacrylate Esters, °C[a]

Methacrylate	Atactic	Syndiotactic	Isotactic
methyl	105	115	45
ethyl	65	66	12
n-propyl	35		
isopropyl	81	85	27
n-butyl	20		−24
isobutyl	53		8
sec-butyl	60		
t-butyl	115	111	7
cyclohexyl	104		110

[a] Refs. 15–16.

Increasing the bulkiness of the side chain also restricts the motion of polymer chains past each other, as evidenced by an increase in the T_g within a homologous series of side-chain isomers. As shown in Table 1, the member of the butyl series with the

Table 3.   Physical Properties of Common Soft Acrylate Polymers

Homopolymer	CAS Registry No.	$T_g{}^a$, °C	Density (at 20°C)^b, g/cm^3	Solubility parameter, (J/cm^3)$^{1/2}$ ^{b,c}	Refractive index, n_D^{25} ^{b}
poly(methyl acrylate)	[9003-21-8]	6	1.22	20.7	1.479
poly(ethyl acrylate)	[9003-32-1]	−21	1.12	19.13	1.464
poly(butyl acrylate)	[9003-49-0]	−55	1.0	18.0	1.474
poly(2-ethylhexyl acrylate)	[9003-77-4]	−85			1.4650

^{a} Refs. 5, 13, 15.
^{b} Refs. 13, 15.
^{c} To convert (J/cm^3)$^{1/2}$ to (cal/in^3)$^{1/2}$, multiply by 2.0.

Table 4.   Comparison of Mechanical Properties of Polyacrylates and Polymethacrylates^a

Ester	Tensile strength, MPa^b		Elongation at break, %	
	Polymethacrylate	Polyacrylate	Polymethacrylate	Polyacrylate
methyl	62	6.9	4	750
ethyl	34	0.2	7	1800
n-butyl	6.9	0.02	230	2000

^{a} Ref. 19.
^{b} To convert MPa to psi, multiply by 145.

bulkiest side chain, poly(t-butyl methacrylate), has a T_g which is almost identical to that of poly(methyl methacrylate); whereas the member of the butyl series with the most flexible side chain, poly(n-butyl methacrylate) has a T_g of 20°C. Further increase in the rigidity and bulk of the side chain further increases the T_g, eg, poly(isobornyl methacrylate) has a T_g of 170°C.

Electromagnetic Spectrum.   Poly(methyl methacrylate) transmits light almost perfectly, ie, 92% compared to 92.3% theoretical, at 360–1000 nm. (The wavelengths of visible light are ca 400–700 nm.) At thicknesses of >2.54 cm, poly(methyl methacrylate) absorbs virtually no visible light (21). Beyond 2800 nm, essentially all ir radiation is absorbed (22). Commercial grades of poly(methyl methacrylate) often contain uv radiation absorbers which block light in the 290–350-nm range (see Uv absorbers). The absorber, acting as a sunscreen, protects the user from sunburn and the polymer against long-term degradation from light (23–24). Poly(methyl methacrylate)'s transparency to x-ray radiation is about the same as that of human flesh or water. Sheets are opaque to α particles and, above 6.35 mm, the polymer is essentially opaque to β-ray radiation; poly(methyl methacrylate) also is used as a transparent neutron barrier (25). Most formulations of colorless sheet have high transmittance within standard broadcast and television waves as well as within most radar bands (25).

Poly(methyl methacrylate) also can serve as a conduit for light (24) (see Fiber optics). A ray of light that moves within the polymer and encounters an air interface at or greater than the critical angle, ie, 42.2° from the normal, is totally reflected back through the plastic at an equal and opposite angle and is trapped within the plastic until it emerges at the opposite end. Image transmission around bends and corners may be accomplished by reflecting periscopically from facets (24). The ability to mold poly(methyl methacrylate) precisely allows complex optical designs to be manufactured easily and inexpensively. Many items, eg, magnifiers, reducers, camera lenses, prisms, and complex reflex lenses, which are used widely in automotive taillights, are made from poly(methyl methacrylate) (see also Filters, optical).

Electrical. Electrical properties of poly(methyl methacrylate) are listed in Table 5. The surface resistivity of poly(methyl methacrylate) is higher than that of most plastic materials. Weathering and moisture affect poly(methyl methacrylate) to a minor degree. High resistance and nontracking characteristics have resulted in its use in high voltage applications. Its excellent weather resistance coupled with its electrical properties have promoted the use of poly(methyl methacrylate) for outdoor electrical applications.

Solution. If a polymer is soluble in a solvent, it is soluble in all proportions and, as the solvent evaporates from the solution, no phase separation or precipitation occurs (5,26–32). With increased evaporation the solution viscosity continues to increase until a coherent film forms. As with the acrylates, the solubility of a methacrylate polymer is affected by the nature of the alcohol-derived side group. Methacrylate polymers which contain short ester side chains are relatively polar and are soluble in polar solvents, eg, ketones, esters, ethers, and alcohols. With increasing length of an alkyl side chain, the polymer becomes less polar and dissolves in relatively nonpolar solvents, eg, aromatic or aliphatic hydrocarbons.

The solubility parameter, which is a measure of the cohesive energy of a molecule, can be used to predict the solubility of a polymer in a solvent (see Table 3). A polymer is soluble in a solvent or tends to be compatible with another polymer if the solubility parameters and polarities of both are similar.

Table 5. Electrical Properties of 6.35-mm Thick Poly(Methyl Methacrylate) Sheet[a]

	Typical values	ASTM method
dielectric strength (short-term test), $V/\mu m^b$	>16.9–20.9	D 149
dielectric constant, V/mm^c		D 150
at 60 Hz	142–154	
at 1,000 Hz	130–134	
at 1,000,000 Hz	83–118	
power factor, V/cm^d		D 150
at 60 Hz	20–24	
at 1,000 Hz	16–20	
at 1,000,000 Hz	8–12	
loss factor, V/cm^d		D 150
at 60 Hz	75–87	
at 1,000 Hz	51–59	
at 1,000,000 Hz	24–31	
arc resistance	no tracking	D 495
volume resistivity Ω/cm	1×10^{14}–6×10^{17}	D 257
surface resistivity, $\Omega/square$	1×10^{17}–2×10^{18}	D 257

[a] Ref. 26.

[b] To convert $V/\mu m$ to V/mil, multiply by 25.4.

[c] To convert V/mm to V/mil, divide by 39.4.

[d] To convert V/cm to V/mil, divide by 394.

The dilute solution properties of isotactic poly(methyl methacrylate) differ from those of a conventional, free-radical-initiated polymer. Isotactic poly(methyl methacrylate) remains more expanded in a thermodynamically good solvent, eg, acetone, than the atactic polymer (33). Mixtures of dimethylformamide solutions of isotactic poly(methyl methacrylate) with syndiotactic poly(methyl methacrylate) result in the rapid formation of a gel which has a specific melting point. This phenomenon is used as a basis for a relative measure of the average syndiotactic sequence length (34).

Chemical Properties

Methacrylate polymers have a greater hydrolytic resistance to both acidic and alkaline hydrolysis than do acrylate polymers; both are far more stable than poly(vinyl acetate) and vinyl acetate copolymers (35–37). There is a marked difference in the chemical reactivity among the noncrystallizable and crystallizable forms of poly(methyl methacrylate) relative to alkaline and acidic hydrolysis. Conventional, ie, free-radical bulk-polymerized, and syndiotactic polymers hydrolyze relatively slowly compared with the isotactic type (38). Polymer configuration is unchanged by hydrolysis (39). Complete hydrolysis of nearly pure syndiotactic poly(methyl methacrylate) in sulfuric acid and reesterification with diazomethane yields a polymer with an nmr spectrum that is identical to that of the original polymers. There also is a high proportion of syndiotactic configuration present in conventional poly(methyl methacrylate). The high proportion of syndiotactic structure in conventional poly(methyl methacrylate) contributes to its high degree of chemical inertness. The chemical resistance of poly-(methyl methacrylate) is summarized in Table 6 (21). The combination of chemical resistance and excellent light stability of poly(methyl methacrylate) compared to two other transparent plastics is illustrated in Table 7 (40).

Methacrylates readily depolymerize with high conversion, ie, 95%, at >300°C (1,41). Methyl methacrylate can be obtained in high yield from mixed polymer materials, ie, scrap (42).

Table 6.   Chemical Resistance of Poly(Methyl Methacrylate)[a]

Not affected by	Attacked by
most inorganic solutions	lower esters, eg, ethyl acetate, isopropyl acetate
mineral oils	aromatic hydrocarbons, eg, benzene, toluene, xylene
animal oils	phenols, eg, cresol, carbolic acid
low concentrations of alcohols	aryl halides, eg, chlorobenzene, bromobenzene
paraffins	aliphatic acids, eg, butyric acid, acetic acid
olefins	alkyl polyhalides, eg, ethylene dichloride, methylene chloride
amines	high concentrations of alcohols, eg, methanol, ethanol, isopropanol
alkyl monohalides	high concentrations of alkalies and oxidizing agents
aliphatic hydrocarbons	
higher esters, ie, >10 carbon atoms	

[a] Ref. 21.

Table 7. Relative Outdoor Stability of Poly(Methyl Methacrylate) [a]

Material	Initial light transmittance, %	Light transmittance after 3 yr outdoor exposure, %	Initial haze, %	Haze after 3 yr outdoor exposure, %
poly(methyl methacrylate)	92	92	1	2
polycarbonate (PC)	85	82	3	·19
cellulose acetate butyrate	89	68	3	70

$$\text{polycarbonate (PC)} = \left[\begin{array}{c} O \\ \parallel \\ OCO- \end{array} \bigcirc \begin{array}{c} CH_3 \\ | \\ C \\ | \\ CH_3 \end{array} \bigcirc \right]$$

[a] Ref. 40.

Manufacture and Processing

The preparation and properties of methacrylic esters is discussed in detail in the article Methacrylic acid and derivatives.

Methacrylate polymers are produced commercially in the form of sheets, rods, tubes, and blocks as well as pellets, solutions, latices, and beads. The physical characteristics of the neat polymers range from soft and flexible to hard and rigid, depending upon the monomer composition and method of polymerization. Most of the commercially available polymers are prepared by free-radical processes involving initiation, propagation, chain transfer, and termination. The type and level of initiators and chain-transfer agents depends upon the polymerization method (see Initiators). In homopolymerizations, the chain-propagation step involves the head-to-tail growth of the polymeric free radical by attack on the double bond of the monomer (eq. 1) (43). Chain termination can occur either by combination or by disproportionation, depending on the process (44–47).

Methacrylate polymerizations are accompanied by liberation of heat and a decrease in volume, as shown in Table 8. Both of these factors strongly influence most manufacturing processes. Excess heat must be dissipated to avoid uncontrolled exothermic polymerizations. Volume changes are particularly important in sheet casting where the mold must compensate for the decreased volume. In general, the percent shrinkage decreases as the size of the alcohol substituent increases; on a molar basis, the shrinkage is constant (49).

Methacrylate polymerizations are markedly inhibited by oxygen; therefore, considerable care is taken to exclude air during polymerization. The inhibitory effect of oxygen results from the copolymerization of oxygen and methacrylate monomer

$$R{-}\left[\begin{array}{c}R\\|\\CH_2C\\|\\CO_2R\end{array}\right]_n \begin{array}{c}R\\|\\CH_2C\cdot\\|\\CO_2R\end{array} \; + \; \begin{array}{c}R\\|\\CH_2{=}C\\|\\CO_2R\end{array} \; \longrightarrow \; R{-}\left[\begin{array}{c}R\\|\\CH_2C\\|\\CO_2R\end{array}\right]_{n+1} \begin{array}{c}R\\|\\CH_2C\cdot\\|\\CO_2R\end{array} \qquad (1)$$

Table 8.   Heats of Polymerization and Volume Shrinkages of Methacrylate Monomers[a]

Monomer	Heat of polymerization[b]		Shrinkage, vol %
	kJ/mol	kJ/g	
methyl methacrylate	57.7	0.576	21.0
ethyl methacrylate	57.7	0.506	18.2
butyl methacrylate	59.4	0.418	14.9

[a] Ref. 48.
[b] To convert kJ to kcal, divide by 4.184.

with the consequential formation of an alternating copolymer (eqs. 2–3) (50–51). If oxygen is present, the reaction expressed by equation 2 is extremely rapid, but the addition of the terminal peroxy radical to the monomer (eq. 3) is much slower than with normal monomer addition (eq. 1). The overall effect, other than the change in the polymer composition, is a decrease in the rate of monomer reaction, in the kinetic chain length, and in the polymer molecular weight (52).

$$
R\left[\begin{array}{c} R \\ | \\ CH_2COO \\ | \\ CO_2R \end{array}\right]_n \begin{array}{c} R \\ | \\ CH_2C\cdot \\ | \\ CO_2R \end{array} \;+\; O_2 \;\longrightarrow\; R\left[\begin{array}{c} R \\ | \\ CH_2COO \\ | \\ CO_2R \end{array}\right]_n \begin{array}{c} R \\ | \\ CH_2COO\cdot \\ | \\ CO_2R \end{array} \tag{2}
$$

$$
R\left[\begin{array}{c} R \\ | \\ CH_2COO \\ | \\ CO_2R \end{array}\right]_n \begin{array}{c} R \\ | \\ CH_2COO\cdot \\ | \\ CO_2R \end{array} \;+\; \begin{array}{c} R \\ | \\ CH_2{=}C \\ | \\ CO_2R \end{array} \;\longrightarrow\; R\left[\begin{array}{c} R \\ | \\ CH_2COO \\ | \\ CO_2R \end{array}\right]_{n+1} \begin{array}{c} R \\ | \\ CH_2C\cdot \\ | \\ CO_2R \end{array} \tag{3}
$$

In methacrylate bulk polymerizations, an autoacceleration begins at 20–50% conversion (53–54). At this point, there is a corresponding increase in the molecular weight of the polymer that is being formed. This acceleration, which continues to high conversion, is known as the Trommsdorff effect and is attributed to the increase in viscosity of the mixture to such an extent that the diffusion rate, and therefore, the termination reaction of the growing radicals is reduced. The reduced termination rate ultimately results in a polymerization rate that is limited only by the diffusion rate of the monomer.

The relative ease of copolymerization of several methacrylate monomers with other common monomers is presented in Table 9. Values above 25 indicate that co-polymers form easily. Lower values often can be compensated for by the proportions of comonomers or by the method of their introduction into the polymerization reaction (55). Copolymers that are prepared with acrylate monomers are most useful. The acrylates soften and improve the flexibility of the normally hard, brittle methacrylates without compromising durability.

Specially functionalized monomers often are copolymerized at low levels with methacrylic monomers in order to modify or improve the properties of the polymer either directly or by providing sites for further cross-linking. These special functional groups include amino, carboxyl, hydroxyl, epoxy, and N-hydroxymethylamide groups.

Table 9. Relative Ease of Copolymer Formation for 1:1 Ratios of Methacrylic Esters and Other Monomers[a], $r_{smaller}{}^{b}/r_{larger}{}^{b} \cdot 100$

| | Monomer 1 | | |
Monomer 2	Methyl methacrylate	Ethyl methacrylate	Butyl methacrylate
acrylonitrile	12	24	19
butadiene	33	8.6	8.1
ethyl acrylate	18	20	18
styrene	88	60	90
vinyl chloride	<0.1	0.6	0.5
vinylidene chloride	9.5	14	16
vinyl acetate	0.1	0.9	0.2

[a] Ref. 48.
[b] r = reactivity ratio.

Bulk Polymerization. The bulk polymerization of monomeric methacrylic esters is used principally to manufacture sheets, rods, tubes, and molding powder (56–61). In terms of volume, sheet casting is the most important process and probably accounts for nearly half of the methacrylate monomer produced. The monomer that is used usually is methyl methacrylate and, often, with minor portions of other monomers, uv absorbers, pigments or dyes, and other additives. The polymerization is a free-radical process that is initiated by heat and radical initiators, eg, peroxides and azo compounds.

Sheets are produced in widths of ca one to several hundred meters, and in thicknesses from 0.16 to 15 cm. Poly(methyl methacrylate) sheet is produced primarily by three methods, ie, batch cell, continuous, and extruded sheet. The batch-cell method is the most common because it is inherently simple and is easily adapted for manufacturing a wide variety of grades, colors, and sizes (62–64). The continuous process is economically more advantageous and is used primarily to manufacture the widely used grades of thinner-gauge sheet. Sheets also are produced by extruding poly(methyl methacrylate) molding powders. For gauges below 0.3 cm and for certain speciality applications, extruded sheets are preferred because of their lower cost.

In the batch-cell method, each sheet is cast in a mold that is assembled from two sheets of highly polished plate glass separated by a flexible spacer. The glass plates are clamped against the spacer, and the tension on the clamping force is such that shrinkage during polymerization is accommodated by movement of the glass faces closer together. The mold is filled from one open corner with exact amounts of monomer or monomer–polymer syrup, initiator, and other additives and then is closed and heated for curing. A programmed temperature cycle is used to effect a reasonable cure time without initiating an uncontrolled exothermic reaction which would result in monomer vaporization and bubble formation. The heat of polymerization is dissipated using high velocity air ovens or water baths. Typical cure times are 10–12 h for thin-gauge sheets to several days for thick sheets. After curing, the molds are cooled and the glass plates are separated from the plastic sheet. The glass plates may be cleaned and reused. The plastic sheet may be used as is or it may be annealed by heating to 140–150°C for several hours to reduce strains and to achieve maximum dimensional stability. Low molecular weight pieces should be annealed at lower temperatures. As a protective measure, a masking paper or a plastic film often is applied to the surface of the finished sheets.

Many of the problems encountered in preparing poly(methyl methacrylate) sheets can be reduced by casting with a monomer–polymer syrup instead of with monomer alone. The use of a monomer–polymer syrup reduces shrinkage and the amount of heat to be dissipated, thus reducing the potential for a runaway reaction; shortens the induction period; and lowers the initiation temperature. The casting syrup typically contains about 20% polymer and is prepared either by partially polymerizing the monomer in a reactor under controlled conditions or by dissolving poly(methyl methacrylate) powder in the monomer.

A process for the continuous casting of poly(methyl methacrylate) sheet has been described (65). This or a similar process is probably used by all U.S. companies that produce poly(methyl methacrylate) sheet by a continuous process. A monomer–polymer syrup is introduced between two parallel stainless steel belts which travel through curing and annealing zones. In the curing zone, the belts are inclined to the horizontal such that the hydrostatic pressure of the syrup maintains the desired spacing between the belts. Spring-mounted rollers on the top side of the upper belt maintain external pressure. The edges between the belts are sealed by a compressible and flexible gasket which travels at the same speed as the belts. Heating in the curing zone is accomplished using a hot-water spray and, in the annealing zone, using electric heaters or hot air. The residence times and temperatures in the curing and annealing zones are ca 45 min and 70°C and ca 10 min and 110°C, respectively. The resulting sheet is covered with a masking paper before being cut into the desired lengths.

One of the most useful properties of poly(methyl methacrylate) plastic sheet is its formability. Because it is thermoplastic, it becomes soft and pliable when heated and can be formed into almost any shape and, as the material cools, it stiffens and retains the shape to which it has been formed (65–68) (see Film and sheeting materials).

Poly(methyl methacrylate) molding powders are available in numerous grades and sizes. They are used by the plastics industry primarily in extrusion- or injection-molding processes. The most common form of molding powder is 3.2-mm pellets, although fine beads and granulated powders also are sold. The pellets are prepared by extruding either melted poly(methyl methacrylate) that has been prepared by a separate bulk polymerization or by extruding a monomer–polymer syrup. The poly(methyl methacrylate) is extruded as a small-diameter rod which is cut into 3.2 mm long segments. Details on molding procedures are available from the manufacturers' technical bulletins (69–70).

Solution Polymerization. The solution polymerization of methacrylic monomers to form soluble methacrylic polymers or copolymers is an important commercial process for the preparation of polymers for use as coatings, adhesives, impregnates, and laminates. Typically, the polymerization is accomplished batchwise by adding the monomer to an organic solvent in the presence of a soluble initiator which usually is a peroxide or an azo compound. This method is suitable for the preparation of polymers having molecular weights of ca 2,000–200,000; higher molecular weight polymers are not only difficult to prepare by solution polymerization but have viscosities that are too high for easy handling. For a review of the quantitative aspects of the solution polymerization of methyl methacrylate, see ref. 71.

In general, the polymethacrylate esters of the lower alcohols are soluble in aromatic hydrocarbons, esters, ketones, and chlorohydrocarbons. They are insoluble or only slightly soluble in aliphatic hydrocarbons, ethers, and alcohols. The polymethacrylate esters of the higher alcohols, ie, $\geq C_4$, are soluble in aliphatic hydrocarbons. Cost, toxicity, flammability, volatility, and chain-transfer activity are the primary considerations in the selection of a suitable solvent. Chain transfer to solvent is an important factor in the determination of the molecular weight of polymers prepared by this method. The chain-transfer constants for poly(methyl methacrylate) radicals with various solvents are listed in Table 10. The transfer constants with thiols and certain chlorohydrocarbons are very large, and these materials generally are used at low levels, eg, 0.1–5%, as chain-transfer agents for the regulation of molecular weight.

The type of initiator that is utilized for a solution polymerization depends on several factors including the solubility of the initiator in the solvent, the rate of decomposition of the initiator, and the use of the polymeric product. The amount of initiator that is used may vary from a few hundredths to several percent of the monomer weight. With decreasing amounts of initiator, the molecular weight of the polymer increases as a result of initiating fewer polymer chains per unit weight of monomer. The initiator concentration often is used to control the molecular weight of the final product. Organic peroxides, hydroperoxides, and azo compounds are the preferred initiators for the preparations of most methacrylic solution polymers and copolymers (see Initiators).

A typical process for the preparation of a 45% methyl methacrylate–45% ethyl methacrylate–10% methacrylic acid polymer as a 50% solution in xylene is given in Table 11.

The solvent is charged to the reactor and is heated with stirring under nitrogen to 136°C. The monomer and initiator charges then are added uniformly over 1.75 h and the temperature is controlled at 136–143°C. After the monomer and initiator feeds are completed, the temperature is maintained at 136–143°C for an additional 5 h and then is cooled to 65°C and the dilution charge is added (75) (see Acrylic ester polymers).

Emulsion Polymerization. The emulsion polymerization of methacrylic esters to form aqueous dispersion polymers is used for the preparation of polymers suitable for applications in the paint, paper, textile, floor polish, and leather industries where they are used principally as coatings or binders. Copolymers of methyl methacrylate with either ethyl acrylate or butyl acrylate are the most common (see Latex technology).

Table 10. Chain-Transfer Constants for Poly(Methyl Methacrylate) Radicals with Common Solvents

Solvent	Chain-transfer constant, $\times 10^5$		Ref.
	at 60°C	at 80°C	
benzene	0.4	0.8	72
toluene	2.6	5.3	72
chlorobenzene	0.74	2.1	72–73
isopropanol	5.8	19.1	72–73
isobutanol	1.0	2.3	72
3-pentanone	8.3	17.3	72
chloroform	4.5	11.3	72
carbon tetrachloride	9.3	24.2	72
2-mercaptoethanol	66,000		74
thiophenol	270,000		74

Table 11.   Preparation of 45% Methyl Methacrylate–45% Ethyl Methacrylate–10% Methacrylic Acid Polymer

	wt %
Reactor charge	
xylene	30.00
Monomer charge	
methyl methacrylate	22.50
ethyl methacrylate	22.50
methacrylic acid	5.00
t-dodecyl mercaptan	0.50
Initiator charge	
di-*t*-butyl peroxide	1.00
xylene	1.83
Dilution charge	
ethanol	16.67

The monomer is polymerized in water in the presence of a surfactant and a water-soluble initiator. The product is an opaque gray or milky white dispersion of high molecular weight, low viscosity polymer at 30–60 wt % in water. The particle size of methacrylic/acrylic copolymer dispersions is ca 0.1–1.0 μm. Difficulties in agitation, heat transfer, and transfer of materials which often are encountered in the handling of viscous polymer solutions are greatly decreased with aqueous dispersions. In addition, the safety hazards and expense of flammable solvents are eliminated. The mechanistic aspects of emulsion polymerizations (76–79) and the quantitative aspects of the emulsion polymerization of methacrylic esters (80–82) have been described (see also Acrylic ester polymers).

Numerous recipes describe the emulsion polymerization of methacrylate homopolymers and copolymers (83–85). A typical process for the preparation of a 50% methyl methacrylate–49% butyl acrylate–1% methacrylic acid terpolymer as ca a 45% dispersion in water is given in Table 12. The monomer emulsion charge is prepared by adding the listed ingredients in given order and maintaining agitation. The reactor charge is heated with agitation under nitrogen to 85°C, then the initiator charge is

Table 12.   Preparation of a 50% Methyl Methacrylate–49% Butyl Acrylate–1% Methacrylic Acid Terpolymer

	wt %
Monomer emulsion charge	
deionized water	15.16
sodium lauryl sulfate	0.12
methyl methacrylate	25.00
butyl acrylate	24.50
methacrylic acid	.50
Initiator charge	
ammonium persulfate	0.26
Reactor charge	
deionized water	34.33
sodium lauryl sulfate	0.12
Total	*99.99*

added to the reactor and the monomer emulsion is fed uniformly for 2.5 h at 85°C. After the addition, the temperature is raised to 95°C to complete the conversion of the monomer. The product is filtered at room temperature.

Suspension Polymerization. Suspension polymerization yields polymethacrylates in the form of tiny beads which are used primarily as molding powders and ion-exchange resins (see Ion exchange). Suspension polymers that are prepared by molding powders generally are poly(methyl methacrylate) copolymers containing up to 20% acrylate for reduced brittleness and improved processibility also are common. Suspension polymers of poly(methyl methacrylate) which are copolymerized with an amino or acid functional monomer and with a di- or trivinyl monomer for cross-linking are useful as ion-exchange resins. Recipes for the suspension polymerization of methacrylate monomers have been reported (86–88).

In a suspension polymerization, the monomer is suspended in water as droplets and is stabilized by protective colloids or suspending agents. Polymerization is initiated by a monomer-soluble initiator and occurs within the monomer droplets. The water serves as both the dispersion medium and as a heat-transfer agent. Particle size is controlled primarily by the rate of agitation and the concentration and type of suspending aids. The polymer is obtained as small beads which are isolated by filtration or centrifugation. Because the polymerization occurs totally within the monomer droplets without any substantial transfer of materials between individual droplets or between the droplets and the aqueous phase, the course of the polymerization is expected to be similar to bulk polymerization (89–91).

A typical process for the preparation of a poly(methyl methacrylate) suspension polymer involves charging a mixture of 24.64 wt % of methyl methacrylate and 0.25 wt % of benzoyl peroxide to a rapidly stirred, 30°C solution of 0.42 wt % of disodium phosphate, 0.02 wt % of monosodium phosphate, and 0.74 wt % of Cyanomer A-370 (polyacrylamide resin) in 73.93 wt % of distilled water. The reaction mixture is heated under nitrogen to 75°C and is maintained at this temperature for 3 h. After being cooled to room temperature, the polymer beads are isolated by filtration, washed, and dried (102).

Graft Copolymerization. Graft copolymers are prepared by attaching one polymer as a branch to the chain of another polymer of a different composition. Graft copolymerization usually is accomplished by generating radical sites on the first polymer onto which the monomer of the second polymer is polymerized. The grafting may be accomplished in either bulk, solution, or dispersion systems. The presence of distinct but chemically bonded segments of two polymers often confers interesting and useful properties. Commercially, the most important methacrylate graft copolymers are the methyl methacrylate–acrylonitrile–butadiene–styrene (MABS) [9010-94-0] and methyl methacrylate–butadiene–styrene (MBS) [25053-09-2] polymers.

The MABS copolymers are prepared by dissolving or dispersing polybutadiene rubber in a methyl methacrylate–acrylonitrile–styrene monomer mixture. The graft copolymerization is accomplished by either a bulk or a suspension process (93–94). The final polymer is a two-phase system in which the continuous phase is a terpolymer of methyl methacrylate, acrylonitrile, and styrene grafted onto the dispersed polybutadiene phase. These polymers are utilized by the plastics industry in applications requiring a tough, transparent, highly impact resistant, and thermally formable material. Except for their transparency, the MABS polymers are similar to the opaque acrylonitrile–butadiene–styrene (ABS) plastics (see Acrylonitrile polymers). Transparency is accomplished by matching the refractive indexes of the two phases; this is the primary function of the methyl methacrylate.

MBS polymers are prepared by grafting methyl methacrylate and styrene onto a styrene–butadiene rubber in an emulsion process (95–96). The product is a two-phase polymer that is useful as an impact modifier for rigid poly(vinyl chloride). The preparation and properties of both MABS and MBS polymers has been reviewed (97).

Detailed investigations of the grafting of methyl methacrylate onto rubbers by a variety of methods include chemical (98–100), photochemical (101), radiation (101–102), and mastication processes (103). The grafting of methacrylates, principally methyl methacrylate, onto a variety of other substrates, eg, cellulose (104), poly(vinyl alcohol) (105), polyethylene (106), polystyrene (107), poly(vinyl chloride) (108), and other alkyl methacrylates (109) also is reported. The synthesis and characterization of numerous graft polymers, including methacrylate graft polymers, has been reviewed (110–111) (see Copolymers).

Ultraviolet-Initiated Curing. The uv curing of coatings, printing inks, and photoresists has become increasingly important in the last decade. Ultraviolet curing requires less energy than thermal curing and involves reduced solvent emissions. Ultraviolet-curable coatings and printing inks typically are composed of a pigment, monomer, polymer, photoinitiator, and inhibitor. The formulation is applied to a substrate as a thin film which is cured (polymerized) rapidly by exposure to uv radiation. The polymers often are unsaturated for co-curing with the monomer. The photoinitiator, which usually is present at about 2%, absorbs the radiation and initiates the free-radical polymerization. Because of their rapid cure rates, high boiling methacrylate monomers, particularly those which are multifunctional, have some use in uv-curable systems; although, in general, the faster-curing acrylates are preferred. Methacrylate functionality also is used as the co-curing unsaturation site on the polymer (112). Methacrylate functional polymers also are used in uv-curing photoresist applications (113) (see Photoreactive polymers). The state of the art in uv curing, including the use of methacrylate monomers, is reviewed in refs. 112, 114–115 (see Photochemical technology).

Anionic Polymerization. Until recently, anionic polymerizations were utilized principally for the preparation of stereoregular and block polymers. Stereoregular forms of several polymethacrylates have been prepared and characterized (116). The physical and chemical properties of the various forms often are different. Methacrylate block polymers formed by the living polymer method also are described in the literature (117). A recent review of the anionic polymerization of methacrylates is available (118). The anionic polymerization of methacrylates to form very low molecular weight polymers has been described (119). Because of their low molecular weight, ie, 500–5000, these polymers have unique properties and a wide variety of commercial applications (120–124). The polymers are prepared in bulk with alkoxide anion initiators in the presence of low levels of chain-regulating alcohols. The molecular weights and molecular weight distributions of these polymers are controlled by the ratio of alcohol to monomer. Methods for functionalizing and cross-linking also are described (122–124). The low solution viscosities of these polymers makes them particularly suited for use in high solids coatings (see also Elastomers, synthetic).

Analytical and Test Methods

Poly(methyl methacrylate) plastic sheet is manufactured in a wide variety of types including clear and colored transparent, clear and colored translucent, and semiopaque colored. Various surface textures also are produced. Grades with improved weatherability, ie, with added uv absorbers; mar resistance; crazing resistance; impact resistance; and flame resistance are available. Standard physical properties of poly(methyl methacrylate) sheet are listed in Table 13.

Methacrylate dispersion polymers are classified as either anionic, cationic, or nonionic, depending upon the type of surfactants and functional monomers used in the synthesis. Typical characterizations include composition, solids content, viscosity, pH, particle-size distribution, glass-transition temperature, and minimum film-forming temperature. Polymer compositions are most readily determined by spectroscopy, pyrolytic gas–liquid chromatography, and refractive-index measurements

Table 13. Typical Properties of Commercial Poly(Methyl Methacrylate) Sheet [a]

Property	Value	ASTM method
sp gr	1.19	D 792-66
refractive index	1.49	D 542-50 (1965)
tensile strength, specimen, MPa[b]		D 638-64T
maximum	72.4	
rupture	72.4	
modulus of elasticity,	3100	
elongation, maximum, %	4.9	
elongation, rupture, %	4.9	
flexural strength, span depth ratio 16, 2.5 mm/min, MPa[b]		D 790-66
maximum	110.3	
rupture	110.3	
modulus of elasticity	3103	
deflection, maximum, cm	1.52	
deflection, rupture, cm	1.52	
compressive strength (5.1 mm/min), MPa[b]		D 695-68T
maximum	124.1	
modulus of elasticity	3103	
compressive deformation under load[c], %		D 621-64
at 14 MPa[b] and 50°C for 24 h	0.2	
at 28 MPa[b] and 50°C for 24 h	0.5	
shear strength, MPa[b]	62.1	D 732-46 (1961)
impact strength		
charpy unnotched, J/cm²[d]	2.94	D 256-56 (1961)
Izod milled notch, J/m[d]	21	D 256-56 (1961)
Rockwell hardness	M-93[e]	
hot-forming temperature, K	417–455	
heat-distortion temperature		D 648-56 (1961)
−15.8°C/min at 1.8 MPa[b]	369[e]	
−15.8°C/min at 0.46 MPa[b]	380[e]	

[a] Ref. 125.

[b] To convert MPa to psi, multiply by 145.

[c] Sheet is conditioned for 48 h at 50°C.

[d] To convert J to cal, divide by 4.184; to convert J/m to ft·lbf/in, divide by 53.38; to convert J/cm² to ft·lbf/in², multiply by 4.758.

[e] Values change with thickness; the reported value is for 16.13 mm.

(126). Typically, the solids content of dispersions are determined by gravimetric procedures, viscosities with a Brookfield viscometer, and glass-transition temperatures by calorimetry. Minimum film-forming temperatures are determined by casting a film on a variable-temperature bar and observing the temperature at which a continuous film is obtained. Details on all of these methods are given in ref. 82. Traditionally, particle-size measurements are performed by light-scattering (127) or electron-microscope techniques (128).

Methacrylate solution polymers generally are characterized by their composition, solids contents, viscosity, molecular weight, glass-transition temperature, and solvent. Where applicable, details of these methods are similar to those described for dispersion polymers. Methods for estimating molecular weights by intrinsic viscosity are described (126,129) (see Rheological measurements).

Methacrylate suspension polymers are characterized by their composition and particle-size distribution. Screen analysis is the most common method for determining particle size. Melt-flow characteristics under various conditions of heat and pressure are important for polymers that are intended for extrusion or injection-molding applications. Suspension polymers that are prepared as ion-exchange resins are characterized by their ion-exchange capacity; density, both apparent and wet; solvent swelling; moisture-holding capacity; porosity; and salt-splitting characteristics (130).

Health and Safety Factors

In general, methacrylate polymers are nontoxic. Various methacrylate polymers are used in the packaging and handling of food, in dentures and dental fillings, and as medicine dispensers and contact lenses (qv). Some care must be exercised because certain low level additives, which are present in various types of methacrylate polymers, can be toxic, eg, some latex dispersions are mild skin and/or eye irritants. This toxicity usually is ascribed to the surfactants in the latex and not to the polymer. Most of the health and safety aspects of methacrylic polymers are involved with their manufacture and fabrication. During manufacture, considerable care is exercised to reduce the potential for violent polymerizations and to reduce exposure to flammable and potentially toxic monomers and solvents. Dust explosions ignited by static discharge are a recognized hazard which is encountered in the handling of poly(methyl methacrylate) powders or in the fabrication of poly(methyl methacrylate) plastic sheet. Methacrylic solution polymers, molding powders, and plastic sheet are treated as flammable materials. Most local building codes allow the use of poly(methyl methacrylate) in residential and commercial building with certain restrictions that are intended to reduce its flammability hazard. Dispersion polymers are nonflammable when in the water-dispersed state. However, the dried powder is flammable and subject to dust explosion hazards.

Economic Aspects

The bulk of the methyl methacrylate monomer (ca 368,000 metric tons produced in the U.S. in 1980) is used captively by the producers: Cy/Ro Industries, Fortier, La; DuPont, Belle, W. Va. and Memphis, Tenn.; and Rohm and Haas, Houston, Tex. The merchant market in the United States is estimated at $(68–91) \times 10^3$ t (3). About 40%

(estd) of methyl methacrylate has gone into cast sheet; and about 20% (estd) into surface coatings. Molding and extrusion compounds have accounted for ca 20% and miscellaneous uses for 20%. Latex accounted for 67% of the methyl methacrylate consumed in the production of surface coatings in 1974 and lacquers accounted for 30%. Polymethacrylates account for 73% of the current consumption of plastics for glazing and polycarbonates account for 20%; the remainder is split among polystyrene, poly(vinyl chloride) and other resins. Japan is the second largest producer of methyl methacrylate. In 1975, Japanese production was 113,400 t. Worldwide polymethacrylate production volumes are not available.

Uses

Glazing. The largest use for polymethacrylates is as a glazing, lighting, or decorative material. The uses are varied and depend upon the polymers' unique combination of light transmittance, light weight, dimensional stability, and formability, and their weather, chemical, impact, and bullet resistance; eg, they are used in bank-tellers' windows, police cars, panels around hockey rinks, storm doors, bath and shower enclosures, sliding partitions, and show cases. In architecture, the polymethacrylates are used for domes over stadia, pools, tennis courts, glazing archways between buildings, curved windows, and geodesic domes. Because of their impact resistance, polymethacrylates are used in the maintenance field for the glazing of windows in schools, housing-authority buildings, and factories. Colored and clear polymethacrylate sheets are used in decorative applications, eg, window mosaics, side glazing, color-coordinated structures, and for solar control in sunscreens by providing temperature and comfort regulation, and by reducing air-conditioning and heating expenses (131–133). Cast sheet and molding pellets have been used for lighting, as a sunscreen, and in signs (21,100,134–136).

Medicine. The polymethacrylates have been used for many years in the manufacture of dentures, teeth, denture bases, and filling materials (see Dental materials). In the orthodontics market, methacrylates have found acceptance as sealants, or pit and fissure resin sealants which are painted over teeth and act as a barrier to tooth decay (137). High binder strength dental and surgical bindings and filling agents which adhere to natural bone or teeth are prepared by mixing specific methacrylate esters and are brought to the appropriate viscosity by mixing with polymethacrylate powder (138). The dimensional behavior of curing bone-cement masses has been reported (139), as has the characterization of the microstructure of a cold-cured acrylic resin (140). Polymethacrylates are used to prepare both soft and hard contact lenses (141–144) (see Contact lenses).

Optics. The preparation of light-focusing plastic fibers by a heat-drawing process and of low attenuation optical fibers has been described (145–146) (see Fiber optics). Methods for the preparation of Fresnel lenses and a Fresnel lens film are reported in refs. 147–148. Compositions and methods for the industrial production of cast plastic eyeglass lenses are given in ref. 149.

Oil Additives. Long-chain polymethacrylates are used as additives to improve the performance of internal-combustion-engine lubricating oils and hydraulic fluids (150–151) (see Lubrication and lubricants; Hydraulic fluids). Long-chain polymethacrylates add little viscosity to oil when it is cold but increase the viscosity of the oil as the engine temperature rises. The proper balance of composition and molecular weight of the polymer allow oils of controlled and constant properties to be formulated (152). In addition to improving the viscosity index, sludge dispersancy and antioxidant qualities can be built into the polymer (153–154) (see Antioxidants and antiozonants). Polyethylene and polypropylene grafts onto linear, long-chain polymethacrylates, eg, *N,N*-dialkylaminoalkyl methacrylate copolymers, are claimed to function as a multipurpose lube-oil additive having viscosity-index-improving properties, pour-point-depressing properties, and detergent-dispersing properties (155–160).

Other.   A discussion of the use of polymethacrylates for the preparation of cultured-marble plastic sanitary fixtures and thermoformed bathtubs is given in refs. 161–162. High performance polymethacrylates are used in the toy industry to meet the requirements of the Child Protection and Toy Safety Act (163–164). Slate that is impregnated with methyl methacrylate monomer followed by polymerization of the monomer improves the water resistance of the slate (165). A method for producing anion-exchange fibers and films by emulsion-spinning techniques has been described (166) (see Ion exchange). The use of opaque and clear methacrylate sheet for the construction of recreational vehicles has been reported (167). Polymethacrylates also are used as electrical insulators (see Insulation, electric). Other comonomer uses are discussed in the article Acrylic ester polymers.

BIBLIOGRAPHY

1. E. H. Riddle, *Monomeric Acrylic Esters*, Reinhold Publishing Corp., New York, 1954.
2. M. Salkind, E. H. Riddle, and R. W. Keefer, *Ind. Eng. Chem.* **51**, 1232, 132B (1959).
3. *Chemical Profile*, Schnell Publishing Co., Inc., July 1, 1979.
4. J. W. C. Crawford, *J. Soc. Chem. Ind.* **68**, 201 (July 1949).
5. D. W. Van Krevelen, *Properties of Polymers*, Elsevier Publishing Co., Inc., Amsterdam, 1976.
6. H. Burrell, *Off. Dig. Fed. Soc. Paint Technol.* **34**, 131 (Feb. 1962).
7. S. Krause, J. J. Gormley, N. Roman, J. A. Shetter, and W. H. Watanabe, *J. Polym. Sci. Part A* **3**, 3573 (1965).
8. W. A. Lee and G. J. Knight, *Br. Polym. J.* **2**, 73 (1970).
9. R. H. Wiley and G. M. Braver, *J. Polym. Sci.* **3**, 647 (1948).
10. R. H. Wiley and G. M. Braver, *J. Polym. Sci.* **3**, 455 (1948).
11. D. H. Klein, *J. Paint Technol.* **42**, 335 (1970).
12. *CM-44*, Rohm and Haas Co., Philadelphia, Pa.
13. *Initial Rates of Homopolymerization and Copolymerization Parameters of Acrylic Monomers Glass Transition Temperature of Acrylic Monomers CM-20*, Rohm and Haas Co., Philadelphia, Pa.
14. O. G. Lewis, *Physical Constants of Linear Homopolymers*, Springer-Verlag, New York, Inc., 1968.
15. J. Brandrup and E. H. Immergut, *Polymer Handbook*, 2nd ed., Wiley-Interscience, New York, 1975.
16. J. A. Shetter, *Polym. Lett.* **1**, 209 (1963).
17. L. S. Luskin in J. Agranoff, ed., *Modern Plastics Encyclopedia*, Vol. 56, McGraw-Hill, Inc., New York, 1979, p. 8.
18. J. L. O'Brien and J. O. Van Hook, *Plasticizer Technology*, Vol. 1, Reinhold Publishing Co., New York, 1965, p. 219.
19. W. H. Brendley, Jr., *Paint Varnish Prod.* **63**, 19 (July 1973).
20. C. E. Rehberg and C. H. Fisher, *Ind. Eng. Chem.* **40**, 1429 (1948).
21. *Plexiglas Molding Pellets*, *PL-926a*, Rohm and Haas Co., Philadelphia, Pa.
22. *Plexiglas Design and Fabrication Data PL-927b*, *Plexiglas Cast-Sheet for Lighting*, Rohm and Haas Co., Philadelphia, Pa.
23. L. S. Luskin and R. J. Myers in N. M. Bikales, ed., *Encyclopedia of Polymer Science and Technology*, Vol. 1, John Wiley & Sons, Inc., New York, 1964, pp. 246–328.
24. *Optics*, *PL-897b*, Rohm and Haas Co., Philadelphia, Pa.
25. *Plexiglas Design and Fabrication Data*, *PL-53i*, Rohm and Haas Co., Philadelphia, Pa.

26. J. H. Hildebrand and R. L. Scott, *The Solubility of Non-Electrolytes*, 3rd ed., Reinhold Publishing Co., New York, 1949.
27. P. A. Small, *J. Appl. Chem.* **3**, 71 (1953).
28. H. Burrell, *Off. Dig. Fed. Soc. Paint Technol.* **27**, 726 (Oct. 1955).
29. J. L. Gardon, "Cohesive-Energy Density" in N. M. Bikales, ed., *Encyclopedia of the Polymer Science and Technology*, Vol. 3, John Wiley & Sons, Inc., 1955, p. 833.
30. J. L. Gardon, *J. Paint Technol.* **38**, 43 (1966).
31. A. Rudin and H. J. Johnson, *J. Paint Technol.* **43**, 39 (1971).
32. B. Vollmert, *Polymer Chemistry*, Springer-Verlag, New York, 1973.
33. S. Krause and E. Cohn-Ginsberg, *J. Phys. Chem.* **67**, 1479 (1963).
34. W. H. Watanabe, C. F. Ryan, P. C. Fleischer, Jr., and B. S. Garrett, *J. Phys. Chem.*, 896 (1961).
35. J. C. Bevington, D. E. Eaves, and R. L. Vale, *J. Polym. Sci.* **32**, 317 (1958).
36. R. F. B. Davies and G. E. J. Reynolds, *J. Appl. Polym. Sci.* **12**, 47 (1968).
37. U.S. Pat. 3,029,228 (Apr. 10, 1962), F. J. Glavis (to Rohm and Haas Co.).
38. F. J. Glavis, *J. Polym. Sci.* **36**, 547 (1959).
39. E. M. Loebl and J. J. O'Niel, *Abstracts*, P-28v, 142nd Meet. Am. Chem. Soc., Atlantic City, N.J., Sept. 1962, p. 466.
40. *Plexiglas Acrylic Plastic Molding Powder*, PL-866, Rohm and Haas Co., Philadelphia, Pa.
41. R. Simha, "Degradation of Polymers" in *Polymerization and Polycondensation Processes*, No. 34, Advances in Chemistry Series, American Chemical Society, Washington, D.C., p. 157, 1962.
42. Ger. Offen. 2,452,309 (May 1975), L. R. Mahoney.
43. B. Baysal, *J. Polym. Sci.* **8**, 529 (1952).
44. J. L. O'Brien, *J. Am. Chem. Soc.* **77**, 4757 (1955).
45. J. C. Buington, *J. Polym. Sci.* **14**, 163 (1954).
46. L. Maduga, *An. Quim.* **65**, 993 (1969).
47. E. P. Bonsall, *Trans. Faraday Soc.* **49**, 686 (1953).
48. *Preparation, Properties, and Uses of Acrylic Polymers*, CM-19 C/Ci, Rohm and Haas Co., Philadelphia, Pa.
49. T. G. Fox and R. Loshock, *J. Am. Chem. Soc.* **75**, 3544 (1953).
50. G. V. Schulz and G. Henrici, *Makromol. Chem.* **18/19**, 437 (1956).
51. F. R. Mayo and A. A. Miller, *J. Am. Chem. Soc.* **80**, 2493 (1956).
52. M. M. Mogilevich, *Russ. Chem. Rev.* **48**, 199 (1979).
53. E. Tramsdorff and co-workers, *Makromol. Chem.* **1**, 169 (1948).
54. C. A. Detrick and co-workers, *Ind. Eng. Chem. Proc. Des. Dev.* **9**, 191 (1970).
55. F. W. Billmeyer, Jr., *Textbook of Polymer Chemistry*, 2nd ed., Wiley-Interscience, New York, 1971.
56. U.S. Pat. 2,471,959 (May 31, 1949), M. Hunt (to DuPont).
57. U.S. Pat. 3,113,114 (Dec. 3, 1963), R. A. Maginn (to DuPont).
58. U.S. Pat. 3,382,209 (May 7, 1968), W. G. Deichert (to American Cyanamid).
59. U.S. Pat. 2,744,886 (May 8, 1956), T. F. Protzman (to Rohm and Haas).
60. Ger. Pat. 673,394 (March 22, 1939), W. Bauer (to Rohm and·Haas G.m.b.H.).
61. U.S. Pat. 2,576,712 (Nov. 27, 1951), J. Boyko (to DuPont).
62. L. S. Luskin, J. A. Sawyer, and E. H. Riddle, "Manufacture of Acrylic Polymers" in W. M. Smith, ed., *Polymer Manufacturing and Processing*, Reinhold Publishing Co., New York, 1964.
63. E. H. Riddle and P. A. Horrigan in P. H. Groggins, ed., *Unit Processes in Organic Synthesis*, 5th ed., McGraw Hill Book Co., Inc., New York, 1958.
64. J. O. Beattie, *Mod. Plast.* **33**, 109 (1956).
65. U.S. Pat. 3,376,371 (Apr. 2, 1968), C. J. Opel and co-workers (to Swedlow, Inc.).
66. *Forming Plexiglas Sheet*, PL-4k, Rohm and Haas Co., Philadelphia, Pa.
67. *Plexiglas Acrylic Sheet*, PL-80M, Rohm and Haas Co., Philadelphia, Pa.
68. *Plexiglas DR Rigidizing Manual*, Rohm and Haas Co., Philadelphia, Pa.
69. *Plexiglas Molding Manual*, PL-710, Rohm and Haas Co., Philadelphia, Pa.
70. *Plexiglas Molding Powder Design Manual*, PL-897, Rohm and Haas Co., Philadelphia, Pa.
71. G. M. Burnett, *IUPAC INT. Syrup. Macromol. Chem. Plenary Main Lecture*, 403 (1971).
72. R. N. Chedha and co-workers, *Trans. Faraday Soc.* **53**, 240 (1957).
73. S. Basn and co-workers, *Proc. Royal Soc. (London)* **A202**, 485 (1950).
74. J. L. O'Brien and F. Gornick, *J. Am. Chem. Soc.* **77**, 4757 (1955).

75. C. L. Sturn and I. Rosen in J. A. Moore, ed., *Macromolecular Synthesis*, Collective Vol. 1, John Wiley & Sons, Inc., New York, 1977, p. 195.
76. W. F. Smith and R. H. Ewart, *J. Chem. Phys.* **16,** 592 (1948).
77. H. Fikentscher and co-workers, *Angew. Chem.* **72,** 856 (1960).
78. J. L. Gordon, *J. Polym. Sci. A-1* **6,** 623 (1968).
79. F. K. Hansen and J. Ugelstad, *J. Polym. Sci. Ed.* **16,** 1953 (1978).
80. K. G. McCurdy and co-workers, *Can. J. Chem.* **42,** 825 (1964).
81. M. Fujii and co-workers, *Kobunski Kaghu* **26,** 163 (1969).
82. *Emulsion Polymerization of Acrylic Monomers*, *CM-104*, Rohm and Haas Co., Philadelphia, Pa.
83. U.S. Pat. 3,458,466 (July 29, 1969), W. J. Lee (to Dow Chemical).
84. U.S. Pat. 3,344,100 (Sept. 26, 1967), F. J. Donat and co-workers (to B. F. Goodrich).
85. U.S. Pat. 2,857,360 (Oct. 21, 1958), S. S. Feuer (to Rohm and Haas).
86. U.S. Pat. 2,264,376 (Dec. 2, 1941), J. R. Hiltner and W. F. Bartol (to Rohm and Haas Co.).
87. U.S. Pat. 3,232,915 (Feb. 1, 1966), J. L. Bush and co-workers (to DuPont).
88. U.S. Pat. 3,450,796 (June 17, 1969), B. P. Griffin (to ICI).
89. G. S. Whitby and co-workers, *J. Polym. Sci.* **16,** 549 (1955).
90. B. N. Rutovshii and co-workers, *J. Appl. Chem. USSR* **26,** 397 (1953).
91. *The Manufacture of Acrylic Polymers*, *CM-107*, Rohm and Haas Co., Philadelphia, Pa.
92. D. P. Hart in J. A. Moore, ed., *Macromolecular Synthesis*, Collective Vol. 1, John Wiley & Sons, Inc., New York, 1977.
93. U.S. Pat. 3,029,223 (Apr. 10, 1962), B. B. Hibbard (to Dow Chemical).
94. U.S. Pat. 3,267,178 (Aug. 16, 1966), L. H. Lee (to Dow Chemical).
95. U.S. Pat. 2,943,074 (June 28, 1960), S. S. Feuer (to Rohm and Haas).
96. U.S. Pat. 3,657,391 (Apr. 18, 1972), D. C. Curfman (to Borg-Warner).
97. T. O. Purcell, Jr. in N. M. Bikales, ed., *Encyclopedia of Polymer Science and Technology*, Suppl. 1, John Wiley & Sons, Inc., New York, 1976, pp. 307–325.
98. W. Kobryner, *J. Polym. Sci.* **34,** 381 (1959).
99. P. M. Swift, *J. Appl. Chem.*, 803 (Dec. 8, 1958).
100. P. W. Allen and co-workers, *J. Polym. Sci.* **36,** 55 (1959).
101. W. Cooper and co-workers, *Appl. Polym. Sci.* **1,** 329 (1959).
102. W. Cooper and co-workers, *J. Poly. Sci.* **34,** 651 (1959).
103. D. J. Angier and co-workers, *J. Polym. Sci.* **20,** 235 (1956).
104. T. Toda, *J. Polym. Sci.* **58,** 411 (1962).
105. U.S. Pat. 3,030,319 (Apr. 17, 1962), S. Kaizirman and G. Mino (to American Cyanamid).
106. A. Chapiro, *J. Polym. Sci.* **29,** 321 (1958).
107. Brit. Pat. 788,175 (Dec. 23, 1957), R. G. Norrish (to Distillers Co., Ltd.).
108. S. P. Rao and co-workers, *J. Polym. Sci. A-1* **5,** 2681 (1967).
109. R. K. Graham and co-workers, *J. Polym. Sci.* **38,** 417 (1959).
110. R. J. Ceresa, *Block and Graft Copolymers*, Vol. 1, Butterworth, Inc., Washington, D.C., 1962.
111. H. A. J. Battaerd and G. W. Tregear, *Polymer Reviews*, *Graft Copolymers*, Vol. 16, Wiley-Interscience, New York, 1967.
112. S. P. Pappas, ed., *U.V. Curing: Science and Technology*, Technology Marketing Corporation, Stamford, Conn., 1978.
113. U.S. Pat. 3,418,295 (Dec. 24, 1968), A. C. Schwenthaler (to DuPont).
114. V. D. McGinnis, *SME Tech. Pap. (Ser.) FC 76-486*, Cleveland, Ohio, 1976.
115. *The Curing of Coatings with Ultra-Violet Radiation*, D 8667 G. C., Tioxide of Canada, Sorel, Quebec.
116. E. Selegny and P. Segain, *J. Macromol. Sci.* **A5,** 603 (1971).
117. M. Szwarc and A. Rembaum, *J. Polym. Sci.* **22,** 189 (1956).
118. Y. Heimei and co-workers, *Appl. Polym. Symp.* **26,** (1975).
119. U.S. Pat. 4,056,559 (Nov. 1, 1977), S. N. Lewis and R. A. Haggard (to Rohm and Haas).
120. U.S. Pat. 4,022,730 (May 10, 1977), S. N. Lewis and co-workers (to Rohm and Haas).
121. U.S. Pat. 4,023,977 (May 17, 1977), A. Mercurio and S. N. Lewis (to Rohm and Haas).
122. U.S. Pat. 4,133,793 (Jan. 9, 1979), S. N. Lewis and R. A. Haggard (to Rohm and Haas).
123. U.S. Pat. 4,064,161 (Dec. 20, 1977), S. N. Lewis and R. A. Haggard (to Rohm and Haas).
124. U.S. Pat. 4,103,093 (July 25, 1978), S. N. Lewis and R. A. Haggard (to Rohm and Haas).
125. *Plexiglas Acrylic Sheet*, *PL 783a*, Rohm and Haas Co., Philadelphia, Pa.
126. P. W. Allen, *Technique of Polymer Characterization*, Butterworths, London, 1959.

127. J. G. Brodnyan, *J. Colloid Sci.* **15**, 76 (1960).

128. E. B. Bradford and J. W. Vanderhoff, *J. Polym. Sci. Part C* **1**, 41 (1963).

129. *Dilute Solution Properties of Acrylic and Methacrylic Polymers*, SP-160, Rohm and Haas Co., Philadelphia, Pa.

130. R. Kunin, *Elements of Ion Exchange*, Robert E. Krieger Publishing Co., Huntington, N.Y., 1971.

131. A. E. Sheer, *SPE J.* **28**, 24 (Nov. 1972).

132. H. J. Gambino, Jr., *Security You Can See Through*, PL-1228, Rohm and Haas Co., Philadelphia, Pa.

133. *Mod. Plast.* (5), 52 (May 1975).

134. *Sun Screen Innovations with Plexiglas*, PL-935, Rohm and Haas Co., Philadelphia, Pa.

135. *Transparent Plexiglas Solar Control Series*, Rohm and Haas Co., Philadelphia, Pa., June 1975.

136. *Plexiglas DR Sign Manual*, PL-1097e, Rohm and Haas Co., Philadelphia, Pa., Feb. 1978.

137. *Chem. Week*, **12**, 47 (Mar. 7, 1973).

138. Jpn. Kokai 74 57,054 (June 1974), E. Masuhara, J. Tarumi, N. Nakabayashi, M. Baba, S. Tanaka, and E. Mochida (to Mochida Pharmaceuticals).

139. J. R. De Wijn, F. C. M. Driessenj, and T. J. J. H. Sloof, *J. Biomed. Mater. Res.* **9**(4), 99 (1975).

140. R. P. Kusy and D. T. Turner, *J. Dent. Res.* **53**, 948 (1974).

141. W. Timmer, *Chem. Tech.* **9**, 175 (March 1979).

142. U.S. Pat. 3,951,528 (Apr. 20, 1976), H. R. Leeds (to Patent Structures, Inc.).

143. U.S. Pat. 3,947,401 (March 30, 1976), P. Stamberger (to Union Optics Corporation).

144. U.S. Pat. 4,239,513 (Feb. 13, 1979), K. Tanaka, K. Takahashi, M. K. Aichi, Sikanome, and T. Nakajima (to Toyo Contact Lens Co., Ltd.).

145. Y. Ohtsuka and Y. Hatanaka, *Appl. Phys. Lett.* **29**, 735 (1976).

146. U.S. Pat. 4,138,194 (Feb. 6, 1979), J. K. Beasley, R. Beckerbauer, H. M. Schleinitz, and F. C. Wilson (to DuPont).

147. I. Kaetsu, K. Yoshida, and H. Okubo, *J. Appl. Polym. Sci.* **24**, 1515 (1979).

148. H. Okubo, K. Yoshida, and I. Kaetsu, *Int. J. Appl. Radiat. Isot.* 30, 209 (1979).

149. U.S. Pat. 4,146,696 (Mar. 17, 1979), H. M. Bond, D. L. Torgersen, and C. E. Ring (to Buckee-Mears Co.).

150. R. J. Kopko, R. L. Stambaugh, *SAE Paper 750693, Fuels and Lubricants Meetings*, Houston, Tex., June 3–5, 1975, Society of Automotive Engineers, Inc., Warrendale, Pa.

151. U.S. Pat. 2,834,733 (May 13, 1950), D. H. Moreton (to Douglas Aircraft Co., Inc.).

152. U.S. Pat. 2,091,627 (Aug. 31, 1937), H. A. Bruson (to Rohm and Haas Co.).

153. U.S. Pat. 3,142,664 (July 28, 1964), L. N. Bauer (to Rohm and Haas Co.).

154. U.S. Pat. 3,147,222 (Sept. 1, 1964), L. N. Bauer (to Rohm and Haas Co.).

155. U.S. Pat. 3,879,304 (Apr. 22, 1975), J. O. Waldbilling (to Texaco, Inc.).

156. A. F. Talbot, *Rheol. Acta* **13**, 305 (1974).

157. H. Pennewiss, *Siefen Oele Fette Wachse* **100**(15–16), 381 (1974).

158. Y. Tamai, M. Masataka, K. Nakajima, and Y. Shibata, *Sekiyu Gakkaishi* **22**(3), 154 (1979).

159. A. K. Misra, G. C. Misra, and A. N. Nandy, *Indian J. Technol.* **11**, 381 (1973).

160. N. V. Messina and H. H. Radtke, *SAE Paper 700053, Automotive Engineering Congress*, Detroit, Mich., Jan. 12–16, 1970, Society of Automotive Engineers, Inc., New York.

161. A. S. Wood, *Mod. Plast.* **52**, 40 (Mar. 1975).

162. *Mod. Plast.* **49**, 44 (Aug. 1972).

163. R. R. MacBride, *Mod. Plast.* **49**, 40 (Aug. 1972).

164. R. R. MacBride, *Mod. Plast.* **56**, 47 (Dec. 1974).

165. H. C. Pyun, B. R. Cho, and S. K. Kwon, *J. Korean Nucl. Soc.* **7**, 9 (1975).

166. U.S. 3,233,026 (Feb. 1, 1966), G. A. Richter, Jr., C. H. McBurney, and B. B. Kine (to Rohm and Haas Co.).

167. R. Martino, *Mod. Plast.* **51**, 62 (Nov. 1974).

General References

E. H. Riddle, *Monomeric Acrylic Esters*, Reinhold Publishing Corporation, New York, 1956.

L. S. Luskin in E. C. Leonard, ed., *Vinyl and Diene Monomers*, Part 1, Wiley-Interscience, New York, 1970.

K. J. Saunders, *Organic Polymer Chemistry*, Chapman and Hall, London, 1973.

H. Warson, *The Applications of Synthetic Resin Emulsions*, Ernest Benn Ltd., London, 1972.

M. B. Horn, *Acrylic Resins*, Reinhold Publishing Corporation, New York, 1960.

Benjamin B. Kine
R. W. Novak
Rohm and Haas Company

METHANE. See Hydrocarbons, C_1–C_6.

METHANOL

Methanol [67-56-1] (methyl alcohol), CH_3OH, is a clear, water-white liquid with a mild odor at ambient temperatures. From its discovery in the late 1600s, methanol has grown to become the 21st largest commodity chemical with over 12×10^6 metric tons annually produced in the world. Methanol has been called wood alcohol (or wood spirit) because it was obtained commercially from the destructive distillation of wood for over a century. However, true wood alcohol contained more contaminants (primarily acetone, acetic acid, and allyl alcohol) than the chemical-grade methanol available today.

For many years the largest use for methanol has been as a feedstock in the production of formaldehyde, consuming almost half of all the methanol produced. In the future, formaldehyde's importance to methanol will decrease as newer uses increase such as the production of acetic acid and methyl *tert*-butyl ether (MTBE, a gasoline octane booster) (see Gasoline). Methanol's direct use as a fuel may be significant in special circumstances (see Fuels, synthetic).

Physical Properties

The physical properties of methanol are given in Table 1. The vapor pressure of methanol from 15.00 to 64.50°C is given by the following equation (6):

$$\ln P = 15.76 - 2.846 \times 10^3 \, T^{-1} - 3.743 \times 10^5 \, T^{-2} + 2.189 \times 10^7 \, T^{-3} \qquad (1)$$

where P = kPa (7.5 mm Hg) and T = temperature in degrees Kelvin. An equation that covers a wider temperature range from −67.4 to 240°C is probably accurate enough for most purposes (11). Density and viscosity values are reported for aqueous methanol from −90 to +50°C (7–8,12). Azeotropic data for 214 compounds with methanol are available (13).

Table 1. Physical Properties of Methanol

Property	Value	Ref.
freezing point, °C	−97.68	1
boiling point, °C	64.70	1
critical temperature, °C	239.43	1
critical pressure, kPa[a]	8096	1
critical volume, mL/mol	118	1
critical compressibility factor z in $PV = znRT$	0.224	1
heat of formation (liquid) at 25°C, kJ/mol[b]	−239.03	2
free energy of formation (liquid) at 25°C, kJ/mol[b]	−166.81	2
heat of fusion, J/g[b]	103	1
heat of vaporization at boiling point, J/g[b]	1129	1
heat of combustion at 25°C, J/g[b]	22,662	1
flammable limits in air		
lower, vol %	6.0	3
upper, vol %	36	3
autoignition temperature, °C	470	4
flash point, closed cup, °C	12	4
surface tension, mN/m (= dyn/cm)	22.6	5
specific heat		
of vapor at 25°C, J/(g·K)[b]	1.370	1
of liquid at 25°C, J/(g·K)[b]	2.533	1
vapor pressure at 25°C, kPa[a]	16.96	6
solubility in water	miscible	
density at 25°C, g/cm^3	0.78663	7
refractive index, n_D^{20}	1.3284	1
viscosity of liquid at 25°C, mPa·s (= cP)	0.541	8
dielectric constant at 25°C	32.7	9
thermal conductivity at 25°C, W/(m·K)	0.202	10

[a] To convert kPa to mm Hg, multiply by 7.5.

[b] To convert J to cal, divide by 4.184.

Chemical Reactions

Methanol undergoes reactions that are typical of alcohols as a chemical class (14). The reactions of particular importance from an industrial standpoint are dehydrogenation and oxidative dehydrogenation to formaldehyde (see Formaldehyde) over silver or molybdenum–iron oxide catalysts and carbonylation to acetic acid catalyzed by cobalt or rhodium (see Acetic acid). Dimethyl ether can be formed by the acid-catalyzed elimination of water. The acid-catalyzed reaction of isobutylene and methanol to form methyl *tert*-butyl ether, an important gasoline-octane improver, has increasing application. Methyl esters of carboxylic acids can be prepared by acid-catalyzed reaction with azeotropic removal of water to force the reaction to completion. Methyl hydrogen sulfate, methyl nitrite, methyl nitrate, and methyl halides are formed by reaction with the appropriate inorganic acids. Mono-, di-, and trimethylamine result from the direct reaction of methanol with ammonia (see Amines).

Manufacturing and Processing

The oldest industrially significant method of methanol manufacture was destructive distillation of wood. Practiced from the mid-19th century to the early 1900s, it is no longer used in the United States. The technology became obsolete with the development of a synthetic route from hydrogen and carbon oxides in the mid-1920s (15–16). Methanol also was produced as one of the products of the noncatalytic oxidation of hydrocarbons, a practice discontinued in the United States since 1973 (see Hydrocarbon oxidation). Methanol also can be obtained as a by-product of Fisher-Tropsch synthesis (see Fuels, synthetic).

Modern industrial-scale methanol production is based exclusively on synthesis from pressurized mixtures of hydrogen, carbon monoxide, and carbon dioxide gases in the presence of metallic heterogeneous catalysts. The required synthesis pressure is dependent on the activity of the particular catalyst. By convention, technology is generally distinguished by pressure as follows: low pressure processes, 5–10 MPa (50–100 atm); medium pressure processes, 10–25 MPa (100–250 atm); and high pressure processes, 25–35 MPa (250–350 atm).

The first commercial synthetic methanol was produced by BASF in Germany in 1923. The process employed a zinc oxide–chromium oxide catalyst system, thus starting the high pressure technology era. In 1927 United States industry, in separate efforts, began commercial high pressure synthesis at plants owned by Commercial Solvents and DuPont. By 1965 a modern, high pressure methanol synthesis plant had a typical capacity of 225–450 t/d (330 d/yr), operated at 35 MPa (350 atm), used reciprocating compressors for pressures above 21 MPa (210 atm), and consumed a net 1300 m^3 natural gas/t methanol (see also High pressure technology).

In the late 1960s medium and low pressure methanol technology came into use with the successful development of highly active, durable copper–zinc oxide catalysts (17). ICI Ltd., UK, began commercial, low pressure synthesis of methanol in late 1966 (18), operating a 400 t/d plant at 5 MPa (50 atm) using all rotating-compression equipment. In 1971 Lurgi, FRG, started an 11 t/d, low pressure demonstration plant using copper-based catalysts (19). Copper catalysts' sensitivity to poisons required careful purification of feed streams. Low and medium pressure technology has advantages of reduced compression power, good catalyst life, larger capacity single-train converter designs, and milder operating pressures. Coupled with large single-train rotating-compression equipment, the result is economy of scale, reduced energy consumption, and improved reliability unmatched by high pressure technology. Since 1970, with few exceptions, new methanol expansions have been based on the low–medium pressure technology. Low pressure technology accounted for 55% of United States methanol capacity in 1980. By 1981, 80% of United States capacity will use low pressure technology as old high pressure plants are being revamped gradually to low pressure technology or shut down. A modern typical low–medium pressure methanol plant built in 1980 would have a capacity of 1000–2000 t/d (350 d/yr), operate at pressures of 8–10 MPa (80–100 atm), feature single-train construction use all rotating-compression equipment, and have a net natural-gas consumption approaching 890 m^3/t methanol. Further information on methanol manufacture is described in references 18–33.

The only substantial process innovation on the horizon at the present time appears to be Chem Systems' three-phase process (34). An inert liquid is used to fluidize the catalyst and remove the heat of reaction. Higher single-pass conversions are claimed to be possible with this design than with the usual two-phase processes.

Raw Materials. Common feedstocks used in producing synthesis gas for methanol manufacture in the United States are natural gas and petroleum residues. Other suitable feedstocks are naphtha and coal. Combined, natural gas, petroleum residues and naphtha account for 90% of worldwide methanol capacity with miscellaneous off-gas sources making up the difference.

Natural Gas. Natural-gas feedstock accounts for 75% of domestic and 70% of worldwide methanol capacity for 1980. A typical schematic of the process steps is shown in Figure 1.

In the modern natural-gas-based methanol process, natural gas (principally methane) is desulfurized (usually to <0.25 ppm (vol) H_2S), mixed with steam and preheated to 425–550°C. The mixture is fed to a reformer where it passes through an arrangement of externally fired tubes containing a nickel-impregnated ceramic catalyst. The following reactions occur:

Steam reforming of CH_4:

$$CH_4 + H_2O \rightleftarrows CO + 3\,H_2 \quad \Delta H_{298\,K} = 206.2\ \text{kJ/mol (195.6 Btu/mol)} \tag{2}$$

Hydrocracking of heavy hydrocarbons:

$$C_nH_{(2n+2)} + (n-1)\,H_2 \rightarrow n\,CH_4 \quad \Delta H_{298\,K}^{C2H6} = -65.07\ \text{kJ/mol } (-61.74\ \text{Btu/mol}) \tag{3}$$

Water gas shift:

$$CO + H_2O \rightarrow CO_2 + H_2 \quad \Delta H_{298\,K} = -41.25\ \text{kJ/mol } (-39.14\ \text{Btu/mol}) \tag{4}$$

The reformed-gas product can be described adequately by assuming reaction 3 goes to completion with exit concentrations of CO, CO_2, H_2, CH_4, and H_2O approaching equilibrium with respect to reactions 2 and 4 at typical exit conditions of 840–880°C and 0.7–1.7 MPa (7–17 atm). The overall reaction is highly endothermic and requires significant amounts of fuel. Reformer waste heat recovered from flue gas and product gas, principally used to generate 4–10 MPa (40–100 atm) steam for meeting driver requirements and distillation loads, is becoming more heavily integrated into the overall process to reduce net energy consumption (25).

The synthesis gas generated by steam reforming of natural gas contains more hydrogen than necessary for the methanol reaction. Methanol synthesis stoichiometrically requires a feed gas, $H_2/(2\,CO + 3\,CO_2)$ ratio near 1.05 whereas steam reforming of natural gas yields a ratio of about 1.4 without CO_2 addition. For low pressure catalysts, the excess hydrogen improves the catalyst effectiveness. Thus, converter costs are reduced and the necessity of shifting and removing excess hydrogen from the synthesis feed gas, as commonly practiced with high pressure technology, is avoided. Excess hydrogen is vented during synthesis and used as fuel in the reforming step. Thus, a high overall energy efficiency is maintained which makes the process economical.

Natural-gas-based low pressure methanol plants can be designed for CO_2 addition, taking advantage of the excess hydrogen to reduce natural-gas usage per ton of methanol as long as the CO_2 is inexpensive (21–22). Sufficient CO_2 addition produces a near stoichiometrically perfect synthesis feed gas similar to that produced from naphtha feedstock. Recovery of carbon dioxide from flue gas rarely is economical.

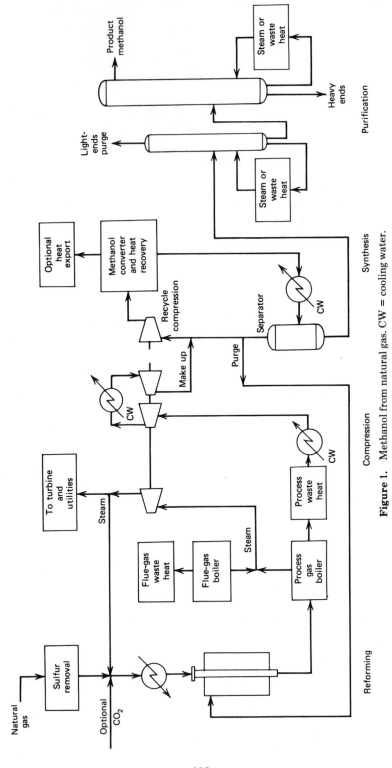

Figure 1. Methanol from natural gas. CW = cooling water.

Petroleum Residues. Methanol produced from residual fuel oil accounts for just under 15% of domestic and worldwide methanol capacity for 1980. Heavy oil is converted to methanol synthesis gas via a partial oxidation (POX) route (19,28). A simplified schematic of the process steps is given in Figure 2.

Petroleum residue is fed to a partial-oxidation reactor where it reacts noncatalytically and adiabatically in a refractory-lined vessel with compressed oxygen to form a mixture of carbon oxides and hydrogen (35). The following partial oxidation reaction takes place initially with the disappearance of oxygen:

$$C_nH_m + \left(n + \frac{m}{4}\right) O_2 \rightarrow n\, CO_2 + \left(\frac{m}{2}\right) H_2O \quad \text{(exothermic)} \tag{5}$$

Sufficient heat is released to cause reforming of the remaining feedstock according to reactions 2, 3, and 4. As with steam reforming of natural gas, the concentrations of CO, CO_2, H_2, CH_4 and H_2O in the partial-oxidation product gas approach equilibrium with respect to reactions 2 and 4 at common exit conditions of 1200–1500°C and pressures of 4–8 MPa (40–80 atm). Oxygen consumption varies according to the level of preheat and exit temperature desired as well as feedstock composition.

Waste heat is recovered from the product gas in a special boiler. Carbon particles (soot), sulfur (hydrogen sulfide and carbonyl sulfide), and heavy metals are removed. To provide a more appropriate hydrogen-to-carbon oxide ratio, part of the carbon monoxide is shifted to carbon dioxide which is then removed.

The overall process is somewhat more complex than producing methanol from natural gas. However, partial oxidation offers more feedstock flexibility than reforming and can be designed to handle feedstocks ranging from natural gas to sour heavy resid (36). Normally, partial oxidation is carried out at sufficient pressure to flow directly into the synthesis system, eliminating the need for a synthesis gas compression step.

Naphtha. Naphtha feedstocks account for just over 5% of the worldwide methanol capacity for 1980. Naphtha is not being used in the United States to make methanol.

The process steps to produce methanol from naphtha are nearly identical to those shown for reforming of natural gas, Figure 1, except that naphtha requires vaporization prior to the desulfurization step. The desulfurization may be slightly more complex than required for natural gas and a different type of reforming catalyst is required (37). The synthesis gas produced by naphtha reforming is nearly of optimum stoichiometry for methanol synthesis.

Coal. Methanol production from coal (qv) accounts for under 2% of worldwide capacity for 1980. No methanol-from-coal plants currently exist in North America. Methanol from coal is, however, attracting serious attention in the United States and Europe as world reserves of oil and natural gas are being depleted and foreign oil cartels continue effectively to influence world energy price and supply. A number of plant designs are currently in the conceptual stages of development and economic evaluation (38–41). Several companies appear committed to building coal-to-methanol plants (42–43).

Reaction Mechanism. The formation of methanol from mixtures of carbon monoxide, carbon dioxide and hydrogen proceeds according to the following reactions:

$$CO + 2\,H_2 \rightarrow CH_3OH \quad \Delta H_{298\ K} = -90.77 \text{ kJ/mol } (-86.12 \text{ Btu/mol}) \tag{6}$$

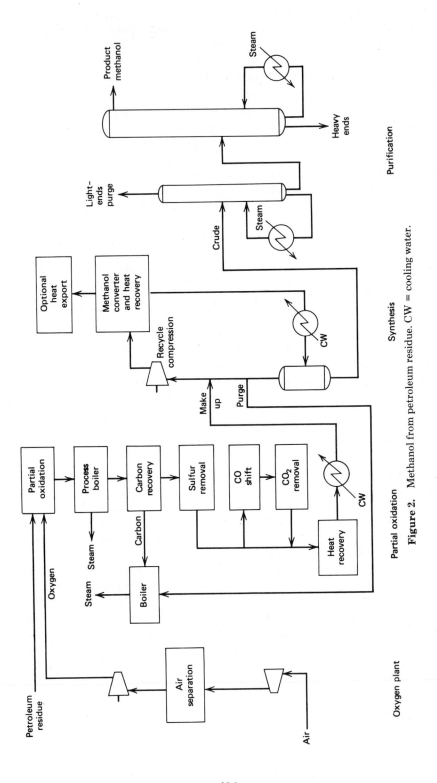

Figure 2. Methanol from petroleum residue. CW = cooling water.

$$CO_2 + 3\,H_2 \rightarrow CH_3OH + H_2O \quad \Delta H_{298\,K} = -49.52\ \text{kJ/mol}\ (-46.98\ \text{Btu/mol}) \tag{7}$$

It is common practice in industry to use the sum of equation 5 and the reverse water-gas shift reaction:

$$CO_2 + H_2 \rightarrow CO + H_2O \quad \Delta H_{298\,K} = 41.25\ \text{kJ/mol}\ (39.14\ \text{Btu/mol}) \tag{8}$$

when describing the methanol synthesis equilibrium instead of equation 7. The equilibrium constants, K_{a6} and K_{a8}, for reactions 6 and 8 are solely a function of temperature and are approximated by the following numerical equations:

CO conversion to methanol (44):

$$K_{a6} = 9.740 \times 10^{-5} \times \exp\left[21.225 + \frac{9143.6}{T} - 7.492 \ln T \right.$$

$$\left. + 4.076 \times 10^{-3}\,T - 7.161 \times 10^{-8}\,T^2\right] \tag{9}$$

Reverse water-gas shift reaction (45):

$$K_{a8} = \exp\left[13.148 \frac{-5639.5}{T} - 1.077 \ln T - 5.44 \times 10^{-4}\,T \right.$$

$$\left. + 1.125 \times 10^{-7}\,T^2 + \frac{49{,}170}{T^2}\right] \tag{10}$$

where T is Kelvins and K_{a8} is in kPa^{-2}. Equilibrium-constant data for CO conversion to methanol, equation 6, are varied with many different results presented in the literature (20,46). Equation 9 is presented as typical—not as absolutely correct.

An equilibrium mixture of methanol-synthesis gases is simultaneously described by reactions 6 and 8 by the following two equations:

CO conversion to methanol:

$$K_{a6} = \left[\frac{N_{CH_3OH} N_T^2}{N_{CO}(N_{H_2})^2 P^2}\right] \times K_{\gamma 6} \tag{11}$$

Reverse water–gas shift reaction:

$$K_{a8} = \left[\frac{N_{CO} N_{H_2O}}{N_{CO_2} N_{H_2}}\right] \times K_{\gamma 8} \tag{12}$$

where N_i = moles of component i in the mixture; N_T = total moles of mixture; $K_{\gamma 6} = (\gamma_{CH_3OH})/[\gamma_{CO}(\gamma_{H_2})^2]$; $K_{\gamma 8} = (\gamma_{CO}\gamma_{H_2O})/(\gamma_{CO_2}\gamma_{H_2})$; γ_i = fugacity ratio or activity coefficient of components i in the mixture; and $P = \text{kPa}$. Values of $K_{\gamma i}$ are significant in the methanol-synthesis reactions. They may be determined experimentally or calculated from generalized correlations (47) assuming ideal solution behavior, and are, therefore, known. A plot of $K_{\gamma 6}$, based on generalized correlations, is given in Figure 3 for the CO reaction to methanol. Similar data have been calculated for $K_{\gamma 6}$ (46–48). A plot of $K_{\gamma 8}$ for the reverse water–gas shift reaction is shown in Figure 4. Again generalized correlations were used. Higher values for $K_{\gamma 8}$ have been calculated at low temperatures and pressures (46).

In terms of a material balance across a methanol synthesis converter, the arguments in equations 11 and 12 can be defined as follows:

$$N_{CH_3OH} = NI_{CH_3OH} + X \tag{13}$$

$$N_{CO} = NI_{CO} - X + Y \tag{14}$$

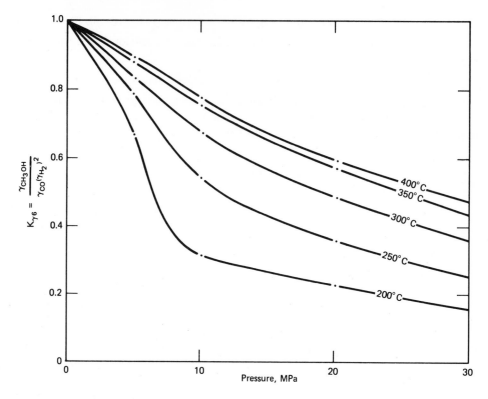

Figure 3. Values of $K_{\gamma 6}$ for the reaction of $CO + 2\,H_2 \rightarrow CH_3OH$ calculated from generalized fugacity tables (47). To convert MPa to atm, divide by 0.101.

$$N_{H_2} = NI_{H_2} - 2\,X - Y \tag{15}$$

$$N_{CO_2} = NI_{CO_2} - Y \tag{16}$$

$$N_{H_2O} = NI_{H_2O} + Y \tag{17}$$

$$N_T = NI_T - 2\,X \tag{18}$$

where X = moles CH_3OH formed (or CO converted); Y = moles H_2O formed (or CO_2 converted); NI_i = moles of component i entering the converter inlet; and N_i = moles of component i exiting the converter. The above terms, when substituted into equations 11 and 12, yield two equations in two unknowns which can be solved numerically to determine equilibrium exit compositions produced from a specific set of converter feed conditions. Alternatively, an approach to reaction equilibrium temperatures may be determined. The percent conversion of CO and CO_2 becomes:

$$\% \text{ CO conversion} = \frac{(X - Y)}{NI_{CO}} \times 100 \tag{19}$$

$$\% \text{ CO}_2 \text{ conversion} = \frac{Y}{NI_{CO_2}} \times 100 \tag{20}$$

Equilibrium conversions and exit methanol concentrations calculated using the

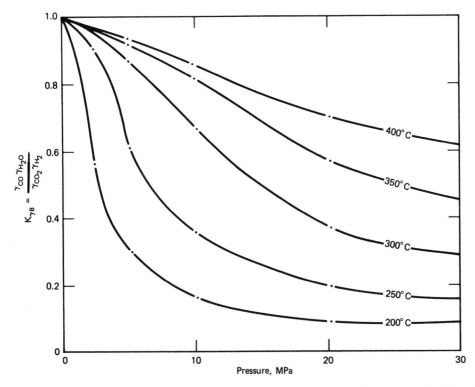

Figure 4.   Values of $K_{\gamma 8}$ for the reaction $CO_2 + H_2 \rightarrow CO + H_2O$ calculated from generalized fugacity tables (47). To convert MPa to atm, divide by 0.101.

above equations are shown in Table 2 for a synthesis gas composition typical of methane reforming.

Catalyst.   Catalyst used in high pressure (25–35 MPa or 250–350 atm) synthesis is zinc oxide–chromium oxide. It is a more robust catalyst than the low pressure copper-based catalyst and can tolerate higher temperature and sulfur levels. The copper–zinc oxide catalyst, however, is more active and can be operated at lower pressure (5–25 MPa or 50–250 atm) and temperature (200–300°C). Both types of catalyst are

Table 2.   Equilibrium CO, CO$_2$ Conversion, and Exit CH$_3$OH Concentration vs Pressure and Temperature [a,b,c]

Temperature, °C	CO conversion, %			CO$_2$ conversion, %			Exit CH$_3$OH, vol %		
	5 MPa[d]	10 MPa[d]	30 MPa[d]	5 MPa[d]	10 MPa[d]	30 MPa[d]	5 MPa[d]	10 MPa[d]	30 MPa[d]
200	95.6	99.0	99.9	44.1	82.5	99.0	27.8	37.6	42.3
250	72.1	90.9	98.9	18.0	46.2	91.0	16.2	26.5	39.7
300	25.7	60.6	92.8	14.3	24.6	71.1	5.6	14.2	32.2
350	−2.3	16.9	73.0	19.8	23.6	52.1	1.3	4.8	21.7
400	−12.8	−7.2	38.1	27.9	30.1	44.2	0.3	1.4	11.4

[a] Typical reformed natural gas composition 15% CO, 8% CO$_2$, 74% H$_2$, 3% CH$_4$.

[b] H$_2$/(2 CO + 3 CO$_2$) = 1.30.

[c] Simplified single-pass operation, no recycle.

[d] To convert MPa to psi, multiply by 145.

usually sold in cylindrical form of ca 5 × 10 mm (dia × height). Bulk density is ca 1.1–1.4 g/cm^3. Both zinc–chromium- and copper-based catalysts require reduction prior to methanol synthesis. Copper-containing catalysts are pyrophoric once reduced.

Commercial-Scale Manufacture. Low pressure methanol manufacture today is distinguished from older, high pressure plants by the reactor design and capacity. All new plants use low pressure technology and many existing plants are converting to it. Above 150 t/d rates, centrifugal compressors generally can be used to compress synthesis gas to methanol-reactor pressure.

Basically, two types of reactors are in commercial use: the shell-and-tube-type reactor and the quench type. The shell-and-tube-type reactor recovers heat of reaction by generation of steam on the shell side. The quench type recovers this heat in a separate heat exchange that preheats boiler feed water or performs a similar duty. Both reaction loops have otherwise similar equipment, ie, recycle compressor, feed–product interchanger, product cooler, and vapor–liquid separator.

Side Reactions. Prior to the commercialization of the low-to-medium pressure process using copper catalysts, the most troublesome side reaction was the reverse of the steam reforming reaction, equation 2. This reaction, known as methanation, occurs in high pressure plants above ca 450°C and causes exit bed temperatures to exceed 600°C. Such runaway temperatures usually require reactor shutdown to prevent catalyst and equipment damage. The low pressure copper-based catalysts operate in a lower temperature range, ie, 200–300°C, where the methanation reaction is unimportant. Hydrocarbons, alkanes C_2–$C_{>20}$, are produced at the ppm level by the Fischer-Tropsch reaction. Concentrations in the crude methanol can vary from 10 to 100 fold, depending on the amount of iron in the system and its catalytic activity.

Alcohols other than methanol are produced in small quantities with ethanol the chief impurity. Formation of the higher alcohols can be suppressed by keeping the reaction temperature as low as possible for the methanol production rate desired. High hydrogen concentration also suppresses the formation of higher alcohols and other by-products. Other by-products produced in small amounts are aldehydes, ketones, ethers and esters.

Purification. Methanol is purified by distillation. The complexity required depends on the desired methanol purity and on the purity of the crude methanol. Sales-grade methanol normally must meet U.S. Federal Grade Specifications A or AA as summarized in Table 3. Typically, grade A methanol requires two-column distillation whereas grade AA requires three columns. Since there is much proprietary knowledge used in methanol purification, various distillation schemes may require greater or fewer columns than typical. Captive use of methanol may require only flashing of dissolved synthesis gases.

Normally, distillation takes the following form: the crude liquid-methanol stream leaving the reaction section is reduced to a lower pressure to flash the dissolved gas, then it is fed to a light-ends removal column that separates compounds with boiling point lower than methanol. Frequently, water is fed near the top of the column to improve the relative volatility of light ends. The bottoms from the first column are sent to a second column where components generally heavier than methanol are removed as side streams and water is removed from the base section. The methanol taken overhead from the second column may be sent to storage or to a third column for

Table 3. Specifications for Pure Methanol

Property	Grade A	Grade AA
methanol content, wt %, min	99.85	99.85
acetone and aldehydes, ppm, max	30	30
acetone, ppm, max		20
ethanol, ppm, max		10
acid (as acetic acid), ppm, max	30	30
water content, ppm, max	1500	1000
specific gravity, 20/20°C	0.7928	0.7928
permanganate time, min	30	30
odor	characteristic	characteristic
distillation range at 101 kPa (760 mm Hg)	1°C, must include 64.6°C	1°C, must include 64.4°C
color, platinum–cobalt scale, max	5	5
appearance	clear–colorless	clear–colorless
residual on evaporation, g/100 mL	0.001	0.001
carbonizable impurities; color platinum–cobalt scale, max	30	30

further refining, depending on purity desired. Light and heavy ends are normally burned for heat recovery or are recovered elsewhere. Steam requirement for distillation has been reduced significantly in new, energy-efficient, methanol plants. Older plants require greater than 2 t steam/t methanol whereas new plants require 0.8–1.2 t steam/t methanol.

The purification of methanol is described in ref. 49 (see also Distillation).

Production and Economic Aspects

World methanol production capacity in 1979 was close to 15×10^6 metric tons. Geographically, 28% of this capacity was in the United States, 23% in Western Europe, and 9% in Japan. Actual methanol production in the same year was about 12×10^6 t with 3.4×10^6 t produced in the United States, 3×10^6 t in Western Europe and 10^6 t in Japan. Table 4 traces production, sales, and published list pricing in the United States over the past twenty years.

Although production has increased almost fourfold over this period, the number of producers and plants has actually decreased. A comparison of current producers and their capacities with a similar list of 1966 producers (Table 5) shows the shift in producers and plant size. These changes were brought about by changes in technology and demand for methanol. As demand for methanol increased, production moved towards larger and more efficient units. In fact, by 1982 none of the 1966 plants will be in operation under the same conditions previously used. The two or three units still operating will be expanded versions of the old units converted to low pressure technology.

Table 4.   United States Methanol Production, Sales and Prices

Year	Production[a]	Sales[a]	$/t[b]
1960	892	500	99.7[c]
1965	1302	633	89.7[c]
1970	2238	845	89.7[c]
1971	2246	1152	83.1[c]
1972	2936	1607	39.9[d]
1973	3205	1743	53.2[d]
1974	3121	1613	119.6[d]
1975	2349	1093	126.3[d]
1976	2832	968	139.6[d]
1977	2928	1647	146.2[d]
1978	2923	1398	146.2[d]
1979	3362	1681	186.1[d]
1980	3176	na	236.1

[a] Thousands of metric tons per year.

[b] All prices are at the end of the respective calendar year.

[c] Delivered in tankcars or 15.1 m³ (4000 gal) minimum tank truck quantities.

[d] FOB Gulf Coast plants, freight collect.

Specifications

Methanol is sold in two U.S. Federal Grades, A and AA (50) (see Fine chemicals). The requirements for each grade are shown in Table 3. Basically, grade AA has a limit on ethanol and acetone content. The tests to be used are ASTM methods:

D 1087	distillation range
D 1209	color (platinum-cobalt scale)
D 1296	odor
D 1353	nonvolatile constituents
D 1363	permanganate time
D 1612	acetone
D 1613	acidity, as acetic acid

ASTM D 1152-77 gives a more general set of specifications for methanol.

Storage and Handling

Methanol may be stored and handled in clean carbon–steel equipment. Storage tanks should be constructed with an internal floating roof with an inert gas pad to minimize vapor emissions. Because of the flammability of the product, tanks are usually enclosed by a dike and protected by a foam-type (either CO_2 or dry chemical) fire-extinguishing system. All applicable federal, state, and local agencies should be contacted for guidance on emission control and fire safety regulations.

All shipping containers (tank cars, tank trucks, and drums) should be of carbon steel and in a clean and dry condition prior to loading. Air pressure should never be used to load or unload methanol. Pumping is preferred but inert gas should be used when pressure loading or unloading.

Each shipping container should bear the DOT red label for flammable liquids and each tank car or tank truck must be provided with DOT flammable placards. In

Table 5. United States Methanol Producers

Company/location (1966)	Capacity[a]	Company/location (1980)	Capacity[a]
Allied Chemical Co.		Air Products & Chemicals, Inc.	
South Point, Ohio	70	Pensacola, Fla.	150
Borden Chemical Co.		Allemania Chemical Co.	
Geismar, La.	240	Plaquemine, La.	300
Celanese Chemical Co., Inc.		Borden Chemical Co.	
Bishop, Texas	225	Geismar, La.	480
Commercial Solvents		Celanese Chemical Co., Inc.	
Sterlington, La.	150	Bishop, Texas	440
Dupont		Clear Lake, Texas	630
Huron, Ohio	90	DuPont	
Orange, Texas	390	Beaumont, Texas	660
Escambia Chemical Corp.		Dear Park, Texas	600
Pensacola, Fla.	60	Georgia Pacific	
Gulf Oil Corp.		Plaquemine, La.	360
Military, Kan.	25	Monsanto	
Hercules Powder Co.		Texas City, Texas	300
Hercules, Calif.	35	Tenneco Chemicals	
Louisiana, Ma.	25	Houston, Texas	240
Monsanto Chemical Co.		*Total*	*4160*
Texas City, Texas	75		
Rohm & Haas			
Dear Park, Texas	70		
Tenneco Chemicals			
Houston, Texas	80		
Union Carbide Corp.			
South Charleston, W. Va.	35		
Texas City, Texas	120		
Total	*1690*		

[a] Thousands of metric tons per year.

addition, a caution label indicating the product to be both flammable and poisonous may be required. Other cautionary labels are prescribed in some states.

Toxicity

The most generally known health hazard associated with methanol is blindness, usually as a result of ingestion. The ingestion of methanol has resulted in a wide range of responses, probably owing to the concurrent intake of varying amounts of ethanol. Ethanol is selectively metabolized by the body, allowing detoxification by respiration to occur to some extent. By mouth, 25–100 mL of methanol is reported to be fatal (4). Initial symptoms vary: weakness, fatigue, headache, dizziness, nausea, and abdominal pain are typical. A latent period of 6–30 h usually follows the initial symptomatic episode. The same responses then occur only in much greater severity; in addition, loss of vision occurs. If death does not result from cardiac or respiratory failure, permanent blindness may be a residual effect.

The recommended maximum exposure level to vapor is 200 ppm TWA per 40 h work week. It is not clear that blindness has resulted from inhalation (51). Table 6 gives estimated exposure limits for other time periods.

First aid should be undertaken if ingestion is suspected, and is performed best

Table 6.   Estimated Tolerance Values for Methanol[a]

Duration	Concentration, ppm
single but not repeated exposure	
1 h	1000
8 h	500
24 h	200
5 × 8 h work d	200
168 h	50
30 d	10
60 d	5
90 d	3

[a] Ref. 52.

by a qualified physician. Ethanol has proved partially effective in the treatment of methanol poisoning for reasons mentioned earlier but this therapy must be monitored closely because of the potential combined central nervous system depression which can result (4). Ordinary sodium bicarbonate in 5–10 g amounts every 15 min can be used to combat acidosis (4). In severe cases, hemodialysis has been reported to be highly effective (4).

Uses

Historically, almost half of all methanol produced has been used to produce formaldehyde (qv), a trend that continues today. In the future, formaldehyde will lose some of this position owing to methanol's use in the production of faster-growing chemicals such as acetic acid, methyl *tert*-butyl ether (MTBE), Oxinol (a methanol–*tert*-butyl alcohol blend for gasoline octane improvement), and other chemical intermediates (see Fuels, synthetic; Gasoline).

Formaldehyde is consumed in the production of amino and phenolic resins (49%), 1,4-butanediol (8%), acetal resins (7%), pentaerythritol (7%), 4,4'-methylenediphenyl diisocyanate (MDI) (3%), and a host of miscellaneous uses (25%). The large share used in amino and phenolic resins is highly dependent on the housing and automotive markets (see Amino resins; Phenolic resins). These resins are used to make plywood, particle board, plastic laminates, foundry resins, and coatings. Since these markets swing dramatically with changes in the economy, it is easy to understand the major gains and losses in methanol production when the economy cycles through high and low periods.

Acetic acid (qv) production via methanol carbonylation (the Monsanto process) is one of the faster-growing uses for methanol. This process enjoys a significant advantage over ethylene-based routes and is expected to capture an increasing share of acetic acid production.

One of the more exciting and faster-growing uses for methanol is in gasoline octane improvers. Methyl *tert*-butyl ether (MTBE) is obtained by the reaction of methanol with isobutylene. Oxinol is a blend of equal parts methanol and *tert*-butyl alcohol. Both are used to raise the octane levels of unleaded gasoline. Together, their uses are projected to grow at rates approaching 20%/year over the next ten years.

Other important methanol markets include the synthesis of methyl methacrylate,

Table 7. United States Methanol Uses[a]

Use	1975	1980[b]	1985[b]
formaldehyde	960	1145	1590
acetic acid	95	340	670
octane improvers		270	995
chemical intermediates	505	710	840
solvent and miscellaneous uses	640	710	865
Total	*2200*	*3175*	*4960*

[a] Thousands of metric tons per year.
[b] Projected.

methylamines, methyl halides, and dimethyl terephthalate (see Methacrylic acid; Methacrylic polymers; Phthalic acid; Polyesters). Methanol is also used as solvent and in various miscellaneous uses. Table 7 summarizes methanol consumption in the United States.

One area of promise for methanol is its direct use in fuels. Potentially it can be used as a replacement for diesel fuel and gasoline or as a gasoline extender. Other fuel uses include use as a clean-burning boiler or turbine fuel to generate electricity. It is very attractive as a fuel for electric peaking turbines in locations where very clean-burning fuels are required. Methanol can also be used to make gasoline in Mobil's MTG (methanol-to-gasoline) process. Although all of these uses are still under study, they could become much larger than the chemical uses for methanol by the late 1980s.

Another potentially large market for methanol is its use in the production of single-cell protein (SCP). SCPs are used as animal feed additives replacing such historical protein supplements as powdered milk, soybean meal, and fish meal (see Foods, nonconventional). The only commercial large-scale facility is ICI's 50,000–75,000 t/yr plant in Billingham, UK. At this time, SCP does not appear to be economically competitive with conventional protein sources but reliability in supply and quality, as well as predictable costs, may make SCP an important market for methanol in the late 1980s.

Methanol is also considered for use in many other areas. These include its use as feedstock to produce olefins (see Feedstocks), as a reducing-gas source for steel mills, to remove nitrogen from sewage sludge (see Water, sewage), and use in fuel cells (see Batteries, fuel cells). At this time, these potential uses are not expected to become a reality until the very late 1980s.

Methanol has kept pace with changing technology in the past and will continue to do so in the future. Because the uses for methanol are also changing and adapting to new technology advances, methanol promises to continue to be one of the main feedstock chemicals for industry in the future.

BIBLIOGRAPHY

"Methanol" in *ECT* 1st ed., Vol. 9, pp. 31–61, by W. W. Yeandle, B. H. Sanders, and H. C. Zeisig, Jr., Spencer Chemical Company; "Methanol" in *ECT* 2nd ed., Vol. 13, pp. 370–398, by H. F. Woodward, Jr., Gulf Oil Corporation.

1. R. C. Wilhoit and B. J. Zwolinski, *J. Phys. Chem. Ref. Data* **2**(Suppl. 1), 40 (1973).
2. S. S. Chen, R. C. Wilhoit, and B. J. Zwolinski, *J. Phys. Chem. Ref. Data* **6**, 105 (1977).

3. *Fire Protection Guide on Hazardous Materials*, 7th ed., National Fire Protection Association, Boston, Mass., 1978.
4. *Chemical Safety Data Sheet*, *Manual Sheet SD-22*, Manufacturing Chemists' Association, Washington, D.C., revised 1970.
5. Y. V. Efremov, *Russ. J. Phys. Chem.* **42**, 10003 (1968).
6. H. F. Gibbard and J. L. Creek, *J. Chem. Eng. Data* **19**, 308 (1974).
7. M. L. Glashan and A. G. Williamson, *J. Chem. Eng. Data* **21**, 196 (1976).
8. S. Z. Mikhail and W. R. Kimel, *J. Chem. Eng. Data* **6**, 533 (1961).
9. P. S. Albright and L. J. Gosting, *J. Am. Chem. Soc.* **68**, 1061 (1946).
10. L. Riedel, *Chem. Ing. Technol.* **23**, 465 (1951).
11. C. L. Yaws, *Physical Properties, a Guide to the Physical, Thermodynamic and Transport Property Data of Industrially Important Chemical Compounds*, McGraw-Hill, New York, 1977, p. 215.
12. T. W. Yegovich, G. W. Swift, and F. Kurata, *J. Chem. Eng. Data* **16**, 222 (1971).
13. L. H. Horsley, *Azeotropic Data III, Advances in Chemistry Series*, Vol. 116, American Chemical Society, Washington, D.C., 1973.
14. J. A. Monick, *Alcohols, Their Chemistry, Properties and Manufacture*, Reinhold Book Corp., New York, 1968, pp. 93–101.
15. Ger. Pat. 293,787 (March 8, 1913), (to Badische Anilin-und Soda-Fabrik).
16. Fr. Pat. 540,543 (Aug. 19, 1921), G. Patart.
17. U.S. Pat. 3,326,956 (June 20, 1967), P. Davies and F. F. Snowdon (to Imperial Chemical Industries Limited).
18. P. L. Rogerson, *Chem. Eng. Prog. Symp. Ser. No. 98* **66**, 28 (1970).
19. H. Hiller, F. Marschner, and E. Supp, *Chem. Econ. Eng. Rev.* **3**, 9 (Sept. 1971).
20. S. Strelzoff, *Chem. Eng. Prog. Symp. Ser. No. 98* **66**, 54 (1970).
21. N. M. Nimo and M. J. Royal, *Hydrocarbon Process.* **48**, 147 (March 1969).
22. D. D. Mehta and D. E. Ross, *Hydrocarbon Process.* **49**, 183 (Nov. 1970).
23. M. J. Pettman and G. C. Humphreys, *Hydrocarbon Process.* **54**, 77 (Jan. 1975).
24. A. Pinto and P. L. Rogerson, *Chem. Eng. Prog.* **73**, 95 (July 1977).
25. A. Pinto and P. L. Rogerson, *Chem. Eng.* **84**, 102 (July 4, 1977).
26. *Hydrocarbon Process.* **58**, 191 (Nov. 1979).
27. H. Hiller and F. Marschner, *Hydrocarbon Process.* **49**, 281 (Sept. 1970).
28. E. Supp, *Chem. Technol.* **3**, 430 (July 1973).
29. *Hydrocarbon Process.* **58**, 192 (Nov. 1979).
30. H. Takahashi and T. Yoshito, *Chem. Econ. Eng. Rev.* **6**, 21 (Nov. 1974).
31. *Hydrocarbon Process.* **58**, 193 (Nov. 1979).
32. *Chem. Econ. Eng. Rev.* **2**, 33 (April 1970).
33. J. H. Prescott, *Chem. Eng.* **78**, 60 (April 5, 1971).
34. *Chem. Week*, 36 (April 16, 1980).
35. du B. Eastman, *1959 Fifth World Petroleum Congress* (Sect. IV), 153 (June 3, 1959); C. J. Kuhre and C. J. Shearer, *Hydrocarbon Process.* **50**, 113 (Dec. 1971).
36. G. E. Weismantel and L. Ricci, *Chem. Eng.* **86**, 57 (Oct. 8, 1979).
37. G. W. Bridger, *Chem. Process Eng.* **53**, 38 (Jan. 1972).
38. J. P. Leonard and M. E. Frank, *Chem. Eng. Prog.* **75**, 68 (June 1979).
39. B. M. Harney and G. A. Mills, *Hydrocarbon Process.* **59**, 67 (Feb. 1980).
40. R. I. Kermode, A. F. Nicholson, and J. E. Jones, Jr., *Chem. Eng.* **87**, 111 (Feb. 25, 1980).
41. E. Supp, *Oil Gas J.* **78**, 73 (June 16, 1980).
42. *Eur. Chem. News* **33**, 48 (Nov. 19, 1979).
43. *Chem. Eng. Prog.* **76**, 89 (March 1980).
44. V. M. Cherednichenko, dissertation, Karpova, Physico Chemical Institute, Moscow, U.S.S.R., 1953.
45. L. Bissett, *Chem. Eng.* **84**, 155 (Oct. 24, 1977).
46. W. Kothowski, *Przem. Chem.* **44**(2), 66 (1965).
47. O. A. Hougen, K. M. Watson, and R. A. Ragatz, *Chemical Process Principles, Part 2*, John Wiley & Sons, Inc., New York, 1959.
48. R. H. Ewell, *Ind. Eng. Chem.* **32**, 149 (1940).
49. D. D. Metha and W. W. Pau, *Hydrocarbon Process.* **50**, 115 (Feb. 1971).
50. *Federal Specification O-M 232 F*, U.S. Government Printing Office, Washington, D.C., June 5, 1975.
51. *Acceptable Concentrations of Methanol*, American National Standards Institute, New York, 1971.

52. *Toxicity Evaluation of Potentially Hazardous Materials, Part III*, National Research Council–National Academy of Sciences, Washington, D.C., 1959, pp. 82–103.

L. E. WADE
R. B. GENGELBACH
J. L. TRUMBLEY
W. L. HALLBAUER
Celanese Chemical Company, Inc.

METHIONINE, $CH_3SCH_2CH_2CH_2CH(NH_2)COOH$.   See Amino acids.

METHYL ACETATE, CH_3COOCH_3.   See Acetic acid; Esters, organic.

METHYLACETYLENE.   See Acetylene.

METHYL ALCOHOL, CH_3OH.   See Methanol.

METHYLAMINES.   See Amines, lower aliphatic.

METHYL SULFATE, $(CH_3)_2SO_4$.   See Sulfuric and sulfurous esters.

MICAS, NATURAL AND SYNTHETIC

Mica [12001-26-2] is the name for a group of complex hydrous aluminum silicate minerals constructed of extremely thin cleavage flakes and characterized by near-perfect basal cleavage, and a high degree of flexibility, elasticity, and toughness. Laminae as thin as 15 μm are obtained readily and utilized.

The various micas, although structurally similar, may vary widely in chemical composition, particularly the rarer types; however, within any one variety, the composition does not vary greatly.

The properties of mica derive from the periodicity of weak chemical bonding alternating with strong bonding. Each ultimate single sheet of mica has one weakly bonded layer (potassium) and three strongly bonded layers (two silicon and one aluminum or magnesium). The ultimate sheets are about 1 nm in thickness.

The minerals of the mica group are shown in Table 1. In general, the Si:Al ratio is about 3:1 (1). The general formula gives the number of ions in the unit cell and is the repeat unit of the inorganic mica structure.

The formula shown for phlogopite in Table 1 is stoichiometric. Because it forms a continuous series with biotite (the boundary between them is arbitrary), phlogopite contains some ferrous iron substituting for some of the magnesium. Natural phlogopite contains both OH and F ions. An all-fluorine natural phlogopite does not seem to exist; however, synthetic fluorine micas of very high purity can be manufactured.

Vermiculite, a product of hydrothermal alteration of phlogopite and biotite, should not be classified commercially as a mica. Some micas are sources of rarer elements, eg, lepidolite for lithium, and roscoelite for vanadium.

All micas crystallize in flat, six-sided monoclinic crystals. Characteristic is the nearly perfect basal cleavage, which permits the crystals, or books, to be split into thin sheets. The ultimate sheet thickness depends almost entirely on mechanical considerations. Strength and elasticity of the sheet are characteristic of the commercial varieties.

Properties of muscovite, phlogopite, and fluorophlogopite are shown in Table 2. Muscovite and phlogopite are the two natural micas, and fluorophlogopite the synthetic mica most widely used in the electrical industry. The suitability of mica film

Table 1. Mica Group Minerals of the General Formula $W_2(X,Y)_{4-6}Z_8O_{20}(OH,F)_4$ [a]

Mineral	CAS Registry No.	Formula
principal types		
muscovite	[1318-94-1]	$K_2Al_4(Al_2Si_6O_{20})(OH)_4$
phlogopite	[12257-58-0]	$K_2Mg_6(Al_2Si_6O_{20})(OH,F)_4$
biotite	[1302-27-8]	$K_2(Mg,Fe)_6(Al_2Si_6O_{20})(OH)_4$
lepidolite	[1317-64-2]	$K_2Li_3Al_3(Al_2Si_6O_{20})(OH,F)_4$
others		
roscoelite	[12271-44-2]	$K_2V_4(Al_2Si_6O_{20})(OH)_4$
fuchsite	[12198-09-3]	$K_2Cr_4(Al_2Si_6O_{20})(OH)_4$
fluorophlogopite	[12003-38-2]	$K_2Mg_6(Al_2Si_6O_{20})F_4$
paragonite	[12026-53-8]	$Na_2Al_4(Al_2Si_6O_{20})(OH)_4$

[a] W usually = K; X,Y = Al, Mg, Fe^{2+}, Fe^{3+}, Li; Z = Si and Al.

Table 2. **Properties of Natural and Synthetic Micas**[a]

Property	Natural		Synthetic fluorophlogopite
	Muscovite	Phlogopite	
density, g/cm^3	2.6–3.2	2.6–3.2	2.8
hardness, Mohs	2.0–3.2	2.5–3.0	3.4
luster	vitreous	pearly to submetallic	vitreous
optical axial angle (2 V)[b]	38–47°	0–10°	14.6 ± 0.5°
orientation of optic plane to plane of symmetry	perpendicular	parallel	parallel
refractive index, n_D			
α	1.552–1.570	1.54–1.63	1.522
β	1.582–1.607	1.57–1.69	1.549
γ	1.588–1.611	1.57–1.69	1.549
maximum temperature at which no decomposition occurs, °C	400–500	850–1000	1200
specific heat at 25°C, J/g[c]	0.049–0.05	0.049–0.05	0.046
thermal conductivity perpendicular to cleavage, W/(m·K)	0.6694	0.6694	
coefficient of expansion per °C			
perpendicular to cleavage			
20–100°C	$(15-25) \times 10^{-6}$	$1 \times 10^{-6} - 1 \times 10^{-3}$	
100–300°C	$(15-25) \times 10^{-6}$	$2 \times 10^{-4} - 2 \times 10^{-2}$	
300–600°C	$(16-36) \times 10^{-6}$	$1 \times 10^{-5} - 3 \times 10^{-3}$	
parallel to cleavage			
0–200°C	$(8-9) \times 10^{-6}$	$(13-14.5) \times 10^{-6}$	$(10-11.5) \times 10^{-6}$
200–500°C	$(10-12) \times 10^{-6}$	$(13-14.5) \times 10^{-6}$	
tensile strength, MPa[d]	225–296	255–296	310–358
modulus of elasticity, MPa[d]	1.723×10^5	1.723×10^5	
water of constitution, %	4.5	3.2	0
melting point, °C	decomposes	decomposes	1387 ± 3
dielectric constant	6.5–9[e]	5–6	6.5
dielectric strength[e] (highest qualities 25–75 μm thick) at 21°C V/μm	235–118	165–83	235–157
power factor at 25°C (highest qualities), %			
at 60 Hz	0.08–0.09		
at 1 MHz	0.01–0.02	0.3	0.02
resistivity, Ω·cm	$10^{12}-10^{15}$	$10^{10}-10^{13}$	$10^{12}-10^{15}$

[a] Ref. 2.
[b] Acute angle between the optic axes of biaxial mineral.
[c] To convert J to cal, divide by 4.184.
[d] To convert MPa to psi, multiply by 145.
[e] Not specified in good quality mica.

for use in capacitors depends on the power factor. Since overheating and damage develop from high power losses, good condenser mica must have a dissipation factor of less than 0.04% at 1 MHz. The Q value, which is the reciprocal of the dissipation factor, commonly is used as a test. For good condensers, it should be at least 2500 (corresponding to a power factor of 0.04%).

Muscovite mica is virtually unaffected by the chemicals usually encountered by electrical insulation. Oil does not attack the surface; however, when combined with water and other liquids, it can enter between the laminae from the edges of the mica by capillary action. Although resistant to combustion gases, muscovite is attacked by hydrofluoric acid, molten alkali hydroxide, warm alkali carbonate, and water containing carbon dioxide. Muscovite does not react with sulfuric or hydrochloric acid; however, some reaction takes place with phosphoric acid.

Synthetic fluorophlogopite and phlogopite are attacked similarly. In addition, the latter is decomposed by sulfuric acid.

All natural micas when heated to a high enough temperature lose some water, but muscovite can be used for electrical purposes up to about 500°C and phlogopite up to 750 to 1000°C, depending on the source and application. The synthetic fluorine micas have no water in their structure and can be used up to ca 1200°C (2).

Natural Micas

Mica is widely distributed as small flakes in many igneous, metamorphic, and sedimentary rocks (2–3). Some schists and kaolin deposits contain sufficient muscovite to be recovered for use as scrap mica (see also Clays).

Sheet muscovite is obtained from coarse-grained igneous rocks called pegmatites. Well-zoned pegmatites are the best source of commercial sheet muscovite. Generally, zones are successive shells, complete or incomplete, that often reflect the shape or structure of a whole pegmatite body. At best development, the zones are concentric around an innermost zone referred to as the core. The accepted nomenclature for the zones, from the outermost inward, is border zone, wall zone, intermediate zones, and core. The best quality and greatest concentration of sheet muscovite usually is found in the wall zone. Less important concentrations occur in the intermediate zones and the core margin. When the pegmatite is poorly zoned, muscovite is scattered throughout the rock (4).

Phlogopite deposits are found in areas of metamorphosed sedimentary rocks intruded by masses of pegmatite-rich granitic rocks. The phlogopite is found as veins or pockets in pyroxenite interlayered with or intersecting marble or gneiss (4).

Throughout the world, the occurrence of high quality and larger-grade sizes of sheet mica is irregular and spotty. Even in the famous Bihar mica area of India, only two or three veins in a hundred are said to be rich enough to permit development on an extensive scale (3). Large runs of book mica in commercially mined pegmatites constitute 2–6% of the rock. Pods of mica may represent up to 40% of some sections of pegmatites, but in large pegmatite bodies, mica rarely accounts for more than 2% of the host rock body (4).

Quality changes greatly from location to location within the same pegmatite. This randomness of quantity and quality makes it very difficult to delineate and quantify actual reserves or resources.

The primary source of United States scrap (flake) mica for grinding purposes is weathered alaskite bodies and pegmatites in North Carolina. The mica is recovered as a coproduct or by-product of feldspar, kaolin, and lithium processing. Flake mica also is recovered from muscovite schists. These schist bodies can range in size from three to several hundred meters thick and several kilometers long. The mica content varies but can range up to 90%.

Mining, Preparation, and Processing. The unique properties of mica and the erratic character in its natural occurrence are such that methods employed in mining and processing other types of minerals are applicable only to a limited extent in the development of mica deposits.

Mining methods for sheet-mica mines tend to be rather primitive small-scale, often underground, operations. Open-pit mining is used when feasible. Care must be taken to avoid drilling through good mica crystals. A low velocity explosive may be needed to blast around a pocket of mica. Several holes are shot at one time to avoid destruction of available mica books.

The nature, extent, and quality of many ores and minerals can be forecast from geologic evidence, but sheet mica may be distributed so unevenly in pegmatites that it is virtually impossible to estimate either the quantity or quality present. Most metallic and nonmetallic minerals can be beneficiated and either reduced or chemically treated to yield highly purified marketable products. The value of sheet mica, however, depends on the size, structure, and purity of the sheets that can be obtained by splitting and trimming the natural mica books. None of these important characteristics of sheet mica can be improved by either metallurgical or chemical processes.

Preparation of sheet mica requires time- and cost-consuming manual labor. The crude mica books are picked out of the broken rock by hand and cobbed to remove adhering dirt, rock, and defective mica. The cobbed mica is then roughly split (rifted or sheeted) into suitable thicknesses, and principal flaws are removed. The mica produced in this manner is broadly classified as untrimmed and scrap mica. It is then trimmed with a knife to a beveled edge, removing broken and ragged edges, loose scales, and other imperfections. Care is taken to provide a final sheet of maximum usable area with minimum waste.

In India, most mica is full-trimmed, ie, all imperfections in the edges are eliminated. The edges are beveled in order to facilitate subsequent splitting into thinner sheets or films. Other types of trimming also are known in the industry. Thumb-trimmed, the usual preparation of punch mica, is confined to breaking off ragged edges by hand. A knife-trimmed block is trimmed closer. Half-trimmed mica is trimmed on two adjacent edges, and in three-quarter trim the imperfections are cut from three sides. After trimming, the mica is classified according to grade (size) and quality. It is either sold as block mica or split further for film or splittings.

Film is split from block mica and other sheets that are too thin to be classified as block. Production of film mica requires skilled manual labor. The worker must be capable of splitting the mica into narrow ranges of thickness.

Splittings are processed from lower quality block and trimmed sheets too thin for block and unsatisfactory for producing film. Again, skilled manual labor is required to split the mica to the required thickness. The splittings are packed three ways: (1) Book-form splittings are laminae that are split to the desired thickness from the same book of mica then dusted with mica dust and restacked in the same form as the original book. (2) Pan-packed splittings have been put in a pan in even layers with each layer separated by a thin sheet of paper and pressed together. The stacks of mica are then bundled, packed, and shipped. (3) Loose-packed splittings are the easiest and cheapest to process. The splittings are screened to separate the different sizes, and then packed loose in shipment cases. In some instances, the splittings are dusted with mica powder to prevent them from adhering to each other (2).

Scrap mica obtained from mining, trimming, or fabrication of sheet mica is bagged and sold to mica grinders without further processing. In the United States, scrap (flake) mica is mined by large-scale quarrying methods. The flake mica, which is recovered from pegmatities, schists, or other rock, is obtained by crushing and screening the host rock and beneficiated by flotation (qv). After passing through a draining bin and sometimes a rotary dryer, the mica is ground.

Colors, Stains, and Associated Secondary Minerals. Muscovite is often called white or ruby mica to distinguish it from phlogophite, which is often called amber mica. In crystals or in thick sheets, the colors of muscovite range through varying shades of red and green, depending primarily on the iron content but also on other ions. Color, in fact, is the basis for the industrial classification of muscovite into ruby (pinkish buff, cinnamon brown, brown) and nonruby (brownish olive, yellowish olive, yellowish green, green). When split into thin films, both muscovite and phlogopite are nearly colorless and transparent, although when viewed against a very white background the green variety usually can be picked out. Staining and visible mineral (clay or other) or vegetable inclusions generally reduce the value of commercial mica.

Inclusions and intergrowths of other minerals in mica books are sometimes broadly referred to as stains. Some minerals, such as magnetite, Fe_3O_4, and hematite, Fe_2O_3, are parallel with the cleavage, but others, such as quartz, SiO_2, and zircon, $ZrSiO_4$, penetrate cleavage planes and produce pinholes.

Stains are classified as primary or secondary. The former are formed during crystallization and include mottling, a kind of stain referred to as vegetable although it is inorganic, and mineral inclusions and intergrowths. Secondary stains are formed between the cleavage planes, chiefly through the action of circulating groundwater, and include air creep; clay, iron, and manganese stains; and the true vegetable stains caused by organic matter. Primary air stains consist of gas trapped beneath cleavage surfaces in flattened pockets, tiny bubbles, or groups of closely spaced bubbles; there is doubt as to their origin. Air creep is formed by air entering at the edges of mica books and penetrating along cleavage planes, and resembles some types of air stain. Air creep is caused by rough handling during preparation of the mica, particularly by trimming with shears or a dull knife.

Primary stains are not related to depth; they are as likely to increase as they are to decrease with depth. Secondary stains occur only at or near the ground surface and are absent from parts of the deposit that are beneath the oxidized zone (2–3).

Mineral stains are the most serious primary impurities. They comprise intergrowths and inclusions of recognizable crystals or crystal groups. Magnetite and hematite are the most common inclusion minerals, but more than thirty other minerals have been identified within books of muscovite. Whereas magnetite inclusions may seriously impair the splitting qualities of muscovite, hematite inclusions appear to have little affect as far as commercial preparation is concerned. Both hematite and magnetite, however, seriously increase the conductance of mica in which they occur, and thus lower its value.

Grades, Qualities, and Specifications of Sheet Mica. Sheet mica is classified according to: preparation, including crude, hand-cobbed, thumb-trimmed, half-trimmed, three-quarter trimmed, and full-trimmed; thickness, ie, block, film, and splittings; size (see Table 3); appearance (for muscovite only), including color, flatness, stain, cracks, air inclusions, waviness, herringbone, sandblast, and others (2); electrical quality (see Table 4); thermal stability (for phlogopite) (high heat phlogopite must withstand 750°C for 30 min with less than 25% increase in thickness after cooling, and with no evident transformation, such as silvery or friable stains)—regular-quality phlogopite fails this test—detailed data on classification may be obtained from ASTM (5), ref. 2, and the National Electrical Manufacturers Association (6), which includes standards for muscovite mica splittings.

The following classifications refer to thickness:

Block mica is at least 0.18 mm thick with a minimum usable area of 6.5 cm².

Film mica is a thin sheet or sheets split from better-quality block mica. Thickness specifications range from 0.032 to 0.10 mm, with a tolerance of 0.0062 mm. The films, usually split from good-quality stained or block mica, should not be confused with the thinner residues of mica processing known as scalings, cleanings, or chillas.

Splittings are thin pieces of irregularly shaped mica with a maximum thickness of 0.03 mm. The desirable thickness is 0.018–0.025 mm. Splittings usually are, although not necessarily, made from lower qualities and smaller sizes of mica, both muscovite and phlogopite.

In addition to these categories, a classification termed thin refers to knife-trimmed mica 0.05–0.18 mm thick. Subdivisions of this category are thick-thins (0.10–0.18 mm) and thins (0.05–0.10 mm).

Clear ruby muscovite is the highest quality natural mica, and the most reliable with regard to performance. Green and dark micas are more variable, particularly with regard to dissipation factor, but may be equally valuable provided they pass the necessary laboratory tests for electrical properties. Iron, both ferric and ferrous, may substitute for Mg or Al, and is the principal constituent that gives rise to inferior electrical properties. In a quick, but destructive, test for iron, the sample is heated in an oxidizing atmosphere in a muffle furnace at 800–1000°C for 5–10 min. A silvery color indicates a low iron concentration, whereas tan to brown colors indicate higher concentrations.

Grading sheet mica for size and quality is a costly hand operation requiring skill and experience. In general, world standards are based on grading established in India, the world's principal producer and exporter of sheet mica; however, the standards are not observed with the same degree of precision everywhere.

Table 3 gives the grades of muscovite block and film mica according to sizes.

Muscovite block mica is classified by ASTM specifications (5), based on the Bengal India System, into 13 quality groups according to color and appearance as well as to stains, waves, cracks, inclusions, etc. The best is V-1, clear, which is hard, flat, or nearly so, of uniform color, and free from all stains, foreign inclusions, cracks, and other similar defects. The poorest is V-12, densely black and red stained, which though usable for some applications, may contain all ordinary defects.

A resonant-circuit test set, known as the Q-meter, tests mica and other insulation. Results of the test are in Q units, and for the best mica should be about 2500. Since Q is in reality the reciprocal of the dissipation or power factor ($Q = 1$ power factor), apparatus that actually determines the dissipation factor may be used to determine

Table 3.   Muscovite Block and Film Mica, Grades by Sizes[a]

Area of min rectangle, cm²	Dimension on side (min), mm	United States domestically produced, cmᶜ	Specifications[b]		
			ASTM and India	Madagascar	Brazil
650	100	20 × 30	OOEE Special	OOOO	OOE Special
520	100	20 × 25	OEE Special	OOO	EE Special
390	100	15 × 25	EE Special	OO	E Special
310	100	15 × 20	E Special	O	Special
235	89	15 × 15	{ ASTM-A1 Special { India Special	A1	A1
155	76	10 × 15	1	1	1
97	51	8 × 13	2	2	2
65	51	8 × 8	3	3	3
		8 × 10			
40	38	5 × 8	4	4	4
20	25	4 × 5	5	5	5
		5 × 5			
15	22	circle	5½	6	5½
6.5	19	punch	6	6	6

[a] Refs. 2, 5.

[b] Indian specifications: O = over, E = extra.

[c] The sizes listed were used during the U.S. government purchase program which ended in 1962. Since that time, U.S. sheet mica production has dropped dramatically with production in recent years being insignificant. Most of the sheet mica currently used by U.S. fabricators is imported from India, Brazil, and Madagascar.

Q. Table 4 gives the Q values for the electrical quality groups. Experience has shown that the Q-value range of ruby and white block mica, regardless of source, is 80–95% E-1, whereas the Q-value range of the various shades of green mica and rum-colored block is 45–90% E-1 (see Insulation, electric).

Scrap and Ground Mica.   Scrap mica comprises mica not usable as sheet (block, film, or splittings). Scrap mica resulting from mining, trimming, or fabricating sheet mica is bagged and sold to mica grinders without further processing. Larger-size better-quality scrap, including waste and trim from punching operations, is processed into reconstituted mica paper and board (see under Special Products).

Table 4. Q Values for Electrical Quality Groups[a]

Electrical quality designation[b]	Mica form	Requirements, Q (min)	
		At 1 MHz	At 1 KHz
E-1 Special	film	2500	1000
	block	2500	
E-1	film	2500	500
	block	1500	
E-2	film	1500	200
	block	200	
E-3	film	200	100

[a] Ref. 2.

[b] Indian.

Even from the best mines, most mica is recoverable only as scrap. The outer parts of some mica books are crushed and tangled and must be cut before they can be rifted. Structural defects render many mica books almost worthless as sources of even small commercial sheets because of buckling or imperfect crystallization. On the average, only about 3–8% of mine-run mica finally is obtained in the form of full-trimmed block mica or its equivalent in film or splittings. Furthermore, when the products are stamped from the sheet mica, much remains as trimmings.

Mine scrap, removed from mine-run mica within or near the mine, may be a sole product from pegmatite or a by-product of sheet-mica production; it varies greatly in quality. Bench scrap or shop scrap is the mica discarded during the processing of hand-cobbed mica into sheet. Factory scrap, usually the material of highest grade and quality, is the waste obtained from cutting and stamping sheet mica into pieces of definite size and shape. Scrap mica that is recovered from schists and from the beneficiation of kaolin, feldspar, and lithium pegmatites is often referred to as flake mica (2).

Ground mica is made from scrap by either dry- or wet-grinding methods. For dry-grinding, high speed hammer mills or pulverizers of the disintegrator type are used most widely. The Micronizer (Sturtevant Mill Co.) uses high pressure steam introduced at the periphery of a cylinder to keep the mica rotating at high speed. It produces mica of $<5\ \mu m$–$20\ \mu m$ (3000–625 mesh) particle size. The Majac mill (Majac, Inc., a subsidiary of Blackstone Corp.) grinds by entraining mica in two directly opposed jet streams of air or steam; particles to micrometer size may be obtained by classification.

Wet-ground mica is produced in chaser-type mills to preserve the sheen or luster. This type of mill consists of a cylindrical steel tank that is lined on the bottom with wooden blocks laid with the end grain up. The central spindle rotates at a rate that causes wooden rollers to revolve at 15–30 rpm. Hopper-fed scrap goes to the mill where water is added slowly to form a thick paste. When the bulk of the mica has been ground and delaminated to the desired size, which requires about 5–8 h, depending on feed material, the charge is washed from the mill into settling bins to separate the gritty impurities. The ground mica overflows to a settling tank and is dewatered by centrifugation and steam drying. The product is screened on enclosed multiple-deck vibrating screens, and bagged for shipment (2).

Specifications for high grade wet-ground mica require 99.5% through a $180\text{-}\mu m$ sieve (80 mesh) and 88% through a $45\text{-}\mu m$ sieve (325 mesh). Specifications for dry-ground mica vary for different uses and for different suppliers. In general, dry-ground mica for roofing passes 840-, 600-, or $420\text{-}\mu m$ sieves (20, 30, or 40 mesh); that used for paint passes 250-, 180-, or 150 μm sieves (60, 80, or 100 mesh); and that for other uses is under 250 μm (60 mesh).

Production of wet-ground mica costs several times as much as that of dry-ground mica. The former is distinguished by its sheen. Dry-ground mica, in contrast, looks more like ordinary flour, as the edges of the flakes are torn and hackly.

Synthetic Fluorine Micas

Compared to the natural-mica industry, the synthetic-mica industry is very small, and because of costs and substitution for mica products in general, will most likely remain so in the future. An excellent summary of the early history of mica synthesis is given in refs. 7 and 8.

Systematic studies in the USSR (9) from 1934 to 1943 provided the background for the modern synthesis of fluorophlogopite mica that aided in industrial application. These studies recognized the role and interchangeability of fluorine and hydroxyl in the mica structure.

About 1938, anticipating the need for mica, the synthesis of a fluorophlogopite mica was investigated in Germany and Japan (10–11). Research in the United States, largely supported by the Federal Government and mainly centered in the U.S. Department of the Interior, Bureau of Mines, resulted in the development of several production processes and synthesis of various types of mica. Currently, the only synthetic mica of any commercial significance is fluorophlogopite (see Fig. 1).

The fluorine mica program since World War II, including government-sponsored and nongovernmental research and development, is reviewed in refs. 12–13.

In synthetic fluorine micas, fluoride replaces the hydroxyl ion. They are true micas, and the potassium varieties have the characteristics of natural mica and can be split as easily. The fluorine micas are formed at atmospheric pressure by melting or solid-state reaction of the proper raw materials. Owing to the absence of hydroxyl ion, fluoromicas do not decompose as readily as natural micas and have a much higher thermal stability. By ionic substitution, a wide variety of fluoromicas may be synthesized, some of which have different or very unusual properties.

Synthetic hydroxyl micas are prepared in a closed system under pressure because they decompose on heating above 500–600°C at atmospheric pressure. Furthermore, the water vapor must be contained to furnish hydroxyl for the reaction. Hydrothermal crystallization presents problems, because of the delicate control of temperature

Figure 1. Crude synthetic fluorophlogopite as grown from an electric-furnace melt. Courtesy of Mycalex, Division of Spaulding Fibre Company, Inc.

gradients and pressures required in the sealed containers. Although many hydroxyl micas have been synthesized in the laboratory under hydrothermal conditions, their particle size and yield have been too small, and their cost too high, for any commercial application. The advantage of completely replacing hydroxyl groups with fluorine is that the fluorine mica can be crystallized from a melt at atmospheric pressure.

Properties. The properties of fluorophlogopite are given in Table 2. In general, the physical, electrical, and chemical properties of fluorophlogopite are similar to those of the better grades of natural muscovite or phlogopite. Single crystals of fluorophlogopite are clear and transparent, and readily cleave parallel to the (001) crystallographic plane. The main differences are the higher heat stability (mp = 1387°C), and the high purity of fluorophlogopite which contains <0.04% Fe_2O_3.

Phase Relations, Batch Compositions. The fluoride ion is essential for the synthesis of fluoromicas. Without a sufficient quantity of fluorine present, mica is formed in lesser amounts, and with some compositions fluoroamphiboles may be formed instead. The theoretical amount of fluorine needed must be present, eg, four F ions for fluorophlogopite, $K_2Mg_6Al_2Si_6O_{20}F_4$. Ordinarily, a slight excess of fluoride compensates for small but inevitable fluorine losses owing to volatility of HF, AlF_3, SiF_4, and KF at the process temperature needed. The HF is derived from reaction of the fluorides with the adsorbed water present on the ground raw materials. A 12 mol % excess of fluoride is usually satisfactory, ie, $F_{4.5}$ in the above formula rather than F_4. Crystalline fluoromica has the theoretical content of F_4 even if excess fluorine is used. Excessive amounts of fluorine in the batch formula give rise to larger amounts of secondary crystalline phases (particularly MgF_2) plus glass.

Fluorophlogopite is prepared easily in small or large amounts from potassium fluorosilicate, feldspar, magnesia, alumina, and quartz sand. The <150-μm (−100 mesh), or finer, raw materials are mixed and fired to temperatures above melting (1387°C). Crystallization in crucibles may be accomplished by stopping heating and letting the furnace cool at its natural rate or by programmed reduction of temperature. Larger crystals are obtained by slow cooling (below 5° and preferably 1°C/h) through the crystallization range. Fluorophlogopite has been grown in sheets up to 7.5 × 12.5 cm.

Potassium carbonate may be used if the sodium from feldspar is not desired. Sodium does not appreciably enter the mica structure if sufficient potassium is present; instead, it forms secondary phases such as $NaMgF_3$.

Typical batch compositions used for the preparation of potassium fluorophlogopite are given in Table 5. Other fluoromicas may be synthesized with batch compositions calculated from the desired formula, usually with a 12.5 mol % excess of fluorine. The mixture is heated above the melting temperature and cooled. The raw materials should not be wet. Raw materials that give off water at elevated temperatures preferably should not be used because fluorine escapes as HF. Impurities, such as Fe, may be dispersed throughout the melt (14) in the form of small globules of metallic alloy and should be avoided. With pure raw materials, large quantities of synthetic fluoromicas can be produced easily in very high and reproducible purity.

Containers and Crucibles. The most satisfactory crucibles for fluoromica melts are made of platinum or graphite. The crystallized mass sticks to platinum but can be dislodged easily from a tapered graphite crucible. The graphite is protected from oxidation by placing it inside a fire-clay crucible which is sealed tightly with a cover. Other crucible construction materials are silicon carbide, silicon nitride, magnesium

Table 5.   Commercial Batch Compositions for K-Fluorophlogopite Production

	For $F_{4.5}$ content		From K_2CO_3, wt %[a]	F_4, theoretical[c], from K_2CO_3, wt %[a]
	From feldspar, wt %[a]	Typical commercial raw materials, wt %[b]		
K_2SiF_6	19.33	19.35	19.23	17.14
feldspar[d]	25.71	25.56		
MgO	28.30	28.62	28.15	28.21
Al_2O_3	7.50	7.45	11.87	11.89
SiO_2	19.17	19.02	36.72	37.38
K_2CO_3			4.02	5.37

[a] Figures are for anhydrous, pure compounds; corrections must be made for both ignition loss and impurities in the actual batch materials used. Feldspar is corrected for impurities but not ignition loss.

[b] Weights of raw materials corrected as follows: K_2SiF_6, 1.00 wt %; feldspar, 0.33 wt % for ignition loss only; MgO, 2.00 wt % for ignition loss and impurities; Al_2O_3, 0.20 wt % for ignition loss and Na_2O; SiO_2, 0.15 wt % for ignition loss.

[c] For comparison only, formula is not used commercially.

[d] Feldspar used here had this chemical analysis in wt %: SiO_2, 68.8; Al_2O_3, 17.2; Na_2O, 2.8; K_2O, 10.7. A significant change in feldspar composition would require alteration of the batch composition.

oxide, or aluminum nitride. However, these must be broken to remove the fluoromica melt. Dense fire-clay crucibles corrode rapidly and may not be used for long firings but are satisfactory for short firing times with reducible substances. Any type of crucible should have a well-fitting cover.

Internal-Resistance Melting.   The fluoromicas are prepared by melting or solid-state reaction of the raw materials at atmospheric pressure. For large-scale production, up to 45 metric tons, an internal-resistance electric furnace is employed (12–13). It consists of a shell to hold the batch, a pair of electrodes positioned to generate heat by electric resistance in the central interior of the batch, and a variable transformer to supply electric current at the appropriate strength.

Both single- and double-wall furnaces are employed; the former is cascade water-cooled. With both vertical and starting electrodes in the position shown in Figure 2, the furnace is filled with a previously mixed batch. When the batch reaches the level of the apex of the starting electrodes, about 150 g powdered graphite or lampblack is so placed as to form a continuous electrical path between the electrodes. The furnace is then filled to about 30 cm above the electrodes. The electrodes range 5–10 cm in diameter. The furnace walls and bottom are constructed from sheets of cold-rolled iron, stainless steel, or Monel (see also Furnaces, electric).

Melting is started by applying 10–15 V and about 100 A of single-phase, a-c power. The voltage is raised until, after several hours, >30 V may be applied. When the melt has reached the main electrodes, and particularly when the starting electrodes have burned, the resistance rises and more power must be applied. At this stage, typical power values for 3–9-t melts are 100 V and 200–300 A. If melting proceeds too rapidly during the early stages, the melt spreads into the upper half of the furnace and does not drop sufficiently to give rapid and maximum melting. When the melt is well underway, power may be increased to 20–50 kW in smaller furnaces, 150–300 kW in larger furnaces.

At this point, more batch mix is fed in increments, either through a hole in the crown, or by caving in the entire crown and adding more raw material on top. Feeding

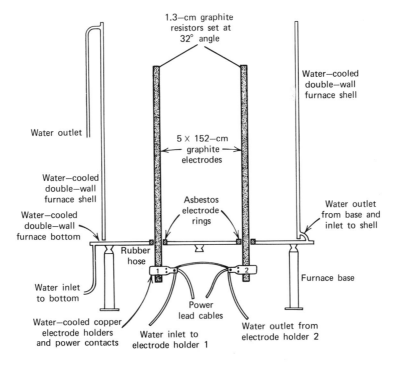

Figure 2. A double-wall, water-cooled, internal-resistance electric furnace.

in increments prevents a high rise in the electrical resistance. The fumes, consisting of the fluorides of K, Al, and Si, must not be inhaled. A liner of sheet aluminum, ca 0.81 mm thick, greatly extends the life of the furnace and prevents contamination of the melt with iron or alloy elements.

To achieve maximum melting, the furnace must be water-cooled. Hot spots in a single-wall furnace should be cooled locally and, if necessary, the power cut off for a few minutes. Melting is continued until the liquid is within 30–40 cm of the bottom. Power is then cut off and the furnace is allowed to cool at its natural rate. The cooling water is continued to prevent furnace damage.

About 900 g fluoromica per kW·h is obtained by this process.

Melts of over 31 t have been made by using three-phase power employing three starting electrodes. No significant operating difficulties were encountered.

Arc-Resistance Melting.   For arc-resistance electric melting, a slightly conical, single-wall furnace is preferred (see Fig. 3). It is water-cooled on the outside by cascade. The two graphite electrodes are inserted from the top and are raised during melting.

At the start, mixed raw materials or fluoromica is placed in the bottom of the furnace in a depth of ca 30 cm. An arc is started and a small pool of melt obtained. Mixed batch is then added in small increments and the power is increased. After start-up, batch mix is added in sufficient amounts to cover the liquid to a depth of as much as 30 cm. A crown is not needed, and if it forms, it is broken, pushed down to the liquid, and more batch is added. Feeding, increase of power, and raising of electrodes are continued until the furnace is nearly full. At this time, the electrodes are

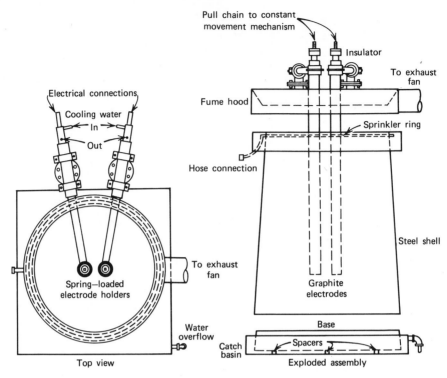

Figure 3. A cascade water-cooled, arc-resistance furnace. For less carbon contamination, the electrodes should be separated as far as possible.

removed from the melt, the batch is placed on top, and the furnace operation discontinued.

Typically, the electrodes are immersed in liquid ca 30–45 cm during the main melting operations. Power input may range 80–140 V at 1000 A or higher. Yield of fluoromica may be 1.5 kg/kW·h or better. If the electrodes are separated as much as possible, the product is nearly white and has less carbon contamination than if the electrodes are closely spaced (13).

Solid-State Reaction. Fluoromicas also may be prepared by solid-state reaction. Fluoromuscovite may be synthesized at temperatures of 600–700°C but is not stable on melting.

Starting material is a mixture of finely divided (75 μm) fluorides, fluorosilicates, oxides, carbonates, silicates, or other compounds in the proper stoichiometric proportions. A slight excess (12.5 mol %) of fluoride is desirable. Although not usually needed, highly reactive solids, such as clay dehydrated at 800°C, and lightly calcined $Al(OH)_3$ or $Mg(OH)_2$, may be used.

Solid-state reactions yielding fluoromicas start at 800°C or less; however, 1000–1100°C is preferred because of a greatly increased rate. Containers may be constructed of any material that withstands such temperatures since reaction with the container is nearly negligible. To prevent fluoride loss, the container is tightly covered. Some raw materials, particularly those commonly used to synthesize fluorophlogopite, expand during solid-state reaction and may break a large rigid crucible, especially if the batch has been tightly packed before firing.

Product Size and Quality. Both the internal- and arc-resistance melting processes give typically clear, crystalline fluoromica with only small amounts of secondary phases in yields of usually better than 97%. Under proper conditions, the product should be nearly white and free of carbon contamination. The secondary phases do not present problems in glass-bonded mica and are ground with the mica without separation.

Single crystals up to 10×13 cm in size have been grown in the melts but their yield is low and unpredictable. However, for special applications, clear crystals, 2.5 cm and larger, sometimes are separated and sold. Most crystals are 2 to 1.3 cm and smaller. Since the mica pig is a tough mass of interlocking crystals, separation of large crystals without damage is difficult. It appears that single crystals grow into the liquid ahead of the main crystallization front. Since in the absence of a strong thermal gradient their growth is more random than parallel, pockets of liquid are left which severely limit the crystal size of the remaining fluoromica.

Milling. The fluoromica pig may be broken with explosives, followed by air hammer and jaw crusher. For glass-bonded mica or other applications, the material may be further treated in a hammer mill, jet-impact mill, wet-ball mill, or vibratory-ball mill, either wet or dry (see Size reduction).

Isomorphic Substitution. A large number of substitutions may be made in the fluoromica structure (12,15). In fluorophlogopite, $K_2Mg_6Al_2Si_6O_{20}F_4$, the potassium ion is in twelvefold coordination with the anions, the Mg^{2+} ion sixfold or octahedral, and $AlSi_3$ fourfold or tetrahedral. Fluorine occupies the position which in natural micas mostly is filled with hydroxyl ion. Substitutions can be made in the following order of decreasing ease of synthesis and stability: for potassium, Ba, Pb, Sr, Na, Li, Rb, Tl, or Cs; for magnesium, Li, Ni, Fe, Al, Co, or Mn; for aluminum, Si, B, Fe, or Be; and for silicon, Ge or Al. Partial substitution of some other ions may be made but only by solid-state reactions. No successful substitution for F has been found. Disilicic, trisilicic, and tetrasilicic fluoromicas have been made.

Special Products

Built-Up Mica. Built-up mica is sheet or board made by alternate layers of over-lapping splittings and suitable binder, consolidated by heat and pressure. Temperatures of up to 150°C or more are used, and pressures from 0.34 to 6.9 MPa (3.4–68 atm).

The manufacture of built-up mica comprises a number of steps, the most critical of which is the distribution of splittings in thin, overlapping layers of uniform thickness. Each successive layer is no more than 25, 50, or 75 μm thick and must be as uniform and homogeneous as possible without bare or thick spots. Formerly this operation was performed largely by hand but today mechanical suction devices are employed to handle the dry, loose splittings.

The type and proportion of binder varies according to the purpose for which the built-up material is intended. The most widely used binders contain shellac, epoxy, polyester, alkyd vinyl, and silicone resins dissolved in suitable solvents for ready distribution.

Glass-Bonded Mica. Glass-bonded mica, a composite material, combines the molding ability of glass with the superior electrical properties of mica. It is machinable, and smooth surfaces with close dimensional tolerances are obtained. By using preheated precision dies, the product may be made by compression or injection molding in a one-shot, rapid process. Glass-bonded mica does not change with age. It does not char or track with an arc, and it is essentially unaffected by oxygen, ozone, or contaminants in the atmosphere. Metallic inserts may be molded in place, and by appropriate techniques, the product may be plated with conducting films.

Glass-bonded mica may be made from natural or synthetic mica, or from combinations of both. The glasses may be lead borate or lead borosilicates for natural mica, and lead aluminum silicate or other silicates for synthetic fluoromicas. The high temperature stability of the fluoromicas allows the use of higher temperature glasses that possess better electrical qualities. Typically, 250-μm mica is mixed thoroughly with 75-μm powdered glass and 5–10% water. The composition of compression molding is generally between 60 wt % mica and 40 wt % glass, whereas for injection molding it is between 40 wt % mica and 60 wt % glass. The mixture is cold-pressed into preforms which are then dried or cured, ie, they are heated slowly in a gas or electric furnace to 50–1100°C, and then transferred rapidly to a mold or die at 340°C. Pressure of 34–69 MPa (4900–10,000 psi) is applied immediately. The molded pieces are then annealed. Possible metal inserts are set in the die before forming. The process is essentially similar to the molding of plastics (see Plastics processing).

Glass-bonded mica may be sawed, drilled, tapped, milled, turned on the lathe, or ground. In fact, practically all machine operations except punching can be performed on glass-bonded mica. Ranges of physical, electrical, and thermal properties for typical glass-bonded mica materials are shown in Table 6.

Phosphate-Bonded Mica. A heat-resistant mica board bonded with soluble aluminum phosphate is made from ground mica by application of elevated temperature and pressure. It may be punched after forming or made into various shapes. Phosphate-bonded mica is used in appliances, electric heaters, and other applications requiring a strong, heat-resistant insulator.

Reconstituted Mica Sheet or Paper. Reconstituted mica sheet or paper is made up of small, thin platelets or flakes of mica. It has substantially the same overlapping structure as built-up mica but on a micro scale. The flakes in reconstituted mica paper are usually in the range of 150–850 μm (100–20 mesh), less than 1 mm in their largest dimension. In most mica paper 150-μm particles are kept to a minimum or eliminated. Although similar in appearance to cellulose paper, the paper before bonding is 100% mica (17–18) (see also Pulp, synthetic).

Mica paper is made from a water slurry on a modified Fourdrinier papermaking machine (see Paper). As ordinarily produced, mica paper is a continuous, uniform, flexible, self-supporting sheet, up to 1 m wide. It has a dielectric strength of 24 V/μm or more, a tensile strength of 6.9–20.7 MPa (1000–3000 psi) or higher, a thickness variation of as little as 10%, resistance to thermal and mechanical shock, nonflammability, low power factor, high resistance to radiation, and high arc, corona, and insulation resistance. In other words, it has the excellent properties of the mica from which it is made. The continuous mica sheet may be as thin as 12–25 μm but thicknesses of 25 to 100 μm are more common and are prepared by lamination.

Before use, most mica paper or laminates are bonded with resins, silicone, glass, or borate–phosphates (19). The product competes with built-up mica and is superior in many respects. It has been used successfully as punchings for electron tubes.

The slurry of mica particles is prepared by one of two general processes (2).

The Heyman or Integrated Mica process takes advantage of the ability of nascent or virgin surfaces of thinly split mica to recohere. Mica is split by high velocity water

Table 6.   Ranges of Physical, Electrical, and Thermal Properties for Typical Glass-Bonded Mica Materials [a]

Properties	Molded materials		Machined materials	
	Synthetic mica	Natural mica	Synthetic mica	Natural mica
dissipation factor at 1 MHz	$(1.5\text{--}2.3)) \times 10^{-3}$	$(1.1\text{--}2.5) \times 10^{-3}$	$(1.3\text{--}1.7) \times 10^{-3}$	1.8×10^{-3}
dielectric constant at 1 MHz	8.8–6.7	9.3–6.6	6.8–6.9	6.7
loss factor at 1 MHz	0.02–0.01	0.01–0.016	0.12–0.009	0.012
dielectric strength, 3 mm thick, V/μm	11–14	14–16	15–16	16
arc resistance, ASTM s	300–250	250–200	325–300	300
endurance, max temp, °C	345–360	345	370–595	370
thermal expansion, 10^{-6}/°C	11.4–9.4	10.7–10.0	11.2–9.4	10.5
thermal conductivity, W/(m·K)	0.628–0.502	0.460–0.418	0.586–0.502	0.418
tensile strength, MPa[b]	34–28	34	41–34	41
compressive strength, MPa[b]	267–172	267–172	267–220	310
flexural strength, MPa[b]	83	96–83	103–83	103
modulus of elasticity, 10^4 MPa[b]	8.27–5.51	5.51–5.17	8.27–7.30	7.58
impact strength, Charpy, J/cm^2 [c]	0.420–0.253	0.315–0.253	0.504–0.231	0.420
water absorption in 24 h at 3 mm thickness, %	nil	nil	nil	nil
specific gravity, g/cm^3	3.8–2.7	3.8–2.6	3.0–2.8	3.0

[a] Ref. 16.
[b] To convert MPa to psi, multiply by 145.
[c] To convert J/cm^2 to lbf/in.^{2}, divide by 0.210.

jets acting like a knife in a high speed disintegrator having no moving mechanical parts. The disintegrator is made up of several chambers one above the other. The mica is fed into the bottom chamber through a funnel by an automatic vibrator. At the sides and top of the chamber, jets give the water a rotating motion. The suspended mica particles are thrown outward by centrifugal force and become oriented parallel to the walls of the chamber. The jets split the whirling mica flakes into finer particles. Thin flakes rise into succeeding higher chambers and, when fine enough, pass out of the system. The mica is fed into a tank where it is allowed to settle and rest evenly on a belt. The belt passes over a suction box and around a vacuum wheel to remove most of the water. The moist sheet then travels through a heater where it is dried. The finished sheet usually is impregnated with a suitable resin.

    In the Samica or Bardet process, clean scrap mica is heated to partial dehydration. The hot mica is quenched in an alkaline solution. After cooling, the excess solution is allowed to drain, and the mica is immersed in a dilute sulfuric acid solution. The reaction between the alkaline and acid solutions generates a gas between the mica laminae, causing the mica to expand greatly. At this point, the mica laminae require only washing and proper agitation to be suitable for mica paper production.

A General Electric process, which is a modification of the Samica process, delaminates the mica less drastically. The scrap mica is heated briefly to remove a small percentage of combined water; this partly exfoliates the mica. It is then chopped in water over a screen with revolving knives in a hammermill-type disintegrator. The mica pulp is removed to a constant-level tank from where it is discharged into the papermaking machine, which produces a continuous sheet of mica paper. The paper is dewatered by passing over suction boxes and gravity draining. The dewatered sheets are passed into a steam-heated drum dryer and rolled onto cores. The mica flakes are held together by the mutual attraction of each flake (2).

Reconstituted mica paper combined with other sheeting materials forms laminates for special applications. Among these are cotton cloth, polyester film, glass cloth, and silicone-varnished cloths (16) (see Laminated and reinforced plastics).

Mica Substitutes. Mica-based substitutes for sheet mica include built-up, reconstituted mica sheets (mica paper), and glass-bonded and phosphate-bonded micas. As yet, no process has been developed to produce large crystals of synthetic mica. Several materials not based on mica are used as a substitute for natural mica, including alumina ceramics, bentonite, glass, fused quartz, talc, and silicone polymers. Organic mica substitutes include polytetrafluoroethylene, nylon, polystyrene, and Mylar. None of these matches all the electrical and thermal properties of natural mica.

Alternative systems and components, such as solid-state devices, ceramic capacitors, and miniaturization, have reduced the need for mica, a trend that is expected to continue into the future. However, owing to some of its unique electrical and thermal properties, sheet mica will remain in some demand for items such as gauge glass and diaphragms and speciality applications.

Economic Aspects

Table 7 shows United States mica production and consumption through 1979. Only small quantities of low grade sheet mica are produced in the United States, generally recovered as a hand-picked by-product of feldspar mining. There is little likelihood that this situation will change.

Figures 4 and 5 show the generalized supply–demand relationship for sheet mica and scrap flake and ground mica in the United States (19).

Table 8 gives the United States consumption of mica splittings, whereas Tables 9 and 10 give the distribution of built-up mica and ground mica, respectively. Distribution of muscovite block and film consumption is shown in Table 11; total United States consumption of phlogopite in 1979 was 5.35 metric tons.

Extensive production data (United States and worldwide) are available in the *Minerals Yearbook*, published annually by the U.S. Department of the Interior, Bureau of Mines.

The Mica Trading Corporation of India (Mitco) plays a vital role in the worldwide marketing of sheet mica. In addition to regulating and setting prices for various mica types and grades, this agency controls 40–50% of India's mica export. Since India supplies an estimated 80% of world demand for sheet mica, Mitco exerts great power over the international mica market.

Table 7.   United States Production and Consumption of Natural Mica, and World Production[a]

	1960–1965[b,c]	1966–1969[b]	1970–1977[b]	1978	1979
Production					
sheet mica, t	192	6	5	d	d
value, 000 $	1,342	2	6	d	d
scrap (flake) mica, 10³ metric tons	98	112	122	126	122
value, 000 $	2,892	3,129	4,908	7,916	7,708
ground mica, 10³ metric tons	102	99	110	112	111
value, 000 $	6,412	6,783	9,448	12,979	14,522
Consumption					
block and film, t	1,187	897	434	112	128
value, 000 $	3,347	2,896	1,809	1,362	1,866
splittings, t	3,101	2,625	2,199	2,512	2,213
value, 000 $	2,899	2,546	2,367	3,031	3,248
exports, 10³ metric tons	4	7	7	8	11
imports, 10³ t	9	5	5	6	9
world production[e], 10³ metric tons	165	152	210	242	239

[a] Ref. 20.
[b] Average.
[c] This was a period of U.S. government stockpiling and price support.
[d] Less than ½ unit.
[e] Estimated.

Table 8.   Consumption of Mica Splittings in the United States, By Source[a]

	India		Madagascar		Total[b]	
Year	Quantity, t	Value, 10³ $	Quantity, t	Value, 10³ $	Quantity, t	Value, 10³ $
1975	2098	2529	54	104	2153	2634
1976	2224	3084	55	142	2279	3226
1977	1805	2525	75	193	1880	2718
1978	2436	2837	75	194	2512	3031
1979	2138	2745	74	503	2212	3248

[a] Ref. 20.
[b] Data may not add to totals shown because of independent rounding.

Prices and Production Costs.   Prices for sheet mica are based upon type of material (muscovite or phlogopite), grade, and quality. Block and film prices range <$1.00–300/kg. Splittings prices range $0.50–10/kg and higher. Prices also reflect the method of packing. More economic data are given in refs. 19–21.

Reported prices for scrap and flake mica range from ca $22–90/t and sometimes higher, depending on quality. Average time–price relationships for flake as well as ground mica can be found in refs. 19–21. As of April, 1980, the average quoted price for dry-ground mica ranged $77–110/t; prices for wet-ground mica ranged $265–400/t, depending on grade (22).

Labor and transportation are the two largest cost items in sheet mica production.

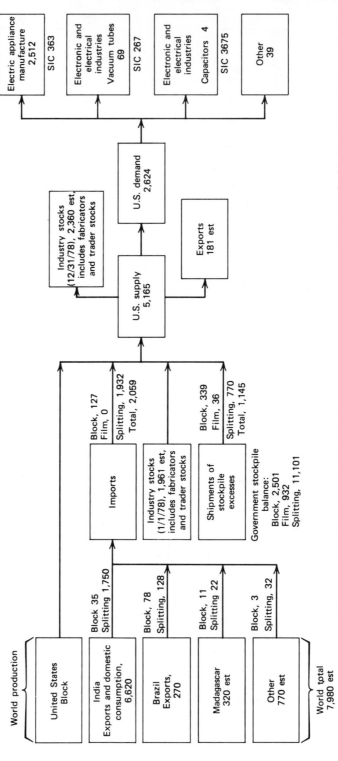

Figure 4. United States supply–demand relationships for sheet mica, 1978, in metric tons (19). SIC = standard industrial classification.

434

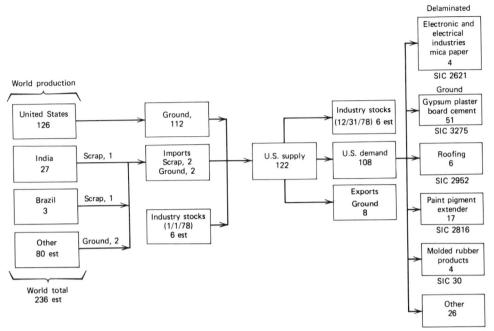

Figure 5. United States supply–demand relationships for scrap and flake mica, 1978, in thousand metric tons (19). SIC = standard industrial classification.

Table 9. Built-Up Mica[a] Sold or Used in the United States, By Product[b]

	1977		1978		1979	
Product	Quantity, t	Value, 10^3 $	Quantity, t	Value, 10^3 $	Quantity, t	Value, 10^3 $
molding plate	557	2,751	653	3,062	703	3,951
segment plate	639	3,787	707	3,892	707	4,423
heater plate	78	249	75	329	76	485
flexible (cold)	280	1,804	262	1,914	288	2,276
tape	351	3,414	398	3,018	337	2,721
other	157	1,246	147	1,301	182	1,801
Total[c]	*2,062*	*13,251*	*2,241*	*13,516*	*2,293*	*15,657*

[a] Consists of alternate layers of binder and irregularly arranged and partly overlapped splittings.
[b] Ref. 20.
[c] Data may not add to totals shown because of independent rounding.

The high cost of skilled labor in the United States has greatly reduced sheet mica production. Increasing energy costs for drying and wet grinding also are reflected in market prices.

Health and Safety Factors

At present, no definite health hazard caused by mica has been established. Its crystal lattice does not produce fibers although some deposits and products may contain fibers of other minerals as impurities. Mica produces bladed or platy particles when ground or milled.

Table 10.   Ground Mica Sold or Used by United States Producers, By Use [a]

Use	1977 Quantity, thousand metric tons	1977 Value, 10^3 $	1978 Quantity, thousand metric tons	1978 Value, 10^3 $	1979 Quantity, thousand metric tons	1979 Value, 10^3 $
roofing	3	200	6	635	W[b]	W[b]
rubber	5	1,202	4	1,044	4	1,177
paint	23	2,703	19	2,367	17	2,233
joint cement	52	4,481	53	4,898	57	6,315
other uses[c]	29	3,320	31	4,034	33	4,796
Total[d]	*111*	*11,906*	*112*	*12,979*	*111*	*14,522*

[a] Ref. 20.

[b] W = Withheld to avoid disclosing company proprietary data; included in other uses.

[c] Included mica used for agricultural products, molded electric insulation, plastics, welding rods, well drilling, textile and decorative coating, wallpaper, and uses indicated by symbol W.

[d] Data may not add to totals shown because of independent rounding.

Table 11.   Distribution of United States Muscovite Consumption, 1979 [a]

Block and film uses	Metric tons
capacitors	2.36
vacuum tubes	76.93
other electronic devices	22.18
gauge glass and diaphragms	2.90
other nonelectronic devices	18.19

[a] Ref. 20.

Granuloma and functionally noticeable fibrogenesis caused by mica dust in humans is reported only after very prolonged exposures to high concentrations (23). Micas have not been identified as carcinogenic, clinically or experimentally (23). An extensive bibliography pertaining to the health effects of mica can be found in ref. 23.

Uses

Sheet Mica.   *Block and Film.* The various uses of sheet mica take advantage of its unique electrical and thermal insulating properties, and of its ability to be cut, punched, or stamped to very fine tolerances. Electronic and electrical industries are the main users of sheet mica; the largest quantity of block and film mica is consumed in the fabrication of vacuum-tube spacers. These spacers serve to position, insulate, and support the tube elements within the envelope. Backing mica is used on the stack top and bottom to give rigidity to capacitors. Block mica is fabricated into washers of various sizes to act as insulators in electronic apparatus.

Transparent and extremely flat pieces of very high quality mica are used to line the gauge glasses of high pressure steam boilers. Diaphragms for oxygen equipment are made of high quality block muscovite. Other uses include marker dials for navigation compasses, quarter-wave plates of optical instruments, pyrometers, thermal regulators, stove windows, lamp chimneys, microwave windows, and hair-dryer elements.

A process has been developed for coating sheets of high quality mica with gold and silver in specific patterns (24). These coated disks and stamped sheets are used in communication devices.

In helium–neon lasers, high quality natural sheet mica is used as retardation plates (see Lasers). Various industries have used mica as a special optical filter (see Filters, optical). Mica washers are used extensively by the computer industry. The small thin mica disks serve as gap separators in recording heads.

The planar surface of the mica used as gap separators and other specialties must be clear, unscratched, and untouched by hands. Disks are punched out of sheets, collected, collated with paper separators, and stacked together. The stacks are then put into a lathe vise and machined to the desired diameter to fit the close tolerances demanded by the customer.

Phlogopite is inferior to muscovite in many electrical properties but replaces muscovite as an electrical insulator where its ability to withstand temperatures up to 1000°C, compared with 500°C for muscovite, is applied.

Splittings and Built-Up Mica. Built-up mica from splittings serves as a substitute for natural sheet in electrical insulation applications. Built-up mica is the largest use of sheet mica and indispensable as an insulating medium in the electrical industry. It is used for segment plate, molding plate, flexible plate, heater plate, and tape.

Segment plate, the largest use of built-up mica, acts as insulation between the copper commutator segments of d-c universal motors and generators.

Molding plate is the sheet from which V-rings are cut and stamped for the insulation of copper segments from the steel shaft at the ends of a commutator. Molding plate is also fabricated into tubes and rings for insulation in transformers, armatures, and motor starters.

Flexible plate is used in electric motor and generator armatures, field-coil insulation, and magnet and commutator core insulation. Heater plate is used where high insulation strength at high temperatures is required.

In some types of built-up mica, the bonded splittings are reinforced with glass cloth, polyester film, Dacron mats, and varnished glass cloth. These products are very flexible and are produced in wide continuous sheets that either are shipped in rolls or first cut into ribbons, tapes, or any desired dimensions.

Scrap, Flake, and Ground Mica. High quality scrap mica is delaminated and used for the production of mica paper. Ground mica is used primarily in gypsum plasterboard cement where the mica acts as a filler and extender in the cement. Ground mica is used in the production of rolled roofing and asphalt shingles. The mica serves as an inert filler and surface coating to prevent sticking of adjacent surfaces. The coating is not absorbed by new roofing because mica has a platy structure and is not affected by the acid in the asphalt or by weathering (see Fillers).

Ground mica is used as a pigment extender which facilitates suspension, reduces checking and chalking, prevents shrinking and shearing of the paint film, increases resistance to water penetration and weathering, and brightens the tone of colored pigments (see Pigments).

In the rubber industry ground mica is used as an inert filler and mold lubricant in the manufacture of molded rubber products such as tires. The uses in the plastics industry are similar, where ground mica also acts as a reinforcing agent (see Fillers).

Other uses include decorative coatings on wallpaper and concrete, stucco, and tile surfaces; as an ingredient in special greases; as a coating for cores and molds in metal castings; and as an ingredient in well-drilling muds.

A possible future use of ground mica may be in the control of aphid infestation. Wet-ground mica is spread around crops and plants. The theory behind this use concerns the aphids' movements in relation to the sky. In wet-ground mica, the reflective properties of the flakes are retained. Aphids see the reflection of the sky on the mica flakes, become disoriented, and move elsewhere. Mica has been proved to be nearly as effective as aluminum flakes for this purpose and is much cheaper. The project is still in an experimental stage.

Glass-Bonded Mica. Glass-bonded mica, produced with natural or synthetic mica, is used in a wide variety of molded and sheet electrical insulation, from time controls for automatic washing machines to the most exotic space applications. Included are strategic applications and where unusual environmental conditions or difficult design problems exist, such as in gyro parts, sonar gear, computers, radar, communications equipment, and nuclear equipment (15).

Reconstituted Mica Sheet or Paper. Many different types of insulating materials based on mica paper are available, ranging from rigid, cured plate for commutator segments through materials for hot- and cold-forming to reinforced flexible tapes and wrappers, and composite sheets for slot cells and vacuum tubes.

Major uses are mica paper tapes for d-c armature coils, a-c stator coils, class 180 transformers; commutator segments—impregnated laminates for punchings—have zero slip and ooze at 138 MPa (20,000 psi), 240°C, and are used in automotive electrical systems; washers, spacers, flat insulation; coil wrappings—impregnated paper or laminate for high temperature uses; coil-slot liners; composites with glass cloth, mica paper, silicone bond—a heavy-duty insulation, useful, eg, in traction motors; sheet insulation for capacitors, including jet engine ignition systems; inorganic bonded sheet mica—in spacers for electron tubes, high temperature insulation for electrical appliances and radomes; and tape, sheet, and board—insulation for small appliances and field and shop service.

BIBLIOGRAPHY

"Mica" in *ECT* 1st ed., Vol. 9, pp. 68–75, by Paul M. Tyler, Consultant; "Mica, Synthetic" in *ECT* 1st ed., Suppl. 2, pp. 480–487, by Alvin Van Valkenburg, National Bureau of Standards; "Micas, Natural and Synthetic" in *ECT* 2nd ed., Vol. 13, pp. 398–424, by H. R. Shell, U.S. Bureau of Mines.

1. L. G. Berry and B. Mason, *Mineralogy—Concepts, Descriptions, Determinations*, W. H. Freeman and Company, San Francisco, Calif., 1959.
2. M. L. Skow, *U.S. Bur. Mines Inf. Circ.* **8125**, (1962).
3. C. M. Rajgarhia, *Mining, Processing and Uses of Indian Mica*, McGraw-Hill Book Co., Inc., New York, 1951.
4. F. G. Lesure, *Mica, U.S. Geol. Survey Prof. Paper 820*, U.S. Government Printing Office, Washington, D.C., 1973, pp. 415–423.
5. *Electrical Insulation, Specifications*, American Society for Testing and Materials, Philadelphia, Pa., 1977, Part 40.
6. *Standards for Manufactured Electrical Mica, Pub. FI-1977*, National Electrical Manufacturers Association, New York, 1977.
7. F. W. Clark, *U.S. Geol. Surv. Bull.* **770**, 396 (1924).
8. C. Doelter, *Neues Jahrb. Mineral. Pt. 1, 2*, 178 (1888).
9. D. P. Grigoriev, *Zentralbl. Mineral. Geol. Palaeontol. Teil 1* **8B**, 219 (1934).
10. W. Eitel, *Synthesis of Fluorine Mica of the Phlogopite Group, FIAT Final Report No. 747, PB 20*, 1946, p. 530.

11. T. Noda, *J. Am. Ceram. Soc.* **38,** 147 (1955).
12. R. A. Hatch, R. A. Humphrey, W. Eitel, and J. E. Comeforo, *U.S. Bur. Mines Rep. Invest.* **5337,** (1957).
13. H. R. Shell and K. H. Ivey, *U.S. Bur. Mines Bull.* **647,** (1969).
14. H. R. Shell and W. Warwich, *U.S. Bur. Mines Rep. Invest.* **6077,** (1960).
15. H. R. Shell, *U.S. Bur. Mines Rep. Invest.* **5667,** (1960).
16. *Introduction to Precision Molded SUPRAMICA Ceramoplastic, and MYCALEX Glass-bonded Mica, High Temperature Insulation Materials,* Mycalex, Division of Spaulding Fibre Co., Clifton, N.J.
17. *Insulation (Libertyville, Ill.)* **12**(6), (1966).
18. R. J. Ketterer, *Insulation (Libertyville, Ill.)* **10,** 24 (1964).
19. A. B. Zlobik, *Mineral Facts and Problems,* 1980 ed., U.S. Dept. of Interior, Bureau of Mines, Washington, D.C., 1981.
20. A. B. Zlobik, *General Yearbook, 1978–1979,* U.S. Dept. of Interior, Bureau of Mines, Washington, D.C., 1980.
21. A. B. Zlobik, *Mineral Commodity Profile—Mica,* U.S. Dept. of the Interior, Bureau of Mines, U.S. Government Printing Office, Washington, D.C., Oct. 1979, 18 pp.
22. *Chem. Mark. Rep.* **215**(16), 52 (April 16, 1980).
23. I. Lusis, *Ind. Miner. (London)* **157,** 45 (Oct. 1980).
24. S. K. Haines, *Min. Eng. (N.Y.)* **27**(2), 1 (1975).

ALVIN B. ZLOBIK
U.S. Bureau of Mines

MICELLULAR FLOODING. See Petroleum, tertiary recovery.

MICROBIAL POLYSACCHARIDES

Polysaccharides of microorganisms occur as intracellular storage amylosaccharides, lipid-associated substances in conjunction with cytoplasmic membranes, structural glycans that impart rigidity to cell walls, both discrete and diffuse capsular slimes that remain attached to the cells, and extracellular products in the media. Only the capsular and extracellular (exo-) polysaccharides can be produced in sufficiently high yields to merit commercial interest. Because physical, chemical, and enzymatic means are required to free the capsular types from cells, these types generally have not been considered for industrial production. Exopolysaccharides that either have been produced or are being considered for industrial production are described in this article.

Microbial extracellular polysaccharides are useful in place of or as supplements to hydrocolloids that are derived from plant and animal sources. From 1955–1978, research at the USDA's Northern Regional Research Center (NRRC) demonstrated that bacteria, yeasts (qv), and molds could produce, in commercially feasible fermentations involving the substrate corn sugar, ie, dextrose [50-99-7], the industrial form of D-glucose, distinctive polysaccharides that display a large variety of component sugars, macromolecular structures, and physical properties (1–2). Xanthan gum [*11138-66-2*] was the first polysaccharide that was derived from fermentative conversion of corn sugar that achieved commercial status (3) (see Gums). A small number of genera capable of exopolysaccharide formation have been examined as sources of possibly useful polysaccharides. When genetic factors regulating formation and release of polysaccharides are more fully understood, powerful tools should become available for manipulating and eliciting production from sources representative of a large segment of the microbial world (4) (see Genetic engineering).

Dextran

Dextrans are α-D-glucans [*9074-78-6*] in which 1→6 linkages predominate; ie, 50% or more of the α-D-glucopyranosyl residues are linked as such. Dextrans are produced from sucrose by bacteria belonging to the genera *Leuconostoc*, *Streptococcus*, and *Lactobacillus*, all of which are in the family *Lactobacillaceae*. The majority of known dextrans are formed by strains of *Leuconostoc mesenteroides*. Because they interfere in the production of sucrose, dextrans were the first extracellular microbial polysaccharides to be investigated. Aside from impeding the filtration and handling of cane and beet sugar juices, dextran causes sucrose to crystallize in the form of impure, elongated needles (see Sugar). This problem still exists in the sugar-processing industry because the causative organisms can enter as contaminants from soil, plants, and air. A process employing dextranase has been devised to prevent formation of elongated crystals (5). Traces of dextran [*9004-54-0*] are present in all samples of commercial sucrose that have been tested serologically (6); less is present in sugar from beets where there is less exposure to infection during harvest or during the early stages of processing.

Biosynthesis of dextran from sucrose was the first direct enzymic polymerization demonstrated for a disaccharide donor substrate (7). The reaction is catalyzed by an inducible enzyme, dextransucrase (sucrose: 1,6-α-D-glucan 6-α-D-glucosyltransferase, E.C. 2.4.1.5), which may be either cellbound or extracellular:

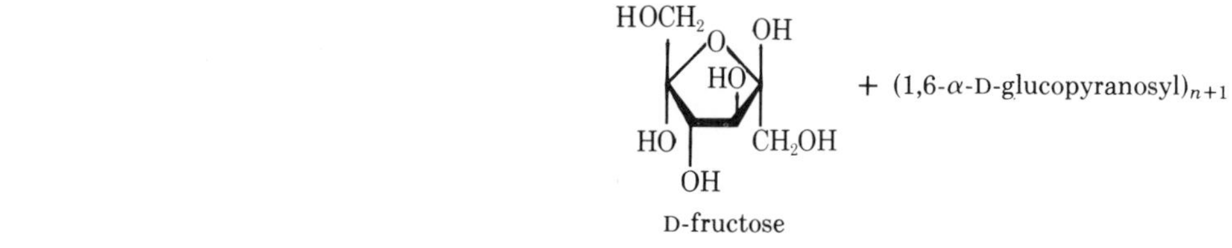

Sucrose is the only donor substrate known and it is the initial acceptor; repetitive α-D-glucopyranosyl transfer occurs so rapidly that high molecular weight products are formed without detectable oligosaccharide intermediates. At high concentrations

(>10–70%), sucrose competes as the acceptor with growing dextran chains, and low molecular weight products are formed (8–9). Other mono- and disaccharide acceptors lead to the formation of oligosaccharides that are related to the isomaltose (6-O-α-D-glucopyranosyl-D-glucose) series (10). Low molecular weight dextrans that are obtained by acid depolymerization of native dextran are efficient primer molecules (11–12), and they are used for controlled enzymic synthesis in which a significant fraction of the product is within a desired range of molecular weights (13).

No exclusively (1→6)-linked dextran has been obtained by either fermentative or enzymatic synthesis. Non-(1→6)-linked residues have an abundance of 4–50% and include the 2-O-, 3-O-, and 4-O-linked α-D-glucopyranosyls (14–15). These non-(1→6)-linkages primarily occur at points of chain branching and involve single, non-reducing glucosyl end groups. Consequently, the presence of branching enzymes has been postulated. However, there is no strong evidence for their existence. The occurrence of unusual dextrans that contain linear sequences composed of alternating (1→3)- and (1→6)-linked residues is noteworthy (16–17).

A number of investigators have concluded that synthesis occurs not through sequential transfer of glucosyl residues to the nonreducing end of a growing dextran chain, but by an insertion mechanism at the reducing end (18). Evidence has been obtained from inhibition of dextransucrase activity by glucoamylase, which exhibits exodextranase activity, for addition of glucosyl groups to the nonreducing end (19).

Whole-culture dextran preparations often contain more than one structural type. A number of these mixtures have been separated by fractional precipitation with ethanol; the fractions are designated S (soluble) and L (less soluble) (20). Some organisms also synthesize a second type of dextran that can be relatively insoluble (14). The insoluble dextrans from certain strains have sequences of contiguous (1→3)-linked residues. The presence of dextransucrases with different catalytic specificities is thought to be responsible. The ultrastructural aspects of dextran formation by strains of *Leuconostoc mesenteroides* have been studied (21). In many cultures where water-soluble and water-insoluble dextrans are formed simultaneously, they are produced by different populations of cells. The insoluble dextrans produced by *L. mesenteroides* are associated with a minority population of capsulate variants that arise from acapsulate parent cells.

Production. In 1945 it was demonstrated that fractions of partially hydrolyzed native dextran could be used as a blood-volume expander (22) and, subsequently, production for this purpose began in pharmaceutical companies in Sweden and in the UK. Production of dextran in the UK for clinical use and for conversion to low molecular weight hydrogen sulfate esters has been described (23) (see Blood).

Production of clinical dextran in the United States was expedited by a large and intensive program at NRRC from 1951–1957. Dextrans from many strains were examined for linkage composition and for correlation of physical properties with non-(1→6)-linkage content (14) and, as a result of this work, *L. mesenteroides* NRRL (Northern Regional Research Laboratory) B-512(F) was selected for dextran production. This organism elaborates copious amounts of an extracellular dextransucrase that catalyzes the synthesis of a dextran with only 5% of non-(1→6)-linkages, all of which are at branching points. In contrast to fractions prepared for clinical use from dextrans having more non-(1→6)-links, those from B-512(F) are substantially less antigenic. The B-512(F) strain is used for dextran production in the United States and Western Europe. Similarly constituted dextrans are produced elsewhere (24).

Because of the wide range of structural types represented in the series, the dextrans characterized in the early work at NRRC have been useful tools for investigating specific interactions with protein components of the immune system, with dextranases, and with plant hemagglutinins (15).

Fermentation conditions were established for attaining maximum yields of dextran and for production of dextransucrase (25). Use of the enzyme leads to a more uniform, easily purified product.

A process was devised for direct enzymatic synthesis of clinical-size dextrans that involved, as the primer, low molecular weight dextran obtained by partial acid hydrolysis of either native dextran or higher molecular weight dextran fractions (13). Although this process is not used, the fundamental information obtained regarding the role of dextran primers permits recycling of a significant portion of the product that would otherwise be lost to the process (8,11). A problem with the synthesis was a drift in molecular weight distribution with successive cycles of primer. Currently available gel permeation chromatographic technology for molecular weight determination of dextrans should allow better control of the process (26).

Although most dextrans are considered to have high molecular weights, ie, values $>10^6$, a dextran with low molecular weight distribution produced by *Streptococcus viridans* B-1351 has been described (27). The organism's dextransucrase is cell-bound rather than extracellular. In a process developed for direct fermentative production of clinical dextran, a fraction of the desired molecular weight distribution could be precipitated from the cell-free culture broth directly by proper choice of alcohol concentrations (28); the process should be more economical than any involving depolymerization by an acid or enzyme. Recent methylation structural analysis of fraction *S* of B-1351 dextran indicates a low degree of branching through 3,6-di-*O*-substituted residues (15). It has been suggested that the dextransucrase has an unusually high affinity for sucrose as an acceptor substrate, which accounts for the relatively high content of D-fructose in certain fractions of the dextran, ie, 0.27–0.37% (27). Failure to detect a tri-*O*-methyl ether other than 2,3,4-tri-*O*-methyl-D-glucose [*4060-09-7*] in the methylated dextran implies that the 6-hydroxyl group belonging to the D-glucosyl moiety of sucrose is the initial D-glucosyl acceptor.

Fermentative. A number of differences in whole-culture processes are described for laboratory-scale production of water-soluble dextrans by *Leuconostoc mesenteroides* NRRL B-512(F) and *L. dextranicum* B-1146 and of water-insoluble dextrans by *L. mesenteroides* B-523 (29). The medium employed for inoculum build-up and production of the *L. mesenteroides* dextrans contains, in addition to 100 g of sucrose per liter of distilled water, a small amount of yeast extract as the nitrogen source with dipotassium phosphate and a trace of magnesium salt; the initial pH is around 7.0. Early stages of inoculum build-up occur under aerobic conditions but the later stages involve the use of occasionally shaken Fernbach flasks and, finally, large, partly filled, unshaken serum bottles. The liquid cultures receive a high percentage of inoculum, ie, 20% of their initial volumes, and incubations are carried out at 25°C for 16–20 h with strain B-512(F) and for about 48 h for the slower growing strain B-523. A critical factor for B-512(F) dextran production under these conditions is the increasing culture acidity. Inocula lose effectiveness if the culture pH drops below 4.8, and a sharp drop in culture viscosity results from a decrease in molecular size of the product. Therefore, careful monitoring of culture pH is required for fermentation production of B-512(F) dextran.

Strain B-523 produces gelatinous, insoluble particles that are suspended in the viscous culture which can be removed with the cells by diluting with water and then centrifuging. As with strain B-512(F), the dextran can be precipitated by adding ethanol. In contrast to the water-soluble B-512(F) dextran, the culture-soluble fraction of B-523 dextran is water-insoluble and requires dilute alkali for redissolution; however, the latter remains in solution after the alkali is neutralized. The culture-insoluble fraction behaves similarly.

Unlike the *L. mesenteroides* strains, *L. dextranicum* B-1146 does not elaborate an extracellular dextransucrase. The organism requires microaerophilic growth conditions and longer incubation at a temperature not exceeding 25°C to produce a water-soluble dextran that is similar in linkage composition to that of B-512(F). Dextran production is doubled by addition of thiamine and niacin to the medium. Although the production medium contains 100 g of sucrose per liter of distilled water, much lower concentrations of sucrose are used during inoculum build-up to maintain the cultures above pH 5 and to avoid transferring highly viscous cultures; 10% inoculum rates are used.

The production of a clinical-size dextran in high yield by direct fermentation also involves a microaerophilic organism, *Streptococcus viridans* B-1351, which forms a cell-bound dextransucrase (28). As carried out in both the laboratory and pilot plant, the process requires only a simple medium composed of sucrose, yeast extract, and dipotassium phosphate. Fermentations are conducted without agitation or aeration at an initial pH of 7.4 and at 37°C. A principal factor is the reduction of the level of inoculum much below that used in fermentations with *Leuconostoc*. A 0.2% inoculum gives better yields, ie, 60–70% theoretical, than a large percentage of inoculum, and it shortens the fermentation time to 48 h. Increasing the sucrose concentration in the medium from 10 to 15% neither lengthens the fermentation time nor alters the percentage yield. During the reduced inoculum fermentation, the pH drops to about 4.5. Because the organism is a potential pathogen, the culture is heated to 100°C to kill the cells and is cooled prior to dextran recovery. A desired clinical fraction is obtained by refractionation of isolated, native dextran with 44–65% methanol in 42–43% theoretical yields (see also Fermentation).

Enzymatic. In the industrial production of dextran, the extracellular enzyme is used. Initially, dextransucrase is produced as in a conventional fermentation. After adjustment of culture pH and removal of cells, the culture fluid is distributed into vessels containing sugar solutions (and, perhaps, low molecular weight dextran primers) where the polymerization reaction takes place. Advantages to this approach are more efficient use of equipment; virtually complete conversion of substrate into product; ease of product recovery and purification; control of reaction conditions, ie, pH, temperature, and primer addition; and the possibility of recovering the co-product D-fructose which otherwise would be consumed by the cells metabolically with concomitant production of lactic acid (30).

Maximum elaboration of dextransucrase by *L. mesenteroides* B-512F occurs in cultures that are maintained at pH 6.7. The enzyme is unstable at that pH but has maximum stability and activity at pH 5.0–5.2. Because the activity of dextransucrase is highly sensitive to temperatures above 25°C, incubation temperatures do not exceed 30°C. Although *Leuconostoc* is microaerophilic, aerated cultures consistently produce more enzyme than still cultures (30). A sucrose level of 2% is optimal for enzyme production, because higher levels lead to formation of an amount of dextran that interferes with removal of cells. There is evidence that calcium ion is an activator of dextransucrase (31). Details of dextransucrase production and its use in the synthesis of dextran have been reported (32–33), and a review of biosynthetic and structural aspects of dextrans has been published (18).

Uses. Unlike most other microbial polysaccharides, the utility of B-512(F) dextran depends much less on its ability to impart high viscosity to aqueous solutions than on its inherent structural features. For example, a 2% soln (polymer wt/soln wt) of the dextran at 25°C gives a viscosity of ca 100 mPa·s (= cP); the same concentration of xanthan gum displays a viscosity of ca 7 Pa·s (70 P) (2) (see Gums). The useful characteristics of B-512(F) dextran derive from its primary structural features, whereas the properties of xanthan gum arise from secondary and tertiary macromolecular structural effects (24). On the basis of methylation–fragmentation analysis, an average repeat unit structure for B-512(F) dextran can be written:

where Glc represents a D-glucopyranosyl residue. However, according to chemical degradation studies, only about 85% of the side chains are 1 or 2 glucosyl residues in length (34). Enzymic degradations indicate that the remaining 15% of the side chains may have an average length of 33 glucosyl residues and may not be distributed uniformly in the molecule (35). This highly irregular structure, if considered with the flexible coil conformation assumed by α-(1→6)-linked glucans in solution, could account for low viscosities even though light-scattering measurements suggest weight-average molecular weight $\overline{M}_w$ values of (35–50) × 10^6 (36–37). Determinations of number-average molecular weight $\overline{M}_n$ values for dextrans give much lower values and suggest that the high values of $\overline{M}_w$ reflect molecular aggregation (38). The type and extent of occurrence of non-(1→6)-linked residues in dextrans have been shown to correlate generally with physical properties (14). Extensive methylation and spectroscopic investigations of dextran structures have been carried out (17,39).

Uses for dextran B-512(F) derive from the high proportion of (1→6)-linkages which imparts flexibility to the polysaccharide chain; minimizes its immunogenicity; affords numerous adjacent equatorial hydroxyl groups at sites for alkylation, esterification, and metal-complex formation; and confers a degree of stability to acid-catalyzed hydrolysis. The latter effect allows facile preparation of fractions with desired ranges of molecular weight by acid hydrolysis (40).

Pharmaceutical. Pharmaceutical uses probably are the main outlets for dextran. Clinical use of dextran as fractions having specific molecular size ranges is based on its compatibility with human tissues and complete metabolic utilization, whether it is ingested or administered parenterally (41). Material of $\overline{M}_w$ 75,000 ± 25,000 is specified for parenteral infusion as a blood-volume expander in treatment of shock, eg, from hemorrhage or burns; the lower limit was set because lower molecular weight material clears too rapidly through excretion from the kidneys. There are indications that dextran fractions of $\overline{M}_w$ above the upper limit interfere with blood clotting. A fraction of $\overline{M}_w$ 40,000 is used to reduce blood viscosity and the possibility of erythrocyte aggregation during surgery (23) (see Blood).

Low molecular weight fractions ($\overline{M}_w$ 2000–3000) are converted directly to sulfate esters derivatives for use as anticoagulant, antilipemic, and antiulcer agents (23–24). Manufacture principally is in Western Europe and Japan. At least one process gives simultaneous sulfation and depolymerization of native dextran to this molecular weight range; whether given orally or parenterally, dextran hydrogen sulfate [9042-14-2] exerts its antilipemic effect through mobilization of lipoprotein lipase. The antiulcer effect results from an interaction of the sulfated polysaccharide with protein and mucus secreted by the stomach, which leads to inactivation of pepsin. An orally administered Na–Al salt of dextran sulfate derived from high molecular weight dextran may be used for treatment of peptic ulcer (42). In the United States, however, the favored treatment is by antihistamine to inhibit acid secretion (see Gastrointestinal agents).

Dextran can form complexes with di- and trivalent cations. Iron dextran [9004-66-4], a soluble, largely nonionic complex, has been used to treat iron-deficiency anemias in humans and livestock (43). Current use primarily is as an injectable preparation for treatment of anemia in baby pigs. A soluble calcium salt of dextran is used to treat hypocalcemia associated with cattle delivery paresis (44).

Industrial. Development in Sweden of the concept of gel filtration has provided a second important outlet for dextran (45). Molecular sieves (qv) consisting of cross-linked dextran gels have been prepared by treatment of alkaline dextran solutions with the cross-linking agent epichlorohydrin (46). Bead forms are obtained by dispersing the alkaline solutions in toluene using poly(vinyl acetate) as the stabilizer prior to the addition of epichlorohydrin (47). These gels can be modified by appending ionic and lipophilic substituents. The various cross-linked preparations are used as fine chemicals and in the food and pharmaceutical industries for desalting protein solutions and for separating proteins and other biological macromolecules. Applications of dextran to gel-precipitation techniques in refining and concentrating metals and to proprietary photographic emulsions have been summarized (24).

Dextran uses and derivatives have been reviewed (48), and an extensive bibliography has been published (49).

Extracellular Polysaccharides as Metabolic Products

Biosynthetic Mechanisms. Unlike dextrans and levans [9013-95-0] (the latter are predominantly (2→6)-linked fructans), whose biosynthesis through transglycosylation reactions is driven by the energy of the glycosidic bond in sucrose, biosynthesis of all other polysaccharides requires activation through conversion of monosaccharide precursors to glycosyl phosphate derivatives. Energy in the form of adenosine triphosphate (ATP), which is derived from the catabolism of simpler organic substances, is used to assimilate substrates in phosphorylative reactions and to synthesize phosphate esters of monosaccharides from precursors entering the pathways of carbohydrate metabolism. The glycosyl phosphates then react with specific nucleoside triphosphates to form sugar nucleotides. These nucleotides may be either ortho- or pyrophosphate diesters. Conversion to precursor monosaccharide components of the ultimate polysaccharide often occurs at the sugar nucleotide level. Assembly of repeat-unit sequences involves transfer of glycosyl groups from the sugar nucleotides to polyisoprenoid alcohol phosphate carriers. The lipophilic carriers transport subunits through the lipoidal cytoplasmic membrane and are the coenzymes that participate directly in the synthesis of polysaccharide chains. Many of the enzymes that are involved are associated with this membrane. Because of their complexity, these reaction sequences have been demonstrated for only two systems: the biosyntheses of the O-antigen moiety of an enterobacterial lipopolysaccharide and of a capsular polysaccharide. In the O-antigen system, chain elongation was shown to occur by insertion of subunits at the reducing end of the growing polysaccharide chain (50). These aspects of polysaccharide biosynthesis, and the evidence for the manner in which competition for sugar nucleotide and polyprenol coenzymes for synthesis of cell wall, lipopolysaccharide, and glycoproteins affects exopolysaccharide biosynthesis have been reviewed (4,51–53). Examples of possible direct biosyntheses from sugar nucleotide precursors also have been reported (4).

Biotechnological Aspects. Elaboration of extracellular polysaccharides in high yields requires a strong oxidative catabolism of part of the substrate to provide sufficient ATP to drive anabolic polymerization. Consequently, polysaccharide production is carried out in vigorously aerated fermentations involving media with high ratios of carbon source to nitrogen source. Conventional fermentation equipment is used. Development of viscosity as the fermentation proceeds limits conversion of the carbon source to the polysaccharide; this effect probably is related to inefficient mechanical dispersal of air bubbles in viscous media. Another common limitation is a lowering of medium pH resulting from the formation of organic acids and acidic polysaccharide. Buffering of the medium or, better, use of pH control can maintain a functional polymerase system associated with the cytoplasmic membrane.

Productivity is not only a function of yield from the substrate but of the length of time that the fermentor is occupied. However, these considerations might not apply to production for pharmaceutical use. For example, a vaccine is currently available that consists of a mixture of capsular polysaccharides from 14 different types of *Streptococcus pneumoniae*, which is a relatively slow-growing species (54). Because polysaccharide fermentation broths are dilute solutions, usually less than 5% (wt/vol), recovery costs are high, and selling prices are in the specialty chemicals class. Consequently, functionality, ie, viscosities of dilute solutions, suspending power, emulsion stabilization, stability to salts, and temperature and pH extremes, is of prime importance and is the principal reason for their use in place of less expensive plant gums and synthetic hydrocolloids (see Vaccine technology).

As with most fermentation processes, careful microbiological, chemical, and physical control is necessary to ensure that conditions do not lead to dominance by less productive, variant strains that often produce an altered polysaccharide. When necessary, excretion of depolymerizing enzymes, which can lower product yields, usually can be dealt with by heating the culture after the fermentation is completed. Pasteurization also is employed in those instances where a plant or animal pathogen is the producing organism. The example of pneumococcal polysaccharides is the only instance of exopolysaccharide production involving human or animal pathogens. Environmental factors involved in production of polysaccharides and control of their properties have been reviewed (55–57).

Sources. Organisms producing useful amounts of extracellular polysaccharide occur among the bacteria, yeasts (qv), and molds. The producers usually belong to taxonomically related groups that form mucoid colonies when grown on solid media. In terms of structure and composition, these polysaccharides often are species-characteristic. There are many examples, however, of similar polysaccharides produced by different species within a genus and by related genera (56). Other species include many strains that make quite dissimilar polysaccharides. Examples of these species are *Klebsiella aerogenes* (*Enterobacter aerogenes*) and *Streptococcus pneumoniae*, which include many strains that form various type-specific capsular heteropolysaccharides (52).

The polysaccharides of commercial interest can be classified as either neutral or anionic. The neutral polysaccharides are D-glucans [*9012-72-0*]. Groups that impart anionic character are carboxylate and may belong to uronic acids or pyruvic acetal and succinic acid half-ester substituents. Yeast D-mannans [*51395-96-1*] and glucogalactans that contain phosphate diesters as the anionic components are known but they have not been produced on an industrial scale (58). (Generally, anionic polysaccharides are recovered as either potassium or sodium salts.) Although suitable cationic and amphoteric types are not known, such groups can be appended chemically (48,57).

Anionic Heteropolysaccharides. *Xanthan Gum.* The extracellular polysaccharide of *Xanthomonas campestris* NRRL B-1459 was the first biopolymer product of a fermentation based on corn sugar that has attained commercial status (59–60). The associated rapid fermentation gives high conversions of carbohydrate into a highly viscous product and the unusual stability and rheological properties of the gum. The FDA has approved xanthan gum for use as a food additive where such use is not precluded by Standards of Identity Regulations (42,57). By the end of 1979 the world capacity for biopolymer production (mainly xanthan gum, but excluding dextran) exceeded 20,000 metric tons (61) (see Gums). The structure of xanthan gum consists of a β-(1→4)-linked D-glucopyranosyl backbone chain, eg, in cellulose:

xanthan gum

To the chain are appended trisaccharide side chains composed of D-mannopyranosyl (Man) and D-glucopyranosyluronic acid (GlcA) residues (62–63). The β-(1→2)-linked mannosyl residues have 6-*O*-acetyl substituents. An average of about half of the β-D-mannosyl end groups bear 4,6-*O*-(1-carboxyethylidene) substituents; ie, 4,6-acetal-linked pyruvic acid. Although the indicated distribution of side chains reflects average values, it is in accordance with evidence for repeat-unit structure in microbial heteropolysaccharides (52) and with the sharp temperature transition in optical rotation displayed by depyruvylated xanthan gum as contrasted with the broad transition displayed by the native gum (64).

Physical Properties. Aqueous dispersions of xanthan gum exhibit several novel and remarkable rheological properties (2,42,57). Low concentrations of the gum have relatively high viscosities which permits its economical use in applications, eg, emulsion stabilization and mobility control of water-flooding fluids in petroleum reservoirs. Xanthan gum solutions are characterized by high pseudoplasticity; ie, over a wide range of gum concentrations, rapid shear-thinning occurs that is instantaneously reversible.

If a salt is present, the viscosity is independent of pH from pH 1.5–13 (57). Viscosity also is stable to heat over a wide temperature range, and this stability is enhanced by salts of mono- and divalent cations (65). In addition to pseudoplastic behavior, xanthan gum solutions of 0.75% concentration or greater display rheological yields that are characteristic of plastic dispersions, ie, a tendency not to flow at ultra-low shear stress (59). This property is exhibited by a number of anionic, microbial, extracellular polysaccharides and is responsible for the ability to suspend heavy particles (2). The yield value for xanthan gum in aqueous dispersion is increased in the presence of a salt (57).

Although xanthan gum is a polyelectrolyte by virtue of ionizable carboxyl groups belonging to D-glucuronic acid residues and pyruvate acetal, the viscosity of ca 0.35% solutions is virtually unaffected by the presence of 0.01–1% salt (KCl) concentrations. The presence of salts slightly increases viscosity for gum concentrations higher than 0.35% and slightly lowers it for concentrations below this amount. Solution viscosities of almost all other polyelectrolytes are sensitive to the presence of salts. Xanthan gum retains its water-binding and rheological properties in dense or saturated salt solutions (42).

Structure and Conformation. The mol wt of xanthan gum is $(2–15) \times 10^6$ (66–67). Many of the properties of xanthan gum dispersions indicate that the molecule assumes a rodlike, ordered secondary structure. A sharp increase in the viscosity of 0.5–1% aqueous dispersions has been noted at 50–60°C (60) and has been confirmed by optical rotation (77), circular dichroism (78), and nmr measurements (78). Increasing the ionic strength raises the temperature at which the transition occurs; such melting out or denaturation is a helix-to-coil conformational transition which is similar to that observed for double-stranded nucleic acids, triple-stranded collagen, and certain polysaccharides (68). X-ray diffraction studies with xanthan gum fibers indicate a helix of fivefold symmetry with side chains folded along the backbone in a hydrogen-bonded structure (69). Weak, noncovalent associations between aligned molecules build up a tenuous gel-like network responsible for the observed yield stress of xanthan gum dispersions. Rapid shear-thinning with instantaneous regaining of viscosity results from progressive breakdown of the gel-like network with increasing shear rate (70).

The increase in transition temperature with increasing salt concentration is indicative of a stabilization of the native conformation by suppressed repulsion among side-chain carboxylate anions (69). Because deacetylation decreases the transition temperature by about 15°C, it is suggested that the acetyl groups stabilize the native conformation (64). The observed increase in viscosity with conformational change is attributed to a spreading of the side chains away from the backbone which results in increased hydrodynamic volume (69,71). In this denatured state, xanthan gum exhibits typical polyelectrolyte behavior; the viscosity collapses in the presence of added salt (see Polyelectrolytes).

Many aspects of xanthan gum denaturation by heat and renaturation by slow cooling resemble the behavior of DNA (deoxyribonucleic acid). A multistranded helix structure has been proposed and electron microscopic evidence has been provided for it (72). Native polysaccharide appears to consist of unbranched, probably double-stranded fibers 4 nm wide and 2–10 μm long. Denaturation, followed by quick cooling to prevent reannealing, gives a single strand 2 nm wide and 0.3–1.8 μm long. In the renatured product, short unravelled regions have 2 or 3 strands arranged in a right-handed twist. The contrary view is that the reversible temperature transitions do not reflect a multichain process but only two characteristic conformations, ie, random and ordered (71). The measurements show a hydrodynamic volume increase of only 10%, which is equivalent to that calculated for the change in axial ratio of the molecule from the helical to the extended conformation (69).

Production. The industrial process for xanthan gum has been outlined (57) (see Gums). Inoculum build-up from a stock culture proceeds through stages to a seed tank, the contents of which 5 vol % are added to the production fermentor directly. An aerated, 2-d fermentation is carried out at pH 6.0–7.5 and at 28–31°C (3). After fermentation, the culture is pasteurized and then treated with isopropyl alcohol to recover the polysaccharide. The precipitate is dried, milled, tested, and packaged. Both the industrial and available food grades contain bacterial cells. Even so, these preparations dissolve in either hot or cold water to produce solutions of comparable viscosity. Meticulous care is taken to ensure that the food-grade gum meets microbiological and chemical specifications.

The wide variety of nitrogen and carbohydrate sources that can be employed to produce xanthan gum have been summarized (55,73–74). The first nitrogen source used in the process was dried distillers' solubles. Additional benefit is obtained by pH control with ammonium hydroxide (75). The conditions for maintaining the specific rate of xanthan gum formation are 28°C, pH 7.0, and dissolved oxygen tension above 20% of air saturation. Polysaccharide formation ceases if the culture is allowed to reach pH 5.0. Viscosities of completed fermentations are 7–12 Pa·s (70–120 P) for xanthan concentrations of 1.4 to nearly 3%.

A synthetic medium in which diammonium phosphate is employed as the nitrogen source has been described (74). The process is based upon conditions for production of gum with a high pyruvic acid content (about 4 wt %). Because low pyruvate gum was found during the early stages of the fermentation, it was concluded that the final product is a mixture of high and low pyruvate types. This conclusion was confirmed when xanthan gums of different pyruvic acid contents were recovered by fractional precipitation of a preparation with alcohol (76). Continuous fermentations for xanthan gum have been reported (77–78); however, it is thought that continuous fermentation entails too great a risk of contamination to merit its production on an industrial scale (3,57).

Polysaccharide production by *Xanthomonas campestris* growing in a chemostat culture on various limiting nutrients has been studied (78). Except for the pyruvate content, the composition of the polysaccharide is unchanged regardless of the limiting nutrient. Phosphate and magnesium limitations severely reduce pyruvate content. Significant polysaccharide production, even under conditions of glucose and phosphorus limitations, suggest that such production is an intrinsic property of the organism and not simply a form of overflow metabolism (79). Studies on batch (80) and continuous fermentations (57) have led to the recognition that *X. campestris* B-1459 tends to form variant substrains that are either less productive or form a polysaccharide having a low content of pyruvate acetal. Procedures for culture maintenance and production, purification, and analysis of xanthan gum have been published (81).

Uses. The many applications of xanthan gum have been tabulated and summarized (42,57). It has been estimated that over half of microbial polysaccharide sales are in to the food industry (61). Food and pharmaceutical applications take advantage of properties, eg, thickening, emulsion stabilization, water-binding, suspending, and salt and acid compatibilities (57). The reversible, shear-thinning property imparts good mouth feel, ie, addition of the gum does not alter the characteristic texture of the food in the mouth (82). Xanthan gum also interacts synergistically with galactomannans, eg, locust bean and guar gums, to produce combined viscosities greater than would be expected from the individual polysaccharides (83–84). The interaction with locust bean gum is much stronger, and appropriate combinations can form a heat-reversible gel. These interactions have been explained as occurring through formation of junction zones between the β-(1→4)-linked glucosyl backbone of xanthan gum and regions of the β-(1→4)-linked D-mannosyl backbones of the galactomannans that do not bear α-D-galactosyl substituents in (1→6)-linkage (83).

Industrially, xanthan gum is used to thicken oil-well drilling muds, frequently, where brines are encountered; as a carrier and suspending agent for agricultural sprays and feeds; in the manufacture of gelled and slurry explosives (qv); in gelled or thickened cleaning compositions; in ceramic glazes; and in textile sizing and printing agents. Xanthan gum also is used in formulation of paints and adhesives. For recovery of natural gas, it is used in water-based fracturing fluids and in sand-bearing propping agents (see Gas, natural). In these applications, the polysaccharide can be removed from rock formations by adding a controlled amount of hypochlorite or hydrochloric acid to the fluid before injection into the well (57). Predicted annual usage of xanthan gum in drilling-mud formulations alone is 3000 metric tons by 1980 (61) (see Petroleum, drilling fluids).

Application of xanthan gum as a mobility-control agent in secondary and tertiary recovery of petroleum from underground formations by flooding with water and surfactant solutions, respectively, requires that cells and other particulate matter be removed to avoid plugging the pores of the oil-bearing rock. Usually, gum solutions are filtered through diatomaceous earth prior to injection. There are companies that ship whole broth cultures for this purpose. A recent patent describes preparation of *Xanthomonas* fermentation broths in which the presence and size of particulates are controlled so as to make whole broth suitable for injection (85). Calcium and other cations that precipitate as insoluble salts are controlled by preparing the fermentation medium with deionized water and adding a chelating agent (see Chelating agents). Storage in the presence of formaldehyde as a bactericide shrinks the shorter dimension of the rod-shaped *Xanthomonas* cells to ≤0.65 μm. Alkaline heat treatment in the presence of a salt with simultaneous deacetylation of the polysaccharide also is used to hydrolyze the bacterial cells (86). Treatment with a proteinase followed by filtration of the gum solution before the cell bodies are completely disintegrated has been proposed (87).

Use of xanthan gum for enhanced oil recovery is being evaluated (88) (see Petroleum, enhanced oil recovery). It represents, by far, the largest potential market for the gum (61). With regard to flow through porous media, pseudoplasticity of xanthan gum at low concentrations imparts desirable low viscosities at the high shear rates encountered during injection but significantly higher viscosities when the waterflood is moving through the formation where there are ultralow shear rates (89). Other advantages attributed to the polysaccharide are high viscosity at low concentrations; resistance to mechanical and chemical degradation; low sensitivity of fluid properties to pH, elevated temperature, salinity, and the presence of divalent metal ions; and low loss from the fluid by retention of the polymer in porous rock (88).

Dispersions of xanthan gum having high pyruvate content (4.0–4.8%) are much more viscous than dispersions prepared from xanthan gum of low pyruvate content (2.5–3.0%) (90). It has been shown that low pyruvate gum can be produced either because of strain variance in *X. campestris* B-1459 cultures (80) or because of the culture conditions (75,78). The increasingly divergent viscosities exhibited at low xanthan gum concentrations, eg, (0.1–0.5%), and particularly in the presence of salt, imply the need for care in selecting strains and fermentation conditions.

Zanflo. Zanflo has a significant market in the paint industry because its dispersions display excellent flow and leveling properties. Viscosity is of the pseudoplastic type but the gum is sensitive to shear. Zanflo is composed of glucose, galactose, glucuronic acid, and fucose in respective molar ratios of 3:2:1.5:1 and has an *O*-acetyl content of 4.5 wt %. The product also contains 3 wt % protein (57). Although glucuronic acid accounts for almost 20% of the weight, Zanflo is not precipitated by cationic dyes as are almost all anionic polysaccharides.

Although designated in the patent as *Erwinia tahitica*, the producing organism has been reclassified as a strain of *Klebsiella pneumoniae* (ATCC 21711) (91). The original soil isolate has been extensively mutated so that the mutant used for production grows at 30°C and only in the presence of added iron (57).

Additional details of fermentative production (pilot plant) are given by the inventors (92). Lactose- and α-amylase-hydrolyzed starch are the preferred carbon sources. Enzyme degraded soy meal and ammonium nitrate are included as nitrogen sources. Viscosity development reaches a maximum after 64 h, at which point >50% of the substrate is converted to polysaccharide. The minimum dissolved oxygen concentration during the first 24 h of the fermentation is 5–10%, and cell population reaches its maximum within 10 h. Product recovery is similar to that for xanthan gum.

Viscosity of Zanflo is higher than that of xanthan gum. Greater differences are found at increasing gum concentrations; in aqueous dispersions containing 1.5% gum, Zanflo exhibits a viscosity that is twice that of xanthan gum (92). Unlike xanthan gum, Zanflo undergoes a reversible, linear decrease in viscosity with an increase in temperature. The linear decrease in viscosity is such that the same concentrations of both polysaccharides give similar viscosity at 80°C. Viscosities of aqueous dispersions of Zanflo decrease sharply outside the range pH 5–10. The viscosity is stable in the presence of inorganic salts (91).

Beijerinckia indica Polysaccharide. A polysaccharide has been described that is produced by a strain of the soil bacterium *Azotobacter indicus* (*Beijerinckia indica*) (93). The polysaccharide is composed of glucose, rhamnose, and galacturonic acid in the weight ratio of ca 6.6:1.5:1; the *O*-acetyl content is 8–10%. Aqueous dispersions of the polysaccharide exhibit higher viscosities than are given by xanthan gum. The viscosity is reversible and pseudoplastic and is stable over a wide temperature and pH. Other important properties are ability to suspend particulate matter, eg, sand and colloidal clays; compatibility with a wide variety of salts; high solubility in sea water and in brine containing 25% salt; and tendency to gel in the presence of cationic or polyvalent ions at high pH (above 10) (94). Trivalent chromium ion causes gelation at pH 9.0–9.5. These properties suggest use in oil-well drilling and fracture-propping fluids; in waterflood systems; in dripless water-based latex paints; in adhesives, and in printing inks (see Inks). A variety of carbohydrates are used as substrates in aerobic fermentations conducted at 30°C for 37–48 h. Complex nitrogen sources seem to be required.

Agarlike Polysaccharide. Development of a proposed agar [9002-18-0] substitute has been reported, and the organism, which is isolated from an aquatic environment, is a new species of the genus *Pseudomonas* (95). Glucose (3%) serves as the substrate in a 3-d aerobic fermentation. The approximate composition of the polysaccharide is rhamnose (46 wt %), glucose (30 wt %), uronic acid (21 wt %), and O-acetyl derivative (3 wt %) (96). Deacetylation by mild alkaline treatment and the presence of cations are required to produce a thermally reversible, brittle, optically clear gel that can be autoclaved for several cycles. Gel strength, melting point, and setting point are controlled by the concentration of cations. Consequently, the amount of polysaccharide used for preparing solid microbiological media (0.5–1.25 wt %) depends on the salt contents of the media (97). Recent wide fluctuations in the price of agar could lead to extensive use of the microbial product.

Bacterial Alginic Acid. Certain strains of *Azotobacter vinelandii* elaborate acidic polysaccharides resembling alginic acid (98). The sugar components, D-mannopyranuronic acid and L-gulopyranuronic acid, occur in varying proportions in (1→4)-linkage, in linear block, and in alternating sequences (99). Except for the presence of the O-acetyl derivative, the bacterial products have the same basic structures as do alginic acids extracted with alkali from the brown algae. Growth in the presence of calcium ions increases the amount of L-guluronic acid (100). An extracellular enzyme that has been isolated catalyzes the partial epimerization of D-mannuronic acid residues to L-guluronic acid residues in the bacterial alginate.

Studies on the biosynthetic steps leading to bacterial alginate have been summarized (55,101). A β-D-mannopyranuronic acid homopolymer [27638-01-3] is the initial product that undergoes subsequent, extracellular partial epimerization. The epimerization is an important reaction because the ability of the alginate to form gels in the presence of calcium ions depends on the content of L-guluronic acid.

Developmental research has resulted in fermentative products that have viscoelastic properties comparable to those of the seaweed alginates (101) (see Gums). Limiting the amount of phosphate in the culture medium results in higher conversions of the substrate, ie, sucrose, that is utilized as well as a product of high average molecular weight and narrow size distribution. Continuous culture experiments under various limiting conditions suggest that improved conversions of sucrose can be obtained by careful manipulation of culture respiration.

Succinoglucan. Succinoglucan [9014-37-3], an acidic polysaccharide, is produced by *Alcaligenes faecalis* var *myxogenes* strain 10C3. The organism was isolated from a soil-enrichment culture that contained ethylene glycol as the only carbon source (102). A variety of sugars in addition to ethylene glycol can be used as substrates in a chemically defined medium. Production and early structural investigations have been reviewed (55). The rheological behavior of succinoglucan in aqueous dispersions is unusual in that the acid and calcium forms show high viscosity but the sodium form exhibits very low viscosity. The viscosity of the calcium form does not change, even in the presence of sodium chloride, over a wide pH range.

According to the most recent component analysis, succinoglucan contains D-glucose (78%), D-galactose (10%), succinic acid (6.3%), and pyruvic acid (5.4%); succinic acid occurs entirely as the half ester and pyruvic acid as acetal (103–104). Examination of extracellular polysaccharides of several *Agrobacterium* strains from *A. radiobacter*, *A. rhizogenes*, and *A. tumefaciens* reveals similar components in addition to small amounts of the *O*-acetyl derivative (105). Except for the presence of succinic acid, succinoglucan and the extracellular polysaccharides of *Rhizobium meliloti* and *A. radiobacter* share the same octasaccharide repeat-unit structure wherein Glc and Gal represent D-glucopyranosyl and D-galactopyranosyl residues, respectively (106):

succinoglucan

It is not known whether commercial production of succinoglucan is contemplated.

Neutral Glucans. *Curdlan.* The production, properties, conformational aspects, and uses of curdlan [54724-00-4], a water-insoluble, linear, (1→3)-linked β-D-glucan are reviewed in refs. 107–108. Although they are water insoluble, heated suspensions of curdlan become clear at about 54°C and form resilient, thermally irreversible gels at higher temperatures. Sharp melting points at 140–160°C depend on polysaccharide concentration and the degree of polymerization (DP) (109). Details of gelation behavior are given in ref. 107. Cooling of aqueous curdlan suspensions that have been swelled by heating from 54 to 60°C leads to formation of soft, low set gels. Hard, high set gels that are formed by heating at 80–90°C have properties which are intermediate between the brittleness of agar and the elasticity of gelatin.

Curdlan was discovered as a contaminant polysaccharide in preparations of succinoglucan. Stable, spontaneous variants and mutants of *Alkaligenes faecalis* var *myxogenes* 10C3 have been obtained that form curdlan exclusively. Curdlan-positive strains of *Agrobacterium* have been detected by utilizing the specific color complex formed between (1→3)-linked β-D-glucans and aniline blue (110). Over 99% of the linkages are β-(1→3) in all of the preparations of curdlan. The curdlan-positive agrobacteria also form the water-soluble succinoglucan (105).

Commercial development of curdlan has reached the pilot-plant stage. High yields are obtained from glucose and many other carbon sources in a simple, defined medium if the pH is maintained at 7 (107). For example, 10% concentrations of glucose are converted in 50% yield. One reason for the high yield is the insolubility of the fibrillar product and its precipitation in the medium as it is formed; there are no secondary effects of high fermentation viscosity upon mixing and dispersion of air bubbles in the fermentor. Results from conformational analyses of (1→3)-linked α-D-glucans suggest an extended ribbon conformation like that assumed by (1→4)-linked β-D-glucan (cellulose) wherein the chains pack together in parallel formation (36). In contrast, (1→3)-linked β-D-glucans assume a helical, coiled-spring conformation as does the (1→4)-linked α-D-glucan amylose [9005-82-7]. Such linear, helical molecules tend to form multihelical junction zones. X-ray diffraction studies indicate the existence of triple-stranded helical structures in fibrous preparations of curdlan (108).

Curdlan is soluble in alkaline but insoluble in neutral and acidic solutions. This unusual behavior greatly simplifies recovery and purification by eliminating the expense of having to recover water-miscible solvent(s) used to precipitate the water-soluble polysaccharides. Addition of sodium hydroxide to the culture dissolves the polysaccharide and permits removal of the cells by filtration. Neutralization of the filtrate with hydrochloric acid precipitates a water-swelled gum that can be collected by centrifugation and then spray-dried.

Studies of gelation of curdlan suspensions in the presence of urea suggest disordering of molecular interactions during swelling and establishment of new interactions during formation of soft-set gels (107). Electron-microscopic studies of assembly and dissociation of fibrils by heating indicate that curdlan fibrils of high DP form a netlike structure when aqueous suspensions are heated to 90°C (111). Assembly of fibrils during formation of high set gel is thought to involve hydrophobic interactions between the macromolecules. This notion is supported by the increasing opacity of gels formed at higher temperatures. Behavior in alkaline dispersions is ascribed to polysaccharide chains in ordered conformation in <0.2 M sodium hydroxide which undergo transition to random coils at higher concentrations of base (112).

Proposed uses of curdlan include preparation of gelled desserts and canned, multiple-layer jelly containing both low and high set gels (107–108). Use of curdlan in the United States requires approval by the FDA. Potential industrial uses are as gelling agents for fish feeds, binding agents for tobacco, and carriers for immobilized enzymes (see Enzymes, immobilized). Clear, flexible films can be cast from alkaline solutions of curdlan by neutralizing with carbon dioxide that diffuses under quiescent conditions and then by drying the thin gel (113).

Scleroglucan. Capsular β-D-glucans [9041-22-9] that are similar to curdlan are produced by a variety of imperfect fungi grown in media containing either glucose or sucrose as carbon sources and complex nitrogen sources supplemented with mineral salts (55,114). The repeat unit structure of scleroglucan [39464-87-4], which is formed by *Sclerotium glucanicum* NRRL 3006, has been determined (115):

$$\begin{array}{c} \underset{\beta}{}\text{Glc} \\ \downarrow^{6} \\ \longrightarrow \underset{\beta}{\overset{3}{}}\text{Glc} \longrightarrow \underset{\beta}{\overset{3}{}}\text{Glc} \longrightarrow \underset{\beta}{\overset{3}{}}\text{Glc} \longrightarrow \end{array}$$

The estimated DP of the polysaccharide from *S. glucanicum* is ca 100. In commercial production, however, a strain of *S. rolfsii* is used to form a product with a DP of ca 800. Aside from the degree of polymerization, the various scleroglucans differ slightly in number and length of side chains.

The preparation of the available grades of scleroglucan has been described (114). During the aerobic, submerged fermentation at 30°C for 60 h, a pelletlike growth develops and is shrouded by a capsule. The capsule is an extended gel phase that is largely immiscible with water. The culture is heated to inactivate glucanase activity and to kill the organism which is a plant pathogen. The culture is homogenized to free the gum from the mycelia and to convert the polysaccharide to a form that is readily dispersible from the dried state. Biopolymer CS-6 is prepared by spray-drying the homogenized culture; Biopolymer CS-11 is prepared by filtration of diluted homogenized culture to remove the mycelia and other particulate matter. The filtrate is concentrated by evaporation and then is precipitated by addition of a water-miscible alcohol. The fibrous precipitate is dehydrated with alcohol and then ground to a powder after removal of residual alcohol.

Scleroglucan was developed and marketed by the Pillsbury Co. as Polytran (116). Worldwide rights are owned by the French company CECA, S.A., which markets the gum as Biopolymer CS.

Refined grades of scleroglucan disperse easily in water to give highly viscous, pseudoplastic solutions that exhibit high thermal stability over a broad pH range. Use is primarily in the manufacture of ceramics (57).

Pullulan. Many strains of the dimorphic fungus *Aureobasidium* (formerly *Pullularia*) *pullulans* elaborate an extracellular α-D-glucan. The polysaccharide usually referred to as pullulan [9057-02-7] is a linear (poly)maltotriose [1109-28-0] linked through α-(1→6) bonds on terminal glucopyranosyl groups of the trisaccharide (117):

$$\rightarrow \, {}^{6}_{\alpha}\text{Glc} \rightarrow {}^{4}_{\alpha}\text{Glc} \rightarrow {}^{4}_{\alpha}\text{Glc} \rightarrow$$

Maltotetraose subunits, however, are present to ca 6% and make the polysaccharide susceptible to partial depolymerization by amylolytic activity late in the fermentation (118). Some pullulans are branched and contain (1→3)-linked glucosyl residues (119–120).

Hayashibara Biochemical Laboratories in Japan is producing pullulan on a pilot scale. Yield, molecular size, and the length of the culture period can be controlled by adjustment of the initial pH and of the phosphate concentration (121). Conversion of glucose to high molecular weight pullulan in a 7-d fermentation involves a medium of an initial pH of 6.5 and that contains 0.2–0.4% phosphate.

Possible uses of pullulan have been discussed (24,122). Pullulan can function as a flocculator of clay slimes in hydrometallurgical processes (123) (see Flocculating agents). The principal use probably will be in the manufacture of compression-molded articles (122). Multilayered molded plastics of pullulan and polymers, paper, or aluminum foil, which exhibit low gas permeability and high strength, have been described (124).

BIBLIOGRAPHY

1. P. A. Sandford, *Carbohydr. Res.* **66**, 3 (1978).
2. A. Jeanes in N. M. Bikales, ed., *Water-Soluble Polymers*, Plenum Press, New York, 1973, pp. 227–242.

3. W. H. McNeely and K. S. Kang in R. L. Whistler and J. N. BeMiller, eds., *Industrial Gums*, 2nd ed., Academic Press, Inc., New York, 1973, pp. 473–497.

4. A. Markovitz in I. Sutherland, ed., *Surface Carbohydrates of the Prokaryotic Cell*, Academic Press, Inc., New York, 1977, pp. 415–445.

5. P. A. Inkerman and G. P. James, *Proc. Queensl. Soc. Sugar Cane Technol.* **43,** 307 (1976).

6. J. M. Neill, J. Y. Sugg, E. J. Hehre, and E. Jaffe, *J. Exp. Med.* **70,** 427 (1939); *Am. J. Hyg.* **34,** 65 (1941).

7. E. J. Hehre, *Science* **93,** 237 (1941).

8. H. M. Tsuchiya, N. N. Hellman, H. J. Koepsell, J. Corman, C. S. Stringer, S. P. Rogovin, M. O. Bogard, G. Bryant, V. H. Feger, C. A. Hoffman, F. R. Senti, and R. W. Jackson, *J. Am. Chem. Soc.* **77,** 2412 (1955).

9. GDR Pat. 21,324 (Aug. 2, 1957), U. Behrens and M. Ringpfeil; *Chem. Abstr.* **56,** 3927 (1962).

10. H. J. Koepsell, H. M. Tsuchiya, N. N. Hellman, A. Kazenko, C. A. Hoffman, E. S. Sharpe, and R. W. Jackson, *J. Biol. Chem.* **200,** 793 (1953).

11. H. M. Tsuchiya, N. N. Hellman, and H. J. Koepsell, *J. Am. Chem. Soc.* **75,** 757 (1953).

12. E. J. Hehre, *J. Am. Chem. Soc.* **75,** 4866 (1953).

13. N. N. Hellman, H. M. Tsuchiya, S. P. Rogovin, B. L. Lamberts, R. Tobin, C. A. Glass, C. S. Stringer, R. W. Jackson, and F. R. Senti, *Ind. Eng. Chem.* **47,** 1593 (1955).

14. A. Jeanes, W. C. Haynes, C. A. Wilham, J. C. Rankin, E. H. Melvin, M. J. Austin, J. E. Cluskey, B. E. Fisher, H. M. Tsuchiya, and C. E. Rist, *J. Am. Chem. Soc.* **76,** 5041 (1954).

15. F. R. Seymour, M. E. Slodki, R. D. Plattner, and A. Jeanes, *Carbohydr. Res.* **53,** 153 (1977).

16. I. J. Goldstein and W. J. Whelan, *J. Chem. Soc.,* 170 (1962).

17. F. R. Seymour, R. D. Knapp, E. C. M. Chen, S. H. Bishop, and A. Jeanes, *Carbohydr. Res.* **74,** 41 (1979).

18. R. L. Sidebotham, *Adv. Carbohydr. Chem. Biochem.* **30,** 371 (1974).

19. M. Kobayashi and K. Matsuda, *Carbohydr. Res.* **66,** 277 (1978).

20. C. A. Wilham, B. H. Alexander, and A. Jeanes, *Arch. Biochem. Biophys.* **59,** 61 (1955).

21. B. E. Brooker, in *Microbial Polysaccharides and Polysaccharidases*, Special Publication of the Society for General Microbiology, Cambridge, UK, 1979, pp. 85–115.

22. A. Groenwall and B. Ingelman, *Acta Physiol. Scand.* **9,** 1 (1945).

23. F. H. Foster, *Process Biochem.* **3**(2), 15 (1968); **3**(3), 55 (1968).

24. A. Jeanes in P. Sandford and A. Laskin, eds., *Extracellular Microbial Polysaccharides*, *ACS Symp. Ser. No. 45*, American Chemical Society, Washington, D.C., 1977, pp. 284–298.

25. H. J. Koepsell and H. M. Tsuchiya, *J. Bacteriol.* **63,** 293 (1952).

26. R. M. Alsop, G. A. Byrne, J. N. Done, I. E. Earl, and R. Gibbs, *Process Biochem.* **12**(12), 15 (1977).

27. E. J. Hehre, *J. Biol. Chem.* **222,** 739 (1956).

28. S. P. Rogovin, F. R. Senti, R. G. Benedict, H. M. Tsuchiya, P. R. Watson, R. Tobin, V. E. Sohns, and M. E. Slodki, *J. Biochem. Microbiol. Technol. Eng.* **2,** 381 (1960).

29. A. Jeanes, *Methods Carbohydr. Chem.* **5,** 118 (1965).

30. H. M. Tsuchiya, H. J. Koepsell, J. Corman, G. Bryant, M. O. Bogard, V. H. Feger, and R. W. Jackson, *J. Bacteriol.* **64,** 521 (1952).

31. W. B. Neely and J. Hallmark, *Nature* **191,** 385 (1961).

32. E. J. Hehre, *Methods Enzymol.* **1,** 178 (1955).

33. Ref. 29, p. 127.

34. O. Larm, B. Lindberg, and S. Svensson, *Carbohydr. Res.* **20,** 39 (1971).

35. G. J. Walker and A. Pulkownik, *Carbohydr. Res.* **29,** 1 (1973).

36. E. R. Morris, D. A. Rees, D. Thom, and E. J. Welsh, *J. Supramol. Struct.* **6,** 259 (1977).

37. F. R. Senti, N. N. Hellman, N. H. Ludwig, G. E. Babcock, R. Tobin, C. A. Glass, and B. L. Lamberts, *J. Polym. Sci.* **17,** 527 (1955).

38. K. H. Ebert and G. Rupprecht, *Makromol. Chem.* **94,** 153 (1966).

39. F. R. Seymour, R. D. Knapp, E. C. M. Chen, A. Jeanes, and S. H. Bishop, *Carbohydr. Res.* **71,** 231 (1979); **75,** 275 (1979).

40. I. A. Wolff, C. L. Mehltretter, R. L. Mellies, P. R. Watson, B. T. Hofreiter, P. L. Patrick, and C. E. Rist, *Ind. Eng. Chem.* **46,** 370 (1954).

41. A. Jeanes in A. Jeanes and J. Hodge, eds., *Physiological Effects of Food Carbohydrates*, *ACS Symp. Ser. No. 15*, American Chemical Society, Washington, D.C., 1975, pp. 336–347.

42. A. Jeanes in K. C. Frisch, D. Klempner, and A. V. Patsis, eds., *Polyelectrolytes*, Technomic Publ. Co., Inc., Westport, Conn., 1976, pp. 207–225.
43. J. S. Cox, R. E. King, and G. F. Reynolds, *Nature* **207**, 1202 (1965).
44. U.S. Pat. 3,262,847 (July 26, 1966), P. G. M. Flodin, J. A. O. Johansson, and H. L. E. Johansson (to Aktiebolaget Pharmacia).
45. A. Tiselius, J. Porath, and P. A. Albertsson, *Science* **141**, 13 (1963).
46. U.S. Pat. 3,042,667 (July 3, 1962), P. G. M. Flodin and B. G. A. Ingleman (to Aktiebolaget Pharmacia).
47. Brit. Pat. 974,054 (Nov. 4, 1964), (to Aktiebolaget Pharmacia).
48. P. T. Murphy and R. L. Whistler in ref. 3, pp. 513–542.
49. A. Jeanes, *Misc. Publ. 1355*, U.S. Department of Agriculture, Washington, D.C., 1978.
50. P. W. Robbins, D. Bray, M. Dankert, and A. Wright, *Science* **158**, 1536 (1967).
51. I. Sutherland in ref. 24, pp. 40–57.
52. I. Sutherland in ref. 4, pp. 27–96.
53. F. A. Troy, II, in M. P. Starr, J. L. Ingram, and S. Raffel, eds., *Annu. Rev. Microbiol.* **33**, 519 (1979).
54. G. Schiffman in A. Voller and H. Friedman, eds., *New Trends & Development in Vaccines*, University Park Press, Baltimore, Md., 1978, pp. 237–243.
55. M. E. Slodki and M. C. Cadmus, *Adv. Appl. Microbiol.* **23**, 19 (1978).
56. P. A. Sandford, *Adv. Carbohydr. Chem. Biochem.* **36**, 265 (1979).
57. K. S. Kang and I. Cottrell in H. J. Peppler and D. Perlman, eds., *Microbial Technology*, 2nd ed., Vol. 1, Academic Press, Inc., New York, 1979, pp. 417–481.
58. M. E. Slodki and J. A. Boundy, *Dev. Ind. Microbiol.* **11**, 86 (1970).
59. S. P. Rogovin, R. G. Anderson, and M. C. Cadmus, *J. Biochem. Microbiol. Technol. Eng.* **3**, 51 (1961).
60. A. Jeanes, J. E. Pittsley, and F. R. Senti, *J. Appl. Polym. Sci.* **5**, 519 (1961).
61. J. Wells in ref. 24, pp. 299–314.
62. P.-E. Jansson, L. Kenne, and B. Lindberg, *Carbohydr. Res.* **45**, 275 (1975).
63. L. D. Melton, L. Mindt, D. A. Rees, and G. R. Sanderson, *Carbohydr. Res.* **46**, 245 (1976).
64. G. Holtzwarth and J. Ogletree, *Carbohydr. Res.* **76**, 277 (1980).
65. G. Holtzwarth, *Biochemistry* **15**, 4333 (1976).
66. F. R. Dintzis, G. E. Babcock, and R. Tobin, *Carbohydr. Res.* **13**, 257 (1970).
67. G. Holtzwarth, *Carbohydr. Res.* **66**, 173 (1978).
68. D. A. Rees, *Biochem. J.* **126**, 257 (1972).
69. R. Moorhouse, M. D. Wilkinshaw, and S. Arnott in ref. 24, pp. 90–102.
70. E. R. Morris in ref. 24, pp. 81–89.
71. M. Milas and M. Rinaudo, *Carbohydr. Res.* **76**, 189 (1979).
72. G. Holtzwarth and E. B. Prestridge, *Science* **197**, 757 (1977).
73. P. Suow and A. L. Demain, *Appl. Environ. Microbiol.* **37**, 1186 (1979).
74. M. C. Cadmus, C. A. Knutson, A. A. Lagoda, J. E. Pittsley, and K. A. Burton, *Biotechnol. Bioeng.* **20**, 1003 (1978).
75. R. A. Moraine and P. Rogovin, *Biotechnol. Bioeng.* **15**, 225 (1973).
76. P. A. Sandford, P. R. Watson, and C. A. Knutson, *Carbohydr. Res.* **63**, 253 (1978).
77. R. W. Silman and P. Rogovin, *Biotechnol. Bioeng.* **12**, 75 (1970); **14**, 23 (1972).
78. I. W. Davidson, *FEMS Microbiol. Lett.* **3**, 347 (1978).
79. O. M. Neijssel and D. W. Tempest, *Arch. Microbiol.* **107**, 215 (1976).
80. M. C. Cadmus, S. P. Rogovin, K. A. Burton, J. E. Pittsley, C. A. Knutson, and A. Jeanes, *Can. J. Microbiol.* **22**, 942 (1976).
81. A. Jeanes, P. Rogovin, M. C. Cadmus, R. W. Silman, and C. A. Knutson, *ARS-NC-51*, U.S. Department of Agriculture, Peoria, Ill., 1976, pp. 1–14.
82. A. S. Szczesniak and E. Farkas, *J. Food Sci.* **27**, 381 (1962).
83. I. C. M. Dea and E. R. Morris in ref. 24, pp. 174–182.
84. T. R. Andrew in ref. 24, pp. 231–241.
85. U.S. Pat. 4,119,546 (Oct. 10, 1978), W. C. Wernau (to Pfizer, Inc.).
86. U.S. Pat. 3,054,689 (Sept. 18, 1962), A. Jeanes and J. H. Sloneker (to U.S. Secretary of Agriculture).
87. U.S. Pat. 4,119,491 (Oct. 10, 1978), S. L. Wellington (to Shell Oil Company).

88. E. I. Sandvik and J. M. Maerker in ref. 24, pp. 242–264.
89. J. H. Elliott in ref. 24, pp. 144–159.
90. P. A. Sandford, J. E. Pittsley, C. A. Knutson, P. R. Watson, M. C. Cadmus, and A. Jeanes in ref. 24, pp. 192–210.
91. U.S. Pat. 3,933,788 (Jan. 20, 1976), K. S. Kang, G. T. Veeder, III, and D. D. Richey (to Kelco Company).
92. K. S. Kang, G. T. Veeder, and D. D. Richey in ref. 24, pp. 211–219.
93. U.S. Pat. 3,960,832 (June 1, 1976), K. S. Kang and W. H. McNeely (to Kelco Company).
94. K. S. Kang and W. H. McNeely in ref. 24, pp. 220–230.
95. T. Kaneko and K. S. Kang, *Abstr. Annu. Meet. Am. Soc. Microbiol.*, 101 (1979).
96. K. S. Kang, G. T. Veeder, and I. W. Cottrell, *Abstr. Annu. Meet. Am. Soc. Microbiol.*, 200 (1979).
97. G. T. Veeder, K. S. Kang, P. J. Mirrasoul, L. Koupal, and E. O. Stapley, *Abstr. Annu. Meet. Am. Soc. Microbiol.*, 200 (1979).
98. P. A. J. Gorin and J. F. T. Spencer, *Can. J. Chem.* **44**, 993 (1966).
99. B. Larsen and A. Haug, *Carbohydr. Res.* **17**, 287 (1971).
100. A. Haug and B. Larsen, *Biochim. Biophys. Acta* **192**, 557 (1969); *Carbohydr. Res.* **17**, 297 (1971).
101. L. Deavin, T. R. Jarmon, C. J. Lawson, R. C. Righelato, and S. Slocombe in ref. 24, pp. 14–26.
102. T. Harada and T. Yoshimura, *Biochim. Biophys. Acta* **83**, 374 (1964).
103. M. Hisamatsu, J. Abe, A. Amemura, and T. Harada, *Carbohydr. Res.* **66**, 289 (1978).
104. T. Harada, *Arch. Biochem. Biophys.* **112**, 65 (1965).
105. M. Hisamatsu, K. Sano, A. Amemura, and T. Harada, *Carbohydr. Res.* **61**, 89 (1978).
106. T. Harada, A. Amemura, P.-E. Jansson, and B. Lindberg, *Carbohydr. Res.* **77**, 285 (1979).
107. T. Harada in ref. 24, pp. 265–283.
108. T. Harada in J. M. V. Blanchard and J. R. Mitchell, eds., *Polysaccharides in Food*, Butterworths Publications, Inc., Woburn, Mass., 1979, pp. 283–300.
109. T. Kuge, N. Suetzugu, and K. Nishiyama, *Agric. Biol. Chem.* **7**, 1315 (1977).
110. I. Nakanishi, K. Kimura, T. Suzuki, M. Ishakawa, I. Banno, T. Sakane, and T. Harada, *J. Gen. Appl. Microbiol.* **22**, 1 (1976).
111. T. Harada, A. Koreeda, S. Sato, and N. Kasai, *J. Electron. Microsc.* **28**, 147 (1979).
112. K. Ogawa, T. Watanabe, J. Tsurugi, and S. Ono, *Carbohydr. Res.* **23**, 399 (1972).
113. U.S. Pat. 4,012,333 (March 15, 1977), G. A. Towle (to Hercules, Inc.).
114. N. E. Rodgers in ref. 3, pp. 499–511.
115. J. Johnson, Jr., S. Kirkwood, A. Misaki, T. E. Nelson, J. V. Scaletti, and F. Smith, *Chem. Ind. (London)*, 820 (1963).
116. U.S. Pat. 3,301,848 (Jan. 31, 1967), F. E. Halleck (to The Pillsbury Company).
117. K. Wallenfels, H. Bender, G. Keilich, and D. Freudenberger, *Biochem. Z.* **341**, 433 (1965).
118. B. J. Catley, *FEBS Lett.* **10**, 190 (1970).
119. W. Sowa, A. C. Blackwood, and G. A. Adams, *Can. J. Chem.* **41**, 2314 (1963).
120. N. P. Elinov and A. K. Matveeva, *Biokhimiya* **37**, 255 (1972).
121. U.S. Pat. 3,912,591 (Oct. 14, 1975), K. Kato and M. Shiosaka (to Hayashibara Biochemical Laboratories, Inc.).
122. S. Yuen, *Process Biochem.* **9**(9), 7 (1974).
123. J. E. Zajac and A. LeDuy, *Appl. Microbiol.* **25**, 628 (1973).
124. U.S. Pat. 3,997,703 (Dec. 14, 1976), S. Nakashio, K. Tsuji, N. Toyota, F. Fujita, and T. Sato (to Sumitomo Chemical Company, Ltd. and Hayashibara Biochemical Laboratories, Inc.).

MOREY E. SLODKI
Northern Regional Research Center
U.S. Department of Agriculture

MICROBIAL TRANSFORMATIONS

Microorganisms are of considerable economic importance in the manufacture of antibiotics (qv), alkaloids (qv), vitamins (qv), amino acids (qv), industrial solvents, organic acids, nucleosides, nucleotides, fermented beverages, and fermented foods (see Solvents, industrial; Carboxylic acids; Beer; Beverage spirits, distilled; Wine; Food processing; Foods, nonconventional; Fermentation; Malts and malting). They also are principal to simple and chemically well-defined reactions involving compounds that are not related to these products. Since some of such reactions can be carried out more economically by microbial means than by a strictly chemical manipulation, the reactions have been included in the processes that yield a number of important products, eg, L-ascorbic acid, steroid hormones, 6-aminopenicillanic acid, various L-amino acids, L-ephedrine, D-fructose, vinegar, and malt (see Hormones, adrenal-cortical; Sugar). In addition, they have been employed in many studies of synthetic, structural, stereochemical, and kinetic problems in organic chemistry to functionalize nonactivated carbon atoms, to introduce centers of chirality into optically inactive substrates, and to carry out optical resolution of racemic mixtures of DL-amino acids and of other compounds. Such properties result from the ability of the microbes to elaborate both constitutive and inducible enzymes that possess a broad substrate specificity and remarkable regio- and stereospecificities.

There are many publications regarding microorganism-catalyzed experimental procedures (1–2). They involve the reactions related to carbohydrates (3); steroids, sterols, and bile acids (4–11); nonsteroid cyclic compounds (12); alicyclic and alkane hydroxylations (13–16); the preparation of medicinals (17); and transformations of antibiotics (18–19). Reviews and discussions of the microbial oxidation of aromatic hydrocarbons (20–21), monoterpenes (22), pesticides (23), lignin (24), and other organic molecules (25–26) also have been published (27–29) (see Terpenoids; Insect control technology; Lignin).

Reactions

Oxidation. The ability of microorganisms to affect organic compounds was generally accepted when Pasteur described the conversion of ethanol to acetic acid by bacteria (30) and when Bertrand showed that polyhydric alcohols were oxidized to their corresponding keto sugars by *Acetobacter xylinum* (31–32). The latter studies led to the formulation of a rule which states that a secondary hydroxyl group of a polyol is oxidized to a ketone only if that group is situated between a primary and another secondary group which is in the cis position with respect to the oxidizable group (33–34). The oxidation of one of the polyols, ie, D-sorbitol, by *A. suboxydans* became an important step in the Reichstein synthesis of L-ascorbic acid 40 years later. D-Sorbitol (1) solution can be converted to L-sorbose (2) within 3 d of incubation with 90–95% yields (35); the low production cost of L-ascorbic acid directly results from this efficiency. The L-sorbose which is obtained is in turn converted to L-ascorbic acid via diacetone-L-sorbose, diacetone-2-keto-L-gulonic acid, and methyl 2-keto-L-gulonate.

Both mono- and polynuclear aromatic hydrocarbons can be oxidized by different

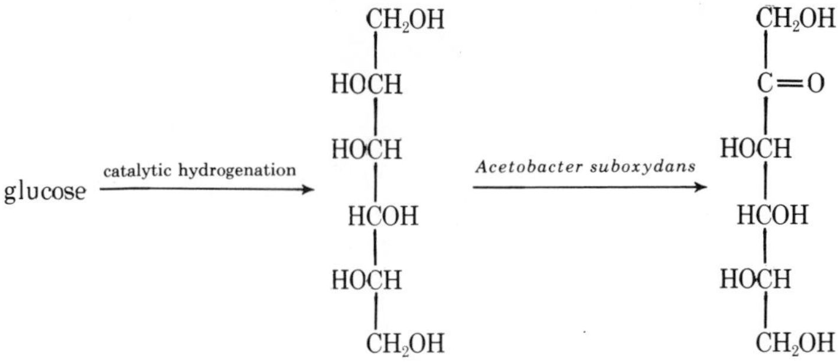

(1)   (2)

microorganisms. *p*-Cymene is converted to cumic acid and *p*-xylene to *p*-toluic acid (36); a high yielding (98%) process has been developed in Japan for the production of salicylic acid (4) from naphthalene (3) (37):

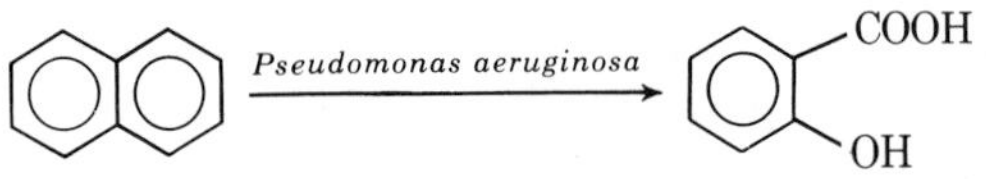

(3)   (4)

Hydroxylation, Dehydrogenation, and β-Oxidation.   When cortisone and hydrocortisone were identified in 1949 as potent antiinflammatory agents and no adequate synthesis existed to meet the sharply increased demand for these compounds, hydroxylation of steroid intermediates at *C*-11 became crucial for their large-scale production. The problem of introducing functionality at that site was solved when it was discovered that progesterone (5) is oxidized to 11α-hydroxyprogesterone (6) by *Rhizopus arrhizus* (38).

In the wake of this finding and the subsequent increased efficiency of this reaction by *Rh. nigricans*, massive programs were launched to modify other sites of the steroid molecules by microorganisms, to develop efficient syntheses of steroid hormones and to find new derivatives with more specific physiological activities than the parent compounds. As a result of this research, stereospecific microbial hydroxylations at practically all available carbon atoms of the steroid molecule were found and other microbial reactions were reported, eg, dehydrogenation, epoxidation, reduction, side-chain cleavage, aromatization of ring A, lactonization of ring D, esterification, hydrolysis, and isomerization (4–6,8–9). Of these, 1-dehydrogenation has assumed practical prominence in the synthesis of prednisone, prednisolone (7), and their derivatives, all of which are more potent and have fewer side effects than the parent hormones (see Steroids).

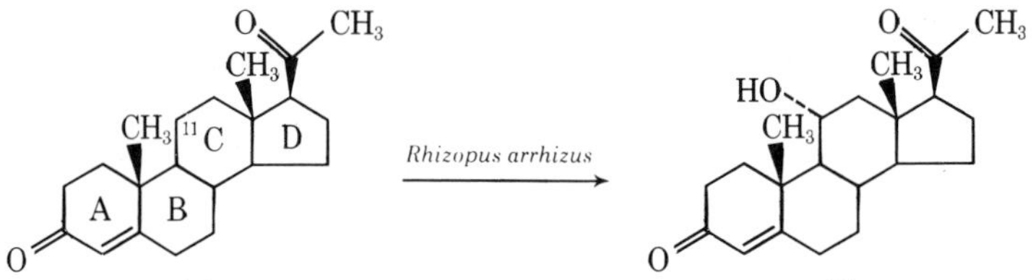

(5)   (6)

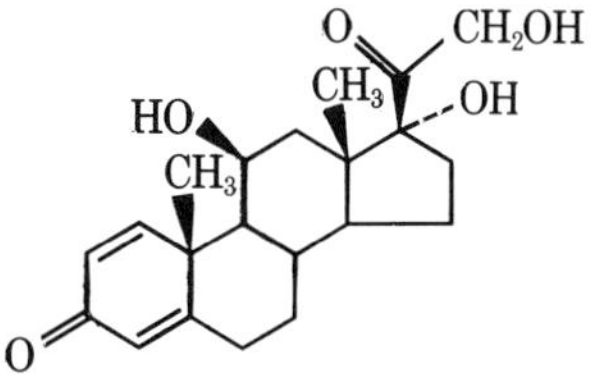

(7)

Bacterial removal of sterol side chains is carried out by a stepwise β-oxidation while the degradation of the perhydrocyclopentanophenanthrene nucleus is prevented by metabolic inhibitors (39), chemical modification of the nucleus (40), or the use of bacterial mutants (10,41). β-Sitosterol (8), a plant sterol which has been a waste product in the manufacture of steroid hormones, is used as a new raw material for the preparation of 4-androstene-3,17-dione (9), 1,4-androstadiene-3,17-dione (10), 9α-hydroxy-4-androstene-3,17-dione (11), and related compounds by means of selected mutants of the β-sitosterol-degrading bacteria (42):

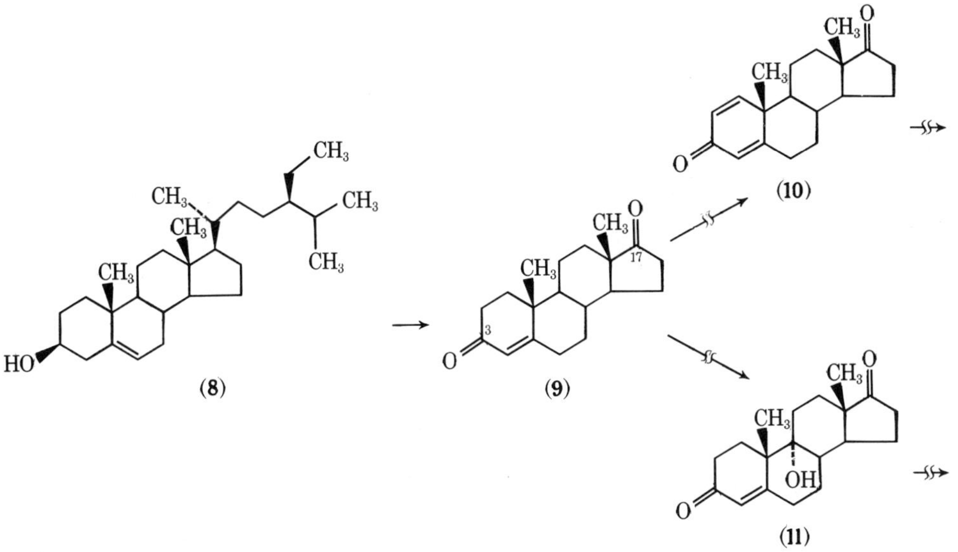

Reduction. Although the reduction of ketonic substrates usually is only partially asymmetric, more stereospecific microbial reduction depends on the size and nature of the substituents in the substrates; eg, the reduction of racemic decalone and hexahydroindanone derivatives and of related di- and tricyclic ketones by *Curvularia falcata* is highly stereospecific. Stereoselective reduction occurs if the ketone is flanked by a large L and small s groups, and yields an alcohol of the *S*-configuration (43):

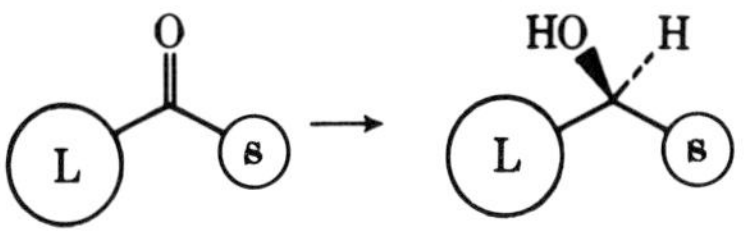

This model may help to predict if the substrate will be reduced to an optically active alcohol, which may be difficult to obtain by purely chemical means. Microorganisms also can be used to implement selective reductions of β-diketones which are important in steroid syntheses (44). Reduction of 4-androstene-3,17-dione (9) to 17β-hydroxy-4-androsten-3-one (testosterone) (12) by yeasts is one of the earliest observed conversions of steroids (45):

Hydrolysis. Microbial hydrolysis of a large number of esters, glycosides, epoxides, lactones, β-lactams, and amides has been described. Acetylated steroids have been hydrolyzed with varying degrees of selectivity by numerous organisms; (−)-14-ace-toxycodeine has been converted to (−)-14-hydroxycodeine (46) and atropine to tropine; the sugar moiety has been removed from heart glycosides and saponins (17); and L-amino acids have been produced from their optically inactive forms. Tartaric acid (15) is produced from glucose by *A. suboxydans* (47) and, more recently, a method has been developed for its preparation from maleic anhydride (13) by hydrolysis of the inter-mediary *cis*-epoxysuccinic acid (14) (99.8% yield) (48):

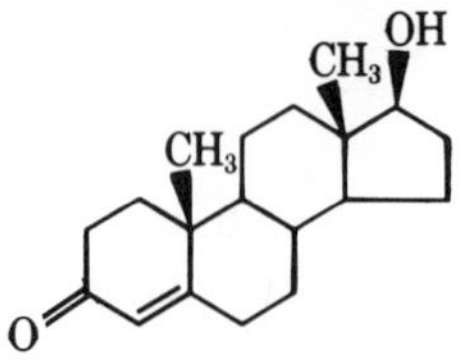

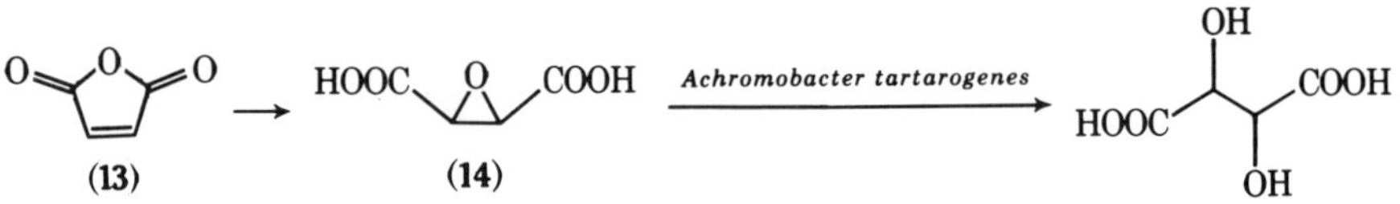

Penicillins are inactivated by many bacteria through hydrolysis of their β-lactam rings. The hydrolysis of the amide bond 6-aminopenicillanic acid (6-APA) (16) is economically valuable since this acid is the principal intermediate in the preparation and manufacture of semisynthetic penicillins. Although 6-APA was originally produced by direct fermentation in the absence of the side-chain precursors, the yields were low and the presence of penicillins which also formed complicated its extraction and purification (49). 6-APA is made on a large scale by selective enzymatic hydrolysis of benzylpenicillin (17) (penicillin G) or phenylmethoxypenicillin (18) (penicillin V) which are produced in high yields by direct fermentation. Penicillin acylases, which carry out this selective hydrolysis, generally are of two types: those which are produced by bacteria and primarily hydrolyze penicillin G, and those which are elaborated by actinomycetes, fungi, and yeasts and which preferentially hydrolyze penicillin V. The hydrolysis usually is favored by high temperatures and an alkaline pH. The reaction can be reversed at an acidic pH whereby 6-APA is acylated to penicillins in the presence of suitable acyl donors. Mutants which have been selected by systematic culture de-velopment hydrolyze high concentrations of penicillin G. The reaction pH, tempera-ture, and ratios of the cells to the substrate must be regulated carefully to force the equilibrium toward complete deacylation and to minimize undesirable degradation, eg, the hydrolysis of the β-lactam ring, of both the substrate and the product. 6-APA has been prepared by conventional batch process, and by using immobilized cells, spores, and immobilized enzymes (qv) (50–52). When the enzymatic hydrolysis has been completed, the released side chain, eg, phenylacetic acid, is extracted with an organic solvent and 6-APA is isolated by crystallization and used for chemical acylation to semisynthetic penicillins as shown in Figure 1.

In a similar way, several cephalosporins have been hydrolyzed to 7-aminodeacetoxycephalosporanic acid (53), and nocardicin C to 6-aminonocardicinic acid (54).

Enzymatic hydrolysis also is used for the preparation of L-amino acids. Amino acids and their acyl derivatives can be obtained chemically as racemic mixtures of their D and L enantiomers which, in turn, are resolved to their natural L forms. A resolution of this kind can be accomplished using aminoacylase from *A. oryzae* which specifically hydrolyzes the L enantiomers of acyl-DL-amino acids. The latter can be separated readily from the unchanged acyl-D form which is recycled by racemization:

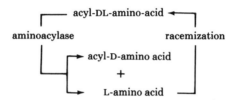

Several L-amino acids, eg, methionine, phenylalanine, tryptophan, and valine, have been manufactured by this process in Japan and their production cost has been reduced by 40% by the application of immobilized cell technology (55).

In a similar way, chlorocyclohexane, which is a by-product in nylon manufacture, is chemically converted to DL-amino-ε-caprolactam (**19**). The racemization and the

Figure 1. Production of semisynthetic penicillins by microbial hydrolysis.

hydrolysis of the L isomer give L-lysine (20) in almost 100% yields at 140 g/L (56):

Condensation. Asymmetric microbial condensation was discovered in 1921 (57) and was utilized in 1934 in the synthesis of the natural (1R,2S)-ephedrine (58). In this process, benzaldehyde (21) is added to the fermenting yeast and reacts with acetaldehyde (22), which is generated from glucose by the organism, and yields (R)-1-phenyl-1-hydroxy-2-propanone (23). The latter undergoes reductive chemical condensation with methylamine and yields the desired (1R,2S)-ephedrine (24):

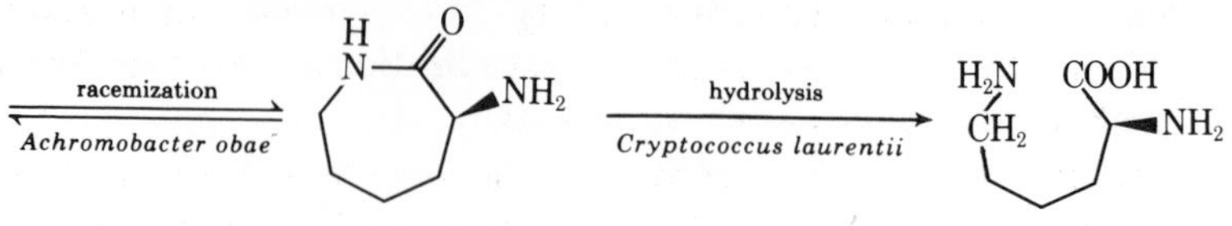

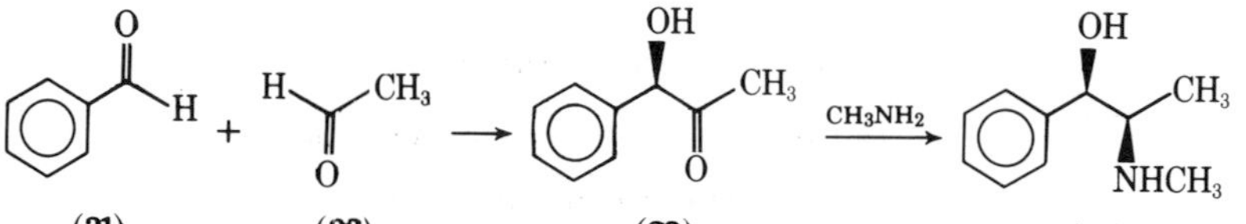

Substituted benzaldehyde derivatives react in the same manner (59).

Amination and Hydration. Optically active products, which correspond to the naturally occurring L isomers, have been obtained by the asymmetric addition of ammonia or water to fumaric acid (25). Thus aspartase-producing bacteria have been used in the manufacture of L-aspartic acid (26) (60). Others with high fumarase activity hydrate (25) to L-malic acid (27) (61–62) in 70% yields, especially when pretreated with detergents, eg, bile extracts, in order to suppress the formation of succinic acid:

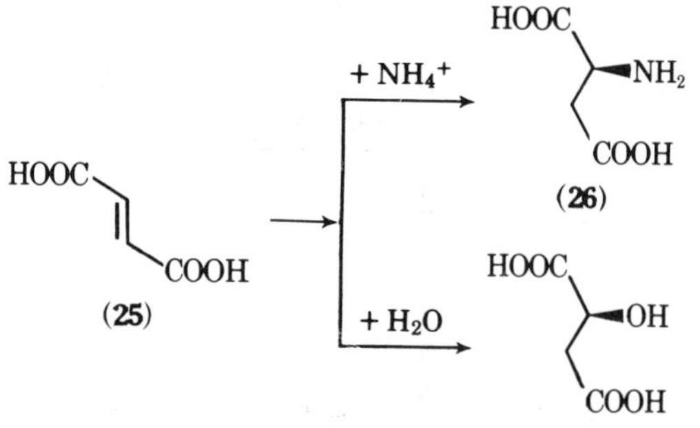

Both reactions have been carried out also by cells that are immobilized in polyacrylamide gel, whereby the half-lives of the enzymes increase from 11 to 120 and 53 d, respectively, and thus, contribute to the economy of both processes. L-Citrulline (28) is manufactured from L-arginine (29) by the same method (61).

Other L-amino acids have been obtained by bacterial enzymes. Thus, by the addition of NH_4^+ and pyruvate to the reaction mixtures, L-tyrosine has been produced from phenol, L-DOPA from catechol, L-tryptophan from indole and 5-hydroxy-L-tryptophan from 5-hydroxyindole. By hydration of DL mixtures, L isomers of a few other amino acids have been generated in 95–100% yields: L-cysteine from DL-2-

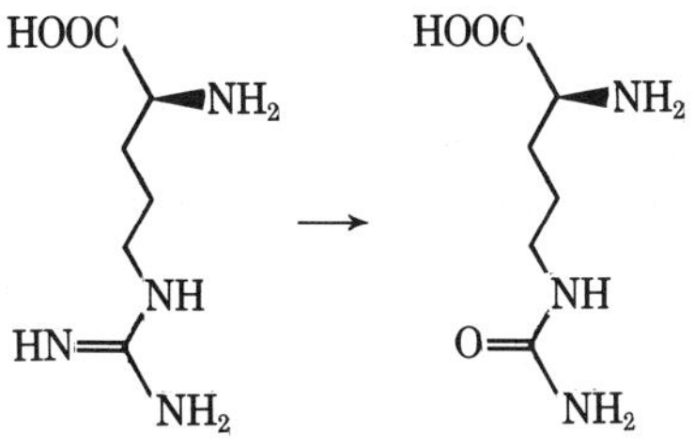

(29)   (28)

aminothiazoline-4-carboxylic acid, L-phenylalanine from DL-phenylalanine-hydantoin, and L-tryptophan from DL-tryptophan-hydantoin (36). Amination is also involved in the production of guanosine-5'-monophosphate (5'-GMP), a compound used as a food seasoning, by means of two mutants of *Brevibacterium ammoniagenes*. The first mutant ferments glucose to xanthosine-5'-monophosphate (5'-XMP) which is converted to 5'-GMP by the second mutant either by sequential operation or by mixed cultures of both mutants (63).

Deamination. Two kinds of deamination that have been observed are hydrolytic, eg, the conversion of L-tyrosine to 4-hydroxyphenyllactic acid in 90% yields (64), and oxidative (12,65–66), eg, isoguanine → xanthine and formycin A → formycin B.

Dehydration. Dehydration of hydroxy fatty acids is quite common; other compounds also undergo the same reaction, eg, elymoclavine to agroclavine, chanoclavine, and other compounds; *cis*-terpine hydrate to α-terpineol; histidine to urocanic acid, or L-phenylalanine to phenylpyruvic, phenylacetic, cinnamic, benzoic, and other acids (17).

N- and O-Demethylation. Because of the high selectivity involved, microbial N- and C-demethylation is preferable, especially of natural products, to chemical treatments which often are too drastic and nonspecific. O-Demethylation of 10,11-[methylenebis(oxy)]aporphine (30) by *Cunninghamella blakesleeana* also proceeds with high regioselectivity and in 100% yields to isoapocodeine (31) (67).

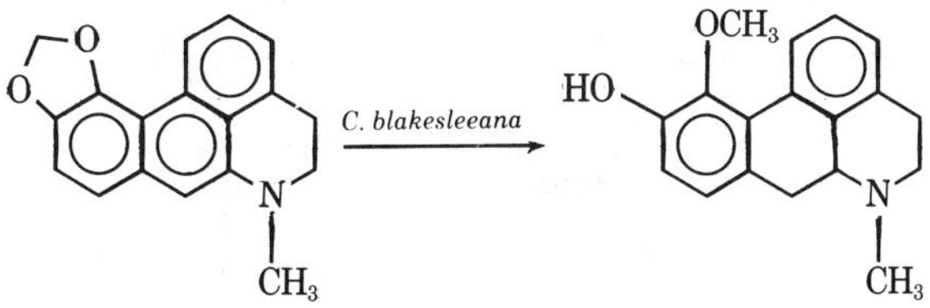

(30)   (31)

Vindoline (32), a monomeric *Vinca* alkaloid which is of interest as a material in the synthesis of antitumor compound, is converted to N-demethylvindoline (33) by *Streptomyces albogriseolus* and other susceptible groups and linkages remain intact (68) (see Alkaloids):

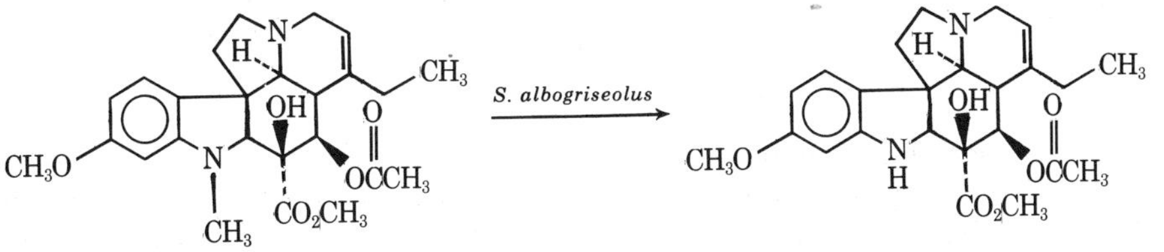

(32)   (33)

Decarboxylation.   Decarboxylation of linear and aromatic carboxylic acids and of amino acids is common and of practical interest. L-Lysine can be synthesized by stereospecific decarboxylation of *meso-*(but not DL)-α,α'-diaminopimelic acid (**34**) (DAP) to L-lysine (**20**). The reaction is catalyzed by *Bacillus sphaericus* and proceeds in quantitative yields (69):

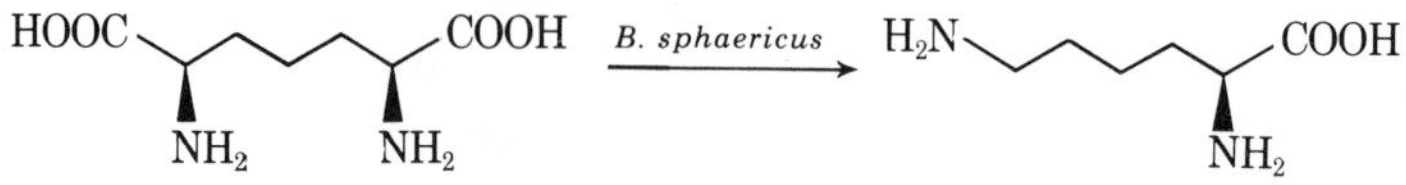

After the recovery of L-lysine, the residual DL-DAP is epimerized to a mixture of the DL and meso isomers, and the latter are subjected to the same decarboxylation step. This reaction is a part of a microbial process in which glucose is fermented by a lysine auxotroph of *E. coli* to *meso*-DAP which accumulates in the medium. As in the preceding case, *meso*-DAP is decarboxylated quantitatively to L-lysine by cell suspensions of *Aerobacter aerogenes* (70). However, L-lysine and practically all L-amino acids are manufactured much more economically in thousands of tons per year in Japan by simplified fermentations directly from glucose, ethanol, acetic acid, glycerol, or *n*-paraffin by means of selected auxotrophic, regulatory, and analogue-resistant bacterial mutants (71–72) (see Amino acids).

N-Acetylation, O-Phosphorylation, and O-Adenylylation.   *N*-Acetylation, *O*-phosphorylation, and *O*-adenylylation provide mechanisms by which therapeutically valuable aminocyclitol antibiotics, eg, kanamycin, gentamicin, sisomicin, streptomycin, neomycin, and spectinomycin, are rendered either partially or completely inactive. Thus, eg, kanamycin B (**35**) is inactivated at several sites by any of the three mechanisms:

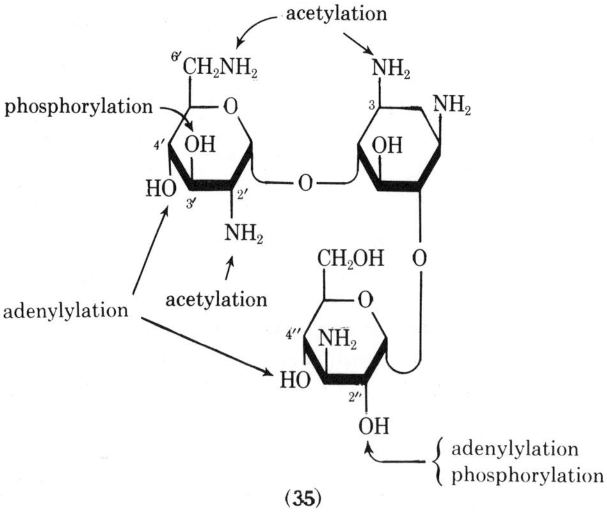

The elucidation of these mechanisms has allowed chemical modification of the sites at which the inactivation occurs. Several such bioactive analogues, eg, dibekacin and amikacin have been prepared and are not subject to the inactivation; hence, they inhibit those organisms against which the parent antibiotics are ineffective (73) (see Antibiotics, aminoglycosides).

Acetylation of the hydroxyl and esterification of the carboxyl groups have been observed in a limited number of cases but, in general, have no preparative advantage over the chemical methods. By comparison, phosphorylation has been useful in the preparation of modified purine and pyrimidine mononucleotides from their corresponding nucleosides, eg, 6-thioguanosine (**36**) (74).

Transglycosylation. Transglycosylation can be expressed by the following equation:

$$R\text{—}O\text{—}R' + R''OH \rightleftharpoons R\text{—}O\text{—}R'' + R'OH$$
$$\text{donor} \quad \text{acceptor} \quad \text{product} \quad \text{by-product}$$

This reaction allows enzymatic preparation of various oligosaccharides that cannot be made readily by strictly chemical means. Transglycosylation was first used in 1944 to prepare sucrose from glucose 1-phosphate and fructose by sucrose phosphorylase of *Pseudomonas saccharophila* (75). Corresponding disaccharides are formed if fructose is replaced in the same system by other sugars, eg, xylose, arabinose, or sorbose.

Because the energy required for these reactions is derived from the high energy phosphate bond and because the glucosidic linkage has a similarly high energy bond, sucrose is used as a glucosyl donor for sorbose and yields glucosylsorbose when incubated with the *Ps. saccharophila* enzyme (76). Tri- and tetrasaccharides also are formed from sucrose, maltose, lactose, and cellobiose by various bacterial, yeast, and fungal enzymes (12) (see Sugar).

Isomerization. Isomerization of steroids is well-known, eg, both the 5-androstene and 5-pregnene structures have been investigated, but is not of practical value (77). Yet, it is of considerable importance in the manufacture of high fructose syrup which is used as a food sweetener. A process has been developed whereby inexpensive carbohydrates are hydrolyzed to glucose which, in turn, is isomerized in 50% solutions by a *Streptomyces* strain. Upon purification and evaporation, syrup is obtained which contains 75 wt % solids consisting of 55 wt % fructose and 45 wt % glucose. The isomerase has been immobilized and is used in high volume continuous processes in the United States and Japan (71). (See Syrups; Sugar, special sugars.)

(**36**)

Methodology

The selection of the organisms that carry out the desired transformation is of paramount importance. However, because only a few systematic examinations have been made of the microbial action on specific classes of organic compounds, there is no assured way to select the organism that will perform the desired reaction with the substrate of interest. In general, fungi and *Streptomycetes* hydrolyze various esters and carry out specific hydroxylations, yeasts reduce carbonyl groups, and certain bacteria oxidize alcohols and aldehydes. Taxonomically related organisms also may carry out the desired reaction if the enzyme has a low substrate specificity. In many cases, suitable organisms have been uncovered by screening randomly selected cultures isolated from various natural sources, eg, soil, decomposing organic material, or spontaneous fermentations. Others have been found by the enrichment method which utilizes the substrate in question as the only source of carbon and nitrogen for growth and energy. Such populations are mutated and the proper mutants are selected (2).

For the transformation, the selected organism usually is grown in aerated flasks until a sufficient amount of cells has been generated. The substrate then is added to the cells, the incubation is continued, and the progress of the transformation is monitored by suitable chromatographic, spectroscopic, or biological methods. When the maximum transformation has been obtained, the reaction is terminated and the product is isolated and identified.

There are many variations of this procedure. The substrate may be dissolved in water or in a relatively nontoxic solvent, eg, ethanol, acetone, dimethylformamide, or dimethylsulfoxide, or be finely dispersed in water. When enzyme-induction is required, some of the substrate is added during early growth. When carried out by constitutive enzymes, the transformation may be added in the late logarithmic or stationary phases of growth or to the washed, ie, nonproliferating, buffered cell suspensions. A number of useful reactions have been carried out by microorganisms, spores, or isolated enzymes that have been immobilized by ionic binding to a water-insoluble ion exchanger, cross-linking with bifunctional reagents, or entrapping into a polymer matrix (see Ion exchange). This immobilization permits reuse of the cells and results in a more economical process.

BIBLIOGRAPHY

1. D. Perlman in J. B. Jones, C. J. Sih, and D. Perlman, eds., *Applications of Biochemical Systems in Organic Chemistry*, John Wiley & Sons, Inc., New York, 1976, Part 1, p. 47.
2. J. R. Norris and D. W. Ribbons, *Methods in Microbiology*, Academic Press, Inc., New York, 1969.
3. J. F. T. Spencer and P. A. J. Gorin, *Progr. Ind. Microbiol.* **7**, 178 (1965).
4. A. Čapek, O. Hanč, and M. Tadra, *Microbial Transformations of Steroids*, Academia, Prague, Czechoslovakia, 1966.
5. H. Iizuka and A. Naito, *Microbial Transformation of Steroids and Alkaloids*, University of Tokyo Press, Tokyo; University Park Press, State College, Pa., 1967.
6. W. Charney and H. L. Herzog, *Microbial Transformations of Steroids*, Academic Press, Inc., New York, 1967.
7. G. K. Skryabin and L. A. M. Golovleva, *Microorganisms in Organic Chemistry*, Nauka Press, Moscow, USSR, 1976.
8. L. L. Smith, *Spec. Period. Rep. Chem. Soc. London* **4**, 394 (1974).
9. K. Kieslich and O. K. Sebek, *Ann. Rep. Ferment. Proc.* **3**, 275 (1979).
10. W. J. Marsheck, *Progr. Ind. Microbiol.* **10**, 49 (1971).
11. S. Hayakawa, *Adv. Lipid Res.* **11**, 143 (1973).

12. K. Kieslich, *Microbial Transformations of Non-Steroid Cyclic Compounds*, G. Thieme, Publ., Stuttgart, FRG, 1976.

13. O. K. Sebek and K. Kieslich, *Ann. Rep. Ferment. Proc.* **1**, 263 (1977).

14. A. J. Markovetz, *Crit. Rev. Microbiol.* **1**, 225 (1971).

15. P. Jurtshuk and G. E. Cardini, *Crit. Rev. Microbiol.* **1**, 239 (1971).

16. B. J. Abbott and W. E. Gledhill, *Adv. Appl. Microbiol.* **14**, 249 (1971).

17. R. Beukers, A. F. Marx, and M. H. J. Zuidweg, *Drug Des.* **3**, 1 (1972).

18. O. K. Sebek, *Lloydia* **37**, 115 (1974); *Acta Microbiol. Acad. Sci. Hung.* **22**, 381 (1975).

19. M. Shibata and M. Uyeda, *Ann. Rep. Ferment. Proc.* **2**, 267 (1978).

20. R. L. Raymond, V. S. Jamison, and J. O. Hudson, *Lipids* **6**, 453 (1971).

21. D. T. Gibson, *Crit. Rev. Microbiol.* **1**, 199 (1971).

22. I. C. Gunsalus and V. P. Marshall, *Crit. Rev. Microbiol.* **1**, 291 (1971).

23. J. M. Bollag, *Crit. Rev. Microbiol.* **2**, 35 (1972).

24. T. K. Kirk, *Ann. Rev. Phytopathol.* **9**, 185 (1971).

25. G. S. Fonken and R. A. Johnson, *Chemical Oxidations with Microorganisms*, Marcel Dekker, Inc., New York, 1972.

26. R. A. Johnson, *Oxidation in Organic Chemistry*, Academic Press, Inc., 1978, Part C, p. 131.

27. L. L. Wallen, F. H. Stodola, and R. W. Jackson, *Type Reactions in Fermentation Chemistry*, ARS-71-13, U.S. Dept. of Agriculture, U.S. Government Printing Office, Washington, D.C., 1959.

28. K. Kieslich, *Synthesis*, 120, 147 (1969); R. Hütter, T. Leisinger, J. Nüesch, and W. Wehrli, eds., *Antibiotics and Other Secondary Metabolites*, Academic Press, Inc., New York, 1978, p. 57.

29. Ch. Tamm, *FEBS Lett.* **48**, 7 (1974).

30. L. Pasteur, *Memoir sur la fermentation acétique*, 1864; *Études sur la vinaigre*, 1868.

31. G. Bertrand, *Compt. Rend.* **122**, 900 (1896).

32. G. Bertrand, *Ann. Chim. Phys.* **3**, 181 (1904).

33. R. M. Hann, E. B. Tilden, and C. S. Hudson, *J. Am. Chem. Soc.* **60**, 1201 (1938).

34. B. Magasanik, R. E. Franzl, and E. Chargaff, *J. Am. Chem. Soc.* **74**, 2618 (1952).

35. P. A. Wells, J. J. Tubbs, L. B. Lockwood, and E. T. Roe, *Ind. Eng. Chem.* **29**, 1385 (1937); **31**, 1518 (1939).

36. Y. Hirose and H. Okada in H. J. Peppler and D. Perlman, eds., *Microbial Technology*, Academic Press, Inc., New York, 1979, p. 211.

37. A. Kitai and A. Ozaki, *J. Ferment. Technol.* **47**, 527 (1969).

38. D. H. Peterson and H. C. Murray, *J. Am. Chem. Soc.* **74**, 1871 (1952).

39. J. M. Whitmarsh, *Biochem. J.* **90**, 23P (1964).

40. C. J. Sih, H. H. Tai, Y. Y. Tsong, S. S. Lee, and R. G. Coombe, *Biochemistry* **7**, 808 (1968).

41. W. J. Marsheck, S. Kraychy, and R. D. Muir, *Appl. Microbiol.* **23**, 72 (1972).

42. O. K. Sebek and D. Perlman in ref. 36, p. 483.

43. W. Acklin and V. Prelog, *Helv. Chim. Acta* **48**, 1725 (1965).

44. H. Kosmol, K. Kieslich, R. Vössing, H. J. Koch, K. Petzoldt, and H. Gibian, *Liebigs Ann. Chem.* **701**, 198 (1967).

45. L. Mamoli and A. Vercellone, *Ber. Deutsch. Chem. Ges.* **70**, 470 (1937).

46. M. Yamada, K. Iizuka, S. Okuda, T. Asai, and K. Tsuda, *Chem. Pharm. Bull.* **11**, 206 (1963).

47. U.S. Pat. 2,314,831 (March 23, 1943), J. Kamlet (to Miles Laboratories).

48. U.S. Pat. 3,957,579 (May 18, 1976), E. Sato and A. Yanai (to Toray Industries, Inc.).

49. T. R. Carrington, *Proc. R. Soc. London Ser. B.* **179**, 321 (1971).

50. B. J. Abbott, *Ann. Rep. Ferment. Proc.* **2**, 91 (1978).

51. I. Chibata, ed., *Immobilized Cells*, Kodansha, Ltd., Tokyo and John Wiley & Sons, Inc., New York, 1978, pp. 182, 184.

52. L. B. Wingard, Jr., E. Katchalski-Katzir, and L. Goldstein, eds., *Applied Biochemistry and Bioengineering*, Vol. 2, Academic Press, Inc., New York, 1979.

53. M. Shimizu, T. Masuike, H. Fujita, K. Kimura, R. Okachi, and T. Nara, *Agr. Biol. Chem.* **39**, 1225 (1975).

54. T. Komori, K. Kunugita, K. Nakahara, H. Aoki, and H. Imanaka, *Agric. Biol. Chem.* **42**, 1439 (1978).

55. I. Chibata, T. Tosa, T. Sato, T. Mori, and T. Matsuo, *Fermentation Technol. Today*, 383 (1972).

56. U.S. Pats. 3,770,585 (Nov. 6, 1973); 3,796,632 (March 12, 1974), T. Fukumura (to Toray Industries, Inc.).

57. C. Neuberg and J. Hirsch, *Biochem. Z.* **115**, 282 (1921).

58. U.S. Pat. 1,956,950 (May 1, 1934), G. Hildebrandt and W. Klavehn (to E. Bilhuber, Inc.).
59. U.S. Pat. 3,338,796 (Aug. 29, 1967), J. W. Rothrock (to Merck and Co., Inc.).
60. I. Chibata, T. Tosa, and T. Sato, *Appl. Microbiol.* **27**, 878 (1974).
61. I. Chibata and T. Tosa, *Adv. Appl. Microbiol.* **22**, 1 (1977).
62. K. Yamamoto, T. Tosa, K. Yamashita, and I. Chibata, *Biotechnol. Bioeng.* **19**, 1101 (1977).
63. A. Furuya, R. Okachi, K. Takayama, and S. Abe, *Biotechnol. Bioeng.* **15**, 795 (1973).
64. F. Ehrlich and K. A. Jacobsen, *Ber. Dtsch. Chem. Gesell.* **44**, 888 (1911).
65. S. Friedman and J. S. Gots, *Arch. Biochem.* **32**, 227 (1951).
66. Y. Sawa, Y. Fukagawa, I. Homma, T. Wakashiro, T. Kakeuchi, M. Hori, and T. Komai, *J. Antibiot.* **21**, 334 (1968).
67. J. P. Rosazza, A. W. Stocklinski, M. A. Gustafson, and J. Adrian, *J. Med. Chem.* **18**, 791 (1975).
68. N. Neuss, D. S. Fukuda, D. R. Brannon, and L. L. Huckstep, *Helv. Chim. Acta* **57**, 1891 (1974).
69. B. S. Gorton, J. N. Coker, H. P. Browder, and C. W. DeFiebre, *Ind. Eng. Chem. Prod. Res. Develop.* **2**, 308 (1963).
70. U.S. Pat. 2,771,396 (Nov. 20, 1956), L. E. Casida, Jr. (to Chas. Pfizer and Co., Inc.).
71. K. Arima, *Dev. Ind. Microbiol.* **18**, 79 (1977).
72. K. Yamada, *Biotechnol. Bioeng.* **19**, 1563 (1977).
73. J. Davies and D. I. Smith, *Ann. Rev. Microbiol.* **32**, 469 (1978).
74. Fr. Pat. 1,388,758 (Feb. 12, 1965), K. Mitsugi.
75. W. Z. Hassid, M. Doudoroff, and H. A. Barker, *J. Am. Chem. Soc.* **66**, 1416 (1944).
76. M. Doudoroff, H. A. Barker, and W. Z. Hassid, *J. Biol. Chem.* **168**, 725, 733 (1947).
77. P. Talalay and V. S. Wang, *Biochem. Biophys. Acta* **18**, 300 (1955).

Oldrich K. Sebek
The Upjohn Company

MICROCHEMISTRY.   See Analytical methods.

MICROENCAPSULATION

"Small is better" would be an appropriate motto for the many people studying microencapsulation, a process in which tiny particles or droplets are surrounded by a coating to give small capsules with many useful properties. In its simplest form, a microcapsule is a small sphere with a uniform wall around it. The material inside the microcapsule is referred to as the core, internal phase, or fill, whereas the wall is sometimes called a shell, coating, or membrane. Most microcapsules have diameters between a few micrometers and a few millimeters, as illustrated in Figure 1.

Many microcapsules, however, bear little resemblance to these simple spheres. The core may be a crystal, a jagged adsorbent particle, an emulsion, a suspension of solids, or a suspension of smaller microcapsules. The microcapsule even may have multiple walls.

In the formation of some microcapsules, a finely divided solid such as a drug is suspended in a polymer solution. Droplets then are formed and solidified. In this case, the wall actually has become the continuous phase of the particle. The solid is now said to be held in a matrix of polymer.

Figure 1.   Microcapsules come in a wide size range. Courtesy of Eurand America, Inc.

The reasons for microencapsulation are countless. In some cases, the core must be isolated from its surroundings, as in isolating vitamins from the deteriorating effects of oxygen, retarding evaporation of a volatile core, improving the handling properties of a sticky material, or isolating a reactive core from chemical attack. In other cases, the objective is not to isolate the core completely but to control the rate at which it leaves the microcapsule, as in the controlled release of drugs or pesticides (see Pharmaceuticals, controlled release). The problem may be as simple as masking the taste or odor of the core, or as complex as increasing the selectivity of an adsorption or extraction process.

Properties Important in Choice of Process

In choosing among processes for a particular application, the following physical properties must be carefully considered.

Core Wettability. Coacervation is the formation of a second polymer-rich liquid phase from a polymer solution, eg, by addition of a nonsolvent. In coacervation coating, the crucial property is the wettability of the core by the coacervate. As long as solid particles are properly wetted, they are frequently easier to coat than liquid cores. If a liquid core material is highly insoluble in the coacervate-forming solution, proper wetting may be difficult. To the dismay of workers in this field, both solids and liquids sometimes present a wettability problem.

In principle, the wettability of a solid in a particular coacervation system is determined easily. However, in practice, a surface of the proper configuration rarely is available for accurate measurement of a wetting angle or spreading coefficient. The wettability usually is determined directly during the microencapsulation process by examining the ability of the coacervate particles to coat the core particles adequately.

Core Solubility. In a coacervation system, it is critical that the core not be soluble in the polymer solvent and that the polymer not partition strongly into a liquid core. In interfacial polymerization systems, determination of the solubilities of the reactants in the phases permits a choice of solvents and polymers.

In some cases, cores can be employed that have some solubility in the external polymer solution. In spray-coating, it is possible to coat a water-soluble solid with an aqueous polymer solution because the water evaporates so rapidly after the spray droplets hit the core that there is little penetration or dissolution of the core.

Wall Permeability and Elasticity.   The polymer permeability indicates whether a core can be isolated or a drug released at the required rate (1) (see Membrane technology). The microcapsules must be able to tolerate handling but may be required to break above a threshold pressure. The wall polymer, capsule size, and wall thickness determine elasticity and friability.

Other Properties.   Other important variables concern the range of concentrations and temperatures over which the wall polymer is sticky or tacky, causing clumping. Stickiness of the wall solution during drying may be the most difficult problem to overcome. Similarly, spray-coating is impeded by stringy polymer solutions.

Melting points, glass-transition temperatures, degree of crystallinity, wall-degradation rate, and many other properties have to be considered.

Techniques

A great many microencapsulation techniques are available and new ones are being developed each year (see General References). Most procedures have seemingly endless variations, depending on wall-polymer solubility, core solubility, particle size, wall permeability, surface free energies, desired release pattern, physical properties, etc.

Formation.   The pan coating process, widely used in the pharmaceutical industry, is among the oldest industrial procedures for forming small coated particles or tablets. The particles are tumbled in a pan or other device while the coating material is applied slowly. The classical example is the sugar-coating of medicinals. The expense of this process and the need to encapsulate fine particles and liquid cores stimulated the development of the new microencapsulation processes (see Coating processes; Pharmaceuticals).

Air-Suspension Coating.   Air-suspension coating of particles by solutions or melts gives better control and flexibility (2). The particles are coated while suspended in an upward-moving air stream, as shown in Figure 2. They are supported by a perforated plate having different patterns of holes inside and outside a cylindrical insert. Just sufficient air is permitted to rise through the outer annular space to fluidize the settling particles (see also Fluidization). Most of the rising air (usually heated) flows inside the cylinder, causing the particles to rise rapidly. At the top, as the air stream diverges and slows, they settle back onto the outer bed and move downward to repeat the cycle. The particles pass through the inner cylinder many times in a few minutes.

As the particles start upward they encounter a fine spray of the coating solution. Only a small amount of solution is applied in each pass. Hence, the solvent is driven off and the particles are nearly dry by the time they fall back onto the outer bed. Particles as large as tablets or as small as 150 μm can be coated easily. Since many thin layers of coating are sprayed onto all surfaces of the randomly oriented particles, a uniform coating is applied, even on crystals or irregular particles.

A great variety of coating materials have been used with this process, including waxes, cellulosic compounds and water-soluble polymers. Development and coating are done by Coating Place, Inc., Verona, Wisc. Equipment of several sizes is sold worldwide by Glatt Industries. The largest unit is 117 cm in diameter and requires

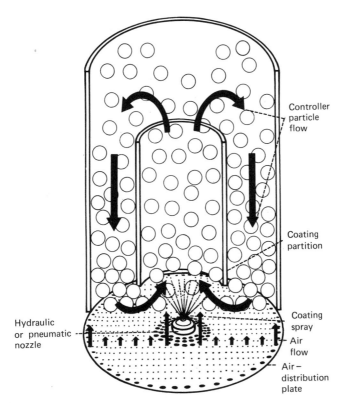

Figure 2. Wurster coating chamber. Perforations in air distribution plate are designed to control direction of particle movement. Courtesy of Coating Place, Inc.

over 283 m³/min (10⁴ ft³/min) of air in a typical operation. Up to 400 kg can be handled in one batch, and cycle time can be as short as 15 min, although it is typically 1–2 h.

Centrifugal Extrusion. Several processes have been patented by the Southwest Research Institute, in which liquids are encapsulated using a rotating extrusion head containing concentric nozzles (3) (Fig. 3). In this process, a jet of core liquid is surrounded by a sheath of wall solution or melt. As the jet moves through the air it breaks, owing to Rayleigh instability, into droplets of core, each coated with the wall solution. While the droplets are in flight, a molten wall may be hardened or a solvent may be evaporated from the wall solution. Since most of the droplets are within ±10% of the mean diameter, they land in a narrow ring around the spray nozzle. Hence, if needed, the capsules can be hardened after formation by catching them in a ring-shaped hardening bath.

This process is excellent for forming particles 400–2000 μm in diameter. Since the drops are formed by the breakup of a liquid jet, the process is only suitable for liquids or slurries. A high production rate can be achieved, ie, up to 22.5 kg of microcapsules can be produced per nozzle per hour; heads containing 16 nozzles are available.

This is an inexpensive process for producing large drops having meltable coatings

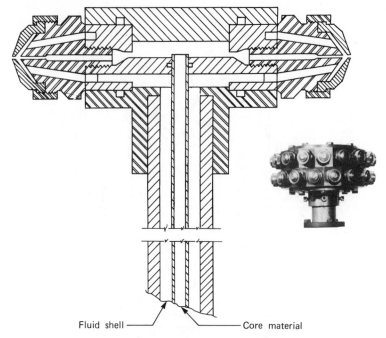

Fluid shell ———   ——— Core material

Figure 3. Rotating head to form biliquid jets. Courtesy of Southwest Research Institute.

such as fats or waxes. However, it also has been used extensively for coating liquids with synthetic polymers and gelatin and to form calcium alginate shells. The only wall materials that are unsuitable are those that are stringy and do not permit the clean breakup of the jet. With this process it is possible to encapsulate aqueous solutions in waxy wall materials, a difficult problem for some processes.

The 3M Company also has a process in which a biliquid column is formed, with subsequent breakup of the jet into coated droplets (4). In this case also, melt systems work well and it is easiest to form 800–4000-μm droplets. The fill phase in this process should have a viscosity of at least 20 mPa·s (= cP).

Vacuum Metallizing. The National Research Corporation has developed vacuum-deposition techniques to coat particles with a wide variety of metals and some nonmetals (5). The particles can be as small as 10 μm. The entire operation is carried out under vacuum, with the particles conveyed slowly down a refrigerated vibrating table while being exposed to a metal-vapor beam from a heated furnace. The beam condenses on the cool small particles, eventually coating them evenly. The process can handle only solids and usually forms a coating that is not completely impervious to gases or liquids. The solidification of the vapors occurs so rapidly that the particles do not tend to agglomerate unless the vaporized material is waxy or sticky. The process forms an even coating on rough or irregular surfaces, since all surfaces of the vibrating particles are exposed equally to the vapor beam (see Film deposition techniques).

Since the coating material must be evaporated and the substrate particles must be kept in vacuum and under refrigeration, the process may be expensive compared to other ways of depositing organic coatings. However, it is unique for depositing metals on fine particles.

Liquid-Wall Microencapsulation. A liquid as the wall of a microcapsule offers several advantages (6). Since the wall can be broken at will, the wall material can be recovered and reused. The core also can be recovered if desired. In addition, the permeabilities of many molecules through a liquid wall are better controlled and predicted than through a solid wall, where morphology can vary greatly.

The liquid core can be a single droplet, an emulsion, or a suspension. It can consist of a variety of chemical reactants, which suggests a wide spectrum of possible uses. The liquid wall is stabilized by addition of a surfactant, such as saponin or Igepal, and through the addition of strengthening agents. It is possible to form thin, aqueous membranes around organic droplets, such as toluene, or oil-based membranes around aqueous solutions. These microcapsules are so stable that they remain suspended with the mildest agitation and coalesce only slowly when stagnant. Droplets of one hydrocarbon, surrounded by a stable aqueous membrane, can be made to pass upward through another hydrocarbon without intermixing. Extraction can be carried out based on the selective permeability of the aqueous membrane. Because the process is inexpensive and the membrane material can be recycled, liquid-membrane microcapsules are attracting attention for a variety of industrial uses such as removal of contaminants from wastewater.

For many applications, the droplets containing reactant (eg, an aqueous solution) can be emulsified into the encapsulating immiscible liquid. This highly stable emulsion can then be gently dispersed into the liquid to be treated. Hydrocarbon droplets containing emulsified aqueous microdroplets are shown in Figure 4.

Spray-Drying. Spray-drying serves as a microencapsulation technique when an active material is dissolved or suspended in a melt or polymer solution and becomes trapped in the dried particle. The main advantage is the ability to handle labile materials because of the short contact time in the dryer; in addition, the operation is economical. In modern spray dryers the viscosity of the solutions to be sprayed can be as high as 300 mPa·s (= cP), meaning that less water must be removed from these concentrated solutions.

Recent examples of advances in spray-drying for controlled release are the

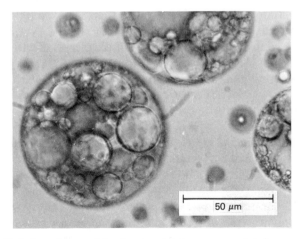

Figure 4. Stable hydrocarbon emulsion drops containing an encapsulated aqueous emulsion. Courtesy of Exxon Research and Engineering Company.

products in which flavor oils are emulsified into a solution of a polymer, such as gum arabic, and spray-dried to make fine particles. Particles of 250 μm (60-mesh) are ideal for inclusion in such products as bath powders. However, for use in an aerosol product, all particles must be below 74 μm (200 mesh) (see Aerosols). With careful gel formulation, the amount of fragrant oil on the surface of the final product may be less than 1% (see Flavors and spices; Perfumes).

When the core material is incorporated into a meltable fat or wax, the process is called spray-chilling, since the emulsion or suspension has only to be chilled below its melting point to form particles. This has been found practical for microencapsulation of such products as citric acid (qv), vitamin C (see Vitamins), and ferrous sulfate.

In a variety of other microencapsulation processes spray-drying could be a final step, since the capsules are in an aqueous dispersion from which they could be isolated economically by spray-drying (see Drying).

Hardened Emulsions.   Since the aim of microencapsulation is the formation of many tiny particles, the first step in a number of processes is the formation of an emulsion or suspension of the core material in a solution of the matrix material (see Emulsions). This emulsion can be emulsified in another liquid, and the droplets hardened.

Emulsions and suspensions can be converted directly to hardened microcapsules by several methods, eg, aspirin is encapsulated by first suspending it in a solution of ethyl cellulose (7). This suspension is dispersed in saturated ammonium sulfate solution and the solvent driven off by heating. This general process also has been used for enzymes (8) and many other core materials. Polystyrene, ethyl cellulose, and silicones are reported to work well as matrix materials.

In another process, a water solution is dispersed in melted waxes or fats. This emulsion is then dispersed in an aqueous solution held at a temperature above the melting point of the wax or fat. The aqueous solution is then cooled to solidify the wax particles (9).

Processes based on emulsion hardening are particularly useful when a drug has high solubility in the polymer solution, preventing formation of a separate polymer wall around the drug by other methods. Initially, the solution of polymer and drug is emulsified in an immiscible liquid. Solvent is then removed by application of heat and/or vacuum. As the organic solvent evaporates, the drug crystallizes inside the polymer solution droplet. Control of the rate of drug crystallization at this point in the process is critical in obtaining particles with good long-term release characteristics. Emulsion-hardening methods are being applied to the encapsulation of various drugs in biodegradable poly(d,l-lactic acid) to obtain injectable particles. Active in this area are Southern Research Institute, Battelle Memorial Research Institute and Washington University (St. Louis).

Liposomes and Surfactant Vesicles.   The incorporation of drugs into liposomes (phospholipid vesicles) and surfactant vesicles has attracted recent attention (10). Liposomes, such as lecithin (qv) and phosphatidylinositol, are smectic mesophases of phospholipids organized into bilayers (see Liquid crystals). Liposomes can be prepared in the laboratory by rotary evaporation of a chloroform solution of the phospholipid and cholesterol, with the subsequent removal of the thin lipid film from the wall of the flask by shaking with aqueous buffer. In order to incorporate a hydrophobic drug into these liposomes, the drug is dissolved in the chloroform solution before evaporation. A water-soluble drug can be incorporated by dissolving it in the buffer solution used to form the liposome. The untrapped drug is removed by gel filtration.

The liposomes formed are typically in the 0.1–0.15 μm range but can be prepared larger. With sonication during the formation step, the liposomes can be as small as 25 nm (see Ultrasonics). The amount of drug that can be entrapped depends on the ratio of the lipid to the trapped aqueous phase and on the solubility of the drug. Some drugs are trapped to the extent of a few percent, but up to 60% bleomycin, eg, can be encapsulated. The liposomes are typically stored below the phase-transition temperature of the solid wall but generally are used above this temperature.

Release rates from the liposomes can be small, in spite of the large area–volume ratio. The release of 8-azaguanine is reported to be around 10% in 20 h. Liposomes can be targeted to specific cells by associating them with immunoglobulins raised against the target cells. This has been accomplished with several drugs. Surfactant vesicles can be formed also from dioctadecyldimethylammonium chloride by sonicating it in a concentrated drug solution and gel-filtering the suspension (11). The amount of drug trapped is many times that trapped in liposomes.

Phase Separation. In several microencapsulation processes, the core material first is suspended in a solution of the wall material. The wall polymer then is induced to separate as a liquid phase, eg, by adding a nonsolvent for the polymer, decreasing the temperature, or adding a phase inducer, another polymer that has higher solubility in the solvent. In the last case, incompatibility between the two polymers causes the first polymer to separate as another phase. When the wall polymer separates as a polymer-rich liquid phase, this phase is called a coacervate and the process is called coacervation.

As the coacervate forms, it must wet the suspended core particles or droplets and coalesce into a continuous coating. The final step is the hardening and isolation of the microcapsules, usually the most difficult step in the process. An understanding of the complex physical chemistry of phase-separation microencapsulation is helpful in designing the process (12).

Complex Coacervation. The first commercially valuable microencapsulation process, developed by the National Cash Register (NCR) Company in Dayton, Ohio, was based on coacervation. The coacervate was formed from the reaction product complex between gelatin and gum arabic (13). The first application was in No-Carbon-Required carbonless copy paper. Carbonless copy systems are still by far the largest market for microencapsulated products. Applications of this technology in the paper industry are now handled by Appleton Papers, Inc. Applications in other industries are handled by Eurand America, Inc., Dayton, Ohio.

In this process, gelatin having a high isoelectric point and gum arabic, which contains many carboxyl groups, are added to a core-containing suspension at pH 2–5 above 35°C to ensure that the coacervate is in the liquid phase. As the gelatin and gum arabic react, viscous liquid microdroplets of polymer coacervate separate. If the core particles are easily wetted by these microdroplets, a wall of the liquid coacervate forms on the core particles. This wall can be hardened by several means; eg, by the addition of formaldehyde. Finally the mixture is cooled to 10°C, the pH adjusted to 9, and the microcapsule suspension filtered. The capsules shown in Figure 1 have walls produced by complex coacervation.

In a modification of this process, two gelatins of differing isoelectric points are used as the reacting species. In this case, the coacervation is reversible before hardening, and distorted capsules can be heated to restore their spherical shape. Such microcapsules can be hardened by the addition of glutaraldehyde.

The complex coacervation process works well in the microencapsulation of solids and oily materials. The walls are polar and have low permeability to nonpolar molecules. Toluene and carbon tetrachloride can be contained so well that only a small fraction of the contents is lost from a thin layer of 1-mm capsules over a period of two years at 21°C at 70% humidity.

Thermal Coacervation. The effect of temperature on solubility is sometimes strong enough to cause a dissolved polymer to form a coacervate upon cooling. However, the temperature effect is nearly always used in conjunction with another method of coacervate formation, as illustrated in the following section.

Polymer–Polymer Incompatibility. This broadly applicable technique, also developed at NCR (14–16), works well for encapsulating many solids and some liquids. A coacervate of the wall material is induced to form from solution by the presence or addition of a phase-inducer. The general phenomenon of polymer–polymer incompatibility is based on the fact that the free energy of mixing for polymers is positive owing to a positive enthalpy change and negligible entropy change.

An example of this process is the microencapsulation of activated carbon in ethyl cellulose using cyclohexane as the solvent. The carbon is slurried in a 2% suspension of ethyl cellulose in cyclohexane, followed by the addition of 2% of low molecular weight polyethylene as a phase-inducer. The entire system is heated to 80°C, where both polymers dissolve. The system then is cooled slowly to room temperature. The change in temperature and the presence of the polyethylene cause the formation of a coacervate containing the ethyl cellulose which coats or wraps around the activated carbon particles. At room temperature, the particles can be separated and dried. Careful attention must be paid to rates of cooling, agitation, etc, to obtain satisfactory, free-flowing microcapsules.

A system also has been developed by NCR to microencapsulate some polar liquids and aqueous solutions (17). In this process, the wall polymer is partially hydrolyzed ethylene–vinyl acetate copolymer which coacervates upon the addition of polyisobutylene. Smooth coherent walls can be formed around glycerol and aqueous solutions of methylene blue or ferrous ammonium sulfate. The walls are nearly impenetrable by ions and all but the smallest polar molecules. The walls can, however, be penetrated at reasonable rates by uncharged organic molecules. A few of these microcapsules containing ferrous ammonium sulfate are shown in Figure 5.

Another NCR variation of this process can be carried out at constant temperature, an advantage in encapsulating temperature-sensitive materials. After the core material is dispersed in the solution of a wall-forming polymer, poly(dimethyl siloxane) is added as the phase-inducer. A possible disadvantage of this process is that a small amount of the silicone is trapped in the polymer wall, making dispersion difficult for some uses.

Nonsolvent Coacervation. A number of patents have issued in which microcapsules are formed or hardened by the addition of nonsolvents for the wall polymer. In a particularly successful example of such a technique (18), the core, typically an aqueous solution, is dispersed in a methylene chloride solution of cellulose acetate butyrate (CAB). Toluene, a nonsolvent for the polymer, is then added slowly to cause coacervation of CAB and the wrapping or coating of the core with the polymer as it

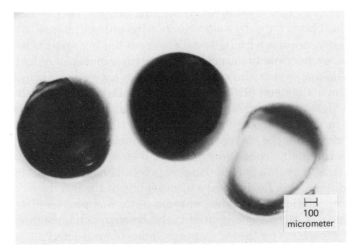

Figure 5. Clear capsule: ferrous ammonium sulfate encapsulated in partially hydrolyzed ethyl-ene–vinyl acetate copolymer. Dark capsules: same after contact with sodium salicylate, showing formation of Fe(III) salicylate complex inside the capsule.

separates from solution. The wall is hardened by the addition of petroleum ether to extract solvent from the walls. Typically, spherical microcapsules are obtained with thin elastic walls. This process has been used to encapsulate some organic compounds by emulsifying them in aqueous solution before coacervation.

Interfacial Coacervation. In another technique, a polymer coacervate forms directly at an interface without the addition of a phase-inducer (19). For example, tris-buffered hemolyzate is emulsified in ice-cooled ethyl ether, saturated with water, which contains 1% Span 85. A solution of collodion is added. Weak microcapsules develop within 45 min. They contain hemolyzate and have a collodion coacervate wall. The wall is hardened with n-butyl benzoate, after which the microcapsules are separated by centrifugation, washed, and stored in Tween 20.

In another process, an aminoplast wall is induced to coat the surface of droplets (20). A water-insoluble fill material is emulsified into an aqueous solution containing a preformed urea–formaldehyde polymer of sufficiently low molecular weight to retain water solubility (see Amino resins). After emulsification, a water-soluble acid such as formic acid is added to lower the pH to 1.5–3, causing the formation of the urea–formaldehyde polymer shell, microencapsulating the emulsion droplets. The microcapsules usually range 10–50 μm and are used in 3M Microfragrance scents for advertising.

Another method of forming microcapsule walls at an interface is based on the observations that some cellulosic polymers are soluble in cold water but not in hot water (21), eg, hydroxypropyl cellulose and ethyl hydroxyethyl cellulose. These materials precipitate from solution when the temperature exceeds a critical value of 40–70°C. To carry out microencapsulation based on this phenomenon, the oil to be encapsulated is emulsified in an aqueous solution of the cellulosic material, at a low temperature where it is still soluble, in the presence of a surfactant. The emulsion is then heated until the cellulosic polymer gels or precipitates from solution, wetting the oil droplets. While the suspension is hot, the cellulosic wall is solidified by reaction with a water-soluble agent such as dimethylolurea, urea–formaldehyde resin, or methoxymethyl-melamine resin. This is typically a slow reaction, resulting in somewhat porous walls. To produce microcapsules with tight walls, a cross-linking agent is included before emulsification, eg, polyfunctional acyl chlorides, isocyanates, or anhydrides. Water-soluble cross-linking agents also can be included in the aqueous solution. Capsules from 1 to 50 μm in diameter are made easily and wall thickness is well-controlled.

Chemical Methods. *Interfacial Polymerization.* In interfacial polymerization, the two reactants in a polycondensation meet at an interface and react rapidly (22). The basis of this method is the classical Schotten-Baumann reaction between an acid chloride and a compound containing an active hydrogen atom, such as an amine or alcohol. Polyesters, polyurea, polyurethane, or polycarbonates may be obtained. Under the right conditions, thin flexible walls form rapidly at the interface.

Microencapsulation of proteins, enzymes, etc, was pioneered by Chang (see General References). An example of interfacial polymerization is the process developed by Pennwalt Corporation for the microencapsulation of pesticides (23). In this process, a solution of the pesticide and a diacid chloride (eg, in toluene) are emulsified in water and an aqueous solution containing an amine and a polyfunctional isocyanate is added. Base is present to neutralize the acid formed during the reaction. Condensed polymer walls form instantaneously at the interface of the emulsion droplets. The isocyanate is not required for wall formation but acts as a cross-linker, giving a tougher, less-permeable wall. Surfactants and suspending agents are used in the process to control particle size. The first commercial product, Penncap-M, contained methyl parathion in microcapsules having a mean diameter of 30 μm. Figure 6 is a photograph of a microcapsule of another insecticide, Knox-out 2FM, trapped on the leg of a cockroach.

Other methods for carrying out similar chemical reactions at the interface have been patented by Moore Business Forms to produce microcapsules for carbonless copy paper (24). A hydrophobic marking fluid plus an oil-soluble reactant are emulsified into water, followed by addition of the other reactant which immediately causes the formation of the condensation polymer at the drop interface.

The processes developed by Pennwalt and Moore have the advantages that they are capable of being run on a large industrial scale and that the reactions are very fast. If the final product can be used in aqueous suspension, the drying step can be eliminated. Hence, the cost is relatively low. Variations in cross-linkers, reactant concentrations, solvent composition, and processing steps permit optimization of the process to obtain different capsule properties.

The importance of carbonless copy paper has stimulated microencapsulation studies by many paper companies. Another application of interfacial polymerization is that of the Mead Corporation (25). Rather than place the reactants of a polycondensation system in each phase, they form an emulsion in which the organic phase contains a monomer and the aqueous phase an initiator that causes polymerization at the droplet surface. This process can be used with monomers such as styrene, methyl methacrylate, vinyl acetate, and acrylonitrile. Initiators (qv) include hydrogen peroxide, acetyl peroxide, zinc peroxide, and alkali metal hydroxides.

Surface polymerization processes to make opacifiers, pigments and dyes for paper coating also have been developed (26). A toluene solution of a reactant, such as toluene diisocyanate, is emulsified in an aqueous solution containing a polymeric film-forming emulsifying agent such as methyl cellulose, starch, or poly(vinyl alcohol). The emul-

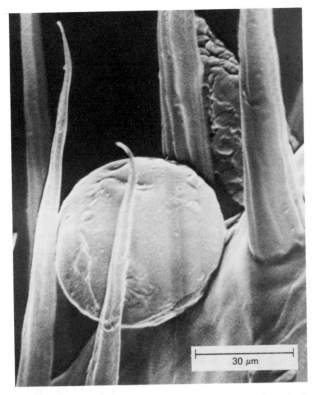

30 μm

Figure 6. A Knox-out 2FM microcapsule trapped on the leg of a cockroach. Courtesy of Pennwalt Corporation.

sifying agent, which is present at the surface of the emulsion droplets, is then cross-linked and insolubilized by reaction with the organic reactant. If the microcapsules are to be used as opacifiers, the internal solvent is driven out and the microcapsules should have diameters <1 μm. Even with typical wall thickness of 5–60 nm, these small capsules tolerate calendering at 3.4 MPa (500 psi).

A different method of microcapsule formation is based on the fact that alginic acid anion forms an insoluble precipitate instantly upon contact with calcium ions. Hence, if droplets of an aqueous solution of sodium alginate are placed in a calcium chloride solution, a membrane quickly forms around the droplets (see Gums). If the initial droplets contain enough alginate and are allowed to remain in the calcium-containing solution long enough, they gel completely.

A method for forming extremely small microcapsules was developed at the Swiss Federal Institute of Technology in Zurich (27). The particles are so small that the process is called nanoencapsulation. Micelles are formed in a water-in-oil emulsion, using polymerizable surface-active agents such as acrylamide (qv). The surfactant is polymerized in various ways such as by x-ray or gamma radiation. Particle sizes range from 80 to 250 nm. It has been possible to suspend up to 40 vol % of water in hexane in this manner. The particles are sufficiently small that they cannot be observed with a microscope but can be seen with the Tyndall effect. Measured pore diameters in the walls appear to be 2–5 nm. The wall is 3–5 nm thick or more. Hence, the particles can be sturdier than liposomes. The final particles, containing a drug, are sustained-release vehicles which are small enough for parenteral use.

Microencapsulation techniques based on surface polymerization at elevated temperatures have been developed (28). In these processes, two film-forming reactive materials are dissolved together with an oily liquid, at low concentration in a volatile solvent, at a temperature sufficiently low to effectively prevent reaction. This oil phase is emulsified in an aqueous phase and the temperature increased, eg, to 90°C, where the solvent evaporates from the droplets and the film formers increase in concentration at the surface of the droplets, reacting to form a wall. A wide variety of polymer walls can be formed in such a process, such as polyols, polyisocyanates, polythiols, and polyamines.

An all-aqueous system for interfacial polymerization has been patented (29). This technique is based on the reaction between polar reactive materials, such as hydroxyethyl cellulose, in one phase and inorganic silicate, such as lithium hectorite clay, suspended in the other phase. When the material to be dispersed, such as a pigment suspension containing the organic material, is mixed with the clay suspension, the reaction is essentially instantaneous, causing the formation of particles of one aqueous suspension in another. This technique is used, eg, for the formation of suspended particles of colored water-dispersible pigment in a latex paint formulation.

In-Situ Polymerization. In a few microencapsulation processes, the direct polymerization of a single monomer is carried out on the particle surface. In one process, eg, cellulose fibers are encapsulated in polyethylene while immersed in dry toluene (30). In the first step, a Ziegler-type catalyst is deposited on the surface of the fiber at 20–30°C by the addition of $TiCl_4$ followed by triethylaluminum. When the catalyst is formed, the surface appears dark brown. The addition of ethylene, propylene, or styrene results in immediate polymerization directly on the surface. The polymerization is typically carried out to a final weight ratio of 50:50 polyethylene–cellulose. This coating has a molecular weight of up to 2×10^6 and a melting point of 56–58°C, vs the theoretical maximum of 61°C. This process also has been applied to coating glass. The coating, typically translucent or opaque, can be clarified by melting after formation (see Olefin polymers).

Unique coatings can be obtained with the Union Carbide Parylene process (31). Di-p-xylylene or a derivative is pyrolyzed at 550°C where it dissociates yielding free radicals in the vapor state. When the vapor is cooled below 50°C, the free radicals polymerize to a high molecular weight polymer:

n-ca 5000

The reaction is typically carried out in a partial vacuum. The p-xylylene is heated in one part of the vessel whereas the material to be coated is cooled in another section. The product is a linear thermoplastic, insoluble in organic solvents up to 150°C, which has many attractive properties. It can be deposited on any solid surface, and has very low permeability to gases and moisture, although it is somewhat permeable to aromatic compounds. The coating retains some ductility at cryogenic temperature, has excellent electrical resistance, and a softening range of 290–400°C (depending on the derivative). Usual deposition rates are about 0.5 μm/min. Coating thickness ranges 0.2–75 μm. The coating is uniform, even over sharp projections, as shown by the coating over the stylus point shown in Figure 7 (see Film deposition techniques).

Matrix Polymerization. In a number of processes, a core material is imbedded in a polymeric matrix during formation of the particles. A simple method of this type is spray-drying, in which the particle is formed by evaporation of the solvent from the matrix material. However, the solidification of the matrix also can be caused by a chemical change.

Matrix particles can be formed by suspending the core in gelatin solution, emulsifying the gelatin solution, and then hardening the droplets with formaldehyde or glutaraldehyde. Solid particles can be obtained in some of the processes mentioned previously if the wall-forming chemicals are present in sufficient quantity and if the reaction time is prolonged.

In silastic microspheres containing a suspended drug in silicone oil, addition of a catalyst such as stannous octanoate polymerizes the silicone prepolymer (32).

An inexpensive process for encapsulating water-insoluble oxidation-resistant liquids, in the form of emulsified droplets in starch xanthate, has been developed by the U.S. Department of Agriculture (33). First, the xanthate derivative of starch is made by reaction with a base and CS_2 (see Starch). A concentration of about 10%

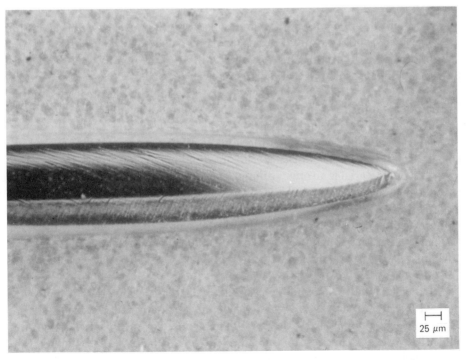

Figure 7. Parylene coating over stylus point. Courtesy of Union Carbide Corporation.

xanthate has the proper consistency for the subsequent steps. A core material such as a pesticide is added with gentle stirring, giving a stable suspension. If the core material is sensitive to base, the pH is reduced from near 11 to 6–7 with acetic acid. After the dispersion is formed, the xanthate is solidified by oxidation and cross-linking with sodium nitrate or hydrogen peroxide. With agitation, solid granules form in a few minutes, and 80% of the water in the granules can be squeezed out physically. The resulting crumbs, containing approximately 50% moisture at this stage, can be ground to the desired particle size, then dried to, eg, 5% moisture. No active ingredient is removed with the water or with the grinding steps. The material appears to be self-sealing. Drying temperature can be as high as 121°C (if the core does not degrade). After the material has been dried, it can no longer be ground without losing active ingredient.

The granules may contain up to 40% active ingredient, although a few percent is more typical for pesticides. A large number of water-insoluble compounds have been encapsulated by this technique. They are extracted with difficulty into nonpolar organic solvents, indicating that such solvents do not penetrate the wall readily. On the other hand, alcohol and water increase the release rate. In the soil, the core is probably released by the action of moisture and bacteria.

The xanthate matrix material can be cross-linked with formaldehyde or epichlorohydrin to decrease its permeability even further.

Process Selection

The range of applicability of the microencapsulation processes described above is extremely wide, and many applications overlap. A slight change in surface properties or solubilities may determine the success or failure of the operation. New materials require experimentation and testing.

There are, nevertheless, some factors that exclude certain processes and favor others. For example, if the material to be encapsulated is a solid and cannot be used as a slurry, liquid-jet methods are not suitable. Phase-separation methods or air-suspension coating may work well if the solid particles are large. If they are very fine, incorporation into a matrix of the coating material may be appropriate.

If the core material is an organic liquid, air-suspension coating cannot be employed but liquid-jet methods may work well for particles larger than a few hundred micrometers. If smaller particles are needed, coacervation techniques or interfacial polymerization may be applicable. If the particles must be hard or have thick walls, interfacial polymerization may be too slow and expensive. If the material forms stable emulsions in a polymer solution or a meltable medium, solidified particles of these emulsions might be the answer.

Microencapsulation of a highly polar liquid is, in general, more difficult but the liquid-jet method, interfacial polymerization, or some of the NCR phase-separation methods may be applicable. If the core and the desired wall material have similar solubilities, phase-separation methods are excluded but matrix polymerization or emulsion solidification may be suitable.

Release Methods and Patterns

Even when the aim of a microencapsulation application is the isolation of the core from its surroundings, the wall must be ruptured at the time of use. Many walls are ruptured easily by pressure or shear stress, as in the case of breaking dye particles during writing to form a copy. Capsule contents may be released by melting the wall, or dissolving it under particular conditions, as in the case of an enteric drug coating. In other systems, the wall is broken by solvent action, enzyme attack, chemical reaction, hydrolysis, or slow disintegration.

Microencapsulation can be used to slow the release of a drug into the body. This may permit one controlled-release dose to substitute for several doses of nonencapsulated drug and also may decrease toxic side effects for some drugs by preventing high initial concentrations in the blood. There is usually a certain desired release pattern. In some cases, it is zero-order, ie, the release rate is constant. In this case, the microcapsules deliver a fixed amount of drug per minute or hour during the period of their effectiveness. This can occur as long as a solid reservoir of dissolving drug is maintained in the microcapsule.

A more typical release pattern is first-order in which the rate decreases exponentially with time until the drug source is exhausted. In this situation, a fixed amount of drug is in solution inside the microcapsule. The concentration difference between the inside and the outside of the capsule decreases continually as the drug diffuses.

In most processes, a small amount of drug is not encapsulated properly. In use, these accessible drug particles release rapidly into the surrounding fluid, causing a burst effect. Sometimes this is desirable, and functions as an initial loading dose to increase the blood concentration toward the therapeutic level. However, if undesirable, the poorly encapsulated particles have to be removed before use.

Typical release patterns for drugs are shown in Figure 8.

An excellent reference on the mathematics of controlled-release systems (34) gives solutions for rates and amounts released as a function of time for the typical cases of constant release rate, exponential release rate, and release from matrices. Many examples and an extensive reference list are provided.

Economic Aspects

Wall polymers may cost $2–40/kg. Solvent cost is substantial, and may represent over half the cost of the entire operation. Hence, installation of solvent-recovery

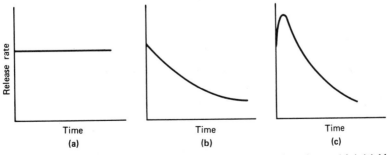

Figure 8.   (a) Constant release; (b) first-order release; (c) first-order release with initial burst.

equipment offers both environmental and economic advantages. The most economical wall materials are waxes and fats, which can be melt-coated. This eliminates all solvent costs.

Wall thickness, particle size, and particle density may have a strong affect on the coating cost of finished product. For small particles, the area to be covered with polymer is high for a given weight, and the rate at which the particles can be produced is limited for processes such as spray-coating, vapor deposition, or jet methods. On the other hand, these considerations are not important in coacervation or interfacial polymerization systems.

As with any manufacturing process, the costs of microencapsulation decrease significantly with increasing volume of production. Melt-coating solid particles at high volume in a trouble-free system may give a product at $0.50/kg, whereas expensive wall materials or time-consuming operations can increase costs to $10–20/kg.

Characterization and Evaluation

Size and Size Distribution. Determination of the size distribution of the capsules, as produced, is an important measurement. It also is critical in studying the effect of the process variables on the final product. A wide range of particle-size-measuring techniques is available (see Size measurement of particles).

Loading Fraction. The amounts of polymer coating and core can be measured directly by isolating the microcapsules, dissolving or crushing the wall, and making a mass balance on the core and polymer. The fraction of the core in the microcapsules or matrix is frequently a function of particle size, and the deviation of the actual percent loading from the design loading is sometimes a function of the loading itself.

Release Properties. In many applications, such as carbonless copy paper, microcapsules are ruptured by pressure. Rupture pressure is measured under controlled conditions similar to those of the application. Most companies have developed their own test apparatus for purposes of quality control and process development.

In a more complex situation, the release of a core drug or the adsorption of an external solute is time-dependent. The kinetics of these processes is usually determined by contacting the microcapsules with a fixed volume of fluid at constant temperature and measuring the change in solution concentration. To avoid the effect of a stagnant fluid layer adjacent to the microcapsules, the bottles are placed in a shaker bath or mounted on a rotating-arm apparatus. Similar results are obtained in both tests. Since solute interactions can have strong affects on distribution coefficients and, hence, on diffusional driving forces, the test fluid should resemble the fluid used in the application. A case in point is drug release into blood or intestinal fluid, where tests in buffers can be highly misleading.

Health and Safety Factors

In any application of microencapsulation in foods, pharmaceuticals or veterinary products, only wall materials approved by the FDA should be used. The GRAS list (generally recognized as safe) can be consulted. Some materials, such as starches, cellulosic compounds, and gums, have been used for many years as food additives or as excipients in drug formulation. However, any process or material not previously approved must be submitted to the FDA. The approval requirement extends to methods of solvent removal, conditions of storage, methods of handling, etc.

Uses

Carbonless Copy Paper.   The most significant application of microcapsules is in business forms permitting copies to be made without the need for carbon paper. Industry-wide production of carbonless-copy business forms is approximately 500,000 t/yr. In carbonless business forms, a dye intermediate, such as crystal violet lactone, is microencapsulated to form particles less than 20 μm in diameter. These are deposited in a thin layer on the underside of the top sheet of paper. The receiving sheet or copy is coated with another reactant such as acidic clay. The microcapsules are designed to resist breakage under normal conditions of storage and handling, but to break under the high local pressure of a pen or pencil point. Upon breaking of the capsules, the two chemicals react, producing a dye copy of the original. Figure 9 is a photomicrograph of NCR microcapsules on the paper produced by Appleton Papers Inc.

Flavors and Essences.   Since many of the flavors and essences in candies, foods, and perfumes are volatile or are affected by oxidation, microencapsulation of these materials offers advantages (see Flavors and spices). Spray-drying has proved to be an excellent, economical way to form particles containing suspended flavors and scents in a polymer matrix. The short contact time of this high-volume process minimizes loss of volatile cores.

Eurand America, Inc. has used coacervation and 3M has used its aminoplast-wall process for new approaches in advertising by microencapsulating essences and coating them onto paper to be used in magazines and promotional literature. The capsules are broken by scratching, releasing the scent of products such as soaps or perfumes. The materials that have been successfully microencapsulated range from lemon oil and essence of roses to the aroma of dill pickles or chocolate mint cookies.

There is much interest in microencapsulating insect pheromones such as *cis*-2-decyl-3(5-methylhexyl)oxirane (disparlur), the scent that guides the male gypsy moth to the female. Disparlur could be used to lure males to traps or, if it was widely dis-

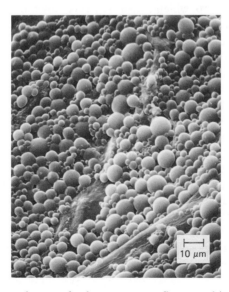

Figure 9.   Microcapsules on carbonless copy paper. Courtesy of Appleton Papers Inc.

persed, to confuse them in their attempt to find females. Microencapsulation greatly increases the time over which the volatile scent is active. Pennwalt Corporation and Zoecon, Inc. are among the companies active in this area.

Pesticides and Herbicides.   Some volatile pesticides and herbicides are highly toxic substances. Microencapsulation decreases their toxicity upon contact and controls their release rate (see Herbicides; Insect control technology). Pennwalt Corporation markets several microencapsulated pesticides, such as Penncap M, which is microencapsulated methyl parathion. The microencapsulation permits early safe entry into a sprayed field, and increases the time of effectiveness from 1–2 d to 5–7 d. Since the tiny capsules (about 30 μm) are in the form of an aqueous suspension, spills can be cleaned easily and the capsules can be flushed easily from the skin. In tests on rabbits, encapsulation lowers the dermal toxicity by a factor of 12. These microcapsules are made by interfacial polymerization using bifunctional acid chlorides and bifunctional amines, cross-linked with a polyfunctional isocyanate. The walls of the capsules are about 1 μm thick.

Pharmaceuticals.   By choice of polymers, particle-size distribution, wall thickness, and processing conditions, the release profile of a drug can be controlled (see Pharmaceuticals, controlled-release).

Probably the largest-selling drug product based on controlled release is Contac, the cold-relief remedy. Its relatively large particles are pan-coated. The coatings dissolve at different times. Another commercial controlled-release drug form is timed-release aspirin, marketed eg by, Sterling Drug Company in the United States and by Rhone-Poulenc S.A. in Europe. Tests show that microencapsulation increases the time of effective action at least by a factor of two. Furthermore, the polymeric coating combined with slower release decreases gastric irritation.

A large number of controlled-release pharmaceuticals have been developed by Eurand S.p.A., the European licensor for National Cash Register technology. The three basic forms (35) are Diffucap, which consists of particles about 0.5 mm in diameter from which the drug is released by diffusion; Chronodrug, in which the membrane disintegrates; and Microcap, which consists of very fine particles that release by both diffusion and disintegration. Among their many controlled-release drugs are ampicillin, indomethacin, potassium chloride, amytryptyline, and phenylbutazone.

An ingenious method for attaining a constant release rate is the Oros osmotic pump (Alza Corporation) shown in Figure 10. In this device, the drug to be released is enclosed, along with a soluble material, in a wall that has a known permeability to water. Upon immersion of the device, water enters slowly by osmosis, causing the internal pressure to increase and dissolving part of the contents. The internal pressure forces the saturated solution through a small hole that has been drilled through the wall. The diameter of the hole is calculated to give a known amount of hydraulic resistance to the flowing drug solution. When the device reaches a steady rate of osmotic water influx, the solution flows at a steady rate, giving a constant rate of drug delivery until the remaining solid contents are dissolved. Then the drug release decreases exponentially. Since the osmotic flux of water depends on the difference in water concentration, there is little affect of external solute concentration or stirring on the drug-release rate, which closely approximates the design release rate.

At present, pharmaceutical uses are the focus of much effort in microencapsulation, and a variety of new controlled-release products should appear in the next decade.

Figure 10.   Oros osmotic pump discharging dyed stream. Courtesy of Alza Corporation.

Medical Applications.   A number of medical problems might be solved by microcapsulation to achieve isolation or controlled release. In the late 1950s, Chang studied the preparation of artificial red blood cells. A procedure was developed for encapsulating aqueous solutions, such as red blood hemolyzate, in a very thin membrane of nylon through the use of interfacial polymerization. Although the original aim of artificial blood cells was put aside, the general technique has been used for microencapsulation of enzyme solutions as a remedy for enzyme deficiency, such as phenylketonuria or acatalasemia. A further study examined the administration of L-asparaginase microcapsules to remove L-asparagine in treating asparagine-dependent tumors, and demonstrated temporary remission of such tumors in mice. The enzyme microcapsules were given to the mice intraperitoneally, where the capsule wall prevented the usual rapid destruction of the enzyme.

The microencapsulation of adsorbents such as activated carbon has been studied for the removal of drugs or waste metabolites from patients. In the case of detoxification by direct perfusion of blood through a bed of activated carbon or other sorbent, a coating is frequently used to increase blood compatibility, giving some control over protein adsorption and platelet interaction (36). Several companies market products in this field, including Smith and Nephew, Ltd. in the U.K., Gambro in Sweden, Asahi in Japan, and Becton-Dickinson, Inc. and Extracorporeal Medical Specialties, Inc. in the United States.

In the intestinal removal of toxins, there is severe competition for sites on the carbon surface by the hydrophobic molecules in the intestinal tract. This competition can be overcome partially by selectively permeable polymer coatings on the carbon. Capacities for creatinine in the fluid from the intestinal tract of the pig were increased nearly fourfold in this manner (37).

Injectable microcapsules can control the long-term release of drugs given intramuscularly or subcutaneously. The capsules must be below 100 μm in diameter to pass through a typical hypodermic needle after suspension in an injection medium. To avoid leaving debris in the body, the microcapsules are made with poly(d,l-lactic acid), which slowly degrades in the body without harmful effect (see Biopolymers). This system can also slowly release narcotic antagonists for eventual use in the treatment of drug addiction (38). A sustained-release form of astiban has been studied as an approach to the treatment of schistosomiasis mansoni, a tropical parasite disease (39). A further intriguing application is the injection of progesterone-loaded microcapsules, to obtain systemic release of the drug for contraception (40).

Droplets of fluids containing islets of Langerhans have been encapsulated in calcium alginate (41). (Islets are the pancreas cells responsible for secreting insulin into the blood in response to increased glucose.) This may be another step toward the use of mammalian islets cells from one species to treat the effects of diabetes in another species. The microcapsule wall would prevent direct contact of the foreign tissue with the lymphocytes and antibodies of the host, slowing or eliminating immunological rejection. Another approach is the direct deposition of a polymeric wall onto the surface of living cells (42) (see Insulin and other antidiabetic agents).

Workers at the University of Chicago School of Medicine incorporated a drug, together with magnetite particles (10–20 nm), in albumin solution (43). The suspension was emulsified in cotton seed oil, sonicated, and the albumin set with aldehydes or heat. The final particles are less than 1.5 μm in diameter, and can pass into blood capillaries. The microspheres can be localized by an external magnetic field to obtain local drug release. Experiments in rats show that the local effects obtained with the microspheres match those obtained with one hundred times as much drug given intravenously. The same workers have coupled protein A (a surface protein from *Staphylococcus aureas* which binds selectively with immunoglobulin G) onto the iron in the microspheres to permit separation of B-lymphocytes from T-lymphocytes. T-cells contain no surface IgG, but B-cells do and are bound to the microspheres.

Among the large number of applications of the Exxon liquid-microcapsule technique are several of medical importance. In one, blood oxygenation is carried out by passing small bubbles of air or oxygen through blood but with the normally deleterious blood-gas interface eliminated by encapsulating the gas in a thin fluorocarbon wall. In another application, aimed at use in kidney failure, urea is removed from the intestinal tract by using oil droplets containing an emulsion of concentrated citric acid solution (44). Uncharged NH_3, generated by enzymatic hydrolysis of urea to ammonia and carbon dioxide, passes through the oil phase into the citric acid where it becomes ionized and can no longer pass through the oil. Hence, the selective permeability of the continuous oil phase allows ammonia to be trapped inside the citric acid droplets.

Veterinary Applications.   A number of drug-release and feed-additive problems in the veterinary field are attracting the attention of workers in microencapsulation (see Veterinary drugs). The U.S. Department of Agriculture has developed the following method for replacing some saturated fat in milk with unsaturated fat, should this be of benefit in the diet (45). Safflower oil is emulsified in 10% sodium caseinate solution and the emulsion is spray-dried. The protein is cross-linked with formaldehyde. When this material is added to the feed of dairy cows, the coating and the unsaturated oil pass through the rumen without the usual enzymatic hydrogenation. In the abomasum, the microcapsules break at the low pH and the unsaturated oil is then incorporated into the milk by the cow. In tests, up to 24% of the milk fat contained linoleic acid. There is no change in the taste of the fresh milk containing the unsaturated fat, but there is sufficient oxidation in three days to affect the taste adversely (see Milk and milk products).

Adhesives. A unique use of microcapsules is the formation of adhesive mixtures which only set when the air concentration is extremely low (see Adhesives). These are referred to as anaerobic adhesives. They are manufactured primarily by Loctite Corporation and by Omni-technik in Munich, FRG. These microencapsulated adhesives can be coated onto bolts or other fasteners. When a nut is screwed onto a coated bolt, the microcapsules break, releasing the adhesive into the void spaces of the fastener. The oxygen concentration is low and the adhesive sets. Such adhesives can be made to set over a controllable time, and to cure at room temperature. The cured assembly has very high breakaway strength and performs better than a toothed lockwasher in vibration tests. A typical formula is based on the polymerization of tetraethylene glycol dimethacrylate, to which has been added cumene hydroperoxide, N,N-dimethyl-p-toluidine, benzoic sulfimide, and p-benzoquinone (46). The first three produce free radicals, but air and the benzoquinone prevent the material from turning solid until the oxygen concentration is low. The solidification is irreversible.

Southwest Research Institute (SWRI) is producing 1.5–1.8 mm microcapsules filled with water to be mixed with a fast-setting gypsum in a novel system for shoring up the ceilings of mines. As mines are dug, support plates, held by rods, must be placed on 1.2–1.5 m centers. The support rods are typically 1.8 m long and 2.2 cm in diameter. They can be set in polymers but these frequently are expensive, difficult to apply, and may require appreciable setting time. In the SWRI system, the water-containing microcapsules are mixed with fast-setting gypsum in a bag the shape of a sausage casing. The bag is inserted into a drilled hole (2.5 cm dia) and the rod rammed into place. This ruptures the capsules, releasing the water. Within a few minutes the bolt withstands a pull of 89 kN (20,000 lbf). The capsules contain 65% water and have a wall of wax, resin, and polymer of low water permeability to give long shelf life. The system is inexpensive and nonflammable.

Visual Indicators. Some liquid-crystal materials change color over a narrow temperature range because of changes in internal structure. Such materials tend to be oily liquids and are hard to handle and keep in place. They are stabilized by microencapsulation. Among the first uses were thermometers on which only the temperatures near 20°C were visible. Several companies are now developing such systems for medical use (see Chromogenic materials; Liquid crystals).

Appleton Paper Company has recently described the Magne-Rite system in which encapsulated metal flakes are suspended in oil. The application of a field of 0.05–0.1 T (500–1000 G), eg, with a magnetic stylus, causes the flakes to align, forming an image within 1–3 ms. A display system has also been developed by Fuji Photo Film Company in Japan.

Outlook. The range of possible applications of microencapsulation is breathtaking. As with all new techniques, microencapsulation will find those situations in the market place where it provides unique products. Cost will limit large-scale industrial uses for all but a few processes but uses in pharmaceuticals and medicine should grow and the markets in specialty products should continue to expand.

BIBLIOGRAPHY

"Microencapsulation" in *ECT* 2nd ed., Vol. 13, pp. 436–456, by James A. Herbig, The National Cash Register Co.

1. J. Crank and G. S. Park, *Diffusion in Polymers*, Academic Press, Inc., New York, 1968.
2. U.S. Pat. 2,648,609 (Aug. 11, 1953); 2,799,241 (July 16, 1957), D. E. Wurster (to Wisconsin Alumni Research Foundation).
3. J. T. Goodwin and G. R. Somerville, *Chem. Tech.* **4**, 623 (1974); J. T. Goodwin and G. R. Somerville in J. E. Vandegaer, ed., *Microencapsulation: Process and Applications*, Plenum Press, New York, 1974, pp. 155–163.
4. U.S. Pat. 3,423,489 (Jan. 21, 1969), R. P. Arens and N. P. Sweeney (to 3M Company).
5. U.S. Pat. 2,846,971 (Aug. 12, 1958), C. W. Baer and R. W. Steeves (to National Research Corporation).
6. N. N. Li, *AIChE J.* **17**, 459 (1971).
7. U.S. Pat. 3,703,576 (Nov. 21, 1972), M. Kitajima, Y. Tsuneoka and A. Kondo (to Fuji Photo Film Co., Ltd.).
8. U.S. Pat. 3,691,090 (Sept. 12, 1972), M. Kitajima, T. Yamaguchi, A. Kondo, and N. Muroya (to Fuji Photo Film Co., Ltd.).
9. U.S. Pat. 3,726,805 (April 10, 1973), Y. Maekawa, S. Miyano and A. Kondo (to Fuji Photo Film Co., Ltd.).
10. J. H. Fendler and A. Romero, *Life Sci.* **20**, 1109 (1977); D. Papahadjopoulos, *Liposomes and Their Use in Biology and Medicine*, New York Academy of Sciences, New York, 1978.
11. A. Romero, C. D. Tran., P. L. Klahn and J. H. Fendler, *Life Sci.* **22**, 1447 (1978).
12. C. Thies, *Polym. Plast. Technol. Eng.* **5**, 1 (1975).
13. U.S. Pat. 2,800,457 (July 23, 1957), B. K. Green and L. S. Schleicher (to The National Cash Register Co.).
14. U.S. Pat. 3,155,590 (Nov. 3, 1964), R. E. Miller and J. L. Anderson (to The National Cash Register Co.).
15. Br. Pat. 1,099,066 (Jan. 10, 1968), (to The National Cash Register Co.).
16. U.S. Pat. 3,415,758 (Dec. 10, 1968), T. C. Powell, M. E. Steinle and R. A. Yoncoskie (to The National Cash Register Co.).
17. U.S. Pat. 3,674,704 (1972), R. G. Bayless, C. P. Shank, R. A. Botham, and D. Werkmeister (to The National Cash Register Co.).
18. D. L. Gardner, R. D. Falb, B. C. Kim, and D. C. Emmerling, *Trans. ASAIO* **17**, 239 (1971).
19. T. M. S. Chang, *Artificial Cells*, Charles C Thomas, Springfield, Ill., 1972, pp. 16–18.
20. U.S. Pat. 3,516,846 (June 30, 1970); 3,516,941 (June 23, 1970), G. W. Matson (to Minnesota Mining and Manufacturing Company).
21. U.S. Pat. 4,025,455 (May 24, 1977), D. R. Shackle (to Mead Corp.).
22. P. W. Morgan, *Condensation Polymers by Interfacial and Solution Methods*, Wiley-Interscience New York, 1965.
23. E. E. Ivy, *J. Econom. Entomol.* **65**, 473 (1972).
24. U.S. Pat. 3,429,829 (Feb. 25, 1969), H. Ruus (to Moore Business Forms, Inc.).
25. M. Gutcho, *Capsule Technology and Microcapsules*, Noyes Data Corp., Parkridge, N.J., 1972, p. 152.
26. U.S. Pat. 3,779,941 (Dec. 18, 1973), M. P. Powell (to Champion International Corporation).
27. P. Speiser in J. R. Nixon, ed., *Microencapsulation*, Marcel Dekker, Inc., New York, 1976, pp. 1–12.
28. U.S. Pat. 3,726,804 (April 10, 1973), H. Matsukawa and M. Kiritani (to Fuji Photo Film Co., Ltd.).
29. U.S. Pat. 3,852,076 (Dec. 3, 1974), S. C. Grasko (to Jack W. Ryan).
30. Fr. Pat. 2,648,609 (April 27, 1960), J. A. Orsino and co-workers (to National Lead Company).
31. W. F. Gorham, *J. Polym. Sci. Part A-1* **4**, 3027 (1966).
32. C. Deng, B. B. Thompson, and L. A. Luzzi in T. Kondo, ed., *Microencapsulation*, Techno Inc., Tokyo, Japan, 1979, p. 149.
33. W. M. Doane, B. S. Shasha, and C. R. Russell, *Am. Chem. Soc. Symp. Ser.* **53**, 74 (1977).
34. R. W. Baker and H. K. Lonsdale in A. C. Tanquary and R. E. Lacey, eds., *Controlled Release of Biologically Active Agents*, Plenum Press, New York, 1974, pp. 15–71.
35. M. Calanchi in ref. 27, pp. 93–102.
36. H. Yatzidis, *Proc. Eur. Dial. Transplant Assoc.* **1**, 83 (1964); T. M. S. Chang, *Can. J. Physiol. Pharmacol.* **47**. 1043 (1969).

37. K. K. Goldenhersh, W. Huang, N. S. Mason, and R. E. Sparks, *Kidney Int.* **10,** S251 (1976).
38. C. Thies in ref. 30, p. 143.
39. P. Gopalratnam, M.S. Thesis, Washington University, St. Louis, Mo., 1978.
40. L. R. Beck, V. Z. Pope, D. R. Cowsay, D. H. Lewis, and T. R. Tice, *Contracept. Del. Syst.* **1,** 79 (1980).
41. F. Lim and A. M. Sun, *Science* **210,** 908 (1980).
42. P. Aegerter, M. S. Thesis, Washington University, St. Louis, Mo., 1980.
43. J. Widder, A. E. Senyei, and D. F. Ranney, *Adv. Pharm. Chemother.* **16,** 213 (1979).
44. W. J. Asher, K. C. Bovèe, T. C. Vogler, R. W. Hamilton, and P. G. Holtzapple, *Clin. Nephrol.* **2,** 92 (1979).
45. J. Bitman, T. R. Wrenn, and L. F. Edmondson in ref. 32, p. 195.
46. U.S. Pat. 3,218,305 (Nov. 16, 1965), V. K. Krieble (to Loctite Corporation).

General References

J. R. Nixon, ed., *Microencapsulation*, Marcel Dekker, Inc., New York, 1976.

T. Kondo, ed., *Microencapsulation*, Techno Inc., Tokyo, Japan, 1979.

J. E. Vandegaer, *Microencapsulation: Processes and Applications*, Plenum Press, New York, 1974.

T. M. S. Chang, ed., *Artificial Cells*, Charles C Thomas, Springfield, Ill., 1972.

M. H. Gutcho, *Microcapsules and Microencapsulation Techniques*, Noyes Data Corp., Park Ridge, N.J., 1976.

A. C. Tanquary and R. E. Lacey, eds., *Controlled Release of Biologically Active Agents*, Plenum Press, New York, 1974.

A. F. Kydonieus, ed., *Controlled Release Technologies: Methods, Theories, and Applications*, CRC Press, Cleveland, Ohio, 1980.

D. R. Paul and F. W. Harris, eds., *Controlled Release Polymer Formulations*, ACS Symposium Series No. *33*, American Chemical Society, Washington, D.C., 1976.

W. Sliwka, "Microencapsulation," *Angew. Chem. Int. Ed.* **14,** 539 (1975).

H. B. Scher, ed., *Controlled Release Pesticides*, ACS Symposium Series No. *53*, American Chemical Society, Washington, D.C., 1977.

ROBERT E. SPARKS
Washington University

MICROPLANTS.   See Pilot plants and microplants.

MICROSCOPY CHEMICAL.   See Analytical methods.

MICROWAVE TECHNOLOGY

The application of electrical or electromagnetic (EM) energy to materials as part of some chemical process is a broad subject and of long history in chemical technology if the whole spectrum of electromagnetic energy is considered. Electrical energy can be delivered to materials through conductive, near-field coupling, or radiative techniques. This article is restricted to a review of the microwave spectrum in its role as a power source for chemical and materials processing.

The electromagnetic spectrum is shown in Figure 1. Microwaves is the loose name applied to the central portion of the nonionizing radiation part of the spectrum which conventionally is defined as ranging from dc to visible light. This portion can be divided into five regions in order of increasing frequency: static, quasistatic, microwave, quasioptical (nanowave), and optical. The term nonionizing was coined (1) with regard to the research and literature on biological effects and refers to that part of the spectrum for which the quantum energy is too low to ionize an atom on a single-event basis. Microwaves are used to ionize gases with sufficient applied power but only through the intermediate process of classical acceleration of plasma electrons to energy values exceeding the ionization potential of molecules in the gas (see also Plasma technology). The term nonionizing radiation is used to distinguish the spectrum below visible light in frequency from the ionizing region well above visible light which exhibits more biological-effect potential whatever the power-flux levels. In fact, the distinction between these two parts of the spectrum is so sharp that for years the traditional literature (2) in health physics used the term radiation hazards without qualification, the relation to the ionizing spectral region being obvious.

In a general sense, the term radio waves has evolved to refer to coherent generation of energy in the nonionizing spectral region and its application for information processes such as communication, broadcasting, and target location (radar). The development of radio-wave technology began historically at lower frequencies where electromechanical techniques, then vacuum-tube techniques, and later solid-state techniques are easiest to apply. The term microwaves was introduced (3) just before World War II when vacuum-tube-technique development had proceeded far enough toward higher frequency to permit significant power levels to be transmitted in hollow pipes

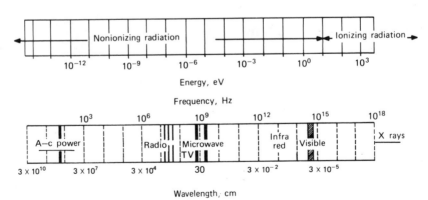

Figure 1. The electromagnetic spectrum in the region of nonionizing radiation.

(waveguides) of practical laboratory size (ca 15 cm diameter) and high antenna gain >20 dB to become feasible with practical rotating-antenna diameters, eg, <5 m. This led to definitions of the microwave spectrum as >1000 MHz. Over the years, these definitions have been variously specified to include frequencies as low as 100 MHz to as high as 3000 GHz. The more scientific meaning of microwaves refers to the principles and techniques applying to electromagnetic systems (4) where the principal dimensions are of order of a wavelength (λ), or more broadly, ca 0.1–10 λ. In this sense, the microwave part of the spectrum lies in the center of the nonionizing spectrum between the quasistatic regime at lower frequencies and quasioptical and optical regimes of higher frequencies.

The distinctive meaning for microwave power applications derives from the fact that for most materials, in particular biological tissue, maximum penetration of the electromagnetic energy irradiating objects of macroscopic size in human commerce occurs in the microwave range. This is illustrated by the example described in Figure 2, ie, penetration into the human body (central internal field relative to external field) as calculated for a simple homogeneous human muscle-tissue model (5).

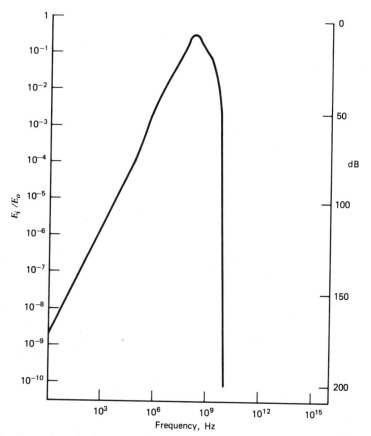

Figure 2. Dependence on frequency of penetration capability of nonionizing radiation in humans (15 cm minimum dimension). Ordinate is ratio of electric field in center of body to incident (external) electric field.

At low frequencies, the body acts as a conductor and the electric field is shunted out, ie, the body has a shielding effect. At high frequencies, the penetration depth (decay by $1/e$ in field strength) rapidly becomes much smaller than body dimensions. Only in the microwave range, here ca 100 MHz, is deep penetration to the core significant.

At low frequencies, penetration of magnetic fields into most materials is efficient. Almost all microwave interactions with materials, however, involve the electric field which penetrates only in the microwave range.

This property of resonance in lossy materials is related in a hybrid manner to the concept of resonant-conducting antennas and resonant low-loss dielectric modes. Most materials of potential interest for microwave processing have dielectric properties that permit interpretation in terms of geometric resonance. A consequence of this is the scaling of processing frequency with some characteristic size of the material or its container being processed.

The other prime reason for interest in microwaves in chemical technology refers to the field of dielectric spectrometry (6) where quantum transitions between rotational states of many molecules are in the microwave region above 100 MHz and are resolvable when the material is in gaseous form. Quantum transitions for electronic states in solids can be detected in the microwave quantum-energy range under special conditions of electron magnetic resonance (7); nmr requires applied magnetic fields and low temperatures (8). Most applications relating to microwave quantum effects in chemical technology are of a diagnostic nature and are not reviewed here (see Analytical methods). This review is restricted to those processes involving microwave power as an important factor in the commercial processing of materials. These processes almost exclusively relate to classical interaction processes of dielectric heating in solids or plasma heating. Other interactions may be possible.

The field of microwave power applications as an application of electronics distinguished from information processing is relatively new. Its development and future possibilities was recognized in a notable review (9). Subsequent interest was shown in U.S. symposia (10), other publications (11), and topical books (12). The International Microwave Power Institute (IMPI) was founded in 1966 as a Canadian Charter Organization. Its offices were at the University of Alberta until it moved to New York in 1979. Its publications include the *Journal of Microwave Power*. The future is likely to see a slow expansion of power uses in the microwave spectrum.

Frequency Allocations for Microwave Power Applications

Under ideal conditions, an optimum frequency should be selected for each application of microwave power. Historically, however, development of the radio spectrum has been predominantly for communications and information-processing purposes (eg, radar or radio location). Thus, within each country, and to some degree through international agreements, a complex list of frequency allocations and regulations on permitted radiated or conducted signals have been generated in order to permit efficient use of the spectrum. Frequency allocations were developed later on a much smaller scale for industrial, scientific, and medical (ISM) applications.

Very recently, the ISM frequency allocations have been revised as a result of the 1979 World Administrative Radio Conference (WARC) (13) of the International Telecommunications Union held in Geneva. A considerable effort was made to increase the number and worldwide uniformity of ISM frequency allocations, but most of these proposals were rejected. The resulting allocations are listed in Table 1.

The three frequencies of 13.56, 27.12, and 40.68 are well-known allocations for r-f heating using quasistatic coupling to loads, such as a capacitor arrangement. The most popular of these assignments is 27.12 MHz for which the broadest bandwidth is authorized. It also overlaps the frequency allocations for citizen's radio bands (CB) in the United States. The techniques used at these frequencies are not generally considered microwave for small loads, but they would be if energy were applied to sufficiently large loads.

The frequency of 433.92 MHz is a harmonic of the three above r-f frequencies and is allocated only in some European countries (see Table 1). It is used mostly for medical diathermy.

The remaining ISM allocations above this frequency are not harmonically related, which is unfortunate with regard to the problem of minimizing radio-frequency interference (rfi) (except for the harmonic relation in the millimeter-wave range).

The frequency bands at ca 915 and 2450 MHz (2375 MHz in eastern Europe) are the most developed microwave bands for power applications. Microwave ovens are almost all at 2450 MHz and many of the industrial heating applications are at 915 MHz. The 1979 WARC indicated that eastern Europe will adopt 2450 MHz in place of 2375.

The higher frequencies at 5,800 and 24,125 MHz have been allocated for some years but have not yet been further developed. At the 1979 WARC, additional ISM allocations were adopted at 61.25, 122.5, and 245 GHz in anticipation of future applications. Actually, until recently, power sources above 30 GHz were not feasible. Therefore, little exploration was made of power applications in that region, called the millimeter waves.

Table 1. Frequency Allocations for ISM Applications [a]

Frequency, MHz	Region	Conditions
6.765–6.795	worldwide	special authorization with CCIR[b] limits; both in-band and out-of-band
13.553–13.567 26.957–27.283 40.66–40.70	worldwide	free radiation bands
433.05–434.79	selected countries in Region 1[c]	free radiation bands
433.05–434.79	rest of Region 1[c]	special authorization with CCIR[b] limits
902–928	Region 2[d]	free radiation band
$2.40–2.50 \times 10^3$	worldwide	free radiation band
5.725–5.875	worldwide	free radiation band
24.0–24.25	worldwide	free radiation band
61.0–61.5 122–123 244–246	worldwide	special authorization with CCIR[b] limits; both in-band and out-of-band

[a] Ref. 13.

[b] CCIR = "International Radio Consultative Committee" of the International Telecommunications Union (ITU).

[c] Region 1 comprises Europe and parts of Asia; the selected countries are the Federal Republic of Germany, Austria, Liechtenstein, Portugal, Switzerland, and Yugoslavia.

[d] Region 2 comprises the Western hemisphere.

In general, ISM applications at allocated frequencies are permitted to freely radiate energy within the allocated band and any other users of this band must tolerate possible interference from such emissions. Emissions outside of these bands, however, must be limited to values specified by regulatory bodies such as the FCC in the United States and verified by specified certification procedures. Typically, field strengths for signals within a 5-MHz band must be below 10 μV/m at a distance of 1.6 km from industrial heating equipment. The FCC rules for ISM equipment are being revised at present (14). An important change is the introduction of rather stringent limits on conducted interference.

In other countries, the rfi regulations must conform to similar limitations. In many cases, the so-called CISPR (Comité international spécial de perturbations radioélectriques) limits, developed by the International Electrotechnical Commission are applicable (15).

Experiments and even production operations can be conducted at any frequency, providing the radiated and conducted signals meet the applicable rfi limits for ISM equipment and tests to certify this are carried out before inception of operations. This implies well-shielded enclosures at high power levels which is expensive but this is justified in certain applications.

Principles of Microwave Power in Processing Materials

In most practical applications of microwave power, the material to be processed is adequately specified in terms of its dielectric permittivity and conductivity. The permittivity is in general taken as complex to reflect loss mechanisms of the dielectric polarization process, whereas the conductivity may be specified separately to designate free carriers. For simplicity, it is common to lump all loss or absorption processes under one constitutive parameter (16) which can be alternatively labeled a conductivity σ or an imaginary part of the complex dielectric constant ϵ_i, as expressed in the following equations for complex permittivity:

$$\epsilon = \epsilon_o(\epsilon_r + j\epsilon_i) = \epsilon_o\left(\epsilon_r + j\frac{\sigma}{\omega\epsilon_o}\right) \tag{1}$$

where ϵ is the complex dielectric permittivity in F/m, $\epsilon_o = 8.86 \times 10^{-12}$ F/m, the permittivity of free space, ϵ_r is the real part of the relative dielectric constant, ϵ_i is the imaginary part of the relative dielectric constant, and σ is the conductivity in S/m (mhos/m) which is equivalent to

$$\epsilon_i = \frac{\sigma}{\omega\epsilon_o}$$

where ω is the assumed radian frequency of the fields. It is convenient to define auxiliary terms like the loss tangent tan δ:

$$\tan \delta = \frac{\epsilon_i}{\epsilon_r} = \frac{\sigma}{\omega\epsilon_r\epsilon_o} \tag{2}$$

From Maxwell's equation (17), the current density $\vec{J}$ in A/m^2 is related to the internal electric field:

$$\vec{J} = (\sigma - j\omega\epsilon_r\epsilon_o)\vec{E}_i \tag{3}$$

Thus, the rate of internal density of absorbed energy, or power, is given by

$$P = r.p.(\vec{J} \cdot \vec{E}*) = \sigma |E_i|^2 \tag{4}$$

or simply

$$P = \omega \epsilon_r \epsilon_o \tan \delta |E_i|^2 \tag{5}$$

This is the practical equation for computing power dissipation in materials and objects of uniform composition adequately described by the simple dielectric parameters.

The internal field is that microwave field which is generally the object for solution when Maxwell's equations are applied to an object of arbitrary geometry and placed in a certain electromagnetic environment. This is to be distinguished from the local field (18) seen by a single molecule which is not necessarily the same. The dielectric permittivity as a function of frequency can be described by theoretical models (19) and measured by well-developed techniques for uniform (homogeneous) materials (20).

The dielectric permittivity as a function of frequency may show resonance behavior in the case of gas molecules as studied in microwave spectroscopy (21) or more likely relaxation phenomena in solids associated with the dissipative processes of polarization of molecules, be they nonpolar, dipolar, etc. (There are exceptional circumstances of ferromagnetic resonance, electron magnetic resonance, or nmr.) In most microwave treatments, the power dissipation or absorption process is described phenomenologically by equation 5, whatever the detailed molecular processes.

The general engineering task in most applications of microwave power to materials or chemicals is to deduce from the geometry of samples, and the EM environment (applicator) the internal field distribution $E_i(r)$ and hence the distribution $P(r)$ of absorbed power. From this, the temperature distribution can be calculated. That allows the materials specialist to deduce processing time and applied power required for the desired result, generally by internally heating the material.

The electromagnetic problem is one of solving Maxwell's equation under various boundary conditions (17). If the object is small, the applied EM field may be little perturbed and perturbation theory is adequate (22). If the object is large in terms of penetration depth, quasioptical radiation calculations are valid. If the object is of the order of a wavelength in dimension, geometric resonance can apply with moderately enhanced absorption cross-section. Many calculations for simple models of biological tissue are available in handbooks (23).

Penetration depth D (at which fields are reduced by a factor of $1/e$) is given by the formula

$$D = \frac{0.225 \lambda}{\sqrt{\epsilon_r} \sqrt{\sqrt{1 + \tan^2 \delta} - 1}} \tag{6}$$

or for low loss materials, $\tan \delta \ll 1$

$$D \cong \frac{0.318 \lambda}{\sqrt{\epsilon_r} \tan \delta} \tag{7}$$

where λ is the free-space wavelength.

For low frequencies or small objects, the components of electrical fields normal to a material boundary are related by

$$\left| \frac{E_i}{E_o} \right| \cong (\omega \epsilon_o / \sigma) \tag{8}$$

where E_i and E_o are the internal and external fields, respectively, the material is characterized by σ, and the outside volume is free space. If the applied field E_o is parallel to the surface, then the internal field is equal to E_o. Thus, if the applied field is parallel to the long axis of an extended object, then E_i is ca E_o. Otherwise, if $\vec{E}_o$ is perpendicular to the long axis, the internal field E_i is given by equation 8 and $E_i \ll E_o$ for even moderate conductivity ca 1 S/m (mho/m).

In a typical application of microwave power, the engineering task is to solve for the internal fields of an object for the given system, compute its heating distribution vs time and resulting change of state, and similar problems. Clearly, the dielectric parameters are key data for such a calculation, particularly their dependence on temperature. Literature references to such dielectric data are found in refs. 24–25. These do not include extensive data on temperature dependence except those for water. Because of the concentrated interest in microwave cooking at 2450 MHz, more data are available at this frequency. Some definitive data on temperature dependence of typical food dielectric properties are shown in Figures 3 and 4 (26). Other references and techniques for measuring dielectric parameters are given in a recent monograph (27).

An examination of Figure 4 indicates a basic problem in microwave-heating applications. If tan δ increases with temperature, this may create runaway conditions in hot spots, although there are possible mitigating factors. Fortunately the properties of water show the opposite behavior above freezing and presumably are conducive to even cooking. The case of ham in Figure 4 is an exception and reflects the effect of salts and increased ion conductivity with temperature.

Below freezing, the loss is much lower for water. Hence, there is a basic conceptual difficulty in designing a process for uniform thawing by microwave heating. Heating to a point short of the melting point, on the other hand, is aided by the negative value of the curve slope and underlies the commercial success of meat tempering.

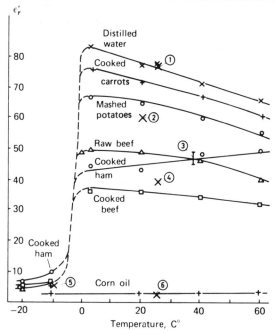

Figure 3. Dielectric constants of some foods near 2.45 GHz as a function of temperature (26).

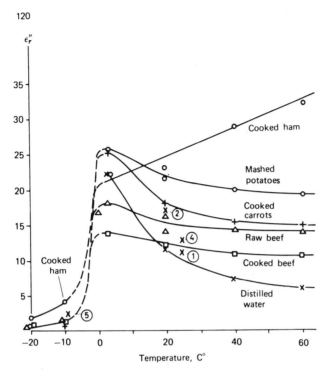

Figure 4. Load factors ($\epsilon'' = \epsilon \tan \delta$) of some foods near 2.45 GHz as a function of temperature. Courtesy of *Journal of Microwave Power*.

Most demonstrated and useful effects of microwave irradiation of materials are explainable as heating effects. The results may be unique to microwave heating because most conventional heating techniques are not associated with deep penetration of the radiant energy.

The literature, particularly that concerning microwave bioeffects, includes speculations on more esoteric effects, sometimes called nonthermal or field-force effects. The only well-demonstrated effort in this category is that of thermoelastic conversion accompanying short pulses or very rapid changes of microwave-radiation levels (28–29). Other speculative mechanisms, but generally very weak, are those of electrostriction (30), dielectropheresis (31), and pearl-chain effects. A review of these mechanisms in biological systems is given in ref. 32. In all these cases the effects are expected to be proportional to $|E|^2$.

The interactions that make use of the magnetic fields are generally of less interest commercially, although of considerable scientific and possibly medical interest. The B and H fields play a key role in weak mechanical forces induced by electromagnetic waves (33). The interactions of esr (7) and nmr (8) are generally only of scientific interest but are being examined for possible medical applications, both diagnostic and therapeutic (34). The most widespread applications of interactions with microwave magnetic fields are those with ferromagnetic materials. The dissipation of microwaves in systems of ferromagnetic materials can be quite complicated in the presence of nonreciprocal transmission properties in biased ferrite systems (35). In the simplest cases, eg, linearly polarized or operations far from ferromagnetic resonance, the localized absorption, in analogy with equation 5, is given by

$$P_{\text{abs}} = \omega\mu''|H|^2 \tag{9}$$

where μ'' is the imaginary part of the permeability and the contributions from the imaginary part of the cross-components of the permeability tensor are negligible (see Magnetic materials).

The dissipation is temperature dependent and disappears above the Curie temperature of the material. This is the basis of some browning dishes developed for microwave-oven use, designed for browning at a Curie temperature of ca 230°C (36).

The interaction of microwaves with ferrites (qv) has many complicating features, eg, low field-loss mechanism (35), nonlinear effects, and losses at high power levels (35,37) as well as dielectric losses.

The application of microwave power to gaseous plasmas is of interest not only in plasma chemistry but for many other reasons. The basic microwave engineering procedure is to calculate first the microwave fields internal to the plasma and then the internal power absorption given the externally applied fields. Again, constitutive dielectric parameters are useful in such calculations. In the absence of d-c magnetic fields, the dielectric permittivity ϵ of a plasma is given by

$$\epsilon = \epsilon_o \left(1 - \frac{Ne^2}{m(\omega^2 + \nu_c^2)\epsilon_o}\right) \tag{10}$$

and the conductivity σ by

$$\sigma = \frac{Ne^2\nu_c^2}{m(\omega^2 + \nu_c^2)} \tag{11}$$

where N is the electron volume density, ω is the microwave radian frequency, ν_c is the collision frequency, and e and m are the electron charge and mass, respectively ($e = 1.602 \times 10^{-19}$ C, $m = 9.11 \times 10^{-31}$ kg) and $\epsilon_o = 8.85 \times 10^{-12}$ F/m, the permittivity of free space.

Maximum power transfer to electrons for a given internal field occurs when $\nu_c = \omega$. The plasma frequency ω_p is the frequency at which $\epsilon = 0$:

$$\omega_p = \left(\frac{Ne^2}{m\epsilon_o}\right)^{1/2} \tag{12}$$

Since the permittivity is negative for $\omega < \omega_p$, transmission through the plasma is cut off and penetration is only by means of evanescent waves.

Various data sources (38) on plasma parameters can be used to calculate conditions for plasma excitation and resulting properties for microwave coupling. Interactions in a d-c magnetic field are more complicated and offer a rich array of means for microwave power transfer (39) (see Plasma technology). The literature offers many data sources for dielectric or magnetic permittivities or permeability of materials (24–25,40). Table 2 gives some data for widely used materials to illustrate typical properties at popular frequency ranges. Because these properties vary considerably with frequency and temperature, the available experimental data are insufficient to satisfy all proposed applications. In these cases, available theories can be applied or the dielectric parameters can be determined experimentally (41).

Table 2.   Some Dielectric Properties of Foods and Other Materials at 2450 MHz and 20°C

Material	ϵ	tan δ
distilled water	78	0.16
raw beef	49	0.33
mashed potatoes	65	0.34
cooked ham	45	0.56
peas	63	0.25
ceramic[a]	8–11	0.0001–0.001
most plastics	2–4.5	0.001–0.02
some glasses[b]	ca 4.0	ca 0.001–0.005
papers	2–3	0.05–0.1
woods	1.2–5.0	0.01–0.1

[a] Alumina.
[b] Pyrex.

Power Sources

The development of electron tubes, including those for the microwave range, has been a mature field (9) for some time. Today it is feasible to generate almost any desired power level for most microwave frequencies of practical interest, limited only by costs. The curves in Figure 5 express the state of the art on power limits for various devices (42). Below 500 MHz large power is typically generated by gridded tubes. Above 1 GHz, the feasible power drops roughly as f^{-5} (f = frequency) because of fundamental limitations on electron-beam current density and circuit losses. The boundary is similar for microwave tubes (magnetrons, klystrons, traveling-wave tubes, backward-wave oscillators) except it is at a frequency roughly thirty times higher.

Power sources in the millimeter-wave range are mostly in the category of extended-interaction klystrons or narrow-band backward-wave oscillators (43). They are quite expensive and suffer from low life and efficiency. Thus, commercial applications in the millimeter-wave range have been hindered, though they are slowly developed including the ISM bands above 50 GHz. Significant power is generated only at frequencies below 300 GHz.

Above this frequency, the expectation has always been that useful laser sources would eventually be developed, albeit with power limits decreasing with decreasing frequency because of fundamental principles (44). Thus, the most difficult region for power generation appears to be that of submillimeter waves or the far infrared, ie, 300–3000 GHz (1000–100 μm). The feasibility of power generation in this frequency range has recently been dramatically changed by the demonstration in the USSR of the gyrotron (45). This is an example of cyclotron-resonance tubes that may make possible a shift of the microwave tube frontier in Figure 5 to the right by a factor of ten in frequency (42). It will be some time, however, before millimeter-wave tubes are available at low cost for commercial applications.

The remaining class depicted in Figure 5 is that of solid state devices, ie, transistors, avalanche diodes, etc. At frequencies below 1 GHz, it is reasonable to expect some widespread inroads of solid-state power sources for certain ISM applications based on today's state of the art (46). For power above 1 kW, however, or for frequencies above 1 GHz, tube sources are likely to prevail for ISM applications for many years to come.

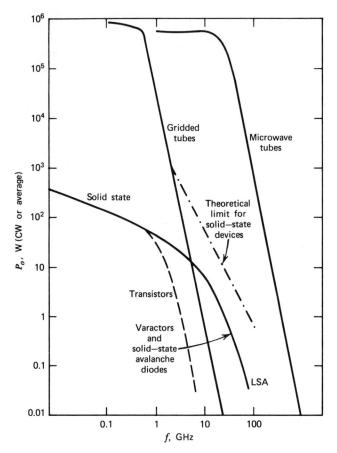

Figure 5. Maximum limits of average output power from various microwave power sources (ca 1975). CW = continuous wave.

The most dramatic evolution of a microwave power source is that of the cooker magnetron for microwave ovens (42). Such tubes for ca 700-W capability were generally bulky (>4.5 kg) and water-cooled 20 years ago. Today, cooker magnetrons are air-cooled and weigh <1.4 kg (see Fig. 6). These tubes generate well over 700 W at 2450 MHz into a matched load and exhibit a tube efficiency on the order of 70%. Their application is enhanced by the availability of comparatively inexpensive microwave power and microwave oven hardware (47).

For many applications at 2450 MHz, it is feasible to utilize a number of such tubes to generate large total powers, eg, 25 or 50 kW. In such cases, multiple feeds or antennas to a heating chamber are used. At 2450 or 915 MHz, it is preferable to use larger power sources at 5, 25, or even 50 kW. Table 3 shows available tubes commonly considered for ISM use at 915 or 2450 MHz. Most of these are magnetrons, although klystrons become competitive at the higher power levels. These tubes are designed to meet the requirements of government agencies on out-of-band spurious emissions. Hence, the use of filter boxes around the high voltage terminals of the tubes.

Microwave tubes for other ISM bands are not commonly available as tubes specifically designed for this use. The available tubes are generally of military and communications types and thus more expensive.

Figure 6.   Typical cooker magnetrons used in 600–700-W consumer microwave ovens.

Applicators and Instrumentation

The basic elements of a microwave power system for materials processing are indicated schematically in Figure 7. A power supply drives the microwave tubes with an applied d-c voltage or even raw rectified voltage (60, 50 Hz). The former may be required for klystrons (48), whereas the latter has been perfected in the form of voltage-doubler power supplies for microwave ovens (49). The microwave tube source is thus generally operated CW (continuous wave, ie, unmodulated signal), except for the possible incidental 60-Hz amplitude modulation for power applications. Only for

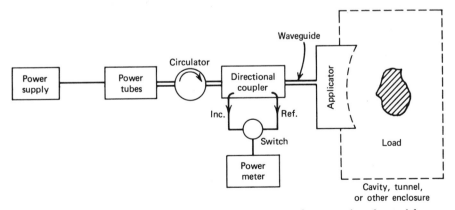

Figure 7.   Basic elements of a microwave power system for processing of materials.

a few applications is a capability of amplitude or frequency modulation interesting and these modifications would entail a significant departure from the present low cost sources. The spectrum of typical microwave power sources is not tightly controlled. Considerable frequency drift and signal broadening is tolerable within the most popular designated ISM bands.

Many power tubes can be operated without the need for a protective ferrite isolator device, particularly the modern cooker magnetrons for consumer microwave ovens (50). At higher power levels, such as 25 kW, it is more common to employ a protective ferrite device, particularly in the form of a circulator (51), as shown in Figure 7. This results in a power loss equivalent to a few percentage points in system efficiency. The ferrite circulator prevents reflected power from returning to the power tube and instead directs it into an auxiliary dummy load. The pulling of tube frequency is thus minimized.

It is common to employ microwave power monitoring by means of a dual-directional coupler in the waveguide transmission system between power tube and useful load. Part of the coupled signals may be used for examinaton with spectrum analyzers, frequency meters, and other microwave instrumentation for special purposes. Generally, this is not necessary in a practical application. Microwave-measurement techniques are described in many available texts (52–53). The availability of microwave components, plumbing, and instrumentation is well described in the trade journals, such as *Microwaves*, *Microwave Journal*, and *Microwave System News*.

At the right of Figure 7 is depicted the useful load, ie, material to be heated by microwave or otherwise affected by exposure to microwave, and some type of applicator, which is the feed (transmitter), or antenna arrangements that couple microwave energy into the load. It is the object to maximize the coupling and this is indicated by zero-reflected power in the waveguide-monitoring system. Not shown in Figure 7 are various waveguide components used to produce an effectively matched applicator, $P_{ref.} = 0$ (54). The power transmitted by the applicator is not necessarily all captured by the load, ie, the circuit efficiency in the heating enclosure is not necessarily 100%, but generally it can be designed to be 90% or better, especially for large loads.

The heating system is calculated by taking into account the power-supply efficiency (generally over 90%); tube efficiency (60–90%); waveguide (including circulator) transmission efficiency (generally over 90%); and circuit efficiency of the heating enclosure. The net efficiency of such systems, ie, useful power into load divided by line input power, varies significantly among systems, but a typical value is ca 50%.

Different applicator types serve various forms, shapes, and sizes of the material being processed or heated. The latter may be gaseous plasma enclosed in a glass tube passing through a cavity or waveguide; thin films passing through a single cavity; or large-size boxes of frozen meat passing in a conveyor through a long metal cavity enclosure. The aim to achieve efficient and uniform heating (or temperature) in the product is at the heart of the design art of microwave power systems, and is reflected in much of the patent literature in this field.

The large variety of possible applicators can be classified in various ways. Most fall into one of the following categories: single-mode cavity; folded guide; slow-wave structures; multimode cavity, batch and conveyor fed; and radiating antenna.

In the single-mode cavity, classical microwave theory and techniques about resonant cavities apply. Generally, the material being heated can be used to calculate the shift in cavity-resonance frequency and change in cavity Q caused by the insertion of the material. This technique is better suited for measuring dielectric properties than most materials processing. It has been applied in special cases for heating (55), particularly of thin fibers passing along the axis of a cavity of cylindrical symmetry, such that the fiber axis is aligned with the maximum E field.

The folded-guide applicator is formed by bending a rectangular guide in its E-plane to form a meander. Longitudinal slots on the broad walls in the center of the meander permit continuous passage of material in sheet form through the successive portions of guide with the sheets in the E-plane for good coupling. The meandering properties permit efficient heating without resonant cavity properties of high E-field and cavity losses. The folded guide is described amply in the literature (56).

Slow-wave structures (57–58) are related to the folded waveguide since the phase velocity of waves along an axis passing through all the guides (ie, through the slots where the sheet material is fed) is less than that on free space, ie, $v < c$. A variety of planar structures, meander lines, interdigital lines, and vane lines, produce a region of fringing field where $EH \propto e^{-\gamma x}$ (γ ca β ca ω/v is the slow-wave propagation constant). In some cases, such structures are suitable for heating sheet material with more space flexibility than the folded guide. Similar slow-wave structures exist in cylindrical form with a fringing field.

Perhaps the most generally used applicator is the multimode cavity, designed for both batch or conveyor-fed processing. The theoretical properties of large cavities with large loads have never been completely established. A few initial studies are reported (59–60). Earlier developments were marked by the use of waveguide apertures (dump feeds) with rotating blades somewhere in the cavity acting as mode stirrers (61–62). More recently, it has been found that rotating antenna feeds achieve a more uniform heating of the load material (63–64).

A rotating turntable is a simple way to improve heating uniformity (65). A related technique is the use of a linearly moving load-bearing surface, ie, a conveyor belt. In carrying materials through a long cavity, each portion of the load undergoes, in principle, a similar history of field pattern and consequent heating. In Figure 8 a practical conveyor-fed cavity, typical of systems operating at 915 MHz, is depicted. In order to accept substantial-size boxes of material, the conveyor belt passes through long entrance and exit tunnels that are designed to minimize microwave leakage by various techniques, including liquid absorbers along the walls of the tunnels.

Finally, in some applications, such as the repair of road surfaces, the load cannot be enclosed in a cavity and radiating applicators must be used. These could be in the form of simple horns (67) or various modified waveguide apertures designed for microwave diathermy (68–69). They can be designed for efficient coupling for a closely spaced load surface as well as minimizing leakage or loss of energy to the sides.

In addition to the measurements of incident and reflected power (Fig. 7), the systems designer may want to make measurements related to field strengths, field patterns, and resulting heating. In simple cavity-waveguide or slow-wave-systems perturbation techniques (52), dielectrics or metals can be used to determine the E-field strengths where the load will be. This is done by measuring cavity-resonance or structure-phase shifts. For multimode cavities the use of conventional E or H probes is laborious and not generally valid because of perturbation of the feed lines for the probes. The E field can be measured with small gas-discharge lamps (70). Various techniques, none completely satisfactory, have been used to determine field patterns in a plane or volume of a multimode cavity in the absence of the load (71–73).

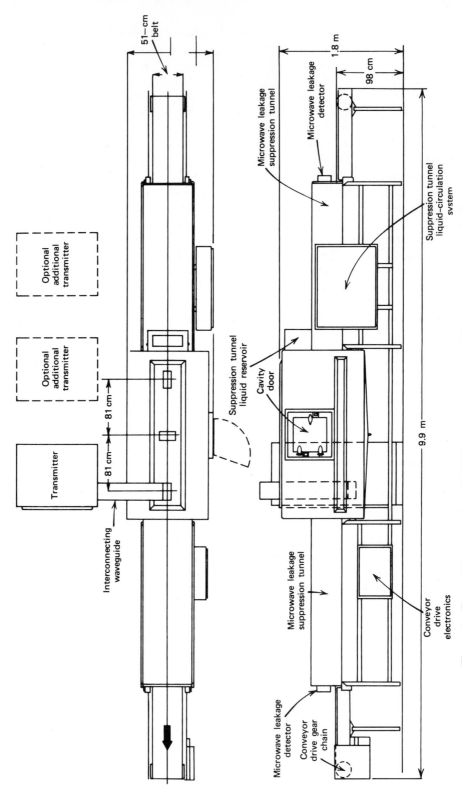

Figure 8. Top and front view of a typical high power conveyor-type industrial microwave equipment (66).

51—cm belt

Optional additional transmitter

Optional additional transmitter

Transmitter

81 cm

81 cm

81 cm

Interconnecting waveguide

Suppression tunnel liquid reservoir

Cavity door

Microwave leakage suppression tunnel

Microwave leakage detector

1.8 m

98 cm

Suppression tunnel liquid-circulation system

9.9 m

Conveyor drive electronics

Microwave leakage suppression tunnel

Microwave leakage detector

Conveyor drive gear chain

The most satisfactory measurement is the determination of fields or temperature distribution in the load. Conventional temperature probes or measurements have been investigated (73) either at the end of microwave heating or during the microwave process by arranging probe leads to be perpendicular to the microwave E field. Because this is in general impossible, nonperturbing probes of various degrees of efficiency and use have been developed. These include probes with fiber-optic leads using liquid-crystal sensors (74), semiconductors (75), and fluorescent materials (76). In addition, small thermocouple and thermistor sensors have been used with newly developed high resistance leads (77). These techniques permit measurement of internal temperatures during microwave heating and, in conjunction with miniature nonperturbing E-field probes (78), offer hope for better design. The determination of surface temperature and temperature patterns can be made noninvasively by infrared pyrometers (79) and thermographic cameras (80), an art which is still being developed (see Infrared technology; Temperature measurement).

Economic Aspects

The cost of microwave power systems is generally considerably greater than that of conventional heating systems. For example, a 50-kW conveyor system in 1980 may cost $150,000–200,000. The acceptance of this technology in industry must follow a sober economic evaluation. The costs of installation, operating power, maintenance (including replacement of power tubes), and financing must be weighed against the potential savings in labor, space, yield, productivity, and energy. An additional benefit of increasing worth is the reduction in chemical pollution which accompanies some microwave applications.

A basic choice is that of operating frequency. In principle, operation can take place at any frequency at the cost of suppression of electromagnetic leakage to regulatory limits on rfi (eg, 25 μV/m at 304 m). This cost is avoided by operating within assigned ISM bands and minimum cost results in bands of considerable use where components are readily available, possibly by economical mass production. In the United States, these popular microwave bands are 915 and 2450 MHz.

The price of cooker magnetrons (ca 700 W) 20 years ago was in the range of several hundred dollars. Today, because of the large microwave-oven market (more than 10^7 units in the United States), the original equipment manufacture (OEM) price of such tubes is well below $30.00. The total sales of microwave ovens, worldwide, now exceed 25×10^6. In Japan (8) and in the United States (81), the markets are over 25 and 15%, respectively, saturated.

In many cases, the alternative of lower-cost r-f systems should be considered, ie, so-called induction and r-f heating systems at 40 MHz and below. These are not reviewed here but have recently been thoroughly discussed (82) (see Furnaces, electric). Refs. 83–85 present more extensive discussions of the economic aspects of microwave systems and payback calculations.

Health and Safety Factors

In addition to the usual mechanical, chemical, thermal, and electrical hazards of power equipment, there are some unique safety considerations in microwave systems.

Microwave voltage breakdown can occur in microwave systems and waveguides at power levels far below theoretical values for ideal systems ie, by a factor of at least 100 below theoretical breakdown (86–87). This is often due to impurities or dirt particles that overheat and cause a breakdown or spurious high Q, resonances in the system which build up high E fields. In addition, the presence of sharp metal objects, accidental small gaps, and other situations often can induce localized arcing or corona which may or may not lead to a basic system breakdown. In this case, the plasma region of the breakdown travels down the feed waveguide toward the source and may cause failure of the tube through cracking of the output window. Therefore, flammable materials should not be processed in microwave systems except with precautions, the more so the greater the degree of flammability.

In most cases, microwave heating, like conventional heating, produces the highest temperatures at or near the surface of load objects, except that microwave heating is more penetrating. In special situations, such as the heating of small objects of a few centimeters diameter, microwave frequencies produce rather unique heating patterns with maximum temperatures at the center of the object (88–89). This would be a genuine heating from the inside out, a common but not generally true characteristic attributed to microwave-oven heating. A superheating of small volumes of water can occur followed by mild explosions. The user of microwave power should be aware of this phenomenon when heating small objects to high temperatures.

Several hazards could result from significant leakage of microwave energy from industrial systems. Such hazards are significant at most frequencies, not just microwave.

The most serious hazard is that from incidental interference (90–91) with other systems. This could be caused by out-of-band radiation, ie, a violation of rfi regulations, or by so-called high power effects where the offending radiation is out of the band of the affected system but is still effectively interfering because of its intense level (92). An example of this would be the incidental interference with cardiac pacemakers (93). This problem is partly caused by insufficient protection from rfi in the pacemaker unit, ie, susceptibility. In the last ten years this susceptibility has been reduced (94) and because of government supervision, this problem appears to be under control (95).

It is known that radiated r-f energy is a hazard to systems with flammable fuel or electronic explosive devices (EED) used for construction blasting. Studies (96) indicate that this possibility is remote but it is recommended that users of large amounts of microwave/r-f energy be aware of guides on safe distances of EEDs from sources of radiated power (97).

The hazard of exposure of personnel to microwave energy is generally recognized as exaggerated in the general press and has been thoroughly reviewed many times in recent years (98–99). Exposure safety limits in Western countries have been set at 10 mW/cm^2 for durations over six min but are now modified to more stringent limits at very high frequency (VHF) bands (body-resonance range of humans) and more relaxed limits below VHF (100). Limits in eastern Europe are far more conservative but there are signs that these are being relaxed upward (101).

In the United States, leakage emission from microwave ovens is regulated to the stringent limit of 5 mW/cm^2 at 5 cm (102). There is no Federal limit on emission from industrial systems but the IMPI has set a voluntary standard which specifies 10 mW/cm^2 at 5 cm (103). The emission values are equivalent to personnel exposures at several meters, well below limits for eastern Europe (see Fig. 9). This was derived for microwave ovens but the conclusions should be valid for all microwave systems (104).

Leakage through door-seal areas is reduced through choke techniques (105) as well as absorbing materials (106) described in manufacturers literature on lossy gaskets. Leakage through holes in viewing screens is kept to acceptable limits by well-known limits on hole sizes (107). Generally, the technology for minimizing microwave leakage is being utilized effectively (108).

Uses

Literature and patents on microwave power applications are extensive, most of which have been limited to research or small-scale production efforts. Applications are usually limited by economic rather than technical considerations. The publications of IMPI are a continuing source of articles in this field including reviews on particular applications. Other recent reviews include special issues in the engineering literature (109).

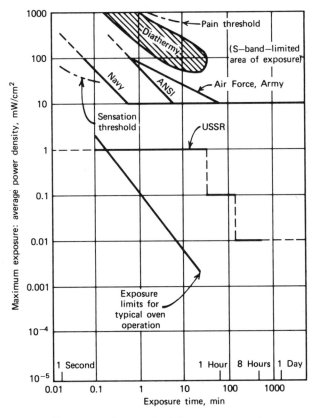

Figure 9. An exposure diagram: maximum potential exposure near microwave oven complying with U.S. emission standard is compared with exposure standards of the United States and the USSR.

Food. The most successful application is that of food processing, cooking, and reheating (see Food processing). The domestic microwave oven has been the economic surprise success, surpassing by far all other microwave power applications.

Essentially all microwave ovens operate at 2450 MHz except for a few U.S. combination-range models that operate at 915 MHz. The success of this appliance was due to the successful development of low cost magnetrons producing over 700 W for oven powers of 500–700 W (Table 3).

As pointed out earlier with reference to Figures 3 and 4, the dielectric properties of food, at least near 2450 MHz, parallel those of water, the principal lossy constituent of food. The dielectric properties of free water are well known (24) and presumably are the basis for absorption in most foods by the interaction of the dipole of the water molecule with the microwave E field. Ice and water of crystallization by comparison absorb very little microwave energy. Adsorbed water, however, can retain its liquid character below 0°C and absorb microwaves (110). Its prevalence in practical situations is not very common.

It was suggested that microwave interaction with foods should be similar to interaction with ionic solutions such as sodium chloride (111–112). Thus, the dielectric loss for a 0.5 N solution below ca 2450 MHz is mostly owing to ionic conductivity, whereas above 2450 MHz the loss is owing mostly to water dipole relaxation. At low frequencies the dielectric loss increases with temperature, whereas at higher frequencies (but still below the water-relaxation frequency, ca 20 GHz), the dielectric loss decreases with temperature. These tendencies are seen in Figures 3 and 4 where the increased ionic conductivity in salted products such as ham tend to show increasing loss with temperature. Generally, the other temperature dependence prevails, however,

Table 3. Microwave Tubes for ISM Applications

Type	Frequency, MHz	Nominal power, W	Voltage, kV	Manufacturer
Magnetrons				
8684	915 ± 3	30×10^3	13	RCA
RM101	915 ± 15	30×10^3	13	Relmag
QKH1838	915 ± 3	30×10^3	13	Raytheon
BM252	915 ± 15 or 896	30×10^3	13	EEV[a]
8684-V1	896 ± 2	30×10^3	13	RCA
RM174	896	30×10^3	13	Relmag
OM72B	2455	875	4.15	Amperex
2M120	2455	2.1×10^3	4.5	Hitachi
2M170	2455	840	4.0	Hitachi
2M156	2455	540	2.5	Hitachi
2M107	2455	840	4.0	Hitachi
2M172	2455	870	4.0	N.E.C.[b]
2M167	2455	840	4.0	Panasonic
2M1725	2455	840	4.0	Toshiba
2M1255	2455	560	2.6	Toshiba
F1123 MCF1327	2450	5×10^3		Thomson-CSF
Klystron				
TH2054	2450	50×10^3	25.5	Thomson-CSF

[a] English Electric Valve Co., Ltd.
[b] Nippon Electric Company.

and is believed to stabilize food heating and avoid runaway heating. The suitability of the higher frequencies for stable heating was one argument in addition to the interest in surface browning put forth by Litton Microwave Cooking, Inc. in its application for a microwave-oven frequency allocation at ca 10.5 GHz (113).

Most systems that are to be heated by microwaves are heterogeneous and, in the case of food, are mixtures of water in either free or bound (adsorbed) states and other materials. The dielectric properties of such heterogeneous systems have been studied and the results suggest only modification but no radical change from the viewpoint described herein (114).

The technology of microwave ovens has developed rapidly in the last decade with advances in techniques for achieving uniform heating (eg, mode stirrers) (115); temperature monitoring (116), humidity monitoring (117), and the use of microprocessors (118) in programming cooking time, defrost, variable power levels, and instructions related to monitor levels. In addition, a wide variety of ceramic, glass, plastic as well as new paperboard products have been developed for utensils, shelves, and other items used in microwave ovens (119). A growing number of accessory products have been developed (120), eg, browning dishes or utensils, popcorn poppers, coffee-makers, etc. Combination ranges have been developed (121) for microwave-electric, microwave-gas, and microwave-convection, the latter in countertop arrangement.

The more general food-processing applications require data on dielectric and thermal properties (122). These define resultant temperature distributions, given imposed E-field distributions in various time profiles in applied energy and surface cooling or heating conditions, eg, convection or infrared.

The microwave oven at 2450 MHz is used mainly for the reheating and cooking of foods; thawing is often included. It may also be used for drying, eg, flowers or food materials.

Of a variety of industrial food-processing applications, only a few are commercialized (123). Some, like the drying of potato chips, were too costly, others, such as freeze-drying (124), encountered technical problems like vacuum breakdown. Microwave ovens have been used extensively for thawing but no effective solution has been yet found for the runaway heating caused by the great increase in dielectric loss at 0°C (123) (see Fig. 4).

The most successful food-processing application is that of meat tempering, particularly at 915 MHz (125), although systems exist at 2450 MHz (126). Microwave power in the range of 25–100 kW is applied to frozen food conveyed in tunnels, such as shown in Figure 8, so as to achieve a fairly uniform temperature in the range of −5 to −2°C where the product can be cut and otherwise processed mechanically before it is returned to a freezer in its final form. It is estimated that over 50 such installations exist in the western hemisphere, some of which process over 7000 metric tons of meat or fish annually.

Generally speaking, nutritional quality of food cooked or heated by microwaves does not differ greatly from food cooked or heated by other means (127–129). Food scorching is minimized in microwave heating, though underheating or undercooking can be avoided only in ovens with superior mode-stirring techniques or by occasional manual rotation.

Microwave heating has been applied to inhibition of mold growth (129), conditioning of wheat (130), poultry cooking (131), sherry making (132), chewing-gum manufacture (133), and pasteurization of beer (134).

Most food-processing applications involve the effects of heating but there may be exceptions using a more sophisticated mechanism. For example, the use of microwaves in oyster-shucking (135) may involve the effects of thermal shock in reducing the mechanical resistance to shucking.

The extensive literature on food processing is partially reviewed in ref. 136.

Biological, Medical, and Agricultural Applications. For at least the last three decades, diathermy at both 27.33 and 2450 MHz has been used extensively in physical therapy (137–138). Recent studies suggest that 915 MHz would be more effective than 2450 MHz because of deeper penetration (139). New diathermy applicators are designed for a more quantitative prediction of tissue-heating patterns and reduced leakage (68–69).

An extension of diathermy-heating techniques has been investigated in the last decade with regard to its use in hyperthermia as an adjunct to cancer therapy (140). The basis for preferential destruction of tumor cells is being studied by hyperthermia alone, or in conjunction with ionizing radiation or chemotherapy. A variety of applicators have been designed, including multiple-focused antennae, injected-probe antennae, and contact applicators.

Other medical applications of microwaves under study include transcutaneous stimulation of nerve-action potentials (141), microwave radiometry for breast-cancer detection (142), pulse and doppler diagnostic techniques (143), microwave imaging (144), and rapid thawing of frozen organs in transplant procedure (145).

A subject of recent interest is the use of microwaves for the rapid inactivation of brain enzymes in rodents (146–147). Microwave power at levels as high as 915 or 2450 MHz is applied by means of a waveguide applicator to achieve a rapid sacrifice of the rodent.

In general, most, if not all, biological effects and potential medical applications are believed to be the results of heating, ie, thermal effects. The phenomenon of microwave hearing (the hearing of clicking sounds when exposed to intense radarlike pulses) is now generally believed to be a thermoelastic effect (148). There is some evidence of an effect of modulated r-f fields on calcium efflux from brain tissue in vitro (149). In addition, based mainly on reports from the USSR on bioeffects in the millimeter-wave range (150), some theoreticians speculate on truly frequency-specific nonlinear coupling to macromolecules as a unique mechanism (151). These matters have not been resolved at this time. Excellent overviews of the field of microwave bioeffects have appeared in recent special issues of some journals (152–153).

Occasionally, reports appear of sterilization against bacteria by some nonthermal microwave effects. It is generally believed, however, that the effect is only that of heating (154). Because microwave heating is often not uniform, studies in this area can be flawed seriously by simplistic assumptions of uniform sample temperature.

The use of microwaves has been investigated to affect plant growth by irradiating seeds or to achieve insect control (155–156). It is generally found, however, that most useful effects result from heating (157).

Plastics Fabrication and Processing. The application of microwave r-f energy in various stages of fabrication of plastic products is perhaps the most widespread industrial application. Most systems operate at low frequency, ca 27 MHz, rather than conventional microwave frequencies. The technology and status of this application is thoroughly reviewed in a U.S. Government report (158). Microwaves are used in curing molded parts, sealing plastic parts, in the fabrication of foamed plastic materials, and for processing industrial coatings. A variety of cavity and folded-wave guide applications have been used on these applications with microwave power as large as 200 kW (see Plastics technology).

Microwave curing of epoxy resin is being extended to glass-fiber reinforced laminates (159) and resin-impregnated wood (160) (see Epoxy resins). The microwave-foaming process is applied to the manufacture of thermally stable polyimide components of low density and high strength (161) (see Polyimides). Microwave heating is used in bonding of plastics and resin adhesives (162) and this is being developed for bonding of wood (163) (see Laminated wood-based composites). Microwave curing is considered for various types of coatings and the drying and application of paint (164–166) (see Coatings).

Textiles. The uses of microwaves in the textile industry have been reviewed recently (167). Crease resistance is imparted to cotton fabrics by microwave curing of fabric impregnated with amino-resin mixtures (168) (see Textiles). Cotton fabrics finished with fire-proofing agents and dried by microwaves, however, are reported to be stiffer and have reduced foaming strength (169). Microwaves are being studied for techniques of reducing endotoxin activity in cotton dust and lints produced in industrial location (170).

Microwaves as heat source facilitate the dyeing of fibers, such as polyester and nylon, as well as cotton and wool (171–172). Cavity techniques offer high coupling efficiency (173).

The microwave drying of textiles appears to attract more interest (174) as energy costs increase.

Drying of Castings and Other Products. The use of microwaves in the curing and drying of foundry cores is well established (158). The best example is the use of microwaves for drying water-based core washes (175); eg, 60 kW at 2450 MHz permits a water removal rate of ca 61 kg/h. Installations up to 150 kW are used in such applications. Similar techniques are being investigated for drying casting molds, including gypsum molds. Microwave techniques for drying inks (176), paper (177), wood (178), and photographic materials (179) have been investigated. The use of microwaves to restore flood-damaged library materials has met with varying success (180). In some cases, lower frequency appears more effective. Expansion of microwave drying into new areas can be expected (181) (see Drying).

Rubber Products. Microwave energy is used in the preheating, curing, and drying of rubber products (158). The most successful industrial applications are those for large products where heating time is greatly reduced because of the penetration properties of electromagnetic energy. Batch-oven units at microwave power over 30 kW at 915 MHz are now used for preheating of giant tires (182) (see Elastomers, synthetic; Rubber, natural).

Polar elastomers heat readily in a microwave field but nonpolar elastomers (including natural rubber) do not heat as readily and must be doped with material such as carbon black to permit microwave heating. A secondary application for lossy rubbers is their use as microwave absorbers in microwave equipment (183). The quality of microwave-cured rubbers appears satisfactory (184); however, economic factors remain an obstacle to such treatment. In addition, it is a problem to develop optimum rubber formulations and optimum microwave applicators, including choice of frequency, for small extruded and molded rubber goods (185–187).

Desulfurization and Waste Treatment. The use of microwave energy (2450 MHz) and NaOH to remove 60–80% inorganic and organic sulfur from U.S. coals has been reported (188). It is believed that water-soluble sulfides (Na_2S, etc) are rapidly formed as a result of deep-heating the coal, although localized discharge sites may play a beneficial role in the process. A similar process is claimed to remove sulfur from crude oils (189). Microwave techniques appear promising for coal-cleaning objectives (190).

The patent literature is extensive on the application of microwave heating and possibly microwave plasma to a variety of waste-treatment problems including sewage, nuclear wastes, nitrogen oxide and regeneration of activated charcoal in wastewater treatment (see Wastes). Studies indicate promise of microwaves for drying of pelleted nuclear waste and vitrification of a calcined waste (191). Microwave-plasma techniques appear to be successful in the detoxification of a red-dye pyrotechnic smoke mixture (192). It is reported that microwave (2450 MHz) heating of PVC resins in the presence of water results in removal of vinyl chloride (193).

Microwave Plasmas. Applications of microwave plasma have been reviewed recently (194). Microwave plasma presents a unique problem in microwave heating because of its high conductivity and frequency-dependent properties. This complicates the task of coupling energy efficiently to the plasma. The problem of single-cavity coupling systems is treated in ref. 195. Slow-wave structure-coupling systems enable much larger plasma volumes to be excited (196). These systems may be applied to organic reactions, polymer modification, and light sources, especially uv. Surface-wave plasmas can be extended as much as 50 cm beyond a 915 MHz coupler, thus forming essentially a microwave plasma torch at atmospheric pressure (197). The principle of the plasma column derives from the excitation of a surface wave that propagates along the interface between a plasma and a quartz tube.

Microwave-plasma discharges produce results similar to other types of discharges, eg, the tendency to produce organic surface films (194). The main distinction is based on differences in electrical coupling efficiency and system simplicity. Still, there are reports of chemical products unique to microwave plasmas which may be owing to a narrower distribution of electron energies (194).

Microwave plasmas may be applied to produce thin oxide films (198), and films of silicon nitride (199) and titanium nitride (200). There is considerable interest in nitrogen discharges, both for synthesis of nitrogen oxide (201) and decomposition of nitrogen monoxide (202).

The objective of microwave heating of nuclear-fusion plasmas is still pursued (203), though interest is shifting from using the lower microwave frequencies to electron-cyclotron resonance heating at millimeter-wave frequencies, perhaps with the gyrotron (see Fusion energy).

Some novel uses of microwave plasma are being studied; for example, the use of plasma excitation by a 2.45 GHz source for the operation of a CO_2 laser (204). The other is the proposal to use microwave heating of the flame plasma in an internal combustion engine to increase combustion efficiency (205) (see Plasma technology).

Other Applications. The use of microwave power is investigated for retorting oil shale, particularly *in situ* (206–207). Optimum frequencies appear to be below the conventional microwave range. The more rapid microwave-heating process may give different products than result from conventional heating of oil shale which is much slower (208). Some interest has also been shown in using microwave heating to extract hydrocarbons from bituminous sands (209) (see Oil shale; Tar sands).

Microwave heating of ceramics to melting point temperatures have been studied (210–211).

The use of microwave heating to simulate meltdown accidents in nuclear reactors (qv) is reported in ref. 212.

An unusual application of the ordinary microwave oven has been the determination of moisture and solid weight of substances like cured hides (213) and detergents (214) by weighing before and after drying.

A host of other applications in scientific experiments, measurements and nondestructive testing exist but are beyond the scope of this article.

Outlook. Based on the past growth, this field is expected to increase with more large-scale production applications as better and cheaper microwave systems are developed. It is interesting to reflect on the fact that since most (perhaps 90%) of the previous studies are at 2450 and 915 MHz, optimum frequencies for all applications have not been found. In particular, uses for 5800 MHz, 24125 and the millimeter-wave frequencies will come in time.

The largest microwave power application ever conceived is that of the solar-power satellite (215–217). If this system comes to reality, it will profoundly change the views and realities with regard to the use of microwave power (see Solar energy).

BIBLIOGRAPHY

1. G. M. Wilkening in *The Industrial Environment—its Evaluation and Control*, U.S. Dept. of HEW, National Institute for Occupational Safety and Health, 1973, Chapt. 28.
2. T. Lauriston, *Legislative History of Radiation Control for Health and Safety Act of 1968*, Testimony before U.S. Congress, DHEW Publication (FDA) 75-8033, Washington, D.C., May 1975, pp. 157, 487, 1025.
3. H. E. M. Barlow, *Microwaves and Waveguides*, Constable, London, Eng., 1948.
4. E. L. Ginzton, *Microwave Measurements*, McGraw-Hill, New York, 1957, pp. viii–ix.
5. J. M. Osepchuk, *Bull. N.Y. Acad. Med.* **55**, 976 (Dec. 1979).
6. W. Gordy, *Rev. Mod. Phys.* **20**, 668 (Oct. 1948).
7. J. E. Wertz, *Chem. Rev.* **55**, 829 (1955).
8. E. M. Purcell, H. C. Torrey, and R. V. Pound, *Phys. Rev.* **69**, 37 (1946).
9. E. W. Herold, *IEEE Spectrum* **2**, 50 (Jan. 1965).
10. E. Okress and co-workers, *IEEE Spectrum* **1**, 76 (Oct. 1964).
11. P. L. Kapitza, *Uspekhi fiz. Nauk* **78**, 181 (1962).
12. E. C. Okress, *Microwave Power Engineering*, Vols. 1 and 2, Academic Press, New York, 1968.
13. *Final Acts of the World Administrative Radio Conference (1979)*, Vols. I and II, International Telecommunications Union (Geneva, Switzerland), Geneva, Dec. 1979.
14. "Industrial, Scientific and Medical Equipment," *Rules and Regulations*, Federal Communication Commission, Vol. II, Part 18, Subpart H.
15. *Limits and Methods of Measurement of Radio Interference Characteristics of Industrial, Scientific, and Medical (ISM) Radio Frequency Equipment*, International Electrotechnical Commission, Special Committee on Radio Interference (CISPR), Publication 11, 1975.
16. R. W. P. King, *Electromagnetic Engineering*, Vol. 1, McGraw-Hill, New York, 1948.
17. J. A. Stratton, *Electromagnetic Theory*, McGraw-Hill, New York, 1941.
18. F. N. H. Robinson, *Macroscopic Electromagnetism*, Pergamon Press, New York, 1973.
19. H. Fröhlich, *Theory of Dielectrics*, Oxford Press, Oxford, 1958.
20. H. Altschuler, "Dielectric Constant" in M. Sacher and J. Fox, eds., *Handbook of Microwave Measurements*, Vol. 2, Polytechnic Press, New York, 1962.
21. C. H. Townes and A. L. Schawlow, *Microwave Spectroscopy*, McGraw-Hill, New York, 1955.
22. G. Birnbaum and J. Franeau, *J. Appl. Phys.* **20**, 817 (1949).
23. C. H. Durney and co-workers, *Radiofrequency Radiation Dosimetry Handbook*, 2nd ed., Report SAM-TR-78-22, USAF, Brooks Air Force Base, Tex., May 1978.

24. A. R. Von Hippel, ed., *Dielectric Materials and Applications*, MIT Press, Cambridge, Mass., 1954.
25. W. R. Tinga and S. O. Nelson, *J. Microwave Power.* **8**(1), 23 (1973).
26. N. E. Bengtsson and P. O. Risman, *J. Microwave Power* **6**(2), 107 (1971).
27. E. H. Grant, R. J. Sheppard, and G. P. South, *Dielectric Behavior of Biological Molecules in Solutions*, Clarendon Press, Oxford, Eng., 1978.
28. R. M. White, *J. Appl. Phys.* **34**, 3559 (1963).
29. K. K. Foster and E. D. Finch, *Science* **185**, 256 (1974).
30. A. W. Guy, C. K. Chou, J. C. Lin, and D. Christensen, *Ann. N.Y. Acad. Sci.* **247**, 194 (1975).
31. H. A. Pohl, *J. Appl. Phys.* **29**, 1182 (1958).
32. H. P. Schwan, *Ann. N.Y. Acad. Sci.* **303**, 198 (Dec. 1977).
33. G. Franceschetti and C. H. Papas, *Appl. Phys.* **23**, 153 (1980).
34. P. C. Lauterbur, *IEEE Trans. Nuclear Sci.* **N5-26**, 2808 (Apr. 1979).
35. J. Helszajn, *Principles of Microwave Ferrite Engineering*, Wiley-Interscience, New York, 1969.
36. E. A. Maguire and D. W. Ready, *J. Am. Ceram. Soc.* **59**, 434 (1976).
37. S. V. Vonsovskii, ed., *Ferromagnetic Resonance*, Pergamon Press, New York, 1966.
38. S. C. Brown, *Base Data of Plasma Physics*, MIT Press, Cambridge, Mass., 1966.
39. A. F. Harvey, *Microwave Engineering*, Academic Press, New York, 1963, p. 897.
40. W. Von Aulock and J. H. Rowen, *Bell System Tech. J.* **36**, 427 (1957).
41. M. A. Hollis, C. F. Blackman, C. M. Wail, J. W. Allis, and D. J. Schaefer, *IEEE Trans.* **MTT-28**, 791 (July 1980).
42. M. M. Osepchuk, *Microwave J.* **21**, 51 (Nov. 1978).
43. J. Wiltse, "Review of Millimeter-Wave Sources," *Digest 1979 IEEE MTT-S Symposium.*
44. N. D. Devyatkov and M. B. Golant, *Radio Eng. Elec. Phys.* **12**, 1835 (1967).
45. V. A. Flyagin, A. V. Gaponov, M. I. Petelin, and V. K. Yulpatov, *IEEE Trans.* **MTT-25**, 514 (June 1977).
46. J. M. Osepchuk and R. W. Bierig, *Digest of IEEE Electro '76*, Session 1, 1976, pp. 1–8.
47. W. Schmidt, *Philips Tech. Rev.* **22**(3), 89 (1960/1961).
48. E. D. Moloney and G. Faillon, *J. Microwave Power* **9**, 231 (Sept. 1974).
49. R. E. Davis and T. Maeda, *Appliance Manufacture* **13**, 87 (Feb. 1979).
50. I. Oguro, *J. Microwave Power* **13**(1), 27 (Mar. 1978).
51. C. Forterre, C. Fournet-Fayas, and A. Priou, *J. Microwave Power* **13**(1), 65 (Mar. 1978).
52. A. F. Harvey, *Microwave Engineering*, Academic Press, New York, 1963, Chapt. 4.
53. J. L. Altman, *Microwave Circuits*, Van Nostrand, New York, 1964.
54. G. L. Ragan, *Microwave Transmission Circuits*, McGraw-Hill, New York, 1948.
55. A. L. Van Koughnett, *J. Microwave Power* **7**(1), 17 (1972).
56. J. Gerling, *IMPI Short Course*, 38 (Nov. 1970).
57. D. A. Dunn, *J. Microwave Power* **2**(1), 7 (1967).
58. A. F. Harvey, *IRE Trans.* **MTT-8**, 30 (1960).
59. T. G. Mihran, *IEEE Trans.* **MTT-26**, 381 (June 1978).
60. M. Watanabe, M. Suzuki, and S. Ohkawa, *J. Microwave Power* **13**(2), 173 (June 1978).
61. D. A. Copson, *Microwave Heating*, Avi Publishing Co., Westport, Conn., 1975, Chapt. 11.
62. P. Bhartia, S. C. Kashyap, and M. A. K. Hamid, *J. Microwave Power* **6**, 221 (1971).
63. W. W. Teich and K. Dudley, *Proc. 1980 Symposium on Microwave Power*, IMPI, New York, pp. 122–124.
64. J. E. Simpson, *Microwave J.* **23**, 47 (Jan. 1980).
65. D. A. Copson and R. V. Decareau, "Ovens" in E. C. Okress, ed., *Microwave Power Engineering*, Vol. 2, Academic Press, New York, 1968.
66. *J. Microwave Power* **10**(4), 333 (1975).
67. J. S. Burgess, "Focused Microwave Energy" in E. C. Okress, ed., *Microwave Power Engineering*, Vol. 2, Academic Press, New York, 1968.
68. G. Kantor and T. C. Cetas, *Radio Science* **12**(6s), 111 (Nov.–Dec. 1977).
69. A. W. Guy, J. F. Lehmann, J. B. Stonebridge, and C. C. Sorenson, *IEEE Trans.* **MTT-26**, 550 (Aug. 1978).
70. J. K. White, *J. Microwave Power* **5**(2), 145 (1970).
71. W. T. Berntsen and B. D. David, *Microwave Eng. Appl. Newsletter* **8**(4), 3 (1975).
72. A. Hiratsuka, H. Inoue, and T. Takagi, *J. Microwave Power* **13**(2), 189 (June 1978).
73. D. A. Copson, *Microwave Heating*, 2nd ed., Avi Publishing Co., Westport, Conn., 1975, Chapt. 18.

74. T. C. Rozzell and co-workers, *J. Microwave Power* **9**, 241 (Sept. 1974).
75. D. A. Christensen, *J. Biomed. Eng.* **1**, 541 (1977).
76. K. A. Wickersheim and R. B. Alves, *Ind. Res./Dev.* **21**, 82 (Dec. 1979).
77. R. R. Bowman, *IEEE Trans.* **MTT-24**, 43 (1976).
78. H. Bassen, P. Herchenroeder, A. Cheung, and S. Neuder, *Radio Sci.* **12**(6), 15 (Nov.–Dec. 1977).
79. C. W. Brice, *IEEE Trans.* **IA-15**(3), 319 (May/June 1979).
80. T. Ohlsson and P. O. Risman, *J. Microwave Power* **13**, 303 (Dec. 1978).
81. *J. Microwave Power* **13**(1), (Mar. 1978).
82. H. Barber and J. E. Harry, *Proc. IEE.* **126**, 1126 (Nov. 1979).
83. J. A. Jolly, *J. Microwave Power* **11**, 233 (1976).
84. T. K. Ishii, *J. Microwave Power* **9**, 355 (Dec. 1974).
85. Y. Kase and K. Ogura, *J. Microwave Power*, **13**(2), 115 (1978).
86. W. Beust and W. L. Ford, *Microwave J.*, 91 (Oct. 1961).
87. Y. P. Raizer, *Soviet Phys. JETP* **34**, 114 (Jan. 1972).
88. H. W. Kritikos and H. P. Schwan, *IEEE Trans. Biomed. Eng.* **BME-22**, 457 (1975).
89. J. C. Lin, A. W. Guy, and G. H. Kraft, *J. Microwave Power* **8**, 275 (1973).
90. W. E. Ours, *Proc. 1971 IEEE EMC Symposium*, pp. 4–7.
91. A. S. McLachlan, *Rad. Elec. Engr.* **46**, 267 (June 1976).
92. M. Skolnik, *Radar Handbook*, McGraw-Hill, New York, 1970, pp. 27-24 to 29-27.
93. W. H. Walter, III and co-workers, *J. Am. Med. Assoc.* **224**, 1628 (1973).
94. J. C. Mitchell and W. D. Hurt, *The Biological Significance of Radiofrequency Radiation Emission on Cardiac Pacemaker Performance*, Report SAM-TR 76-4, USAF School of Aerospace Medicine, San Antonio, Tex., 1976.
95. R. Reis, *Bull. NY Acad. Med.* **55**, 1216 (Dec. 1979).
96. *Radio Electron. Eng.* **49**(6), 264 (1979).
97. *Safety Guide for the Prevention of Radio-Frequency Hazards in the Use of Electric Blasting Caps*, ANSI C95.4, 1971.
98. *Brit. Med. J.*, (July 12, 1980).
99. S. M. Michaelson, *Proc. IEEE* **68**, 40 (Jan. 1980).
100. *Safety Level of Electromagnetic Radiation with Respect to Personnel*, ANSI C95.1 Standard (original version 1966, Revised 1974, 1980).
101. P. Czerski, *Abstracts for Papers, 146th National Meeting AAAS*, Jan. 3–8, 1980, p. 65.
102. "Performance Standard for Microwave Ovens," from Part 1030 of *Regulations for Administration and Enforcement of the Radiation Control for Health and Safety Act of 1968*, U.S. Dept. of HHS/FDA, DHHS Publication No. FDA 76-8035, Jan. 1976.
103. *IMPI Performance Standard on Leakage from Industrial Microwave Systems*, IMPI Publication IS-1, Aug. 1973.
104. J. M. Osepchuk, R. A. Foerstner, and D. R. McConnell, "Computation of Personnel Exposure in Microwave Leakage Fields and Comparison with Personnel Exposure Standards," *Digest of the 1973 Microwave Power Symposium*, IMPI, New York, 1973.
105. J. M. Osepchuk, J. E. Simpson, and R. A. Foerstner, *J. Microwave Power* **8**, 295 (1973).
106. *Ferrite Absorber*, TDK Electronics Co., Ltd., Japan, 1973.
107. T. Y. Otoshi, *IEEE Trans.* **MTT-20**, 235 (Mar. 1972).
108. J. M. Osepchuk, *J. Microwave Power* **13**(1), 13 (Mar. 1978).
109. *Proc. IEEE* **62**, (Jan. 1974).
110. H. P. Schwan, *Ann. N.Y. Acad. Sci.* **125**, 344 (Oct. 1965).
111. S. A. Goldblith, ed., "Principles and Applications of Radiofrequency Energy to Food Preservation with Particular Reference to Concentration and Dehydration," *Freeze Drying Advanced Food Technology*, Academic Press, London, 1975.
112. B. D. Roebuck, S. A. Goldblith, and W. B. Westphal, *J. Food Sci.* **37**(2), 199 (1972).
113. *In the Matter of Allocation of the 10,500 to 10,700 MHz Band for the Use of Microwave Ovens*, Litton Microwave Cooking Products; Petition to FCC for Rule Making, Nov. 1976.
114. L. K. H. Van Beek, "Dielectric Behavior of Heterogeneous Systems" in J. B. Birk, ed., *Progress in Dielectrics*, Vol. 7, CRC Press, Cleveland, Oh., 1967.
115. S. M. Bakanowski and L. H. Belden, *Proc. 1979 Mic. Power Symposium*, IMPI, New York, p. 14.
116. K. Sato and co-workers, *Proc. 1979 Mic. Power Symposium*, IMPI, New York, pp. 14–16.
117. S. Nagamoto and co-workers, *Proc. 1979 Microwave Power Symposium*, IMPI, New York, pp. 17–19.

118. D. Winstead, *J. Microwave Power* **13**(1), 7 (Mar. 1978).
119. R. J. Boutin, *J. Microwave Power* **13**(1), 47 (Mar. 1978).
120. R. F. Bowen and G. Freedman, *Proc. 1980 Mic. Power Symposium*, IMPI, New York, pp. 37–39.
121. H. J. Lehmann, *Proc. 1979 Mic. Power Symposium*, IMPI, New York, pp. 15–16.
122. R. W. Dickerson, Jr. "Thermal Properties of Foods" in *The Freezing Preservation of Foods*, 4th ed., Vol. 2, Avi Publishing Co., Westport, Conn., 1967, Chapt. 2.
123. N. E. Bengtsson and T. Ohlsson, *Proc. IEEE* **62**, 44 (Jan. 1974).
124. J. W. Gould and E. M. Kenyon, *J. Microwave Power* **6**(2), 151 (1971).
125. A. F. Bezanson, *Food Technol.*, 34 (Dec. 1976).
126. M. N. Meisel, *Microwave Energy Appl. Newsletter* **5**(3), 3 (1972).
127. A. G. Sorbier and co-workers, *Ann. Nutr. Aliment* **32**(2–3), 437 (1978).
128. Jorg Augustin and co-workers, *J. Food Sci.* **45,** 814 (1980).
129. K. Ikawa and co-workers, *Hokusuishi Geppo*, **33**(9), 17 (1976).
130. C. Doty and C. W. Baker, *J. Agric. Food Chem.* **25,** 815 (1977).
131. N. A. Dunn and J. L. Heath, *J. Food Sci.* **44,** 339 (1979).
132. Sh. A. Abramov, *Pishch. Prom-st. Ser.* **1**(1), 10 (1980).
133. Jpn. Kokai 80 29,982 (Mar. 3, 1980), K. Ogawa.
134. Ya. D. Kadaner, *Elektron. Obrab. Mater.* (4), 69 (1976).
135. J. M. Mendelsohn and co-workers, *Fishery Ind. Res.* **4,** 241 (1969).
136. S. A. Goldblith and R. V. Decareau, *An Annotated Bibliography on Microwaves: Their Properties, Production, and Applications to Food Processing*, MIT Press, Cambridge, Mass., 1973.
137. S. Licht, ed., *Therapeutic Heat and Cold*, Licht, New Haven, Conn., 1965.
138. J. F. Lehmann, "Diathermy" in F. H. Krusen, F. J. Kottke, and P. M. Elwood, eds., *Handbook of Physical Medicine and Rehabilitation*, Saunders, Philadelphia, Pa., 1971.
139. A. W. Guy, J. F. Lehmann, and J. B. Stonebridge, *Proc. IEEE* **62**, 55 (Jan. 1974).
140. *IEEE Trans.* **MTT-26**(8), (Aug. 1978).
141. C. C. Johnson and A. W. Guy, *Proc. IEEE* **60**, 692 (June 1972).
142. A. E. Barrett, P. C. Myers, and N. L. Sadowsky, *Radio Sci.* **12**(6S), 1675–1715 (1977).
143. M. F. Iskander and C. H. Durney, *Proc. IEEE* **68**, 126 (Jan. 1980).
144. J. H. Jacobi and L. E. Larsen, *Med. Phys.* **7**(1), 1 (Jan./Feb. 1980).
145. R. V. Rajotte, J. B. Dossetor, W. A. G. Voss, and C. R. Stiller, *Proc. IEEE* **62**, 76 (Jan. 1974).
146. W. B. Stavinoha, *Psychopharmacol. Bull.* **15**(4), 63 (1979).
147. T. Moroji and co-workers, *J. Microwave Power* **12**, 273 (1977).
148. J. C. Lin, *Proc. IEEE* **68**, 67 (Jan. 1980).
149. W. R. Adey, *Proc. IEEE* **68**, 119 (Jan. 1980).
150. N. D. Devyatkov, *Radiotekh. Elektron.* **23**(9), 1882 (1978).
151. H. Fröhlich, "The Biological Effects of Microwaves and Related Questions," L. Marton and C. Marton, eds., *Advances in Electronics and Electronic Physics*, Vol. 53, Academic Press, New York, 1980.
152. *Bull. NY Acad. Med.* **55**(11), (Dec. 1979).
153. *Proc. IEEE* **68**(1), (Jan. 1980).
154. R. V. Lechowich, L. R. Beuchat, K. I. Fox, and F. H. Webster, *Appl. Microbiol.* **17**(1), 106 (1969).
155. G. R. Hooper and co-workers, *J. Am. Soc. Hortic. Sci.* **103**(2), 173 (1978).
156. K. J. Lessman and A. W. Kirleis, *Crop Sci.* **19**(2), 189 (1979).
157. S. O. Nelson and L. E. Stetson, *IEEE Trans.* **MTT-22**, 1303 (Dec. 1974).
158. A. F. Readdy, Jr., *Plastics Fabrication by Ultraviolet Infrared, Induction, Dielectric and Microwave Radiation Methods*, PLASTEC Report R43, Ricatinny Arsenal, Dover, N.J., Apr. 1976.
159. L. K. Wilson and J. P. Salerno, *Gov. Rep. Announce Index (U.S.)* **79**(18), 133 (1979).
160. K. Matsuda, *Kagoshima Daigaku Kyoikugakubu Kenkyu Kiyo Shizen Kagaku Hen* **30,** 61 (1979).
161. J. Gagliani and D. E. Supkis, *Adv. Astronaut. Sci.* **38**(1), 193 (1979).
162. K. Saito, *Nippon Setchaku Kyokaishi* **14,** 347 (1978).
163. K. Matsuda and co-workers, *Kagoshima Daigaku Kyoikugakubu Kenkyu Kiyo, Shizen Kagaku Hen* **27,** 59 (1975); K. Matsuda, *ibid.* **27,** 69 (1975).
164. P. D. Francis, *Measurements at 2.45 GHz of the Loss Factor and Energy Deposition in Thermosetting Metal Coating Lacquers*, Electr. Counc. Res. Cent., (Memo), ECRC/M1121, 1978.
165. H. Jullien and co-workers, *Angew. Makromol. Chem.* **62**, 241 (1977).
166. K. Ohshima, *Toso To Toryo*, **236,** 65 (1974).
167. M. Manoury, *Teintex* **45**(3), 7 (1980).

168. K. Furuya and co-workers, *Sen'i Kobunshi Zairyo Kenkyusho Kenkyu Happyokai Sanko Shiryo* **51,** 55 (1976).

169. A. B. Pepperman, Jr. and co-workers, *Text. Chem. Color.* **9**(7), 137 (1977).

170. J. J. Fischer and co-workers, *Proc. Spec. Sess. Cotton Dust Res. Beltwide Cotton Prod. Res. Conf.*, 3rd, 1979, pp. 13–14.

171. H. L. Needles and co-workers, *Book Pap., Nat'l Tech. Conf., AATCC,* 1977, pp. 201–207.

172. R. S. Berns and H. L. Needles, *J. Soc. Dyers Colour* **95,** 207 (1979).

173. A. Metaxas and co-workers, *J. Microwave Power* **13,** 341 (1978).

174. J. Chabert and co-workers, *Ind. Text* **1081,** 533 (1978).

175. J. Crowley and J. Apelbaum, *Electronic Progress*, Vol. XVIII, Raytheon Co., Lexington, Mass., 1976, pp. 13–18.

176. A. R. H. Tawn, *Farbe Lack* **80,** 416 (1974).

177. N. Anderson and co-workers, *Sv. Papperstidn.* **75,** 663 (1972).

178. R. Morrow, *Australas. Conf. Heat Mass Transfer, 2nd,* 1977, pp. 437–444.

179. E. P. Ertsgaard, *Res. Discl.* **154,** 10 (1977).

180. D. J. Fischer, *Adv. Chem. Ser.* **164,** 124 (1977).

181. R. M. Perkin, *J. Sep. Process Technol.* **1**(1), 14 (1979).

182. *Microwaves* **9**(9), 44 (Sept. 1970).

183. E. J. Zachariah and co-workers, *Indian J. Pure Appl. Phys.* **18**(3), 216 (1980).

184. H. F. Schwarz and co-workers, *J. Microwave Power* **8**(3–4), 303 (1973).

185. U. Schindler, *Elastomerics* **110**(5), 42 (1978).

186. J. Ippen, *Plastichem* **9,** 93 (1979).

187. S. A. Sofratec, *Caucho* **115,** 68 (1979).

188. P. D. Zavitsanos, *Gov. Rep. Announce. Index* (*U.S.*) **78**(26), 106 (1978).

189. Jpn. Kokai 72 25,206 (Oct. 19, 1972), S. Tsutsumi and T. Takihara (to Sanyo Electric Trading Co. Ltd.).

190. G. Y. Contos and co-workers, *Gov. Rep. Announce. Index* (*U.S.*) **79**(8), 65 (1979).

191. S. J. Priebe and co-workers, *INIS Atomindex* **10**(19), Abstr. No. 480850 (1979).

192. L. J. Bailin, *Gov. Rep. Announce. Index* (*U.S.*) **28**(25), 111 (1978).

193. U.S. Pat. 4,117,220 (Sept. 26, 1978), C. H. Worman, Jr. (to Air Products & Chemicals, Inc.).

194. J. P. Wightman, *Proc. IEEE* **62,** 4 (Jan. 1974).

195. J. Asmussen, Jr., R. Mallararpu, J. R. Hamann, and H. C. Park, *Proc. IEEE* **62,** 109 (Jan. 1974).

196. R. G. Bosisio, C. F. Weissfloch, and M. C. Westheimer, *J. Microwave Power* **7,** 325 (Dec. 1972).

197. M. Moisan, R. Pantel, J. Habert, and E. Bloyet, *J. Microwave Power* **14**(1), 57 (Mar. 1979).

198. G. Loncar and co-workers, *Czech. J. Phys. B* **30,** 688 (1980).

199. M. Shibagaki and co-workers, *Proc. Conf. Solid State Devices* **9,** 215 (1978).

200. V. N. Troitskii and co-workers, *Fiz. Khim. Obrab. Mater* (2), 173 (1979).

201. M. Locqueneux and co-workers, *Conf. Proc. Int. Symp. Plasma Chem., 4th,* Vol. 2, 1979, pp. 366–371.

202. S. Suzuki and co-workers, *Nippon Kagaku Kaishi* (7), 1037 (1978).

203. D. B. Batchelor, *INIS Atomindex* **10**(6), Abstr. No. 434525 (1979).

204. K. G. Handy and J. E. Brandelik, *J. Appl. Phys.* **49,** 3753 (1978).

205. M. A. V. Ward and co-workers, *J. Microwave Power* **12**(3), 187 (1977).

206. E. T. Wall and co-workers, *Adv. Chem. Ser.* **183,** 329 (1979).

207. *EOS* **61**(4), 33 (Jan. 22, 1980).

208. C. L.J. Hu, *IEEE Trans.* **MTT-27,** 38 (Jan. 1979).

209. J. L. Cambon and co-workers, *Can. J. Chem. Eng.* **56,** 735 (1978).

210. P. A. Haas, *Am. Ceram. Soc. Bull.* **58**(9), 873 (1979).

211. P. Colomban and J. C. Badod, *Ind. Ceram.* **725,** 101 (1979).

212. H. Makowitz and T. Ginsberg, *Trans. Am. Nucl. Soc.* **34,** 548 (1980).

213. W. E. Kallenberger and R. M. Lollar, *J. Am. Leather Chem. Assoc.* **74,** 458 (1979).

214. S. Yamaguchi and co-workers, *J. Am. Oil Chem. Soc.* **54,** 539 (1977).

215. P. E. Glaser, *Science* **162,** 857 (Nov. 22, 1968).

216. W. C. Brown, *Proc. IEEE* **62,** 11 (Jan. 1974).

217. V. A. Vanke, V. M. Lopubkin, and V. L. Savvin, *Sov. Phys. Uspekhi* **20,** 989 (Dec. 1977).

General References

Refs. 3–4, 12, 16–21, 23–24, 35, 37–39, 52–54, 114, and 136–137 are also general references.

Journals

Journal of Microwave Power.
Microwaves.
Microwave Journal.
Microwave System News.

JOHN M. OSEPCHUK
Raytheon Company

MILK AND MILK PRODUCTS

Milk has served as a food for humans since the beginning of recorded history. Before urbanization each family depended on its own animals for milk. Later, dairy farms were developed close to the cities. The milk industry became a commercial enterprise when methods for the preservation of fluid milk were introduced. Like many other early industries, small in size, large in number, and associated with a local production area, the dairy industry developed into large production units with large processing plants, often far from the production areas. The success in evolution of the dairy industry from small to large units of production, ie, the farm to the dairy plant, depended on sanitation of product and equipment; cooling facilities; health standards for animals and workers; transportation systems; construction materials for process machinery and product containers; pasteurization methods; containers for distribution; and refrigeration for products in stores and homes.

Composition and Properties

Milk consists of 85 to 89% water and 11 to 15% total solids (Table 1). The latter comprises solids-not-fat (SNF) and fat. Milk with a higher fat content also has higher solids-not-fat content with an increase of SNF of 0.4% for each 1% fat increase. The principal components of solids-not-fat are protein, lactose, and minerals (ash). The fat content and other constituents of the milk varies with the species, and for the milk cow, with the breed. Likewise, the composition of milk varies with feed, stage of lac-

Table 1.   Constituents of Milk from Various Mammals, Average, wt %

Species	Water	Fat	Protein	Lactose	Ash	Nonfat solids	Total solids
human	87.4	3.75	1.63	6.98	0.21	8.82	12.57
cows							
Holstein	88.1	3.44	3.11	4.61	0.71	8.43	11.87
Ayrshire	87.4	3.93	3.47	4.48	0.73	8.68	12.61
Brown Swiss	87.3	3.97	3.37	4.63	0.72	8.72	12.69
Guernsey	86.4	4.5	3.6	4.79	0.75	9.14	13.64
Jersey	85.6	5.15	3.7	4.75	0.74	9.19	14.34
goat	87.0	4.25	3.52	4.27	0.86	8.65	12.90
buffalo (India)	82.76	7.38	3.6	5.48	0.78	9.86	17.24
camel	87.61	5.38	2.98	3.26	0.70	6.94	12.32
mare	89.04	1.59	2.69	6.14	0.51	9.34	10.93
ass	89.03	2.53	2.01	6.07	0.41	8.49	11.02
reindeer	63.3	22.46	10.3	2.50	1.44	14.24	36.70

tation, health of animal, udder position at withdrawal, and seasonal and environmental conditions.

The nonfat solids, fat solids, and moisture relationships are well established and can be used as a basis for detecting adulteration with water.

The physical properties of milk are given in Table 2.

Table 2.   Physical Properties of Milk

Property	Value
density at 20°C of milk with 3–5% fat, average, g/cm^3	1.032
weight at 20°C, kg/L[a]	1.03
density at 20°C of milk serum, 0.025% fat, g/cm^3	1.035
weight at 20°C of milk serum, 0.025% fat, kg/L[a]	1.03
freezing point, °C	−0.540
boiling point, °C	100.17
maximum density at °C	−5.2
electrical conductivity, S ($= \Omega^{-1}$)	$(45–48) \times 10^{-8}$
specific heat at 15°C, kJ/(kg·K)[b]	
skim milk	3.94
whole milk	3.92
40% cream	3.22
fat	1.95
relative volume 4% milk at 20°C = 1, volume at 25°C	1.002
40% cream 20°C = 1.0010 at 25°C	1.0065
viscosity at 20°C, mPa·s (= cP)	
skim milk	1.5
whole milk	2.0
whey	1.2
surface tension of whole milk at 20°C, mN/m (= dyn/cm)	50
acidity, pH	6.3 to 6.9
titratable acid, %	0.12 to 0.15
refractive index at 20°C	1.3440–1.3485

[a] To convert kg/L to lb/gal, multiply by 8.34.

[b] To convert kJ/(kg·K) to Btu/(lb·°F), divide by 4.183.

Nutritional Content. To assure that milk provides the necessary nutrients, milk may be fortified with vitamins. Vitamin D milk has been sold since the 1920s. The milk was fortified with vitamin D by irradiation or by feeding irradiated yeast to the cows. Ergosterol is converted to vitamin D by uv irradiation. Vitamin D is now added directly to the milk to provide 400 USP units per liter. Vitamin A may be added to low fat or skim milk to provide 1000 RE (retinol equivalents) per liter. Multivitamin, mineral-fortified milk is available to meet the recommended daily requirements. The vitamin content of milk from various mammals is given in Table 3 (see Vitamins). The daily nutritional needs for an adult, and the constituents of milk are given in Table 4.

Fat. Milk fat is a mixture of triglycerides and diglycerides (see Fats and fatty oils). The triglycerides are short chain, C_{24}–C_{46}; medium chain, C_{34}–C_{54}; and long chain, C_{40}–C_{60}. Milk fat contains more fatty acids than vegetable fats. In addition to being classified according to the number of carbon atoms, the fatty acids in milk may be classified as saturated and unsaturated as well as soluble and insoluble. The fat carries

Table 3.   Vitamin Content of Milk from Various Mammals, mg/L [a]

Species	A, RE[b]	B_6	B_{12}	C	Thiamine	Riboflavin	Nicotinic acid	Panto-thenic acid	Biotin	Folic acid
cow	312	0.48	0.0056	16	0.42	1.57	0.85	3.50	0.035	0.0023
goat	415	0.07	0.0006	15	0.40	1.84	1.87	3.44	0.039	0.0024
sheep	292		0.0064	43	0.69	3.82	4.27	3.64	0.093	0.0024
horse	160	0.21	0.0012	100	0.30	0.33	0.58	3.02	0.022	0.0012
human	380	0.10	0.0003	43	0.16	0.36	1.47	1.84	0.008	0.0020
pig	207	0.40	0.0016	140	0.70	2.21	8.35	5.28	0.014	0.0039
whale	1439	1.10	0.0085	70	1.16	0.96	20.40	13.10	0.050	

[a] Ref. 1.
[b] Vitamin A is reported as retinol equivalents per liter (RE = 5 IU = 1.25 μg vitamin A or 3 μg β carotene).

Table 4.   Nutritional Content (for Adults) of Cow Milk [a]

Nutrient	Recommended daily allowance	Supplied by 1 L, %
energy, kJ (kcal) [b]	11,720 (2,800)	96 (23)
protein, g	56	49
calcium, g	0.8	155
phosphorus, g	0.8	115
iron, mg	10	4.5
vitamin A, RE [c]	1,000	31
thiamine, mg	1.4	30
riboflavin, mg	1.7	92
niacin, mg	18.5	5
ascorbic acid, mg	60	27
vitamin D, IU	200	200 [d]

[a] Ref. 2.
[b] To convert kJ to kcal, divide by 4.184; food calorie = 1 kcal.
[c] RE, retinol equivalent, is the new standard for vitamin A; 1 RE = 5 IU.
[d] Fortified milk.

numerous lipids (see Table 5), and vitamins A, D, E, and K, which are fat soluble. Tables 6–8 give the acids content of milk fat.

Milk is an emulsion of fat in water (serum). The emulsion is stabilized by phospholipids which are absorbed on the fat globules. In treatments, such as homogenization and churning, the emulsion is broken.

Processing

The processing operations for fluid milk or manufactured milk products include centrifugal sediment removal and cream separation, pasteurization and sterilization, homogenization, and packaging, handling, and storing.

Table 5. Composition of Lipids in Cow Milk[a]

Class of lipid	Range of occurrence
triglycerides of fatty acids, %	97.0–98.0
diglycerides, %	0.25–0.48
monoglycerides, %	0.016–0.038
keto acid glycerides, %	0.85–1.28
aldehydrogenic glycerides, %	0.011–0.015
glyceryl ethers, %	0.011–0.023
free fatty acids, %	0.10–0.44
phospholipids, %	0.2–1.0
cerebrosides, %	0.013–0.066
sterols, %	0.22–0.41
free neutral carbonyls, ppm	0.1–0.8
squalene, ppm	70
carotenoids, ppm	7–9
vitamin A, ppm	6–9
vitamin D, ppm	0.0085–0.021
vitamin E, ppm	24
vitamin K, ppm	1

[a] Ref. 3.

Table 6. Fatty Acids in Samples of Milk Fat for Cows Fed Normal Rations

Fatty acid	Acid content[a]	
	Range	Average
butyric (4:0)[b]	2.4–4.23	2.93
hexanoic (6:0)	1.29–2.40	1.90
octanoic (8:0)	0.53–1.04	0.79
decanoic (10:0)	1.19–2.01	1.57
lauric (12:0)	4.53–7.69	5.84
myristic (14:0)	15.56–22.62	19.78
oleic (18:1)	25.27–40.31	31.90
palmitic (16:0)	5.78–29.0	15.17
stearic (18:0)	7.80–20.37	14.91

[a] Percent of total acids.
[b] A shorthand designation for fatty acids is used in this table. For example, 18:0 = saturated C_{18}; 18:1 = C_{18} acid with one double bond; 18:2 = C_{18} acid with two double bonds; 18:0 br = branched-chain saturated C_{18} acid; etc (see Carboxylic acids).

Table 7. Saturated Acids as % of Total Acids of Milk Fat[a]

Even		Odd	
Acid[b]	%	Acid[b]	%
4:0	2.79	5:0	0.01
6:0	2.34	7:0	0.02
8:0	1.06	9:0	0.03
10:0	3.04	11:0	0.03
12:0	2.87	13:0	0.06
14:0	8.94	13:0 br	0.04
14:0 br	0.10	15:0	0.79
16:0	23.80	15:0 br A[c]	0.24
16:0 br	0.17	15:0 br B[c]	0.38
18:0	13.20	17:0	0.70
18:0 br	trace	17:0 br A[c]	0.35
20:0	0.28	17:0 br B[c]	0.25
20:0 br	trace	19:0	0.27
22:0	0.11	21:0	0.04
24:0	0.07	23:0	0.03
26:0	0.07	25:0	0.01

[a] Ref. 4.
[b] See footnote b, Table 6.
[c] A and B designate isomers.

Table 8. Unsaturated Acids as % of Total Acids of Milk Fat[a]

Even				Odd	
Acid	%	Acid	%	Acid	%
10:1[b]	0.27	20:2	0.05	15:1	0.07
12:1[c]	0.14	20:3	0.11	17:1	0.27
14:1[c]	0.76	20:4	0.14	19:1	0.06
16:1[d]	1.79	20:5	0.04	21:1	0.02
18:1[d]	29.60	22:1	0.03	23:1	0.03
18:2	2.11	22:2	0.01		
18:2 c,t conj.[e]	0.63	22:3	0.02		
18:2 t,t conj.[e]	0.09	22:4	0.05		
18:3	0.50	22:5	0.06		
18:3 conj.	0.01	24:1	0.01		
20:1	0.22				

[a] Ref. 4.
[b] Terminal double bond.
[c] Includes cis, trans, and terminal double-bond isomers.
[d] Includes cis and trans isomers.
[e] c,t = cis-trans isomer; t,t = trans-trans isomer; conj. = conjugated.

Cooling. After removal from the cow by a mechanical milking machine, usually at ca 34°C, the milk should be cooled as rapidly as possible to 4.4°C or below to maintain quality. At this low temperature, enzyme activity and growth of micro-organisms are minimized. Commercial dairy operations usually consist of a milking machine, a pipeline to convey the milk directly to the tank, and a refrigerated bulk milk tank in which the milk is cooled and stored. A meter may be in the line to measure the quantity of milk from each cow. Development of rancidity must be avoided by preventing air from passing through the warm milk as a result of excessive air flow, air leaks, and long risers in the pipeline. The pipelines, made of glass or stainless steel, are usually cleaned by a CIP (cleaning-in-place) process as described later.

Centrifugation. Centrifugal devices include clarifiers for removal of sediment and extraneous particulates and separators for removal of fat from milk (see Centrifugal separation). Modifications include a standardizing clarifier that removes fat to provide a certain fat content of the product while removing sediment; a clarifixator that partially homogenizes while separating the fat; and a high speed clarifier that removes bacteria cells in a bactofuge process (see below).

Clarification. Clarifiers have replaced filters in the dairy plant for removing sediment, although the milk may have been previously strained or filtered on the farm. A clarifier has a rotating bowl with conical disks between which the product is forced. The sediment is forced to the outside of the rotating bowl where the sludge or sediment remains. Some clarifiers have dislodging devices to flush out the accumulated material. The clarified milk leaves through a spout or outlet.

Clarification is usually performed at 4.4°C, although a wide range of temperatures is permitted. The clarifier may be used in numerous positions in the processing system, depending on the temperature, standardization procedure, flow rate, and use of the clarified product. The clarifier may be between the bulk-milk tanker and raw-milk storage tank; the raw-milk receiving tanks and raw-milk storage tank; the storage tank and standardizing tank; the standardizing tank and HTST pasteurizer; the preheater or regenerative heater for raw milk and the heating sections of the HTST pasteurizer; or the regenerative cooler for the pasteurized-milk side and the final cooling sections of the HTST pasteurizer (rarely because of possible postpasteurization contamination).

To avoid sediment following homogenization, the clarifier is used generally before homogenization. It clarifies the cold incoming raw milk. Clarification at this point provides a milk ready for pasteurization, particularly if standardized; permits longer operation of clarifier without stopping or cleaning, because sediment builds up more rapidly on a warm product; and when used as an operation independent from pasteurization, does not interfere with the pasteurization if maintenance is necessary.

Bactofugation. Bactofugation is a specialized process of clarification in which two high velocity centrifugal devices (bactofuges) operate at 20,000 rpm in series. The first device removes 90% of the bacteria. The second removes 90% of the remaining bacteria, providing a 99% bacteria-free product.

The milk is heated to 77°C to reduce the viscosity. There is a continous discharge from the centrifugal bowl of bacteria and some nonfat portion of the milk (1 to 1.5%), which is the constituent with the highest density.

This process is not used for ordinary fluid milk but for sterile milk or cheese.

Separation. Continuous-flow centrifugal cream separators using cone disks in a bowl were introduced in 1890. Originally, the cream separators were the basic plant equipment, and dairy plants were then known as creameries. Today's separators are pressure- or force-fed sealed air-tight units, in contrast to the original gravity-fed units that incorporated air to produce foam. Separators develop 5,000–10,000 times the force of gravity to separate the fat (cream) from the milk. Originally the remaining skim milk was discarded or returned to the farm as animal feed. The cream was used for butter and other fat-based dairy products. Today, the separator removes all or a portion of the fat and the skim milk or reduced-fat milk is sold as a beverage.

Separation is performed between 32°C and 38°C; temperatures as high as 71°C are acceptable. Cold milk separators, which have less capacity at the lower temperatures, may be used in processing systems in which the milk is not heated.

In the separation of fat globules from milk serum, the action is proportional to the difference in densities, the square of the radius of the fat globule, and the centrifugal force; and inversely proportional to the resistance to flow of the fat globule in the serum, the viscosity of the product through which the fat globule must pass, and the speed of flow through the separator.

The ease of the separated products leaving the bowl determines the richness of the fat. The fluid whole milk enters the separator at the center under pressure with a positive displacement pump or centrifugal pump with flow control (see Fig. 1). The fat is separated and moves toward the center of the bowl, while the skim milk passes to the outer space. There are two spouts or outlets, one for cream and one for skim milk. Cream, a mixture of fat and milk serum, leaves the center of the bowl, with the percentage of fat, usually 30–40%, controlled by the adjustment of either a cream screw or skim milk screw. This screw is a type of valve that controls the flow of the products leaving the field of centrifugal force and thus affects the separation.

Standardization. Standardization is the process of adjusting the ratio of butterfat and solids-not-fat to meet legal or industry standards. Adding cream of a high butterfat into the serum of low butterfat milk might result in a product with low SNF. Thus, careful control must be exercised.

A standardizing clarifier and a separator are employed equipped with two discharge spouts. The higher fat product is removed at the center and the lower fat product at the outside. The standardizing clarifier removes sediment and a smaller portion of the fat than the conventional separator which leaves only 0.25% of the fat behind. The fat in the milk discharge of a standardizing clarifier is only slightly less than that of the entering milk; the reduction is ca 10% from 4.0 to 3.6% fat. Accurate standardization is performed by sampling a storage tank of milk and adding appropriate fat or solids, or by putting the product through a standardizing clarifier and then into a tank for adjustment of contents.

Homogenization. Homogenization is the process by which a mixture of components is treated mechanically to give a uniform product that does not separate. In milk, the fat globules are broken up into small particles that form a more stable emulsion in the milk. In homogenized milk, the fat globules do not rise by gravity to form a creamline. The fat globules in raw milk are 1–15 μm in diameter; they are reduced to 1–2 μm by homogenization. The U.S. Public Health Service defines homogenized milk as "milk that has to be treated to insure the breakup of fat globule to such an extent

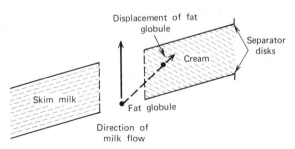

Figure 1. Diagrammatic representation of a fat separation in a centrifugal separator (5).

that, after 48 h of quiescent storage at 45°F [7°C] no visible cream separation occurs in the milk . . ." (6). Today, most fluid milk is homogenized. Homogenization is an integral part of the continuous HTST pasteurization process (see below).

Milk is homogenized in an homogenizer or viscolizer. The milk is forced at high pressure through small openings of a homogenizing valve formed by a valve or a seat or a disposable compressed stainless steel conical valve in the flow stream (see Fig. 2). The globules are broken up as a result of shearing, impingement on the wall adjacent to the valve, and perhaps to some extent by the effects of cavitation and explosion after the product passes through the valve. In a two-stage homogenizer, the first valve is at a pressure of 10.3–17.2 MPa (1500–2500 psi) and the second valve at ca 3.5 MPa (500 psi). The latter functions primarily to break-up clumps of homogenized fat particles, and is particularly applicable for cream and products with more than 6–8% fat.

A homogenizer is a high pressure, positive pump with 3, 5, or 7 pistons, driven by a motor, and equipped with adjustable homogenizing valves. Smoother flow and

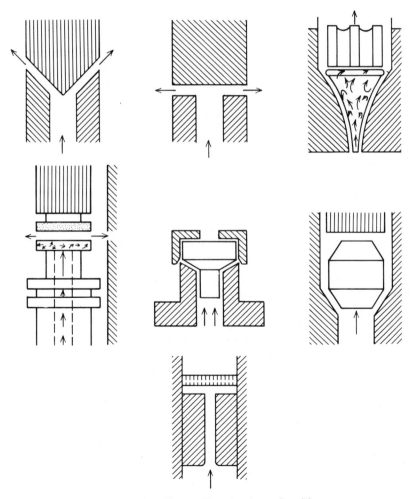

Figure 2.   Types of homogenizer valves (7).

greater capacity are obtained with more pistons. The pistons force the product into a chamber which feeds the valve. In design and operation, it is desirable to minimize the power requirements for obtaining an acceptable level of homogenization. At 17.2 MPa (2500 psi) and 0.91 metric ton per hour, a 56-kW (75-hp) motor is required.

The following factors have to be considered:

Before homogenization, milk must be heated to 54–60°C to break-up the fat globules, and prevent undesirable lipase activity.

It is difficult to separate the cream from homogenized milk to make butter.

As the temperature of the milk is increased, the size of globules decreases.

The viscosity of fluid milk is not greatly influenced by homogenization, whereas the viscosity of cream is increased.

Clarification before or after homogenization prevents the formation of sediment which otherwise adheres to the fat.

The homogenizer must be placed appropriately in the system to assure proper temperature of incoming product, provide for clarification, and avoid air incorporation which would cause excessive foaming. The homogenizer also may be used as a pump in the pasteurization circuit.

Pasteurization. Pasteurization is the process of heating milk to kill yeasts, molds, and pathogenic bacteria, and most other bacteria and to inactivate certain enzymes, without greatly altering the flavor. The principles were developed by and named after Louis Pasteur and his work on wine in 1860–1864 in France. Since then, stringent codes have been developed. The basic regulations are included in the U.S. Public Health Milk Ordinance (6) which has been adopted by most local and state jurisdictions. However, since the quality of milk depends on the care of the animals, the environment on the farm, and the care of the product throughout, pasteurization cannot substitute for quality.

Pasteurization may be carried out by batch or continuous-flow processes. In the batch process, each particle of milk must be heated to at least 62.8°C and held continuously at or above 62.8°C for at least 30 min. In the continuous process, the milk is heated to at least 71.7°C for at least 15 s. This is known as the HTST (high temperature short-time) pasteurization. For milk products with a fat content above that of milk or with added sweeteners, 65.6°C is required for the batch process and 74.4°C for the HTST process. For either method, following pasteurization, the product should be cooled quickly to 7.2°C or less. Other time–temperature relationships have been established for other products. Ice-cream mix is pasteurized at 79–81°C and held for at least 25 s.

Another continuous pasteurization process is known as ultrahigh temperature (UHT), in which shorter time, 1–2 s, and higher temperatures, 87–132°C, are employed. The UHT process approaches aseptic processing (Fig. 3).

The enzyme phosphatase is inactivated by pasteurization. Thus, the degree of pasteurization is determined by measuring the phosphatase present.

Batch Holding. The milk in the batch holding tanks is heated by a hot-water spray on the tank liner, a large-diameter coil which circulates in the milk through which the hot water is pumped, a flooded tank around which hot water or steam is circulated, or by coils surrounding the liner through which the heating medium is pumped at a high velocity. Table 9 gives the overall heat-transfer coefficients for these methods.

Design and operating features of batch units include:

Time–temperature exposure is recorded on a chart. The chart must be kept for

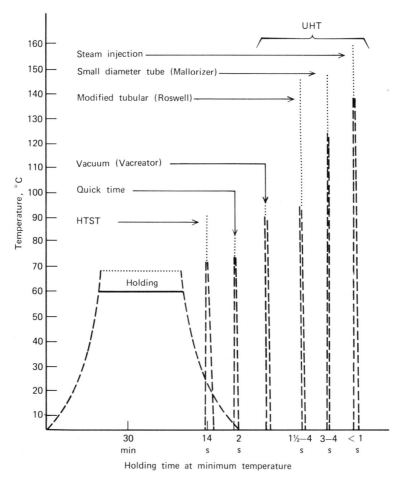

Figure 3. Pasteurization by various methods (8).

Table 9.   U-Values[a] for Holding Methods of Batch Pasteurization

Method	kW/(m²·K)[b]	Remarks
water-spray	0.350	heat from 10–63°C in 25 min, hot water at 71°C
coil vat	0.350	coils turns at 130 rpm, water through coil at 100 rpm
flooded	0.350	gravity circulation, agitator
high velocity	0.525	requires more energy to pump heating fluid

[a] Overall heat-transfer value.

[b] To convert kW/(m²·K) to Btu/(h·ft²·°F), multiply by 571.2.

proof of treatment. If the lid is opened, the exposure is interrupted and pasteurization must be started again.

A self-acting regulator closely controls the temperature of water which is usually heated with steam.

An air space heater ejects clean steam into the air space above the product and into the foam, maintaining a temperature at least 3°C above the holding temperature.

Valves are mounted so that the plug of the valve is flush with the tank to avoid a pocket of unpasteurized milk. Furthermore, a leak-detector valve permits drainage of the milk trapped in the plug of the valve.

Covers, piping, and tubing must drain away from the pasteurizer.

Agitators provide adequate mixing, without churning, assist in heat transfer by sweeping the milk over the heated surface, and help assure that all particles are properly pasteurized.

High Temperature Short-Time Pasteurizers. The principal continuous-flow process is the high temperature short-time method. The product is heated to at least 72°C and held at that temperature for not less than 15 s. Other features are nearly the same as in the holding method.

The following equipment is needed: balance tank, regenerative heating unit, positive pump, plates for heating to pasteurization temperature, tube or plates for holding for the specified time, flow-diversion valve (FDV), regenerative cooling unit, and cooling unit (see Fig. 4). Other devices often incorporated in the HTST circuit are clarifier, standardizing–clarifier, homogenizer, and booster pump.

The balance or float tank collects the milk entering the unit, receives the milk returned from the flow-diversion valve which has not been adequately heated, and maintains a uniform product elevation on the pasteurizer intake.

The heat-regeneration system partially heats the incoming cold product and partially cools the outgoing pasteurized product. The regenerator is a stainless steel plate heat exchanger, usually of the product-to-product type. The configuration is so arranged that the outgoing pasteurized product is at a higher pressure to avoid contamination. A pump in the circuit moves the milk from the raw milk side and the discharge to the final heater. Heat regenerators are usually 80–90% efficient. The regeneration efficiency may be improved by increasing the number of regenerator plates which increases the energy for pumping and cost of heat-exchanger plates.

The final heater increases regeneration temperature (ca 60°C) to pasteurization temperature (at least 72°C) with hot water. The hot water is ca 1–2°C above the highest product temperature (73°C). From 4 to 6 times as much hot water is circulated as the amount of product circulated on the opposite side of the plates.

The holder or holding tube is at the discharge of the heater. Its length and diameter assure that the product has been exposed to the minimum time–temperature, which for milk is 72°C for 15 s. Glass or stainless steel tubing, or plate heat exchangers may be used for holders.

On the outlet of the holder tube, the flow-diversion valve directs the pasteurized product to the regenerator and then to the final cooling section, known as forward flow. Alternatively, the product, which is below the temperature of pasteurization, is diverted back to the balance tank (diverted flow). The FDV is controlled by the safety thermal-limit recorder.

The final cooling section is usually a plate heat exchanger cooled by chilled water. Brine or compression refrigeration may be used. Milk leaves the regenerator and enters the cooler at ca 18–24°C and is cooled to 4.4°C by water circulating at 1°C. The relationship of regenerator, heater, and cooler for flow, number of plates, and pressure drop is given in Table 10.

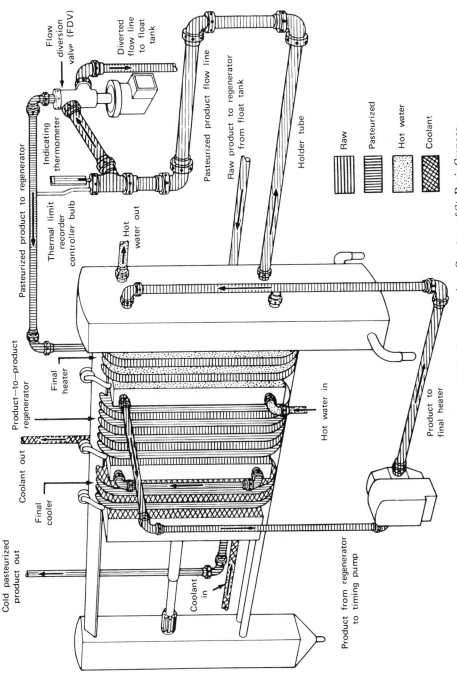

Figure 4. Flow through typical HTST plate pasteurizer. Courtesy of St. Regis Crepaco.

Flow diversion valve (FDV)

Diverted flow line to float tank

Indicating thermometer

Pasteurized product flow line

Raw product to regenerator from float tank

Holder tube

Pasteurized product to regenerator

Thermal limit recorder controller bulb

Hot water out

Raw

Pasteurized

Hot water

Coolant

Product—to—product regenerator

Final heater

Hot water in

Coolant out

Final cooler

Product to final heater

Cold pasteurized product out

Coolant in

Product from regenerator to timing pump

533

Table 10.   Representative Capacities of HTST Plate Pasteurizers[a]

Capacity, L/h	3,800	7,600	11,360	15,140	18,930
Regenerator, 84%; up temperature, 4–65°C;					
down temperature, 77–16°C					
plates, number	31	51	71	91	111
pressure drop milk, kPa	62	90	103	103	117
Heater, milk temperature, 65–77°C;					
water temperature, 79–77°C					
plates, number	9	15	21	29	33
water, L/min	261	522	587	787	492
pressure drop milk, kPa[b]	55	76	76	69	96
pressure drop water, kPa[b]	83	117	76	69	165
Cooler, milk temperature, 16–3°C;					
water temperature, 1–4°C					
plates, number	9	17	31	41	49
water, L/min	326	662	462	643	772
pressure drop milk, kPa[b]	55	55	117	117	145
pressure drop water, kPa[b]	131	131	165	165	179
Total, 84% regeneration					
plates, number	49	83	123	161	193
pressure drop, milk, kPa[b]	172	221	296	289	358
size of frame, m	1.22	1.52	1.83	2.13	2.13
Total, 90% regeneration					
plates, number	73	109	147	189	239
pressure drop, milk, kPa[b]	131	200	214	221	207
size of frame, m	1.52	1.83	1.83	2.13	2.44

[a] Courtesy of Crepaco, Inc.

[b] To convert kPa to mm Hg, multiply by 7.5.

The heat-transfer sections of the HTST pasteurizer, ie, regenerator, heater, and cooler, are usually stainless steel plates made of 302 series, with a dull 2D finish and ca 0.635 mm (0.025 in.) to 0.91 mm (0.036 in.) thick. The plates for the different sections are separated by a terminal that includes the piping connections to direct the product into and out of the plates. The plates hang on a support from above and can be moved, along with the terminals, for inspection or for closing the unit. A screw assembly can be operated, manually or mechanically, to hold the plates together during operation. The plates are mounted and connected in such a manner that the product can flow through ports connecting alternate plates. The heat-transfer medium flows between every other set of plates.

The pasteurized product drains from the unit avoiding pockets of unpasteurized or pasteurized products. The stainless steel plates are separated by nonabsorbant vulcanized gaskets, to maintain a space of ca 3 μm between plates. Various profiles and configurations provide a rapid, uniform heat-transfer surface for the plates. Raised knobs, crescents, channels, and diamonds, depending on the manufacturer, make up the plate surfaces.

During operation the plates must be pressed together to provide a seal, and mounted and connected in such a manner that the product drains from the plates without opening, and air is eliminated.

Various arrangements and configurations are available for the HTST pasteurizer.

For regeneration, the milk-to-milk regenerator is most common. A heat-transfer medium, usually water, provides a milk-to-water-to-milk system. Both sides may be closed (see Fig. 5) or the raw milk supply may be open.

A timing or metering pump provides a positive, fixed flow through the pasteurization system. A homogenizer or a rotary positive pump is used (Fig. 6). The pump is placed ahead of the heater and holding section. Various control drives assure that the pasteurized side of the heat exchanger is at a higher pressure than the opposite side.

The homogenizer can be used as a timing pump as it is homogenizing the product (Fig. 7). When both the timing pump and homogenizer are used in the same unit, appropriate connections and relief valves must be provided to permit the product to by-pass one unit if that unit is not operating.

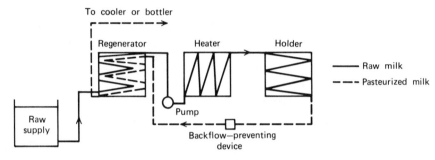

Figure 5. Milk-to-milk regenerator with both sides closed to atmosphere (9). Courtesy of U.S. Department of Health, Education, and Welfare.

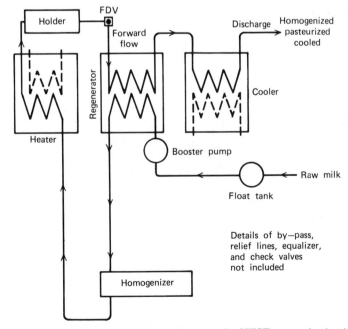

Figure 6. Homogenizer used as a timing pump for HTST pasteurization (10).

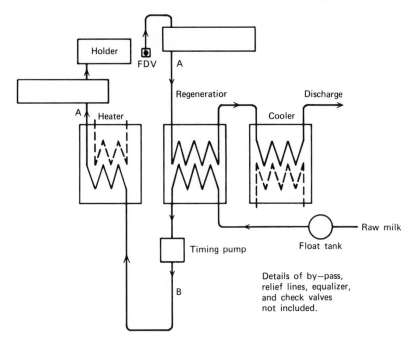

Figure 7. Homogenization of regenerated milk (10). A = after HTST heat treatment. B = before HTST pasteurization.

Booster Pump. Use of a centrifugal booster pump avoids a low intake pressure, particularly for large, high volume units. A low pressure of over 26.6 kPa (200 mm Hg) on the intake of a timing pump could cause vaporization of the product. The booster pump is in the circuit ahead of the timing pump, but not necessarily immediately preceding it (Fig. 8). The booster pump is connected to operate only when the FDV is in forward flow, the metering pump is in operation, and the pasteurized product is at least 7 kPa (1 psi) above the maximum pressure developed by the booster pump.

Clarifier. The incoming milk is clarified as it moves into storage. The clarifier also may be a part of the HTST circuit, usually placed ahead of the homogenizer. The milk flows from the regenerator (57°C), timing pump, clarifier, homogenizer, heater, etc. A standardizing clarifier is used in a similar manner.

Separator. The fat is normally separated from the milk before the HTST. However, in one system the air-tight separator is placed after the FDV, after the product is pasteurized. A restriction is placed in the line after the FDV to maintain a constant flow. Several control combinations ensure that flow is maintained, that vacuum does not develop in line, that the timing pump stops if the separator stops, and that the legal holding time is met.

Control System. For quality control, a complete record of the control and operation of the HTST is kept with a safety thermal-limit recorder–controller (Fig. 9). The temperature of product leaving the holder tube (ahead of the FDV) is recorded. The forward or diverted flow of the FDV is determined. Various visual indicators (green and red lights), operator temperature-calibration records, and thermometers also are provided.

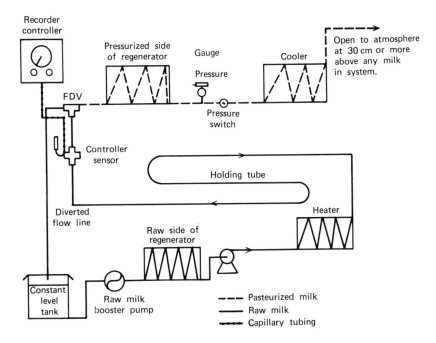

Figure 8.   Booster pump for milk-to-milk regeneration (11).

Utilities.   Electricity, water, steam refrigeration, and compressed air must be provided to the pasteurizer for water for heating, cooling, and cleaning. Water is heated with steam and cooled with a direct-expansion refrigeration system. It may be cooled directly or over an ice bank formed by direct-expansion refrigeration. The compressed air should be clean and relatively dry and supplied at about 138 kPa (20 psi) to operate the valves and controls.

The water is heated by steam injection or an enclosed heating and circulating unit. The controller, sensing the hot-water temperature, permits heating until the preset temperature is reached, usually 1–2°C above the pasteurization temperature.

A diaphragm valve, directed by the controller, maintains the maximum temperature of the hot water by control of the steam.

Other Heat Treatments.   Today, the trend is toward higher treatment temperatures usually with shorter times, approaching one second or less. The product has to be cooled to prevent deleterious effects. The various pasteurization heat treatments include quick time, vacuum treatment (vacreator), modified tubular, small diameter tube, and steam injection. The latter three methods are UHT processes (see Fig. 3).

Vacuum Treatment.   Milk may be exposed to a vacuum to remove low boiling substances. Certain weeds and feeds, such as onions, garlic, and some silage, may impart off-flavors to the milk, particularly the fat portion. These may be removed by vacuum treatment. A three-stage vacuum unit, known as a vacreator, produces pressures of 17, 51–68, and 88–95 kPa (127, 381–508, and 660–711 mm Hg). A vacuum can be incorporated in the HTST system. The continuous vacuum unit may consist of one or two chambers and be heated by live steam (with an equivalent release of water by evaporation) or flash steam carrying off the volatiles. If live steam is used, it must be

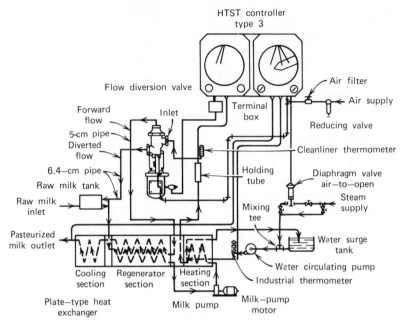

Figure 9.   HTST control system. Courtesy of Taylor Instrument Co.

culinary steam or steam from culinary water. Culinary steam is produced by heating potable water with an indirect heat exchanger. Dry saturated steam is desired for food-processing operations.

Pasteurization Unit.   Equivalent heat treatment for destruction of microorganisms or inactivation of enzymes can be represented by plotting the logarithm of the time versus the temperature. These relationships were originally developed for sterilization of food at 121.1°C. The time to destroy the microorganism is the F_0 value at 121.1°C or F_0 value at 250°F which, for the example shown in Figure 10, is 10 min. The slope of the curve is z, the temperature span is one log cycle, which for the example is 10°C. The heat treatment at 131°C for 1 min is equivalent to 121.1°C for 10 min. (Note: Nomenclature in Europe, and being considered by U.S. scientists and engineers, is to use F_0 at 120°C, with $z = 10$°C, as being more convenient).

The pasteurization unit, P.U., is based on similar relationships, except that the equivalent heat treatment is based on a lower temperature than sterilization (Fig. 11). The P.U. is based on a heat treatment at 60°C. A 1-min treatment is 1 P.U., where $z = 10$°C. The P.U. $= e^{0.23 (T - 60)}$, with T in °C.

The z value is related to the destruction of organisms or enzymes; larger values represent longer heat treatment; in SI units, z has a value of 5 to 7 for organisms and 11 to 28 for enzymes of milk. Pasteurization is measured on the basis of killing *Mycobacterium tuberculosis* and *Microccus* MS 102.

Irradiation.   Milk can be pasteurized or sterilized by beta rays produced by an electron accelerator or gamma rays produced by cobalt-60. The bacteria and enzymes in milk are more resistant to irradiation than higher life forms. For pasteurization, 5000–7500 Grey (Gy) (500,000–750,000 rad) are required and for inactivating enzymes at least 20,000 Gy (2,000,000 rad). A much lower level of irradiation, about 70 Gy (7000 rad), causes an off-flavor.

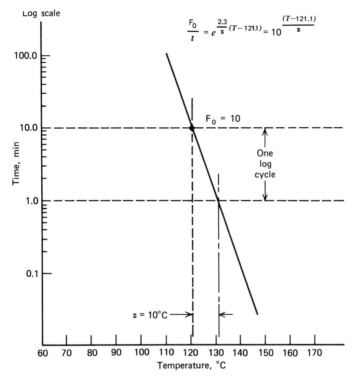

$$\frac{F_0}{t} = e^{\frac{2.3}{z}(T-121.1)} = 10^{\frac{(T-121.1)}{z}}$$

Figure 10. Representation of z and F-values (12). F_0 is the zero point for identifying the sterilization value at 121.1°C. In degrees Fahrenheit: $F_0/t = e^{2.3/z\,(T-250)} = 10^{(T-250)/z}$; and F_0 is the zero point for identifying the sterilization value at 250°F.

Equipment. *3A Sanitary Standards.* Equipment is designed according to 3A Sanitary Standards, which are established by the 3A Committee of users, manufacturers, and sanitarians in the food industry. It is the objective of the committee to provide interchangeability of parts and equipment, establish standards for inspection, and provide knowledge of acceptable design and materials, primarily to fulfill sanitary requirements. The features of sanitary equipment design are as follows:

Material of construction should be 18-8 stainless steel, with a carbon content not more than 0.12%. Equally corrosion-resistant material is acceptable.

The metal gauge for various applications is specified.

Surfaces fabricated from sheets shall have a No. 4 finish or equivalent.

Weld areas shall be substantially as corrosion resistant as the parent material, with an equivalent finish. Soldered joints are not acceptable.

Minimum radii are often specified; eg, for a storage tank, 0.62 cm for inside corners of permanent attachments; and for tank, 1.86 cm.

No threads shall be in contact with food.

Threads should be acme threads (acme threads are flat-headed instead of V-shaped).

Materials of Construction. *Stainless Steel.* An important development in materials for milk and dairy equipment during the past 25 years is the use of stainless steel (qv) for flat surfaces, tubing, coils, and castings. Previously metal-coated materials

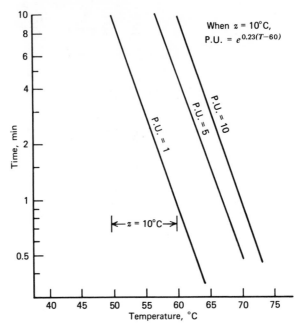

Figure 11. Representation of pasteurization unit, P.U. (13). For $z = 18°F$, P.U. $= e^{0.128\,(T-140)}$. For $z = 10°C$, P.U. $= e^{0.23(T-60)}$.

such as tinned copper were used for most applications and copper alloys were used for castings and fittings. Today, the contact surfaces of equipment with milk are primarily stainless steel, which permits cleaning-in-place (CIP), automation, continuous operations, and aseptic processing. Glass and glass-coated surfaces also have become popular.

Many types of stainless steels are available. The type most widely used in the dairy industry is 18-8 (18% chromium, 8% nickel, plus iron). Small amounts of silicon, molybdenum, manganese, carbon, sulfur, and phosphorus might be included to obtain characteristics desired for specific applications.

The most important series of stainless steel are the 200-, 300-, and 400-series (see Steel). The 300-series is used in the dairy industry, mostly the 302, 304, and 316, whereas the 400-series is used for special applications, such as pump impellers, plungers, cutting blades, scrapers, and bearings (see Table 11). Surface finishes are specified from No. 1 to No. 8 (highly polished); the No. 4 finish is most commonly used.

Stainless steel develops a passive protective layer of chromium oxide up to 5 nm thick. This layer must be maintained or be permitted to rebuild after it is removed by product flow or cleaning, since it protects the surface. The passive layer may be removed by electric current flow across the surface as a result of dissimilar metals being in contact. Corrosion may occur in welds, between dissimilar materials, at points under stress, and in places where the passive layer is removed. The creation of an electrolytic cell with subsequent current flow and corrosion has to be avoided in construction. Corrosion also is caused by food material, residues, cleaning solutions, and brushes.

Table 11. Stainless Steels Used in Food-Processing Equipment

Identification	Alloy content		Characteristics	Uses
	Chromium	Nickel		
300-Series[a]				
301	16–18 wt %	6–8 wt %	ductile; lower resistance to corrosion, particularly as temperature increases	
302	17–19 wt %	8 wt %	good corrosion resistance; can be cold worked and drawn; anneal following welding to avoid intergranular corrosion in corrosive environment	general purpose, used widely
304	18–20 wt %	8–12 wt %	better corrosion resistance than 302	most widely used for food
310	24–26 wt %	19–22 wt %	scale resisting properties at elevated temperatures	high temperature applications
316	16–18 wt %, 2–3% molybdenum	10–14 wt %	superior corrosion resistance of all the stainless steels	in contact with brine and various acids; gaining importance in food industry
400-Series[b]				
410	11.5–13.5 wt %, 0.15% carbon	0.15 wt %	basic martensitic alloy hardenable by heat treatment	roofing; siding; blades on freezers
416	12–14 wt % carbon	0.15 wt % carbon	easily machinable	valve stems, plugs, and gates
420	12–14 wt %		hardenable by heat treatment	cladding over steel; high spring temper
430	14–18 wt %		nonhardenable, good corrosion resistance	trim, structural, and decorative purposes
440	16–18 wt % carbon	0.60 wt % carbon	harder than others; generally not recommended for welding	pumps, plungers, gears, seal rings, cutlery, bearings

[a] Nonmagnetic or slightly magnetic.
[b] Magnetic.

Aluminum. Aluminum is also a widely used material for processing and handling equipment. For dairy equipment a high purity cast or aluminum sheet metal is used. Impurities increase susceptibility to corrosion, which is prevented by a protective oxide layer formed on the surface.

Cleaning. The equipment is cleaned to prevent contamination of subsequent dairy-processing operations and damage to the surface. In cleaning stainless steel, the surface contaminants are removed which otherwise destroy the protective passive layer. The surface is dried and exposed to air to rebuild the protective passive chromium oxide layer. Metal adhering to the stainless steel surface should be removed with the least abrasive material. After cleaning, the surface should be washed with hot water and left to dry. The equipment should be sanitized (often incorrectly called sterilized) with 200 ppm chlorine solution within 30 min before use, not necessarily after cleaning, to avoid corrosion resulting from chlorine on the surface for an extended period of time. For cleaning-in-place (CIP) (see below), the velocity of the cleaning solution over surfaces should be 1.5 m/s or less. Excessive velocities cause erosion of the surface and reduction of the protective layer. Excessive time of contact of the cleaning solution may cause corrosion, depending upon the strength of the cleaning solution.

Piping and Tubing.   Piping has thicker walls than tubing, and its size is designated by a nominal rather than an exact inside diameter (see Piping systems). Thus, a pipe of 2.5 cm diameter could have an inside diameter slightly more or less than 2.5 cm, depending upon the wall thickness. Tubing size is designated by the outside diameter. A tube of 2.5 cm diameter would have an outside diameter of 2.5 cm and as the thickness of the tubing increases the inside diameter decreases. Both piping and tubing have fixed outside diameters, and standard fittings can be used with different wall thicknesses.

The food industry uses stainless steel and glass tubing or piping extensively for moving food products. Conventional steel piping, cast iron, copper, plastic, glass, aluminum, and other alloys are used for utilities.

Most piping and tubing systems are designed for in-place cleaning. Classification is based on the type of connections for assembly: welded joints for permanent connections, ground joints with acme threads and hexagon nuts with gaskets for connections that are opened daily or periodically; and clamp-type joints.

Corrosion between the support device and the pipeline must be avoided. Drainage is provided by the pipeline slope, which is normally 0.48–0.96 cm/m of length. Gaskets must be nonabsorbent and of a type that does not affect the food product.

Fittings connect different pipes and provide for attachment of equipment or change of the flow direction. They must be easily cleaned inside and out, have no pipe threads exposed, and if of the detachable type, have an appropriate gasket. The fittings are constructed of the same or similar materials as the pipeline. For glass pipelines, stainless steel or other alloys can be used. Fittings are installed on tubing with a device that expands the tubing against the fitting to make a firm connection. This type has in general replaced soldered fittings. Welded connections are used for permanent installations. Standard shapes and sizes, specified by the 3A Standards Committee are made by many equipment manufacturers.

An air valve, sometimes called the air-activated valve, is now widely used for automated food-handling operations. These valves can be operated remotely. Although electronic or electric control boxes may be a part of the system, the valve itself generally is air activated, and is more reliable and positive than other types considering the usual moist environment. Air-operated valves are used for in-place cleaning systems, and transfer and flow control of various products. Air-operated valves may be piston- or diaphragm-operated compression valves, or operated with O-ring product pistons.

Pipelines and joints are welded in an atmosphere of inert gas. The welding electrode is moved over the work area at a uniform speed, and the finished welding is inspected by a device such as a borescope. The appropriate welding rod must be used to avoid the creation of corrosive cells.

Glass and Plastic.   Glass piping is used extensively for food plants and particularly for in-place cleaning. These lines can be inspected easily without disassembly. The pipe size is designated by the nominal inside diameter. Appropriate slope, proper mounting, and sanitary connections are very important.

Plastic tubing is used more for farm-to-receiving operations than for permanent food-handling installations. It is widely used for the transport of water for cleaning and sanitizing.

Pumps. The flow of fluids through a dairy processing plant is maintained by either a centrifugal (nonpositive) or a displacement (positive) pump. Positive displacement pumps are either of the piston or plunger type, usually equipped with multiple pistons, or of the rotary positive type. The pump is selected on the basis of the quantity of product to be moved against a specified head. The material of construction should not affect the characteristics of the product. Generally, a hardenable 400-series stainless steel is used for the moving parts. These parts chip easily and must be handled very carefully during disassembly, cleaning, and assembly.

Centrifugal Pump. The centrifugal pump consists of a directly connected impeller, which operates in a casing at high speed. Fluid enters the center and is discharged at the outer edge of the casing. The centrifugal pump is used either for moving products against low discharge heads or where it is necessary to regulate the flow of product through a throttling valve or restriction. It is designed with an external spring-loaded seal around the shaft and uses an O-ring. The spring serves as a compression unit to hold the sealed surfaces together with a force of ca 22–31 N (5–7 lbf). Pumps for a CIP system include self-cleaning diaphragm contacts in which air or fluid disengages the seal during flushing, cleaning, or sanitizing.

Positive Pumps. Positive pumps employed by the food industry have a rotating cavity between two lobes, two gears (which rotate in opposite directions), or a crescent or a stationary cavity and a rotor. Rotary positive pumps operate at relatively low speed. Fluid enters the cavity by gravity flow or, in some cases, from a centrifugal pump. The positive pump also may use a reciprocating cavity, and may be a plunger or piston pump. Sealing is often achieved by the product itself. These pumps are not truly positive with respect to displacement, but are considered positive displacement pumps and are used for metering product flow. For a particular product and speed, almost regardless of head within the capacity of the pump, the flow is positive. These pumps generally are constructed with a pressure-relief valve to permit the surge of fluid if the discharge is closed.

Speed Devices. Many positive displacement pumps are connected by variable-speed drives or devices to vary the speed. When these pumps are used as a timing device on a homogenizer, the setting is fixed, that is, the maximum speed is limited in order to meet the requirements of pasteurization. Cavitation is possible in high speed pumps. Pump sizes should be selected to operate in the middle of the output curve to avoid noisy operation.

Pump Suction. The net positive suction head required (NPSHR) affects the resistance on the suction side of the pump. If it drops to or near the vapor pressure of the fluid being handled, cavitation and loss of performance occurs (14). The NPSHR is affected by temperature and barometric pressure and is of most concern on evaporator CIP units where high cleaning temperatures might be used. A centrifugal booster pump may be installed on a homogenizer or on the intake of a timing pump to prevent low suction pressures.

Cleaning Systems. Both manual and automatic methods of cleaning food-processing equipment are used. Even in a plant with advanced cleaning equipment, some manual cleaning is involved.

Cleaning In Place. Today in most plants the equipment surfaces are cleaned in place, generally at least once every 24 h. Continuously operated equipment may be cleaned every few days, depending upon the product and the potential build-up of residue on the surfaces.

Cleaning-in-place systems evolved from recirculating of cleaning solutions in pipelines and equipment to a highly automatic system with valves, controls, and timers. In the early circulation systems, considerable manual operation was required in the assembly and disassembly of units. Homogenizers and heat exchangers were cleaned manually or by a system of circulating solutions.

The results of cleaning in place are influenced by equipment surfaces and time of exposure, and the temperature and concentration of the solution being circulated. Cleaning is a mechanical–chemical operation.

In the CIP procedure, a cold or tempered aqueous prerinse is followed by the circulation of a cleaning solution for 10 min to 1 h at 54–82°C. Hot water rinses may harden the food product on the surface to be cleaned. The temperature of the cleaning solution should be as low as possible but high enough to avoid an excess of cleaning chemicals. A wide variety of cleaning solutions may be used, depending upon the food product, the hardness of the water, and the equipment. Alkali or chlorinated acid cleaners are preferred. A chlorinated alkaline cleaner may be used separately or in combination with an acid detergent. The best combination of chemical, timing, and temperature is determined experimentally. A CIP system includes pipelines interconnected with valves to direct the fluid to appropriate locations. The control circuit consists of interlines to control the valves which direct the cleaning solutions and water through the lines, and air lines which control and move the valves. A programmer controls the timing and the air flow to the valves on a set schedule. The 3-A Standards for CIP components, equipment, and installation have been developed. A simple CIP system circuit is shown in Figure 12.

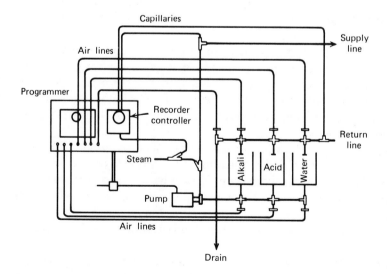

Figure 12. Simple circuit for CIP system.

Economic Aspects

In 1977, U.S. milk production was 55.8×10^6 metric tons from nearly 11×10^6 cows, a decrease from a high of 57.6×10^6 t in 1964. The world production in 1977 was 198.2×10^6 t from nearly 50×10^6 cows. About 60% of the animals are in herds of ≥ 50 cows (see Table 12). Table 13 gives the 1977 production of fluid milk and milk products. Table 14 gives the distribution of milk production to milk products, and Table 15 total U.S. and per capita consumption. Table 16 gives the 1976 total U.S. and per capita consumption of frozen desserts. With the exception of cheese and ice cream, U.S. per capita consumption of dairy products has decreased.

The leading states in milk production in decreasing order are Wisconsin, Minnesota, New York, California, and Pennsylvania. These states produce ca 45% of the U.S. milk supply. Less than 5% of the total production is used on farms with the remainder sold for commercial purposes. Whereas milk and cream were shipped formerly in 19-, 30-, or 38-L cans from the farm to the plant, commercial production is now moved in bulk from the cows to refrigerated farm tanks to insulated bulk truck tankers and to the manufacturing plant. The investment in equipment and the cost of hired labor are associated with large, capital-intensive dairy production centers.

Storage, Cooling, Shipping, and Packaging

Bulk Milk Tanks. Commercial dairy production enterprises generally employ tanks in which the milk is cooled and stored. In some operations, the warm milk is first cooled

Table 12. Size of Dairy Herds in the United States, 1977[a]

Size of operations	Number of operations	Cattle
50 and larger	64,140	6,540,000
30 to 49	67,936	2,616,000
1 to 29	247,454	1,744,000

[a] Ref. 15.

Table 13. U.S. Production of Milk and Milk Products[a]

Product	Quantity, 10^5 metric tons[b]	Total manufacturing establishments	Employees[c]
fluid milk	233	2,507	126,000
butter	4.4	231	4,000
cheese, total	22.4	872	25,200
nonfat dry milk	4.2	225	
evaporated, condensed	4.06	283	12,300
ice cream	16.68	697	21,100

[a] Refs. 16–17.
[b] 1977.
[c] 1972.

Table 14. U.S. Milk Production and Utilization, 1976[a]

	10^5 metric tons
Milk	
produced on farms	545.53
fed to calves	6.96
consumed as fluid milk or cream, on farms	5.88
used on farms	12.83
Milk sold to plants and dealers	
as whole milk	537.14
as farm-separated cream	0.92
sold directly to consumers	6.82
Total production	*544.88*
Milk products[b]	
creamery butter	99.5
cheese	130.46
cottage cheese, creamed	4.76
canned milk	8.58
bulk condensed whole milk	2.9
dry whole milk	2.5
ice cream and frozen products	62.24
other manufactured products	3.3
Total manufactured products	*292.35*
Total fluid milk sales	*233.62*

[a] Ref. 16.
[b] Milk equivalent.

Table 15. U.S. Consumption of Dairy Products[a]

Product	Total consumption, 10^5 metric tons		Per capita consumption, kg		
	1966	1976	1966	1976	1978
fluid milk and cream	264.9	239	134.71	111.13	108
butter (farm and factory)	5.2	4.29	2.6	2	2.1
cheese (except cottage)	8.73	15.57	4.4	7.2	7.5
condensed and evaporated milk	9.7	5.5	4.4	2.3	2.0
ice cream	16.35	17.8	8.2	8.2	8.1
dry whole milk	0.41	0.34	0.14	0.06	0.11
nonfat dry milk	7.34	4.07	2.2	1.6	1.5
cottage cheese, estimate (manufactured, several types)	5.72	7.80	2.1	3.6	4.2

[a] Ref. 16.

and then stored in a tank. 3-A Standards have been established for their design and operation. Among other requirements, the milk must be cooled to 4.4°C within 2 h after milking. In some areas, the temperature must not be permitted to increase above 10°C when warm milk from the following milking is placed in the tank. Bulk milk tanks are classified according to: method of refrigeration—direct expansion (DX) or ice bank (IB); pressure in tank—atmospheric or vacuum; regularity of pickup—everyday (ED) or every-other-day (EOD); capacity—in liters, when full or amount which can be re-

Table 16.   Production and per Capita Consumption of Various Frozen Desserts, 1976

Product	Production, 10^4 metric tons	Per capita consumption, kg
ice cream	148.55	8.2
ice milk[b]	12.9	0.72
sherbet	2.24	0.12
mellorine-type	4.3	0.24

[a] Ref. 16.

[b] Ice milk does not include water ices.

ceived per milking; shape—cylindrical, half-cylindrical, or rectangular; position—vertical or horizontal; and method of cooling—by water, air, or both.

Cooling.   A compression refrigeration system, driven by an electric motor, supplies the cooling for either the direct expansion (DX) or the ice-bank (IB) systems (see Fig. 13). In the former, the milk is cooled by the evaporator (cooling coils) on the bulk-tank liner opposite the milk side of the liner. The compressor must have the capacity to cool the milk as rapidly as it enters the tank.

In the ice-bank system, ice is formed over the evaporator coils. Water is pumped over the ice bank and circulated over the inner liner of the tank to cool the milk. The

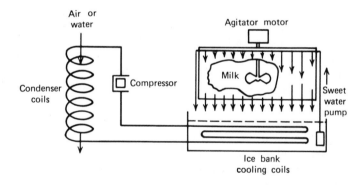

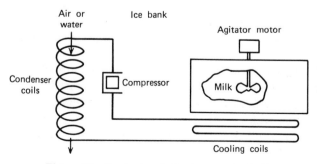

Figure 13.   Ice bank and direct-expansion bulk tank (18).

water is returned to the ice-bank compartment. This system provides a means of building refrigeration capacity for later cooling. Therefore, a smaller compressor and motor can be used, although the unit operates 2–3 times as long as a direct expansion system for the same cooling capacity. Off-peak electricity might be used for the ice-bank system, reducing operating costs.

Important features of bulk milk tanks are measuring devices, generally equipped with a calibrated rod or meter; cleaning and sanitizing facilities; and stirring with an appropriate agitator to cool and maintain cool temperatures.

Surface Coolers. Milk obtained from the cow may be rapidly cooled over a stainless steel surface cooler before entering a bulk tank or milk cans. The cooler may use compression refrigeration or may have two sections, one using cold water followed by a section using compression refrigeration.

Can Coolers. Cans, usually 38 L in capacity, may be cooled with spray coolers that spray cold water over the cans, or placed in tanks in which they are surrounded by refrigerated water. These tanks are known as immersion coolers.

Shipping. Bulk milk is normally hauled to the processing plant in insulated tanks to maintain the right temperature. Truck tankers or trailer tankers are used. Receiving operations were formerly performed at the processing plant but are now the responsibility of the tank-truck driver. These include quantity determination, sampling for quality determination, and sensual evaluation (appearance and odor). The milk is transferred from the bulk tank to the tanker with a positive or centrifugal-type pump.

For routes of some distance, every-other-day (EOD) pick-up reduces hauling costs.

Receiving Operations. Bulk milk receiving operations are simple, consisting primarily of transfering the milk from the tanker to a storage tank in the plant. The handling of milk in 38-L cans requires provisions for equipment and space, for quantity and quality check of the product, washing of cans, and conveyors for moving the cans. Small plants utilize a rotary can washer (primarily a manual operation) with a capacity of 3–6 cans per minute. Large plants use a straight-away can washer (cans move in straight line through unit), appropriately connected to conveyors for nearly automatic operation, with a capacity from 6 to 14 cans per minute. Hot water, steam, and electricity must be furnished for these operations.

Packaging. The pasteurized product is placed in a glass or plastic bottle, or a wax- or plastic-coated paper carton. The filling operation must be carried out without contamination of the product. The cost of paper cartons and plastic bottles per liter or two liters is greater than the per-trip cost of glass containers. The total cost of handling glass containers includes the return of the bottles and washing and handling in the plant. Plastic bottles and paper cartons may be made at the dairy plant. The containers must be closed and capped without contamination and be able to withstand the further handling.

Following filling and closing, the bottles or cartons are placed in a case. The containers are stacked on pallets and moved to a refrigerated area (2–4°C). These operations are mechanized, with continuous operator supervision.

Aseptic Packaging. Aseptic packaging has developed in conjunction with high temperature processing and has contributed to make sterilized milk and milk products a commercial reality. Early processes consisted of heat treatment in the container, usually metal.

The objective in packaging cool sterilized product is to maintain the product under aseptic conditions, to sterilize the container and its lid, and to place the product into the container and seal it without contamination. The Tetrapak and the Dole systems are the most widely used (Figs. 14–15).

Contamination of the head space between the product and the closure is avoided by the use of superheated steam, maintaining high internal pressure, spraying the container surface with a bactericide such as chlorine, irradiation with a bactericidal lamp, and filling the space with an inert sterile gas such as nitrogen.

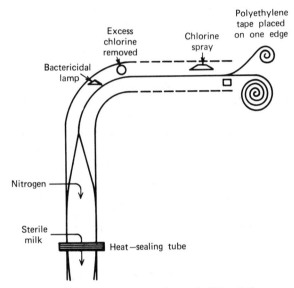

Figure 14.   Tetrapak aseptic filling (19).

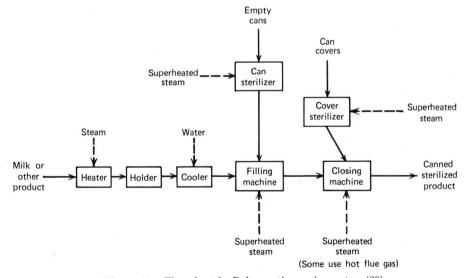

Figure 15.   Flow chart for Dole aseptic canning system (20).

Analysis and Testing

Milk and its products can be subjected to a variety of tests to determine compositions, microbial quality, adequacy of pasteurization, contamination with antibiotics or pesticides, and radioactivity (21).

Microbial Quality. The microbial quality of dairy products is related to the number of viable organisms present. A high number of microorganisms in raw milk suggests it was produced under unsanitary conditions or that it was not adequately cooled after removal from the cow. If noncultured dairy products contain excessive numbers of bacteria, in all likelihood postpasteurization contamination occurred or the product was held at a temperature permitting substantial microbial growth. Raw milk (and some milk products) is commonly examined for its concentration of microorganisms by dye-reaction tests (methylene blue or resazurin), the agar plate test, or the direct microscopic method.

The methylene blue and resazurin reduction methods indirectly measure bacterial densities in milk and cream in terms of the time interval required, after starting incubation, for a dye–milk mixture to change color (methylene blue, from blue to white; resazurin, from blue through purple and mauve to pink). In general, reduction time is inversely proportional to bacterial content of the sample when incubation starts.

The agar plate method consists of adding a known quantity of sample (usually 0.1 mL or less, depending on the concentration of bacteria) to a sterile Petri plate and then mixing the sample with a sterile nutrient medium. After the agar medium solidifies, the Petri plate is incubated at 32°C for 48 h after which the bacterial colonies are counted and the number expressed in terms of 1 mL of sample. This procedure measures the number of viable organisms present and able to grow under test conditions.

The direct microscopic count determines the number of viable and dead microorganisms in a milk sample. A small amount (0.01 mL) of milk is spread over a 1.0 cm^2 area on a microscope slide and allowed to dry. After staining with an appropriate dye (usually methylene blue), the slide is examined with the aid of a microscope (oil-immersion lens). The number of bacterial cells and clumps of cells per microscopic field is determined and, by proper calculations, is expressed as the number of organisms per milliliter of sample.

Coliform Bacteria. Pasteurized products are tested for numbers of coliform bacteria in order to detect significant bacterial recontamination from improper processing, damaged or poorly sanitized equipment, condensate dripping into pasteurized milk, and direct or indirect contamination of equipment by insects or by hands or garments of workers. Coliform bacteria are detected by using the agar plate method and a selective culture medium (violet-red bile agar). A liquid medium (brilliant green lactose bile broth) can also be used to detect this group of organisms. Coliform bacteria are not present in properly processed products that have not been recontaminated.

Thermoduric, Thermophilic, and Psychrophilic Bacteria. Thermoduric bacteria survive but do not grow at pasteurization temperatures. They are largely nonspore-forming, heat-resistant types that develop on surfaces of unclean equipment. These bacteria are determined by subjecting a sample to laboratory pasteurization and then examining it by the agar plate method.

Thermophilic bacteria are able to grow at 55°C. They are sporeforming bacilli which can enter milk from a variety of farm sources. Thermophiles grow in milk held at elevated temperatures. Their presence in milk is determined by means of the agar plate method and an incubation at 55°C.

Psychrophilic bacteria can grow relatively rapidly at low temperatures, commonly within a range of 2–10°C. They are particularly important in the keeping quality of products held at refrigeration temperatures, and their growth is associated with the development of fruity, putrid, and rancid off-flavors. These bacteria can be detected and counted by the agar plate method and an incubation at 7°C for 10 d.

Inhibitory Substances. When antibiotics or other chemicals appear in milk, starter culture growth in such milk may be inhibited. To test for the presence of such chemicals, an agar medium is inoculated with spores of *Bacillus subtilis*. A thin layer of the medium is poured into a Petri dish and allowed to harden. Filter disks (1.25 cm in diameter) are dipped into milk samples and then placed on the surface of the agar medium. After appropriate incubation, plates are examined for a zone of growth inhibition surrounding the disks; the presence of such a zone suggests that the milk contained an antibiotic or other inhibitory agent.

Sediment. The sediment test consists of filtering a definite quantity of milk through a white cotton sediment test disk and observing the character and amount of residue. Efficient use of single-service strainers on dairy farms has reduced the value of sediment tests on milk as delivered at receiving plants. Although the presence of sediment in milk indicates unsanitary production or handling, its absence does not prove that sanitary conditions always existed.

Phosphatase Test. The phosphatase test is a chemical method for measuring the efficiency of pasteurization. All raw milk contains phosphatase and the thermal resistance of this enzyme is greater than that of pathogens over the range of time and temperature of heat treatments recognized for proper pasteurization. Phosphatase tests are based on the principle that alkaline phosphatase is able, under proper conditions of temperature and pH, to liberate phenol from a disodium phenyl phosphate substrate. The amount of liberated phenol, which is proportional to the amount of enzyme present, is determined by the reaction of liberated phenol with 2,6-dichloroquinone chloroimide and colorimetric measurement of the indophenol blue formed. Underpasteurization as well as contamination of a properly pasteurized product with raw product can be detected by this test.

Pesticides. Chlorinated hydrocarbon pesticides often are found in feed or water consumed by cows (22–23); subsequently they may appear in the milk. At present, low-level residues of some of these chemicals are permitted in milk, and hence milk and its products must sometimes be tested for chlorinated hydrocarbon pesticides. Such tests, which are seldom carried out on a routine basis in the dairy plant, are most often conducted in regulatory laboratories or private specialized laboratories. Examining milk for insecticide residues involves extraction of fat (since the insecticide is contained in the fat), partitioning with acetonitrile, cleanup (Florisil column) and concentration, saponification if necessary, and determination by means of paper, thin-layer, microcoulometric gas, or electron capture gas chromatography (see Trace and residue analysis).

Fat Content of Milk. Raw milk as well as many dairy products are routinely analyzed for their fat content by means of the Babcock test (or one of its modifications). This test employs bottles with an extended and calibrated neck, milk plus sulfuric acid (to digest the protein), and a centrifuge (to concentrate the fat into the calibrated neck). The percentage of fat in the milk is read directly from the neck of the bottle by means of a divider or caliper.

Other tests for measuring fat in milk and dairy products include the Mojonnier method, which employs thermostatically controlled vacuum drying ovens and hot plates together with desiccators whose temperature is controlled by circulating water; the Gerber test, developed by a Swiss chemist and used extensively in Europe, which employs sulfuric acid to dissolve solids other than fat, amyl alcohol to prevent charring of fat, and centrifuging to separate the fat into the calibrated neck of the Gerber test bottle; the DPS detergent test, which is based on the principle that the selected detergent(s) dissolves readily in both the fat and water phases of milk and then go out of solution upon application of heat and/or salt, thereby liberating the accumulated fat for measurement; and the TeSa test, which employs a protein-solubilizing agent, two dispersing agents, supplementary alkaline buffering and agitant agents, and test bottles fitted with a side arm and plunger; the test is essentially a chemical extraction method and is applicable to a variety of animal and vegetable fat products. These fat tests are described in detail in ref. 24.

Protein Content. The protein content of milk can be determined by a variety of methods including gasometric, Kjeldahl, titration, colorimetric, and optical procedures. Since most of the techniques are too cumbersome for routine use in a dairy plant, payment for milk has seldom been made on the basis of its protein content. Recently, dye-binding tests have been applied to milk for determination of its protein content; they are relatively simple to perform and hence could be carried out in dairy-plant laboratories. Undoubtedly these procedures will be utilized in the future to assess milk for its nutritional value and results of the tests may replace those of the fat test as a basis for payment.

In dye-binding tests, milk is mixed with an excess of an acidic dye solution; the protein binds the dye in a constant ratio and form a precipitate. This reaction can be explained as follows:

At an acidic pH value, the protein micelles are electropositively charged since dissociation of the acidic carboxyl groups of the polypeptide chains is suppressed, whereas the basic amino groups are positively charged and are in equilibrium with the small negative ions from solution.

The dyestuffs such as amido black and Orange G contain sulfonic acid groups, and negatively charged dyestuff ions are formed in solution which replace the smaller anions on the protein surface.

The complexes thus formed are less dissociated and tend more to aggregation and precipitation.

After the dye–protein interaction has taken place, the mixture is centrifuged and the optical density of the supernatant is determined. Utilization of dye is thus measured and from it the protein content is calculated. Several methods for application of dye-binding techniques to milk are given in refs. 25–26.

Health and Safety Factors

Milk may be a carrier of diseases from animals or other sources to man. To avoid contamination before pasteurization, healthy animals should be separated from sick animals or those with infected udders. The animals should be clean, kept in clean housing with clean air, and handled by workers and equipment under strictly sanitary conditions. Postpasteurization contamination can occur as a result of improper handling, due to exposure to contaminated air, improperly sanitized equipment, or an infected worker.

Proper refrigeration should be practiced from production to consumption. It prevents the growth of some microorganisms, such as Salmonella, and the production of toxins, such as from some types of Staphylococcus, particularly *Staphylococcus aureus*. The growth of bacteria which cause diarrhea, *Escherichia coli* and *Bacillus subtilus*, is substantially checked by proper cooling and handling of milk. Table 17 lists the diseases transmitted by cows to humans. Pasteurization is the best means of prevention.

Manufactured Products

In the United States, 55% of the fluid milk production is used for manufactured products, mainly cheese, evaporated and sweetened condensed milk, nonfat dry milk, and ice cream. Evaporated and condensed milk and dry milk are made from milk only; other ingredients are added to ice cream and sweetened condensed milk.

Evaporated and Condensed Milk. Evaporated milk is produced by removing moisture from milk under a vacuum, followed by packaging and sterilizing in cans (see Fig. 16). The milk is condensed to half its volume in single- or multiple-effect evaporators. The final product has a fat to solids-not-fat ratio of 1:2.2785, and is standardized before and after evaporation. It must have at least 7.9% fat and 25.9% of total milk solids, including fat. The process for evaporated skim milk is similar. A key operation is sterilization in the container, at 116–118°C for 15–20 min. Subsequent cooling with

Table 17. Diseases Transmitted by Milk to Humans

Disease	Microorganism	Carrier
Direct transmission		
tuberculosis (cow)	*Mycobacterium bovis*	udder and manure of infected cows
brucellosis	*Brucella abortus*	milk
foot-and-mouth	virus	blood to udder
milk sickness	white snakeroot	in forage
anthrax	*Bacillus anthracis*	udder by systemic disease; organisms live in soil
Q fever	*Rickettsiae burneti;* also called *Coxiella burneti*	spread by ticks and inhalation
mastitis	*Streptococcus agalactiae*, plus several other bacteria	udder
gastroenteritis	*Escherichia coli,* *Bacillus subtilus*, and salmonella of many types	manure, soil, forage, udder
Indirect transmission		
tuberculosis, human	*Mycobacterium tuberculosis*	sputum, breath droplets
typhoid fever	*Salmonella typhi*	human excreta, flies, polluted water
paratyphoid fever	*Salmonella paratyphi*	feces and urine
scarlet fever	hemolytic streptococcus	udder infection
salmonellosis	salmonella of many types	water, milk, feces, other animals
staphylococcal infections	*Staphylococcus aureus*	udder, human infection
diphtheria	*Corynebacterium diphtheriae*	throat, nose, tonsils[a]
dysentery, bacillary	*Shigella dysenteriae*	bowel discharge[a]
dysentery, amoebic	*Entamoeba histolytica*	bowel discharge[a]

[a] From humans.

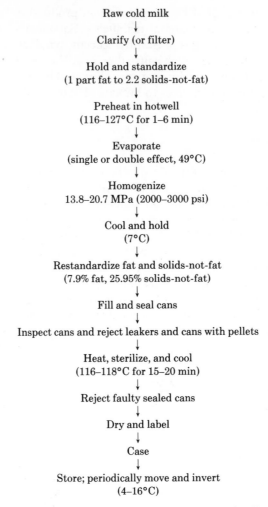

Raw cold milk
↓
Clarify (or filter)
↓
Hold and standardize
(1 part fat to 2.2 solids-not-fat)
↓
Preheat in hotwell
(116–127°C for 1–6 min)
↓
Evaporate
(single or double effect, 49°C)
↓
Homogenize
13.8–20.7 MPa (2000–3000 psi)
↓
Cool and hold
(7°C)
↓
Restandardize fat and solids-not-fat
(7.9% fat, 25.95% solids-not-fat)
↓
Fill and seal cans
↓
Inspect cans and reject leakers and cans with pellets
↓
Heat, sterilize, and cool
(116–118°C for 15–20 min)
↓
Reject faulty sealed cans
↓
Dry and label
↓
Case
↓
Store; periodically move and invert
(4–16°C)

Figure 16.   Steps in producing evaporated milk.

cold water should be completed in 15 min. The cans are continuously turned and moved through the sterilizing unit. Sterilization in the can imparts a distinct cooked flavor to the product. Higher temperatures and shorter treatment (UHT) lessen this effect. Tables 18 and 19 give the standards and definitions for evaporated and condensed milk, as set by the WHO Food and Agriculture Organization (27).

Vitamin A (845 RE per liter) and vitamin D (913 IU per liter) may be added to fortify evaporated milk. Other possible ingredients are sodium citrate, disodium phosphate, and salts of carrageenin. Phosphate ions maintain an appropriate salt balance to prevent coagulation of the protein (casein) during sterilization. The amount of phosphate to be added depends on the amount of calcium and magnesium present.

Large quantities of evaporated milk are used to manufacture ice cream, bakery products, and confectionary products. When used for manufacturing other foods, the evaporated milk is not sterilized, but placed in bulk containers, refrigerated, and used

Table 18.   Standard for Evaporated Milk and Evaporated Skimmed Milk[a]

Definitions

Evaporated milk is a liquid product, obtained by the partial removal of water
only from milk.

Evaporated skimmed milk is a liquid product, obtained by the partial removal
of water only from skimmed milk.

Essential composition and quality factors

Evaporated milk

minimum milk-fat content, mol %	7.5
minimum milk-solids content, mol %	25.0

Evaporated skimmed milk

minimum milk-solids content, mol %	20.0

Food additives

Stabilizers, max

sodium, potassium, and calcium salts of

hydrochloric acid	2000 mg/kg singly
citric acid	3000 mg/kg in
carbonic acid	combination
orthophosphoric acid	expressed as an-
polyphosphoric acid	hydrous substances
carrageenin, mg/kg	150

Labeling

In addition to sections 1, 2, 4, and 6 of the General Standards for the Labeling
of Prepackaged Foods (Ref. No. CAC/RS 1-1969), the following specific
provisions apply:

The name of the food

The name of the product shall be "Evaporated milk" or "Evaporated whole
milk" or "Evaporated full cream milk" or "Unsweetened condensed whole
milk" or "Unsweetened full cream condensed milk," or "Evaporated
skimmed milk," or "Unsweetened condensed skimmed milk" as appropriate.

Where milk other than cow's milk is used for the manufacture of the product
or any part thereof, a word or words denoting the animal or animals from
which the milk has been derived should be inserted immediately before
or after the designation of the product except that no such insertion need
be made if the consumer would not be misled by its omission.

[a] Ref. 27.

fresh. This is called condensed milk. Skim milk may be used as a feedstock to produce
evaporated skim milk.

The moisture content of other liquid milk products can be reduced by evaporation
to produce condensed whey, condensed buttermilk, and concentrated sour milk.

Sweetened Condensed Milk. Sweetened condensed milk, unlike evaporated milk,
is not sterilized. Sugar is added as a preservative which replaces sterilization as a means
of maintaining keeping quality. The equipment is similar to that used for evaporated
milk, except that sugar is added in a hot well before condensing (evaporating) the
liquid. Preheating pasteurizes the product and no sterilizer is needed. According to
standards, sweetened condensed milk must contain 8.5% fat min and 28.0% min total
milk solids, including fat (fat to solids-not-fat ratio = 1:2.294). The final product
contains 43–45% sugar. Sweetened condensed skim milk has not less than 24% total
milk solids, but up to 50% sugar may be added.

Age-thinning and age-thickening defects occur in sweetened condensed products,

Table 19.   Standard for Sweetened Condensed Milk and Skimmed Sweetened Condensed Milk [a]

Definitions

  Sweetened condensed milk is a product obtained by the partial removal of water only from milk, with the addition of sugars.

  Skimmed sweetened condensed milk is a product obtained by the partial removal of water only from skimmed milk with the addition of sugars.

Essential Composition and Quality Factors

  Sweetened condensed milk

minimum milk-fat content, mol %	8.0
minimum milk solids content, mol %	28.0

  Skimmed sweetened condensed milk

minimum milk solids content, mol %	24.0

Food additives

  Stabilizers, max

sodium, potassium, and calcium salts of	
hydrochloric acid	2000 mg/kg singly
citric acid	3000 mg/kg in
carbonic acid	combination
orthophosphoric acid	expressed as an-
polyphosphoric acid	hydrous substances

Labeling

  In addition to sections 1, 2, 4 and 6 of the General Standard for the Labeling of Prepackaged Foods (Ref. No. CAC/RS 1-1969), the following specific provisions apply:

  The name of the food

  The name of the product shall be "Sweetened condensed milk" or "Sweetened condensed whole milk" or "Sweetened full cream condensed milk," or "Skimmed sweetened condensed milk" or "Sweetened condensed skimmed milk," as appropriate.

  Where milk other than cow's milk is used for the manufacture of the product or any part thereof, a word or words denoting the animal or animals from which the milk has been derived should be inserted immediately before or after the designation of the product except that no such insertion need be made if the consumer would not be misled by its omission.

  When one or several sugars are used, the name of each sugar shall be declared on the label.

[a] Ref. 27.

and their density changes. These defects are related to the preheating temperature before evaporation of the water. A low temperature can result in thinning, a high temperature in thickening. The optimum preheating temperature is in the range of 60 to 81°C.

Dry Milk.   Dry milk reduces transportation costs, provides long-term storage, and supplies a product that can be used for food-manufacturing operations. Dry milk is generally made using a drum by the so-called roller or spray process. These processes generally follow condensation in an evaporator where the water removal is less costly.

The moisture content for nonfat dry milk, the principal dry milk product, is ≤5.0% for standard grade and ≤4.0% for extra grade. Dry whole milk contains ≤3.0% moisture. Other drying methods include the use of foam sprays, foam mats, jet sprays, freeze-drying, and tall towers.

Clarification and homogenization precede evaporating and drying. Homogenization of whole milk at 63–74°C with pressures of 17–24 MPa (2500–3500 psi) is particularly desirable for reconstitution and the preservation of quality.

The properties of dry milk are given in Table 20 and the WHO standards in Table 21.

Dry whole milk should be vacuum or gas packed to maintain the quality under storage. Products with milk fats deteriorate in the presence of oxygen, giving oxidation off-flavor. Several factors may be involved in oxidative deterioration: preheating of product; storage temperature; presence of metallic ions, particularly copper and iron; presence of oxygen (air) in product; and light. Antioxidants of many kinds have been used with various degrees of success, but a universally acceptable antioxidant which meets the requirements for food additives has so far not been found.

Dry milk has been used primarily for manufactured products, but is now used to a much greater extent for beverage products. Nonfat dry milk is the principal dry-milk product.

Drum Drying. The drum or roller dryers used for milk operate on the same principles as for other products. A thin layer or film of product is dried over an internally steam-heated drum with steam pressures up to 620 kPa (90 psi) and 149°C. Approximately 1.2 to 1.3 kg steam are required per kg water removed. The dry film is scraped from the surface with a knife called a doctor blade, moved from the dryer by conveyor, and pulverized, sized, cooled, and containerized.

Table 20. Properties of Dry Milk[a,b]

moisture content, nonfat, wt %	4–5
apparent or bulk density, including voids, g/cm^3	
drum dried, nonfat	0.3–0.5
spray dried, nonfat	0.5–0.6
true density without voids	
dry milk	1.31–1.32
nonfat dry milk	1.44–1.46
coefficient of friction at 20°C, 5 wt % fat	0.64
porosity, spray dried, nonfat, wt %	0.482
solubility index, spray process	1.2
vapor pressure	
5% moisture, nonfat, 38°C, kPa^c	1.17
5% moisture, 13% fat, 38°C, kPa^c	0.75
threshold radiation level to produce off-flavor	
dry whole milk, Gy^d	590
dry nonfat milk, Gy^d	1280
titratable acidity, wt %	0.15
specific heat, kJ/(kg·K)^e	1.04
thermal conductivity, k, W/(m·K)^f	0.05
thermal conductivity, k, W/(m·K)^f	
4.2%, 40°C	0.05
and at 65°C	0.06

[a] Approximate values.
[b] Atomization of one liter of condensed product to an average particle size of 50 μm dia equals 341,000 cm^2 surface.
[c] To convert kPa to mm Hg, multiply by 7.5.
[d] To convert Gy to rad, multiply by 100.
[e] To convert kJ/(kg·K) to Btu/(lb·°F), divide by 4.184.
[f] To convert W/(m·K) to Btu/(h·ft·°F), multiply by 1.874.

Table 21.   Standard for Whole Milk Powder, Partly Skimmed-Milk Powder, and Skimmed-Milk Powder[a,b]

Scope

This standard applies exclusively to dried milk products as defined, having a fat content of not more than 40 mol %.

Definitions

Milk powder is a product obtained by the removal of water only from milk, partly skimmed milk or skimmed milk.

Essential composition and quality factors

Whole milk powder

minimum milk-fat content, mol %	26
maximum milk-fat content, mol %	<40
maximum water content, mol %	5

Partly skimmed-milk powder

minimum milk-fat content, mol %	>1.5
maximum milk-fat content, mol %	<26
maximum water content, mol %	5

Skimmed-milk powder

maximum milk-fat content, mol %	1.5
maximum water content, mol %	5

Food additives

Stabilizers, max

sodium, potassium and calcium salts of

hydrochloric acid	5000 mg/kg singly
citric acid	or in combination
carbonic acid	expressed as an-
orthophosphoric acid	hydrous substances
polyphosphoric acid	

Emulsifiers in instant milk powders only

monoglycerides and diglycerides	2500 mg/kg
lecithin	5000 mg/kg

Anticaking agents in milk powders intended to be dispensed in vending machines

tricalcium phosphate	
silicates of aluminum, calcium, magnesium, and sodium–aluminum	
silicon dioxide (amorphous)	10 g/kg singly or
calcium carbonate	in combination
magnesium oxide	
magnesium carbonate	
magnesium phosphate	

[a] Ref. 27.

[b] Dry milk was referred to as milk powder until the mid-1960s, when this designation was changed by the American Dry Milk Institute to dry milk in the U.S.

The operating variables for a drum or roller dryer include condensation of incoming product, temperature of incoming product, steam pressure (temperature) in drum, speed of drum, and height of product over drum.

The capacity of the dryer is increased by increasing the steam pressure, the temperature of the milk feed, the height of milk over the drums, the gap between drums (double), and the speed of rotation. Increasing the capacity is limited by the effect on the product quality.

Drum-dried products are more affected by heat than spray-dried products. Drying in a vacuum chamber decreases the temperature and thus the heat effect on the product, although the atmospheric dryers are used more widely.

Drum-dried products, mostly nonfat, make up only ca 5–10% of the dried milk products. Because of the high temperature and longer contact time, considerable protein denaturation occurs. Drum-dried products are identified as high heat dry milk and as such have a lower solubility index, lower protein nitrogen content, and a darker color.

Spray Drying. The spray dryer provides a chamber in which the milk or milk product is atomized in a heated air stream that removes moisture. The dry product is separated from the air stream and removed from the chamber. A flow sheet of the process is given in Figure 17.

In spite of the higher energy requirements, the spray dryer has gained in popularity because of the reduced heat effect on the product as compared to the drum dryer. Modifications such as foam spraying are being developed to reduce the heat effect further.

In the manufacture of dry milk by the spray process, a condensed product is pumped to an atomizer to produce a large surface area to enhance drying. A high pressure nozzle or centrifugal device, such as a rotating disk or wheel, is used for atomization. The air is filtered, heated to 149–260°C, and moved over the atomizing product, and then, saturated with water, exhausted from the dryer. The dry product is centrifuged in a separator and filtered outside the drying chamber. In order to

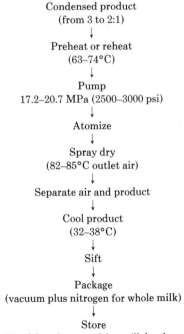

Figure 17.   Manufacture of dry milk by the spray process.

minimize heat effects, the dried product is removed as rapidly as possible from the drying chamber and cooled. Considerable variation exists in the operation of spray dryers, depending on the product and the dryer. A low heat, nonfat dry milk product is obtained by minimizing heating before and after drying and including quick cooling.

Foam-spray drying consists of forcing gas, usually air or nitrogen, into the product stream at 1.38 MPa (200 psi) ahead of the pump in the normal spray dryer circuit. This method improves the characteristics of dried products, such as dispersibility, bulk density, and uniformity. The foam-spray dryer can accept a condensed product with 60% total solids, as compared to 50% without the foam process. The usual neutralization of acid whey is avoided with the foam-spray dryer.

Agglomeration. The process of treating dried products, particularly nonfat products, in order to increase the speed and ability to reconstitute those products is known as "instantizing" or agglomeration. The particles are agglomerated into larger particles which dissolve more easily than small particles. The particle surface is first wetted, followed by agglomeration and drying. Instantized products can also be obtained by foam-spray drying.

Instantized products have a lower density, are more fragile than conventional products, and must be handled with extra care. They are of particular importance to the fast-food market. The process is used for various beverage and milk products.

Packaging. Dry milk is packaged in large bulk or small retail containers. A suitable container keeps out moisture, light, and air (oxygen). For dry whole milk, oxygen is removed by vacuum, and an inert gas, such as nitrogen, is inserted in the heat space. An oxygen level of 2.0% or below is required by U.S. standards for premium quality.

Cream. Cream is a high fat product which is secured by gravity or mechanical separation through differential density of the fat and the serum. Fat content may range from 10 to 40%, depending on use and Federal and state laws. The U.S. Public Health Service (6) milk ordinance defines creams as products that contain not less than 18% milk fat. Whipping cream has a fat content of 34–40%, table, coffee, or light cream has a fat content of 20–25%. Half-and-half, suggesting a mixture of cream and milk, has not less than 10.5% milk fat, and in some states up to 12%.

Cream is standardized in the same manner as milk, following separation. The addition of whole milk rather than serum is preferred.

The sale of fresh cream as a table item for serving has decreased greatly over the past twenty years, primarily as a result of changing customer demand based on diet. A variety of cream and fat substitutes are available for spreads, toppings, whiteners, and cooking (see Synthetic dairy products).

In the early days of the dairy industry, cream was an important product. It was separated on the farm and handled with less care and cooling than needed to maintain beverage quality of whole milk.

Anhydrous Milk Fat; Butter Oil. A high milk fat material is butter oil, which is 99.7% fat, also called anhydrous milk fat or anhydrous butter oil, if less than 0.2% moisture is present. Although the terms are used interchangeably, anhydrous butter oil is made from butter and anhydrous milk fat is made from whole milk. For milk and cream there is an emulsion of fat-in-serum. For butter oil and anhydrous milk fat there is an emulsion of serum-in-fat, such as with butter. It is easier to remove moisture in the final stages to make anhydrous milk fat with the serum-in-fat emulsion.

Butter. In the United States, about 10% by weight of the edible fats used are butter. Butter is defined as a product that contains 80% milk fat with not more than 16% moisture. It is made of cream with 25–40% milk fat. The process is primarily a mechanical one in which the cream, an emulsion of fat-in-serum, is changed to butter, an emulsion of serum-in-fat. It is an inverting or breaking of the emulsion. The process is accomplished by churning or, in recent years, by a continuous operation with automatic controls. Some of the physical properties are given in Table 22 (see also Emulsions).

Buttermilk. Buttermilk is drained from the butter (churn) after the butter granules are formed. As such, it is the fluid other than the fat which is removed by churning. Buttermilk may be used as a beverage, or may be dried and then used for baking. Buttermilk from churning is ca 91% water and 9% total solids. The total solids includes lactose (4.5%), nitrogeneous matter (3.4%), ash (0.7%), and fat (0.4%). Table 23 gives the U.S. specifications.

Most of the beverage buttermilk is produced by fermentation of skim milk, often with some cream added, known as cultured buttermilk. Two principal fermentation organisms used for fermentation are *Streptococcus cremoris* or *Streptococcus lactis*,

Table 22.   Representative Values of Physical Properties of Milk Fat and Butter[a]

fat content, wt %	80
size of fat globules, μm	1–20
melting point of milk fat, °C	31–36
solidification of milk fat, °C	19–24
apparent specific heat, kJ/(kg·K)[b]	
0°C	2.14
15°C	2.20
40°C	2.32
60°C	2.42
density of milk fat, g/cm^3	
34°C (just above melting point)	0.91–0.95
60°C	0.896
viscosity of milk fat, mPa·s (= cP)	
30°C	25.8
50°C	12.4
70°C	7.1
viscosity of butter 21°C, mPa·s (= cP)	
(Brookfield at 1 rpm)	3.1×10^5
iodine number, normal butter	30.5
melting point of butter, °C	33.3
spreadibility,	good at 21°C
	desirable at 7–16°C
	difficult at 4°C
ratio of firmness to butter/firmness of butterfat	
in summer	1.97:1
in winter	1.48:1
coefficient of expansion of liquid pure butterfat	0.00076 (30–60°C)
free acidity, fresh butterfat	0.05–0.10%

[a] Refs. 28–29.
[b] To convert kJ/(kg·K) to Btu/(lb·°F), divide by 4.184.

Table 23.   U.S. Specifications for Dry Buttermilk and Dry Whey[a]

| | Spray process DBM | | Roller process DBM | | Dry whey[b], extra |
	Extra	Standard	Extra	Standard	
moisture, wt %, not more than	5.0	5.0	5.0	5.0	5.0
milk fat, wt %, not less than	4.5	4.5	4.5	4.5	1.25% (not more than)
solubility index, mL, not more than	≤1.25	2.0	15.0	15.0	1.25
scorched particles, mg, not more than	15.0	22.5	22.5	32.5	15.0
titratable acidity					
wt %, not more than	0.18	0.20	0.18	0.18	0.16
wt %, not less than	0.10	0.10	0.10	0.10	
bacteria count, per gram, not more than	50,000	200,000	50,000	200,000	50,000
ash alkalinity, mL of 0.1 N HCl/100 g, not more than	125	125	125	125	125

[a] Ref. 30.
[b] Not applicable to cottage cheese whey.

and *Leuconostoc citrovorum*. The effect of the high processing temperature and the lactic acid provide an easily digestable product.

Dried buttermilk is made by either the drum or the spray process. The buttermilk is usually pasteurized before drying, even though the milk was previously pasteurized before churning. Dried buttermilk is used primarily for baking, confectionary, and dairy products.

Cheese.   The making of cheese is based on the coagulation of casein from milk, or to a minor extent, of the proteins of whey. The casein is precipitated by acidification, which can be accomplished by natural souring of milk. The procedures for making cheese vary greatly and cheese products are countless.

Considerable art is involved in making cheese. The composition and handling of the original milk, bacterial flora, and starter culture are the basic variables, which along with heat treatments, flavoring, salting, and forming of the final product, affect the final product.

Membrane Separation.   The separation of components of liquid milk products can be accomplished with semipermeable membranes by either ultrafiltration (UF) or hyperfiltration, also called reverse osmosis (RO) (31). With ultrafiltration, the membrane selectively prevents the passage of large molecules such as protein. In reverse osmosis, different low molecular weight molecules are separated which are small in size like water. Both procedures require pressure to be maintained and energy needed is a cost item. The materials from which the membranes are made is similar for both processes and include cellulose acetate, poly(vinyl chloride), nylon, and polyamide (see Membrane technology).

Ultrafiltration.   Membranes are used which are capable of selectively passing large molecules (molecular weights >500) (see Ultrafiltration). Pressures of 0.1 to 1.4 MPa (up to 200 psi) are exerted over the solution to overcome the osmotic pressure, although providing an adequate flow through the membrane for use. Ultrafiltration has been particularly successful for the separation of whey from cheese. It separates the protein from the lactose and mineral salts, the protein being the concentrate. Ultrafiltration also is used to obtain a protein-rich concentrate from skim milk from which cheese is made. The whey protein obtained by ultrafiltration is 50–80% protein which can be spray dried. Ultrafiltration is used for removing minerals from whey and buttermilk.

Reverse Osmosis (Hyperfiltration).   Membranes are used for the separation of smaller components (molecular weights <500). They have smaller pore space and are tighter than those used for ultrafiltration (see Reverse osmosis). High pressure pumps, usually of the positive-piston type or multistage centrifugal type, provide pressures up to 4.14 MPa (600 psi).

Reverse osmosis is used for the separation of lactose from whey. Treatment of the permeate from the ultrafiltration of whey by reverse osmosis removes the lactose. By appropriate selection of the membrane, the minerals are removed from the lactose. Reverse osmosis can be used to concentrate the water in a dairy plant, giving a product with 18% solids. Concentration of the rinse water gives a product with 4–5% total solids.

Proper maintainance of the membrane can prolong its life up to 2 years. Membranes are available for use up to 100°C with pH ranges from 1 to 14; the usual temperature range is 0–50°C.

Cheddar Cheese.   Milk is heated to 30°C and a lactic acid-producing starter is added. The milk is held for about one hour, during which time the acidity increases. Rennet extract is mixed with the milk, which produces a curd in ca 30 min. The curd is cut into cubes and the whey is expressed. The curd solidifies, and is stirred and heated slowly. The heating is continued until the curd becomes completely firm. The whey is drained and separated by forming channels. With the action of the lactic acid and the removal of the whey, the curd becomes a solid mass. The curd is cut with the pieces moved to continue the removal and drainage of whey. The whey increases from 0.1% acid at the time of cutting of the curd, to 0.5% acid at the end of drainage. Cheddared cheese is put through a curd mill to reduce the curd sizes.

Cottage Cheese.   Cottage cheese is made from skim milk. As compared to most other cheeses, cottage cheese has a short shelf-life, and must be refrigerated to maintain quality, usually at 4.4°C or below to provide a shelf-life of 2 weeks. Cottage cheese is a soft uncured cheese which contains not more than 80% moisture.

Cottage cheese is made by several procedures. In general, pasteurized skim milk is innoculated with lactic acid culture and rennet starter to coagulate the protein. The coagulated material is divided or cut and the resulting curd cooked to remove the whey. The whey is drained and the curd washed with water. Mechanized operations are used for large-scale production. The conditions of manufacture are given in Table 24.

Horizontal vats are employed for manual and mechanized operations. The starter

Table 24.   **Manufacture of Cottage Cheese**

Conditions	Value
amount of starter, wt %	0.5–5
setting temperature, °C	21–32
coagulating time, h	4–12
size of curd cubes, cm^3	0.25–2.00
cooking temperature, °C	49–60
rennet extract, g/500 kg milk	0.5–1.0

may be blended with the incoming product or added in the vat. The setting temperature of the treated whey is typically 30°C and it is held for 4.5–5 h. The curd is cut when the titratable acidity is 0.52% for lactic acid milk with 9.0% nonfat milk solids, or a pH of 4.6–4.7. The acidity controls the calcium level of the casein which determines the characteristics of the curd. A low acidity causes a rubbery curd; a high acidity causes a tender curd which shatters easily. The curd is cut by moving a knife first horizontally, then vertically, and finally crosswise through the vat. The cut curd is cooked about 15 min after the cutting is finished. The temperature gradually increased in increments of 0.5–1°C every 3–5 min to avoid the formation of a hardened protein layer that would inhibit moisture removal. The curd is washed successively with cooler water, pasteurized or treated with chlorine, with the final rinse at 4.4°C to firm the curd. Curd pumps move the curd to the blender where salt and stabilizer may be added. Creamed cottage cheese, which has a fat content of at least 4%, is produced by mixing in 12–14% fat cream.

Yogurt.   Yogurt is a fermented milk product that is rapidly increasing in consumption in the United States. Milk is fermented with *Lactobacillus bulgaricus* and *Streptococcus thermophilus* organisms producing lactic acid. Usually some cream or nonfat dried milk is added to the milk in order to obtain a heavy-bodied product.

Yogurt is manufactured similarly to buttermilk. Milk with a fat content of 1–5% and SNF content of 11–14%, is heated to ca 82°C and held for 30 min. After homogenization, the milk is cooled to 43–46°C and inoculated with 2% culture. The product is incubated at 43°C for 3 h in a vat or in the final container. The yogurt is cooled and held at ≤4.4°C. The cooled product should have a titratable acidity of 1.0–1.2% and a pH of 4.3–4.4. The titratable acidity is expressed in terms of percentage of lactic acid which is determined by the amount of 0.1 N NaOH/100 mL required to neutralize the substance. Thus, 10 mL of 0.1 N NaOH represents 0.10% acidity. Yogurts with ≤2% fat are popular. Fruit-flavored yogurts are also common; 30–50 g of fruit is placed in the carton before or with the yogurt.

Frozen Desserts.   Ice cream is the principal frozen dessert produced in the United States. It is known as the American dessert, and was first sold in New York City in 1777. The composition and nutrient contents of various frozen desserts are given in Tables 25 and 26.

Ice Cream.   Ice cream is a frozen-food dessert prepared from a mixture of dairy ingredients (16–35%), sweeteners (13–20%), stabilizers, emulsifiers, flavoring, and fruits and nuts. Ice cream has 8–20% milk fat and 8–15% nonfat solids with a total of 38.3% (36–43%) total solids. These ingredients can be varied. The dairy ingredients are milk or cream, and milk fat, which is supplied by milk, cream, butter, or butter oil, as well as solids-not-fat, which is supplied by condensed whole or nonfat milk or dry milk. The quantities of these products are specified by standards. The milk fat provides the characteristic texture and body in ice cream. Sweeteners are a blend of cane or beet sugar and corn-syrup solids. The quantity of these vary depending on the sweetness desired and the cost.

Stabilizers to improve the body of the ice cream include gelatin, sodium alginate, pectin, and guar gum. Emulsifiers such as lecithin, monoglycerides, and diglycerides, assist the incorporation of air and improve the whipping properties. The mixture of components for making ice cream is called ice-cream mix. Dry ice-cream mix is available. Properties of ice cream are given in Table 27.

Table 25. Composition of Frozen Desserts[a], %

Component	Ice cream Premium[b]	Average	Ice milk	Sherbet	Ice	Soft-serve
milk fat[c]	16.0	10.5	3.0	1.5		6.0
milk solids, nonfat	9.0	11.0	12.0	3.5		12.0
sucrose	16.0	12.5	12.0	19.0	23.0	9.0
corn syrup solids		5.5	7.0	9.0	7.0	6.0
stabilizer[d]	0.1	0.3	0.3	0.5	0.3	0.3
emulsifier[d]		0.1	0.15			0.2
total solids, kg/L	41.1	39.9	34.45	33.5	30.3	33.5
	1.09	1.12	1.13	1.14	1.13	1.11
draw from freezer:						
% overrun	65–70	95–100	90–95	50	10	40
approx kg/L	0.64	0.55	0.57	0.74	1.01	0.77

[a] Frozen desserts containing vegetable fat (mellorine type) are permitted in some states. A wide variation of composition exists depending on individual state standards.
[b] To be classified as custard or French—must contain not less than 1.4% egg yolk solids.
[c] Milk fat content regulated by individual state.
[d] Usage level as recommended by manufacturer of stabilizer and emulsifier.

Table 26. Nutrient Composition per 100 g of Various Frozen Desserts

Analysis	Vanilla ice cream		Vanilla ice milk	Orange sherbet	Orange ice
energy, kJ[a]	845	895	598	561	602
fat, %	10.2	12.1	3.0	1.3	trace
protein, %	4.1	3.7	4.8	1.2	0.01
carbohydrate, %	23.5	23.2	24.8	31.4	35.4
total solids, %	39.0	40.0	34.5	34.5	37.0
calcium, mg	155	138	182	53	1
phosphorus, mg	115	106	139	35	2
sodium, mg	66	61	78	23	trace
vitamin A, RE[b]	86.4	101.6	25.2	16.2	2.8
thiamine, mg	0.05	0.04	0.05	0.02	0.01
riboflavin, mg	0.22	0.20	0.27	0.07	trace
niacin, mg	0.11	0.10	0.13	0.08	0.02
vitamin D, IU	4	5	1	trace	0

[a] To convert kJ to kcal, divide by 4.184; food calorie = 1 kcal.
[b] See footnote b in Table 3.

Preceded by a blending operation, the ingredients are mixed in a freezer which whips the mix to incorporate air and freezes a portion of the water. Freezers may be of a batch or continuous type. Commercial ice cream is produced mostly in continuous operation.

The incorporation of air decreases the density and improves the consistency. If one half of the final volume is occupied by air, the ice cream is said to have 100% overrun, and 4-L weight of 2.17 kg. The ice cream from the freezer is at ca −5.5°C with one-half of the water frozen, preferably in small crystals.

Containerized ice cream is hardened on plate-contact hardeners or by convection air blast as the product is carried on a conveyor or through a tunnel. Air temperatures for hardening are −40 to −50°C. The temperature at the center of the container as

Table 27.   Properties of Ice Cream and Ice-Cream Mix, Approximate Values

structural constituents[a], particle diameter, μm	
ice crystals	45–56
air cells	110–185
unfrozen materials	6–8
average distance between air cells	100–150
lamellae thickness	30–300
lactose crystals (when apparent to tongue feel)	16–30
individual fat globules	0.5–2.0
small fat globules	$\leq$20
agglomerated fat	$\leq$25
coalesced fat	$\geq$25
weight per 3.9 L, kg, 100% overrun	2.04
specific gravity (at 100% overrun), g/cm^3	0.54
specific heat, kJ/(kg·K)[b]	
ice cream	1.88
ice-cream mix	3.35
fuel values, kJ[b]	8.70
overrun, %	60–100
temperature at which freezing begins, °C	3.3
at −5° to −6°C, water in ice cream which is frozen	50%
at −30°C, water in ice cream which is frozen	90%
ice-cream mix	
pH	6.3
acidity, %	0.19
specific gravity	1.054–1.123
surface tension, N/m (= dyn/cm)	50×10^{-3}
composition of solids-not-fat of mix	
protein, %	36.7
lactose, %	55.5
minerals, %	7.8

[a] Ref. 32.
[b] To convert kJ to Btu, divide by 1.054.

well as the storage temperature should be −23°C or below. Approximately one-half of the heat is removed at the freezer and the remainder in the hardener process.

Other Frozen Desserts.   Although ice cream is by far the most important frozen dessert, other frozen desserts such as bice milk, sherbet, and mellorine-type products are also popular.

Ice milk is a frozen product which has less fat (2–5%) and slightly more nonfat milk solids than ice cream. Stabilizers and emulsifiers are added. About half of the ice milk is made as a soft-serve dessert, produced in freezers with an overrun of 30–60%.

Sherbets have a low fat content (1–2%), low milk solids (2–5%), and a sweet but tart flavor. Ice-cream mix and water ice can be mixed to obtain a sherbet. The overrun in making sherbets is about 25–40%. A sherbet-type product in which vegetable fat is used is called sherbine.

Mellorine is similar to ice cream except that the milk fat is replaced with a vegetable fat (6% min). The total solids in mellorine are 35–39%, of which there are 10–12% milk solids.

Table 28. Composition and Concentrations of Milk Protein[a]

Component	Whole	Casein α-	Casein β-	Casein γ-	β-Lactoglobulin	α-Lactalbumin	Blood serum albumin	Euglobulin	Pseudoglobulin[b] (a)	(b)
Concentration, g/100 mL	2.23–8.84	1.4–2.3	0.5–1.0	0.06–0.22	0.20–0.42	0.07–0.15	0.02–0.05	0.03–0.06	0.02–0.05	
Composition, g/100 g										
N	15.63	15.53	15.33	15.40	15.60	15.86	16.07	16.05	15.29	15.9
amino N	0.93	0.99	0.72	0.67	1.24		0.78			
amide N	1.6	1.6	1.6	1.6	1.07					
P	0.86	0.99	0.61	0.11	0.00	0.02	0.00	0.00	0.00	
S	0.80	0.72	0.86	1.03	1.60	1.91	1.92	1.01	1.00	1.1
hexose								2.93	2.96	
hexosamine								1.58	1.45	
Gly[c]	2.7	2.8	2.4	1.5	1.4	3.2	1.8			
Ala	3.0	3.7	1.7	2.3	7.4	2.1	6.2			
Val	7.2	6.3	10.2	10.5	5.8	4.7	5.9	10.4	9.6	8.7
Leu	9.2	7.9	11.6	12.0	15.6	11.5	12.3	10.4	9.6	8.5
Ile	6.1	6.4	5.5	4.4	6.1	6.8	2.6	3.0	3.0	4.2
Pro	11.3	8.2	16.0	17.0	4.1	1.5	4.8			10.0
Phe	5.0	4.6	5.8	5.8	3.5	4.5	6.6	3.6	3.9	3.9
Cys_2	0.34	0.43	0.0–0.1	0.0	2.3	6.4	5.7	3.3	3.0	
Cys	0.0	0.0	0.0	0.0	1.1	0.0	0.3	0.0	0.0	
Met	2.8	2.5	3.4	4.1	3.2	1.0	0.8	0.9	0.9	1.3
Trp	1.7	2.2	0.83	1.2	1.9	7.0	0.7	2.4	2.7	3.2
Arg	4.1	4.3	3.4	1.9	2.9	1.2	5.9	5.1	3.3	5.6
His	3.1	2.9	3.1	3.7	1.6	2.9	4.0	2.0	2.1	2.3
Lys	8.2	8.9	6.5	6.2	11.4	11.5	12.8	6.3	7.1	6.1
Asp	7.1	8.4	4.9	4.0	11.4	18.7	10.9			9.4
Glu	22.4	22.5	23.2	22.9	19.5	12.9	16.5			12.3
Ser	6.3	6.3	6.8	5.5	5.0	4.8	4.2			
Thr	4.9	4.9	5.1	4.4	5.8	5.5	5.8	10.6	10.3	9.0
Tyr	6.3	8.1	3.2	3.7	3.8	5.4	5.1			6.7

[a] Ref. 33.
[b] From milk (a), from colostrum (b).
[c] See Amino acids, survey for abbreviations.

567

Other frozen desserts are parfait, souffle, ice-cream pudding, punch, and mousse. These are often classified with the sherbets and ices.

By-Products From Milk. Milk is a source for numerous by-products resulting from separation or alteration of the components. These components may be used in other so-called nondairy manufactured foods, dietary foods, pharmaceuticals, and as a feedstock for numerous industries, such as casein for glue.

Lactose. Lactose [63-42-3] or milk sugar, $C_{12}H_{22}O_{11}.H_2O$ makes up about 5% of cow's milk. Compared to sucrose, lactose has about one-sixth the sweetening strength (see Sugar). Because of its low solubility, lactose is limited in its application. However, it is soluble in milk serums and can be removed from the whey. Upon fermentation by bacteria, lactose is converted to lactic acid, and is therefore of particular importance in producing fermented or cultured dairy products, such as cultured buttermilk, cheeses, and yogurt.

The ratio of α-lactose and β-lactose in dry milk and whey vary according to the speed and temperature of drying. An aqueous solution at equilibrium at 25°C contains 35% alpha and 63% beta lactose. The latter is more soluble and sweeter than DL-lactose. It is obtained by heating an 80% DL-lactose solution above 93.5°C, followed by drying on a drum or roller dryer. Lactose is used for foods and pharmaceutical products.

Casein. Milk contains proteins and essential amino acids lacking in many other foods. Casein, a mixture of several proteins (qv), is the principal protein in the skim milk (nonfat) portion of milk (3–4% of the weight). After it is removed from the liquid portion of milk, whey remains. It can be denatured by heat treatment of 85°C for 15 min. Various protein fractions are identified as α-, β-, γ-, and δ-casein; α-actalbumin; β-, γ-lactoglobulin; and blood-serum albumin, which have specific characteristics for various uses. Table 28 gives the concentration and composition of milk proteins.

Casein is used to fortify flour, bread, and cereals. Casein also is used for glues and microbiological media. Calcium caseinate is made from a pressed casein, by rinsing, treating with calcium hydroxide, heating, and mixing followed by spray drying. A product of 2.5–3% moisture is obtained.

Casein hydrolyzates are produced from dried casein. With appropriate heat treatment, addition of alkalies and enzymes, digestion proceeds. Following pasteurization, evaporation, and spray drying a dried product of 2–4% moisture is obtained.

Many so-called nondairy products such as coffee cream, topping, and icings utilize caseinates (see Synthetic dairy products). In addition to fulfilling a nutritional role, the caseinates impart creaminess, firmness, smoothness, and consistency of products. Imitation meats, soups, etc, use caseinates as an extender and to improve moistness and smoothness.

Nomenclature

CIP	= cleaning-in-place
DBM	= dry buttermilk
DX	= direct expansion
EOD	= every other day
FDV	= flow-diversion valve
HTST	= high temperature short-time
IB	= ice bank
NPSHR	= net positive suction head required
RO	= reverse osmosis

SNF   = solids-not-fat
UF   = ultrafiltration
UHT   = ultrahigh temperature

BIBLIOGRAPHY

"Dairy Products" in *ECT* 1st ed., Vol. 4, pp. 774–846, by Arnold H. Johnson, National Dairy Research Laboratories, Inc.; "Milk and Milk Products" in *ECT* 2nd ed., Vol. 13, pp. 506–576, by E. H. Marth, University of Wisconsin, and R. V. Hussong, L. F. Cremers, J. H. Guth, L. D. Hilker, H. W. Jackson, O. J. Krett, E. G. Stimpson, R. A. Sullivan, and L. Tumerman, National Dairy Products Corporation.

1. B. H. Webb, A. H. Johnson, and J. A. Alford, *Fundamentals of Dairy Chemistry*, Avi Publishing Co., Westport, Conn., 1974, p. 396.
2. *Recommended Dietary Allowances*, 8th ed., National Research Council, National Academy of Sciences, Washington, D.C., 1980.
3. Ref. 1, p. 125.
4. S. F. Herb, P. Magidman, F. E. Luddy, and R. W. Riemenschneidet, *J. Am. Oil. Chem. Soc.* **39**, 142 (1962).
5. W. J. Harper and C. W. Hall, *Dairy Technology and Engineering*, Avi Publishing Co., Westport, Conn., 1976, p. 413.
6. *Grade A Pasteurized Milk Ordinance*, United States Department of Health, Education, and Welfare, Public Health Service Publication No. 299, 1965.
7. Ref. 5, p. 426.
8. C. W. Hall and G. M. Trout, *Milk Pasteurization*, Avi Publishing Co., Westport, Conn., 1968, p. 51.
9. Ref. 8, p. 64.
10. C. W. Hall, G. M. Trout, and A. L. Rippen, *Michigan Arg. Exp. Stn. Q. Bull.* **43**, 634 (1961).
11. Ref. 8, p. 73.
12. Ref. 8, p. 125.
13. Ref. 8, p. 128.
14. Ref. 5, pp. 411–412.
15. United States Department of Agriculture, ESCS, Crop Reporting Board Report, January 30, 1979.
16. *Agricultural Statistics*, United States Department of Agriculture, Washington, D.C., 1979, pp. 363, 380–386.
17. *Census of Manufacturers*, United States Department of Commerce, Vol. 1, Washington, D.C., 1972, pp. SR 3-4 to SR 3-6.
18. C. W. Hall and D. C. Davis, *Processing Equipment for Agricultural Products*, Avi Publishing Co., Westport, Conn., 1979, p. 49.
19. T. I. Hedrick and C. W. Hall, *Dairy Eng.* **78**, 248 (1961).
20. I. J. Pflug, C. W. Hall, and G. M. Trout, *Dairy Eng.* **76**, 328 (1959).
21. *Standard Methods for the Examination of Dairy Products*, American Health Public Health Association, 12th ed., New York, 1967.
22. E. H. Marth, *J. Milk Food Technol.* **25**, 72 (1962).
23. E. H. Marth and B. E. Ellickson, *J. Milk Food Technol.* **22**, 112, 145 (1959).
24. *Laboratory Manual, Methods of Analysis of Milk and Its Products*, The Milk Industry Foundation, Washington, D.C., 1959.
25. R. M. Dolby, *J. Dairy Res.* **28**, 43 (1961).
26. R. W. Weik, M. Goehle, H. A. Morris, and R. Jenness, *J. Dairy Sci.* **47**, 192 (1964).
27. *Code of principles concerning milk and milk products*, FAO/WHO, Food and Agriculture Organization of the United Nations, Rome, Italy, 1973, pp. 27–32.
28. C. W. Hall, A. W. Farrall, and A. L. Rippen, *Encyclopedia of Food Engineering*, Avi Publishing Co., Westport, Conn., 1971, pp. 102–103.
29. F. H. McDowell, *The Buttermakers Manual*, New Zealand Univ. Press, Wellington, N.Z., 1953, pp. 51–58.
30. C. W. Hall and T. I. Hedrick, *Drying of Milk and Milk Products*, 2nd ed., Avi Publishing Co., Westport, Conn., 1971, pp. 212–213.
31. R. F. Madsen in S. A. Goldblith, L. Rey, and W. W. Rothmayr, eds., *Freeze Drying and Advanced Food Technology*, Academic Press, Inc., New York, 1975, pp. 575–587.
32. C. W. Hall, A. W. Farrall, and A. L. Rippen, *Michigan Agr. Exp. Stn. Q. Bull.* **43**, 433 (1961).
33. R. Jenness and S. Patton, *Principles of Dairy Chemistry*, Chapman & Hall, Ltd., London, Eng., 1959, p. 125.

General References

W. S. Arbuckle, *Ice Cream*, Avi Publishing Co., Westport, Conn., 1977, 410 pp.
J. G. Brennan and co-workers, *Food Engineering Operations*, 2nd ed., Applied Science Publishers, Ltd., London, Eng., 1976, 532 pp.
J. R. Campbell and R. T. Marshall, *The Science of Providing Milk for Man*, McGraw-Hill Book Co., New York, 1975, 801 pp.
A. W. Farrall, *Engineering for Dairy and Food Products*, John Wiley & Sons, Inc., New York, 1963, 674 pp.
C. W. Hall, A. W. Farrall, and A. L. Rippen, *Encyclopedia of Food Engineering*, Avi Publishing Co., Westport, Conn., 1971, 755 pp.
C. W. Hall and T. I. Hedrick, *Drying of Milk and Milk Products*, 2nd ed., Avi Publishing Co., Westport, Conn., 1971, 234 pp.
W. J. Harper and C. W. Hall, eds., *Dairy Technology and Engineering*, Avi Publishing Co., Westport, Conn., 1976, 631 pp.
J. L. Henderson, *The Fluid Milk Industry*, 3rd ed., Avi Publishing Co., Westport, Conn., 1971, 677 pp.
M. Loncin, *Food Engineering*, Academic Press, Inc., New York, 1979, 494 pp.
B. H. Webb, A. H. Johnson, and J. A. Alford, *The Fundamentals of Dairy Chemistry*, Avi Publishing Co., Westport, Conn., 1976, 929 pp.
E. H. Marth, ed., *Standard Methods for the Examination of Dairy Products*, 14th ed., American Public Health Association, Washington, D.C., 1978.
Dairy Science Abstracts.
Food Science and Technology Abstracts (England).

CARL W. HALL
Washington State University

MINERAL NUTRIENTS

Mineral nutrients are involved in the most fundamental processes of life. The oxygen that humans breathe is utilized with the aid of two metal complexes, ie, iron-containing hemoglobin and zinc-containing carbonic anhydrase. With the evolution of life from a reducing to an oxidizing atmosphere, mechanisms involving enzymes were developed by organisms in order to protect the cells from high levels of oxygen. One such class of protective enzymes is the superoxide dismutases which contain metals, eg, manganese, copper, zinc, and iron (1).

The human skeleton is composed of calcium and phosphorus and traces of other ions, eg, magnesium and sodium embedded in an organic matrix. The regulation of body-fluid volume and acid–base balance requires sodium, potassium, and chloride. Neuromuscular excitability and blood coagulation occurs in the presence of calcium. Metabolic energy, cellular homeostasis, and most enzyme activities are dependent on phosphorus. The electron-transport chain requires copper and iron. Several vitamins contain sulfur and one contains cobalt. Hormones contain iodine, sulfur, and zinc. Each cell contains complex sets of enzymes, many of which require metal ions, either as part of the basic structure or as activators (see Enzymes; Hormones; Vitamins).

As with other biological substances, a state of dynamic equilibrium exists for the mineral nutrients, and mechanisms exist whereby the system can adjust to varying amounts of minerals in the diet. In the form usually found in foods, and under circumstances of normal human metabolism, most nutrient minerals are not toxic when ingested orally, even in amounts considerably greater than the Recommended Dietary Allowances (RDAs) of the National Academy of Science (see Table 1) (2).

Some elements that are found in body tissues have no apparent physiological role and have not been shown to be toxic, eg, rubidium, strontium, titanium, niobium, boron, germanium, and lanthanum. Other elements are toxic when found in greater than trace amounts, and sometimes in trace amounts, eg, arsenic, mercury, lead, cadmium, silver, zirconium, beryllium, and thallium. Numerous elements that have been used in medicine in nonnutrient roles include lithium, bismuth, antimony, bromine, platinum, and gold (see Fig. 1).

Important trends of investigation in the field of mineral nutrients are mineral-nutrient interactions, mineral–vitamin interactions, and interactions of mineral nutrients with toxic elements (3).

The amount of each element that is required in a daily dietary intake varies with the individual bioavailability of the mineral nutrient which depends on body need as determined by the absorption and excretion patterns of the element and by general solubility and absence of substances that cause formation of insoluble products, eg, $Ca_3(PO_4)_2$. In some cases, there are additional requirements for transport substances and a further requirement for a specific chemical compound of the element, eg, a

Table 1. Characteristics of the Mineral Nutrients

Element	Body content, mg/kg body wt	Daily requirement, mg/d[a]
Major elements		
sodium	1,500–1,600	1,100–3,300[b]
potassium	2,000–3,500	1,875–5,625[b]
magnesium	270–500	300[c,d,e]; 350[d,f]
calcium	14,000–20,000	800[d,e]
phosphorus	11,000–12,000	800[d,e]
sulfur	1,600–2,500	[g]
chlorine	1,200–1,500	1,700–5,100[b]
Trace elements		
copper	1.0–2.5	2.0–3.0[b]
zinc	33–50	15[d,e]
selenium	0.2–0.3	0.05–0.2[b]
chromium	0.06–0.2	0.05–0.2[b]
molybdenum	0.1–0.5	0.15–0.5[b]
fluorine	37	1.5–4.0[b]
iodine	0.2–0.4	0.15[d,e]
manganese	0.2–4.0	2.5–5.0[b]
iron	60–66	10[d,f]; 18[c,d,e]
cobalt	0.02	0.003[h]

[a] Ref. 2.
[b] Estimated safe and adequate daily intake, adults.
[c] Female.
[d] RDA, adults.
[e] Increased amounts required during pregnancy and lactation.
[f] Male.
[g] Adequate intake with adequate intake of protein.
[h] As vitamin B_{12}.

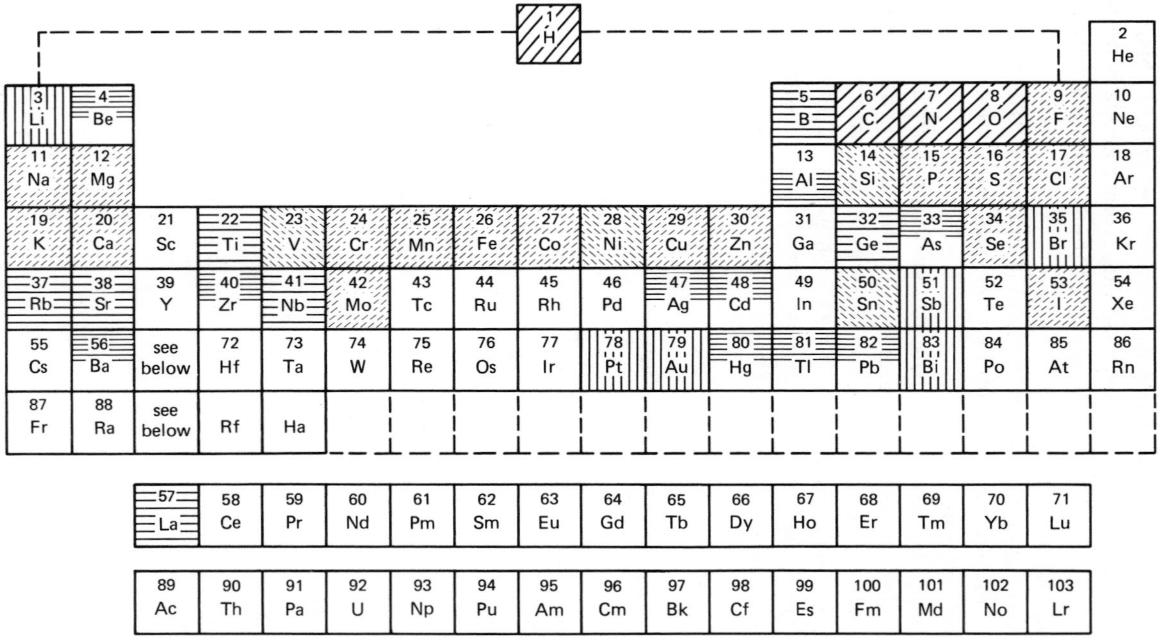

Figure 1. Periodic table showing elements of importance in biological systems. ▨ = Principal element of bioorganic compounds. ▨ = Essential mineral nutrient (humans, animals). ▨ = Essential mineral nutrient (animals, possibly humans). ▨ = Present in body, not known to be a nutrient or toxic element. ▥ = Element present in body, possibly toxic. ▤ = Element used in medicine. ▥ = Element generally poisonous. ▤ = Present in body, possibly toxic.

calcium-binding protein for transport or an intrinsic factor for vitamin B_{12} (cobalt) uptake.

The essential mineral nutrients are grouped as major elements and trace elements; the distinction is the relative amounts in the dietary requirement. Certain characteristics of the mineral nutrients that can be summarized uniformly are presented in Tables 2–6. Basic sources of information for this article were standard treatises (4–20).

Table 2.   Normal Blood Plasma or Serum Values for Selected Mineral Nutrients [a]

Major elements		Concentration, mg/100 mL
sodium	serum	310–340
potassium	serum	14–20
magnesium	serum	1–3
calcium	serum	9.0–10.6
phosphorus	serum	3.0–4.5
sulfur (as sulfate)	plasma or serum	0.5–1.5 [b]
chlorine (as chloride)	serum	360–375
Trace elements		Concentration, µg/100 mL
copper	serum	100–200
zinc	serum or plasma	72–120
	whole blood	408–1170
selenium	whole blood	0.13–0.34
chromium	serum	0.5–3.1
molybdenum	whole blood	1.35–1.59
fluorine	whole blood	280
iodine	serum (bound)	4–8
manganese	whole blood	2.4–6.9
iron	serum	65–175
cobalt	whole blood	0.35–6.3

[a] Refs. 2–20.
[b] Meq/L.

The Major Elements

Sodium and Potassium.   Sodium ion is the most abundant cation in the extracellular fluid; most of the potassium ion is present in the intracellular fluid although a small amount is required in the extracellular fluid to maintain normal muscle activity. Cation and anion composition of extracellular and intracellular fluids is given in Figure 2.

Metabolic Functions.   Sodium ion acts in concert with other electrolytes, in particular K^+, to regulate the osmotic pressure and to maintain the appropriate water balance of the body and the acid–base balance (pH); homeostatic control of these functions is accomplished by the lungs and kidneys interacting by way of the blood (21–22). Sodium is essential for glucose absorption and transport of other substances across cell membranes and is involved, as is K^+, in transmitting nerve impulses and in muscle relaxation. Potassium ion acts as a catalyst in the intracellular fluid, in energy metabolism, and is required for carbohydrate and protein metabolism.

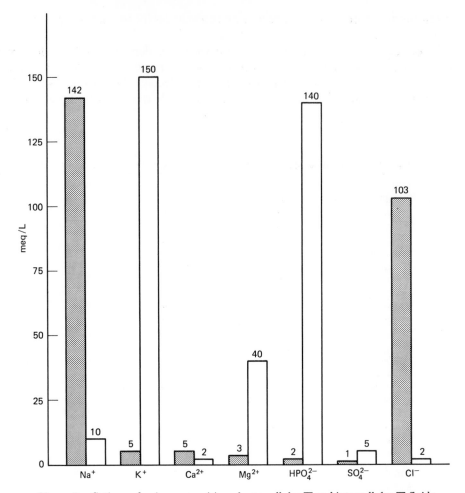

Figure 2.   Cation and anion composition of extracellular ▨ and intracellular ☐ fluids.

Active Transport: The Na+/K+ Pump.   Maintenance of the appropriate concentrations of K^+ and Na^+ in the intra- and extracellular fluids involves active transport, ie, a process requiring energy (23). Sodium ion in the extracellular fluid (0.136–0.145 M Na^+) diffuses passively and continuously into the intracellular fluid (< 0.01 M Na^+) and must be removed. Sodium ion is pumped from the intracellular to the extracellular fluid and K^+ is pumped from the extracellular (ca 0.004 M K^+) to the intracellular fluid (ca 0.14 M K^+) (23–25). The energy for these processes is provided by hydrolysis of adenosine triphosphate (ATP) which requires the enzyme Na^+/K^+ ATPase, a membrane-bound enzyme which is widely distributed in the body. In some cells, eg, brain and kidney, 60–70 wt % of the ATP is used to maintain the required Na^+–K^+ distribution.

Sodium and potassium ions are actively absorbed from the intestine. As a consequence of the electrical potential caused by transport of these ions, an equivalent

Table 3. Primary Sites of Absorption and Excretion of Mineral Nutrients

Nutrient	Absorption[a]	Excretion[b]
Major elements		
sodium	large intestine, ileum (a); jejunum (f); stomach, skin (p); some two-step absorption in parts of small intestine (p/a)	kidney and intestine; some from skin as perspiration
potassium	small intestine (p)	kidney and skin
magnesium	ileum (a)	kidney and skin; very small amount from intestine
calcium	duodenum and jejunum (a, f)	kidney and intestine, from the latter as digestive juices
phosphorus	small intestine (a)	kidney
sulfur	small intestine (a, f transport of S-containing amino acids)	kidney and intestine, from the latter as bile acids
chlorine	absorbed with Na^+, K^+, and Ca^{2+}	kidney, intestine, and skin
Trace elements		
copper	stomach and upper intestine with low pH (p, f)	intestine as bile and pancreatic enzymes; kidney
zinc	duodenum (f)	intestine as bile and pancreatic juices; skin as perspiration; almost none from kidney
selenium	duodenum (possibly (p))	kidney; intestine as bile and pancreatic juices; lungs (in expired air) if excess is ingested
chromium	small intestine (possibly (p))	kidney
molybdenum	small intestine (possibly (f))	kidney and skin (as perspiration)
fluorine	stomach (p); possibly intestine	kidney
iodine	small intestine, entire gastro-intestinal tract as I^- (p)	kidney
manganese	small intestine in two-step mechanism (p/a)	intestine as bile and pancreatic juices
iron	duodenum and jejunum (a, f)	no significant excretion mechanism; skin as perspiration, exfoliation of cells, eg, intestinal, etc
cobalt	as B_{12} in ileum (f); as inorganic Co (a, f)	kidney; skin as perspiration

[a] (a) = active; (p) = passive; (f) = facilitated.
[b] Usually some fraction of the mineral nutrient that is ingested is not absorbed and passes into the feces. Modes of excretion listed in the table pertain only to the fraction of the mineral nutrient that is absorbed.

quantity of Cl^- is absorbed; the resulting osmotic effect causes absorption of water (26).

Excretion and Reabsorption of Na^+ and K^+. Selective excretion and reabsorption of Na^+ and K^+ are accomplished by means of the kidney tubular cell membranes (21,25). Water, Na^+, and Cl^- passively diffuse into the proximal tubular cells. Potassium ion is pumped into the cells and Na^+ is pumped out by the Na^+/K^+ pump. In the extracellular fluid, Na^+ and K^+ account for 90–92 wt % and 3 wt %, respectively, of the cations; in the intracellular fluid, the distribution is ca 70–80 wt % K^+ and 6 wt % Na^+. In the adult, ca 160 L of fluid per day is filtered. Urinary volume is 0.5–2.5 L/d so that ca 99 wt % of the filtered water is reabsorbed.

Table 4. Some Food Sources Rich in Mineral Nutrients [a]

		Mineral	
	Wt or vol of food	mg	μg
Major elements			
Sodium			
Canadian bacon	113 g (4 oz)	2144	
table salt	5 mL (1 tsp)	2000	
corned beef	113 g (4 oz)	1474	
tuna, canned	113 g (4 oz)	988	
pickle	large	1428	
ham, cured	113 g (4 oz)	861	
cheese, process American	28 g (1 oz)	406	
vegetables, canned (beans, carrots, peas)	113 g (4 oz)	267	
Potassium			
split peas, cooked	275 mL (1 cup)	1790	
sunflower seeds	275 mL (1 cup)	1334	
raisins	275 mL (1 cup)	1259	
prunes, dried	275 mL (1 cup)	1117	
peanuts (shelled)	275 mL (1 cup)	1009	
soybeans, cooked	275 mL (1 cup)	972	
potato, baked	large	782	
kidney beans (canned)	275 mL (1 cup)	670	
avocado	0.5 average	600	
banana	average (150 g)	550	
orange juice	227 g (8 oz)	503	
Magnesium			
wheat germ	275 mL (1 cup)	336	
wheat bran	275 mL (1 cup)	279	
corn meal, whole grain	276 mL (1 cup)	125	
chocolate, baking	28 g (1 oz)	82	
molasses, blackstrap	15 mL (1 tbsp)	52	
oysters	113 g (4 oz)	36	
shrimp	113 g (4 oz)	47	
Calcium			
sockeye salmon, canned with bones	275 mL (1 cup)	570	
Parmesan cheese	28 g (1 oz)	336	
whole milk	275 mL (1 cup)	290	
mustard greens, cooked	275 mL (1 cup)	284	
Swiss cheese	28 g (1 oz)	272	
turnip greens, cooked	275 mL (1 cup)	267	
Cheddar cheese	28 g (1 oz)	211	
kale, cooked	275 mL (1 cup)	206	
Muenster cheese	28 g (1 oz)	203	
broccoli, cooked	275 mL (1 cup)	136	
Phosphorus			
sunflower seeds	275 mL (1 cup)	1214	
almonds	275 ml (1 cup)	791	
black walnuts	275 mL (1 cup)	713	
peanuts, shelled	275 mL (1 cup)	586	
split peas, cooked	275 mL (1 cup)	536	
brains	113 g (4 oz)	354	
soybeans, cooked	275 mL (1 cup)	322	
salami	113 g (4 oz)	321	
whitefish	113 g (4 oz)	306	
chicken, meat only	113 g (4 oz)	300	

Table 4 (*continued*)

	Wt or vol of food		Mineral	
			mg	μg
Sulfur[b]				
turkey	113 g (4 oz)	1120		
chicken	113 g (4 oz)	920		
fish	113 g (4 oz)	880		
liver	113 g (4 oz)	810		
veal	113 g (4 oz)	770		
lamb	113 g (4 oz)	760		
cottage cheese	140 mL (0.5 cup)	680		
whole grain flour	275 mL (1 cup)	590		
buckwheat flour	275 mL (1 cup)	420		
egg	large	350		
Chloride[c]				
table salt	5 mL (1 tsp)	3500		
pickle	large	2204		
tuna, canned	113 g (4 oz)	1525		
vegetables, canned (beans, carrots, peas)	113 g (4 oz)	412		
hamburger	113 g (4 oz)	133		
Trace elements				
Copper				
veal liver	113 g (4 oz)	9.0		
lamb liver	113 g (4 oz)	6.35		
beef liver	113 g (4 oz)	3.17		
sunflower seeds	275 mL (1 cup)	2.57		
lobster	113 g (4 oz)	2.49		
bakers' yeast	28 g (1 oz)	1.96		
beef consomme	275 mL (1 cup)	1.85		
crab, steamed	113 g (4 oz)	1.47		
oysters	113 g (4 oz)	1.36		
wheat germ	275 mL (1 cup)	1.3		
Zinc				
oysters	113 g (4 oz)	85.0		
crab, steamed	113 g (4 oz)	5.0		
beef	113 g (4 oz)	4.0		
pork	113 g (4 oz)	4.0		
almonds	275 mL (1 cup)	4.0		
walnuts and pecans	275 mL (1 cup)	3.0		
lobster	113 g (4 oz)	2.1		
chicken	113 g (4 oz)	2.0		
tuna, canned	275 mL (1 cup)	1.8		
mango	average	1.4		
Selenium[d]				
coconut				
onions				
brewer's yeast				
grains grown on seleniferous soils (see text)				
liver (pork, calf, lamb)				
herring				
Chromium[d]				
brewer's yeast				
corn				
wheat germ				

Table 4 (*continued*)

	Wt or vol of food	Mineral	
		mg	μg
whole-grain cereals			
clams			
paprika			
Molybdenum[d]			
legumes			
liver			
dark green vegetables			
whole-grain cereals			
milk			
Fluorine[d]			
tea			
fluoridated water			
seafood			
Iodine[d]			
seafoods			
kelp			
iodized salt			
Manganese			
chestnuts	275 mL (1 cup)	5.9	
filberts	275 mL (1 cup)	5.7	
Brazil nuts, raw	275 mL (1 cup)	3.9	
barley, dry	275 mL (1 cup)	3.4	
brown rice	275 mL (1 cup)	3.2	
almonds, raw	275 mL (1 cup)	2.7	
sunflower seeds	275 mL (1 cup)	2.6	
sesame seeds	275 mL (1 cup)	2.4	
peanuts, roasted	275 mL (1 cup)	2.2	
carrot	large	2.2	
Iron[e]			
millet, whole grain, dry	275 mL (1 cup)	15.5	
lamb liver	113 g (4 oz)	12.4	
bran flakes, fortified	275 mL (1 cup)	12.4	
prune juice	275 mL (1 cup)	10.5	
veal liver	113 g (4 oz)	10.0	
wheat germ, raw	275 mL (1 cup)	9.4	
chicken liver	113 g (4 oz)	9.0	
wheat bran, raw	275 mL (1 cup)	8.5	
apricots, dried	275 mL (1 cup)	7.2	
wild rice, raw	275 mL (1 cup)	6.7	
Cobalt[f]			
lamb liver	113 g (4 oz)		118.0
beef liver	113 g (4 oz)		90.8
turkey liver	113 g (4 oz)		68.0
veal liver	113 g (4 oz)		68.0
chicken liver	113 g (4 oz)		28.2
liverwurst	113 g (4 oz)		15.8
brains	113 g (4 oz)		4.5
lamb	113 g (4 oz)		2.0
beef, lean	113 g (4 oz)		2.0
chicken breast	113 g (4 oz)		0.5
Silicon[d]			
high fiber grains (eq, oats)			
Barbados brown sugar			
organ meats			

Table 4 (*continued*)

| | Wt or vol of food | Mineral | |
		mg	μg
Tin[d]			
foods preserved in tin cans			
(eq, fruit juices, vegetables, fish)			
grains			
Vanadium[d]			
seafoods			
grains and cereals			
seeds			
milk powder			
Nickel[d]			
grains			
vegetables			
oysters			

[a] Refs. 16, 17, and 20.
[b] Calculated as sulfur-containing amino acids, methionine plus cystine.
[c] Calculated from sodium.
[d] Accurate quantitative data not yet available; best reported sources listed.
[e] Not all iron from different sources is equally absorbed; see text for discussion.
[f] Cobalt as μg of vitamin B_{12}.

The volume of extracellular fluid is directly related to the Na^+ concentration which is closely controlled by the kidneys. Homeostatic control of Na^+ concentration depends on the hormone aldosterone (see Hormones, adrenal-cortical hormones). The kidney secretes a proteolytic enzyme, rennin, which is essential in the first of a series of reactions leading to aldosterone. In response to a decrease in plasma volume and Na^+ concentration, the secretion of rennin stimulates the production of aldosterone which results in increased sodium retention and increased volume of extracellular fluid (21,25).

Sodium and Hypertension. Salt-free or low salt diets often are prescribed for hypertensive patients (27–28). However, sodium chloride increases the blood pressure in some individuals but not in others; conversely, restriction of dietary NaCl lowers the blood pressure of some hypertensives but not of others. Genetic factors and other nutrients, eg, Ca^{2+} and K^+, may be involved. The optimal intakes of Na^+ and K^+ remain to be established (29).

Magnesium. In the adult human, 50–70% of the magnesium is in the bones associated with calcium and phosphorus. The rest is widely distributed in the soft tissues and body fluids; in these, most of the Mg^{2+}, like K^+, is located in the intracellular fluid in which it is the most abundant divalent cation. Magnesium ion is efficiently retained by the kidney when the plasma concentration of Mg^{2+} falls; in this respect it resembles Na^+. The functions of Na^+, K^+, Mg^{2+}, and Ca^{2+} are interrelated so that a deficiency of Mg^{2+} affects the metabolism of the other three ions (30).

Metabolic Functions. Magnesium is essential in numerous metabolic processes because it is the activator of many enzymes, eg, alkaline phosphatases and the phosphokinases, pyrophosphatases, and thiokinases (31–33). Because the phosphokinases are required for the hydrolysis and transfer of phosphate groups, magnesium is es-

Table 5. Some Disorders Associated with Deficiency, Excess[a], and/or Faulty Utilization of the Mineral Nutrients[b]

Nutrient	Deficiency	Excess	Faulty utilization
Major elements			
sodium	muscle weakness; nausea	hypertension	Addison's disease; Cushing's disease
potassium	muscle weakness		Addison's disease; Cushing's disease
magnesium	neuromuscular irritability; convulsions; muscle tremors; mental changes (confusion, disorientation, hallucinations); heart disease; kidney stones	Mg intoxication (drowsiness, stupor, coma)	kidney stones
calcium	hypocalcemia; tremor; rickets; osteomalacia; osteoporosis; muscle spasm (tetany); possibly heart disease		rickets; osteomalacia; osteoporosis; muscle spasm (tetany); possibly heart disease; Paget's disease
phosphorus	rickets; osteomalacia; osteoporosis		rickets; osteomalacia; osteoporosis, Paget's disease; renal rickets (vitamin D-resistant rickets)
sulfur			homocystinuria
chlorine[c]	impaired growth in infants		
Trace elements			
copper	impaired elastin formation; impaired hemopoiesis		Menkes' kinky-hair syndrome; Wilson's disease
zinc	mental retardation; delayed sexual maturity; dwarfism; sterility; slow wound healing; hypogeusia		acrodermatitis enteropathica
selenium[d]	heart disease; increased cancer		
chromium	impaired glucose tolerance; possibly atherosclerosis		
molybdenum[d]			
fluorine		fluorosis; mottled teeth	
iodine	hypothyroidism; cretinism; myxedema; goiter		hyperthyroidism; Grave's disease
manganese	impaired mucopolysaccharide synthesis		
iron	anemia	hemochromatosis; Bantu siderosis	hemochromatosis
cobalt	pernicious anemia	polycythemia	pernicious anemia

[a] Excess of a dietary nature is considered; industrial toxicities are not included in the table.
[b] Refs. 4–20.
[c] Abnormalities of chlorine metabolism usually accompany those of sodium metabolism.
[d] For toxicity see text.
[e] No known deficiency disease.

Table 6. Toxicities of the Mineral Nutrients [a]

Nutrient	Compound	Animal	LD$_{50}$ (oral) mg/kg body wt	LD$_{Lo}$[b] (oral) mg/kg body wt
Major elements				
sodium	sodium acetate	rat	3,530	
		mouse	4,960	
	sodium chloride	human		500
		rat	3,000	
		mouse	4,000	
	sodium bicarbonate	human		500
		rat	4,220	
	sodium carbonate	rat		4,000
potassium	potassium acetate	rat	3,250	
	potassium chloride	human		500
		infant		938
		rat		2,430
		guinea pig	2,500	
	potassium carbonate	rat	1,870	
magnesium	magnesium chloride	human		500
		rat	2,800	
	magnesium chloride hexahydrate	rat	8,100	
	magnesium sulfate	mouse		5,000
		rabbit		3,000
calcium	calcium acetate monohydrate	rat	4,280	
	calcium chloride	rat	1,000	
		rabbit		1,384
	calcium carbonate	human		5,000
phosphorus	sodium monohydrogen phosphate heptahydrate	rat	12,930	
chlorine	sodium chloride (see sodium)			
	potassium chloride (see potassium)			
Trace elements				
copper	copper(II) chloride	rat	140	
		mouse	190	
		guinea pig	31	
	copper sulfate	child		200
		human		50
		rat	300	
		duck		600
	copper sulfate pentahydrate	human		1,088
		rat	300	
		pigeon		1,000
zinc	zinc acetate dihydrate	rat	2,460	
	zinc chloride	human		50
		rat	350	
		mouse	350	
		guinea pig	200	
	zinc sulfate	human		50
		rat		2,200
		rabbit		2,000
	zinc sulfate heptahydrate	rat		2,200
		rabbit		1,914
selenium	sodium selenite	dog		4
		rabbit	2.2	7
		rat	7	
		mouse	7	
		guinea pig	5.1	

581

Table 6 (*continued*)

Nutrient	Compound	Animal	LD$_{50}$ (oral) mg/kg body wt	LD$_{Lo}$[b] (oral) mg/kg body wt
Trace elements (*continued*)				
	sodium selenate	human		5
		rat	2.5	
		rabbit	4	
chromium	chromium(III) chloride	rat	1,870	
	chromium(III) chloride hexahydrate	rat	101	
molybdenum	calcium molybdate	rat	101	
	sodium phosphomolybdate	rat	3,900	
fluorine	sodium fluoride	human		75
		rat	180	
		mouse		97
		dog		75
		rabbit		100
	potassium fluoride	rat	245	
	calcium fluoride	mouse		28
	sodium hexafluorosilicate	human		50
		rat	125	
		rabbit		125
	potassium hexafluorosilicate	guinea pig	500	
	magnesium hexafluorosilicate hexahydrate	guinea pig	200	
	zinc fluorosilicate	rat	200	100
	sodium fluoroaluminate	rat	200	
		rabbit		9,000
iodine	sodium iodide	human		500
		rat	4,340	
		mouse		1,650
	potassium iodide	human		500
		mouse		1,862
manganese	manganese acetate	rat	2,940	
	manganese acetate tetrahydrate	rat	3,730	
iron	iron(III) chloride heptahydrate	rat		900
	iron(II) chloride tetrahydrate	rat	984	
		rabbit		890
	iron(II) sulfate	child		390
		woman		200
		rat	319	
		mouse	979	
	iron(II) sulfate heptahydrate	rat		1,989
		mouse	1,520	
silicon	sodium silicate (2:1)	rat	1,300	
	sodium silicate (3:1)	rat	1,600	
	sodium metasilicate	human		500
		dog		250
		pig		250
vanadium	vanadyl chloride	rat	140	
	ammonium vanadate	rat	18	
	monosodium vanadate	rat		200
		rabbit		200
	trisodium vanadate	rabbit		100
tin	tin(II) chloride	rat	700	
		mouse	1,200	
		dog		500
		rabbit		40

Table 6 (*continued*)

Nutrient	Compound	Animal	LD_{50} (oral) mg/kg body wt	$LD_{Lo}{}^{b}$ (oral) mg/kg body wt
Trace elements (*continued*)				
nickel	nickel acetate	rat	350	
		mouse	410	

[a] Ref. 158.

[b] The lowest lethal dose, other than that for LD_{50}, of a substance introduced by any route other than inhalation over any period of time and reported to have caused death when introduced in one or more divided portions.

sential in glycolysis and in oxidative phosphorylation (see Phosphorus). The thiokinases are required for the initiation of fatty-acid degradation. Magnesium is also required in systems in which thiamine pyrophosphate is a coenzyme.

As an activator of the phosphokinases, magnesium is essential in energy-requiring biological processes, eg, activation of amino acids, acetate, and succinate; synthesis of proteins (qv), fats, coenzymes, and nucleic acids; generation and transmission of nerve impulses; and muscle contraction (33) (see Amino acids, survey; Fats and fatty oils).

Regulation of Serum Mg^{2+} Concentration. Regulation of serum Mg^{2+} appears to result from a balance among intestinal absorption, renal reabsorption, and excretion (31,34). The controlling factor is probably the renal threshhold (35). In the normal adult, intestinal absorption is, to a large extent, proportional to the Mg^{2+} supplied in the diet. The system responds to a wide range of dietary intake by increasing or decreasing urinary excretion; the plasma Mg^{2+} concentration varies only within the normal range. Although exchange of bone salt Mg^{2+} with blood Mg^{2+} is slow, as compared to exchange of Ca^{2+}, the bone salts serve as a reservoir of Mg^{2+} to buffer depletions developing over long periods, ie, weeks or months (36). Parathyroid hormone may be involved in mobilizing bone Mg^{2+} and increasing tubular reabsorption of Mg^{2+} but not to as great an extent as with Ca^{2+}; it may also increase absorption of Mg^{2+} from the intestine.

Magnesium Deficiency. A severe magnesium deficiency in humans is seldom encountered except as a secondary effect resulting from numerous disease states, eg, chronic alcoholism with malnutrition, acute or chronic renal disease, long-term Mg^{2+}-free parenteral feeding, protein-calorie malnutrition, and hyperthyroidism. In these situations, it is difficult to attribute specific clinical manifestations to magnesium deficiency (31).

Magnesium ion is essential for normal Ca^{2+} and K^+ metabolism. In acute experimental magnesium deficiency in humans, hypocalcemia occured despite adequate calcium intake and absorption and despite normal renal and parathyroid functions. Negative K^+ balance was also observed. All biochemical and clinical abnormalities disappear with restoration of adequate amounts of magnesium to the diet (31).

Magnesium supplements (as MgO) have been used successfully in the treatment of patients with a history of calcium oxalate stone formation (37–40). Marginal magnesium deficiencies may occur in areas where food crops are grown on magnesium-deficient soil. One such area is a narrow strip of the Atlantic coast of the United States extending from Pennsylvania to Florida, sometimes called the "stone belt" because of the high incidence of calcium oxalate kidney stones (37). According to some authorities, the stone formation is the consequence of the low magnesium intake. Evidently a high Mg^{2+} concentration in the kidney increases the solubility of calcium oxalate; no completely satisfactory explanation for this has been given.

Calcium.   Calcium is the most abundant mineral element in mammals comprising 1.5–2.0 wt % of the adult human body; over 99 wt % of the calcium that is present occurs in bones and teeth (41). The normal calcium content of blood is 9–11 mg Ca/100 mL. About 48% of the serum calcium is ionic; ca 46% is bound to blood proteins; the rest is present as diffusible complexes, eg, of citrate (42). The calcium ion level must be maintained within definite limits.

Metabolic Functions.   Bones act as a reservoir of certain ions, in particular Ca^{2+} and PO_4^{3-}, which readily exchange between bones and blood. Bone structure comprises a strong organic matrix combined with an inorganic phase which is principally hydroxyapatite, $3Ca_3(PO_4)_2.Ca(OH)_2$. Bones contain two forms of hydroxyapatite. The less soluble crystalline form contributes to the rigidity of the structure. The crystals are quite stable but, because of their small size, they present a very large surface area available for rapid exchange of ions and molecules with other tissues. There is also a more soluble intercrystalline fraction. Bone salts also contain small amounts of magnesium, sodium, carbonate, citrate, chloride, and fluoride (41).

Calcium is necessary for blood-clot formation; Ca^{2+} stimulates release of blood-clotting factors from platelets (41) (see Blood, coagulants and anticoagulants). Neuromuscular excitability depends on the relative concentrations of Na^+, K^+, Ca^{2+}, Mg^{2+}, and H^+ (30). With a decrease in Ca^{2+} concentration (hypocalcemia), excitability increases; if this condition is not corrected, the symptoms of tetany appear (muscular spasm, tremor, even convulsions). Too great an increase in Ca^{2+} concentration (hypercalcemia) may impair muscle function to such an extent that respiratory or cardiac failure may occur.

Contraction of muscle follows increase of Ca^{2+} in the muscle cell as a result of nerve stimulation and this initiates processes causing the proteins myosin and actin to be drawn together making the cell shorter and thicker. The return of the Ca^{2+} to its storage site, the sarcoplasmic reticulum, by an active pump mechanism allows the contracted muscle to relax (43). Calcium ion is also a factor in the release of acetylcholine on stimulation of nerve cells; it influences the permeability of cell membranes; activates enzymes, eg, ATPase, lipase, and some proteolytic enzymes; and facilitates intestinal absorption of vitamin B_{12} (44).

Blood Ca^{2+} Level.   In the normal adult, the blood Ca^{2+} level is established by: equilibrium between the blood Ca^{2+} and the more soluble intercrystalline calcium salts of the bone; and by a subtle and intricate feedback mechanism (responsive to the Ca^{2+} concentration of the blood) that involves the less soluble crystalline hydroxyapatite. Also participating in Ca^{2+} control are the thyroid and parathyroid glands, the liver, kidney, and intestine. The salient features of the mechanism are summarized in Figure 3 (45–47).

Major factors that control calcium homeostasis are calcitonin, parathyroid hormone (PTH), and a vitamin D metabolite. Calcitonin (a polypeptide of 32 amino acid residues, mol wt ca 3600) is synthesized by the thyroid gland; its release is stimulated by small increases in the blood Ca^{2+} concentration. Parathyroid hormone (a polypeptide of 83 amino acid residues, mol wt 9500) is produced by the parathyroid glands.

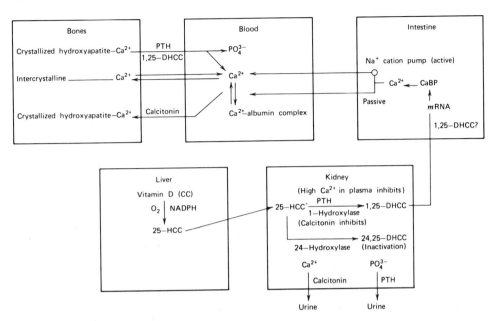

Figure 3. Homeostatic control of blood Ca^{2+} level. PTH = parathyroid hormone. CC = cholecalciferol (vitamin D_3). HCC = hydroxycholecalciferol. DHCC = dihydroxycholecalciferol. CaBP = calcium-binding protein. NADPH = protonated nicotinamide-adenine dinucleotide phosphate.

Its release is activated by a decrease of blood Ca^{2+} level below normal. The sites of action of calcitonin are the bones and kidneys. It increases bone calcification and, thereby inhibits resorption. In the kidney, it inhibits Ca^{2+} reabsorption and increases Ca^{2+} excretion in urine. Calcitonin operates via a cAMP (cyclic adenosine monophosphate) mechanism. Parathyroid hormone increases blood Ca^{2+} concentration by increasing resorption of bone, renal reabsorption of calcium, and absorption of calcium from the intestine. A cAMP mechanism is also involved in the action of PTH. Parathyroid hormone also induces formation of 1-hydroxylase in the kidney, required in formation of the active metabolite of vitamin D.

Metabolites of vitamin D (cholecalciferol, CC) are essential in maintaining the appropriate blood level of Ca^{2+}. The active metabolite is 1,25-dihydroxycholecalciferol (1,25-DHCC) which is synthesized in two steps. In the liver, CC is hydroxylated to 25-hydroxycholecalciferol (25-HCC) which in combination with a globulin carrier, is transported to the kidney where it is converted to 1,25-DHCC, a step that requires 1-hydroxylase whose formation is induced by PTH. This may be the controlling step in regulating Ca^{2+} concentration. The sites of action of 1,25-DHCC are the bones and the intestine. Formation of 1,25-DHCC is limited by an inactivation process, ie, conversion of 25-HCC to 24,25-DHCC which is catalyzed by 24-hydroxylase.

Calcium is absorbed from the intestine by facilitated diffusion and active transport. In the former, Ca^{2+} moves from the mucosal to the serosal compartments along a concentration gradient. The active transport system requires a cation pump. In both processes, a calcium-binding protein, CaBP, is thought to be required for the transport. Synthesis of CaBP is activated by 1,25-DHCC. In the active transport, release of Ca^{2+} from the mucosal cell into the serosal fluid requires Na^+.

Paget's Disease of Bone.   Paget's disease (osteitis deformans) occurs mainly in people over 40; about twice as many men as women are affected. The disease may be mild and asymptomatic and require little or no treatment. The clinical signs are high alkaline phosphatase and high urine hydroxyproline as well as abnormal bone structure which is usually unrecognized until discovered accidentally by routine x-ray examination (48).

About 10% of the cases are symptomatic, ie, considerable disability, even crippling may occur. In these cases, the disease is generalized; it affects many bones including the long bones of the legs, the pelvis and the skull. The bones soften and buckle, the skull may become enlarged, height may decrease (if the spine is involved) and a bent-over stooped posture may result. In addition there may be severe bone pain and other neurologic complications such as deafness.

One method of treatment is to inject calcitonin, which decreases blood Ca^{2+} concentration and increases bone calcification (49). Another is to increase the release of calcitonin into the blood by increasing the blood level of Ca^{2+} (50). This is accomplished by increasing Ca^{2+} absorption from the intestine which requires dietary calcium supplements and avoidance of high phosphate diets, which decrease Ca^{2+} absorption by precipitation of insoluble calcium phosphate.

Phosphorus.   Eighty-five percent of the phosphorus in the body occurs in bones and teeth (51). There is constant exchange of calcium and phosphorus between bones and blood but very little turnover of ions in teeth (41) (see Calcium). Phosphorus is the second most abundant element in the human body; the ratio of Ca:P in bones is constant at ca 2:1. Every tissue and cell contains phosphorus, generally as a salt or ester of mono-, di-, or tribasic phosphoric acid. Phosphorus is involved in a large number and wide variety of metabolic functions, eg, carbohydrate metabolism (52–53), ATP from fatty acid metabolism (54), and oxidative phosphorylation (52,55).

Energy-Rich Compounds.   Reactions of energy-rich compounds are required to drive the many endergonic metabolic processes, eg, active transport, muscle contraction, and biosynthesis of fats and macromolecules, eg, nucleic acids and proteins. Energy-rich compounds contain "high energy" bonds. The negative free energy for breaking these bonds, eg, a P—O bond, by hydrolysis is large. Most of the high energy compounds are phosphates, eg, adenosine triphosphate (ATP).

Two and twelve moles of ATP are produced respectively per mole of glucose consumed in the glycolytic pathway and each turn of the Krebs (citrate) cycle. In fat metabolism, many high energy bonds are produced per mole of fatty ester oxidized, eg, 129 high energy phosphate bonds per mole of palmitate. Oxidative phosphorylation has a remarkable efficiency of 75% (three moles of ATP per transfer of two electrons compared to the theoretical four). The process occurs via a series of reactions involving flavoproteins, quinones, eg, coenzyme Q, and cytochromes (see Iron).

Carbohydrate Metabolism.   The formation of phosphate esters is the essential initial process in carbohydrate metabolism. The glycolytic (anaerobic, Embden-Meyerhof) pathway comprises a series of nine such esters. The phosphogluconate pathway, starting with glucose, comprises a succession of twelve phosphate esters.

Phospholipids.   Phospholipids are components of every cell membrane and are active determinants of membrane permeability. They are sources of energy, components of certain enzyme systems, and are involved in lipid transport in plasma. Because of their polar nature, they can act as emulsifying agents (56). The structure of most phospholipids resembles that of triglycerides except that one fatty acid radical has been replaced by a radical derived from phosphoric acid and a nitrogen base.

Nucleic Acids. Phosphorus is an essential component of the nucleic acids which are polymers consisting of chains of nucleosides, ie, sugar plus a nitrogenous base, joined by phosphate groups (57–58). In ribonucleic acids (RNA), the sugar is D-ribose; in deoxyribonucleic acids (DNA), the sugar is 2-deoxy-D-ribose. Both contain four different bases and three are the same for both sugars, ie, adenine (A), cytosine (C), and guanine (G). The fourth is thymine (T) in RNA and uracil (U) in DNA.

Other Metabolic Functions. The ester, cyclic adenosine monophosphate (cAMP), which is produced from ATP, is involved in a large number of cellular reactions including glycogenolysis, lipolysis, active transport of amino acids, and synthesis of protein (59). Inorganic phosphate ions are involved in controlling the pH of blood (21). The principal anion of intracellular fluid is HPO_4^{2-} (see Fig. 2) (21).

Sulfur. Sulfur is present in every cell in the body, primarily in proteins in the amino acids methionine, cystine, and cysteine. Inorganic sulfates and sulfides occur in small amounts relative to the total body sulfur, but the compounds that contain them are very important to metabolism (60–61). Sulfur intake is thought to be adequate if protein intake is adequate and sulfur deficiency has not been reported.

Sulfur is part of several vitamins and cofactors, eg, thiamine, pantothenic acid, biotin, and lipoic acid. Mucopolysaccharides, eg, heparin and chondroitin sulfate, contain a monoester of sulfuric acid with an HSO_3^- group. Sulfur-containing lipids isolated from brain and other tissues usually are sulfate esters of glycolipids. The sulfur-containing amino acid, taurine, is conjugated to bile acids (60). Labile sulfur is attached to the nonheme iron in stoichiometric amounts in the respiratory chain where it is associated with the flavoproteins and cytochrome b (see Iron) (62).

Disulfides. The A and B chains of insulin are connected by two disulfide bridges and there is an intrachain cyclic disulfide link on the A chain (see Insulin and other antidiabetic agents). Vasopressin and oxytocin also contain disulfide links (63). Oxidation of thiols to disulfides and reduction of the latter to thiols are quite common and important in biological systems, eg, cysteine to cystine, reduced lipoic acid to oxidized lipoic acid (see Sulfur compounds). Many enzymes depend on free SH groups for activation–deactivation reactions. The oxidation–reduction of glutathione (glu-cys-gly) depends on the sulfhydryl group from cysteine (see Selenium).

Sulfur in Fat Metabolism. Although sulfur is in the same periodic group (VIA) as oxygen because of similar outermost electronic structure, it functions much more like its neighbor in Group VA, phosphorus, in biological systems. In fat metabolism, sulfur plays a key role analogous to that of phosphorus in carbohydrate metabolism. Fatty acid synthesis and degradation begin and end with the same compound, acetyl-*S*-coenzyme A (acetyl-SCoA) (64).

Detoxification. Detoxification systems in the human body often involve reactions that utilize sulfur-containing compounds, eg, reactions in which sulfate esters of potentially toxic compounds are formed, rendering them less toxic or nontoxic, and acetylation reactions involving acetyl-SCoA (60). Another important compound is *S*-adenosylmethionine (SAM), the active form of methionine, which acts as a methylating agent, eg, in detoxification reactions such as the methylation of pyridine derivatives, and in the formation of choline, creatine, carnitine, and epinephrine (65) (see also Radioprotective agents).

Chlorine. *Metabolic Functions.* The chlorides are essential in the homeostatic processes maintaining fluid volume, osmotic pressure, and acid-base equilibrium (5). Most of the chloride is in the body fluids and a little is in bone salts. It is the major anion accompanying Na^+ in the extracellular fluid. Less than 15 wt % of the Cl^- is associated with K^+ in the intracellular fluid in which the major inorganic anions are HPO_4^{2-} and HCO_3^-. Chloride passively and freely diffuses between intra- and extracellular fluids through the cell membrane.

If chloride diffuses freely, but most of it remains in the extracellular fluid, it follows that there is some restriction on the diffusion of phosphate; the nature of this restriction does not appear to be conclusively established. There may be a transport device (66). Moreover, cell membranes may not be very permeable to phosphate ions so that their loss from intracellular fluid is minimized (67).

Some of the blood Cl^- is used for formation in the gastric glands of HCl, which is required for digestion (68). Hydrochloric acid is secreted into the stomach where it acts with the gastric enzymes in the digestive processes. The chloride is then reabsorbed with other nutrients into the blood stream. Chloride is actively transported in gastric and intestinal mucosa. In the kidney, chloride is passively reabsorbed in the thin ascending loop of Henle and actively reabsorbed in the thick segment of the ascending loop, ie, the distal tubule.

In the chloride shift, Cl^- plays an important role in the transport of CO_2. In the plasma, CO_2 is present as HCO_3^- which is produced in the erythrocytes from CO_2. The diffusion of HCO_3^- requires the counterdiffusion of another anion to maintain electrical neutrality. This function is performed by Cl^- which readily diffuses into and out of the erythrocytes (see Fig. 4).

Fruit and vegetable juices high in potassium have been recommended to correct hypokalemic alkalosis in patients on diuretic therapy but apparently the efficacy of this treatment is questionable. A possible reason for ineffectiveness is the low Cl^- content of most of these juices; Cl^- is high only in juices in which Na^+ is high, which of course, have to be excluded (69).

Trace Elements

Copper. All human tissues contain copper. The highest amounts are found in the liver, brain, heart, and kidney (70). In blood, plasma and erythrocytes contain almost equal amounts of copper, ie, ca 110 and 115 μg per 100 mL, respectively.

Metabolic Functions. In plasma, ca 90 wt % of the copper is in the metalloprotein ceruloplasmin (a_2-globulin; mol wt 151,000) which contains 8 atoms of copper per molecule (71). Ceruloplasmin has been identified as a ferroxidase(I) which catalyses the oxidation of aromatic amines and of Fe^{2+} to Fe^{3+} (72–73). The ferric ion is then incorporated into transferrin which is necessary for the transport of iron to tissues involved in the synthesis of iron-containing compounds, eg, hemoglobin (see Iron Compounds). Lowered levels of ceruloplasmin interfere with hemoglobin synthesis.

Erythrocuprein, which contains about 60 wt % of the erythrocyte copper, hepatocuprein, and cerebrocuprein act as superoxide dismutases. Each contains 2 atoms of copper per molecule (mol wt ca 34,000). The superoxide ion and peroxide are the two main toxic by-products of oxygen reduction in the body (1,74). Superoxide dismutase catalyses the dismutation, ie, the simultaneous oxidation, reduction, and decarboxylation, of the superoxide free-radical anion and, thus, protects the cell from oxidative damage.

The oxidation of the ϵ-amino groups of lysine is required for the cross-linking of polypeptide chains of collagen and elastin. The catalyst for this reaction is the copper

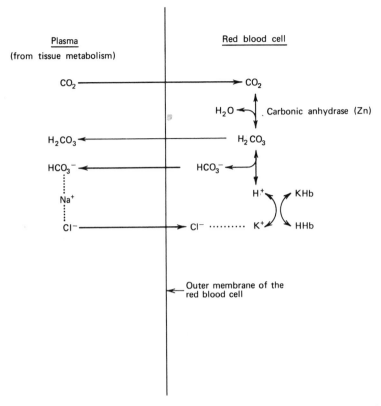

Figure 4. Chloride shift (9). KHb = potassium hemoglobin. HHb = acid hemoglobin.

metalloenzyme lysyl oxidase (71). Copper deficiency is characterized by poorly formed collagen which leads to bone fragility and spontaneous bone fractures in animals. Anemia, neutropenia, and bone disease have been reported in children with protein calorie malnutrition (PCM) and accompanying hypocupremia (70). Some other copper metalloproteins are cytochrome c oxidase, dopamine β hydroxylase, urate oxidase, tyrosinase, and ascorbic acid oxidase. Most copper enzymes are involved in redox reactions (71).

Genetic Disease. At least two genetic diseases involving copper are known. Wilson's disease is an autosomal recessive disease, usually detected in adulthood, in which there is a toxic increase in copper storage with neurological and liver damage, but a decrease in the amount of circulating copper due to decreased ceruloplasmin (71,72,75). Menkes' kinky-hair syndrome is an X-linked defect of copper transport out of the intestinal cell that results in lowered activity of several copper-dependent enzymes, lowered copper levels in the serum, progressive mental deterioration, defective keratinization of hair, and degenerative changes in the aorta (71,72,76,77).

Dietary Copper. Recent analytical data indicate that many diets contain less than the RDA for copper (78).

Zinc. The 2–3 g of zinc in the human body is widely distributed in every tissue and tissue fluid (79). About 90 wt % is in muscle and bone; unusually high concentrations are in the choroid of the eye and in the prostate gland (80). Almost all of the zinc in the blood is associated with carbonic anhydrase in the erythrocytes (81). Zinc is concentrated in nucleic acids but its function there is not clear (79). Zinc is also found in the nuclear, mitochondrial, and supernatant fractions of all cells that have been examined by ultracentrifugation.

Metabolic Functions. Zinc is essential for the function of many enzymes, either as a nondialyzable component in numerous metalloenzymes or as a dialyzable activator in various enzyme systems (82–83). Well-characterized zinc metalloenzymes are the carboxypeptidases A and B, thermolysin, neutral protease, leucine amino peptidase, carbonic anhydrase, alkaline phosphatase, aldolase (yeast), alcohol dehydrogenase, superoxide dismutases, and aspartate transcarbamylase. Other enzymes reported to contain zinc include RNA and DNA polymerases and a number of dehydrogenases, eg, lactic, malic, and glutamic (82–83). Generally, it is thought that enzymes that contain zinc in one species contain zinc in other species; this is true across species for carbonic anhydrase, but it is not true for aldolase which is a zinc enzyme in yeast but not in mammalian muscle (82–83). In addition to its role in the various enzyme activities, zinc is a membrane stabilizer and a participant in electron-transfer processes (80).

Zinc–hormone interactions include hormonal influence on absorption, distribution, transport, and excretion of zinc and zinc influence on synthesis, secretion, receptor binding, and function of numerous hormones. Zinc enhances pituitary activity by increasing circulating levels of growth hormone, thyroid-stimulating hormone, luteinizing hormone, follicle-stimulating hormone, and adrenocorticotropin (80) (see Hormones). The role of zinc in insulin action is recognized but not well understood (80). Zinc is required for maintenance of normal plasma concentrations of vitamin A and for normal mobilization of vitamin A from the liver (80,84–85) (see Vitamins).

Zinc Deficiency. Zinc was confirmed as essential for man in 1954 (81,83–86). The size of the human fetus is correlated with zinc concentration in the amniotic fluid and habitual low zinc intake in the pregnant female is thought to be related to several congenital anomalies in humans (83–84). Low zinc intakes result in hypogonadism, dwarfism, low serum and red blood cell zinc in humans and animals and retarded growth and teratogenic effects on the nervous system in rats.

In children suffering from marginal zinc deficiency, impaired taste acuity, poor appetite, and suboptimal growth were reversed with zinc supplementation (81,85). Accelerated wound healing occurs in humans with zinc supplementation (84–86); this suggests that marginal zinc deficiency in humans may be more widespread than has been thought. Zinc supplementation was also effective in alleviating symptoms of active rheumatoid arthritis in clinical trials (87). Acrodermatitis enteropathica, a hereditary disease that involves aberrant zinc metabolism, responds to oral zinc supplementation (81,85) (see Table 4).

Selenium. Selenium is thought to be widely distributed throughout the body tissues, and animal experiments suggest that greater concentrations are in the kidney, liver, and pancreas and lesser amounts are in the lungs, heart, spleen, skin, brain, and carcass (88).

Metabolic Functions. The most clearly documented role for selenium is as a necessary component of glutathione peroxidase (see Fig. 5) (89–91). Glutathione peroxidase reduces hydrogen peroxide, which is formed by free-radical oxidant-stressor reactions, to H_2O and reduces organic peroxides, eg, those formed by the peroxidation

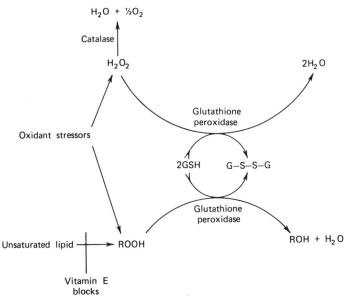

Figure 5. Glutathione peroxidase system (90). Glutathione peroxidase is a selenium enzyme.

of unsaturated fatty acids, to alcohols and H_2O (89–91). Selenium is also involved in the functions of additional enzymes, eg, leukocyte acid phosphatase and glucuronidases (92). Recent studies suggest a role for selenium in electron transfer (88,93) and involvement in nonheme iron proteins (88,91). Selenium and vitamin E appear to be necessary for proper functioning of lysosomal membranes (94) (see Vitamins).

Toxicity and Medicinal Aspects. Historically, interest in the biological effects of selenium developed with the recognition that alkali disease and blind staggers of grazing livestock in the western United States were the result of selenium poisoning. There are unusually high concentrations of selenium in certain plants, because of their selenium-accumulating properties, or in ordinary plants growing on highly seleniferous soils. Since no treatment for this type of poisoning is known, excess selenium in the animal diet must be avoided (91). Prolonged ingestion of up to 600 μg daily of selenium does not produce toxic effects in humans (95).

Selenium may have anticarcinogenic effects possibly because of the antioxidant properties of selenium compounds (88,91). Two types of data support this latter conclusion, ie, epidemiological investigations (88,91), and laboratory-feeding experiments (88,91). Animals involved in early carcinogenicity studies lived significantly longer if they were fed supplementary selenium. The role of selenium in the prevention of cancer and other chronic diseases, eg, heart conditions, as an antiaging and antimutagenic agent are being studied (96) (see Memory-enhancing agents and antiaging drugs).

Chromium. The history of the investigations establishing the essentiality of chromium has been reviewed (97). An effect of brewer's yeast in preventing or curing impaired glucose tolerance in rats was revealed, and the active factor was identified as a Cr(III) organic complex, ie, glucose tolerance factor (GTF).

Metabolic Functions. Chromium(III) potentiates the action of insulin; it may be considered a cofactor for insulin. In *in vitro* tests of epididymal fat tissue of chromium-deficient rats, Cr(III) increases the uptake of glucose only in the presence of insulin. The interaction of Cr(III) with insulin also is demonstrated by experimental results indicating an effect of Cr(III) in translocation of sugars into cells, ie, at the first step of sugar metabolism. These results and the data from other studies have led to the hypothesis that chromium forms a complex with insulin and insulin receptors (97).

There appears to be a chromium pool in individuals who are not chromium deficient (97). When there is an increase in level of circulating insulin, in response to a glucose load, an increase in circulating chromium occurs over a period of 0.5–2 h and this is followed by a decline; and, excretion of chromium in urine increases. Chromium deficiency is indicated when no increase or a small increase in blood chromium level or urine chromium occurs.

Test Results with Humans. In some studies with elderly people and mildly diabetic patients, significant improvement in the glucose tolerance test (GTT) was observed in 40–50% of the patients given chromium supplementation of 150–200 μg/d as chromium chloride (97–100); in other tests, positive results were not obtained (101). The results may indicate that not all subjects are capable of utilizing inorganic chromium to the same extent and that some might require a preformed GTF. Chromium chloride supplementation has been effective in normalizing impaired glucose tolerance in malnourished children and in patients receiving total parenteral nutrition for a long time (102–103). The most available form of chromium is GTF obtained from brewer's yeast. In human studies in which GTF was administered, one of the most significant results was normalization of the exaggerated insulin responses to glucose loads (104).

Attempts to isolate GTF from brewer's yeast have resulted in production of very active concentrates but the substance is too labile to be obtained in the solid state (97) (see Yeast). However, it has been shown that GTF is a Cr(III) complex containing two coordinated nicotinate radicals and other amino acid anions (105). Active preparations containing similar complexes have been synthesized (97,106).

Chromium(III) Chemistry. The most characteristic reactions of Cr(III) in aqueous solution at >4 pH, eg, in the intestine and blood, are hydrolysis and olation (107). As a consequence, inorganic polymeric molecules are formed that probably are not able to diffuse through membranes; this may be prevented by ligands capable of competing for coordination sites on Cr(III) (107). Thus it is not expected that any large fraction of ingested Cr(III) would be absorbed; this is the case with ordinary Cr(III) compounds, eg, chromium chloride. Chromium(III) in the form of GTF may be more efficiently absorbed.

Molybdenum. Molybdenum is a component of the metalloenzymes xanthine oxidase, aldehyde oxidase, and sulfite oxidase in mammals (108). Two other molybdenum metalloenzymes present in nitrifying bacteria have been characterized; nitrogenase and nitrate reductase (108). The molybdenum in xanthine oxidase, aldehyde oxidase, and sulfite oxidase is involved in redox reactions; the heme iron in sulfite oxidase also is involved in the electron transfer (109).

Xanthine oxidase (mol wt ca 275,000) is present in milk, liver, and intestinal mucosa (108), and is required in the catabolism of nucleotides. The free bases guanine and hypoxanthine from the nucleotides are converted to uric acid with xanthine as the intermediate. Xanthine oxidase catalyzes oxidation of hypoxanthine to xanthine and xanthine to uric acid. In these processes and in the oxidations catalyzed by aldehyde oxidase, molecular oxygen is reduced to H_2O_2 (110–111). Xanthine oxidase is also involved in iron metabolism. Release of iron from ferritin requires reduction of Fe^{3+} to Fe^{2+} and reduced xanthine oxidase participates in this conversion (111).

Copper–Molybdenum Antagonism. A Cu–Mo antagonism involving sulfate occurs in animals. Large amounts of Mo and sulfate can depress copper absorption (108). Cattle grazing on pasturage of high Mo content succumb to teart or peat scours, which is characterized by diarrhea and general wasting. Control involves increasing the copper intake, eg, with $CuSO_4$. The Cu–Mo antagonism has been observed in humans (108,112). Significant increases in urinary Cu excretion have been observed with increasing Mo intake (113).

Fluorine. Fluoride is present in the bones and teeth in very small quantities. Human ingestion is 0.7–3.4 mg/d from food and water. Evidence for the essentiality of fluorine was obtained by maintaining rats on a fluoride-free diet, which resulted in decreased growth rate, decreased fertility, and anemia; these impairments were remedied by supplementing the diets with fluoride (114–115).

Fluoridation and Dental Caries. In the opinion of many authorities, fluoridation of the public water supply is an effective means of significantly reducing the incidence of dental caries (116–119) (see Water, municipal water treatment). The view is not universally accepted and some have expressed concern regarding the narrow range of safety between effective and toxic concentrations (see also Dentifrices) (120–121). Assertions that fluoridation of water supplies increases the incidence of cancer have not been substantiated (122).

Other Effects of Fluoride. Excess fluoride ingestion damages developing teeth, causing mottling, chalky-white coloration and pitting (117,123). Adding fluoride to animal feed leads to fragile and brittle teeth and bones (123). Fluoride is an inhibitor of enzymes, especially enolase in the glycolytic pathway (123). The reports of the effects of fluoride on osteoporosis have not been fully established (117). Fluoride supplementation results in poorly mineralized bone unless very high pharmacologic levels, ie, 50,000 IU of vitamin D and 900 mg of calcium are included in the regimen (124).

Iodine. Of the 10–20 mg of iodine in the adult body, 70–80 wt % is in the thyroid gland; it is present in all tissues. The essentiality of iodine depends solely on its utilization by the thyroid gland to produce thyroxin and related compounds. Well-known consequences of faulty thyroid function are hypothyroidism, hyperthyroidism, and goiter.

Metabolic Functions. The functions of the thyroid hormones and, thus, of iodine are control of energy transductions (125). These hormones increase oxygen consumption and basal metabolic rate by accelerating reactions in nearly all cells of the body. A part of this effect is attributed to increase in activity of many enzymes. Protein synthesis is affected by the thyroid hormones (125–126).

Thyroid Hormones. Iodine, which is absorbed as I^-, is oxidized in the thyroid and is bound to a thyroglobulin. The resultant glycoprotein (mol wt 670,000) contains 120 tyrosine residues of which ca two thirds are available for binding iodine, which is bound to the tyrosine residues in several different ways. Proteolysis introduces the active hormones 3,5,3′-triiodothyronine (T_3) and 3,5,3′,5′-tetraiodothyronine (T_4, thyroxine) in the ratio (T_4/T_3) of 4/1 (125–126).

Only small amounts of free T_4 are present in plasma; most of it is bound to the specific carrier, ie, thyroxine-binding protein. T_3, which is very loosely bound to protein, passes rapidly from blood to cells, and accounts for 30–40% of total thyroid hormone activity (125). According to some authorities, not much of the T_3 is synthesized in the thyroid; most of it may be produced by conversion of T_4 at the site of action of the hormone, ie, T_4 may be a prehormone requiring conversion to T_3 to exert its metabolic effect (127) (see Thyroid and antithyroid preparations).

Thyroid-stimulating hormone (TSH, thyrotropin) influences thyroid activity by a feedback mechanism (125–127). TSH, which is secreted by the pituitary gland, stimulates the thyroid to increase production of thyroid hormones when the blood hormone level is low, with a resulting increase in size of thyroid cells. If this continues, the increase in size of the thyroid becomes noticeable as simple goiter. In many parts of the world, simple goiter is endemic and usually results from dietary iodine deficiency (125). In technologically advanced countries, the problem of iodine deficiency has been minimized by the use of iodized salt (125,128–129) (see also Hormones).

Manganese.   The adult human body contains ca 10–20 mg of manganese (129–130). Manganese is widely distributed throughout the body with the largest concentration in the mitochondria of the soft tissues, especially in the liver, pancreas, and kidneys (130–131). Manganese concentration in bone varies widely with dietary intake (131).

Metabolic Functions.   Manganese is essential for normal body structure, reproduction, normal functioning of the central nervous system, and activation of numerous enzymes. Synthesis of the mucopolysaccharide chondroitin sulfate involves a series of reactions where manganese is required in at least five steps (132). These reactions are responsible for formation of polysaccharides and linkage between the polysaccharide and proteins that form the mucopolysaccharide of cartilage (130). In addition to the glycosyl transferases of mucopolysaccharide synthesis that require manganese, a number of metalloenzymes contain Mn^{2+} (131–132). Superoxide dismutase contains 2 atoms of manganese per molecule and pyruvic carboxylase contains 4 atoms of manganese per molecule. Most enzymes that require magnesium, eg, kinases, can use manganese in *in vitro* reactions. It is not known if Mn^{2+} can substitute for Mg^{2+} *in vivo*.

Manganese Deficiency.   In animals manganese deficiency results in wide-ranging disorders, eg, impaired growth, abnormal skeletal structure, disturbances of reproduction, and defective lipid and carbohydrate metabolism (131). The common denominator appears to be the impairment of activity of enzymes that have a specific requirement for manganese. Although overt manganese deficiency in man does not appear to be widespread, some forms of epilepsy in man and animals and a decrease in glucose tolerance in animals have been linked to low levels of manganese in the tissues (133–134).

Iron.   The total body content of iron, ie, 3–5 g, is recycled more efficiently than other metals. There is no mechanism for excretion of iron and what little iron is lost daily, ie, ca 1 mg in the male and 1.5 mg in the menstruating female, is lost mainly through exfoliated mucosal, skin, or hair cells, and menstrual blood (135–137).

Metabolic Functions.   A large percentage of the iron in the human body is in hemoglobin: 85 wt % in the adult female, 60 wt % in the adult male (137). The remainder is present in other iron-containing compounds that are involved in basic metabolic functions or in iron transport or storage compounds. Myoglobin, the cytochromes, catalase, sulfite oxidase, and peroxidase are heme iron enzymes. NADH-dehydrogenase, succinate dehydrogenase, α-glycerophosphate dehydrogenase, monoamine oxidase, xanthine oxidase, and alcohol dehydrogenase are nonheme metalloflavoproteins that contain iron. Aconitase and microsomal lipid peroxidase do not contain iron but do require it as a cofactor. The transport and storage proteins for iron are transferrin, ferritin, and hemosiderin (see Fig. 6) (135,137–138). The iron in transferrin in the blood is a combination of freshly absorbed iron and recycled iron.

The hemoglobin molecule (mol wt 65,000) is a tetramer that contains four iron atoms (138–139). The iron remains in the ferrous (2+) state and reversibly binds oxygen, allowing the hemoglobin molecule to function in its role as an oxygen carrier to the tissues (139). Myoglobin is a monomer containing one iron atom which functions in the muscle by accepting oxygen from the blood and storing it for use during muscle contraction. The iron in myoglobin also remains in the ferrous (2+) state (138).

The ability of iron to exist in two stable oxidation states, ie, the ferrous (2+) and ferric (3+), in aqueous solutions is important to its role as a biocatalyst. Although the cytochromes of the electron-transport chain are porphyrins like hemoglobin and myoglobin, they differ in that the iron atom is involved in redox reactions ($Fe^{2+} \rightleftharpoons Fe^{3+}$) (138,140). Catalase is a tetramer which contains four atoms of iron and peroxidase is a monomer with one atom of iron per molecule; the iron in these enzymes also is oxidized and reduced (141).

Homeostatic Control of Iron Levels. Absorption of iron from food to maintain homeostasis is tightly controlled and increases in instances of increased demands, eg, during pregnancy, lactation, and iron-deficiency states which are the result of blood loss or iron-deficiency anemia that is the result of previous inadequate intake (136–137). Iron absorption is greatly reduced in the normal individual when iron stores are adequate or excessive. Absorption is enhanced by acid conditions and reducing agents; heme iron from animal sources is absorbed more readily than nonheme iron from cereals and vegetables (136–137).

A system of internal iron exchange exists which is dominated by the iron required for hemoglobin synthesis. For formation of red blood cells, iron stores can furnish 10–40 mg of iron per day, as compared to 1–3 mg from dietary sources (136); only ca 10 wt % of ingested iron actually is absorbed. Transferrin is essential for movement of iron and without it, as in genetic absence of transferrin, iron overload occurs in tissues and is coupled with iron-deficiency anemia.

Iron Deficiency and Toxicity. Iron deficiency is a major worldwide nutritional problem and cause of anemia (136,142). Insufficient dietary iron intake; iron losses, eg, bleeding and parasite infestation, and malabsorption of iron are the principal causes. The groups at greatest risk for developing iron-deficiency anemia are menstruating females, pregnant or nursing females, and young children.

Iron toxicity resulting from excess absorbable iron ingestion is rare except in Africa where fermented beverages are made in large iron pots with levels of iron approaching 80 mg/L in a brew where the pH is very low. This results in siderosis which can result in hemochromatosis, ie, damage to various organs from excessive storage of iron, causing numerous disease states, eg, hepatic fibrosis and diabetes in 80% of the cases of idiopathic hemochromatosis patients (136–137). Iron overload is frequently a complication of repeated blood transfusions in anemias, eg, thalassemia (136).

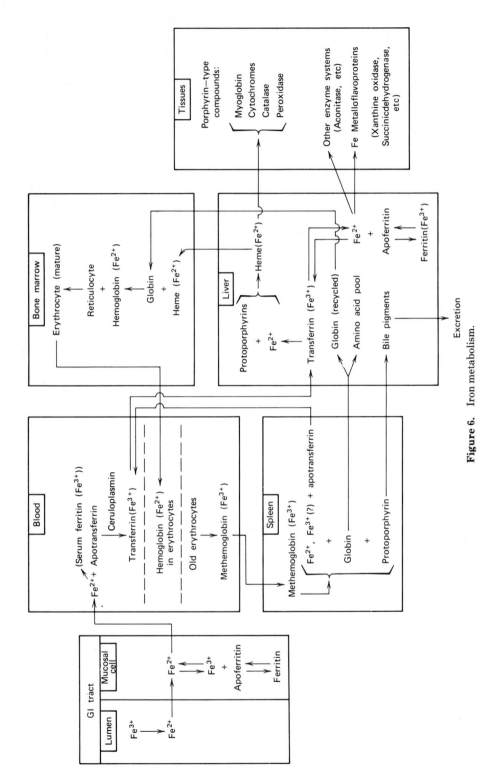

Figure 6. Iron metabolism.

Cobalt. Cobalt is nutritionally available only as vitamin B_{12} (143). Although Co^{2+} can function as a replacement *in vitro* for other divalent cations, in particular Zn^{2+}, no *in vivo* function for inorganic cobalt is known for humans (144). In ruminant animals, B_{12} is synthesized by bacteria in the rumen (143).

Occurrence and Structure of B_{12} Vitamins. In foods, vitamin B_{12} occurs only in animal products (143). A particularly rich source is liver from which it was originally isolated following the successful use of liver in treatment of pernicious anemia, a form of megaloblastic anemia not responsive to iron supplementation. It is produced commercially by bacterial fermentation in the form of cyanocobalamin for use as a dietary supplement and for injection in the treatment of pernicious anemia (see Vitamins). In vitamin B_{12}, the cobalt atom is at the center of the corrin ring to which it is attached by coordination to the nitrogen atoms of the four five-membered heterocyclic rings (see Vitamins, vitamin B_{12}). In cyanocobalamin, the source of the cyanide is apparently contamination from the reagents used since naturally occurring forms of B_{12} contain little or no cyanide (145–146). In coenzyme B_{12} the 5'-C of the deoxyadenosyl group is bonded to the cobalt atom:

(partial structure of Vitamin B_{12})

The chemical and biological activity of coenzyme B_{12} are dependent on this carbon–metal bond (146).

Metabolic Functions of Vitamin B_{12} and Coenzyme B_{12}. In pernicious anemia, the bone marrow fails to produce mature erythrocytes as a result of defective cell division, which is a consequence of impaired DNA synthesis which requires vitamin B_{12}. If the disease is untreated, extensive neurological damage, eg, irreversible degeneration of the spinal cord by demyelinization, may occur because of faulty fatty acid metabolism (146).

Coenzyme B_{12} is required in mammalian enzyme systems involving methylmalonyl CoA and methyl transferase (transmethylase). Methylmalonyl CoA is an intermediate in the metabolism of propionate which it converts to succinyl CoA. Transmethylase is required for methylation of homocysteine to methionine (143,145,147).

Intrinsic Factor. Vitamin B_{12} deficiency commonly is caused by inadequate absorption resulting from a lack of or insufficient intrinsic factor (IF) (148). Intrinsic factor is a glycoprotein (mol wt ca 50,000) which binds in a 1:1 molar ratio with vitamin B_{12}. The B_{12}–IF complex, formed in the stomach, is absorbed in the ileum; absorption in this part of the intestine occurs because of the specific characteristics of the cells of the microvilli (brush border) of the ileum (148). The IF remains in the intestine attached to the epithelial cells. Transport of B_{12} into the blood stream requires Ca^{2+}.

In the blood, B_{12} is bound to transcobalamin II (transport protein). Whatever bound B_{12} that is not utilized immediately is stored in the liver. With increasing quantities of dietary B_{12}, the fraction that is absorbed decreases. Generally, the vitamin is excreted in the urine, but with large intake, some is excreted in the bile.

Essential Nutrients in Animals

Silicon. Silicon is present in the body as a result of ingestion of silicates, primarily from vegetables, and normal blood serum levels are ca 1 mg/100 mL regardless of the intake because of efficient kidney excretion of excess (149). Silicon is necessary for calcification, growth, and as cross-linking material in mucopolysaccharide formation (149). Essentiality of silicon in rats and chicks has been established (121,149–151).

Tin. The widespread use of canned foods results in a daily intake of tin that is ca 1–17 mg for an adult male (152). At this level it has not been shown to be toxic. Essentiality has been shown only for the rat and has been evidenced by greatly enhanced growth rate resulting from tin supplementation to low tin diets (121).

Vanadium. Vanadium is essential in rats and chicks (121,153). Estimated human intake is less than 4 mg/d. In animals, deficiency results in impaired growth, reproduction, and lipid metabolism (154). The levels of coenzyme A and coenzyme Q_{10} in rats are reduced and monoamine oxidase activity is increased when rats are given excess vanadium (154).

Nickel. There is considerably more evidence for the essentiality of nickel than for silicon, tin, or vanadium. Various pathological manifestations of nickel deficiencies have been observed with chicks, rats, and pigs (155–156). *In vitro* studies have shown nickel to be an activator of several enzymes. Nickel stabilizes RNA and DNA against thermal denaturation and may have a role in membrane structure or metabolism (155). A nickel metalloprotein has been isolated from human serum (155). Average intake is ca 0.3–0.6 mg/d.

Arsenic. Arsenic is under consideration for inclusion as an essential element, but no clear role has been established (157).

Health and Safety Factors, Toxicology

Under unusual circumstances, toxicity may arise from ingestion of excess amounts of minerals, although this is uncommon except in the cases of five of the trace elements, ie, fluorine, molybdenum, selenium, copper, and iron (see Table 5). Toxicosis also results from exposure to industrial forms of some of the minerals, eg, chromic acid (CrO_3), manganese(IV) oxide, organophosphates, chlorine (Cl_2), fluorine (F_2), and bromine (Br_2).

Because of the efficient homeostatic controls of the mammalian organism, it is not likely that toxicity occurs from ingestion of the mineral nutrients except under unusual conditions far removed from those of nutritional significance or with individuals suffering from some pathological conditions. Because of their very low concentrations in foods and with few exceptions, eg, selenium and iron, the trace elements are not toxic under normal nutritional conditions.

The data of Table 6 pertaining to the toxicities of compounds of the essential mineral elements were obtained from *Registry of Toxic Effects of Chemical Substances* published (and revised annually) by the National Institute for Occupational

Health and Safety (NIOSH) (158). The scientists responsible for producing this compendium regard everything as potentially toxic and nutritional aspects are not their primary concern. They include the following warning: "Under no circumstances can the toxic dose values presented with these chemical substances be considered definitive values for describing safe versus toxic doses for human exposure." Nevertheless, the values listed permit an evaluation of relative toxicities as may be seen by comparing the values for sodium and potassium chloride with those for sodium selenite and sodium selenate. Data for cobalt and sulfur are not included since no values for vitamin B_{12}, cysteine, cystine, and methionine (oral administration) were found in the NIOSH compendium.

BIBLIOGRAPHY

1. I. Fridovich, *Am. Sci.* **63,** 54 (1975).
2. *Recommended Dietary Allowances*, Food and Nutrition Board, National Academy of Sciences–National Research Council, Washington, D.C., 1980.
3. *Conference on Micronutrient Interactions: Vitamins, Minerals and Hazardous Elements*, The New York Academy of Sciences, Feb. 20–22, 1980.
4. A. L. Lehninger, *Biochemistry*, 2nd ed., Worth Publishers, Inc., New York, 1975.
5. R. Montgomery, R. L. Dryer, T. W. Conway, and A. A. Spector, *Biochemistry—A Case-Oriented Approach*, 2nd ed., The C. V. Mosby Company, St. Louis, Mo., 1977.
6. J. M. Orten and O. W. Neuhaus, *Human Biochemistry*, 9th ed., The C. V. Mosby Company, St. Louis, Mo., 1975.
7. A. White, P. Handler, and E. L. Smith, *Principles of Biochemistry*, 5th ed., McGraw-Hill Book Co., New York, 1973.
8. A. C. Guyton, *Textbook of Medical Physiology*, 5th ed., W. B. Saunders Company, Philadelphia, Pa., 1976.
9. H. A. Harper, *Review of Physiological Chemistry*, Lange Medical Publications., Los Altos, Calif., 1975.
10. B. L. Oser, ed., *Hawk's Physiological Chemistry*, McGraw-Hill Book Company, New York, 1965.
11. H. S. Mitchell, H. J. Rynbergen, L. Anderson, and M. V. Dibble, *Nutrition in Health and Disease*, J. B. Lippincott Co., Philadelphia, Pa., 1976.
12. R. L. Pike and M. L. Brown, *Nutrition: An Integrated Approach*, 2nd ed., John Wiley & Sons, Inc., New York, 1975.
13. *Present Knowledge in Nutrition*, 4th ed., The Nutrition Foundation, Inc., Washington, D.C., 1976.
14. F. P. Antia, *Clinical Dietetics and Nutrition*, 2nd ed., Oxford University Press, London, Eng., 1973.
15. H. A. Guthrie, *Introduction to Nutrition*, 2nd ed., The C. V. Mosby Company, St. Louis, Mo., 1971.
16. J. D. Kirschmann, *Nutrition Almanac*, McGraw-Hill Book Co., New York, 1979.
17. C. F. Adams, *Nutritive Value of American Foods in Common Units, Agriculture Handbook No. 456*, Agricultural Research Service, U.S. Department of Agriculture, Washington, D.C., 1975.
18. H. J. M. Bowen, *Trace Elements in Biochemistry*, Academic Press, Inc., New York, 1966.
19. H. A. Schroeder, *The Trace Elements and Man*, The Devin-Adair Co., Old Greenwich, Conn., 1973.
20. E. J. Underwood, *Trace Elements in Human and Animal Nutrition*, 4th ed., Academic Press, Inc., New York, 1977.
21. Ref. 5, p. 157.
22. Ref. 11, pp. 180, 190.
23. Ref. 5, p. 198.
24. Ref. 12, p. 378.
25. C. R. Martin, *Textbook of Endocrine Physiology*, Oxford University Press, Inc., New York, 1976, p. 118.
26. Ref. 8, p. 881.

27. G. R. Meneely and H. D. Batterbee in Ref. 13, p. 259.
28. H. C. Trowell, *Executive Health* **XVI**(2), (Nov. 1979).
29. "Research Needs for Establishing Dietary Guidelines for Sodium" in *Research Needs for Establishing Dietary Guidelines for the U.S. Population*, The National Research Council, National Academy of Sciences, Washington, D.C., 1979.
30. J. B. Peterson, *Limestone*, (Fall 1980).
31. M. E. Shils in Ref. 13, p. 247.
32. Ref. 12, p. 185.
33. W. E. C. Wacker, *Ann. N.Y. Acad. Sci.* **162,** 717 (1969).
34. W. E. C. Wacker and A. F. Parisi, *New Eng. J. Med.* **278,** 658, 712, 772 (1968).
35. F. W. Heaton, *Ann. N.Y. Acad. Sci.* **162,** 775 (1969).
36. B. A. Barnes, *Ann. N.Y. Acad Sci.* **162,** 786 (1969).
37. I. Melnick, R. R. Landes, A. A. Hoffman, and J. F. Burch, *J. Urol.* **105,** 119 (1971).
38. P. F. De Albuquerque and M. Tuma, *J. Urol.* **87,** 504 (1962).
39. E. L. Prien and S. F. Gershoff, *J. Urol.* **112,** 509 (1974).
40. S. N. Gershoff and E. L. Prien, *Am. J. Clin. Nutr.* **20,** 393 (1967).
41. Ref. 11, p. 52.
42. M. Walser, *J. Clin. Invest.* **40,** 723 (1961).
43. Ref. 12, p. 660.
44. Ref. 12, p. 182.
45. Ref. 6, pp. 430, 533.
46. Ref. 25, p. 155.
47. H. F. DeLuca, *J. Steroid Biochem.* **11,** 35 (1979).
48. S. Wallach, ed., *Paget's Disease of Bone*, Armour Pharmaceutical Co., Phoenix, Ariz., 1979.
49. A. Avramides, *Clin. Orthop.* **127,** 78 (1977); W. C. Sturtridge, J. E. Harrison, and D. R. Wilson, *Can. Med. Assoc.* **117,** 1031 (1977).
50. R. A. Evans, *Aust. N.Z. J. Med.* **7,** 259 (1977).
51. R. S. Goodhart and M. E. Shils, *Modern Nutrition in Health and Disease*, 6th ed., Lea & Febiger, 1980, Philadelphia, Pa., p. 305; Ref. 12, p. 184.
52. Ref. 6, p. 173.
53. T. P. Bennett and E. Frieden, *Modern Topics in Biochemistry*, The MacMillan Co., New York, 1966, p. 81.
54. Ref. 53, p. 117.
55. Ref. 53, p. 70.
56. Ref. 12, p. 42.
57. Ref. 53, p. 120.
58. Ref. 6, p. 29.
59. Ref. 12, p. 74.
60. Ref. 12, p. 191.
61. O. H. Muth and J. E. Oldfield, eds., *Symposium: Sulfur in Nutrition*, The Avi Publishing Co., Inc., Westport, Conn., 1970.
62. Ref. 9, p. 181.
63. Ref. 6, p. 390.
64. Ref. 12, p. 113.
65. Ref. 6, p. 340; Ref. 9, p. 370; Ref. 5, p. 428.
66. H. White, P. Handler, and E. L. Smith, *Principles of Biochemistry*, 3rd ed., McGraw-Hill Book Company, New York, 1964, p. 789.
67. H. Netter, *Theoretical Biochemistry*, Wiley-Interscience Division, John Wiley & Sons, Inc., New York, 1969, p. 784.
68. Ref. 12, p. 189.
69. S. A. Miller, P. A. Roche, P. Srinavasan, and V. Vertes, *Am. J. Clin. Nutr.* **32,** 1757 (1979).
70. Ref. 12, p. 196.
71. B. L. O'Dell in Ref. 13, p. 302.
72. Ref. 20, p. 56.
73. W. G. Hoekstra, J. W. Suttie, H. E. Ganther, and W. Mertz, eds., *Trace Element Metabolism in Man—2 (Tema-2)*, University Park Press, Baltimore, Md., 1974; E. Frieden, *ibid.*, p. 105.
74. E. M. Gregory and I. Fridovich in Ref. 73, p. 486.
75. Ref. 11, p. 466.

76. Ref. 11, p. 66.
77. N. A. Holtzman, *Fed. Proc.* **35,** 2276 (1976).
78. L. M. Klevay, S. J. Reck, R. A. Jacob, G. M. Logan, J. M. Munoz, and H. H. Sanstead, *Am. J. Clin. Nutr.* **33,** 45 (1980).
79. Ref. 11, p. 65; Ref. 12, p. 206.
80. *Zinc*, Subcommittee on Zinc, Committee on Medical and Biologic Effects of Environmental Pollutants, National Research Council, National Academy of Sciences, University Park Press, Baltimore, Md., 1979, p. 123.
81. Ref. 20, p. 196.
82. Ref. 80, p. 211.
83. H. H. Sandstead in Ref. 13, p. 290.
84. Ref. 80, p. 173.
85. Ref. 80, p. 225.
86. R. E. Burch and J. F. Sullivan in R. E. Burch and J. F. Sullivan, eds., *The Medical Clinics of North America*, Vol. 60, No. 4 (Symposium on Trace Elements), W. B. Saunders Co., Philadelphia, Pa., 1976, p. 675.
87. P. A. Simkin, *Lancet ii*, 539 (1976); *Prog. Clin. Biol. Res.* **14,** 343 (1977).
88. *Selenium*, Subcommittee on Selenium, Committee on Medical and Biologic Effects of Environmental Pollutants, National Research Council, National Academy of Sciences, Washington, D.C., 1976, p. 51.
89. R. F. Burk in Ref. 13, p. 310.
90. W. G. Hoekstra in Ref. 73, p. 61.
91. Ref. 20, p. 302; G. N. Schrauzer, D. A. White, and C. J. Schneider, *Bioinorg. Chem.* **8,** 387 (1978); R. J. Shamberger, *Executive Health*, **XIV**(12), (Sept. 1978).
92. J. R. Chen and J. M. Anderson, *Science* **206,** 1426 (1979).
93. T. C. Stadtman, *Science* **183,** 915 (1974).
94. Ref. 12, p. 542.
95. G. N. Schrauzer and D. A. White, *Bioinorg. Chem.* **8,** 303 (1978).
96. G. N. Schrauzer, private communication, Jan. 30, 1980.
97. W. Mertz in D. Shapcott and J. Hubert, eds., *Chromium in Nutrition and Metabolism, Developments in Nutrition and Metabolism*, Vol. 2, Elsevier-North Holland Biomedical Press, Amsterdam, 1979, p. 1.
98. R. A. Levine, D. H. P. Streeten, and R. J. Doisy, *Metabolism* **17,** 114 (1968).
99. L. L. Hopkins, Jr. and M. G. Price, *Proceedings Western Hemisphere Nutrition Congress*, *Puerto Rico*, Vol. II, Puerto Rico, 1968, p. 40.
100. H. Schroeder, *Am. J. Clin. Nutr.* **21,** 230 (1968).
101. L. Sherman, J. A. Glennon, W. J. Brech, G. H. Klomberg, and E. S. Gordon, *Metabolism* **17,** 439 (1968).
102. L. L. Hopkins, Jr., O. Ransome-Kuti, and A. S. Majaj, *Am. J. Clin. Nutr.* **21,** 203 (1968); C. T. Gursen and G. Saner, *Am. J. Clin. Nutr.* **24,** 1313 (1971); *Am. J. Clin. Nutr.* **26,** 988 (1973).
103. K. N. Jeejeebhoy, R. C. Chu, E. B. Marliss, G. R. Greenberg, and A. Bruce-Robertson, *Am. J. Clin. Nutr.* **30,** 531 (1977); H. Freund, S. Atamian, and J. E. Fisher, *J. Am. Med. Assoc.* **241,** 496 (1979).
104. R. J. Doisy, D. H. P. Streeten, I. M. Freiberg, and A. J. Schneider, *Trace Elements in Human Health and Disease*, Vol. II, Academic Press, New York, 1976, p. 79; V. J. K. Liu and J. S. Morris, *Am. J. Clin. Nutr.* **31,** 972 (1978).
105. E. W. Toepfer, *Fed. Proc.* **33,** 695 (1974).
106. C. L. Rollinson, E. Harte, and A. Ruth, unpublished work.
107. C. L. Rollinson in *Comprehensive Inorganic Chemistry*, Vol. 3, Pergamon Press, Oxford, 1973, p. 676; A. J. Gould, ed., *Radioactive Pharmaceuticals*, U.S. Atomic Energy Commission, Division of Technical Information, Oak Ridge, Tenn., 1966, p. 429; C. L. Rollinson and E. W. Rosenbloom in S. Kirschner, ed., *Coordination Chemistry: Papers Presented in Honor of Prof. John C. Bailar, Jr.*, Plenum Press, New York, 1969, p. 108.
108. Ref. 20, p. 109.
109. P. D. Boyer, ed., *The Enzymes*, Academic Press, Inc., New York, 1970.
110. Ref. 5, p. 546.
111. Ref. 12, p. 202.
112. A. Galli, *Ann. Biol. Clin.* **26,** 976 (1968).
113. Y. G. Doestahale and C. Gopalan, *Br. J. Nutr.* **31,** 351 (1974).

114. K. Schwarz in Ref. 73, p. 355.
115. H. H. Messer, W. D. Armstrong, and L. Singer in Ref. 73, p. 425.
116. H. H. Messer and L. Singer in Ref. 13, p. 325.
117. Ref. 20, p. 347.
118. H. J. Sanders, *Chem. Eng. News*, 30 (Feb. 25, 1980).
119. B. A. Bart, *Chem. Eng. News*, 56 (Oct. 22, 1979).
120. G. I. Waldbott, *Fluoridation: The Great Dilemma*, Coronado Press, Lawrence, Kansas, 1978.
121. G. I. Waldbott, P. A. Coleman, and M. B. Schacter, *Chem. Eng. News*, 2, 3 (Dec. 17, 1979); J. R. Lee, *Chem. Eng. News*, 4 (Jan. 28, 1980).
122. D. R.Taves in H. H. Hiatt, J. D. Watson, and J. A. Winsten, eds., *Origins of Human Cancer*, Vol. 4, Cold Spring Harbor Conferences on Cell Proliferation, Cold Spring Harbor Laboratory, 1977, p. 357.
123. Ref. 6, p. 549.
124. J. Jowsey, B. L. Riggs, P. J. Kelly, and D. L. Hoffman, *Am. J. Med.* **53,** 43 (1972).
125. Ref. 12, p. 198.
126. Ref. 25, p. 189.
127. Ref. 5, p. 615.
128. Ref. 15, p. 150; Ref. 20, p. 271.
129. Ref. 12, p. 201.
130. L. S. Hurley in Ref. 13, p. 345.
131. Ref. 20, p. 170.
132. M. F. Utter in Ref. 86, p. 713.
133. Y. Tanaka, C. Dupont, and E. R. Harpur, *Abstracts of 174th American Chemical Society Meeting*, Abstract No. 130, Chicago, Ill., Aug. 28–Sept. 2, 1977.
134. G. J. Everson and R. E. Schrader, *J. Nutr.* **94,** 89 (1968).
135. Ref. 20, p. 13.
136. C. A. Finch in Ref. 13, p. 280.
137. *Iron*, Subcommittee on Iron, Committee on Medical and Biologic Effects of Environmental Pollutants, National Research Council, National Academy of Sciences, University Park Press, Baltimore, Md., 1979, p. 79.
138. Ref. 12, p. 192.
139. Ref. 6, pp. 92, 770.
140. E. Frieden, *Nutr. Rev.* **31,** 21 (1973).
141. Ref. 9, p. 72.
142. Ref. 137, p. 107.
143. Ref. 13, p. 132.
144. B. L. Vallee in S. K. Dhar, ed., *Advances in Experimental Medicine and Biology*, Vol. 40, Plenum Press, New York, 1973, p. 1.
145. Ref. 12, p. 122.
146. Ref. 5, p. 22.
147. Ref. 5, p. 262.
148. Ref. 12, p. 256.
149. Ref. 20, p. 398.
150. E. M. Carlisle in Ref. 13, p. 337.
151. E. M. Carlisle in Ref. 73, p. 407.
152. Ref. 20, p. 449.
153. L. L. Hopkins, Jr. in Ref. 73, p. 397.
154. Ref. 20, p. 388.
155. Ref. 20, p. 159.
156. F. H. Nielsen in Ref. 73, p. 381.
157. Ref. 20, p. 426.
158. R. J. Lewis, Sr. and R. L. Tatkin, eds., *Registry of Toxic Effects of Chemical Substances*, 8th ed., 1978, National Institute for Occupational Safety and Health, Public Health Service Center for Disease Control, U.S. Government Printing Office, Washington, D.C., 1979.

General References

G. L. Eichhorn, ed., *Inorganic Biochemistry*, Vols. 1 and 2, Elsevier Scientific Publishing Co., Amsterdam, 1973.
D. J. D. Nicholas and A. R. Egan, eds., *Trace Elements in Soil–Plant–Animal Systems*, Academic Press, Inc., New York, 1975.
E. I. Hamilton, *The Chemical Elements and Man*, Charles C Thomas, Springfield, Ill., 1979.
S. Ochiai, *Bioinorganic Chemistry*, Allyn and Bacon, Inc., Boston, Mass., 1977.
M. Kirchgessner, ed., *Trace Elements Metabolism in Man and Animals—3 (Tema-3). Arbeitskreis für Tiernährungsforschung*, Weihenstephen, West Germany, 1978.
H. A. Lowenstam, "Minerals Formed by Organisms," *Science* **211**, 1126 (1981).

CARL L. ROLLINSON
MARY G. ENIG
University of Maryland

MINERAL WOOL.   See Refractory fibers.

MINIUM, Pb$_3$O$_4$.   See Lead compounds.

MISCH METAL.   See Cerium.

MIXING AND BLENDING

Mixing, an important operation in the chemical process industries, can be divided into five areas: liquid–solid dispersion, gas–liquid dispersion, liquid–liquid dispersion, the blending of miscible liquids, and the production of fluid motion. Mixing performance is evaluated by two criteria. The first is physical uniformity, ie, a physical relationship is required in terms of samples of uniformity in various parts of the mixing vessel. Specifications describe this requirement. The other criterion is based on mass transfer or chemical reaction. The elements of mixer design are: (*1*) process design— fluid mechanics of impellers, fluid regime required by process, scale-up, and hydraulic similarity; (*2*) impeller power characteristics—relate impeller power, speed, and diameter; and (*3*) mechanical design—impellers, shafts, and drive assembly. The curves in Figure 1 give data for a wide variety of impeller types and systems (1). The power consumption curves shown in Figure 1 are completely independent of process performance.

Homogeneous chemical reactions require a knowledge of the overall concentration in the tank, primarily on a microscale mixing level. A study showed that the root-mean-square (RMS) velocity fluctuation value correlated well with the direction in competitive consecutive second-order reactions. This indicates that this type of microscale mixing can affect certain kinds of chemical reaction (2). The speed of a chemical reaction is related to the blend time on both macro- and microscale levels (3–4). Thus, mixing may affect the progress of a homogeneous chemical reaction.

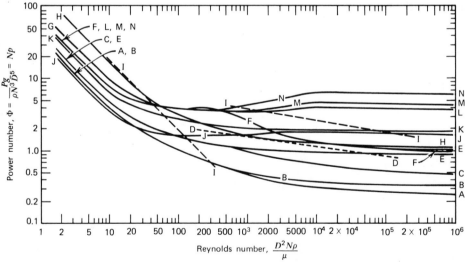

Figure 1.   Power characteristics of mixing impellers. When no baffles (for N_{Re} over 300), $\Phi = (P_g/\rho N^3 D^6)(g/N^2 D)(a - \log N_{Re}/b)$. A, propeller, pitch equal to diameter, no baffles; B, propeller, pitch equal to diameter, four baffles each 0.1 T; C, propeller, pitch twice the diameter, no baffles; D, paddle, no baffles; E, propeller, pitch twice the diameter, four baffles each 0.1 T; F, flat six-blade turbine, no baffles; G, shrouded six-blade turbine, four baffles each 0.1 T; H, shrouded six-blade turbine, stator ring with 20 blades; I, paddle, no baffles; J, flat paddle, two-blade, four baffles each 0.1 T; K, fan turbine, eight-blade, four baffles each 0.1 T; L, arrowhead six-blade turbine, four baffles each 0.1 T; M, curved six-blade turbine, four baffles each 0.1 T; and N, flat six-blade turbine, four baffles each 0.1 T.

Fluid Mechanics

Mixer power P produces a pumping capacity Q expressed in kg/s, and a specific velocity work term H expressed in J/kg (5).

$$P = QH$$

The term H is related to the square of the velocity and, therefore, to fluid shear rates. In low and medium viscosity, the pumping capacity is related to speed and diameter of the impeller (6):

$$Q \propto ND^3$$

The power drawn by the impeller is proportional to N^3D^5:

$$(P \propto \rho N^3 D^5)$$

These relationships can be combined to show that at constant power input, the flow-to-velocity work ratio is related to the impeller diameter:

$$(Q/H)_P \propto D^{8/3}$$

This is sometimes expressed as the flow-to-fluid shear ratio which is correct conceptually but is not in terms of the mathematical equation above. This equation also does not hold as a constant on scale-up, and is, therefore, used primarily to evaluate the effect of geometric variables. It does not have a ready evaluation in terms of the actual numbers when comparing large to small tanks.

Most mixing applications are sensitive primarily to fluid-pumping capacity. Thus, if the pumping capacity of different impellers is compared (7) or the overall flow pattern in the tank considered, process results are proportional in the same fashion to the actual circulating capacity of the impeller and in the tank. The equations relate to pumping capacity of the impeller itself and do not include the entrainment provided as the jet from the impeller circulates through the tank.

Most open impellers, propellers, turbines, and axial-flow turbines are normally limited to an upper range of about 0.6–0.7 D/T, because at that point, no further entrainment is possible to provide additional flow in the mixing tank.

If a process is dependent primarily upon pumping capacity, the fluid velocities and the individual shearing rates, both on a macro- and a microscale, are above a certain minimum level to allow other process requirements to proceed unhindered. If the pumping capacity is increased and some of the other velocity and shear rate values are decreased below some minimum, then fluid shear stress enters into the overall design.

Other process applications sensitive to fluid shear rates include fermentation (qv), crystallization (qv), solids dispersion, polymerization, and many others. Shear rates should be considered if they are part of the overall process requirement or mechanism. However, oversimplification of the complex mixing phenomenon involved can result in serious errors in analysis and interpretation. Thus, high pumping capacity and low impeller velocity work can be obtained by large impeller diameters running at slow speeds for a given power level. Low pumping capacities and high impeller velocity work are obtained by running small impellers at high speeds.

In a general way, a minimum circulation rate in the tank sets the entire volume of fluid in motion. This is usually a minimum value to make the processing vessel a full participant in the process. Furthermore, the level of specific impeller velocity work or fluid shear rate which is needed to carry out the blending, dispersion, or diffusion required by the process objective has to be determined.

Most manufacturers supply portable mixers with either direct drive or gear drive. Direct-drive portable mixers have small impellers at high speeds whereas the gear-drive unit has large impellers and low speeds. The former are used for applications that require high shear rates, whereas the latter are used for applications requiring lower shear rates and higher pumping capacity.

Figure 2 shows the velocity profile of the blade of a flat-blade turbine radial-flow impeller. It was determined by measuring the average velocity at a point in a baffled tank (Fig. 3). By taking the slope at any point on this profile, the velocity gradient dv/dy is obtained which is the definition of fluid shear rate. Thus, the maximum shear rate around the impeller zone and the average shear rate can be evaluated (8–9).

The shear rate multiplied by the viscosity gives the shear stress, ie, shear stress = μ (shear rate). The shear stress carries out the dispersion and diffusion required in a process. Even at low viscosities, eg, 1–5 mPa·s (= cP), the shear stress increases five times from the same shear rate around the impeller.

Increasing the viscosity increases shear stress at a given shear rate. Thus, an increase in viscosity might improve the mixing performance. This is correct in terms of the actual shear stress produced but the material has to circulate through the shear zone and throughout the entire tank. Increasing the viscosity may introduce additional requirements for pumping capacity, blending, and other variables in the overall tank, hence the net result of the total energy required in the system depends on the viscosity. However, it is easier to disperse materials into a high viscosity liquid since the impeller has the ability to generate high shear stresses, assuming that the material is being circulated through the impeller shear zone.

A high frequency-response velocity probe that could pick the instantaneous velocity fluctuations in low viscosity turbulent-flow systems (Fig. 4) would give a plot as shown in Figure 5. This plot permits the calculation of the RMS velocity fluctuation which is a measure of microscale-mixing shear rates. Thus, in a mixing tank large shear stresses are generated by the average velocities at a point and the resulting shear stress (10) whereas macroscale mixing is generated by the existing high frequency turbulent fluctuations.

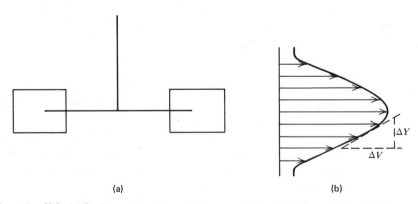

(a)

(b)

Figure 2.   Velocity discharge from (**a**) a radial-flow impeller and (**b**) the definition of fluid shear rate. Shear rate = $\Delta V/\Delta Y$.

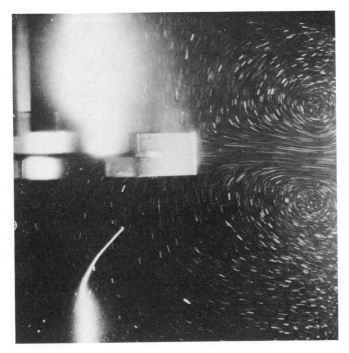

Figure 3.   Radial-flow impeller in a baffled tank in the laminar region. A thin plane of light is passed through the center of the tank.

To evaluate the effect of macroscale shear rates, the average point velocity is used. Four values give a good approximation to describe the shear rate profile, maximum and average impeller-zone shear rates, and average and minimum tank-zone shear rates.

Microscale shear rates are used as velocity fluctuations at a point and can be expressed as $\sqrt{(\mu')^2}$.

The power put into the mixer has to be dissipated as heat through the mechanism of viscous shear—regardless of the viscosity—in order to obtain shear rates small enough to dissipate their velocity energy in terms of the viscous shear rates. It has been estimated that the transition size is ca 500 μm between particles subject primarily to macroscale shear rates and microscale shear rates (11–12).

Impellers

Impellers are either radial flow or axial flow (13). Figure 6 illustrates a radial-flow disk turbine. Flow patterns in baffled and unbaffled tanks are shown in Figures 7 and 8.

The flat-blade turbine is normally placed 0.5–1 impeller diameter off bottom and has a coverage of 1–2 impeller diameters. The spacing between multiple impellers is somewhere between 1.5 and 3 impeller diameters.

Axial-flow impellers include the square-pitch marine-type propeller shown in Figure 9 and the axial-flow turbine shown in Figure 10. The former has a variable angle

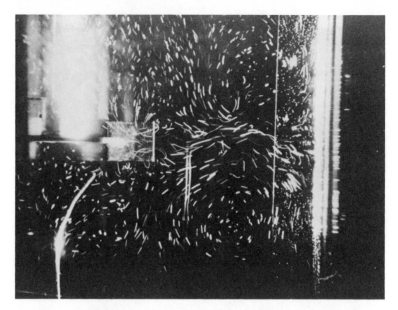

Figure 4.   Flow patterns in a mixing tank in the turbulent region. A thin plane of light is passed through the center of the tank.

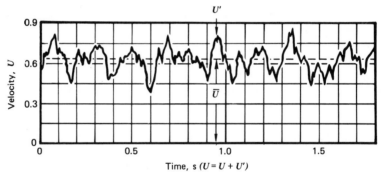

Figure 5.   Typical velocity fluctuation pattern obtained from high frequency velocity probe placed at a point in the mixing vessel.

and, therefore, an approximately constant pitch across the impeller face. This gives the most uniform flow pattern across the impeller periphery, and the propeller tends to have the highest pumping capacity per unit power.

   The axial-flow turbine has a constant blade angle and, therefore, a variable pitch across the surface. It has an effective flow pattern but is not quite as efficient as the propeller. Axial-flow turbines are used in large-size equipment primarily because of their low cost. The performance of axial-flow impellers in baffled tanks is illustrated in Figure 11. In general, for applications requiring primarily pumping capacity, such as blending and solid suspension, axial-flow turbines are the choice. For applications requiring gas–liquid or liquid–liquid mass transfer, or in multistage columns, radial-flow turbines are preferred.

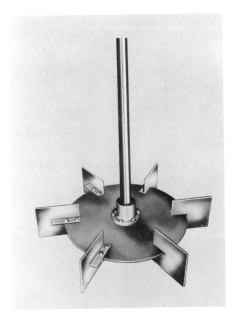

Figure 6.   Radial-flow, flat-blade, disk turbine.

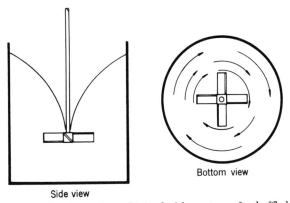

Bottom view

Side view

Figure 7.   Vortexing flow pattern obtained with any type of unbaffled impeller.

The fluid force acting in a radial direction on the impeller tends to deflect the shaft owing to the many variables of impeller design, tank geometry, and flow patterns. The manufacturer may supply relationship estimates.

Various types of mixer drives are illustrated in Figures 12–15. Portable mixers may have direct or gear drive, usually operated with propellers (Figs. 12–13). They are commonly used in angular, off-center position to achieve a good top-to-bottom flow pattern without the use of tank baffles. The fixed-mounted portable mixer has a mounting base and may either be used in open tanks or in a stuffing box for closed tanks. Since the mounting base is rigid, the shaft's critical speed and length must be designed with care.

For large equipment, heavy-duty fixed-mounted mixer drives are used (top-

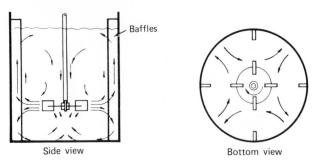

Figure 8.   Typical baffled flow pattern of radial-flow impeller.

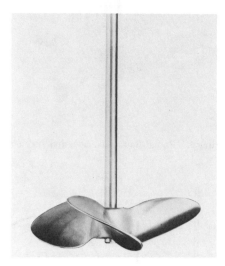

Figure 9.   Square-pitch marine-type impeller.

entering, Fig. 14). They must operate below the first critical speed of the mixer shape and most commonly use axial-flow or radial-flow turbines (Figs. 10 and 6, respectively).

The side-entering mixer (Fig. 15) normally uses propellers, has relatively short shafts, and is an extremely cost-efficient device for blending homogeneous fluids. The shaft enters the tank via the mechanical seal or stuffing box. Severe abrasive or corrosive conditions impede operation and cause leakage.

Bottom-entering drives shown in Figure 16 also may be used. The mixer is easily accessible for maintenance but leakage at the seal or stuffing box may have serious consequences. These drives are more suitable for single impellers near the bottom of the tank, since long shaft extensions into the upper part of the tank cause design problems.

In general, large impellers running at slower speeds need less power. However, depending upon the exponential relations of power and D/T, the torque required for the mixer drive typically increases. Torque is the ratio of power to impeller speed. It is the main criterion of cost of the mixer drive assembly. Mixer drives are normally rated on their torque capacity, and output energy is a function of output speed.

A mixer shaft has a natural vibration frequency. When it is rotated at that speed,

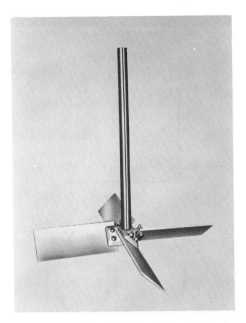

Figure 10.   Typical 45° axial-flow turbine.

shaft deflection becomes infinite, resonance occurs, and it may be destroyed. Figure 17 shows the general relationship. The mixer shaft must be operated at a speed below its critical speed.

It is theoretically possible to run mixer shafts above natural frequency, but when combined with long overhung shafts with bearing support only at the upper end, they cannot be used on large equipment.

Portable mixers, such as shown in Figure 12, run above and below the critical speed because of the instability of many portable installations.

Mixing

The mixing requirement may be expressed as pumping or circulating capacity. The impeller flow is defined through the impeller peripheral-discharge area, or total flow in the tank (14). In draft-tube circulators, shown in Figure 18, the impeller is enclosed in a draft tube. Head, flow, and pumping efficiency are measured as in a pump.

In an unbaffled tank (see Fig. 7), a swirl and a vortex are obtained which can be useful for certain kinds of mixing processes. However, the swirl has a tendency to suck gases into the vortex and is troublesome for many mixing applications. Baffles (see Figs. 10–11) give the top-to-bottom flow pattern which is often desired but require more power than the unbaffled tank. If the swirling flow pattern, with or without a detrimental accompanying vortex, is satisfactory for the process, it gives motion in the mixing tank at the lowest power requirement. Baffles give a better overall result, but lead to higher fluid shear rates in the tank. The improvement in process result has to justify the additional energy required.

In square and rectangular tanks, the effect of the corners and the shape provides

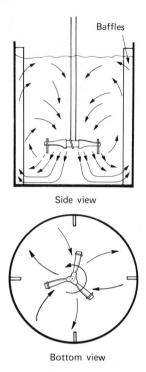

Side view

Bottom view

Figure 11. Typical baffled-flow pattern of axial-flow impeller.

a certain amount of baffle action (see Fig. 19). Therefore, at low power levels, eg, in storage, paint-blending, or milk tanks, a satisfactory flow pattern may be obtained without the use of baffles. The shape of the bottom is not of great importance. There are minor differences within the general classification of flat bottom, ASME dish bottom, or shallow cone but they can be handled on an individual mixing-specification basis.

When concerned with elliptical heads, spherical heads, or deep cones, it is necessary to consider the insertion of some kind of baffle in the tank bottom to prevent a residual swirling action.

Power Consumption. The power drawn by an impeller at a given speed and diameter is independent of process performance. The power number N_P is a function of the Reynolds number $ND^2\rho/\mu$:

$$N_P = P/(\rho N^3 D^5)$$

This results in the series of curves shown in Figure 1 which are for standard baffles defined as four baffles each one twelfth of the tank diameter in width.

In turbulent or viscous areas special equations are employed. In the turbulent region, power varies with $\rho N^3 D^5$, whereas in the viscous area, it varies with $\mu N^2 D^3$. Figure 1 gives curves for a variety of different impellers, and interpolation to other impeller shapes can be made readily.

Solid–Liquid Contacting. The settling velocity of solid particles is a critical factor in the process design of many mixing applications. Figure 20 gives the approximate settling velocity for spheres in water. It is, however, desirable to obtain the settling

Figure 12. Portable propeller mixer.

Figure 13. Fixed-mounted propeller mixer.

velocity experimentally by dropping individual particles of various sizes into a graduated cylinder and timing the settling velocities.

In a free-settling system, the settling velocity is above 300 mm/min and is a main factor in the design of the equipment. If the settling velocity is less than 300 mm/min, there is a fair degree of uniformity in the tank, once there is motion.

There are three types of solid suspension (15–21): complete motion on the tank bottom; all the particles have an upward motion (off-bottom suspension); or complete

Figure 14.   Top-entering drive with axial-flow turbine.

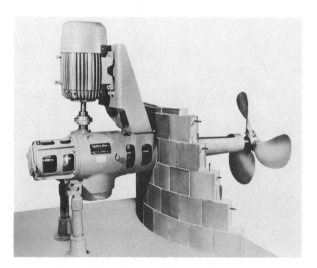

Figure 15.   Side-entering propeller mixer.

uniformity. The last is a relative term since the particles have a vertical settling velocity and the top of the tank has a horizontal velocity across the surface. It is, therefore, not possible to obtain complete uniformity in the upper layers of the tank. The relationship between these three types depends upon the settling velocity of the system (see Table 1). High concentrations of solids require increased power. Since the settling velocity of the particle is lower in the hindered settling range, an inverse function is needed to correlate the percentage of solids and the effect of settling velocity on power.

Figure 16.　Bottom-entering drive.

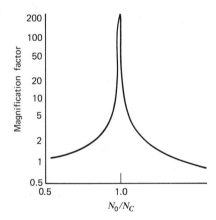

Figure 17.　Shaft deflection has a ratio to total indicated run-out, TIR, as a function of the operating speed to critical speed ratio.

Figure 21 gives an approximate selection of the power required for a solid suspension as a function of percent solids and settling velocity of the heaviest particle in the system. This relationship is not suitable when precise design is required for a given application.

The effect of power on solid-liquid mass transfer also can be measured by considering various suspension definitions (22). In the range up to complete off-bottom

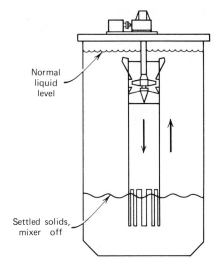

Figure 18.   Draft-tube circulator.

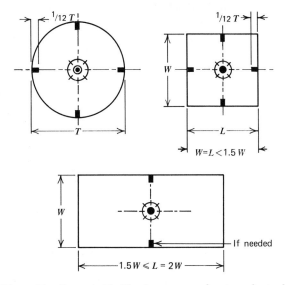

Figure 19.   Suggested baffles for square and rectangular tanks.

suspension, the effect of power is typically given by a slope of 0.2 or 0.3 on a logarithmic plot. As shown on Figure 22, above that point, the slope changes to about 0.1. In this range, only the slip velocity of the particles affects the film coefficient. Much greater gains for a given power level are expended up to that point than up to complete uniformity.

In general, axial-flow impellers are superior to radial-flow impellers for solid suspension. However, for gas–liquid–solid systems, the effect of the upward gas velocity on the downward pumping capacity of the axial-flow turbine almost completely neutralizes the impeller's flow pattern, and causes the power consumption for a given

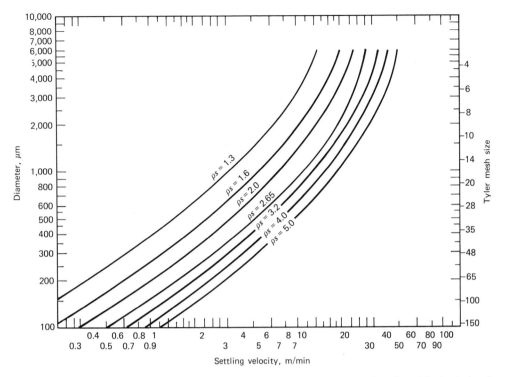

Figure 20. Settling velocity of spheres of various sizes in water as a function of particle size (ps) and specific gravity of the solids.

Table 1. Process Power Ratios

Process type	N^a	Power ratio Settling velocity		
		4.9–18.3[b]	1.2–2.4[b]	0.03–0.18[b]
complete uniformity	2.9	25	9	2
off-bottom suspension	1.7	5	3	2
on-bottom motion	1.0	1	1	1

[a] Impeller rotational speed, rps.
[b] Settling velocity is in m/min.

degree of suspension to be much higher with axial flow than with the radial flow (see also Extraction, liquid–solid).

Gas–Liquid Operations. Gas–liquid processes are affected by changes in power, speed, pumping capacity, and shear rate (23). Specification should give the process conditions required, such as mass-transfer and reaction rates. The dispersion specification alone, eg, 1000 m³/min, is not sufficient.

For any gas–liquid process selection, the superficial gas velocity is needed. This is defined as the average volumetric gas-flow rate in and out of the vessel divided by its cross-sectional area at the temperature and pressure at the midpoint of the tank expressed in meters per second. Normally a value of about 0.1 m/s is the boundary between normal mixer applications and those which must be carefully designed for

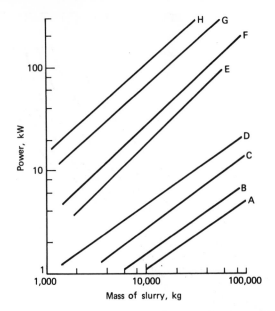

Figure 21. Approximate selection of mixer power, D/T ratio of 0.4, for two different settling velocities, two different percent solids concentration, as a function of mass of mother liquor in the tank. A, 2% off bottom, 5 m/min; B, 30% off bottom, 5 m/min; C, 2% uniform, 5 m/min; D, 30% uniform, 5 m/min; E, 2% off bottom, 15 m/min; F, 30% off bottom, 15 m/min; G, 2% uniform, 15 m/min; H, 30% uniform, 15 m/min.

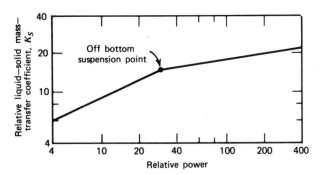

Figure 22. Effect of liquid–solid mass transfer.

process considerations such as liquid interface foaming and splashing, and fluid forces on mechanical equipment.

Intimate dispersions are controlled by the mixer flow patterns accompanied by power levels on the order of three times more energy in the mixer than in the expanding gas stream.

Minimum dispersion occurs at a point where the energy content of the mixer and gas stream are nearly equal. This usually results in a gas-controlled flow pattern in which uniform bubble dispersion leaves the area of the impeller. However, the overall flow pattern is governed by the rising expanding gas stream. Below that point are the areas where gas tends to form geysers and large splashing brings bursts of liquid to the surface.

As a general rule, the flow pattern from axial-flow impellers is governed by the upward velocity of the gas stream. Thus, axial-flow impellers do not operate satisfactorily unless their energy input is five to ten times higher than that of the gas stream.

Radial-flow impellers are most commonly used and also disk-type impellers. The latter prevent the gas from rising through the low shear zone around the hub.

Sparge rings or open inlets permit the admission of the gas at a somewhat smaller diameter than the impeller periphery below. Thus, the gas can rise into the maximum impeller-zone shear rate.

Liquid–Liquid Dispersion. Stable emulsions include household products, cosmetics, pharmaceuticals, and a wide variety of other combinations (see Emulsions). A minimum of fluid shear rate is usually required for the production of a uniform stable emulsion. When this fluid shear rate is produced by the mixer, the composition can either maintain its dispersion or the particles coalesce. If, on the other hand, the mixer is not capable of producing the shear rate required, the product will not be satisfactory. However, when the mixer produces the emulsion at high impeller shear rate, eventually all particles achieve the minimum particle size necessary for a satisfactory product.

The mixing time depends on the pumping capacity of the unit and on the number of times the particles pass through the high shear zone of the impeller and the length of time each particle spends in that shear zone.

Thus a variable known as the shear work, which is the product of shear rates and the time of contact, gives a measure of the total amount of energy expended into the dispersion.

Figure 23 shows a multistage mixer column in which the dispersed phase travels countercurrent to a continuous phase, either heavy or light (24). Mixer columns are similar in performance to packed, spray, or plate columns but can handle solids in either one or both phases. Since the dispersion produced is a function of the impeller relationships, scale-up is extremely reliable (see also Extraction, liquid–liquid).

Blending

Miscible Liquids. When blending miscible liquids, two distinct mechanisms are involved. In general, one material is run into the vessel with the mixer, whereas the second is injected into the tank. The uniformity of some particle or a physical or chemical property is then measured. A sampling point has to be chosen and the uniformity required for blending has to be defined.

Larger impellers at slow speeds reduce blend times. For a propeller, the blending time is proportional to D^{-2}, whereas for a turbine the blending time is proportional to D^{-1}. To obtain the same circulating time on scale-up, the ratio P/V increases with the square of the tank diameter. However, this relationship cannot be used for design. Therefore, an increase in blending time is incorporated into the overall plant design (25).

In another blending method, the tank is initially stratified. After mixing is started, the time required to eliminate the stratified condition is measured. This is typical of large petroleum storage tanks (26–28). Using hot and cold water and the associated temperature differences, the blending equation is

$$\theta \propto P^{-1}(D/T)^{-2.3}\left(\frac{\Delta\rho}{\rho}\right)^{0.9}$$

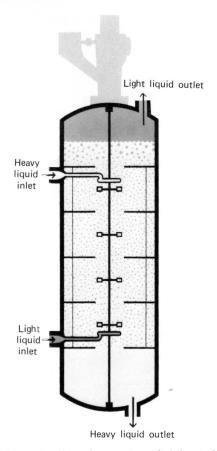

Figure 23.   Oldshue-Rushton extraction column, using radial-flow turbines in each compartment.

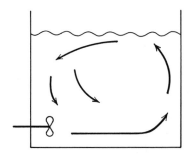

Figure 24.   Typical flow pattern of properly positioned side-entering drive.

Other injection-technique studies give different exponents. In fact, blend time is proportional to power and not speed. It takes several times more power to achieve blending in a stratified tank than in an injection tank, but this difference is reduced on scale-up since P/V is more typically constant in the case of stratified blending.

For homogeneous fluids, side-entering mixers offer the most economic combination of capital and operating costs. Their effectiveness is based on the fact that the

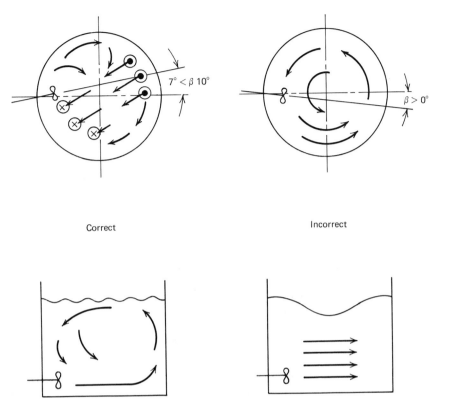

Correct   Incorrect

Figure 25.   Flow pattern when proper angle of entry of the impeller shaft is or is not maintained.

fluid in the tank, as shown in Figure 24, makes a gentle series of 90° turns, and not the sharp 180° turns occurring in top-entering systems. However, there are quiet zones at the sides of these tanks, 90° away from the impeller position. In large gasoline and petroleum storage tanks this can cause the deposition of so-called bottom sediment and water. To alleviate this condition the angle of the mixer or the rotation is changed to keep the solids deposition moving around the tank at an acceptably controllable level. Figure 25 illustrates the flow pattern with and without proper angle of the impeller shaft.

Viscous Fluids.   There are several areas of viscous blending. In large industrial tanks, low viscosity is defined as 5 Pa·s (50 P). From 5 to 50 Pa·s (50–500 P), which is the medium-viscosity region, either open-axial or radial-flow turbines may be considered or the close-clearance anchor or helical impellers.

The area above 50 Pa·s is defined as high viscosity mixing, in which typically an anchor or helical impeller should be used (see Figs. 26–27). Figure 28 gives typical data obtained in measuring the circulation time of helical-flow mixers in both Newtonian and non-Newtonian fluids (3,29).

Anchor impellers do not have any tendency for top-to-bottom flow and, therefore, do not provide effective blend time or achieve effective temperature uniformity in heat transfer applications. However, for many processes anchor impellers are used.

The helical impeller has a strong axial-flow component and can give effective blending and circulation times in both Newtonian and non-Newtonian fluids. Typical

Figure 26.   Anchor impeller.

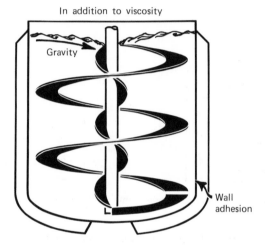

Figure 27.   Helical impeller.

are a pitch of 0.5 and a blade width of $\frac{1}{12}$ to $\frac{1}{6}$ T. The inner flight is only effective on non-Newtonian fluids but gives superior performance over noninner flight.

A single outer flight gives the lowest torque and is completely satisfactory for large equipment. However, if a double outer flight is desired, it can be easily produced but results in a lower operating speed and a larger mixer drive for the same process requirement.

In the plant, the nominal clearance of 25–100 mm between the blade and the tank wall may be 25–100 mm, which requires no machining of either component. On the other hand, a clearance of 5–25 mm requires manufacture of the tank first. A template is prepared and the impeller machined to suit. Scraper blades also can be used which may be product loaded or spring loaded.

Scraper blades normally increase both power consumption and the heat-transfer coefficient by a factor of two.

Heat input from the mixer may be an appreciable part of the total heat load in the process.

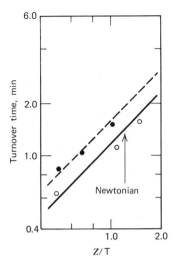

Figure 28. Turn-over time vs helical impeller speed Z for a variety of viscosities. $N = 20$ rps; pitch ratio = 0.5; 46 cm tank; 43 cm helix. Non-Newtonian fluid: O, single outer, single inner; ●, single outer.

In-Line Mixers. Figures 29–31 illustrate several of the in-line mixing elements used in pipelines (see Pipelines). For viscous fluids, elements are preferred that can twist and cut the streams or actually force materials through channels or tubes. Mixer performance is expressed as the width of disbursement, ie, the number of cuts per element which can be typically either 2, 3, or 4, raised to the nth power where n is the number of elements. A prediction can be made of the size of the final dispersion produced in the unit. Pressure-drop data generally are available.

In low viscosity operations, fluid turbulence and shear stresses are introduced by the elements as the flow passes through the unit. For a certain pressure drop, the same degree of shear rate and mixing effect are obtained. Thus, different types of elements require different lengths, depending on their specific pressure drop relationship.

In-line mixer elements provide primarily transverse uniformity but not time-interval uniformity (see Fig. 32). The latter requires a type of mixing volume that is accomplished either by a vessel in the pipeline, or a suitable mixing tank of the required size. In general, to effect time-interval uniformity, the residence time has to be about the same order of magnitude as the length of time it takes to blend the elements.

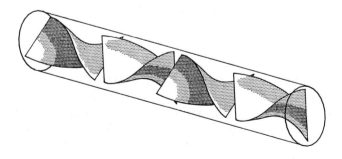

Figure 29. Kenics in-line mixer.

Figure 30.   Koch in-line mixer.

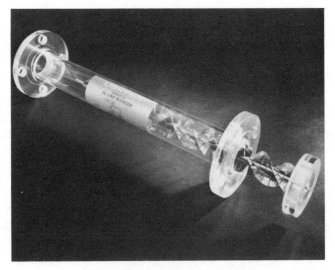

Figure 31.   Lightnin in-line mixer.

Scale-Up

Table 2 shows the change of typical mixing parameters on a scale of 100–12,500 L capacity. The linear-scale ratio is 5:1. In calculating scale-up, these parameters are considered in correlation. Another consideration is geometric similarity, which has been used to derive Table 2. Figure 33 shows the maximum impeller-zone shear rate tends to increase, whereas the average impeller shear rate tends to decrease with increasing scale. Thus, the distribution of shear rates in a large mixing tank differs from

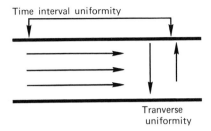

Figure 32.   In-line mixers can provide transverse uniformity but not time-interval uniformity.

Table 2.   Properties of a Fluid Mixer on Scale-Up

Property	Pilot scale, 0.1 m³	Plant scale, 12.5 m³			
P	1.0	125	3125	25	0.2
P/V	1.0	1.0	25	0.2	0.0016
N	1.0	0.34	1.0	0.2	0.04
D	1.0	5.0	5.0	5.0	5.0
Q^a	1.0	42.5	125	25	5.0
Q/V	1.0	0.34	1.0	0.2	0.04
ND^b	1.0	1.7	5.0	1.0	0.2
$\dfrac{ND^2\rho}{\mu}$	1.0	8.5	25.0	5.0	1.0

[a] Impeller pumping capacity.
[b] Tip speed.

Table 3.   Dry Mixers

Type of mixing	Mixers used[a]
bakery mixes, foodstuffs	A, B, E
dry chemicals, crystalline, moderately fine to coarse	A, B, C, D, F, G, H
dry chemicals, crystalline, fine powders	A, B, C, D, F, G, H
pharmaceuticals	A, D
fertilizers	A, D, G
glue chips or powder	A, B, C, D, F, G, H
resin chips or powder	A, B, C, D, F, G, H
powdered metals	A, D, G
tobacco	A, D, G
feed and grain	A, B, C, G
foundry sand mixes	A, F

[a] A, tumble type, drum, container, V, cone; B, ribbon; C, paddle; D, centrifugal;
   E, planetary; F, pan, wheel, plow; G, spiral elevator; H, fluid bed.

those in a small tank. For a dynamic, heterogeneous system, in which bubbles and drops are dispersed and coalesced, the particle size distribution in a large tank differs from that in a small tank. This might or might not affect the overall process objective.

It is difficult to combine the two shear rates in big tanks as in small tanks. How-

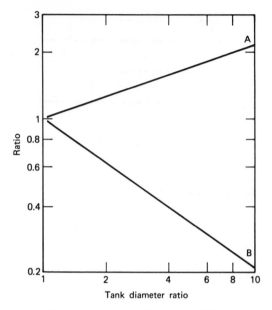

Figure 33. Maximum impeller-zone shear rates, A, increase while average impeller-zone shear rates, B, decrease when using geometric similarity and constant power per unit volume on scale-up.

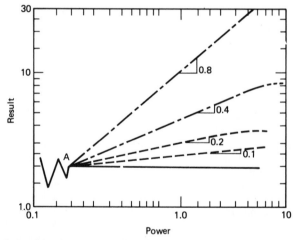

Figure 34. Relationship of process result and impeller power consumption, obtained by varying impeller speed. The slope can be used as an indicator of controlling factors.

ever, it is possible to control the increase or decrease of either of these shear rates by nongeometric scale-up or scale-down techniques.

In general, a small pilot-scale mixing tank designed to perform a particular plant-scale process is not a good model for a full-scale process because the pumping capacity and the maximum impeller-zone shear rate are too high, and the blend time is too short.

Geometric, kinematic, and dynamic similarity govern many scale-up consider-

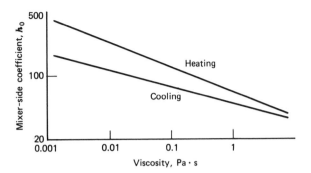

Figure 35. Design curves for the mixer-side coefficient as a function of viscosity for heating and cooling. To convert Pa·s to P, multiply by 10.

ations (30–31); eg, dynamic similarity requires that the ratio of the four main fluid forces in the tank, ie, the inertia force put in by the mixer F_i; the opposing force of viscosity F_v; the surface tension F_σ; and the gravity F_g in both the model M and the prototype P be equal to a common constant ratio R shown below. μ = viscosity, g = force of gravity, σ = surface or interfacial tension.

Geometric:

$$\frac{X_M}{X_P} = X_R$$

Dynamic:

$$\frac{(F_i)_M}{(F_i)_P} = \frac{(F_v)_M}{(F_v)_P} = \frac{(F_g)_M}{(F_g)_P} = \frac{(F_\sigma)_M}{(F_\sigma)_P} = F_r$$

Force ratios:

$$\frac{F_i}{F_v} = N_{Re} = \frac{ND^2\rho}{\mu}$$

$$\frac{F_i}{F_g} = N_{Fr} = \frac{N^2D}{g}$$

$$\frac{F_i}{F_\sigma} = N_{We} = \frac{N^2D^3\rho}{\sigma}$$

Unless the fluid in the model and the prototype is changed, dynamic similarity of all four fluid forces cannot be obtained. The viscosity is considered the main force but the Reynolds number and the ratio of inertia force to viscous force are often used in mixing correlations.

It is difficult to write an appropriate process group in dimensional terms which corresponds in principle to the dynamic similarity ratio of the Reynolds number. The usual requirements of pumping capacity, shear rates, diffusion rates, and many other quantitative and qualitative factors preclude the use of a single dimensionless group as a suitable indicator of process result. Exceptions are the Nusselt number for heat transfer and the blend number for well-defined blending criteria.

For scale-up, the following method is used: Some experimental data can give an approximate estimation of the principal controlling process steps; and the pertinent mixing parameters relating to these controlling steps have to be identified.

In general, the power in the pilot-plant experiment should be varied in order to change flow and fluid shear rates markedly. This is normally accomplished by changing

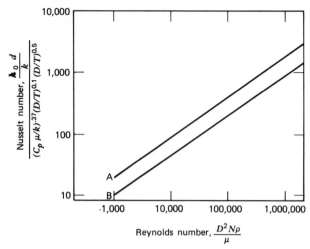

Figure 36. Correlation of Nusselt no. vs Reynolds no. for A, radial-flow flat-blade turbines, 6 blades; and B, propellers, 3 blades, pitch ratio 1.0. Tank, 1.2 m; liquid, 1.2 m level; baffles, 10 × 10 cm wide at wall.

the speed. Figure 34 is a quantitative plot of the process result, although it can be a qualitative estimation. A zero slope indicates a homogeneous chemical reaction mechanism, and mixing does not have a decisive effect on the process. On the other hand, a slope, on the order of 0.5 to 0.8, typically indicates gas–liquid or liquid–liquid type of mass-transfer-controlled processes (see Mass transfer).

A pilot process can determine the controlling factor, and thus indicate the best scale-up technique. However, most mixing-process designs do not require a pilot-plant model. Mostly, sufficient data are available to permit designers and vendors to select the equipment required. If not, the proper scale-up technique can be determined on the basis of some operating data (see Pilot plants and microplants).

In order to maintain equal blend time on scale-up, the ratio P/V must be increased with the square of the tank size. This usually is not feasible and, therefore, blending times are allowed to increase in larger tanks.

Constant P/V is usually too conservative for many processes. It is, however, a good criterion for homogeneous chemical reactions, and for scale-up over volume ranges of 5:1 or less. Parameters of big tanks are different in many respects and must be evaluated as to their ultimate effect.

Constant tip speed equals constant torque per unit volume, and is normally not predictable on scale-up. The value P/V drops in direct proportion to scale ratio, and this creates problems with macro- and micro-scale shearing effects as well as large increases in blend and circulation time. It does, however, work well for draft-tube circulators.

Heat-transfer correlations with Nusselt and Reynolds numbers allow accurate prediction of performance on different scales.

Heat Transfer. The source of turbulence affecting the heat-transfer coefficient in a mixing vessel is the fluid flow around and across the heat-transfer surface. The impeller affects only the flow that actually reaches the heat-transfer surface.

For design purposes, usually only a gross overall approximate coefficient is needed

(32). Figure 35 shows typical heat-transfer coefficients as a function of viscosity for organic fluids with various viscosities. The material with 1 mPa·s (= cP) viscosity is an aqueous solution.

In general, propellers and turbines give similar heat-transfer performance. For example, Figure 36 shows a correlation of Nusselt vs Reynolds numbers with other dimensionless groups (33). However, the propeller draws only 3% of the power of the flat-blade turbine at the same Reynolds number. Therefore, when it is accelerated to draw the same power, it has about the same heat-transfer coefficient.

The slope of heat-transfer coefficient vs mixer power in the forced convection region is very low, $h \propto P^{0.22}$. Typically, once forced convection is achieved, the heat-transfer coefficient is not raised by increasing power. Usually it is calculated, and the overall heat-transfer performance is modified by design changes (see Heat-transfer technology).

Mixing of Dry Solids

Table 3 gives an overview of dry-mixing applications. Bulk blenders are used for the complete gamut of dry-solids blending, from coating of solids to paste mixing.

Tumbling. Gentle mixing by a tumbling action causes materials to cascade from the top of the rotating vessel. Common types offer various vessel configurations including drum, container, V, and cone. Intake and discharge of materials take place through an opening in the vessel end. Dry and partly dry powders, granules, and crystalline substances are readily mixed in such equipment. Liquid feed and intensive blending are possible with the addition of a liquid-feed or high speed mix bar.

Ribbon Type. Spiral or other blade styles transfer materials from one end to the other or from both ends to the center for discharge. This mixer can be used for dry materials or pastes of heavy consistency. It can be jacketed for heating or cooling.

Paddle Type. This type is similar to the ribbon type except interrupted flight blades or paddles transfer materials from one end to the other, or from both ends to the center for discharge. The paddle-type mixer can be used for dry materials or pastes of heavy consistency. It can be jacketed for heating or cooling.

Centrifugation. Materials are passed through a rotor consisting of disks spaced ca 2.5 cm apart and held together by rod-type supports that act as impacters. Centrifugal forces throw the material against the rods and inside walls or the housing, from which it drops by gravity to the discharge (see Centrifugal separation).

Planetary Type. Paddles or whips of various configurations are mounted in an off-center head that moves around the central axis of a bowl or vessel. Material is mixed locally and moved inward from the bowl side, causing intermixing. This mixer handles dry materials or pastes.

Pan Type. Mulling action of the wheel type is similar to the action of a mortar and pestle. Scrapers move the materials from the center and sides of a pan into the path of rotating wheels where mixing takes place. The pan may be of the fixed or rotating type. Discharge is through an opening in the pan. The flow type uses rotating plows in a rotating pan to locally mix and intermix by the rotation of plows and pan, respectively (Fig. 37).

Spiral Elevator. Materials are moved upward by the centrally located spiral-type conveyor in a cylindrical or cone-shaped vessel. Blending occurs by the downward movement at the outer walls of the vessel. The vessel serves the dual purposes of blending and storage.

Figure 37.   Model 507 Clearfield mixer.

Fluidized-Bed Type.   Particles suspended in a gas stream behave like a liquid. They are mixed by turbulent motion and intimate contact of components (see Fluidization). This mixer is used for mixing and drying, or mixing and reaction.

Equipment for Pastes and Viscous Materials

Blending of viscous materials and pastes can be achieved either by the age-old art of dividing and recombining, or by layering the various components to some predefined striation thickness until adequate uniformity is attained. Moving agitators may have to come close to walls or stationary baffles in order to provide the high shear required to separate agglomerates and to reduce the size of regions occupied by one component. The energy requirements may be very high because of the work involved in dividing and shearing the material. Allowable heat rise frequently imposes a limit at which power can be applied.

Mixing machinery is selected according to its capacity to shear material at low speed and to wipe, smear, fold, stretch, or knead the mass to be handled. Mixers with intermeshing blades are sometimes required to keep the material from clinging un-mixed to the lee side of the blade. Wiping of heat-transfer surfaces promotes addition or removal of heat.

Batch Mixers.   Batch rather than continuous mixing is still preferred when batch identity must be maintained, eg, in pharmaceutical preparations; frequent product changes would require too much off-specification production between products, eg, in dye and pigment manufacture; a multitude of ingredients is required, each of which can be accurately weighed and charged without relying on absolute constancy of flowmeters, eg, in some adhesives and caulking compounds; various changes of state may be involved, eg, the pulling of a vacuum on the end product to drive off volatiles after reaction of a granular material which would not permit a fluid seal in a continuous mixer; and very long mixing and reacting times are required.

Change-Can Mixers.   In change-can mixers one or more blades cover all regions of the can either by a planetary motion of the blades or a rotation of the can (see Fig.

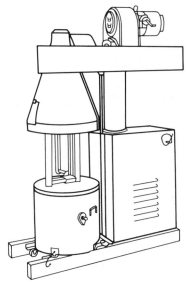

Figure 38. Change-can mixer. Courtesy of Chas. Ross & Son.

38). The blades may be lowered into the can or the can may be raised to the mixing head. Separate cans allow the ingredients to be measured carefully before the mixing operation begins, and can be used to transport the finished batch to the next operation while the next batch is being mixed.

Stationary-Tank Mixers. Stationary-tank mixers are recommended when the particular advantages of the change-can mixer are not required. The agitator may be particular to a specific industry, like the soap crutcher, or for general use as with anchor mixers (Fig. 26) or gate mixers. In the latter, some part of the agitator moves in close proximity to the vessel walls or stationary bar baffles.

The impeller may also consist of a single- or double-helical blade (Fig. 27) to promote top-to-bottom turnover while minimizing the amount of hardware that must be moved through the viscous mass.

Double-Arm Kneading Mixers. These mixers have been used for a long time (see Fig. 39). The material is carried by two counterrotating blades over the saddle section of a W-shaped trough. Randomness is introduced by the difference in blade speed and end-to-end mixing by differences in the length of the arms on the Σ-shaped blades. Other blade shapes are used for specific end purposes, such as smearing or cutting edges on the blade faces. Discharge is usually by tilting the trough or by a door in the bottom of the trough. Double-arm kneading mixers also are available with a screw centrally located to discharge the contents (Fig. 40).

Intensive Mixers. Intensive mixers, such as the Banbury (Fig. 41), are similar in principle to the double-arm kneading mixers, but are capable of much higher torques. Used extensively in the rubber and plastics industries, the Banbury mixer is operated with a ram cover so that the charge can be forced into the relatively small volume mixing zone. The largest of these mixers holds only 500 kg but is equipped with a 2000-kW motor.

Roll Mills. When dispersion is required in exceedingly viscous materials, the large surface area and small mixing volume of roll mills allows maximum shear to be

Figure 39.   Sigma-blade mixer. Courtesy of Baker Perkins Inc.

maintained as the thin layer of material passing through the nip is continuously cooled. The rolls rotate at different speeds and temperatures to generate the shear force with preferential adhesion to the warmer roll.

Ribbon Blenders.   Ribbon blenders (Fig. 42) provide end-to-end mixing and overall lifting of the blender contents by the pitch of the helical blades. Variations include closed vessels with a plow-shaped head on a horizontal rotor operating at relatively high speed to scrape the trough wall as well as to hurl the contents throughout the free space.

Similar applications also are handled in vertically mounted mixers equipped with a high speed blade. This blade scoops material from the bottom of the vessel, hurling it upward along the vessel walls, where baffles may be located to provide variation in turnover. Such mixers may be used for dispersion in thin slurries, or in wetting or coating of granular materials.

Cone-and-Screw Mixer.   A more gentle blender is the Nauta-type cone-and-screw mixer (Fig. 43). This mixer uses an orbiting screw to provide bottom-to-top circulation. Reversing the screw aids discharge. Other variants are available that provide either an epicyclic orbit or two screws to provide greater volumetric coverage. The mixing action of cone-and-screw mixers is independent of the degree of fill.

Figure 40.   Sigma-blade mixer with screw discharged. Courtesy of B. P. Guittard.

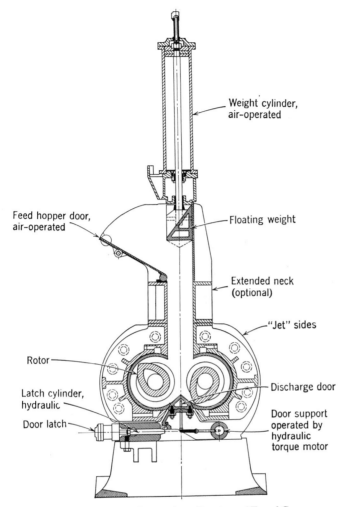

Weight cylinder,
air-operated

Feed hopper door,
air-operated

Floating weight

Extended neck
(optional)

"Jet" sides

Rotor

Discharge door

Latch cylinder,
hydraulic

Door latch

Door support
operated by
hydraulic
torque motor

Figure 41.   Banbury mixer. Courtesy of Farrel Corp.

Pan Muller. A mixer similar to mortar and pestle is the pan muller, in which plows bring material into the path of the rolling mullers. Such mixers are used for applications where the final blend is neither very fluid nor very sticky, such as foundry sand, clay, or chocolate.

Continuous Mixers. In most continuous mixers one or more screw or paddle rotors operate in an open or closed trough. Discharge may be restricted at the end of the trough to control holdup and degree of mixing. Some ingredients may be added stagewise along the trough or barrel. The rotors may be cored to provide additional heat-transfer area. The rotors may have interrupted flights to permit interaction with pins or baffles protruding inward from the barrel wall.

Single-Screw Extruders. Single-screw extruders (Fig. 44) incorporate ingredients, such as antioxidants, stabilizers, pigment, and other fillers into plastics and elastomers. In order to provide a uniform distribution of these additives, the polymer is brought to a fusion state primarily by the work energy imparted in the extruder, rather than

Figure 42.   Spiral-ribbon mixer. Courtesy of Teledyne Read Co.

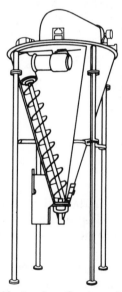

Figure 43.   Nauta mixer. Courtesy of J. H. Day Co.

by heat transfer through the barrel wall. In addition to melting the polymer, the extruder is used as a melt pump to generate pressure for extrusion through a die, shaping the molten product in a specific profile, strands, or pellets. The extruder screw drags the polymer through the barrel, generating shear between screw and barrel. In addition, some axial mixing occurs in response to a back flow along the screw channel caused by the pressure required to get through the die.

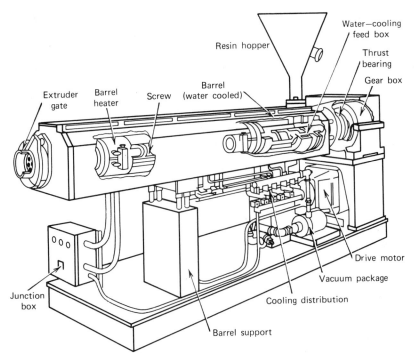

Figure 44. Sectional view of a typical extruder.

Figure 45. Twin-screw mixer. Courtesy of Baker Perkins Inc.

Mixing in a single-screw extruder can be enhanced by interrupting the flow pattern within the screw flight channel. This can be done by variations in the screw flight-channel width or depth, or by causing an interchange with spiral grooves in the barrel, or by interengaging teeth.

Twin-Screw Mixers.   Twin-screw continuous mixers provide more radial mixing by interchange of material between the screws, rather than just acting as screw conveyors.

Intermeshing co-rotating twin-screw mixers (Fig. 45) have the additional advantage that the two rotors wipe each other as well as the barrel wall. This action eliminates any possiblity of dead zones or unmixed regions. In addition to variations in screw helix angles, these mixers can be fitted with kneading paddles that interactively cause a series of compressions and expansions to increase the intensity of mixing. Such mixers are used for a variety of pastes and doughs, as well as in plastics compounding.

Nomenclature

a	= constant
b	= constant
C	= impeller off bottom distance
D	= impeller diameter, m
d	= tube diameter, m
F	= superficial gas velocity, m/s
H	= velocity head, J/kg
h_0	= mixer side coefficient
K_S	= mass-transfer coefficient
N	= impeller rotational speed, rps
N_{Fr}	= Froude number
N_P	= power number
N_{Re}	= Reynolds number
N_{We}	= Weber number
P	= power, W
Q	= impeller flow; pumping capacity, kg/s
T	= tank diameter, m
TIR	= total indicated run-out
U	= liquid velocity, m/s
V	= volume
Y	= distance, m
Z	= helical impeller speed
θ	= blend time
μ	= viscosity
Φ	= power number
ρ	= fluid density
σ	= surface tension

BIBLIOGRAPHY

"Mixing and Agitating" in *ECT* 1st ed., Vol. 9, pp. 133–166, by J. H. Rushton, Illinois Institute of Technology, R. D. Boutros, Mixing Equipment Co., Inc., and C. W. Selheimer, Illinois Institute of Technology; "Mixing and Blending" in *ECT* 2nd ed., Vol. 13, pp. 577–613, by J. H. Rushton, Purdue University, and R. D. Boutros, Mixing Equipment Co., Inc.

1. R. L. Bates, P. L. Fondy, and J. G. Fenic in V. W. Uhl and J. B. Gray, eds., *Mixing*, Vol. I, Academic Press, Inc., New York, 1966, Chapt. 3.
2. E. L. Paul and R. E. Treybal, *AIChE J.* **17,** 718 (1971).
3. J. R. Bourne, W. Knoepfli, and R. Riesen, *Proc. 3rd European Conf. on Mixing, April 1979*, BHRA Fluid Engineering, **I** (1979).

4. J. R. Bourne and H. L. Toor, *AIChE J.* **23** (1977).

5. J. Y. Oldshue, *CHEMECA '70, Proceedings, Australia*, Butterworth, Reading, Mass., 1970.

6. I. Fort, *Coll. Czech. Chem. Commun.* **44,** (1979).

7. J. Gray in ref. 1.

8. L. A. Cutter, *AIChE J.* **12,** 35 (1966).

9. J. Y. Oldshue, *Biotech. Bioeng.* 8(1), 3 (1966).

10. A. B. Metzner and R. E. Otto, *AIChE J.* **3**(3), (1957).

11. J. Y. Oldshue and O. B. Mady, *Chem. Eng. Prog.*, 103 (Aug. 1978).

12. *Ibid.* 72 (May 1979).

13. J. Y. Oldshue and J. H. Rushton, *Chem. Eng. Prog.* **49,** 161, 267 (1953).

14. J. P. Sachs and J. H. Rushton, *Chem. Eng. Prog.* **50,** 597 (1954).

15. J. Y. Oldshue, *Hydrometallurgy*, Gordon & Breach Ltd., London, Eng., 1961.

16. L. E. Gates, J. R. Morton, and P. L. Fondy, *Chem. Eng.*, 144 (May 24, 1976).

17. J. Y. Oldshue, *Ind. Eng. Chem.* **61,** 79 (1969).

18. J. R. Connolly and R. L. Winter, *Chem. Eng. Prog.* **65**(8), (1969).

19. E. J. Lyons in V. W. Uhl and J. B. Gray, eds., *Mixing*, Vol. II, Academic Press, Inc., New York, Chapt. 3.

20. A. W. Nienow, *Chem. Eng. Sci.* **23,** 1453 (1968).

21. A. W. Nienow, *Chem. Eng.* **15,** 13 (1978).

22. J. Y. Oldshue in R. H. Perry and C. H. Chilton, eds., *Perry's Handbook*, 5th ed., McGraw-Hill Book Company, New York, 1973, Chapter 19.

23. J. Y. Oldshue and G. L. Connelly, *Chem. Eng. Prog.* (Mar. 1977).

24. J. Y. Oldshue and J. H. Rushton, *Chem. Eng. Prog.* **48,** (1952).

25. J. C. Middleton in ref. 1.

26. J. Y. Oldshue, A. T. Gretton, and H. E. Hirschland, *Chem. Eng. Prog.* **52,** 481 (1956).

27. E. A. Fox and V. E. Gex, *AIChE J.* **2,** 539 (1956).

28. R. W. Hicks, J. R. Morton, and J. G. Fenic, *Chem. Eng.*, (April 26, 1976).

29. C. K. Coyle, H. E. Hirschland, B. J. Michel, and J. Y. Oldshue, *Can. J. Chem. Eng.* **48** (1970).

30. J. H. Rushton, *Chem. Eng. Prog.* **47**(9), 485 (1951).

31. R. R. Corpstein, R. A. Dove, and D. S. Dickey, *Chem. Eng. Prog.*, 66 (Feb. 1979).

32. J. Y. Oldshue and A. T. Gretton, *Chem. Eng. Prog.* **50,** 615 (1954).

33. J. Y. Oldshue, *Chem. Process. Eng.*, 183 (1966).

General References

F. A. Holland and F. S. Chapman, *Liquid–Liquid Mixing in Stirred Tanks*, Reinhold, New York, 1966.

S. Nagata, *Mixing Principles and Applications*, John Wiley & Sons, Inc., New York, 1975.

Z. Sterbacek and P. Tausk, "Mixing in the Chemical Industry," *International Series of Monographs in Chemical Engineering*, Vol. 5, Pergamon Press, New York, 1965.

V. W. Uhl and J. B. Gray, eds., *Mixing*, Vols. I–II, Academic Press, Inc., New York, 1966–1967.

JAMES Y. OLDSHUE
Mixing Equipment Co., Inc.

DAVID B. TODD
Baker Perkins, Inc.

MOLASSES. See Syrups.

MOLD-RELEASE AGENTS. See Abherents.

MOLECULAR SIEVES

Molecular-sieve zeolites are crystalline aluminosilicates of group IA and group IIA elements such as sodium, potassium, magnesium, and calcium. Chemically, they are represented by the empirical formula:

$$M_{2/n}O.Al_2O_3.ySiO_2.wH_2O$$

where y is 2 or greater, n is the cation valence, and w represents the water contained in the voids of the zeolite. Structurally, zeolites are complex, crystalline inorganic polymers based on an infinitely extending framework of AlO_4 and SiO_4 tetrahedra linked to each other by the sharing of oxygen ions. This framework structure contains channels or interconnected voids that are occupied by the cations and water molecules. The cations are mobile and ordinarily undergo ion exchange. The water may be removed reversibly, generally by the application of heat, which leaves intact a crystalline host structure permeated by micropores which may amount to 50% of the crystals by volume. In some zeolites, dehydration may produce some perturbation of the structure such as cation movement and some degree of framework distortion (1).

The structural formula of a zeolite [1318-02-1] is based on the crystal unit cell, the smallest unit of structure, represented by

$$M_{x/n}[(AlO_2)_x(SiO_2)_y].wH_2O$$

where n is the valence of cation M, w is the number of water molecules per unit cell, x and y are the total number of tetrahedra per unit cell, and y/x usually has values of 1–5. However, recently high silica zeolites have been prepared in which y/x is 10 to 100 or even higher and, in one case, a molecular-sieve silica has been prepared (2–3).

Zeolites were first recognized as a new type of mineral in 1756. The word zeolite was derived from two Greek words meaning to boil and a stone. Several properties of zeolite minerals have been studied, including adsorption and ion exchange. This led to the preparation of amorphous aluminosilicate ion exchangers for use in water softening. Studies of the gas-adsorption properties of dehydrated natural-zeolite crystals more than 60 years ago led to the discovery of their molecular-sieve behavior. As microporous solids with uniform pore sizes that range from 0.3 to 0.8 nm, these materials can selectively adsorb or reject molecules based on their molecular size. This effect, with obvious commercial overtones leading to novel processes for separation of materials, inspired attempts to duplicate the natural materials by synthesis. Many new crystalline zeolites have been synthesized and several fulfill important functions in the chemical and petroleum industries and consumer products such as detergents. More than 150 synthetic zeolite types and 40 zeolite minerals are known. The nomenclature of zeolite minerals follows established procedures. No practical system of nomenclature for the synthetic materials and their many modifications has yet been devised. Consequently, a system based on trivial symbols is used to denote the synthetic zeolite in terms of its composition and structure.

Mineral Zeolites

Zeolite minerals are formed over much of the earth's surface, including the sea bottom (1,4). Until about twenty years ago, zeolite minerals were considered as typically occurring in cavities of basaltic and volcanic rocks. During the last 20–25 years, however, the use of x-ray diffraction for the examination of very fine-grained sedimentary rocks has led to the identification of several zeolite minerals which were formed by the natural alteration of volcanic ash in alkaline environments. More common types include clinoptilolite, mordenite, chabazite, and erionite (see Table 1). A few occur in a state of high purity and are used without any beneficiation treatment in several commercial applications.

Of 40 known zeolite minerals, chabazite, erionite, mordenite, and clinoptilolite occur in quantity and reasonably high purity and are available as commercial products (1,4). In general, these zeolites, originated by the alteration of aluminosilicate ash of volcanic origin, occur in two types of deposits:

Closed-system deposits, where volcanic ash was deposited in the Cenozoic lakes of the western United States, for example, and over long periods of time converted to zeolites. Owing to hydrolysis of the alkaline constituents of the volcanic ash, the water became salty and alkaline and the ash crystallized to zeolites. The pH may have reached 9.5. The resulting zeolites were produced as readily accessible flat-lying beds.

The open-system type refers to the deposition of sediments on land in thick beds and subsequent conversion to zeolite by the downward percolation of surface water. The city of Naples is underlain by a zeolitic deposit some 200 km^2 (77 mi^2) in area which is only 5 to 10 thousand years old.

In general, a high grade zeolite ore is mined. It is processed by crushing, drying, powdering, and screening. Beneficiation techniques have been investigated, but are not applied to commercial processes (2).

Table 1. Zeolite Compositions

Zeolite	CAS Registry No.	Typical formula
Natural		
chabazite	[12251-32-0]	$Ca_2[(AlO_2)_4(SiO_2)_8].13H_2O$
mordenite	[12173-98-7]	$Na_8[(AlO_2)_8(SiO_2)_{40}].24H_2O$
erionite	[12150-42-8]	$(Ca,Mg,Na_2,K_2)_{4.5}[(AlO_2)_9(SiO_2)_{27}].27H_2O$
faujasite	[12173-28-3]	$(Ca,Mg,Na_2,K_2)_{29.5}[(AlO_2)_{59}(SiO_2)_{133}].235H_2O$
clinoptilolite	[12321-85-6]	$Na_6[(AlO_2)_6(SiO_2)_{30}].24H_2O$
Synthetic		
zeolite A		$Na_{12}[(AlO_2)_{12}(SiO_2)_{12}].27H_2O$
zeolite X		$Na_{86}[(AlO_2)_{86}(SiO_2)_{106}].264H_2O$
zeolite Y		$Na_{56}[(AlO_2)_{56}(SiO_2)_{136}].250H_2O$
zeolite L		$K_9[(AlO_2)_9(SiO_2)_{27}].22H_2O$
zeolite omega		$Na_{6.8}TMA_{1.6}[(AlO_2)_8(SiO_2)_{28}].21H_2O$ [a]
ZSM-5	[58339-99-4]	$(Na,TPA)_3[(AlO_2)_3(SiO_2)_{93}].16H_2O$ [b]

[a] TMA = tetramethylammonium.
[b] TPA = tetrapropylammonium.

Structure

Of the many synthetic and mineral zeolites, 34 structure types are known, of which ten are synthetic. There are three structural aspects: the basic arrangement of the individual structural units in space, which defines the framework topology; the location of charge-balancing metal cations; and the channel-filling material, which is water, as the zeolite is formed. After the water is removed, the void space can be used for adsorption of gases, liquids, salts, elements, and many other substances. The current concepts of zeolite structures were developed by Pauling in 1930 (5). Modern tools such as x-ray crystallography have provided a very detailed description of many structures.

There are two types of structures, one provides an internal pore system comprised of interconnected cagelike voids; the second provides a system of uniform channels which, in some instances, are one-dimensional and in others intersect with similar channels to provide two- or three-dimensional channel systems. The preferred type has two- or three-dimensional channels to provide rapid intracrystalline diffusion in adsorption and catalytic applications.

In most zeolite structures, the primary structural units—the tetrahedra—are assembled into secondary building units which may be simple polyhedra such as cubes, hexagonal prisms, or octahedra. The final structure framework consists of assemblages of the secondary units. Models of the structures are often constructed of skeletal tetrahedra. Packing models are more realistic but are very difficult to construct. Figure 1 illustrates these models.

Zeolite Minerals. Crystal structures of zeolite minerals are illustrated by the zeolites chabazite and mordenite. The structure of chabazite is hexagonal and the framework consists of double six-membered rings of $(Si,Al)O_4$ tetrahedra arranged in parallel layers in an AABBCC sequence. These tetrahedra are cross-linked by 4-membered rings, as shown in Figure 2, resulting in cavities, 0.67×1.0 nm, each of which is entered by elliptical apertures 0.44×0.31 nm. Exchangeable metal ions, such as calcium, occupy positions within or near the double six-membered rings. The mineral mordenite is more complex, as illustrated in Figure 3, and provides for a one-dimensional channel of about 0.6×0.7 nm. The framework itself is built from chains of 5-membered rings which are cross-linked. The mineral does not exhibit adsorption properties commensurate with the channel size, apparently because of occluded material.

Synthetic Zeolites. Examples of important synthetic zeolites are shown in Table 1. These include zeolites A, X, Y, and Zeolon H [53569-61-2], a synthetic form of mordenite. In addition, high silica synthetic zeolites are also known, such as ZSM-5 and ZSM-11 (2). A high purity silica has been synthesized with a framework structure resembling ZSM-5 (3). These materials have very low alumina contents and are prepared using a templating ion (see below) (6).

The secondary structure unit in zeolites A, X, and Y is the truncated octahedron shown in Figure 1f. These polyhedral units are linked in three-dimensional space through four- or six-membered rings. The former produces the zeolite A structure, and the latter the topology of zeolites X and Y and of the mineral faujasite. In zeolite A, the internal cavity is 1.1 nm in diameter and is entered by six circular apertures 0.42 nm in size. The interlinked cavities form three-dimensional, unduloid channels with a 0.42 nm minimum diameter. The sodium ions lie in the 6-membered rings and near the 8-membered rings (7).

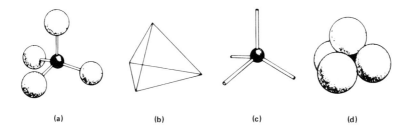

(a)   (b)   (c)   (d)

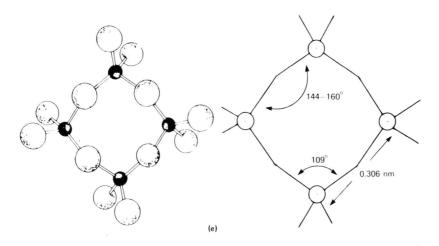

(e)

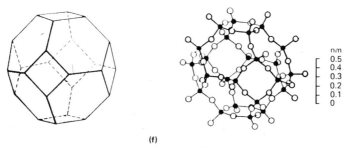

(f)

Figure 1. (**a–d**) Methods for representing SiO_4 and AlO_4 tetrahedra by means of ball-and-stick model, solid tetrahedron, skeletal tetrahedron, and space filling of packed spheres (1). (**e**) Linking of four tetrahedra in a four-membered ring. (**f**) Secondary building unit called the truncated octahedron as represented by a solid model, left, and ball-and-stick model, right.

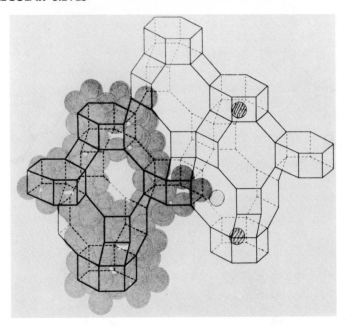

Figure 2. Structure of the mineral zeolite chabazite is depicted by packing model, left, and skeletal model, right. The silicon and aluminum atoms lie at the corners of the framework depicted by solid lines. In this Figure, and Figure 1, the solid lines do not depict chemical bonds. Oxygen atoms lie near the midpoint of the lines connecting framework corners. Cation sites are shown in three different locations referred to as sites I, II, and III. Courtesy of *Scientific American*.

The faujasite-type structure consists of a tetrahedral arrangement of the truncated octahedra by joining of the six-membered rings. The resulting cages are 1.3 nm in size and each is entered by a 12-membered ring, 0.8 nm in diameter. This is the largest known pore size in zeolites. Figure 4 illustrates several types of cation positions; some lie within the smaller polyhedra and some are exposed on the internal surface (1).

Silicalite, a silica molecular sieve, topologically resembles ZSM-5 and contains the same type of building unit. After synthesis and calcination, the unit cell of 96 SiO_2 tetrahedra has a pore volume of 0.32 cm^3/cm^3. There are no exchangeable metal cations, and by definition it is not a zeolite. The structure of silicalite consists of sheets of hexagonal SiO_4 rings, 3 layers wide extending in one direction. They are cross-linked by two SiO_4 units, as shown in Figure 5. The structure provides for two sets of intersecting channels which give a pore size of 0.52–0.57 nm (3).

Structure Modification. Several types of structural defects or variants can occur which figure in adsorption and catalysis:

Surface defects due to termination of the crystal surface and hydrolysis of surface cations.

Structural defects due to imperfect stacking of the secondary units, which may result in blocked channels.

Ionic species may be left stranded in the structure during synthesis (eg, OH^-, AlO_2^-, Na^+, SiO_3^{2-}).

The cation form, acting as the salt of a weak acid, hydrolyzes in water suspension to produce free hydroxide and cations in solution.

Hydroxyl groups in place of metal cations may be introduced by ammonium ion

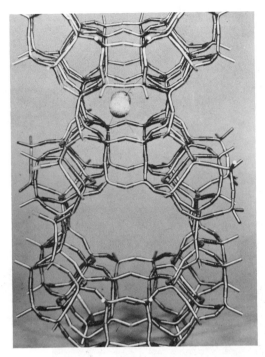

Figure 3.   Model of the crystal structure of the mineral mordenite showing the main channel formed by 12-membered ring and small channels which contain some of the sodium cations. Synthetic types of mordenite exhibit the adsorption behavior of a 12-membered ring, whereas the mineral does not, probably owing to the channel blocking.

exchange followed by thermal deammoniation. These impart acidity to the zeolite, which is important in hydrocarbon-conversion reactions.

Tetrahedral aluminum atoms can be removed from the framework by internal hydrolysis to produce $Al(OH)_3$ when heated in steam. Chemical treatment with acids or chelating agents may also be used to carry out dealumination, but this may cause severe structural damage.

Properties

Adsorption.   Although several types of microporous solids are used as adsorbents for the separation of vapor or liquid mixtures (see Adsorptive separation), the distribution of pore diameters does not enable separations based on the molecular-sieve effect, that is, separations caused by differences in the molecular size of the materials to be separated. The most important molecular-sieve effects are shown by dehydrated crystalline zeolites. Zeolites selectively adsorb or reject molecules based upon differences in molecular size, shape, and other properties such as polarity. The sieve effect may be total or partial.

Activated diffusion of the adsorbate is of interest in many cases. As the size of the diffusing molecule approaches that of the zeolite channels, the interaction energy becomes of increasing importance. If the aperture is small relative to the molecular size, then the repulsive interaction is dominant and the diffusing species needs a specific activation energy to pass through the aperture. Similar stereospecific effects

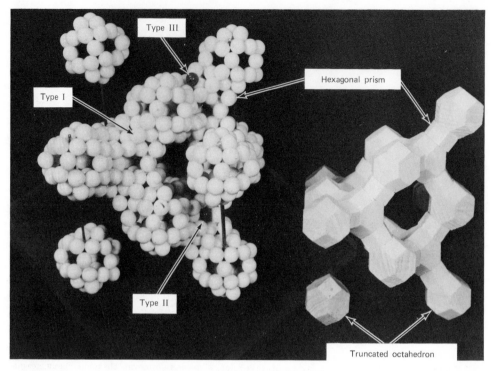

Figure 4.   Model of the crystal structure of zeolites X, Y, and the mineral faujasite. At the right is shown the tetrahedral arrangement of truncated octahedra surrounding one large cavity. On the left the packing model of zeolite X is shown containing three types of Na$^+$ cations.

are shown in both catalysis and ion exchange, two important applications of these materials (8) (see also Clathration).

During the adsorption or occlusion of various molecules, the micropores fill and empty reversibly. Adsorption in zeolites is a matter of pore filling, and the usual surface area concepts are not applicable. The pore volume of a dehydrated zeolite and other microporous solids which have type I isotherms may be related by the Gurvitch rule.

The quantity of material adsorbed is assumed to fill the micropores as a liquid having its normal density. The total pore volume V_p is given by

$$V_p = x_s/d_a$$

where d_a = the density of the liquid adsorbate in g/cm^3, x_s = amount adsorbed at saturation in g/g; and V_p in cm^3/g.

The total void volume in a dehydrated zeolite is usually calculated from the amount of adsorbed water based on the assumption that the water is present as the normal liquid. It is also possible to calculate the micropore volume from the crystal structure when known. Measured void volumes can be correlated with the structurally-derived volumes (9).

The channels in zeolites are only a few molecular diameters in size, and overlapping potential fields from opposite walls result in a flat adsorption isotherm which is characterized by a long horizontal section as the relative pressure approaches unity

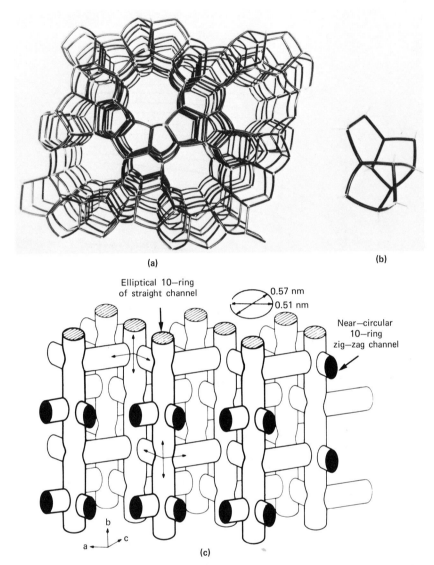

Figure 5. (a) Framework structure showing the topology of the molecular-sieve silicalite viewed in the direction of a main channel. (b) The 12-tetrahedra secondary building unit. (c) Idealized channel system in silicalite (3). Courtesy of *Nature*.

(Fig. 6). The adsorption isotherms do not exhibit hysteresis as do those in many other noncrystalline, microporous adsorbents. Adsorption and desorption are reversible, and the contour of the desorption isotherm follows that of adsorption.

In order to utilize the absorption properties of the synthetic zeolite crystals (1–5 μm) in processes, the commercial materials are prepared as pelleted agglomerates containing a high percentage of the crystalline zeolite together with an inert binder. The formation of these agglomerates introduces macropores in the pellet which may result in some capillary condensation at high adsorbate concentrations.

In commercial materials, the macropores contribute diffusion paths. However,

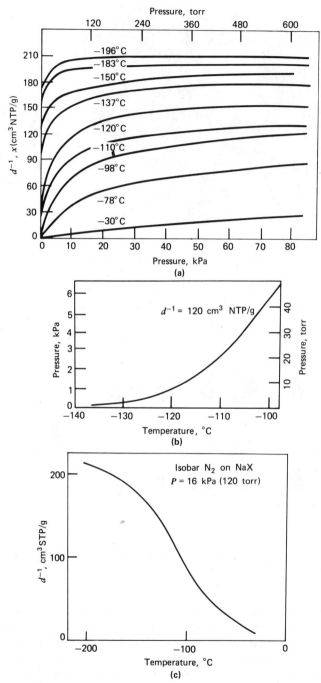

Figure 6. (a) Family of adsorption isotherms for adsorption of nitrogen on zeolite X at temperatures of $-30°$ to $-196°C$. (b) Adsorption isobar at constant $P = 16$ kPa (120 torr) by plotting $d^{-1} = f(T)$ where d = density. (c) Adsorption isostere obtained from (a) showing $P = f(T)$ at constant $d^{-1} = 120$ cm^3 N$_2$ STP/g (1).

the main part of the adsorption capacity is contained in the voids within the crystals.

Zeolites are high capacity, selective adsorbents because they separate molecules based upon the size and configuration of the molecule relative to the size and geometry of the main apertures of the structures; and they adsorb molecules, in particular those with a permanent dipole moment which show other interaction effects, with a selectivity that is not found in other solid adsorbents.

Separation may be based upon the molecular-sieve effect or may involve the preferential or selective adsorption of one molecular species over another. These separations are governed by several factors discussed below.

The basic framework structure, or topology, of the zeolite determines the pore size and the void volume.

The exchange cations, in terms of their specific location in the structure, their population or density, their charge and size, affect the molecular-sieve behavior and adsorption selectivity of the zeolite. By changing the cation types, and number, one can tailor or modify within certain limits the selectivity of the zeolite in a given separation.

The cations, depending upon their locations, contribute electric field effects that interact with the adsorbate molecules.

The effect of the temperature of the adsorbent is pronounced in cases involving activated diffusion.

Sieving by dehydrated zeolite crystals is based on the size and shape differences between the crystal apertures and the adsorbate molecule. The aperture size and shape in a zeolite may change during dehydration and adsorption because of framework distortion or cation movement.

In some instances, the aperture is circular, such as in zeolite A. In others, it may take the form of an ellipse such as in dehydrated chabazite. In this case, subtle differences in the adsorption of various molecules result from a shape factor.

Some typical molecular dimensions are shown in Figure 7, based on the Lennard-Jones (7–8,10–14) potential function (1).

As shown in Figure 7, the calcium-exchanged form of zeolite A has a pore diameter of 0.42 nm. This compares well with the value of 0.43 nm for the kinetic diameter for normal paraffin hydrocarbons and 0.44 nm for dichloromethane. The apparent pore size, therefore, varies from 0.42 to 0.44 nm. This molecular sieve is referred to as 5A.

For the sodium A zeolite, because of the higher cation population, the apparent pore diameter is 0.36–0.40 nm, depending on temperature, for reasons explained before. This is referred to as 4A.

The potassium form of zeolite A [68989-21-9] (KA) when highly exchanged, adsorbs some carbon dioxide and, at lower degrees of exchange, ethylene. A pore diameter of 0.33 nm is appropriate to these results, and KA is referred to as 3A.

The zeolite sodium X [68989-23-1] (type 13X) has a crystallographic aperture of 0.74 nm. This compares well with the adsorbate value of 0.81 nm. Zeolite calcium X exhibits a smaller apparent pore size of 0.78 nm. This difference is probably due to some distortion of the aluminosilicate framework upon dehydration and calcium-ion migration.

When two or more molecular species involved in a separation are both adsorbed,

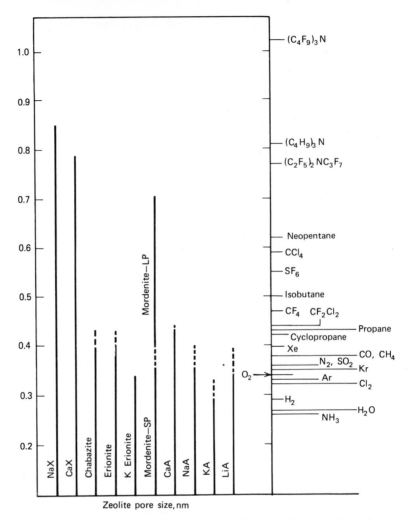

Figure 7.   Molecular dimension and zeolite pore size. Chart showing a correlation between effective pore size of various zeolites over temperatures of 77 to 420 K (indicated by - - -) with the kinetic diameters of various molecules (1). M–A is a cation–zeolite–A system. M–X is a cation–zeolite–X system.

selectivity effects become important because of interaction between the zeolite and the adsorbate molecule. These interaction energies include dispersion and short-range repulsion energies (ϕ_D and ϕ_R), polarization energy (ϕ_P), and components attributed to electrostatic interactions.

Types of Separations.   The first type of adsorption separation is based on differences in the size and shape of molecules (1,11).

The molecular-sieve separation of hydrocarbons by the zeolite calcium A [68989-22-0] (5A) is used in commercial processes for the recovery of normal paraffins from hydrocarbon feedstocks (15). Paraffin isomers and cyclic hydrocarbons are too large to be adsorbed and are excluded. The recovered n-paraffins are utilized, for example, in the manufacture of biodegradable detergents.

All zeolites have a high affinity for water and other polar molecules and could be used for drying gases and liquids (see Drying agents). However, in many instances, secondary reactions such as polymerization of a coadsorbed olefin may take place. This is avoided by using the potassium-exchanged form of zeolite A (3A) for the removal of water from unsaturated hydrocarbon streams. Its effective pore size excludes all hydrocarbons including ethylene.

The molecular-sieve effect for water removal is also utilized in the drying of refrigerants. In this instance, zeolite A (4A) is employed because the size of the refrigerant molecules, such as refrigerant-12 with a kinetic diameter of 0.44 nm, is too large and precludes adsorption.

The second type of adsorption separation is based upon differences in the relative selectivity of two or more coadsorbed gases or vapors. An outstanding example of this type is the production of oxygen from air by the selective adsorption of nitrogen at ambient temperatures on various molecular sieve zeolites including calcium A, calcium X, and types of mordenite.

Many separations based on relative selectivity range from simple drying processes to the separation of sulfur compounds from natural gas, and of aromatics from saturated hydrocarbons.

Alteration and Tailoring of Zeolite Adsorption Selectivities. It is possible to tailor the zeolite adsorption characteristics in terms of size selectivity or the selectivity caused by other interactions, including cation exchange; cation removal or decationization; the presorption of a very strongly held polar molecule, such as water; pore-closure effects, that is, effects which alter the size of the openings to the crystal; and the introduction of various defects such as removal of framework aluminum and an increase in the silicon/aluminum ratio (12).

When synthetic mordenite is dealuminated by acid treatment, the SiO_2/Al_2O_3 ratio can increase to about 100 with the result that water-adsorption capacity is essentially eliminated and the zeolite becomes hydrophobic (12). The limit is attained in the all-silica molecular sieve, silicalite. This material is capable of removing organic compounds from water (3).

Adsorption Kinetics. In zeolite adsorption processes the adsorbates migrate into the zeolite crystals. First, transport must occur between crystals contained in a compact or pellet, and second, diffusion must occur within the crystals. Diffusion coefficients are measured by various methods, including the measurement of adsorption rates and the determination of jump times as derived from nmr results. Factors affecting kinetics and diffusion include: channel geometry and dimensions; molecular size, shape, and polarity; zeolite cation distribution and charge; temperature; adsorbate concentration; impurity molecules; and crystal-surface defects.

Binary Mixtures. In a two-component mixture, the separation factor is derived from a plot of the equilibrium gas-phase composition versus the composition of the adsorbed phase (Fig. 8). The separation factor α is given by

$$\alpha = \frac{Y_a . X_g}{Y_g . X_a}$$

where X_a and Y_a are the mol fractions of the two adsorbates X and Y in the adsorbed phase and X_g and Y_g mol fractions in the gas phase. Some progress has been made in predicting the behavior of gas mixtures from isotherms but the experimental approach is still preferred.

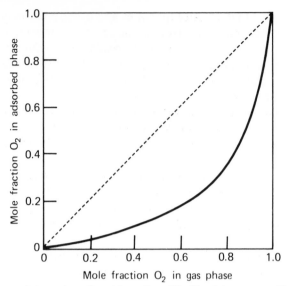

Figure 8.   Binary mixture adsorption. Adsorption of O_2–N_2 mixtures on zeolite NaX at $-78°C$, 100 kPa (1).

Catalytic Properties.   In zeolites, catalysis takes place within the intracrystalline voids (see also Catalysis). The aperture size and channel system affect catalytic reactions caused by diffusion of reactants and products. Activity and selectivity are achieved or altered by modifying the zeolite in several ways. In hydrocarbon reactions in particular, the zeolites with the largest pore sizes are preferred (6). These include mordenite and zeolites Y, L, and omega. Modification techniques include ion exchange, composition in terms of Si/Al ratio, hydrothermal dealumination or stabilization which produces Lewis acidity, introduction of acidic groups such as OH which impart Brønsted acidity, and introducing dispersed metal phases such as the noble metals. In addition, the zeolite framework structure determines shape-selective effects. Several types have been demonstrated including reactant selectivity, product selectivity, and restricted transition-state selectivity.

Acid Sites.   Acidic zeolites have outstanding catalytic activity. Acidity is introduced by the decomposition of the NH_4^+ ion-exchanged form, by hydrogen-ion exchange, or by hydrolysis of a zeolite containing multivalent cations during dehydration. For example:

$$NH_4Z \rightarrow NH_3 + HZ$$
$$H^+ + NaZ \rightarrow HZ + Na^+$$
$$Ce^{3+}Z + H_2OZ \rightarrow CeOH^{2+}Z + HZ$$

where Z = zeolite. The number and acid strength of these Brønsted acid sites are both important. Cation hydrolysis is attributed to the ionization of water by the polar environment of the zeolite internal surface. The hydroxyl groups thus generated have been well characterized by several techniques such as ir spectroscopy. It has been demonstrated, for example, that the acid hydroxyl groups have a characteristic frequency that is determined by the charge density of the framework (13).

It was furthermore originally proposed that Lewis acid sites occur in zeolites as the result of dehydroxylation of two hydroxyl groups, which produces a trigonal coordinated aluminum atom and a positive-charged silicon atom. Recent evidence favors Lewis sites that are generated by the formation of hexacoordinated aluminum atoms at cationic positions in extra-lattice positions (13).

Dispersed Metals. Bifunctional zeolite catalysts, primarily zeolite Y, are used in commercial processes such as hydrocracking. These are acidic zeolites containing dispersed metals such as platinum or palladium. The metals are introduced by cation exchange of the ammine complexes, followed by a reductive decomposition (8) (see Petroleum).

$$NaY + Pt(NH_3)_4^{2+} \rightarrow Pt(NH_3)_4^{2+}Y + Na^+$$

$$2\,Pt(NH_3)_4^{2+}Y + H_2 \xrightarrow{+4\,e} 2\,Pt^0.HY + 8\,NH_3$$

Although it was originally conceived that the platinum was atomically dispersed, it now appears that a bidisperse system involving agglomerates in the supercages and some crystallites at the external surface is present.

Other transition metal ions such as Cd, Zn, Ni, and Ag are introduced by ion exchange followed by reduction with hydrogen. Agglomeration and migration to the external surface can also occur with these metals.

Dehydrated zeolites can be loaded with metals by adsorption of neutral compounds such as carbonyls, followed by thermal decomposition. Molybdenum, ruthenium, and nickel have been loaded by this method into large-pore zeolites (14).

Stabilized Zeolites. Thermal and hydrothermal stability of certain zeolites, in particular zeolite Y, is necessary in many catalytic applications. The stability increases with Si/Al ratio and by exchange with polyvalent cations such as rare earths. Mixed rare-earth-exchanged zeolite Y is used in cracking catalysts. Increased stability is achieved by hydrothermal treatment of the ammonium or rare-earth-exchanged form. When heated at high temperatures in the presence of water vapor, dealumination occurs with the formation of extra-framework Al(OH) species that hydrolyze to Al(OH)$_3$ and dimeric Al(OH) species located within the framework, imparting additional structural stability (1,13). The acidic hydroxyl groups are also maintained. At least one commercial cracking catalyst contains a hydrothermally stabilized zeolite Y as the active component.

In shape-selective catalysis the pore size of the zeolite is important. For example, the ZSM-5 framework contains 10-membered rings with 0.6 nm pore size. This material is used in xylene isomerization, ethylbenzene synthesis (see Xylenes and ethylbenzene), and the conversion of methanol to liquid hydrocarbon fuels (8) (see Gasoline).

The zeolites used are primarily modified forms of zeolite Y, acid forms of synthetic mordenite, silicalite, and ZSM-5. Smaller-pore-size zeolites such as zeolite T are used in shape-selective catalysis (16).

Some current and possible future zeolite catalyst applications are listed below.

alkylation	dehydration
cracking	methanol to gasoline
hydrocracking	organic catalysis
isomerization	inorganic reactions
hydrogenation and dehydrogenation	H$_2$S oxidation
hydrodealkylation	NH$_3$ reduction of NO
methanation	H$_2$O $\rightarrow$ $\frac{1}{2}$ O$_2$ + H$_2$
shape-selective reforming	CO oxidation

Ion Exchange. The exchange behavior of nonframework cations in zeolites (selectivity, degree of exchange) depends upon the nature of the cation (the size and charge of the hydrated cation), the temperature, the concentration and, to some degree, the anion species. Cation exchange may produce considerable changes in other properties such as thermal stability, adsorption behavior, and catalytic activity.

The ion-exchange process is represented by:

$$z_A B^{z+}_{B}(z) + z_B A^{z+}_{A}(s) \rightleftharpoons z_A B^{z+}_{B}(s) + z_B A^{z+}_{A}(z)$$

where z_A and z_B are the ionic charge of cations A and B and (z) and (s) represent the zeolite and solution.

The ion-exchange isotherm (Fig. 9) is constructed by plotting A_z versus A_s, where A_z and A_s represent the mol fractions of cation A in the zeolite and solution respectively. Similarly, with B_z and B_s representing the mol fraction of cation B in the zeolite and solution, the preference of the zeolite for ion A is given by the separation factor:

$$\alpha^A_B \equiv \frac{A_z B_s}{B_z A_s}$$

Ion-exchange isotherms assume different shapes depending on the selectivity factor and the variations in A_s with the level of exchange A_z. The rational selectivity coefficient K^A_B includes the ionic charge and is given by

$$K^A_B \equiv \frac{A_z^{z_B} B_s^{z_A}}{B_z^{z_A} A_s^{z_B}}$$

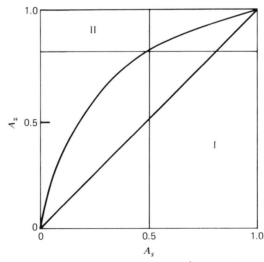

Figure 9.   Ion-exchange isotherm. The separation factor α^A_B is given by the ratio of area I/area II

$$\alpha^A_B = \frac{A_z B_s}{B_z A_s}$$

where A_z and A_s are mol fractions of cation A in zeolite and solution, and B_z and B_s are mol fractions of cation B in zeolite and solution, respectively (1).

Typical exchange isotherms are given in Figure 10 for zeolite X. The exchange capacity of various zeolites is given in Table 2. In many cases, complete exchange does not take place, such as for dipositive and tripositive ions in zeolite Y because of non-occupancy of cation site type I located in the double six-membered ring units. This corresponds to a maximum level of 0.68. Similarly, the level of exchange diminishes with the size and volume of the cation since the intracrystalline volume available does not permit full cation-site occupancy.

Kinetics. Ion-exchange rates in zeolites are controlled by ion diffusion within the zeolite structure, and are affected by particle radii, ionic diffusion coefficients, and temperature. For example, in zeolite NaX [*68989-73-1*], diffusion of sodium ions occurs by migration from type II sites (in the 6-membered rings) into the large cages, followed by diffusion to the surface through the large channels. As a result, complete replacement of sodium by calcium can be accomplished only at elevated temperatures.

Selectivity. Cation sieving in zeolites is attributed to cation size, distribution of charge in the zeolite structures, and size of the hydrated ion in aqueous solution. Solvation influences exchange since, for the ion to diffuse into the crystal, exchange of solvent molecules such as H_2O must always occur.

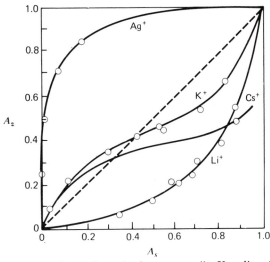

Figure 10. Ion-exchange isotherms on zeolite X, sodium form.

Table 2. Ion-Exchange Capacity of Various Zeolites[a]

Zeolite	Si/Al ratio	meq/g
chabazite	2	5
mordenite	5	2.6
erionite	3	3.8
clinoptilolite	4.5	2.6
zeolite A	1	7.0
zeolite X	1.25	6.4
zeolite Y	2.0	5.0

[a] Anhydrous powder basis.

The selectivity coefficient K_B^A varies with the Si/Al ratio of the zeolite. In zeolites A (Si/Al = 1) and X (Si/Al = 1.2) the selectivity series for unipositive ions is Na > K > Rb > Cs. In zeolite Y (Si/Al = 2.8), the selectivity series is Cs > Rb > K > Na > Li. Even in unipositive–dipositive ion exchange, the normal preference for the dipositive ion shown by zeolites A and X is reversed in zeolites with Si/Al ratio of about 3.

Manufacture

Zeolites are formed under hydrothermal conditions, defined here in a broad sense to include zeolite crystallization from aqueous systems containing various types of reactants. Most synthetic zeolites are produced under nonequilibrium conditions, and must be considered as metastable phases in a thermodynamic sense.

Although more than 150 synthetic zeolites have been reported, many important types have no natural mineral counterpart. Conversely, synthetic counterparts of many zeolite minerals are not yet known. The conditions generally used in synthesis are reactive starting materials such as freshly coprecipitated gels, or amorphous solids; relatively high pH introduced in the form of an alkali metal hydroxide or other strong base, including tetraalkylammonium hydroxides; low temperature hydrothermal conditions with concurrent low autogeneous pressure at saturated water vapor pressure; and a high degree of supersaturation of the gel components leading to the nucleation of a large number of crystals.

A gel is defined as a hydrous metal aluminosilicate prepared from either aqueous solutions, reactive solids, colloidal sols, or reactive aluminosilicates such as the residue structure of metakaolin and glasses.

The gels are crystallized in a closed hydrothermal system at temperatures varying from room temperature to about 200°C. The time required for crystallization varies from a few hours to several days. When prepared, the aluminosilicate gels differ greatly in appearance, from a stiff, translucent appearance to opaque gelatinous precipitates and heterogenous mixtures of an amorphous solid dispersed in an aqueous solution. The alkali metals form soluble hydroxides, aluminates, and silicates. These materials are well suited for the preparation of homogeneous mixtures.

The gel preparation and crystallization is represented schematically using the $Na_2O–Al_2O_3–SiO_2–H_2O$ system as an example (1).

$$NaOH(aq) \ + \ NaAl(OH)_4(aq) \ + \ Na_2SiO_3(aq)$$

$$\downarrow 25° \ C$$

$$[Na_a(AlO_2)_b(SiO_2)_c . NaOH . H_2O] \ gel$$

$$\downarrow 25\text{-}175° \ C$$

$$Na_x[(AlO_2)_x(SiO_2)_y] . m H_2O \ + \ solution$$

Typical gels are prepared from aqueous solutions of reactants such as sodium aluminate, NaOH, and sodium silicate; other reactants include alumina trihydrate ($Al_2O_3.3H_2O$), colloidal silica, and silicic acid. Some synthetic zeolites prepared from sodium aluminosilicate gels are given in Table 3.

Table 3. Some Synthetic Zeolites Prepared from Sodium Aluminosilicate Gels

Zeolite type	Typical composition, mol/mol Al_2O_3			Reactants	Reactant temp, °C	Zeolite product composition, mol/mol Al_2O_3		
	Na_2O	SiO_2	H_2O			Na_2O	SiO_2	H_2O
A	2	2	35	$NaAlO_2$ NaOH sodium silicate	20–175	1	2	4.5
X	3.6	3	144	$NaAlO_2$ NaOH sodium silicate	20–120	1	2.0–3.0	6
Y	8	20	320	$NaAlO_2$ colloidal SiO_2 NaOH	20–175	1	3.0–6.0	9
mordenite, Zeolon	6.3	27	61	$NaAlO_2$ diatomite sodium silicate	100	1	9–10	6.7
omega	5.60[a]	20	280	colloidal SiO_2 $Al(OH)_3$ TMAOH[c] NaOH	100	0.71 0.36 TMA	7.3	6.3
ZSM-5	10[b]	27.7	453	$NaAlO_2$ SiO_2 TPAOH[d]	150	0.89	31.1[e]	2.0

[a] Also 1.4 TMA_2O.
[b] Also 8.6 TPA_2O.
[c] TMA = tetramethylammonium.
[d] TPA = tetrapropylammonium.
[e] After calcination at 1000°C.

When the reaction mixtures are prepared from colloidal silica sol or amorphous silica, additional zeolites may form which do not readily crystallize from the homogeneous sodium silicate–aluminosilicate gels.

The temperature strongly influences the crystallization time of even the most reactive gels; for example, zeolite X crystallizes in 800 h at 25°C and 6 h at 100°C.

Synthesis mechanisms of the typical low silica zeolites, such as A, X, and Y, are apparently different from the high silica zeolites such as ZSM-5. In the low silica zeolites, nuclei are formed consisting of alkali metal-ion complexes of the aluminosilicate species. Structural units consisting of 4-membered rings, 6-membered rings, and cages coordinated with cations are involved in the nucleation and crystallization. In the high silica zeolites, the mechanism appears to be a templating type (6) where an alkylammonium cation complexes with silica by hydrogen bonding. These complexes cause the structures to replicate by hydrogen bonding of the organic cation with framework oxygen atoms (17–19).

Processes. Processes for the manufacture of commercial molecular-sieve products may be classified into three groups, as shown in Table 4 (1): the preparation of molecular-sieve zeolites as high purity crystalline powders or as preformed pellets from

Table 4.   Processes For Molecular-Sieve Zeolites

Process	Reactants	Products
hydrogel	reactive oxides	high purity powders
	soluble silicates	gel preform
	soluble aluminates	zeolite in gel matrix
	caustic	
clay conversion	raw kaolin	
	meta-kaolin	
	calcined kaolin	low to high purity powder
	acid-treated clay	binderless, high purity preform
	soluble silicate	zeolite in clay-derived matrix
	caustic	
	sodium chloride	
other	natural SiO_2	
	amorphous minerals	low to high purity powder
	volcanic glass	zeolite on ceramic support
	caustic	binderless preforms

reactive aluminosilicate gels or hydrogels; the conversion of clay minerals into zeolites, either in the form of high purity powders or as binderless high purity preformed pellets; and processes based on the use of other naturally occurring raw materials.

The hydrogel and clay conversion processes may also be used to manufacture products that contain the zeolite as a major or a minor component in a gel matrix, a clay matrix, or a clay-derived matrix. Powdered products are often bonded with inorganic oxides or minerals into agglomerated particles for ease in handling and use.

Hydrogel Processes.   The first commercial process for preparing synthetic zeolites on a large scale was based on a laboratory synthesis using amorphous hydrogels. The typical starting materials included an aqueous solution of sodium silicate, sodium aluminate, and sodium hydroxide. Hydrogel processes are based either on homogenous gels, that is, hydrogels prepared from solutions of soluble reactants, or on heterogeneous hydrogels which are prepared from reactive alumina or silica in a solid form, for example, solid amorphous silica powder.

In gel-preform processes the reactive aluminosilicate gel is first formed into a pellet which reacts with sodium aluminate solution and caustic solution. The zeolite crystallizes *in situ* within an essentially self-bonded pellet, or as a component in an unconverted amorphous matrix.

A process flow sheet for the manufacture of types 4A, 13X, and Y as high purity crystalline zeolite powders is shown in Figure 11.

A typical material balance and chemical compositions are given in Table 5. The raw materials are metered into the makeup tanks in the proper ratios. Crystallization takes place in a separate crystallizer. An intermediate aging step at ambient temperature may be required for the synthesis of certain high purity zeolites.

The process appears to be simple in terms of equipment and experimental conditions; however, because of the metastability of zeolite species formed from typical reactant systems, problems may arise when large-scale synthesis is attempted. Generally, the crystallization temperature is near the boiling point of water; in some instances, such as in the synthesis of mordenite-type zeolites, higher temperatures are required. After the digestion period, the slurry of crystals in the mother liquor is filtered in a rotary filter.

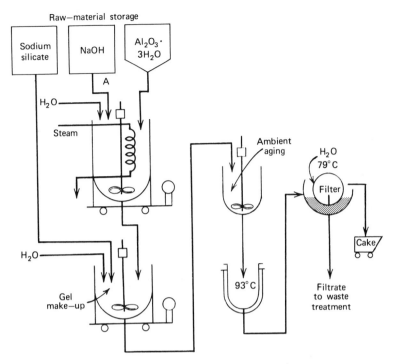

Figure 11. Hydrogel process. Process flow sheet for the manufacture of zeolite type 4As, 13X, and Y from reactant hydrogels (1).

Clay Conversion. The starting material for this process is kaolin, which usually must be dehydroxylated three or four times a day to meta-kaolin by air calcination (see Clays). At 500–600°C, meta-kaolin forms, followed by a mullitized kaolin at 1000–1050°C.

$$Al_2Si_2O_5(OH)_4 \xrightarrow{550°C} Al_2Si_2O_7 + 2\,H_2O$$
$$\text{kaolin} \qquad\qquad \text{meta-kaolin}$$

$$3\,Al_2Si_2O_7 \xrightarrow{1050°C} Si_2Al_6O_{13} + 4\,SiO_2$$
$$\text{meta-kaolin} \qquad\qquad \text{mullite}\ \ \text{cristobalite}$$

The zeolites are prepared as essentially binderless preformed particles. The kaolin is shaped in the desired form of the finished product and is converted *in situ* in the pellet by treatment with suitable alkali hydroxide solutions. Preformed pellets of zeolite A are prepared by this method. In another process, crystalline zeolite powder in the form of a filter cake is mixed with the clay, blended, extruded into pellets, and dried. The raw kaolin component is then calcined to meta-kaolin, and crystallized to zeolite A to form a preformed pellet. These pellets may be converted by ion exchange to other forms such as molecular-sieve Type 5A (1).

Agglomeration of Synthetic Zeolite Powders. High purity zeolite crystals used in adsorption processing must be formed into agglomerates having high physical strength and attrition resistance. The crystalline powders are formed into agglomerates by the addition of an inorganic binder, generally a clay, in a wet mixture. The blended

Table 5.   Typical Material Balance for Hydrogel Process, kg [a]

Raw materials	Zeolite		
	A	X	Y
sodium silicate [b]	1350	2000	
SiO_2 powder [c]			1450
alumina trihydrate [d]	575	500	340
caustic, 50% NaOH	870	1600	1400
water	3135	7687	5300
Gel composition, mol ratio			
Na_2O	2.04	4.09	4.0
Al_2O_3	1	1	1
SiO_2	1.75	3.0	10.6
H_2O	70	176	161

[a] To produce 1000 kg, dry basis.
[b] 9.4% Na_2O, 28.4% SiO_2.
[c] 95% SiO_2.
[d] 65% Al_2O_3, 35% H_2O.

clay–zeolite mixture is extruded into cylindrical-type pellets or formed into beads that are subsequently calcined to form a strong composite (see also Size enlargement).

The preparation of zeolite–binder agglomerates as spheres or cylindrical pellets which have high mechanical attrition resistance is not difficult. However, in order to utilize the zeolite in a process of adsorption or catalysis, the diffusion characteristics must not be unduly affected. Consequently, the binder system must permit a macroporosity which does not increase unduly diffusion resistance. The problem, therefore, is to optimize the binder–zeolite combination to achieve a particle of maximum density (to produce a high volumetric adsorption capacity) with maximum mechanical attrition resistance and minimum diffusion resistance.

Economic Aspects

The synthetic zeolite markets have grown from the million (10^6) dollar range in the late 1950s to an estimated 40×10^6 in 1970, and to 250×10^6 in 1979 (20). The market for zeolite builders in the detergent market is reported to be 25×10^6 in 1980 and growing to 100×10^6 in 1982 (21). The worldwide consumption of zeolites in catalytic cracking is estimated at 31,800–40,800 t per year (22).

Estimates of worldwide sales of natural zeolites have grown from 10,900 metric tons in 1965, to 72,600 t in 1970, to 254,000 t in 1979. Over 90% is in bulk mineral markets (20). Japan has a mordenite and clinoptilolite production of 5000–6000 t per month. A mid-1970s estimate indicated that about 22,000 t synthetic zeolite, mainly zeolite Y, was produced annually to meet worldwide demands (23).

Analytical Procedures

Identification.   Each zeolite has a characteristic x-ray powder diffraction pattern which is used for identification and determination of the purity or quality of zeolite present in a composite such as a catalyst. Generally, powder patterns are determined over a 2θ range of 56° to 4° since these materials have large unit cells and, correspondingly, exhibit the strongest lines at low diffraction angles. However, peak intensities and, to some extent, positions vary with dehydration and/or cation exchange. Suitable procedures for x-ray analysis have been developed by ASTM.

Other procedures are based on infrared spectroscopy, thermal analysis, and standard chemical analyses (24).

Adsorption. The BET (Brunauer-Emmett-Teller) method for surface-area measurement commonly employed to characterize adsorbents and catalysts is not relevant for zeolites since adsorption occurs by a pore-filling mechanism.

Oxygen adsorption at low temperature ($-183°C$) is employed as a method for determining zeolite content, utilizing an appropriate reference. Since the structure of the zeolite is known, the void volume and oxygen capacity can be calculated as a reference value (see above under Adsorption Properties). Nitrogen could be used but, because cation–nitrogen interactions would contribute to additional adsorption capacity, oxygen is preferred. A gravimetric microbalance of the McBain-Bakr type is used. Before adsorption, the zeolite sample is outgassed at 350–450°C under a reduced pressure of 1.3 mPa (10^{-5} mm Hg) for 9–16 h. Complete isotherms should be measured in order to determine deviation from the Type I contour.

Health and Safety Factors; Toxicology

Zeolites have applications in food, drugs, cosmetic products, and detergents. Thus, extensive toxicological and environmental studies have been carried out (25). In single oral-intubation studies, rats have survived a single massive dose equivalent to 32 g/kg of bodyweight (powder form of type 4A, 5A, 13X, and Y). Feeding of 5.0 g/kg of bodyweight for seven days produced no ill effect (26). There is no contraindication to the use of zeolite A (Sasil) in detergents. No negative effect on biological wastewater treatment was found. In toxicity studies using algae, macroinvertebrates, and fish, zeolite A showed no evidence of acute toxicity to four species of fresh-water fish. No mortality was found for either cold- or warm-water fish exposed to suspensions of 680 mg/L.

Uses

Some commercially available molecular-sieve products and related materials are shown in Table 6, classified according to the basic zeolite structure types. In most cases, the water content of the commercial product is below 1.5–2.5 wt %; certain products, however, are sold as fully hydrated crystalline powders.

Adsorption. During the last 20 years, molecular-sieve adsorbents have become firmly established as a means of performing difficult separations, including gases from gases, liquids from liquids, and solutes from solutions (15). They are supplied as pellets, granules, or beads, and occasionally powders. The adsorbents may be used once and discarded, or more commonly, regenerated and used for many cycles. They are stored generally in cylindrical vessels through which the stream to be treated is passed. For regeneration, two or more beds are usually employed with suitable valving, in order to obtain a continuous process.

As a unit operation, adsorption is unique in a number of respects. In some cases, one separation is equivalent to hundreds of mass-transfer units. In others, the adsorbent allows the selective removal of one component from a mixture, based on molecular size differences, which would be nearly impossible to perform by any other

Table 6.   Commercial Molecular Sieve Products

Zeolite type	Designation	Cation	Pore size, nm	Forms available[a]	Manufacturer[b]
A	3A	K	0.3	pwd, ext, bd	L, D
	4A	Na	0.4	pwd, ext, bd	L, D
	5A	Ca	0.5	pwd, ext, bd	L, D
X	13X	Na	0.8	pwd, ext, bd	L, D
	10X	Ca	0.7	pwd, ext	L
Y	LZY-52	Na	0.8	pwd, ext	L
	LZY-62	NH₄	0.8	pwd, ext	L
	LZY-72	H	0.8	pwd, ext	L
	LZY-82	NH₄	0.8	pwd, ext	L
mordenite					
small port	AW-300, Zeolon 300	Na, mixed cations	0.3–0.4	pwd, ext	L, N
large port	Zeolon 100, 200, 900	H or Na	0.8	pwd, ext	N
chabazite	AW-500, Zeolon 500, Davison 714	mixed cations	0.4–0.5	pwd, ext, bd	L, D, N
F[c]	Ionsiv F80	K, Na	0.4	pwd	L
W[d]	Ionsiv W85	K, Na	0.4	pwd, ext	L

[a] Pwd—powder; ext—pellet; bd—bead.

[b] L—Linde Division, Union Carbide Corporation; D—Davison Chemical Division, W. R. Grace & Co.; N—Norton Co., Chemical Process Products Division.

[c] Zeolite F.

[d] Zeolite W.

means. In addition, contaminants can be removed from fluid streams to attain virtually undetectable impurity concentrations. Adsorbents are used in applications requiring a few grams to several tons (see Adsorptive separation).

Process Cycles.   Regenerative adsorption units can be operated by a thermal-swing cycle, pressure-swing cycle, displacement-purge cycle, or inert-purge stripping cycle. Combinations of these are frequently employed. Brief descriptions are given below with reference to Figure 12.

Thermal-Swing Cycle.   If a carrier stream containing an impurity at concentration X_1 is brought into contact with the adsorbent at a temperature T_1, the equilibrium loading is defined by the isotherm at point A (loading L_1). If the temperature is raised to T_2, still in the presence of the carrier, desorption occurs and the equilibrium loading will be defined by the isotherm at temperature T_2 at point B (loading L_2). The usable delta loading would be $(L_1 - L_2)$ per cycle. The attainable effluent product purity on the subsequent adsorption cycle, after cooling to T_1, is defined by the isotherm T_1 at point C, which is X_2.

Pressure-Swing Cycle.   This cycle employs a pressure differential to perform the adsorption–desorption cycle. If a carrier gas containing an impurity at a concentration of X_1 is brought into contact with the adsorbent at temperature T_1, the equilibrium loading is defined by the isotherm T_1, at point A as L_1. However, the concentration X_1, in vapor-phase operation, is a function of the amount of impurity present and the pressure. Thus, by reducing system pressure the concentration in the

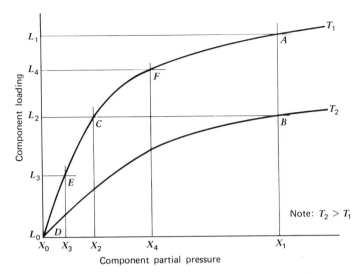

Figure 12.   Illustration of adsorption-process cycles.

voids surrounding the adsorbent is reduced, for example to X_4. This results in an equilibrium loading reduction, as defined by the isotherm T_1 at the point F, to L_4. The usable delta loading is $(L_1 - L_4)$. Upon repressurization with feed at the higher pressure, the adsorbent loads (adsorbs) to L_1.

Displacement-Purge Cycle.   This cycle is accomplished by purging the adsorbent with a material which is equally or more strongly sorbed than the adsorbate. The adsorbate can be completely removed and replaced with the desorbent. In the subsequent adsorption stroke, the desorbent is driven off and replaced with the component being removed from the feed. Starting at A, the loading on the adsorbent follows the path $AFCED$. The theoretical delta loading is $(L_1 - L_0)$.

Isothermal-Purge (Stripping) Cycle.   Starting at point A, it is possible to perform a desorption by purging the adsorbent with a fluid that contains no adsorbate. This results in a reduction in concentration of adsorbate in the voids surrounding the adsorbent. Consequently, the system moves toward equilibrium by desorption of adsorbate into the voids. Since the purge gas carries away the adsorbate in the voids, the desorption continues. The path of the loading follows the isotherm T_1 along path $AFCED$. If the stripping is continued to point E, an equilibrium loading L_3 is achieved and the usable delta loading is $(L_1 - L_3)$. The effluent purity attainable when the next adsorption step is performed is defined by the residual loading left on the adsorbent at E (X_3).

Purification.   Purification refers to separations wherein the feed stream is upgraded by the removal of a few percent or even traces of a contaminant (see Fig. 13). A heated purge gas is usually employed for this purpose.

Water.   The dehydration of natural gas and air were the first gas-purification applications for molecular sieves. Because of their high adsorptive selectivity for water and high capacity at low water partial pressures, molecular sieves were an obvious choice for water removal from natural gas and air before cryogenic extraction of helium and cryogenic separation of oxygen, nitrogen, and the rare gases, respectively (see Cryogenics; Gas, natural). Molecular-sieve dehydration is used in the cryogenic pro-

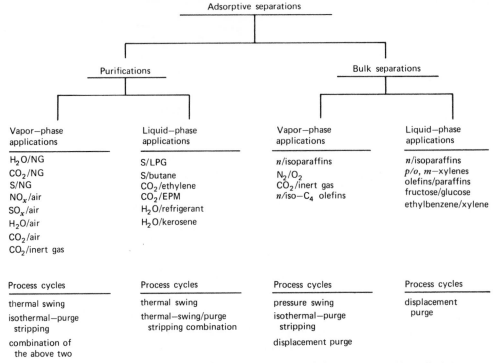

Figure 13.   Classification of adsorptive separations. NG = natural gas. S = sulfur.

duction of liquefied natural gas (LNG), for small peak-demand-type storage facilities and giant base-load facilities. In addition, molecular sieves have proved to be the most effective dehydration technique for the cryogenic recovery of ethane and heavier liquids from natural gas (see Liquefied petroleum gas).

Molecular sieves have had increasing use in the dehydration of cracked gases in ethylene plants before low temperature fractionation for olefin production. Type 3A molecular sieve is size selective for water molecules and does not coadsorb the olefin molecules.

The unique features of molecular sieves are demonstrated in the removal of water from natural-gas streams containing high percentages of acid gases (eg, H_2S and CO_2). Other dry-bed adsorbents degrade in highly acidic environments. However, acid-resistant molecular sieves have been developed which maintain dehydration capacities over long periods of on-stream use. They also are used to dehydrate gas streams containing corrosive components such as chlorine, sulfur dioxide, and hydrogen chloride (see Drying).

Molecular sieves also are used widely in the dehydration of liquid streams. Both batch-type and continuous processes have been developed for drying a variety of hydrocarbon and chemical liquids.

For nonregenerative drying, the molecular sieve is designed for the lifetime of the unit. A typical example is refrigerant drying and purification. A suitable-size cartridge of the proper molecular sieve, installed in the circulating refrigerant stream, adequately protects the refrigeration system from freeze-ups and corrosion for the life of the unit by adsorbing water and the acidic decomposition products of the refrigerant.

Another nonregenerative drying application for molecular sieves is their use as an adsorbent for water and solvent in dual-pane insulated glass windows. The molecular sieve is loaded into the spacer frame used to separate the panes. Once the window has been sealed, low hydrocarbon and water dew points are maintained within the enclosed space for the lifetime of the unit. Consequently, no condensation or fogging occurs within this space, which would cloud the window.

Gas and liquid dehydrators employing molecular sieves provide product gas streams of <0.1 ppm(v) water and product liquid streams routinely to <10 ppm(v) water. Applicable pressures range from less than one to several hundred times atmospheric pressure. Temperatures range from subzero to several hundred °C. Processing units range in capacity from as little as 10 m^3/h to as much as 10^8 m^3/d in multiple-train units.

Carbon Dioxide. Molecular sieves are used to purify gas streams containing carbon dioxide in cryogenic applications where freeze-out of CO_2 would cause fouling of low temperature equipment. They also are employed for air purification in cryogenic air-separation plants where one front-end purifier unit can be used for the simultaneous removal of both water and CO_2. Peakshaving natural-gas liquefaction is employed by utilities to store LNG during the summer. Molecular-sieve adsorbents remove water and CO_2 before liquefaction.

Commercial processes are available for the removal of CO_2 from air, natural gas, ethylene, ethane–propane mix, and synthesis gases. Operations cover wide ranges of pressure and temperature.

Sulfur Compounds. Various gas streams are treated by molecular sieves to remove sulfur contaminants. In the desulfurization of wellhead natural gas, the unit is designed to selectively remove sulfur compounds, but not carbon dioxide, which would occur in liquid-scrubbing processes. Molecular-sieve treatment offers advantages over liquid-scrubbing processes in reduced equipment size because the acid gas load is smaller; in production economics because there is no gas shrinkage (leaving CO_2 in the residue gas); and in the fact that the gas is also fully dehydrated, alleviating the need for downstream dehydration.

Molecular sieves are being used to treat refinery hydrogen streams containing trace amounts of H_2S. A single molecular-sieve unit may be designed to remove trace water and H_2S in the recycle hydrogen loop of a catalytic reformer to protect the catalyst from poisoning. Then, during catalyst regeneration, the same unit acts as a dryer for treating the inert gas used in regenerating the reforming catalyst.

A large use of molecular sieves in the natural-gas industry is LPG sweetening. Sweetening and dehydration are combined in one unit and the problem associated with the disposal of caustic wastes from liquid-treating systems is eliminated. The regeneration medium is typically natural gas. In LPG sweetening, H_2S and other sulfur compounds are removed. Commercial plants are processing as little as ca 30 m^3/d (200 bbl/d) to over 8000 m^3/d (50,000 bbl/d).

Bulk Separation. The adsorptive separation of process streams into two or more main components is termed bulk separation (see Fig. 13). The development of processes and products is complex. Consequently, these processes are proprietary and are purchased as a complete package under licensing agreements. High purities and yields can be achieved.

Separation of Normal and Isoparaffins.   The recovery of normal paraffins from mixed refinery streams was one of the first commercial applications of molecular sieves. Using Type 5A molecular sieve, the n-paraffins can be adsorbed and the branched and/or cyclic hydrocarbons rejected. During the adsorption step, the effluent contains isoparaffins. During the desorption step, the n-paraffins are recovered. Isothermal operation is typical.

Regeneration is carried out by a pressure-swing process for a process separating light hydrocarbons ($\leq C_7$). For heavier streams, a displacement-purge cycle employing lighter n-paraffins is used. The n-paraffins are separated by distillation from the regeneration effluent and recycled.

There are seven commercial processes in operation; six operate in the vapor phase. The Universal Oil Products process operates in the liquid phase and is unique in the simulation of a moving bed. The adsorption unit consists of one vessel segmented into sections with multiple inlet and outlet ports. Flow to the various segments is accomplished by means of a rotary valve which allows each bed segment to proceed sequentially through all the adsorption/desorption steps.

The normal paraffins produced are raw materials for the manufacture of biodegradable detergents, plasticizers, alcohols, and synthetic proteins. Removal of the n-paraffins upgrades gasoline, improves the octane number of the branched fraction (see Adsorptive separation, liquids).

Xylene Separation. p-Xylene is separated from mixed xylenes and ethylbenzene by means of the Universal Oil Products Company's Parex process. A proprietary adsorbent and process cycle are employed in a simulated moving-bed system. High purity p-xylene is produced.

Olefin Separation.   Olefin-containing streams are separated either by Union Carbide Corporation's OlefinSiv process (n-butenes from isobutenes in the vapor phase) or Universal Oil Product's Olex process (liquid phase).

Oxygen from Air.   Recently, a demand has developed for oxygen for processes needing from a few to about 50 t/d. In this size range, a pressure-swing adsorption process is often competitive with the conventional cryogenic separation route. Oxygen of 95% purity can be obtained; the main impurities are the inert gases found in air.

Catalysts.   At the present time, zeolite-based catalysts are employed in catalytic cracking, hydrocracking, isomerization of paraffins and substituted aromatics, and in a process converting methanol to hydrocarbons (16) (see also Catalysis).

Catalytic Cracking.   The addition of relatively small amounts of hydrothermally stable acidic zeolites to conventional cracking catalyst formulations significantly increases both the yield and the quality of the products from fluidized-bed and moving-bed cracking reactors. At present, catalytic cracking is the largest-scale industrial process employing zeolite catalysts. In general, the rare-earth-exchanged form is used, but some catalysts based on NH_4–Y are also marketed. The commercial catalysts comprise 5–40% zeolite dispersed in a matrix of synthetic silica–alumina, semisynthetic clay-derived gel, or natural clay. Such composites can be prepared either by blending a synthetic zeolite with a binder, or by chemical treatment of suitable clays to produce the zeolite component *in situ*.

Zeolite-promoted cracking catalysts offer the advantage of high rates of intermolecular hydrogen transfer coupled with extremely high intrinsic cracking activity, and the high thermal stability of the zeolite cracking catalyst. The stability often can be further improved via rare-earth ion exchange. Zeolite catalysts increase the yield of light cycle oil, the yield and octane of the gasoline product fraction, and decrease the production of coke. The high cracking rates obtained with zeolite catalysts result in greatly reduced contact times for given conversion levels, thus further increasing liquid-product yields. Additional advantages are increased tolerance to poisons and greater operating flexibility. The relationships between catalyst properties, feedstock composition, and reactor operating conditions are very complex (27).

Hydrocracking for Fuels Production. Hydrocracking is catalytic cracking in the presence of hydrogen using a dual-function catalyst possessing both cracking and hydrogenation–dehydrogenation activity. At present, it is the second largest use for zeolite-containing catalysts. In general, such catalysts consist of an acidic, hydro-thermally stable, large-pore zeolite loaded with a small amount of a noble metal, or admixed with a relatively large amount of an active hydrogenation system such as NiO + MoO_3 or WO_3.

Although several proprietary hydrocracking technologies are in use, the Union Oil Company Unicracking process exemplifies the value of zeolite catalysts in broadening the range of feedstocks that can be handled and in simplifying hydro-cracker design and operation.

The most elaborate and versatile Unicracking process scheme is the two-stage configuration shown in Figure 14. Feedstock is mixed with hydrogen and admitted to reactor R-1, a conventional hydrotreater in which it is substantially freed of nitrogen and sulfur compounds. The product then enters the first-stage Unicracker, R-2, in which it is hydrocracked, typically at a per-pass conversion of 40–70%. The zeolite catalysts developed for the Unicracking process can operate stably and efficiently in the presence of hydrogen sulfide and ammonia. Thus, separation of these substances from the R-2 feed is not necessary. The fractionator bottoms are recycled through a second-stage reactor, R-3. The recycled gas which is mixed with the R-3 feed is es-

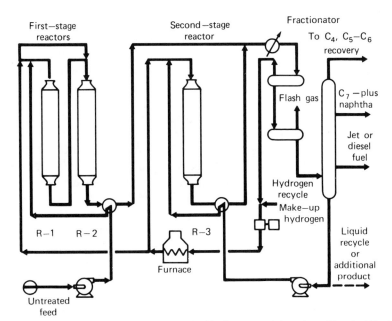

Figure 14.   Two-stage Unicracking unit (8). Courtesy of American Chemical Society.

sentially ammonia-free; thus, the reactor can be operated efficiently at comparatively low temperatures and pressures while maintaining conversion at 50–80% per pass.

In a single-stage Unicracking, the R-2 product is treated in liquid–gas separators and fractionated.

In hydrocracking, as with all multipurpose refinery processes, the relationships between feedstock properties, catalyst type, operating conditions, and product yield and character, are complex. With zeolite catalysts, a variety of feedstocks is converted into a range of fuels, including LPG, medium-octane unleaded gasoline, and jet, diesel, and heating oils, or into feedstocks suitable for catalytic reforming or petrochemical manufacture. Zeolite hydrocracking catalysts are noted for permitting long periods (2–6 yr) of highly efficient reactor operation at moderate conditions. Following such service, some of these catalysts can be completely restored to their original levels of activity (see Petroleum, refinery processes).

Shape-Selective Catalysts for Special-Purpose Hydrocracking. The shape-selectivity of certain zeolite molecular sieves is exploited in two special types of hydrocracking processes.

The Selectoforming process of the Mobil Corporation employs a metal-loaded offretite–erionite zeolite to selectively hydrocrack the normal paraffin components of catalytic reformate. The process thus allows either an increase in reformate octane at a particular operating severity, or an increase in reformate yield at a specified octane. The former is the usual mode of operation, with plant experience indicating a typical gain of 2–5 octane numbers.

The second type is catalytic dewaxing, which has been commercialized by several companies, and typically employs a metal-loaded tubular-pore zeolite catalyst such as mordenite. The main purpose of this process is usually pour-point reduction. With such a catalyst, long-chain normal paraffin waxes can be selectively hydrocracked to propane, isobutane, and isopentane. Uses for the process include upgrading of diesel and heating oils, conversion of low sulfur, waxy crudes directly to fuel oils, and general improvement of lubricating oils.

Paraffin Isomerization. The only other well-established commercial process which employs zeolite catalysts at present is the isomerization of normal paraffins into higher octane, branched isomers. The catalyst for the well-known Hysomer process of the Shell Oil Company is dual functional, and consists of a highly acidic, large-pore zeolite loaded with a small amount of a noble-metal hydrogenation component. This catalyst possesses the same hydrogenation–dehydrogenation and acid functions as hydrocracking catalysts. However, since hydrocracking of the Hysomer feedstock, which is usually a light, straight-run naphtha containing principally pentanes and hexanes, is not desired because it represents a direct reduction of gasoline yield, the process operating conditions are more stringently defined than those for hydrocracking. In general, reaction conditions are adjusted to allow the lowest possible reactor temperature consistent with economical production rates since lower temperatures favor higher equilibrium concentrations of branched isomers, and, the lowest hydrogen pressure consistent with economy of operation. In this way, yields of the highest octane highly branched isomers are maximized, and hydrocracking is minimized.

The Hysomer process produces an increase of about 12 octane numbers in suitable naphtha feedstocks. The process can be operated in conjunction with the Union Carbide Corporation's Isosiv process for the separation of normal and isoparaffins, achieving complete isomerization of a C_5/C_6 stream. The combined process is trade named TIP (Total Isomerization Process), and results in increases in octane numbers of about 20 rather than the 12 obtained with a once-through Hysomer treatment.

Catalysis of Aromatic Reactions. In the 1960s, zeolite-syntheses which included organic cations such as quaternary ammonium species in addition to the traditional alkali metal ions were developed, leading to a number of new molecular sieves. Mobil Corporation laboratories developed, among others, the zeolites ZSM-5, ZSM-11, ZSM-12, and ZSM-21. This group is characterized by an unusual pore structure, the openings of which are 10-membered rings, and highly siliceous frameworks, with SiO_2/Al_2O_3 molar ratios of 20–100. The ZSM-5 zeolites have been used to prepare some unique catalysts for a variety of processes.

Xylene Isomerization. The objective of C_8-aromatics processing is the conversion of the usual four-component feedstream (ethylbenzene and the three xylenes) into an isomerically pure xylene. Although the bulk of current demand is for p-xylene isomer for polyester fiber manufacture, significant markets for the other isomers exist (see Polyester fibers). The primary problem is separation of the 8–40% ethylbenzene that is present in the usual feedstocks, a task that is complicated by the closeness of the boiling points of ethylbenzene and p-xylene. In addition, the equilibrium concentrations of the xylenes present in the isomer separation train raffinate have to be reestablished to maximize the yield of the desired isomer.

In the most readily adopted C_8-aromatics process based on ZSM-5 and related zeolites, the catalyst is an acid form of the sieve to which is added a group VIII metal hydrogenation–dehydrogenation component. Such catalysts can be used in existing plants in place of the Pt/silica–alumina materials originally developed for the process. They have the advantages of higher throughput, owing to the ability to operate at low H_2/hydrocarbon ratios, and longer run life between regenerations.

Ethylbenzene Synthesis. The synthesis of ethylbenzene for styrene production is another process in which ZSM-5-type catalysts are employed. Although some ethylbenzene is obtained directly from petroleum, about 90% is synthetic. In general, benzene is alkylated with high purity ethylene in liquid-phase slurry reactors with promoted $AlCl_3$ catalysts or the vapor-phase reaction of benzene with a dilute ethylene-containing feedstock with a BF_3 catalyst supported on alumina (see Alkylation; Friedel-Crafts reactions). Both of these catalysts are corrosive and their handling presents problems.

An acidic ZSM-5 catalyst has been employed successfully to produce ethylbenzene using both pure and dilute ethylene sources. In both cases, the alkylation is accomplished under vapor-phase conditions of about 425°C, 1.5–2 MPa (15–20 atm), 300–400 kg benzene per kg catalyst per hour, and a benzene:ethylene feed ratio of about 30. With both types of ethylene sources, raw material efficiency exceeds 99%, and heat recovery efficiency is high (see Xylenes and ethylbenzene).

The Methanol-to-Gasoline Process. Mobil Corporation laboratories have reported the conversion of methanol and some other oxygenated organic compounds to hydrocarbons using an acidic ZSM-5 zeolite catalyst. In the case of methanol, the highly exothermic reaction sequence includes the formation of dimethyl ether. The process is quite selective in producing an aromatics-rich gasoline-range mixture of hydrocarbons free of oxygen compounds (see Gasoline). It has been proposed that the highly charged zeolite framework induces carbene formation, and that these species react with each other, methanol, or dimethyl ether to form C_2–C_5 olefins which are subsequently converted to the final product. The hydrocarbon product contains no compounds with more than 10 carbon atoms, and this, in conjunction with the lower-than-predicted concentrations of certain multipolyalkylated benzenes, is construed as evidence of shape selectivity owing to the moderate pore size of the catalyst.

Ion Exchange. Crystalline molecular-sieve ion exchangers do not follow the typical rules and patterns exhibited by organic and other inorganic ion exchangers. Many provide combinations of selectivity, capacity, and stability superior to the more common cation exchangers. Their commercial utilization has been based on these unique properties (28) (see Ion exchange).

Cesium and Strontium Radioisotopes. Because of their stability in the presence of ionizing radiation and in aqueous solutions at high temperatures, molecular-sieve ion exchangers offer significant advantages in the separation and purification of radioisotopes (see Radioisotopes). Their low solubility over wide pH ranges, together with their rigid frameworks and dimensional stability and attrition resistance, have endowed zeolites with properties which generally surpass those of the other inorganic ion exchangers. The high selectivities and capacities of several zeolites for cesium and strontium radioisotopes resulted in the development of processes currently used by nuclear processing plants.

Ammonium Ion Removal. A fixed-bed molecular-sieve ion-exchange process has been commercialized for the removal of ammonium ions from secondary wastewater-treatment effluents. This application takes advantage of the superior selectivity of molecular-sieve ion exchangers for the ammonium ion. The first plants employed clinoptilolite because of its availability in natural deposits as a potentially low cost material. The bed is regenerated with a lime-salt solution that can be reused after the ammonia is removed by pH adjustment and air stripping. The ammonia is subsequently removed from the air stream by acid scrubbing. Currently, three units are in operation with throughputs ranging from 0.6 to 12.5 t/d.

Detergent Builders. The prime function of phosphates in detergents is to reduce the activity of the hardness ions, Ca^{2+} and Mg^{2+}, in the wash water by complexing. Zeolite ion exchangers in powder form replace Ca^{2+} and Mg^{2+} in the solution with soft ions such as Na^+. Heavy-duty detergents employ the sodium form of Type A zeolite for this purpose in low or zero-phosphate formulations. The zeolite powder is incorporated into the detergent powder during formulation. Large amounts of zeolites are used in this application (see Silicon compounds, synthetic inorganic silicates; Surfactants and detersive systems).

BIBLIOGRAPHY

1. D. W. Breck, *Zeolite Molecular Sieves, Structure, Chemistry, and Use*, John Wiley & Sons, Inc., New York, 1974.
2. D. H. Olson and W. M. Meier, *Nature* **272,** 437 (1978).
3. E. M. Flanigen, J. M. Bennett, R. W. Grose, J. P. Cohen, R. L. Patton, R. M. Kirchner, and J. V. Smith, *Nature* **271,** 512 (1978).
4. L. B. Sand and F. A. Mumpton, *Natural Zeolites, Occurrence, Properties, Use*, Pergamon, Oxford, Eng., 1978.
5. L. Pauling, *Proc. Nat. Acad. Sci.* **16,** 453 (1930).
6. L. D. Rollman, *Adv. Chem. Ser.* **173,** 387 (1979).
7. J. V. Smith, *ACS Monogr.* **171,** 3 (1976).
8. J. A. Rabo, ed., *Zeolite Chemistry and Catalysis*, ACS Monograph 171, American Chemical Society, Washington, D.C., 1976.
9. D. W. Breck and R. W. Grose, *Adv. Chem. Ser.* **121,** 319 (1973).

10. W. M. Meier and D. H. Olson, *Atlas of Zeolite Structure Types*, Structure Commission of the International Zeolite Association, Polycrystal Book Service, Pittsburgh, Pa., Juris Druck and Verlag AG, Zurich, Switz., 1978.

11. R. M. Barrer, *Zeolites and Clay Minerals as Sorbents and Molecular Sieves*, Academic Press, London, Eng., 1978.

12. N. Y. Chen, *J. Phys. Chem.* **80,** 60 (1976).

13. P. A. Jacobs, *Carboniogenic Activity of Zeolites*, Elsevier Scientific Publishers Co., New York, 1977.

14. J. B. Uytterhoeven, *Acta Phys. Chem.* **24**(1–2), 53 (1978).

15. R. A. Anderson, *ACS Symp. Ser.* **40,** 637 (1977).

16. J. A. Rabo, R. D. Bezman, and M. L. Poutsma, *Acta Phys. Chem.* **24**(1–2), 39 (1978).

17. E. M. Flanigen, *Adv. Chem. Ser.* **121,** 119 (1973).

18. S. P. Zhdanov, *Adv. Chem. Ser.* **101,** 20 (1971).

19. C. L. Angell and W. H. Flank, *ACS Symp. Ser.* **40,** 194 (1977).

20. D. W. Leonard, *preprint of Paper Presented at Soc. of Mining Engineers of A.I.M.E.*, Tuscon, Ariz., Oct. 17–19, 1978.

21. *Chem. Week*, 29 (Jan. 2, 1980).

22. D. P. Burke, *Chem. Week*, 42 (Mar. 28, 1979).

23. J. S. Magee, *ACS Symp. Ser.* **40,** 650 (1977).

24. A. P. Bolton in R. B. Anderson and P. T. Dawson, eds., *Experimental Methods in Catalysis Research*, Vol. 2, Academic Press, Inc., New York, 1976, pp. 1–42.

25. P. Berth, *J. Am. Oil Chem. Soc.* **55,** 52 (1978).

26. *Union Carbide Adsorbent Data Sheets*, Nos. 3797, 4172, 4174, 4175, New York.

27. P. B. Venuto and E. T. Habib, Jr., *Fluid Catalytic Cracking with Zeolite Catalysts*, Marcel Dekker, New York, 1979.

28. J. D. Sherman, *A.I.Ch.E. Symp. Ser. No. 179*, **74,** 98 (1978).

General References

Refs. 1 and 8 are also general references.

R. P. Townsend, ed., *The Properties and Applications of Zeolites*, No. 33, The Royal Society of Chemistry, London, Eng., 1980.

D. Fraenkel, "Encapsulate Hydrogen," *Chemtech*, 60 (Jan. 1981).

D. W. Breck
R. A. Anderson
Union Carbide Corporation

MOLLUSCICIDES.   See Poisons, economic.

MOLYBDENUM AND MOLYBDENUM ALLOYS

Molybdenum [7439-98-7] first was identified as a discrete element in 1778 by the Swedish chemist, Carl Wilhelm Scheele. The metal remained a laboratory curiosity until the late 1880s when French metallurgists produced a molybdenum-containing armor-plate steel, followed by applications as an additive to tool steels and in dyes. The first uses of any importance were developed during World War I when molybdenum was employed in armor-plate steels, tool steels, and high strength steels for aircraft engines (see High temperature alloys; Laminated and reinforced metals).

After World War I, the demand for molybdenum dropped to very low levels. The success of molybdenum steels for war materiel had, however, encouraged metallurgical research (1), and the Society of Automotive Engineers adopted the 4100 series of chromium–molybdenum steels (see Steel).

Over one half of the world supply of molybdenum comes from mines where its recovery is the primary objective of the operation. The balance is recovered as a by-product of copper mining. The most abundant mineral, and the only one of commercial significance, is molybdenite, MoS_2. Powellite, $Ca(MoW)O_4$, and wulfenite, $PbMoO_4$, also are known. Primary ore bodies in the western hemisphere contain ca 0.2–0.4% molybdenum and give a recovery of 2–4 kg per metric ton of ore.

Molybdenite is concentrated by first crushing and grinding the ore, and passing the finely ground material (called pulp) through a series of flotation cells (see Flotation). Operations recovering molybdenum as a by-product of copper mining produce a concentrate containing both metals. Molybdenite is separated from the copper minerals by differential flotation.

Molybdenite concentrate contains about 90% MoS_2. The remainder is primarily silica. The concentrate is roasted to remove the sulfur and convert the sulfide to oxide. Molybdenum is added to steel in the form of this oxide, known as technical molybdic oxide. In modern molybdenum-conversion plants, the sulfur formed by roasting MoS_2 is converted to sulfuric acid instead of being discharged as SO_2 into the atmosphere (2).

A mixture of technical molybdic oxide and iron oxide can be reduced to ferromolybdenum [11121-95-2] by silicon plus aluminum in a thermite reaction (3) (see Aluminum). The resulting product contains about 60% molybdenum and 40% iron. Foundries generally use ferromolybdenum for adding molybdenum to cast iron and steel, and steel mills may prefer it to technical molybdic oxide for some types of steels.

A small portion of molybdenite concentrate production is purified to yield lubricant-grade molybdenum disulfide.

Several high purity molybdenum compounds are produced for applications where trace impurities cannot be tolerated. The most common are pure molybdic oxide and ammonium molybdate. The pure oxide is produced either by subliming technical-grade oxide or by calcining ammonium molybdate.

Molybdenum

Physical Properties. Molybdenum, Mo, is in group VIB, between chromium and tungsten vertically and niobium and technetium horizontally in the periodic table. It has a silvery-gray appearance. The most stable valence state is +6; lower, less stable valence states are +5, +4, +3, +2, and 0.

Molybdenum is a typical transition element with the maximum number of five unpaired $4d$ electrons, which account for its high melting point, strength, and modulus of elasticity. There are many similarities between molybdenum and its horizontal and vertical neighbors in the periodic system. Physical properties are given in Table 1.

The melting point of molybdenum is about 2626°C, 1100°C above that of iron. Only two commercially significant elements—tungsten and tantalum—have higher melting points than molybdenum. As a result of its high melting temperature, molybdenum metal has strength characteristics at temperatures where most metals are in the molten state, and some applications, such as furnace parts, rocket nozzles, welding tips, thermocouples, glass melting electrodes, dies, molds, etc, are based on this property.

Chemical Properties. Resistance to corrosion is one of the most valuable properties of molybdenum metal. It has particularly good resistance to corrosion by mineral acids, provided oxidizing agents are not present. Molybdenum metal also offers excellent resistance to attack by some liquid metals. The approximate temperature limits where molybdenum can be considered for long-time service while in contact with nine metals in the liquid state are given in Table 2.

In addition, molybdenum has high resistance to a number of alloys of these metals and also to copper, gold, and silver. Among the molten metals that severely attack molybdenum are tin (at 1000°C), aluminum, nickel, iron, and cobalt. Molybdenum has moderately good resistance to molten zinc but a molybdenum–30% tungsten alloy is practically completely resistant to molten zinc at temperatures up to 800°C. Molybdenum metal is substantially resistant to many types of molten glass and to most nonferrous slags.

Above ca 600°C, unprotected molybdenum oxidizes so rapidly in air or oxidizing atmospheres with formation of volatile MoO_3 that extended use under these conditions is impractical. Since no molybdenum-base alloy that combines high oxidation resistance and good high temperature properties has been discovered, protective coatings seem to be the answer where oxidation is a problem. Various coatings, differing in maximum time–temperature capabilities and in physical and mechanical characteristics, are available. The most widely used coatings depend on the formation of a thin coating of $MoSi_2$ on the molybdenum metal part. This compound has outstanding oxidation resistance at temperatures up to about 1650°C.

In vacuum, uncoated molybdenum metal has an unlimited life at high temperatures, as proved by numerous furnace applications. This is also true under the vacuumlike conditions of outer space. Pure hydrogen, argon, and helium atmospheres are completely inert to molybdenum at all temperatures, whereas water vapor, sulfur dioxide, and nitrous and nitric oxides have an oxidizing action at elevated temperatures. Molybdenum is relatively inert in carbon dioxide, ammonia, and nitrogen atmospheres up to about 1100°C; a superficial nitride film may be formed at higher temperatures in the last two gases. Hydrocarbons and carbon monoxide may carburize molybdenum above 1100°C.

Table 1. Physical Properties of Molybdenum[a]

Property	Value
Atomic	
atomic number	42
isotopes	
natural	92, 94, 95, 96, 97, 98, 100
artificial	90, 91, 93, 99, 101, 102, 105
atomic weight	95.94
atomic radius, coordination number 8, nm	0.136
ion radius, nm	
trivalent	0.092
hexavalent	0.062
atomic volume, cm^3/mol	9.41
lattice	
type	body-centered cubic
constant at 25°C	3.1405 kX
fast-neutron-absorption cross section, 10/250 keV, m^2 [b]	9×10^{-28}
ionization potential, eV	7.2
apparent positive-ion work function, eV	8.6
apparent electron work function, eV	4.2
Thermal	
melting point, °C	2626 ± 9
heat of fusion (estimated), kJ/mol[c]	28
boiling point, °C	5560
heat of vaporization, kJ/mol[c]	491
entropy of crystals, $S^\circ_{298.16}$, $J/(mol \cdot K)$[c]	28.6
vapor pressure, Pa[d]	
at 1725°C	3.95×10^{-9}
at 2225°C	4.35×10^{-6}
at 2610°C	1.72×10^{-4}
at 2725°C	4.05×10^{-4}
at 3225°C	8.71×10^{-3}
at 3725°C	8.41×10^{-2}
at 4225°C	4.76×10^{-1}
at 4725°C	1.82
at 5225°C	5.57
rate of evaporation, $g/(cm^2 \cdot s)$ log (rate of evaporation) = $17.11 - 38,600\,T - 1.76 \log T$	
diffusivity, cm^2/s	
at 200°C	0.43
at 540°C	0.40
at 870°C	0.38
specific heat at 100°C, $kJ/(kg \cdot K)$[c]	0.27
coefficient of linear expansion, %	
at 0–400°C	0.23
at 0–800°C	0.46
at 0–1200°C	0.72
thermal conductivity, $W/(m \cdot K)$	
at 500°C	122
at 1000°C	101
at 1500°C	82

Table 1 (*continued*)

Property	Value
Electrical and magnetic	
electrical resistivity, nΩ·m	
at 0°C	50
at 1000°C	320
at 2000°C	610
Franz-Wiedemann or Lorenz constant	2.72
hydrogen overpotential (1×10^{-2} A/cm^2), V	0.44
electrochemical equivalent (hexavalent), mg/C	0.1658
minimum arcing amperage, A	
24 V	10
110 V	1.5
220 V	1.0
magnetic susceptibility (κ_p), m^3/kg	
at 25°C	0.93×10^{-6}
at 1825°C	1.11×10^{-6}
paramagnetic	
Optical and emissivity	
optical reflectivity, %	
at 500 nm	46
at 10,000 nm	93
total hemispherical emissivity	
at 1200 K	0.104
at 1600 K	0.163
at 2000 K	0.209
at 2400 K	0.236
thermionic emission in high vacuum, mA/cm^2	
at 1600°C	ca 0.7
at 2000°C	85
spectral emissivity	
at 390 nm	ca 0.43
at 6700 nm	0.40
radiation for 550 nm at 20°C	54% of black-body radiation
total radiation, W/cm^2	
at 527°C	ca 0.2
at 1127°C	3.0
at 1727°C	19
at 2327°C	68
Other	
density at 20°C, g/cm^3	10.22
coefficient of friction vs steel at HRC^e 44	
dry	
static	0.271
dynamic	0.370
humid	
static	0.405
dynamic	0.465
compressibility at 293°C, cm^2/kg	3.6×10^{-7}
velocity of sound at 2.25 MHz, cm/s	

Table 1 (*continued*)

Property	Value
longitudinal wave (V_L)	$6.37 \pm 0.02 \times 10^5$
shear wave (V_S)	$3.41 \pm 0.06 \times 10^5$
thin rod (V_0)	5.50×10^5
surface tension at melting point, mN/m (= dyn/cm)	2240

[a] Refs. 4–8.
[b] To convert m^2 to barn, divide by 10^{-28}.
[c] To convert J to cal, divide by 4.184.
[d] To convert Pa to mm Hg, multiply by 0.0075.
[e] HRC = Rockwell C hardness.

Table 2.   Temperature Limits for Molybdenum in Contact with Other Metals

Metal	Limit, °C	Metal	Limit, °C
bismuth	1425	mercury	600
gallium	400	potassium	1100
lead	1200	sodium (liquid	
lithium	925	and vapor)	1500
magnesium	700	sulfur	440

In a reducing atmosphere, molybdenum is resistant at elevated temperatures to hydrogen sulfide, which forms a thin adherent sulfide coating. In an oxidizing atmosphere, however, molybdenum is rapidly corroded by sulfur-containing gases. Molybdenum has excellent resistance to iodine vapor up to about 800°C. It is also resistant to bromine up to about 450°C and to chlorine up to about 200°C. Fluorine, the most reactive of the halogens, attacks molybdenum at room temperature.

Manufacture.   The technical-grade MoO$_3$ formed by roasting molybdenite concentrate can be sublimed to produce high purity molybdenum trioxide or it can be leached with dilute ammonia to give ammonium molybdate. Ammonium molybdate or molybdenum trioxide is reduced to molybdenum powder by hydrogen at 500–1150°C. Boat- and tube-type furnaces are usually selected for this purpose.

In the conventional powder-metallurgy process, molybdenum powder is compacted hydraulically in dies at a pressure of about 207–276 MPa (30,000–40,000 psi) to bars which are sintered electrically at 2200–2300°C in hydrogen-atmosphere bells. Sintering currents supply about 90% of the current required to fuse the bar. In an alternative process, adaptable to larger sections, the bars are sintered in a hydrogen-atmosphere muffle furnace for about 16 h at 1600–1700°C. The longer time at the lower temperatures in the muffle furnace accomplishes the same densification produced by bell sintering. Ingots weighing ≥2000 kg have been produced by these methods.

The production of molybdenum ingots by consumable electrode melting in vacuum was developed during the late 1940s. In the original arc-cast method, molybdenum powder is fed from a hopper into a compacting device, which forms a continuously compacted bar. The bar is used as a consumable molybdenum electrode. The arc is struck by touching the bar against a stub ingot in the bottom of a water-cooled copper sleeve, and transfers molybdenum continuously from the electrode to the molten pool, where it solidifies. The entire operation is carried out in vacuum. Carbon is added to the charge for deoxidation. Ingots ≤ 816 kg have been produced by this method. Arc-cast molybdenum ingots have the theoretical density. They may be extruded or forged to break the coarse solidification structure. Once this has been accomplished, the arc-cast molybdenum may be handled by the same procedures as those used for processing powder-metallurgy products.

Rods and bars of molybdenum are made by rolling, forging, or swaging the sintered ingot or extruded billet at 1200–1400°C. As the working operation proceeds, lower temperatures are used. Molybdenum wire is drawn from rod, using tungsten carbide dies for heavy-gauge wire and diamond dies for fine-gauge wire. Drawing at the coarser sizes is accomplished at elevated temperatures with graphite or molybdenum disulfide suspensions for lubrication. Final drawing of fine wire is usually performed at ambient temperatures. The wire may be annealed, normally at 800–850°C, to soften it between cold-drawing operations. To prevent embrittlement care must be taken to avoid overannealing. Sheet is produced by rolling forged or extruded slabs at 1200–1400°C into plates which then are rolled to sheet at lower temperatures of 1000–1250°C. At last, fine-gauge sheet or foil may be rolled, cold or warm, with intermediate annealing. As molybdenum produced by powder-metallurgy methods becomes worked to greater and greater degrees, its density increases to the theoretical value.

Molybdenum metal can be mechanically worked by almost any commercial process with suitable practices such as forging, extrusion, rolling, forming, bending, punching, stamping, deep drawing, spinning, and power-roll forming. Except for fine wire and sheet, at least a moderate amount of heating is recommended for all working operations. The mechanical properties of unalloyed molybdenum depend to a large degree on the working below the recrystallization temperature. For optimum ductility, parts should be given at least a 50% reduction in area. Fully recrystallized molybdenum gives lower strengths and flows more easily in working. Under those conditions, however, molybdenum has low bend and impact characteristics, although the tensile elongation is higher than in the wrought and stress-relieved condition. Fully recrystallized molybdenum can be used for the fabrication of parts where bending or deep-drawing are not involved, mechanical properties after working are not of concern, or subsequent processing involves sufficient working to produce the required properties.

With proper tool angles, molybdenum metal can be machined without difficulty. In turning with sintered-carbide tools, speeds as high as 182 surface m/min commonly are used.

Although brazing and mechanical methods have been the accepted means of joining molybdenum, satisfactory fusion welding of molybdenum has been accomplished on a commercial basis under closely controlled conditions. Wrought arc-cast molybdenum usually can be welded with less difficulty than conventionally processed powder-metallurgy molybdenum products. The ductility of molybdenum is adversely affected by even minute amounts of oxygen or nitrogen, not only in the parent metal but also on the surface and in the welding atmosphere.

Thorough cleaning of faying surfaces is essential for optimum ductility in welds, probably because it eliminates surface layers that have become contaminated during processing. For optimum ductility, fusion welding must be carried out in a closely controlled atmosphere. Commercial argon, helium, and hydrogen may contain sufficient oxygen to cause porosity and cracking unless specific purification measures are taken. Molybdenum can be welded to itself by a number of different processes including arc, electric-resistance, and electron-beam welding. The weldment must be warm-worked to produce ductilities as high as those of the parent metal. It also can be welded to any other metallic material with which it readily alloys.

Economic Aspects. Most of the supply for the noncommunist nations comes from the United States (60%), Canada (18%), and Chile (12%). Demand has exceeded supply through much of the 1970s. Economic statistics (1973–1979) are given in Table 3. Installed capacities of the principal molybdenum mines are given in Table 4. Climax Molybdenum Company, a division of AMAX Inc., is the leading producer.

Table 3.   Economic Statistics for Molybdenum, Noncommunist Countries, 1000 Metric Tons[a]

Statistic	1973	1976	1979
demand[b]	82	80	91
mine production			
primary	37	42	48
by-product from copper mines	35	36	41
•*Total*	72	78	89
deficit	10	3	2
GSA[c] releases	3	1	0

[a] Ref. 9.
[b] Includes shipments to communist nations.
[c] GSA = General Services Administration (a U.S. government agency).

Historically, primary and by-product production shares the market on an almost equal basis. However, the contribution of by-product molybdenum did not change significantly over 1973–1979, because of weakness in the copper market, a situation that provided no incentive for the development of new copper mines. Although the supply of primary molybdenum increased significantly over this period, it could not keep up with demand.

Supply deficits in the early to mid-1970s were made up either from the United States government stockpiles or from industrial inventory. The last of the government stockpile material was released in 1976. Industry stocks were reduced by 14,000 metric tons in the 1973–1976 period. Producer inventories were at very low levels in 1979 and can no longer be counted on to make up future supply deficits. Because of this situation, producer prices increased more than fourfold during the 1970s, whereas speculative or trader prices have increased to as much as four times the producer prices in 1979. As a consequence, mine evaluation and development activities were greatly accelerated. Current prices (Oct. 1980) are shown in Table 5.

Analytical Methods; Specifications; Health and Safety. Commercial molybdenum metal is usually of a high order of purity. Most trace metals present can be measured by standard chemical, x-ray fluorescent, or spectrographic means. For the determination of potassium or sodium, either standard chemical or spectrographic methods may be used. The carbon content of arc-cast molybdenum is determined by heating a sample in a stream of pure oxygen; the carbon dioxide evolved is absorbed in a barium hydroxide solution, where the change in electrical conductivity gives a measure of the amount of barium carbonate formed. Oxygen and hydrogen may be determined by vacuum fusion, nitrogen by the micro-Kjeldahl or diffusion-extraction methods. The composition limits listed in ASTM specifications for unalloyed molybdenum are shown in Table 6.

Specifications. The following ASTM specifications cover molybdenum metal:

molybdenum and molybdenum alloy billets for reforging	B 385-74
molybdenum and molybdenum alloy forgings	B 384-74
molybdenum and molybdenum alloy strip, sheet, foil, and plate	B 386-74
molybdenum and molybdenum alloy bar, rod, and wire	B 387-74
chemical analysis of molybdenum	E 315-71 T
molybdenum strip for electron tubes	F 49-68
molybdenum wire under 510 μm (20 mil) in diameter	F 289-60
round wire for winding electron-tube grid laterals	F 290-62 T
molybdenum flattened wire for electron tubes	F 364-73 T

Other specifications have been established by other technical organizations and by private companies.

Health and Safety. Industrial poisoning by molybdenum has not been reported. Lethal doses for rats of molybdenum and several of its compounds are shown in Table 7 (see Mineral nutrients).

Uses. Table 8 gives the present distribution of uses of molybdenum in the United States. Approximately 85% of all molybdenum consumption is as an alloying additive to steels and irons (14–16).

Molybdenum metal is the most widely used furnace resistance element where temperatures beyond the limits of ordinary resistance alloys are required. Such furnaces are generally used for temperatures up to about 1650°C, but some are in successful operation at 2200°C. The elements, which may be wire, rod, ribbon, or expanded sheet, must be protected from oxidation by a reducing or inert atmosphere, or vacuum; hydrogen is commonly employed. Under these conditions, molybdenum has a long life and seldom is the limiting factor in the durability of the furnace. The furnace industry also consumes sizable amounts of molybdenum sheet for susceptors in high frequency units, as well as radiation shields, baffles, structural supports, muffle liners, skids, hearths, boats, and firing trays in all types of high temperature vacuum and controlled-atmosphere furnaces (see Furnaces).

The same properties that make molybdenum metal effective in high temperature furnace applications make it useful as support wires for tungsten filaments in incandescent light bulbs and as targets in x-ray tubes.

The glass industry has become an important user of large molybdenum parts. The advantages of molybdenum include its high melting point, high strength at elevated temperatures, good electrical conductivity, resistance to attack by most molten glasses, and the fact that any oxide formed is colorless under normal operating conditions. The main application is as electrodes which may be either of the plate or rod type. Stirrers, pumps, bowl liners, wear parts, and molds are also made from molybdenum metal.

Because of its high modulus of elasticity, molybdenum is used in machine-tool accessories such as boring bars and grinding quills.

Molybdenum metal has good thermal-shock resistance because of its low coefficient of thermal expansion combined with high thermal conductivity. This combination accounts for its use in casting dies and in some electrical and electronic applications.

Table 4.   Principal Molybdenum Mines in the World, 1978[a]

Country	Company	Mine	Installed capacity, t/yr
Bulgaria	government	Medet	180
Canada	Brenda Mines Ltd. (Noranda Mines, Ltd.)	Brenda	3,800
	Lomex Mining Corp. Ltd. (Rio Algom Ltd.)	Lomex	1,700
	Noranda Mines Ltd.	Boss Mountain	1,000
		Copper Mountain	na
		Needle Mountain	na
	Placer Development Ltd.	Endako	6,600
		Gibraltar	140
	Utah Mines Ltd. (Utah International Inc.)	Island Copper	850
Chile	Corporacion Nacional del Cobre de Chile	Chuquicamata	7,500
	(CODELCO)	El Teniente	2,500
		El Salvador	1,400
		Andina	350
Peru	Southern Peru Copper Corp. (SPCC)	Toquepala	820
United States	ASARCO Inc.	Mission	200
		Silver Bell	na
	Anamax Mining Co.	Twin Buttes	1,700
	Cities Service Co.	Pinto Valley	325
	Climax Molybdenum Co. (AMAX Inc.)	Climax	28,000
		Henderson	22,600
	Cyprus Mines Corp.	Bagdad	900
		Pima	750
	Duval Corp. (Pennzoil Co.)	Esperanza	1,700
		Mineral Park	1,250
		Sierrita	6,500
	Inspiration Consolidated Copper Co.	Inspiration	na
	Kennecott Copper Corp.	Bingham	3,500
		Chino	300
		Nevada Mines	na
		Ray	330
	Magma Copper Co. (Newmont Mining Corp.)	Magma	na
	Molycorp Inc. (Union Oil Co. of California)	Questa	5,500
	Union Carbide Corp.	Pine Creek	200
U.S.S.R.	government	Kalmakyr	1,000
		Balkhash (Kounrad)	500
		Umaltinsk	na
		Tyrny Auz	3,000
		Kadzharan	1,600

[a] Refs. 10–11.

Molybdenum-Base Alloys

Currently, there are two molybdenum-base alloys of commercial significance. Their designations and nominal compositions are:

Alloy	Zirconium, %	Titanium, %	Tungsten, %	Molybdenum
TZM	0.06–0.12	0.40–0.55		balance
Mo-30W			27–33	balance

The TZM alloy was the result of an extensive development effort to produce a material

Table 5. Prices of Molydenum Products, $/kg[a]

Product	Price, $/kg
molybdenite concentrate	20.30
ferromolybdenum	23.40
technical molybdic oxide	
bags	21.40
briquettes	21.60
molybdenum powder	36.70–39.60
miscellaneous molybdenum metal products	44–187

[a] Prices from Oct. 1980.

Table 6. Specified Composition for Unalloyed Molybdenum Metal According to ASTM Designation B 385-74

Impurity, %	Arc-cast material	Powder-metallurgy material
C, max	0.010–0.040	0.010
O, max	0.0030	0.0070
N, max	0.0010	0.0020
Fe, max	0.010	0.010
Ni, max	0.005	0.005
Si, max	0.010	0.010
Mo, min[a]	99.931–99.961	99.956

[a] By difference.

Table 7. Toxicity of Molybdenum and Several of its Compounds in Rats

Element or compound	LD_{50}, mg/kg		Ref.
	Peros (oral)	Intraperitoneal	
molybdenum, Mo		114–117	12
molybdenum trioxide, MoO_3	125		13
calcium molybdate, $CaMoO_4$	101		13
ammonium molybdate, $(NH_4)_6Mo_7O_{24}$	333		13

that could withstand higher temperatures and exhibit greater toughness than molybdenum metal. The relatively small amount of alloying elements present achieve these objectives without sacrificing castability or workability. The TZM alloy has a much higher recrystallization temperature than unalloyed molybdenum, thus the beneficial effects of work hardening can be retained to higher operating temperatures.

Typical tensile strength properties for the TZM alloy and for unalloyed molybdenum are given in Table 9.

The Mo-30W composition was originally developed to take advantage of the high melting point of tungsten (3380°C) to produce an alloy with a higher melting point than molybdenum. However, the most significant application for this alloy is now related to the fact that it is not attacked chemically by molten zinc. As a result, Mo-

Table 8.   United States Use Distribution of Molybdenum

Use	%
alloy steels	47
stainless steels	20
tool steels	9
cast irons	7
specialty alloys	3
chemicals	9
molybdenum metal	4
miscellaneous	1

Table 9.   Tensile Strength of Rolled Bars—Molybdenum and TZM, MPa[a,b]

Temperature, °C	Unalloyed molybdenum	TZM alloy
ambient	790	830
200	750	790
400	660	730
600	570	670
800	480	600
1000	370	530
1200	180	450

[a] To convert MPa to psi, multiply by 145.
[b] Ref. 17.

30W is used extensively in zinc handling equipment for parts such as impellers, tundishes, valves, and thermocouple protection tubes.

Molybdenum as an Alloying Element.   Molybdenum is an unusually versatile alloying element. Its extensive consumption in all industrial countries is attributed to the numerous beneficial effects it imparts to irons and steels and to some alloy systems based on cobalt, nickel or titanium (15–16).

Addition of molybdenum to iron and steels enhances their hardenability (16). The hardenability of a given steel may be measured in various ways. For example, a series of cylinders of increasing diameters is quenched from the austenitizing temperature and the hardness determined along a diameter of a cross section. The bar with the largest diameter that hardens all the way through provides the measure of hardenability.

Hardenability can be evaluated by time–temperature-transformation (TTT) diagrams and continuous-cooling-transformation (CCT) diagrams. These diagrams predict the behavior of steels over a broad range of cooling rates and identify the microconstituents that occur when the steels are cooled at different rates.

The concentrations of molybdenum usually fall in the range 0.15–0.30% in engineering steels used for carburized gears, pinions, ball and roller-bearing races, and journals and for homogeneous rods, shafts, tool joints, casing, tubing, grinding balls, and grinding rods. A few specialty steels contain 0.30–0.60%. Molybdenum is most effective when combined with other alloying elements, such as manganese, chromium, nickel, and boron. The best-known alloy steels are, in fact, chrome–moly and chrome–nickel–moly grades. The abbreviation moly, for molybdenum, is used throughout the metal industry. The molybdenum-containing steels have been standardized by the Society of Automotive Engineers and the American Iron and Steel Institute.

Many engineering alloy steels, particularly those containing chromium and chromium–nickel combinations, are susceptible to a loss of toughness if they are slowly cooled following tempering >600°C or if they are exposed to ca 500°C for extended periods of time. This susceptibility to loss of toughness is called temper brittleness. The addition of molybdenum eliminates or reduces temper brittleness in steels. The mechanism of this phenomenon is not completely understood.

An early application of molybdenum was its substitution for tungsten in the high speed steels popular before World War I. The red hardness of a steel containing 9% molybdenum is as high as that of a steel containing 18% tungsten. Subsequently, a series of molybdenum-type high speed steels containing 3.75–9.5% molybdenum was developed. It is estimated that today 92% of the hacksaws and twist drills produced in the United States are manufactured from molybdenum-type high speed steels. About 95% of all high speed steel tool shipments in the United States in 1976 comprised molybdenum types.

High strength, low alloy (HSLA) steels and dual-phase steels often contain 0.10–0.30% molybdenum. The former exhibit toughness at low temperatures and good weldability. They are used, eg, for the large-diameter pipelines transporting gas and oil through remote Arctic areas.

Dual-phase steels are a new form of low carbon steels for automotive structural parts that are produced by stamping or forming. They deform plastically with a smooth, continuous stress–strain curve in contrast to the discontinuous yielding of most other steels. As a result, complex parts can be formed without visible surface evidence of yielding. Dual-phase steel allows vehicle parts to be made in lighter gauges than the usual carbon steel. As a result, the part can be lighter in weight although retaining adequate strength.

Molybdenum improves the corrosion resistance of stainless steels that are alloyed with 17–29% chromium. The addition of 1–4% molybdenum results in high resistance to pitting in corrosive environments. Stainless steel containing 1% molybdenum is used for automotive trim. The presence of molybdenum increases the corrosion resistance of wheel covers, body trim, and windshield-wiper components.

A simple, low cost steel for high temperature service in electric power generation is the C–0.50% Mo steel known as carbon-half moly which was widely used for many years. Today, the power industry and oil refineries depend upon the 1.25% Cr–0.5% Mo and 2.25% Cr–1% Mo steels for high stress and high temperature service, since these steels have improved resistance to graphitization and oxidation and higher creep and rupture strength (see High temperature alloys).

In cast irons, addition of molybdenum increases hardness and strength. Such irons have been used for crankshafts in large diesel engines for power generation, in ship propulsion and pipeline pumps, and in some special railroad castings.

Abrasion-resistant high alloy cast irons are used in ore processing, in the crushing and pulverizing of coal, and in the grinding of cement clinker. These white irons are alloyed with 15–28% chromium and 2–3% molybdenum to provide effective wear resistance. Although the highly alloyed irons are sometimes heat treated to develop the desired structure, recent research has shown that castings often develop a suitable matrix. They may, therefore, go into service without heat treatment, providing substantial savings in fuel and handling costs. The operating effect of molybdenum in these castings is its ability to stabilize the structure so effectively that it does not transform during a relatively slow cooling to room temperature.

BIBLIOGRAPHY

"Molybdenum and Molybdenum Alloys" in *ECT* 1st ed., Vol. 9, pp. 191–199, by R. I. Jaffee, Battelle Memorial Institute; "Molybdenum and Molybdenum Alloys" in *ECT* 2nd ed., Vol. 13, pp. 634–644, by J. Z. Briggs, Climax Molybdenum Co.

1. H. W. Gillett and E. L. Mack, *Molybdenum, Cerium and Related Alloy Steels*, Chemical Catalogue Co., New York, 1925.
2. W. W. Duecker and J. R. West, *The Manufacture of Sulfuric Acid*, Robert E. Krieger Publishing Co., Inc., Huntington, N.Y., 1971, p. 203.
3. A. Sutulov, *International Molybdenum Encyclopaedia 1778–1978*, Vol. II, Intermet Publications, Santiago, Chile, 1979, p. 296.
4. *Molybdenum Metal*, Climax Molybdenum Company, Ann Arbor, Mich., 1960.
5. *Metals Handbook*, 9th ed., Vol. 2, American Society for Metals, Metals Park, Ohio, 1979, p. 771.
6. *Aerospace Structural Metals Handbook*, Mechanical Properties Data Center, Department of Defense, Belfour Stulen, Inc., Code 5301, March 1963, p. 5.
7. V. E. Peletskii and V. P. Druzhinin, *High Temp. High Pressures* **2**, 69 (1970).
8. M. M. Kenisarin, B. Ya. Berezin, and V. Ya. Chekhovskoi, *High Temp. High Pressures* **4**, 707 (1972).
9. J. W. Goth, "Molybdenum in the Eighties," *Second International Ferro-Alloys Conference, Copenhagen, Denmark, Oct. 1979.*
10. J. T. Kummer, *Molybdenum, Mineral Commodity Profiles*, U.S. Dept. of the Interior, Bureau of Mines, 1979.
11. Ref. 3, Vol. I, 1978, p. 305.
12. F. Maresh, M. J. Lustok, and P. P. Cohen, *Proc. Soc. Exp. Biol. Med.* **45**, 576 (1940).
13. L. T. Fairhall, R. C. Dunn, N. E. Sharpless, and E. A. Pritchard, *U.S. Public Health Serv. Publ.* **293**, (1945).
14. W. F. Distler, *Eng. Min. J.* **181**(3), 99 (1980).
15. Ref. 3, Vol. III, 1980, p. 41.
16. R. S. Archer, J. Z. Briggs, and C. M. Loeb, *Molybdenum Steels–Irons–Alloys*, Climax Molybdenum Company, New York, 1948.
17. *Molybdenum Mill Products*, AMAX Specialty Metals Division, Greenwich, Conn., 1971.

ROBERT Q. BARR
Climax Molybdenum Company

MOLYBDENUM COMPOUNDS

The chemical applications of molybdenum have increased substantially over the past fifteen years. New applications have been developed in catalysis, lubrication, corrosion inhibition, protective coatings, inhibitive pigments, and flame and smoke retardants. Molybdenum compounds are used in the transportation and petroleum and petrochemical industries; molybdenum catalysts are prime candidates in the conversion of coal to liquid fuels (see Fuels, synthetic).

The chemistry of molybdenum is extremely complex because it forms compounds in which it has the valence states of 0, +2, +3, +4, +5, or +6; molybdenum compounds may disproportionate to mixtures of compounds in which molydenum occurs in different valence states; shifts between the several coordination numbers (4, 6, and 8) of molybdenum can result from changes in reaction conditions; and molybdenum has a strong tendency to form complex compounds.

Halides and Oxyhalides

Molybdenum forms halogen compounds of widely different degrees of stability (see Table 1). The hexafluoride and pentachloride deposit molybdenum metal in the vapor phase (1) as a result of reaction with hydrogen.

The highest member of each series (MoF_6, $MoCl_5$, $MoBr_4$, and MoI_3) can be made by direct halogenation of molybdenum metal. The hexafluoride, pentafluoride, pentachloride, oxytetrafluoride, oxytetrachloride, and dioxydichloride are volatile at moderate temperatures. The entire series exhibits a wide range of stability, color, and volatility. Few of the compounds are monomeric in their normal state. Generally, the lower halides are prepared by reduction of the highest member of the series with molybdenum metal, hydrogen, or a hydrocarbon. By refluxing in benzene, the pen-

Table 1. Halogen Compounds of Molybdenum

Compound	CAS Registry No.	Melting point, °C	Compound	CAS Registry No.	Melting point, °C
Fluorides			$MoOCl_3$	[13814-74-9]	302[b]
MoF_6	[7783-77-9]		$MoOCl$	[41004-72-2]	
MoF_5	[13819-84-6]	64	MoO_2Cl	[20770-33-6]	
MoF_4	[23412-45-5]		$MoOCl_2$	[24989-40-0]	subl
MoF_3	[20193-58-2]	dec	MoO_2Cl_2	[13637-68-8]	184[b]
MoF_2	[20205-60-1]		Mo_2OCl_8	[77727-64-1]	
$MoOF_4$	[14459-59-7]	98	*Bromides*		
$MoOF \cdot 4H_2O$	[77727-63-0]		$MoBr_4$	[13520-59-7]	dec
MoO_2F_2	[13824-57-2]	270 subl	$MoBr_3$	[13446-57-6]	dec
Chlorides			$MoBr_2$	[13446-56-5]	
$MoCl_5$	[10241-05-1]	201	$MoOBr_3$	[13596-04-8]	
$MoCl_4$	[13320-71-3]	dec	MoO_2Br_2	[13595-98-7]	
$MoCl_3$	[13478-18-7]	[a]	*Iodides*		
$MoCl_2$	[13478-17-6]	dec	MoI_3	[14055-75-5]	
$MoOCl_4$	[13814-75-0]	102	MoI_2	[14055-74-4]	

[a] Disproportionates above 410°C.
[b] Sealed tube.

tachloride is converted to $MoCl_4$; $MoOCl_4$ is converted to $MoOCl_3$ (2). The tetrachloride also can be made by the reaction of MoO_2 with octachlorocyclopentane (3).

The halides are extremely reactive toward moisture and oxygen in the air and must, therefore, be handled under an inert atmosphere. Thus, air exposure converts the pentachloride to the oxytetrachloride. The reduction of $MoCl_5$ in the presence of air gives $MoOCl_3$, thermodynamically the most stable oxychloride. Molybdenum tetrachloride sublimes when heated, but upon cooling disproportionates to $MoCl_3$ and $MoCl_5$ (3). At 500°C, under dry helium, $MoCl_3$ decomposes to $MoCl_2$ and $MoCl_4$. At 700°C, $MoCl_2$ disproportionates to metallic molybdenum and $MoCl_4$. In acid solution, $MoCl_4$ dissociates to a mixture of trivalent and pentavalent compounds.

Molybdenum pentachloride is dimeric in the solid state; the pentafluoride is a tetramer. The trihalides appear to be polymeric, generally trimers. The dihalides have been shown to be hexamers that contain the complex cation $(Mo_6X_8)^{4+}$ and are preferably written as $(Mo_6X_8)X_4$ (4).

Oxides and Hydroxy Compounds

Molybdenum Trioxide. Molybdenum trioxide, MoO_3, is a white crystalline powder. Most molybdenum compounds are prepared directly or indirectly from it. Physical constants are given in Table 2.

Technical-grade MoO_3 is prepared commercially by oxidizing molybdenite (MoS_2) in a multiple-hearth roaster. Air flow and temperature are carefully controlled to provide a low sulfur product with minimum MoO_2 content. High purity chemical grades are prepared either by sublimation of the technical grade or calcination of crystallized ammonium dimolybdate (ADM). Impurity concentrations of the two high purity products differ slightly.

Molybdenum trioxide reacts with strong acids (notably conc sulfuric) to form the complex molybdenyl, MoO_2^{2+}, and molybdyl, MoO^{4+}, cations which form soluble compounds. Alkali solutions react readily with molybdenum trioxide to form molybdates. Fused alkalies and many metal oxides react directly with the trioxide to form normal molybdates, such as M_2MoO_4, $MMoO_4$, or $M_2(MoO_4)_3$, or polymolybdates, depending upon the stoichiometry of the system.

Reduction of molybdic trioxide with hydrogen at 300–400°C proceeds smoothly to yield molybdenum dioxide. At higher temperatures (ca 500°C), the reduction proceeds to metallic molybdenum.

The chemistry of molybdenum suggests that a series of oxides should exist, each having a valence number from two to six. Actually, the molybdenum–oxygen system is complex and not completely defined. The oxides which have been reported are listed

Table 2. Physical Properties of Molybdenum Trioxide, MoO₃

Property	Value
melting point, °C	795
boiling point, °C	1155
sublimation point, °C	ca 700
crystal form	orthorhombic
a:b:c, nm	0.39628:1.3855:0.36964

in Table 3 (5). Many of these compounds are metastable and are prepared by prolonged heating of stoichiometric mixtures of oxides and metal, followed by quenching. Reduction of the trioxide does not appear to go stepwise through the oxides listed in Table 3.

Halogens and hydrohalides react with MoO_3 to form oxyhalides. Reaction with thionyl chloride, $SOCl_2$, gives molybdenum oxytetrachloride or pentachloride, depending on the reaction time (6).

Molybdenum Dioxide. Molybdenum dioxide is a relatively inert compound that is not attacked by nonoxidizing acids, alkalies, or molten salts. At 1777°C, in the absence of air, MoO_2 dissociates to form volatile MoO_3 and metallic molybdenum:

$$3\ MoO_2 \rightarrow 2\ MoO_3 + Mo$$

In the presence of air, MoO_2 readily oxidizes at elevated temperatures to form MoO_3. Chlorine adds directly to MoO_2 at 300°C to yield the dioxydichloride, MoO_2Cl_2. With chlorinated hydrocarbons at elevated temperatures molybdenum dioxide gives molybdenum tetrachloride, $MoCl_4$ (7).

Hydroxides. The structures of molybdenum blues and the series of hydrated molybdenum oxides are intimately related to those of the oxides and are characterized as genotypic (8). The series of hydrated oxides, listed in Table 4, was prepared by the

Table 3. Molybdenum Oxides[a]

Name	CAS Registry No.	Formula Empirical	Formula Molecular	Crystal structure	Color
trioxide	[1313-27-5]	MoO_3	MoO_3	rhombic	white
ζ-oxide	[12136-82-0]	$MoO_{2.89}$	$Mo_{18}O_{52}$	triclinic	blue black
β′-oxide	[12163-89-2]	$MoO_{2.89}$	Mo_9O_{26}	monoclinic	violet
β-oxide	[12058-34-3]	$MoO_{2.84}$	Mo_8O_{23}	monoclinic	violet
θ-oxide	[12438-84-5]	$MoO_{2.80}$	Mo_5O_{14}	tetragonal	blue violet
κ-oxide	[12777-01-4]	$MoO_{2.78}$	$Mo_{17}O_{47}$	orthorhombic	red blue
η-oxide	[12033-38-4]	$MoO_{2.75}$	Mo_4O_{11}	monoclinic	wine red
γ-oxide		$MoO_{2.75}$	Mo_4O_{11}	orthorhombic	wine red
dioxide	[18868-43-4]	MoO_2	MoO_2	tetragonal	brown black
sesquioxide	[1313-29-7]	$MoO_{1.5}$	Mo_2O_3	amorphous	black
	[12136-80-0]	$MoO_{0.33}$	Mo_3O	cubic	metallic

[a] Ref. 5.

Table 4. Molybdenum Hydroxides

Compound	CAS Registry No.	Color
$MoO(OH)_3$	[27845-91-6]	white
$Mo_4O_{10}(OH)_2$	[12507-76-5]	blue
$Mo_2O_4(OH)_2$[a]	[55004-54-1]	blue
$Mo_5O_8(OH)_8$[b]	[66588-05-4]	brown
$Mo_5O_7(OH)_8$[c]	[66589-56-8]	red
$Mo_5O_5(OH)_{10}$	[66588-05-4]	green

[a] From $Mo_5O_8(OH)_8$ by oxidation.
[b] From $Mo_5O_5(OH)_{10}$ with KOH.
[c] From $Mo_5O_8(OH)_8$ by reduction.

carefully controlled stepwise reduction of MoO_3 with either atomic hydrogen, zinc and hydrochloric acid, molybdenum metal powder and water, or lithium aluminum hydride.

Mild reduction of an acid solution of a molybdate gives a strong blue color, molybdenum blue [1313-29-7], as the first step in the reduction reaction. The exact composition of this compound is uncertain. There is evidence that it is colloidal and a mixture of several compounds sensitive to oxidation-reduction. Most probably, it is a mixture of low valent, hydrous molybdenum oxides.

Molybdenum blue is used in several colorimetric methods, including the determination of trace quantities of phosphorus or phosphates by the preferential reduction of molybdophosphoric acid [12026-57-2] and a number of other methods employing molybdophosphoric acid as a reducing reagent for compounds such as phenols, tyrosine, or tryptophan. When molybdenum blue suspensions are made alkaline by the addition of ammonia or sodium hydroxide, a hydrated oxide of pentavalent molybdenum precipitates; it easily loses water and forms $MoO(OH)_3$.

Acids. Molybdic(VI) acid monohydrate [13462-95-8], $H_2MoO_4 \cdot H_2O$ or $MoO_3 \cdot 2H_2O$ is prepared by acidifying a molybdate solution with nitric acid. It takes several weeks to separate the yellow solid from the mother liquor. The white crystalline acid, H_2MoO_4 [7782-91-4], may be prepared by careful dehydration of the yellow monohydrate.

Molybdates

Isopolymolybdates. The isopolymolybdates, a large class of inorganic compounds, consist of a cation and a condensed molybdate anion. The naming of isopolymolybdates is straightforward: they are called simply di, tri, etc, molybdates depending upon the degree of condensation. The compositions of isopolymolybdate salts that can be isolated from aqueous solutions are listed in Table 5 (9).

Ammonium Dimolybate. Ammonium dimolybdate [27546-07-2], $(NH_4)_2Mo_2O_7$, ADM, is obtained by crystallization from aqueous solutions of MoO_3 containing excess NH_3. An evaporative crystallizer is used at 100°C. Ammonium dimolybdate is a high purity source for the preparation of molybdenum metal (powder, sheet, pellets, and wire) and for specialty catalysts.

Ammonium Heptamolybdate. Ammonium heptamolybdate [12027-67-7], $(NH_4)_6Mo_7O_{24} \cdot 4H_2O$, AHM, is prepared by crystallization from aqueous solutions of MoO_3 containing excess NH_3. It is isolated in a cooling crystallizer at low temper-

Table 5.   Composition of Isopolymolybdates[a]

Name	Formula[b]
dimolybdates	$M_2O.2MoO_3$ or $M_2Mo_2O_7$
trimolybdates	$M_2O.3MoO_3$ or $M_2Mo_3O_{10}$
heptamolybdates (paramolybdates)	$3M_2O.7MoO_3.xH_2O$ or $M_6Mo_7O_{24}.xH_2O$
octamolybdates (tetramolybdates)	$M_2O.4MoO_3.xH_2O$ or $M_4Mo_8O_{26}.xH_2O$

[a] Ref. 9.
[b] M represents a univalent cation.

atures. This compound is used for the preparation of specialty catalysts and wherever a highly aqueous soluble form of molybdate is required free from metal cations. A phase study of the MoO_3–H_2O–NH_3 system has been made (10).

Normal Molybdates. Normal or orthomolybdates, $M_2O.MoO_3.x H_2O$ or $M_2MoO_4.x H_2O$, are prepared by direct combination of the oxides, neutralization of slurries of MoO_3 with MOH or M_2CO_3, or by precipitation from molybdate solution by salts of the desired metal. The properties of normal molybdates are given in Table 6 (11).

Molybdenum is frequently recovered from spent catalysts and various metal

Table 6. Properties of Normal (Ortho) Molybdates

Molybdate	CAS Registry No.	Color	Density, g/cm^3	Crystal structure	Melting point, °C	Soly °C, g solute per 100 g water
Na_2MoO_4	[7631-95-0]	white	3.28^{18}	$H1_1$, spinel $a_0 = 0.912$ nm	686	39.38^{25}
$CaMoO_4$	[7789-82-4]	white	4.28	C_{4h}^6, scheelite $c/a = 2.19$	965 (dec)	0.0050
$ZnMoO_4$	[13767-32-3]	white		tetragonal pyramids	700 (900)	0.5
$CoMoO_4$	[13762-14-6]	violet rose	3.6 (α) 4.5 (β) 4.1 (γ)		1040	
$NiMoO_4$	[12673-58-4]	green	3.5 (α) 4.9 (β) (γ)	monoclinic	970	
$FeMoO_4$	[13718-70-2]	dark brown, yellow		monoclinic	850 (1068)	0.0076
$CuMoO_4$	[13767-34-5]	light green			820 (dec)	0.038
$PbMoO_4$	[10190-55-3]	white	6.811	C_{4h}^6, scheelite $c/a = 2.23$	1065 (1075)	0:000012
$CdMoO_4$	[13972-68-4]	light yellow	5.347	C_{4h}^6, scheelite $c/a = 2.174$	ca 900 (dec)	0.0067
Li_2MoO_4	[13568-40-6]	white	2.66	C_{3i}^2, phenacite $c/a = 1.153$	702	44.81^{25}
K_2MoO_4	[13446-49-6]	white	2.342	isomorphous with K_2SO_4, K_2CrO_4	919	64.57^{25}
Rb_2MoO_4	[13718-22-4]	white			958 (919)	67.88^{18}
Cs_2MoO_4	[13597-64-3]	white			936 (929)	67.07^{18}
$SrMoO_4$	[13470-04-7]	white	4.662	C_{4h}^6, scheelite $c/a = 2.23$	≤1040 (dec)	ca 0.003
$BaMoO_4$	[7787-37-3]	white	4.975	C_{4h}^6, scheelite $c/a = 2.29$	1480	0.0055
$MnMoO_4$	[14013-15-1]	yellow		monoclinic		
Ag_2MoO_4	[13765-74-7]	white, pale yellow[a]		O_h^7, spinel[a] cubic	483	0.00386
Tl_2MoO_4	[34128-09-1]	white[b], yellow[c]		$a_0 = 0.926$ nm	red heat	insoluble

[a] After fusion.
[b] When precipitated.
[c] During fusion.

scraps by digesting the products with caustic. The molybdate can be recovered as the water-soluble sodium molybdate or it can be precipitated by lime addition and reused in the steel industry as water-insoluble $CaMoO_4$.

Normal sodium and potassium molybdates behave as typical 1-2 electrolytes in water. In concentrated solutions, appreciable association has been observed by cryoscopy. The degree of association for the potassium salt was less than that for the sodium salt.

Sodium Molybdate.   Sodium molybdate, Na_2MoO_4, SMC, is prepared by evaporative crystallization of aqueous solutions of molybdic oxide and sodium hydroxide. Upon crystallization, the dihydrate [10102-40-6], $Na_2MoO_4.2H_2O$, is obtained; the anhydrous sodium salt is formed by heating at 100°C. Sodium molybdate is used in the pigment and metal-finishing industries, as a soil additive, and for aqueous corrosion-inhibition applications.

Molybdate Acidification.   Acidification of molybdate ions in alkaline solution, in which they exist as $(MoO_4)^{2-}$, results in the formation of polynuclear species. The MoO_4^{2-} ion is stable above pH 6.5. Acidification leads to the first polymeric species, the heptamolybdate ion $[Mo_7O_{24}]^{6-}$:

$$7 \, MoO_4^{2-} + 8 \, H^+ \rightleftharpoons [Mo_7O_{24}]^{6-} + 4 \, H_2O$$

This reaction is practically complete at pH = 4.5. At higher acidity (between pH 2.9 and 1.5), the octamolybdate ion forms, $[Mo_8O_{26}]^{4-}$. At H^+ to MoO_4^{2-} acidification ratios higher than 1.5, more aggregated species may form but the equilibria are established very slowly. At pH 0.9, the isoelectric point of molybdic acid is reached and molybdic acid precipitates. At still lower pH, this precipitate redissolves to form cationic molybdenyl species such as MoO_2^{2+}.

Heteropolymolybdates.   The heteropolymolybdates form a large family of salts and free acids with each member containing a complex and high molecular weight anion. These anions contain two to eighteen hexavalent molybdenum atoms around one or more central atoms. The heteropolymolybdates are all highly oxygenated, eg, $[PMo_{12}O_{40}]^{3-}$, $[As_2Mo_{18}O_{62}]^{6-}$, and $[TeMo_6O_{24}]^{6-}$, where the P^{5+}, As^{5+}, and Te^{6+} are the central atoms.

Approximately 36 different elements have been reported to function as central atoms in distinct heteropoly anions. Many of these elements can act as central atoms in more than one series of heteropoly anions. Thus, P^{5+} occurs in six distinct stable species and P^{3+} occurs in two others; the total number of possible acids and salts is very large.

Heteropolymolybdates fall into distinct series with properties that differ from one series to another. Table 7 gives the principal species of heteropolymolybdates (12). Heteropolymolybdates are always made in solution, generally after acidifying and heating quantities of reactants. When the central atom is a transition metal, a simple salt of the element may be mixed hot with a soluble molybdate solution of appropriate pH. Free acids may be prepared by extraction with ether from acidified aqueous solutions of the simple acids, by double decomposition of salts, or by ion exchange from heteropoly salts.

Heteropolymolybdates show the following general properties: high molecular weights ranging up to 4000 for inorganic electrolytes; solubility in water up to 85%, several are soluble in organic solvents as well; high degree of hydration, a single acid or salt often forms several solid hydrates; the intense colors range through the spectrum

Table 7.   Principal Species of Heteropolymolybdates

Number of atoms, X:Mo	Principal central atoms	Typical formulas
1:12[a]	Series A: N^{5+} (?), P^{5+}, As^{5+}, Ge^{4+}, Sn^{4+} (?), Ti^{4+}, Si^{4+}, Zr^{4+}	$[X^{n+}Mo_{12}O_{40}]^{(8-n)-}$
	Series B: Ce^{4+}, Th^{4+}, U^{4+}	$[X^nMo_{12}O_{42}]^{(12-n)-}$
1:11[b]	P^{5+}, As^{5+}, Ge^{4+}	$[X^{n+}Mo_{11}O_{39}]^{(12-n)-}$ (possibly dimeric)
1:10[b]	P^{5+}, As^{5+}, Pt^{4+} (?)	$[X^{n+}Mo_{10}O_x]^{(2x-60-n)-}$
1:9[a]	Mn^{4+}, Ni^{4+}	$[X^{n+}Mo_9O_{32}]^{(10-n)-}$
1:9[a]	P^{5+}	$[X^{n+}M_9O_{31}(OH)_3]^{(11-n)-}$
1:6[a]	Series A: Te^{6+}, I^{7+}	$[X^{n+}Mo_6O_{24}]^{(12-n)-}$
	Series B: Co^{3+}, Al^{3+}, Cr^{3+}, Rh^{3+}, Ga^{3+}, Fe^{3+}, Ni^{2+}	$[X^{n+}Mo_6O_{24}H_6]^{(6-n)-}$
2:10[a]	Co^{3+}	$[X_2^{n+}Mo_{10}O_{38}H_4]^{(12-2n)-}$
2:17[b]	P^{5+}, As^{5+}	$[X_2^{n+}Mo_{17}O_x]^{(2x-102-2n)-}$
2:5[a]	P^{5+}	$[X_2^{n+}Mo_5O_{23}]^{(16-2n)-}$
1m:6m[b]	Co^{2+}, Mn^{2+}, Cu^{2+}, Se^{4+}, As^{3+}, P^{5+}, P^{3+}	$[X^{n+}Mo_6O_x]_m^{(2x-36-n)-m}$
4:12[a]	As^{5+}	$[H_4As_4Mo_{12}O_{50}]^{4-}$
1:1[a]	As^{3+}	$[(CH_3)_2AsMoO_{14}OH]^{2-}$

[a] Structure known by x-ray determination.
[b] X-ray structure not determined.

and occur in many shades of green, yellow, orange, red, and blue (other compounds in this group are colorless); they are strong oxidizing agents for organic substrates, the reduction products show an intense deep color; the free heteropoly acids are strong acids, dissociation constants are usually in the range of 10^{-1} to 10^{-3}; the heteropolymolybdate anions are decomposed by strongly basic solutions, the products are simple molybdate ions and either an oxyanion or a hydrous metal oxide of the central atom; over specific ranges of pH and other conditions, most solutions of heteropolymolybdates appear to contain one distinct species of anion.

Sulfides, Selenides, and Tellurides

Molybdenum forms a series of homologous compounds with sulfur, selenium, and tellurium. The disulfide [1317-33-5], MoS_2, diselenide [12058-18-3], $MoSe_2$, and ditelluride [12058-20-7], $MoTe_2$, are isomorphous. These chalcogenides occur as shiny gray plates and may be prepared by the direct combination of the elements or by heating MoO_3 and the appropriate element in potassium carbonate to a high temperature.

The sesquisulfide [12033-33-9], Mo_2S_3, sesquiselenide [12325-96-1], Mo_2Se_3, and sesquitelluride [12325-97-2], Mo_2Te_3, have been prepared by the direct reaction of the elements in sealed evacuated tubes at elevated temperatures (13). All three compounds are inert to attack by nonoxidizing acids.

Molybdenum Disulfide.   Molybdenum disulfide (molybdenite) is the principal molybdenum-containing mineral mined, and the commercial source of molybdenum and its compounds. It is stable under ordinary conditions but, under strongly oxidizing conditions, is oxidized to molybdic trioxide, MoO_3. Molybdenum disulfide is hexagonal with sixfold symmetry with two molecules per unit cell. Each sulfur atom is equidistant from three molybdenum atoms and each molybdenum atom is surrounded by six equidistant sulfur atoms at the corners of a trigonal prism with altitude 0.317 ± 0.01 nm and edge 0.315 ± 0.002 nm. The Mo–S distance is 0.241 ± 0.006 nm. A rhombohedral form of MoS_2 has been prepared synthetically; a few isolated natural occurrences have also been reported. Here the a axis is the same as in the hexagonal form but the c axis is one and one half that in the hexagonal form resulting in three molecules per unit cell. The hexagonal, layer–lattice structure of MoS_2 results in easy cleavage (sliding) of the crystal. This phenomenon is responsible for the inherent lubricity and low coefficient of friction of MoS_2 (see Lubrication and lubricants).

In the absence of air, water vapor, or carbon dioxide, MoS_2 is stable to about 1200°C. Heated *in vacuo* at 1370°C, molybdenite begins to dissociate very slowly to sulfur and Mo_2S_3. At 1600°C *in vacuo*, the products are molybdenum metal and free sulfur. In the presence of air, MoS_2 begins to oxidize slowly at 315°C; the rate increases with increasing temperatures.

Molybdenum disulfide has a specific gravity of 4.8 and an HV of 29 (1–1.5 Mohs). As a naturally occurring material, the electrical properties of commercially available molybdenite are erratic. Molybdenum sulfide is diamagnetic and a linear photoconductor and semiconductor which exhibits either p- or n-type conductivity. Rectifier and transistor actions have been observed.

Molybdenum Trisulfide. Molybdenum trisulfide [12033-29-3], MoS_3, is a brown-to-black amorphous powder formed when solutions of ammonium tetrathiomolybdate [15060-55-6], $(NH_4)_2MoS_4$, are acidified or heated. It is reconverted to the thiomolybdate by reaction with ammonium sulfide.

Thiomolybdates. Thiomolybdates resemble normal molybdates and undergo many similar reactions in which the oxygen and sulfur are interchangeable. Dark maroon ammonium tetrathiomolybdate may be prepared by treating a solution of ammonium molybdate [12027-67-7] with H_2S. An orange dithiomolybdate [16150-60-0], $(NH_4)_2MoO_2S_2$, also has been prepared by treating a cooled solution of ammonium molybdate with hydrogen sulfide. Ammonium dithiomolybdate decomposes thermally to yield an amorphous material, molybdenum oxydisulfide [58797-84-5], with the composition $MoOS_2$. A comprehensive review of the Mo–S system has been prepared (14).

Cyanides, Thiocyanates, and Carbonyls

Cyanides and thiocyanates form complex anions with molybdenum in a lower valence state (see Table 8). In most of these, molybdenum atoms have a coordination number of 8. These compounds are highly colored. The cyanides and thiocyanates are light-sensitive, but are generally more stable in air than the complex halides.

The red thiocyanate complex is formed by adding a solution of a soluble thiocyanate to a reduced acid solution of molybdenum. It is utilized in the colorometric determination of molybdenum. The compound contains a thiocyanate to molybdenum ratio of 3; the Mo is pentavalent.

Molybdenum Hexacarbonyl. Molybdenum hexacarbonyl [13939-06-5], $Mo(CO)_6$, may be prepared from molybdenum metal powder and carbon monoxide under high pressure; the molybdenum is zero valent in this structure. Generally, $Mo(CO)_6$ is prepared from $MoCl_5$, zinc dust, and CO in ether at a pressure of 9–12 MPa (90–120

Table 8. Complex Molybdenum Cyanides and Thiocyanates

Formula	CAS Registry No.	Crystal form and color	Solubility
$K_4[Mo(CN)_7H_2O].H_2O^a$	[76900-85-1]	crystalline	sol H_2O
$K_3MoS(CN)_4.2H_2O$	[76900-86-2]	deep-blue microcrystalline powder	soln in H_2O turns violet
$K_4Mo(CN)_8.2H_2O^b$	[17457-89-5]	amber-yellow rhombic bipyramids and plates^c	easily sol in H_2O, 0.017 g/L abs alc; insol ether
$K_4Mo(OH)_4(CN)_4$	[21978-12-1]	blue	
$K_4Mo(OH)_4(CN)_4.xH_2O^d$	[76550-45-3]	red–violet feathery rhombic needles or plates	sol H_2O
$K_4Mo(OH)_3(CN)_5.3H_2O$	[76550-42-2]	red–violet crystals	sol H_2O
$K_3Mo(CN)_8^e$	[19442-23-0]	yellow needles	sol H_2O
$K_3Mo(SCN)_6.4H_2O$	[14649-96-8]	small amber-yellow rhombic (pseudohexagonal) crystals	soln in H_2O is dark red, becoming yellow
$KSCN.K_2O.4MoO_3.5H_2O$	[76721-95-4]	yellow needles or prisms	sol H_2O and dilute acids (HCl)

^{a} Forms the following colored precipitates with: Pb, red–brown; Mn, gray–green; Co, dark gray; Ni, green; Cu, purple–brown; Fe, bright blue.

^{b} Forms the following colored precipitates with: Tl, red–yellow; Au and Cd, yellow; Co, reddish yellow; Mn, bright yellow; Cu, violet–brown amorphous; Ag, light yellow amorphous.

^{c} Sp gr 2.337.

^{d} $x = 2, 4, 6,$ or 8.

^{e} Highly sensitive to light. The copper(II) salt is a green precipitate, turning purple–red in light.

atm). Some of the pertinent properties of $Mo(CO)_6$ are listed in Table 9 (see Carbonyls).

Molybdenum hexacarbonyl reacts with a variety of organic chemicals to yield organomolybdenum compounds; the literature on this subject is vast (18). Among these are nitrogen bases; cyclopentadiene and cycloheptatriene; aromatic hydrocarbons; nitriles and isonitriles; organic phosphine, arsine, stibine and bismuthine compounds and organic sulfides and ethers. Metallic molybdenum coatings can be deposited onto steel substrates by the thermal decomposition of $Mo(CO)_6$ (19).

Organomolybdenum Compounds

In addition to the substituted molybdenum hexacarbonyl compounds, organomolybdenum–sulfur compounds containing phosphorus or nitrogen have been prepared. Many of these compounds are oil-soluble and, thus, the low friction benefits of the molybdenum–sulfur interaction can be imparted to engine and gear oils. Typical structures for these compounds can be assigned as:

$$Mo_2S_2O_2^{2+} \left[S \cdots P \underset{OR}{\overset{S}{\diagdown}} \right]_2^{-} \quad \text{and} \quad R_2NC \underset{S}{\overset{S}{\diagdown}} Mo \underset{S}{\overset{O}{\diagdown}} \underset{S}{\overset{O}{\diagdown}} Mo \underset{S}{\overset{S}{\diagdown}} CNR_2$$

For oil-solubility, the R groups are generally C_8 or larger alkyls or aryls.

Organomolybdenum–sulfur–phosphorus compounds can be prepared by dis-

Table 9.   Properties of Molybdenum Hexacarbonyl, $Mo(CO)_6$

Property	Value or description
molecular weight	264.01
appearance	white, shiny, highly refracting crystals
specific gravity	1.96
melting point, °C[a]	150–151
absolute boiling point, °C	156.3[b] 155.1[c]
crystal structure[d]	orthorhombic
a, nm[e]	1.20
b, nm[e]	0.64
c, nm[e]	1.12
space group[f]	C_{2v}^9[c]
molecular structure[g]	
bond distance, nm	
Mo—C	0.19–0.21
C—O	0.12
magnetic properties	diamagnetic

[a] Determined in sealed, inert-gas filled tube. Decomposes around 150°C in air.
[b] Ref. 15.
[c] Ref. 16.
[d] Unit cell contains four molecules.
[e] Ref. 17.
[f] Identification of $Mo(CO)_6$ by x-ray diffractometry.
[g] Six CO molecules are distributed octahedrally about molybdenum.

solving MoO_3 in a caustic solution and then acidifying with an amount of sulfuric acid equivalent to the hydroxide. An organophosphorodithioic acid reactant may be prepared separately by treating a monohydric alcohol with P_2S_5 in the mol ratio of 4:1. The resultant reactant is added to the aforementioned molybdate solution with subsequent stirring at reflux temperature (85–100°C); reaction time is generally 2 h (20).

The organomolybdenum–sulfur–nitrogen compounds are prepared in good yield by acidification of a solution of sodium molybdate and sodium dialkyldithiocarbamate (21).

Various molybdenum-containing xanthates, naphthenates, acetylacetonates, and ethoxylated amines have been prepared and evaluated as lubricant additives. Thus far, these compounds have had limited applicability.

Health and Safety Factors

No occurrence of significant toxicity of molybdenum compounds has been observed in humans, nor has any cumulative effect been noticed (see Molybdenum and molybdenum alloys). In some areas where soil molybdenum is high, and, consequently, the molybdenum content of forage crops is high, animals develop a toxic reaction called molybdenosis, which can be overcome by injections or feed implementations of copper salts.

Molybdenum in Biological Processes.   Molybdenum is an important component of the redox catalytic activity of enzymes, the high molecular weight proteins that fix nitrogen or result in the assimilation of nitrogen (nitrate metabolism) (see Mineral nutrients; Nitrogen fixation). Thus, molybdenum takes part in the life processes of plants and animals. Table 10 contains a list of the six known enzymes in which molybdenum is found (22). The most important enzyme is nitrogenase in which molybdenum undergoes catalytic multielectron redox and may serve as an intermediate MoN_2 or $Mo.N_2.Mo$ in the overall reaction:

$$N_2 + 2\,H^+ \xrightarrow{2e} N_2H_2$$

$$N_2H_2 + 2\,H^+ \xrightarrow{2e} N_2H_4$$

$$N_2H_4 + 2\,H^+ \xrightarrow{2e} 2\,NH_3$$

Iron–sulfur centers in the protein supply electrons; water is the proton source.

Molybdenum-deficient soils respond (in some cases dramatically) to spray applications of sodium molybdate; in certain areas the molybdate is added as part of the seed treatment. Cauliflower, cotton, corn, tomatoes, peas, and even trees respond favorably to the application of sodium molybdate.

Uses

Catalysts. Molybdenum compounds have widespread use in catalytic applications (24). Large-scale industrial installations in the petroleum and petrochemical industries have employed molybdenum compounds for the following processes:

Desulfurization. Sulfur compounds can be removed from petroleum feedstocks by CoO—MoO_3/γ-Al_2O_3 catalysts. The MoO_3 is generally deposited on the γ-Al_2O_3 substrate from an ammoniacal solution. The molybdenum catalyst is used in a guard chamber to protect the platinum catalyst from sulfur poisoning in naphtha reforming reactions (for octane improvement). Variants of this catalyst are used in the hydrodesulfurization (HDS) of distillates, heater oils, lubricating oils, and residium. Air-

Table 10.   Molybdenum Enzymes [a,b]

Enzyme	CAS Registry No.	Metal composition [c]	Remarks
nitrogenase	[9013-04-1]	2(?) Mo, 24 Fe–S	elementary source of nitrogen, $N_2 \rightarrow NH_3$
nitrate reductase	[9013-03-0]	2 Mo, 6 Fe–S? Haem?	source of nitrogen, $NO_3^- \rightarrow NO_2^- \rightarrow (NH_3)$
xanthine oxidase [d]	[9002-17-9]	2 Mo, 8 Fe–S, 2 Flavin (FAD)	nitrogen assimilation (fungi) and purine degradation
aldehyde oxidase	[9029-38-3]	2 Mo, 8 Fe–S, 2 Flavin (FAD)	
sulfite oxidase	[9029-07-6]	2 Mo, 2 Haem (Fe)	removal of sulfite?
formate dehydrogenase	[9028-85-7]	1 Mo, 1 Se, 1 Haem, 12 Fe–S	oxidation of formate

[a] Ref. 22.
[b] Detailed structures are given in ref. 23.
[c] Per protein unit.
[d] Purine metabolism.

quality standards require the continued lowering of the sulfur content of liquid fuels. Thus, the use of molybdenum HDS catalysts is expected to increase markedly during the 1980s (25) (see Petroleum).

Denitrification.   NiO–MoO$_3$/γ-Al$_2$O$_3$ systems are used to remove nitrogen-containing compounds from petroleum fractions. Nitrogen compounds can adversely affect the storage stability and octane value of naphthas and may poison downstream catalysts. Nitrogen removal improves air quality to some extent, since it lowers the potential for NO$_x$ formation during the subsequent fuel combustion. Denitrification catalysts are expected to be widely used for oil shale (qv) and tar sands (qv) processing.

Acrylonitrile Synthesis.   A complex molybdenum-containing catalyst is used for the direct ammoxidation of propylene to synthesize acrylonitrile (qv), a 3.1×10^6 t/yr monomer.

Propylene Oxide.   Molybdenum compounds are used for the homogeneous catalytic epoxidation (qv) of propylene to propylene oxide (qv).

Formaldehyde.   An iron molybdate catalyst is used to prepare formaldehyde (qv) from methanol. Formaldehyde is used in a resin in the plywood and construction industries.

Coal Conversion.   Scheduled on-stream in 1980, are two large-scale pilot plants for coal liquefaction using Mo-containing catalysts. Molybdenum compounds appear to function catalytically in the hydrogen transfer reaction (to increase the H–C ratio and, thus, provide the basis for liquefaction). It is anticipated that coal-conversion catalysts will be a large-volume application for molybdenum compounds in the late 1980s (see Fuels, synthetic).

Pigments.   Molybdenum Orange.   Molybdenum Orange [12656-85-8] is formed by the coprecipitation of the chromate and molybdate of lead; lead sulfate may also be added for some applications (see Pigments, inorganic). Moly-orange is a popular pigment, widely used to prepare red to orange hues for automotive and appliance applications; current production volume is about 8900 metric tons of pigment per year.

Inhibitive Pigments.   Calcium and zinc molybdates are white compounds of low aqueous solubility that function as inhibitive pigments in primer paints for steel substrate applications. These compounds have been commercialized successfully and are enjoying unusual growth in industrial-maintenance-paint applications (26) (see Paint; Pigments).

Colors, Toners, and Dyes.   Basic dyes, color lakes and certain acid dyes are precipitated from solution using molybdophosphoric acid (MPA). Dyes such as Methyl Violet [8004-87-3] and Victoria Blue [1325-94-6] offer color and light stability at reasonable cost; commercial acceptance of the MPA dyes is widespread (see Dyes, application and evaluation).

Aqueous Corrosion Inhibitors.   Sodium molybdate is used widely at low concentrations as a corrosion inhibitor for cooling-tower applications (27) (see Water, industrial water treatment; Corrosion).

Molybdates provide corrosion inhibition in ethylene glycol antifreeze (qv) solutions; sodium molybdate is effective on steel, cast iron, and aluminum. With the advent of aluminum-containing engines and radiators, the use of sodium molybdate is expected to increase (28).

In cutting-fluid applications, the use of sodium molybdate–sodium nitrate blends offer the potential of equal inhibition performance but without the danger of forming carcinogenic nitrosamines (29).

Flame and Smoke Suppressants. Molybdenum complexes containing calcium and zinc molybdate are used to lower the smoke and provide flame retardancy to plastics (see Flame retardants). Plasticized poly(vinyl chloride) containing molybdenum complexes is used in the automotive, wire and cable, and construction industries. It is anticipated that this use will increase as smoke standards become more stringent (30).

Lubrication. Molybdenum disulfide is an important solid lubricant. It is used extensively in automotive lubrication, in ball joints, constant-velocity universals, and wheel bearings. Molybdenum disulfide also is used in the cold- and warm-forming of metals and as an aid to the assembly and disassembly of close-fitting parts. It is used in gear oils, cutting fluids, plastics, and wire-drawing lubricants, and in the aerospace and aircraft industries. Molybdenum disulfide generally is used under conditions of boundary lubrication, ie, high loads and low speeds. Depending on the specific use, it can be applied dry or in a carrier (oil, grease, bonded coating, plastics or paste) (31).

Organomolybdenum compounds have been used in lubricants for many years as antiwear or extreme-pressure additives. Recent data indicate that compounds of this type can function as friction-modifier additives to provide improved fuel economy with motor oils (32). Because of rising energy costs, it is expected that the oil-soluble organomolybdenum compounds will become increasingly popular in fuel-efficient motor oils (see Lubrication and lubricants).

Analytical Reagents. Molybdenum compounds are used widely as reagents to test for phosphorus, silicon, and arsenic. These tests employ the insolubility of ammonium molybdophosphate in nitric acid solution as a means to separate phosphate ion from interfering compounds and ions (especially iron, aluminum, and calcium). A color reaction from the reduction of molybdophosphate to blue is used for low phosphate concentrations.

Other Applications. Sodium molybdate continues to be used as a soil additive in certain parts of the world. The molybdate is frequently added as part of the seed treatment; it is especially effective on legumes.

Molybdenum compounds are added to ceramic glazes and enamels to provide increased adhesion of the vitreous enamels on steel. Molybdates lower the surface tension of the melt and increase the spreading and wetting power.

Molybdates frequently are added to steel phosphatizing baths to improve the coating weight and thickness. They also are added to black oxide baths to improve the color and coating adhesion to the steel substrate (see Steel).

Various tests have indicated that molybdenum can be codeposited electrochemically from specific baths (33) with nickel, zinc, or iron. These deposits have shown some promise as antiwear surfaces but additional data are required to determine the ultimate utility (34).

A few experiments have indicated that leather can be tanned by using nitric acid solutions of sodium molybdate. Additional work is required to determine if the process is competitive with trivalent chromium.

Several molybdenum compounds are photochromic (see Chromogenic materials). So far, no satisfactory way has been found to utilize this effect in sunglasses, paints, wall-coverings, etc; generally, the reversible process is too slow to be useful.

BIBLIOGRAPHY

"Molybdenum Compounds" in *ECT* 1st ed., Vol. 9, pp. 199–214, by A. Linz, Climax Molybdenum Co.; "Molybdenum Compounds" in *ECT* 2nd ed., Vol. 13, pp. 645–659, by J. C. Bacon, Climax Molybdenum Co.

1. P. L. Raymond, *J. Electrochem. Soc.* **106**, 444 (1959).
2. M. L. Larson and F. W. Moore, *Inorg. Chem.* **3**, 285 (1964).
3. D. E. Couch and A. Brenner, *J. Res. Natl. Bur. Stand.* **63A**, 185 (1959).
4. *Inorg. Synth.* **12**, 165 (1970).
5. L. Kihlborg, *Ark. Kemi* **21**, 443, 471 (1964).
6. H. J. Seifert and H. P. Quak, *Angew. Chem.* **73**, 621 (1961).
7. A. B. Bardawil, F. N. Collier, Jr., and S. Y. Tyree, Jr., *J. Less Common Metals* **9**(1), 20 (1965).
8. O. Glemser and G. Lutz, *Z. Anorg. Allgem. Chem.* **285**, 173 (1956).
9. G. A. Tsigdinos and C. J. Hallada, *Isopoly Compounds of Molybdenum, Tungsten, and Vanadium, Bulletin Cdb-14*, Climax Molybdenum Company, Greenwich, Conn., Feb. 1969.
10. C. J. Hallada, "Phase Diagram of the NH_3–MoO_3–H_2O System," *Proceedings of the Climax First International Conference on the Chemistry and Uses of Molybdenum, Reading, Eng., Sept. 1973*, p. 52.
11. *Properties of the Simple Molybdates, Bulletin Cdb-4*, Climax Molybdenum Company, Greenwich, Conn., 1962.
12. G. A. Tsigdinos, "Heteropoly Compounds of Molybdenum and Tungsten," *Topics in Current Chemistry*, Vol. 76, Springer-Verlag, Heidelberg, FRG, 1978.
13. A. A. Opalovskii and V. E. Fedorov, *Dokl. Akad. Nauk. SSR* **163**, 1163 (1965); *Chem Abstr.* **63**, 15652f (1965).
14. G. A. Tsigdinos, *Inorganic Molybdenum-Sulfur Compounds, Bulletin Cdb-17*, Climax Molybdenum Company, Greenwich, Conn., March 1973.
15. W. Hieber and E. Romberg, *Z. Anorg. Allg. Chem.* **221**, 321 (1935).
16. T. N. Rezukhina and V. V. Shuyrev, *Vestn. Mosk. Univ. Ser. Fiz. Mat. Estestv. Nauk* **7**(4), 41 (1952).
17. W. Rudorff and U. Hofmann, *Z. Phys. Chem. Abt. B* **28**, 351 (1935).
18. M. L. Larson, *Organic Complexes of Molybdenum, Bulletin Cdb-9*, Climax Molybdenum Company, Greenwich, Conn., 1956, *Suppl. 3–8*, 1966–1971.
19. *Properties of Molybdenum Hexacarbonyl, Bulletin Cdb-13a*, Climax Molybdenum Company, Greenwich, Conn., April 1970.
20. U.S. Pat. 3,494,866 (Feb. 10, 1970), E. V. Rowan and H. H. Farmer (to R. T. Vanderbilt Co.).
21. M. L. Larson and F. W. Moore, *Inorg. Chem.* **2**, 881 (1963).
22. R. J. P. Williams, *The Biological Role of Molybdenum*, Climax Molybdenum Company, London, Eng., 1978.
23. R. C. Bray and J. C. Swann, *Struct. Bonding (Berlin)* **11**, 107 (1972).
24. *Molybdenum Catalyst Bibliography*, bulletin, Climax Molybdenum Company, Greenwich, Conn., 1952–1964; *Suppl. No. 1*, 1964–1967; *Suppl. No. 2*, 1968–1969; *Suppl. No. 3*, 1970–1972; *Suppl. No. 4*, 1973–1974; *Suppl. No. 5*, 1975–1976; *Suppl. No. 6*, 1977–1978.
25. B. Delmon, "Recent Approaches to the Anatomy and Physiology of Cobalt Molybdenum Hydrodesulfurization Catalysts," *Proceedings of the Climax Third International Conference on the Chemistry and Uses of Molybdenum, Ann Arbor, Mich., Aug. 19–23, 1979*, p. 73–84.
26. W. J. Banke, *Mod. Paint Coat.* (Feb. 1980).
27. R. J. Lipinski and R. F. Weidner, *Heat. Piping Air Cond.* (Jan. 1978).
28. R. R. Wiggle, V. Hospadaruk and E. A. Styloglou, "The Effectiveness of Engine Coolant Inhibitors for Aluminum," *NACE Paper No. 69*, NACE, Chicago, Ill., March 1980.
29. M. S. Vukasovich, "Sodium Molybdate Corrosion Inhibition of Synthetic Metalworking Fluids," *ASLE Preprint No. 80-AM-1C-1*, ASLE, Anaheim, Calif., May 1980.
30. D. A. Church and F. W. Moore, *Plast. Eng.* **31**(12) (Dec. 1975).
31. H. F. Barry and J. P. Binkelman, *NLGI Spokesman* (May 1966).
32. C. A. Passaut and R. E. Kollman, *SAE Spec. Publ.* **780601** (June 1978).

33. H. S. Myers, "The Electrodeposition of Molybdenum," Ph.D. thesis, Columbia University, New York, 1941.

34. H. Kei, F. Hisaaki, O. Hidetsuga, and A. Tetsuya, *Kinzoku Hyomen Gijutsu* **27**, 590 (1976).

General References

S. H. Killefer and A. Linz, *Molybdenum Compounds*, Interscience Publishers, New York, 1952.

Proceedings of the Climax International Conference on the Chemistry and Uses of Molybdenum, Vol. 1, Reading, Eng., 1973; Vol. 2, Oxford, Eng., 1976; Vol. 3, Ann Arbor, Mich., 1979.

E. J. Stiefel, "The Coordination and Bioinorganic Chemistry of Molybdenum" in S. J. Lippard, ed., *Progress in Inorganic Chemistry*, Vol. 22, John Wiley & Sons, Inc., New York, 1977.

H. F. BARRY

Climax Molybdenum Company of Michigan

MONAZITE, (Th, Ce, La, Di, Si)PO₄.   See Cerium; Rare-earth metals; Thorium.

MONOSODIUM GLUTAMATE, NaOOCCH₂CH₂CH(NH₂)COOH.   See Amino acids (MSG).

MORPHOLINE.   See Amines—cyclic.

MOUTHWASHES.   See Dentifrices.

MUCILAGES.   See Gums.

MUNTZ METAL.   See Copper alloys (wrought).

MUSCLE RELAXANTS.   See Neuroregulators; Psychopharmacological agents.

N

NAPHTHALENE

This article deals mainly with naphthalene [91-20-3]. The hydrogenated naphthalenes, the alkylnaphthalenes (particularly methyl- and isopropylnaphthalenes), and acenaphthene also are discussed (see also Naphthalene derivatives).

Properties

The accepted configuration of naphthalene, ie, two fused benzene rings sharing two common carbon atoms in the ortho position, was established in 1869 and was based on its oxidation product, phthalic acid (1). Based on its fused-ring configuration, naphthalene is the first member in a class of aromatic compounds with condensed nuclei. Its numbering is shown in structure (1). The use of α and β, shown in structure (2), is used rarely by theoretical chemists.

(1)   (2)

Naphthalene is a resonance hybrid having three contributing forms (3–5):

(3)   (4)   (5)

In chemical reactions, naphthalene usually acts as though the bonds were fixed in the positions as shown in (3). For most purposes, the conventional formula in (1) is adequate; the numbers represent the carbon atoms with attached hydrogen atoms.

The two carbons that bear no numbers are common to both rings and carry no hydrogen atoms. From the symmetrical configuration of the naphthalene molecule, it should be possible for only two isomers to exist when one hydrogen atom is replaced by another atom or group. Therefore, positions 1, 4, 5, and 8 are identical and often are designated as α positions; likewise, positions 2, 3, 6, and 7 are identical and are designated as β positions, as shown in (**2**). Some selected chemical and physical properties of naphthalene are given in Table 1.

Selected values from the vapor pressure–temperature relationship for naphthalene are listed in Table 2.

Selected viscosity–temperature relationships for liquid naphthalene are listed in Table 3.

Naphthalene forms azeotropes with several compounds; some of these mixtures are listed in Table 4.

Naphthalene is very slightly soluble in water but is appreciably soluble in many organic solvents, eg, 1,2,3,4-tetrahydronaphthalene, phenols, ethers, carbon disulfide, chloroform, benzene, coal-tar naphtha, carbon tetrachloride, acetone, and decahydronaphthalene. Selected solubility data are presented in Table 5.

The ir, uv, mass, nmr, and ^{13}C-nmr spectral data for naphthalene and other related hydrocarbons have been reported (7–11). Additional information regarding the properties of naphthalene has been published (3,6,12–13).

Table 1.   Chemical and Physical Properties of Naphthalene

Property	Value
mol wt	120.1732
mp, °C	80.290[a]
normal bp (at 101.3 kPa[c]), °C	217.993[b]
triple point (t_{tp}), °C	80.28[b]
critical temperature (t_c), °C	475.2[b]
critical pressure (p_c), kPa[c]	4051[b]
flash point (closed cup), °C	79[d]
ignition temperature, °C	526[d]
flammable limits, vol %	
upper	5.9[e]
lower	0.9[e]
heat of vaporization, kJ/mol[f]	43.5[b]
heat of fusion at triple point, kJ/mol[f]	18.979[b]
heat of combustion (at 15.5°C and 101.3 kPa[c]), kJ/mol[f]	−5158.41[b]
heat capacity (at 15.5°C and 101.3 kPa[c]), J/(mol·K)[f]	159.28[b]
heat of formation (at 25°C), kJ/mol[f]	
solid	78.53[b]
gas	150.58[b]
density (at 25°C), g/cm³	1.175[b]
density (at 90°C), g/cm³	0.97021[b]

[a] Ref. 2.
[b] Ref. 3.
[c] To convert kPa to atm, divide by 101.3.
[d] Ref. 4.
[e] Ref. 5.
[f] To convert J to cal, divide by 4.184.

Table 2.   Selected Values of Vapor Pressure–Temperature Relationship for Naphthalene[a]

Temperature, °C	Pressure, kPa[b]
0	0.0008
10	0.003
20	0.007
40	0.043
87.6	1.33
119.1	5.33
166.3	26.66
191.3	53.33
214.3	93.33
218.0	101.33
230.5	133.32
250.6	199.98

[a] Ref. 6.
[b] To convert kPa to mm Hg, multiply by 7.5.

Table 3.   Selected Values of Viscosity–Temperature Relationship for Liquid Naphthalene[a]

Temperature, °C	Viscosity, mPa·s (= cP)
80.3	0.96
90	0.846
100	0.754
110	0.678
120	0.616
150	0.482
180	0.394
220	0.320

[a] Ref. 3.

Reactions

Substitution.   Substitution products retain the same nuclear configuration as naphthalene. They are formed by the substitution of one or more hydrogen atoms with other functional groups. Substituted naphthalenes that are important commercially have been obtained through the use of sulfonation, sulfonation and alkali fusion, alkylation, nitration and reduction, and chlorination.

The hydrogen atoms in the 1 positions of naphthalene can be substituted somewhat more easily than hydrogen atoms in benzene, and they tend to do so under mild conditions. For example, 1-chloronaphthalene can be formed by direct substitution with little or no catalyst and 1-nitronaphthalene can be prepared using dilute nitric acid. Sulfonation of naphthalene also occurs readily in the 1 position but can be influenced by temperature. When a second group substitutes, its position also is influenced by the nature and position of the first group. In the case of the substitution of the second identical group during nitration or sulfonation, it predominately attaches to the unsubstituted ring. With substitution of the second identical group, there are ten possible disubstituted isomers. If the two substituting groups are different, 14 disubstitution isomers are possible. The number of possibilities is still larger when

Table 4.   Azeotropes of Naphthalene[a]

Compound	bp of compound (at 101.3 kPa[b]), °C	Naphthalene, wt %	bp of azeotrope (at 101.3 kPa[b]), °C
water	100	16	98.8
dodecane	216.3	60.5	140.2
dipropylene glycol	231.8	87.6	142.9
ethylene glycol	197.4	49	183.9
benzyl alcohol	205.2	40	204.1
p-ethylphenol	218.8	55	215.0
p-chlorophenol	219.8	63.5	216.3
diethylene glycol	245.5	78	216.6
3,4-dimethylphenol	226.8	84	217.6
benzoic acid	249.2	95	217.7

[a] Ref. 3.

[b] To convert kPa to mm Hg, multiply by 7.5.

Table 5.   Naphthalene Solubility Data[a]

Solvent	Solubility (at 25°C), mol fraction
ethylbenzene	0.2926
benzene	0.2946
toluene	0.2920
cyclohexane	0.1487
carbon tetrachloride	0.2591
n-hexane	0.1168
water	0.18×10^{-5}

[a] Ref. 3.

the substituents are different and three or more hydrogen atoms are replaced. The number of possible isomeric substitution products from naphthalene has been calculated and reported (14). The use of the numbered positions of the appropriate carbon atoms are used when two or more hydrogen atoms are replaced. Positions are designated numerically according to the lowest position number for the first substituent, eg (6),

(**6**) 1,3-dinitronaphthalene

Confusion is avoided by applying the rule that the sum of the position numbers shall be the lowest possible, eg (7),

(**7**) 1,4,5-trinitronaphthalene (not 1,4,8-trinitronaphthalene)

Sulfonation. Sulfonation of naphthalene with sulfuric acid produces mono-, di-, tri-, and tetranaphthalenesulfonic acids (see Naphthalene derivatives). All of the naphthalenesulfonic acids form salts with most bases. Naphthalenesulfonic acids are important starting materials in the manufacture of organic dyes (15) (see Azo dyes). They also are intermediates used in reactions, eg, caustic fusion to yield naphthols, nitration to yield nitronaphthalenesulfonic acids, etc.

Nitration. Naphthalene is easily nitrated with mixed acids, eg, nitric and sulfuric, at moderate temperatures to give mostly 1-nitronaphthalene and small quantities, ca 3–5%, of 2-nitronaphthalene. 2-Nitronaphthalene is not made in substantial amounts by direct nitration and must be produced by indirect methods, eg, the Bucherer reaction starting with 2-naphthalenol (2-naphthol). However, the 2-naphthylamine made using this route is a carcinogen; thus, the Bucherer method is seldom used in the United States.

Halogenation. Under mild catalytic conditions, halogen substitution occurs, and all of the hydrogen atoms of the naphthalene molecule can be replaced. The only commercially significant halogenated naphthalene products are the mixed chlorinated naphthalenes. Naphthalene is chlorinated readily by introducing gaseous chlorine into molten naphthalene at ambient pressure and temperatures of 100–220°C in the presence of small amounts of a catalyst, eg, ferric chloride. The chlorination of molten naphthalene gives a mixture of mono-, di-, and polychloronaphthalenes; the degree of chlorination is controlled by monitoring the sp gr or mp of the crude reaction product.

The commercial products are mixtures ranging from liquids, eg, mono- and mixed mono- and dichloronaphthalenes, to various waxlike solids, which contain di-, tri- and polychloronaphthalenes, with high melting points, ie, 90–185°C. Chloronaphthalenes are flame resistant and have high dielectric constants. Their high degree of chemical stability is indicated by their resistance to most acids and alkalies as well as to dehydrochlorination. They also are resistant to attack by fungi and insects. The U.S. demand for chloronaphthalenes has declined steadily in recent years. Koppers Company, Inc., the sole U.S. producer, ceased manufacturing their chloronaphthalene products (Halowax) in 1977.

The toxicity of chloronaphthalenes requires that special attention and caution be used during their manufacture and use; acne is the most common result of excessive skin exposure to them, and the most frequently affected areas are the face and neck (16). Liver damage has occurred in workers who have been exposed repeatedly to vapors, particularly to those of penta- and hexachloronaphthalene (17–18). Uses for the chlorinated naphthalenes include solvents, gauge and instrument fluids, capacitor impregnants, components in electrical insulating compounds, and electroplating stop-off compounds (see Insulation, electric).

Alkylation. Naphthalene can be easily alkylated using various alkylating agents, eg, alkyl halides, olefins, or alcohols in the presence of a suitable catalyst (19–20). In vapor-phase alkylations, phosphoric acid on kieselgur and silica/alumina catalysts are useful. In liquid-phase reactions, acid catalysts, eg, sulfuric acid, hydrofluoric acid, and phosphoric acid, are used widely when olefins are the alkylating agent. Aluminum chloride and other metal halides, eg, iron and zinc chlorides, also are active alkylating catalysts; however, their use often involves reactions that yield undesirable resinous by-products. Sulfuric acid is the preferred catalyst for alkylations with alcohols since the active aluminum chloride forms a complex with alcohols and must be used in prohibitively large quantities.

Isopropylnaphthalenes produced by alkylation of naphthalene with propylene have gained commercial importance as chemical intermediates, eg, 2-isopropylnaphthalene [2027-17-0], and as multipurpose solvents, eg, mixed isopropylnaphthalenes. Alkylation of naphthalene with alkyl halides (except methyl halides), acid chlorides, and acid anhydrides proceeds in the presence of anhydrous aluminum chloride by the Friedel-Crafts reaction (qv). The products are alkylnaphthalenes or alkyl naphthyl ketones, respectively (see Alkylation).

Chloromethylation. The reactive intermediate, 1-chloromethylnaphthalene, has been produced by the reaction of naphthalene in glacial acetic acid and phosphoric acid with formaldehyde and hydrochloric acid. Heating of these ingredients at 80–85°C at 101.3 kPa (1 atm) with stirring for ca 6 h is required. The potential hazard of such chloromethylation reactions, which results from the possible production of small amounts of the powerful carcinogen methyl chloromethyl ether, has been reported (21).

Addition. Addition reactions are characterized by the introduction of an element onto two or more of the adjacent carbon atoms of the naphthalene nucleus. The most important addition products of naphthalene are the hydrogenated compounds. Of less commercial significance are those made by the addition of chlorine.

Hydrogenation. Hydrogen is added to the naphthalene nucleus by reagents that do not affect benzene. It is possible to obtain products into which two, four, six, eight, or ten hydrogen atoms have been added. Of these, only the tetra- and decahydronaphthalenes are commercially significant. In addition to the commercially important 1,2,3,4-tetrahydronaphthalene, the 1,4,5,8-isomer has been reported. A review of the other hydronaphthalenes is presented in ref. 22. Some chemical and physical properties of 1,2,3,4-tetrahydronaphthalene and the decahydronaphthalenes are listed in Table 6.

Table 6.   Chemical and Physical Properties of Tetra- and Decahydronaphthalenes

	1,2,3,4-Tetrahydronaphthalene	Decahydronaphthalene *cis*	*trans*
mol wt	132.2048	138.2522	
mp, °C	−35.749[a]	−42.98[b]	−30.38[b]
normal bp (at 101.3 kPa[c]), °C	207.62[b]	195.815[b]	187.310[b]
density (at 25°C), g/cm³	0.9659[a]	0.8929[b]	0.8660[b]
viscosity (at 25°C), mPa·s (= cP)	2.012[a]	2.99[b]	1.936[b]
flash point (closed cup), °C	71[d]	57.8[b,e]	
ignition temperature, °C	385[d]	250[b,e]	
flammable limits, vol %			
upper	5[d,f]	4.9[d,e,g]	
lower	0.8[d,g]	0.7[d,e,g]	

[a] Ref. 23.
[b] Ref. 24.
[c] To convert kPa to mm Hg, multiply by 7.5.
[d] Ref. 4.
[e] Mixed isomers.
[f] At 150°C.
[g] At 100°C.

1,2,3,4-Tetrahydronaphthalene.   1,2,3,4-Tetrahydronaphthalene (Tetralin) is a water-white liquid that is insoluble in water, slightly soluble in methyl alcohol, and completely soluble in other monohydric alcohols, ethyl ether, and most other organic solvents. It is a powerful solvent for oils, resins, waxes, rubber, asphalt, and aromatic hydrocarbons, eg, naphthalene and anthracene. Its high flash point and low vapor pressure make it useful in the manufacture of paints, lacquers, and varnishes; for cleaning printing ink from rollers and type; in the manufacture of shoe creams and floor waxes; as a solvent in the textile industry; and for the removal of naphthalene deposits in gas-distribution systems (25). The commercial product typically has a tetrahydronaphthalene content ≥97 wt % with some decahydronaphthalene and naphthalene as the principal impurities.

1,2,3,4-Tetrahydronaphthalene reacts similarly to an alkylbenzene since its structure contains only one aromatic nucleus. It can be sulfonated readily, nitrated, oxidized, and hydrogenated. Sulfonation occurs first in the 6 and, to some extent, in the 5 position. Nitration with mixed acid yields the 5- and 6-nitro compounds in the cold; 5,6- and 5,7-dinitro compounds are formed at 35–40°C (22). 1,2,3,4-Tetrahydronaphthalene is oxidized to a hydroperoxide by passing air or oxygen through the warm liquid (see Peroxides) (26–27).

Further hydrogenation under pressure in the presence of a nickel catalyst gives a mixture of *cis*- and *trans*-decahydronaphthalene (Decalin). 1,2,3,4-Tetrahydronaphthalene dehydrogenates to naphthalene at 200–300°C in the presence of a catalyst; thermal dehydrogenation takes place at ca 450°C and is accompanied by cracking to compounds, eg, toluene and xylene.

Tetrahydronaphthalene is produced by the catalytic treatment of naphthalene with hydrogen. Various processes have been used, eg, vapor-phase reactions at 101.3 kPa (1 atm) as well as higher pressure liquid-phase hydrogenation where the conditions are dependent upon the particular catalyst used. Nickel or modified nickel catalysts generally are used commercially; however, they are sensitive to sulfur and only naphthalene that has very low sulfur levels can be used. Thus, many naphthalene producers purify their product to remove the thionaphthene which is the major sulfur compound present. Sodium treatment and catalytic hydrodesulfurization processes have been used for the removal of sulfur from naphthalene; the latter treatment is preferred because of the hazardous nature of sodium treatment.

1,2,3,4-Tetrahydronaphthalene is not a highly toxic compound. A threshold limit value of 25 ppm or 135 mg/m^3 has been suggested for Tetralin. Tetralin vapor is an irritant to the eyes, nose, and throat and dermatitis has been reported in painters working with it (28). The single-dose oral toxicity LD$_{50}$ for rats is 2.9 g/kg (29).

Decahydronaphthalene.   Decahydronaphthalene (Decalin) is the product of complete hydrogenation of naphthalene. Like Tetralin, it is a clear, colorless liquid with excellent solvent properties. It is produced commercially by the catalytic hydrogenation of naphthalene or 1,2,3,4-tetrahydronaphthalene and consists of a mixture of cis and trans isomers (22,30–32). The commercial product typically has a decahydronaphthalene content ≥97 wt % with the major impurity being 1,2,3,4-tetrahydronaphthalene. Decahydronaphthalene can be converted to naphthalene by heating with platinum, palladium, or nickel catalyst at 300°C (33).

Decahydronaphthalene is slightly to moderately toxic (29). The vapors are irritating to the eyes, nose, and throat. Excessive exposure to high concentrations causes numbness, nausea, headache, and vomiting. Dermatitis has been noted among painters handling decahydronaphthalene. No serious cases of industrial poisoning have been reported (34). A threshold limit value has not been established, although a value of 25 ppm has been suggested (28).

The uses of decahydronaphthalene are similar to those of 1,2,3,4-tetrahydrona-phthalene. Mixtures of the two are used for certain applications where a synergistic solvency effect is noted.

Some selected chemical and physical properties of 1,2,3,4-tetrahydronaphthalene and the decahydronaphthalenes are listed in Table 6.

Chlorine Addition. Chlorine addition and some chlorine substitution occurs at normal or slightly elevated temperatures in the absence of catalysts. The chlorination of molten naphthalene under such conditions yields a mixture of naphthalene tetra-chlorides, a monochloronaphthalene tetrachloride, and a dichloronaphthalene tet-rachloride, as well as mono- and dichloronaphthalenes (35). Sunlight or uv radiation initiates the addition reaction of chlorine and naphthalene resulting in the production of the di- and tetrachlorides (36). These addition products are relatively unstable and, at ca 40–50°C, they decompose to form the mono- and dichloronaphthalenes.

Oxidation. Naphthalene may be oxidized directly to 1-naphthalenol (1-naphthol) and 1,4-naphthoquinone, but yields are not good. Further oxidation beyond 1,4-naphthoquinone results in the formation of o-phthalic acid, which can be dehydrated to form phthalic anhydride. The manufacture of phthalic anhydride is the largest single use for naphthalene. Catalytic vapor-phase oxidation of naphthalene is preferred over liquid-phase oxidation. All vapor-phase processes involve the use of a catalyst based on vanadium pentoxide. In the United States, all phthalic anhydride units currently operating from naphthalene feed utilize fluidized catalyst-bed reactors. The naph-thalene can be fed as a liquid injected directly into the reactor at the bottom of the catalyst bed. Rapid dispersion of the naphthalene within the catalyst bed occurs as it is vaporized and oxidized to phthalic anhydride. Since all of the naphthalene is ox-idized in the reactor, there is no need for a naphthalene vaporizor, which is a re-quirement for fixed-bed reactors. Also, the rapid movement of the catalyst within the bed serves to dissipate the heat of the reaction so as to provide a relatively narrow temperature differential across the catalyst bed; this allows good control of the reaction with little possibility of a runaway reaction. A typical range of bed temperatures is 340–380°C. The catalyst in a fluid-bed reactor, as opposed to a fixed-bed reactor, readily carries the heat of reaction away from the reactants (37).

Naphthalene-based fixed-bed phthalic anhydride plants are in use outside of the United States. All phthalic anhydride plants built in the United States in the late 1960s and 1970s utilized fixed-bed technology, primarily the BASF process, based on o-xylene rather than naphthalene as the raw material. However, the rapidly escalating prices of petroleum products coupled with the possibility that demand for xylenes as octane boosters in the gasoline pool is expected to increase, may signal renewed interest in naphthalene, particularly the coal-tar-derived product, as a phthalic an-hydride feedstock (see Phthalic acids).

The quality of naphthalene required for phthalic anhydride manufacture varies depending upon the process and catalyst being used. Petroleum naphthalene generally can be used because of its low sulfur content (<10 ppm) and high purity (>99%). However, a typical commercial coal-tar naphthalene (freezing point, 78°C) has a purity of ca 96% and a sulfur content of ca 0.5%, with other impurities dependent upon the source. Further purification is required for fluid-bed phthalic processes. Hydrodesulfurization is a commonly used technique for the purification of coal-tar naphthalene in the United States (38).

Fixed-bed phthalic anhydride processes utilizing naphthalene, eg, the Von Heyden process, can operate on a 78–79°C freezing point coal-tar material obtained from a crude fraction by conventional fractional distillation after removal of the tar acids. Such fixed-bed reactors typically operate under slightly positive pressure with a reaction-zone temperature of ca 385°C.

Manufacture

Two sources of naphthalene exist in the United States: coal tar and petroleum. Coal tar was the traditional source until the late 1950s when it was in short supply. In 1960, the first petroleum-naphthalene plant was brought on stream and, by the late 1960s, petroleum naphthalene accounted for over 40% of total naphthalene production. The availability of large quantities of o-xylene at competitive prices during the 1970s affected naphthalene's position as the prime raw material for phthalic anhydride. In 1971, 45% of U.S. phthalic anhydride capacity was based on naphthalene as compared to only 29% in 1979. Since 1970, naphthalene production has decreased at an average rate of 3.3%/yr; coal-tar naphthalene decreased 6.5% and petroleum naphthalene dropped 1.5% (39). As a result of this decline, in the United States there are only three petroleum-naphthalene producers and three coal-tar naphthalene producers with a capacity distribution at ca 60% coal tar/40% petroleum.

Coal Tar. Coal tar is condensed and separated from the coke-oven gases formed during the high temperature carbonization of bituminous coal in coke plants (see Coal, coal-conversion processes). Although some naphthalene is present in the oven gases after tar separation and is removed in subsequent water cooling and scrubbing steps, the amounts are of minor importance. The largest quantities of naphthalene are obtained from the coal tar that is separated from the coke-oven gases. A typical dry coal tar obtained in the United States contains ca 10 wt % naphthalene. The naphthalene content of the tar varies somewhat depending upon the coal source, the coke-oven operating temperature, and the coking cycle times: the higher the coking temperature and rate, the higher the naphthalene content of the tar.

The coal tar first is processed through a tar-distillation step where ca the first 20 wt % of distillate, ie, chemical oil, is removed. The chemical oil, which contains practically all the naphthalene present in the tar, is reserved for further processing and the remainder of the tar is distilled further to remove additional creosote oil fractions until a coal-tar pitch of desirable consistency and properties is obtained.

The chemical oil contains ca 50 wt % naphthalene, 6 wt % tar acids, 3 wt % tar bases, and numerous other aromatic compounds. The chemical oil is processed to remove the tar acids by contacting with dilute sodium hydroxide and, in some few cases, next is treated to remove tar bases by washing with sulfuric acid.

Principal U.S. producers obtain their crude naphthalene product by fractional distillation of the tar acid-free chemical oil. This distillation may be accomplished in either a batch or continuous fashion. One such method for the continuous recovery of naphthalene by distillation is shown in Figure 1. The tar acid-free chemical oil is charged to the system where most of the low boiling components, eg, benzene, xylene, and toluene, are removed in the light-solvent column. The chemical oil next is fed to the solvent column, which is operated under vacuum, where a product containing the

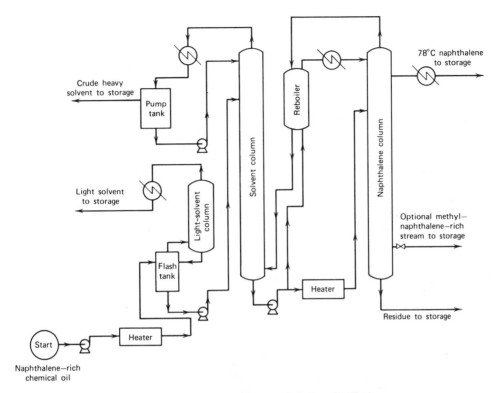

Figure 1. Typical coal-tar naphthalene distillation.

prenaphthalene components is taken overhead. This product, which is called coal-tar naphtha or crude heavy solvent, typically has a boiling range of ca 130–200°C and is used as a general solvent and as a feedstock for hydrocarbon-resin manufacture because of its high content of resinifiables, eg, indene and coumarone (see Hydrocarbon resins). The naphthalene-rich bottoms from the solvent column then are fed to the naphthalene column where a 78°C freeze-point product (78°C naphthalene) is produced. The naphthalene column is operated at near atmospheric pressure to avoid difficulties, eg, naphthalene-filled vacuum jets and lines, which are inherent to vacuum distillation of this product. A side stream which is rich in methylnaphthalenes may be taken near the bottom of the naphthalene column. The crude 78°C naphthalene that is produced by distillation is not of suitable quality for many applications and must be further upgraded.

Various techniques have been used in the past for the purification of crude 78°C coal-tar naphthalene and include solvent extraction with methanol, sodium treatment, washing with various aqueous solutions, sulfuric acid washing, hot pressing, and sublimation. The principal commercial processes for purifying crude coal-tar naphthalene have been discussed (40). One of the common processes involves treatment with concentrated sulfuric acid followed by neutralization and redistillation to give a product with a freezing point of over 79°C and a reduced sulfur content, ie, total sulfur reduced from ca 0.5 wt % in the starting material to ca 0.1 wt % after treatment. This refined product also can be sublimed into a colorless material which is suitable for manufacture of flakes or pellets for insecticidal use, ie, mothballs or flakes (see Insect control technology).

However, since the naphthalene produced from petroleum is of high purity and quality, the production of refined naphthalene by such chemical treatments essentially has ceased in the United States. Not only are such treatments expensive, but they also generate a significant amount of waste sludge which creates additional costs for appropriate waste-disposal facilities.

The main impurity in crude 78°C-coal-tar naphthalene is sulfur which is present in the form of thionaphthene (1–3%). Methyl- and dimethylnaphthalenes also are present (1–2 wt %) with lesser amounts of indene, methylindenes, tar acids, and tar bases.

Because thionaphthene is not completely removed by acid washing and since sulfur compounds are undesirable impurities for fluid-bed phthalic anhydride processes and those for tetrahydronaphthalene and decahydronaphthalene, much crude coal-tar naphthalene is desulfurized by hydrodesulfurization (see Petroleum). Hydrogenolysis of crude coal-tar naphthalene over cobalt molybdate-on-alumina catalysts at ca 400°C and 1.4 MPa (200 psi) readily converts the thionaphthene to hydrogen sulfide and ethylbenzene. Other impurities in the crude naphthalene are converted to lower-boiling hydrocarbons and are removed by distillation. The resulting desulfurized naphthalene has a sulfur content of <0.05 wt %. Depending upon the extent of fractionation achieved during the distillation, a product with freezing points from 77.5°C to >79°C can be obtained by hydrogenation of naphthalene, with the major impurity being 1,2,3,4-tetrahydronaphthalene.

Purification of crude coal-tar naphthalene also can be achieved by fractional crystallization techniques. Equipment, eg, the Brodie Purifier (Union Carbide Australia, Ltd.) is capable of producing a refined colorless product with a freezing point of 80°C or higher and sulfur contents of ca 0.05 wt % (41). The Brodie Purifier utilizes a long horizontal and somewhat tapered body in which crystals grow as they pass countercurrently against the flow of liquid. There are three general process zones: recovery, refining, and purification (see Fig. 2). In the recovery zone, the product is continuously cooled by the shell jacket; crystals form as the liquid volume diminishes. Crystals move from the recovery zone toward the intermediate refining section, picking up mass from liquid coming from the purification zone. In the vertical purification zone, the crystals settle by gravity to form a moving bed that is subjected to a coun-

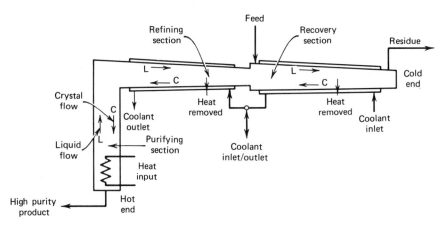

Figure 2.   Typical form of Brodie Purifier (41).

tercurrent liquid-reflux stream. Near the base of this zone, the crystals completely melt to form a liquid reflux and the final molten product. The manufacturer claims low utility consumption and operating costs with high thermal efficiency. There are no Brodie Purifiers using naphthalene in the United States, although such units are in commercial operation in Europe and elsewhere.

Petroleum. The production of naphthalene from petroleum involves two principal steps. The first is the production of an aromatic oil in the naphthalene–alkylnaphthalene boiling range by hydroaromatization or cyclization. The second step is the dealkylation of such oils either thermally or catalytically. The naphthalene that is produced is recovered as a high quality product and usually by fractional distillation (42–43).

The dealkylation process involves the following steps: hydrogen and the alkylnaphthalene feed are combined and then heated in a furnace to reaction temperature; the reaction takes place in a catalytic reactor; and the reactor effluent is separated in a series of distillation columns.

In the Unidak process (Union Oil Company of California), the feed, which may be catalytic reformate bottoms, enters a fractionator where a naphthalene concentrate, a middle product of alkylnaphthalenes which is treated further, and a small amount of bottoms which is used as fuel oil (see Fig. 3) are produced. The alkylnaphthalenes, which are mixed with make-up hydrogen at 3.5 MPa (500 psig) and some water to moderate the reaction, are transferred to a fixed-bed reactor. The reaction product enters a high pressure separator where 80 wt % hydrogen is evolved and recycled. A low pressure separator releases a small amount of fuel gas and the dealkylated product is sent to the feed. The naphthalene concentrate is purified from a by-product which can be used in gasoline.

The Hydeal process (Universal Oil Products) is shown schematically in Figure

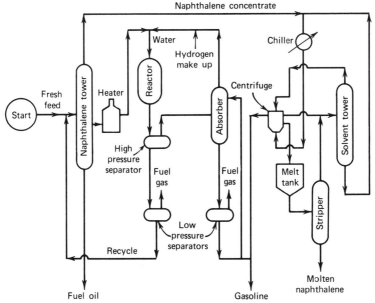

Figure 3. Unidak process.

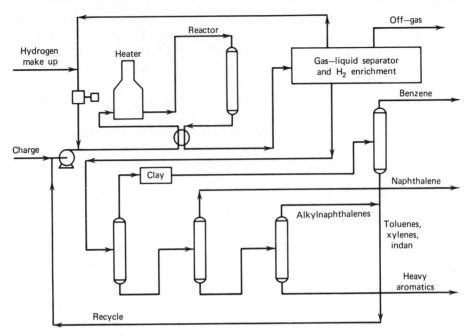

Figure 4.   Hydeal process (simplified process flow diagram).

4. A feedstock which is derived from heavy reformate is processed with excess hydrogen over a chromia–alumina catalyst of high purity and low sodium content. Naphthalene is recovered by distillation after treatment with clay to remove traces of olefins.

Economic Aspects

The total nameplate capacity for all U.S. naphthalene producers in 1979 was 324,000 metric tons with 206,000 t produced from coal tar and 118,000 t from petroleum (44). Naphthalene production from 1968 to 1979 is listed in Table 7 and is shown graphically in Figure 5.

Table 7.   U.S. Naphthalene Production 1968–1979[a], 10^3 Metric Tons

Year	Coal tar	Petroleum	Total
1968	206	171	377
1969	208	162	370
1970	188	132	320
1971	170	117	287
1972	181	105	286
1973	163	109	272
1974	170	91	261
1975	159	50	209
1976	159	49	208
1977	150	68	218
1978	141	78	219
1979	152	74	226

[a] Ref. 45.

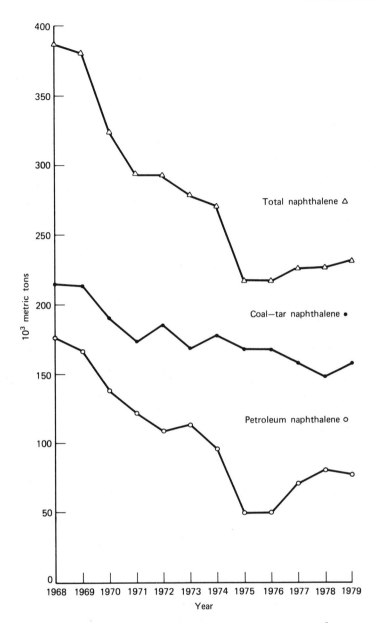

Figure 5. U.S. naphthalene production 1968–1979, 10^3 t.

The decline in naphthalene production in 1973 primarily resulted from competition with o-xylene as the feedstock for phthalic anhydride. Periods of feedstock shortages and the loss of one major producer also affected petroleum naphthalene output.

Naphthalene imports provided about 10–20% of the material consumed in the United States until ca 1963 when that percentage dropped to and leveled at less than 5%.

The economics of naphthalene recovery from coal tar can vary significantly depending upon the particular processing operation used. A major factor is the cost of the coal tar. As the price of fuel oil increases, the value of coal tar also increases. The economics of naphthalene production from petroleum have been affected to the greatest extent by the value of the alkyl naphthalene raw material which must be used in cost calculations (43). The price history of naphthalene from 1969 to 1979 is shown in Table 8. The prices shown are ca mid-year list prices; whenever a range of prices is given in the source, the lowest is listed in the table.

The higher price of the petroleum product results from its higher quality, ie, higher purity, lower sulfur content, etc. The price of naphthalene primarily is associated with that of *o*-xylene, its chief competitor as phthalic anhydride feedstock. Pressure from the phthalic anhydride manufacturers has made it difficult for naphthalene producers to operate their units profitably, and, in the case of the petroleum naphthalene producers, has initiated searches for other ways to utilize their dealkylation equipment.

It should be mentioned that potentially large quantities of naphthalene could become available as a result of the utilization and refinement of by-product streams from coal-gasification processes (46).

Specifications and Test Methods

Naphthalene usually is sold commercially according to its freezing or solidification point, because there is a correlation between the freezing point and the naphthalene content of the product; the correlation depends on the type and relative amount of impurities which are present (47). Because the freezing point can be changed appreciably by the presence of water, values and specifications are listed on a dry, wet, or as-received basis using an appropriate method agreed upon between buyer and seller, eg, ASTM D 1493.

Gas–liquid chromatography is used extensively to determine the naphthalene content of mixtures. Naphthalene can be separated easily from thionaphthene, the methyl- and dimethylnaphthalenes, and other aromatics. Analysis of the various other impurities may require the use of high resolution capillary columns.

Other tests that are routinely performed on commercial grades of naphthalene include: evaporation residues (ASTM D 2232), APHA color (ASTM D 1686), water

Table 8. Naphthalene Price History 1969–1979, ¢/kg

	Crude coal tar (77.5–78°C)	Petroleum (phthalic grade)
1969	14.3	16.5
1970	14.3	14.9
1971	14.3	14.9
1972	14.3	14.9
1973	12.1	13.5
1974	12.1	13.5
1975	19.8	30.0
1976	26.5	30.9
1977	26.5	30.9
1978	26.5	30.9
1979	26.5	30.9

(ASTM D 95), and acid-wash color (ASTM D 2279). Three methods used to measure sulfur content are the oxygen-bomb combustion method (ASTM D 129), the lamp-combustion method (ASTM D 1266), and the Raney nickel-reduction technique (48). Some typical specifications of commercially available grades of coal-tar and petroleum naphthalene and the ASTM specifications for refined naphthalene are listed in Table 9.

Table 9. Naphthalene Specifications

Analysis	Crude coal tar[a]	Coal tar low sulfur[a]	ASTM refined[b]	Petroleum refined[c]
solidification point (min), °C	78.0	77.5	79.80	80.0
assay (min), wt %	95.9	95.0		
sulfur (max), wt %	0.9	0.05		0.0010
color (max), APHA			100	10
acid wash color (max)			10	1
nonvolatiles (max), wt %	0.25	0.25	0.10	
water (max), wt %	0.25	0.25		

[a] Koppers Company, Inc., Pittsburgh, Pa.
[b] ASTM D 2431-68.
[c] Ashland Chemical Company, Columbus, Ohio.

Health and Safety Factors, Toxicology

Handling. Naphthalene generally is transported in molten form in tank trucks or tank cars that are equipped with steam coils. Depending upon the transportation distance and the insulation on the car or truck, the naphthalene may solidify and require reheating before unloading. Without inert-gas blanketing and at the temperature normally used for the storage of molten naphthalene, ie, 90°C, the vapors above the liquid are within the flammability limits. Thus, storage tanks containing molten naphthalene have a combustible mixture in the vapor space and care must be taken to eliminate all sources of ignition around such systems. Naphthalene dust also can form explosive mixtures with air which necessitates care in the design and operation of solid handling systems. Perhaps the greatest hazard to the worker is the potential for operating or maintenance personnel to be accidentally splashed with hot molten naphthalene while taking samples or disassembling process lines (see ASTM D 3438). Molten naphthalene tank vents must be adequately heated and insulated to prevent the accumulation of sublimed and solidified naphthalene. A collapsed tank can result easily from pumping from a tank with a plugged vent.

Toxicology. The acute toxicity of naphthalene is low with LD_{50} values for rats from 1780 to 2500 mg/kg (49). The inhalation of naphthalene vapors may cause headache, nausea, confusion, and profuse perspiration and, if exposure is severe, vomiting, optic neuritis, and hematuria may occur (28). Rabbits that received one to two grams per day of naphthalene either orally or hypodermically developed changes in the lens after a few days followed by definite opacity of the lens after several days (49). Rare cases of such corneal epithelium damage in humans have been reported (28). Naphthalene can be irritating to the skin and hypersensitivity does occur.

There are few reports with respect to animal experimentation of the chronic toxicity of naphthalene and no chronic toxic effects have been reported as a result of industrial exposure to naphthalene (28,49). In the *Salmonella* microsome mutagenicity (Ames) test, naphthalene is nonmutagenic (50–51). Naphthalene shows no biological activity in chemical carcinogen tests, indicating little carcinogenic risk (52). Since naphthalene vapors can cause eye irritation at concentrations of 15 ppm in air and since continued exposure may result in eye effects, a threshold limit value of 10 ppm (50 mg/m^3) has been set by the ACGIH (53). The amount is about 10% of the air-saturation value at 27°C.

Uses

The U.S. naphthalene consumption by markets for 1978 is listed in Table 10.

The production of phthalic anhydride by vapor-phase catalytic oxidation has been the main use for naphthalene. Although its use has declined in favor of *o*-xylene through the 1970s, naphthalene is expected to maintain its present share of this market, ie, ca 29%, or make slight gains through the 1980s in view of increasing prices of petroleum-based *o*-xylene. Both petroleum naphthalene and desulfurized coal-tar naphthalene can be used for phthalic anhydride manufacture. U.S. phthalic anhydride capacity was 597,000 metric tons in 1979 (54).

The second large use of naphthalene is as a raw material for the manufacture of 1-naphthyl-*N*-methylcarbamate (8) (carbaryl, Sevin). Crude or semirefined coal tar or petroleum naphthalene can be used for carbaryl manufacture. Carbaryl is used extensively as a replacement for DDT and other products that have become environmentally unacceptable (see Insect control technology).

$$\begin{array}{c} O \\ \parallel \\ OCNHCH_3 \end{array}$$

(8)

Another major outlet for naphthalene is for the manufacture of 2-naphthalenol (2-naphthol). The current U.S. producer of 2-naphthalenol uses a process that requires

Table 10.   U.S. Naphthalene Consumption, 1978[a]

Use	10^3 t	% of Total
phthalic anhydride	141	66.5
insecticides	35	16.5
2-naphthol	16	7.6
synthetic tanning agents	12	5.7
moth repellants	4	1.9
surfactants	2	0.9
miscellaneous	2	0.9
Total	212	100.0

[a] Refs. 39, 43.

high quality petroleum naphthalene. The technique involves alkylation to obtain 2-isopropylnaphthalene which then is oxidized to the hydroperoxide which in turn is cleaved, using an acid catalyst, to yield 2-naphthalenol and acetone (see also Acetone; Cumene). 2-Naphthalenol is an intermediate used in the manufacture of a variety of dyes, pigments, and rubber-processing chemicals. The 2-naphthalenol market is considered mature with a growth rate of about 4%/yr which translates to a forecast naphthalene requirement of ca 20,400 t in 1983 (39).

Approximately 5.7% of the U.S. naphthalene supply is consumed in the manufacture of tanning agents (see Leather). These are derivatives of naphthalenesulfonic acids, their salts, and the sodium salts of the reaction products of the sulfonic acid and formaldehyde. Both petroleum and purified coal-tar naphthalene are used in these processes.

Naphthalenesulfonates represent another small outlet for naphthalene, ie, ca 1 % of supply. The products are used as wetting agents and dispersants in paints and coatings and in a variety of pesticides and cleaner formulations. Their application as surfactants is expected to continue as a low growth item, although recent use of these products as concrete additives, ie, plasticizers (qv), may alter this pattern (see Surfactants).

Miscellaneous uses include several organic compounds and intermediates, eg, 1-naphthalenol, 1-naphthylamine, 1,2,3,4-tetrahydronaphthalene, decahydronaphthalene, and chlorinated naphthalenes. One new use of naphthalene which could develop to significant volumes is the manufacture of anthraquinone (9) (qv). Almost all anthraquinone worldwide is presently made by the oxidation of anthracene, an expensive chemical recovered from coal tar. However, commercial production of anthraquinone from naphthalene has been planned. In this process, coal-tar naphthalene is oxidized with air to naphthoquinone (10) and phthalic anhydride using a vanadium oxide catalyst. Without prior separation, the naphthoquinone reacts with butadiene to form tetrahydroanthraquinone which is dehydrogenated to anthraquinone. Subsequent separation and purification give the main product, anthraquinone, and phthalic anhydride as a by-product. This process is much more economical than the anthracene route; world capacity of anthraquinone is estimated at about 40,000 t/yr which satisfies a world demand of ca 36,000 t/yr (55). Recent developments in the use of anthraquinone in the wood-pulp industry could increase the world demand for anthraquinone to 100,000 t/yr or greater (56–57). Such demand would require ca 115,000 t/yr of additional naphthalene if the naphthalene route were used. Bayer and Ciba-Geigy have proposed the building and operation of a 12,000–15,000 t/yr plant (58) and Kawasaki Kasei Chemicals in Japan has a 1000 t/yr anthraquinone-from-naphthalene plant in operation (1000 t/yr capacity).

(9)   (10)

Alkylnaphthalenes

Methyl- and dimethylnaphthalenes are contained in coke-oven tar and in certain petroleum fractions in significant amounts. A typical high temperature coke-oven coal tar, for example, contains ca 3 wt % of combined methyl- and dimethylnaphthalenes (6). In the United States, separation of individual isomers seldom is attempted; instead, a methylnaphthalene-rich fraction is produced for commercial purposes. Such mixtures are used as solvents for pesticides, sulfur, and various aromatic compounds. They also can be used as low freezing, stable heat-transfer fluids. Mixtures that are rich in monomethylnaphthalene content have been used as dye carriers (qv) for color intensification in the dyeing of synthetic fibers, eg polyester. They also are used as the feedstock to make naphthalene in dealkylation processes. Phthalic anhydride also can be made from methylnaphthalene mixtures by an oxidation process that is similar to that used for naphthalene.

A mixed monomethylnaphthalene-rich material can be produced by distillation and can be used as feedstock for further processing. By cooling this material to about 0°C, an appreciable amount of 2-methylnaphthalene crystallizes leaving a mother liquor consisting of ca equal quantities of 1- and 2-methylnaphthalene. Pure 2-methylnaphthalene [91-57-6] (bp 341.1°C; mp 34.58°C) is used primarily as a raw material for the production of vitamin K preparations. Oxidation produces 2-methyl-1,4-naphthoquinone (menadione, vitamin K_3) which itself and in the form of water-soluble sodium hydrogen sulfite adducts show similar antihemorrhagic effects similar to the natural vitamin K_1. Other compounds of the vitamin K series can be prepared from menadione (59) (see Vitamins).

1-Methylnaphthalene [90-12-0] (bp 244.6°C; mp −30.6°C) can be used as a general solvent because of its low melting point. It also is used as a test substance for the determination of the cetane number of diesel fuels (see Gasoline). By side-chain chlorination of 1-methylnaphthalene to 1-chloromethylnaphthalene and formation of naphthaleneacetonitrile, it is possible to produce 1-naphthylacetic acid which is a growth regulator for plants, a germination suppressor for potatoes, and an intermediate for drug manufacture (59) (see Plant growth substances).

Of the individual dimethylnaphthalenes, 2,6-dimethylnaphthalene [28804-88-8] has been of particular interest as a precursor to 2,6-naphthalenedicarboxylic acid, a potentially valuable monomer for polyesters.

Isopropylnaphthalenes can be prepared readily by the catalytic alkylation of naphthalene with propylene. 2-Isopropylnaphthalene [2027-17-0] is an important intermediate used in the manufacture of 2-naphthol (see Naphthalene derivatives). The alkylation of naphthalene with propylene, preferably in an inert solvent at 40–100°C with an aluminum chloride, hydrogen fluoride, or boron trifluoride/phosphoric acid catalyst, gives 90–95% wt % 2-isopropylnaphthalene; however, a considerable amount of polyalkylate also is produced. Preferably, the propylation of naphthalene is carried out in the vapor phase in a continuous manner, over a phosphoric acid on kieselguhr catalyst under pressure at ca 220–250°C. The alkylate, which is low in di- and polyisopropylnaphthalenes, then is isomerized by recycling over the same catalyst at 240°C or by using aluminum chloride catalyst at 80°C. After distillation, a product containing >90 wt % 2-isopropylnaphthalene is obtained (60).

Mixtures containing various concentrations of mono-, di-, and polyisopropylnaphthalenes have been prepared by treating molten naphthalene with concentrated sulfuric acid and propylene at 150–200°C. followed by distillation (61). Products comprised of such isomeric mixtures have extremely low pour points, ie, ca −50°C, are excellent multipurpose solvents, and have been evaluated as possible liquid-phase heat-transfer oils. Of the higher alkylnaphthalenes, those of importance are the amyl-, diamyl-, polyamyl-, nonyl-, and dinonylnaphthalenes. These alkylnaphthalenes are used in sulfonated form as surfactants and detergent products.

Acenaphthene. Acenaphthene [*83-32-9*] (**11**) is a hydrocarbon ($C_{12}H_{10}$) present in high temperature coal tar (6). Acenaphthene may be halogenated, sulfonated, and nitrated in a manner similar to naphthalene (62). Oxidation first yields acenaphthenequinone and then 1,8-naphthalenedicarboxylic acid anhydride, an important intermediate for dyes, pigments, fluorescent whiteners, and pesticides. Acenaphtylene [*208-96-8*] (**12**) is formed upon catalytic dehydrogenation of acenaphthene (63). Acenaphthene can be isolated and recovered from a tar-distillation fraction by concentrating it by fractional distillation followed by crystallization to give ca 40% recovery of 98–99% pure acenaphthene. This material can be further purified by recrystallization from a suitable solvent, eg, ethanol (64).

(11)   (12)

BIBLIOGRAPHY

"Naphthalene" in *ECT* 1st ed., Vol. 9, pp. 216–231, by George Riethof and Abbot Pozefsky, Gulf Research & Development Co.; "Naphthalene" in *ECT* 2nd ed., Vol. 13, pp. 670–690, by Gilbert Thiessen, Koppers Company, Inc.

1. J. R. Partington, *A Short History of Chemistry*, Macmillan & Company, Ltd., London, Eng., 1957, pp. 293, 316.
2. J. P. McCullough and co-workers, *J. Phys. Chem.* **61**, 1105 (1957).
3. "Naphthalene" *Amer. Petrol. Inst. Monograph Series*, API Publication 707, Washington, D.C., Oct. 1978.
4. *Fire Protection Guide on Hazardous Materials*, 7th ed., National Fire Protection Assoc., Boston, Mass., 1978, pp. 49–212, 213, 225M-61, 173.
5. G. W. Jones and G. S. Scott, *U.S. Bur. Mines Rep. Invest.* **3881**, (1946).
6. *The Coal Tar Data Book*, 2nd ed., The Coal Tar Research Association, Gomersal, Leeds, U.K., 1965, B-2, p. 63; A-1, pp. 3, 4; B-2, p. 74.
7. B. J. Zwolinski and co-workers, "Selected Infrared Spectral Data" *American Petroleum Institute Research Project 44*, Thermodynamics Research Center, Texas A&M University, College Station, Texas (loose leaf sheets), 1954.
8. B. J. Zwolinski and co-workers, "Selected Ultraviolet Spectral Data" *American Petroleum Institute Research Project 44*, Thermodynamics Research Center, Texas A&M University, College Station, Texas (loose leaf sheets), 1961.
9. B. J. Zwolinski and co-workers, "Selected Mass Spectral Data" *American Petroleum Institute Research Project 44*, Thermodynamics Research Center, Texas A&M University, College Station, Texas (loose leaf sheets), 1949.
10. B. J. Zwolinski and co-workers, "Selected Nuclear Magnetic Resonance Spectral Data" *American Petroleum Institute Research Project 44*, Thermodynamics Research Center, Texas A&M University, College Station, Texas (loose leaf sheets), 1962.
11. B. J. Zwolinski and co-workers, "Selected ^{13}C Nuclear Magnetic Resonance Spectral Data" *American Petroleum Institute Research Project 44*, Thermodynamics Research Center, Texas A&M University, College, Station, Texas (loose leaf sheets), 1974.

12. H. C. Anderson and W. R. K. Wu, *U.S. Bur. Mines Bull.* **606,** 304 (1963).
13. C. L. Yaws and A. C. Turnbough, *Chem. Eng.* **82,** 107 (Sept. 1).
14. R. F. Evans and W. J. LeQuesne, *J. Org. Chem.* **15,** 19 (1950).
15. H. E. Fierz-David and L. Blangley, *Fundamental Processes of Dye Chemistry*, Interscience Publishers, Inc., New York, 1949, pp. 125, 440–453.
16. "Chloronaphthalenes" in *Hygienic Guide Series*, American Industrial Hygiene Assoc., Jan.–Feb., 1966.
17. F. B. Flinn and N. E. Jarvik, *Proc. Soc. Exptl. Biol. Med.* **35,** 118 (1936).
18. M. Mayers and A. Smith, *N.Y. Ind. Bull.* **21,** 30 (Jan. 1942).
19. A. N. Sachanen, *Conversion of Petroleum*, 2nd ed., Reinhold Publishing Corp., New York, 1948, pp. 550–565.
20. G. A. Olah, *Friedel-Crafts and Related Reactions*, Vols. 1–4, Wiley-Interscience, New York, 1963–1965.
21. *The Bureau of National Affairs, Current Report*, Nov. 22, 1979, 589; C. C. Yao and G. Miller, NIOSH Contract No. 210-75-0056 (Research Study on Bis(Chloromethyl)Ether), Jan. 1979.
22. N. Donaldson, *The Chemistry and Technology of Naphthalene Compounds*, Edward Arnold Publishers, London, U.K., 1958, pp. 455–473.
23. "Tetralin" *Amer. Petrol. Inst. Monograph Series*, API Publication 705, Oct. 1978.
24. "*cis* and *trans*-Decalin," *Amer. Petrol. Inst. Monograph Series*, API Publication 706, Oct. 1978.
25. *Tetralin and Decalin Solvents*, Bulletin, E. I. du Pont de Nemours & Co., Inc., Organic and Chemicals Division, Wilmington, Del., Rev. 1176.
26. J. S. Bogen and G. C. Wilson, *Pet. Refiner* **23,** 118 (1944).
27. U.S. Pat. 2,462,103 (Feb. 22, 1949), R. Johnson (to Koppers Company, Inc.).
28. E. E. Sandmeyer in G. D. Clayton and F. E. Clayton, *Patty's Industrial Hygiene and Toxicology*, 3rd rev. ed., Vol. II, Wiley-Interscience, New York, 1981, Chapt. 46.
29. N. I. Sax, *Dangerous Properties of Industrial Materials*, 4th ed., Van Nostrand Reinhold Co., New York, 1975, pp. 348, 599, 948–949, 1152.
30. Millard, *Annales de l'Office National des Combustibles Liquides* **9,** 1013 (1934); *Annales de l'Office National des Combustibles Liquides* **10,** 95 (1935).
31. R. Baker and R. Schultz, *J. Am. Chem. Soc.* **69,** 1250 (1947).
32. W. Seyer and R. Walker, *J. Am. Chem. Soc.* **60,** 2125 (1938).
33. Wessely and Grill, *Montatsh.* **77,** 282 (1947).
34. C. Marsden and S. Mann, *Solvents Guide*, 2nd ed., Wiley-Interscience, New York, 1963, p. 161.
35. Faust and Saame, *Ann. Chem.* **160,** 67 (1871).
36. A. Leeds and E. Everhart, *J. Am. Chem. Soc.* **2,** 205 (1880).
37. J. J. Graham, *Chem. Eng. Prog.* **66,** 54 (1970).
38. G. Gilbert, R. Weil, and R. Hunter, *Ind. Eng. Chem.* **53,** 993 (1961).
39. R. Gerry, "Product Review on Naphthalene," *Chemical Economics Handbook*, Stanford Research Institute, Menlo Park, Calif., Dec. 1978.
40. D. McNeil, *Gas World Coking Suppl.* **149,** 43 (Mar. 1959).
41. *Introducing the Brodie Purifier*, Pamphlet RJB-1000, Union Carbide Australia Ltd., Sydney, Australia, 1971.
42. N. Ockerbloom, *Hydrocarbon Process.* **50,** 101 (Dec. 1971).
43. R. Stobaugh, *Hydrocarbon Process.* **45,** 149 (1966).
44. *Chem. Mark. Rep.* (Apr. 1, 1979).
45. *Marketing Estimates*, Chemical Division, Koppers Company, Inc., Pittsburgh, Pa., 1979.
46. R. Serrurier, *Hydrocarbon Process.* **55,** 253 (Sept. 1976).
47. W. Kirby, *J. Soc. Chem. Ind.* **59,** 168 (1940).
48. L. Granatelli, *Anal. Chem.* **31,** 434 (1959).
49. M. Kawai, *Aromatics* **31**(7–8), 26 (1979).
50. J. McCann and co-workers, *Proc. Natl. Acad. Sci. U.S.A.* **72,** 5135 (1975).
51. B. Commoner, *Reliability of Bacterial Mutagenesis Techniques to Distinguish Carcinogenic and Noncarcinogenic Chemicals*, EPA Report No. 600/1-76-022, Apr. 1976.
52. I. Purchase and co-workers, *Br. J. Cancer* **37,** 873 (1978).
53. *Documentation of the Threshold Limit Values*, 3rd ed., ACGIH, Cincinnati, Ohio, 1971.
54. *Chem. Mark. Rep.* (Dec. 31, 1979).
55. *Chem. Eng. Int. Ed.* **85,** 18 (Aug. 14, 1978).
56. *Chem. Week* **123,** 22 (July 5, 1978).

57. U.S. Pat. 4,012,280 (Mar. 15, 1977), H. Holton (to Canadian Industries).

58. H. Bretscher, G. Eigemann, and E. Plattner, *Chimia* **32**(5), 180 (1978).

59. H. Franck and G. Collin, *Erzeuginisse aus Steinkohlenteer*, Springer Verlag, Berlin, FRG, 1968, p. 173.

60. K. Handrick, *Erdoel and Kohle*, *Compendium 76/77*, p. 308.

61. U.S. Pat. 3,962,365 (May 28, 1975), R. M. Gaydos and co-workers (to Koppers Company, Inc.).

62. A. E. Everest, *The Higher Coal Tar Hydrocarbons*, Longmans Green & Company, 1927, pp. 1–57.

63. M. Kaufman and A. Williams, *J. Appl. Chem.* **1**, 489 (1951).

64. G. Markus and T. Kraatsover, *Coke and Chem. USSR* **5**, 37 (1971).

R. M. GAYDOS
Koppers Company, Inc.

NAPHTHALENE CARBOXYLIC ACIDS.   See Naphthalene derivatives.

NAPHTHALENE DERIVATIVES

Much of the chemical knowledge concerning naphthalene derivatives has grown out of dye-intermediates research (see Dyes and dye intermediates). Naphthalene derivatives are of growing importance as intermediates for agricultural chemicals, drugs, and rubber chemicals.

Several systems of nomenclature have been used for naphthalene and many trivial and trade names are well established. In this article, the usage of the *Chemical Abstracts Index Guide* is employed. The numbering of the naphthalene nucleus is shown in (**1**); older practices are given in (**2**) and (**3**).

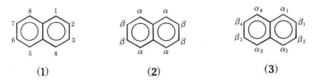

(1)   (2)   (3)

Substituents in the 1,8 or 2,6 positions of naphthalene are in the peri or amphi positions, respectively.

The number of naphthalene derivatives is very large since the number of positional isomers is large: 2 for monosubstitution, 10 for disubstitution/same substituent, 14 for disubstitution/different substituents, 14 for trisubstitution/same substituent, 42 for trisubstitution/two different substituents, 84 for trisubstitution/three different substituents, and so on with multiplying complexity. The commercially important compounds are described below. Detailed information is given in refs. 1–7.

Naphthalenesulfonic Acids

Naphthalenesulfonic acids are important chemical precursors for dye interme- diates, wetting agents and dispersants, naphthols, and air-entrainment agents for concrete (see Azo dyes; Plasticizers). The production of many intermediates used for making azo, azoic, and triphenylmethane dyes involves naphthalene sulfonation and one or more unit operations, eg, caustic fusion, nitration, reduction, or amination.

Generally, the sulfonation of naphthalene leads to a mixture of products. Naphthalene sulfonation as an electrophilic substitution reaction, the kinetics and mechanism of sulfonation with the various sulfonating agents, substituent orientation in naphthalene sulfonation, and the isomerization of the sulfonic acids are reviewed in ref. 8. Analyses of sulfonation products includes the use of nmr, gas chromatogra- phy/mass spectroscopy, tlc, paper and column chromatography, and especially, high pressure liquid chromatography (9–11).

Naphthalene sulfonation at less than ca 100°C is kinetically controlled and pro- duces predominantly 1-naphthalenesulfonic acid. Sulfonation of naphthalene at above ca 150°C provides thermodynamic control of the reaction and 2-naphthalenesulfonic acid as the main product. In naphthalene polysulfonation, a staged program of acid addition or control of acid concentration and time–temperature control often is used to obtain a desired product mix.

Naphthalenesulfonic acid production technology used to be all batch operations but now emphasizes continuous processes, removal of excess sulfonating agent by stripping under vacuum, and the use of chlorosulfonic acid or sulfur trioxide to min- imize the need for excess sulfuric acid. The improved analytical methods have con- tributed to the success of process optimization.

A demonstration of the paths of naphthalene sulfonation, with selected conditions for each step, is given in Figure 1; alternative paths are possible. A list of naphth- alenesulfonic acids and some of their properties is given in Table 1.

Production figures for the sulfonic acids are not available, but an indication of scale is the report that, in 1979, Schelde-Chemie (a joint venture of Bayer and CIBA-GEIGY) was building a plant in The Federal Republic of Germany with a ca- pacity of 11,000–14,000 metric tons per year of naphthalenesulfonic acid derivatives (alphabet acids).

1-Naphthalenesulfonic Acid.   The sulfonation of naphthalene with excess 96 wt % sulfuric acid at <80°C gives >85 wt % 1-naphthalenesulfonic acid; the balance is mainly the 2-isomer. An older German commercial process is based on the reaction of naphthalene with 96 wt % sulfuric acid at 20–50°C (12). The product can be used unpurified to make dyestuff intermediates by nitration or can be sulfonated further. The sodium salt of 1-naphthalenesulfonic acid is required, for example, for the con- version to 1-naphthalenol (1-naphthol) by caustic fusion. In this case, the excess sul- furic acid first is separated by the addition of lime and is filtered to remove the in- soluble calcium sulfate; the filtrate is treated with sodium carbonate to precipitate calcium carbonate and leave the sodium 1-naphthalenesulfonate [130-14-3] in solution. The dry salt then is recovered, eg, by spray drying the solution.

The older methods have been replaced by methods which require less, if any, excess sulfuric acid. For example, sulfonation of naphthalene can be carried out in tetrachloroethane solution with the stoichiometric amount of sulfur trioxide at no greater than 30°C, followed by separation of the precipitated 1-naphthalenesulfonic

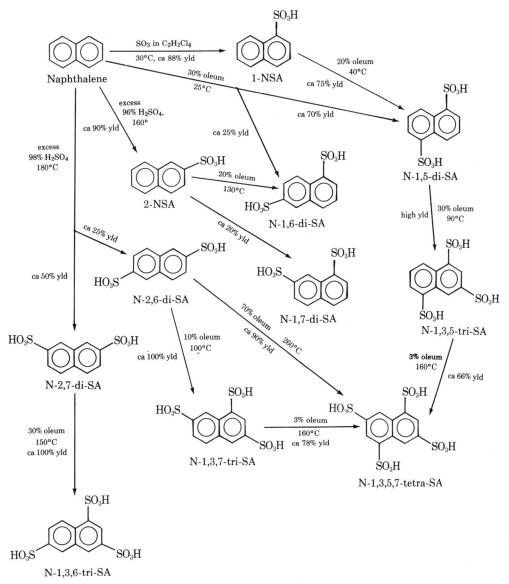

Figure 1. Selected paths to naphthalenesulfonic acids (N = naphthalene, SA = sulfonic acid, yld = yield).

acid; the filtrate can be reused as the solvent for the next batch (13). The purification of 1-naphthalenesulfonic acid by extracting or washing the cake with 2,6-dimethyl-4-heptanone (diisobutyl ketone) or a C_{1-4} alcohol has been described (14–15). The selective insoluble salt formation of 1-naphthalenesulfonic acid in the sulfonation mixture with 2,3-dimethylaniline has been patented (16).

1-Naphthalenesulfonic acid can be converted to 1-naphthalenethiol [529-36-2] (4) by reduction of the related sulfonyl chloride; this product has some utility as a dye intermediate, and is converted, by reaction with alkyl isocyanates, to S-naphthyl-

Table 1.   Melting Points of Naphthalenesulfonic Acids

Compound	CAS Registry No.	Trivial name	mp, °C	mp of corresponding sulfonyl chloride, °C
1-naphthalenesulfonic acid	[85-47-2]	α-acid	139–140	68
1-naphthalenesulfonic acid dihydrate	[6036-48-2]		90	
2-naphthalenesulfonic acid	[120-18-3]	β-acid	139–140	76
2-naphthalenesulfonic acid hydrate	[76530-12-6]		124–125	
2-naphthalenesulfonic acid trihydrate	[17558-84-8]		83	
1,2-naphthalenedisulfonic acid	[25167-78-6]			160
1,3-naphthalenedisulfonic acid	[6094-26-4]			137.5
1,4-naphthalenedisulfonic acid	[46859-22-7]		240–245 dec	162
1,5-naphthalenedisulfonic acid	[81-04-9]	Armstrong acid	125 dec	183
1,6-naphthalenedisulfonic acid	[525-37-1]	Ewer-Pick acid		129
1,7-naphthalenedisulfonic acid	[5724-16-3]			123
2,6-naphthalenedisulfonic acid	[581-75-9]	Ebert-Merz β-acid	199 dec	228–229
2,7-naphthalenedisulfonic acid	[92-41-1]	Ebert-Merz α-acid		159.5
1,3,5-naphthalenetrisulfonic acid	[6654-64-4]			146
1,3,6-naphthalenetrisulfonic acid	[86-66-8]			194–197
1,3,7-naphthalenetrisulfonic acid	[85-49-4]			165–166
1,4,5-naphthalenetrisulfonic acid	[60913-37-3]			156–157
1,3,5,7-naphthalenetetrasulfonic acid	[6654-67-7]			261–262

(4)

N-alkylthiocarbamates which have pesticidal and herbicidal activity (17) (see Herbicides).

2-Naphthalenesulfonic Acid.   The standard manufacture of 2-naphthalenesulfonic acid involves the batch reaction of naphthalene with 96 wt % sulfuric acid at ca 160°C for ca 2 h (12). The product contains the 1- and 2-isomers in ca a 15/85 ratio. Because of its faster rate of desulfonation, 1-naphthalenesulfonic acid can be hydrolyzed selectively by dilution of the charge with water and agitating for 1 h at 150°C; the naphthalene that is formed can be removed by steam distillation.

Sulfonation can be conducted with naphthalene/92 wt % H_2SO_4 in a 1:1.1 mol ratio with staged acid addition at 160°C over 2.5 h to give a 93% yield of the desired product (18). Continuous monosulfonation of naphthalene with 96 wt % sulfuric acid in a cascade reactor at ca 160°C gives 2-naphthalenesulfonic acid and small amounts of by-product naphthalenedisulfonic acids (19). The purification of 2-naphthalenesulfonic acid by hydrolysis of the 1-isomer can be done in continuous manner (20).

In the manufacture of 2-naphthalenol, 2-naphthalenesulfonic acid must be converted to its sodium salt; this can be done by adding sodium chloride to the acid, and by neutralizing with aqueous sodium hydroxide or neutralizing with the sodium sulfite by-product obtained in the caustic fusion of the sulfonate. The crude sulfonation product, without isolation or purification of 2-naphthalenesulfonic acid, is used to make 1,6-, 2,6-, and 2,7-naphthalenedisulfonic acids and 1,3,6-naphthalenetrisulfonic acid by further sulfonation; by nitration, 5- and 8-nitro-2-naphthalenesulfonic acids, [89-69-1] and [117-41-9], respectively, are obtained, which are intermediates for Cleve's acid. All are dye intermediates. The crude sulfonation product can be condensed with formaldehyde and/or alcohols or olefins to make valuable wetting, dispersing, and tanning agents.

1,5- and 1,6-Naphthalenedisulfonic Acids. 1,5 and 1,6-Naphthalenedisulfonic acids are co-products in the low temperature disulfonation of naphthalene. A typical process involves the sulfonation of naphthalene with an excess of 20–30 wt % oleum at not over 25°C (21). The sulfonation mass is diluted with water and the sodium salt of the disulfonic acid is formed by reaction with sodium sulfate. On cooling, the dihydrate of the disodium 1,5-naphthalenedisulfonate [76758-30-0] precipitates (ca 70% yield) and is recovered by filtration or centrifugation. The filtrate contains sodium 1,6-naphthalenedisulfonate [1655-43-2] which can be recovered by lime addition to precipitate sulfate as calcium sulfate and by filtration and evaporation of the filtrate. The sulfonation of naphthalene at −5 to 40°C in a chlorinated alkane solvent with SO_3 or first with chlorosulfonic acid, followed by sulfur trioxide, gives 1,5-naphthalenedisulfonic acid in excellent yields and purity (22).

2,6- and 2,7-Naphthalenedisulfonic Acids. The sulfonation of naphthalene with excess 98 wt % sulfuric acid at 135–180°C gives a mixture of 2,6- and 2,7-naphthalenedisulfonic acids; the mixture is diluted with water and converted to the sodium salt by the addition of NaCl and/or Na_2SO_4 (12). At first, sodium 2,6-naphthalenedisulfonate [1655-45-4] precipitates and is recovered by filtration in ca 25% yield; further salt addition and cooling precipitates sodium 2,7-naphthalenedisulfonate [1655-35-2] in ca 50% yield.

A naphthalene sulfonation product that is rich in the 2,6-isomer and low in sulfuric acid is formed by the reaction of naphthalene with excess sulfuric acid at 125°C and by passing the resultant solution through a continuous wiped-film evaporator at 245°C at 400 Pa (3 mm Hg) (23). The separation in high yield of 99% pure 2,6-naphthalenedisulfonate as its anilinium salt from a crude sulfonation product has been claimed (24).

2,7-Naphthalenedisulfonic acid is partially isomerized in sulfuric acid at 160°C to 2,6-naphthalenedisulfonic acid. The reaction takes place by a desulfonation–resulfonation mechanism.

1,3,5- and 1,3,6-Naphthalenetrisulfonic Acids. The sulfonation of 1,5-naphthalenedisulfonic acid with oleum at 90°C gives 1,3,5-naphthalenetrisulfonic acid in good yield (25). 1,3,6-Naphthalenetrisulfonic acid can be made by sulfonation of 1,6- or 2,7-naphthalenedisulfonic acid but this is a fairly costly process (26). A more acceptable manufacturing method involves the time–temperature–acid concentration-programmed sulfonation of naphthalene with sulfuric acid and oleum (27).

Alkylnaphthalenesulfonic Acids and Naphthalenesulfonic Acid–Formaldehyde Condensates. The alkylnaphthalenesulfonic acids can be made by sulfonation of alkylnaphthalenes, eg, with sulfuric acid at 160°C, or by alkylation of naphthalenesulfonic acids with alcohols or olefins. These products, as the acids or their sodium salts, are commercially important as textile auxiliaries, surfactants, wetting agents, dispersants, and emulsifying aids, eg, for dyes, wettable powder pesticides, tars, clays, and hydrotropes (see Surfactants and detersive systems). A wide variety of products can be made, eg, naphthalenemono- or -disulfonic acid (mostly as the sodium salts) with one or two methyl, isopropyl, butyl, or nonyl groups. In 1977, the U.S. production of such naphthalenesulfates was 5,100 metric tons with an average unit value of $1.41/kg; 820 t of sodium diisopropylnaphthalenesulfonate [sodium di-(1-methylethyl)naphthalenesulfonate] [1322-93-6] was produced with a unit value of $2.01/kg (28).

The sodium salts of the condensation products of naphthalenesulfonic acid with formaldehyde constitute the most important class of synthetic tanning agents for hides (see Leather). In 1977, U.S. production of these materials was 16,000 t with a unit value of $0.95/kg (28). The naphthalenesulfonic acid–formaldehyde condensation products are also used as plasticizers (qv) for concrete (29).

Hydronaphthalenesulfonic Acids. Sodium tetralinsulfonate [37837-69-7] (sodium 1,2,3,4-tetrahydronaphthalenesulfonate) is marketed by DuPont under the trade name Alkanol S as a dispersing and solubilizing agent. Poly(dihydronaphthalene)sulfonates have been proposed for use as ion-exchange (qv) resins and antistatic agents (qv) for thermoplastics (30).

Nitronaphthalenes and Nitronaphthalenesulfonic Acids

The nitro group does not undergo migration on the naphthalene ring during the usual nitration procedures. Therefore, mono- and polynitration of naphthalene is similar to low temperature sulfonation. The first nitro group enters the 1-position and a second nitro group tends to enter the 8-position to give 1,8- and 1,5-dinitronaphthalene in a 60/40 ratio. The nitronaphthalenes and some of their physical properties are listed in Table 2. Many of these compounds are not accessible by direct nitration of naphthalene but are made by indirect methods, eg, nitrite displacement of diazonium halide groups in the presence of a copper catalyst, decarboxylation of nitronaphthalenecarboxylic acids, or deamination of nitronaphthalene amines.

1-Nitronaphthalene. 1-Nitronaphthalene is manufactured by nitrating naphthalene with nitric and sulfuric acids at ca 40–50°C (31). The product is obtained in very high yield and contains ca 3–10 wt % 2-nitronaphthalene and traces of dinitronaphthalene; the product can be purified by recrystallization from alcohol. 1-Nitronaphthalene is important for the manufacture of 1-naphthaleneamine.

2-Nitronaphthalene is metabolized to the carcinogenic 2-naphthylamine in the human body (32). Respirators, protective clothing, proper engineering controls, and medical monitoring programs for workers involved in making by-product 2-nitronaphthalene should be used.

1,5- and 1,8-Dinitronaphthalenes. 1,5- and 1,8-dinitronaphthalenes (see (5) and (6), respectively) can be made by nitration of 1-nitronaphthalene in ca a 40/60 ratio. Similar results are obtained by the direct dinitration of naphthalene with mixed acid at 40–80°C and separation of isomers by fractional crystallization from ethylene dichloride (12).

1,5-Dinitronaphthalene is the starting material for 1,5-naphthalenediisocyanate [3173-72-6]. 1,8-Dinitronaphthalene is reduced to 1,8-diaminonaphthalene, which is an intermediate for making phthaloperinone (7), an orange colorant for plastics.

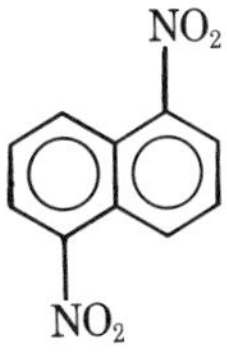

(5)

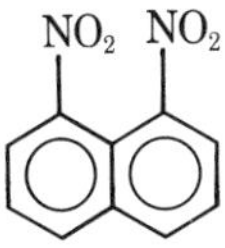

(6)

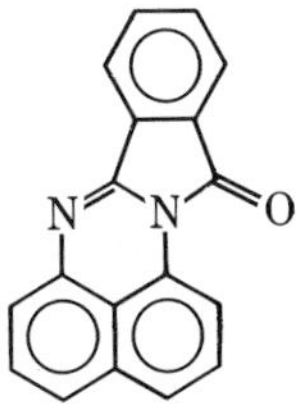

(7) [6925-69-5]

Table 2. Melting Point of Nitronaphthalenes

Compound	CAS Registry No.	mp, °C
1-nitronaphthalene	[86-57-7]	52[a]; 57.8[b]
2-nitronaphthalene[c]	[581-89-5]	78.7[d]
1,2-dinitronaphthalene[c]	[24934-47-2]	161–162
1,3-dinitronaphthalene[c]	[606-37-1]	148
1,4-dinitronaphthalene[c]	[6921-26-2]	134
1,5-dinitronaphthalene	[605-71-0]	219
1,6-dinitronaphthalene[c]	[607-46-5]	166.5[e]
1,7-dinitronaphthalene[c]	[24824-25-7]	156
1,8-dinitronaphthalene	[602-38-0]	172
2,3-dinitronaphthalene[c]	[1875-63-4]	174.5–175
2,6-dinitronaphthalene[c]	[24824-26-8]	279
2,7-dinitronaphthalene[c]	[24824-27-9]	234
1,2,3-trinitronaphthalene[c]	[76530-13-7]	190
1,2,4-trinitronaphthalene[c]	[76530-14-8]	258
1,3,5-trinitronaphthalene	[2243-94-9]	122
1,3,6-trinitronaphthalene[c]	[59054-75-0]	186
1,3,8-trinitronaphthalene	[2364-46-7]	218
1,4,5-trinitronaphthalene	[2243-95-0]	149
1,3,5,7-tetranitronaphthalene[c]	[60619-96-7]	260
1,3,5,8-tetranitronaphthalene	[2217-58-5]	194–195
1,3,6,8-tetranitronaphthalene	[28995-89-3]	203
1,4,5,8-tetranitronaphthalene	[4793-98-0]	340–345 dec

[a] Metastable form.

[b] bp 304°C (169°C at 1.6 kPa (12 mm Hg)).

[c] Made by indirect methods, not by the direct nitration of naphthalene or naphthalene-nitration products.

[d] bp 312.5°C at 97.8 kPa (733 mm Hg) and 165°C at 2.0 kPa (15 mm Hg).

[e] bp 370°C (235°C at 1.3 kPa (9.75 mm Hg)).

Recent process studies involve improvements in the separation of 1,5- and 1,8-dinitronaphthalenes by solvent extraction (33). The analysis of the mono- and dinitronaphthalenes can be done by glc.

Nitronaphthalenesulfonic Acids. Nitronaphthalenesulfonic acids can be obtained either by the sulfonation of 1-nitronaphthalene or by the nitration of 1- or 2-naphthalenesulfonic acid. Thus, the sulfonation of 1-nitronaphthalene with oleum at ca 25°C gives mainly 5-nitro-1-naphthalenesulfonic acid [17521-00-5]. The mononitration of 1-naphthalenesulfonic acid gives mainly 5- and 8-nitro-1-naphthalenesulfonic acid [117-41-9] and mononitration of 2-naphthalenesulfonic acid gives mainly 5-nitro-2-naphthalenesulfonic acid [86-69-1] and 8-nitro-2-naphthalenesulfonic acid [18425-74-6]. These compounds seldom are isolated; usually, the nitro group is reduced to the amino group to obtain dye intermediates.

Naphthaleneamines and Naphthalenediamines

Selected physical properties are listed in Table 3.

1-Naphthaleneamine. 1-Naphthaleneamine (1-naphthylamine; α-naphthylamine) can be made from 1-nitronaphthalene by reduction with iron–dilute HCl, or by catalytic hydrogenation. 1-Naphthaleneamine is toxic, LD_{50} (dogs) = 400 mg/kg, and a suspected human carcinogen, which mandate appropriate precautions in manufacture and use. 1-Naphthaleneamine is a dye intermediate and is used as the starting material in the manufacture of the rodenticide, Antu (8), 1-naphthalenethiourea; the rubber antioxidant, N-phenyl-1-naphthaleneamine (9) made by the condensation of 1-naphthaleneamine or 1-naphthalenol with aniline; the insecticide and miticide, Nissol (10); and the herbicide, Naptalam (11) Alanap or Dyanap, or N-1-naphthylphthalamic acid.

A tetrahydronaphthaleneamine derivative, 2-(5,6,7,8-tetrahydro-1-naphthaleneamine)-2-imidazoline (12), is used in the hydrochloride form as an adrenergic agent, ie, tramazoline hydrochloride, also called KB 227 and Rhinaspray.

Table 3. Physical Properties of Naphthalenamines and Naphthalenediamines

Compound	CAS Registry No.	mp, °C	Density	Other
1-naphthaleneamine	[134-32-7]	50	1.13_4^{14}	flash pt, 157°C; sol 0.496 g/L H_2O; vol with steam; bp 301°C (160°C at 1.6 kPa[a])
2-naphthaleneamine	[91-59-8]	111–113	1.061_4^{98}	sol hot water; vol with steam; bp 306°C (175.8°C at 2.7 kPa[a])
1,2-naphthalenediamine	[938-25-0]	96–98		sol hot water, alc, ether; bp at 0.01 kPa[a] 150–151°C
1,4-naphthalenediamine	[2243-61-0]	120		sl sol hot water
1,5-naphthalenediamine	[2243-62-1]	189.5		sol hot water, alc
1,6-naphthalenediamine	[2243-63-2]	78	$1.147_4^{99.4}$	sol hot water, alc
1,7-naphthalenediamine	[2243-64-3]	117.5		sol alc
1,8-naphthalenediamine	[479-27-6]	66.5	$1.127_4^{99.4}$	sol alc, ether; bp at 1.6 kPa[a] 205°C
2,3-naphthalenediamine	[771-97-1]	191		sol alc, ether
2,6-naphthalenediamine	[2243-67-6]	216–218		sparingly sol alc, ether
2,7-naphthalenediamine	[613-76-3]	159		

[a] To convert kPa to mm Hg, multiply by 7.5.

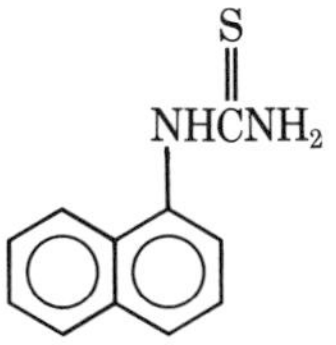

(8) [86-88-4] 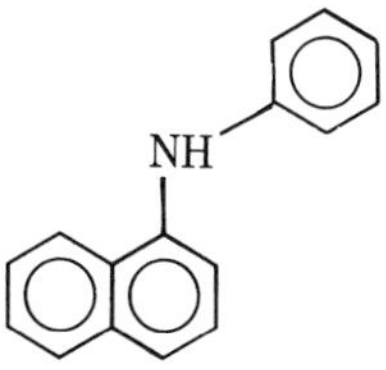   (9) [90-30-2] 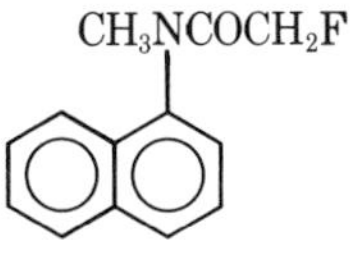   (10) [5903-13-9] 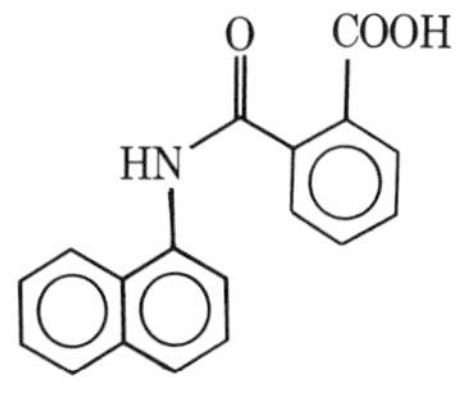   (11) [132-66-1]

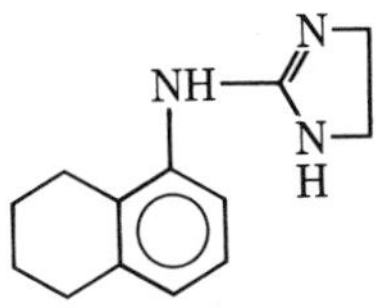

(12)

2-Naphthaleneamine. 2-Naphthaleneamine (2-naphthylamine, β-naphthylamine) has been recognized as a human carcinogen, producing bladder cancer on prolonged exposure. Thus, 2-naphthaleneamine as such is no longer commercially produced or used in the United States. Before its commercial demise, 2-naphthaleneamine was used in the production of dyes. An important derivative is the rubber antioxidant, N-phenyl-2-naphthaleneamine (PBNA) (13) which is made by the condensation of 2-naphthol and with aniline in the presence of an acid catalyst. In 1967, U.S. production of PBNA was 2500 t. N-Phenyl-2-naphthaleneamine is metabolized in the human body to 2-naphthaleneamine; NIOSH has published recommendations for working with this product (30).

1,5-Naphthalenediamine. 1,5-Naphthalenediamine (14) (1,5-naphthylenediamine) is manufactured by metal–acid, eg, Fe–CH$_3$COOH, reduction or by catalytic hydrogenation of 1,5-dinitronaphthalene (34). Aside from its possible use as an intermediate for azo dyes, 1,5-naphthalenediamine is used for the manufacture of 1,5-naphthalene diisocyanate (15) by the phosgenation route (see Isocyanates). This diisocyanate is used for making high grade urethane cast elastomers (see Urethane polymers).

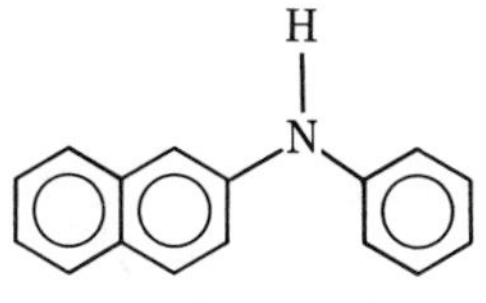

(13) [135-88-6]

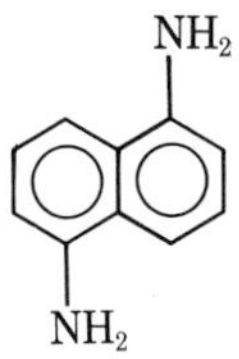

(14) 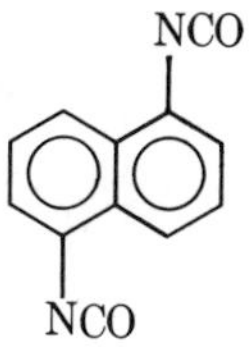   (15) [3173-72-6] 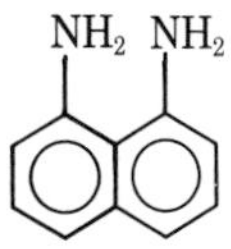   (16)

1,8-Naphthalenediamine. 1,8-Naphthalenediamine (**16**) (1,8-naphthylenediamine) is produced by metal–acid reduction or by catalytic hydrogenation of 1,8-dinitro-naphthalene (34). The most important use of 1,8-naphthalenediamine is for the manufacture, by condensation with phthalic anhydride, of phthaloperinone (**7**), an orange colorant for plastics used in automobile turn-signal and warning-light lenses.

Aminonaphthalenesulfonic Acids

Many aminonaphthalenesulfonic acids are important in the manufacture of azo dyes or are used to make intermediates for azo acid dyes, direct, and fiber-reactive dyes. Usually, the aminonaphthalenesulfonic acids are made by either the sulfonation of naphthalenamines, the nitration–reduction of naphthalenesulfonic acids, the Bucherer-type amination of naphtholsulfonic acids, or the desulfonation of an amino-naphthalenedi- or trisulfonic acid. Most of these processes produce by-products or mixtures which often are separated in subsequent purification steps. A summary of commercially important aminonaphthalenesulfonic acids is given in Table 4.

7-Amino-1,3-naphthalenedisulfonic acid (**17**) is now made by the Bucherer amination route (35). A mixture of dipotassium 7-hydroxy-1,3-naphthalenedisulfonate [842-18-2], excess aqueous ammonia, and ammonium sulfite is heated slowly in an autoclave to 185°C and is maintained at this temperature for ca 16 h. Aqueous (50 wt %) sodium hydroxide is added to the charge to liberate ammonia which is recovered for recycling. The charge is neutralized with a mineral acid. The yield of Amino G-salt is 97% of theoretical yield.

Another example of manufacture in this series is the sulfonation of an aminona-phthalenesulfonic acid, followed by selective desulfonation, to make 6-amino-1,3-naphthalenedisulfonic acid (**18**). Thus, 2-amino-1-naphthalenesulfonic acid made by amination of 2-hydroxy-1-naphthalenesulfonic acid is added to 20 wt % oleum at ca 35°C. At this temperature, 65 wt % oleum is added and the charge is stirred for 2 h, then is slowly heated to 100°C and is maintained for 12 h to produce 6-amino-1,3,5-naphthalenetrisulfonic acid. The mass is diluted with water and maintained for 3 h at 105°C to remove the sulfo group adjacent to the amino group. After cooling to ca 20°C and filtration, 6-amino-1,3-naphthalenedisulfonic acid is obtained in 80% yield (36).

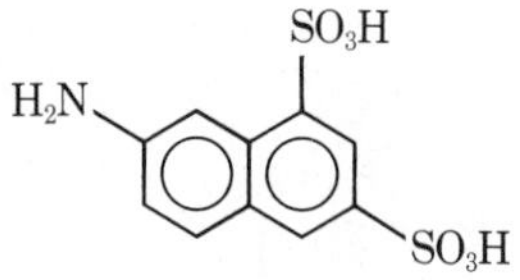

(**17**)

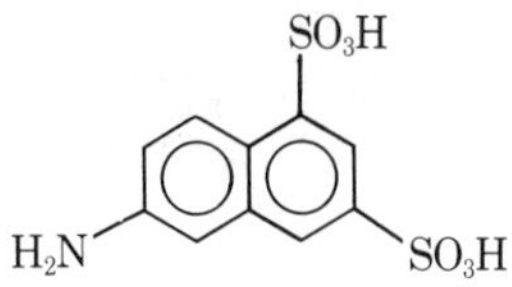

(**18**)

Table 4. Manufacture, Production, and Application Data for Selected Aminonaphthalenesulfonic Acids

Compound	Trivial name	CAS Registry No.	Manufacturing method[a-e]	U.S. production, metric tons	Intermediate for
1-amino-2-naphthalenesulfonic acid		[81-06-1]	a or e using naphthionic acid		azo dyes, eg, CI Direct Violet 11
4-amino-2-naphthalenesulfonic acid	Piria's acid; naphthionic acid	[134-54-3]	d		4-hydroxy-2-naphthalenesulfonic acid
4-amino-1-naphthalenesulfonic acid		[84-86-6]	a	80.7 (1969)	azo dyes, eg, CI Acid Red 88 and Acid Brown 14; 4-hydroxy-1-naphthalenesulfonic acid
5-amino-1-naphthalenesulfonic acid	Laurent's acid	[84-89-9]	a or b with Peri acid as the co-product		azo dyes, eg, CI Acid Black 24 and Mordant Brown 1; 5-hydroxy-1-naphthalenesulfonic acid; 5-amino-1,3-naphthalenedisulfonic acid; M-acid; 5-amino-1-naphthalenol
5-amino-2-naphthalenesulfonic acid	1,6-Cleve's acid	[119-79-9]	b with 1,7-Cleve's acid as the co-product	29 (1967)	azo dyes, eg, CI Direct Blue 120; 5-amino-8-acetamino-2-naphthalenesulfonic acid
8-amino-2-naphthalenesulfonic acid	1,7-Cleve's acid	[119-28-8]	b with 1,6-Cleve's acid	44 (1967)	azo dyes, eg, CI Direct Green 51; 8-amino-2-naphthalenol; 4-amino-1,6-naphthalenedisulfonic acid
5- and 8-amino-2-naphthalenesulfonic acids	Cleve's acid (mixed)		b	23 (1970)	azo dyes, eg, CI Direct Brown 62
8-amino-1-naphthalenesulfonic acid	Peri acid	[82-75-7]	b with Laurent's acid as the major by-product	84.8 (1968)	azo dyes, eg, CI Acid Black 35; 8-hydroxy-1-naphthalenesulfonic acid; 4-amino-1,5-naphthalenedisulfonic acid; 1,8-naphthosultam; 4-amino-1,3,5-naphthalenetrisulfonic acid
8-phenylamino-1-naphthalenesulfonic acid	Phenyl Peri acid	[82-76-8]	by condensation of aniline with Peri acid	58.5 (1970)	azo dyes, eg, CI Acid Blue 113
2-amino-1-naphthalenesulfonic acid	Tobias acid	[81-16-3]	c		pigments, eg, CI Pigment Red 49; 6-amino-1-naphthalenesulfonic acid; 6-amino-1,3-naphthalenedisulfonic acid
6-amino-1-naphthalenesulfonic acid	Dahl's acid	[81-05-0]	d using Tobias acid		azo dyes, eg, CI Acid Green 12, one of six listed; 6-hydroxy-1-naphthalenesulfonic acid

Table 4 (*continued*)

Compound	Trivial name	CAS Registry No.	Manufacturing method[a-e]	U.S. production, metric tons	Intermediate for
6-amino-2-naphthalenesulfonic acid	Broenner's acid	[93-00-5]	c	41.0 (1970)	azo dyes, eg, CI Direct Red 4
7-amino-2-naphthalenesulfonic acid	F-acid	[92-40-0]	c		azo dyes, eg, CI Direct Red 22
7-amino-1-naphthalenesulfonic acid	Badische acid	[86-60-2]	c		azo dyes, eg, CI Direct Green 33
1-amino-2,7-naphthalenedisulfonic acid	Kalle's acid	[486-54-4]	d		triphenylmethane dye
4-amino-2,7-naphthalenedisulfonic acid	1,3,6-Freund's acid	[6521-07-6]	b with 1,3,7-Freund's acid		azo dyes, eg, CI Acid Black 7; 5-hydroxy-2,7-naphthalenedisulfonic acid
4-amino-2,6-naphthalenedisulfonic acid	1,3,7-Freund's acid	[6362-05-6]	b 1,3,6-Freund's acid is co-product		azo dyes, eg, CI Direct Orange 49
8-amino-1,6-naphthalenedisulfonic acid	Amino-ε-acid	[129-91-9]	b		4-amino-2-naphthalenesulfonic acid; 8-hydroxy-1,6-naphthalenedisulfonic acid; 4-hydroxy-2-naphthalenesulfonic acid
4-amino-1,7-naphthalenedisulfonic acid	Dahl's acid II	[85-74-5]	a using naphthionic acid or 1,6-Cleve's acid		azo dyes, eg, CI Direct Orange 69
4-amino-1,6-naphthalenedisulfonic acid	Dahl's acid III	[85-75-6]	a as by-product of Dahl's acid II		azo dyes, eg, CI Direct Orange 49; 4-hydroxy-1,6-naphthalenedisulfonic acid
8-amino-1,5-naphthalenedisulfonic acid		[117-55-5]	a or b		8-hydroxy-1,5-naphthalenedisulfonic acid; 4-amino-5-hydroxy-1-naphthalenesulfonic acid
5-amino-1,3-naphthalenedisulfonic acid		[13306-42-8]	d		4-hydroxy-8-amino-2-naphthalenesulfonic acid
3-amino-2,7-naphthalenedisulfonic acid	Amino-R-acid	[92-28-4]	c		azo dyes, eg, CI Direct Orange 13; 6-hydroxy-7-amino-2-naphthalenesulfonic acid

Compound	Trivial name	CAS Registry Number	Method / remarks	US production, t (year)	Related products
3-amino-1,5-naphthalenedisulfonic acid	Cassella acid	[131-27-1]	b		azo dyes, eg, CI Direct Red 15; 7-hydroxy-1,5-naphthalenedisulfonic acid
6-amino-1,3-naphthalenedisulfonic acid	Amino J-acid	[118-33-2]	d	225 (1970)	J acid (3-hydroxy-6-amino-2-naphthalenesulfonic acid)
7-amino-1,3-naphthalenedisulfonic acid	Amino G-acid	[86-65-7]	c	171 (1970)	azo dyes, eg, CI Direct Orange 74; γ-acid (4-hydroxy-6-amino-2-naphthalenesulfonic acid)
4-amino-1,3,5-naphthalenetrisulfonic acid (as the sultam)	Chicago acid	[76530-15-9]	a		Chicago acid (4-amino-5-hydroxy-1,3-naphthalenedisulfonic acid)
8-amino-1,3,6-naphthalenetrisulfonic acid	Koch's acid	[117-42-0]	b		H-acid; chromotropic acid; 8-hydroxy-1,3,6-naphthalenetrisulfonic acid
8-amino-1,3,5-naphthalenetrisulfonic acid	B-acid	[17894-99-4]	b		K-acid (4-amino-5-hydroxy-1,7-naphthalenedisulfonic acid)
6-amino-1,3,5-naphthalenetrisulfonic acid		[55524-84-0]	a		6-amino-1,3-naphthalenedisulfonic acid; 2-amino-5-hydroxy-1,7-naphthalenedisulfonic acid
7-amino-1,3,6-naphthalenetrisulfonic acid	2R amino acid	[118-03-6]	a		3-amino-5-hydroxy-2,7-naphthalenedisulfonic acid
6,8-di(phenylamino)-1-naphthalenesulfonic acid	Diphenyl-ε-acid	[129-93-1]	from aniline and 8-amino-1,6-naphthalenedisulfonic acid		safranine dyes, eg, CI Acid Blue 61

[a] By sulfonation of the appropriate naphthaleneamine or aminonaphthalenesulfonic acid.
[b] By nitration/reduction of the appropriate naphthalene(poly)sulfonic acid.
[c] By amination of the appropriate hydroxynaphthalenesulfonic acid.
[d] By the desulfonation of an aminonaphthalenedi- or trisulfonic acid.
[e] By rearrangement of another aminonaphthalenesulfonic acid.

Naphthalenols and Naphthalenediols

Naphthalenols, naphthalenediols, and their sulfonated derivatives are important intermediates for dyes, agricultural chemicals, drugs, perfumes (qv), and surfactants. The methods of manufacture include sulfonation–caustic fusion of naphthalene, hydrolysis of 1-naphthaleneamine, oxidation–aromatization of tetralin, and hydroperoxidation of 2-isopropylnaphthalene [2027-17-0]. As the toxic hazard of the 1-naphthaleneamine was recognized, it's commercial use was minimized. The sulfonation–caustic fusion process is more difficult to operate than in the past because of increasing difficulties posed by product purity requirements, high investment and replacement costs, and by-product–effluent handling problems. In the United States, the naphthalenols are made by hydrocarbon oxidation routes.

The chemical properties of the naphthalenols are similar to those of phenol and resorcinol, with added reactivity and complexity of substitution because of the condensed ring system (see Hydroquinone, resorcinol, and catechol). Some of the naphthols and naphthalenediols are listed with some of their physical properties in Table 5.

1-Naphthalenol. 1-Naphthalenol (1-naphthol, α-naphthol: 1-hydroxynaphthalene) forms colorless crystals which tend to become colored on exposure to air or light. Of the three commercial methods of manufacture, the pressure hydrolysis of 1-naphthaleneamine with aqueous sulfuric acid at 180°C has been abandoned, at least in the United States. The caustic fusion of sodium 1-naphthalenesulfonate with 50 wt % aqueous sodium hydroxide at ca 290°C followed by neutralization gives 1-naphthalenol in ca 90% yield. However, the starting material is not readily available

Table 5. Properties of Naphthalenols and Naphthalenediols

Compound	CAS Registry No.	mp, °C	Density	Other
1-naphthalenol	[90-15-3]	95.8–96.0	1.224_4^4 1.099_4^{99}	sublimes; sol 0.03 g/100 mL H_2O at 25°C; readily sol alc, ether, benzene; bp 280°C (158°C at 2.6 kPa[a])
2-naphthalenol	[135-19-3]	122	1.078_4^{130} 1.22_4^{25}	sublimes; sol 0.075 g/100 ml H_2O at 25°C; readily sol alc, ether, benzene; flash pt 161°C; bp 295°C (161.8°C at 2.6 kPa[a])
1,2-naphthalenediol	[574-00-5]	103–104		
1,3-naphthalenediol	[132-86-5]	124		
1,4-naphthalenediol	[571-60-8]	195		heat of combustion 4.77 MJ[b]
1,5-naphthalenediol	[83-56-7]	258		sublimes; sparingly sol water; readily sol ether, acetone
1,6-naphthalenediol	[575-44-0]	137–138		
1,7-naphthalenediol	[575-38-2]	181		
1,8-naphthalenediol	[569-42-6]	144		
2,3-naphthalenediol	[92-44-4]	159		
2,6-naphthalenediol	[581-43-1]	222		
2,7-naphthalenediol	[582-17-2]	194		sol boiling water

[a] To convert kPa to mm Hg, multiply by 7.5.

[b] To convert MJ to kcal, divide by 4.184×10^{-3}.

in a purity adequate to make the high purity 1-naphthalenol that is required for the synthesis of the insecticide, 1-naphthyl methylcarbamate (Sevin) (**19**). Therefore, 1-naphthalenol is made in the United States by Union Carbide Corp. by the oxidation of tetralin [*119-64-2*] (1,2,3,4-tetrahydronaphthalene) in the presence of a transition-metal catalyst, presumably to 1-tetralol/1-tetralone by way of the 1-hydroperoxide, and dehydrogenation of the intermediate ie, 1-tetralol to 1-tetralone and aromatization of 1-tetralone to 1-naphthalenol, using a noble-metal catalyst (37). U.S. production of 1-naphthalenol is ca 35,000 t/yr (28).

The manufacturer's literature on 1-naphthalenol describes it as a moderately toxic and highly irritating chemical. The dust causes eye injury and skin irritation. Molten 1-naphthalenol is available in tank wagons, as a cast solid in drums, or as a powder. The price for the molten material in 1979 was $3.09/kg.

Several agricultural chemicals are made from 1-naphthalenol, ie, Sevin (**19**) (1-naphthyl methylcarbamate, Carbaryl) (38); Stauffer's Devrinol (**20**) (2-(1-naphthoxy)-*N*,*N*-diethylpropionamide, napropamide); and 1-naphthoxyacetic acid (**21**).

Several drugs are derived from 1-naphthalenol; the magnesium salt of 3-(4-methoxy-1-naphthoyl)propionic acid (**22**), ie, Hepalande, is used as a choleretic; propranolol (**23**) (Inderal; 1-isopropylamino-3-(1-naphthoxy)-2-propanol) is an important adrenergic blocking agent used in the treatment of angina and cardiac arrhythmias; and 1-naphthyl salicylate (**24**) (Alphol) has been used as an antiseptic and antirheumatic (see Cardiovascular agents; Disinfectants).

1-Naphthalenol also is used in the preparation of azo, indigoid, and nitro, eg, 2,4-dinitro-1-naphthol, dyes, and in making dye intermediates, eg, naphtholsulfonic acids, 4-chloro-1-naphthalenol, and 1-hydroxy-2-naphthoic acid. 1-Naphthalenol is an antioxidant for gasoline and some of its alkylated derivatives are stabilizers for plastics and rubber (39).

(**19**) [*63-25-2*]   (**20**) [*16299-99-7*]   (**21**) [*2976-75-2*]

(**22**) [*16643-66-6*]   (**23**) [*3506-09-0*]   (**24**) [*550-97-0*]

2-Naphthalenol. 2-Naphthalenol (2-naphthol; β-naphthol; 2-hydroxynaphthalene) forms colorless crystals of characteristic, phenolic odor which darken on exposure to air or light. The older method of manufacture was by fusion of sodium 2-

naphthalenesulfonate with sodium hydroxide at ca 325°C, acidification of the drowned fusion mass, which was quenched in water, isolation and water-washing of the 2-naphthalenol, and vacuum distillation and flaking of the product; a continuous process of this type has been patented (40). The newer method of manufacture involves the oxidation of 2-isopropylnaphthalene in the presence of a few percent of 2-isopropyl-naphthalene hydroperoxide [6682-22-0] as the initiator, some alkali, and perhaps a transition-metal catalyst, with oxygen or air at ca 90–110°C to ca 20–40% conversion to the hydroperoxide; the oxidation product is cleaved, using a small amount of ca 50 wt % sulfuric acid as the catalyst at ca 60°C to give 2-naphthalenol and acetone in high yield (41). The yields of both 2-naphthalenol and acetone from the hydroperoxide are 90% or better.

A process variation of the extraction of 2-isopropylnaphthalene hydroperoxide from the crude oxidation product with an alkylene glycol has been patented (42). The 2-naphthalenol plant of American Cyanamid, which is using the hydroperoxidation process, has a 13,600 t/yr capacity (43). The product is shipped molten ($2.23/kg) and flaked ($2.27/kg) (1979 manufacturer's prices). The manufacturer's literature gives the following toxicity information for 2-naphthalenol: single oral dose LD_{50} (young male albino rats) 4 g/kg; 24-h closely clipped skin contact LD_{50} (male albino rats) >10 g/kg. Based on plant experience, the product is not a primary irritant and lacks any appreciable capacity as a skin sensitizer. In ordinary use, 2-naphthalenol should not present a health hazard.

The major uses for 2-naphthalenol are in the dyes and pigments industries, eg, as a coupling component for azo dyes, and to make important intermediates, eg, 3-hydroxy-2-naphthalenecarboxylic acid (25), 2-naphtholsulfonic acids, aminonaph-tholsulfonic acids, and 1-nitroso-2-naphthol (26). 2-Naphthalenolsulfonic acid–formaldehyde condensates are used in tanning agents. Other uses of 2-naphthalenol are in the manufacture of perfuming agents, eg, 2-naphthyl methyl ether (27) (Yara Yara) and 2-naphthyl ethyl ether (27) (nerolin, Bromelia); an antioxidant for polyolefins, ie, thio-1,1-bis(2-naphthol) (28); the intestinal antiseptic, 2-naphthyl lactate (29) (Lactol, Lactonaphthol); a gastrointestinal and genitourinary anti-infective, ie, 2-naphthyl salicylate (30) (Betol, Salinaphthol); a semisynthetic penicillin, sodium

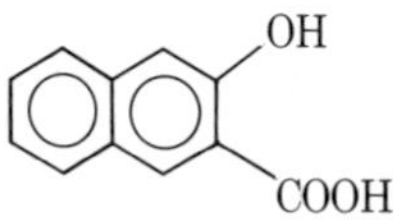

(25)

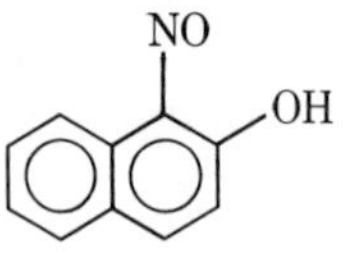

(26) [131-91-9]

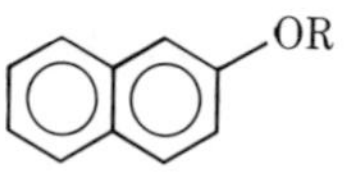

(27) R = CH₃ [93-04-9];
      R = C₂H₅ [93-18-5]

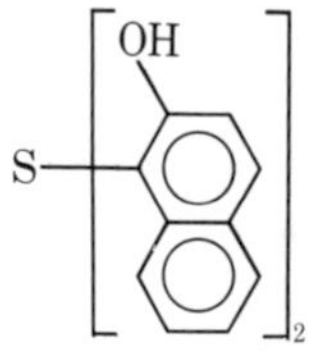

(28) [17096-15-0]

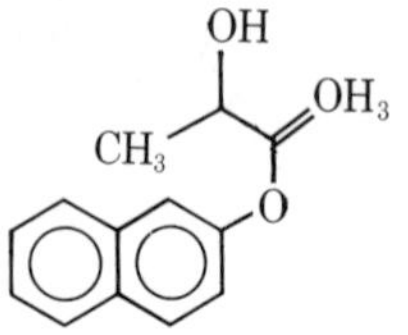

(29) [93-43-6]

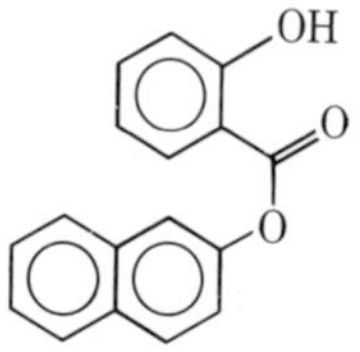

(30) [613-78-5]

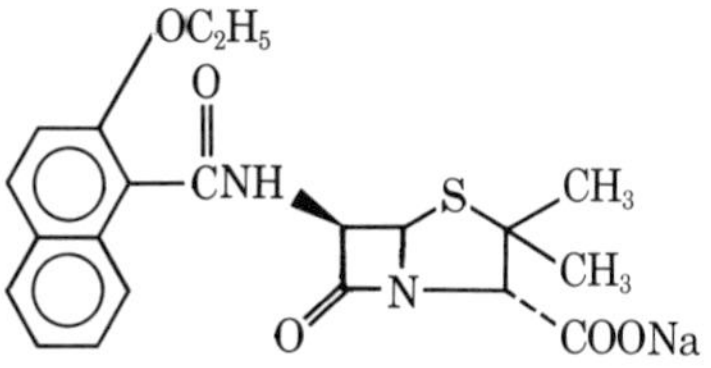

(31) [985-16-0]

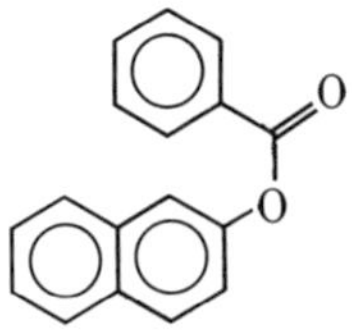

(32) [93-44-7]

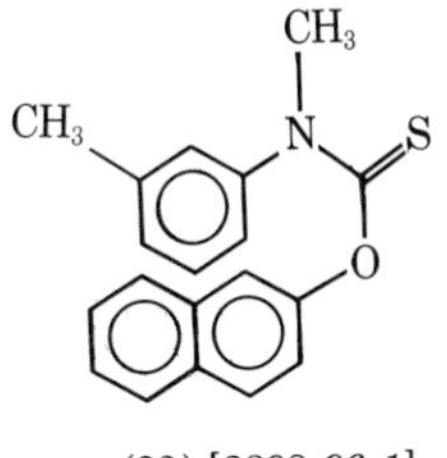

(33) [2398-96-1]

6-(2-ethoxy-1-naphthamido)penicillanate (**31**) (nafcillin sodium, Naptopen, Unipen); an intestinal antiseptic 2-naphthyl benzoate (**32**) (Lintrin, Haertolan); and a topical antifungal agent tolnaftate (**33**) (*m,N*-dimethylthiocarbanilic acid *O*-naphthyl ester, Tinactin, Focussan, Sporiline, Tinaderm) (see Antioxidants and antiozonants).

1,4-Naphthalenediol. 1,4-Naphthalenediol (**34**) can be prepared by the chemical or catalytic reduction of 1,4-naphthoquinone. This diol and the quinone are of interest because of their relation to the vitamin K family (see Vitamins).

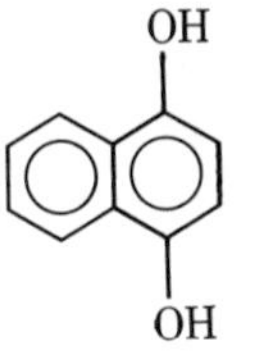

(34)

1,5-Naphthalenediol. 1,5-Naphthalenediol (1,5-dihydroxynaphthalene; Azurol) is a colorless material which darkens on exposure to air. It is manufactured by the fusion of disodium 1,5-naphthalenedisulfonate with sodium hydroxide at ca 320°C in high yield. 1,5-Naphthalenediol is an important coupling component, giving *ortho*-azo dyes which form complexes with chromium. The metallized dyes produce fast black shades on wool. 1,5-Naphthalenediol can be aminated with ammonia under pressure to give 1,5-naphthalenediamine.

1,8-Naphthalenediol. 1,8-Naphthalenediol darkens rapidly in air. It can be made by the fusion of the sultone of 8-hydroxy-1-naphthalenesulfonic acid with 50 wt % sodium hydroxide at 200–230°C or by the hydrolytic desulfonation of 1,8-dihydroxy-4-naphthalenesulfonic acid. The diol also reacts with ammonia to give 1,8-naphthalenediamine.

2,3-Naphthalenediol. 2,3-Naphthalenediol is made by the hydrolytic desulfonation of 2,3-naphthalenediol-6-sulfonic acid at ca 180°C. It is used as a coupler forming azo dyes which are applied in reprographic processes.

2,7-Naphthalenediol.   2,7-Naphthalenediol is made by the fusion of sodium 2,7-naphthalenedisulfonate with molten sodium hydroxide at ca 280–300°C in ca 80% yield. A formaldehyde resin prepared from this diol has excellent erosion resistance, strength, and chemical inertness; it is used as an ablative material in rocket-exhaust environments (44) (see Ablative materials).

Hydroxynaphthalenesulfonic Acids

Hydroxynaphthalenesulfonic acids are important as intermediates either for coupling components for azo dyes or azo components and for synthetic tanning agents. Hydroxynaphthalenesulfonic acids can be manufactured either by sulfonation of naphthols or hydroxynaphthalenesulfonic acids, by acid hydrolysis of aminonaphthalenesulfonic acids, by fusion of sodium naphthalenepolysulfonates with sodium hydroxide, or by desulfonation or rearrangement of hydroxynaphthalenesulfonic acids. In Table 6, a summary of 19 products in this class is given.

In the production of sodium 3-hydroxy-2,7-naphthalenedisulfonate (**35**) (R-salt), 2-naphthol is stirred with excess 98 wt % sulfuric acid at 60°C, sodium sulfate is added, and the mixture is stirred and heated for 36 h at 105–122°C (45). The charge is diluted with water and salted out with ca 15 wt % sodium chloride at 60°C to give R-salt in 68% yield.

Another example of manufacture of a hydroxynaphthalenesulfonic acid is a caustic fusion process to make 7-hydroxy-2-naphthalenesulfonic acid (**36**) (46). Sodium 2,7-naphthalenedisulfonate, which is made by mixing the 40 wt % disulfonic acid paste with ca 70 wt % caustic, is fused with excess sodium hydroxide in an agitated autoclave at 230–265°C for 10 h. The charge is drowned in water, brought to pH 8 with hydrochloric acid, diluted with water, boiled and treated with carbon, and filtered hot. The product is isolated by filtration after cooling to 30°C. Additional product can be obtained by adding sodium chloride to the filtrate to give a combined yield of 90% of the theory of sodium 7-hydroxy-2-naphthalenesulfonate [135-55-7].

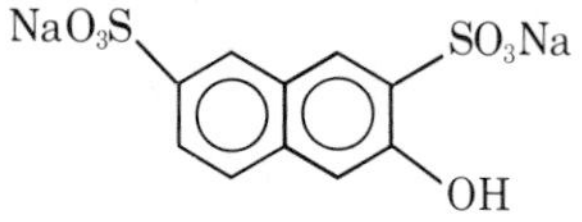

(**35**) [135-59-3]

(**36**)

Aminonaphthols and Aminonaphtholsulfonic Acids

The aminonaphthols are of minor use but the aminohydroxynaphthalenesulfonic acids are intermediates for dyes, eg, fiber-reactive azo dyes and plain and metallized azo dyes. In Table 7, a selection of these products is presented.

4-Hydroxy-6-amino-2-naphthalenesulfonic acid (**37**) is manufactured as follows (44): An aqueous solution of the alkali salt of 7-amino-1,3-naphthalenedisulfonic acid

(Amino G-acid) is added slowly, ie, during ca 16 h, to 70 wt % sodium hydroxide at 190°C; the reactor temperature is maintained at 183–186°C. The charge is held at this temperature for another 2 h, is cooled slowly, and is diluted with water and acidified with dilute sulfuric acid. 4-Hydroxy-6-amino-2-naphthalenesulfonic acid is obtained in ca 85% of the theoretical yield.

The manufacture of 3-hydroxy-4-amino-1-naphthalenesulfonic acid (**38**) involves the nitrosation of 2-naphthalenol, bisulfite addition, and reduction of the nitroso to the amino group by sulfur dioxide generated *in situ* (47). 3-Hydroxy-4-amino-1-naphthalenesulfonic acid is obtained in 80% yield.

A number of *N*-acyl-, *N*-alkyl-, and *N*-arylaminonaphthalenolsulfonic acids are used as couplers for azo dyes. Examples of such intermediates are shown in Table 8.

Naphthalenecarboxylic Acids

Physical properties for naphthalenemono, di-, tri-, and tetracarboxylic acids are summarized in Table 9. Sketches of individual compounds are restricted to naphthalenecarboxylic acids of use or prolonged commercial interest. Most of the other naphthalenedi- or polycarboxylic acids have been made by simple routes, eg, the oxidation of the appropriate di- or polymethylnaphthalenes, or by complex routes, eg, the Sandmeyer reaction of the selected aminonaphthalenesulfonic acid to give a cyanonaphthalenesulfonic acid followed by fusion of the latter with an alkali cyanide with simultaneous or subsequent hydrolysis of the nitrile groups.

1- and 2-Naphthalenecarboxylic Acids. The acids are prepared readily by the oxidation of 1- or 2-alkylnaphthalenes with dilute nitric acid, chromic acid, or permanganate. Recently, the oxygen or air oxidation of alkylnaphthalenes in an alkanoic acid solvent in the presence of a Co- or Mn-containing catalyst and a Br-containing catalyst gives good results (48). The direct carboxylation of naphthalene with CO and oxygen in the presence of a Pd-carboxylate catalyst also has been patented (49).

4-Alkyl-*N*,*N*-dialkyl-1-naphthalenecarboxamides are useful herbicides (qv) (50) and the 2,2-dimethylhydrazide of 1-naphthalenecarboxylic acid has been patented as a plant growth regulator (51) (see Plant growth substances). 2-Propynyl-2-naphthalenecarboxylate [*53548-27-9*] and similar esters are insecticides (52) (see Insect control technology).

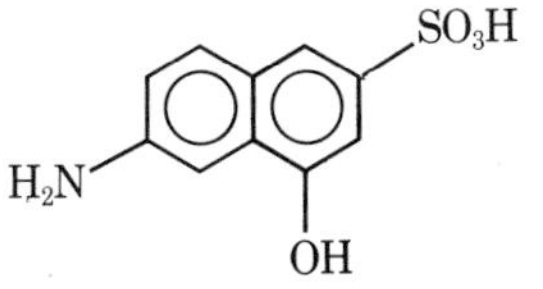

(**37**)

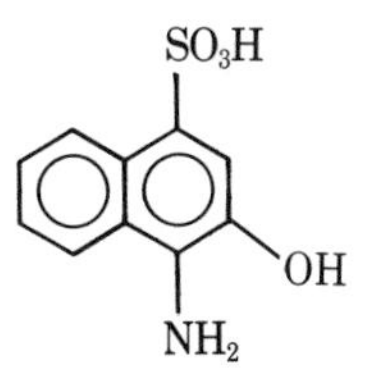

(**38**)

Table 6. Manufacture, Production, and Application Data for Selected Hydroxynaphthalenesulfonic Acids

Compound	Trivial name	CAS Registry No.	Manufacturing method[a-f]	U.S. production[g], kg/yr	Intermediate for
4-hydroxy-2-naphthalenesulfonic acid	Armstrong & Wynne's acid; 1,3-oxy-acid	[3771-14-0]	a,b		azo dyes, eg, CI Direct Blue 127
4-hydroxy-1-naphthalenesulfonic acid	Nevile-Winther acid; 1,4-oxy-acid	[84-87-7]	c,d		azo dyes, eg, CI Acid Red 14; 3-amino-4-hydroxy-1-naphthalenesulfonic acid; tanning agents
5-hydroxy-1-naphthalenesulfonic acid	L-acid	[117-59-9]	d,e		azo dyes and pigments, eg, CI Pigment Red 54, toner; 1,5-naphthalenediol
8-hydroxy-1-naphthalenesulfonic acid		[117-22-6]	f		metallized o,o'-dihydroxyazo dyes, eg, CI Acid Blue 58; 1,8-naphthalenediol
2-hydroxy-1-naphthalenesulfonic acid	oxy-Tobias acid	[567-47-5]	c		Tobias acid (2-amino-1-naphthalenesulfonic acid); J-acid (7-amino-4-hydroxy-2-naphthalenesulfonic acid)
6-hydroxy-2-naphthalenesulfonic acid	Schaeffer's acid	[93-01-6]	c	289,000[b] (1972)	azo dyes, eg, CI Acid Orange 12; synthetic tanning agents
7-hydroxy-2-naphthalenesulfonic acid	F-acid	[92-40-0]	e		azo dyes, eg, CI Direct Blue 128; 2,7-naphthalenediol; 7-amino-2-naphthalenesulfonic acid
7-hydroxy-1-naphthalenesulfonic acid	Crocein acid; Baeyer's acid	[132-57-0]	c with fractional precipitation of the sodium salt		azo dyes, eg, CI Acid Red 70
4,5-dihydroxy-1-naphthalenesulfonic acid	dioxy S acid	[83-65-8]	e		azo dyes, eg, CI Direct Blue 26
6,7-dihydroxy-2-naphthalenesulfonic acid	dioxy R acid	[92-27-3]	e		2,3-dihydroxynaphthalene
5-hydroxy-2,7-naphthalenedisulfonic acid	RG acid; violet acid	[578-85-8]	e		azo dyes, eg, CI Acid Red 99
8-hydroxy-1,6-naphthalenedisulfonic acid	ε-acid; Andresen's acid	[117-43-1]	f		azo dyes, eg, CI Direct Blue 98

738

Compound	Common name	CAS Registry Number		Production, t (yr)	Dyes and derivatives
4-hydroxy-1,6-naphthalenedisulfonic acid	Dahl's acid; D-acid	[6361-37-1]	a,d		nitro coloring matter, eg, CI Acid Yellow 1
4-hydroxy-1,5-naphthalenedisulfonic acid	Schoellkopf's acid; CS acid; δ acid	[82-75-7]	f		azo dyes, eg, CI Acid Blue 169; 4,5-dihydroxy-1-naphthalenesulfonic acid
3-hydroxy-2,7-naphthalenedisulfonic acid	R-acid	[148-75-4]	c	609,000[h] (1970)	azo dyes, eg, CI Acid Red 115, Acid Red 26; 3-amino-2,7-naphthalenedisulfonic acid; 6,7-dihydroxy-2-naphthalenesulfonic acid
7-hydroxy-1,3-naphthalenedisulfonic acid	G-acid	[118-32-1]	c	439,000 (1970)	azo dyes, eg, CI Acid Red 73; triphenylmethane dyes; 4,6-dihydroxy-2-naphthalenesulfonic acid
4,5-dihydroxy-2,7-naphthalenedisulfonic acid	chromotropic acid	[148-25-4]	a,e		azo dyes, eg, CI Acid Violet 3
8-hydroxy-1,3,6-naphthalenetrisulfonic acid	oxy-Koch's acid	[3316-02-7]	a		azo dyes, eg, CI Direct Blue 27; chromotropic acid (4,5-dihydroxy-2,7-naphthalenedisulfonic acid)
7-hydroxy-1,3,6-naphthalenetrisulfonic acid		[6259-66-1]	c		azo dyes, eg, CI Acid Red 41; 7-amino-1,3,6-naphthalenetrisulfonic acid

[a] By the hydrolysis of the corresponding aminonaphthalenesulfonic acid.
[b] By the desulfonation of 8-hydroxy-1,6-naphthalenedisulfonic acid.
[c] By the sulfonation of the appropriate (1- or 2-) naphthalenol.
[d] By the Bucherer reaction (with sulfite) of the appropriate aminonaphthalenesulfonic acid.
[e] By the alkali fusion or alkaline hydrolysis under pressure of the appropriate naphthalenedisulfonic or naphthalenetrisulfonic acid or hydroxynaphthalenedisulfonic acid.
[f] By the alkaline hydrolysis of the sultone formed on boiling an aqueous solution of the diazonium salt of 8-amino-1-naphthalenesulfonic acid or its appropriate derivatives.
[g] Ref. 28.
[h] As the sodium salt.

Table 7. Selected Aminonaphthalenols and Aminohydroxynaphthalenesulfonic Acids

Compound	Trivial name	CAS Registry No.	Manufacturing method[a-c]	U.S. production, t/yr	Intermediate for
5-amino-1-naphthalenol	Purpurol	[83-55-6]	a		azo dyes, eg, CI Acid Blue 70; sulfur dyes
7-amino-2-naphthalenol	Cyanol	[118-46-7]	a		azo dyes, eg, CI Mordant Brown 65
3-hydroxy-4-amino-1-naphthalenesulfonic acid	1,2,4-acid; Boeniger acid	[116-63-2]	b	26 (1970)	azo dyes, eg, CI Acid Red 186, Mordant Red 7; chrome complex dyes
5-amino-6-hydroxy-2-naphthalenesulfonic acid	Amino-Schaeffer acid	[5639-34-9]	c		photographic developer; rarely used for dyes
4-hydroxy-8-amino-2-naphthalenesulfonic acid	M acid	[489-78-1]	a		azo dyes, eg, CI Direct Green 42
4-hydroxy-7-amino-2-naphthalenesulfonic acid	J acid	[87-02-5]	a	272 (1969)	azo dyes, eg, CI Direct Blue 71, Direct Red 16; direct dyes using N-phenyl-J-acid and J-acid imide
4-hydroxy-6-amino-2-naphthalenesulfonic acid	γ-acid	[90-51-7]	a	202 (1969)	azo dyes, eg, CI Direct Black 22
4-amino-5-hydroxy-2,7-naphthalenedisulfonic acid	H acid	[90-20-0]	a		azo dyes, eg, CI Direct Black 19, Direct Blue 15
4-amino-5-hydroxy-1,3-naphthalenedisulfonic acid	Chicago acid; SS acid; 2S acid	[82-47-3]	a		azo dyes, eg, CI Acid Blue 42
4-amino-5-hydroxy-1,7-naphthalenedisulfonic acid	K acid	[130-23-4]	a		azo dyes, eg, Sulfon Acid Blue G, CI 13400
3-amino-5-hydroxy-2,7-naphthalenedisulfonic acid	RR acid; 2R acid	[90-40-4]	a		azo dyes, eg, CI Direct Brown 31

[a] By the alkali fusion or hydrolysis of the appropriate aminonaphthalenesulfonic acid.
[b] By the nitrosation of 2-naphthalenol and the reaction of the nitroso compound with sodium bisulfite.
[c] By nitrosation/reduction of 6-hydroxy-2-naphthalenesulfonic acid.

Tetrahydrozoline (**39**) (2-(1,2,3,4-tetrahydro-1-naphthyl)2-imidazolin; Tysine, Visine), a sympathomimetic and nasal decongestant, is made by the condensation of 1,2,3,4-tetrahydro-1-naphthoic acid or its methyl ester with 1,2-ethylenediamine.

1,8-Naphthalenedicarboxylic Acid. 1,8-Naphthalenedicarboxylic acid (naphthalic acid) readily dehydrates on heating to 1,8-naphthalenedicarboxylic acid anhydride, (**40**) (naphthalic anhydride). The anhydride and its imide (**41**) (R = H; naphthalimide) are intermediates for important dyes, pigments, optical bleaches, and biologically active compounds (see Bleaching agents).

The anhydride can be made by the liquid-phase oxidation of acenaphthene [83-32-9] with chromic acid in aqueous sulfuric acid or acetic acid (53). A postoxidation of the crude oxidation product with hydrogen peroxide or an alkali hypochlorite is advantageous (54). An alternative liquid-phase oxidation process involves the reaction of acenaphthene, molten or in alkanoic acid solvent, with oxygen or aid at ca 70–200°C in the presence of Mn resinate or stearate or Co and/or Mn salts and a bromide. An addition of an aliphatic anhydride accelerates the oxidation (55).

(**39**) [84-22-0]

(**40**) [81-84-5]   (**41**) [81-83-4]

The anhydride of 1,8-naphthalenedicarboxylic acid is obtained in ca 95–116 wt % yield by the vapor phase air-oxidation of acenaphthene at ca 330–450°C, using unsupported or supported vanadium oxide catalysts, with or without modifiers (56).

The anhydride of 1,8-naphthalenedicarboxylic acid has fungicidal properties (57) (see Fungicides). This anhydride has been commercially introduced, under the trade name, Protect, as a seed treatment (eg, for corn) to prevent injury to the seed by thiocarbamate herbicides.

4-Halogenated and 4,5-halogenated derivatives of 1,8-naphthalenedicarboxylic acid anhydride are useful intermediates for dyes, pigments, and fluorescent whiteners for polymers.

Imides of 1,8-naphthalenedicarboxylic acid are used as drugs; an anthelmintic for animals, eg, Naphthalophos; and as rodenticides (58–59) (see Poisons, economic; Veterinary drugs). Other imides are useful fluorescent whiteners for polyesters and acrylonitrile polymers (60). The imide of 1,8-naphthalenedicarboxylic acid gives, by oxidative alkali fusion, the diimide of 3,4,9,10-perylenetetracarboxylic acid, the parent compound of an important class of red dyes and pigments for plastics and coatings with high color fastness (see Brighteners, fluorescent). The diimides are useful elec-

Table 8. Selected N-Substituted Aminohydroxynaphthalenesulfonic Acids

Compound	Structure	Trivial name	CAS Registry No.	Intermediate for
7,7'-ureylene-bis-4-hydroxy-2-naphthalenesulfonic acid		J-acid urea	[137-47-4]	azo dyes, eg, CI Direct Orange 26
7-benzamido-4-hydroxy-2-naphthalene-sulfonic acid		N-benzoyl J-acid	[132-87-6]	azo dyes, eg, CI Direct Red 81
7-phenylamino-4-hydroxy-2-naphthalene-sulfonic acid		N-phenyl J-acid	[119-40-4]	azo dyes, eg, CI Direct Violet 7
7,7'-imino-bis-4-hydroxy-2-naphthalene-sulfonic acid		Di-J-acid; J-Acid Imide	[87-03-6]	azo dyes, eg, CI Direct Red 149
6-phenylamino-4-hydroxy-2-naphthalene-sulfonic acid		N-phenyl γ-acid	[119-40-4]	azo dyes, eg, CI Mordant Brown 40
4-acetamido-5-hydroxy-2,7-naphthalene-disulfonic acid		N-acetyl-H-acid	[134-34-9]	azo dyes, eg, CI Acid Violet 7
4-benzamido-5-hydroxy-1,7-naphthalene-disulfonic acid		N-benzoyl-K-acid	[6361-49-5]	azo dyes, eg, CI Acid Red 133

742

Table 9. Selected Properties of Naphthalenecarboxylic Acids

Compound	CAS Registry No.	mp, °C	Other
1-naphthalenecarboxylic acid	[86-55-5]	162	sol ethanol; sparingly sol water; $K_a = 2.04 \times 10^{-4}$ at 25°C; bp at 6.7 kPa[a] 231°C
2-naphthalenecarboxylic acid	[93-09-4]	184–185	sol ethanol, ether, chloroform; $K_a = 6.78 \times 10^{-5}$ at 25°C; bp >300°C
1,2-naphthalenedicarboxylic acid	[2088-87-1]	175 dec	sol ethanol, ether, acetic acid; mp anhydride 168–169°C
1,3-naphthalenedicarboxylic acid	[2089-93-2]	267–268	
1,4-naphthalenedicarboxylic acid	[605-70-9]	309	sol ethanol; insol boiling water
1,5-naphthalenedicarboxylic acid	[7315-96-0]	315–320 dec	insol common solvents
1,6-naphthalenedicarboxylic acid	[2089-87-4]	310	sol hot ethanol, acetic acid
1,7-naphthalenedicarboxylic acid	[2089-91-0]	308	sol common organic solvents
1,8-naphthalenedicarboxylic acid	[518-05-8]	on heating, converts to anhydride (mp 274°C)	sol warm ethanol; bp anhydride at 440 Pa[b] 215°C
2,3-naphthalenedicarboxylic acid	[2169-87-1]	239–241 dec	sol hot ethanol; mp anhydride 246°C
2,6-naphthalenedicarboxylic acid	[1141-38-4]	310–313 dec	sol aq alc
2,7-naphthalenedicarboxylic acid	[2089-89-6]	>300	sol ethanol
1,2,5-naphthalenetricarboxylic acid	[36439-99-3]	270–272	sol methanol
1,3,8-naphthalenetricarboxylic acid	[36440-24-1]		mp 1,8-anhydride 289–290°C
1,4,5-naphthalenetricarboxylic acid	[28445-09-2]	on heating, forms the anhydride (mp undefined)	mp 4,5-anhydride 274°C
1,2,4,5-naphthalenetetracarboxylic acid	[22246-61-3]	263	mp dianhydride 263°C
1,4,5,8-naphthalenetetracarboxylic acid	[128-97-2]	on heating, forms the anhydride	sol acetone; dianhydride sublimes >300°C

[a] To convert kPa to mm Hg, multiply by 7.5.
[b] To convert Pa to mm Hg, divide by 133.3.

trical conductors and semiconductors, eg, in solar cell systems (61). Some benzimidazole derivatives of 1,8-naphthalenedicarboxylic acid are excellent thickening agents for high temperature greases (62).

2,6-Naphthalenedicarboxylic Acid. 2,6-Naphthalenedicarboxylic acid (**42**) can be condensed with diols to give polyester fibers and plastics with superior properties. Potentially useful methods of manufacture for 2,6-naphthalenedicarboxylic acid are

(42)

based on a Henkel-type process involving either the isomerization under pressure of 1,8-naphthalenedicarboxylic acid at ca 400–500°C in the presence of a CdO catalyst and K_2CO_3 or the carboxylation–isomerization of a naphthalenecarboxylic acid in a similar system. Also, 2,6-naphthalenedicarboxylic acid can be prepared in high selectivity by the reaction of sodium 2-naphthalenecarboxylate with carbon monoxide, sodium carbonate, and sodium formate at ca 300°C and 2.1–4.8 MPa (300–700 psi) under CO_2 (63). Manufacturing methods for the diacid are reviewed in ref. 64.

Teijin Ltd. has produced polyester film with good elevated-temperature resistance based on 2,6-naphthalenedicarboxylic acid. The product is marketed for use in magnetic tape (qv) and electrical insulation applications (see Insulation, electric). Teijin's process for the diacid involves the carboxylation/isomerization of a naphthalenecarboxylic acid (65).

1,4,5,8-Naphthalenetetracarboxylic Acid. Traditionally, the tetracarboxylic acid (43) has been manufactured by oxidation of the coal-tar component, pyrene, eg, with chromic acid, or by a chlorination–oleum hydrolysis–oxidation sequence. Alternative processes start with acylation of acenaphthene in the 5,6 position followed by an oxidation of naphthalene ring substituents. For example, 5,6-acenaphthenedicarboximide (44) can be prepared in 84% yield by the reaction of acenaphthene with excess sodium cyanate in anhydrous HF (46). The intermediate can be oxidized to the tetracarboxylic acid.

The dianhydride of the acid (43) has been of research interest for the preparation of high temperature polymers, ie, polyimides (qv). The condensation of the dianhydride with o-phenylenediamines gives vat dyes and pigments of the benzimidazole type.

(43) [81-30-1]   (44)

Hydroxynaphthalenecarboxylic and Aminonaphthalenecarboxylic Acids

Some properties of selected hydroxynaphthalenecarboxylic acids are presented in Table 10. 3-Hydroxy-2-naphthalenecarboxylic acid is of commercial importance as a dye intermediate. The compounds of technical interest are prepared by the Kolbe-Schmitt reaction, ie, the carboxylation of alkali naphthoxides with CO_2. Less direct syntheses have to be resorted to for the other isomers, eg, the oxidation of hydroxyaldehydes or acylated naphthols, the Sandmeyer reaction of appropriately substituted naphthaleneamines, alkali fusion of sulfonated naphthalenecarboxylic acids, etc.

Table 10.   Selected Properties of Hydroxynaphthalenecarboxylic Acids

Position —OH	—COOH	CAS Registry No.	mp, °C	Other
2–	1–	[2283-08-1]	157–159	sparingly sol water; readily sol alc, benzene
3–	1–	[19700-42-6]	248–249	
4–	1–	[7474-97-7]	188–189	
5–	1–	[2437-16-3]	236	
6–	1–	[2437-17-4]	213	
7–	1–	[2623-37-2]	256–257	
8–	1–	[1769-88-6]	169	lactone, mp 108°C
1–	2–	[86-48-6]	200	0.55 wt % sol boiling water; sol alc, ether, benzene
3–	2–	[92-70-6]	222–223	0.1 wt % sol water at 25°C; sol alc, ether, benzene, chloroform
4–	2–	[1573-91-7]	225–226	
5–	2–	[2437-18-5]	215–216	
6–	2–	[16712-64-4]	245–248	
7–	2–	[613-17-2]	274–275	
8–	2–	[5776-28-3]	229	

2-Hydroxy-1-Naphthalenecarboxylic Acid.   2-Hydroxy-1-naphthalenecarboxylic acid (2-hydroxy-1-naphthoic acid) (45) is manufactured by a Kolbe-type process, ie, by reaction of the thoroughly dried potassium or sodium 2-naphthalenolate with CO_2 at ca 115–130°C in an autoclave at ca 300–460 kPa (3.0–4.5 atm) for 10–16 h. It decarboxylates readily, eg, in water starting at ca 50°C.

1-Hydroxy-2-Naphthalenecarboxylic Acid.   1-Hydroxy-2-naphthalenecarboxylic acid (1-hydroxy-2-naphthoic acid) (46) is made similarly to (45) by reaction of dry sodium 1-naphthalenolate with CO_2 in an autoclave at ca 125°C. It has been used in making triphenylmethane dyes and metallizable azo dyes. Alkylamides and arylamides of 1-hydroxy-2-naphthalenecarboxylic acid are cyan couplers, ie, components used in indoaniline dye formation in color films (see Color photography).

3-Hydroxy-2-Naphthalenecarboxylic Acid.   3-Hydroxy-2-naphthalenecarboxylic acid (3-hydroxy-2-naphthoic acid; β-oxynaphthoic acid; BON; BONA; Developer 8) (47) is the principal commercial product among the hydroxynaphthalenecarboxylic acids. It is manufactured by the Kolbe-type carboxylation of sodium 2-naphtholate (65). American Cyanamid's commercial product is a lemon to greenish yellow, odorless powder, mp 218°C min, purity 98% min. It is shipped in 36-kg paper bags or 102-kg fiber drums. The manufacturer's price (1979) is ca $5.18/kg (66).

The oral LD_{50} for young male albino rats of 3-hydroxy-2-naphthalenecarboxylic acid is ca 1.77 g/kg. The compound is irritating to rabbit eyes but not to rabbit skin.

The condensation of 3-hydroxy-2-naphthalenecarboxylic acid with aromatic

(45)   (46)   (47)

amines in the presence of phosphorus trichloride produces amides of the acid in high yield. These amides (48) which are of the Naphthol AS type, are important coupling components that are applied to fiber, eg, cotton. They then react with a diazo component on the fiber to produce insoluble azo dyes of high washfastness and lightfastness. A wide range of arylamides of 3-hydroxy-2-naphthalenecarboxylic acids and diazo components is available: azo pigments of similar structure are made from Naphthol AS by coupling a diazo component with a 3-hydroxy-2-naphthalenecarboxamide (see Dyes, application and evaluation).

8-Amino-1-Naphthalenecarboxylic Acid. In most methods of 8-amino-1-naphthalenecarboxylic acid (49) manufacture, the lactam (50) (naphthostyril) is obtained. For example, 8-amino-1-naphthalenesulfonic acid is converted to 8-cyano-1-naphthalenesulfonic acid in the Sandmeyer reaction and the nitrile is treated with concentrated alkali at 185°C to form the lactam. The lactam can be hydrolyzed to the amino acid by treatment with dilute alkali at 100°C (67). A potentially useful process is the reported treatment of the imide (41) (R = H) of 1,8-naphthalenedicarboxylic acid with sodium hypochlorite to give the lactam (50) in 57% yield (68). A practical method of manufacture is the preparation of the imide by the reaction of 1-naphthaleneisocyanate with anhydrous aluminum chloride followed by hydrolysis (69).

8-Amino-1-naphthalenecarboxylic acid can be converted, by diazotization and treatment with ammoniacal cuprous oxide, to 1,1'-naphthalene-8,8'-dicarboxylic acid (1,1'-binaphthalene-8,8'-dicarboxylic acid) (51); the latter is treated with concentrated sulfuric acid to yield anthranthrone. The dihalogenated anthranthrones are valuable vat dyes.

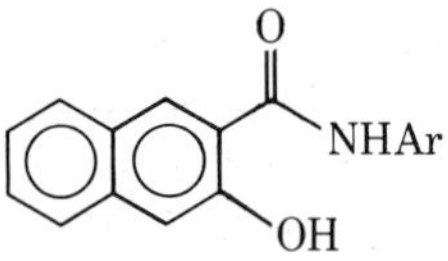

(48)

The (N-alkylated) lactam of 8-aminonaphthalenecarboxylic acid (50) also is a valuable dye intermediate, eg, for cyclomethine type dyes used for dyeing polyacrylonitrile fibers and other synthetics.

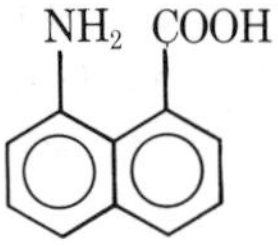

(49)

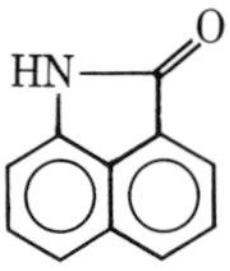

(50) [130-00-7]

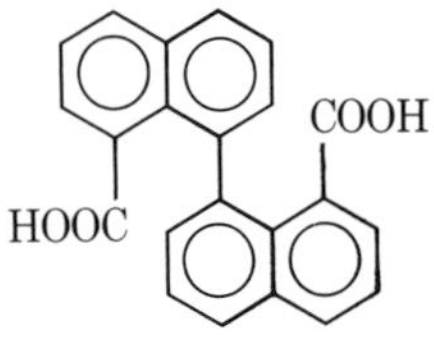

(51) [29878-91-9]

BIBLIOGRAPHY

"Amino Naphthols and Amino Naphtholsulfonic Acids" in *ECT* 1st ed., Vol. 1, pp. 730–737; by J. Werner and A. W. Dawes, General Aniline Works Div., General Aniline & Film Corp.; "Naphthalenesulfonic Acids," "Naphthols and Naphthosulfonic Acids," and "Naphthylamines and Naphthylaminesulfonic Acids" in *ECT* 1st ed., Vol. 9, pp. 232–240, 248–258, and 258–270, by J. Werner, General Aniline Works Div., General Aniline.

1. F. Radt, ed., *Elsevier's Encyclopaedia of Organic Chemistry*, Sec. III, Vols. 12B, Elsevier Publ. Co., Amsterdam, 1949–1955.
2. N. Donaldson, *The Chemistry and Technology of Naphthalene Compounds*, E. Arnold Ltd., London, Eng., 1958.
3. E. Müller, ed., *Methoden Der Organischen Chemie*, (Houben-Weyl), 4th ed., George Thieme Verlag, Stuttgart, FRG, 1958 to present.
4. *Ullmann's Encyklopadie Der Technischen Chemie*, 4th ed., Verlag Chemie, Weinheim/Bergstr., FRG, 1972 to present.
5. *Color Index*, 3rd ed., The Soc. Dyers & Colorists Publ., Bradford, England and Research Triangle Park, N.C., 1971; Rev. 3rd ed., 1975.
6. H. A. Lubs, *The Chemistry of Synthetic Dyes and Pigments*, ACS Monograph Series No. 127, Reinhold Publ. Corp., New York, 1955.
7. K. Venkataraman, ed., *The Chemistry of Synthetic Dyes*, 8 vols., Academic Press, New York, 1952–1978.
8. H. Cerfontain, *Mechanistic Aspects in Aromatic Sulfonation and Desulfonation*, Interscience Publishers, a Division of John Wiley & Sons, Inc., New York, 1968.
9. N. A. Korneeva, O. M. Prokhorova, and V. V. Kozlov, *Izv. Vyssh. Uchebn. Zaved. Khim. Khim. Tekhnol.* **19,** 171 (1976).
10. K. Kaufmann and F. Wolf, *Z. Chem.* **11,** 352 (1971).
11. H. Bretscher, G. Eigemann, and E. Plattner, *Chimia* **32**(5), 180 (1978).
12. *B.I.O.S. Final Report No. 1152*, Item No. 22, London, Eng.
13. Neth. Pat. 138,100 (Feb. 15, 1973), (to Sandoz A.G.).
14. Ger. Offen. 2,337,395 (Feb. 13, 1975), (to Farbwerke Hoechst).
15. Jpn. Pat. 75 13,792 (Mar. 23, 1970), to A. Ito and H. Hiyama.
16. Jpn. Pat. 53 071,046 (June 24, 1978), (to Sugai Kagaku Kogyo).
17. U.S. Pat. 4,059,609 (Nov. 22, 1977), J. K. Rinehart (to PPG Industries).
18. B. V. Passet, V. A. Kholodnov, and A. V. Matveev, *Zh. Prikl. Khim.* **51,** 1606 (1978).
19. U.S. Pat. 4,110,365 (Aug. 29, 1978), S. Bildstein, R. Lademann, S. Pietzsch, and G. Schaeffer (to Hoechst AG).
20. U.S. Pat. 3,655,739 (Nov. 4, 1972), H. Clasen (to Farbwerke Hoechst AG).
21. *FIAT Final Rep.* **1016,** 45 (1947).
22. Jpn. Pat. 73 38,699 (Nov. 19, 1973), (to Sumitomo Chem. Co.); USSR Pat. 596,565 (Sept. 27, 1977), (to Leningrad Chem. Pharm.).
23. U.S. Pat. 3,546,280 (Dec. 8, 1970), H. Dressler and K. G. Reabe (to Koppers Company).
24. Jpn. Kokai 78 50,149 (May 8, 1978), H. Fujii, T. Nagashima, and H. Oguri (to Sugai Chem. Ind.).
25. F. Allison, G. Brunner, and H. E. Fierz-David, *Helv. Chim. Acta* **35,** 2139 (1952).
26. Jpn. Pat. 54 467,559 (Apr. 12, 1979), (to Sugai K. K.).
27. Belg. Pat. 866,223 (Oct. 23, 1978), (to Bayer A.G.).
28. *United States International Trade Commission Report No. 833*, Washington, D.C., 1976.
29. *Chem. Eng. News* **54,** 11 (June 23, 1975); *Chem. Age*, 3 (Feb. 13, 1976).
30. USSR Pat. 267,889 (July 24, 1970), (to Phys. Org. Chem. Inst. Acad. Sci. Beloruss. SSR); USSR Pat. 443,885 (June 5, 1975), T. I. Vasilenok, B. A. Konoplev, V. N. Lagumova, and co-workers.
31. *B.I.O.S. Final Report 1143*, London, Eng.
32. *Chem. Eng. News* **55,** 7 (Jan. 3, 1977).
33. U.S. Pat. 4,053,526 (Apr. 19, 1975), H. U. Blank, F. Duerholz, and G. Skipka (to Bayer A.G.); Jpn. Pat. 54 016,460 (Feb. 2, 1979), (to Nippon Synth. Chem. Indust.).
34. Ger. Offen. 2,523,351 (Dec. 9, 1976), H. U. Blank, F. Duerholz, and G. Skipka (to Bayer A.G.); Jpn. Pat. 50 0970,954 (July 26, 1975), (to Mitsui Toatsu Chem. Co.).
35. *B.I.O.S. Documents FD 4637/47*, frames 58–60, London, Eng.

36. *PB Report 74197*, Washington, D.C.

37. V. I. Trofimov, Ya. L. Levkov, and A. M. Yakubson, *Sov. Chem. Ind.* (3), 168 (1973).

38. *Chem. Eng. News* **53**, 25 (July 28, 1975).

39. S. P. Starkov and Ye. I. Mostyaev, *Sov. Chem. Ind.* (5), 302 (1973).

40. USSR Pat. 340,270 (Jan. 31, 1973), (to Novosibirsk Org. Chem. Inst. Siberian Acad. Sci. USSR).

41. Brit. Pat. 654,035 (May 30, 1951), W. Webster and D. C. Quin (to the Distillers Co., Ltd.); U.S. Pat. 3,804,723 (Apr. 16, 1974), J. P. Dundon, H. R. Kemme, and E. J. Scharf (to American Cyanamid Co.); U.S. Pat. 3,848,001 (Nov. 12, 1974), J. P. Dunden (to American Cyanamid Co.); U.S. Pat. 3,939,211 (Feb. 17, 1976), R. H. Spector and R. K. Madison (to American Cyanamid Co.); U.S. Pat. 4,021,495 (May 3, 1977), H. Hosaka, K. Tamaka, and Y. Ueda (to Sumitomo Chem. Co,); U.S. Pat. 4,049,720 (Sept. 20, 1977), H. Hosaka, K. Tamimoto, and H. Yamachika (to Sumitomo Chem. Co.); Ger. Offen. 2,207,915 (Aug. 23, 1973), M. Elstner, H. Hoever, and M. Fremery (to Union Rheinische Braunkohlen Kraftstoff AG); Ger. Offen. 2,517,591 (Oct. 23, 1975), M. Takahashi, T. Yamanchi, and T. Imura (to Kureha Chem. Ind.); USSR Pat. 498,292 (Jan. 5, 1976), M. I. Ferberov, B. N. Bychkov, G. N. Shustovskaya, and G. D. Mantyukov (to Yaroslavl Polytechnic Inst.); USSR Pat. 593,729 (Mar. 9, 1976); M. I. Ferberov, B. N. Bychkov, G. N. Shustovskaya, and G. D. Mantyukov (to Yaroslavl Polytechnic Inst.).

42. Jpn. Kokai 75 0076,054 (June 21, 1975), K. Kobayaschi, I. Dogane, and Y. Nagao (to Sumitomo Chem. Ind.).

43. *J. Commerce* **325** (June 26, 1972).

44. U.S. Pat. 3,391,117 (July 2, 1968), N. Bilow and L. J. Miller (to U.S. Dept. of the Air Force); *Chem. Week*, 84 (Aug. 21, 1965).

45. *FIAT Final Report No. 1016*, 22–23.

46. Ger. Offen. 2,304,873 (Aug. 8, 1974), K. Eiglmeier and H. Luebbers (to Farbwerke Hoechst AG).

47. *B.I.O.S. Final Report 986*, Pt. I, London, Eng.

48. Fr. Pat. 1,543,144 (Oct. 18, 1968), G. J. Rolman (to Ashland Oil); USSR Pat. 225,249 (Oct. 29, 1970), N. D. Rusyanova and N. S. Mulyaeva (Eastern Scientific Research Inst. of Coal Chem.); Jpn. Pats. 73 43,891 and 73 43,893 (Dec. 21, 1973), G. Yamashita and K. Yamamoto (to Teijin Ltd.).

49. U.S. Pat. 3,920,734 (Nov. 18, 1975), Y. Ichikawa and T. Yamaji (to Teijin Ltd.).

50. Ger. Offen. 2,028,555 (Dec. 17, 1970), E. Arsuro and co-workers (to Montecatini Edison S.p.A.).

51. U.S. Pat. 3,855,289 (Dec. 17, 1974), G. H. Alt (to Monsanto Co.).

52. Ger. Offen. 2,824,988 (Dec. 21, 1978), M. Suchy (to F. Hoffmann-LaRoche and Co.).

53. USSR Pat. 517,579 (Sept. 6, 1976), L. A. Kozorez and co-workers (to Voroshilovgrad Mech. Inst.).

54. Brit. Pat. 1,224,418 (March 10, 1971), (to Mitsubishi Chem. Ind.); U.S. Pat. 3,646,069 (Feb. 29, 1972), H. Okada and co-workers (to Sandoz Ltd.); Jpn. Kokai 76 11,749 (Jan. 30, 1976), Y. Wakisada and M. Sumitami (to Nippon Kayaku Co.).

55. Ger. Pat. 1,262,268 (Mar. 7, 1968), A. Marx and co-workers (to Rutgerswerke AG); USSR Pat. 291,910 (Jan. 6, 1971), V. L. Plakin and co-workers; Ger. Offen. 2,520,094 (Nov. 20, 1975), H. Okushima, I. Nitta, and R. Furuya (to Mitsui Petrochem. K.K.); Jpn. Pat. 53 141,253 (Dec. 12, 1978), (to Nippon Kayaku K.K.).

56. Belg. Pat. 525,660 (June 8, 1956), (to Rutgerswerke A.G.); Jpn. Pat. 71 13,737 (Apr. 13, 1971), Y. Shimada, H. Yoshizumi, and H. Namikiri (to Japan Catalytic Chem. Ind. Co.); U.S. Pat. 3,708,504 (Jan. 2, 1973), O. Kratzer, H. Suter, and F. Wirth (to Badische Anilin and Soda Fabrik A.G.); Jpn. Kokai 75 39,292 (Apr. 11, 1975), Y. Nanba and co-workers (to Japan Catalytic Chem. Ind. Co.); USSR Pat. 535,306 (Dec. 20, 1976), D. Kh. Sembaev and co-workers (to AS Kazakistan Chem. Sci.).

57. U.S. Pat. 3,860,720 (Jan. 14, 1975), M. F. Covey (to Gulf Research & Dev. Co.).

58. U.S. Pats. 3,940,397 and 3,940,398 (Feb. 24, 1976); 3,947,452 (Mar. 30, 1976); 3,959,286 (May 25, 1976); 3,996,363 (Dec. 7, 1976); 4,007,191 (Feb. 8, 1977); 4,051,246 (Sept. 27, 1977); 4,062,953 (Dec. 13, 1977); 4,070,465 (Jan. 24, 1977), P. C. Wade and B. R. Vogt (to E. R. Squibb & Sons, Inc.).

59. Can. Pat. 1,030,147 (Apr. 25, 1978), R. Martinez and co-workers (to Laboratories Made S.A.).

60. Ger. Offen. 2,507,459 (Sept. 2, 1976), (to Badische Aniline and Soda Fabrik A.G.).

61. Ger. Offen. 2,636,421 (Feb. 16, 1978), (to Badische Aniline and Soda Fabrik A.G.).

62. U.S. Pat. 4,040,968 (Aug. 9, 1977), H. A. Harris (to Shell Oil Co.).

63. U.S. Pat. 3,718,690 (Feb. 27, 1973), R. D. Bushick, O. L. Norman, and H. J. Spinnelli.

64. I. I. Kiiko, *Khim. Prom.* **1**, 13 (1975).

65. *Chem. Week* **112**, 43 (Apr. 25, 1973).

66. *B.I.O.S. Final Rep. No. 986*, Item 22, pp. 234–250, London, Eng.

67. *Product Bulletin IC-4*, American Cyanamid Co., July 1978.

68. *B.I.O.S. Final Rep. No. 1152*, pp. 118–123, London, Eng.

69. T. Maki, S. Hashimoto, and K. Kamada, *J. Chem. Soc. Jpn. Ind. Chem. Sect.* **55**, 4835 (1952).
70. Ger. Pat. 2,635,693 (Apr. 26, 1979), (to Bayer A.G.).

HANS DRESSLER
Koppers Company, Inc.

NAPHTHENIC ACIDS

Naphthenic acids [*1338-24-5*] are the carboxylic acids that are derived from petroleum during the refining of the various distilled fractions and are predominately monocarboxylic acids (see Carboxylic acids). The main distinguishing structural feature of naphthenic acids is a hydrocarbon chain consisting of single or fused cyclopentane rings, which are alkylated in various positions with short aliphatic groups. Other acids, eg, aliphatic, dicarboxylic, and those containing the cyclohexane ring, also are present in moderate quantities.

The commercial grades of naphthenic acids vary widely in properties and impurities, depending on their source and the refining method. All contain 5–25 wt % hydrocarbons whose composition is the same as the petroleum fraction from which the naphthenic acids are derived. The average molecular weight is higher for acids that are extracted from higher boiling fractions. All contain acidic impurities, eg, phenols, mercaptans, and thiophenols, in small quantities. The content of the odorous impurities is reduced and the color is improved markedly for distilled grades of naphthenic acids as compared with the corresponding crudes. A comprehensive review of naphthenic acid technology is given in ref. 1.

Composition

The predominant acids that are present in naphthenic acids are described by structure (**1**):

$$H\text{---}\left[\underset{R}{\overbrace{}}\right]_n\text{---}(CH_2)_m\text{---}COOH$$

(**1**)

The number of fused rings, n, may range from 1 to 5 and a small fraction of these rings may be cyclohexyl. R consists of small aliphatic groups eg, methyl, and m is greater than one.

The preponderance of monocarboxylic acids was determined by identification of the low molecular weight acids and by research involving high molecular weight acids (2–9). Monocarboxylic acids (90 mol %) have been reported for naphthenic acids that are derived from diesel fuel and have an average molecular weight of 260 (10).

The carboxyl group seldom is attached directly to a ring. Studies of distilled fractions of naphthenic acids having an average molecular weight of 243 reveal that $m \geq 4$ in structure (1) (11–12). Both cyclopentyl and cyclohexyl moieties are contained in the hydrocarbon chain. The structure of an acid containing the cyclohexyl group has been isolated and determined (13). However, most of the acids are based on cyclopentane. The trimethylsilyl esters of naphthenic acids have been prepared and analyzed by ir and mass spectroscopy (10):

Acids	%
aliphatic	10.3
monocyclic, $n = 1$	38.5
bicyclic, $n = 2$	34.8
tricyclic, $n = 3$	8.6
tetracyclic, $n = 4$	5.4
pentacyclic, $n = 5$	2.4

Naphthenic acids in a given crude oil probably are derived from the same substituted cycloparaffins (naphthenes) and other types of hydrocarbons which constitute the neutral oil or unsaponifiable fraction of the acids. In one study, the acids from a Venezuelan crude oil were converted to hydrocarbons which were similar to the hydrocarbons obtained from the same crude oil (14). However, analysis is extremely difficult, as evidenced by the determination of ca 120 different compounds in an Austrian crude oil (15).

Properties

Naphthenic acids are viscous liquids with a characteristic odor resulting from the phenols and sulfur compounds which are extracted with the acids; these impurities are very difficult to remove. Various chemical treatments, such as mild oxidation or reaction with aldehydes followed by distillation, only partially reduce the odor and no economically feasible general treatment has been devised. Undistilled acids are very dark and distilled acids typically are light yellow but quickly change to yellow or light amber. Attempts to synthesize acids that have the composition and properties of the naturally occurring acids by oxidation of cycloparaffinic hydrocarbons have been unsuccessful (16–17).

The product of refractive index and density can be used to distinguish naphthenic acids from other compounds (5):

Compounds	$n_D^{20} \times d_4^{20}$	
	Possible	Typical
aliphatic acids	1.28–1.35	1.30–1.31
naphthenic acids	1.39–1.47	1.41–1.44
phenols	>1.50	
hydrocarbons	1.30	1.28

Naphthenic acids undergo the same chemical reactions as other saturated carboxylic acids. The most important commercial reactions are amidation, esterification (qv) and the formation of metal soaps (see Driers and metallic soaps). The utility of the metal soaps, especially of heavy metals, results from their high solubility in hydrocarbons. These soaps allow the required concentrations of metals to be dissolved in solvent-based paint as driers, in lubricants, in solvents as preservatives, and in various organic systems as catalysts (see Lubrication and lubricants; Catalysis). The amides of naphthenic acids are used as corrosion inhibitors and antimicrobials (see Corrosion and corrosion inhibitors; Industrial antimicrobial agents).

The properties of one crude and two distilled commercial grades of naphthenic acid are given in Table 1.

Table 1. Properties of Commercial Naphthenic Acids

Property	Crude	NA 170	NA 210
acid number, mg KOH/g	160–170	170	210
acid number (oil-free)	200–220	200–215	220–235
unsaponifiables (neutral oil), wt %	20–25	15–20	5–10
water content, wt %	1	0.2	0.05
color	black	6–8[a]	6–8[a]
sp gr at 16°C	0.95–0.96	0.94–0.95	0.94–0.95
viscosity at 38°C, mPa·s (= cP)	40–80	100–120	90–130

[a] Gardner.

Occurrence

All sources of crude oil contain naphthenic acids but the concentrations vary widely. In the distilled fractions from which the acids are recovered, the naphthenic acid content is greater in the higher boiling cuts. Typically, the 300–400°C fraction contains the maximum concentration of acids. Much of the acid formerly recovered by extraction to free the distillates of acidic components is no longer available because modern purification procedures destroy the acids.

Manufacture and Processing

Naphthenic acids are recovered as by-products from the refining of straight-run petroleum distillates. Most of the commercial acids are derived from the kerosene and gas-oil fractions and some higher molecular weight acids are derived from the light lubricating-oil cuts (see Petroleum).

The distillates are extracted with caustic soda solutions to form the sodium soap of the naphthenic acids. This soap stock contains varying amounts of emulsified oil which ultimately become the unsaponifiables in the recovered acids. Phenols and sulfur compounds also are extracted by the caustic. The soap stock may be brought into contact with naphtha to extract and replace some of the emulsified oil. The stock is treated with sulfuric acid and the crude naphthenic acids are separated from the aqueous salt layer. The phenols and sulfur compounds also are released during the acid treatment and become part of the crude naphthenic acids.

Further refining by vacuum distillation in stainless-steel equipment greatly improves the color of the acids. The vacuum distillation reduces the unsaponifiables slightly if a forerun is taken. Water is removed almost totally and any salts that remain from the separation of the aqueous layer are left in the heel. Odiferous impurities distill with the acids unless chemical treatment is used to reduce them.

Other methods of recovering naphthenic acids have been practiced, eg, distillation of the crude oil from caustic soda, bringing the petroleum vapors into contact with molten alkali, ion exchange (qv), and extraction from crude oil with solvents. In one process, oil containing 3 wt % naphthenic acids is treated with ammonia and the salt is separated and decomposed during distillation to recover the acids (18). The ammonia is recycled to the process. Attempts to recover naphthenic acids from crude oil lead to intractable emulsions.

Economic Aspects

In 1977, United States production of naphthenic acids was ca 9400 metric tons and the unit value was $0.30/kg (19). The 1980 production of naphthenic acids of all types is expected to be ca 9100 t with an additional 3600 t imported to meet the demand of 12,700 t (19).

The scarcity of naphthenic acid feedstocks and the increased prices of crude oil have caused price levels to rise to $0.40–0.49/kg for crude grades of naphthenic acids and to $0.99–1.21/kg for refined grades having acid values of 190–220. Future availability may depend largely on the development of crude oil sources, which have high naphthenic acids content, in South America, Mexico, and the Caribbean. The long-term growth of naphthenic acids will be governed primarily by availability rather than by use potential. The decline in oil-based paints that contain naphthenate salts as driers has stabilized. The use of naphthenate salts as oxidation and polymerization catalysts and in corrosion inhibitors has grown.

Uses

Naphthenates are used primarily in the manufacture of paint driers, corrosion inhibitors, lubricants, catalysts, and preservatives. In 1977, paint driers accounted for ca 75% of the United States production with a volume of 6800 t (19). The naphthenate salts of lead [61790-14-5], calcium [61789-36-4], cobalt [61789-51-3], zinc [12001-85-3], and manganese [1336-93-2] salts of naphthenic acids in paint and varnishes function as catalysts in the oxidation and polymerization of drying oils and resins. Salts of the heavy metals promote the solidification of applied films of paint, varnish, and other oleoresinous coatings. The organic portion of these compounds serves as a carrier for the catalytic metal and provides the necessary solubility in the drying oil. The salts of naphthenic acids have superior solubility characteristics and greater stability in concentrated form than the older metallic resinates and linoleates.

Lead naphthenate has been used widely in lubricants for its anticorrosion and extreme-pressure properties. Lead naphthenate also enhances antiseize, oiliness, and film-strength characteristics. Its use, however, is expected to decline because of the concern over the toxicity of lead compounds. The aliphatic amine and diamine salts of naphthenic acids are used as corrosion inhibitors for ferrous metals.

The use of metallic naphthenates as catalysts has grown steadily. Cobalt naphthenate has been employed in conjunction with peroxide catalysts to accelerate the curing of unsaturated polyester resins in the building of glass-fiber laminates. It also has been used as a catalyst in the multistep conversion of unsaturated hydrocarbons to monohydric alcohols, ie, the oxo process (qv).

Copper and zinc naphthenates are well known for their fungicidal activity. They provide long-term protection against fungus, mold, mildew, and marine parasites. Common uses include protecting canvas, burlap, tents, awnings, lumber, and utility poles (see Fungicides).

BIBLIOGRAPHY

"Naphthenic Acids" in *ECT* 1st ed., Vol. 9, pp. 241–246, by V. L. Shipp, Socony-Vacuum Oil Co.; "Naphthenic Acids" in *ECT* 2nd ed., Vol. 13, pp. 727–734, by S. E. Jolly, Sun Oil Co.

1. H. L. Lochte and E. R. Littman, *Petroleum Acids and Bases*, Chemical Publishing Co., New York, 1955, pp. 9–279.
2. K. Hancock and H. L. Lochte, *J. Am. Chem. Soc.* **61,** 2488 (1939).
3. C. D. Nenitzescu and T. A. Volrap, *Ber.* **71B,** 2056 (1938).
4. Ref. 2, p. 2448.
5. H. Schutze, B. Shive, and H. L. Lochte, *Ind. Eng. Chem. Anal. Ed.* **12,** 262 (1940).
6. W. Schive and co-workers, *J. Am. Chem. Soc.* **65,** 770 (1943).
7. U.S. Pat. 1,694,461 (Dec. 11, 1928), G. Alleman (to Sun Oil Co.).
8. R. W. Harkness and J. H. Brunn, *Ind. Eng. Chem.* **32,** 499 (1940).
9. G. E. Goheen, *Ind. Eng. Chem.* **32,** 503 (1940).
10. V. G. Lebedevskaya and co-workers, *Neftekhimiya* **17**(5), 765 (1977).
11. J. von Braun, *Science of Petroleum*, Vol. II, Oxford University Press, New York, 1938, p. 1007.
12. P. D. Shikhalizade and co-workers, *Izv. Vyssh. Uchebn. Zaved. Neft. Gaz* **19**(10), 60 (1976).
13. J. Cason and K. L. Liauw, *J. Org. Chem.* **30,** 1763 (1965).
14. J. Knotnerus, *J. Inst. Pet.* **43,** 307 (1957).
15. R. Bock and K. Behrends, *Z. Anal. Chem.* **208**(5), 338 (1965).
16. A. A. Akhundov and co-workers, *Vopr. Neftekhim.* **9,** 3 (1977).
17. B. K. Zeinalov and co-workers, *Sb. Tr. Inst. Neftekhim. Protessov. Akad. Nauk B SSR* **4,** 38 (1973).
18. D. I. Allakhverdiev and co-workers, *Azerb. Neft. Khoz.* **6,** 41 (1977).
19. *United States International Trade Commission Report No. 920*, U.S. Government Printing Office, Washington, D.C.

W. E. Sisco
W. E. Bastian
E. G. Weierich
CPS Chemical Company

NAVAL STORES.   See Terpenoids; Rosin and rosin derivatives.

NEODYMIUM.   See Rare-earth elements.

NEPTUNIUM.   See Actinides and transactinides.

NEUROREGULATORS

Although the concept of chemical neurotransmission in the peripheral nervous system was established about 60 years ago, the existence of chemical transmission in the CNS has been generally accepted only for the past 25 years (1). It had been widely believed that only electrical transmission could provide the rapid transfer of information needed in the CNS (2–4); however, in the 1950s, results from microelectrode recordings of single cells (5–6) and microiontophoretic application of compounds onto neurons (7–8) were not explicable in terms of electrical junctions. In addition, the occurrence and functional importance of noradrenaline in the brain provided further evidence of chemical neurotransmission in the CNS (9).

The number of endogenous substances that have been isolated and shown to have a neurochemical, neurophysiological, or behavioral effect in animals or humans has increased enormously (10). Investigators are endeavoring to accommodate these new substances into a general framework that attempts to classify endogenous substances that affect neuronal communication on a much broader basis than the older term neurotransmitter does (10–11). It has been suggested that the preoccupation with neurotransmission and the identification of neurotransmitters may have tended to overshadow other possibly valuable areas of neurochemical research and potentially interesting new substances (11–13). Neuroregulator has been suggested as a generic name for these endogenous compounds, and the term includes the neurotransmitter and neuromodulator groups (10–11). A neurotransmitter is a substance that conveys a transient and unilateral signal across a specialized synapse. A neuromodulator alters neuronal activity by mechanisms which may or may not involve a synapse. Neuromodulators may be classified as hormonal neuromodulators, which provide direct, short- or long-lasting modulation of neurons in areas far removed from the site of release; and synaptic neuromodulators, which act indirectly by modulating neurotransmitter function (11). Various criteria for establishing the identity of neurotransmitters and neuromodulators are listed below (10,11):

Neurotransmitter

(*1*) The substance must be present in presynaptic elements of neuronal tissue, possibly in an uneven distribution throughout the brain.

(*2*) Precursors and synthetic enzymes must be present in the neuron, usually in close proximity to the site of presumed action.

(*3*) Stimulation of afferents should cause release of the substance in physiologically significant amounts.

(*4*) Direct application of the substance to the synapse should produce responses that are identical to those of stimulating afferents.

(*5*) Specific receptors should interact with the substance and should be in close proximity to presynaptic structures.

(*6*) Interaction of the substance with its receptor should induce changes in postsynaptic membrane permeability leading to excitatory or inhibitory postsynaptic potentials.

(*7*) Specific inactivating mechanisms, which stop interactions of the substance with its receptor in a physiologically reasonable time frame, should exist.

(*8*) Stimulation of afferents and direct application of the substance should be equally responsive to and similarly affected by interventions involving postsynaptic sites or inactivating mechanisms.

Neuromodulator

(*1*) The substance does not act as a neurotransmitter, in that it does not act transsynaptically.

(*2*) The substance must be present in physiological fluids and have access to the site of potential modulation in physiologically significant concentrations.

(*3*) Alterations in endogenous concentrations of the substance should affect neuronal activity consistently and predictably.

(*4*) Direct application of the substance should mimic the effect of increasing its endogenous concentrations.

(*5*) The substance should have one or more specific sites of action through which it can alter neuronal activity.

(*6*) Inactivating mechanisms, which account for the time course of effects of endogenously or exogenously induced changes in concentrations of the substance, should exist.

(*7*) Interventions, which alter the effects on neuronal activity of increasing endogenous concentrations of the substance, should act identically when concentrations are increased by exogenous administration.

Unlike neurotransmitters, which are receptor-specific, neuromodulators may act by affecting neurotransmitter synthesis, release, uptake, metabolism, or receptor interaction. However, because the term neurotransmitter has been used for a long time and because there is general agreement as to what constitutes a reasonable set of criteria for this term, it is easier to classify such substances as acetylcholine, dopamine, noradrenaline, serotonin, glycine, and γ-aminobutyric acid (GABA) as being probable neurotransmitters in the CNS than it is to designate, with any degree of certainty, certain other compounds as being neuromodulators (11).

The value but complexity of studying the effects of neuroregulators on animal behavior at the neurochemical level has been outlined (10). The difficulties involved in attempting to relate the activity of various neuroregulators to certain types of mental illness also has been reviewed (10,14). Various criteria by which a neuroregulator may be related to severe mental illness are listed below (10):

(*1*) Each neuroregulator must be an endogenously formed substance for which receptors exist in nerve cells or which alters function.

(*2*) A characteristic pattern of neuroregulator activity should occur in relation to the psychiatric disorder.

(*3*) Alteration of the activity of neuronal systems involving one or more neuroregulators or alteration of the balance between them should affect the psychiatric disorder.

(*4*) Appropriate manipulation of the neuroregulatory system or systems should induce the psychiatric disorder.

(*5*) Unless the physiological process is irreversible, appropriate restoration of neuroregulator activity or balance should ameliorate the psychiatric disorder.

In this article, owing to limitations of space, no attempt has been made to cover all the recently described naturally occurring neuroactive substances. However, most of the more important candidates for a role as neuroregulators are discussed. In addition to the publications of Barchas and co-workers (10–11) on neuroregulators, the reader is referred to the previously mentioned publications (12–13) for general details of neurotransmitters and neurotransmission and to the works of Copper, Bloom, and Roth (15) and Lipton, DiMascio, and Killam (16) for further general reading in neurochemistry and neuropharmacology.

Acetylcholine

Properties. Acetylcholine [51-84-3] occurs as the bromide [66-23-9] (Pragmoline) and the chloride [60-31-1] (Acecoline). The chloride is a hygroscopic, crystalline powder. It is very soluble in cold water and alcohol but is practically insoluble in diethyl ether. It is decomposed by hot water and alkalies.

Acetylcholine bromide can be prepared by direct reaction of trimethylamine and β-bromoethyl acetate in benzene (17).

$$
\underset{\overset{\underset{\displaystyle CH_3}{|}}{\underset{\displaystyle CH_3}{|}}{H_3C-N} \; + \; BrCH_2CH_2OCCH_3 \; \longrightarrow \; \underset{\overset{\underset{\displaystyle CH_3}{|}}{\underset{\displaystyle O}{}}}{H_3C-N^+CH_2CH_2OCCH_3} \; + \; Br^-
$$

acetylcholine bromide

Acetylcholine is the product of the reaction between choline and acetyl coenzyme A in the presence of choline acetylase (15).

$$
Acetyl\ CoA \; + \; (CH_3)\overset{+}{N}CH_2CH_2OH \; \xrightarrow[acetylase]{choline} \; (CH_3)\overset{+}{N}CH_2CH_2O\overset{O}{\overset{\|}{C}}CH_3 \; + \; CoA
$$

acetylcholine

Pharmacology. Acetylcholine is a neurotransmitter at the neuromuscular junction in autonomic ganglia and at postganglionic parasympathetic nerve endings. In the CNS, the motor-neuron collaterals to the Renshaw cells are cholinergic (15). In the rat brain, acetylcholine occurs in high concentrations in the interpeduncular and caudate nuclei (16). The LD_{50} (subcutaneous) of the chloride in rats is 250 mg/kg.

Adenosine

Properties. Adenosine [58-61-7], mp 234–235°C, $[\alpha]_D^{11}$ −61.7° (c = 0.706 in water), occurs as crystals from water. Its uv absorption maximum is 260 pm (ϵ 15,100). It is quite soluble in water, especially on warming, and it is practically insoluble in alcohol. Adenosine forms a picrate [6128-21-8], mp 195–198°C.

The first reported synthesis of adenosine involves condensation of the silver salt of 2,8-dichloroadenine with acetochloro-D-ribofuranose followed by deacetylation which gives 2,8-dichloroadenosine (18). Hydrogenation over a palladized barium sulfate catalyst yields adenosine.

adenosine

Adenosine can, in principle, be formed via a number of pathways, eg, dephosphorylation of 5′-adenylic acid, from S-adenosylhomocysteine formed in the process of methylation, from 3′-adenylic acid, from adenosine end groups in ribonucleic acid (RNA), and by the reaction of adenine and ribose-1-phosphate (19).

Pharmacology. The role of the purinergic nerves in the functioning of the peripheral nervous system has been thoroughly reviewed, and there has been considerable interest in exploring the possible role of such systems in the CNS (20–22). Adenosine can be released from the intact brain and from brain slices and synaptosomal preparations following electrical stimulation. Certain central neurons are depressed following the iontophoretic application of adenosine. The brain also has a specific adenosine-sensitive, adenylate cyclase system. ^{3}H-Adenosine is taken up readily by certain neurons (see Radioactive drugs). Intraventricular administration of adenosine to cats leads to sedation.

(−) Adrenaline (Epinephrine)

Properties. The (±) free base [329-69-7] is sparingly soluble in water and alcohol. The (±) HCl salt [329-63-5] has a mp of 157°C; it is readily soluble in water but is sparingly soluble in absolute alcohol. The (+) free base [150-05-0] occurs as crystals, mp 211°C. The naturally occurring (−) form [51-43-4] (free base) has a mp of 211–212°C and an $[\alpha]_D^{25}$ not less than −50° and not more than −53.5° (USP). The latter form is sparingly soluble in water and is insoluble in alcohol, chloroform, ether, acetone, and oils. It is readily soluble in aqueous solutions of mineral acids and of sodium or potassium hydroxide but not in aqueous solutions of ammonia and of the alkali carbonates. Neutral and alkaline solutions of adrenaline are very susceptible to oxidation in the presence of oxygen (see Epinephrine). (−)-Adrenaline base has the following trade names: Adnephrine, Adrenal, Adrenamine, Adrenine, Adrin, Chelafrin, Epinephran, Epirenan, Hemisine, Hemostasin, Hemostatin, Hypernephrin, Levorenine, Nephridine, Nieraline, Paranephrin, Renaglandin, Renaleptine, Renalina, Renoform, Renostypticin, Renostyptin, Scurenaline, Styptirenal, Supracapsulin, Supranephrane, Suprarenaline, Suprarenin, Surrenine, Takamina, Vasoconstrictine, and Vasotonin.

(−)-Adrenaline also occurs as the crystalline (+)-bitartrate [51-42-3], $(C_9H_{13}NO_3 \cdot C_4H_6O_6)$, mp 147–154°C. It darkens slowly on exposure to air and light. It is soluble in water (1 g/3 mL) and is slightly soluble in alcohol.

Catechol reacts with chloroacetyl chloride in the presence of phosphorus oxychloride to yield 4-chloroacetylcatechol. This compound is treated with methylamine and the resulting ketone is reduced to (−)-adrenaline catalytically or with aluminum amalgam (23).

Adrenaline is formed from the amino acid tyrosine via 3-(3,4-dihydroxyphenyl)-alanine (DOPA), dopamine, and noradrenaline (norepinephrine) with the aid of the appropriate enzymes, as outlined below (15):

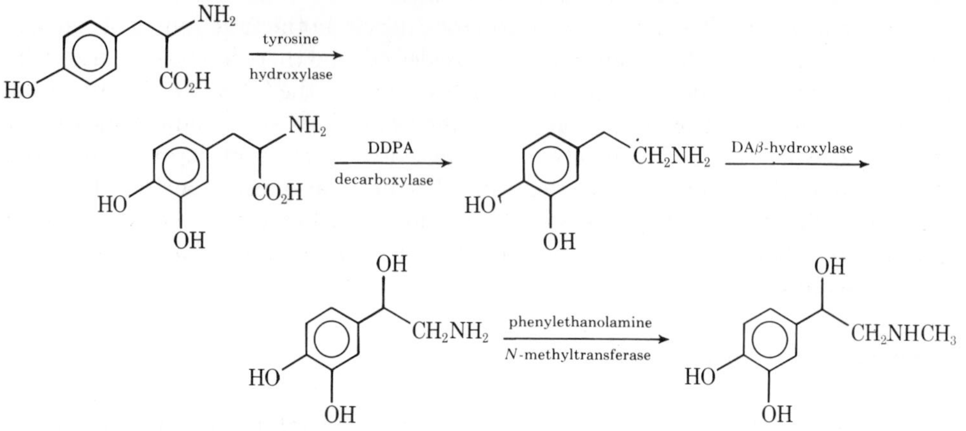

Pharmacology. Adrenaline is the major hormone of the adrenal medulla (see Hormones). The most important influences of endogenous adrenaline probably are those on various metabolic processes. It elevates blood glucose and free fatty acid concentrations and inhibits insulin secretion via an action on α receptors (24). In the CNS the adrenaline-containing cell bodies are intermingled with various noradrenergic cells in the lateral tegmentum and dorsal medulla (15).

Medically, (−)-adrenaline is used as a sympathomimetic, vasoconstrictor, cardiac stimulant, and bronchodilator. It can cause anxiety, headache, tremors, dizziness, weakness, palpitations, and insomnia (see Epinephrine and norepinephrine).

Adrenocorticotropic Hormone (ACTH)

The structure of human ACTH [*12279-41-3*] is

H−Ser−Tyr−Ser−Met−Glu−His−Phe−Arg−Trp−Gly−Lys−Pro−Val−Gly−Lys−Lys−Arg−Arg−
              5   10   15

Pro−Val−Lys−Val−Tyr−Pro−Asn−Gly−Ala−Glu−Asp−Glu−Ser−Ala−Glu−Ala−Phe−Pro−Leu−
    20   25   30   35

Glu−Phe−OH.

Bovine [*39319-42-1*] and ovine ACTH [*39319-42-1*] both have a 33-Gln instead of Glu.

Properties (Bovine ACTH). Bovine ACTH [*39139-42-1*] is a white powder and is freely soluble in water. The aqueous solutions are stable to heat; thus, a solution at pH 7.5 may be placed in a boiling water bath for 120 min. It is quite soluble in 60–70% alcohol or acetone. It is partly precipitated at the isoelectric point (pH 4.65–4.80). It is precipitated by 2.5% trichloracetic acid, 20% sulfosalicyclic acid, and 5% lead acetate solutions. In 0.1 M HCl, a 0.2% solution retains its activity when maintained at 100°C for 1 h; however, in 0.1 M NaOH and under the same conditions, potency is lost within 30 min. Its activity is not destroyed by digestion with pepsin.

Chemical synthesis of human ACTH [*12279-41-3*] has been carried out by fragment condensation in solution and by solid-phase synthesis employing symmetrical anhydride coupling (25). Biosynthesis of ACTH occurs in the anterior lobe of the pituitary.

Pharmacology. ACTH is necessary for the maintenance and growth of the cells of the adrenal cortex and for the regulation of their secretory activity. It stimulates the synthesis of steroids in the adrenal glands which, in humans, results primarily in an increased secretion of cortisol and corticosterone. ACTH also affects areas remote from those of the adrenal glands; thus, in amphibians and reptiles it causes dispersal of melanin as does melanocyte-stimulating hormone (MSH).

ACTH, its analogues, and its sequential fragments have been studied in connection with their effects on memory and behavior (see Memory-enhancing agents and antiaging drugs). An ACTH-like material has been detected by both radioimmunoassay and bioassay in rat brain and its presence may account for the behavioral activity of ACTH and its analogues (16). ACTH is the first peptide hormone of which a modified synthetic product (ACTH 1–24, Synacthen, CIBA-GEIGY) has become commercially available, it is used as a test of adrenocortical function.

Androgens

For a discussion of testosterone, see ref. 26 (see also Hormones, sex hormones).

testosterone

Angiotensins

Properties. The angiotensins are very stable (see Hormones, brain oligopeptides). They are hydrolyzed by strong acids and bases above pH 9.5. They are soluble in aqueous solutions at pH 5–8 and in some organic solvents. At high dilution, they can be absorbed by glass.

H–Asp–Arg–Val–Tyr–Ile–His–Pro–Phe–His–Leu–OH

―――――――angiotensin I [484-42-4] (peptitensin)――――|

|――――― angiotensin II [4474-91-3]――|

|―angiotensin III [13602-53-4]―|

For a review of the methods of chemical synthesis, see ref. 27.

Angiotensin was the original name for a vasoconstricting decapeptide formed in blood from an α-2 hepatic globulin by the proteolytic enzyme, renin. This decapeptide, angiotensin I, is converted enzymatically to an octapeptide angiotensin II which has a similar pharmacological profile but of higher intrinsic activity. It also appears that there is an angiotensin III which may be even more potent and may be able to pass through biomembranes more readily. All of the substrates and enzymes required for these conversions of angiotensin occur in the brain (15).

Pharmacology. Angiotensin II is a very potent inducer of drinking behavior in animals following administration of picomolar amounts into the third ventricle of the rat brain (15–16). The neurons in the supraoptic nucleus and certain neuronal-like cells of the subfornical organ are excited by angiotensin II.

Angiotensin II-like nerve terminals occur as moderately dense networks in several hypothalamic nuclei, including the dorsomedial nucleus, the ventral basal hypothalamus, the median eminence, and the central amygdaloid nucleus. Cell bodies have been observed in the paraventricular nucleus and the perifornical area of the hypothalamus (16).

L-Aspartic Acid

Properties. L-Aspartic acid [56-84-3] occurs as orthorhombic bisphenoidal leaflets or rods: mp 270–271°C, $d^{12.5}$ 1.661 g/cm^3, $[\alpha]_D^{20}$ +25° (c = 1.97 in 6 N HCl), pK_1 1.88, pK_2 3.65, pK_3 9.60. It has the following solubility in water: 1 g/376.3 mL at 0°C, 1 g/256.4 mL at 10°C, 1 g/222.2 mL at 20°C, 1 g/57.4 mL at 60°C and 1 g/18.6 mL at 100°C; 100 mL of water dissolves 5.37 g at 97°C. L-Aspartic acid readily forms supersaturated solutions. It is more soluble in salt solutions, it is also soluble in acids and alkalies but is insoluble in alcohol. Rotation of the D form is $[\alpha]_D^{27}$ −23.0° (c = 2.30 in 6 N HCl).

A mixture of the magnesium and potassium salts is marketed as Aspara, Asparstat, and Trophicard. Potassium aspartate salt (L form) is called Aspara K. Ferrous aspartate is known as Sideryl or Spartocine. Calcium aspartate is marketed as Calciretard.

Reaction of diethyl sodium phthalimidomalonate and ethyl chloroacetate yields triethyl α-phthalimidoethane-α,α,β-tricarboxylate. This is hydrolyzed with a mixture of acetic and hydrochloric acids to DL-aspartic acid (28).

Aspartic acid is formed from the corresponding α ketoacid by a transamination reaction, in which glutamate is the donor of the α amino group, in the presence of pyridoxal phosphate as the coenzyme.

Pharmacology. Aspartic acid is one of the nonessential amino acids. It produces excitatory responses from various CNS neurons and is present in abundant quantities in the mammalian CNS. It has a similar action and distribution in the CNS to glutamic acid. Aspartate may be concentrated in the interneurons of the polysynaptic reflex arc. For further details of its neurochemistry, see ref. 29.

Corticosteroids—Corticosterone, Cortisone, and Hydrocortisone (Cortisol)

Properties. Corticosterone occurs as trigonal plates from acetone: mp 180–182°C,

corticosterone [50-22-6]: R = OH, R′ = H
cortisone [53-06-5]: R = O, R′ = OH
hydrocortisone [50-23-7]: R = OH, R′ = OH

$[\alpha]_D^{15}$ +223° (c = 1.1 in alcohol), absorption max 240 nm. It is insoluble in water, but soluble in many organic solvents (see Hormones, adrenal-cortical). Treatment with concentrated H_2SO_4 gives an orange-yellow solution which is strongly fluorescent.

Cortisone occurs as rhombohedral platelets from 95% alcohol: mp 220–224°C, $[\alpha]_D^{25}$ +209° (c = 1.2 in 95% alcohol), absorption max 237 nm (ϵ = 1.4 × 10⁴). It is fairly soluble in cold methanol, ethanol, and acetone but is much less soluble in ether, benzene, and chloroform. It is slightly soluble in water: 28 mg/100 mL at 25°C. In concentrated H_2SO_4, it yields an intense green fluorescent solution.

The C-21-acetate is marketed as Cortistab, Cortelan, Cortisyl, or Artriona. It occurs as flat needles from acetone: mp 235–238°C; $[\alpha]_D^{25}$ +164° (c = 0.5 in acetone); absorption max 238 nm (ϵ = 1.58 × 10⁴); solubility: in water, 2.2 mg/100 mL at 25°C; in propylene glycol, 44 mg/100 mL; in chloroform, 182 mg/g; in ether, 1.9 mg/g; and in acetone, 17.4 mg/g. It reduces ammoniacal silver nitrate solution at room temperature. It is soluble in sulfuric acid giving a nonfluorescent yellow solution.

Hydrocortisone occurs as crystalline, striated blocks from absolute ethanol: mp 217–220°C; $[\alpha]_D^{22}$ + 167° (c = 1 in absolute ethanol); absorption max 242 nm; absorbance, 445. The solubility at 25°C: water 0.28 mg/mL, ethanol 15.0 mg/mL, acetone 9.3 mg/mL; chloroform 1.6 mg/mL, propylene glycol 12.7 mg/mL, and ether 0.35 mg/mL. In concentrated H_2SO_4, it produces an intense green fluorescence. It reduces an alkaline silver solution at room temperature.

One synthesis of cortisone involves 27 stages and results in an overall yield of ca 1%; the starting material is 6-methoxy-1-tetralone (30):

Cortisone is produced on an industrial scale by C-11 microbiological hydroxylation of progesterone to yield 11α-hydroxyprogesterone which can be converted to hydrocortisone and cortisone in about 8 and 10 steps, respectively (31). Cortisone can be converted easily to corticosterone (32).

The corticosteroids are synthesized from cholesterol by the adrenal cortex. Pregnenolone is formed from cholesterol by a mitochondrial cytochrome P-450 enzyme system; pregnenolone is converted by a 3-β-hydroxydehydrogenase and isomerase to progesterone. Progesterone is hydroxylated at C-17 by an enzyme in the microsomes of the adrenals, ovaries, or testes to yield 17-hydroxyprogesterone. This is hydroxylated at C-21 in adrenal microsomes to yield 11-desoxycortisol. The latter intermediate is hydroxylated to hydrocortisone by an 11β-hydroxylase in adrenal mitochondria. Hydrocortisone is oxidized to cortisone; however, this reaction is readily reversible (33).

progesterone   17-hydroxyprogesterone   11-desoxycortisol

hydrocortisone   cortisone

Corticosterone is biosynthesized in a manner similar to cortisone from progesterone via 11- and 12-hydroxylation (33).

Pharmacology. The corticosteroids produce increased gluconeogenesis and a small effect on salt and water balance. They also inhibit the response of tissues to inflammation, and they stimulate and promote fat synthesis. They are used in the treatment of collagen diseases, anaphylaxis, asthma, hay fever, serum sickness, adrenal insufficiency as occurs in Addison's disease, and various skin and eye disorders.

They may act as hormonal neuromodulators in the CNS (11). These steroids influence the steady-state levels of tyrosine hydroxylase in the brain which, in turn, affects the activity of catecholaminergic neurons (10). Radioactive corticosteroids have been used to map high-affinity binding sites in the rat brain; high levels of binding are found in the hippocampus and septum (26,34) (see Radioactive drugs and tracers). Cortisone and hydrocortisone have been identified in human cerebral spinal fluid (CSF) and the latter hormone is present in the human brain (26). Corticosterone levels also have been determined in various areas of rat brain (26). The behavioral effects of corticosterone on extinction of avoidance behavior and suppression of paradoxical sleep, and sensory perception have been reviewed (35). Administration of hydrocortisone often has been complicated by the development of both manic and depressive symptoms (16). The circadian control of hydrocortisone secretion is abnormal in many depressed patients (16). The role of the other steroid hormones, eg, the estrogens, androgens, and progestins in the CNS, has been reviewed (26).

N,N-Dimethyltryptamine (DMT)

Properties. The crystalline compound has a mp of 44.6–46.8°C and a pK_a of 8.68 (ethanol–water). It is freely soluble in dilute acetic acid and dilute mineral acids. It forms a picrate [20203-37-6], mp 169.5–170°C, and a methiodide [13558-34-4], mp 216–217°C.

The methods of synthesizing N,N-dimethyltryptamine (DMT) have been reviewed (36). By one method, the methyl ester of indole-3-acetic acid and dimethylamine react and the resulting amide is reduced with lithium aluminum hydride (37).

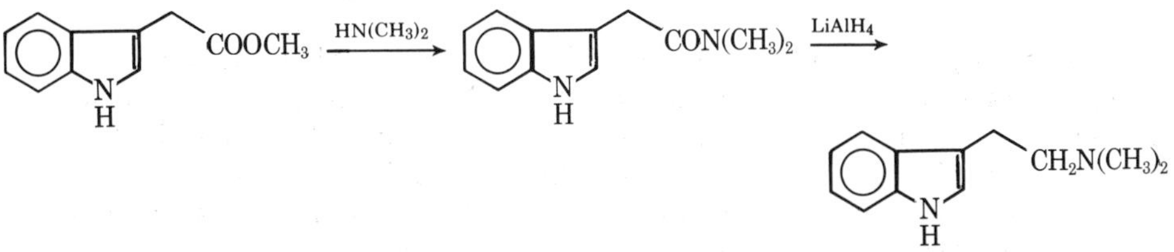

N,N-dimethyltryptamine [61-50-7]

DMT occurs in the seeds of the leguminous shrubs *Piptadenia peregrina* and *Piptadenia macrocarpa*. Methyl transferases that are capable of synthesizing DMT, have been described in human lung, brain, and blood (3). The postulated route is shown below.

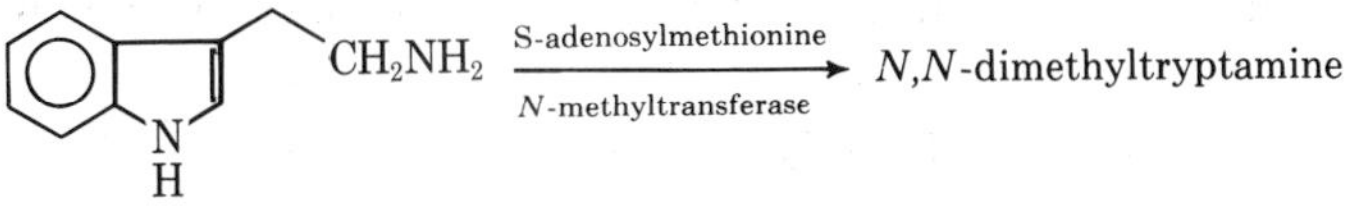

Pharmacology. DMT produces hallucinogenic effects in humans at a dose of 0.7 mg/kg injected intramuscularly (38). It has a rapid action and a short duration of ca 1 h. It has been suggested as a possible schizotoxin, ie, an endogenous compound capable of causing schizophreniclike symptoms (16). Although it has been detected in both normal and schizophrenic patients, the concentrations do not differ significantly (38–39). Although these findings may reflect endogenous synthesis, they also could result from diet, bacterial production, or laboratory error. Thus, its status as a possible neuroregulator must be viewed with a great deal of caution.

Dopamine

Properties. The free base occurs as stout prisms, and it is very sensitive to oxidation. The hydrochloride [62-31-7] forms rosettes of needles from water, dec 241°C. It is freely soluble in water, methanol, and hot 95% ethanol but it is practically insoluble in ether, petroleum ether, chloroform, benzene, and toluene. It also is soluble in aqueous solutions of alkali hydroxides. It occurs as a crystalline hydrobromide [645-31-8], dec 210–214°C, and as a picrate [75802-62-9] appearing as minute brownish-yellow crystals from water, dec 189°C.

The possible synthetic routes to dopamine have been reviewed (40); one is shown here (41):

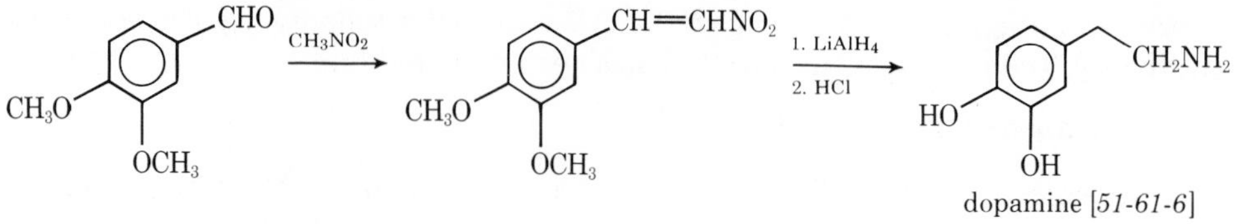

dopamine [51-61-6]

Dopamine is synthesized as an intermediate in the biosynthesis of noradrenaline and adrenaline, ie, from tyrosine (15).

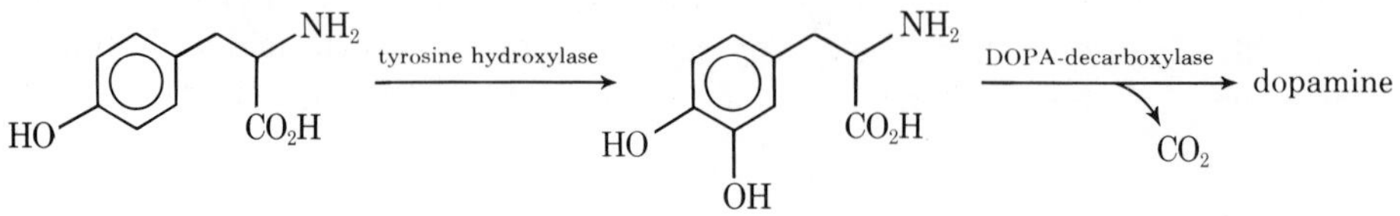

Pharmacology.   Second to acetylcholine, it is arguable that dopamine is one of the best-characterized neurotransmitters in the CNS (40). Its neuronal pathways in the CNS have been mapped by the use of fluorescence histochemistry. The concentrations of dopamine in 101 brain areas have been determined and its neurophysiology and neuropharmacology have been investigated extensively. The involvement of dopamine pathways in Parkinsonism is well documented and has led to the clinical use of the dopamine precursor, L-DOPA, as an effective treatment. Much attention has been focused on the possible role of dopaminergic pathways in schizophrenia. One important reason for this is the fact that the drugs that are most effective in the treatment of schizophrenia, the neuroleptics, are dopamine-receptor blockers.

Endorphins and Enkephalins

Ovine β-Lipotropin and its neurotropic subunits are shown below:

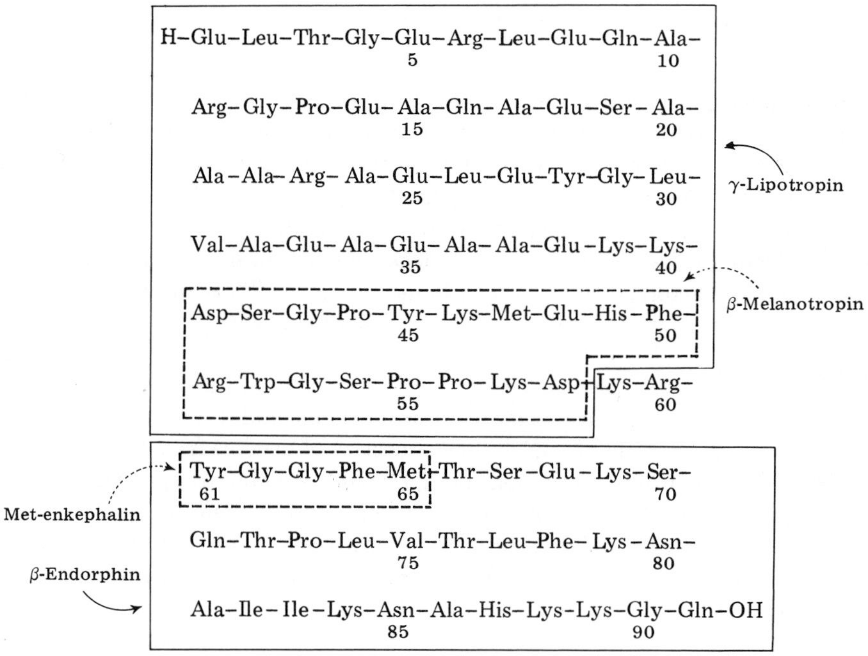

Amino acid sequence of ovine β-lipotropin

61–65 = 5-methionine–enkephalin [58569-55-4]; 61–77 = γ-endorphin [61512-77-4]; 41–58 = α-MSH (melanocyte-stimulating hormone); 61–76 = α-endorphin [59004-96-5]; and 61–91 = β-endorphin [61214-51-5]. 5-Leucine–enkephalin [58822-25-6] has a leucine residue at the C-terminal end instead of a methionine group.

Properties. β-Endorphin occurs as a white powder, $[\alpha]_D^{24}$ $-76.6°$ (c = 0.3 in 0.5 N HCl). Methionine–enkephalin occurs as needles from methanol, mp 196–198°C, $[\alpha]_{589}^{22}$ $-21.9°$ (c = 1 in DMF), +14.1° (c = 1 in N HCl). Leucine–enkephalin occurs as a white crystalline solid from methanol, mp 206°C, $[\alpha]_{589}^{22}$ $-23.4°$ (c = 1 in DMF), + 18.3° (c = 1 in N HCl).

Both of the enkephalin pentapeptides have been synthesized by conventional solution methods employing a (3+2) fragment-coupling approach (42). Human β-endorphin has been synthesized by several groups using either solution (43) or solid-phase techniques (44).

Although all the various endorphins and methionine–enkephalin are contained in the amino acid sequence of β-lipotropin, the precise biosynthetic pathways involved in the formation of these molecules is uncertain. Two types of systems exist in the brain; by way of immunohistochemical techniques, it has been shown that the enkephalins are distributed throughout the brain stem with numerous cell groups and short neuronal tracts. In contrast, β-endorphin and its possible precursor, β-lipotropin, are present both in the pituitary and the brain. Systems containing the latter two peptides are apparently located in a single set of hypothalamic neurons with axons projecting throughout the brain stem. Of the various endorphins, it appears that the major naturally stored product is β-endorphin, with very high concentrations in the intermediate lobe of the pituitary. Specific lesions can affect the enkephalin system without substantially altering the β-lipotropin/β-endorphin system and *vice versa* (45).

Pharmacology. The isolation and structure elucidation of the enkephalins, endogenous compounds with morphinelike activity, was followed by the isolation of the endorphins, pituitary peptides with opiatelike properties (46–47). It appears that these substances resemble each other and the opiates in their ability to produce analgesia and specific motor effects and in their induction of tolerance and dependence (10) (see Analgesics). Most are blocked by the opiate antagonist, naloxone. These endogenous, opiatelike molecules do differ, however, in their potencies; thus, the effects of the enkephalins are short-lived because of the rapid metabolic degradation, whereas the endorphins produce effects that are longer lasting.

The physiological role of these compounds is not clear. High concentrations of both enkephalins are found in brain areas involved in pain transmission, respiration, motor activity, endocrine control, and mood. An interconnection between ACTH and the endorphins has been suggested from the results of studies that show an increase in the brain and blood concentration of the endorphins following stress. Various lines of evidence also indicate a possible interaction between the catecholamine systems and the endogenous opiatelike substances (10). Various attempts have been made to link these systems to human behavioral and disease states but their possible role in schizophrenia and depression is unclear. For further details see ref. 48.

The possible existence of an endogenous neuroleptic-like agent, Des-Tyr¹-γ-endorphin (ie, residues 62–77 of β-LPH), has been claimed (49–50). Initial clinical trials in schizophrenics who were resistant to normal neuroleptic therapy were positive. See ref. 51 for a review of various peptides, including the enkephalins, as possible neurotransmitters.

Estrogens

See ref. 26 (see also Hormones, sex hormones).

17β-estradiol [50-28-2]

γ-Aminobutyric Acid (GABA)

Properties. γ-Aminobutyric acid [56-12-2] is freely soluble in water but is insoluble or poorly soluble in other solvents. It crystallizes as leaflets from a mixture of methanol and ether and as needles from a mixture of water and alcohol. It has a mp of 202°C; $K_a = 3.7 \times 10^{-11}$ and $K_b = 1.7 \times 10^{-10}$ at 25°C. It decomposes on melting to form pyrrolidinone and water. It also occurs as the HCl salt [5959-35-3], mp 135–136°C. The ethyl ester is a liquid, $bp_{1.6 \text{ kPa (12 mm Hg)}}$ 76°C.

GABA can be prepared in various ways, one of the best being the reaction of γ-chlorobutyronitrile and potassium phthalimide to form γ-phthalimidobutyronitrile followed by heating with concentrated H_2SO_4 and prolonged boiling after addition of water (52).

GABA is formed by the α decarboxylation of L-glutamic acid catalyzed by glutamic acid decarboxylase, an enzyme that apparently occurs uniquely in the mammalian CNS and retinal tissue (15).

$$NH_2 \cdot CH(CH_2)_2COOH \xrightarrow[\text{decarboxylase}]{\text{glutamic acid}} GABA$$
$$|$$
$$COOH$$

Pharmacology. The function of GABA in the CNS is not certain but there are strong indications that it functions as an inhibitory transmitter (15,29). GABA has been implicated both directly and indirectly in the pathogenesis of Huntington's disease, Parkinsonism, epilepsy, schizophrenia, and senile dementia (15,29). It occurs in high concentrations in the substantia nigra and globus pallidus in the human and Rhesus monkey brains (15).

L-Glutamic Acid

Properties. The L form occurs as orthorhombic, bisphenoidal crystals from aqueous alcohol, d_4^{20} vac 1.538 g/cm^3, dec 247–249°C, $[\alpha]_D^{22.4}$ +31.4° (c = 1.00 in 6 N HCl); pK_1' 2.19; pK_2' 4.25; pK_3' 9.67. It's solubility in water at 25°C = 8.64 g/L, at 50°C = 21.86 g/L, at 75°C = 55.32 g/L, and at 100°C = 140.00 g/L. It is quite insoluble (0.07 g/L at 25°C) in methanol or ethanol. It is practically insoluble in ether, acetone, or cold glacial acetic acid. The hydrochloride [138-15-8] is marketed under the names: Acidulin, Acidoride, Hypochylin, Antalka, Aciglumin, Pepsdol, Glutamidin, Acidogen, Aclor, Gastuloric, Glutan-HCl, Glutasin, Hydrionic and Muriamic. It forms orthorhombic bisphenoidal plates, dec 214°C, $[\alpha]_D^{22}$ +24.4° (c = 6 in water). Glutamic acid also occurs as sodium [142-47-2] and calcium [19238-49-4] salts and as a calcium chloride complex [34427-83-3]. DL-Glutamic acid occurs as orthorhombic crystals from water, d_D^{20} 1.4601 g/cm^3, dec 225–227°C. Its solubility in water at 25°C = 20.54 g/L and at 100°C = 284.9 g/L. It is sparingly soluble in alcohol and ether.

The various methods of synthesizing and resolving this amino acid have been reviewed (53). The synthesis from malonic ester is detailed below.

$$CH_2(COOC_2H_5)_2 \xrightarrow{HNO_2} HON{=}C(COOC_2H_5)_2 \xrightarrow[Zn/CH_3COOH]{(CH_3CO)_2O}$$

$$CH_3CONHCH(COOC_2H_5)_2 \xrightarrow[NaOC_2H_5]{Cl(CH_2)_2COOC_2H_5}$$

$$\underset{\underset{CH_3CONH}{|}}{C_2H_5OOC(CH_2)_2C(COOC_2H_5)_2} \xrightarrow{HCl} \underset{\underset{NH_2}{|}}{HOOC(CH_2)_2CHCOOH}$$

L-glutamic acid

In brain tissue, L-glutamic acid is formed via a variation in the citric acid cycle. Acetyl-CoA and oxaloacetic acid are condensed in the usual way to form citric acid which then is converted to α-ketoglutaric acid. The latter is transformed to L-glutamic acid either by direct amination or by transamination, the amino donor being γ-aminobutyric acid (GABA). GABA is formed by decarboxylation of glutamic acid (see GABA) and is catabolized via transamination to succinic semialdehyde followed by oxidation to succinate and oxaloacetate. If the two transamination steps are linked, a complete cycle occurs, the GABA shunt (see Fig. 1).

Pharmacology. The sodium salt often is used to impart a meat flavor to food and the hydrochloride has been used to improve the taste of beer. The L-form is the active agent in these applications. L-Glutamic acid also has been used as an antiepileptic. The hydrochloride is used to correct gastric hypo- or achlorhydria. In dogs, it is used to treat chronic intestinal flatulence, to change the pH of the urine, and to treat gastric HCl deficiency.

It occurs in high concentration in the brain and is known to be an excitatory amino acid when applied iontophoretically to nerve cells. Its role as an excitatory neuromuscular transmitter is better established in the invertebrate nervous system than in the mammalian brain. As a result of its unequal distribution in the mammalian spinal cord, ie, it occurs in higher concentrations in dorsal regions than in ventral regions, it is thought that it may act as an excitatory transmitter that is released from primary afferent nerve endings. The status of glutamate as a neurotransmitter/neu-

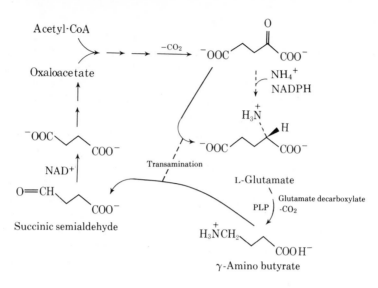

Figure 1.   The γ-aminobutyrate shunt. PLP = pyridoxal phosphate.

roregulator in the mammalian CNS is weaker than that for acetylcholine, the cate-cholamines, 5-HT, GABA, or glycine (29) (see Amino acids, monosodium gluta-mate).

Glycine

Properties.   Glycine [56-40-6] occurs as monoclinic prisms from alcohol. It starts to decompose at 233°C and it is completely sintered at 290°C; d 1.1607 g/cm^3; pK_1' 2.34; pK_2' 9.60. The pH of a 0.2 M solution in water = 4.0. Its solubility in 100 mL of water at 25°C = 25.0 g, at 50°C = 39.1 g, and at 100°C = 67.2 g; 100 g of absolute eth-anol dissolve 0.06 g glycine. It is soluble in 164 parts of pyridine and is almost insoluble in ether. There are two forms of the hydrochloride; glycine.HCl [6000-43-7] and (glycine)$_2$.HCl [13059-60-4]; the former has a mp of 182°C and the latter has a mp of 189°C. It also occurs as a sodium salt [6000-44-8].

Glycine is readily prepared by hydrolysis of methyleneaminoacetonitrile with hydrobromic acid (54).

$$CH_2{=}NCH_2CN \xrightarrow{HBr/H_2O} NH_2CH_2COOH$$
$$\text{glycine}$$

Although the biosynthesis of glycine has been extensively studied in various tissues, its metabolism in nervous tissue is not understood. It can be formed from serine by a reversible folate-dependent reaction catalyzed by the enzyme, serine transhy-droxy-methylase. Serine is formed in nerve tissue from glucose via 3-phosphoglycerate and 3-phosphoserine. It also is possible that glycine may be formed from glyoxylatic acid via a transaminase reaction with glutamic acid.

$$\underset{\text{serine}}{HOCH_2\overset{\overset{\displaystyle NH_2}{|}}{C}HCOOH} \longrightarrow \underset{\text{glycine}}{NH_2CH_2COOH}$$

tetrahydrofolic   methylene
acid (THF)   THF

Pharmacology. Glycine is the simplest amino acid and is classified as nonessential for growth. It has a unique distribution in the mammalian CNS; it is most highly concentrated in the spinal cord and medulla. Electrophysiological and autoradiographic studies indicate that glycine may serve as an inhibitory neurotransmitter in the spinal cord at synapses on motor neurones, spinal interneurones, in dorsal column nuclei, and the medullary reticular formation (15). It also may serve as a transmitter or neuroregulator in the retina. Apart from spasticity, there is little evidence directly linking glycine to any other specific neurological or psychiatric disorders (29).

Histamine

Properties. Histamine occurs as deliquescent needles from chloroform, mp 83–84°C, $bp_{2.4 \text{ kPa (18 mm Hg)}}$ 209–210°C. It is freely soluble in water, alcohol, and hot chloroform but is only sparingly soluble in ether. The LD_{50} (intraperitoneal) in mice is 13.0 g/kg.

Histamine also occurs as the salicylate [75802-62-9] (Cladene) and as the dihydrochloride [56-92-8] mp 244–246°C; histamine monopicrate [5990-88-5], mp 160–162°C; histamine dipicrate [6026-98-8], dec 241°C; and histamine dipicrolonate [6026-99-9], mp 262–264°C.

Histamine can be prepared from diaminoacetone as outlined below (55).

Histamine is formed from the amino acid, histidine, by enzymatic decarboxylation.

Pharmacology. Histamine appears to occur in two types of cells in the brain, ie, mast cells and certain unidentified hypothalamic neurons (15). There is not a great deal of direct neurophysiological evidence that indicates the existence of specific histamine receptors in the brain. Intraventricular histamine changes the cortical electroencephalogram (EEG) and produces sedation. Iontophoretic application of histamine to unidentified hypothalamic spinal and brain stem neurons produces depressant activity but also excites certain hypothalamic and cortical cells (see Histamine and antihistamine agents). The direct evidence for altered histamine metabolism in schizophrenia is unconvincing (16). Histamine is used clinically as a diagnostic agent for gastric secretion and pheochromocytoma (see Medical diagnostic reagents). A review of histamine's possible neuroregulator role in the CNS is given in ref. 56.

4-Hydroxybutyric Acid

Properties. The sodium salt of 4-hydroxybutyric acid, sodium oxybate [502-85-2], occurs as a crystalline solid from alcohol mp 145–146°C.

The sodium salt can be prepared from γ-butyrolactone and sodium hydroxide (57).

$$\text{(γ-butyrolactone)} + NaOH \longrightarrow HO(CH_2)_3CO_2Na$$

In the rat brain, 4-hydroxybutyric acid arises *in vivo* and *in vitro* from GABA. Its natural occurence in the rat brain has been established by gas chromatography–mass spectroscopy techniques, and the acid content is 2 nmol/g. High concentrations of the acid occur in the human substantia nigra and in areas of the basal ganglia (15).

$$H_2N(CH_2)_3COOH \xrightarrow{\text{brain}} HO(CH_2)_3COOH$$
$$\text{4-hydroxybutyric acid}$$

Pharmacology. 4-Hydroxybutyric acid [300-85-4] is used clinically as an anesthetic in the form of its sodium salt. In animals, doses that produce sedation cause a large increase in brain dopamine content but have little effect on brain noradrenaline, serotonin, or GABA (15).

5-Hydroxytryptamine (Serotonin)

Properties.   Serotonin [50-67-9] occurs as the hydrochloride [153-98-0], as light-sensitive, hygroscopic crystals, mp 167–168°C. It is soluble in water and the aqueous solutions are stable at pH 2–6.4. It forms a picrate [6106-56-5], mp 191–193°C.

It also forms a complex with creatinine sulfate occurring as minute plates containing 1 mol H_2O, dec 215°C. It has an absorption max of 275 nm (water at pH 3.5, specific extinction coefficient = 1.5 × 10^4). Its solubility in water at 27°C = 20 mg/mL. It is soluble in glacial acetic acid but only very sparingly soluble in methanol and 95% ethanol. It is insoluble in acetone, pyridine, chloroform, ether, and benzene.

Serotonin has been prepared from 5-benzyloxyindole, as outlined below (58):

serotonin

Serotonin is present in various nonneuronal cells, eg, platelets, mast cells, and enterochromaffin cells. Only 1–2% of the endogenous serotonin occurs in the brain. Tryptophan from the diet is actively taken up by brain cells and is hydroxylated to 5-hydroxytryptophan which is decarboxylated to serotonin.

Pharmacology. Since its isolation from beef in 1948 (59), serotonin has played and continues to play a key role in the neuropharmacology of the CNS. Serotonin appears to play an important role as a neurotransmitter in the CNS (15). Serotonin-containing neurons are restricted to nine clusters of cell bodies lying in or near the midline or raphe regions of the pons and upper brain stem (15). Among other areas, the cerebral cortex and caudate nucleus are innervated by serotonin-containing nerve terminals. In the mammalian CNS, serotonin produces inhibitory effects upon iontophoretic administration. Psychotropic drugs in which a role for serotonin has been implicated are the reserpine type, the tricyclic antidepressants, and the hallucinogens, eg, lysergic acid diethylamide (LSD) (15–16) (see also Hypnotics, sedatives, and anticonvulsants; Psychopharmacological agents).

Luteinizing Hormone Releasing Hormone (LHRH)

The structure of luteinizing hormone releasing hormone [9034-40-6] is H-(Pyro) Glu–His–Trp–Ser–Tyr–Gly–Leu–Arg–Pro–Gly–NH$_2$ (see Hormones, brain oligopeptides).

Properties. The pituitary hormone-releasing action of LHRH is destroyed by chymotrypsin, papain, subtilisin, and thermolysin. Rotational data for LHRH include $[\alpha]_D^{25}$ −50° in 1% acetic acid.

The chemical synthesis of LHRH by solid-phase methods has been reported (60).

Pharmacology. LHRH stimulates the secretion of the pituitary hormones, luteinizing hormone (LH) and follicle-stimulating hormone (FSH) which produce biochemical changes that result in the inducement of ovulation. LHRH occurs in a high concentration in the hypothalamus, particularly near the median eminence. Iontophoretic application of LHRH to neurons in the hypothalamus result in either excitation or inhibition. In ovariectomized, estrogen-primed female rats that are given injections of LHRH into the medial-basal hypothalamus, mating behaviors are activated (15); see also ref. 61.

Melatonin (*N*-Acetyl-5-Methoxytryptamine)

Properties. Melatonin [73-31-4] crystallizes from benzene as pale-yellow leaflets, mp 116–118°C. Its absorption max is 223 nm (ϵ = 27,550); 278 nm (ϵ = 6,300).

Melatonin can be prepared from 5-methoxyindole as shown below (62).

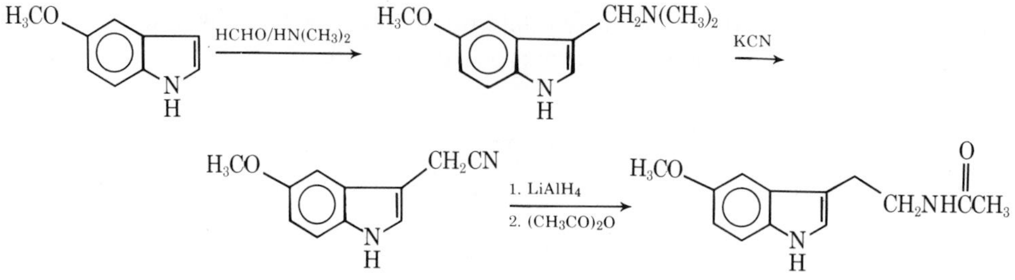

Melatonin occurs in the pineal gland. It is formed from serotonin by acetylation of the side chain amino group followed by O-methylation of the resulting phenol.

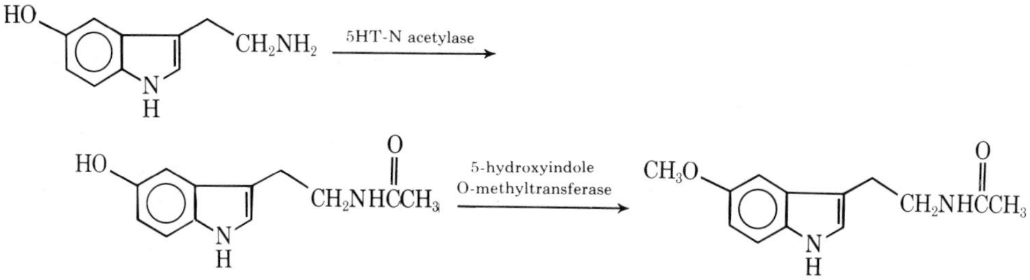

Pharmacology. Melatonin, the pineal factor, lightens skin color by reversing the darkening effect of MSH, the melanocyte-stimulating hormone (see also ref. 63). The melatonin content of the pineal gland and its action in supressing the female gonads is increased by environmental light and lowered by darkness. This cyclic daily rhythm is brought about through a sympathetic innervation, since the adrenergic receptors of the pineal gland are of the β type. Melatonin has been reported to induce sedation and, sometimes, sleep in cats and humans and it potentiates hexobarbital sleeping time in mice (63) (see Hypnotics, sedatives, anticonvulsants).

Neurotensin

The structure of neurotensin [39379-15-2] is

H–Pyr–Leu–Tyr–Glu–Asn–Lys–Pro–Arg–Arg–Pro–Tyr–Ile–Leu–OH.

Properties. Neurotensin occurs as a fluffy white powder, $[\alpha]_D^{23}$ −65.6° (c = 0.5, in H_2O).

Bovine neurotensin [55508-42-4] has been prepared using conventional methods (64). Very little is known regarding the biosynthesis and metabolism of this polypeptide.

Pharmacology. Although it has various biological actions, the normal physiological role of neurotensin is unclear. It induces vasodilation, hypotension, and hyperglycemia. Radioimmunoassays have shown that high levels of immunoreactivity occur in the hypothalamus and basal ganglia (65). Immunohistochemical mapping of this peptide in the rat CNS has been reported (65). Its immunohistochemical distribution resembles that of enkephalin. Central administration of neurotensin produces a decrease in rat body temperature and locomotor activity. An intracisternally injected, 2-fmol dose in mice produces analgesia; thus, it is ca 1000 times more potent than enkephalin as an analgesic (66) (see Analgesics).

(−)-Noradrenaline (Norepinephrine)

Properties. (±) Noradrenaline occurs as crystals, dec 191°C. It is sparingly soluble in water but readily soluble in dilute acids and sodium hydroxide solution. It is only very slightly soluble in alcohol and ether. The racemic mixture can be resolved through the formation of the (+) acid tartrates. The naturally occurring (−)-form [51-41-2] is known as Levarterenol, Adrenor, Aktamin, Binodrenal, and Levophed. It forms microcrystals, dec 216.5–218°C, $[\alpha]_D^{25}$ −37.3° (c = 5 in water with 1 equivalent HCl). The (−) form also occurs as the hydrochloride [55-27-6], mp 145.2–146.4°C, $[\alpha]_D^{25}$ −40° (c = 6). It is freely soluble in water, the solutions are readily oxidized under the influence of oxygen and light. The (−) isomer is available also as the d-bitartrate [51-40-1] mp 102–104°C, $[\alpha]_D^{25}$ −10.7° (c = 1.6 in H_2O); it is freely soluble in water.

Synthetic aspects of the chemistry of noradrenaline have been reviewed (23). Friedel-Crafts acylation of catechol with chloroacetyl chloride yields an α-haloacetophenone which, after reaction with hexamethylenetetramine and catalytic reduction, gives noradrenaline.

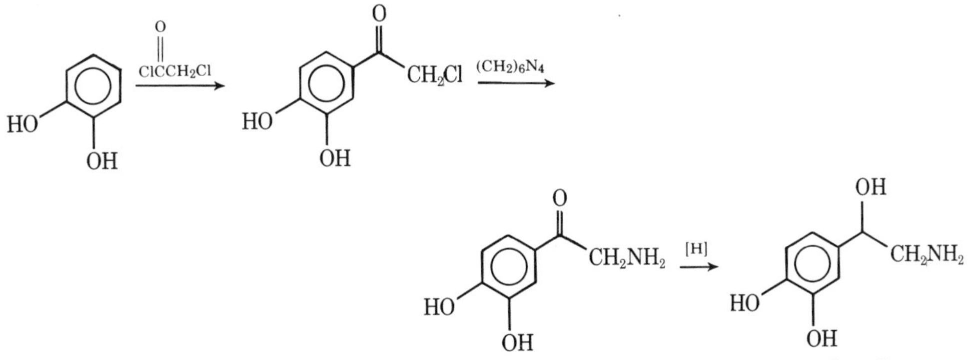

noradrenaline

(−)-Noradrenaline is formed from dopamine with the aid of the enzyme, dopamine β-hydroxylase, which is localized primarily in the membrane of the amine storage granules.

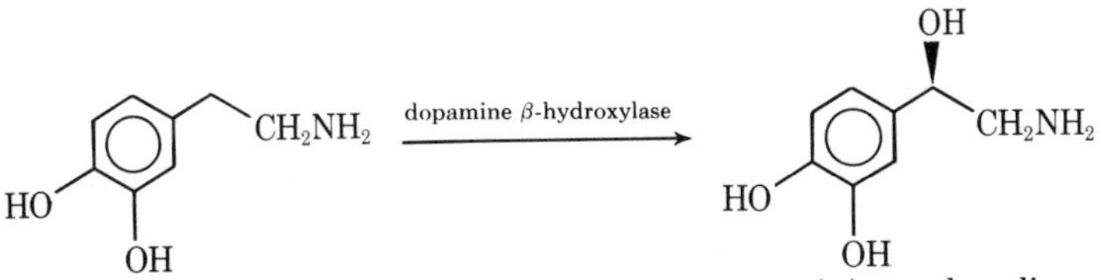

(−)-noradrenaline

Pharmacology. Noradrenaline is synthesized and stored in sympathetic neurons, is released in significant amounts in response to sympathetic nerve stimulation, produces effector-organ responses identical to those resulting from sympathetic nerve stimulation, and drugs blocking the action of applied noradrenaline also block the effect of sympathetic nerve stimulation. It is generally accepted that this amine is the neurotransmitter in the mammalian peripheral sympathetic nervous system (15). It is highly likely that noradrenaline also plays a neurotransmitter role in the CNS. As with dopamine and serotonin, fluorecence histochemical methods have facilitated the mapping of the neuronal pathways containing noradrenaline. There are two main groups of noradrenaline cell bodies; one is located in the locus coeruleus and the other in the lateral ventral tegmental fields (15).

The noradrenergic system in the CNS is thought to be involved in the mode of action of the tricyclic antidepressants, stimulants, eg, the amphetamine group, the reserpine alkaloids and analogues, the monoamine oxidase inhibitors, and the minor tranquilizers. The catecholamine theory of affective disorders suggests that depression and mania may result from an impaired function of the noradrenergic systems in the CNS (16).

Octopamine

Properties. The DL form of octopamine [104-14-3] occurs as a hydrochloride [770-05-8], crystals dec 170°C, and is sold in ampules under the names Norden or Norphen. It is freely soluble in water.

Octopamine may be prepared by reaction of phenol and aminoacetonitrile followed by catalytic reduction (67).

Octopamine is formed from tyramine by the action of dopamine β-hydroxylase. Tyramine arises from tyrosine by decarboxylation.

Pharmacology. Octopamine was first discovered in the salivary glands of the octopus. It occurs in large quantities in the nerve tissue of many invertebrates and in sympathetically innervated and CNS tissue of mammals, including man.

It has been suggested that octopamine is a false transmitter because, following MAO inhibition, it is taken up, stored, and released from catecholamine-containing nerve terminals. However, it has been shown to be normally synthesized, stored, and released from nerve endings; it is enzymatically degraded; and it can stimulate the relatively specific adenylate cyclase system in invertebrate nerve tissue (68). Its possible role as a neurotransmitter/neuroregulator is best characterized in diverse invertebrate systems (69).

Phenylethanolamine

Properties. Phenylethanolamine [7568-93-6] occurs as pale yellow crystals, mp 56–57°C, $bp_{2.3 \text{ kPa (17 mm Hg)}}$ 157–160°C. It is freely soluble in water and it forms a sulfate, $2 \text{ } C_8H_{11}NO.H_2SO_4$, Apophedrin [7568-93-6], crystals mp 275–276°C. Apophedrin is water-soluble.

Phenylethanolamine has been prepared by catalytic reduction of ω-aminoacetophenone (70).

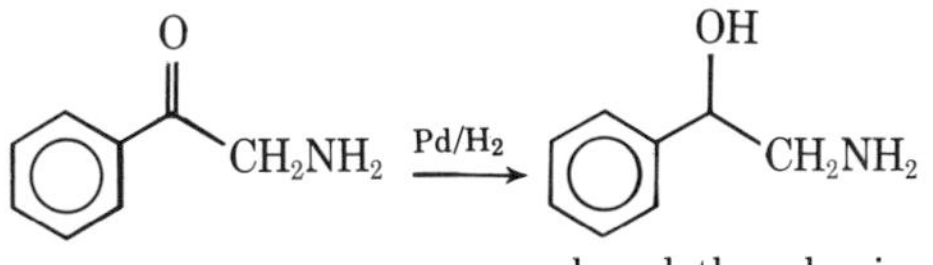

phenylethanolamine

Phenylethanolamine is formed biosynthetically from phenylethylamine [64-04-0] via β-hydroxylation

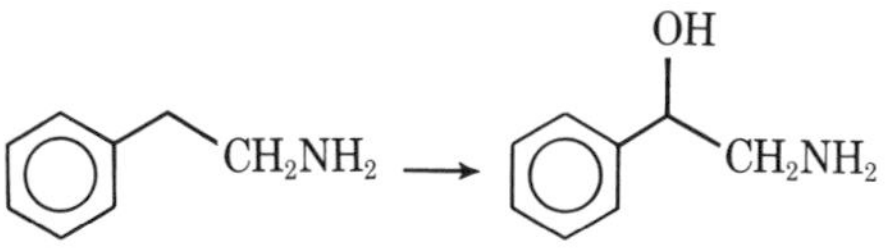

Pharmacology. By means of radioenzymatic and gas chromatography–mass spectroscopy techniques, it has been shown that this amine occurs in trace amounts in various areas of the rat brain (71). In general, its concentrations are lower than those of phenylethylamine. It also has been claimed to occur in human brain (72).

Phenylethylamine

Properties. Phenylethylamine [64-04-0] is a liquid, bp 194.5–195°C, d_4^{25} 0.9640 g/cm^3. It is a strong base, is soluble in water, and is freely soluble in alcohol and ether. It also occurs as a hydrochloride [156-28-5], which appears as orthorhombic bipyramidal platelets from absolute alcohol, mp 217°C. At 15°C, 100 parts of water dissolve 80 parts of the salt. The salt is soluble in alcohol but is insoluble in ether.

Phenylethylamine is readily prepared by the reduction of benzyl cyanide with either sodium and ethanol or with Raney nickel (73).

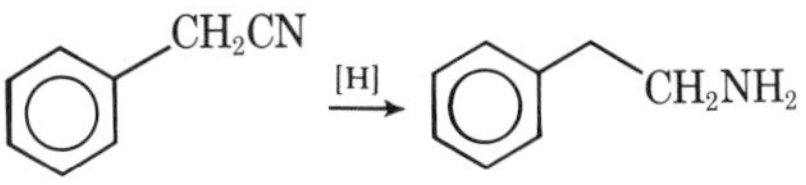

phenylethylamine

Phenylethylamine is formed from phenylalanine by enzymatic decarboxylation.

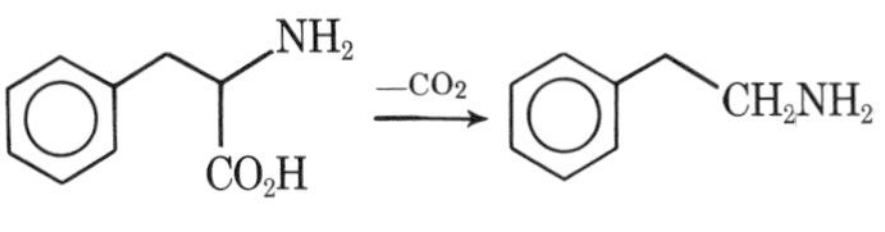

Pharmacology. Phenylethylamine has been detected in low concentrations in the brain and peripheral tissues of humans and animals. It produces an increase in spontaneous motor activity in animals (74). On iontophoretic application, it has a variable effect on the firing rate of optic cortex neurons (75). It is excreted at an abnormal rate from patients suffering from phenylketonuria. It may play a role in schizophrenia and migraine (16,76). The LD_{50} (intraperitoneal) of phenylethylamine hydrochloride is 366 mg/kg; see also ref. 77.

Piperidine

Properties. Piperidine [110-89-4] is a liquid with a characteristic odor. It solidifies at -13 to $-17°C$, mp $-7°C$, $bp_{101.3 \text{ kPa (1 atm)}}$ $106°C$, $bp_{2.7 \text{ kPa (20 mm Hg)}}$ $18°C$, d_4^{20} 0.8622 g/cm^3, n_D^{20} 1.4534. It is a strong base, pK at $25°C$ = 2.80. It is miscible with water and is soluble in alcohol, benzene, and chloroform. Piperidine also occurs as a water-soluble hydrochloride [6091-44-7], mp $247°C$; nitrate [6091-45-8], mp $110°C$; and picrate [6091-49-2], dec $150°C$. The picrate is freely soluble in acetone, ethyl acetate, and hot water.

Piperidine usually is prepared by the electrolytic reduction of pyridine.

piperidine

Piperidine is derived from pipecolic acid by decarboxylation (78). Pipecolic acid is an intermediary metabolite of lysine.

Pharmacology. Piperidine is found in the brain, skin, and urine of animals and in the brain, cerebrospinal fluid, and urine of humans. It occurs in high concentrations in the cerebellum of the dog. It has potent nicotinelike actions on the peripheral and central nervous systems. In the CNS, it produces synaptic stimulation followed by depression. It counteracts experimentally induced aggression in mice and rats and is said to have a tranquilizing action on schizophrenic patients (78). It has transmitterlike properties in the CNS of the mollusk. The levels of piperidine in the mouse brain increase during periods of sleep (79). The LD (subcutaneous) in rabbits is 500 mg/kg.

Polyamines

Properties. Putrescine [110-60-1], $H_2N(CH_2)_4NH_2$, occurs as crystals and has a strong odor, mp $27–28°C$. It is water soluble and readily absorbs CO_2. It forms a dihydrochloride as needles from water, mp $290°C$. Putrescine dihydrochloride [323-93-7] is very soluble in water but is insoluble in methanol. Spermidine [124-20-9], $H_2N(CH_2)_3NH(CH_2)_4NH_2$, occurs as a trihydrochloride [334-50-9], mp $256–258°C$;

as a phosphate [*1945-32-0*], (base)$_2$(H$_3$PO$_4$)$_3$.6H$_2$O, mp 207–208°C; and as a tripicrate, mp 211°C. Spermine [*71-44-3*], H$_2$N(CH$_2$)$_3$NH(CH$_2$)$_4$NH(CH$_2$)$_3$NH$_2$, is a liquid, bp$_{67 \text{ Pa (0.5 mm Hg)}}$ 141–142°C. It readily absorbs CO$_2$. It forms a tetrahydrochloride [*306-67-2*], mp 312–314.5°C and a tetrapicrate [*38693-36-6*], needles, dec 247–248°C.

Putrescine can be prepared as outlined below (80).

$$C_2H_5O_2C(CH_2)_4CO_2C_2H_5 \xrightarrow{\text{N}_2\text{H}_4} H_2NHNCO(CH_2)_4CONHNH_2$$

$$\xrightarrow{\text{HNO}_2} N_3CO(CH_2)_4CON_3 \xrightarrow{\Delta} OCN(CH_2)_4NCO \xrightarrow{\text{HCl/H}_2\text{O}} H_2N(CH_2)_4NH_2$$
$$\text{putrescine}$$

Spermidine is synthesized in two steps from γ-aminobutyronitrile (81).

$$H_2N(CH_2)_3CN + CH_2{=}CHCN \rightarrow NC(CH_2)_2NH(CH_2)_3CN \xrightarrow{\text{Pt/H}_2} H_2N(CH_2)_3NH(CH_2)_4NH_2$$
$$\text{spermidine}$$

Spermine has been synthesized from tetramethylenediamine (putrescine) (82).

$$H_2N(CH_2)_4NH_2 + 2\,Br(CH_2)_3OC_6H_5 \rightarrow C_6H_5O(CH_2)_3HN(CH_2)_4NH(CH_2)_3OC_6H_5$$

$$\xrightarrow[\text{2. NH}_3-\text{C}_2\text{H}_5\text{OH}]{\text{1. HBr}} H_2N(CH_2)_3NH(CH_2)_4NH(CH_2)_3NH_2$$
$$\text{spermine}$$

Putrescine is formed from ornithine by ornithine decarboxylase and is converted to spermidine by a specific synthase, putrescine 3-aminopropyl transferase, which reacts with decarboxylated *S*-adenosylmethionine, as a propylamine donor. With the same donor a spermine synthase converts spermidine to spermine (83).

$$H_2N(CH_2)_3CHCOOH \xrightarrow[\text{ornithine decarboxylase}]{-\text{CO}_2} \text{putrescine} \xrightarrow[\text{synthase}]{\text{SAM}} \text{spermidine} \xrightarrow[\text{synthase}]{\text{SAM}} \text{spermine.}$$
$$|$$
$$NH_2$$
$$\text{ornithine}$$

Pharmacology. The polyamines have been found in the brains of all species that have been examined to determine polyamine presence. The putrescine content of the adult mammalian brain is only a few nmol/g, ie, only 2–3% of the spermidine concentration. During early brain development in the rat, both spermidine and spermine concentrations fall during the first three postnatal weeks. The highest concentrations of spermidine, approaching 100 nmol/g are found in the brain stem and spinal cord. In most species, the spermine content of the brain is lower than that of spermidine. The degradation of spermidine and spermine in the CNS is apparently very slow (83).

A possible neuromodulator role for spermine is more likely than for the other two amines (83). Central administration of spermine produces convulsions in mice and rabbits. The polyamines, in general, produce a depression of the neuronal firing rate. In mice that are made aggressive by isolation, an increase in brain spermidine content parallels the development of the aggression (83).

Prostaglandins

Prostaglandins (qv) are derivatives of the hypothetical parent compound, prostanoic acid:

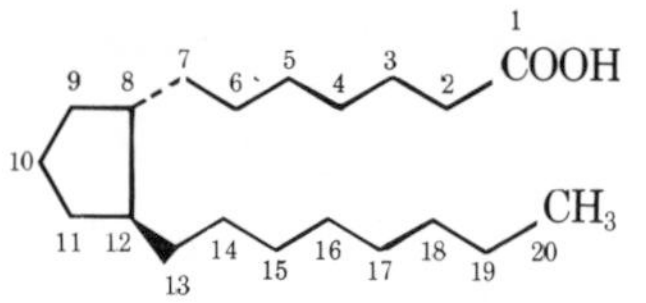

They are classified into various groups, depending on the nature of the cyclopentane ring system:

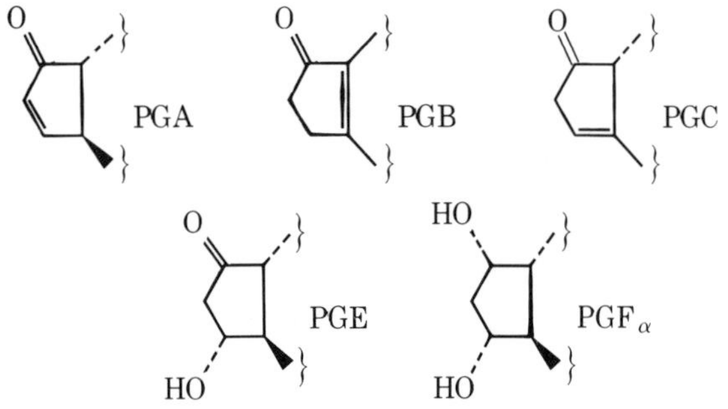

Prostaglandins of the E and F groups are the primary prostaglandins as they are the primary products of the PG synthetase systems. Two prostaglandins, PGE_2 and $PGF_{2\alpha}$, are illustrated below:

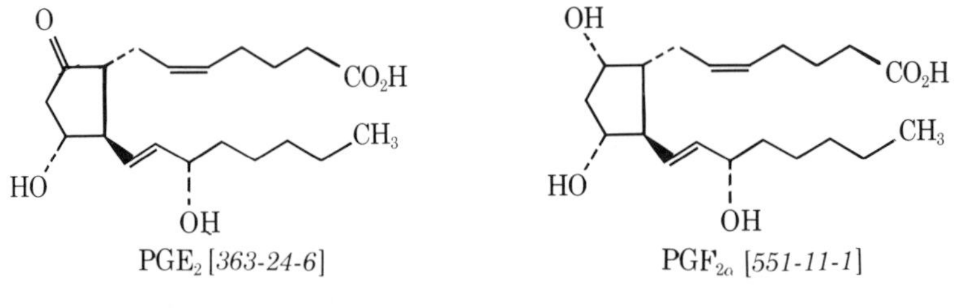

PGE_2 [363-24-6] $PGF_{2\alpha}$ [551-11-1]

The subscript 2 indicates that two double bonds are present in the side chain. The α designation shows that the hydroxyl group at C_9 is below the plane of the cyclopentane ring.

Syntheses.   The total synthesis of various prostaglandins has been reported by several groups (84). One of the most versatile and elegant is for the synthesis of PGE_2 and $PGF_{2\alpha}$, as illustrated in Figure 2 (85). From a common intermediate, both PGE_2 and $PGF_{2\alpha}$ are obtained.

The prostaglandins are formed from arachidonic acid which is made from the essential polyunsaturated fatty acids that occur in the diet. A scheme for the biosynthesis of the primary prostaglandins is shown in Figure 3. A key step in the process is the incorporation of the oxygen atoms at C-9, C-11, and C-15 to produce the endoperoxide, PGG_2 (86).

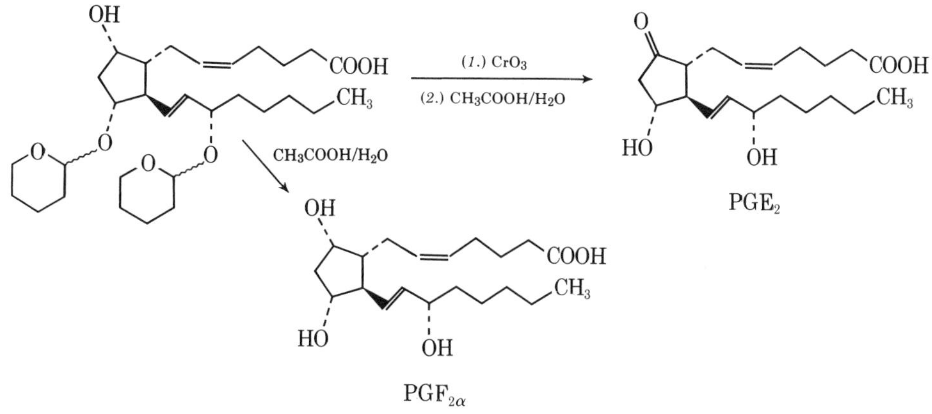

Figure 2. Synthesis of PGE₂ and PGF₂α.

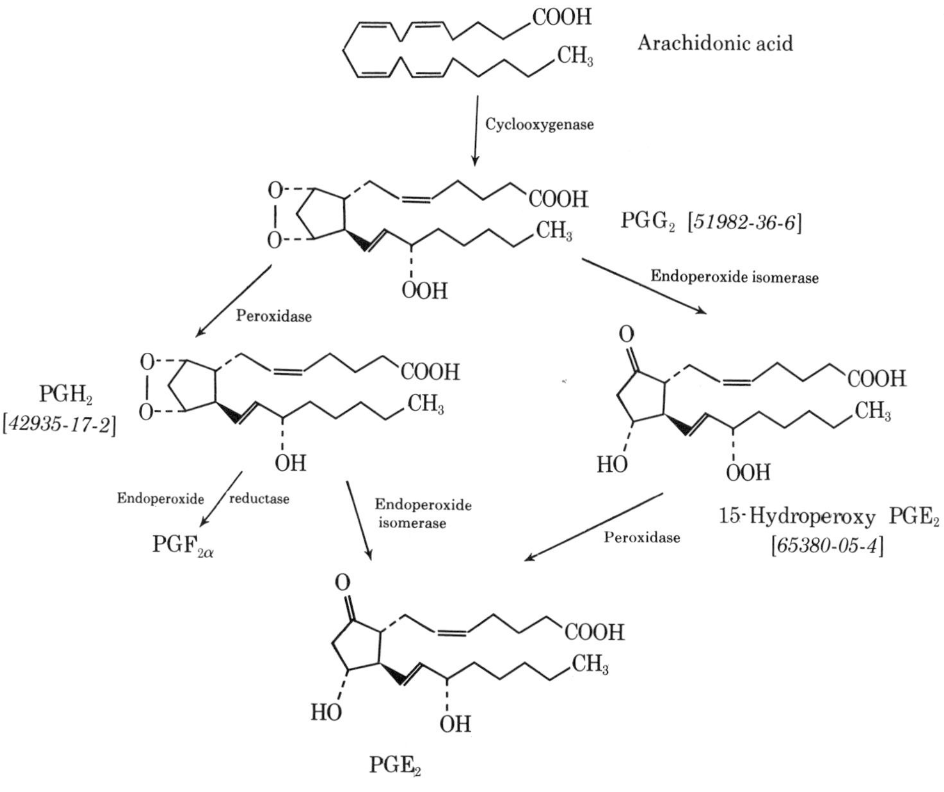

Figure 3. Biosynthesis of the primary prostaglandins.

Pharmacology. Prostaglandins occur throughout the CNS. Although, under resting conditions, the amount of PGs in the CNS is small, they can be synthesized in large amounts on demand. The CNS differs from most other tissues in that stimulation releases increased amounts of PGF$_{2\alpha}$ and PGE$_2$. Rather than limiting their central actions by conversion to metabolites, the CNS excretes prostaglandins into the general circulation. The firing rate of the stimulated neurons occurs on direct application of prostaglandins of the E series to brain-stem cells or the spinal cord. Administration to animals leads to symptoms of sedation. The PGEs enhance the sedative action of barbiturates. Nerve stimulation, analeptic drugs, low blood flow states, trauma, and serotonin all cause a release of prostaglandins in the CNS (86). It is suggested that PGs may exercise a neuromodulator role (86).

Somatostatin

The structure of somatostatin [51110-01-0] is:

H–Ala–Gly–Cys–Lys–Asn–Phe–Phe–Trp–Lys–Thr–Phe–Thr–Ser–Cys–OH

Synthesis. The synthesis of somatostatin has been reported (87–91). Solid-phase synthesis usually yields 15–35% of the linear dithiol form. High dilution techniques are used to obtain cyclic somatostatin in ca 60% yield. Somatostatin also has been prepared by several solution procedures using various fragment condensation schemes (87–91).

Pharmacology. Somatostatin inhibits the release of growth hormone (GH) from the pituitary and the release of thyrotropin and prolactin. It is present in the stomach, intestine, and pancreas. The hypothalamus contains only about 25% of the total body somatostatin content, with the highest concentration in the median eminence. Iontophoretic application leads to both excitatory and inhibitary neuronal responses (65). Intracerebroventricular injections in animals produce decreases in spontaneous motor activity, reduced sensitivity to barbiturates, loss of slow wave and REM (rapid eye movement) sleep, and increased appetite (65). Somatostatin or its analogues may be useful in the treatment of acromegaly, diabetes mellitus, bleeding ulcers, and pancreatis (92).

Substance P

The structure of substance P [11035-08-8, 12769-48-1] is

H–Arg–Pro–Lys–Pro–Gln–Gln–Phe–Phe–Gly–Leu–Met–NH_2.

Properties. The biological activity of substance P is destroyed by trypsin and chymotrypsin. Aqueous solutions of highly purified material lose biological activity in a few minutes; however, this loss can be prevented by storage at low pH, under nitrogen, or with various antioxidants (see Antioxidants and antiozonants). The addition of Tween 80, gelatin, human plasma albumin, or bovine γ-globulin increases the stability of aqueous solutions.

The amino acid sequence of bovine substance P and its solid-phase synthesis have been reported (93–94). The biosynthetic pathways leading to the formation of substance P are unclear.

Pharmacology. Substance P was isolated in 1931 in impure form as an acetone powder extract from the brain and intestine (15). It causes contractions of the isolated rabbit jejunum. Substance P occurs in small neuronal systems in many areas of the CNS. There are high concentrations in neurons that project into the substantia gelatinosa of the spinal cord from the dorsal-root ganglia and it may be the neurotransmitter for primary afferent sensory fibers. The levels of substance P are very high in the substantia nigra, caudate-putamen, amygdala, hypothalmus, and cerebral cortex. Iontophoretic application to neurons in these regions indicate that it has an excitatory action. In patients suffering from Huntington's Chorea, the levels of substance P in the substantia nigra are reduced (15). Of all of the neuropeptides, this substance is thought to be one of the best candidates for a possible neurotransmitter role in the CNS (95–96).

Taurine

Properties. Taurine [107-35-7] occurs as large monoclinic prismatic rods, dec 300°C, pK_1' 1.5, pK_2' 8.74. It is soluble in 15.5 parts of water at 12°C, and 100 parts of 95% alcohol dissolve 0.004 parts at 17°C. It is insoluble in absolute alcohol.

Taurine may be prepared by reaction of β-bromoethylamine with sodium sulfite:

$$Br(CH_2)_2NH_2 \cdot HBr + Na_2SO_3 \rightarrow H_2N(CH_2)_2SO_3H + 2\ NaBr$$
taurine

Taurine is formed in the brain by the oxidation of hypotaurine which is derived from cysteine sulfinic acid via decarboxylation:

$$H_2NCHCH_2SO_2H \xrightarrow{-CO_2} H_2N(CH_2)_2SO_2H \xrightarrow{[O]} H_2N(CH_2)_2SO_3H$$
$$\underset{\displaystyle COOH}{|}$$

Pharmacology. Taurine is found in quite high concentrations throughout the mammalian CNS and in the retina and heart. Total brain taurine content increases as the fetus develops, peaking at weaning and decreasing slowly thereafter. Taurine can be accumulated and released by brain tissue. Its action on CNS neurons is inhibitory. Significant alterations in taurine levels may be associated with retinitis pigmentosa, epilepsy, mongolism, and possibly heart disease. Systemic administration of taurine may be useful in treating epilepsy (97).

Thyrotropin-Releasing Hormone (TRH)

Properties. Thyrotropin is highly soluble in absolute methanol, partially soluble in chloroform, and completely insoluble in pyridine. It is inactivated by diazotized sulfanilic acid and by plasma, serum, or whole blood *in vitro*. It is resistant to inactivation by proteolytic enzymes.

TRH has been synthesized by both classical and solid-phase methods (98–99). An outline of one method is shown below:

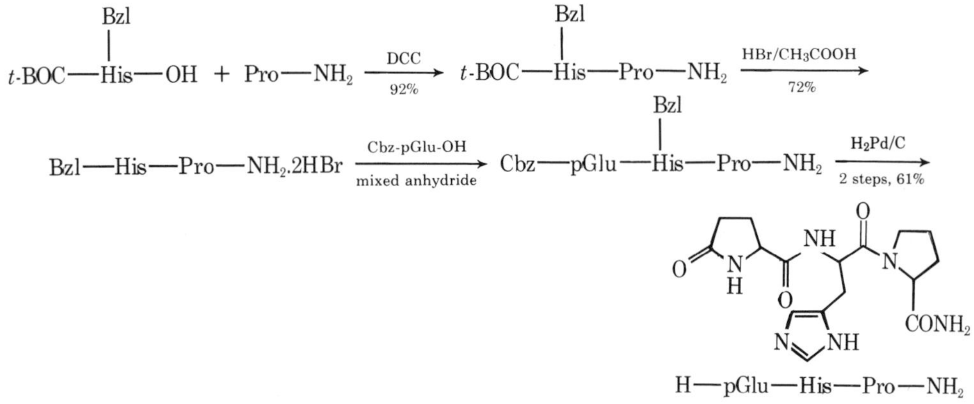

$$\text{H—pGlu—His—Pro—NH}_2$$

TRH is synthesized in the hypothalamus and is released from nerves into portal vessels. Very little is known about the biosynthesis of TRH. It has been shown, however, that guinea pig hypothalamic cultures and whole newt brain can incorporate ^{3}H-proline into TRH but it is unknown whether TRH is synthesized via a mRNA template.

Pharmacology. TRH stimulates the release of thyrotropin from the anterior pituitary via the hypophyseal portal system. In the rat brain, only about 20% of TRH-like immuno reactivity is in the hypothalamus (65). Other areas of importance are the thalamus (34%), cerebrum (27%), and the brain stem (19%). By far the highest concentration of TRH in the hypothalamus occurs in the median eminence. Iontophoretically applied TRH increases the firing of spinal motor neurons and decreases the firing rate in several supraspinal regions (65). Administration of TRH directly into the brain produces antagonism of sedation and of hypothermia induced by barbiturates and ethanol, increased locomotor activity, and anorexia (65). TRH may have antidepressant activity (65).

Tryptamine

Properties. Tryptamine [61-54-1] occurs as needles from petroleum ether, mp 118°C. It is soluble in ethanol and acetone but is practically insoluble in water, ether, benzene, and chloroform. The uv absorption maxima in ethanol are 222, 282, and 290 nm (log ϵ 4.56, 3.78, 3.71). It also occurs as the hydrochloride [343-94-2], mp 248°C and the picrate [6159-31-5], dec 242°C.

Tryptamine syntheses have been reviewed (100) (see also Alkaloids). Gramine is the starting material used in one method (101).

Tryptamine is formed from tryptophan by enzymatic decarboxylation.

Pharmacology. Tryptamine occurs throughout the rodent CNS, the highest brain concentrations being in the hypothalamus and striatum; the highest overall concentrations are in the spinal cord (102). It also has been found in the human brain (103).

Although there is some evidence that tryptamine may act as an excitatory neurotransmitter (104) and that it can produce an LSD-like state after parenteral administration, there is no evidence of abnormalities in tryptamine excretion or formation in schizophrenia (39).

Tyramine

Properties. Tyramine [51-67-2] occurs as crystals from benzene and alcohol, mp 164–165°C, $bp_{3.33\ kPa\ (25\ mm\ Hg)}$ 205–207°C, $bp_{267\ Pa\ (2\ mm\ Hg)}$ 166°C. At 15°C, 1 g tyramine dissolves in 95 mL water and 10 mL boiling alcohol. It is sparingly soluble in benzene and xylene. It forms a crystalline hydrochloride [60-19-5] from alcohol and ether, mp 269°C.

Tyramine has been prepared by reducing p-hydroxyphenylacetonitrile with sodium in ethanol (105).

Tyramine is formed from tyrosine by enzymatic decarboxylation.

Pharmacology. Tyramine is an indirectly acting sympathomimetic amine, ie, it can release catecholamines from storage sites. Its concentration in the brain is much lower than that of the catecholamines and serotonin (106). Oral doses precipitate migraine attacks in susceptible patients. Urinary excretion of the unconjugated amine is abnormal in patients suffering from Parkinson's disease and schizophrenia (107).

Vasoactive Intestinal Polypeptide (VIP)

The structure of VIP [40077-57-4] is H–His–Ser–Asp–Ala–Val–Phe–Thr–Asp–Asn–Tyr–Thr–Arg–Leu–Arg–Lys–Gln–Met–Ala–Val–Lys–Lys–Tyr–Leu–Asn–Ser–Ile–Leu–Asn–NH_2. It is selectively cleaved by kallikrein (108).

Synthesis. VIP has been synthesized using the solution method (108). It was first isolated from the porcine small intestine in 1970 (65). Its amino acid sequence and biological actions are similar to those of secretin and glucagon. Precise details of its biosynthesis are unclear.

Pharmacology. Vasoactive intestinal polypeptide has a broad spectrum of biological activity, including vasodilation, stimulation of glycogenolysis, lipolysis and insulin secretion, stimulation of secretion by the pancreas and small intestine, and inhibition of gastric acid production (65). VIP occurs in the CNS (65), and radioimmunoassay shows that the highest levels occur in the cortex. It also is present in peripheral nerves that innervate the intestinal wall (65). Binding studies with ^{125}I-VIP in guinea pig brain membranes have demonstrated the occurrence of specific, saturable, and reversible binding to receptors that are regulated by guanine nucleotides. It is thought that VIP may act via a cyclic AMP system (65).

Nomenclature

AMP, cyclic	=	adenosine monophosphate [60-92-4]
Bzl	=	benzyl ($C_6H_5CH_2$—)
t-BOC	=	t-butoxycarbonyl (($CH_3)_3CCO$—)
Cbz	=	carbobenzoxy ($C_6H_5CH_2OCO$—)
DA	=	dopamine [51-61-6]
DCC	=	dicyclohexylcarbodiimide [538-75-0]
GABA	=	γ-aminobutyric acid [56-12-2]
5-HT	=	5-hydroxytryptamine [50-67-9]
MAO	=	monoamine oxidase [9001-66-5]
MSH	=	melanocyte-stimulating hormone [9002-79-3]
PLP	=	Pyridoxal 5-phosphate [54-47-7]
SAM	=	S-adenosylmethionine [29908-03-0]

BIBLIOGRAPHY

1. O. Loewi, *Arch. Ges. Physiol.* **189**, 239 (1921).
2. H. H. Dale, *Pharmacol. Rev.* **6**, 7 (1954).
3. J. C. Eccles, *Ann. Rev. Physiol.* **10**, 93 (1948).
4. K. Krnjevic, *Physiol. Rev.* **54**, 418 (1974).
5. P. Fatt and B. Katz, *J. Physiol. (London)* **115**, 320 (1951).
6. J. W. Woodbury and H. D. Patton, *Cold Spring Harbour Symp. Quart. Biol.* **17**, 185 (1952).
7. J. Del Castillo and B. Katz, *J. Physiol. (London)* **128**, 157 (1955).
8. D. R. Curtis and R. M. Eccles, *J. Physiol. (London)* **141**, 446 (1958).
9. M. Vogt, *J. Physiol. (London)* **123**, 451 (1954).
10. J. D. Barchas, H. Akil, G. R. Elliott, R. B. Holman, and S. J. Watson, *Science* **200**, 964 (1978).
11. G. R. Elliott and J. D. Barchas in D. DeWied and P. A. van Kep, eds., *Hormones and the Brain*, M.T. Press Ltd., Lancaster, Eng., in press.
12. R. Werman, *Comp. Biochem. Physiol.* **18**, 745 (1966).
13. H. McClennan, *Synaptic Transmission*, 2nd ed., Sanders, London, 1970.
14. E. Usdin, D. A. Hamburg, and J. D. Barchas, eds., *Neuroregulators and Psychiatric Disorders*, Oxford University Press, New York, 1977.
15. J. R. Cooper, F. E. Bloom, and R. H. Roth, *The Biochemical Basis of Neuropharmacology*, 3rd ed., Oxford University Press, New York, 1978.
16. M. A. Lipton, A. DiMascio, and K. F. Killam, eds., *Psychopharmacology—A Generation of Progress*, Raven Press, New York, 1978.
17. E. Fourneau and J. Page, *Bull. Soc. Chim. France* **15**, 544 (1914).
18. J. Davoll, B. Lythgoe, and A. R. Todd, *J. Chem. Soc.*, 967 (1948).
19. H. P. Baer and G. I. Drummond, eds., *Physiological and Regulatory Functions of Adenosine and Adenine Nucleotides*, Raven Press, New York, 1979.
20. G. Burnstock, *Pharmacol. Rev.* **24**, 509 (1972).
21. G. Burnstock in E. Usdin, D. A. Hamburg and J. D. Barchas, eds., *Neuroregulators and Psychiatric Disorders*, Oxford University Press, New York, 1977, pp. 470–477.
22. H. McIlwain, *Neuroscience* **2**, 357 (1977).
23. H. Loewe, *Arzneimittel Forsch.* **4**, 583 (1954).
24. L. S. Goodman and A. Gilman, eds., *The Pharmacological Basis of Therapeutics*, 5th ed., Macmillan, New York, 1975, pp. 488–489.
25. K. Koyama, H. Watanabe, H. Kawatani, J. Iwai, and H. Yajima, *Chem. Pharm. Bull. (Tokyo)* **24**, 2558 (1976).
26. R. E. Zigmond in L. L. Iversen, S. Iversen and S. H. Snyder, eds., *Handbook of Psychopharmacology*, Vol. 5, Plenum Press, New York, 1975, pp. 239–328.
27. M. Bodanszky and M. A. Ondetti, *Peptide Synthesis*, John Wiley & Sons, Inc., New York, 1966, pp. 215–223.
28. A. C. Cope, ed., *Organic Syntheses*, Vol. 30, John Wiley & Sons, Inc., New York, 1950, pp. 7–10.
29. S. J. Enna in F. H. Clarke, ed., *Ann. Rep. Med. Chem.*, Vol. 14, Academic Press, New York, 1979, pp. 41–50.

30. A. A. Akhrem and Y. A. Titov, *Total Steroid Synthesis*, Plenum Press, New York, 1970.
31. D. Lednicer and L. A. Mitscher, *The Organic Chemistry of Drug Synthesis*, John Wiley & Sons, Inc., New York, 1977.
32. U.S. Pat. 2,927,108 (1960), Oliveto and co-workers (to Schering Corp.).
33. E. Heftmann, *Steroid Biochemistry*, Academic Press, Inc., New York, 1970.
34. B. S. McEwen, *Prog. Brain Res.* **39,** 87 (1973).
35. B. S. McEwen in P. Dell, ed., *Neuroendocinologie de l'axe corticotrope (Brain-Adrenal Interactions),* Les Colloques de l'Institut National de la Sante et de la Recherche Medicale, INSERM, Paris, 1973, pp. 41–62.
36. D. F. Downing in M. Gordon, ed., *Psychopharmacological Agents*, Vol. 1, Academic Press, New York, 1964, pp. 558–559.
37. M. S. Fish, N. M. Johnson, and E. C. Horning, *J. Am. Chem. Soc.* **77,** 5892 (1955).
38. J. C. Gillin, J. Kaplan, R. Stillman, and R. J. Wyatt, *Am. J. Psychiat.* **133,** 203 (1976).
39. J. R. Smythies, *Lancet ii*, 136 (1976).
40. A. S. Horn, J. Korf, and B. H. C. Westerink, eds., *The Neurobiology of Dopamine*, Academic Press, London, 1979.
41. G. A. Swan and D. Wright, *J. Chem. Soc.*, 381 (1954).
42. J. D. Bower, K. P. Guest, and B. A. Morgan, *J. Chem. Soc. Perkin I* 2488 (1976).
43. M. Kubota, T. Hirayama, O. Nagase, and H. Yajima, *Chem. Pharm. Bull.* **26,** 2139 (1978).
44. C. H. Li, D. Yamashiro, L. F. Tseng, and H. H. Loh, *J. Med. Chem.* **20,** 325 (1977).
45. S. J. Watson, H. Akil, P. A. Berger, and J. D. Barchas, *Arch. Gen. Psychiat.* **36,** 35 (1979).
46. J. Hughes, T. W. Smith, H. W. Kosterlitz, L. A. Fothergill, B. Morgan, and H. R. Morris, *Nature, Lond.* **258,** 577 (1975).
47. E. Casta and E. M. Traburchi, eds., *The Endorphins, Advances in Biochemical Psychopharmacology,* Vol. 18, Raven Press, New York, 1978.
48. E. Usdin and W. E. Bunney, Jr., eds., *Endorphins in Mental Illness*, MacMillan Press, London, 1978.
49. D. De Wied, B. Bohus, J. M. van Ree, G. L. Kovacs, H. M. Greven, *Lancet i*, 1046 (1978).
50. J. P. H. Burbach, J. G. Loeber, J. Verhoef, V. M. Wiegant, E. R. de Kloet, and D. de Wied, *Nature* **283,** 96 (1980).
51. S. H. Snyder, *Science* **209,** 976 (1980).
52. *Organic Syntheses*, Coll. Vol. II, John Wiley & Sons, Inc., New York, 1943, p. 25.
53. C. W. Huffman and W. G. Skelly, *Chem. Rev.* **63,** 625 (1963).
54. *Organic Syntheses*, Vol. IV, John Wiley & Sons, Inc., New York, 1925, p. 31.
55. K. K. Koessler and M. T. Hanke, *J. Am. Chem. Soc.* **XL,** 1716 (1918).
56. J. C. Schwartz, *Ann. Rev. Pharmacol. Toxicol.* **17,** 325 (1977).
57. C. S. Marvel and E. R. Birkhimer, *J. Am. Chem. Soc.* **51,** 260 (1929).
58. M. E. Speeter and W. C. Anthony, *J. Am. Chem. Soc.* **76,** 6209 (1954).
59. M. M. Rapport, A. A. Green, and I. H. Page, *Science* **108,** 329 (1948).
60. M. Monahan, J. Rivier, R. Burgus, M. Amoss, R. Blackwell, W. Vale, and R. Guilleman, *C. R. Acad. Sci.* **273,** 205 (1971).
61. R. Guilleman, *Science* **202,** 390 (1978).
62. J. Szmuszkovicz, W. C. Anthony, and R. V. Heinzelman, *J. Org. Chem.* **25,** 857 (1960).
63. K. P. Minneman and R. J. Wurtman, *Ann. Rev. Pharmacol. Toxicol.* **16,** 33 (1976).
64. K. Kitagawa, T. Akita, T. Segawa, M. Nakano, N. Fujii, and H. Yajima, *Chem. Pharm. Bull. (Tokyo)* **24,** 2692 (1976).
65. S. H. Snyder and R. B. Innis, *Ann. Rev. Biochem.* **48,** 755 (1979).
66. B. V. Clineschmidt and J. C. McGuffin, *Eur. J. Pharmacol.* **46,** 395 (1977).
67. M. Asscher, *Rec. Trav. Chim.* **68,** 960 (1949).
68. A. J. Harmar and A. S. Horn, *Molec. Pharmacol.* **13,** 512 (1977).
69. T. P. Hicks, *Can. J. Physiol. Pharmacol.* **55,** 137 (1977).
70. M. C. Rebstock, G. W. Moersch, A. C. Moore, and J. M. Vandenbelt, *J. Am. Chem. Soc.* **73,** 3666 (1951).
71. J. Willner, H. F. LeFevre, and E. Costa, *J. Neurochem.* **23,** 857 (1974).
72. E. E. Inwang, A. D. Mosnaim, and H. C. Sabelli, *J. Neurochem.* **20,** 1469 (1973).
73. J. Robinson and H. Snyder, *Organic Syntheses*, Coll. Vol. III, John Wiley & Sons, Inc., New York, 1955, p. 720.
74. D. M. Jackson, *J. Pharm. Pharmacol.* **24,** 383 (1972).

75. W. J. Giardina, W. A. Pedemonte, and H. C. Sabelli, *Life Sci.* **12**, 153 (1973).
76. M. Sandler, M. B. H. Youdim, and E. Hanington, *Nature* **250**, 335 (1974).
77. A. D. Mosnaim and M. E. Wolf eds., *Noncatecholic Phenylethylamines, Phenylethylamine: Biological Mechanisms and Clinical Aspects*, Part I, Marcel Dekker, New York, 1978.
78. Y. Kasé, Y. Okano, T. Miyata, M. Kataoka, and N. Yonehara, *Life Sci.* **14**, 785 (1974).
79. M. Stepita-Klauco, H. Dolezalova, and R. Fairweather, *Science* **183**, 536 (1974).
80. *Organic Syntheses*, Coll. Vol. IV, John Wiley & Sons, Inc., 1963, p. 819.
81. M. Danzig and H. P. Schultz, *J. Am. Chem. Soc.* **74**, 1836 (1952).
82. H. W. Dudley, O. Rosenheim, and W. W. Starling, *Biochem. J.* **20**, 1082 (1926).
83. G. G. Shaw, *Biochem. Pharmacol.* **28**, 1 (1979).
84. S. M. M. Karim, ed., *Prostaglandins: Chemical and Biochemical Aspects*, University Press, Baltimore, Md., 1976.
85. E. J. Corey, N. M. Weinshenker, T. K. Schaaf, and W. Huber, *J. Am. Chem. Soc.* **91**, 5675 (1969).
86. R. T. Buckler and D. L. Garling.in M. E. Wolff, ed., *Burgers Medicinal Chemistry*, 4th ed., Part II, John Wiley & Sons, Inc., New York, 1979, pp. 1133–1199.
87. N. Fujii and H. Yajima, *Chem. Pharm. Bull. (Tokyo)* **23**, 1596 (1975).
88. A. M. Felix, M. H. Jimenez, T. Mowles, and J. Meienhofer, *Int. Peptide Protein Res.* **11**, 329 (1978).
89. D. Sarantakis and W. A. McKinley, *Biochem. Biophys. Res. Commun.* **54**, 234 (1973).
90. H. U. Immer, K. Sestanj, V. R. Nelson, and M. Götz, *Helv. Chim. Acta* **57**, 730 (1974).
91. D. F. Veber, R. G. Strachan, S. J. Bergstrand, F. W. Holly, C. F. Homnick, R. Hirschmann, M. L. Torchiana, and R. Saperstein, *J. Am. Chem. Soc.* **98**, 2367 (1976).
92. J. Meienhofer in ref. 86, pp. 769-774.
93. M. M. Chang, S. E. Leeman, and H. D. Niall, *Nature New Biology* **232**, 86 (1971).
94. G. W. Tregear, H. D. Niall, J. T. Potts, Jr., S. E. Leeman, and M. M. Chang, *Nature New Biology* **232**, 87 (1971).
95. L. L. Iversen, *Chem. Eng. News*, 30 (Nov. 19, 1979).
96. S. E. Leeman and E. A. Mroz, *Life Sci.* **15**, 2033 (1974).
97. A. Barbeau, N. Inoue, Y. Tsukada, and R. F. Butterworth, *Life Sci.* **17**, 669 (1976).
98. G. Flouret, *J. Med. Chem.* **13**, 843 (1970).
99. J. K. Chang, H. Sievertsson, C. Bogentoft, B. Currie, and K. Folkers, *J. Med. Chem.* **14**, 481 (1971).
100. J. E. Saxton in R. Manske, ed., *The Alkaloids*, Vol. 7, Academic Press, New York, 1960, pp. 5–143.
101. J. Thesing and F. Schülde, *Chem. Ber.* **85**, 324 (1952).
102. S. R. Snodgrass and A. S. Horn, *J. Neurochem.* **21**, 687 (1973).
103. J. M. Saavedra and J. Axelrod, *J. Pharmacol. Exp. Ther.* **185**, 523 (1973).
104. J. A. Bell and W. R. Martin, *J. Pharmacol. Exp. Ther.* **190**, 492 (1974).
105. G. Barger, *J. Chem. Soc.* **95**, 1123 (1909).
106. S. Axelsson, A. Björklund, and N. Seiler, *Life Sci.* **13**, 1411 (1973).
107. A. A. Boulton, R. J. Pollitt, and J. R. Majer, *Nature* **215**, 132 (1967).
108. M. Bodanszky, Y. S. Klausner, C. Yang Lin, V. Mutt, and S. I. Said, *J. Am. Chem. Soc.* **96**, 4973 (1974).

ALAN S. HORN
University of Groningen

NEUTRON ACTIVATION.   See Radioactive tracers; Analytical methods.

NICKEL AND NICKEL ALLOYS

The first reported use of nickel [7440-02-0] was in a nickel–copper–zinc alloy produced in China in the Middle Ages or earlier; alloys of nickel also may have been used in prehistoric times. In the mid 1700s, Cronstedt was the first to isolate the metal for analytical study. He named it nickel, which derives from the German word *kupfernickel*, or false copper.

Nickel, iron, and cobalt occur in the transition group VIII of the Periodic Table. Some physical properties of nickel are given in Table 1 (1–4). It is a high melting element with a ductile crystal structure and with chemical properties that allow it to be combined with other elements to form many alloys.

In the United States in 1978, 44% of the nickel was used in stainless steels and alloy steels, 33% in nonferrous and high temperature alloys (qv), 17% in electroplating, and the remaining 6% was consumed primarily as catalysts, in ceramics, in magnets, and as salts (5–6). The markets for nickel alloys in 1978 were ca 23% in the transportation and aircraft industries, 15% in the chemical industry, 13% in the electrical equipment industry, 9% each as construction and fabricated metal products, and 31% in other uses. In recent years, these proportions have remained quite constant with transportation and aircraft usage increasing slightly. The total United States consumption of nickel remained approximately constant for the past decade with the 1978 consumption being ca 240,000 metric tons. Of this amount, about 181,000 t was primary nickel and 59,000 t was recycled nickel (see Recycling). The world mine production of nickel in 1978 was almost 740,000 t. Hence, the United States now consumes about 24.5% of the world's primary nickel production.

Properties

Select chemistries and properties of nickel and its alloys are given in Tables 2 and 3 (1–3,7–21). Nickel-base alloys provide excellent mechanical properties from cryogenic temperatures to temperatures in excess of 1000°C.

Nickel

Nickel metal is available in many wrought forms and usually is designated as Nickel 200 or Nickel 201, 205, and 270. Nickel 200 is the general-purpose nickel used in ambient-temperature applications in food-processing equipment, chemical containers, caustic-handling equipment and plumbing, electromagnetic parts, and aerospace and missile components. Nickel 201 has much lower trace carbon content than 200 and is more suitable for elevated-temperature applications where the lower carbon content prevents elevated-temperature stress-corrosion cracking. Nickel 205 is low in carbon but contains trace amounts of magnesium and Nickel 270 is one of the purest, ie, 99.98 wt %, commercial nickels. Duranickel alloy 301, which contains about 4.5 wt % aluminum and 0.5 wt % titanium, can be aged to form very fine γ'-(Ni$_3$Al) [12003-81-5] precipitates. This type of alloy combines high strength and hardness with the excellent corrosion resistance that is characteristic of Nickel 200. Various of these nickel metals also are used as welding electrodes for joining ferritic

Table 1. Physical Constants of Nickel[a]

Property	Value
atomic weight	58.71
crystal structure	fcc
lattice constant at 25°C, nm	0.35238
melting point, °C	1453
boiling point (by extrapolation), °C	2732
density at 20°C, g/cm^3	8.908
specific heat at 20°C, kJ/(kg·K)[b]	0.44
av coefficient of thermal expansion × 10^{-6} per °C	
at 20–100°C	13.3
at 20–300°C	14.4
at 20–500°C	15.2
thermal conductivity, W/(m·K)	
at 100°C	82.8
at 300°C	63.6
at 500°C	61.9
electrical resistivity at 20°C, $\mu\Omega$·cm	6.97
temperature coefficient of resistivity at 0–100°C, ($\mu\Omega$·cm)/°C	0.0071
Curie temperature, °C	353
saturation magnetization, T[c]	0.617
residual magnetization, T[c]	0.300
coercive force, A/m[d]	239
initial permeability, mH/m[e]	0.251
max permeability, mH/m[e]	2.51–3.77
modulus of elasticity, × 10^3 MPa[f]	
tension	206.0
shear	73.6
Poisson's ratio	0.30
reflectivity, %	
at 0.30 μm	41
at 0.55 μm	64
at 3.0 μm	87
total emissivity, μW/m^2 [g]	
at 20°C	45
at 100°C	60
at 500°C	120
at 1000°C	190
thermal neutron cross section (for neutron velocity of 2200 m/s), m^2 [h]	
absorption	4.5 × 10^{-28}
reaction cross section	17.5 × 10^{-28}

[a] Refs. 1–4.
[b] To convert J to cal, divide by 4.184.
[c] To convert T to G, multiply by 1.0 × 10^{-4}.
[d] To convert A/m to Oe, divide by 79.58.
[e] To convert mH/m to G/Oe, multiply by 795.8.
[f] To convert MPa to psi, multiply by 145.
[g] To convert μW/m^2 to erg/(s·cm^2), multiply by 10^{-3}.
[h] To convert m^2 to barn, divide by 1.0 × 10^{-28}.

Table 2. Nominal Chemical Composition of Some Nickel Alloys[a], wt %

Alloy	CAS Registry No.	Ni	Fe	Cr	Cu	Mo	Mn	Si	C	Al	Ti	Other
Nickel 200	[12671-92-0]	99.5	0.15		0.05		0.25	0.05	0.06			
Monel Alloy 400	[11105-19-4]	66.5	1.25		31.5		1.0	0.25	0.15			
Monel Alloy K-500	[11105-28-5]	65.0	1.0		29.5		0.6	0.15	0.15	2.8	0.5	
Nichrome	[12605-70-8]	77.0	0.5	20.0			1.0	1.0	0.06			
Inconel Alloy 600	[12606-02-9]	76.0	8.0	15.5			0.5	0.2	0.08			
Incoloy Alloy 800	[11121-96-3]	32.5	46.0	21.0			0.8	0.5	0.05	0.4	0.4	
Hastelloy Alloy B-2	[61608-60-4]	65.4	2.0	1.0		28.0	1.0	0.1	0.02			2.5 Co
Hastelloy Alloy G	[39367-32-3]	42.0	19.5	22.0	2.0	6.5	1.5	1.0	0.05			2.5 Co, 2.0 (Cb + Ta), 1.0 W
Hastelloy Alloy C-276	[12604-59-0]	55.4	5.0	16.0		16.0	1.0	0.08	0.02			2.5 Co, 4.0 W
Inconel Alloy 718	[12606-10-9]	52.5	18.5	19.0		3.0	0.2	0.2	0.04	0.5	0.9	5.1 Cb
Udimet 200[b]	[12616-78-3]	42.7	34.0	13.5		6.2	0.4	0.4	0.05	0.2	2.5	
B-1900	[12773-54-5]	64.0		8.0		6.0			0.1	6.0	1.0	10.0 Co, 4.0 Ta, 0.015 B, 0.1 Zr
Mar-M200	[12604-85-2]	60.0		9.0					0.15	5.0	2.0	10.0 Co, 12.0 W, 1.0 Cb, 0.015 B, 0.05 Zr
Waspaloy	[11068-93-2]	58.0		19.5		4.3			0.08	1.3	3.0	13.5 Co, 0.006 B, 0.06 Zr
Udimet 500	[11068-87-4]	54.0		18.0		4.0			0.08	2.9	2.9	18.5 Co, 0.006 B, 0.05 Zr
Udimet 700	[11068-91-0]	53.0		15.0		5.2			0.08	4.3	3.5	18.5 Co, 0.03 B
Nimonic Alloy 80A	[11068-71-6]	76.0		19.5			0.3	0.3	0.06	1.4	2.4	0.003 B, 0.06 Zr
Nimonic Alloy 115	[51204-21-8]	60.0		14.3		3.3			0.15	4.9	3.7	13.2 Co, 0.16 B, 0.04 Zr
René 41	[11068-84-1]	55.0		19.0		10.0			0.09	1.5	3.1	11.0 Co, 0.005 B
Inconel Alloy 754	[62112-97-4]	78.0		20.0					0.05	0.3	0.5	0.6 Y$_2$O$_3$

[a] Monel, Duranickel, Inconel, Incoloy and Nimonic are trademarks of INCO companies; Hastelloy is a trademark of the Cabot Corporation; Udimet is a trademark of the Special Metals Corporation; Mar M is a trademark of the Martin Marietta Corporation; René 41 is a trademark of Teledyne Allvac; and Waspaloy is a trademark of United Technologies Corporation.

[b] Also known as Incoloy 901.

Table 3. Typical Properties of Some Nickel Alloys[a]

Alloy	Melting range, °C	Yield strength[b], MPa[c]				100-h rupture strength, MPa[c]		
		20°C	538°C	760°C	982°C	649°C	812°C	982°C
Nickel 200	1435–1446	103–931		(139 at 316°C)				
Monel alloy 400	1299–1349	172–1173		(179 at 316°C)				
Monel alloy K-500	1316–1349	241–1380		(648 at 316°C)				
Nichrome	1399	345–1311						
Inconel alloy 600	1355–1415	285	220	180	41	160	55	19
Incoloy alloy 800	1355–1385	250	180	150		240	63	21
Hastelloy alloy B-2[d]	1320–1350	412						
Hastelloy alloy C-276	1323–1371	356	233					
Hastelloy alloy G[d]	1260–1343	319	226	217		310	97	26
Inconel alloy 718	1260–1335	1125	1020	800		724		
Udimet 200	1229–1416	890	759	455		607	152	
B-1900[e]	1275–1300	825	870	808	415		505	170
Mar-M200	1315–1370	840	880	840	470		495	179
Waspaloy	1330–1355	795	725	675	140	760	275	45
Udimet 500	1300–1395	840	795	730	230	930	305	83
Udimet 700	1205–1400	965	895	830	305	828	400	110
Nimonic alloy 80A	1360–1390	620	530	505	62	595	195	14
Nimonic alloy 115		865	795	800	240		400	110
René 41	1315–1370	1060	1020	940	260	635	275	
Inconel alloy 754	1320–1390	662	504	262	166		(131 at 1093°C)	62

[a] Refs. 1–3, 7–21.
[b] Where two numbers appear, the first refers to the annealed condition, the second to the condition when maximum strength is achieved by cold working and/or aging. Otherwise the number refers to the alloy being heat-treated for optimum strength.
[c] To convert MPa to psi, multiply by 145.
[d] 3.18-mm sheet.
[e] As cast.

or austenitic steels to high nickel-containing alloys and for welding the clad side of nickel-clad steels (see Welding).

Nickel has excellent corrosion-resistance properties (see Corrosion and corrosion inhibitors). Nickel and nickel alloys are useful in reducing environments and under some oxidizing conditions in which a passive oxide film is developed. In general, nickel is very resistant to corrosion in marine and industrial atmospheres and outdoors and in distilled and natural waters and flowing sea water. Nickel has excellent resistance to corrosion by caustic soda and other alkalies. In nonoxidizing acids, nickel does not readily discharge hydrogen; hence, it has fairly good resistance to sulfuric acid, hydrochloric acid, organic acids, and other acids but has poor resistance to strongly oxidizing acids like nitric acid. Nickel has excellent resistance to neutral and alkaline salt solutions. Nonoxidizing acid salts are moderately corrosive, and oxidizing acid salts and oxidizing alkaline salts generally are corrosive to nickel. Nickel also is resistant to corrosion by chlorine, hydrogen chloride, fluorine, and molten salts.

Wrought and cast nickel anodes are used widely for nickel electrodeposition onto many base metals. Nickel also can be plated by an electroless process (see Electroless plating). Nickel plating provides resistance to corrosion for many commonly used articles, eg, pins, paper clips, scissors, keys, fasteners, etc, as well as for materials used in food processing, the paper and pulp industries and the chemical industries which often are characterized by severely corrosive environments. Nickel plating is used in conjunction with chromium plating to provide decorative finishes and corrosion resistance to numerous articles. Nickel plating is used to salvage worn, corroded, or incorrectly machined parts. Nickel electroforming, in which nickel is electrodeposited onto a mold which subsequently is separated from the deposit, is used to form complex shapes, eg, phonograph-record stampers, printing plates, tubing, nozzles, screens, and grids (see Electroplating).

Porous nickel electrodes made from nickel powder are used in storage batteries and fuel cells (see Batteries, secondary). Nickel–cadmium batteries have attractive properties including long operating life and long storage life, high rate discharge capability, high rate charge acceptance, and high and low temperature capability.

Nickel also is an important industrial catalyst (see Catalysis). The most extensive use of nickel as a catalyst is in the food industry in connection with the hydrogenation or dehydrogenation of organic compounds to produce edible fats and oils (see Fats and fatty oils).

Alloys

Nickel–Copper. In the solid state, nickel and copper form a continuous solid solution. The nickel-rich, nickel–copper alloys are characterized by a good compromise of strength and ductility and are resistant to corrosion and stress corrosion in many environments, in particular, water and sea water, nonoxidizing acids, neutral and alkaline salts, and alkalies. These alloys are weldable and are characterized by elevated and high temperature mechanical properties for certain applications. The copper content in these also ensure improved thermal conductivity for heat exchange (see Heat exchange technology). Monel 400 is a typical nickel-rich, nickel–copper alloy in which the nickel content is ca 66 wt %. Monel alloy K-500 is Monel 400 with small additions of aluminum. Aging of the K-500 alloy results in very fine γ' precipitates and increased strength.

Typical applications for the nickel–copper alloys are in industrial plumbing and valves, marine equipment, petrochemical equipment, and water-fed heat exchangers. The age-hardened alloys are used as pump shafts and impellers, valves, drill parts, and fasteners. Nickel–copper alloys also are used as coated electrodes or filler alloys for welding purposes. Coinage nickel is an alloy of 75 wt % Cu and 25 wt % Ni.

Copper and nickel can be alloyed with zinc to form nickel silvers. Nickel silvers are ductile, easily formed and machined, have good corrosion resistance, can be worked to provide a range of mechanical properties, and have an attractive white color. These alloys are used for ornamental purposes, as silverplated and uncoated tableware and flatware, in the electrical industry as contacts, connections, and springs and as many formed and machined parts (see also Electrical connectors).

Nickel–Chromium.   Nickel and chromium form a solid solution up to 30 wt % chromium. Chromium is added to nickel to enhance strength, corrosion resistance, oxidation, hot corrosion resistance, and electrical resistivity. In combination, these properties result in the nichrome-type alloys used as electrical furnace heating elements. The same alloys also provide the base for alloys and castings which can withstand hot corrosion in sulfur and oxidative environments, including those with vanadium pentoxides which are by-products of petroleum combustion in fossil-fuel electric power plants and in aircraft jet engines. Alloy additions to nickel–chrome usually are ca 4 wt % aluminum and ca ≤ 1 wt % yttrium. Without these additions, the nichrome-type alloys provide hot oxidation or hot corrosion resistance through the formation of surface nickel–chromium oxides. Aluminum provides for surface Al_2O_3 formation and the yttrium or other rare earth additions improve the adherence of the protective oxide scales to the nickel–chromium–aluminum substrates.

Nickel–Molybdenum.   Molybdenum in solid solution with nickel strengthens the latter metal and improves its corrosion resistance, eg, the Hastelloy alloys. Hastelloy alloy B-2 is noted for its superior resistance to corrosion by hydrochloric acid at all concentrations up to the boiling point, by other nonoxidizing acids, such as sulfuric and phosphoric, and by hot hydrogen chloride gas. Other nickel–molybdenum alloys contain chromium which improves the resistance to corrosion and, especially, to oxidation. Hastelloy alloy C-276, eg, has excellent resistance to corrosion by oxidizing environments, oxidizing acids, chloride solutions, and other acids and salts. Hastelloy alloy N, developed for use in the nuclear industry, contains moderate additions of chromium and molybdenum and is particularly noted for its resistance to corrosion by molten fluoride salts.

Another set of nickel alloys, which have a high chromium content, a moderate molybdenum content, and some copper, are the Illium alloys. These are used in the cast form and generally are wear and erosion resistant and highly resistant to corrosion by acids and alkalies under both oxidizing and reducing conditions.

Nickel–Iron–Chromium.   A large number of industrially important materials are derived from nickel–iron–chromium alloys. These alloys are within the broad austenitic, gamma-phase field of the ternary Ni–Fe–Cr phase diagram and are noted for good resistance to corrosion and oxidation and good elevated-temperature strength. Examples are the Inconel alloys which are based on the Inconel alloy 600 composition. Alloy 600 is a solid solution alloy with good strength and toughness from cryogenic to elevated temperatures and good oxidation and corrosion resistance in many media. In addition, the alloy is easily fabricated and joined. Many modifications of alloy 600 have been made to produce other alloys with different characteristics. For example,

Inconel alloy 601 [12631-43-5] contains aluminum for improved high temperature oxidation resistance, and Inconel alloy 625 [12682-01-8] contains molybdenum and niobium in solid solution for better strength. Other Inconel alloys are specially made for particular corrosive environments at high temperatures. Several age-hardenable alloys contain additions of aluminum and titanium. For example, Inconel alloys 718 [12606-10-9] and X750 [11145-80-5] have higher strength and better creep and stress rupture properties than alloy 600 and maintain the same good corrosion and oxidation resistance.

Another class of nickel–iron–chromium alloys is exemplified by the Incoloy alloys, which are based on the Incoloy alloy 800 composition. Alloy 800 is resistant to hot corrosion, oxidation, and carburization and has good elevated-temperature strength. The modifications of alloy 800 impart different strength or corrosion-resistance characteristics. For example, Inconel alloy 801 [12605-97-9] contains more titanium which, with appropriate heat treatments, can age-harden the alloy and provide increased resistance to intergranular corrosion; Inconel alloy 802 [51836-04-5] contains more carbon which provides improved high temperature strength through carbide strengthening. Incoloy alloy 825 [12766-43-7] and the Hastelloy alloy G contain molybdenum and copper and other additions and are exceptionally resistant to attack by aggressively corrosive environments. Udimet 200, an alloy 901-type material, which contains significant amounts of titanium and molybdenum, has good forgeability and can be age-hardened to impart excellent elevated-temperature mechanical properties.

The corrosion- and heat-resistant alloys, eg, Inconel alloy 600 and Incoloy alloy 800, are used extensively in heat-treating equipment, nuclear and fossil-fuel steam generators, heater-element sheathing and thermocouple tubes, and in chemical and food-processing equipment. Inconel alloy 625 and Incoloy alloy 825 are used in chemical processing, pollution control, marine and pickling equipment, ash-pit seals, aircraft turbines and thrust reversers, and radiation waste-handling systems. The age-hardened Inconel and Incoloy alloys are used in gas turbines, high temperature springs and bolts, nuclear reactors, rocket motors, spacecraft, and hot-forming tools. There are also nickel–iron–chromium alloys used as welding electrode and filler metals.

Nickel-Base Superalloys. Superalloys, which are critical to gas-turbine engines because of their high temperature strength and superior creep and stress rupture resistance, basically are nickel–niobium that is alloyed with a host of other elements (see High temperature alloys). The alloying elements include the refractory metals tungsten, molybdenum, or columbium for additional solid-solution strengthening especially at higher temperatures and aluminum in appropriate amounts for the precipitation of γ' for coherent particle strengthening. Titanium is added to provide stronger γ', and niobium reacts with nickel in the solid state to precipitate the γ'' phase; γ'' is the main strengthening precipitate in the 718-type alloy discussed above. Cobalt generally is present in many superalloys in large amounts, ie, ≥ 10 wt %. It probably enhances strength and oxidation and hot-corrosion resistance which also is provided by the chromium in the alloy. Small excess amounts of carbon usually are present in superalloys for intentional carbide precipitation at grain-boundaries which, as discrete and equiaxed particles, can provide obstacles for grain boundary sliding and motion, thus suppressing creep at high temperatures. Small or trace amounts of elements, eg, zirconium, boron, and hafnium, may be present and they enhance grain-boundary strength and improve ductility. The strength and elevated-temperature properties of a superalloy are dependent upon the volume fraction of the fine γ' precipitates, which can be increased to ca 60 wt %, depending on the aluminum and titanium content. Besides precipitation control at the grain boundaries, improved heat resistance can result from either the elimination of grain boundaries or through the growth of aligned grains with minimum grain boundaries perpendicular to the principal applied stress direction, eg, in turbine-blade applications.

Because of their constitutional complexity, the exact chemistries of nickel-base superalloys must be controlled carefully in order to avoid the precipitation after long-term high temperature exposure of deleterious topologically close-packed (TCP) phases and extraneous carbides. Heat treatment schedules and thermomechanical treatments in the case of wrought alloys also are important to provide optimum strength and performance.

Oxide-Dispersion-Strengthened (ODS) Alloys.   Through mechanical alloying and other powder-metallurgical techniques, highly hot-oxidation- and corrosion-resistant-nickel–chromium matrices are strengthened by very fine dispersions of somewhat chemically inert oxide particles. These oxide dispersoids replace γ' as the main strengthening agent and provide strength benefits close to the melting temperatures. Gamma-prime precipitation strengthening usually begins to decline above 800°C. The ODS nickel-base and iron- and cobalt-base alloys are used mainly in the sheet form in combustion chambers and as exhaust hardware in very high temperature applications; ODS alloys also are used as high temperature turbine hardware.

Nickel–Iron.   A large amount of nickel is used in alloy and stainless steels and in cast irons (see Steel; Iron). Nickel is added to ferritic alloy steels to increase the hardenability and to modify ferrite and cementite properties and morphologies and, thus, to improve the strength, toughness, and ductility of the steel. In austenitic stainless steels, the nickel content is 7–35 wt %. Its major roles are to stabilize the ductile austenite structure and to provide, in conjunction with chromium, good corrosion resistance. Nickel is added to cast irons to improve strength and toughness.

Many nickel–iron alloys have useful magnetic characteristics and are used in a wide range of devices in the electronics and telecommunication fields. Some nickel–iron alloys are magnetically soft and have attractive properties of high initial permeability, high maximum magnetization and low residual magnetization, low coercive force, and low hysteresis and eddy-current losses. These properties are sensitive to alloying and to precipitate and grain morphologies. Important soft magnetic alloys are based on compositions of 78 wt % Ni–22 wt % Fe, 65 wt % Ni–35 wt % Fe, and 50 wt % Ni–50 wt % Fe and often include a few weight percent of molybdenum, copper, or chromium (see Magnetic materials).

The majority of permanent magnets are made from magnetically hard alloys of nickel and iron that are characterized by high values of residual magnetization and coercive force. The many Alnico alloys, consisting of (14–28) wt % Ni–(5–35) wt % Co–(6–12) wt % Al–(0–6) wt % Cu–(0–8) wt % Ti—balance iron, are precipitation-strengthened, hard, brittle alloys in which the magnetic properties are sensitive to heat treatments which determine precipitate and grain morphologies.

Some nickel–iron alloys known as Invar have anomalously low thermal-expansion coefficients within certain temperature ranges. This behavior results from a balance between the normal thermal expansion and a contraction caused by magnetostriction. These alloys, eg, nickel–iron alloys with 36 wt %, 42 wt %, or 50 wt % nickel, and a 29 wt % Ni–17 wt % Co–54 wt % Fe alloy, are used as glass-to-metal joints and in metrology equipment, thermostats and thermometers, cryogenic structures and devices, and many other electrical and engineering applications.

Another anomalous property of some nickel–iron alloys, which are called constant-modulus alloys, is a positive thermoelastic coefficient which occurs in alloys with 27–43 wt % nickel; the elastic moduli in these alloys increase with temperature. Usually and with additions of chromium, molybdenum, titanium, or aluminum, the constant-modulus alloys are used as watch hair springs, in precision weighing machines, measuring devices, and oscillating mechanisms.

Reserves and Resources

Nickel comprises ca 3% of the earth's composition and is exceeded in abundance by iron, oxygen, silicon, and magnesium. However, although nickel comprises ca 7% of the earth's core, nickel ranks 24th in order of abundance in the earth's crust of which it comprises only about 0.009%. Fortunately, ore forms amenable to economic mining exist. The 1978 reserve and resource quantities, the countries or geopolitical regions of occurrence, the ore grades, and the mine production rates are listed in Table 4.

The trend of total world mine production rates from 1966 to 1978 is given in Table 5. The average actual price of nickel in U.S. dollars and the prices based on 1976 U.S. dollars for the same period also are given. The 1980 price of nickel is ca $7.50/kg.

Canada and New Caledonia have the largest reserves. In the other market-economy countries, Indonesia, the Philippines, Australia, and the Dominican Republic also have sizable reserves. The USSR and Cuba have by far the majority of the central-market-economy countries' reserves. As shown in Table 4, the United States has less than 0.4% of the world's estimated reserves. Canada, New Caledonia, and the other market-economy countries produced about 70% of the world's nickel in 1978, and the USSR and Cuba produced most of the rest. The United States produced only 2% of the world's nickel in the same year. From 1973 to 1979, the net import reliance of the United States as a percent of apparent consumption remained about 70% until 1978 when it jumped to 77%.

Table 4   World Nickel Reserves and Resources, Ore Grades, and Mine Production in 1978[a], Thousand Metric Tons

Location	Reserves	Ore grades of reserves, %	Resources	Mine production
Market-economy countries				
Canada	8,600	1.5–3.0	12,500	140.5
New Caledonia	15,000	1.0–3.0	31,000	72.8
United States	200	0.8–1.3	14,900	14.4
other (primarily Australia, Indonesia, South Africa, Dominican Republic, and the Philippines)	28,000	0.2–4.0	82,700	293.5
Central-economy countries				
Cuba	3,400	1.4	14,200	40.0
other (primarily the USSR)	4,800	0.4–4.0	12,800	177.2
Seabed nodules			760,000	

[a] Refs. 5–6, 22.

Table 5.   World Nickel Mine Production and Nickel Prices from 1966 to 1979[a]

Year	World mine production, thousand metric tons	Average annual price, $/kg	Average price[b], $/kg
1966	475	1.87	3.26
1967	480	2.07	3.51
1968	545	2.27	3.68
1969	530	2.82	4.37
1970	685	2.82	4.14
1971	706	2.93	4.08
1972	708	3.09	4.12
1973	726	3.37	4.28
1974	826	3.84	4.43
1975	900	4.48	4.72
1976	886	4.85	4.85
1977	851	4.78	4.28
1978	738	4.59	4.03
1979[c]	746	6.61	

[a] Ref. 22.
[b] Price based on 1976 U.S. dollar.
[c] Estd.

The world's mine production rates between 1966 and 1978 show large fluctuations from year to year. A 12-yr averaging shows ca a 4% growth in annual production rates. The average price of nickel also has varied from year to year, and the actual price has doubled from 1968 to 1978. However, the inflation-compensated price has increased only by about 2.2% annually.

The world nickel reserves are estimated at 6×10^7 metric tons. At the 1979 world rate of mine production, the reserves are expected to last over 80 yr. If, however, annual mine production increases at a rate that reflects a predicted increase in the world primary nickel consumption of 3.6% annually, the identified nickel reserves will be depleted in less than 40 yr (6,23–24).

In addition to the reserves, there are substantial other nickel resources which will be amenable to mining and refining when the appropriate technology becomes available and commercially feasible. By far the single largest resource is the seabed or ocean nodules which contain ca 1% nickel and which represent an estd 7.6×10^8-t resource. Ocean-floor mining is being debated internationally with respect to mining rights and other legal, political, and environmental aspects. If all of the nickel reserves and resources are considered and at the current rate of mine production, the reserves and resources should last well over a thousand years. If the annual world growth in primary nickel consumption is 3.6%, the world's nickel reserves and resources will be depleted in about a century (see Ocean raw materials).

Nickel Ores

There are two types of nickel ore which can be mined economically and which are classified as sulfide and lateritic (25). The sulfide deposits currently account for most of the nickel that is produced in the world. The most common nickel sulfide is pentlandite, $(Ni,Fe)_9S_{16}$, which is almost always found in association with chalcopyrite, $CuFeS_2$, and large amounts of pyrrhotite, Fe_7S_8. Other much rarer nickel sulfides include millerite, NiS, heazlewoodite, Ni_3S_2, and the sulfides of the linnaeite series, $(Fe,Co,Ni)_3S_4$. The nickel sulfides were formed thousands of meters below the surface of the earth by the reaction of sulfur with nickel-bearing rocks. These sulfides generally are found in northern regions where glacial action has planed away much of the overlying weathered surface rock; important deposits are found in Canada, the USSR, and Finland.

In contrast to the sulfide ores, the lateritic ores were formed over long periods of time as a result of weathering of exposed nickel-containing rocks. The lateritic weathering process resulted in nickel solutions that were redeposited elsewhere in the form of oxides or silicates. One type of laterite is limonitic or nickeliferous iron laterite which consists primarily of hydrated iron oxide in which the nickel is dispersed in solid solution. The other type of laterite is nickel silicate in which nickel is contained in solid solution in hydrated magnesium–iron minerals, eg, garnierite. Lateritic ores occur primarily in tropical regions, eg, New Caledonia, or in regions which were once at least subtropical for extended periods, eg, Oregon. In such regions, the weathering that produces limonitic and silicate ores is active. These deposits are distributed widely and constitute the largest nickel reserves.

Extraction, Refining, and Alloying

The treatments used to recover the nickel from sulfide and lateritic ores differ considerably because of the ores' different physical characteristics. The sulfide ores, in which the nickel, iron, and copper occur in a physical mixture as distinct minerals, are amenable to initial concentration by mechanical methods, eg, flotation (qv) and magnetic separation (qv). The lateritic ores are not susceptible to these physical processes of beneficiation and chemical means must be used to extract the nickel. The nickel concentration processes that have been developed are not as effective for the lateritic ores as for the sulfide ores (see also Extractive metallurgy).

Sulfide Ores. *Pyrometallurgical Processes.* Sulfide ores first undergo crushing and milling operations to reduce the material to the necessary degree of fineness for separation. Froth flotation or magnetic separation processes separate the sulfides from the gangue. Most sulfide ores then undergo a series of pyrometallurgical processes consisting of roasting, smelting, and converting. Roasting, in which much of the iron is oxidized and a large portion of the sulfur is removed as sulfur dioxide, is carried out in multihearth furnaces, fluidized-bed roasters, or rotary kilns. The material then is smelted in reverberatory furnaces or blast furnaces or by flash smelting or arc-furnace smelting. During smelting, a siliceous slag with iron oxide and other oxide compounds is removed and the sulfur content is further reduced yielding an impure copper–nickel–iron–sulfur matte. In the converting or Bessemerizing stage, the matte is charged into a horizontal-type converter and the molten matte with added silica is blown with air, which removes virtually all the remaining iron in a slag and more of the sulfur as sulfur dioxide and yields a sulfur-deficient copper–nickel matte.

The matte can be treated in different ways, depending on the copper content and on the desired product. In some cases, the copper content of the Bessemer matte is low enough to allow the material to be cast directly into sulfide anodes for electrolytic refining. Usually it is necessary first to separate the nickel and copper sulfides. The copper–nickel matte is cooled slowly for ca 4 d to facilitate grain growth of mineral crystals of copper sulfide, nickel sulfide, and a nickel–copper alloy. This matte is pulverized and the nickel and copper sulfides are isolated by flotation and the alloy is extracted magnetically and refined electrolytically. The nickel sulfide is cast into anodes for electrolysis or, more commonly, is roasted to nickel oxide and further reduced to metal for refining by electrolysis or by the carbonyl method. Alternatively, the nickel sulfide may be roasted to provide a nickel oxide sinter that is suitable for direct use by the steel industry.

Electrolytic Refining.   The electrolytic refining process generally is carried out in a divided cell using anodes which are cast from the crude metal or from nickel sulfides. The electrolyte is pumped continuously through the cell and impure anolyte, which forms by solution of the anode, is pumped out of the electrolyzing tank and through a purification train to remove soluble impurities. The impure anolyte is prevented from coming into direct contact with the cathode by the use of a porous diaphragm. The nickel cathode starting sheets are made by deposition onto stainless steel blanks from which the nickel sheets are stripped after two days and then built up in ca 10 d. The final nickel metal that is obtained has a purity exceeding 99.9 wt %. The electro-refining process also facilitates the recovery of precious metals and other metals of value, eg, cobalt, that remain in the insoluble anode residues which can be collected during the nickel refining.

Carbonyl Process.   Crude nickel also can be refined to very pure nickel by the carbonyl method. The crude nickel and carbon monoxide react at ca 100°C to form a nickel carbonyl, $Ni(CO)_4$, which upon further heating to ca 200–300°C, decomposes to nickel metal and carbon monoxide. The process is highly selective because, under the operating conditions of temperature and atmospheric pressure, carbonyls of other elements that are present, eg, iron and cobalt, are not readily formed. The nickel carbonyl is decomposed in the presence of small nickel pellets upon which the nickel is deposited. The process yields nickel pellets typically about 0.8 cm dia and of better than 99.9 wt % purity (see Carbonyls).

In a variation of the carbonyl process, a high pressure is used during the formation of the nickel carbonyl which facilitates its subsequent condensation to liquid. The liquid is purified, vaporized, and rapidly heated to ca 300°C which results in the decomposition of the vapor to carbon monoxide and a fine high purity nickel powder with a particle size <10 μm. This product is useful for powder metallurgical applications (see Powder metallurgy).

Hydrometallurgical Processes.   Hydrometallurgical refining also is used to extract nickel from sulfide ores. Sulfide concentrates can be leached with ammonia to dissolve the nickel, copper, and cobalt sulfides as amines. The solution is heated to precipitate copper, the nickel and cobalt solution is oxidized to sulfate, and then reduced with hydrogen at a high temperature and pressure to precipitate the nickel and cobalt. The nickel is deposited as a 99 wt % pure powder.

Lateritic Ores.   *Pyrometallurgical Processes.*   Nickel oxide ores are processed by pyrometallurgical or hydrometallurgical methods. In the former, oxide ores are smelted with a sulfiding material, eg, gypsum, to produce an iron–nickel matte that can be treated similarly to the matte obtained from sulfide ores. The iron–nickel matte may be processed in a converter to eliminate iron, and the nickel matte then can be cast into anodes and refined electrolytically. A different type of nickel product is obtained by roasting the nickel matte to the oxide, grinding and compacting the oxide, and re-

Table 6.   Commercial Forms of Primary Nickel

Type	Approximate nickel content, wt %	Uses
electrolytic (cathode)	>99.9	alloy production, electroplating
carbonyl pellets	>99.7	alloy production, electroplating
briquettes	99.9	alloy production
rondelles	99.25	alloy production
powder	99.74	sintered parts, battery electrodes
nickel oxide sinter	76.0	steel and ferrous alloy production
ferronickel[a]	20–50	steel and ferrous alloy production
nickel salts		electroplating, catalysts
nickel chloride	24.70[b]	
nickel nitrate	20.19[b]	
nickel sulfate	20.90[b]	

[a] Different grades of ferronickel are produced, and the nickel content denoted includes 1–2 wt % Co.
[b] Nickel content is theoretical.

ducing to metal with charcoal in a muffle furnace. The metal sinters to form rondelles that contain ca 99.3 wt % nickel. Alternatively, the nickel oxide ore may be smelted without a sulfiding agent and reduced with coke in an electric furnace to produce ferronickel. Ferronickel generally contains 20–30 wt % nickel but the nickel content may be higher.

Hydrometallurgical Processes.   The hydrometallurgical treatments of oxide ores involve leaching with ammonia or with sulfuric acid. In the ammoniacal leaching process, the nickel oxide component of the ore first is reduced selectively. Then the ore is leached with ammonia which removes the nickel into solution from which it is precipitated as nickel carbonate by heating. A nickel oxide product used in making steel is produced by roasting the carbonate.

In the acid-leaching process, the oxide ore is leached with sulfuric acid at elevated temperature and pressure which causes nickel and sulfur, but not iron, to enter solution. The leach solution is purified and then it reacts with hydrogen sulfide with subsequent precipitation of nickel and cobalt sulfides. The nickel sulfide is refined by conversion to a sulfate solution and reduction with hydrogen to produce a high purity nickel powder.

Available Forms of Nickel.   Nickel is available in many forms suited to meet the various demands of industry. The main forms of nickel that are marketed are listed in Table 6. The very pure unwrought nickels, primarily the electrolytic cathodes and pellets and, to a lesser degree, the briquettes and rondelles, are used for production of alloys in which contamination by undesired elements must be minimized in order to obtain desired properties, eg, in nickel-base superalloys and magnetic materials. The pure nickel powder is utilized in the production of porous plates for batteries and in the production of powder-metallurgy parts. A large amount of nickel is available as nickel oxide sinter and ferronickel, which is widely used in the steel and foundry industries as an economical nickel source.

Alloying.   Nickel is alloyed into low alloy steels, ferritic alloy steels, and austenitic stainless steels through the conventional steelmaking processes, eg, open hearth, basic oxygen conversion, and the argon–oxygen degassing (AOD) processes. The AOD process is used to produce a substantial quantity of the stainless steels in the world. It is a highly productive process that yields cleaner products at lower operating and materials costs as compared to the older conventional electric-arc-furnace steelmaking practice. Electroslag remelt (ESR) processing also is used to further refine stainless steels.

Alloys that are heavily alloyed with nickel, including the nickel-base and iron-base superalloys, almost always are produced by vacuum-induction melting (VIM). In this process, the melt, alloying, and melt treatments and the casting are carried out under vacuum. Industrial furnaces generally can process 1-t to 10-t batches. For further alloy refinement, VIM castings are used as electrodes and are remelted by the vacuum arc argon. In certain countries, eg, the People's Republic of China and the USSR, where there is a shortage of argon, ESR remelting often is done instead of VAR. Investment castings of the chemically complex nickel-base alloys, especially those containing the reactive elements aluminum and titanium, also are carried out under vacuum. In recent years, directional solidification techniques, in which the heat is extracted directionally through a controlled solidification rate and temperature gradient, are used to produce either monocrystalline nickel-base superalloys or polycrystalline structures with long columnar grains. Vacuum powder-atomizing techniques, which involve VIM master melts, also are used routinely to produce fine nickel-base powders for subsequent powder metallurgical consolidation of net-shape components.

BIBLIOGRAPHY

"Nickel and Nickel Alloys" in *ECT* 1st ed., Vol. 9, pp. 271–288, by W. A. Mudge and W. Z. Friend, The International Nickel Company, Inc.; "Nickel and Nickel Alloys" in *ECT* 2nd ed., Vol. 13, pp. 735–753, by J. B. Adamec and T. E. Kihlgren, The International Nickel Company, Inc.

1. S. J. Rosenberg, *Natl. Bur. Stand. Monogr.* **106,** (1968).
2. *Metals Handbook*, 9th ed., Vol. 3, American Society for Metals, Metals Park, Ohio, 1980.
3. W. Betteridge, *Nickel and Its Alloys, Industrial Metals Series*, MacDonald and Evans, Ltd., London, Eng., 1977.
4. R. C. Weast, *Handbook of Chemistry and Physics*, 61st ed., Chemical Rubber Publishing Co., Boca Raton, Fla., 1980–1981.
5. N. A. Matthews in *Mineral Commodity Summaries 1979*, U.S. Bureau of Mines, Washington, D.C., 1979, p. 104.
6. N. A. Matthews, *Nickel, Mineral Commodity Profiles*, U.S. Bureau of Mines, Washington, D.C., May 1979.
7. F. B. White-Howard, *Nickel, An Historical Review*, D. Van Nostrand Co. Inc., New York, 1963.
8. J. L. Everhart, *Engineering Properties of Nickel and Nickel Alloys*, Plenum Press, New York, 1971.
9. *Nickel Alloys*, Huntington Alloy Products Division, The International Nickel Co., Inc., Huntington, W. Va., 1972.
10. *Monel Alloys*, Huntington Alloy Products Division, The International Nickel Co., Inc., Huntington, W. Va., 1978.
11. *INCONEL Alloy 600*, Huntington Alloy Products Division, The International Nickel Co., Inc., Huntington, W. Va., 1978.
12. *INCOLOY Alloys*, Huntington Alloy Products Division, The International Nickel Co., Inc., Huntington, W. Va., 1973.
13. *Resistance to Corrosion*, Huntington Alloy Products Division, The International Nickel Co., Inc., Huntington, W. Va., 1970.
14. *High Temperature, High Strength Nickel Base Alloys*, The International Nickel Co., Inc., New York, 1977.

15. *1979 Databook, Metal Progress*, American Society of Metals, Metals Park, Ohio, June 1979.
16. *Alloy Performance Data*, Special Metals Corp., New Hartford, N.Y.
17. *Hastelloy Alloy B-2*, Cabot Corporation, Stellite Division, Kokomo, Ind., 1977.
18. *Hastelloy Alloy C-276*, Cabot Corporation, Stellite Division, Kokomo, Ind., 1978.
19. *Hastelloy Alloy G*, Cabot Corporation, Stellite Division, Kokomo, Ind., 1976.
20. S. Purushothaman and J. K. Tien, "Metallurgy of High Temperature Alloys" in *Properties of High Temperature Alloys*, Electrochemical Society, Inc., Princeton, N.J., 1976.
21. B. H. Kear, D. R. Muzyka, J. K. Tien, and S. T. Wlodek, eds., *Superalloys: Metallurgy and Manufacture*, Claitor's Publishing Division, Baton Rouge, La., 1976.
22. "Nickel," *Mineral Commodity Summaries, 1967–1979*, U.S. Bureau of Mines, Washington, D.C., 1967–1979.
23. J. K. Tien, R. M. Arons, and R. W. Clark, *J. Met.* **28**(12), 26 (1976).
24. J. K. Tien and R. M. Arons, *AIChE Symp. Ser.* **73**(170), (1977).
25. J. R. Boldt, Jr., and P. Queneau, *The Winning of Nickel*, D. Van Nostrand Co., Inc., New York, 1967.

John K. Tien
Timothy E. Howson
Columbia University

NICKEL COMPOUNDS

Inorganic Compounds

Nickel has a $3d^8 4s^2$ electronic configuration and forms compounds in which the nickel atom has oxidation states of -1, 0, $+1$, $+2$, $+3$, and $+4$. Ni(II) represents the bulk of all known nickel compounds. Examples of stable crystalline derivatives of Ni(III) and Ni(IV) are the fluoride anions $(NiF_6)^{3-}$ [32698-29-6] (1) and $(NiF_6)^{2-}$ [23712-86-9] (2). Examples of the binuclear and diamagnetic species of Ni(I) are the cyanonickelates $K_4[Ni_2(CN)_6]$ [40810-33-1] and $K_6[Ni_2(CN)_8]$ (3–4). Nickel(I) dihydrohexacarbonyl [12549-35-8], $H_2Ni_2(CO)_6$, is an example of nickel in the -1 oxidation state (5). Many Ni(0) compounds have been prepared; nickel[0] carbonyl [13463-39-3], $Ni(CO)_4$, is the most common (6). Other types of Ni(0) compounds are $[(C_6H_5)_3P]_2Ni(CCl_2CHCl)$, $(C_6H_5NC)_4Ni$ [23411-45-2], $Ni[P(OCH_3)_3]_4$ [14881-35-7], and the anion $[Ni(CN)_4]^{4-}$ [15453-80-2] (7–10).

The octahedral configuration, in which nickel has a coordination number of 6, is the most common structural form for nickel compounds. The square-planar and tetrahedral configurations, in which nickel has a coordination number of 4, are the other structural forms and are less common. Generally, the latter tend to be reddish brown, whereas the octahedral forms are almost always green.

Nickel Oxides. *Properties.* Nickel oxide [1313-99-1], NiO, is a green cubic crystalline compound, mp 2090°C, density 7.45 g/cm^3. The properties of nickel oxide are related to its method of preparation. Green nickel oxide is prepared by firing a mixture of water and pure nickel powder in air at 1000°C or by firing a mixture of high purity nickel powder, nickel oxide, and water in air (11–12). The latter provides a more rapid reaction than the former method. Single whiskers of green nickel oxide have been made by the closed-tube transport method from oxide powder formed by the decomposition of nickel sulfate [17786-81-4] using HCl as the transport gas (13). Green nickel oxide also is formed by thermal decomposition of nickel carbonate [3333-67-3] or nickel nitrate [13138-45-9] at 1000°C. Green nickel oxide is an inert and refractory material.

Black nickel oxide, NiO, is a microcrystalline form resulting from calcination of the carbonate or nitrate at 600°C. This incomplete calcination product typically has more oxygen than its formula indicates, ie, 76–77 wt % nickel compared to the green form which has 78.5 wt % nickel. Black nickel oxide compositions are reactive chemically and form the simple salts of nickel when heated with mineral acids. Both black and green nickel oxide can be converted to the metal by heating with carbon, carbon monoxide, or hydrogen. Either green or black nickel oxide fuses with potassium hydroxide at 700°C to form potassium nickelate [50811-97-7], K_2NiO_2 (14). Other nickel oxides, eg, Ni_2O_3 [1314-06-3], NiO_2 [12035-36-8], Ni_3O_4 [12137-09-6], and NiO_4 [37194-86-8], have been reported but there is insufficient evidence for their characterization (15–16).

Manufacture. Several nickel oxides are manufactured commercially. A sintered form of green nickel oxide is made by smelting a purified nickel matte at 1000°C (17). A powder form of green nickel oxide is made by the desulfurization of nickel matte (18). Black nickel oxide is made by the calcination of nickel carbonate at 600°C (19). The carbonate results from an extraction process whereby impure reduced nickel metal is oxidized in the presence of ammonia and CO_2 to nickel hexammine carbonate [67806-76-2] (20). Nickel oxides also are made by the calcination of nickel carbonate or nickel nitrate that were made from a pure form of nickel. A high purity, green nickel oxide is made by firing a mixture of nickel powder and water in air (11).

Uses. The sinter oxide form is used as charge nickel in the manufacture of alloy steels and stainless steels (see Steel). The oxide furnishes oxygen to the melt for decarburization and slagging. In 1979, >50,000 metric tons of nickel contained in sinter oxide was shipped to the world's steel industry. Nickel oxide sinter is charged as a granular material to an electric furnace with steel scrap and ferrochrome; the mixture is melted and blown with air to remove carbon as CO_2. The melt is slagged, poured into a ladle, the composition is adjusted, and the melt is cast into appropriate shapes. A modification of the use of sinter oxide is its injection directly into the molten metal (21–22).

Green nickel oxide powder, which is used in the refining of nickel, is agglomerated to a particular shape and then is reduced to metal in a furnace. Green and black nickel oxides are used in the ceramic industry for making frit, ferrites, and inorganic colors (see Ceramics; Colorants; Ferrites). Black nickel oxide is used for the manufacture of nickel salts and the specialty ceramics. Green and black nickel oxides are used for nickel catalyst manufacture by admixing, usually when wet, with a powdered ceramic support material. The mixture is formed into a suitable shape and then reduced with hydrogen to form the finished catalyst.

Nickel Sulfate.   Properties. Nickel sulfate hexahydrate [10101-97-0], $NiSO_4.6H_2O$, is a monoclinic emerald-green crystalline salt that dissolves easily in water and ethanol. When heated, it loses water and, above 800°C, it decomposes into nickel oxide and SO_3. Its density is 2.03 g/cm^3.

Manufacture. Much nickel sulfate is made commercially by adding nickel powder to hot dilute sulfuric acid. Adding sulfuric acid to nickel powder in hot water enhances a competing reaction whereby the acid is reduced to hydrogen sulfide. Nickel sulfate also is made by the reaction of black nickel oxide with hot dilute sulfuric acid or of dilute sulfuric acid and nickel carbonate. One possibility for the potential continuous large-scale manufacture of nickel sulfate is the reaction of nickel carbonyl, SO_2, and oxygen in the gas phase at 100°C (23). Another possibility for the continuous manufacture of nickel sulfate is the gas-phase reaction of nickel carbonyl with nitric acid, recovering the solid product in sulfuric acid, and continuous removal of nickel sulfate from the acid mixture (24). In the latter scheme, nickel carbonyl and sulfuric acid are fed into a closed-loop reactor with the production of nickel sulfate and carbon monoxide; CO is recycled to form nickel carbonyl.

Uses. The principal use for nickel sulfate is as an electrolyte for the metal-finishing application of nickel electroplating (qv). Nickel sulfate also is used as the electrolyte for nickel electrorefining and in electroless nickel plating (see Electroless plating). Nickel sulfate and a reducing agent, eg, sodium hypophosphite, are brought together in hot water in the presence of the workpiece to be plated (25). Another application for nickel sulfate is as a nickel strike solution, which is used for replacement coatings or nickel flashing on steel that is to be porcelain enameled. Nickel sulfate is used as an intermediate in the manufacture of other nickel chemicals and as a catalyst intermediate.

Nickel Nitrate. *Properties and Preparation.* Nickel nitrate hexahydrate [13478-00-7], $Ni(NO_3)_2 \cdot 6H_2O$, is a green monoclinic deliquescent crystal (mp 56°C, density 2.05 g/cm^3). It is extremely soluble in water. Nickel nitrate hexahydrate loses water on heating and eventually decomposes forming nickel oxide. The loss of the individual waters of hydration upon heating the hexahydrate can be studied and the existence of the anhydrous covalent compound $Ni(NO_3)_2$ [13138-45-9], before it decomposes, can be observed using dta and tga techniques. The latter compound is prepared by the reaction of red fuming nitric acid and nickel nitrate hexahydrate. Nickel nitrate hexahydrate can be prepared by the reaction of dilute nitric acid and nickel carbonate.

Manufacture. Nickel nitrate hexahydrate is made commercially by several methods. Nickel metal reacts vigorously with nitric acid and, if the reaction is not closely controlled, excess heating occurs and causes breakdown of the nitric acid. Nickel powder is added slowly to a stirred mixture of nitric acid and water to yield nickel nitrate. Adding nitric acid to nickel powder results in the formation of nickel ammonium nitrate [22026-79-5]. Nickel nitrate also is made by employing a two-tank reactor system: one tank contains a solid nickel form, eg, 10- X 10- X 1-cm electrolytic nickel squares; the other tank contains nitric acid and water. The tanks are connected at the bottom with a tube that passes through a pump which circulates the water/nitric acid mixture into the second tank where the reaction occurs. The solution then overflows to the first tank by a tube connection near the top. The reaction rate is controlled by the pump. Another method is by the addition of nitric acid to a mixture of black nickel oxide powder and hot water. The reaction is controlled by using a cooling coil or cold-water condenser because the reaction is highly exothermic.

Uses. Nickel nitrate is an intermediate in the manufacture of nickel catalysts, especially those that are sensitive to sulfur and that, therefore, preclude the use of the less expensive nickel sulfate. Nickel nitrate also is an intermediate in loading active mass in nickel–cadmium batteries of the sintered-plate type (see Batteries, secondary). Typically, hot nickel nitrate syrup is impregnated in the porous sintered-nickel positive plates. Subsequently, the plates are soaked in potassium hydroxide solution whereupon nickel hydroxide [12054-48-7] precipitates within the pores of the plate.

Nickel Halides. *Properties.* Nickel forms anhydrous as well as hydrated halides. The properties of the anhydrous salts are given in Table 1.

Nickel chloride hexahydrate [7791-20-0] is formed by the reaction of nickel powder or nickel oxide with a hot mixture of water and HCl. Nickel fluoride [13940-83-5], $NiF_2.4H_2O$, is prepared by the reaction of hydrofluoric acid on nickel carbonate. Nickel bromide [18721-96-5], $NiBr_2.6H_2O$, is made by the reaction of black nickel oxide and HBr. The reaction of hydriodic acid with nickel carbonate yields nickel iodide [7790-34-3], $NiI_2.6H_2O$.

Uses. Nickel chloride hexahydrate is an important material in nickel electroplating. It is used with nickel sulfate in the conventional Watts plating bath (28). Nickel chloride is an intermediate in the manufacture of certain nickel catalysts, and it is used to absorb ammonia in industrial gas masks. Nickel bromide has limited use in nickel electroplating. The reaction of nickel chloride or nickel bromide with dimethoxyethane yields ether-soluble $NiX_2.2C_2H_4(OCH_3)_2$ compounds which are useful as nickel-containing reagents for a variety of reactions used to form coordination compounds of nickel (29).

Nickel Carbonate. Nickel carbonate [3333-67-3], $NiCO_3$, is a light-green, rhombic crystalline salt that is very slightly soluble in water (density of 2.6 g/cm^3). The addition of sodium carbonate to a solution of a nickel salt precipitates an impure basic nickel carbonate. The commercial material is the basic salt $2NiCO_3.3Ni(OH)_2.4H_2O$ [29863-10-3]. Nickel carbonate is prepared best by the oxidation of nickel powder in ammonia and CO_2. Boiling away the ammonia causes precipitation of pure nickel carbonate.

Nickel carbonate is used in the manufacture of catalysts, in the preparation of colored glass, in the manufacture of certain nickel pigments, and as a neutralizing compound in nickel electroplating solutions. It also is used in the preparation of many specialty nickel compounds.

Nickel Hydroxide. Nickel hydroxide [12054-48-7], $Ni(OH)_2$, is a light-green, microcrystalline powder (density 4.15 g/cm^3). It decomposes into nickel oxide and water when heated at 230°C, and it is extremely insoluble in water. A solution of nickel sulfate which is treated with sodium hydroxide yields the gelatinous nickel hydroxide which, when neutralized, forms a fine precipitate that can be filtered. Another in-

Table 1.   Properties of Anhydrous Nickel Halides[a]

Compound	CAS Registry No.	mp, °C	Density, g/cm^3	Color	Solubility, in 100 cm^3 water at 0°C
nickel difluoride	[10028-18-9]	1000 (sublimes)	4.63	light green	4
nickel dichloride	[7718-54-9]	1001	3.56	yellow	64
nickel dibromide	[13462-88-9]	963	5.10	orange	113
nickel diodide	[13462-90-3]	797	5.83	black	124

[a] Refs. 26–27.

dustrial route for the manufacture of nickel hydroxide is by electrodeposition at an inert cathode using metallic nickel as the anode and nickel nitrate as the electrolyte. High purity crystalline nickel hydroxide can be made from nickel nitrate solution and potassium hydroxide by subsequently extracting the gelatinous precipitate with hot alcohol (30). Nickel hydroxide is an intermediate in the manufacture of nickel catalysts. The major use for nickel hydroxide is in the manufacture of nickel–cadmium batteries. Nickel hydroxide can be formed electrochemically within the sintered positive-electrode plaque if the porous plaque is used as the cathode (31). A variation of this method involves two porous electrodes and a d-c pole-reversing technique (32).

Nickel Fluoroborate. Fluoroboric acid reacts with nickel carbonate to form nickel fluoroborate [14708-14-6], $Ni(BF_4)_2.6H_2O$; upon crystallization, the high purity product is obtained (33). Nickel fluoroborate is used as the electrolyte in specialty high speed nickel plating. It is available commercially as a concentrated solution.

Nickel Cyanide. Nickel cyanide tetrahydrate [20427-77-4], $Ni(CN)_2.4H_2O$, forms apple-green plates which are, like other metal cyanides, highly poisonous. When the tetrahydrate is heated to 200°C, anhydrous nickel cyanide [557-19-7], $Ni(CN)_2$, forms. Further heating causes decomposition. Nickel cyanide is made by the reaction of potassium cyanide and nickel sulfate. Nickel cyanide is highly insoluble in water and precipitates from the reaction medium. Nickel cyanide is soluble in aqueous alkali cyanides as well as in other bases, including ammonium hydroxide and alkali metal hydroxides. The latter yield the orange tetracyanonickelates(II) which are water-soluble. A water solution of $K_2[Ni(CN)_4]$ [14220-17-8] does not yield a nickel sulfide precipitate when treated with hydrogen sulfide. Nickel cyanide has been used in the Reppe process for the conversion of acetylene to butadiene (qv) and other products (34) (see Acetylene-derived chemicals).

Nickel Sulfamate. Nickel sulfamate [13770-89-3], $Ni(SO_3NH_2)_2.4H_2O$, commonly is used as an electrolyte in nickel electroforming systems where low stress deposits are required. As a crystalline entity for commercial purposes, nickel sulfamate never is isolated from its reaction mixture. It is prepared by the reaction of fine nickel powder or black nickel oxide with sulfamic acid in hot-water solution. Care must be exercised in its preparation and the reaction should be completed rapidly because sulfamic acid hydrolyzes readily to form sulfuric acid.

Nickel Sulfide. Nickel, like iron and cobalt, forms similar monosulfides which may show considerable deviation from stoichiometry without exhibiting heterogeneity. Nickel sulfide [1314-04-1], NiS, occurs naturally as the mineral millerite, and has a trigonal crystalline form and a yellow metallic luster; density 5.65 g/cm^3, mp 797°C. It is insoluble in water. Nickel sulfides often are thought of as binary alloys of sulfur and nickel because the metallic appearance of the sulfides resembles alloys more than chemical compounds. Other nickel sulfides include two subsulfides Ni_2S [12137-08-5], Ni_3S_2 [12035-72-2]; the latter is found as the mineral heazlewoodite. Another naturally occurring sulfide is polydymite [12137-12-1], Ni_3S_4.

Nickel sulfide, NiS, can be prepared by the fusion of nickel powder with molten sulfur or by precipitation using hydrogen sulfide treatment of a buffered solution of a nickel(II) salt. The behavior of nickel sulfides in the pure state and in mixtures with other sulfides is of interest in the recovery of nickel from ores, in the high temperature sulfide corrosion of nickel alloys, and in the behavior of nickel-containing catalysts.

Nickel Arsenate.   Nickel arsenate [7784-48-7], $Ni_3(AsO_4)_2.8H_2O$, is a yellowish-green powder (density 4.98 g/cm^3). It is highly insoluble in water but is soluble in acids, and it decomposes on heating to form As_2O_5 and nickel oxide. Nickel arsenate is formed by the reaction of a water solution of arsenic anhydride and nickel carbonate. Nickel arsenate is a selective fat-hardening hydrogenation catalyst.

Nickel Phosphate.   Trinickel orthophosphate [14396-43-1], $Ni_3(PO_4)_2.7H_2O$, exists as apple-green plates which decompose upon heating. It is prepared by the reaction of nickel carbonate with hot dilute phosphoric acid. Nickel phosphate is a component of a mixture used to coat steel prior to its being painted (see Metal surface treatments).

Double Salts.   Nickel ammonium chloride [16122-03-5], $NiCl_2.NH_4Cl.6H_2O$, nickel ammonium sulfate [15699-18-0], $NiSO_4.(NH_4)_2SO_4.6H_2O$, and nickel potassium sulfate [10294-65-2], $NiSO_4.K_2SO_4.6H_2O$, are prepared by crystallizing the individual salts from a water solution. They have limited use as dye mordants and are used in metal-finishing compositions.

Nickel Ammine Complexes.   Differential thermal analysis (dta) and thermogravimetric studies (tga) have been used to obtain a better understanding of the ammine complexes of nickel salts (36). The thermal stability of the nickel hexaammine halides increases with the size of the halide. The decomposition temperatures at 13 kPa (100 mm Hg) for $Ni(NH_3)_6Cl_2$ [10534-88-0], $Ni(NH_3)_6Br_2$ [13601-55-3], and $Ni(NH_3)_6I_2$ [13859-68-2] are 398°C, 433°C, and 450°C, respectively. The thermal decomposition of nickel tetrakispyridine dichloride [14076-99-4] has been studied and, at 110°C, 2 mol of pyridine are lost (37). Further heating to 190°C causes the loss of an additional mole of pyridine and, at 270°C, the last mole of pyridine is eliminated with the formation of anhydrous nickel chloride. Likewise, the stepwise elimination of single moles of ethylenediamine (en) from the bidendate complex $[Ni(en)_3]Cl_2$ [13408-70-3] has been studied, and decomposition temperatures of 250°C, 350°C, and 420°C have been reported (38) (see Coordination compounds).

Organic Compounds

Contributions to the understanding of the organic chemistry of nickel include the Reppe discovery of acetylene polymerization employing nickel salts as catalysts to form cyclooctatetraene (39), the reduction of nickel halides with sodium cyclopentadienide to form nickelocene [1271-28-9] (40–41), the synthesis of cyclododecatriene nickel [39330-67-1] (42), and formation from elemental nickel powder and other reagents of tetrakis ligand nickel(0) complexes that are catalysts for oligomerization and hydrocyanation reactions (43) (see also Organometallics).

Nickel Carbonyl.   *Properties.*   Nickel carbonyl, $Ni(CO)_4$, is a colorless liquid (bp 42.6°C; crystallization −25°C; density at 17°C = 1.3186 g/cm^3; vapor pressure at 0°C, 10°C, and 20°C = 19.2, 28.7, and 44.0 kPa (144, 215, and 330 mm Hg), respectively; critical t = 200°C). The thermodynamic properties of nickel carbonyl are documented (44–45). The vapor density of nickel carbonyl is ca four times that of air. Nickel carbonyl is miscible in all proportions with most organic solvents and is practically insoluble in water. It reacts slowly with HCl and H_2SO_4 but reacts vigorously with nitric acid and the halogens to form the corresponding nickel salts with concurrent liberation of CO. Bromine water is a useful reagent for the controlled decomposition of nickel carbonyl and for destroying residual amounts of it in a chemical apparatus. The vibrational spectrum of nickel carbonyl has been studied (46–47) (see Carbonyls).

Manufacture. Nickel carbonyl can be prepared by the direct combination of carbon monoxide and metallic nickel (48). The presence of sulfur, the surface area, and the surface activity of the nickel affect the formation of nickel carbonyl (49). The thermodynamics of formation and reaction are documented (50–54). Two commercial processes are used for large-scale production of nickel carbonyl. An atmospheric method, whereby carbon monoxide is passed over freshly reduced nickel metal, is used in the United Kingdom to produce pure nickel carbonyl (55). The second method, which is used in Canada, involves high pressure CO in the formation of the iron and nickel carbonyls; the two are separated by distillation (55). Very high pressure CO permits the formation of cobalt carbonyl and a method has been described where the mixed carbonyls are scrubbed with ammonia or an amine and the cobalt is extracted as the ammine carbonyl (56). A discontinued commercial process in the United States involved the reaction of carbon monoxide with nickel sulfate solution (57).

Uses. Nickel refining (involving nickel carbonyl as an intermediate) capacity at Inco Metals Division of Inco Ltd. is >100,000 t/yr. High purity nickel pellets for melting and dissolving are a product of the carbonyl-refining process. Nickel powders useful in nickel chemical synthesis, and for making nickel alkaline-battery electrodes and powder metallurgical parts are derived from the carbonyl-refining process. (58–61). Nickel carbonyl also is used in a carbonylation reaction in the synthesis of acrylic and methacrylic esters from acetylene and alcohols (62) (see Acrylic acid; Methacrylic acid). Nickel carbonyl has been proposed as catalyst or as an addition agent for a variety of organic reactions including catalysis, polymerization, and other carbonylation reactions (63–69).

Substituted Nickel Carbonyl Complexes. The reaction of trimethyl phosphite with nickel carbonyl yields the monosubstituted colorless oil, $(CO)_3NiP(OCH_3)_3$ [*17099-58-0*], the disubstituted colorless oil, $(CO)_2Ni[P(OCH_3)_3]_2$ [*16787-28-3*], and the tri-substituted white crystalline solid, $(CO)Ni[P(OCH_3)_3]_3$ [*17084-87-6*] (mp 98°C) (70). Liquid complexes result from the reaction of trifluorophosphine with nickel carbonyl yielding $(CO)_3Ni(PF_3)$ [*14264-32-5*], $(CO)_2Ni(PF_3)_2$ [*13859-78-4*], and $(CO)Ni(PF_3)_3$ [*14219-40-0*] (71–72). The substituted arsine complexes, $(CO)_3NiAs(C_6H_5)_3$ [*37757-32-7*] (a white solid, mp 105°C), and $(CO)_2Ni[As(C_6H_5)_3]_2$ [*15709-52-1*], which is isolated as a cream, have been prepared (73). The substituted stibine derivative, $(CO)_2Ni[Sb(OC_6H_5)_3]_2$ [*28042-59-3*] is a white solid, mp 132°C (74). A bidendate substituted nickel carbonyl $((CO)_2Ni[o\text{-}C_6H_4[P(CH_3)_2]_2]$ [*76404-14-3*]) is a white crystalline compound, mp 123°C (75).

π-Cyclopentadienyl Nickel Complexes. Nickel bromide dimethoxyethane [*29823-39-9*] forms bis(cyclopentadienyl)nickel [*1271-28-9*] upon reaction with sodium cyclopentadienide (76). This complex, known as nickelocene, $\pi\text{-}(C_5H_5)_2Ni$, is an emerald-green crystalline sandwich compound, mp 173°C, density 1.47 g/cm^3. It is paramagnetic and it slowly oxidizes in air. A number of derivatives of nickelocene are known, eg, methylnickelocene [*1292-95-4*] (green, mp 37°C) and bis(π-indenyl)nickel [*52409-46-8*] (red, mp 150°C) (77–78).

Substituted derivatives of nickelocene, where one ring has been replaced, include the complex cyclopentadienylnickel nitrosyl [*12071-73-7*], $(\pi\text{-}C_5H_5)NiNO$ (a red liquid, mp −41°C) (79). It has been claimed to be a hydrocarbon fuel additive but has not been successful commercially (80). The dimer complex of cyclopentadienylnickel carbonyl [*12170-92-2*], $(\pi\text{-}C_5H_5NiCO)_2$, is made in reversible reaction from nickel carbonyl and nickelocene (81). The complex is a red-violet solid and is diamagnetic, and spectroscopic studies show the presence only of bridging carbonyl groups.

Tetrakis Nickel(0) Complexes.   Tetrakis nickel(0) complexes are made by several methods, one of which is the substitution of CO molecules in nickel carbonyl. Tetrakistrichlorophosphinenickel(0) [36421-86-0], $Ni(PCl_3)_4$ (yellow, mp 120°C dec) is an example of the CO substitution synthesis (82). Another method of preparation involves the substitution of more powerful donor ligands in other tetrakisligand nickel(0) complexes and can be used to prepare complexes that cannot be made by the CO substitution method. The red solid tetrakistriphenylphosphinenickel(0) [15133-82-1], mp 125°C, and the yellow solid tetrakistrimethylphosphinenickel(0) [28069-69-4], mp 185°C dec, also are examples (83–84). A third method of preparation involves the direct reaction of nickel powder with ligands that have halogens on the donor atom. Examples of complexes prepared by this method are $Ni(PF_3)_4$ [13859-65-9] and $Ni(CH_3PCl_2)_4$ [76404-15-4] (85–86). A variation of this method involves only minor amounts of the halogenated ligand in the presence of the nonhalogenated ligands, eg, tetrakis(tri-m-) and (tri-p-tolyl phosphite)nickel(0) [35884-66-3] has been prepared in excellent yield with only 0.1 wt % of the ligand quantity present as (m- and p-$CH_3C_6H_4O)_2PCl$ (42). A fourth method of preparation of NiL_4 complexes involves the reaction of nickel halides with ligands in the presence of a reducing agent, eg, zinc metal. The mixed ligand complex $Ni[P(o-CH_3C_6H_4)_3]_3$ [NCCH$_2$CH=CHCH$_3$] [41686-95-7] has been prepared from nickel chloride [7718-54-9], 3-pentenenitrile, and tri-o-tolylphosphine (87). Tetrakis ligand nickel(0) complexes have tetrahedral structures. Their detailed electronic structure and conformational analysis have been studied (88). Quantitative equilibria measurements of the ligands in tetrakis ligand nickel(0) complexes imply a dominant role for the steric effects when they are employed in catalytic applications (84).

Tetrakis ligand nickel(0) complexes catalyze the reaction of ethylene with butadiene to give 1,4-hexadiene (89), the isomerization of 1-butene to 2-butene (90), and hydrocyanation of butadiene to form adiponitrile (91–92). The thermal decomposition of tetrakis(triorganophosphite)nickel(0) complexes in high boiling solvents is a method for depositing a high purity coating of nickel on steel (93–94).

Other Complexes.   Several other classes of organonickel complexes are known. Allyl bromide reacts with nickel carbonyl to give a member of the π-allyl system [12012-90-7], [(π-C_3H_5)NiBr]$_2$ (95). Allylmagnesium bromide reacts with anhydrous $NiBr_2$ to form cis and trans isomers (96) of bis(π-allyl)nickel [12077-85-9], (π-$C_3H_5)_2Ni$ (97).

1,2,3,4-Tetramethylcyclobutadienenickel dichloride [76404-16-5] can be prepared by the reaction of nickel carbonyl with 3,4-dichlorotetramethylcyclobutene (CBD) in polar solvents (98). The complex is black-violet, mp 270°C dec.

The reaction of a mixture of 1,5,9-cyclododecatriene (CDT), nickel acetylacetonate [3264-82-2], and diethylethoxyaluminum in ether gives red, air-sensitive, needle crystals of (CDT)Ni [12126-69-1] (41). Crystallographic studies indicate that the nickel atom is located in the center of the 12-membered ring of (CDT)Ni (99). The latter reacts with 1,5-cyclooctadiene (COD) readily to yield bis(COD) nickel [1295-35-8], (COD)$_2$Ni (yellow crystals, fairly air stable, mp 142°C dec). Bis(COD)nickel also can be prepared by the reaction of 1,5-COD, triethylaluminum, and nickel acetylacetonate (100).

Another class of compounds involves nickel in the anion, eg, tetraethylammonium tetrachloronickelate [5964-71-6], [($C_2H_5)_4N]_2$[NiCl$_4$] (101) and tetraethylammonium triphenylphosphinetribromonickelate [41828-60-8], [($C_2H_5)_4N][(C_6H_5)_3PNiBr_3]$ (102).

Nickel salts form coordination compounds with many ligands. Examples of these salts are bis(tri-n-butylphosphine)nickel dibromide [15242-92-9], [(n-C$_4$H$_9$)$_3$P]$_2$NiBr$_2$ (103), nickel amminocyanide, Ni(NH$_3$)(H$_2$O)(CN)$_2$ (104), and bis(triphenylphosphine)nickel nitrosyl bromide [14586-72-2] (105). These complexes and many others are used for syntheses in preparative organonickel chemistry.

Reduction of compounds of the type LNiX$_2$ and L$_2$NiX$_2$ (L = ligand) yields hydride complexes, ie, LNiHX and L$_2$NiHX (see Hydrides). The former generally are stable only at low temperatures whereas the latter are more stable. A high degree of stability can result when bulky ligands are employed. Bis(tricyclohexylphosphine)hydridonickel chloride [25703-57-5], HNi[P(cyclo-C$_6$H$_{11}$)$_3$]$_2$Cl, mp 150°C dec, is an example of L$_2$NiHX (106).

The presence of powerful electron-donating ligands has a large effect on the synthesis of alkyl and aryl nickel compounds. Dimethylnickel [54836-89-4], Ni(CH$_3$)$_2$, cannot be isolated, even at −130°C; whereas bis(tricyclohexylphosphine)dimethylnickel [36427-03-9] [(cyclo-C$_6$H$_{11}$)$_3$P]$_2$Ni(CH$_3$)$_2$, is stable at ambient temperature (107). Bis(trityl)nickel [7544-48-1], [(C$_6$H$_5$)$_3$C]$_2$Ni (violet crystals, mp 120°C dec) is a ligand-free nickel aryl complex (108). It can be prepared by the reduction of nickel salts in the presence of hexaphenylethane. Bis(triphenylphosphine)phenylnickel chloride [38415-93-3], [(C$_6$H$_5$)$_3$P]$_2$NiC$_6$H$_5$Cl (mp 122°C) is a bis ligand aryl nickel halide (109). The bis ligand dialkylnickel complex [60802-48-4], [(CH$_3$)$_3$P]$_2$Ni(CH$_3$)$_2$, reacts further with the ligand to yield the tris ligand dialkylnickel complex [42725-08-4], [(CH$_3$)$_3$P]$_3$Ni(CH$_3$)$_2$ (orange-red solid, mp 49–51°C) (110).

π-Complexes of alkylnickel and arylnickel also have been prepared; strong electronic effects impart stability. Methyl Grignard reagent reacts with π-allylnickel [12077-85-9] to yield the dimer of π-allylmethylnickel, a complex which cannot be isolated above −78°C. However, tricyclohexylphosphine π-allylmethylnickel [76422-11-2], π-CH$_2$CHCH$_2$NiCH$_3$[P(cyclo-C$_6$H$_{11}$)$_3$], is a stable yellow solid, mp 50°C dec (111).

Tri-n-butylphosphine π-cyclopentadienylmethylnickel [7298-70-0], P(C$_4$H$_9$)$_3$-(π-C$_5$H$_5$)NiCH$_3$, is a π-cyclopentadienyl alkylnickel complex. It can be prepared by the reaction of a cyclopentadienyl ligand nickel halide with methyl Grignard reagent. It is a greenish brown solid, mp 29–30°C (112).

Arene–nickel complexes are formed from nickel carbonyl or nickel bromide in the presence of aromatics. Polymeric and intractable solids also have been reported. The dark red tricyclohexylphosphine 1,3,5-triphenylbenzenenickel [76404-17-6] probably has been isolated from the reaction of phenyllithium and tricyclohexylphosphine triphenylboranenickel in the presence of sym-triphenylbenzene (113).

Nickel Salts and Chelates. Nickel salts of simple organic acids can be prepared by reaction of the organic acid with nickel carbonate or hydroxide, reaction of the acid with a water solution of a simple nickel salt and, in some cases, reaction of the acid with fine nickel powder or black nickel oxide.

Nickel acetate [6018-89-9], Ni(C$_2$H$_3$O$_2$).4H$_2$O, is a green powder which has an acetic acid odor, density 1.74 g/cm^3. When heated, it loses its water of crystallization and then decomposes to form nickel oxide. Nickel acetate is used as a catalyst intermediate, as an intermediate in the formation of other nickel compounds, as a dye mordant, as a sealer for anodized aluminum, and in nickel electroplating.

Nickel formate [15694-70-9], Ni(HCOO)$_2$.2H$_2$O, is a green monoclinic crystalline compound which melts with decomposition to nickel oxide at 180°C; density 2.15

g/cm^3. Nickel formate is used in the preparation of fat-hardening nickel hydrogenation catalysts.

Other simple nickel salts of organic acids include the oxalate [20543-06-0], oleate [68538-38-5], and stearate [2223-95-2]. The latter two have been used as oil-soluble nickel forms in the dying of synthetic polyolefin fibers (see Driers). Nickel oxalate has been used as a catalyst intermediate.

Nickel acetylacetonate [3264-82-2], Ni(C$_5$H$_7$O$_2$)$_2$, is a green powder which can be made by the aqueous reaction of a soluble nickel salt with 2,4-pentanedione. It is the simplest of the bidentate coordination compounds of nickel. Its use is primarily in preparative organonickel chemistry. Other well-known chelates include nickel ethylenediaminebisacetylacetonate [42948-35-6], nickel phthalocyanine [14045-02-8], and nickel dimethylglyoxime [13478-93-8] (114–115). The latter two have been studied as pigments. Nickel also forms derivatives with organic sulfur compounds. Some examples include triammino nickel trithiocarbonate [39282-88-7], (NH$_3$)$_3$NiCS$_3$, nickel dimethyldithiocarbamate [15521-65-0], Ni[(CH$_3$)$_2$NCS$_2$]$_2$, and nickel ethyl xanthate [52139-56-7], Ni(C$_2$H$_5$OCS$_2$)$_2$ (116–118).

Economic Aspects

An estimate of the annual usage of nickel in compound form in noncommunist countries for 1978 and 1979 is

Compound	Quantity of nickel used, 10^3 metric tons
nickel oxide sinter	50
green nickel oxide	10
black nickel oxide	5
nickel sulfate	7–9
nickel nitrate	2
nickel chloride	2
nickel carbonate	1
others	1

The approximate annual usage of nickel chemicals, other than for steel and nickel refining, by noncommunist nations for 1978 and 1979 is

Use	Quantity of nickel used, 10^3 t
catalysts	8–10
plating salts	7–9
specialty ceramics	2–3
other specialty chemicals	2–3

Prices for nickel compounds in January 1980 were:

Compound	$/kg
nickel oxide sinter (as NiO)	5.14
green nickel oxide (as NiO)	5.36
black nickel oxide (as NiO)	5.36
nickel sulfate (as NiSO$_4$.6H$_2$O)	2.43
nickel chloride (as NiCl$_2$.6H$_2$O)	3.22

nickel nitrate (as $Ni(NO_3)_2.6H_2O$)	2.60
nickel carbonate (as $NiCO_3$)	7.39
nickel acetate (as $Ni(C_2H_3O_2).4H_2O$)	4.70

Analytical Methods

Nitric acid can be used for the dissolution of nickel from many inorganic substances. In some cases perchloric acid is used in combination with nitric acid. Simple organic forms of nickel also can be dissolved in nitric acid. In the case of complicated structural organic forms of nickel, oxidation calorimetry must be used to decompose the substances. Analytical determination of nickel in solution usually is made by atomic absorption spectrophotometry and, often, by x-ray fluorescence spectroscopy. Nickel also is determined by a volumetric method employing ethylenediaminetetraacetic acid as a titrant. The classical gravimetric method employing dimethylglyoxime to precipitate nickel as a red complex is used as the most precise analytical technique (115). A colorimetric method employing dimethylglyoxime also is available. Electrodeposition is a commonly employed technique to separate nickel in the presence of other metals, notably copper. X-ray diffraction often is used to identify nickel in crystalline form.

Health and Safety Factors, Toxicology

Eye and Skin Contact.   Some aqueous solutions of nickel are irritating on contact with the eye, eg, nickel sulfate and nickel chloride. Face shields, eye protection, and safety showers should be used where there is a possibility of splashes with nickel-containing solutions.

Nickel and nickel compounds, particularly in aqueous solution, may cause allergic dermatitis in some workers. Nonoccupational dermatitis has occurred in individuals and has resulted from skin contact with nickel in jewelry and in nickel-plated articles. Health studies have demonstrated that 4–13% of patients with skin disease are sensitive to nickel on patch testing (119). In contrast to nonoccupational exposure, allergic dermatitis in nickel-exposed workers is not a significant problem. Nevertheless, care should be taken to prevent skin contact with aqueous solutions containing nickel by means of protective clothing. Surface wounds that may be contaminated with nickel should be cleansed with mild soap and water before treatment.

Inhalation.   The permissible exposure level in the United States for all forms of nickel and its inorganic compounds is 1 mg/m³ air (120). The ACGIH-recommended TLVs for nickel are (121)

	TLV mg/m³ as Ni
nickel, metal and insoluble compounds	1
nickel, soluble compounds	0.1
nickel sulfide roasting, fume and dust	1.0

A monograph on nickel has been published (122). NIOSH recommends a permissible exposure limit of 0.015 mg/m³ for nickel and its inorganic compounds based on the NIOSH conclusion that nickel and its inorganic compounds are carcinogens (123). Airborne concentrations should be maintained below the currently applicable permissible exposure levels. Nickel and nickel oxide powders should be used with good ventilation; local mechanical exhaust ventilation is preferred. Personal respiratory protective equipment may be used for intermittent or short-term exposures in which dust levels exceed the applicable exposure limit.

Nickel Carbonyl.   Nickel carbonyl is an extremely toxic gas. The permissible exposure limit in the United States is 1 ppb in air (120). The ACGIH TLV for an 8-h, time-weighted average concentration is 50 ppb (121).

Nickel carbonyl may form wherever carbon monoxide and finely divided nickel are brought together. Its occurrence has been speculated but never demonstrated in some industrial operations, eg, welding of nickel alloys.

Concentrations as low as 30 ppm in air for 30 min may be lethal for humans. Individuals exposed to these high concentrations show immediate symptoms of dizziness, headache, shortness of breath, and vomiting. These early symptoms generally disappear in fresh air but delayed symptoms may develop 12–36 h later. These latter symptoms include shortness of breath, cyanosis, chest pain, chills, and fever. In severe exposure cases, death results from pneumonitis.

Nickel carbonyl should be used in totally enclosed systems or under good local exhaust. Plants and laboratories where nickel carbonyl is used should make use of air-monitoring devices and alarms in case of accidental leakage. Appropriate personal respiratory protective devices should be readily available for emergency uses. Monitoring of urinary nickel levels is useful to help determine the severity of exposure and the appropriate treatment measures. Some large-scale users of nickel carbonyl maintain a supply of sodium diethyldithiocarbamate, a therapeutic agent, on hand for use in case of overexposure.

Uses

Catalysts.   Nickel is an important hydrogenation catalyst because of its ability to chemisorb hydrogen. One important nickel catalyst is Raney nickel (124–127). Raney nickel catalyst is used widely in laboratory and industrial hydrogenation processes. It is the most active and the least specific of the nickel catalysts. Raney nickel catalyst has been used in a continuous hydrogenation process by filling a tube with chunks of the nickel–aluminum alloy and activating the surface by passing a solution of caustic over it, thereby removing some of the surface aluminum. Periodic flushes with caustic enables the catalyst to be reactivated in place.

A number of variations of the nickel–aluminum catalyst have been developed. One involves rolling nickel and aluminum foil at 630°C followed by leaching with caustic (128). Nickel–aluminum alloy has been flame sprayed on the inside of steel tubes followed by leaching (129) and has been suggested for a continuous reactor for the conversion of synthesis gas to methane (130) (see Fuels, synthetic). Another method is the electroplating of nickel on the inside wall of stainless steel tubing, followed by aluminizing the nickel surface and activating with a caustic leach (131). Other alloying compositions have been devised including a nickel–iron–aluminum alloy which, upon caustic activation, is used as a catalyst for the selective hydrogenation of organic nitro compounds (see Amines by reduction) (132). A composition with nickel and silicon yields NiSi [12035-57-3] and NiSi$_2$ [12201-89-7] which, upon caustic leaching, activates a nickel surface (133). Nickel–boron alloy [12007-00-0], when activated with caustic, has been claimed as a more reactive hydrogenation catalyst than the nickel–aluminum catalyst (134).

Supported nickel catalysts of the precipitated and impregnated types are used for methanation, steam–hydrocarbon reforming, petrochemical hydrogenation, and fat hardening. The nickel compound and the ceramic carrier in a dry reduction technique are heated in a stream of inert gas to decompose the nickel compound to nickel oxide which then is reduced with hydrogen to nickel metal. Precipitated catalysts generally are made from nickel carbonate and nickel hydroxide. Nickel nitrate and, to a minor degree, nickel acetate solutions are used for impregnation. Supported catalysts of nickel on alumina or nickel on zirconia have been suggested as a heterogeneous methanation system for the hydrogenation of CO. The latter catalyst has been reported to be quite active in the presence of H_2S (135).

Nickel is used with other elements for special types of hydrogenation catalysts. Nickel sulfide [16812-54-7] and nickel tungsten sulfide catalysts are used when high concentrations of sulfur compounds are present in the hydrogenation of petroleum distillates. Nickel–molybdenum catalysts are used to denitrogenate petroleum fractions that are high in nitrogen-containing components (see Petroleum).

Efficient NO_x reduction in automobile exhaust has been achieved using nickel–iron and nickel–copper oxides (136). Black nickel oxide enhances the activity of three-way catalysts containing rhodium, platinum, and palladium (137). Three-way catalysts oxidize hydrocarbons and CO and reduce NO_x (see Exhaust control, automotive).

Nickel and other transition metals function as solvent–catalysts for the transformation of carbon species into the diamond allotrope. At temperatures high enough to melt the metal or metal–carbon mixture and at pressures high enough for diamond to be stable, the latter forms by what is probably an electronic mechanism (see Carbon, diamond, synthetic).

Electroplating. The second largest application for nickel chemicals is as electrolytes in nickel electroplating (qv). In ordinary plating systems, nickel in the electrolyte never forms on the finished workpiece but results from dissolution and transfer from nickel anodes. Decorative nickel plating is used for automobile bumpers and trim, appliances, wire products, flatware, jewelry, and many other consumer items. A comprehensive review of nickel electroplating has been compiled (138).

Specialty Ceramics. Nickel compounds, especially black or green nickel oxide, are used extensively in the ceramic industry. Nickel oxide is added to frit compositions which are used for porcelain enameling of steel. Nickel enhances the adhesion of glass to steel (139–140). Nickel oxides also are used in the manufacture of the magnetic nickel–zinc ferrites (qv) which are used in electric motors, antennas, and television yokes. Nickel silicides, ie, Ni_3Si [12059-22-2], Ni_5Si_2 [12059-27-7], and Ni_2Si [12059-14-2], are electroconductive materials that are useful in resistors and resistance heating elements (141). Nickel phosphorus compounds also have been studied for their electroconductive nature (142–143). Nickel selenite [15060-62-5], nickel phosphate, nickel tungstate [14177-51-6], nickel chromite [12018-18-7], potassium nickel molybdate [59228-72-7], nickel oxide, and nickel nitrate have been used as glass colorants (see Colorants). Nickel oxide also is employed as a colorant in ceramic body stains used in ceramic tile, dishes, pottery, and sanitary ware. Nickel oxide imparts avocado green and grey colors in ceramic glazes. When fired, nickel oxide, antimony oxide, and titanium dioxide produce the yellow chalking pigment, nickel (antimony) titanate [11118-07-3]. This pigment is used extensively in exterior house paint and in vinyl house siding because of its good weatherability. Nickel cobalt aluminate has been proposed as a useful fade-resistant blue pigment for exterior paint application (143) (see Ceramics; Pigments).

Plastics Additives.   Many claims have been made for the use of nickel chemicals as additives to various resin systems. One of the important applications is as uv quenchers in polyolefins (144–146). Among the useful nickel complexes in these systems are nickel dibutyldithiocarbamate [13927-77-0], nickel thiobisphenolates, nickel amide complexes of bisphenol sulfides, and nickel complexes of pyrazolones (147–148). Many claims have been made for the use of nickel compounds as light stabilizers in poly(vinyl chloride). Among these are nickel derivatives of organic phosphites (149), bishydrazones (149), thiobisphenolate alkoxides (150), methylenebisphenols (151), trithiocarbonates (152), salicylates (153), and methylenebissalicylates (154). Nickel amino thiobisphenolates have been claimed as light stabilizers in ABS graft copolymers (155). Nickel dialkylhydroxyphenylalkylphosphonate imparts light stability to ABS terpolymers and polyurethanes (156) (see Uv stabilizers; Antioxidants).

Nickel dialkyldithiocarbamates stabilize vulcanizates of epichlorhydrin–ethylene oxide against heat aging (156). Nickel dibutyldithiocarbamate [56377-13-0] has been used as an oxidation inhibitor in synthetic elastomers. Nickel chelates of substituted acetylacetonates are flame retardants for epoxy resins (157). Nickel dicycloalkyldithiophosphinates have been proposed as flame retardant additives for polystyrene (158) (see Flame retardants).

Organic Dyes and Pigments.   A number of nickel pigments have been reported, eg, the nickel disazomethine complex [61312-95-6] which is prepared from 2-hydroxy-1-naphthaldehyde (161), the water-soluble nickel azo–azomethine complex (162), and nickel chelates of azines and bisazines (163). Several nickel azo pigments have had commercial success, including nickel azo yellow [51931-46-5], nickel azo gold, and a nickel azo red made as the nickel salt of the condensation product of β-oxynaphthoic acid and diazotized o-chloro-m-toluidine-p-sulfonic acid (164–165) (see Azo dyes). The lightfastness in other pigment systems, eg, the quinacridones and iron blues, have been improved by the addition of 1–2 wt % nickel in soluble salt form.

Nickel also has been used as a dye site in polyolefin polymers, particularly fibers. When a nickel compound, eg, the stearate or bis(p-alkylphenol)monosulfide, is incorporated in the polyolefin melt which is subsequently extruded and processed as a fiber, it complexes with certain dyes upon solution treatment to yield bright fast-colored fibers which are useful in carpeting and other applications (166–167).

Agricultural Chemicals.   Many claims exist for the use of nickel chemicals as nemotocides, miticides, and other pesticides (168–169). However, extensive testing of many classes of nickel complexes in insecticide, fungicide, nemotocide, and herbicide programs leaves little doubt that, except for a few selected fungus organisms, nickel chemicals afford little more efficacy than the nonnickel-containing derivatives (see Insect control technology; Herbicides; Poisons, economic). The application of nickel ion, particularly the halides, as a commercial fungicide to control blister blight of tea is practiced in the Far East (170). Tea is one of a few botanical species that naturally contains nickel (171). Nickel sulfate is used to control rust in blue grass seed crops in northwestern United States (172). Nickel sulfate has been thoroughly explored for the control of cereal rusts, but no commercial application has resulted (173–177). Nickel sulfate has been proposed as an additive to wood chips against fungus attack in long-term chip storage piles (178–179). Evidence exists for the possible essential requirement of nickel ion in growing chicks (180) (see Mineral nutrients).

Other Specialty Chemicals. In fuel-cell technology, nickel oxide cathodes have been demonstrated for the conversion of synthesis gas and the generation of electricity (181) (see Batteries and electric cells, primary). Nickel salts have been proposed as additions to water-flood tertiary crude-oil recovery systems (see Petroleum, enhanced oil recovery). The salt forms nickel sulfide, which is an oxidation catalyst for H_2S, and provides corrosion protection for downwell equipment. Sulfur-containing nickel complexes have been used to limit the oxidative deterioration of solvent-refined mineral oils (182–183).

Nickel salts and soaps have been used in electrosensitive copy paper for image development. They also have been used for color stabilization of color copy paper (see Electroplating).

Nickel phosphate complexes with ammonia have been used for high speed photographic image amplification (184), and nickel-chelated quenching compounds, which stabilize image dyes in photographic film, also have been used (see Photography) (185).

BIBLIOGRAPHY

"Nickel Compounds" in *ECT* 1st ed., Vol. 9, pp. 289–304, by J. G. Dean, The International Nickel Company, Inc.; "Nickel Compounds" in *ECT* 2nd ed., Vol. 13, pp. 753–765, by D. H. Antonsen and D. B. Springer, The International Nickel Company, Inc.

1. H. Bode and E. Voss, *Z. Anorg. Allgem. Chem.* **269,** 165 (1952).
2. R. Hoppe, *Rec. Trav. Chim.* **75,** 569 (1956).
3. R. Nast and W. Pfab, *Naturwissenschaften* **39,** 300 (1952).
4. R. Nast and H. Kasperl, *Chem. Ber.* **92,** 2135 (1959).
5. H. Brehrens and F. Lohofer, *Z. Naturforsch.* **86,** 691 (1953).
6. L. Mond, C. Langer, and F. Quincke, *J. Chem. Soc.* **57,** 749 (1890).
7. U.S. Pat. 3,395,165 (July 30, 1968), R. D. Feltham (to The International Nickel Co., Inc.).
8. W. Hieber, *Z. Naturforsch.* **56,** 129 (1950).
9. Ger. Pat. 1,618,167 (Mar. 6, 1975), T. Kruck and M. Hoefler (to BASF A.G.).
10. A. A. Blanchard, *Science* **94,** 311 (1941).
11. U.S. Pat. 4,053,578 (Oct. 11, 1977), B. Hill and W. H. Elwood, Jr. (to The International Nickel Co., Inc.).
12. Can. Pat. 1,035,545 (Aug. 1, 1978), R. Sridhar and H. Davies (to Inco Ltd., Canada).
13. S. Saito, K. Kurosawa, and S. Takemoto, *Bull. Univ. Osaka Prefect Ser. A.* **24**(1), 123 (1975).
14. D. M. Mathews and R. Kuth, *Ind. Eng. Chem.* **49,** 55 (1957).
15. G. D. Parkes, ed., *Mellor's Modern Inorganic Chemistry*, Longmans, Green and Co. Ltd., London, Eng., 1951.
16. N. V. Sidgwick, *Chemical Elements*, Oxford University Press, New York, 1950, pp. 1430–1451.
17. *Nickel Oxide Sinter 75*, International Nickel Company, Inc., Product Data Sheet A-789, New York, Dec. 1977.
18. *Green Nickel Oxide*, Product Brochure, Société Metallurgique Le Nickel, Paris, France, 1976.
19. *Inco Nickel Oxide*, International Nickel Company, Inc., Product Data Sheet A-1083, New York, June 1975.
20. Can. Pat. 828,670 (Dec. 2, 1969), A. Illis, H. J. Koehler, and B. J. Brandt (to The International Nickel Company of Canada, Ltd.).
21. C. R. Quail, Jr. and Q. R. Skrabec, *Met. Producing* **33,** 54 (Apr. 1979).
22. H. Gorges and co-workers, *paper presented at the Sixth International Vacuum Metallurgy Conference on Special Melting and Metallurgical Coatings*, San Diego, Calif., Apr. 23–27, 1979.
23. U.S. Pat. 3,869,257 (Mar. 4, 1975), H. P. Beutner and G. Flick (to The International Nickel Co., Inc.).
24. U.S. Pat. 3,857,926 (Dec. 31, 1974), H. P. Beutner and C. E. O'Neill (to The International Nickel Co., Inc.).

25. U.S. Pat. 2,658,839 (Nov. 10, 1953), P. Palney and W. J. Crehan (to GATX).
26. R. Weist, ed., *Handbook of Chemistry and Physics*, 61st ed., Chemical Rubber Company, Boca Raton, Fla., 1980–1981.
27. M. Windholz, ed., *Merck Index*, 9th ed., Merck, Rahway, N.J., 1976.
28. G. A. DiBari, *Metal Finishing Guidebook Directory*, 47th Annual ed., Hackensack, N.J., 1979, p. 270.
29. L. G. L. Ward, D. P. Jordan, and D. H. Antonsen, *paper presented at the 157th American Chemical Society Meeting*, Minneapolis, Minn., Apr. 1969.
30. U.S. Pat. 2,602,070 (July 1, 1952), W. J. Kirkpatrick (to The International Nickel Company, Inc.).
31. U.S. Pat. 3,827,911 (Aug. 6, 1974), D. F. Pickett (to United States of America, Air Force Materiels Laboratory).
32. Ger. Offen 2,653,984 (June 1, 1978), G. Crespey, P. Matthey, and M. Gutjahr (to Daimler-Benz A.G.).
33. Brit. Pat. 814,638 (June 10, 1959), (to Schering A-G).
34. J. W. Copenhaver and M. H. Bigelow, *Acetylene and Carbon Monoxide Chemistry*, Reinhold, Princeton, N.J., 1949.
35. J. R. Boldt, Jr. and P. Queneau, *The Winning of Nickel*, Longmans Canada Ltd., Toronto, Can., 1967, p. 7.
36. D. H. Brown, R. H. Nuttall, and D. W. A. Sharp, *J. Inorg. Nucl. Chem.* **25,** 1067 (1963).
37. A. K. Majumdar and co-workers, *J. Inorg. Nucl. Chem.* **26,** 2177 (1964).
38. T. D. George and W. W. Wendlad, *J. Inorg. Nucl. Chem.* **25,** 395 (1963).
39. W. Reppe and co-workers, *Justus Liebigs Ann. Chem.* **560,** 1 (1948).
40. E. O. Fisher and R. Jira, *Z. Naturforsch* **86,** 217 (1953).
41. G. Wilkensen, *J. Am. Chem. Soc.* **76,** 209 (1954).
42. G. Wilke, *Angew. Chem.* **72,** 581 (1960).
43. U.S. Pat. 3,903,120 (Sept. 2, 1975), H. E. Shook and J. B. Thompson (to E. I. du Pont de Nemours & Co., Inc.).
44. F. D. Rossini and co-workers, *Nat. Bur. Stand. Circ.* **500,** 245, 690, 910, 1104 (1952).
45. K. A. Walsh, *U.S. Atomic Energy Commission Report LA-1649*, Washington, D.C., 1953, p. 14.
46. L. H. Jones, *J. Chem. Phys.* **23,** 2448 (1955).
47. L. H. Jones, *J. Chem. Phys.* **28,** 1215 (1958).
48. U.S. Pat. 455,230 (June 30, 1891), L. Mond.
49. U.S. Pat. 1,909,762 (May 16, 1933), C. M. W. Grieb (to The International Nickel Company, Inc.).
50. J. E. Spice, L. A. K. Staveley, and G. A. Harrow, *J. Chem. Soc.*, 100 (1955).
51. K. W. Sykes and S. C. Townshend, *J. Chem. Soc.*, 2528, 1955.
52. A. K. Fisher, F. A. Cotton, and G. Wilkenson, *J. Am. Chem. Soc.* **79,** 2044 (1957).
53. F. A. Cotton, A. K. Fisher, and G. Wilkenson, *J. Am. Chem. Soc.* **81,** 800 (1959).
54. A. L. Rotingan and D. G. Katsman, *Zh. Neogr. Khim.* **5,** 237 (1960).
55. J. R. Boldt, Jr. and P. Queneau, *The Winning of Nickel*, Longmans Canada Ltd., Toronto, Can., 1967.
56. U.S. Pat. 3,252,791 (May 24, 1966), G. R. Frysinger and H. P. Beutner (to The International Nickel Co., Inc.).
57. G. N. Dobrokhotov, *Zh. Prikl. Khim.* **32,** 757 (1959).
58. A. J. Catotti and co-workers, *Nickel Cadmium Battery Application Engineering Handbook*, General Electric Company, Battery Business Department, Gainesville, Fla., 1975, p. 2-2.
59. V. A. Tracey, *I&EC Prod. Res. Dev.* **18,** 234 (Sept. 1979).
60. *Sintering Data for Inco Nickel Powder A-1293*, The International Nickel Co., Inc., New York, Nov. 1979.
61. W. E. Trout, Jr., *J. Chem. Ed.* **14,** 453 (1937).
62. M. Salkind, E. H. Riddle, and R. W. Keefer, *Ind. Eng. Chem.* **51,** 1232 (1959).
63. J. W. Reppe, *Mod. Plast.* **23**(3), 162 (1945).
64. C. McKinley, *Ind. Eng. Chem.* **44,** 995 (1952).
65. L. Cassar and M. Foa, *Chim. Inc.* (*Miean*) **51,** 673 (1969).
66. E. J. Corey and L. S.Hegedus, *J. Am. Chem. Soc.* **91,** 1233 (1969).
67. M. Ryang and co-workers, *J. Organometal. Chem.* **46,** 375 (1972).
68. I. Hashimoto and co-workers, *J. Org. Chem.* **35,** 3748 (1970).
69. P. Diversi and R. Ross, *Synthesis* **5,** 258 (1971).
70. M. Bigorne and H. Zelwer, *Bull. Soc. Chim. Fr.* **5,** 1986 (1960).

71. R. J. Clark and E. O. Brimm, *Inorg. Chem.* **4**, 651 (1965).
72. H. J. Emeleus and J. D. Smith, *J. Chem. Soc.*, 527 (1958).
73. F. T. Delbeke and G. P. van der Kelen, *J. Organometal Chem.* **21**, 155 (1970).
74. U.S. Pat. 3,247,269 (Apr. 19, 1966), C. D. Storrs and R. F. Clark (to Columbian Carbon Co.).
75. J. Chatt and F. A. Hart, *J. Chem. Soc.*, 1378 (1960).
76. P. L. Pauson, *Quart. Rev. (London)* **9**, 391 (1955).
77. M. F. Retlig and R. S. Drago, *J. Am. Chem. Soc.* **91**, 1361 (1969).
78. H. P. Fritz, F. H. Kohler, and K. E. Schwarzhans, *J. Organometal. Chem.* **19**, 449 (1969).
79. T. S. Piper, F. A. Cotton, and G. Wilkenson, *J. Inorg. Nucl. Chem.* **1**, 165 (1955).
80. U.S. Pat. 3,086,034 (Apr. 16, 1963), J. E. Brown, E. G. deWitt, and H. Shapiro (to Ethyl Corp.).
81. E. O. Fischer and C. Palm, *Chem. Ber.* **91**, 1725 (1958).
82. W. C. Smith, *Inorg. Chem.* **6**, 201 (1960).
83. H. Behrens and A. Muller, *Z. Anorg. Allg. Chem.* **341**, 124 (1965).
84. C. A. Tolman, *J. Am. Chem. Soc.* **92**, 2956 (1970).
85. T. Kruck, K. Bauer, and W. Lang, *Chem. Ber.* **101**, 138 (1968).
86. L. D. Quin, *J. Am. Chem. Soc.* **79**, 3681 (1957).
87. U.S. Pat. 3,846,461 (Nov. 5, 1974), H. E. Shook, Jr. (to E. I. du Pont de Nemours & Co., Inc.).
88. J. M. Savarialt and J. F. Larre, *Theor. Chim. Acta.* **42**(3), 207 (1976).
89. C. A. Tolman, *J. Am. Chem. Soc.* **92**, 6785 (1970).
90. C. A. Tolman, *J. Am. Chem. Soc.* **94**, 2994 (1972).
91. U.S. Pat. 3,631,191 (Dec. 28, 1971), N. J. Kane (to E. I. du Pont de Nemours & Co., Inc.).
92. C. W. Westan, *Inorg. Chem.* **16**(6), 1313 (1977).
93. U.S. Pat. 3,529,989 (Sept. 22, 1970), D. P. Jordan and D. H. Antonsen (to The International Nickel Co., Inc.).
94. D. P. Jordan and D. H. Antonsen, *Ind. Eng. Chem. Prod. Res. Dev.* **8**, 208 (June 1969).
95. W. R. McClellan and co-workers, *J. Am. Chem. Soc.* **83**, 1601 (1961).
96. H. Bonnemann, B. Bogdanovic, and G. Wilke, *Angew. Chem.* **79**, 817 (1967).
97. G. Wilke and B. Bogdanovic, *Angew. Chem.* **73**, 756 (1961).
98. R. Criegee and G. Schroder, *Justus Liebigs Ann. Chem.* **623**, 1 (1959).
99. G. N. Schrauzer and H. Thyret, *J. Am. Chem. Soc.* **82**, 6420 (1960).
100. B. Bogdanovic, M. Kravic, and G. Wilke, *Justus Liebigs Ann. Chem.* **699**, 1 (1966).
101. N. S. Gill and R. S. Nyholm, *J. Chem. Soc.*, 3997 (1959).
102. F. A. Cotton and D. M. L. Goodgame, *J. Am. Chem. Soc.* **82**, 2967 (1960).
103. K. A. Jensen, P. H. Nielsen, and C. T. Pedersen, *Acta. Chem. Scand.* **17**, 1115 (1963).
104. E. E. Aynsley and W. A. Campbell, *J. Chem. Soc.*, 4137 (1957).
105. R. D. Feltham, *J. Inorg. Nucl. Chem.* **14**, 307 (1960).
106. K. Jonas and G. Wilke, *Angew. Chem.* **81**, 534 (1969).
107. P. W. Jolly and G. Wilke, *The Organic Chemistry of Nickel*, Vol. 1, Academic Press, New York, 1974, p. 151.
108. G. Wilke and H. Schott, *Angew. Chem.* **78**, 592 (1966).
109. M. Hidai and co-workers, *J. Organometal. Chem.* **30**, 279 (1971).
110. H. F. Klein and H. H. Karsch, *Chem. Ber.* **105**, 2628 (1972).
111. B. Bogdanovic, H. Bonnemann, and G. Wilke, *Angew Chem.* **78**, 591 (1966).
112. M. D. Rausch, Y. F. Chang, and H. B. Gordon, *Inorg. Chem.* **8**, 1355 (1969).
113. P. W. Jolly and G. Wilke, *The Organic Chemistry of Nickel*, Vol. 1, Academic Press, New York, 1974, p. 490.
114. G. T. Morgan and J. D. M. Smith, *J. Chem. Soc.*, 912 (1926).
115. O. Brunck, *Z. Angew Chem.* **20**, 834 (1907).
116. G. Gattow, *Naturwissenschaften* **46**, 72 (1959).
117. F. A. Cotton and J. A. McCleverty, *Inorg. Chem.* **3**, 1398 (1964).
118. M. Delepine and L. Compin, *Bull. Soc. Chim.* **27**, 469 (1920).
119. F. H. Nielsen, *Advances in Modern Toxicology*, Vol. 2, Hemisphere Publishing Corporation, Washington, 1977, pp. 129–146.
120. *OSHA General Industry*, Title 29, Chapt. XVII, Part 19.10 of CFR.
121. *TLV for Chemical Substances and Physical Agents in the Workroom Environment, with Intended Changes for 1980*, ACGIH, Table Z.1.
122. *Nickel*, National Academy of Sciences, Committee on Medical and Biologic Effects of Environmental Pollutants, Washington, D.C., 1975.

123. *NIOSH Criteria for a Recommended Standard, Occupational Exposure to Inorganic Nickel*, U.S. Department of Health, Education and Welfare Publication No. 77-164, May 1977.
124. U.S. Pat. 1,563,587 (Dec. 1, 1925), M. Raney.
125. U.S. Pat. 1,628,190 (May 10, 1927), M. Raney.
126. U.S. Pat. 1,915,473 (June 27, 1933), M. Raney.
127. M. Raney, *Ind. Eng. Chem.* **32,** 1199 (1940).
128. J. Yasumura and T. Yoshino, *I&EC Prod. Res. Dev.* **7,** 252 (1968).
129. U.S. Pat. 3,271,326 (Sept. 6, 1966), A. J. Forney and J. J. Demeter (to United States of America).
130. J. H. Field and co-workers, *I&EC Prod. Res. Dev.* **3,** 150 (June 1964).
131. U.S. Pat. 3,846,344 (Nov. 5, 1974), F. E. Larsen and E. Snape (to The International Nickel Co., Inc.).
132. Ger.Offen. 2,713,374 (Sept. 28, 1978), H. J. Becker and W. Schmidt (to Bayer A.G.).
133. A. B. Fasman, B. K. Almashev, and B. Usenov, *Kinet. Katal.* **17,** 1353 (1976).
134. Y. Nitta, T. Imanaka, and S. Teranishi, *Nypan Kagaku Kaishi* **9,** 1362 (1976).
135. R. A. Dalla Betta, A. G. Piken, and M. Shelef, *J. Catal.* **40,** 173 (1975).
136. G. L. Bauerle and K. Nobe, *I&E Chem. Prod. Res. Dev.* **13**(3), 185 (1974).
137. U.S. Pat. 4,012,485 (Mar. 15, 1977), G. H. Maguerian, E. H. Hirschberg, and F. W. Rakowosky (to Standard Oil Company, Indiana).
138. F. A. Lowenheim, ed., *Modern Electroplating*, John Wiley & Sons, Inc., New York, 1963, Chapt. 12.
139. N. S. Smirnov, *Vitreous Enameller* **25**(4), 52 (1974).
140. A. I. Nedeljkovic, *Vitreous Enameller* **27**(3), 45 (1976).
141. Jpn. Kokai 78, 53798 (May 16, 1978), M. Hattori and co-workers (to Matushita Electrical Industrial Co. Ltd.).
142. P. A. Vityaz, R. Kisling, and A. Ronnquist, *Material Symposium*, Moscow, USSR, 1968, p. 358.
143. U.S. Pat. 3,748,165 (July 24, 1973), B. Hill (to The International Nickel Co., Inc.).
144. H. J. Heller, *Eur. Polym. J.* **5,** 105 (1969).
145. A. Zweig and W. A. Hendersen, Jr., *J. Polym. Sci. Polym. Chem. Ed.* **13,** 717 (1975).
146. Swiss Pat. 581,090 (Oct. 29, 1976), L. Avar and K. Hofer (to Sandoz).
147. Can. Pat. 862,132 (Jan. 26, 1971), J. H. Bright (to American Cyanamid Co.).
148. Ger. Offen. 2,519,594 (Nov. 27, 1975), L. Avar and K. Hofer (to Sandoz).
149. Brit. Pat. 1,066,349 (Apr. 26, 1967), (to Farbwerke Hoechst A.G.).
150. Ger. Offen. 2,539,034 (Mar. 25, 1976), L. Vuitel, F. L'Eplattenier, and A. Pugin (to CIBA-GEIGY A.G.).
151. U.S. Pat. 3,632,825 (Jan. 4, 1972), D. P. Jordan (to The International Nickel Co., Inc.).
152. U.S. Pat. 3,627,798 (Dec. 14, 1971), L. G. L. Ward (to The International Nickel Co., Inc.).
153. U.S. Pat. 3,689,517 (Sept. 5, 1972), J. T. Carriel (to The International Nickel Co., Inc.).
154. U.S. Pat. 3,651,110 (Mar. 21, 1972), L. G. L. Ward (to The International Nickel Co., Inc.).
155. U.S. Pat. 3,624,116 (Nov. 30, 1971), L. G. L. Ward (to The International Nickel Co., Inc.).
156. A. Zweig and W. A. Henderson, Jr., *J. Polym. Sci. Poly. Chem. Ed.* **13,** 993 (1975).
157. U.S. Pat. 3,310,575 (Mar. 21, 1967), J. D. Spivak (to Geigy Chem. Corp.).
158. U.S. Pat. 4,006,119 (Feb. 1, 1977), H. C. Beadle and I. Gibbs (to R. T. Vanderbilt Co., Inc.).
159. Ger. Offen. 2,631,475 (Jan. 19, 1978), N. Buhl, W. Kleeberg, and R. Wiedenmann (to Siemans A.G.).
160. U.S. Pat. 3,812,080 (May 21, 1974), A. M. Feldman (to American Cyanamid Co.).
161. Ger. Offen. 2,610,308 (Sept. 15, 1977), T. Papenfuhs and H. Volk (to Hoechst A.G.).
162. Ger. Offen. 2,533,958 (Feb. 17, 1977), T. Papenfuhs (to Hoechst A.G.).
163. U.S. Pat. 2,877,252 (Sept. 15, 1961), D. Hein and co-workers (to American Cyanamid Co.).
164. V. C. Vesce, *Offic. Dig.* **31,** 419 (Part 2), 58 (1959).
165. U.S. Pat. 3,338,938 (Aug. 29, 1967), A. S. Matlack (to Hercules, Inc.).
166. U.S. Pat. 4,066,388 (Jan. 3, 1978), R. Botros (to American Color & Chemical).
167. R. R. Haynes, J. H. Mathews, and G. A. Heath, *Mod. Text.* **51,** 64 (Jan. 1970).
168. *Nickel Compounds as Fungicides ICB-39*, International Nickel Co., Inc., New York, 1964.
169. *Nickel Compounds as Fungicides, 5351*, International Nickel Co., Inc., New York, 1971.
170. C. S. V. Ram, *Curr. Sci. (India)* **30,** 57 (1961).
171. K. E. Burke and C. H. Albright, *J. Assoc. Off. Anal. Chem.* **53,** 531 (1970).
172. J. R. Hardisen, *Phytopathology* **53,** 209 (Feb. 1963).
173. H. L. Keil, H. P. Frohlich, and J. O. VanHook, *Phytopathology* **48,** (Dec. 1958).

174. B. Peturson, F. R. Forsyth, and C. B. Lyon, *Phytopathology* **48**, (Dec. 1958).

175. H. L. Keil, H. P. Frolich, and C. E. Glassick, *Phytopathology* **48**, (Dec. 1958).

176. F. R. Forsyth and B. Peturson, *Phytopathology*, **49**, (Jan. 1959).

177. U.S. Pat. 2,971,880 (Feb. 14, 1961), H. L. Keil and H. P. Frohlich (to Rohm & Haas Co.).

178. A. Assarsson, I. Croan, and E. Frisk, *Svens. Papperstidn.* **73**, 493 (1970).

179. *Ibid.*, 528 (1970).

180. F. H. Neilson and H. E. Sauberlich, *Proc. Soc. Exp. Biol. Med.* **134**, 845 (1970).

181. J. M. King, Jr., *Advanced Technology Fuel Cell Programs*, EPRI EE-335, Research Project 114-1, Final Report, Power Systems Division, United Technologies Corp., South Windsor, Conn. Oct. 1976.

182. U.S. Pat. 2,813,076 (Nov. 12, 1957), L. E. Edelman (to The International Co., Inc.).

183. U.S. Pat. 4,090,970 (May 23, 1978), M. Braid (to Mobil Oil Co.).

184. H. E. Spencer and J. E. Hill, *Photogr. Sci. Eng.* **20**, 260 (1976).

185. U.S. Pat. 4,050,938 (Sept. 27, 1977), W. F. Smith, Jr. and G. A. Reynolds (to Eastman Kodak Co.).

General References

W. W. Wendlandt and S. P. Smith, *The Thermal Properties of Transition-Metal Ammine Complexes*, Elsevier Publishing Co., New York, 1967.

J. R. Boldt, Jr. and P. Queneau, *The Winning of Nickel*, Longmans Canada Ltd., Toronto, 1967.

J. F. Thompson and N. Beasley, *For The Years To Come*, G. P. Putnam's Sons, New York; Longmans, Green & Company, Toronto, Can., 1960.

P. W. Jolly and G. Wilke, *The Organic Chemistry of Nickel*, Vol. I, Organometallic Complexes, Academic Press, New York, 1974.

P. W. Jolly and G. Wilke, *The Organic Chemistry of Nickel*, Vol. II, Organic Synthesis, Academic Press, New York, 1975.

J. G. Dean, *Ind. Eng. Chem.* **51**, 48 (Oct. 1959).

A. F. Cotton and G. Wilkensen, *Advanced Inorganic Chemistry*, Wiley-Interscience, New York, 1966, pp. 878–893.

Y. T. Shah and A. J. Perrotta, *Ind. Eng. Chem. Prod. Res. Dev.* **15**(2), 123 (1976).

P. W. Silwood, *J. Catal.* **42**, 148 (1976).

O. B. J. Fraser, *Trans. Electrochem. Soc.* **71**, 425 (1937).

R. B. Pannell, K. S. Chung, and C. H. Bartholomew, *J. Catal.* **46**, 340 (1977).

R. B. King, *Transition-Metal Organometallic Chemistry, An Introduction*, Academic Press, New York 1969, Chapt. VII.

F. G. Ciapetta and G. J. Plank "Catalyst Preparation" in *Catalysis*, Vol. 1, Reinhold Publishing Corp., New York, 1954, Chapt. 7.

Symposium on Advances in Fischer-Tropsch Chemistry, Division of Petroleum Chemistry, Preprints, American Chemical Society, Anaheim Meeting, Calif., Mar. 1978.

C. Y. Wu and H. E. Swift, *Symposium on Homogeous Catalysis*, Vol. 25, No. 2, Division of Petroleum Chemistry, Preprints, American Chemical Society, Houston, Tex., Mar. 1980, pp. 372–381.

D. H. ANTONSEN
The International Nickel Company, Inc.

NICKEL SILVER.   See Copper alloys.

NICOTINAMIDE, $C_5H_4NCONH_2$.   See Vitamins.

NICOTINE, $C_{10}H_{14}N_2$.   See Alkaloids; Insect control technology.

NIOBIUM AND NIOBIUM COMPOUNDS

Niobium

Properties. Elemental niobium [7440-03-1], Nb, which also is called columbium, Cb, in the United States, has a cosmic abundance of 0.9 relative to silicon $\equiv 10^6$ (1), an average value of 24 ppm in the earth's crust (2), and a comparable value for the lunar surface (3). Niobium is a monoisotopic element, although a search for residual radionuclides from the formation of the solar system has established the natural abundance of ^{92}Nb [13982-37-1] ($T_{1/2} = 1.7 \times 10^8$ yr) to be $1.2 \times 10^{-10}\%$ (4) (see Isotopes). In addition, minute amounts of ^{94}Nb [14681-63-1] ($T_{1/2} = 2.03 \times 10^4$ yr) and ^{95}Nb [13967-76-5] ($T_{1/2} = 35$ d) occur in nature: ^{94}Nb from neutron capture by the stable isotope and ^{95}Nb as the daughter of ^{95}Zr in the fission products of ^{235}U. Niobium-93 has a nuclear spin of 9/2 and a thermal neutron-capture cross section of $1.1 \pm 0.1 \times 10^{-28}$ m^2 (1.1 barns), which makes it of much interest to the nuclear industry.

Niobium, like vanadium, undergoes no phase transitions from room temperature to the melting point, is a steel-grey, ductile, refractory metal with a higher melting point than molybdenum and a lower electron work function than tantalum, tungsten, or molybdenum. Niobium closely resembles tantalum in its properties; the former is only slightly more chemically reactive. The metal is unattacked by most gases below 200°C, but is air oxidized at 350°C with the development of an oxide film of increasing thickness which changes from pale yellow to blue to black at 400°C. Absorption of hydrogen at 250°C and nitrogen at 300°C occurs to form interstitial solid solutions which greatly affect the mechanical properties. Niobium is attacked by fluorine and gaseous hydrogen fluoride and is embrittled by nascent hydrogen at room temperature. It is unaffected by aqua regia and mineral acids at ordinary temperatures, except hydrofluoric acid which dissolves it. Niobium is attacked by hot concentrated hydrochloric and sulfuric acids, dissolving at 170°C in concentrated sulfuric acid, and by hot alkali carbonates and hydroxides, which cause embrittling.

The most common oxidation state of niobium is +5, although many anhydrous compounds have been made with lower oxidation states, notably, +4 and +3, and niobium +5 can be reduced in aqueous solution to +4 by zinc. The aqueous chemistry primarily involves halo- and organic acid anionic complexes but virtually no cationic chemistry because of the irreversible hydrolysis of the cation in dilute solutions. Metal–metal bonding is common with extensive formation of polymeric anions. Niobium resembles tantalum and titanium in its chemistry; thus, its separation from these elements is difficult. In the solid state, niobium has the same atomic radius as tantalum and essentially the same ionic, ie, +5, radius as Ta^{5+} (68 pm), Ti^{4+} (68 pm), and Li^+ (69 pm). Some properties of niobium are listed in Table 1; corrosion data are presented in Table 2.

Occurrence. Niobium and tantalum usually occur together. Niobium never occurs in the free state; sometimes it occurs as a hydroxide, silicate, or borate; and most often it is combined with oxygen and another metal, forming a niobate or tantalate in which the niobium and tantalum isomorphously replace each other with little change in physical properties except density (see Tantalum and tantalum compounds). Ore concentrations of niobium usually occur as carbonatites and are associated with tantalum in pegmatites and alluvial deposits. Principal niobium-bearing minerals can be divided into two groups:

Table 1.   Some Properties of Niobium

Property	Value	Ref.
atomic no.	41	
atomic wt	92.906	
atomic vol, cm^3/mol	10.8	
atomic radius, nm	0.147	
electronic configuration	$Kr4d^45s^1$	
ionization potential, V	6.77	
crystal structure	bcc	
lattice constant at 0°C, nm	0.33004	
density at 20°C, g/cm^3	8.66	
mp, °C	2468 ± 10	
bp, °C	5127	
latent heat of fusion, kJ/mol[a]	26.8	5
latent heat of vaporization, kJ/mol[a]	697	5
heat of combustion, kJ/mol[a]	949	6
heat capacity, J/(mol·K)[a]		
at 298 K	24.7	5
at 1500 K	29.7	5
at 3000 K	33.5	5
entropy, J/(mol·K)[a]		
at 298 K	36.5	5
at 1500 K	79.6	5
at 3000 K	111.6	5
vapor pressure at 2573 K, mPa[b]	22	7
evaporation rate at 2573 K, $\mu g/(cm^2 \cdot s)$	1.9	7
thermal conductivity at 298 K, W/(m·K)	52.3	
coefficient of linear thermal expansion (291–373 K), per °C	7.1×10^{-6}	8
volume electrical conductivity, % IACS[c]	13.3	9
electrical resistivity, $\Omega \cdot cm$	13–16×10^{-6}	
temperature coefficient of resistivity, per °C	3.95×10^{-3}	
work function, eV	4.01	
secondary emission (primary δ_{max} = 400 V), eV	1.18	
positive ion emission, eV	5.52	

[a] To convert J to cal, divide by 4.184.
[b] To convert mPa to μm Hg, divide by 133.3.
[c] IACS = International Annealed Copper Standard.

(*1*) Titano-niobates consist of the salts of niobic and titanic acids; the important minerals of this group are pyrochlore, loparite, koppite, and others. Pyrochlore, the most important of the minerals, is complex and of varying composition; the general formula for a typical Canadian pyrochlore is $(Na,Ca)_2(Nb,Ti)_2O_6[F,OH]$ and that of a typical Brazilian pyrochlore is $(Ba,Ca)_2(Nb,Ti,Ce)_2O_6[O,OH]$. The color of pyrochlore ranges from dark grey to brown to orange–brown. Pyrochlore occurs in carbonatite complexes primarily in Brazil and Canada as well as in Kenya, Uganda, Nigeria, Zaire, Norway, and the United States (14). It also occurs with calcite, dolomite, apatite, magnetite, and some silicates. The density of pyrochlore is ca 4.0–4.4 g/cm^3 and the tantalum content usually is low, ca 0.1–0.3% on a metal basis.

(2) Tantalo-niobates consist of the salts of niobic and tantalic acids. The general formula for this group is $(Fe,Mn)(Nb,Ta)_2O_6$, and the minerals consist of isomorphic mixtures of the four possible salts; the compound is a columbite if niobium is predominant and a tantalite if tantalum is predominant. These minerals are brown to black and usually contain titanium, tin, tungsten, and other impurities. The density

Table 2.   Corrosion of Niobium Metal[a]

Medium	Temperature, °C	Concentration, wt %	Corrosion rate, μm/yr
sulfuric acid[b]	23	96	0.5
	50	40	5
	100	20	0.5
	145	96	5,000
	bp	70	dissolves
	190[c]	1–10	slight
	250[c]	1–10	slight
	250[c]	20	250
	250[c]	30	1,300
hydrochloric acid	bp	1	nil
	bp	5	nil
	bp	10	100
	bp	15	450
	bp	20	1,000
	190[c]	5	30
	190[c]	10	500
	190[c]	15	14,000
hydrofluoric acid	23	all	very high
nitric acid	bp	70	nil
	190[c]	70	nil
	250[c]	70	nil
phosphoric acid	23	85	0.5
	100	85	80
aqua regia	23		0.5
	55		20
sodium hydroxide	23	10	20
	23	40	30
potassium hydroxide	23	40	90
zinc chloride	bp	40	nil
ferric chloride	23	10	nil
formic acid	bp	10	nil
acetic acid	bp	≤99.7	nil
oxalic acid	bp	10	20
citric acid	bp	10	20
lactic acid	bp	10	10
	bp	85	2.5
trichloroacetic acid	bp	50	nil
trichloroethylene	bp	99[d]	nil

[a] Refs. 10–13.
[b] Hydrogen embrittlement at higher temperatures.
[c] In sealed tubes.
[d] 1% water present.

of columbite is ca 5 g/cm^3 and that of tantalite is ca 8 g/cm^3, with a regular gradation between these limits in proportion to the tantalum content. Columbite–tantalite minerals usually are finely disseminated in granitic rocks and associated pegmatites or occur as enriched concentrations in alluvial (placer) deposits. The main sources of columbites are Nigeria, Zaire, and Malaysia, and a number of smaller occurrences in other parts of the world. Columbites frequently are associated with cassiterite (SnO_2) deposits and the niobium occurs in the tin slags during processing. These slags from Thailand and Malaysia are processed for their niobium and tantalum content.

Extraction, Refining, and Metallurgy. The process of extracting and refining niobium consists of a series of consecutive operations, frequently with several steps combined as one: the upgrading of ores by preconcentration (15); an ore-opening procedure to disrupt the niobium containing matrix; preparation of a pure niobium compound; reduction to metallic niobium; and refining, consolidation, and fabrication of the metal.

The stability of niobium ores is reflected in their occurrence as the residues of advanced weathering processes, eg, the placer deposits from decomposition of pegmatites and granitic rock and the pyrochlore found in lateritic (decomposed) carbonatites. Thus, stringent conditions are required to render the niobium extractable.

The most straightforward process is the direct conversion of the niobium concentrate to the metallic form. The primary method is the aluminothermic reduction of a pyrochlore and iron–iron oxide mixture (16). A single batch usually contains from 1–10 metric tons of pyrochlore with the necessary aluminum powder; iron scrap and/or iron oxide; frequently small amounts of lime or fluorspar as fluxing agents; and sometimes a small quantity of a powerful oxidizer, eg, sodium chlorate, which provides additional reaction heat. A typical reactor consists of a refractory-lined steel shell, and sometimes a floor of slag from previous reduction reactions is used. Typically, a small amount of a starter mixture, eg, aluminum powder and sodium chlorate, is ignited electrically to start the reaction. In another variation, aluminum powder and barium peroxide are used, and a small water spray initiates the aluminum-peroxide combustion. In some cases, the reaction is started with a small quantity of reactants and additional pyrochlore–aluminum–iron mixture is fed into the reaction mixture by means of a chute until the reactor vessel is filled with molten metal and slag.

In the batch reactions, the reaction time varies from 2–3 min to ca 25 min, primarily depending upon the size of the reaction. At the completion of the reaction, the molten ferroniobium is at the bottom of the reactor and the slag floats on top. Most of the impurities go into the slag and some easily reduced metals go into the ferroniobium. Typical commercial-grade ferroniobium has the following wt % composition:

Nb	62–67
Fe	28–32
Si	1–2.5
Al	0.5–2.0
Ti	0.1–0.4
P	0.05–0.15
S	0.05–0.1
C	0.05–0.1

Minor amounts of tantalum, tin, lead, bismuth, and other elements also occur in the ferroniobium. After cooling for 12–30 h, the metal is separated from the slag and is crushed and sized for shipment. The recovery of niobium in the aluminothermic reaction is 87–93%, with larger reactions generally giving better recoveries.

Ferroniobium also is produced in an electric-furnace procedure and essentially the same reactants are used as in the aluminothermic method. Since the electric furnace provides additional energy input, the quantity of aluminum can be substantially reduced and partly substituted by other reducing agents, eg, ferrosilicon. Since the total heat input can be better controlled using the electric furnace procedure, the recovery of niobium generally is better than with the aluminothermic method. The metal and slag which is produced can be tapped or skimmed with standard electric furnace procedures. Generally, the production volume of ferroniobium is not sufficient to take advantage of the control and semicontinuous operation offered by the electric furnace.

In addition to the standard ferroniobium, there is a lesser but significant demand for high purity niobium alloys, mainly high purity ferroniobium and nickel niobium. These high purity alloys are used in the fabrication of nickel- and cobalt-based superalloys which primarily are used in jet engine and aerospace applications (see High temperature alloys). These alloys are produced by reducing a pure niobium oxide with iron or nickel in an aluminothermic reaction with carefully controlled conditions and raw materials (17). In some cases, the reactions are carried out in water-cooled copper reactors to avoid contamination by refractory materials. The key raw material for these alloys is a pure (99 wt %) niobium oxide, which can be produced only by chemical procedures.

Direct attack by hot 70–80 wt % hydrofluoric acid, sometimes with nitric acid, is effective for processing columbites and tantalo–columbites; yields are >90 wt %. This method was used in the first commercial separation of tantalum and niobium and is used commercially as a lead-in to solvent extraction procedures. It is not suited to direct processing of pyrochlores because of the large alkali and alkaline earth oxide content, ie, ca 30 wt %, and the corresponding high consumption of acid.

Concentrated sulfuric acid (97 wt %) at 300–400°C has been used to solubilize niobium from columbite and pyrochlore (18–19). The exothermic reaction is performed in iron or silicon–iron crucibles to yield a stable sulfato complex, which is filtered free of residue and is hydrolyzed by dilution with water and boiling to yield niobic acid which is removed by filtration as a white colloidal precipitate.

Fusion with caustic soda at 500–800°C in an iron crucible is an effective method for opening pyrochlores and columbites (20). The reaction mixture is flaked and leached with water to yield an insoluble niobate which can be converted to niobic acid in yields >90 wt % by washing with hydrochloric acid.

The reaction of chlorine gas with a mixture of ore and carbon at 500–1000°C yields volatile chlorides of niobium and other metals which can be separated by fractional condensation (21–23). This method is used on columbites but is less suited to the chlorination of pyrochlore because of the formation of nonvolatile alkali and alkaline earth chlorides which remain in the reaction zone as a residue. The chlorination of ferroniobium, however, is used commercially. The product mixture of niobium pentachloride, iron chlorides, and chlorides of other impurities is passed through a heated column of sodium chloride pellets at 400°C to remove iron and aluminum by formation of a low melting eutectic compound which drains from the bottom of the column. The niobium pentachloride passes through the column and is selectively condensed; the more volatile chlorides pass through the condenser in the off-gas. The niobium pentachloride then can be processed further.

The reaction of finely ground ores with an excess of carbon at high temperatures produces a mixture of metal carbides. The reaction of pyrochlore with carbon starts at 950°C and proceeds vigorously. After being heated to 1800–2000°C, the cooled friable mixture is acid-leached leaving an insoluble residue of carbides of niobium, tantalum, and titanium. These may be dissolved in HF or they may be chlorinated or burned to oxides for further processing.

Once the niobium ore has been opened, the niobium must be separated from the tantalum and/or impurities. The classical method of doing this is the addition of an excess of potassium fluoride to hydrofluoric acid solutions of niobium ores to precipitate the complex fluorides and oxyfluorides of niobium, tantalum, and titanium. These are redissolved in dilute hydrofluoric acid, a 3 wt % HF solution containing $K_2NbOF_5.H_2O$ [19200-74-9], $K_2TiF_6.H_2O$, and K_2TaF_7 with respective solubilities at 15°C of ca 77 g/L, 12 g/L, and 5 g/L. The hydrate of K_2TiF_6 is isomorphous with $K_2NbOF_5.H_2O$ but K_2TaF_7 is not. Dipotassium tantalum heptafluoride is stable in 3 wt % HF, its solubility increases with HF concentration and temperature, and it is fifty times more soluble at 85°C than at 15°C. The species changes to $KTaF_6$ at 45% HF. The solubility of K_2NbOF_5 is high in 3 wt % HF and increases with temperature. At 6–40 wt % HF, the occurring species are K_2NbF_7 [36354-32-2] and $KNbF_6$ [16919-14-5] at >45 wt % HF. Increasing the concentration of KF depresses the solubility of all the complex salts. Repeated recrystallization can produce K_2TaF_7 with less than 0.01 wt % niobium content, but the niobium remains in the mother liquor and, hence, still contains the titanium impurity; thus, it limits the compound's usefulness in the economical production of pure niobium.

The recrystallization of complex fluoride salts has been replaced completely by solvent extraction techniques which are used extensively. Many acid–solvent combinations have been reported in the literature, eg, HF–MIBK (methyl isobutyl ketone) (24), HF–HNO$_3$–MIBK (25), HF–HCl–MIBK (26), and HF–H$_2$SO$_4$–MIBK (27). Commercial processes involve the use of various acids in combination with HF and either MIBK or TBP (tributyl phosphate), generally starting with an upgraded feed material. The dissolution in HF is performed in rubber or polyethylene-linked tanks to form the fluoride feed solution. The separation then may be effected by two methods: either the tantalum is extracted first by contacting a low acidity aqueous feed with the organic phase and subsequently raising the acidity of the aqueous phase and extracting the niobium with fresh organic, or both niobium and tantalum are extracted into the organic phase from a strongly acidic medium and the niobium is back-extracted from the organic with dilute acid. The metal values are back-extracted from the organic phase with a low acidity aqueous phase from which they are precipitated as oxides by addition of ammonia or as double fluorides by addition of potassium fluoride.

Another solvent extraction scheme uses the mixed anhydrous chlorides from a chlorination process as the feed (28). The chlorides, which are mostly of niobium, tantalum and iron, are dissolved in an organic phase and are extracted with 12N hydrochloric acid. The best separation occurs from a mixture of MIBK and DIBK (diisobutyl ketone). The tantalum transfers to the hydrochloric acid, leaving the niobium and iron. The latter enhances the separation factor in the organic phase. They are stripped with hot 14–20 wt % H$_2$SO$_4$ which is boiled to precipitate niobic acid, leaving the iron in solution.

Another method of purifying niobium is by distillation of the anhydrous mixed chlorides (29). Niobium and tantalum pentachlorides boil within about 15°C of one another which makes control of the process difficult. Additionally, process materials must withstand the corrosion effects of the chloride. The system must be kept meticulously anhydrous and air free to avoid plugging resulting from the formation of niobium oxide trichloride (NbOCl$_3$). Distillation has been used commercially in the past.

Once purification of the niobium has been effected, the niobium can be reduced to the metallic form. The double fluoride salt with potassium, K_2NbF_7, can be reduced with sodium metal. The reaction is carried out in a cylindrical iron vessel that is filled with alternating layers of K_2NbF_7 and oxygen-free sodium:

$$K_2NbF_7 + 5\,Na \rightarrow Nb + 2\,KF + 5\,NaF$$

Use of excess sodium drives the reaction to completion; the reaction usually is done under an argon or helium blanket. After cooling, the excess sodium is leached with alcohol and the sodium and potassium fluorides are extracted with water, leaving a mass of metal powder. The metal powder is leached with hydrochloric acid to remove iron contamination from the crucible.

Fused-salt electrolysis of K_2NbF_7 is not an economically feasible process because of the low current efficiency (30). However, electrowinning has been used to obtain niobium from molten alkali halide electrolytes (31). The oxide is dissolved in molten alkali halide and is deposited in a molten metal cathode, either cadmium or zinc. The reaction is carried out in a ceramic or glass container using a carbon anode; the niobium alloys with the cathode metal, from which it is freed by vacuum distillation, and the niobium powder is left behind.

Niobium pentoxide can be reduced with carbon in a two-step process (Balke process); formation of the carbide is the first step. The oxide is mixed with the stoichiometric amount of lamp black, placed in a carbon crucible, and heated in vacuum to 1800°C:

$$Nb_2O_5 + 7\,C \rightarrow 2\,NbC + 5\,CO$$

The carbide then is remixed with a stoichiometric amount of oxide, is compacted into chunks, and is refired to >2000°C under reduced pressure:

$$5\,NbC + Nb_2O_5 \rightarrow 7\,Nb + 5\,CO$$

The product chunks are hydrided, crushed, and dehydrided and the resultant powder is blended and pressed into bars which are purified by high temperature sintering. The sintering removes all of the carbon and most of the oxygen and is followed by consolidation by either arc or electron-beam melting.

Niobium pentoxide also is reduced to metal commercially by the aluminothermic process. The finely ground powder is mixed with atomized aluminum and an accelerator compound which gives extra heat during reaction, and is ignited. The reaction is completed quickly and, after cooling, the slag is broken loose to free the metal derby which is purified by electron-beam melting.

The pentachloride $NbCl_5$ can be reduced with hydrogen to yield a metal powder of high purity which is comparable to electron-beam-melted metal from other reduction processes. However, the large excess of hydrogen which is required and the attending safety problems make this route undesirable. Niobium pentachloride also can be reduced by the Kroll process, ie, decomposition of the halide with magnesium, and by reduction with oxygen-free sodium to yield niobium sponge, which must be consolidated.

Powder and sponge may be compacted at 0.7 GPa (6900 atm) into bars which are presintered at 1400–1500°C in vacuum. The bars then are resistance-heated in high vacuum to slightly below the melting point. After cooling, the bars are rolled to consolidate the pores and are resintered at 2300°C to yield a fabricable metal product of 98% theoretical density.

Arc melting also can be used to consolidate bars of metal which are pressed from powder or sponge and which are used as consumable electrodes in a low voltage, high current arc. The bar is suspended vertically and the molten metal falls from the bottom of the bar onto a water-cooled copper crucible, from which it is removed as an ingot.

The most common method in commercial use is electron-beam melting. The furnace essentially is a large thermionic vacuum tube. A high current, ie, >10 A, beam of electrons at >20 kV is focused magnetically on the bottom of a suspended metal compact bar. The molten metal globules fall into a pool on top of the ingot which is contained in a water-cooled copper cylinder. The ingot is retracted at the rate that it is formed. Impurities are boiled out of the molten pool and either are pumped away or deposited on the furnace walls as slag. The final ingot is 20–30 cm in diameter and 1.5–2 m long. Scrap niobium can be recycled by the hydride–dehydride process. Heating niobium under hydrogen pressure results in the formation of niobium hydride, which is brittle enough to crush and size. The crushed hydride can be reheated in a vacuum to form niobium-metal powder.

Niobium metal is available as ingot, sheet, rod, and wire and can be fabricated and formed by most metallurgical and engineering techniques (32). Cold working is necessary to avoid embrittlement which results from absorption of oxygen and nitrogen. Niobium can be reduced by 50–90% without intermediate annealing. When work-hardening occurs, annealing may be done in an inert atmosphere or at <10 mPa (7.5×10^{-5} mm Hg) at 1300–1400°C. The recrystallization temperature is 1050°C.

Economic Aspects. Discussions of the economic aspects of niobium and its compounds are given in refs. 33–35. The first large-scale continuous mining and concentrating of pyrochlore ore was started at Oka, Quebec, Canada, in 1961 by the St. Lawrence Columbium and Metals Corp. which ceased production in February, 1976. Canadian production, since 1976, is by Niobec Inc., a joint venture in which Société Québécoise d'Exploration Minière (SOQUEM) and Teck Corporation have equal interests. The Canadian output of pyrochlore in 1978 was about 4090 metric tons of which ca 1750 t was contained niobium. The Companhia Brazileria de Metalurgia e Mineracao started pyrochlore production at Araxa, Minas Gerais, Brazil, in late 1961 and has exported continuously since 1965. Brazil is the world's largest niobium-producing country, with a 1978 production of about 20,400 t of pyrochlore containing about 8280 t of niobium; this represents 80% of the world output of about 10,300 t of niobium. World niobium reserves are at least 14.5×10^6 t of contained niobium, with the following known distribution in t of contained niobium: Brazil, 12×10^6; USSR, 0.68×10^6; Canada, 0.59×10^6; Zaire, 0.41×10^6; Nigeria, 0.32×10^6; and Uganda, 0.32×10^6. There was no mine production of niobium minerals in the United States in 1979, although there are known deposits in Colorado, ca 63,000 t of which is contained Nb; Idaho, ca 9000 t contained Nb; Arkansas, ca 54,000 t contained Nb; and Oklahoma, ca 9000 t contained Nb. The principal world niobium ore producers are Companhia Brazileria de Metalurgia e Mineracao (CBMM) and Mineracao Catalao de Goias S.A. in Brazil and Niobec Inc. in Canada. Other producers of mineral concentrates that contain niobium are Greenbushes Tin in Australia, Companhia Estanno Sao Joao del Rei and Companhia Industrial Fluminese in Brazil, Datuk Keramet Smelting, Straits Trading and BEH Minerals SDN Berhad in Malaysia, Sociedade Mineira de Marropino Ltd. in Mozambique, Makeri Smelting and ATMN Ltd. in Nigeria, Hochmetals Africa (Pty) Ltd. in the Republic of South Africa, Thailand Smelting and Refining

Co., Ltd. (Thaisarco) in Thailand. Production also takes place in the USSR, Zaire, Zairetain, Sominki, and Somirwa.

Standard-grade ferroniobium was priced in North America at $11.29/kg of niobium contained at the beginning of 1979. The price increased to $11.95/kg of niobium in mid 1979, and increased to $14.00/kg of niobium in April 1980. Pyrochlore concentrate prices increased correspondingly, on an fob mine basis, from $5.53/kg of niobium oxide contained, to $6.08/kg in mid 1979, and to $7.12/kg niobium oxide in April 1980.

High purity niobium oxide reached a price of $35.00–37.00/kg in late 1979 but is available at less than $22.00/kg (June 1980) delivered in the United States. High purity ferroniobium prices increased from ca $35.00/kg of niobium in early 1979 to a current (June 1980) level of $68.00–70.55/kg niobium.

Domestic consumption of niobium in 1979 exceeded 2700 t, which is a 10% increase from 1978. The steel industry accounts for about 80% of this consumption, mostly for high strength, low alloy steels and carbon steels, and most of the remainder is used in superalloys and other alloys. The United States maintained a 1270 t stockpile of niobium in 1979, mostly as concentrates and ferroniobium, and has a stockpile goal of 1400 t. The main domestic niobium producing and processing companies and their products are: Cabot Corp. (Kawecki Berylco Ind., Inc.), metal, oxide, nickel–niobium, and ferroniobium; Mallinckrodt, Inc., oxide; Metallurg, Inc. (Shieldalloy Corp.), nickel–niobium and ferroniobium; H. K. Porter Co. (Fansteel, Inc.), metal and oxide; Reading Alloys, Inc., nickel–niobium and ferroniobium; Teledyne, Inc. (Teledyne Wah Chang Albany), metal, carbide, oxide, ferroniobium, niobium–tungsten, and niobium–titanium.

Analytical and Test Methods. The lack of stable niobium compounds resulting from a pronounced tendency for hydrolysis to colloidal suspensions of the hydrated oxide has given rise to an extensive body of literature (36–39). The analysis of niobium involves solubilizing the sample; separating gross interferences by selective precipitation, solvent extraction, ion exchange, or another chromatographic technique; and determination by colorimetric, spectrometric, or gravimetric methods. There are three methods for dissolution of niobium metal, alloys, oxide, ores, and minerals:

(1) Addition of nitric acid to hydrofluoric acid aids in dissolving metal and alloy samples. Hydrofluoric acid solutions are used in anion-exchange separations, colorimetric determinations, and solvent-extraction separations.

(2) Fusion of niobium ores with potassium hydroxide or potassium carbonate followed by water leaching of the cooled melts dissolves the niobium and tantalum as niobates and tantalates, respectively. Addition of sodium chloride to saturate the alkaline solution precipitates almost quantitatively the sparingly soluble sodium salts. Acidification of a suspension of the sodium salts precipitates the mixed hydrated oxides for further processing.

(3) Fusion with potassium bisulfate, $KHSO_4$, or pyrosulfate, $K_2S_2O_7$, is the most widely used route to solubilize the ignited oxides or ores. The sample is fused in a quartz crucible with an eight- to tenfold excess of reagent to a clear melt at 650–800°C. The cooled melt is dissolved in a saturated solution of ammonium oxalate or 20 wt % tartaric acid or citric acid. A large number of precipitation reactions for the determination of niobium and tantalum are based on these acid solutions of oxalato or tartrato complexes. It is necessary to maintain a large excess of oxalic or tartaric acid for stability of the complex. The oxalic acid solutions are more stable but the tartrato complexes can be made alkaline without precipitation of niobic and tantalic acids.

Of the gravimetric procedures, precipitation with tannin from a slightly acidic oxalate solution is probably the best known. The hot oxalate extract of a potassium pyrosulfate fusion is adjusted with ammonia to pH 3.7–4.0 using Bromothymol Blue. Dropwise addition of a 2 wt % solution of tannin precipitates the lemon-yellow complex of tantalum. Several grams of ammonium chloride is added to prevent peptization, and the solution is digested and filtered. Further addition of tannin and ammonia to the clear filtrate precipitates the vermilion-colored complex of niobium. Many other organic precipitants have been used, including Cupferron, N-benzoyl-N-phenylhydroxylamine, 8-quinolinol (Oxine), phenylarsonic acid, pyrogallol, and others.

Solvent extraction techniques are very useful in the quantitative analysis of niobium. The fluoro complexes are amenable to extraction by a wide variety of ketones. Some of the water-insoluble complexes with organic precipitants are extractable by organic solvents and colorimetry is performed on the extract. An example is the extraction of the niobium–oxine complex with chloroform (40). The extraction of the niobium–pyrocatechol violet complex with tridodecylethylammonium bromide and the extraction of niobium–pyrocatechol–sparteine complex with chloroform are examples of extractions of water-soluble complexes with colorimetry performed on the extract (41–42). Colorimetry also may be performed directly on the water-soluble complex, eg, with ascorbic acid and 5-nitrosalicylic acid (43–44).

Chromatographic methods play a prominent role in the clean separation of niobium from tantalum and other metals. The use of methyl ethyl ketone as the eluent for the cellulose column separation of niobium and tantalum in dilute hydrofluoric acid–ammonium fluoride solutions was a major advance from the tannin precipitations. The anionic-exchange separation of niobium from tantalum and other metals from a hydrochloric–hydrofluoric acid medium is accomplished by elution with ammonium chloride–hydrofluoric acid (45).

Once niobium is isolated, it may be determined gravimetrically by precipitation as an insoluble complex or the earth acid, by firing to the oxide at 800–1000°C, and by weighing. Alternatively, the precipitate may be redissolved and determined colorimetrically in aqueous solution or in an organic extract. The arc and spark spectra of niobium allow its determination in oxide mixtures and alloys. X-ray fluorescence analyses have become the principal method for niobium analysis and are used for routine work by all of the major producers of niobium concentrates (46–47). With careful attention to matrix effects, the precision and accuracy of x-ray fluorescence analyses are at least equal to those of the gravimetric and ion-exchange methods.

Health and Safety Factors (Toxicology). Toxicity data on niobium and its compounds are sparse. The most common materials, eg, niobium concentrates, ferroniobium, niobium metal, and niobium alloys, appear to be relatively inert biologically. Limited animal experiments show high toxicity for some salts which are related to disturbance of enzyme action. The hydride has moderate fibrogenic and general toxic action with recommended maximum allowable concentrations of 6 mg/m^3 (48). Recommended maximum permissible concentration of Nb in reservoir water is 0.01 mg/L. Rats receiving 0.005 mg/kg daily over 9 mo show changes in cholinesterase activity but no effects are observed with doses of 0.0005 mg/kg. The threshold for affecting clarity and BOD (biological oxygen demand) is 0.1 mg/L (49).

Unstable niobium isotopes that are produced in nuclear reactors or similar fission reactions have typical radiation hazards: ^{93m}Nb, $T_{1/2} = 14$ yr, decays by 0.03 MeV gamma emission to stable ^{93}Nb; ^{95}Nb, $T_{1/2} = 35$ d, a fission product of ^{235}U, decays to stable ^{95}Mo by emission of 0.16 MeV β^- and 0.77 MeV γ radiation; and ^{97}Nb, $T_{1/2}$ = 72 min, decays to stable ^{97}Mo by emission of 1.27 MeV β^- and 0.66 MeV γ radiation. Inhalation experiments on mice have been performed with aerosols of radioniobium ^{95}Nb, which was prepared at 100, 250, 600, and 1100°C to give particles of different chemical compositions (50). The activity was restricted to the lungs for the 600°C and 1100°C aerosols but was translocated partially to the skeleton for the 100°C and 250°C aerosols. At the lower temperatures, the highest radiation dose was to the lungs, skeleton, and liver; whereas, the radiation dose was delivered exclusively to the lungs at the higher temperatures.

There is a moderate fire and explosion hazard from niobium dust, either by exposure to flame or by chemical reaction; the dust also reacts with oxidizing agents.

Uses. Niobium, as ferroniobium, is used extensively in the steel industry as an additive in the manufacture of high strength, low alloy (HSLA) and carbon steels. It acts as a grain refiner to increase yield and tensile strength with additions as low as 0.02 wt %; although normal usage is 0.03–0.1 wt %. The most important application of niobium HSLA or microalloyed steels is in oil- and gas-pipeline steels, particularly those which may experience operating conditions below −25°C. The niobium microalloyed steels also are widely used in automobiles, buildings, bridges, ships, towers, concrete reinforcing bars, etc, where the strength-to-weight and cost-per-unit-of-strength ratios are particularly advantageous.

Addition of niobium to austenitic stainless steels inhibits intergranular corrosion by forming niobium carbide with the carbon that is present in the steel. Without the niobium addition, chromium precipitates as a chromium carbide film at the grain boundaries and, thus, depletes the adjacent areas of chromium and reduces the corrosion resistance. An amount of niobium equal to ten times the carbon content is necessary to prevent precipitation of the chromium carbide.

Niobium also is important in nonferrous metallurgy. Addition of niobium to zirconium reduces the corrosion resistance somewhat but increases the mechanical strength. Because niobium has a low thermal-neutron cross section, it can be alloyed with zirconium for use in the cladding of nuclear fuel rods. A Zr–1 wt % Nb [11107-78-1] alloy has been used as primary cladding in the USSR and a Zr–2.5 wt % Nb [11135-92-5] alloy has been used as a structural material in the USSR and Canada. Recently, Zr–2.5 wt % Nb has been used to replace Zircaloy-2 as the cladding in Candu-PHW reactors and has resulted in a 20% reduction in wall thickness of cladding (51) (see Nuclear reactors).

Niobium is a common additive to the nickel- and cobalt-based superalloys. Addition levels are typically 1–2.5 wt % in the cobalt-based alloys and 2–5 wt % in the nickel-based alloys. Niobium-based alloys with tungsten, titanium, and zirconium have superior strength and corrosion resistance up to 1200°C; eg, C-103, a Nb–Hf–Ta–Ti–W–Zr alloy, is used for rocket nozzles in spacecraft and missiles and for turbine blades used in extreme conditions.

In addition, the corrosion resistance of niobium and its high electrical conductivity and ductility make it a valuable structural material for chemical and metallurgical applications. The heat-transfer coefficient of niobium is more than twice that of titanium and three times higher than zirconium and stainless steels. Niobium is very corrosion-resistant to most media, with the exception of hydrofluoric acid and hot concentrated hydrochloric and sulfuric acids. The pickling solution for removal of normal surface oxides is 1 part nitric acid, 1 part sulfuric acid, 2 parts hydrofluoric acid, and four parts water (by volume). Niobium also shows good corrosion resistance to sulfidizing atmospheres of low oxygen potential, which may be useful in the production of substitute natural gas from sulfur-containing materials (52). Liquid sodium, potassium, sodium–potassium alloys, or lithium have little effect on niobium up to 1000°C, and its resistance to many other liquid metals is good.

Niobium is used as a substrate for platinum in impressed–current cathodic protection anodes because of its high anodic breakdown potential (100 V in seawater), good mechanical properties, good electrical conductivity, and the formation of an adherent passive oxide film when it is anodized. Other uses for niobium metal are in vacuum tubes, high pressure sodium-vapor lamps, and in the manufacture of catalysts.

Niobium carbide is used as a component of hard metals, eg, mixtures of metal carbides that are cemented with cobalt, iron, and nickel. Along with tantalum carbide, niobium carbide is added to impart toughness and shock and erosion resistance. The spiraling rise in the price of tantalum has spurred the development of a hafnium carbide–niobium carbide substitute for tantalum carbide (53). These cemented carbides are used for tool bits, drill bits, shovel teeth, and other wear-resistant components; turbine blades; and as dies in high pressure apparatus (see Carbides).

Niobium and many of its alloys exhibit superconductivity, ie, the lack of electrical resistance at very low temperatures; thus, they are of great interest for power generation, propulsion devices, fusion research, electronic devices, and other applications. Niobium becomes superconducting at 9.15 K. Other niobium compounds and their transitional temperatures are: NbTi [12384-42-8], 9.5 K; Nb_3In [12030-07-8], 9 K; Nb_3Sn [12035-04-0], 18 K; Nb_3Al [12003-75-7], 18.8 K; Nb_3Ga [12024-05-4], 20 K; and Nb_3Ge [12025-22-8], 23 K. Most superconducting devices use niobium–titanium because of ease of its fabrication into magnet wire, which is its most common application. Where very high magnetic fields are necessary, niobium–tin Nb_3Sn conductors are used, even though the intermetallic nature of Nb_3Sn makes fabrication difficult. Improved methods of fabrication will lead to wider use of Nb_3Sn and to commercial application of niobium–aluminum and niobium–aluminum–germanium superconductors (see Superconducting materials).

Niobium Compounds

Niobium Boride. A number of niobium boride phases have been described in the literature, ie, Nb_2B [12344-74-0], Nb_3B [56450-58-9], Nb_3B_4 [12045-89-5], NbB, and NbB_2. Only the last two, the monoboride and the diboride, melt congruently; NbB_2 decomposes at the melting point to NbB and boron (54). Some of the properties of the niobium borides and other niobium compounds are listed in Table 3.

The most common methods of preparation have been hot-pressing, sintering, or remelting powdered mixtures of elemental boron with niobium or niobium hydride (88–89). Other methods are the reduction of a mixture of Nb_2O_5 and B_2O_3 with aluminum, silicon, or magnesium (90); carbon reduction at 2000°C of Nb_2O_5 and B_2O_3 (91); reaction of carbon with B_4C and Nb_2O_5 (92); electrolysis of molten mixtures of Nb_2O_5 with alkali metal and alkaline earth metal borates and fluorides to produce NbB_2 (93–94); chemical vapor deposition (CVD) onto a hot substrate by a mixture of niobium halide, boron halide, hydrogen, and argon (61,95); and CVD of boron on niobium accompanied or followed by diffusion into the substrate (96) (see Film de-

Table 3. Properties of Some Niobium Compounds

Compound	CAS Registry No.	Lattice	Lattice constant, pm	Density, g/cm³	Hardness	Mp, °C	Bp, °C	Specific resistivity, μΩ·cm	Thermal conductivity, W/(m·K)	Refs.
niobium boride, NbB	[12045-19-1]	orthorhombic	a = 329.8 b = 872.4 c = 316.6	7.5		2000		64.5 at 25°C		55-58
niobium diboride, NbB₂	[12007-29-3]	hexagonal	a = 308.9 c = 330.3	6.9	Mohs 8+	3050				55
diniobium carbide, Nb₂C	[12011-99-3]	hcp	a = 312.7 c = 497.2	7.80		3090		65 at 25°C	17 at 23°C	56, 59-61 62
niobium carbide, NbC	[12069-94-2]	fcc	a = 447.1	7.788	Mohs 9+	3600	4300	180 max	14 at 23°C	60, 63-68
niobium pentafluoride, NbF₅	[7783-68-8]	monoclinic	a = 963 b = 1443 c = 512 β = 96.1°	3.54		79	234			69-70
niobium fluorodioxide, NbO₂F	[15195-33-2]	cubic	a = 390.2							71
niobium pentachloride, NbCl₅	[10026-12-7]	monoclinic	a = 183.0 b = 1798 c = 588.8 β = 90.6°	2.74		208.3	248.2			72-73

Compound	Registry No.	Crystal system	Lattice constants	Density	Hardness	mp, °C	bp/dec, °C	Thermal property	Refs
niobium trichloromonoxide, NbOCl₃	[13597-20-1]	tetragonal	a = 1087	3.72		vacuum sublimes at ca 200			74–75
niobium pentabromide, NbBr₅	[13478-45-0]	orthorhombic	c = 396 a = 612.7 b = 1219.8 c = 1855	4.36		254	365		76–77
niobium tribromomonoxide, NbOBr₃	[14459-75-7]					vacuum sublimes at 180	ca 320 dec		
niobium pentaiodide, NbI₅	[13779-92-5]	monoclinic	a = 1058 b = 658 c = 1388 β = 109.1°			ca 200 dec			78
niobium hydride, NbH	[13981-86-7]	bcc		6–6.6					79–80
diniobium nitride, Nb₂N	[12033-43-1]	hcp	a = 305.6–304.8 c = 495.6	8.08		2050			
niobium nitride, NbN	[24621-21-4]	fcc	a = 438.2–439.2	8.4	Mohs 8+			200 at 25°C, 450 at mp	80–82
niobium oxide, NbO	[12034-57-0]	cubic	a = 421.08	7.30					83
niobium dioxide, NbO₂	[12034-59-2]	tetragonal	a = 1371 c = 598.5	5.90					84
α-niobium pentoxide, α-Nb₂O₅	[1313-96-8]	monoclinic	a = 2116 b = 382.2 c = 1935	4.55		1491 ± 2			85–87

position techniques). Niobium diboride generally is a gray powder; is unattacked by hydrochloric acid, nitric acid, or aqua regia; is attacked slowly by hot sulfuric or hydrofluoric acid; and is dissolved rapidly by molten alkali, hydroxides, carbonates, bisulfates, and sodium peroxide. It is oxidized in air at red heat.

Niobium Carbide. Apparently three solid single-phase regions exist in the niobium–carbon system, ie, a solid solution of carbon in niobium (bcc), Nb_2C (hexagonal), and NbC (fcc) (62). The compositional range of Nb_2C is very limited, whereas NbC varies from $NbC_{0.7}$ to nearly stoichiometric NbC. Thermodynamic values for these phases have been reported (66,97–98). Industrial preparation utilizes Nb_2O_5 and carbon as starting materials. The reaction starts at ca 675°C but temperatures of 1800–2000°C are needed for completion of the reaction. Heating the elemental powders also produces NbC if a sufficiently high final temperature is used. Chemical vapor deposition can be used to deposit NbC on a hot surface by reaction of $NbCl_5$ with hydrogen and hydrocarbons. Niobium carbide powder has a gray metallic color up to a composition of $NbC_{0.9}$; the color changes to lavender upon addition of carbon up to $NbC_{0.99}$. NbC is very unreactive and resists boiling aqua regia; a mixture of HNO_3 and HF is needed for dissolution. NbC burns on heating in air to >1100°C and can be converted to the nitride by heating in nitrogen or ammonia.

Niobium Halides and Oxyhalides. All possible halides of pentavalent niobium are known and preparations of lower-valent halides generally start with the pentahalide. Ease of reduction decreases from iodide to fluoride.

Niobium Pentafluoride. Niobium pentafluoride is prepared best by direct fluorination of the metal with either fluorine or anhydrous hydrofluoric acid at 250–300°C. The volatile NbF_5 is condensed in a pyrex or quartz cold trap, from which it can be vacuum-sublimed at 120°C to yield colorless monoclinic crystals. It is very hygroscopic and reacts vigorously with water to give a clear solution of hydrofluoric acid and H_2NbOF_5 [12062-01-0]. This acid also is formed by dissolving niobium metal or niobic acid in hydrofluoric acid and, at high acid concentrations, it is converted to H_2NbF_7. Addition of potassium fluoride to a solution of H_2NbOF_5 precipitates $K_2NbOF_5 \cdot H_2O$, which is very soluble in hot water and can be recrystallized from a saturated solution to give large monoclinic platelets. The high solubility of $K_2NbOF_5 \cdot H_2O$ in water was the basis for early separations of tantalum and niobium since the corresponding tantalum salt K_2TaF_7 is twelve times less soluble in 1 wt % HF at 20°C.

Niobium Dioxide Fluoride. Niobium dioxide fluoride, NbO_2F, is formed on dissolution of niobium pentoxide in 48 wt % aqueous hydrofluoric acid, evaporation of the solution to dryness, and heating to 250°C.

Niobium Pentachloride. Niobium pentachloride can be prepared in a variety of ways but most easily by direct chlorination of niobium metal. The reaction takes place at 300–350°C. Chlorination of a niobium pentoxide–carbon mixture also yields the pentachloride; however, generally the latter is contaminated with niobium oxide trichloride. The pentachloride is a lemon-yellow crystalline solid that melts to a red–orange liquid and hydrolyzes readily to hydrochloric acid and niobic acid. It is soluble in concentrated hydrochloric and sulfuric acids, sulfur monochloride, and many organic solvents.

Niobium Oxide Trichloride. Niobium oxide trichloride, $NbOCl_3$, also can be prepared in a variety of ways, ie, oxidation of the pentachloride by air, reaction of the pentoxide with the pentachloride, reaction of carbon tetrachloride or HCl gas with the pentoxide at 400–700°C, and as a intractable impurity in most preparations of the pentachloride. It is a white solid which sublimes at ca 200°C, and it is thermally unstable and forms the pentoxide and the pentachloride at higher temperatures.

Niobium Pentabromide. Niobium pentabromide is most conveniently prepared by reaction of bromine with niobium metal at ca 500°C. It is a fairly volatile yellow–red compound which is hygroscopic and readily hydrolyzes. It is soluble in water, alcohol, and ethyl bromide.

Niobium Oxide Tribromide. Niobium oxide tribromide, $NbOBr_3$, is a yellow–brown solid which is readily hydrolyzed by moist air. It is prepared by reaction of bromine with a mixture of niobium pentoxide and carbon at 550°C. It decomposes in vacuum to the pentabromide and pentoxide at 320°C.

Niobium Pentaiodide. Brass-yellow crystals of niobium pentaiodide are formed by direct reaction of excess iodine with niobium metal in a sealed tube (99). It is thermally unstable and decomposes to the tetraiodide [13870-21-8] at 206–270°C in vacuum (100).

Niobium Hydride. Hydrogen reacts exothermically with niobium to form a stable interstitial solid solution. In a gas-phase reaction at 300–1500°C and hydrogen pressures of 0–101 kPa (0–1 atm), the lattice parameter and heat of solution increase with hydrogen content up to the composition $NbH_{0.85}$ (101). X-ray studies show a linear relation between atomic volume and the hydrogen content (102). The absorption of hydrogen is proportional to the square root of the hydrogen pressure, which indicates dissociation of molecular hydrogen at the metal surface and diffusion of hydrogen atoms (103). Although the hydride is stable at room temperature, heating to 500°C at 0.67 kPa (5 mm Hg) decomposes the hydride to hydrogen and niobium metal. The expansion on absorption and contraction on desorption of hydrogen by the metal lattice usually leaves the metal in powder form after a hydride–dehydride operation, a process of commercial value for production of metal powder and recovery of metal scrap. An unstable dihydride NbH_2 [13981-96-9] has been prepared by the cathodic hydrogenation of niobium foil in $6N$ sulfuric acid (104). It is very unstable and decomposes in vacuum or in air to the monohydride (see Hydrides).

Niobium Nitrides. The uptake of nitrogen by niobium metal proceeds by the exothermic formation of an interstitial solid solution of nitrogen atoms in the bcc lattice of the metal. The solubility of nitrogen in the metal is proportional to the square root of the nitrogen partial pressure until the formation of the nitride phase Nb_2N. This relation holds from 1200–2400°C and over almost ten orders of magnitude of the pressure (105). At the solubility limit of the solid solution α phase, the hcp β phase appears, which has a composition of $NbN_{0.4}$ to $NbN_{0.5}$. Further absorption of nitrogen leads to the formation of a fcc δ phase with a homogeneity of $NbN_{0.88}$ to $NbN_{0.98}$; the δ phase is stable only above 1230°C. Niobium nitride can be prepared by heating the metal in nitrogen or ammonia to 700–1100°C (80), by heating the pentoxide and carbon to 1250°C in the presence of nitrogen (106), and by CVD using $NbCl_5$, H_2, and N_2 (96,107). The nitride is a light gray powder with a yellow cast; it is insoluble in HCl, HNO_3, and H_2SO_4; and it is attacked by hot caustic, lime, or strong alkalis with the evolution of ammonia. It reacts to the pentoxide when heated in air and is accompanied by the liberation of nitrogen; Nb_2N is resistant to acids but reacts with strong alkali to liberate nitrogen rather than ammonia (see Nitrides).

Niobium Oxides. The solubility of oxygen in niobium obeys Henry's law to the solubility limit of the first oxide phase of 850–1300°C (108). The amount of oxygen in solution in niobium is 1.3 at. % at 850°C and nearly 2 at. % at 1000°C (109). Only three clearly defined anhydrous oxides of niobium have been obtained in bulk, ie, NbO, NbO_2, and Nb_2O_5. Niobium monoxide, NbO, is obtained by hydrogen reduction of the pentoxide at 1300–1700°C or by heating a compressed mixture of the metal powder with NbO_2 in argon at 1700°C. It has a gray metallic appearance. Niobium dioxide, NbO_2, also can be obtained by hydrogen reduction of the pentoxide at 800–1300°C, by heating a properly proportioned mixture of the pentoxide and the metal, or by thermal dissociation of the pentoxide at 1150°C in an argon sweep. Niobium dioxide is black with a bluish cast, is a strong reducing agent in the dry state, and is converted to the pentoxide on ignition in air.

The considerable confusion existing in the literature regarding the polymorphism of Nb_2O_5 seems to have been resolved. Three distinct phases have been identified: a low temperature phase T, a middle temperature phase M, and a high temperature phase H (110). With regard to the amorphous oxide produced from the hydrolysis of niobic acid, conversion to the crystalline T form occurs at 500°C and is followed by transformation to the M form at 1000°C and to the H form at 1100°C. These phases were renamed γ (T), β (M), and α (H) in a later study in which the transition of amorphous to γ was found to occur at 440°C (111). Heating at 830°C irreversibly converts the γ phase to a mixture of β and α forms and further heating to 1095°C irreversibly transforms this mixture to the pure α form. It was concluded that the β form is an imperfectly crystallized α phase. Subsequent work has demonstrated the existence of another metastable phase: ϵ (112). It appears that three crystalline forms of Nb_2O_5 are detectable at atmospheric pressure and that the α is the only stable structure and is monoclinic with 14 formula units in the unit cell (87,113). The α phase can be prepared by heating the metal, carbide, nitride, or niobic acid in air at >1100°C and has been prepared in the form of large single crystals (114). It generally is an egg-shell-white powder, and it turns yellowish on heating because of the formation of oxygen vacancies in the lattice. It is insoluble in acids, except hydrofluoric, and can be dissolved by fused alkali pyrosulfates, carbonates, or hydroxides.

Niobic Acid and Salts. *Niobic Acid.* Niobic acid, $Nb_2O_5 \cdot x\,H_2O$, includes all hydrated forms of niobium pentoxide, where the degree of hydration depends on the method of preparation, age, etc. It is a white insoluble precipitate formed by acid hydrolysis of niobates that are prepared by alkali pyrosulfate, carbonate, or hydroxide fusion; base hydrolysis of niobium fluoride solutions; or aqueous hydrolysis of chlorides or bromides. When it is formed in the presence of tannin, a voluminous red complex forms. Freshly precipitated niobic acid usually is colloidal and is peptized by water washing; thus, it is difficult to free from traces of electrolyte. Its properties vary with age and reactivity is noticeably diminished on standing for even a few days. It is soluble in concentrated hydrochloric and sulfuric acids but is reprecipitated on dilution and boiling and can be complexed when it is freshly made with oxalic or tartaric acid. It is soluble in hydrofluoric acid of any concentration.

Niobates. Niobic acid is amphoteric and can act as an acid radical in several series of compounds, which are referred to as niobates. Niobic acid is soluble in solutions of the hydroxides of the alkali metals to form niobates. Fusion of the anhydrous pentoxide with alkali metal hydroxides or carbonates also yields niobates. Most niobates are insoluble in water with the exception of those alkali metal niobates having a base-to-acid ratio that is greater than one. The most well-known water-soluble niobates are the 4:3, and the 7:6 salts (base:acid), with empirical formulas $M_8Nb_6O_{19}$(aq) and $M_{14}Nb_{12}O_{37}$(aq), respectively. The hexaniobate is hydrolyzed in aqueous solution according to the pH-dependent reversible equilibria (115).

$$(Nb_6O_{19}\,aq)^{8-} \underset{}{\overset{H_2O}{\rightleftharpoons}} (HNb_6O_{19}\,aq)^{7-} + OH^-; \quad pH \cong 13$$

$$(HNb_6O_{19}\,aq)^{7-} \underset{H_2O}{\rightleftharpoons} [(Nb_6O_{18}\,aq)^{6-}]_n + OH^-; \quad pH \cong 9$$

The 7:6 salts are the acid salts of the normal 4:3 hexaniobates, and their formulas can be written $M_7H(Nb_6O_{19})(aq)$. Further hydrolysis can take place with the irreversible precipitation of niobic acid:

$$H_6(Nb_6O_{18}) \rightarrow 3\,Nb_2O_5 + 3\,H_2O; \quad pH \cong 4.5$$

The potassium salts are the most soluble and other salts usually are precipitated by addition of the appropriate metal chloride to a solution of the corresponding potassium salt. The metaniobates $MNbO_3$ and orthoniobates $MNbO_4$ generally are prepared by fusion of the anhydrous mixed oxides. The metaniobates crystallize with the perovskite structure and are ferroelectric (116) (see Ferroelectrics). The orthoniobates are narrow-band-gap semiconductors (qv) (117).

Sodium metaniobate [67211-31-8] (1:1), $Na_2O.Nb_2O_5.7H_2O$ or $Na_2Nb_2O_6.7H_2O$, separates as colorless triclinic crystals as a result of concentrating the mother liquor from the preparation of the 7:6 sodium niobate by spontaneous evaporation. It also can be obtained by fusion of the anhydrous pentoxide in sodium hydroxide or carbonate.

Potassium niobate [12502-31-7] (4:3), $4K_2O.3Nb_2O_5.16H_2O$ or $K_8Nb_6O_{19}.16H_2O$, is obtained by dissolving niobic acid in a concentrated solution of potassium hydroxide. The large monoclinic crystals are separated by concentrating the solution. The salt is very soluble; at room temperature a saturated solution contains 425 g/100 g of water. It is much more soluble in hot water and is prone to form supersaturated solutions.

Sodium niobate [12201-59-1] (7:6), $7Na_2O.6Nb_2O_5.31H_2O$ or $Na_{14}Nb_{12}O_{37}.31H_2O$, forms a crystalline precipitate when a hot solution of a soluble niobium compound is added to a hot concentrated sodium hydroxide solution. It is insoluble in the presence of excess sodium hydroxide but is sparingly soluble in pure water. It also can be formed by addition of sodium hydroxide or chloride to a solution of the 4:3 potassium niobate.

Lithium niobate [12031-63-9], $Li_2O.Nb_2O_5$ or $LiNbO_3$, is prepared by the solid-state reaction of lithium carbonate with niobium pentoxide. After being separately predried at 150–200°C, the stoichiometric amounts of the oxides are carefully mixed and heated to 600°C in a platinum crucible. The temperature is increased slowly for 12 h from 600–800°C and is maintained at 800°C another 12 h. The mixture is cooled, crushed, and reheated to 900°C for 12 h. The product may not be completely homogeneous.

BIBLIOGRAPHY

"Columbium" in *ECT* 1st ed., Vol. 4, pp. 314–324, by C. W. Balke, Fansteel Metallurgical Corp.; "Niobium and Niobium Compounds" in *ECT* 2nd ed., Vol. 13, pp. 766–784, by P. A. Butters, Murex Ltd.

1. G. G. Goles, "Cosmic Abundances" in *Handbook of Geochemistry*, Vol. 1, Springer-Verlag Berlin, New York, 1969.
2. M. R. Krishnadev and A. Galibois, *Proc. 3rd Interamerican Conf. Mater. Technol.*, 581 (1972).
3. W. Von Engelhardt and R. Stengelin, *Earth Planet. Sci. Lett.* **42**, 213 (1979).

4. K. E. Apt, J. D. Knight, D. C. Camp, and R. W. Perkins, *Geochim. Cosmochim. Acta* **38,** 1485 (1974).

5. D. R. Stull and G. C. Sinke, *Thermodynamic Properties of the Metals*, American Chemical Society, Washington, D.C., 1956.

6. G. L. Humphrey, *J. Am. Chem. Soc.* **76,** 978 (1954).

7. R. Speiser, P. Blackburn, and H. L. Johnston, *J. Electrochem. Soc.* **106,** 52 (1959).

8. C. R. Tottle, *Nucl. Eng.* **3,** 212 (1958).

9. J. R. Darnell and L. F. Yntema in B. W. Gonser and E. M. Sherwood, eds., *The Technology of Columbium (Niobium)*, John Wiley & Sons, Inc., New York, 1958.

10. D. F. Taylor, *Ind. Eng. Chem.*, 639 (April 1950).

11. C. R. Bishop, *Corrosion* **19**(9), 308t (1963).

12. D. L. Macleary, *Corrosion* **18,** 67t (1962).

13. *ACS Monogr.* **158,** (1963).

14. D. P. Gold, M. Vallée, and J. P. Charette, *Can. Mining Metall. Bull.* **60,** 1131 (1967); M. Vallée and F. Dubuc, *Can. Min. Metall. Bull.* **63,** 1384 (1970); C. Carbonneau and J. C. Caron, *Can. Min. Metall. Bull.* **58,** 281 (1965).

15. M. Robert, *Can. Min. J.*, 38 (March 1978).

16. H. Stuart, O. de Souza Paraiso, and R. de Fuccio, *Iron Steel*, 11 (May 1980).

17. F. Perfect, *Trans. Metall. Soc. AIME* **239,** 1282 (1967).

18. F. J. Kelly and W. A. Gow, *Can. Mining Metall. Bull.* **58,** 843 (1965).

19. U.S. Pat. 3,607,006 (Sept. 21, 1971), E. P. Stambaugh (to Molybdenum Corporation of America).

20. U.S. Pat. 4,182,744 (Jan. 8, 1980), R. H. Nielsen and P. H. Payton (to Teledyne Industries, Inc.).

21. S. L. May and G. T. Engel, *U.S. Bur. Mines Rep. Invest.* **6635,** (1965).

22. F. Habashi and I. Malinsky, *CIM Bull.* **68**(761), 85 (1975).

23. U.S. Pat. 3,153,572 (Oct. 20, 1964), W. E. Dunn, Jr. (to E. I. du Pont de Nemours & Co., Inc.).

24. D. J. Soissan, J. J. McLafferty, and J. A. Pierret, *Ind. Eng. Chem.* **53**(11), 861 (1961).

25. C. H. Faye and W. R. Inman, *Research Report MD210*, Dept. Mines and Technical Surveys, Canada, 1956.

26. J. R. Werning, K. B. Higbie, J. T. Grace, B. F. Speece, and H. L. Gilbert, *Ind. Eng. Chem.* **46,** 644 (1954).

27. C. W. Carlson and R. H. Nielson, *J. Met.* **12,** 472 (June 1960).

28. J. R. Werning and K. B. Higbie, *Ind. Eng. Chem.* **46,** 2491 (1954).

29. B. R. Steele and D. Geldart, "Extraction and Refining of the Rarer Metals," *Proc. Symp. Institute of Mining and Metallurgy, London, Eng.*, 1956, pp. 287–309.

30. T. K. Mukherjee and C. K. Gupta, *Trans. SAEST* **11**(1), 127 (1976).

31. U.S. Pat. 3,271,277 (Sept. 6, 1966), L. F. Yntema.

32. *Tool. Prod.* **45**(2), 76 (1979).

33. H. E. Stipp, *Columbium, Bulletin MCP-10, Mineral Commodity Profiles*, U.S. Bureau of Mines, Washington, D.C., Jan. 1978.

34. *Niobium—Survey of World, Production, Consumption and Prices*, 2nd ed., Roskill Information Services Ltd., Oct. 1975.

35. *Columbium and Tantalum in 1979, Mineral Industry Surveys*, U.S. Bureau of Mines, Washington, D.C., annual.

36. W. R. Schoeller, *The Analytical Chemistry of Tantalum and Niobium*, Chapman and Hall, London, Eng., 1937.

37. W. R. Schoeller and A. R. Powell, *The Analysis of Minerals and Ores of the Rarer Elements*, 3rd ed., Griffin, London, Eng., 1955.

38. R. W. Moshier, *Analytical Chemistry of Niobium and Tantalum*, Pergamon, New York, 1964.

39. I. M. Gibalo, translated by J. Schmorak, *Analytical Chemistry of Niobium and Tantalum*, Ann Arbor–Humphrey Science Publ., Ann Arbor, Mich., London, Eng., 1970.

40. J. L. Kassner, A. G. Porrata, and E. L. Grove, *Anal. Chem.* **27,** 493 (1955).

41. Y. Shijo, *Bull. Chem. Soc. Jpn.* **50,** 1011 (1977).

42. A. G. Ward and O. Borgen, *Talanta* **24,** 65 (1977).

43. R. N. Gupta and B. K. Sen, *J. Inorg. Nucl. Chem.* **37,** 1548 (1975).

44. G. C. Shivahare and D. S. Parmar, *Indian J. Chem.* **13,** 627 (1975).

45. S. Kallman, H. Oberthin, and R. Liu, *Anal. Chem.* **34,** 609 (1962).

46. K. Fujimori and F. DiGiorgi, *Metal. ABM* **30**(204), 751 (1974).

47. H. L. Giles and G. M. Holmes, *X-Ray Spectrom.* **7**(1), 2 (1978).

48. G. A. Shkurko, *Gig. Tr.* **9,** 74 (1973); *Chem. Abstr.* **85,** 129779v (1976).
49. L. A. Sazhina and L. N. Elńichnyky, *Gig. Sanit.* **6,** 8 (1975); *Chem. Abstr.* **83,** 188928z (1975).
50. R. G. Thomas, S. A. Walker, and R. O. McClellan, *Proc. Soc. Exp. Biol. Med.* **138**(1), 228 (1971).
51. B. A. Cheadle, W. J. Langford, and R. I. Coote, *Nucl. Eng. Int.* **24**(289), 50 (1979).
52. K. N. Straffod and J. R. Bird, *J. Less Common Met.* **68,** 223 (1969).
53. P. H. Booker and R. E. Curtis, *Cutting Tool Eng.*, 18 (Sept.–Oct. 1978).
54. F. Fairbrother, *The Chemistry of Niobium and Tantalum*, Elsevier Publishing Co., New York, 1967.
55. L. H. Anderson and R. Kiessling, *Acta Chem. Scand.* **4,** 160 (1950).
56. *Tech. Data Sheet No. 4-B*, Borax Consolidated Ltd., London, Eng.
57. G. V. Samsonov and L. Ya. Markovski, *Usepkhi. Khim.* **25**(2), 190 (1958).
58. F. W. Glaser, *J. Met.* **4,** 391 (1952).
59. B. Post, F. W. Glaser, D. Moskowitz, *Acta Metall.* **2,** 20 (1954).
60. S. J. Sindeband and P. Schwartzkopf, "The Metallic Nature of Metal Borides," *97th Meeting of the Electrochemical Society, Cleveland, Ohio, 1950; Powder Metall. Bull.* **5/3,** 42 (1950).
61. K. Moers, *Z. Anorg. Chem.* **198,** 243 (1931).
62. E. K. Storms and N. H. Krikorian, *J. Phys. Chem.* **64,** 1461 (1960).
63. E. K. Storms, N. H. Krikorian, and C. P. Kempter, *Anal. Chem.* **32,** 1722 (1960).
64. E. K. Storms, *The Refractory Carbides*, Academic Press, Inc., New York, 1967, p. 70.
65. R. Kieffer and F. Kölbl, *Powder Met. Bull.* **4,** 4 (1949).
66. R. Hultgren, P. D. Desai, D. T. Hawkins, M. Gleiser, and K. K. Kelley, *Selection Values of the Thermodynamic Properties of Binary Alloys*, American Society for Metals, Metals Park, Ohio, 1973, p. 500.
67. I. E. Campbell, ed., *High Temperature Technology*, John Wiley & Sons, Inc., New York, 1956.
68. H. Takeshita, M. Miyaki, and T. Sano, *J. Nucl. Mat.* **78,** 77 (1978).
69. Ref. 54, p. 76.
70. A. J. Edwards, *J. Chem. Soc.*, 3714 (1964).
71. L. K. Frevel and H. W. Rinn, *Acta Cryst.* **9,** 626 (1956).
72. J. H. Canterford and R. Colton, *Halides of the Second and Third Row Transition Elements*, John Wiley & Sons, Inc., New York, 1968.
73. D. R. Sadoway and S. W. Flengas, *Can. J. Chem.* **54**(11), 1692 (1976).
74. D. E. Sands, A. Zalkin, and R. E. Elson, *Acta Cryst.* **12,** 21 (1959).
75. Ref. 54, p. 103.
76. S. S. Berdonosova, A. V. Lapitskii, D. G. Berdonosova, L. G. Vlasov, *Russ. J. Inorg. Chem.* **8,** 1315 (1963).
77. S. S. Berdonosova, A. V. Lapitskii, and E. K. Bakov, *Russ. J. Inorg. Chem.* **10,** 173 (1965).
78. W. Littke and G. Brauer, *Z. Anorg. Allg. Chem.* **325,** 122 (1963).
79. G. Brauer, *Z. Elektrochem.* **46,** 39 (1949).
80. N. Schönberg, *Acta Chem. Scand.* **8**(2), 208 (1954).
81. Ref. 54, p. 190.
82. R. Kieffer and P. Schwartzkopf, *Hartstoffe and Hartmetalle*, Springer, Vienna, Austria, 1953.
83. Ref. 54, p. 23.
84. Ref. 54, p. 24.
85. Ref. 54, p. 26.
86. A. Reisman and F. Holtzberg in A. M. Alper, ed., *High Temperature Oxides*, Part II, Academic Press, Inc., New York, 1970, p. 220.
87. B. M. Gatehouse and A. D. Wadsley, *Acta Cryst.* **17,** 1545 (1964).
88. R. Kiessling, *J. Electrochem. Soc.* **98**(4), 166 (1957).
89. L. Brewer, D. L. Sawyer, D. H. Templeton, and C. H. Dauben, *J. Am. Ceram. Soc.* **34,** 173 (1951).
90. U.S. Pat. 2,678,870 (1950), H. S. Cooper.
91. P. M. McKenna, *Ind. Eng. Chem.* **28,** 767 (1936).
92. G. A. Meerson and G. V. Samsonov, *Zh. Prikl. Khim.* **27,** 1115 (1954).
93. A. Andrieux, *Compt. Rend.* **189,** 1279 (1929).
94. J. T. Norton, H. Blumenthal, and S. J. Sindeband, *J. Met.* **185,** 749 (1949).
95. S. Motojima, K. Sugiyama, and Y. Takahashi, *J. Cryst. Growth* **30,** 233 (1975).
96. C. F. Powell, J. H. Oxley, and J. M. Blocher, Jr., *Vapor Deposition*, John Wiley & Sons, Inc., New York, 1966, p. 346.
97. Ref. 64, p. 72.

98. JANAF Thermochemical Tables, 1975 Supplement, *J. Phys. Chem. Ref. Data* **4**(1), 51 (1975).

99. R. F. Rolsten, *J. Am. Chem. Soc.* **79**, 5409 (1957).

100. J. D. Corbett and P. W. Seabaugh, *J. Inorg. Nucl. Chem.* **6**, 207 (1958).

101. S. Komjathy, *J. Less Common Met.* **2**, 466 (1960).

102. H. Wenzl, *J. Physique, Colloque C7*, **38**(Suppl. 12), C7 221 (1977).

103. A. Sieverts and H. Moritz, *Z. Anorg. Chem.* **247**, 124 (1941).

104. G. Brauer and H. Müller, *J. Inorg. Nucl. Chem.* **17**, 102 (1961).

105. G. Hörz, *Electrochem. Soc. Proc. Symp. Prop. High Temp. Alloys* (77-1), 753 (1976).

106. E. Friederich and L. Sittig, *Z. Anorg. Chem.* **143**, 293 (1925).

107. T. Takehashi, H. Itoh, and T. Yamaguchi, *J. Crystal Growth* **46**, 69 (1979).

108. W. Nickerson and C. J. Altstetter, *Scr. Metall.* **7**, 229 (1973).

109. R. Lauf and C. Altstetter, *Scr. Metall.* **11**, 983 (1977).

110. G. Brauer, *Z. Anorg. Allg. Chem.* **248**, 1 (1941).

111. F. Holtzberg, A. Reisman, M. Berry, and M. Berkenblit, *J. Am. Chem. Soc.* **79**, 2039 (1957).

112. A. Reisman and F. Holtzberg, *J. Am. Chem. Soc.* **81**, 3182 (1959).

113. Ref. 86, p. 222.

114. I. Shindo and H. Komatsu, *J. Cryst. Growth* **34**(1), 152 (1976).

115. G. Jander and D. Ertel, *J. Inorg. Nucl. Chem.* **14**, 77 (1960).

116. A. Räuber in K. Kaldis, ed., *Current Topics in Materials Science*, Vol. 1, North Holland Publ. Co., Amsterdam, The Netherlands, 1978, p. 481.

117. G. G. Kasimov, E. G. Vovkotrub, E. I. Krylov, and I. G. Rozanov, *Inorg. Mat.* **11**(6), 891 (1975).

General References

G. L. Miller, *Metallurgy of the Rarer Metals—6 Tantalum and Niobium*, Butterworths Scientific Publications, London, Eng., 1959.

R. J. H. Clark and D. Brown, *The Chemistry of Vanadium, Niobium and Tantalum*, Pergamon Press, Elmsford, N.Y., 1975.

D. L. Douglass and F. W. Kunz, eds., "Columbium Metallurgy," *Proc. Symp., New York, June 9–10, 1960*, Interscience Publishers, New York, 1961.

B. W. Gonser and E. M. Sherwood, eds., *The Technology of Columbium (Niobium)*, John Wiley & Sons, Inc., New York, 1958.

F. T. Sisco and E. Epremian, *Columbium and Tantalum*, John Wiley & Sons, Inc., New York, 1963.

C. T. Wang, "Composition, Properties and Applications of Niobium and Its Alloys" in *Metals Handbook*, 9th ed., American Society for Metals, 1980.

PATRICK H. PAYTON
Teledyne Wah Chang Albany

NITRATION

Nitration involves the reaction between an organic compound and a nitrating agent, eg, nitric acid (qv), either to introduce a nitro group into the hydrocarbon or to produce a nitrate. In C-nitration, a nitro group, —NO_2, is attached to a carbon atom:

$$\text{>CH} + HNO_3 \longrightarrow \text{>CNO}_2 + H_2O$$

O-nitration results in the formation of a nitrate and is regarded as an esterification (qv):

$$\text{>COH} + HNO_3 \longrightarrow \text{>CONO}_2 + H_2O$$

N-nitration is the attachment of a nitro group to a nitrogen atom, eg, in the production of nitramines:

$$\text{>NH} + HNO_3 \longrightarrow \text{>NNO}_2 + H_2O$$

In each of the above examples, a nitro group is substituted for a hydrogen atom; however, nitro groups may be substituted for other atoms or groups of atoms. In the Victor Meyer reaction, which involves the use of silver nitrate, a halide atom, eg, I or Br, is replaced by a nitro group. In a modification of this method, sodium nitrite is employed. Nitro compounds also can be produced by certain addition reactions, eg, the reaction of nitric acid or nitrogen dioxide with unsaturated compounds, eg, olefins or acetylenes.

Nitrations are highly exothermic, ie, ca 126 kJ/mol (30 kcal/mol). However, the heat of reaction varies with the hydrocarbon that is nitrated. The mechanism of a nitration depends on the reactants and the operating conditions. The reactions usually are either ionic or free-radical. Ionic nitrations are used commonly for aromatics; many heterocyclics; hydroxyl compounds, ie, simple alcohols, glycols, glycerol, cellulose, etc; and amines. Nitration of paraffins, cycloparaffins, and olefins frequently involve a free-radical reaction. Aromatic compounds and other hydrocarbons sometimes can be nitrated by free-radical reactions but, generally, such reactions are less successful (see Cellulose derivatives, esters; Explosives; Nitrobenzene and nitrotoluenes; Nitroparaffins).

Ionic Reactions

Most ionic nitrations involve mixed acids, which nitrate the hydrocarbon. The mixture usually contains nitric acid plus a strong acid, eg, sulfuric acid, perchloric acid, selenic acid, hydrofluoric acid, boron trifluoride, and ion-exchange resins containing sulfonic acid groups. Usually, water also is present in the acid mixture, and some is formed during nitration. Industrially, sulfuric acid is the most frequently used acid because it is highly effective and is the least expensive.

Most ionic nitrations are conducted at 0–120°C. In many nitrations of aromatics, there is an organic phase and an acid phase. Sufficient pressure, which usually is slightly above atmospheric, is provided to maintain the liquid phases. In a two-phase system, most nitrohydrocarbons collect in the hydrocarbon phase and water in the acid phase. The site of nitration usually is at or close to the interface between the phases. Regardless of the site of the main nitration reactions, mass transfer (or diffusion) of reactants and products is important.

Chemistry.   The nitronium-ion mechanism has been accepted almost universally since ca 1950 as the mechanism for nitration of most aromatic hydrocarbons, glycerol (qv), and various other hydrocarbons in which either mixed acids or highly concentrated nitric acid are used (1–3). Nitronium ions (NO_2^+) attack the aromatic hydrocarbon (ArH) as follows:

$$ArH + NO_2^+ \longrightarrow \left[Ar \begin{array}{c} H \\ \diagdown \\ NO_2 \end{array} \right]^+ \longrightarrow ArNO_2 + H^+$$

For an alcohol, glycol, or glycerol, or for amines, nitronium ions react as follows:

$$ROH + NO_2^+ \longrightarrow \left[RO \begin{array}{c} H \\ \diagdown \\ NO_2 \end{array} \right]^+ \longrightarrow RONO_2 + H^+$$

$$RNHR' + NO_2^+ \longrightarrow \left[\begin{array}{c} NO_2 \\ | \\ R\overset{+}{N}H \\ | \\ R' \end{array} \right] \longrightarrow RN(NO_2)R' + H^+$$

Nitronium ions are produced from nitric acid by use of strong acids, eg, sulfuric acid:

$$H_2SO_4 + HNO_3 \rightleftarrows NO_2^+ + HSO_4^- + H_2O$$

$$H_2SO_4 + H_2O \rightleftarrows HSO_4^- + H_3O^+$$

or highly concentrated nitric acid:

$$2\,HNO_3 \rightleftarrows NO_2^+ + NO_3^- + H_2O$$

and

$$HNO_3 + H_2O \rightleftarrows H_3O^+ + NO_3^-$$

These reactions are considered to be very rapid and equilibrium concentrations of NO_2^+ are thought to be present at all times in the acid phase during nitrations. The equilibrium concentrations vary depending on several variables; the most important is the composition of the acid mixture (4–5). The composition for various mixtures of nitric acid, sulfuric acid, and water were measured and those measured at 20°C are shown in Figure 1. The NO_2^+ concentrations generally increase with decreased amounts of water. The highest concentrations occur at essentially equal molar amounts of nitric and sulfuric acid. At high concentrations of H_2SO_4, almost all of the nitric acid is ionized to produce NO_2^+. For pure HNO_3 with no water present, ca 3 wt % is ionized

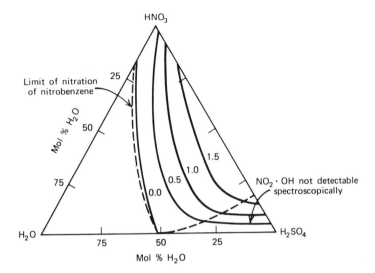

Figure 1. Concentration of NO_2^+ in mol/1000 g solution.

to NO_2^+. At high water concentrations where no NO_2^+ is detected, benzene, toluene, and other aromatics that nitrate moderately easily often can be nitrated. Two explanations have been proposed: first, another ion, perhaps a hydronium ion, $H_2NO_3^+$ (in essence a hydrated nitronium ion), is the nitrating species (6–7); second, nitronium ions are present but are in concentrations that are too low for detection by the Raman spectra used in the Figure 1 studies.

In addition, NO_2^+ concentrations at 40°C are ca 10–20% less than those at 20°C (5). Nitronium-ion values can be predicted for a given temperature within experimental accuracy by use of calculated equilibrium constants for the four ionization equations reported above.

Ipso attack (at a substituent position) of NO_2^+ on substituted aromatics, eg, alkyl benzene, often is important when strong acids are used (8). A cyclohexadienyl cation is formed as an unstable intermediate ion in which the nitro group can migrate on the ring to a carbon atom that is attached to a hydrogen. Loss of a proton results in a stable nitroaromatic. The intermediate also can decompose to the original aromatic hydrocarbon. A significant reduction in production of m-nitrotoluene occurs when toluene is nitrated at −40 to −10°C using a mixture of nitric acid, sulfuric acid, and sulfur trioxide (9). The reduction is of considerable interest in trinitrotoluene (TNT) processes where the meta isomer leads to the eventual production of undesired TNT isomers and to oxidation products.

Mixtures of nitrogen dioxide and sulfuric acid or another strong acid sometimes are employed as the nitrating agents. Details of a process yielding high conversions for either the mononitration or dinitration of benzene or toluene have been reported (10). A process for production of dinitrotoluenes involves the use of toluene, nitrogen dioxide, and oxygen as feedstocks; nitric acid is a by-product (11). Acetic acid and acetic anhydride sometimes are added to nitric acid to form mixtures that are effective for easy nitrations. Acetic anhydride and nitric acid also react to form acetic acid and acetyl nitrate, which is a nitrating agent (12).

$$(CH_3CO)_2O + HNO_3 \rightarrow CH_3COOH + CH_3CONO_3$$

Such mixtures are effective for the commercial N-nitration of hexamethylene-tetramine, $(CH_2)_6N_4$, in a glass tubular reactor to form cyclotrimethylenetrinitramine (RDX), $(CH_2)_3(NNO_2)_3$ (see Explosives). Acetic anhydride and acetic acid also tend to increase the solubility of the two phases in each other.

The nitronium-ion mechanism predicts reasonably well most aspects of nitration including the following: (1) Isomer distribution, eg, for mononitration of toluene (13). However, distribution is difficult to explain when the mixed acids contain ≥ 90 wt % sulfuric acid. (2) The kinetics of nitration. The rates of nitration of nitrobenzene generally parallel the acidity of the mixed acids and the NO_2^+ concentration. The limit for nitrating nitrobenzene is essentially identical to the zero concentration profile for NO_2^+, as shown in Figure 1. However for nitrations involving two liquid phases, it is difficult to determine whether the rates of nitration are directly proportional to the NO_2^+ concentration, since resistances to mass transfer or diffusion usually are significant. Such resistances vary with at least the levels of agitation, the acid composition, and the ratios of two liquid phases. (3) The relative rates of nitration when mixtures of aromatics are used (13). Substituents on the aromatic ring affect the reactivity of the compound because of polar and steric factors. Aromatics with attached alkyl groups, eg, toluene, ethylbenzene, or cumene (qv), are more reactive than benzene. The alkyl group increases the electron density of the ring at the ortho and para positions. As a result, NO_2^+ reacts predominately at these positions of the ring. However, aromatics with attached nitro groups, eg, mononitrobenzene or mononitrotoluene, are nitrated with considerably more difficulty than the unnitrated aromatics. Attached nitro groups reduce the electron density of the ring and are meta directing. As a general rule, an aromatic compound is completely mononitrated before being di- or trinitrated, since more severe conditions are required for the subsequent nitration steps than for the initial one.

Nitrous acid or nitrite salts may be used to catalyze the nitration of very easily nitratable aromatic hydrocarbons, eg, phenol or phenolic ethers. It has been suggested that a nitrosonium ion (NO^+) attacks the aromatic resulting in an initial nitrosation reaction as follows (14):

$$ArH + NO^+ \longrightarrow \left[Ar \begin{array}{c} H \\ \diagup \\ \diagdown \\ NO \end{array} \right]^+ \longrightarrow ArNO + H$$

Oxidation of the nitrosoaromatic then occurs:

$$ArNO + HNO_3 \rightarrow ArNO_2 + HNO_2$$

The nitrosonium ion is produced from nitrous acid and nitric acid

$$HNO_2 + HNO_3 \rightleftharpoons NO^+ + NO_3^- + H_2O$$

However, some investigators believe that NO_2^+ is involved. Some of the difficulties in explaining the isomer distribution of mononitrophenols, based on the nitroso-nium-ion mechanism, have been discussed (15).

Many by-products have been detected in the product stream for ionic nitrations (16–20). Oxidations involving primarily nitric acid often are important. In the case of toluene nitrations, two types of oxidations occur, ie, oxidation of the methyl group and ring decomposition yielding gaseous products (21). The oxidizing power of nitric acid in mixed acids correlates well with the kinetics of nitration over the range of acid compositions, which induces both weak and strong acids (22).

Oxidations occur both in the hydrocarbon and acid phases (21). Nitric acid but not sulfuric acid is relatively soluble in nitrotoluenes (16). More nitric acid is dissolved in the hydrocarbon phase for trinitrations as compared to mononitrations because higher concentrations of nitric acid are used in the acid phase, and higher concentrations of dinitrotoluenes and trinitrotoluenes exist and they dissolve more nitric acid. By-products that are formed during trinitrations include benzaldehyde and nitrobenzaldehydes, nitrophenols, nitrocresols, nitrobenzoic acids, and tetranitromethane; nitrous acid is a major by-product. Some by-products are condensation products (17–18).

Ring decompositions or oxidations occur in the acid phase (21). Ring decomposition is the more significant of the two types of oxidations in terms of product losses.

Kinetics of Aromatic Nitrations. The kinetics of commercial nitration vary greatly; reaction times may be several seconds to many hours. Temperature, acid composition, and degree of agitation are principal factors. Agitation affects the mixing of phases and the physical transfer steps in the two-phase systems. Claims have been analyzed that sufficient agitation can eliminate the resistances to mass transfer so that the kinetics of nitration are controlled chemically (23–24). However, when acids are used with compositions comparable to those of industrial importance, the claims seem unwarranted, but with low strength acids and with vigorous agitation, the claims may be correct. The problems involved in eliminating mass-transfer resistances have been presented (6,23).

Temperature. The energy of activation for various nitrations is 59–75 kJ/mol (14–18 kcal/mol). Yet, the kinetic rate constants for the various chemical steps increase with temperature, and the solubilities in the acid phase of both unnitrated and nitrated aromatics probably increase with temperature. Such solubilities are important since the main reactions probably occur in the zone of the acid phase that is adjacent to the interface between phases. Solubility data for the hydrocarbons in the acid and the acids in the hydrocarbons has been reported for various systems (25–28). At high concentrations of H_2SO_4 or $H_2S_2O_7$, solubility of nitric acid in the organic phase is reduced. In addition, viscosities decrease and diffusivity coefficients increase with temperature. Temperature also affects the interfacial surface tensions of the two phases. Hence, the interfacial area between the phases changes and, probably, increases with temperature so that resistances to mass transfer between the phases change (16). The concentrations of NO_2^+ and other ions in the acid also change with temperature (5). These reversible reactions are assumed to be essentially at equilibrium but, at some temperatures, this assumption may not be correct.

Agitation. Increased agitation generally tends to promote transfer of reactants in the two-phase system and, hence, increased rates of reaction. Droplet size and internal circulation in the droplets in the dispersed phase also are affected by the impeller speed. With very small droplets, increased impeller speeds may decrease the rate of mass transfer of a reactant to or from the dispersed phase (23) (see also Mixing and blending).

Phase Composition. Compositions of the acid and organic phases affect the concentrations of the reactants. In addition, concentrations affect the mutual solubilities of the two phases in each other as well as the viscosities, diffusion coefficients, and surface tension between phases.

Ratio of Acid-to-Organic Phases. The ratio of acid-to-organic phases is important in terms of the type of emulsion formed and of the interfacial area between phases. The shear forces caused by agitation are expended to a greater extent in the continuous phase, which probably is the acid phase in most commercial nitrations, than in the dispersed phase.

The kinetics of toluene nitration for a system in which the two phases are in laminar flow have been reported, and it was concluded that the main reactions occur in the zone of the acid adjacent to the interface between phases (25,29–30).

Processes for Ionic Nitrations. Nitrations of aromatic hydrocarbons, glycerol, and many other organic materials are hazardous and require careful control of the kinetics and temperature. If the temperature becomes too high, the rates of reaction and the degree of exothermicity for the reaction may become too high to maintain adequate temperature control. In such a case, runaway reactions, excessive side reactions, or explosions may occur.

Ideally, it is desirable to contact the aromatic hydrocarbon feed with dilute or used mixed acids and to contact the partially nitrated hydrocarbon with the stronger or feed acids. If a continuous flow process is used, then the organic and acid phases should flow countercurrent to each other. In batch systems, the nitrations often are conducted in stages. When nitration is partially completed, the phases are separated and a stronger acid is used for further nitrations.

Nitrations that require only relatively small production capacities are normally made in batch nitrators that are provided with vigorous agitation. Adequate heat-transfer surfaces are required in each reactor, and often they are provided as a jacket, internal coils, cooled baffles, or an external heat exchanger. Completion of the nitration reactions in batch runs occurs in one to several hours. Often, two to five additional hours are required in each run to load the reactor and preheat the reactants and then to unload and clean the reactor.

Continuous-flow processes have been widely adopted in the last 20–30 yr, especially for large-scale production of TNT, DNT (dinitrotoluenes), nitroglycerin, and nitrobenzene. They offer important advantages as compared to batch processes in terms of safer operation; better-controlled conditions; and lower operating costs, eg, labor and energy costs. Several continuous-flow processes have been adopted and, generally, the organic and acid flows are countercurrent.

Several reactors or reactor compartments that are connected in a series and that have an efficient agitator in each reactor are used (11,21,26,31–32). Such a process can be adapted for production of either TNT or nitroglycerin. Variations in design are relative to the flow patterns in the reactors. For example, some reactors are equipped with draft tubes and the reaction emulsion is recirculated internally many times through the draft tube. However, internal recirculation techniques are not provided in all nitration reactors. However, large internal heat-transfer surfaces always are provided, and the emulsions are separated into acid and organic phases between some or all of the reactors to provide countercurrent flow of reactants.

At least two processes use a loop reactor system which is a modification of the internal-recirculation reactors (33). Recirculation occurs through a loop that consists of a pump which provides the mixing and agitation of the reactants; a heat exchanger, in which most of the reactions occur and in which the heat of reaction is transferred; and a return line, which is used to recirculate most of the emulsion. Flow sheets for loop-reactor processes are similar to those for well-stirred-reactor processes. Loop reactors are used for the mononitration of benzene and toluene.

Tubular reactors are used in the Hercules process and the Bofors process (34–35).

In the former, the two liquid feeds are mixed by use of a mixing tee; in the Bofors process, rapid mixing of the feeds is accomplished using a centrifugal pump. A shell-and-tube heat exchanger is used in both processes and provides excellent heat-transfer characteristics. The velocities of the emulsion in the tubes are high in order to provide sufficient turbulence so that the two phases remain emulsified in the tubes. The emulsion that leaves the tubular reactor is separated into acid and hydrocarbon phases. In the case of the Bofors process, part of the acid is recirculated by means of the pump through the heat exchanger. Apparently, little or no hydrocarbon phase is recirculated in either process. Both processes are recommended for production of both TNT and nitroglycerin. In the Hercules process for TNT, two reactors are used in series. In the first reactor, all mononitration and most, if not all, dinitration occurs in ca ≤10 s. Trinitration occurs in the second reactor in ca 2–6 min.

A tubular reactor system or a series of several stirred reactors also is used for production of either mononitrobenzene or mononitrotoluene (12). In this process, a slight excess of the aromatic hydrocarbon is used and the reactors are operated adiabatically at ca 80–120°C and at a slightly elevated pressure to prevent any vaporization. This process developed by CLR, Inc., and American Cyanamid is used in the world's largest nitrobenzene plant that has an annual capacity of 159×10^6 kg. The desired product contains only trace amounts of by-products, and there is almost complete consumption of the nitric acid. The sulfuric acid that leaves the reactor can be reconstituted easily by use of a vacuum evaporator and then is recycled. Total residence time for the reactants in the reactor are 0.5–12 min. One commercial process involves the use of methylene chloride as a diluent and an extractive agent for mono- and di-nitroaromatics (36).

Both batch and continuous-flow processes are employed for the O-nitration of cellulose to produce cellulose nitrates (37). Mixtures of nitric and sulfuric acids are employed as nitrating agents. Three hydroxyl groups are present for each repeating unit which has the formula $C_6H_{10}O_5$. Each hydroxyl group can be esterified, and the degree of reaction can be varied by changing the operation conditions for the reaction and the type of reactants used. Increasing the temperature from 0 to 80°C results in the following changes: the reaction rate increases; above 40°C, the degree of nitration, ie, the nitrogen content of the product, decreases but remains relatively constant from 60 to 80°C; increased decomposition and chain shortening of the cellulosic material occurs and by-products include hydrocellulose and oxycellulose; and the solubility of the cellulose in the acid phase increases and lower viscosity mixtures are obtained (38). The rate of nitration of the cellulose depends on the kinetics of the nitration or esterification steps and the rate of transfer or diffusion of the mixed acids into and between the cellulose fibers (see Cellulose derivatives).

Sulfuric acid is by far the most common strong acid used industrially; however, it has the disadvantage of becoming diluted by the by-product water, and there is no simple method for the complete separation of the water and sulfuric acid. Partial separation of the water from the sulfuric acid can be accomplished by stripping the waste used acids with steam to remove the unreacted nitric acid and dissolved hydrocarbons. The remaining acid is concentrated to ca 96% by vacuum stripper or 93 wt % acid with combustion gases and is reconstituted to 98 wt % or higher by addition of sulfur trioxide. Concentrated acid that is sufficient for the nitration reaction is recycled and the additional acid, which is equivalent to the amount of sulfur trioxide that was added, is withdrawn from the system.

Free-Radical Reactions

Vapor Phase. Vapor-phase nitrations are performed at sufficiently high temperatures and low pressures to obtain gaseous mixtures of hydrocarbons, nitrating agents, products, and inert vapors. When nitric acid is used as the nitrating agent, temperatures vary from ca 350 to 450°C. A residence time, perhaps of up to several minutes, is required at lower temperatures. At higher temperatures, side reactions become more important and a short time, eg, as low as a fraction of a second, is needed to complete the reaction. Many laboratory nitrations have been carried out at atmospheric pressure, but pressures of 0.8–1.2 MPa (8–12 atm) usually are used in commercial reactors to expedite condensation and recovery of the nitroparaffin product; normal cooling water results in almost complete condensation of the nitroparaffin products.

In addition to the replacement of an H atom with a nitro group, C—C bonds are broken so that nitroparaffins with fewer carbons than the original paraffin are formed. When propane is nitrated, the following nitroparaffins are produced: 1-nitropropane, 2-nitropropane, nitroethane, and nitromethane. The ratio of the four compounds varies with the operating conditions employed. If nitric acid is used as the nitration agent, a typical analysis for the nitroparaffin product (at ca 410–420°C) is 25, 40, 10, and 25 wt %, respectively (39). The nitroparaffin product from nitration with nitrogen dioxide with a feed stream at 280°C is 14, 52, 9, and 25 wt %, respectively (40). In the nitration of n-butane, each of the above four nitroparaffins plus 1-nitrobutane and 2-nitrobutane is produced. Nitrations at lower temperatures, eg, those involving nitrogen dioxide, result in higher molecular weight nitroparaffins than those at higher temperatures.

Vapor-phase nitrations of paraffins using nitric acid result in conversions of nitric acid to various nitroparaffins from ca 20% for methane to ca 40% for n-butane. The remaining nitric acid is converted primarily to nitrogen oxides. In general, conversions of nitric acid to nitroparaffins increase as the molecular weight of the paraffin increases. Methane and ethane, which contain only primary carbon atoms, are more difficult to nitrate than propane and heavier normal paraffins which contain secondary carbon atoms. Nitric acid conversions to nitroparaffins essentially level off for normal paraffins higher than propane. Conversions per pass of nitrogen dioxide to nitroparaffins usually are less than those using nitric acid. Theoretical conversions for a feed mixture of paraffin (RH) and nitrogen dioxide are 66.7%:

$$2\,RH + 3\,NO_2 \rightarrow 2\,RNO_2 + H_2O + NO$$

When propane is nitrated with nitrogen dioxide, conversions are as high as 27%, but conversions are considerably less for methane or ethane.

Increased conversions of nitric acid or of nitrogen dioxide to form nitroparaffins can be achieved by adding oxygen or ozone and/or halogens, eg, chlorine or bromine, to the reaction mixture; these additives increase production of alkyl free radicals (41–44). Oxygen is most beneficial when ca one to two moles of oxygen is added per mol of nitric acid. With bromine and chlorine, ca 0.003 and 0.06 mol/mol nitric acid, respectively, should be used. Conversions based on nitric acid are increased to 50–70% if these additives are used either singly or in combinations. However, additives, eg, oxygen and halogens, are not used industrially. In the case of oxygen, significant amounts of oxygenated products are formed, yet the quantities are not sufficient for economical recovery. The halogens result in the production of corrosive hydrogen halides and in recovery problems (45) (see also Hydrocarbon oxidation).

Considerable amounts of by-products are produced in addition to the nitration products. Nitric acid is an oxidizing agent and produces aldehydes, particularly formaldehyde; carbon monoxide; carbon dioxide; water; lighter paraffins; olefins; and small amounts of alcohols and ketones. When no additives are used, less than half of the nitric acid is converted to a nitroparaffin. The remainder of the nitric acid forms nitric oxide plus lesser amounts of nitrogen dioxide, nitrous oxide, and nitrogen. In commercial plants, the nitric oxide and nitrogen dioxide are recovered and are reformed into nitric acid so that ca 60–80% of the nitric acid is converted into nitroparaffins; yields based on the paraffins also are ca 60–80%.

In the production of nitrocyclohexane, the desired reaction is limited to the substitution of a nitro group for a hydrogen atom and C—C breakage is minimized. The following conditions are used: low temperature nitrations involving nitrogen dioxide (46), halogens as additives (oxygen increases C—C breakage), and careful temperature control of the reaction mixture.

The following procedures have been used to minimize loss of temperature control in nitration reactors. (*1*) Use excess amounts of hydrocarbons, eg, a 4:1 or higher ratio of hydrocarbons to nitrating agent. (*2*) Add an inert gas, eg, steam or nitrogen; normally 60–70% nitric acid is used. (*3*) Use effective heat-transfer surfaces, eg, small-diameter aluminum tubes and molten salt baths as the reaction media (45). Fluidized-bed reactors also provide excellent temperature control. (*4*) Employ the heat of reaction to vaporize the nitric acid and to heat it to reaction temperatures, eg, with spray nitrators (47) (see Nitroparaffins).

The process of Société Chemique de la Grande Paroisse for production of nitroparaffins using propane, nitrogen dioxide, and air as feedstocks is promising (40). The yields of nitroparaffins based on both propane and nitrogen dioxide seem to be high. Recovery of the nitrogen oxides in the product stream is effective; nitric oxide is oxidized to nitrogen dioxide which is adsorbed in nitric acid. Next, the nitric dioxide is stripped from the liquid phase and is recirculated.

Liquid-Phase Nitrations of Paraffins. Liquid-phase nitrations of paraffins occur predominantly by free-radical reactions. Highly ionized mixed acids are ineffective with relatively nonpolar hydrocarbons; however, nitric acid that is unmixed with any other acid is an effective nitrating agent. Temperatures required for the reaction are ca 100–200°C, and sufficient pressure is required to maintain most, if not all, of the mixture of reactants and products of each of the two phases which normally are present in the liquid state. High pressures are needed to maintain the by-product nitric oxide and nitrogen gases dissolved in the liquids. Specified pressures are ca 0.4–20.3 MPa (4–206 atm) (45). At lower pressures, both liquid and vapor phases are present during certain or all portions of some nitrations. The pressure required to maintain only liquids depends on the volatility of the paraffin that is nitrated. Pressures higher than those required to maintain liquid phases have little or no effect on the reaction. Corrosion seems to be more of a problem when both a vapor and a liquid phase are present in the reactor.

The molar ratios of hydrocarbons to nitrating agent vary from ca 2:1 to 6:1. Both nitric acid, usually 60–70 wt % concentrations, and nitrogen dioxide have been used as nitrating agents. With nitric acid, two liquid phases form; thus, the degree of agitation and the character of the emulsion form is important. However, nitrogen dioxide is relatively soluble in hydrocarbons so that only one liquid phase may be present until a water phase forms. The major reactions probably occur in the hydrocarbon phase, since nitrogen dioxide probably is the principal nitrating species even if nitric acid is used. At the temperatures that are used in most liquid-phase nitrations, residence

Table 1. Reaction Steps for Vapor-Phase Nitration of Paraffins

A. Reactions of nitric acid

    1. $HNO_3 \rightarrow \cdot OH + NO_2$ (the NO_2 is a free radical existing as either $\cdot NO_2$ or $\cdot ONO$)

    2. $HNO_3 + NO \rightarrow HNO_2(\cdot OH + NO) + NO_2$

B. Reactions of paraffin (RH)

    1. $RH + \cdot OH \rightarrow R\cdot + H_2O$

    2. $RH + NO_2 \rightarrow R\cdot + HNO_2 \rightarrow R\cdot + \cdot OH + NO$

    3. $RH + O_2 \rightarrow R\cdot + HOO\cdot$

Oxygen might be obtained by the decomposition of nitric acid or nitrogen dioxide, eg, $NO_2 \rightleftharpoons NO + \frac{1}{2} O_2$.

The overall initiation reaction to produce alkyl radicals when nitric acid is used could be

$$3\,RH + HNO_3 \rightarrow 3\,R\cdot + 2\,H_2O + NO$$

The above equation is obtained by adding equations A.1., B.1. (twice), and B.2.

The overall initiation reaction to produce alkyl radicals when nitrogen dioxide is used could be

$$2\,RH + NO_2 \rightarrow 2\,R\cdot + H_2O + NO$$

The above equation is obtained by adding reactions B.1. and B.2.

C. Reactions of alkyl radical (R·)

    1. $R\cdot + HNO_3 \rightarrow RNO_2 + \cdot OH$

    2. $R\cdot + HNO_3 \rightarrow RO\cdot + HNO_2 \rightarrow RO\cdot + \cdot OH + NO$

    3. $R\cdot + \cdot NO_2 \rightarrow RNO_2$

    4. $R\cdot + \cdot ONO \rightarrow RONO$

    5. $R\cdot + \cdot OH \rightarrow ROH$ (probably minor reaction)

    6. $R\cdot + \cdot OH \rightarrow$ olefin $+ H_2O$ (there may be other reactions for abstracting a hydrogen atom from an alkyl radical)

    7. $R\cdot + R' \rightarrow RR'$ (higher alkanes may be formed thus, particularly at wall)

    8. $R\cdot + O_2 \rightarrow ROO\cdot$ (alkyl peroxy radicals react to form oxygenated compounds)

    9. $R\cdot + NO \rightarrow RNO$

D. Decomposition of alkyl nitrites

    1. $RONO \rightarrow RO\cdot + NO$

    (Alkyl nitrites are very unstable especially at higher temperatures. This reaction is probably important for the production of alkoxy radicals at 350–450°C, ie, the temperature required when nitric acid is the nitrating agent. When nitrogen dioxide is used at 200–300°C, some alkyl nitrites may be undecomposed. Significant quantities of cyclohexyl nitrite are obtained in nitrations of cyclohexane with nitrogen dioxide at 200–240°C.)

E. Reaction of alkoxy radicals

Production of oxygenated compounds, eg, aldehydes, alcohols, ketones, etc. The C—C bond frequently is broken to form lower alkyl radicals. Propyl radicals are used in several examples given below:

    1. $CH_3CH_2CH_2O\cdot \longrightarrow CH_3CH_2\cdot + CH_2O$

    2. $CH_3\overset{\overset{\bullet}{O}}{\underset{|}{C}}HCH_3 \longrightarrow CH_3\cdot + CH_3CHO$

    3. $CH_3\overset{\overset{\bullet}{O}}{\underset{|}{C}}HCH_3 + \cdot OH \longrightarrow CH_3\overset{\overset{O}{\parallel}}{C}CH_3 + H_2O$

    (The cyclohexyloxy radical probably reacts similarly to form cyclohexanone which is obtained in significant quantities when cyclohexene is nitrated. Possibly NO_2, ONO, etc, radical also abstracts the hydrogen atom.)

    4. $RO\cdot + RH$ (or aldehyde or other compound) $\rightarrow ROH + R\cdot$

    5. $RO\cdot + NO_2 \rightarrow RONO_2$ (cyclohexyl nitrate is formed in significant quantities during nitration of cyclohexane at 200–240°C)

    6. $RO\cdot + O_2 \rightarrow$ oxygenated products

The aldehydes, ketones, alcohols, etc, continue to oxidize to acids, carbon oxides, water, etc.

times in tubular reactors are ca 1–4 min; 10–20% conversion of the nitrating agent to nitroparaffins is achieved. In general, side reactions become much more prevalent at higher conversions.

Chemistry. The chemistry of free-radical nitrations is complicated since both nitration and oxidation reactions are involved. A chain mechanism frequently occurs in nitrations with nitric acid. A set of the more important reaction steps for the nitration of paraffins or cycloparaffins is given in Table 1 (45).

When nitric acid is used in vapor-phase nitrations, reaction A.1. is the main initiating step. Temperatures > ca 350°C are required to obtain a significant amount of initiation, and A.1. is the rate-controlling step of the overall reaction. Reactions B.1. and C.1. are chain-propagating steps (see also Initiators).

When nitrogen dioxide is used, the main reaction steps are B.2. and C.3. These reactions occur readily at as low as 200°C. The exact temperature depends on the specific hydrocarbon that is nitrated and, presumably, B.2. is the rate-controlling step. Reaction C.3. is of minor importance in nitrations with nitric acid, as indicated by kinetic information (45).

An important side reaction in all free-radical nitrations is reaction C.4. in which an unstable alkyl nitrite is formed. It decomposes to form nitric oxide and an alkoxy radical. The latter radical, as shown in reactions E.1.–E.6., forms oxygenated compounds and low molecular weight alkyl radicals which can form low molecular weight nitroparaffins by reactions C.1. or C.3. The oxygenated hydrocarbons often react further to produce even lighter oxygenated products, carbon oxides, and water.

Additives, eg, halogens or oxygen, promote production of alkyl radicals. The increased concentrations of these radicals result in higher conversions to nitroparaffins with each pass of nitric acid.

Reactions B.2. and C.3. are important steps for the liquid-phase nitration of paraffins. The nitric oxide which is produced is oxidized with nitric acid to reform nitrogen dioxide which continues the reaction. The process is complicated by the frequent presence of two liquid phases; consequently, the nitrogen oxides must transfer from one phase to another.

Reaction C.9. is important in liquid-phase nitrations. The nitroso compound that is formed can rearrange to form an oxime which reacts with nitrogen dioxide to form dinitroparaffins in which both nitro groups are attached to the same carbon atom. 2,2-Dinitropropane can be produced readily by a liquid-phase process but not in high temperature vapor-phase nitrations. If dinitroparaffins are formed, they decompose rapidly at high temperatures (38). However, it also is possible that reaction C.9. is less likely at high temperatures or in the gas phase when the concentrations of the various reactants are low as compared to those in liquid-phase processes.

Health and Safety Factors

Nitrohydrocarbons generally are hazardous; some are explosive, most are highly flammable, and many are toxic. Dangers of explosion usually increase with the degree of nitration, eg, trinitroaromatics are much more hazardous as compared to mononitroaromatics. Some of the most explosive compounds are TNT; picric acid; glycerol trinitrate (nitroglycerin); and highly nitrated cellulose, which often is used as gunpowder. Safety hazards involved in nitrohydrocarbon manufacture have been reviewed (21,27,38,48–51) (see Explosives). Nitroaromatics and some polynitrated paraffins are highly toxic when inhaled or upon direct contact with the skin (see Nitrobenzene and nitrotoluenes).

BIBLIOGRAPHY

"Nitration" in *ECT* 1st ed., Vol. 9, pp. 314–330, by Willard deC. Crater, Hercules Powder Company; "Nitration" in *ECT* 2nd ed., Vol. 12, pp. 784–796, by Lyle F. Albright, Purdue University.

1. J. G. Haggett, R. B. Moodie, J. R. Penton, and K. Schofield, *Nitration and Aromatic Reactivity*, Cambridge University Press, 1971.
2. E. D. Hughes, C. K. Ingold, and R. J. Reed and co-workers, *J. Chem. Soc.*, 2400 (1950).
3. G. A. Olah, *ACS Symposium Series #22*, Washington, D.C., 1976, Chapt. 1, pp. 1–47.
4. J. Chedin, *Ann. Chem.* **8**, 295 (1937).
5. M. B. Zaman, *Nitronium Ions in Nitrating Acid Mixtures*, Ph.D. Thesis, University of Bradford, Bradford, U.K., 1972.
6. L. F. Albright, *Chem. Eng.* **73**(9), 169 (1966); (10), 161 (1966).
7. H. M. Brennecke and K. A. Kobe, *Ind. Eng. Chem.* **48**, 1298 (1956).
8. P. C. Myhre, *ACS Symposium Series #22*, Washington, D.C., 1976, Chapt. 4, pp. 87–94.
9. M. E. Hill and co-workers, *ACS Symposium Series #22*, Washington, D.C., 1976, Chapt. 17, pp. 253–271.
10. U.S. Pat. 4,123,466 (Oct. 31, 1978), C. Y. Lin, F. A. Stuber, and H. Ulrich (to the Upjohn Co.).
11. U.S. Pat. 4,028,425 (June 7, 1977), E. Gilbert (to United States of America).
12. U.S. Pats. 4,021,498 (May 3, 1977), and 4,091,042 (May 23, 1978), V. Alexanderson, J. B. Trecek, and C. Marsden (to American Cyanamid Co.).
13. K. L. Nelson and H. C. Brown in B. T. Brooks and co-eds., *The Chemistry of Petroleum Hydrocarbons*, Vol. 3, Reinhold Publishing Corporation, New York, 1955, p. 465.
14. C. A. Bunton, E. D. Hughes, C. K. Ingoll, and co-workers, *J. Chem. Soc.*, 2628 (1950).
15. D. S. Ross, G. P. Hum, and W. G. Blucher, *Chem. Commun.*, in press.
16. L. F. Albright, D. Schiefferle, and C. Hanson, *J. Appl. Chem. Biotechnol.* **26**, 522 (1976).
17. W. T. Bolleter, *TNT Process Characterization Studies*, Joint US/UK Seminar on TNT Chemistry and Manufacture, Tech. Report 106, Oct. 1971.
18. C. Hanson, T. Kaghazchi, and M. W. T. Pratt, *ACS Symposium Series #22*, Washington, D.C., 1976, Chapt. 8, pp. 132–155.
19. D. S. Ross and N. A. Kirshen, *ACS Symposium Series #22*, Washington, D.C., 1976, Chapt. 7, pp. 114–131.
20. H. Suzuki, *Synthesis*, 217 (Apr. 1977).
21. E. Gilbert, "TNT Production" in S. M. Kaye, ed., *Encyclopedia of Explosives and Related Items*, Vol. 9, U.S. Army Armament Research and Development Command, Dover, N.J. (in press).
22. D. S. Ross, private communications, 1980.
23. C. Hanson, J. G. Marsland, and G. Wilson, *Chem. Ind. (London)*, 675 (1966).
24. A. C. Strachan, *ACS Symposium Series #22*, Washington, D.C., 1976, Chapt. 13, pp. 210–218.
25. C. Hanson and H. A. M. Ismail, *J. Appl. Chem. Biotechnol* **25**, 319 (1975).
26. H. J. Klaasen, and Humphrys, *Chem. Eng. Prog.* **49**, 641 (1953).
27. C. M. Mason, R. W. Van Dolah, and J. Ribovich, *J. Chem. Eng. Data* **10**, 173 (1965).
28. D. F. Schiefferle, C. Hanson, and L. F. Albright, *ACS Symposium Series #22*, Washington, D.C., 1976, Chapt. 11, pp. 176–189.
29. C. Hanson and H. A. M. Ismail, *J. Appl. Chem. Biotechnol.* **26**, 111 (1976).
30. C. Hanson and H. A. M. Ismail, *Chem. Eng. Sci.* **32**, 775 (1977).
31. F. Meissner, G. Wannschaff, and D. F. Othmer, *Ind. Eng. Chem.* **46**, 718 (1954).
32. E. Thomas, *The ROF TNT Process*, Joint US/UK Seminar on TNT Chemistry and Manufacture, Tech. Report 106, Oct. 1971.
33. U.S. Pat. 3,092,671 (1963), S. B. Humphrey and D. R. Smoak (to U.S. Rubber Co.).
34. Bofors Nobel Chematur, personal communication, 1980.
35. U.S. Pats. 2,951,746 and 2,951,877 (Sept. 6, 1960), D. L. Kouba, J. T. Paul, and F. S. Stow (to Hercules Powder Co.).
36. U.S. Pat. 4,005,102 (Jan. 25, 1977), N. C. Cook (to General Electric Co.).
37. S. M. Kaye, *Encyclopedia of Explosives and Related Items*, Vol. 8, U.S. Army Armament Research and Development Command, Dover, N.J., 1978, pp. N40–N88.
38. T. Urbanski, *Chemistry and Technology of Explosives*, Vols. 1 and 2, The MacMillan Co., New York, 1964, 1965.
39. G. B. Bachman, L. M. Addison, J. V. Hewett, L. Kohn, and A. Millikan, *J. Org. Chem.* **17**, 906 (1952).

40. U.S. Pat. 3,780,115 (Dec. 18, 1973), P. L'honore, G. Cohen, and B. Jacquinot (to Societe Chimique de la Grande Paroisse).

41. G. B. Bachman, H. B. Hass, and L. M. Addison, *J. Org. Chem.* **17,** 914 (1952).

42. G. B. Bachman, H. B. Hass, and J. V. Hewett, *J. Org. Chem.* **17,** 928 (1952).

43. G. B. Bachman, J. V. Hewett, and A. Millikan, *J. Org. Chem.* **17,** 935 (1952).

44. G. B. Bachman and L. Kohn, *J. Org. Chem.* **17,** 942 (1952).

45. L. F. Albright, *Chem. Eng.* **73**(12), 149 (1966).

46. R. Lee and L. F. Albright, *Ind. Eng. Chem. Proc. Des. Dev.* **4,** 411 (1965).

47. U.S. Pat. 2,418,241 (Apr. 1, 1947), L. A. Stengel and R. G. Edly (to Commercial Solvents Corp.).

48. G. S. Biasutti, *Histoire des accidents dans l'industrie des explosifs*, Dr. Ing. Mario Biazzi, Soc. An., Vevey, Switzerland, 1978; *ACS Symposium Series #22*, Washington, D.C., 1976, Chapt. 22, pp. 320–326.

49. R. C. Dartnell and T. A. Ventrone, *Chem. Eng. Prog.* **67**(6), 58 (1971).

50. E. J. Fritz, *Loss Prev.* **3,** 41 (1969).

51. T. A. Ventrone, *Loss Prev.* **3,** 38 (1969).

LYLE F. ALBRIGHT
Purdue University

NITRIC ACID

Nitric acid [*7697-37-2*], HNO_3, became a major industrial chemical with the development of the explosives and dyestuffs industries at the end of the nineteenth century. This early growth was dwarfed after World War II, when the enormous expansion in the use of synthetic fertilizers (qv) increased demand for nitric acid to an extent that placed its annual rate of production among the ten largest chemicals produced in the United States. The peak production years in the 1970s were 1973 and 1979 with 7.62×10^6 and 7.77×10^6 metric tons produced, respectively. The anticipated growth of the fertilizer, fiber, and plastic industries assures a continuing increase in demand for nitric acid.

The earliest description of a method for making nitric acid generally is attributed to eighth-century Arabian sources which described the distillation of a mixture of Cyprus vitriol, saltpeter, and alum to yield a liquor with "high solvent action," later called *aqua fortis* (1). In 1798, Milner reported the successful oxidation of ammonia vapor over red-hot manganese dioxide to yield nitrogen oxides and acid. In 1824, Henry demonstrated that ammonia reacts directly with oxygen at elevated temperatures in the presence of spongy platinum. During the nineteenth and early twentieth centuries, hundreds of catalysts were tested for use in the latter reaction which was to become the foundation of the modern nitric acid industry. In 1784, Cavendish synthesized the acid by the action of electric sparks in humid air, and in 1816 Gay-Lussac and Berthollet established the acid's composition.

Until 1900, nitric acid was produced commercially from potassium nitrate and then, with the development of the Chile saltpeter deposits in South America, from sodium nitrate by reaction with sulfuric acid. The latter process was supplanted in 1903 with the operation in Norway of the first successful plant to produce nitric acid directly from nitrogen and oxygen in an electric-arc furnace. More than a dozen electric-arc plants were installed in Norway where inexpensive electricity was available (2–3) (see Furnaces, electric).

The catalytic oxidation of ammonia in air in the presence of platinum catalyst was studied in detail by Ostwald starting in about 1901, and by 1908 a commercial plant was built near Bochum, Germany, which produced 3 t/d of nitric acid. With the demonstration of the Haber-Bosch ammonia synthesis process at Oppau, Germany, in 1913, the future of ammonia oxidation was assured, and additional nitric acid plants were installed in Germany. The first full-size plant for the production of nitric acid by ammonia oxidation in the United States was installed at Muscle Shoals, Alabama, by Chemical Construction Co. in 1917. After World War I, additional plants were erected, and now virtually all nitric acid manufacture is by the oxidation of ammonia. Variations in plant design have developed principally in relation to capacity and operating pressure; modern plants produce 1000 t/d in single trains operating at 1 MPa (10 atm).

In the ammonia oxidation process, nitric acid generally is produced as an aqueous solution of 50–70 wt %. These grades are suitable for direct use in the production of fertilizers but, for most organic reactions, more highly concentrated acid is required. The two major routes for concentrating nitric acid have been by extractive distillation with sulfuric acid or magnesium nitrate and by reaction with additional nitrogen oxides in direct strong nitric (DSN) processes. Variations of DSN processes now produce weak and strong acids simultaneously.

Physical Properties

Nitric acid is extremely difficult to prepare as a pure liquid because of its tendency to decompose and, thereupon, release nitrogen oxides. When produced by vacuum distillation of a mixture of sodium nitrate and concentrated sulfuric acid with condensation of the liquid at just above its freezing point, a colorless liquid (freezing pt −41.59°C) can be collected. Crystals of the pure acid are quite stable, but the liquid degenerates to a limited extent at any temperature above the melting point, and turns yellow within an hour at room temperature.

Nitric acid is completely miscible with water and generally is known and used as an aqueous solution and sometimes with the addition of dissolved nitrogen oxides at high concentrations. Two hydrates may be crystallized from acid solutions, ie, a monohydrate ($HNO_3.H_2O$), corresponding to 77.77 wt % acid (mp −37.62°C) and a trihydrate ($HNO_3.3H_2O$), corresponding to 53.83 wt % acid (mp −18.47°C) (4). With local maxima at the freezing points of the crystal forms, the freezing-point curve for nitric acid solutions is irregular. Local minima occur at ca 32, 71, and 91 wt % acid, as shown in Table 1. The density of nitric acid at any temperature increases with acid concentration: a detailed tabulation of densities is given in ref. 6.

Nitric acid forms a maximum boiling-point azeotrope with water at 68.8 wt % (bp ca 122°C). The partial pressures of acid and water vapor over solutions of nitric acid have been determined but with varying results, since decomposition of nitric acid at high temperatures and concentrations can introduce considerable uncertainty into careful measurements (8–9). Many of the published data are questionable when subjected to strict thermodynamic tests (4). Although data for the total pressure over the solutions have errors of up to 10% at pressures below ca 0.7 kPa (5 mm Hg) at 0°C, the error drops to ca 0.3% at 27 kPa (200 mm Hg) at 80°C (10). For all temperatures,

Table 1. Physical Properties of Nitric Acid Solutions[a]

wt % HNO_3	Density (at 20°C), g/cm³	Freezing point, °C	Bp, °C	Specific heat (at 20°C), J/(g·K)[b]	Partial pressures (at 20°C), Pa[c]		Viscosity (at 20°C), mPa·s (= cP)	Thermal conductivity (at 20°C), W/(m·K)[d]
					HNO_3	H_2O		
0.0	0.99823	0	100.0	4.19		2333	1.0	0.61
10.0	1.0543	−7	101.2	3.73		2266	1.1	0.57
20.0	1.1150	−17	103.4	3.39		2026	1.2	0.54
30.0	1.1800	−36	107.0	3.18		1760	1.4	0.50
40.0	1.2463	−30	112.0	3.01		1440	1.6	0.47
50.0	1.3100	−20	116.4	2.85	27	1053	1.9	0.43
60.0	1.3667	−22	120.4	2.64	120	653	2.0	0.40
70.0	1.4134	−41	121.6	2.43	387	347	2.0	0.36
80.0	1.4521	−39	116.6	2.22	1400	120	1.9	0.35
90.0	1.4826	−60	102.0	1.97	3600	27	1.4	0.31
100.0	1.5129	−42	86.0	1.76	6000	0	0.9	0.28

[a] Refs. 5–10.
[b] To convert J/(g·K) to cal/(g·°C), divide by 4.184.
[c] To convert Pa to atm, divide by 1.013×10^5.
[d] To convert W/(m·K) to Btu/(h·ft·°F), divide by 1.7307.

viscosities of nitric acid solutions reach a maximum at ca 68 wt % acid (10). At 20°C, this maximum is 2.0 mPa·s (= cP), the viscosity falling from this value to ca 0.9 mPa·s for 100 wt % acid and 1.0 mPa·s for pure water.

Fuming nitric acid is concentrated nitric acid that contains dissolved nitrogen dioxide. The density and vapor pressure of such solutions increase with the percentage of NO_2 present. Acid containing ca 45 wt % NO_2 and 55% HNO_3 has a vapor pressure of 101 kPa (760 mm Hg) at 25°C and a density of 1.64 g/cm³. Thermodynamic data for nitric acid solutions are given in Table 2.

Chemical Properties

Nitric acid is a strong, monobasic acid. It reacts readily with alkalies, oxides, and basic materials with the formation of salts. The reaction with ammonia, forming ammonium nitrate for use as a fertilizer, is by far the largest single industrial outlet for nitric acid.

Table 2. Thermodynamic Properties of Nitric Acid and Its Hydrates[a]

	HNO_3	$HNO_3 \cdot H_2O$	$HNO_3 \cdot 3H_2O$
nitric acid, wt %	100.0	77.77	53.83
freezing point, °C	−41.59	−37.62	−18.47
heat of formation (at 25°C), kJ/mol[b]	−173.35	−472.07	−888.45
free energy of formation (at 25°C), kJ/mol[b]	−79.97	−328.29	−810.99
entropy (at 25°C), kJ/(mol·K)[b]	155.71	217.00	347.17
heat of fusion, kJ/mol[b]	10.48	17.52	29.12
heat of vaporization (at 20°C), kJ/mol[b]	39.48		

[a] Ref. 11.
[b] To convert J to cal, divide by 4.184.

Nitric acid is a strong oxidant. Organic material, eg, turpentine, charcoal, and charred sawdust, are violently oxidized and alcohol may react explosively with concentrated nitric acid. Furfuryl alcohol, aniline, and other organic chemicals have been used with nitric acid in rocket fuels (see Explosives and propellants). Most metals, except the platinum metals and gold, are attacked by nitric acid; some are converted into oxides, eg, arsenic, antimony, and tin, but most others are converted into nitrates. Exceptions are aluminum and chromium steel which may become passivated by the acid and show very little attack. The oxides, sulfides, etc, of the lower oxidation states of most elements can be oxidized readily to higher oxidation levels by concentrated nitric acid.

The activity of nitric acid as an oxidizing agent apparently is dependent upon the presence of free nitrogen oxides. Pure nitric acid does not attack copper, but when oxides of nitrogen are introduced, the reaction is at first slow and then proceeds rapidly and violently. The reduction products of nitric acid vary with the concentration of the acid and the strength of the reducing agent with which it reacts. A mixture of oxides generally is produced; dilute nitric acid tends to give a predominance of nitric oxide and concentrated acid produces a mixture that is richer in nitrogen dioxide. The reaction of very dilute acid with a strong reducing agent, eg, metallic zinc, produces a mixture of ammonia and hydroxylamine.

Nitric acid also undergoes reactions with organic compounds where the acid serves neither as an oxidizing agent nor as a source of hydrogen ions. The formation of organic nitrates by esterifications, ie, O-nitration, involves reaction with the hydroxyl group:

$$ROH + HONO_2 \rightarrow RONO_2 + H_2O \tag{1}$$

Esterification with nitric acid includes the industrially important reactions with glycerol to form nitroglycerin and with cellulose to form nitrocellulose.

C-nitration does not involve the hydroxyl group but is a reaction with an aliphatic or aromatic hydrocarbon or a substituted derivative to produce compounds, eg, the nitroparaffins and the nitrotoluenes:

$$RH + HONO_2 \rightarrow RNO_2 + H_2O \tag{2}$$

These reactions normally take place with a mixture of concentrated nitric and sulfuric acid; the sulfuric acid removes the water that is formed (see Nitration).

Manufacture and Processing

Virtually all of the nitric acid that is manufactured commercially is obtained by the ammonia oxidation process. Despite the many variations in operating details among the plants producing nitric acid, three essential steps are common, ie, oxidation of ammonia to nitric oxide, NO, oxidation of the nitric oxide to the dioxide, NO_2, and absorption of nitrogen oxides in water to produce nitric acid with the release of additional nitric oxide. Each of these reactions has been studied but little agreement has been reached on the mechanisms involved in each step.

Oxidation of Ammonia. Ammonia is oxidized with excess of oxygen over a catalyst to form nitric oxide and water:

$$NH_3 + 1.25\,O_2 \rightarrow NO + 1.5\,H_2O \qquad \Delta H_{298\,K} = -226\ \text{kJ/mol}\ (-54.1\ \text{kcal/mol}) \tag{3}$$

The reaction is extremely rapid and goes almost to completion. The major competing reaction yields nitrogen

$$NH_3 + 0.75\,O_2 \rightarrow 0.5\,N_2 + 1.5\,H_2O \qquad \Delta H_{298\ K} = -317 \text{ kJ/mol } (-75.7 \text{ kcal/mol}) \qquad (4)$$

Under normal operating conditions, 95–98% of the ammonia feed follows equation 3. Although various metallic oxides have been used, the almost universally preferred catalyst consists of platinum wire which is woven into fine mesh gauze. Five to ten weight percent rhodium normally is added to the platinum to increase its strength, and up to 5 wt % palladium is used to reduce cost.

A theory has been postulated for the mechanism of the reaction on platinum and is based upon the intermediate formation of a nitroxyl compound, HNO (12–13). An alternative scheme, which is based upon the existence of an imide (NH) as the principal intermediate, also has been reported (14). Reviews of these theories and of those based upon the formation of nitrous acid and hydroxylamine have been published (15). The reaction may be directly between ammonia and oxygen to form nitric oxides on the platinum wire without benefit of intermediates (16–17). Mass transfer is significant in limiting the reaction, since contact times are ca 10^{-3} s.

Despite the lack of agreement on the mechanism of the oxidation reactions, significant data have been accumulated on the effect of operating variables. The efficiency of the reaction, measured as the percent of the ammonia feed which produces nitric oxide, increases with temperature at the gauze. The efficiency first appeared to reach a maximum of ca 97% at 850°C at atmospheric pressure and, at 0.4 MPa (4 atm), a maximum of ca 93% at 900°C (18). Laboratory experiments, which show that the efficiency is independent of pressure, have been reported (19). Industrial experience does not show a marked difference in efficiencies based on pressure, although values of 98% for atmospheric pressure plants at 850°C and of 96% for plants at 0.8 MPa (8 atm) and 900°C are accepted values. Other variables, including linear, space, and mass velocities; reactor design; and, particularly, freedom from impurities, appear to have greater significance than pressure in determining actual efficiencies. The advantage of higher temperatures in improving yields has been demonstrated but, to some extent, this advantage is offset by a rapid increase in the loss of the precious-metal catalyst.

Oxidation of Nitric Oxide. Nitric oxide undergoes a slow homogeneous reaction with oxygen to yield nitrogen dioxide:

$$2\,NO + O_2 \rightleftharpoons 2\,NO_2 \qquad \Delta H_{298\ K} = -114 \text{ kJ/mol } (-27.3 \text{ kcal/mol}) \qquad (5)$$

The equilibrium constant for the reaction strongly favors the production of NO_2 at lower temperatures, so that below 150°C almost all nitric oxide combines with any oxygen that is present if sufficient residence time is allowed. The reaction rate is slow and the rate constant decreases with increasing temperature. Plant operating conditions generally are sufficiently far from equilibrium so that the reverse reaction can be neglected and the change in the partial pressure of nitric oxide can be expressed as (20–21)

$$d(P_{NO})/dt = -k(P_{NO})^2(P_{O_2}) \qquad (6)$$

The rapid decrease in k with increasing temperature means that there is a major advantage to using lower temperatures to speed the reaction. The kinetic equation demonstrates the even more dramatic effect of pressure upon the oxidation. Since residence time and partial pressure are proportional to the total pressure, the volume required to effect the oxidation of a given percentage of nitric oxide in a gas mixture is inversely proportional to the cube of the pressure.

Nitrogen dioxide dimerizes almost instantaneously to an equilibrium mixture with dinitrogen tetroxide:

$$2\,NO_2 \rightleftharpoons N_2O_4 \qquad \Delta H_{298\,K} = -57.4\;kJ/mol\;(-13.7\;kcal/mol) \qquad (7)$$

Lower temperatures and increasing pressures shift the reaction to the production of the tetroxide.

Absorption of Nitrogen Oxides. In the absorption of nitrogen oxides in water, there are uncertainties about the reaction mechanism, and complexities resulting from mass diffusion in a vapor and in a liquid phase are involved. The overall reaction usually is shown as if only the nitrogen dioxide that is present in the gas reacts with liquid water:

$$3\,NO_2(g) + H_2O(l) \rightleftharpoons 2\,HNO_3(aq) + NO(g) \qquad \Delta H_{298\,K} = -135.6\;kJ/mol\;(-32.4\;kcal/mol) \qquad (8)$$

Extensive measurements of the reaction equilibrium have been made at various temperatures and acid concentrations (22). Theories that were formulated regarding the possible mechanisms of absorption were based upon gas-phase reaction or liquid-phase reaction and control by gas-film diffusion or liquid-film diffusion as well as upon the various molecular species present, eg, NO_2, N_2O_4, and N_2O_5. A summary of some of the studies and support for the assumption that N_2O_4 is the reactant species is presented in ref. 23. All the reported equilibrium data can be fairly well correlated by $P_{NO}/(P_{N_2O_4})^{1.5}$ as a function of acid concentration independent of temperature. Ten possible mechanisms for absorption have been compared and it was concluded that, at high concentrations of nitrogen oxide in the vapor, the controlling mechanism is the solution of N_2O_4 and its hydrolysis to HNO_3 plus HNO_2; whereas, at low concentration, the control is either diffusion of NO_2 through the liquid film or absorption of HNO_2 (24).

Principles of equipment design can be deduced from equation 8 and the available equilibrium data. Equipment sizes can be decreased to produce a given amount and concentration of acid, or more of a higher concentration of acid can be produced by making the following changes: reduce operating temperatures, increase operating pressures, increase proportion of NO_2/NO in the feed gas, and increase the reaction volume after vapor–liquid contact to permit reoxidation of released NO. A detailed example of a method for designing an absorption tower for nitric acid as well as additional data on the reactions are given in ref. 25.

Plant Design. Initial developments in nitric acid plant design principally involved modifications to the converter design and included the use of fine-mesh platinum gauze in the reactor. In 1917, at the Muscle Shoals plant in the United States, 696 converters were installed, each of which were 35×70 cm in cross section and included a single sheet of 0.17-mm (80 mesh) gauze, for a capacity of 400 kg/d of nitric acid per unit. In such a plant, the ammonia oxidation reaction takes place at ca 800°C, with the hot gas effluent preheating the feed gas and then generating steam in low pressure boilers before final cooling and absorption. A typical flow sheet for an early atmospheric plant is shown in Figure 1a.

The only materials of construction suitable for absorption of nitric acid in these early plants were stoneware and acidproof brick; however, these materials precluded operation at pressures in excess of 0.1 MPa (1 atm). In the 1920s, the development of stainless iron and steel alloys permitted the industrial application of the laboratory-demonstrated advantages of nitric acid production at high pressures. The in-

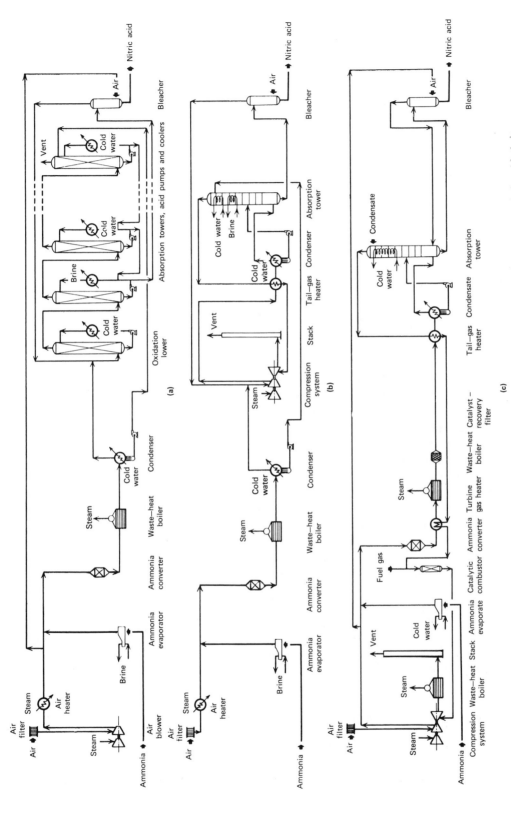

Figure 1. (a) Atmospheric nitric acid plant. (b) Split-pressure nitric acid plant. (c) High pressure nitric acid plant.

859

troduction of pressure processing followed two different routes: in the United States, the high pressure system involved a single pressure throughout; whereas, in Europe, the split-pressure system operated with the ammonia burner at atmospheric pressure, followed by compression of the gas containing nitric oxide for improved absorption (see also High pressure technology).

High Pressure Process. The high pressure process first was used commercially by the DuPont Company in the late 1920s (26). The process essentially is the same as the atmospheric process, but operation at the higher pressure permits dramatic reduction in equipment size. Improved designs were standardized and dozens of identical systems were installed at U.S. Army Ordinance Works in the early 1940s in conjunction with installation of explosives facilities. Although most of these plants have been replaced with more efficient units, some are in operation.

In the Army plants, air is compressed to ca 0.86 MPa (110 psig), preheated to ca 250°C, and mixed with ammonia vapor. The mixture, which contains ca 10 vol % ammonia, flows down through a pack of flat gauzes and produces nitric oxide at ca 95% efficiency at ca 930°C. The hot gas leaving the gauze is cooled by exchange with the feed air and in a tail-gas reheater before flowing to a water-cooled condenser. Weak acid that is produced in the condenser is pumped to an intermediate tray in the absorption tower and the uncondensed process gas flows to the bottom. The absorption tower consists of a series of bubble-cap trays provided with cooling coils for removing the heat of reaction (see Absorption). As the gas flows up the tower countercurrent to the acid flow, nitrogen dioxide dissolves in water forming nitric acid and releasing nitric oxide; the latter is reoxidized in the space between the trays by the excess oxygen which is present. Steam condensate is added to the top of the tower as the absorbent; dissolved nitrogen oxides are removed from the product acid by contact with secondary air in a bleaching tower. The tail gas leaving the absorption tower is reheated to about 250°C by exchange with the process gas and then is expanded through a gas engine which provides ca 40% of the power required for driving the reciprocating air compressors. The DuPont plants were designed to produce 25 t/d of equivalent 100 wt % nitric acid but ultimately reached a capacity of ca 50 t/d per stream.

Variations in design details for nitric acid plants built during World War II included: medium pressure plants operating at about 0.45 MPa (50 psig) and involving centrifugal compressors to supply air, the provision of boilers following the ammonia burner for recovery of heat, and absorption in a series of stacked horizontal drums or a series of vertical packed towers with acid circulated to external coolers. In the 1950s, the application of high temperature designs permitted greatly improved power recovery in the nitric acid plants. In 1954, Chemical Construction Corporation installed a 225-t/d plant which includes a catalytic combustion unit to heat the tail gas to 670°C before expansion; thus, all the power that is required for compression is recovered and the need for a make-up driver is eliminated. With the increase in demand for acid and cost of utilities, similar plants were designed to produce up to 1000 t/d of nitric acid in one stream. A typical high pressure plant with a catalytic combustor is illustrated in Figure 1c. The process scheme is essentially the same for the atmospheric pressure plant except for the use of a single absorber and the addition of the power and heat recovery system.

In the 1960s, use of the catalytic combustor was further developed to destroy simultaneously nitrogen oxides that are present in the tail gas as a fume abatement measure (see Exhaust control). However, the increased cost and uncertainty of natural-gas supply in the 1970s negated the earlier advantage of the combustor, so it is now offered in only unusual circumstances. The most recent plants incorporated more and larger heat exchangers for improved energy recovery, and use much larger absorption towers to reduce emissions to acceptable level.

Split-Pressure Process. Although many European plants were built at a single low pressure, the preferred approach was to maintain atmospheric or a slight negative pressure on the ammonia burner, to remove the generated heat in a waste-heat boiler and water-cooled condenser, and then to compress the process gas to ca 0.34 MPa (35 psig) in a stainless steel compressor. The compressed gas is contacted with water in a tower or towers for production of the acid, and the tail gas is reheated to about 200°C for power recovery in an expander. Figure 1**b** is an illustration of a typical split-pressure plant.

The capacity of the atmospheric ammonia burner used in such a plant is limited to ca 100 t/d of equivalent acid so that, for a large installation, several burners and waste-heat boilers are required. The greatly increased cost of such multiple or large units, when combined with the extra cost of a stainless steel compressor, requires an investment for split-pressure plants from one and one half to two times the amount for equivalent high pressure plants. This difference in cost was justified by the slight possible improvement in ammonia efficiency and the savings resulting from a reduction in loss of platinum catalyst; no such plants were built in the United States. In the late 1950s, a compromise was developed: an air compressor was installed before the ammonia converter so that ammonia oxidation took place at ca 0.3 MPa (29 psig) and absorption at 0.6 MPa (72 psig) with intermediate oxidation efficiencies. In the 1960s, the French company Grande Paroisse supplied a dozen large plants in Eastern Europe with this design, and two such medium split-pressure plants were built for U.S. companies. In the 1970s, Chemico added a booster compressor to their high pressure plant design, which permitted split-pressure plants with ammonia oxidation at 1 MPa (130 psig) and absorption at 1.7 MPa (232 psig).

Latest Process Developments. Changes that have taken place in nitric acid technology in recent years have resulted from governmental restrictions on gaseous emissions and from the skyrocketing increase in the cost of fuels in the 1970s. In 1971, New Source Performance Standards for nitric acid plants were promulgated by the EPA and resulted in limiting emissions to 1.5 kg of equivalent nitrogen dioxide per metric ton of equivalent 100 wt % nitric acid produced. This quantity corresponds to a concentration of about 230 ppm by volume of nitrogen oxides in the vented tail gas or 5–10% of the amount typically discharged in earlier facilities.

For new plants, the earliest means of abatement, catalytic combustion, became uneconomical with the dramatic increase in natural-gas cost, and extended absorption, ie, the provision of much larger absorption towers to reduce emitted oxides, became the rule. In order to minimize the size of these new towers, two approaches, which previously had proven to be uneconomical, were introduced. A refrigeration system can be provided to circulate chilled water to the upper coils of the absorber which results in a lower operating temperature which substantially reduces the tower volume requirement. In addition, for large plants, incorporation of a booster compressor for absorption at very high pressures is justified.

Process Licensers. All commercial nitric acid plants use essentially the same process technology, and variations among them relate chiefly to operating pressures and details of equipment design. In the United States, much of the recent progress can be attributed to the efforts of the engineering contractors, eg, Chemical Construction Corp., C&I/Girdler, and D. M. Weatherly, in the United States. In Europe, most producing companies have developed or improved their own nitric acid plants and offer such technology directly. Licenses for a variety of very similar plants can be obtained from Dutch State Mines; Grande Paroisse; Estabs. Kuhlmann; Montecatini; Pechiney-St. Gobain; Pintsch-Bamag; Societé Belge de l'Azote; Uhde Hoechst; and their successor companies.

Equipment. In a modern, high pressure nitric acid plant, almost one half of the equipment cost involves the air-compression system. This typically consists of an intercooled, multistage compressor on a common shaft with a high temperature gas expander and a make-up steam turbine or electric motor. Air required for a nitric acid plant is ca 170 Nm^3/h per metric ton per day [100 SCF/(min) per t/d] of acid, so that, for plants with capacities of 100–500 t/d, centrifugal compressors normally are supplied. At higher capacities, an axial plus a centrifugal compressor or two axial compressors are required.

The type of gas expander that is provided depends upon the desired operating temperature. Below 540°C, ordinary steam-turbine design, modified only for the difference in gas properties, is acceptable. At higher temperatures, however, designs and materials based upon gas-turbine technology are necessary. Although several plants operate with all of the required compressor power supplied by the expander, it is preferred to limit the temperature to the expander so that ca 10% of the power is supplied by the make-up driver. The latter unit must be supplied in any case in order to start up the system, unless large quantities of air or gas are available from other sources. In large plants, when a steam turbine is provided, sufficient steam can be generated in the process waste-heat boilers to provide all that is required for normal operation plus a substantial amount for other uses.

In a split-pressure plant, which involves oxidation under vacuum, the nitric acid compressor and its direct auxiliaries must be constructed of stainless steel. Tail gas normally is not heated to temperatures higher than ca 200°C for power recovery, because the cost of high temperature exchangers at the low allowable pressure drops after an atmospheric burner is prohibitive. The most modern plants combine both technologies: steel compressors discharging air at 1.0 MPa (10 atm), a stainless steel booster compressing the nitric gas to 1.7 MPa (17 atm), and the total system driven by an expander operating on tail gas at 670°C and by a small steam turbine.

Practically all nitric acid plants built since 1940 use a catalyst consisting of multiple layers of fine-mesh platinum–rhodium wire gauze, which is supported in a vessel of proprietary design. Atmospheric pressure converters use three or four layers of gauze and have been made as large as four meters in diameter. In a typical converter design, the gauze is mounted directly above the tubes of a boiler with the gas passing downwards through the gauze so that the hot gas is cooled rapidly by generating steam and the walls are protected by a water film. Other designs provide for upflow of the ammonia–air mixture so as to simplify the support problems. Pressure converters generally are downflow units with 20–30 gauzes in a pack. The pressure walls are protected by internal insulation or are cooled with an external gas or water jacket.

Most important to the proper operation of any ammonia converter is the provision of suitable means of filtering and mixing the feed gases. In order to minimize the possibility of a preliminary reaction of the ammonia with air before they reach the gauze, the two streams generally are preheated and filtered separately, and mixed thoroughly just prior to their entering the converter.

The ammonia oxidation reaction is maintained at about 850°C in an atmospheric burner and at 900–930°C at higher pressures. With air and ammonia supply temperatures held constant, the gauze temperature is adjusted by slight changes in the proportion of ammonia to air. Since the explosive limit of ammonia is approached at concentrations greater than 12 mol %, plant operation normally is maintained at 9.5–10.5 mol %. For the high pressure plant, therefore, a preheat temperature of 225–275°C is required, which is ca 50°C more than for an atmospheric burner.

The higher operating temperatures and the higher mass velocities which normally are used in high pressure plants lead to appreciably greater losses of catalyst gauze, ie, 250–500 mg/t of nitric acid produced, depending upon design, as compared to 50 mg in an atmospheric plant. Since the cost of the catalyst is so high, a recovery filter usually is provided in a high pressure plant and about half of the lost catalyst is recovered. In addition, proprietary support materials can be provided by catalyst suppliers to reduce the gross loss even in the less efficient plants to ca 250 mg/t.

The original atmospheric-pressure absorption towers consisted of a series of packed stoneware towers and cooled acid was circulated over each. Similar towers in stainless steel later were used at intermediate pressures. In the 1920s, DuPont introduced the bubble-cap plate column with internal cooling coils, and this design is the most popular. The original diameter limit of 1.5 m has been extended so that field-assembled towers that are 5.5 m in diameter are common. Variations in tower details include the substitution of external cooling for some of the internal coil surface and the use of sieve trays instead of bubble-cap trays.

Fume Abatement. Until recent governmental intervention, absorption towers typically were designed to reduce the concentration of nitrogen oxides in the tail gas to 0.2–0.3 vol %, since the cost to increase absorber sizes could not be justified by the additional recovered acid. Nitrogen tetroxide is a dark-red gas so that, even at low concentrations, the vented tail gas from a nitric acid plant, if untreated, has a distinct yellow to red color which often is regarded as a major nuisance. In the late 1960s, environmental concerns required the reduction of total NO_x, ie, $NO + NO_2$, emissions from nitric acid plants, and fume-abatement studies were initiated.

The first successful means of reducing emissions to less than 200 ppm in full-sized plants was by an extension of the use of catalytic combustors. If a sufficient amount of fuel gas to react with all oxygen present is added to the tail gas flowing to the catalyst, any nitric oxide that is present is reduced to nitrogen by reaction with a small amount of excess fuel. Because of the high ignition temperature and large temperature rise accompanying the reaction when natural gas is used as a fuel, a single-stage combustor must be followed by a waste-heat boiler or a water quench, or a dual-combustor system with intermediate quenching is necessary to protect the gas expander.

Although new nitric acid plants were designed with catalytic combustors for abatement, this approach did not lend itself easily to installation in existing plants. Adsorption, absorption, and selective combustion methods have proven successful as retrofits to specific plants. Nitrogen oxides in the tail gas are adsorbed by passing the gas through a bed of acid-resistant molecular-sieve pellets (see Molecular sieves). When the bed is saturated, the gas is switched to a fresh bed and the oxides are desorbed with hot secondary air and are returned to the acid plant for recovery as acid (see Adsorptive separation). In absorption, the tail gas leaving the absorber is passed through a second absorber where it is brought into contact with either water or a solution containing ammonia, urea, or sodium hydroxide. Where the absorbent is water, the weak acid that forms is used as feed to the primary absorber, which directly increases acid recovery. Where other absorbants are used, the oxides are recovered as nitrite/nitrate solution and typically are mixed with nitrogen fertilizer solutions in adjoining facilities. In selective combustion, as an alternative to the high temperature nonselective catalytic combustion system which requires removal of all oxygen in the tail gas before nitric oxide is reduced, special catalysts have been developed which cause the selective reaction of ammonia with the nitrogen oxides that are present in the tail gas at low temperatures. Although the system results in a reduction in acid yield compared to the two preceding approaches, because the oxides are destroyed rather than recovered and additional ammonia is used as fuel, the smaller investment cost has made this an economical solution in many cases (27).

Since the requirement for modifying existing facilities varied from state to state and plant to plant, the methods chosen for abatement also have varied. In 1971, the Federal government set guidelines limiting NO_x emission from new nitric acid plants to a maximum concentration of ca 230 ppm. Although the nonselective catalytic combustion system has provided excellent abatement results where installed, the increased cost and lack of availability of natural gas has effectively eliminated it from consideration for new plants in the United States (28). Extended absorption, therefore, has become the standard design for the new generation of nitric acid plants (see also Air pollution).

Other Processes

The Wisconsin Process. A thermal process for the direct fixation of atmospheric nitrogen was developed at the University of Wisconsin during World War II. Air is heated to a temperature >2000°C by contact with a bed of fuel-heated magnesia pebbles and then is quickly chilled by contact with a second, cold, bed, yielding 1–2 wt % nitric oxide. A unique method of recovering the dilute oxides was developed later. The gas was contacted in three silica-gel beds in series: the first to dehumidify the gas, the second to oxidize catalytically the nitric oxide present to the dioxide, and the third to adsorb the nitrogen dioxide which then could be stripped in a highly concentrated form (29–30) (see Nitrogen fixation).

Further studies and tests were conducted and a small plant was built for the U.S. government in 1953 to produce acid by this process. The plant was operated for eighteen months and the data that were obtained, although proving the process technologically feasible, indicated that it could not be justified economically at the time (31).

The Nuclear Process. Appreciable yields of nitrogen oxides have been produced by irradiating air in a nuclear reactor. Although potentially an economical source of nitric acid in huge amounts, particularly as a possible by-product of nuclear power generation, little progress towards commercialization of the process has been made and much further research will be necessary before this system can become competitive (32).

Other Manufacturing Processes. Several other processes to produce nitric acid have been studied in the laboratory. Generally, these involve different methods of heating air to extremely high temperatures in order to form some nitric oxide and then cooling the nitric oxide rapidly to prevent its decomposition. Procedures to attain the required high temperatures include the use of shock waves and magnetohydrodynamic power generators (33) (see Coal, magnetohydrodynamics).

An alternative approach to acid production evolves from attempts to recover directly the energy released by the oxidation of ammonia to nitric oxide in a fuel cell. Early trials with low and medium temperature cells proved unsuccessful, because all the ammonia degraded to nitrogen. Recent studies of cells operating at ca 850°C report conversion of more than 60 wt % ammonia to nitric oxide, which indicates exciting possibilities for the commercial cogeneration of electricity and nitric acid (34) (see Batteries and electric cells, primary).

Nitric Acid Concentration. Nitric acid is produced by the standard ammonia-oxidation processes as an aqueous solution at a concentration of 50–70 wt %. Such concentrations are suitable for the production of ammonium nitrate but, for use in organic nitrations, anhydrous nitric acid is required. Since nitric acid forms an azeotrope with water at 68.8 wt %, the water cannot be separated from the acid by simple distillation. Two industrial methods for accomplishing the concentration are extractive distillation and reaction with additional nitrogen oxides. The latter are the direct strong nitric (DSN) processes.

Extractive Distillation. The principal method used for producing 98–99 wt % nitric acid involves distillation of weak nitric acid with concentrated sulfuric acid; the latter serves as a dehydrating agent. Typically, 60 wt % nitric acid is mixed with 93 wt % sulfuric acid in a packed tower which is provided with a steam-heated reboiler. The nitric acid vapor is distilled and condensed, and the sulfuric acid and water leave the bottom as ca 70 wt % H_2SO_4. Water is removed from the sulfuric acid in a sulfuric acid concentrator and the 93 wt % acid is recycled to the process. In producing 99 wt % nitric acid by this method, decomposition is minimized and a product containing less than 0.1 wt % of lower oxides is recovered. Sulfuric acid and nitric acid losses are <0.1 wt % of the nitric acid produced (see Azeotropic and extractive distillation).

An alternative extraction medium is a 72 wt % solution of magnesium nitrate in water. The nitrate solution typically leaves the distillation column at ca 68 wt % and is reconcentrated by flashing to a steam-heated vacuum drum. The use of magnesium nitrate is most economical in small plants, particularly where existing sulfuric acid concentrating facilities are not available, or where a sulfate-free acid is required. For large-capacity plants, sulfuric acid generally is preferred.

Direct Strong Nitric Processes. In Europe, modifications of a concentration process that was described in 1932 have been used widely (35). Nitrogen tetroxide is separated from the process gases that leave the ammonia converter by refrigeration or by absorption in concentrated nitric acid. The tetroxide then reacts with weak nitric acid and air or oxygen to yield a 98 wt % product. Despite improvements in the process resulting from the development of continuous rather than batch processing, high investment costs and the production of a sidestream of very dilute acid have prevented its commercial application in the United States. More recent modifications of DSN processes that yield strong and weak acid simultaneously have received much attention in the literature but their installation has been limited (36–38,39).

Process Licensers. Nitric acid concentration processes using extractive distillation with sulfuric acid were perfected in the United States by DuPont, Hercules, and Chemico. The substitution of magnesium nitrate as the extractive agent was developed separately and offered by Hercules, Tennessee Eastman, and Chemico. Direct strong nitric acid processes were perfected by Bamag (now owned by Davy). New strong-acid plants producing weak acid as a sidestream have been built by Bamag, Sumitomo, and Espendiso.

Production and Shipment

Production. The production of nitric acid in the United States has been rapid since World War II. Although figures generally are not available for production in government-owned plants, private production of nitric acid increased from 406,000 t of equivalent 100 wt % nitric acid in 1945 to 1,080,000 t in 1947 and to 4,294,000 t in 1964. The annual production of nitric acid from 1965–1979 is listed in Table 3. More than 90% of the nitric acid produced in the United States is for captive consumption. Listed prices for all technical grades of acid remained constant from 1929 until 1952, and they increased only 10–15% to ca $0.15/kg (100 wt % basis) for 60 wt % acid through 1973. Current prices (1980) are $0.10–0.14/kg for 58–68 wt % acid and $0.26/kg for 95 wt % acid.

Shipment. Nitric acid is classified as a hazardous material for purposes of transportation by the Department of Transportation and so may not be offered or accepted for movement in commerce unless it is properly packaged and labeled and unless specified shipping papers and vehicle marking are provided. Acid in concentrations up to 40 wt % is classified as a corrosive material and higher concentration acid is defined as an oxidizer.

All concentrations of nitric acid may be shipped in stainless steel drums or stainless steel tank trucks or cars. Acid in concentrations of 80 wt % or higher may be shipped in aluminum drums or aluminum tank cars. All concentrations may be shipped in glass bottles of up to 2.4 L capacity, individually enclosed and cushioned in tightly closed metal cans, and packed in wooden boxes or barrels. Concentrations below 72 wt % may be shipped in glass carboys that are packed in cushioned boxes. Every drum and box of acid must bear the appropriate corrosive or oxidizer label, and tank cars and railroad cars carrying one or more containers of acid must bear the corresponding placard. The detailed requirements for packing and shipping nitric acid are given in ref. 41.

Specifications and Standards

The ACS specifies the concentration of reagent-grade nitric acid to be 69.0–71.0 wt % HNO_3 and that it be colorless and free from suspended matter and that it contain limited maxima of chlorides, sulfate, arsenic, heavy metals, and iron. Fuming nitric acid must contain no less than 90 wt % HNO_3, no more than ca 0.1 wt % dissolved ox-

Table 3.   U.S. Production of Nitric Acid[a], Thousand Metric Tons

Year	Production	Year	Production
1965	4445	1973	7621
1966	5004	1974	7368
1967	5864	1975	6830
1968	5773	1976	7162
1969	6554	1977	7215
1970	6899	1978	7306 estd
1971	6931	1979	7771 estd
1972	7242		

[a] Ref. 40. Shown as equivalent 100% HNO_3.

ides, and similar but slightly less stringent limits on impurities. The USP definitions are almost identical to those of the ACS (see also Fine chemicals).

The U.S. Department of Defense military standard MS26047 lists the ACS grades and ACS fuming grades as standard for use as analytical reagents. In addition, it lists a technical grade for use in photoengraving and metal cleaning and for general chemical use. The technical grade is specified to be 61 wt % HNO_3 min and to contain no more than 0.5 wt % chlorides as HCl or 0.5 wt % residual acid as H_2SO_4. The Joint Army–Navy Specifications for nitric acid for ordnance use (JAN-A-183) provides specifications and methods of testing for five classes of acid for the manufacture of specific explosives.

Analytical and Test Methods

Qualitative Analysis. Nitric acid may be detected by the classical brown-ring test, the copper-turnings test, reduction of nitrate to ammonia by active metal or alloy, or the nitrogen precipitation test. Nitrites or nitrous acid interfere with most of the above tests, but such interference may be eliminated by acidifying with sulfuric acid, adding ammonium sulfate crystals, and evaporating to a low volume.

In the brown-ring test, concentrated sulfuric acid is carefully poured down the side of an inclined test tube to form a separate layer beneath the solution to be tested. After cooling of and without mixing the two layers, a few drops of a ferrous sulfate solution are placed inside the inclined tube and drain down to the interface. The appearance within a few minutes of a brown layer of ferronitrososulfate $[Fe(NO)]SO_4$ at the junction of the two liquids shows the presence of nitrates. Nitric acid or nitrates may be detected, after the removal of nitrous acid, by warming a mixture of the material to be tested, a 1:1 solution of sulfuric acid, and copper turnings in a test tube. The evolution of brown fumes indicates the presence of nitrates. Nitrates are easily reduced by active metals or alloys in alkaline solution to give ammonia. Suitable reducing agents are aluminum, zinc, or Devarda's alloy (50 wt % Cu, 45 wt % Al, 5 wt % Zn).

Nitric acid may be precipitated by nitron (4,5-dihydro-1,4-diphenyl-3,5-phenylimino-1,2,4-triazole). The yellow precipitate may be seen at dilutions as low as 1:60,000 at 25°C or 1:80,000 at 0°C. To prevent nitrous acid from interfering with the test results, it may be removed by treating the solution with hydrazine sulfate, sodium azide, or sulfamic acid.

Quantitative Analysis. The total acidity of nitric acid solution may be determined by conventional titration using phenolphthalein as the indicator. Other acidic impurities in commercial nitric acid can be determined as follows:

Sulfuric Acid. The sample is evaporated to dryness on a steam bath. The residue is removed with water and evaporation is repeated until the sample is free from nitric acid fumes. The residue is diluted with water and titrated with standardized sodium hydroxide solution using phenolphthalein as an indicator.

Hydrochloric Acid. Hydrochloric acid is determined gravimetrically as silver chloride.

Nitrous Acid. Nitrous acid and lower oxides of nitrogen generally are determined as reported as NO_2. In the absence of organic matter or other reducing agents, the determination may be made by titrating with potassium permanganate solution. A sharp end point is achieved by rapidly adding the permanganate solution to the solution containing the oxide of nitrogen or nitrous acid. If the addition of permanganate solution is slow, some oxidation of the sample by dissolved air occurs resulting from the dilution of the solution with water.

In general, for nitric acid in mixed acid or for nitrates that are free from interferences, the ferrous sulfate titration, the nitrometer method, and Devarda's method give excellent results.

The determination of nitric acid and nitrates in mixed acid is based on the oxidation of ferrous sulfate by nitric acid and may be subject to interference by other materials that reduce nitric acid or oxidize ferrous sulfate. Small amounts of sodium chloride, potassium bromide, or potassium iodide may be tolerated without serious interference, as can nitrous acid up to 50% of the total amount of nitric acid present. Strong oxidizing agents, eg, chlorates, iodates, and bromates, interfere by oxidizing the standardized ferrous sulfate.

Possible interferences and variation of results resulting from modified techniques can be avoided by titrating the sample in exactly the same way and by employing approximately the same amounts of materials as in the initial standardization of the ferrous sulfate against a known quantity of nitric acid. The ferrous sulfate solution is added in a thin stream until the initially yellowish solution turns brown. The titration is complete when the faint brownish tinged end point is reached.

The nitrometer method also is used to determine nitric acid or nitrates in mixed acid or in oleum, and it involves the measurement of the volume of NO gas that is liberated when mercury is oxidized by nitric acid. The method is based on the reaction represented by the following equation:

$$2 \, HNO_3 + 3 \, H_2SO_4 + 3 \, Hg \rightarrow 3 \, HgSO_4 + H_2O + 2 \, NO \tag{9}$$

Nitrogen in nitrates or nitric acid also may be determined by the Kjeldahl method or by Devarda's method. The latter is both convenient and accurate when no organic nitrogen is present. The nitrate is reduced by Devarda's alloy to ammonia in an alkaline solution. The ammonia is distilled and titrated with standard acid.

Health and Safety Factors

Nitric acid vapors and nitrous oxide fumes or oxides of nitrogen (nitric oxide and nitrogen dioxide) are highly toxic and capable of producing severe injury or death if improperly handled. The extent of injury, the signs and symptoms of poisoning, and the nature of the treatment required depend on the concentration of several toxic substances, the time of exposure, and the susceptibility of the individual. The liquid form of the acid is very corrosive and can destroy skin, respiratory mucosa, and gastrointestinal tissue. The extent of the destruction is proportional to the time of contact and the concentration of the solution. Symptoms of attack are smarting, burning, and yellow discoloration. Continued contact may result in severe burns, followed by chronic ulceration with permanent scarring. Acid in contact with skin or mucous membranes should be removed as quickly as possible by washing with large quantities of cold water. If nitric acid is swallowed, the victim should drink large amounts of lukewarm water, soapy water, or milk. Vomiting should be induced if it does not occur naturally. A suitable demulcent, such as thin flour paste, raw egg white, or corn starch paste should be administered.

Nitrogen oxides are readily produced by the action of nitric acid on organic material, or from the decomposition of nitrates or organic nitro compounds. Severe pulmonary signs and symptoms may set in as late as 5–48 h after breathing as little as 25 ppm throughout an 8-h period; breathing 100–150 ppm for 0.5–1 h may produce serious pulmonary edema, and a few breaths of 200–700 ppm may produce fatal pulmonary edema within 5–8 h. The victim may feel no pain or discomfort at the time of breathing the fumes, and the inhalation may be followed by an asymptomatic period of several hours.

Most cases of nitrous-fume poisoning can be classified into the following categories: immediate fatalities after inhaling very heavy fumes, which is rare and usually is caused by some serious accident; delayed symptoms occurring within 48 h after breathing light nitrous oxide fumes. This form of poisoning occurs most frequently in industry, and mild immediate effects from which recovery is apparently complete, but after which pneumonia eventually follows.

In a typical case of nitrogen oxide poisoning, the sequence of events may be a few breaths of apparently harmless gas; only slight discomfort with the worker continuing his job; five to eight hours after exposure, the victim's lips and ears become cyanotic; and increasing difficulty in breathing follows, accompanied by choking, dizziness, and irregular respiration. Severe untreated cases frequently terminate fatally from excessive pulmonary congestion or suffocation.

Remedial measures which should be taken as soon as possible after any indication that poisoning has occurred are: (1) The patient should be moved to uncontaminated atmosphere and no physical exertion should be permitted. (2) The patient should breathe 100% oxygen for 30 min out of every hour for 6 h. If after this time breathing is normal, oxygen inhalation may be discontinued. (3) During the period of oxygen inhalation, the patient should exhale against a positive pressure of ca 390 Pa (4 cm of water), unless there is an indication or history of cardiovascular failures. This is intended to prevent the development of pulmonary edema. (4) Bed rest should be enforced for 24 h with patient under observation of physician or other personnel capable of ensuring resumption of oxygen inhalations should respiratory symptoms reappear.

Uses

The largest use of nitric acid has continued to be in the production of ammonium nitrate for use as fertilizer (see Ammonium compounds). Such use initially was in the form of solid ammonium nitrate granules, ie, prills, but increasing quantities have been mixed with excess ammonia and/or urea and shipped as aqueous nitrogen solutions for direct application to the soil or for the manufacture of mixed fertilizer (see Fertilizers). In the mid-1950s, the use of ammonium nitrate prills mixed with fuel oil was accepted for direct use as an explosive (see Explosives). Although this shift from specially compounded explosives did not materially increase the use of nitric acid, it has sharply shifted the manufacturing sources for explosives.

Nitric acid is used in the manufacture of cyclohexanone (see Cyclohexanol and cyclohexanone) which is the raw material for adipic acid (qv) and caprolactam, which are monomers used in producing nylon (see Polyamides). From 5 to 10% of nitric acid production in recent years has gone into this use. Similar quantities are used in a large number of other organic syntheses; the most rapidly growing is toluene diisocyanate (TDI) which is used for polyurethane production (see Isocyanates; Urethane polymers).

Relatively small quantities of acid are employed for stainless steel pickling, metal etching, as a rocket propellant, and for nuclear-fuel processing.

The use of nitric acid to directly acidify phosphate rock in producing mixed fertilizers has been relatively common in Europe, but not practiced widely in the United States. Similarly, the production of sodium and potassium nitrates has remained insignificant in the United States.

BIBLIOGRAPHY

"Nitric Acid" in *ECT* 1st ed., Vol. 9, pp. 340–344, by M. D. Barnes, Lion Oil Company; "Nitric Acid" in *ECT* 2nd ed., Vol. 13, pp. 796–814, by S. Strelzoff and D. J. Newman, Chemical Construction Corporation.

1. J. W. Mellor, *Comprehensive Treatise on Inorganic and Theoretical Chemistry*, Vol. VIII, Longmans, Green and Co., Ltd., London, Eng., 1928, pp. 555–558.
2. *Ibid.*, pp. 558–562.
3. J. T. Partington and L. H. Parker, *The Nitrogen Industry*, Constable & Co. Ltd., London, Eng., 1922, pp. 233–329.
4. W. R. Forsythe and W. F. Giauque, *J. Am. Chem. Soc.* **64,** 48 (1942).
5. *International Critical Tables of Numerical Data, Physics, Chemistry, and Technology*, Vol. IV, McGraw-Hill Book Co., Inc., New York, 1928, p. 255.
6. R. H. Perry and C. H. Chilton, eds., *Chemical Engineers' Handbook*, 5th ed., McGraw-Hill Book Co., Inc., New York, 1973.
7. *International Critical Tables*, Vol. III, McGraw-Hill Book Co., Inc., New York, 1928, p. 309.
8. *Ibid.*, p. 304–305.
9. G. L. Wilson and F. D. Miles, *Trans. Faraday Soc.* **36,** 356 (1940).
10. T. R. Bump and W. L. Sibbitt, *Ind. Eng. Chem.* **47,** 1665 (1955).
11. F. D. Rossini and co-workers, *Nat. Bur. Stand. (U.S.) Circ.* **500,** 53 (1952).
12. L. Andrussow, *Z. Angew. Chem.* **39,** 321 (1926).
13. M. Bodenstein, *Z. Angew. Chem.* **40,** 174 (1927).
14. F. Raschig, *Z. Angew. Chem.* **40,** 1183 (1927).
15. J. K. Dixon and J. E. Longfield in P. J. Emmett, ed., *Catalysis*, Vol. VII, Rheinhold Publishing Corp., New York, 1960, pp. 281–300.
16. E. K. Nowak, *Chem. Eng. Sci.* **21,** 19 (1966).
17. Y. M. Fogel and co-workers, *Kinet. Katal.* **5,** 154 (1964).
18. G. Fauser, *Chem. Met. Eng.* **37,** 604 (1930).
19. V. I. Atroschenko, *J. Appl. Chem. (USSR)* **19,** 1214 (1946).
20. M. Bodenstein, *Z. Phys. Chem.* **100,** 68 (1920).
21. *Nat. Bur. Stand. (U.S.) Circ.* **510,** (1957).
22. C. L. Burdick and E. S. Freed, *J. Am. Chem. Soc.* **43,** 518 (1921).
23. J. J. Carberry, *Chem. Eng. Sci.* **14,** 189 (1959).
24. S. P. S. Andrew and D. Hanson, *Chem. Eng. Sci.* **14,** 105 (1961).
25. T. H. Chilton, *Chem. Eng. Prog. Monogr. Ser. #3*, AIChE, New York, 1960.
26. T. H. Chilton, *Strong Water*, The M.I.T. Press, Cambridge, Mass., 1968.
27. *Chem. Week*, 33 (July 28, 1976).
28. *A Review of Standards of Performance for New Stationary Sources*, E.P.A., Washington, D.C., EPA-450/3-79-013 (Mar. 1979).
29. N. Gilbert and F. Daniels, *Ind. Eng. Chem.* **40,** 1719 (1948).
30. E. G. Foster and F. Daniels, *Ind. Eng. Chem.* **43,** 986, 992 (1951).
31. E. D. Ermenc, *Chem. Eng. Prog.* **42,** 149, 488 (1956).
32. P. Harteck and S. Dondes, *Science* **146,** 30 (1964).
33. *Nitrogen* **31,** 35 (Sept. 1964).
34. C. G. Vayenas and R. D. Farr, *Science* **208,** 593 (May 9, 1980).
35. G. Fauser, *Chem. Met. Eng.* **39,** 430 (1932).
36. D. J. Newman and L. A. Klein, *Chem. Eng. Prog.* **68,** 62 (Apr. 1972).
37. L. Hellmer, *Chem. Eng. Prog.* **68,** 67 (Apr. 1972).

38. T. Ohrul, M. Okuba, and O. Imal, *Hydrocarbon Process.* **57,** 163 (Nov. 1978).
39. L. M. Marzo and J. M. Marzo, *Chem. Eng.* **87,** 54 (Nov. 3, 1980).
40. *Census of Manufacturers Nitrogenious Fertilizers SIC 2873*, Bureau of Census, Washington, D.C.
41. *Code of Federal Regulations Title 49.*

DANIEL J. NEWMAN
Barnard and Burk, Inc.

NITRIDES

At elevated temperatures and pressures, nitrogen combines with most of the elements to form nitrogen compounds; with metals and semimetals, it forms nitrides. Atomic nitrogen reacts much more readily with the elements than molecular nitrogen. Atomic nitrogen forms nitrides with elements that do not react with molecular nitrogen even at very high pressures. The nitrides of the high melting transition metals, eg, TiN, ZrN, and TaN, are characterized by high melting points, hardness, and resistance to corrosion and are referred to as refractory hard metals. The nonmetallic compounds, eg, BN, Si_3N_4, and AlN, are corrosion- and heat-resistant, ceramiclike industrial materials with semiconductor properties (see Abrasives; Semiconductors). The binary compounds of nitrogen are given in Table 1. These compounds may be classified, according to their chemical and physical properties, into four groups, ie, saltlike, metallic, nonmetallic or diamondlike, and volatile nitrides.

An alphabetical list of nitrides mentioned in the text with their CAS Registry Numbers is given at the end of the article.

Properties

Saltlike Nitrides. The nitrides of the electropositive metals of groups IA, IIA, and IIIB form saltlike nitrides with predominantly heteropolar (ionic) bonding and are regarded as derivatives of ammonia. The composition of these nitrides is determined by the valency of the metal, eg, Li_3N, Ca_3N_2, and ScN. The thermodynamic stability of the saltlike nitrides increases with increasing group number; for example, the nitrides of the alkali metals are only marginally or not at all stable, whereas the rare-earth metals are very effective nitrogen scavengers in metals and alloys. The saltlike nitrides generally are electrical insulators or ionic conductors, eg, Li_3N. The nitrides of the group IIIB metals are metallic conductors or at least semiconductors and, thus, represent a transition to the metallic nitrides. The saltlike nitrides are characterized by their sensitivity to hydrolysis; they react readily with water or moisture and give ammonia and the metal oxides or hydroxides.

Lithium nitride can be prepared by the reaction of lithium metal and nitrogen. Lithium nitride crystals are dark red and melt at 845°C. The electrical ionic conductivity of the bulk material is 6.6×10^{-4} S/cm at 25°C and 8.3×10^{-2} S/cm at 450°C. The ionic conductivity of single crystals of Li_3N is extremely anisotropic; eg, parallel to the c-axis: 1×10^{-5} S/cm at 20°C; perpendicular to the c-axis: 1.2×10^{-3} S/cm.

Table 1. Binary Compounds of Nitrogen in the Periodic System

IA	IIA	IIIB	IVB	VB	VIB	VIIB	VIII	VIII	VIII	IB	IIB	IIIA	IVA	VA	VIA	VIIA	VIIIA
H_3N																	He
Li_3N	Be_3N_2											BN	$(CN)_2$	N_2	ON O_4N_2	F_3N	Ne
Na_3N	Mg_3N_2											AlN	Si_3N_4	PN	SN	Cl_3N	Ar
K_3N	Ca_3N_2	ScN	Ti_3N TiN	V_2N VN	Cr_2N CrN	Mn_4N Mn_2N Mn_3N_2	Fe_4N Fe_2N	Co_3N Co_2N	Ni_3N	Cu_3N	Zn_3N_2	GaN	Ge_3N_4	AsN	SeN	Br_3N	Kr
Rb_3N	Sr_3N_2	YN	ZrN	Nb_2N Nb_4N_3 NbN NbN_{1+x}	Mo_2N MoN	Tc?	Ru	Rh	Pd	Ag_3N	Cd_3N_2	InN	Sn_3N_2 Sn_3N_4	SbN	TeN	I_3N	Xe
Cs_3N	Ba_3N_2	LaN*	Hf_3N_2 HfN	Ta_2N TaN Ta_3N_5	W_2N WN	Re_2N	Os	Ir	Pt	Au_3N	Hg_3N_2	TlN	Pb_3N_2 Pb_3N_4	BiN	Po?	At?	Rn
Fr?	Ra?	Ac?**	Rf?	Ha?	106?												

Lanthanide series IIIA *

CeN	PrN	NdN	Pm?	SmN	EuN	GdN	TbN	DyN	HoN	ErN	TmN	YbN	LuN

Actinide series IIIA **

ThN Th_3N_4	PaN_2	UN U_2N_3 UN_{2-x}	NpN	PuN	AmN	CmN	BkN	CfN?	Es?	Fm?	Md?	No?	Lr?

Legend:

① Saltlike nitrides
①a Compounds representing the transition from saltlike to metallic nitrides
② Metallic nitrides
③ Diamondlike nitrides
④ Gaseous or volatile nitrides
No formation of nitrides
? Nitride formation is possible but so far not proved

Metallic Nitrides. Properties of metallic nitrides are listed in Table 2. The nitrides of the transition metals of groups IVB–VIIB generally are termed metallic nitrides because of their metallic conductivity and luster and their general metallic behavior. They are characterized by a wide range of homogeneity, high hardness, high melting points, and good corrosion resistance. They are grouped with the carbides (qv), borides (see Boron compounds), and silicides (see Silicon and silicides) as refractory hard metals. They crystallize in highly symmetrical, metallike lattices; the small nitrogen atoms occupy the interstitial voids within the metallic host lattice forming interstitial alloys similar to the generally isotypic carbides. Metallic nitrides can be alloyed with other nitrides and carbides of the transition metals to give solid solutions. Complete solid solubility has been demonstrated for a great number of combinations (2). At high temperatures, all pseudobinary systems between cubic mononitrides and monocarbides of the IVB and VB metals show complete miscibility, with the exception of the pairs ZrN–VN and HfN–VN, ZrN–VC and HfN–VC, and ZrC–VN and HfC–VN. TaN has a cubic high temperature modification which is completely miscible with all other cubic monocarbides and mononitrides (3–4) (see Fig. 1).

Metallic nitrides are wetted and dissolved by many liquid metals and can be precipitated from metal baths. The stoichiometry is determined not by the valency of the metal, but by the number of interstitial voids per host atom. The metallic nitrides are very stable against water and all nonoxidizing acids except hydrofluoric acid. The thermodynamic stability decreases with increasing group number from the nitrides of the IVB metals. The nitrides of Mo and W can be prepared only by the action of nitrogen under high pressure or by reaction with atomic nitrogen or dissociating ammonia. The same is true for the nitrides of the iron group metals (5–6). No nitrides of the platinum-group noble metals are known.

Nonmetallic (Diamondlike) Nitrides. Some properties of nonmetallic nitrides are listed in Table 3. The nitrides of some elements of the IIIA and IVA groups, eg, BN, Si_3N_4, AlN, GaN, and InN, are characterized by predominantly covalent bonding. They are very stable chemically, have high degrees of hardness, eg, cubic BN, high melting points, and are nonconductive or semiconductive. The structural elements of diamondlike nitrides are tetrahedral, M_4N, which are structurally related to diamond. Although the most common graphitelike form of BN does not contain these structural elements, boron nitride is considered a diamondlike nitride because of the existence of a diamondlike form at high pressures and its chemical and physical behavior. Diamondlike nitrides have stoichiometric compositions with no homogeneity range and, as a rule, do not form solid solutions with each other. For the preparation and properties of hexagonal BN, see Boron compounds, refractory).

Silicon nitride can be heated in air up to 1450–1550°C. In nitrogen, inert gas, or reducing atmosphere, Si_3N_4 can be heated up to 1750°C. Above 1750°C, decomposition and sublimative evaporation become severe. When in the presence of carbon, however, Si_3N_4 stability depends on temperature and pressure. The equilibrium temperature for the reaction $Si_3N_4 + 3\,C \rightarrow 3\,SiC + 2\,N_2$ at normal pressure is 1700°C. Under reduced pressure of 130 mPa (10^{-3} mm Hg), the decomposition temperature is <1100°C and at 3.0 MPa (30 atm), the decomposition temperature is >1900°C (7).

Volatile Nitrides. The nitrogen compounds of the nonmetallic elements generally are not very stable; they decompose at elevated temperatures. Some are explosive and decompose upon shock. They form distinct molecules similar to organic compounds, and they behave like organic molecules at low temperatures: they are gaseous, liquid,

Table 2. Properties of Metallic Nitrides[a]

Nitride	Color	Structure	Lattice parameter, RT, nm	Density, g/cm³	Hardness[b]	Mp, °C	Heat conductivity, W/(m·K)	Coefficient of thermal expansion, × 10⁻⁶	Electrical resistivity, μΩ·cm	Transition temperature, K
TiN	golden yellow	fcc NaCl type	0.4246	5.43	HM 2000	2950	29.1	9.35	25	4.8
ZrN	pale yellow	fcc NaCl type	0.4577	7.3	HM 1520	2980	10.9	7.24	21	9
HfN	greenish yellow	fcc NaCl type	0.4518	14.0	HM 1640	3330	11.1	6.9	33	
VN	brown	fcc NaCl type	0.4139	6.10	HM 1500	2350	11.3	8.1	85	7.5
NbN	dark gray	fcc NaCl type	0.4388	8.47	HM 1400	2630 dec	3.8	10.1	78	15.2
ε-TaN	dark gray	hexagonal B 35	a: 0.5191 c: 0.2906	14.3	HM 1100	2950 dec	9.54		128	1.8
δ-TaN	yellowish gray	fcc NaCl type	0.4336	15.6	HM 3200	2950 dec				17.8
CrN	gray	fcc NaCl type	0.4150	6.14	HM 1090	1080 dec[c]	11.7		640	not superconductive
Mo₂N	gray	fcc	0.416	9.46	HM 1700	790 dec[d]		6.7		5.0
W₂N	gray	fcc	0.412	17.7		dec				
ThN	gray	fcc NaCl type	0.5159	11.9	HM 600	2820			20	
UN	dark gray	fcc NaCl type	0.4890	14.4	HK 580	2800	15.5	8.0	176	
PuN	dark gray	fcc NaCl type	0.4907	14.4		2550				

[a] Ref. 1.
[b] HM = microhardness; HK = Knoop hardness.
[c] At 0.1 MPa (1 bar).
[d] At 0.7 MPa (7 bar).

	TiN	ZrN	HfN	VN	NbN	TaN (cub)
ZrN	●					
HfN	●	●				
VN	●	○	○			
NbN	●	●	●	●		
TaN (cub)	●	●	●	●	●	
TaN (hex)	◐	◐	◐	◐	◐	◐

(a)

	TiC	ZrC	HfC	VC	NbC	TaC
TiN	●	●	●	●	●	●
ZrN	●	●	●	○	●	●
HfN	●	●	●	○	●	●
VN	●	○	○	●	●	●
NbN	●	●	●	●	●	●
TaN (cub)	●	●	●	●	●	●
TaN (hex)	◐	◐	◐	◐	◐	◐

(b)

Figure 1.   Solubilities in (**a**) the nitride–nitride and (**b**) the nitride–carbide systems. ● Completely miscible. ◐ Partially miscible in the cubic phase. ○ Not at all or very slightly miscible.

or easily volatile solids. $(SN)_x$ is polymeric and chemically stable with semimetallic properties. $(PNCl_2)_x$ has attracted some scientific interest as inorganic rubber (see Inorganic high polymers). None of the volatile nitrides has attained any substantial industrial application except ammonia (hydrogen nitride) and nitrogen oxide (oxygen nitride). Gaseous nitrogen fluorides are explosive; Cl_3N, a dark-yellow liquid, evaporates somewhat on heating and explodes. $I_3N.NH_3$ [15823-38-8] detonates at the slightest touch.

Preparation

Nitriding Metals or Metal Hydrides.   Metals or metal hydrides may be nitrided using nitrogen or ammonia. The use of pure metal powders or of pure metal hydride powders yield products that are nearly as pure as the prematerials.

The nitrides of groups IVB and VB elements form at ca 1200°C. The nitrides of magnesium and aluminum form at 800°C. Aluminum nitride is obtained by heating aluminum powder with ammonia or nitrogen at 800–1000°C and is formed as a white to grayish-blue powder. A grade of especially high purity results from the decomposition of $AlCl_3$–NH_3 vapor mixtures.

The nitrides of the alkaline earth metals form at 300–400°C and lithium nitride can form at room temperature. Raising the temperature shortens the reaction time

Table 3. Properties of Nonmetallic (Diamondlike) Nitrides

Nitride	Structure	Lattice parameter, RT, nm	Density, g/cm³	Hardness	Temperature stability, up to °C	Heat conductivity, W/(m·K)	Coefficient of thermal expansion, $\beta \times 10^{-6}$
BN	hex	a: 0.2504 c: 0.6661	2.3	like graphite	3000	15	7.51
	fcc Zn blende	a: 0.3615	3.4	approaching diamond			
AlN	hex wurtzite	a: 0.311 c: 0.4975	3.05	HM 1230[a]	2200	30	4.03
GaN	hex wurtzite	a: 0.319 c: 0.518	5.0		600[b]		
Si_3N_4	hex α	a: 0.7748 c: 0.5618	3.2	HM 3340[a]	1900	17	2.75
	hex β	a: 0.7608 c: 0.2911					

[a] HM = microhardness.
[b] *In vacuo.*

and promotes a more complete reaction. Nitrides of metals that do not react or react slowly with molecular nitrogen at normal pressure may require pressures up to 100 MPa (1000 bar) or more (5). Even these high pressures do not suffice for the preparation of thermodynamically unstable nitrides, eg, the nitrides of the iron group metals (Fe, Co, Ni), rhenium nitride Re_2N, and the higher nitrides of Mo and W (6). In these cases, nitriding in a stream of purified ammonia at 600–1000°C leads to the formation of the desired nitrides. Atomic nitrogen is an even more powerful nitriding agent and can be produced by ionizing molecular nitrogen by the action of electrical discharges (8).

Metal Oxides. A process based on the reaction of metal oxides with nitrogen or ammonia in the presence of carbon is economical and has possibilities for large-scale production because less expensive metal oxides can be used in place of metal powders. However, the products are not very pure as they usually contain oxygen and carbon. Removal of residual amounts of oxygen and carbon is difficult, especially in cases where carbon and oxygen atoms are taken into solid solution within the nitride lattice. This is usually the case with nitrides of the transition metals. Low reaction temperatures generally favor the formation of nitrides and high temperatures promote stable carbides as compared to nitrides and favor contamination with carbon.

Metal Compounds. Many nitrides, eg, BN, AlN, TiN, ZrN, HfN, CrN, Re_2N, Fe_2N, Fe_4N, and Cu_3N, may be prepared by the reaction of the corresponding metal halides with ammonia; an intermediate step in this method is the thermal decomposition of the ammonia–halide complex that is formed. Nitrides also may be obtained by the reaction of oxygen-containing compounds, eg, oxyhalides ($VOCl_3$, CrO_2Cl_2), ammonium–oxo complexes (NH_4VO_3, NH_4ReO_4), or oxides (GeO_2, B_2O_3, V_2O_3, and ferrous metal oxides), with ammonia. These nitrides, however, are not very pure and may contain residual oxygen and halogen. On nitriding pure carbides at 3–30 MPa (30–300 bar) and 1100–1700°C, the carbides of the IVB and VB metals are transformed into carbonitrides and free carbon, with the exception of TaC, which is very stable. The carbides of the VIA metals react to form $Cr_3(C,N)_2$ and $Mo(C,N)$; like TaC, WC is very stable under these conditions (9).

Precipitation from the Gas Phase. The van Arkel gas deposition process gives especially pure nitrides and nitride layers which, under certain conditions, may precipitate as single crystals; the nitrides include TiN, ZrN, HfN, VN, NbN, BN, and AlN. In this process, a gaseous reaction mixture consisting of a volatile metal halide, nitrogen, and hydrogen is conducted over a hot substrate, eg, tungsten wire; the metal halide decomposes and the resulting nitride deposits on the wire. Because the deposition temperature is from 1000–1500°C, this procedure is limited by the thermal stability of the nitride. Nevertheless, the formation of nitrides from gaseous halides and ammonia in the plasma torch is possible (see Plasma technology).

Other Methods of Preparation. The nitrides, Si_3N_4, Ge_3N_4, Zn_3N_2, Cd_3N_2, and Ni_3N, also may be produced by thermal decomposition of the corresponding metal amide or imide; Rb_3N and Cs_3N are obtained by decomposition of the azides. Nitrides low in nitrogen also can be synthesized from nitrides with a higher nitrogen content by decomposition in a vacuum or by reduction with hydrogen; for example: UN from U_2N_3, Co_3N from Co_2N, Ta_2N from TaN, etc. Nitrides and complex nitrides of ferrous metals and of high melting transition metals present in steels and superalloys can be isolated by electrolysis (10).

Silicon nitride occurs in two forms, α-Si_3N_4 and β-Si_3N_4. Pure Si_3N_4 is white, but the colors of commercial materials may be tan, gray, or black because of residual silicon or impurities. Si_3N_4 may be prepared by nitriding silicon powder at 1200–1400°C or, for extremely fine-grained Si_3N_4, by the reactions of $SiCl_4$ or SiH_4 with N_2 or NH_3.

Nitride-Containing Layers. Besides case hardening, the hardening and, thus, the increase in nitrogen content, that is achieved by nitriding special alloy steels is technologically significant in the heat treatment of high quality parts, eg, gears (see Metal surface treatments, case hardening).

Hardness properties are imparted by the resulting layers of needle-shaped precipitates of the nitrides and carbonitrides of iron, aluminum, chromium, molybdenum, etc. The nitriding steels (nitroalloy steels) that are developed especially for this process contain ca 0.4% C, 1% Al, 1.5% Cr, 0.2% Mo, 1% Ni, and trace amounts of other elements. In gas nitriding, the parts made from such steels are annealed in ammonia at ca 500°C for up to 100 h, whereby 0.1–1 mm thick layers form. The hardness of these layers exceeds that of the precipitation-hardened parts by ca 30%.

In nitriding or carbonitriding with condensed materials, molten cyanides are used at ca 570°C; this method produces fairly thick layers of nitrides or carbonitrides after ca one hour. Another process involves the use of an atmosphere containing activated nitrogen atoms or ions that are formed by the action of an electrical glow discharge; the temperature is ca 560°C and the time required for completion of the ion nitriding is 10–12 h (8). The advantage of all of these methods is the lack of distortion during surface hardening, unlike quench hardening, which usually results in at least small changes in dimensions and, at worst, in distortions.

Wear-resistant layers can be deposited on the surface of nearly every kind of material, eg, steel, cast iron, cemented carbides, etc, by a chemical vapor deposition (CVD) process (11–13) (see Film deposition techniques). Passing a stream of a mixture of gases containing, for example, $TiCl_4$ vapor, H_2, and N_2, over the surface of heated metallic or nonmetallic bodies, results in the deposition of a thin uniform layer of TiN. Optimum temperatures for this process are 900–1100°C. The layer is extremely hard and wear resistant and improves the cutting performance of carbide tips for machining steel and long chipping materials. Carbonitride Ti(C,N) can be deposited analogous to TiN by feeding N_2, $TiCl_4$, H_2, and CH_4 into the gas stream. Whereas the color of TiN is pure golden yellow, the color of $Ti(C_x N_{1-x})$ can be varied from red golden (ca $Ti(C_{0.1}N_{0.9})$) to a deep purple (ca $TiC_{0.4}N_{0.6}$). These layers can be polished to a very good finish with consequent excellent luster.

One disadvantage of CVD is the necessary high temperature for adherent and pore-free coatings. Deposition of TiN by a sputtering process obviates the use of high temperatures and results in substrate temperatures below 200°C; but, consequently, deposition rates generally are low (<1 μm/h). Ion plating is a process performing simultaneously the evaporation of titanium atoms, reaction with very dilute nitrogen gas in the gas phase, and deposition of these highly active atomic clusters onto a solid metallic surface; thus, higher deposition rates occur without a substrate temperature increase (14). Cemented carbides also can be treated by this process, eg, TiC–$Mo_2C(Mo)$–Ni or WC–TiC–TaC–Co. If the latter is high in TiC, it can be nitrided at 1150–1350°C in an autoclave. After ca 20 h at 4–8 MPa (40–80 bar) N_2, a 10 μm thick layer that is high in Ti(C,N) is produced (15).

Manufacture and Processing

Nitride Layers. Carbide tips coated with titanium nitride or titanium carbonitride usually are manufactured by the CVD process in equipment like that illustrated in Figure 2. Most of the large carbide producers, eg, Metallwerk Plansee, Austria; Sandvik AB, Sweden; Carboloy, Div. of General Electric, U.S.; Teledyne Firth Stirling, U.S.; Krupp Widia, FRG; etc, manufacture nitride-coated carbide tips. The portion of coated throw-away tips are as high as 50–70%; a substantial part of these tips are nitride coated. Teledyne Firth Stirling is manufacturing HfN-coated tips besides TiN- and Ti(C,N)-coated tips (16).

To obtain an adherent, uniform nitride layer, the lapped tips are positioned on grids made of heat-resistant wire inside the hot-wall reactor. The reactor is thoroughly

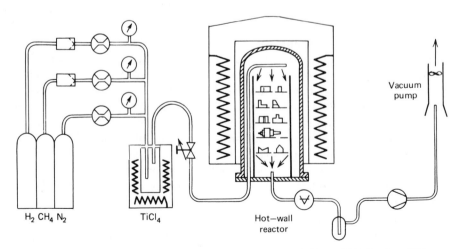

Figure 2.   Typical device for coating of cemented carbides with TiN or Ti(C,N) via a CVD-process. Courtesy of Berna-Bernex AG, Olten, Switzerland.

flushed with nitrogen and hydrogen, the temperature is brought to ca 950–1100°C by moving the preheated furnace over the reactor, and $TiCl_4$ vapor is fed to the gas stream. The pressure within the furnace usually is kept below atmospheric but it can be maintained at ambient pressure with equally good results. After completion of the coating cycle, the reactor is cooled by removing the furnace. The tips are removed after cooling and may be conditioned by sand blasting or tumbling to round the cutting edges.

Silicon Nitride. Silicon nitride is manufactured either as a powder as a prematerial for the production of hot-pressed parts or as self-bonded, reaction-sintered, silicon nitride parts. α-Silicon nitride, which is used in the manufacture of Si_3N_4 intended for hot pressing, can be obtained by nitriding Si powder in an atmosphere of N_2, H_2, and NH_3. Reaction conditions, eg, temperature, time, and atmosphere, have to be controlled closely. Special additions, eg, Fe_2O_3 to the prematerial, act as catalysts for the formation of predominantly α-Si_3N_4. Silicon nitride is ball-milled to a very fine powder and is purified by acid leaching. Silicon nitride can be hot pressed to full density by adding 1–5% MgO.

Self-bonded reaction-sintered Si_3N_4 is manufactured according to the flow sheet in Figure 3. Silicon powder is ball-milled to <63 μm, eventually is purified by acid leaching to remove abraded iron particles, and is conditioned by adding an organic agent as a plasticizer or deflocculant. The mixture can be cold pressed, isostatically pressed, or extruded. Slip casting or pressure slip casting has been applied successfully to obtain complex preforms.

The organic binder or the deflocculant is removed by heating the preforms and prenitriding them at 1000–1250°C. The prenitriding treatment imparts a strength level to the preforms that is sufficient to allow them to be handled, machined, milled, or drilled to the required tolerances with conventional tools. During the final nitriding stage at 1300–1500°C, all the silicon powder particles are converted to Si_3N_4. No shrinkage takes place either during the presintering process or during the final sintering stage.

Economic Aspects

Small amounts of TiN, HfN, and other metallic nitrides are produced on a pilot-plant scale. Titanium nitride is sold for $25–80/kg, depending on purity and grain size. Prices for HfN are ca $300/kg.

Annual production of BN in noncommunist nations is ca 80–100 metric tons per year and its cost is $50–70/kg. The price of cubic boron nitride is similar to that of synthetic diamond bort.

Annual production of aluminum nitride is 50–100 t and it is sold for ca $35/kg. Extra high purity, ie, high heat conductive aluminum nitride is sold for $50–70/kg.

Annual production of silicon nitride is ca 50–100 t. The reaction-sintered parts are sold for $120 to 300/kg, depending on complexity of shape. Hot-pressed, fully dense Si_3N_4 parts are priced 5–10 times higher than reaction-sintered parts.

Analytical and Test Methods

The physical properties of nitrides may be determined by procedures similar to those used for carbides (see Carbides). The same is true for chemical assay methods,

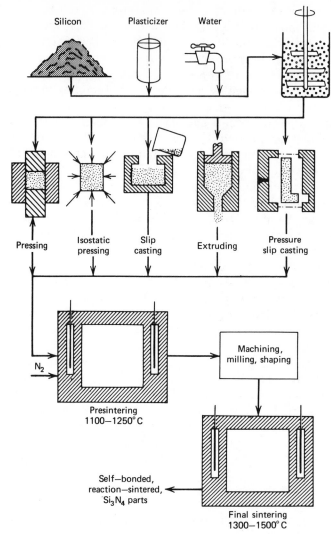

Figure 3. Flow sheet for the manufacture of self-bonded, reaction-sintered silicon nitride. Courtesy of Annawerk, Ceranox, Roedental, FRG.

although the stability of many nitrides to acids creates difficulties, especially in the Kjeldahl determination of nitrogen content. Consequently, modifications of the Dumas combustion method have been developed (17). Reproducible nitrogen values also are obtained by direct gas determination according to modified high temperature extraction processes (18). Nearly all of the nitrides, including the very stable ones, can be dissolved in polytetrafluoroethylene (PTFE) bombs with a mixture of hydrofluoric acid and perchloric acid at 190°C under pressure (19). The resulting solution contains all of the nitrogen in the form of NH_4^+ ions. Boron nitride and silicon nitride can be analyzed by a method that makes use of the reaction of nitrides with molten alkalies (20).

Health and Safety Factors, Toxicology

The saltlike nitrides decompose with water or moisture with resulting formation of ammonia, which can irritate the respiratory organs. The metallic nitrides are very stable chemically. Very fine powder or dust of the nitrides of the transition metals might be pyrophoric, especially the nitrides of the actinide metals, UN, ThN, and PuN, which ignite in air upon shock and during comminutive operations. The nitrides of the actinide metals are carcinogenic. The diamondlike nitrides, especially as dust, might irritate the lungs. Volatile nitrides and the nitrides of the IB and IIB metals have to be handled with extreme care because of their instability and high degree of toxicity. The toxic behavior of the nitrides of the IB and IIB metals is determined by their reaction products with moisture, eg, ammonia and caustic hydroxides (21).

Uses

Metallic Nitrides.   Recently, the classic liquid-phase sintering was successfully applied to nitrides and carbonitrides which normally have poor wetting properties. The oxygen content of the prematerial is kept below 0.05% and suitable binder alloys based on Ni–Mo and Ni–Mo–C are introduced (22–23). This binder composition is especially effective in wetting the nitride or carbonitride particles during sintering and in removing the adhering oxide layers. Cemented titanium carbonitrides with a nickel–molybdenum binder compare well with cemented carbides of this ISO-P series corresponding to C5 to C8 grades of the U.S. Industry Code for cemented carbides. TiN is scarcely soluble in solid iron and is much less so than TiC; hence, TiN has favorable frictional qualities and little tendency towards local welding and seizing during cutting operations. Sealing rings of cemented carbonitride have proved very satisfactory in difficult chemical environments.

Quaternary carbonitride alloys based on (Ti,Mo)(C,N) tend to separate into two isotypic phases; one, eg, Ti(C,N), is rich in nitrogen and poor in molybdenum and the other, eg, (Ti,Mo)C, contains nearly all the molybdenum but little nitrogen. This spinodal decomposition reaction has been applied to the production of cemented carbonitride alloys, ie, spinodal alloys, that show superior cutting performance and very little wear, especially when used to machine cast iron and chilled cast iron (24). More recent work has been carried out with regard to TaN(HfN)–TaC(HfC) alloys and alloys containing the cubic TaN (25–26). Hot-pressed cemented hard metals based on TaN–ZrB$_2$ are far more suitable for the machining of high melting metals and superalloys than cemented carbides are (27).

The nitride coating of cemented carbide tips, ie, throw-away tips, has increased greatly. Layers of TiN, Ti(C,N), and HfN, that are 3–6 μm thick, are applied using CVD (11–13). HfN–containing layers also are produced. In a U.S. process, HfN coatings are manufactured by treating Hf sponge with elementary chlorine, thus introducing HfCl$_4$ *in situ* (16).

An important European development features multilayers on carbide tips. The carbide surface first is coated with TiC, then with Ti(C,N), and finally by TiN layers; thus, the excellent flank wear of TiC coatings is combined with the excellent crater wear of TiN coatings (28). The golden-colored TiN layers are used in the jewelry industry, mainly as the coating of scratch-free watch cases. In Japan, Ti(C,N,O) layers with up to 1.60% O$_2$ are used instead of carbide or nitride or carbonitride layers (29).

Within the system Mo–C–N, a molybdenum–carbonitride phase Mo(C,N) with the WC-type structure can be prepared (9). This phase is completely miscible with WC. Solid solutions (Mo,W)(C,N) between Mo(C,N) and WC behave similarly to WC when sintered with a cobalt or nickel binder and give cemented carbonitrides with properties that are similar to those based on WC–Co. In these cemented carbonitrides, a substantial part of tungsten can be replaced by molybdenum, with two parts of tungsten equivalent to one part of molybdenum. In situations of tungsten shortage, eg, in the mid-1970s, this replacement is expected to gain great technical and economical significance (30).

Transition-metal nitrides improve hardness properties. Nitrogen-alloyed chromium steels, as compared to nitrogen-free austenitic steels, are characterized by increased strength without loss of toughness (31). The corrosion resistance does not decrease if the chromium content is increased to compensate for the amount of chromium that is bound by nitrogen. The industrial preparation of these steels takes place in pressure-melt equipment at ca 2.0 MPa (20 atm) nitrogen, whereby as much as 1.8% N may go into solution. As an alloying constituent, nitrogen can be applied not only by melting under pressure but by employing highly nitrided ferrochromium. This is obtained by treating 50–70% ferrochromium with nitrogen at 950–1150°C in calcium cyanamide pusher furnaces, whereby 3.5–4.5% N is taken up. Research on the ternary systems with IVB, VB and VIB metals as bases for future nitride applications has been reported (22,32). Cubic NbN and Nb(C,N) are superconductors with high transition temperatures (NbN 17.3 K, Nb (C,N) 17.8 K). There have not been technical applications because of the extreme brittleness of these compounds (see Superconducting materials).

The Actinide Nitrides: UN, PuN, and ThN. Use of the nitrides of uranium-235 and thorium as fuels and breeders in high temperature reactors has been proposed (see Nuclear reactors). However, the compounds most frequently used for this purpose are the oxides and carbides. Nitrides could be useful in high temperature breeder reactors because of their stability when in contact with molybdenum as a cladding material. Only those mononitrides that are prepared by powder metallurgy (qv), eg, by hot pressing into sintered parts, and those that are high melting, chemically stable, and metallic as well as the solid solutions of such nitrides with monocarbides (possibly in the form of dispersions in stainless steels), may be suitable for use as nuclear-fuel materials (33). The properties of interest for use in reactors were studied in detail and are included among the data shown in Table 2. The higher nitrides of uranium, ie, β-U_2N_3 and α-UN_{2-x}, and of thorium, ie, Th_3N_4, which might be formed during the reaction of the metals with ammonia or pressurized nitrogen, can be decomposed easily by treating the compounds in vacuum at high temperatures.

The rare-earth nitrides do not have any technical applications; they are high melting compounds but are hydrolyzed easily by moisture and are not stable under normal atmospheric conditions.

Saltlike Nitrides. Of the saltlike nitrides, only Li_3N has attracted technical interest. Lithium nitride has an uncommonly high ionic conductivity in the solid state and, hence, is considered a suitable electrolyte for lithium–sulfur batteries, conferring a favorable capacity as compared with conventional batteries (see Batteries and electric cells). Lithium nitride is used as a solvent catalyst in the production of cubic BN.

Nonmetallic (Diamondlike) Nitrides. *Boron Nitride.* Hexagonal boron nitride is a soft white powder and resembles graphite in crystal structure, in texture, and in many other properties, except that it is an electrical insulator. It is used in the refractories industry as a mold-facing and release agent. Structural parts made of BN are manufactured by hot pressing. Their characteristics include low density, easy workability, good heat resistance with especially good thermal conductivity and stability to thermal shock, excellent corrosion resistance and the ability to provide electrical insulation. Boron nitride-based structural parts are used as wall liners in plasma-arc devices, eg, gas heaters, arc-jet thrusters, and high temperature magnetohydrodynamic devices (34) (see Coal conversion processes, magnetohydrodynamics). BN also is used as a crucible material for reactive metal melts because of its nonwetting properties. Composites, eg, $BN-Si_3N_4$ and BN–SiC are very stable to molten zinc and to covering salts, eg, borax. As an additive, BN adjusts the electrical conductivity of sintered TiB-TiC boats that are used in vacuum evaporation of aluminum. Boron nitride also is used as a solid dopant source for semiconductors.

Cubic Boron Nitride (Borazon). The graphitic form of BN can be converted into a high pressure diamondlike form by applying pressures of 5–9 GPa (50–90 kbar) and temperatures of 1500–2000°C in the presence of catalysts, eg, alkali metals, alkaline earth metals, antimony, tin, and lead. Lithium nitride seems to be the best catalyst. Pure cubic BN is colorless, but the technical product always contains either excess boron and is brown or black or, if it contains excess lithium nitride, the product is yellow. The cubic boron nitride is stable in air at atmospheric pressure to 2000°C. Cubic boron nitride approaches the hardness of diamond and has been introduced successfully as an abrasive for steel, especially high alloy steel and superalloys (see Abrasives). Cubic BN is not used for grinding cemented carbides. During the last ten years the commercial use of cubic BN grit for grinding wheels has grown steadily. The material competes with diamond bort (35) especially for grinding operations involving high alloy steels.

Aluminum Nitride. Aluminum nitride is characterized by covalent bonding; thus, it is a semiconductor with a relatively wide band gap. Structural parts made from it exhibit high degrees of hardness (see also Fig. 1). Structural parts made of AlN can be prepared in the same way as ceramics, ie, by powder metallurgy from nitride powders or by sintering mixtures of Al_2O_3 and C in the presence of nitrogen. These parts are stable to oxidizing and carbon-containing furnace gases up to 800°C. Above 800°C, they are stable only under a protective gas or in a vacuum. Because of their good chemical and heat stability and their stability to changes in temperature, sintered parts made of AlN have been proposed for use in nozzles, thermocouple-protecting tubes, crucibles, and boats, eg, in the manufacture of semiconductors (36). However, AlN is hydrolyzed slowly by water vapor and moisture.

Gallium Nitride and Indium Nitride. The mononitrides of the group IIIA metals Ga and In have semiconductor and electroluminescence properties (see also Table 3). They can be prepared by solid-state reactions of Ga_2O_3 or In_2O_3 with ammonia. The utilization of the luminescence properties of GaN and InN is, however, limited by the thermal decomposition of these compounds which occurs above 600°C.

Silicon Nitride. Sintered parts of Si_3N_4 may be manufactured from Si_3N_4 powder, or from silicon powder by means of cold pressing or slip casting with subsequent sintering in the presence of nitrogen. Silicon preforms that have been presintered under nitrogen at 1200°C have sufficient strength to be machined with conventional carbide

Table 4. Alphabetical List of Nitrides Referred to in Text

Compound	CAS Registry Number
aluminum nitride	[24304-00-5]
americium nitride	[12296-96-1]
ammonia	[7664-41-7]
antimony	[12333-57-2]
arsenic nitride	[26754-98-3]
barium nitride	[12047-79-9]
berkelium nitride	[56509-31-0]
beryllium nitride	[1304-54-7]
bismuth nitride	[12232-97-2]
boron nitride	[10043-11-5]
bromine nitride	[15162-90-0]
cadmium nitride	[12380-95-9]
californium nitride	[70420-93-8]
calcium nitride	[12013-82-0]
carbon nitride	[12069-92-0]
cerium nitride	[25764-08-3]
cesium nitride	[12134-29-1]
chlorine nitride	[10025-85-1]
chromium nitride	[24094-93-7]
chromium nitride (2:1)	[12053-27-9]
cobalt nitride (2:1)	[12259-10-8]
cobalt nitride (3:1)	[12432-98-3]
copper nitride	[1308-80-1]
curium nitride	[56509-28-5]
cyanogen	[2074-87-5]
dinitrogen tetroxide	[10544-72-6]
dysprosium nitride	[12019-88-4]
erbium nitride	[12020-21-2]
europium nitride	[12020-58-5]
fluorine nitride	[13967-06-1]
gadolinium nitride	[25764-15-2]
gallium nitride	[25617-97-4]
germanium nitride	[12065-36-0]
gold nitride	[13783-74-9]
hafnium nitride (1:1)	[25817-87-2]
hafnium nitride (3:2)	[12508-69-9]
holmium nitride	[12029-81-1]
indium nitride	[25617-98-5]
iodine nitride	[21297-03-1]
iron nitride (2:1)	[12023-20-0]
iron nitride (4:1)	[12023-64-2]
lanthanum nitride	[25764-10-7]
lead nitride (3:2)	[58572-21-7]
lead nitride (3:4)	[75790-62-4]
lithium nitride	[26134-62-3]
lutetium nitride	[12125-25-6]
magnesium nitride	[12057-71-5]
manganese nitride (2:1)	[12163-53-0]
manganese nitride (3:2)	[12033-03-3]
manganese nitride (4:1)	[12033-07-7]
mercury nitride	[12136-15-1]

Table 4 (*continued*)

Compound	CAS Registry Number
molybdenum nitride	[12033-19-1]
molybdenum nitride (2:1)	[12033-31-7]
neodymium nitride	[25764-11-8]
neptunium nitride	[12058-90-1]
nickel nitride	[12033-45-3]
niobium nitride (2:1)	[12033-63-5]
niobium nitride (4:3)	[12163-98-3]
niobium nitride	[11092-17-4]
nitrogen	[7727-37-9]
nitrous oxide	[10024-97-2]
phosphorus nitride	[17739-47-8]
plutonium nitride	[12033-54-4]
potassium nitride	[29285-24-3]
praseodymium nitride	[25764-09-4]
protactinium nitride	[75733-54-9]
rhenium nitride	[12033-55-5]
rubidium nitride	[12136-85-5]
samarium nitride	[25764-14-1]
scandium nitride	[25764-12-9]
selenium nitride	[12033-59-9]
silver nitride	[20737-02-4]
α-silicon nitride ⎫ β-silicon nitride ⎭	[12033-89-5]
sodium nitride	[12136-83-3]
strontium nitride	[12033-82-8]
sulfur nitride	[28950-34-7]
tantalum nitride	[12033-62-4]
tantalum nitride (2:1)	[12033-63-5]
tantalum nitride (3:5)	[12033-94-2]
tellurium nitride	[59641-84-8]
terbium nitride	[12033-64-6]
thallium nitride	[12033-67-9]
thorium nitride	[12033-65-7]
thorium nitride (3:4)	[12033-90-8]
thulium nitride	[12033-68-0]
tin nitride (3:2)	[75790-61-3]
tin nitride (3:4)	[75790-62-4]
titanium nitride (1:1)	[25583-20-4]
titanium nitride (2:1)	[12169-08-3]
tungsten nitride	[12058-38-7]
tungsten nitride (2:1)	[12033-72-6]
uranium nitride	[25658-43-9]
uranium nitride (2:3)	[12033-85-1]
vanadium nitride	[24646-85-3]
vanadium nitride (2:1)	[12209-81-3]
ytterbium nitride	[24600-77-9]
yttrium nitride	[25764-13-0]
zinc nitride	[1313-49-1]
zirconium nitride	[25658-42-8]

tools to close dimensional tolerances. Thus, even complicated forms, eg, holes, grooves, threads, etc, can be prepared. During final sintering at 1200–1400°C, no change in dimensions occurs but hardness is increased to a level where only grinding with diamond wheels is effective. Parts with good strength may be obtained; however, residual porosities are 15–30%. Such parts are characterized by low thermal expansion, good thermal shock resistance, resistance to creep, and high electrical resistivity. They also exhibit good resistance to corrosion by acids but are less resistant toward hydrofluoric acid and alkali–hydroxide solutions. They also are stable toward reactive gases and nonferrous melts (37–38).

Sintered parts of Si_3N_4 can be made into tubes, crucibles, boats, nozzles, etc. The tubes can be used for protecting thermocouples or for measuring the temperature in nonferrous melts or gas temperatures in the steel industry. Other parts are suitable as linings of cooling and purification towers for SO_2 roasting gas, cyclone dust separators, spray nozzles, and combustion nozzles, or they may serve as parts in pumps for melting light metals and in linings for Wankel engines. An extensive research effort is being made to develop reaction-sintered or hot-pressed silicon nitride turbine blades as a high temperature material for use in automotive engines (39).

Silicon nitride is one of the few nonmetallic nitrides that is able to form alloys with other refractory compounds. Numerous solid solutions of β-Si_3N_4 with Al_2O_3, $BeSiO_4$, and other oxides have been identified and are referred to as nitrogen ceramics (40–42). One of these, Sialon, which is the solid solution between Si_3N_4 and Al_2O_3, has gained technical interest. During the last 10 yr, many companies have begun to mass-produce reaction-sintered and hot-pressed Si_3N_4 parts.

Table 4 is a list of nitrides mentioned in text.

BIBLIOGRAPHY

"Nitrides" in *ECT* 1st ed., Vol. 9, pp. 345–352, by L. S. Foster, Watertown Arsenal; "Nitrides" in *ECT* 2nd ed., Vol. 13, pp. 814–825, by Friedrich Benesovsky, Metallwerk Plansee A.G., Reutte, and Richard Kieffer, Technische Hochschule, Vienna.

1. R. Kieffer and P. Ettmayer, *High Temp. High Pressures* **6**, 253 (1974).
2. P. Duwez and F. Odell, *J. Electrochem. Soc.* **97**, 299 (1950).
3. R. Kieffer, H. Nowotny, P. Ettmayer, and G. Dufek, *Metall.* **26**, 701 (1972).
4. F. Gatterer, G. Dufek, P. Ettmayer, and R. Kieffer, *Mh. Chem.* **106**, 1137 (1975).
5. P. Ettmayer, H. Priemer, and R. Kieffer, *Metall.* **23**, 307 (1969).
6. D. J. Jack and K. H. Jack, *Mat. Sci. Eng.* **11**, 1 (1973).
7. H. Rassaerts and A. Schmidt, *Planseeber. Pulvermetall.* **14**, 110 (1966).
8. J. Kläusler, *Fachber. Oberflächentech.* **6**, 201 (1968).
9. R. Kieffer, H. Nowotny, P. Ettmayer, and M. Freudhofmeier, *Metall.* **25**, 1335 (1971).
10. W. Koch, *Metallkundliche Analyse*, Stahleisen Düsseldorf, 1965.
11. R. Kieffer, D. Fister, H. Schoof, and K. Mauer, *Powder Metall. Int.* **4**, 1 (1973).
12. U.S. Pat. 3,717,496 (1970), (to Deutsche Edelstahlwerke Krefeld FRG).
13. W. Schintlmeister, O. Pacher, K. Pfaffinger, and T. Raine, *J. Electrochem. Soc.* **123**, 924 (1976).
14. H. K. Pulker, R. Buhl, and E. Moll, "Ion Plating," *paper presented at the 9th Plansee Seminar*, 1977.
15. O. Rüdiger, H. Grewe, and J. Kolaska, *paper no. 31 presented at 9th Plansee-Seminar 1977; Wear* **48**, 267 (1978).
16. E. Rudy, B. F. Kieffer, and E. Baroch, *Planseeber. Pulvermetall.* **26**, 105 (1978).
17. J. Rottmann and H. Nickel, *Fresenius Anal. Chem.* **247**, 208 (1969).
18. G. Paesold, K. Müller, and R. Kieffer, *Fresenius 2. Anal. Chem.* **232**, 31 (1967).
19. W. Werner and G. Tölg, *Fresenius 2. Anal. Chem.* **276**, 103 (1975).

20. H. Puxbaum and A. Vendl, *Fresenius 2. Anal. Chem.* **287**, 134 (1977).
21. N. I. Sax, *Dangerous Properties of Industrial Materials*, Van Nostrand Reinhold Co., New York, 1975.
22. R. Kieffer and P. Ettmayer, *High Temp. High Pressures* **6**, 253 (1974).
23. R. Kieffer, P. Ettmayer, and M. Freudhofmeier, *Metall.* **25**, 1335 (1971); R. Kieffer, P. Ettmayer, and M. Freudhofmeier in H. H. Hausner, ed., *Modern Developments in Powder Metallurgy*, Vol. 5, Plenum Press, New York, 1971, pp. 201–214.
24. E. Rudy, *J. Less Common Met.* **33**, 43 (1973).
25. R. Kieffer, G. Dufek, P. Ettmayer, and R. Ducreux, *papers 5-7, presented at the IV European Symposium for Powder Metallurgy*, Grenoble, Fr., 1975.
26. M. Komac, T. Kosmac, and F. Thümmler, *Planseeber. Pulvermetall.* **25**, 101 (1977).
27. V. Murata and E. D. Whitney, *Am. Ceram. Soc. Bull.* **46**, 643 (1967); *Am. Ceram. Soc. Bull.* **48**, 698 (1969); *Am. Ceram. Soc. Bull.* **47**, 617 (1968).
28. W. Schintlmeister and O. Pacher, *Metall.* **28**, 690 (1974); *Planseeber. Pulvermetall.* **23**, 260 (1975).
29. T. Sadahiro, S. Yamaya, K. Shibuki, and N. Ujiie, *Wear* **48**, 291 (1978).
30. R. Kieffer, P. Ettmayer, and B. Lux, *paper 33 presented at the Recent Advances in Hardmetal Production Conference*, Loughborough, UK, 1979.
31. J. Frehser and C. Kubisch, *Berg-u. Huettenmaenn. Monatsh.* **108**, 369 (1963).
32. A. Vendl, *Planseeber. Pulvermetall.* **26**, 233 (1968).
33. S. J. Paprocki and co-workers, *Rep. Battelle Mem. Inst. BMI*, 1365 (1959).
34. J. Frederickson and W. H. Redanz, *Met. Prog.* **87**(2), 97 (1965).
35. L. Coes, Jr. "Abrasives" in *Applied Mineralogy*, Vol. 1, Springer, Vienna, 1971.
36. G. Long and L. M. Forster, *J. Electrochem. Soc.* **109**, 1176 (1962).
37. J. F. Collins and R. W. Gerby, *J. Met.* **7**, 612 (1955).
38. A. M. Sage and J. H. Histed, *Powder Met.* **8**, 196 (1961).
39. E. Gugel and G. Leimer, *Ber. Dtsch. Keram. Ges.* **50**, 151 (1973).
40. K. H. Jack and W. J. Wilson, *Nature Phys. Sci.* **283**, 28 (1972).
41. W. Wruss, R. Kieffer, E. Gugel, and B. Willer, *Sprechsaal* **108**, (1975).
42. L. J. Gauckler, H. L. Lukas, and G. Petzow, *J. Am. Ceram. Soc.* **58**, 346 (1975).

General References

R. Kieffer and F. Benesovsky, *Hartstoffe*, Springer, Vienna, 1963.
P. Schwarzkopf and R. Kieffer, *Refractory Hard Materials*, Macmillan, New York, 1953.
H. Goldschmidt, *Interstitial Alloys*, Butterworths, London, 1967.
L. E. Toth, *Transition Metal Carbides and Nitrides*, Academic Press, New York, 1971.
G. V. Samsonov, *Nitridij*, Naukova Dumka, Kiev, USSR, 1969.
E. K. Storms, *A Critical Review of Refractory Nitrides*, U.S. At. Energy Comm., LAMS-2674, Part II, 1962.

F. Benesovsky
Metallwerk Plansee, Reutte, Austria

R. Kieffer
P. Ettmayer
Technical University, Vienna

NITRIDING.   See Metal-surface treatments.

NITRILE RUBBER.   See Elastomers, synthetic.

NITRILES

Nitriles are cyano-substituted organic compounds, ie, organic derivatives of hydrogen cyanide. Since the cyano group can be readily hydrolyzed to a carboxylic acid, nitriles commonly are named as the derivatives of the corresponding carboxylic acid, eg, cyanomethane (methyl cyanide) is named acetonitrile, and 1,4-dicyanobutane is referred to as adiponitrile [111-69-3] since it can be hydrolyzed to adipic acid. However, in many cases, the names of nitriles that include two or more functional groups are more clearly designated as cyano-substituted compounds, eg, cyanoacetic acid or cyanoacetamide. Nitriles that are produced by the reaction of hydrogen cyanide with a ketone to form a compound containing a cyano and a hydroxy group on the same carbon or with an epoxide to form a compound with cyano and hydroxy groups on adjacent carbons are commonly named cyanohydrins, eg, acetone cyanohydrin (2-hydroxy-2-methylpropionitrile) from acetone, and ethylene cyanohydrin (3-hydroxypropionitrile) from ethylene oxide. According to *Chemical Abstracts*, aliphatic nitriles are named as derivatives of the longest carbon chain and the carbon of the nitrile is included.

One of the most commercially important nitriles is acrylonitrile (qv). It is produced worldwide because of its high reactivity and the desirable properties of the products made from it. Acetonitrile, a by-product of acrylonitrile manufacture, has achieved commercial importance in an increasing number of uses, eg, solvent extraction, reaction media, and an intermediate in the preparation of pharmaceuticals (qv) and other organic chemicals (see Extraction, liquid–liquid extraction). Propionitrile [107-12-0], which is a by-product of the electrodimerization of acrylonitrile to adiponitrile, is used as a chemical intermediate. Thousands of metric tons of adiponitrile are manufactured worldwide each year. Adiponitrile is used almost exclusively by the manufacturers and, therefore, is not sold widely and is not considered a commodity chemical. Other nitriles that are produced in thousands of metric tons per year are acetone cyanohydrin and fatty acid nitriles. Acetone cyanohydrin is an intermediate for the preparation of methyl methacrylate, which is a monomer used for acrylic resins, eg, Lucite and Plexiglas; and azobis(isobutyronitrile), which is a widely used polymerization initiator, eg, Vazo 64. The fatty acid nitriles are intermediates in the production of a large variety of commercial amines and amides.

The nitriles that are commercially available (1980) and their manufacturers and approximate prices are listed in Table 1. Most of the nitrile manufacturers outside the United States are listed in Table 2.

Reactions

Nitriles are exceedingly versatile reactants that can be used to prepare amines, amides, amidines, carboxylic acids and esters, aldehydes, ketones, large-ring cyclic ketones, imines, heterocycles, orthoesters, and other compounds. Amides and acids are produced by hydrolysis under either acidic or basic conditions:

$$RCN + H_2O \rightarrow RCONH_2$$

$$RCN + 2\,H_2O \rightarrow RCOOH + NH_3$$

If the water is replaced with alcohols, either esters or substituted amides are obtained. Anhydrides, carboxylic acids, and olefins also react with nitriles to produce substituted amides (1).

Table 1. Commercially Available Nitriles

Nitriles	CAS Registry No.	U.S. Manufacturers (1980)	Prices (1980)[a], $/kg
acetone cyanohydrin	[75-86-5]	DuPont, Monsanto	0.70
acetonitrile	[75-05-8]	DuPont, Vistron	0.80
acrylonitrile	[107-13-1]	American Cyanamid, DuPont, Monsanto, Vistron	0.77
2,2'-azobis(isobutyronitrile)	[78-67-1]	DuPont, Eastman	6.49
2,2'-azobis(2,4-dimethylvalero-nitrile)	[4419-11-8]	DuPont	16.60
benzonitrile	[100-47-0]	Sherwin-Williams	1.90
cyanoacetic acid	[372-09-8]	Kay-Fries	5.25
ethyl cyanoacetate	[105-56-6]	Kay-Fries	5.50
ethylene cyanohydrin	[109-78-4]	Thiokol	6.67
fatty acid nitriles		Armak, Tomah	
isophthalonitrile	[6260-60-5]	Sherwin-Williams	3.41
2-methyl-2-butenenitrile	[4403-61-6]	DuPont	0.33
2-methyl-3-butenenitrile	[16529-56-9]	DuPont	1.10
methyl cyanoacetate	[105-34-0]	Kay-Fries	6.50
2-methylglutaronitrile	[4553-62-2]	DuPont	0.55
2-pentenenitrile	[13284-42-9]	DuPont	0.55
3-pentenenitrile	[4635-87-4]	DuPont	1.32

[a] Bulk prices listed.

The reaction of nitriles with alcoholic ammonium chloride leads to substituted amidines,

$$RCN + NH_4Cl \longrightarrow R\overset{\overset{\displaystyle NH}{\|}}{C}NH_2 \cdot HCl$$

and addition of hydrogen sulfide produces thioamides:

$$RCN + H_2S \longrightarrow R\overset{\overset{\displaystyle S}{\|}}{C}NH_2$$

Amidoximes result from the reaction of nitriles with hydroxylamine:

$$RCN + NH_2OH \longrightarrow R\overset{\overset{\displaystyle NOH}{\|}}{C}NH_2$$

Catalytic reduction with hydrogen leads to amines (see Amines by reduction). The proper conditions and catalyst can control the reaction to yield predominantly primary, secondary, or tertiary amines:

$$RCN \xrightarrow[\text{catalyst}]{H_2} RCH_2NH_2, (RCH_2)_2NH, \text{ or } (RCH_2)_3N$$

Raney cobalt is an excellent catalyst for this reduction.

Table 2. Non-U.S. Nitrile Producers

Nitriles	CAS Registry No.	Country	Producer
acetone cyanohydrin		Austria	Meta-Chemie
		France	Societe Norsolor
		FRG	Degussa
		Italy	Montedison
		The Netherlands	Roha
		Spain	Energia e Industris
			PAULAR
		UK	ICI
acetonitrile		Austria	Chemie Linz
		France	Rhone-Poulenc
			Societe Norsolor
		FRG	Degussa
			Erdolchemie
			SKW Trostberg
		Italy	Anic
			Montedison
			Rumianca
		Japan	Asahi
			Mitsubishi
			Nitto
acrylonitrile		Austria	Chemie Linz
		Brazil	ACRINOR
		Bulgaria	Burgas[a]
			Pleven[a]
		People's Republic of China	Fangshan[a]
			Jinshan[a]
			Lanzhou[a]
		GDR	VEB[b]
		FRG	Degussa
			Erdolchemie
			SKW Trostberg
		India	Indian Petrochemicals
		Italy	Anic
			Montedison
			Rumianca
		Japan	Asahi
			Mitsubishi
			Nitto
			Showa Denka
			Sumitomo
			Toyo Chemies
		People's Republic of Korea	Pyong Yang[a]
		Republic of Korea	Tong Suh Petrochemical
		The Netherlands	DSM
		Poland	Tarnow[a]
		Romania	Pitesti[a]
			Savinesti[a]
		Spain	PAULAR
		Republic of China (Taiwan)	China Petrochemical
		Turkey	Petkin Petrokimyo
		UK	Borden
			Monsanto

890

Table 2 (*continued*)

Nitriles	CAS Registry No.	Country	Producer
		USSR	Novopolotsk[a]
			Saratov[a]
2,2-azobis(isobutyronitrile)		France	Société Française d'Organo-syntheses
		FRG	Bayer
		The Netherlands	Akzo Chemie
		UK	Fisons
2,2-azobis(2,4-dimethylvaler-onitrile)		The Netherlands	Akzo Chemie
benzonitrile		FRG	BASF
			SKW Trostberg
		UK	Hopkins & Williams
			Koch-Light Labs
cyanoacetic acid		FRG	Boehringer Ingelheim
			Knoll
		Switzerland	Lonza
		UK	May & Baker
decanonitrile	[1975-78-6]	Switzerland	Suchema
ethyl cyanoacetate		France	Rhone-Poulenc
		FRG	Bayer
			Knoll
		Switzerland	Lonza
		UK	CIBA-GEIGY
ethylene cyanohydrin		UK	Koch-Light Labs
malononitrile	[109-77-3]	Switzerland	Lonza
		UK	Koch-Light Labs
2-methyl-2-butenenitrile		France	Butachemie
methyl cyanoacetate		France	Rhone-Poulenc
		FRG	Bayer
			Boehringer Ingelheim
		Switzerland	Lonza
		UK	Koch-Light Labs
2-methylglutaronitrile		France	Butachemie
2-pentenenitrile		France	Butachemie

[a] State complex.
[b] Government owned.

Iminoethers are produced by the Pinner Synthesis.

$$RCN + HCl + R'OH \longrightarrow R\underset{\underset{O}{\|}}{\overset{\overset{NH \cdot HCl}{\|}}{C}}CR'$$

The reaction of the iminoether with additional alcohol leads to the production of orthoesters (2):

$$RC(=NH.HCl)COR' + 2 R'OH \rightarrow RC(OR')_3 + NH_4Cl$$

Partial reduction of nitriles with stannous chloride to an aldimine followed by hydrolysis leads to aldehydes (3):

$$\text{RCN} \xrightarrow{\text{H}_2,\ \text{SnCl}_2} (\text{RCH}=\text{NH.HCl})_2\text{SnCl}_4 \xrightarrow{\text{H}_2\text{O}} \text{RCHO}$$

Ketones (qv) are formed from the reaction of nitriles with phenols (Hoesch Synthesis), by the reaction of a Grignard reagent followed by hydrolysis, and by reaction with α-bromoesters and zinc (Blaise reaction) (4):

$$\text{RCN} + \text{C}_6\text{H}_4(\text{OH})_2 \xrightarrow{\text{HCl}} \xrightarrow{\text{H}_2\text{O}} \text{RCOC}_6\text{H}_3(\text{OH})_2$$

$$\text{RCN} + \text{R}'\text{MgBr} \xrightarrow{\text{H}_2\text{O}} \text{RCOR}'$$

$$\text{RCN} + (\text{CH}_3)_2\text{C(Br)COOCH}_3 \xrightarrow{\text{Zn}} \xrightarrow{\text{H}_2\text{O}} \text{RCOC(CH}_3)_2\text{COOCH}_3$$

Cyclic ketones are produced by the condensation, hydrolysis, and decarboxylation of dinitriles (5). The number of carbon atoms in the product is one less than that in the starting dinitrile. The smallest ring that can be produced is cyclopentanone. A seventeen-membered ring has been prepared by a high dilution technique (6).

The activating effect of the nitrile group allows base-catalyzed substitutions of hydrogens on the carbon that is adjacent to the nitrile, eg, nitriles may be alkylated with alkyl iodides in the presence of sodium or sodamide:

$$\text{RCH}_2\text{CN} + \text{R}'\text{I} \xrightarrow{\text{Na}} \text{RR}'\text{CHCN} + \text{NaI}$$

Nitriles react with aromatic aldehydes to produce cyanoolefins (Knoevenagel condensation),

$$\text{RCH}_2\text{CN} + \text{C}_6\text{H}_5\text{CHO} \xrightarrow{\text{NaOC}_2\text{H}_5} \text{C}_6\text{H}_5\text{C} = \text{CRCN}$$

and to produce salts of oximinonitriles by nitrosation with amyl nitrite:

$$\text{RCH}_2\text{CN}\ +\ 2\,\text{C}_5\text{H}_{11}\text{ONO} \xrightarrow{\text{KOCH}_2\text{CH}_3} \overset{\overset{\text{NOK}}{\|}}{\text{RCCN}}$$

A few of the nitrogen heterocycles that can be prepared from nitriles include triazines by trimerization or by reaction with dicyandiamide, and oxazoles from cyanohydrins and aldehydes (Fischer oxazole synthesis) (7–8).

Preparation

The first nitrile to be prepared was propionitrile, which was obtained in 1834 by distilling barium ethyl sulfate with potassium cyanide. This is a general preparation of nitriles from sulfonate salts and is referred to as the Pelouze reaction (9). The reaction of alkyl halides with sodium cyanide to produce nitriles also is a general reaction with wide applicability:

$$\text{RX} + \text{NaCN} \rightarrow \text{RCN} + \text{NaX}$$

where X = Cl, Br, or I. If dimethyl sulfoxide is used as solvent, high yields of nitriles can be obtained with both primary and secondary alkyl chlorides (see Sulfoxides). This method also may be used for preparing dinitriles. Reaction times usually are less than one hour (10).

The first commercial method for preparing nitriles was the dehydration of an amide:

$$RCONH_2 \rightarrow RCN + H_2O$$

Eventually, preparation of the amide and its dehydration were combined into an overall reaction which permits the direct preparation of a nitrile from a carboxylic acid or an ester by the reaction of a mixture of the acid or ester with ammonia over a dehydration catalyst.

Ammoxidation can be used to produce nitriles from compounds that contain activated methyl groups, eg, propylene, isobutylene, toluene, or xylenes, to produce acrylonitrile, methacrylonitrile, benzonitrile, or tolunitriles, respectively (11).

$$RCH_3 + NH_3 + O_2 \xrightarrow{\text{catalyst}} RCN + H_2O$$

Ammoxidation is a vapor-phase reaction of the hydrocarbon with ammonia and air to provide oxygen. However, processes have been developed whereby the oxygen is supplied from the crystal lattice of a metal-oxide catalyst (12) (see Acrylonitrile; Methacrylic acid and derivatives).

Another general method for preparing nitriles is the addition of hydrogen cyanide to aldehydes or ketones, epoxides, olefins, and acetylenes (13–15). The catalyzed addition of hydrogen cyanide to acetylene was the basis for an early industrial acrylonitrile process. Hydrogen cyanide can be added selectively to almost any olefin.

There are many other methods for preparing nitriles by utilizing reagents, eg, Grignard reagents and diazonium salts; or reactions, eg, Friedel-Crafts reactions (qv) and Michael addition (cyanoethylation (qv)). Nitrile syntheses are reviewed in ref. 16.

Acetone Cyanohydrin

Acetone cyanohydrin (2-hydroxy-2-methylpropanenitrile, 2-methyllactonitrile), $(CH_3)_2C(OH)CN$, is a colorless liquid. It is completely miscible in water and most organic solvents but has low solubility in aliphatic hydrocarbons and carbon disulfide (see Table 3 for properties). Commercial acetone cyanohydrin is stabilized with sulfuric acid because the commercial product decomposes to hydrogen cyanide and acetone under basic conditions. In strong sulfuric acid, it hydrolyzes and dehydrates to methacrylamide sulfate. Methyl methacrylate is made by the reaction of this intermediate with methanol:

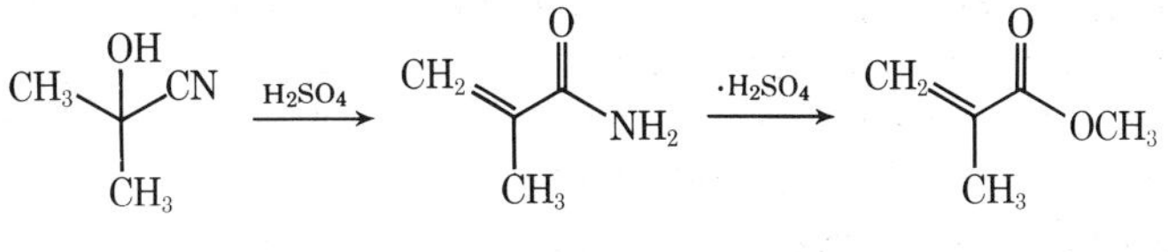

The reaction of acetone cyanohydrin with ammonia produces aminoisobutyronitrile [96-16-2], which is an intermediate in the preparation of azobis(isobutyronitrile) (17).

Acetone cyanohydrin is produced by the reaction of acetone with hydrogen cyanide. The reaction is rapid and exothermic so that cooling is required to maintain the desired reaction temperature of 25°C or less.

Table 3.   Acetone Cyanohydrin Properties

mol wt	85.10
bp (at 3 kPa[a]), °C	81.7
mp, °C	−19
density (at 19°C), g/cm^3	0.932
n_D^{20}	1.3996
flash (COC[b]), °C	74
flammable limits (in air), vol %	
lower	2.2
upper	12.0

[a] To convert kPa to mm Hg, multiply by 7.5.

[b] COC = Cleveland open cup.

Analysis, Shipping, and Storage.   The typical analysis of acetone cyanohydrin is acetone cyanohydrin, 98 wt %; hydrogen cyanide, 0.35 wt %; acetone, 0.80 wt %; water, 0.40 wt %; and stabilizer, eg, H_2SO_4, 0.20 wt %.

The DOT classifies acetone cyanohydrin as a class-B poison. The compound is shipped in 41.6-m^3 (11,000 gal) tank cars. It should be unloaded in well-ventilated areas and, preferably, in enclosed systems. Carbon steel storage tanks, piping, and valves are satisfactory, provided that temperatures are maintained under 50°C. For temperatures above 50°C, Type 316 stainless steel equipment is recommended. A conservation tank vent should be used. Gaskets can be of Teflon TFE (polytetrafluoroethylene) fluorocarbon resin, Buna-S, bonded asbestos, or compressed asbestos. Pumps should be of type 316 stainless steel; mechanical pump seals are preferred but R-54 braided packing of Teflon also can be used. Storage should be in a cool, dry, well-ventilated location away from any fire hazard areas and alkaline materials. Diking is required around the area. Acetone cyanohydrin should not be stored for long periods.

Health and Safety Factors (Toxicology).   The following toxicities have been reported: oral LD$_{50}$ (rats), 17 mg/kg; skin LD$_{50}$ (rabbits), 17 mg/kg; and inhalation LC$_{Lo}$ (rats), 63 ppm/4 h (18). Effects of acetone cyanohydrin overexposure in humans is similar to that for hydrogen cyanide, ie, it is highly toxic by inhalation, skin contact, and ingestion. Exposure to high concentrations of the vapor may result in instantaneous loss of consciousness and death.

First aid and medical procedures associated with acetone cyanohydrin should be established before the compound is used. All persons handling acetone cyanohydrin should be thoroughly familiar with *Chemical Safety Data Sheet SD-67* (hydrocyanic acid), since these safety precautions and handling instructions also apply to acetone cyanohydrin (19). It should be used only in well-ventilated areas. Protective equipment should include rubber gloves and safety goggles. Contact of acetone cyanohydrin with alkaline materials must be avoided since decomposition results in the release of hydrogen cyanide. It is flammable and decomposes when heated to 120°C. Dry chemical or carbon dioxide extinguishers should be used in case of fire. Fire fighters must wear breathing apparatus and complete body protection. Spills should be diked and pumped into drums for disposal. Personnel who must enter the spill area also must wear full protective clothing, including boots and breathing apparatus.

Uses. The major use of acetone cyanohydrin is as an intermediate in the manufacture of methyl methacrylate but it also is used for the manufacture of insecticides, pharmaceuticals, foaming agents, and polymerization initiators. It can be used as a source of hydrogen cyanide for producing cyanohydrins of other ketones, ie, in trans-cyanohydrination, and has been suggested as a means of transporting hydrogen cyanide since it is readily decomposed to its starting materials.

Acetonitrile

Acetonitrile (ethanenitrile), CH_3CN, is a colorless liquid with a sweet, ethereal odor; its properties are listed in Table 4. It is completely miscible with water and its high dielectric strength and dipole moment make it an excellent solvent for both inorganic and organic compounds including polymers. Some representative inorganic compounds which are soluble in acetonitrile are $CaCl_2$, $CuCl$, $FeCl_2$, $FeCl_3$, $KSCN$, $KMnO_4$, $AgNO_3$, and $ZrCl_2$. Many gases also are highly soluble in acetonitrile, eg, olefinic hydrocarbons, halides, HCl, SO_2, and H_2S. It forms low boiling azeotropes with many organics and high boiling azeotropes with BF_3, $SiCl_4$, and $(CH_3)_4Pb$ (20). Some of its azeotropes are listed in Table 5.

Although acetonitrile is one of the more stable nitriles, it undergoes typical nitrile reactions and is used to produce many types of nitrogen-containing compounds, eg, amides (22), amines (23–24); higher molecular weight mono- and dinitriles (25–26); halogenated nitriles (27); ketones (28); isocyanates (29); heterocycles, eg, pyridines

Table 4.   Acetonitrile Properties

mol wt	41.05
bp (at 101.3 kPa = 1 atm), °C	81.6
freezing pt, °C	−45.7
density (at 20°C), g/cm³	0.786
n_D^{20}	1.3441
viscosity (at 20°C), mPa·s (= cP)	0.35
heat of vaporization (at 80°C), J/kg[a]	72.7×10^4
heat of fusion (at −45.7°C), J/kg[a]	21.8×10^4
heat of combustion (at 25°C), J/kg[a]	31.03×10^6
heat capacity (liquid at 20°C), J/(kg·K)[a]	22.59×10^2
surface tension, mN/m (= dyn/cm)	29.3
coefficient of expansion (at 20°C per °C)	1.37×10^{-3}
specific conductance (at 25°C), S	$(5–9) \times 10^{-8}$
dipole moment, C·m[b]	10.675×10^{-30}
dielectric constant	
at 0°C	42.0
at 20°C	38.8
at 81.6°C	26.2
evaporation rate (butyl acetate = 100)	579
flash pt (COC[c]), °C	6
flammable limits (in air), vol %	
lower	4.4
upper	16.0

[a] To convert J to cal, divide by 4.184.
[b] To convert C·m to D, divide by 3.336×10^{-30}
[c] COC = Cleveland open cup.

Table 5.   Acetonitrile Binary Azeotropes[a]

Component	Bp (at 101.3 kPa = 1 atm)	Wt % acetonitrile
benzene	73	34
carbon tetrachloride	65	17
1,2-dichloroethane	79	79
ethanol	73	44
ethyl acetate	75	23
methanol	63	81
water	77	84

[a] Ref. 21.

(30), and imidazolines (31). It can be trimerized to s-trimethyltriazine (32) and has been telomerized with ethylene (33) and copolymerized with α-epoxides (34).

All of the acetonitrile that was produced commercially in the United States in 1980 was isolated as a by-product from the manufacture of acrylonitrile by propylene ammoxidation. The amount of acetonitrile produced in an acrylonitrile plant depends on the ammoxidation catalyst that is used, but the ratio of acetonitrile:acrylonitrile usually is ca 1:35. The acetonitrile is recovered as the water azeotrope, dried, and purified by distillation (35). U.S. capacity (1980) from DuPont and Vistron is ca 8000 t/yr.

Specifications.   Specifications for commercial acetonitrile are given in Table 6. The principal organic impurity in commercial acetonitrile is propionitrile; although small amounts of allyl alcohol also may be present.

Shipping and Storage.   The DOT classifies acetonitrile as a flammable liquid, ie, a red label is required. Its freight class is chemical (NOI) and, for international shipments, IMCO requires that poison labels and flammability labels be affixed to all containers. For storage and piping at normal temperatures and pressures, aluminum or carbon steel may be used. Centrifugal pumps are preferred because the solvency of acetonitrile may affect the lubricant in positive-displacement pumps. All tanks, piping, valves, and pumps should be electrically grounded. Fire protection equipment can be water spray, alcohol foam, CO_2, or dry chemical.

Table 6.   Commercial Acetonitrile Specifications

Specification	Value
sp gr (at 20°C)	0.783–0.787
distillation range, °C	
initial min	80.5
end pt, max	82.5
purity (min), wt %	99.0
acidity (max), wt %	0.05
copper (max), ppm	0.5
iron (max), ppm	0.5
water (max), wt %	0.3
color (max), Pt–Co	15

Health and Safety Factors (Toxicology). The following toxicities for acetonitrile have been reported: oral LD_{50} (rats), 3030–6500 mg/kg; skin LD_{50} (rabbits), 3884–7850 mg/kg; and inhalation LC_{50} (rats), 7500–17,000 ppm (18). Humans can detect the odor of acetonitrile at 40 ppm. Exposure for 4 h at up to 80 ppm has not produced adverse effects. However, exposure for 4 h at 160 ppm results in reddening of the face and some temporary bronchial tightness.

Although acetonitrile has a low order of acute toxicity by ingestion, inhalation, and skin absorption, it can cause severe eye burns. In case of eye contact, eyes should be immediately flushed with water for at least 15 min and a physician should be consulted. In the event of a spill or leak, the spill should be contained, flooded with water, and disposed of according to local regulations. Acetonitrile is flammable (see Table 4) and must be kept away from excessive heat, sparks, and open flame. Associated fires can be extinguished using water spray, alcohol foam, CO_2, or dry chemical extinguishers. OSHA requires that an employee's exposure to acetonitrile in any 8-h shift does not exceed a time-weighted average of 40 ppm (70 mg/m^3) in air (36).

Uses. Because of its good solvency and relatively low boiling point, acetonitrile is used widely as a recoverable reaction medium, particularly for the preparation of pharmaceuticals. Its largest use is for the separation of butadiene from C_4 hydrocarbons by extractive distillation (see Azeotropic and extractive distillation) (37). It also has been proposed for the separation of other olefins, eg, propylene, isoprene, allene, and methylacetylene from hydrocarbon streams (38–40). It is a superior solvent for polymers and can be used as a solvent for spinning fibers and for casting and molding plastics. It is used widely in spectrophotometry and electrochemistry. Since pure acetonitrile does not absorb uv light, a growing use is as a solvent in high pressure liquid chromatography (hplc) for the detection of materials, eg, residual pesticides, in the ppb range. If used in hplc, acetonitrile must be extensively purified; however, hplc-grade acetonitrile can be obtained from suppliers of highly purified solvents (41).

Acetonitrile also is used as a catalyst and as an ingredient in transition-metal complex catalysts (42–43). There are many uses for it in the photographic industry and for the extraction and refining of copper and by-product ammonium sulfate (44–46). It also is used for dyeing textiles and in coating compositions (47–48). It is an effective stabilizer for chlorinated solvents, particularly in the presence of aluminum, and it has some application in the manufacture of perfumes (qv) (49–50). It also is used as a reagent for the preparation of a wide variety of compounds.

Adiponitrile

Adiponitrile (hexanedinitrile), $NC(CH_2)_4CN$, is manufactured principally for use as an intermediate for hexamethylenediamine (1,6-diaminohexane), which is a principal ingredient for nylon-6,6. The first large-scale production of adiponitrile was by DuPont in 1939 and, since World War II, adiponitrile manufacturing plants have been built worldwide (see Polyamides). Pure adiponitrile is a colorless liquid and has no distinctive odor; some properties are shown in Table 7. It is soluble in methanol, ethanol, chloroalkanes, and aromatics but has low solubility in carbon disulfide, ethyl ether, and aliphatic hydrocarbons. At 20°C, the solubility of adiponitrile in water is ca 8 wt %; the solubility increases to 35 wt % at 100°C. At 20°C, adiponitrile dissolves ca 5 wt % water.

Adiponitrile undergoes the typical nitrile reactions, eg, hydrolysis to adipamide

Table 7.   Adiponitrile Properties[a]

mol wt	108.14
bp, °C	
at 101.7 kPa[a]	295
at 1.3 kPa[a]	154
freezing pt, °C	2.49
density (at 20°C), g/cm^3	0.965
n_D^{20}	1.4343
viscosity, mPa·s (= cP)	
at 20°C	9.1
at 70°C	2.6
heat of vaporization, J/kg[b]	70.4×10^4
heat of fusion (at 1°C), J/kg[b]	21.3×10^4
heat of combustion, J/kg[b]	40.4×10^6
critical temperature, °C	507
critical pressure, MPa[c]	2.8
specific conductance, S	3.5×10^{-8}
flash pt (closed cup), °C	159
autoignition temperature, °C	550
flammable limits (in air), vol %	
lower	1.7
upper	5.0

[a] To convert kPa to mm Hg, multiply by 7.5.
[b] To convert J to cal, divide by 4.184.
[c] To convert MPa to atm, divide by 0.101.

and adipic acid and alcoholysis to substituted amides and esters. The most important industrial reaction is the catalytic hydrogenation to hexamethylenediamine.

$$NC(CH_2)_4CN \xrightarrow[\text{catalyst}]{H_2} H_2N(CH_2)_6NH_2$$

A variety of catalysts are used for this reduction including cobalt–nickel (51), cobalt–manganese (52), cobalt boride (53), copper cobalt (54), and iron oxide (55).

Adiponitrile is made commercially by at least three different processes, and each is based on a different hydrocarbon. The original process, which was first commercialized by DuPont in the late 1930s and is the basis for a number of adiponitrile plants, starts with adipic acid (qv) which is produced from oxygenated derivatives of cyclohexane. However, the adipic acid process was abandoned by DuPont in favor of two processes based on butadiene. During the 1960s, Monsanto and Asahi developed routes to adiponitrile by the electrodimerization of acrylonitrile (qv), which is made by the ammoxidation of propylene (see Electrochemical processing).

In the adipic acid process to adiponitrile, adipic acid reacts with ammonia over a catalyst in either the liquid or vapor phase. The most widely used catalysts are based on phosphorus-containing compounds, but boron compounds and silica gel also have been patented for this use (56–60). Vapor-phase processes involve the use of fixed catalyst beds; whereas, in liquid–gas processes, the catalyst is added to the feed. The reaction temperature of the liquid-phase processes is ca 300°C and most vapor-phase processes run at 350–400°C. Both operate at atmospheric pressure. Yields of adipic acid to adiponitrile are as high as 95% (61).

In the now-obsolete furfural process, furfural was decarboxylated to furan which was then hydrogenated to tetrahydrofuran (THF). Reaction of THF with hydrogen chloride produced dichlorobutene. Adiponitrile was produced by the reaction of sodium cyanide with the dichlorobutene. The overall yield from furfural to adiponitrile was around 75% (see Furan derivatives).

In one butadiene-based adiponitrile process, the following reactions occur:

$$CH_2{=}CHCH{=}CH_2 + Cl_2 \rightarrow ClCH_2CH{=}CHCH_2Cl \text{ (ca 66\%)} + CH_2{=}CHCHClCH_2Cl \text{ (ca 33\%)}$$

cis- and trans-1,4-dichloro-2-butene   3,4-dichloro-1-butene

$$\text{dichlorobutene} + NaCN \xrightarrow[Cu_2Cl_2]{Cu} NCCH_2CH{=}CHCH_2CN$$

cis- [2141-58-4], trans- [2141-59-5]
1,4-dicyano-2-butene
+
$NCCH{=}CHCH_2CH_2CN$
cis- [1119-85-3], trans- [18715-38-3]
1,4-dicyano-1-butene

$$\text{dicyanobutenes} + H_2 \xrightarrow{\text{catalyst}} \text{adiponitrile}$$

Butadiene (qv) is chlorinated in the vapor phase in high yield at 160–250°C. The 1,4-dichlorobutenes react with aqueous sodium cyanide in the presence of copper catalysts to produce the isomeric 1,4-dicyanobutenes; yields are as high as 95% (62). Adiponitrile is produced by the hydrogenation of the dicyanobutenes over a palladium catalyst in either the vapor phase or the liquid phase (63–64). The yield in either case is 95% or better.

Another butadiene-to-adiponitrile route is based on direct addition of HCN to the double bonds of butadiene, ie, olefin hydrocyanation. In the first step, HCN is added to butadiene in the presence of a nickel catalyst according to the following equation:

$$CH_2{=}CHCH{=}CH_2 + HCN \xrightarrow{\text{catalyst}}$$
$$CH_3CH{=}CHCH_2CN + CH{=}CHCH(CH_3)CN$$

3-pentenenitrile   2-methyl-3-butenenitrile

The reaction is carried out at ca 100°C at a pressure sufficient to maintain the reactants in the liquid phase (13). 2-Methyl-3-butenenitrile [16529-56-9] is isomerized to linear 3-pentenenitrile [4635-87-4] and 4-pentenenitrile [592-51-8] and the mononitriles are fed into the second hydrocyanation unit (14). After the second hydrocyanation, the by-products, ie, 2-pentenenitrile [13284-42-9], 2-methyl-3-butenenitrile, 3-pentenenitrile, and methylglutaronitrile are separated from adiponitrile by distillation.

The Monsanto dimerization of acrylonitrile was first commercialized in 1965 (65). The process involves the following reaction, which takes place at the cathode in an electrolytic cell:

$$2\,CH_2{=}CHCN + 2\,H^+ + 2\,e \rightarrow NC(CH_2)_4CN$$

Small amounts of propionitrile and bis(cyanoethyl) ether are formed as by-products. The hydrogen ions are formed from water at the anode and pass to the cathode through a membrane. The catholyte that is continuously recirculated in the cell consists of a mixture of acrylonitrile, water, and a tetraalkylammonium salt; the anolyte is recirculated aqueous sulfuric acid. A quantity of catholyte is continuously removed for recovery of adiponitrile and unreacted acrylonitrile; the latter is fed back to the catholyte with fresh acrylonitrile. Oxygen that is produced at the anodes is vented and water is added to the circulating anolyte to replace the water that is lost through electrolysis. The operating temperature of the cell is ca 50–60°C. Current densities are 0.25–1.5 A/cm^2 (see Electrochemical processing).

A typical composition of the catholyte is adiponitrile, 15 wt %; acrylonitrile, 15 wt %; quaternary ammonium salt, 39 wt %; water, 29 wt %; and by-products, 2 wt %. Such a solution is extracted with acrylonitrile and water which separates the organics from the salt which can be returned to the cell. The acrylonitrile is distilled from the extract and the resultant residue consists of ca 91 wt % adiponitrile, which is purified further by distillation. The overall yield of acrylonitrile to adiponitrile is 92–95%.

Production and Shipment. Estimated adiponitrile production capacities are shown in Table 8. It is probable that there will be a worldwide capacity in excess of 10^6 metric tons by the mid-1980s.

Adiponitrile is a nonregulated commodity when transported on land. Its freight classification is chemicals NOIBN, and its Coast Guard classification is combustible liquid, Grade E. Barges used to transport adiponitrile must have a valid U.S. Coast Guard certificate of inspection for transport of Grade E combustible liquid.

Approved materials of construction for shipping, storage, and associated transportation equipment are carbon steel and type 316 stainless steel. Either centrifugal or positive displacement pumps may be used, and gasketing should be of white asbestos. Carbon dioxide or chemical-foam fire extinguishers should be used.

There are no specifications for commercial adiponitrile; the typical composition is 99.5 wt % adiponitrile. Impurities that may be present depend on the method of manufacture and, thus, vary depending on the source.

Health and Safety Factors (Toxicology). The following toxicities for adiponitrile have been reported: oral LD$_{50}$ (rats), 300 mg/kg; oral approx lethal dose (rats), 450 mg/kg; skin approx lethal dose (rats), 2000 mg/kg; and inhalation approx lethal concentration (rats), 3.0 mg/0.25 h. Inhalation may cause nausea, vomiting, irritation of mucosal membranes, and dizziness.

Adiponitrile should be used only with adequate ventilation; breathing of the vapor should be avoided. A person who has inhaled the vapors should be moved to an uncontaminated environment, and cardiopulmonary resuscitation should be adminis-

Table 8.   Estimated Worldwide Adiponitrile Production Capacity^a, Thousand Metric Tons

	1976	1977	1978	1979	1980
U.S.	160	175	420	420	420
North and South America (other than U.S.)	53	53	53	53	54
U.K.	135	170	225	225	225
Western Europe	190	170	205	205	205
Japan	28	28	28	28	28
China					20
Total	*566*	*596*	*931*	*931*	*952*

^{a} Based on data of hexamethylenediamine production capacities collected from various published sources (see Amines by reduction).

tered. If the victim breathes with difficulty, oxygen should be given. In case of eye contact, flush with copious water for at least 20 min and call a physician. In case of ingestion, induce vomiting and call a physician. For skin contact, wash with plenty of soap and water. All personnel should be kept upwind of a spill or leak. Spills should be contained and collected by absorbing with sawdust and incinerating, or by absorbing with earth or clay and disposing in accordance with local requirements. The spill area should be washed with soap and water which should be collected for disposal. All personnel should be far removed and upwind from an adiponitrile-related fire, and firefighters should use CO_2 or chemical foam extinguishers and wear self-contained breathing apparatus.

Uses.   The principal use of adiponitrile is for hydrogenation to hexamethylenediamine. Adipoguanamine, which is prepared by the reaction of adiponitrile with dicyandiamide [461-58-5] (cyanoguanidine), may have uses in melamine–urea amino resins (qv) (see Benzonitrile, Uses). Its typical liquid nitrile properties suggest its use as an extractant for aromatic hydrocarbons.

2,2′-Azobis(isobutyronitrile) and 2,2-Azobis(2,4-dimethylvaleronitrile)

2,2′-Azobis(isobutyronitrile) [78-67-1] (Vazo 64) and 2,2′-azobis(2,4-dimethylvaleronitrile) [4419-11-8] (Vazo 52) are efficient sources of free radicals for vinyl polymerizations and chain reactions, eg, chlorinations (see Initiators). Both compounds decompose in a variety of solvents at nearly first-order rates to give free radicals with no evidence of induced chain decomposition. They can be used in bulk, solution, and suspension polymerizations and, because no oxygenated residues are produced, they are suitable for use in pigmented or dyed systems that may be susceptible to oxidative degradation.

Vazo 64 and Vazo 52 are crystalline solids that are produced by hypochlorite oxidation of 2-aminonitriles (66).

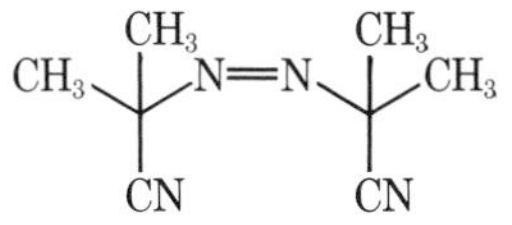

2,2′-azobis(isobutyronitrile) 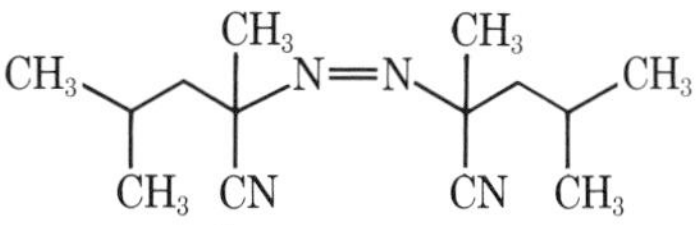   2,2′-azobis(2,4-dimethylvaleronitrile)

These compounds are essentially insoluble in water, sparingly soluble in aliphatic hydrocarbons, and soluble in functional compounds and aromatic hydrocarbons. Some solubility data are listed in Table 9.

In solution, the azonitriles decompose on heating to form two free radicals with the liberation of nitrogen:

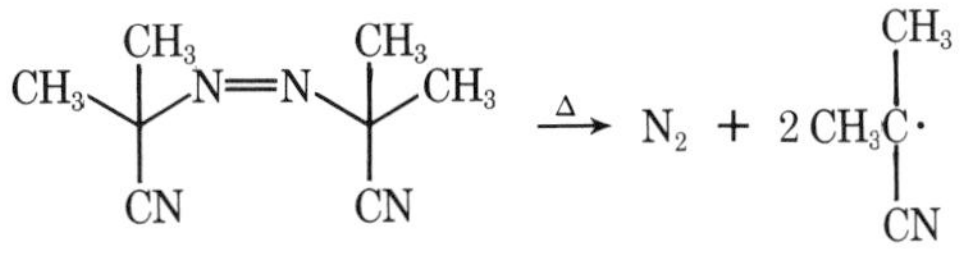

Because the decomposition is first order, the rate of free-radical formation can be controlled by regulating the temperature. Half-lives in toluene as functions of tem-

Table 9.   Solubilities of Vazo 64 and Vazo 52 Initiators, g/100 g Solvent [a]

Solvent	Vazo 64		Vazo 52	
	at 0°C	at 25°C	at 0°C	at 25°C
acetone	12		39	75
acrylonitrile	20	38		
benzene	<1	11	49	108
carbon tetrachloride		0.7		
chloroform	17	25	54	100
dichloromethane	21.5	40		
diisobutyl ketone	2	6		
N,N-dimethylacetamide	7	21		
dioxane	<1			
ethanol (absolute)		2.7	9	20
ethanol 2B		2.7		
ethyl acetate	5	14	35	100
ethyl ether		4		
ethylene chloride	4	23		
formamide		0.5		
Freon 113 (trichlorotrifluoroethane)	<1			
heptane		0.5	4	6
Isopar K [b]			2	6
isopropyl acetate	4	9		
isopropyl alcohol		0.7	5	22
methanol (abs)	2.5	7.5	11	39
methyl isobutyl ketone	3	13		
methyl ethyl ketone	13	30		
methyl formate	10	26		
methyl methacrylate	4	10	28	85
methyl methoxyacetate	8.5	20		
Stoddard solvent			6	18
styrene	3	8	27	96
toluene	2	7	35	82
vinyl acetate			28	85
VM & P naphtha			5	18
xylene		2.5	27	79
water	<0.01	0.04		0.009
white mineral oil (Nujol) [c]			0.2	0.3
mixtures (50 wt %-50 wt %)				
benzene + N,N-dimethylacetamide	8	21		
benzene + methyl formate	15	33		
benzene + methyl methoxyacetate	8	20		

[a] Ref. 67.
[b] Exxon Co.
[c] Plough, Inc.

perature are shown in Figure 1. These decomposition rates are essentially independent of the solvent (68), and activation energies are ca 130 kJ/mol (31 kcal/mol) for both compounds.

Specifications.   Specifications for azobisnitriles are given in Table 10.

Shipping and Storage.   According to the DOT, azobisnitriles are flammable solids and their containers must be affixed with red-and-white-striped hazard labels. Their freight classification is chemical, NOIBN. Azobisnitriles are shipped in 4.5-kg and 22.7-kg fiber drums. Shipping of Vazo 52 by private truck or common carrier must

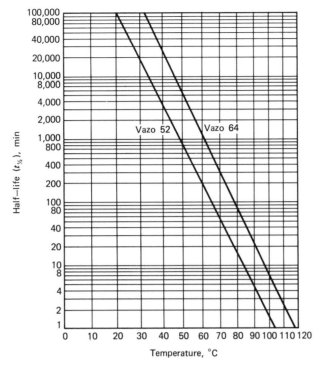

Figure 1. Thermal decomposition of Vazo polymerization initiators half-life in toluene solution (67). Vazo 52: log $(t_{1/2})$ = 6767 $(1/T)$ − 18.037. Vazo 64: log $(t_{1/2})$ = 7142 $(1/T)$ − 18.355. T = K.

Table 10. Specifications for Commercial Azobisnitriles

	Value	
Property	Vazo 64	Vazo 52
assay, wt %	98 min	98 min
water, wt %	0.5 max	0.5 max
iron	20 max	20 max
acetone insolubles, wt %	0.1 max	0.3 max
α color (as 2 wt % soln in dimethylformamide)	15 max	25 max
bulk density, g/cm^3	ca 0.4	ca 0.4

be according to Bureau of Explosives permit BA-1852 which must accompany the bill of lading. Vazo polymerization initiators must be stored out of the sun and away from heat in a cool, dry place. Since decomposition produces nitrogen and, consequentially, a pressure increase, these compounds should not be stored in glass or in any tightly closed containers other than the original shipping container. The maximum storage temperature is 10°C for Vazo 52 and 24°C for Vazo 64.

Health and Safety Factors, (Toxicology).   Vazo 52 and Vazo 64 cause mild transient irritation to the eyes. Neither compound is a skin sensitizer but Vazo 52 is a slight skin irritant. The approx lethal dose (ALD) of Vazo 52 in rats is greater than 5000 mg/kg and the ALD for Vazo 64 is 670 mg/kg. In the absence of a polymerizable vinyl polymer, tetramethylsuccinonitrile [3333-52-6] (TMSN) is the principal decomposition product of Vazo 64. TMSN has an oral ALD of 60 mg/kg and is highly toxic by inhalation. OSHA regulations require that an employee's exposure to TMSN in any 8-h shift does not exceed an 8-h time-weighted average of 0.5 ppm in air (= 3 mg/m^3). Because both TMSN solid and vapor are capable of penetrating the skin and mucous membranes, control of vapor inhalation alone may not be sufficient to prevent absorption of an excessive dose.

All operations should be carried out with good ventilation and contact with eyes and skin should be avoided. In case of eye contact, the eyes should be flushed with water for 15 min and a physician should be consulted. Soap and water should be used to wash azobisnitriles from skin. If these compounds are inhaled, particularly the decomposition products of Vazo 64, the victim should be removed to fresh air and oxygen should be administered. If the victim is not breathing, cardiopulmonary resuscitation should be administered. A physician should be called in either case.

Small quantities of waste Vazo should be disposed of by incineration, preferably by dissolving first in a waste liquid. Large quantities of waste Vazo should be returned to the manufacturer for disposal.

Uses. The azobisnitriles have been used for bulk, solution, emulsion, and suspension polymerization of all of the common vinyl monomers, including ethylene, styrene, vinyl chloride, vinyl acetate, acylonitrile, and methyl methacrylate. The polymerizations of unsaturated polyesters and copolymerizations of vinyl compounds also have been initiated by these compounds.

Benzonitrile

Benzonitrile [100-47-0], C$_6$H$_5$CN, is a colorless liquid with a characteristic almondlike odor. Its physical properties are listed in Table 11. It is miscible with acetone, benzene, chloroform, ethyl acetate, ethylene chloride, and other common organic solvents but is immiscible with water at ambient temperatures and soluble to ca 1 wt % at 100°C. It distills at atmospheric pressure without decomposition but slowly discolors in the presence of light.

Table 11.   Properties of Benzonitrile

mol wt	103.12
bp (at 101.3 kPa = 1 atm), °C	191.1
freezing pt, °C	−12.8
density (at 20°C), g/cm^3	1.3441
n_D^{20}	1.5289
viscosity, mPa·s (= cP)	1.24
heat of vaporization, mJ/kg^a	3.66
surface tension (at 20°C), mN/m (= dyn/cm)	39.0
coefficient of expansion (per °C)	9×10^{-4}
dielectric constant (at 25°C)	25.2
dipole moment (at 22°C), C·m^b	13.144×10^{-30}
evaporation rate (butyl acetate = 100)	13
flash pt (COC^c), °C	79

^{a} To convert J to cal, divide by 4.184.

^{b} To convert C·m to D, divide by 3.336×10^{-30}.

^{c} COC = Cleveland open cup.

Like acetonitrile, benzonitrile is a powerful solvent for many inorganic and organic materials including some polymers. Inorganic salts that are soluble in benzonitrile at 25°C include aluminum chloride, ferric chloride, and silver nitrate. In chemical reactions, benzonitrile displays characteristics of both the nitrile group and the aromatic nucleus. It can be converted to a large number and variety of derivatives by simple syntheses; eg, by hydrolysis, it can be converted to either benzoic acid or benzamide. Hydrolysis in the presence of $ZnCl_2$ and HCl produces benzaldehyde, whereas alcoholysis leads to the formation of benzoic acid esters. Reduction produces benzyl- or dibenzylamine and reductive alkylation produces N-substituted benzamides. The aromatic ring can be halogenated or nitrated without the nitrile group being affected. Benzonitrile can be trimerized in the presence of a catalytic amount of base to 2,4,6-triphenyl-1,3,5-triazine (69). The most important reaction is with dicyandiamide to produce 2,4-diamino-6-phenyl-1,3,5-triazine (benzoguanamine) (70):

Benzonitrile is produced commercially and in high yield by the vapor-phase catalytic ammoxidation of toluene:

$$C_6H_5CH_3 + NH_3 + 1.5\,O_2 \xrightarrow{\text{catalyst}} C_6H_5CN + 3\,H_2O$$

An older route is based on just toluene and ammonia in the absence of air with separate catalyst regeneration (71). The preparation of benzonitrile by the catalytic reaction of toluene with nitric oxide has been reported but has not proceeded beyond the research stage (72).

$$2\,C_6H_5CH_3 + 4\,NO \rightarrow 2\,C_6H_5CN + 3\,H_2O + N_2O$$

A method for making benzonitrile by dehydrogenation of the Diels-Alder adduct of butadiene and acrylonitrile also has been described (73). Benzonitrile also can be made on a small scale by the dehydration of benzamide in an inert solvent with phosphorus oxychloride or benzenesulfonyl chloride and an organic amine (74–75).

Specifications.   Sherwin Williams is the only U.S. producer of benzonitrile. Specifications for the commercial compound are listed below:

frz pt, °C	−14.0
color, APHA	50 max
water content, wt %	0.40 max
gc purity, wt %	99.5 min

Shipping and Storage.   There is no DOT hazard classification for benzonitrile; its freight classification is chemical, NOIBN. Carbon-steel drums and tanks may be used for storage.

Health and Safety Factors (Toxicology).   The following toxicities for benzonitrile have been reported: oral LD_{Lo} (rats), 720 mg/kg; skin LD_{50} (rats), 1200 mg/kg; and inhalation LC_{50} (rats), 950 ppm/8 h. Benzonitrile is a very toxic organic cyanide (76). Toxicity symptoms resulting from industrial accidents include shortness of breath, unconsciousness, and toxic spasms of face and arm muscles (77).

Good ventilation should be used to minimize exposure to vapors. Contact with skin and, particularly eyes, should be avoided. Affected skin areas should be washed immediately with water and mild soap. Eyes that have been contaminated should be flushed immediately with large volumes of water for at least 15 min and a physician should be consulted. In case of ingestion, vomiting should be induced immediately and the victim should be treated for cyanide poisoning. Oxygen or cardiopulmonary resuscitation, if necessary, should be administered in case of unconsciousness or overexposure to vapors and the victim should be treated for cyanide poisoning.

Benzonitrile spills should be controlled by absorbing the liquid with vermiculite or sawdust and then by placing the wet absorbants in a solution of sodium hypochlorite (Clorox) which converts the nitrile to an isocyanate. After several hours, the slurry can be disposed of according to local regulations.

Uses.   The most important commercial use for benzonitrile is in the synthesis of benzoguanamine, which is a derivative of melamine and is used in protective coatings and molding resins (see Amino resins; Cyanamides). Other uses for benzonitrile are as an additive in nickel-plating baths, for separating naphthalene and alkylnaphthalenes from nonaromatics by azeotropic distillation (qv), as a jet-fuel additive, in cotton bleaching baths, as a drying additive for acrylic fibers, and in the removal of titanium tetrachloride and vanadium oxychloride from silicon tetrachloride.

Isophthalonitrile

Isophthalonitrile [626-17-5] (IPN), (1,3-dicyanobenzene), is a white solid which melts at 161°C and sublimes at 265°C. It is slightly soluble in water but readily dissolves in dimethyl formamide or N-methylpyrrolidinone and hot aromatic solvents. IPN undergoes the reactions expected of an aromatic nitrile, eg, hydrogenation to m-cyanobenzylamine [10406-24-3], 1,3-di(aminomethyl)benzene (m-xylylenediamine), or 1,3-di(aminomethyl)cyclohexane, and chlorination to tetrachloroisophthalonitrile [1897-45-6]. Isophthalonitrile is prepared by the vapor-phase ammoxidation of m-xylene. Safe handling practices require minimal exposure. The oral LD_{50} for rats is greater than 5 mg/kg. Its principal use appears to be as an intermediate to the amines noted above. As a reagent, IPN can be used to convert aromatic acids to nitriles in near quantitative yields (78).

Cyanoacetic Acid

The physical properties of cyanoacetic acid [372-09-8] and its commercial derivatives are listed in Table 12. The parent acid is a strong organic acid with a dissociation constant at 25°C of 3.36×10^3. It is hygroscopic and is highly soluble in alcohols and ethyl ether but is insoluble in both aromatic and aliphatic hydrocarbons. It undergoes typical nitrile and acid reactions but the presence of the nitrile and carboxylic acid groups on the same carbon cause the hydrogens on C-2 to be replaced readily. The resulting malonic acid derivative decarboxylates to a substituted acrylonitrile:

cinnamonitrile [4360-47-8]

Table 12.   Properties of Cyanoacetic Acid and Esters [a]

	Cyanoacetic acid, $NCCH_2COOH$	Methyl ester, $NCCH_2COOCH_3$	Ethyl ester, $NCCH_2COOC_2H_5$
mol wt	85.06	99.09	113.11
bp, °C	108[b]	206	208–210
mp, °C	67		−22.5
n_D^{20}		1.419	1.4177
flash pt, °C	107	110	110

[a] Ref. 79.
[b] At 2 kPa (15 mm Hg).

Cyanoacetic acid can be used for the preparation of heterocyclic ketones. It is prepared by the reaction of chloroacetic acid with sodium cyanide.

There are no shipping regulations for cyanoacetic acid.

Health and Safety Factors (Toxicology).   The following toxicities have been reported for cyanoacetic acid: interperitoneal LD_{50} (mice), 200 mg/kg; subcutaneous LD_{Lo} (rabbit), 2000 mg/kg; and subcutaneous LD_{Lo} (frog); 2000 mg/kg. Toxicities in humans are unknown.

Uses.   Although cyanoacetic acid can be used in applications requiring strong organic acids, its principal use is in the preparation of malonic esters and other reagents used in the manufacture of pharmaceuticals, eg, barbital, caffeine, and B vitamins (see Alkaloids; Hypnotics; Vitamins).

Cyanoacetate Esters

Physical properties for methyl [105-34-0] and ethyl cyanoacetate [105-56-6] are listed in Table 12. Both esters are slightly soluble in water but are completely miscible in most common organic solvents including aromatic hydrocarbons. The esters, like the parent acid, are highly reactive, particularly in reactions involving the central carbon atom; however, the esters tend not to decarboxylate. They are prepared by esterification of cyanoacetic acid and are used principally as chemical intermediates. There are no regulations associated with their shipment.

Health and Safety Factors (Toxicology).   The following toxicities for ethyl cyanoacetate have been reported: interperitoneal LD_{50} (mice), 750 mg/kg; subcutaneous LD_{Lo} (rabbits), 1500 mg/kg; and subcutaneous LD_{Lo} (frogs), 4000 mg/kg. Toxicities in humans are unknown.

Table 13.   Fatty Acid Nitrile Physical Properties

Nitrile	CAS Registry No.	bp (at 101.3 kPa (1 atm)), °C	Freezing pt, °C	Iodine value
oleyl	[112-91-4]	>325	5	85 min
coco	[61789-53-5]			15 max
tallow	[61790-28-1]	330–360	4	44–60
stearyl	[638-65-3]	330–360	39	4 max

Fatty Acid Nitriles

Some of the physical properties of fatty acid nitriles are listed in Table 13 (see also Carboxylic acids).

Fatty acid nitriles are produced as intermediates for a large variety of amines and amides. Estimated U.S. production capacity (1980) was >140,000 t/yr. Fatty acid nitriles are produced from the corresponding acids by a catalytic reaction with ammonia in the liquid phase. They have little use other than as intermediates but could have some utility as surfactants (qv), rust inhibitors, and plasticizers (qv).

BIBLIOGRAPHY

"Nitriles" in *ECT* 2nd ed., Supplement, pp. 590–603, by R. W. Ingwalson, Velsicol Chemical Corp.

1. Brit. Pat. 1,341,104 (Dec. 17, 1973), D. I. Hoke (to Lubrizol Corp.).
2. J. Poupaert and co-workers, *Synthesis* **3,** 622 (1972).
3. F. E. King and co-workers, *J. Chem. Soc.*, 352 (1936).
4. J. Cason and co-workers, *J. Org. Chem.* **18,** 1594 (1953).
5. S. R. Best and J. F. Thorpe, *J. Chem. Soc.* **95,** 1901 (1909).
6. K. Ziegler and co-workers, *Annalen* **504,** 94 (1933).
7. R. H. Wiley, *Chem. Rev.* **37,** 410 (1945).
8. V. Migrdichian, *The Chemistry of Organic Cyanogen Compounds*, A.C.S. Monograph 105, Reinhold, New York, 1947.
9. D. T. Moury, *Chem. Rev.* **42,** 192 (1948).
10. U.S. Pat. 2,915,455 (Nov. 10, 1959), R. A. Smiley (to E. I. du Pont de Nemours & Co., Inc.).
11. U.S. Pat. 2,481,826 (Sept. 13, 1949), J. N. Cosby (to Allied Chem.).
12. M. C. Sze and A. P. Gelbein, *Hydrocarbon Process.* **55,** 103 (Feb. 1976).
13. U.S. Pat. 3,496,215 (Feb. 17, 1970), W. C. Drinkard, Jr. and R. V. Lindsey, Jr. (to E. I. du Pont de Nemours & Co., Inc.).
14. U.S. Pat. 3,536,748 (Oct. 27, 1970), W. C. Drinkard, Jr. and R. V. Lindsey, Jr. (to E. I. du Pont de Nemours & Co., Inc.).
15. C. Y. Wu and H. E. Swift, *179th American Chemical Society Meeting, Div. of Petroleum Chemistry Symposia Preprint*, Houston, Texas, Mar. 1980, p. 372.
16. D. H. R. Barton and W. D. Ollis, *Comprehensive Organic Chemistry*, Vol. 2, Pergamon Press, Oxford, Eng., 1979, pp. 528–562.
17. R. A. Jacobson, *J. Am. Chem. Soc.* **68,** 2628 (1946).
18. R. L. Lewis, Sr., *Registry of Toxic Effects of Chemical Substances*, U.S. Dept. of HEW, NIOSH, Jan. 1979.
19. *Chemical Safety Data Sheet SD-67*, Chemical Manufacturers' Association, Washington, D.C., 1961.
20. U.S. Pat. 3,362,889 (Jan. 9, 1968), J. F. Hannan (to E. I. du Pont de Nemours & Co., Inc.).
21. *Advances in Chemistry Series*, Azeotropic Data, No. 6 and No. 35, ACS, Washington, D.C., 1952 and 1962.
22. U.S. Pat. 3,822,313 (July 2, 1974), J. R. Norell (to Phillips Petroleum Co.).
23. A. Buzas and co-workers, *Rev. Chim.* **22,** 656 (1971).
24. N. S. Kozlov and co-workers, *Dokl. Akad. Nauk BSSR* **2**(4), 326 (1977).
25. G. W. Gokel and co-workers, *Tetrahedron Lett.* **39,** 3495 (1976).
26. U.S. Pat. 3,810,935 (May 14, 1974), W. Leimgruber and M. Weigele (to Hoffmann-La Roche Inc.).
27. U.S. Pat. 3,825,581 (July 23, 1974), W. A. Gay and D. F. Gavin (to Olin Corp.).
28. H. Felkin and G. Roussi, *C.R. Acad. Sci. Paris Ser. C* **266,** 1552 (1968).
29. U.S. Pat. 3,920,718 (Nov. 18, 1975), J. E. Nottke (to E. I. du Pont de Nemours & Co., Inc.).
30. U.S. Pat. 3,829,429 (Aug. 13, 1974), R. A. Clement (to E. I. du Pont de Nemours & Co., Inc.).
31. Fr. Pat. 2,121,106 (Sept. 22, 1972), (to Ajinomoto Co., Inc.).
32. T. Cairns and co-workers, *J. Am. Chem. Soc.* **74,** 5633 (1952).
33. I. A. Gunevich and co-workers, *Tr. Mosk. Khim. Technol. Inst.* **86,** 5 (1975).
34. A. A. Durgaryan and co-workers, *Arm. Khim. Zh.* **25,** 401 (1972).

35. U.S. Pat. 3,281,450 (Oct. 24, 1966), P. J. Horvath (to B. F. Goodrich).
36. *OSHA Standard 1910*, June 27, 1974.
37. H. D. Evans and D. H. Sarno, *World Petrol. Congr. Proc.*, *7th* **5**, 259 (1967).
38. E. Ger. Pat. 91,480 (July 20, 1972), G. Hauthal and co-workers.
39. M. Enomoto and co-workers, *Kavgaku Kogaku* **35**, 437 (1971).
40. USSR Pat. 439,143 (Mar. 15, 1976), M. E. Aerov and co-workers.
41. "Organic Solvents" in J. A. Riddick and W. B. Bunger, *Techniques of Chemistry*, Vol. II, John Wiley & Sons, Inc., 1970, pp. 798–805.
42. A. Krause and T. Weiman, *Rocznki Chem.* **40**, 1173 (1967).
43. Ger. Offen. 2,237,704 (Feb. 15, 1973), L. W. Gosser and C. A. Tolman (to E. I. du Pont de Nemours & Co., Inc.).
44. Brit. Pat. 1,312,573 (Apr. 4, 1973), N. T. Notley.
45. D. M. Muir and A. J. Parker, *Adv. Entr. Metall. Int. Symp. 3rd*, 191 (1977).
46. U.S. Pat. 3,607,136 (Sept. 21, 1971), R. A. Smiley and J. A. Vernon (to E. I. du Pont de Nemours & Co., Inc.).
47. L. I. Primak and co-workers, *Tekst. Promst. (Moscow)* (2), 60-1 (1977).
48. Fr. Pat. 1,471,321 (Mar. 3, 1967), (to Dunlop Co., Ltd.).
49. Jpn. Kokai 76 04,107 (Jan. 14, 1976), T. Kita and M. Ishii.
50. W. S. Brud and co-workers, *Int. Cong. Essent. Oils*, *6th*, 73 (1974).
51. Ger. Pat. 848,498 (Sept. 4, 1952), K. Adam and co-workers (to BASF).
52. Fr. Pat. 1,483,300 (June 2, 1967), K. Adam and co-workers (to BASF).
53. Belg. Pat. 763,109 (Aug. 17, 1971), (to ICI).
54. U.S. Pat. 2,284,525 (May 26, 1942), A. W. Larchar and co-workers (to DuPont).
55. U.S. Pat. 3,696,153 (Oct. 3, 1972), B. J. Kershaw and co-workers (to E. I. du Pont de Nemours & Co., Inc.).
56. Brit. Pat. 568,941 (Apr. 26, 1945), (to DuPont Co.).
57. U.S. Pat. 3,299,116 (Jan. 17, 1967), R. Romani and M. Ferri (to Societa Rodiatoce).
58. U.S. Pat. 3,153,084 (Oct. 13, 1964), T. M. Veazey and co-workers (to E. I. du Pont de Nemours & Co., Inc.).
59. Brit. Pat. 893,709 (Apr. 11, 1962), (to Chemstrand).
60. U.S. Pat. 3,242,204 (Mar. 22, 1966), W. A. Lozier (to E. I. du Pont de Nemours & Co., Inc.).
61. U.S. Pat. 3,242,204 (Mar. 22, 1966), M. Decker and co-workers (to BASF).
62. U.S. Pat. 2,680,761 (June 8, 1954), R. H. Hallwell (to E. I. du Pont de Nemours & Co., Inc.).
63. U.S. Pat. 2,749,359 (June 5, 1956), W. H. Calkins and co-workers (to E. I. du Pont de Nemours Co., Inc.).
64. U.S. 2,532,312 (Dec. 5, 1950), L. E. Romilly (to E. I. du Pont de Nemours & Co., Inc.).
65. M. M. Baizer, *J. Electrochem. Soc.* **111**, 215 (1964).
66. U.S. Pat. 2,711,405 (June 21, 1955), A. W. Anderson (to E. I. du Pont de Nemours & Co., Inc.).
67. *Vazo® Polymerization Initiators*, Technical Brochure, E. I. du Pont de Nemours & Co., Inc., Wilmington, Del., 1979.
68. C. Walling, *J. Polym. Sci.* **14**, 214 (1954).
69. U.S. Pat. 2,598,811 (June 3, 1952), J. E. Mahan and S. D. Turk (to Phillips Petroleum).
70. Can. Pat. 545,630 (Sept. 3, 1957), R. T. Corkum and J. M. Salsburg (to American Cyanamid).
71. W. I. Denton and co-workers, *Ind. Eng. Chem.* **42**, 796 (1950).
72. U.S. Pat. 2,736,739 (Feb. 28, 1956), D. C. England and G. V. Mock (to E. I. du Pont de Nemours & Co., Inc.).
73. Brit. Pat. 968,752 (Sept. 2, 1964), M. H. Richmond (to Distillers Co.).
74. Can. Pat. 722,712 (Nov. 30, 1965), C. Herschmann (to Lonza Ltd.).
75. Fr. Pat. 1,170,116 (Jan. 9, 1959), C. R. Stephens (to Chas. Pfizer & Co.).
76. T. L. Junod, *NASA Technical Memorandum 73866*, Cleveland, Ohio, 1978.
77. *Technical Bulletin 164*, Sherwin-Williams Co., Cleveland, Ohio, 1979.
78. W. G. Toland and L. L. Ferstandig, *J. Org. Chem.* **23**, 1350 (1958).
79. Cyanoacetic Esters, *Technical Brochure*, Kay-Fries, Inc., Montvale, N.J.

ROBERT A. SMILEY
E. I. du Pont de Nemours & Co., Inc.

NITRO ALCOHOLS

A nitro alcohol is formed when an aliphatic nitro compound with a hydrogen atom on the carbon that bears the nitro group reacts with an aldehyde in the presence of a base (see Alcohols). Many such compounds have been synthesized, but only those formed by the condensation of formaldehyde and the lower nitroparaffins are marketed commercially (see Nitroparaffins). The condensation may occur one to three times, depending on the number of hydrogen atoms on the carbon with the nitro group, and yield nitro alcohols with one to three hydroxyl groups. In addition to the mononitro compounds, monohydric and dihydric dinitro alcohols have been prepared but are not available commercially.

$$RR'CHNO_2 + CH_2O \overset{OH^-}{\rightleftharpoons} RR'C(CH_2OH)NO_2$$

where R and R′ = alkyl groups.

$$RCH(CH_2OH)NO_2 + CH_2O \overset{OH^-}{\rightleftharpoons} RC(CH_2OH)_2NO_2$$

where R = alkyl and R′ = H.

$$HC(CH_2OH)_2NO_2 + CH_2O \overset{OH^-}{\rightleftharpoons} C(CH_2OH)_3NO_2$$

where R and R′ = H. The formation, properties, and reactions of nitro alcohols are reviewed in refs. 1–2.

Physical Properties

The physical properties of the commercially available nitro alcohols are given in Table 1. When pure, these nitro alcohols are white crystalline solids except for nitrobutanol. They are thermally unstable above 100°C and purification by distillation is a hazardous procedure.

The nitro alcohols generally are soluble in water and in oxygenated solvents, eg, alcohols. The monohydric nitro alcohols are soluble in aromatic hydrocarbons; the diols are only moderately soluble even at 50°C; at 50°C, the triol is insoluble.

Chemical Properties

The nitro alcohols can be reduced to the corresponding alkanolamines (qv). Commercially, reduction is accomplished by hydrogenation of the nitro alcohol in methanol in the presence of Raney nickel. Convenient operating conditions are 30°C and 6900 kPa (1000 psi). Production of alkanolamines constitutes the largest single use of nitro alcohols.

Nitro alcohols form salts upon mild treatment with alkalies. Acidification causes separation of the nitro group as N_2O from the parent compound and results in the formation of carbonyl alcohols, ie, hydroxy aldehydes from primary nitro alcohols and α-ketols from secondary nitro alcohols.

$$2\,RCHOHCH{=}N^+\!\!\begin{array}{c}O^-\\\backslash\,O^-\end{array}M^+ \xrightarrow{H^+} 2\,RCHOHCH{=}N^+\!\!\begin{array}{c}O^-\\\backslash\,O^-\end{array}H^+ \longrightarrow$$

$$2\,RCHOHCHO \;+\; N_2O \;\longleftarrow$$

Table 1. Physical Properties of Nitro Alcohols[a]

Compound	CAS Registry No.	Structural formula	Mol wt	Mp, °C	Bp, °C	Solubility in water at 20°C, g/100 mL
2-nitro-1-butanol (NB)	[609-31-4]	$\begin{array}{c} H \\ | \\ CH_3CH_2CCH_2OH \\ | \\ NO_2 \end{array}$	119.12	−47 to −48	105[b]	54
2-methyl-2-nitro-1-propanol (NMP)	[76-39-1]	$\begin{array}{c} CH_3CCH_2OH \\ | \quad | \\ NO_2 \; CH_3 \end{array}$	119.12	90	94[c]	350
2-methyl-2-nitro-1,3-propanediol (NMPD)	[77-49-6]	$\begin{array}{c} HOCH_2CCH_2OH \\ | \\ NO_2 \\ | \\ C_2H_5 \end{array}$	135.12	ca 160	dec	80
2-ethyl-2-nitro-1,3-propanediol (NEPD)	[597-09-1]	$\begin{array}{c} HOCH_2CCH_2OH \\ | \\ NO_2 \\ | \\ CH_2OH \end{array}$	149.15	56	dec	400
2-hydroxymethyl-2-nitro-1,3-propanediol	[126-11-4]	$\begin{array}{c} HOCH_2CCH_2OH \\ | \\ NO_2 \end{array}$	151.12	175–176	dec	220

[a] Ref. 3.
[b] At 1.3 kPa (10 mm Hg).
[c] At 1.95 kPa (15 mm Hg).

911

The hydroxyl groups in polyhydric nitro alcohols readily react with aldehydes to form cyclic acetals. For example, formaldehyde reacts with 2-methyl-2-nitro-1,3-propanediol in the presence of p-toluenesulfonic acid (PTSA) to yield 5-methyl-5-nitro-1,3-dioxane.

$$CH_2O + HOCH_2\overset{\overset{\displaystyle CH_3}{|}}{\underset{\underset{\displaystyle NO_2}{|}}{C}}CH_2OH \xrightarrow{PTSA} \text{[cyclic structure]} + H_2O$$

Similar reactions have been observed with the more reactive ketones. Noncyclic acetals have been prepared from aldehydes and two molecules of monohydric nitro alcohols.

Chloromethyl ethers may be obtained by the reaction of nitro alcohols with formaldehyde and hydrogen chloride.

$$CH_3\overset{\overset{\displaystyle CH_3}{|}}{\underset{\underset{\displaystyle NO_2}{|}}{C}}CH_2OH + CH_2O + HCl \longrightarrow CH_3\overset{\overset{\displaystyle CH_3}{|}}{\underset{\underset{\displaystyle NO_2}{|}}{C}}CH_2OCH_2Cl + H_2O$$

Because of the high reactivity of the chlorine atom, such chloromethyl ethers are convenient intermediates for synthesis.

Nitro alcohols react with amines to form nitro amines:

$$CH_3\overset{\overset{\displaystyle CH_3}{|}}{\underset{\underset{\displaystyle NO_2}{|}}{C}}CH_2OH + HNRR' \longrightarrow CH_3\overset{\overset{\displaystyle CH_3}{|}}{\underset{\underset{\displaystyle NO_2}{|}}{C}}CH_2NRR' + H_2O$$

Such a reaction can be carried out with a wide variety of primary and secondary amines, both aliphatic and aromatic; a basic catalyst is required if aromatic amines are involved. The products of reactions between dihydric nitro alcohols and amines are nitro diamines, many of which are good fungicides (qv).

Dihydric nitro alcohols, primary amines, and formaldehyde react to yield nitrohexahydropyrimidines (4):

$$RC(CH_2OH)_2 + 2\,R'NH_2 + CH_2O \longrightarrow \text{[ring structure]} + 3\,H_2O$$

Nitrohexahydropyrimidines can be reduced to the corresponding amines, some of which are good fungicides or bactericides, eg, hexetidine [5-amino-1,3-bis(2-ethylhexyl)-5-methylhexahydropyrimidine].

Esters of nitro alcohols with primary alcohol groups can be prepared from the nitro alcohol and the organic acid, but nitro alcohols with secondary alcohol groups can be esterified only through the use of the acid chloride or anhydride.

gem-Dinitro alcohols can be prepared from either monohydric or dihydric mononitro alcohols by reaction with sodium nitrite and silver nitrate in the presence of sodium hydroxide. Alternatively, *gem*-dinitro alcohols can be obtained from nitroalkanes in three steps: (*1*) treatment with sodium hydroxide and a halogen to yield a halonitroalkane; (*2*) treatment of the halonitroalkane with a metal nitrite to form a *gem*-dinitroalkane (*5*); (*3*) reaction of the dinitroalkane with formaldehyde to yield the corresponding dinitro alcohol. Dinitro alcohols are esterified readily by acid chlorides in pyridine.

The nitrate esters of the nitro alcohols are obtained easily by treatment with nitric acid. The resulting products have explosive properties but are not used commercially.

Other reactions involving the hydroxyl functionality of nitro alcohols include the conversion of —OH to —Cl by phosphorus pentachloride, and preparation of phosphate esters by the action of phosphorus oxychlorides.

Primary nitro alcohols react with strong acids in the presence of water to produce hydroxylammonium salts and substituted organic acids. For example, 1-nitro-2-propanol yields hydroxylammonium acid sulfate and lactic acid when treated with sulfuric acid.

$$CH_3CH(OH)CH_2NO_2 + H_2SO_4 + H_2O \rightarrow CH_3CH(OH)CO_2H + [NH_3OH]HSO_4$$

On dehydration, nitro alcohols yield nitroolefins. The ester of the nitro alcohol is treated with caustic or is refluxed with a reagent, eg, phthalic anhydride or phosphorus pentoxide. A milder method involves the use of methanesulfonyl chloride to transform the hydroxyl into a better leaving group. Yields up to 80% after a reaction time of 15 min at 0°C have been reported (*6*).

Nitro alcohols react with carbon disulfide to form xanthates.

Monohydric nitro alcohols are converted to chloronitroparaffins by thionyl chloride; with nitrodiols or 2-hydroxymethyl-2-nitro-1,3-propanediol, the nitroalkyl sulfite is obtained (*7–8*).

In aqueous solution, nitro alcohols decompose at pH >7.0 with the formation of formaldehyde. One mol of formaldehyde is released per mol of monohydric nitro alcohol, and two mol of formaldehyde are released by the nitrodiols. However, 2-hydroxymethyl-2-nitro-1,3-propanediol gives only two mol of formaldehyde instead of the expected three mol. The rate of release of formaldehyde increases with the pH or the temperature or both.

Manufacture and Processing

The nitro alcohols that are available in commercial quantities are manufactured by the condensation of nitroparaffins with formaldehyde. These condensations are equilibrium reactions, and potential exists for the formation of polymeric materials. Therefore, reaction conditions, eg, reaction time, temperature, mol ratio of the reactants, catalyst level, and catalyst removal, must be carefully controlled in order to obtain the desired nitro alcohol in good yield (9). Paraformaldehyde can be used in place of aqueous formaldehyde. A wide variety of basic catalysts, including amines, quaternary ammonium hydroxides, and inorganic hydroxides and carbonates, can be used. After completion of the reaction, the reaction mixture must be made acidic, either by addition of mineral acid or by removal of base by an ion-exchange resin in order to prevent reversal of the reaction during the isolation of the nitro alcohol (see Ion exchange).

The purification of liquid nitro alcohols by distillation should be avoided because violent decompositions and, in at least one case, detonation have occurred when distillation was attempted. However, if the distillation of a nitro alcohol cannot be avoided, the utmost caution should be exercised. Reduced pressure should be utilized, ie, ca 0.1 kPa (1 mm Hg). The temperature of the liquid should not exceed 100°C; hot water should be used as the heating bath. A suitable explosion-proof shield should be placed in front of the apparatus. At any rise in pressure, the distillation should be stopped immediately. The only commercially produced liquid nitro alcohol, 2-nitro-1-butanol, is not distilled because of the danger of decomposition. Instead, it is isolated as a residue after the low boiling impurities have been removed by a vacuum treatment at a relatively low temperature.

Economic Aspects

2-Nitro-1-butanol is available as a developmental chemical and only in drums containing 226.8 kg net wt. The price (1980) per drum is $1.09/kg. 2-Methyl-2-nitro-1-propanol is available in a pelletized form containing stearic acid as a binder. It is available in 22.7-kg bags at $2.75/kg. 2-Ethyl-2-nitro-1,3-propanediol and 2-methyl-2-nitro-1,3-propanediol are available in developmental quantities for $0.86/kg. 2-Hydroxymethyl-2-nitro-1,3-propanediol is available as the Tris Nitro brand of tris(hydroxymethyl)nitromethane [126-11-4] either as the crystalline material or as a 50 wt % aqueous concentrate. The crystals are available in 59-kg fiber drums at $6.60/kg. The concentrate is available in bulk as well as in drums. The truckload price is $1.09/kg and the single-drum price is $1.25/kg.

Health and Safety Factors (Toxicology)

Acute oral LD_{50} data for nitro alcohols in mice are: 2-nitro-1-butanol, 1.2 g/kg; 2-methyl-2-nitro-1-propanol, 1.0 g/kg; 2-methyl-2-nitro-1,3-propanediol, 4.0 g/kg; 2-ethyl-2-nitro-1,3-propanediol, 2.8 g/kg; and 2-hydroxymethyl-2-nitro-1,3-propanediol, 1.9 g/kg (3). Because of their low volatility, the nitro alcohols present no vapor inhalation hazard. They are nonirritating to the skin and, except for 2-nitro-1-butanol, are nonirritating when introduced as a 1 wt % aqueous solution in the eye of the rabbit. When 0.1 mL of 1 wt % commercial-grade 2-nitro-1-butanol in water is introduced into the eyes of rabbits, severe and permanent corneal scarring results. This anomalous behavior may be caused by the presence of a nitroolefin impurity in the unpurified commercial product.

Uses

The nitro alcohols are useful as intermediates for chemical synthesis. In particular they are used to introduce a nitro functionality and, by reduction of the resultant intermediate, an amino functionality.

Antimicrobials. In slightly alkaline aqueous solutions, nitro alcohols are useful for the control of microorganisms, eg, in cutting fluids, cooling towers, oil-field flooding, drilling muds, etc (10–17) (see Petroleum; Industrial antimicrobial agents). 2-Hydroxymethyl-2-nitro-1,3-propanediol, 2-ethyl-2-nitro-1,3-propanediol, and 2-methyl-2-nitro-1,3-propanediol are registered pesticides with the EPA.

Polymers. All nitro alcohols are sources of formaldehyde for cross-linking operations with polymers of urea, melamine, phenols, resorcinol, etc (see Amino resins). Nitrodiols and 2-hydroxymethyl-2-nitro-1,3-propanediol can be used as polyols to form polyester or polyurethane products (see Polyesters; Urethane polymers). 2-Methyl-2-nitro-1-propanol is used in tires to promote the adhesion of rubber to tire cord (qv). Nitro alcohols are used as hardening agents in photographic processes, and 2-hydroxymethyl-2-nitro-1,3-propanediol is a cross-linking agent for starch adhesives, urea resins, or wool and in tanning operations (18–25).

Stabilizers. Nitro alcohols can be used to prevent the decomposition of p-phenylenediamine color-developing agents (26). 2-Hydroxymethyl-2-nitro-1,3-propanediol and 2-nitro-1-butanol have been used as additives for the stabilization of 1,1,1-trichloroethane.

Other. 2-Nitro-1-butanol is an excellent solvent for many polyamide resins, cellulose acetate butyrate, and ethyl cellulose. It can be utilized in paint removers for epoxy-based coatings. 2-Hydroxymethyl-2-nitro-1,3-propanediol is useful for control of odors in chemical toilets. Its slow release of formaldehyde ensures prolonged action to control odor, and there is no reodorant problem which sometimes is associated with the use of free formaldehyde. 2-Hydroxymethyl-2-nitro-1,3-propanediol solutions are effective preservative and embalming fluids. The slow liberation of formaldehyde permits thorough penetration of the tissues before hardening.

BIBLIOGRAPHY

"Nitro Alcohols" in ECT 1st ed., Vol. 7, pp. 375–381, by E. B. Hodge, Commercial Solvents Corp.; "Nitro Alcohols" in ECT 2nd ed., Vol. 13, pp. 826–834 by R. H. Dewey, Commercial Solvents Corporation.

1. H. B. Hass and E. F. Riley, Chem. Rev. **32**, 373 (1943).
2. B. M. Vanderbilt and H. B. Hass, Ind. Eng. Chem. **32**, 34 (1940).
3. NP Series, Technical Data Sheet No. 15, International Minerals & Chemical Corp., Terre Haute, Ind., Jan. 1977.
4. M. Senkus, J. Am. Chem. Soc. **68**, 10 (1946).
5. Fr. Pat. 1,326,923 (May 10, 1963), E. E. Hamel (to Aerojet-General Corp.).
6. J. Melton and J. E. McMurry, J. Org. Chem. **40**, 2138 (1975).
7. U.S. Pat. 2,397,358 (March 26, 1946), S. P. Lingo (to Commercial Solvents Corporation).
8. U.S. Pat. 2,471,274 (May 24, 1949), S. P. Lingo (to Commercial Solvents Corporation).
9. W. E. Noland, "2-Nitroethanol" in H. E. Baumgarten, ed., Organic Syntheses, Coll. Vol. 5, John Wiley & Sons, Inc., New York, 1973, pp. 833–838.

10. H. O. Wheeler and E. O. Bennett, *Appl. Microbiol.* **4,** 122 (1956).
11. E. O. Bennett and H. N. Futch, *Lubr. Eng.* **16,** 228 (1960).
12. U.S. Pat. 3,001,936 (Sept. 26, 1961), E. O. Bennett and E. B. Hodge (to Commercial Solvents Corporation).
13. U.S. Pat. 3,789,008 (Jan. 29, 1974), D. W. Young (to Elco Chemicals Inc.).
14. Ger. Pat. 2,530,522 (Jan. 27, 1977), P. Voegele (to Henkel & Cie G.m.b.H.).
15. U.S. Pat. 3,542,553 (Nov. 25, 1970), M. S. Beach (to Eastman Kodak Co.).
16. U.S. Pat. 4,113,444 (Sept. 12, 1978), P. M. Bunting and co-workers (to Gulf Research & Development Co.).
17. U.S. Pat. 3,592,893 (July 13, 1971), H. G. Nosler and co-workers (to Henkel & Cie G.m.b.H.).
18. U.S. Pat. 3,897,583 (July 29, 1975), C. Bellamy (to Uniroyal, S.A.).
19. M. L. Happich and co-workers, *J. Am. Leather Chem. Assoc.* **65**(3), 135 (1970).
20. U.S. Pat. 3,809,585 (May 7, 1974), H. L. Greenberg (to U.S. Dept. of the Navy).
21. Jpn. Pat. 74 94,748 (Sept. 9, 1974), Y. Hori and co-workers (to Dai-ichi Kogyo Seiyaku Co. Ltd.).
22. U.S. Pat. 3,982,993 (Sept. 28, 1976), R. L. Fife (to Georgia-Pacific Corp.).
23. Ger. Pat. 1,958,914 (Aug. 6, 1970), J. Delmenico and co-workers (to Commonwealth Scientific and Ind. Research Organization).
24. U.S. Pat. 3,475,383 (Oct. 28, 1969), F. D. Stewart (to B. F. Goodrich Co.).
25. U.S. Pat. 4,039,495 (August 2, 1977), J. H. Hunsucker (to IMC Chemical Group, Inc.).
26. Brit. Pat. 1,468,015 (March 23, 1977), R. Cowell and co-workers (to May and Baker Ltd.).

ROBERT H. DEWEY
ALLEN F. BOLLMEIER, JR.
International Minerals & Chemical Corporation

NITROBENZENE AND NITROTOLUENES

Nitrobenzene

Nitrobenzene [*98-95-3*] (oil of mirbane), $C_6H_5NO_2$, is a pale yellow liquid with an odor that resembles bitter almonds. Depending upon the compound's purity, its color varies from pale yellow to yellowish brown.

Nitrobenzene was first synthesized in 1834 by treating benzene with fuming nitric acid (1), and it was first produced commercially in England in 1856 (2). The relative ease of aromatic nitration has contributed significantly to the large and varied industrial applications of nitrobenzene, other aromatic nitro compounds, and their derivatives.

Physical Properties.   Nitrobenzene is soluble readily in most organic solvents and is miscible in all proportions with diethyl ether and benzene. Nitrobenzene is only slightly soluble in water with a solubility of 0.19 parts per 100 parts of water at 20°C and 0.8 pph at 80°C. Nitrobenzene is a good organic solvent and, because aluminum chloride dissolves in it, it is used as a solvent in Friedel-Crafts reactions. The physical properties of nitrobenzene are summarized in Table 1.

Table 1.   Physical Properties of Nitrobenzene

Property	Value	Reference
mp, °C	5.85	3
bp, °C		
at 101 kPa[a]	210.9	3
at 53 kPa[a]	184.5	3
at 13 kPa[a]	139.9	3
at 6.7 kPa[a]	120.2	3
at 4.0 kPa[a]	108.2	3
at 1.3 kPa[a]	85.4	3
at 0.13 kPa[a]	53.1	3
density, g/cm^3		
$d_4^{1.5}$ (solid)	1.344	3
d_4^0 (supercooled liquid)	1.223	3
d_4^{10}	1.213	3
d_4^{25}	1.199	4
refractive index, n_D^{20}	1.55296	3
viscosity (at 15°C), mPa·s (= cP)	2.17×10^{-2}	5
surface tension (at 20°C), mN/m (= dyn/cm)	43.35	5
dielectric constant (at 25°C)	34.89	5
specific heat (at 30°C), J/g[b]	1.418	6
latent heat of vaporization, J/g[b]	331	6
latent heat of fusion, J/g[b]	94.1	5
heat of combustion (at constant volume), MJ/mol[b]	3.074	3
flash point (closed cup), °C	88	4
autoignition temperature, °C	482	7
explosive limit (at 93°C), vol % in air	1.8	7
vapor density (air = 1)	4.1	4

[a] To convert kPa to mm Hg, multiply by 7.5.

[b] To convert J to cal, divide by 4.184.

Chemical Properties.   Nitrobenzene reactions involve substitution in the aromatic ring and reactions involving the nitro group. Under electrophilic conditions, the substitution occurs at a slower rate than for benzene, and the nitro group promotes meta substitution. Nitrobenzene can undergo halogenation, sulfonation, and nitration but it does not undergo Friedel-Crafts reactions. Under nucleophilic conditions, the nitro group promotes ortho and para substitution. The reduction of the nitro group to yield aniline is the most commercially important reaction of nitrobenzene. Usually the reaction is carried out by the catalytic hydrogenation of nitrobenzene, either in the gas phase or in solution, or by using iron borings and dilute hydrochloric acid, eg, in the Béchamp process. Depending upon the conditions, the reduction of nitrobenzene can lead to a variety of products. The series of reduction products is shown in Figure 1 (see Amines by reduction).

Nitrosobenzene, *N*-phenylhydroxylamine, and aniline are primary reduction products. Azoxybenzene is formed by the condensation of nitrosobenzene and *N*-phenylhydroxylamine in alkaline solutions, and azoxybenzene can be reduced to form azobenzene and hydrazobenzene. The reduction products of nitrobenzene under various conditions are given in Table 2.

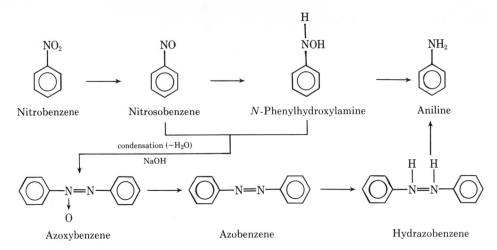

Figure 1.   Reduction products of nitrobenzene.

Table 2.   Reaction Products of the Reduction of Nitrobenzene Under Different Conditions

Reagents	Name
Fe, Zn, or Sn + HCl	aniline
H$_2$ + metal catalyst + heat (gas phase or solution)	aniline
SnCl$_2$ + acetic acid	aniline
Zn + NaOH	hydrazobenzene
	azobenzene
Zn + H$_2$O	N-phenylhydroxylamine
Na$_3$AsO$_3$	azoxybenzene
LiAlH$_4$	azobenzene
Na$_2$S$_2$O$_3$ + Na$_3$PO$_4$	sodium phenylsulfamate, C$_6$H$_5$NHSO$_3$Na

Manufacture and Processing.   Nitrobenzene is manufactured commercially by the direct nitration of benzene using a mixture of nitric and sulfuric acids, which commonly is referred to as mixed acid or nitrating acid. Since two phases are formed in the reaction mixture and the reactants are distributed between them, the rate of nitration is controlled by mass transfer between the phases as well as by chemical kinetics (8). The reaction vessels are acid-resistant, cast-iron or steel vessels that are equipped with efficient agitators. By vigorous agitation, the interfacial area of the heterogenous reaction mixture is maintained as high as possible, thereby enhancing the mass transfer of reactants. The reactors contain internal cooling coils which control the temperature of the highly exothermic reaction (see Explosives and propellants; Nitration).

Nitrobenzene can be produced by either a batch or a continuous process. With a typical batch process, the reactor is charged with benzene, and then the nitrating acid (56–60 wt % H$_2$SO$_4$, 27–32 wt % HNO$_3$, and 8–17 wt % H$_2$O) is added slowly below the surface of the benzene. The temperature of the mixture is maintained at 50–55°C by adjusting the feed rate of the mixed acid and the amount of cooling. The temperature can be raised to ca 90°C toward the end of the reaction to promote completion of reaction. The reaction mixture is fed into a separator where the spent acid settles to the bottom and is drawn off to be refortified. The crude nitrobenzene is drawn from the top of the separator and is washed in several steps with dilute sodium carbonate and then water. Depending upon the desired purity of the nitrobenzene, the product may be distilled. Based on a yield of 1000 kg of nitrobenzene, material requirements for the process are as follows (9)

Material	Quantity, kg
benzene	650
sulfuric acid	720
nitric acid	520
water	110
sodium carbonate	10

Usually a slight excess of benzene is used to ensure that little or no nitric acid remains in the spent acid. The batch reaction time generally is 2–4 h, and typical yields are 95–98 wt % based on the benzene charged.

A continuous nitration process generally offers lower capital costs and more efficient labor usage than a batch process; thus, most, if not all, of the nitrobenzene producers use continuous processes. The basic sequence of operations for a continuous process is the same as that for a batch process; however, for a given rate of production, the size of the nitrators is much smaller in a continuous process. A 0.114-m³ (30-gal) continuous nitrator has roughly the same production capacity as a 5.68-m³ (1500-gal) batch reactor. In contrast to the batch process, a continuous process typically utilizes a lower nitric acid concentration and, because of the rapid and efficient mixing in the smaller reactors, higher reaction rates are observed.

A typical continuous process for the production of nitrobenzene is given in Figure 2. Benzene and the nitrating acid (56–65 wt % H_2SO_4, 20–26 wt % HNO_3, and 15–18

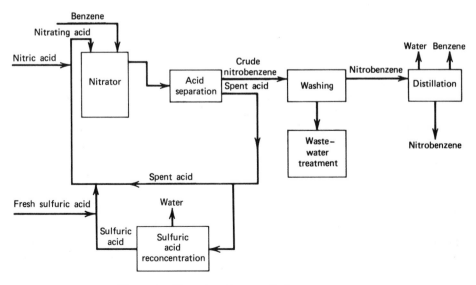

Figure 2. Typical continuous nitrobenzene process.

wt % water) are fed into the nitrator, which can be a stirred cylindrical reactor with internal cooling coils or a cascade of such reactors. The nitrator also can be designed as a tubular reactor (eg, a tube-and-shell heat exchanger with appropriate cooling) involving turbulent flow (10). Generally, with a tubular reactor, the reaction mixture is pumped through the reactor in a recycle loop and a portion of the mixture is withdrawn and fed into the separator. A slight excess of benzene usually is fed into the nitrator to ensure that the nitric acid in the nitrating acid is consumed to the maximum possible extent and to minimize the formation of dinitrobenzene. The temperature of the nitrator is maintained at 50–100°C by varying the amount of cooling. The reaction mixture flows from the nitrator into a separator or centrifuge where it is separated into two phases. The aqueous phase or spent acid is drawn from the bottom and is concentrated in a sulfuric acid reconcentration step or is recycled to the nitrator where it is mixed with nitric and sulfuric acid immediately prior to being fed into the nitrator. The crude nitrobenzene flows through a series of washer-separators where residual acid is removed by washing with a dilute sodium carbonate solution followed by final washing with water. The product then is distilled to remove water and benzene, and, if required, the nitrobenzene can be refined by vacuum distillation. Reaction times of 10–30 min are typical, and theoretical yields are 96–99%. The nitration process is unavoidably associated with the disposal of wastewater from the washing steps. This water principally contains nitrobenzene, some sodium carbonate, and inorganic salts from the neutralized spent acid which was present in the product. Generally, the wastewater is extracted with benzene to remove the nitrobenzene, and the benzene that is dissolved in the water is stripped from the water prior to the final waste treatment.

An adiabatic process for the nitration of benzene has been developed which produces very low levels of the by-product dinitrobenzene (11). The necessity of removing the heat of reaction with extensive cooling is eliminated, and the heat can be used for the reconcentration of the sulfuric acid. The concept is applicable to both batch and continuous processes. The nitrating acid is composed of 60–70 wt % H_2SO_4, 5–8.5 wt % HNO_3, and not less than 25% water. A slight molar excess of benzene is used to promote the complete denitration of the spent acid. The initial temperature is 60–75°C, and the final temperature is 105–145°C. Reaction times are 0.5–7.5 min and vigorous agitation is required.

In the azeotropic nitration of benzene, the need to reconcentrate the sulfuric acid in a separate step is partially or fully eliminated (12–13). The nitration reaction is carried out at 120–160°C, with the excess water being distilled from the nitrator as an azeotrope with benzene. After being separated from the product, the sulfuric acid is recycled to the nitrator without being concentrated. Typical concentrations for the sulfuric acid in the nitrator are 55–70 wt % relative to the sulfuric acid plus the water that is present. An excess of benzene is used to facilitate the removal of water as the azeotrope. After the azeotrope is condensed, water is separated and the benzene is returned to the nitrator; the benzene must be partially or completely vaporized for temperature control of the nitrator. In a duplex process, the nitration is carried out in an azeotropic first stage which is followed by a lower temperature, mixed-acid second stage (13). All or part of the benzene resulting from the azeotrope of the first stage is taken to the second stage for further nitration; thus, the necessity of vaporizing the benzene is avoided (see also Azeotropic and extractive distillation).

A number of other techniques have been investigated for the nitration of benzene.

Processes in which only nitric acid is used have been described. One such process involves the utilization of 40–68 wt % HNO_3 and temperatures of 50–90°C (14). Molar ratios of HNO_3:benzene are 6:1 to 15:1, and it is necessary to reconcentrate the nitric acid by distillation. Another technique involves the use of a perfluorosulfonic acid polymeric membrane which separates the benzene and the nitric acid (15). The use of methylene chloride as a moderating solvent in the nitration of benzene with nitric and sulfuric acid mixtures results in efficient extraction of the undegraded product and improved yields (16).

Economic Aspects. Factors affecting the economic aspects for nitrobenzene manufacture are closely related to the process that is used and the recovery of the spent acid; however, the cost of the raw materials typically accounts for at least 85% of the manufacturing cost. Annual production, sales, and prices in the United States are summarized in Table 3. In January 1980, the price of double-distilled nitrobenzene in tanks was $0.68/kg.

In bulk, nitrobenzene is sold in 0.208-m^3 (55 gal) drums and in tank trucks and tank cars. The major manufacturers of nitrobenzene in the United States are American Cyanamid Company; E. I. du Pont de Nemours & Company, Inc.; First Mississippi Corporation; Mobay Chemical Corporation; and Rubicon Chemicals, Inc.

Specifications. Specifications for double-distilled nitrobenzene are listed in Table 4.

Analytical and Test Methods. Several qualitative spot tests are applicable to nitrobenzene and depend upon a characteristic color developed by its reaction with certain reagents (18). However, these tests are not specific for nitrobenzene since other aromatic nitro compounds yield colored products which are similar or only slightly different in color. One example of such a test is the heating or fusing of nitrobenzene with diphenylamine, which yields a reddish-yellow color. In general, colorimetric methods also are subject to interferences from other aromatic nitro compounds. Certain colorimetric methods are based upon the nitration of nitrobenzene to *m*-dinitrobenzene and subsequent determination by the generation of a red-violet color with acetone and alkali. A general titrimetric method for the determination of aromatic nitro compounds is based upon reduction with titanium(III) sulfate or chloride in acidic

Table 3. Annual Production, Sales, and Prices of Nitrobenzene [a]

Year	Production, metric tons	Sales, metric tons	Price[b], $/kg
1960	73,600	2,800	0.24
1962	90,500	4,100	0.24
1964	108,500	4,300	0.21
1966	148,300	6,200	0.21
1968	180,500	5,200	0.19
1970	248,400	7,600	0.19
1972	250,000	5,700	0.19
1974	229,800	8,800	0.21
1975	187,900	8,700	0.42
1976	185,500	8,800	0.51
1977	250,500	8,700	0.51

[a] Ref. 17.

[b] Prices are for double-distilled material in tanks. Prices for 1960–1974 include freight.

Table 4.   Specifications of Double-Distilled Nitrobenzene

Property	Value
purity, %	≥99.8
color	clear, light yellow to brown
freezing point, °C	≥5.13
distillation range[a] (first drop), °C	≥207
dry point[b], °C	212
moisture, %	<0.1
acidity (as HNO$_3$), %	<0.001

[a] 95% boiling at 207–210°C.
[b] Temperature at which no liquid remains.

solution followed by a back titration of the excess titanium(III) ions with standard ferric alum solution (19). Instrumental methods, eg, polarography, gas chromatography, and high speed liquid chromatography, can be used to provide a nitrobenzene analysis not subject to the interferences noted above. In industry, the freezing point and distillation range of nitrobenzene are used as indicators of purity.

Health and Safety Factors (Toxicology).   Nitrobenzene is a very toxic substance; the maximum allowable concentration for nitrobenzene is 1 ppm or 5 mg/m^3 (20). It is readily absorbed by contact with the skin and by inhalation of the vapor. If a worker were exposed for 8 h to 1 ppm nitrobenzene in the working atmosphere, ca 25 mg of nitrobenzene would be absorbed, of which about one third would be by skin absorption and the remainder by inhalation. The primary effect of nitrobenzene is the conversion of hemoglobin to methemoglobin; thus, the conversion eliminates hemoglobin from the oxygen-transport cycle. Exposure to nitrobenzene may irritate the skin and eyes. Nitrobenzene affects the central nervous system and produces fatigue, headache, vertigo, vomiting, general weakness, and in some cases, unconsciousness and coma. There generally is a latent period of 1–4 h before signs or symptoms appear. Nitrobenzene is a very powerful methemoglobin former, and cyanosis appears when the methemoglobin level reaches 15%. Chronic exposure can lead to spleen and liver damage, jaundice, and anemia. Alcohol ingestion tends to increase the toxic effects of nitrobenzene; thus, alcohol in any form should not be ingested by the victim of nitrobenzene poisoning for several days after the nitrobenzene poisoning or exposure. Impervious protective clothing should be worn in areas where risk of splash exists. Ordinary work clothes that have been splashed should be removed immediately, and the skin should be washed thoroughly with soap and warm water. In areas of high vapor concentration (>1 ppm), full facemasks with organic-vapor canisters or air-supplied respirators should be used. Clean work clothing should be worn daily, and showering after each shift should be made mandatory.

With respect to the hazards of fire and explosion, nitrobenzene is classified as a moderate hazard when exposed to heat or flame. Nitrobenzene is classified by the ICC as a Class-B poisonous liquid.

Uses.   The most significant use of nitrobenzene is in the manufacture of aniline (see Amines, aromatic). Approximately 97–98 wt % of the nitrobenzene produced in the United States is converted to aniline. Fifty percent of this aniline is used in the production of 4,4'-methylenebis(phenyl isocyanate) (MDI) and polymeric MDI, 27% in rubber chemicals, 6% in dyes and intermediates, 5% in hydroquinone production,

3% in drugs, and 9% in other uses, eg, herbicides (qv) and fibers (see Isocyanates, organic). Nitrobenzene also is used as a solvent, eg, in Friedel-Crafts reactions (qv) and in the refining of certain lubricating oils.

DERIVATIVES

Mononitrochlorobenzenes. *Properties.* The physical properties of the ortho, meta, and para isomers of nitrochlorobenzene are summarized in Table 5.

o-Nitrochlorobenzene crystallizes in light yellow, monoclinic needles. It is insoluble in water and is very soluble in benzene, diethyl ether, and hot ethanol.

o-Nitrochlorobenzene reactions involve the nitro group, chlorine atom, and aromatic ring. The nitro group can be partially reduced to the corresponding intermediate or fully to the amino group. The aromatic ring can be nitrated, leading to the formation of 2,4-dinitro-1-chlorobenzene [97-00-7] and 2,6-dinitro-1-chlorobenzene [606-21-3], or it can be sulfonated, yielding 3-nitro-4-chlorobenzenesulfonic acid [121-18-6]. The chlorine atom can be displaced easily by nucleophilic attack by —OH, —OCH$_3$, —OC$_6$H$_6$, —NH$_2$, etc. Treatment of *o*-nitrochlorobenzene with aqueous sodium hydroxide at 130°C results in the formation of *o*-nitrophenol [88-75-5], and with aqueous methanolic potassium hydroxide at high temperature and pressure, *o*-nitroanisole [91-23-6] is formed. When *o*-nitrochlorobenzene is treated with aqueous ammonia under high temperature and pressure, *o*-nitroaniline [88-74-4] is formed. *o*-Nitrochlorobenzene condenses with aniline to form 2-nitrodiphenylamine [119-75-5].

m-Nitrochlorobenzene is a pale yellow crystalline solid which can exist as a stable or labile form in the solid state. It is insoluble in water, very soluble in benzene and diethyl ether, and soluble in acetone, chloroform, and hot ethanol. Unlike the ortho and para isomers, the chlorine atom of *m*-nitrochlorobenzene is not activated for nucleophilic substitution.

p-Nitrochlorobenzene crystallizes in light yellow monoclinic prisms. It is insoluble in water and very soluble in benzene, diethyl ether, and hot ethanol. *p*-Nitrochlorobenzene undergoes the same reactions described for the ortho isomer to yield the

Table 5. Physical Properties of Mononitrochlorobenzenes

Property	Values		
	p-Nitrochlorobenzene [88-73-3]	*m*-Nitrochlorobenzene [121-73-3]	*p*-Nitrochlorobenzene [100-00-5]
melting point, °C	32.5[a]	46 (stable)[b] 24 (labile)[b]	83[b]
boiling point, °C$_{kPa}$[c]	245.5$_{100}$[a] 119$_{1.1}$[b]	235.6$_{101}$[a]	242$_{101}$[b] 113$_{1.1}$[b]
density, g/cm^3	1.368$_4^{22}$[d] 1.305$_4^{80}$[a]	1.534$_4^{20}$[d] 1.343$_4^{50}$[a]	1.520$_4^{22}$[d] 1.298$_4^{91}$[a]
flash point, closed cup, °C	127[e]	127[e]	

[a] Ref. 6.
[b] Ref. 3.
[c] To convert kPa to mm Hg, multiply by 7.5.
[d] Ref. 7.
[e] Ref. 21.

analogous para derivatives. Tin(II) chloride and hydrochloric acid convert p-nitro-chlorobenzene to p-chloroaniline. The aromatic ring of the para isomer can undergo additional substitution by nitration to yield 2,4-dinitro-1-chlorobenzene, by chlorination to yield 1,2-dichloro-4-nitrobenzene [99-54-7], or by sulfonation to yield 2-chloro-5-nitrobenzene sulfonic acid [96-73-1]. The chlorine atom is activated and, as with the ortho isomer, can be easily displaced by nucleophilic attack. Treatment with aqueous ammonia at elevated temperature and pressure results in the formation of p-nitroaniline [100-01-6], and with aqueous sodium hydroxide under pressure, p-nitrophenol [100-02-7] is formed. p-Nitrochlorobenzene reacts with sodium disulfide to form 4,4'-dinitrodiphenyl disulfide [100-32-3] which is an intermediate in the preparation of sulfanilamide derivatives (see Chemotherapeutics).

Manufacture and Processing. Chlorobenzene can be nitrated at 40–70°C with a nitrating acid consisting of 52.5 wt % H_2SO_4, 35.5 wt % HNO_3, and 12 wt % H_2O. The technique and equipment are similar to that described for the nitration of benzene. The resulting product is a mixture of isomers containing ca 34 wt % o-nitrochlorobenzene, 65 wt % p-nitrochlorobenzene, and 1 wt % m-nitrochlorobenzene. The mixture is cooled to a temperature slightly above its freezing point, ie, ca 15°C, and a large portion of the para isomer slowly crystallizes and is separated from the mother liquor. The liquid mixture of isomers is separated by a combination of fractional distillation and crystallization. Other methods of preparing the mononitrochlorobenzenes are chlorination of nitrobenzene, the diazotization of nitro anilines and replacement by chlorine (Sandmeyer reaction), and the reaction of phosphorus pentachloride with the nitrophenols. These reactions are used on a laboratory scale but are not of commercial interest.

Economic Aspects. U.S. production of mononitrochlorobenzenes is ca 70,000 metric tons per year of which ca 25,000 t is the ortho isomer and 45,000 t is the para isomer. The meta isomer is not isolated in U.S. production. The bulk, fob prices of o-nitrochlorobenzene and p-nitrochlorobenzene are $1.37/kg and $1.46/kg, respectively (January, 1980). The mononitrochlorobenzenes are manufactured by E. I. du Pont de Nemours & Company, Inc. and Monsanto Chemical Co.

Health and Safety Factors (Toxicology). The mononitrochlorobenzenes are very toxic substances which may be absorbed through the skin and lungs, giving rise to methemoglobin. Their toxicity is about the same as or greater than that of nitrobenzene. The para isomer is less toxic than the ortho isomer, and the maximum allowable concentration which has been adopted for p-nitrochlorobenzene is 1 mg/m^3 (0.15 ppm) (20). The mononitrochlorobenzenes are moderate fire hazards when they are exposed to heat or flame. They are classified by the ICC as Class-B poisons. The same handling precautions should be used for these compounds as is used for nitrobenzene.

Uses. o-Nitrochlorobenzene is used in the synthesis of azo dye intermediates, eg, o-chloroaniline (Fast Yellow G Base), o-nitroaniline (Fast Orange GR Base), o-anisidine (Fast Red BB Base), o-phenetidine, and o-aminophenol (see Azo dyes). It also is used in corrosion inhibitors, pigments, and agricultural chemicals. p-Nitrochlorobenzene is used principally in the production of intermediates for azo and sulfur dyes. Other uses include pharmaceuticals, photochemicals, rubber chemicals, and insecticides (see Insect control technology). Typical intermediates manufactured from the para isomer are p-nitroaniline (Fast Red GC Base), p-anisidine, p-aminophenol, p-nitrophenol, p-phenylenediamine, 2-chloro-p-anisidine (Fast Red R Base), 2,4-dinitro-1-chlorobenzene, and 1,2-dichloro-4-nitrobenzene.

Other Nitrochlorobenzenes.   *2,4-Dinitrochlorobenzene.*   2,4-Dinitrochloro-benzene [97-00-7], $ClC_6H_3(NO_2)_2$, is a yellow solid which can exist in three forms, one stable and two labile. The stable α form crystallizes in yellow rhombic crystals from diethyl ether; mp α 53.4°C, β 43°C, γ 27°C; bp at 101 kPa (= 1 atm) 315°C with slight decomposition; d_4^{22} α 1.697 g/cm^3, d_4^{20} β 1.680 g/cm^3; flash point (closed cup) 194°C; vapor density (air = 1) 6.98 (6–7). 2,4-Dinitrochlorobenzene is insoluble in water and soluble in benzene, hot ethanol, diethyl ether, and carbon disulfide.

2,4-Dinitrochlorobenzene can be manufactured by either dinitration of chloro-benzene in fuming sulfuric acid or nitration of p-nitrochlorobenzene with mixed acids. Further substitution on the aromatic ring is very difficult because of the deactivating effect of the chlorine atom, but the chlorine is very reactive and is displaced even more readily than in the mononitrochlorobenzenes.

U.S. production of 2,4-dinitro-1-chlorobenzene amounts to 600–700 t/yr. It is used primarily in the manufacture of azo dyes; other areas include the manufacture of fungicides, rubber chemicals, and explosives. It is produced by Martin Marietta Chemicals, Sodyeco Division, and is sold in two grades: 44°C mp and 47°C mp. The prices of the 44°C and 47°C grades in 272-kg drums are $1.68/kg and $1.74/kg, re-spectively (January, 1980).

2,4-Dinitro-1-chlorobenzene is more toxic than nitrobenzene. It is an extremely powerful skin irritant and must be handled with great care.

1,2-Dichloro-4-nitrobenzene.   1,2-Dichloro-4-nitrobenzene [99-54-7], O_2N-$C_6H_3Cl_2$, crystallizes in needles from ethanol and has both a stable and a labile form. The stable α form has a melting point of 42–43°C, and the β form is a liquid which changes to the α form at 15°C (6); bp (at 101 kPa or 1 atm), 225–256°C; d_4^{75} 1.4558 g/cm^3; vapor density (air = 1), 6.6 (3,6–7). It is insoluble in water and soluble in diethyl ether, hot ethanol, and benzene. It can be prepared by the chlorination of p-nitro-chlorobenzene or by the nitration of o-dichlorobenzene. The isomers are separated by fractional distillation and crystallization. 1,2-Dichloro-4-nitrobenzene is produced by E. I. du Pont de Nemours & Co., Inc. and Blue Spruce Co., and its bulk, fob plant site price is $2.23/kg.

1,4-Dichloro-2-nitrobenzene.   1,4-Dichloro-2-nitrobenzene [89-61-2], O_2N-$C_6H_3Cl_2$, crystallizes in prisms or plates from ethanol; mp 56°C; bp (at 101 kPa or 1 atm) 267°C; d_4^{75} 1.4390 g/cm^3; vapor density (air = 1), 6.6 (3,7). It is insoluble in water and soluble in hot ethanol, diethyl ether, and benzene. It is prepared by the nitration of p-dichlorobenzene and is used extensively in the manufacture of dyestuff inter-mediates. Annual U.S. production is several hundred metric tons, and the price is $1.94/kg for truckloads of drums. It is produced by E. I. du Pont de Nemours & Co., Inc. and ICC Industries, Inc.

Nitrotoluenes

MONONITROTOLUENES

The mononitration of toluene results in the formation of a mixture of the ortho, meta, and para isomers of nitrotoluene, $O_2NC_6H_4CH_3$. The presence of the methyl group on the aromatic ring facilitates the nitration, but it also increases the ease of oxidation.

Properties. *o-Nitrotoluene.* *o*-Nitrotoluene [88-72-2] is a clear yellow liquid. The solid is dimorphous and the melting points of the α and β forms are $-9.55°C$ and $-3.85°C$, respectively. *o*-Nitrotoluene is infinitely soluble in benzene, diethyl ether, and ethanol. It is soluble in most organic solvents and only slightly soluble in water, 0.065 g/100 g of water at 30°C. The physical properties of *o*-nitrotoluene are listed in Table 6.

The strong electron-acceptor action of the nitro group in *o*-nitrotoluene confers increased reactivity on the methyl group; thus, the methyl group is easily oxidized. Oxidization with potassium permanganate or potassium dichromate causes the formation of *o*-nitrobenzoic acid [552-16-9]. When boiled with a sodium hydroxide solution, *o*-nitrotoluene exhibits the phenomena of autooxidation and reduction and yields anthranilic acid. When the oxidation is carried out with manganese dioxide and sulfuric acid, *o*-nitrobenzoic acid or *o*-nitrobenzaldehyde [552-89-6] is formed, depending on the reaction conditions. *o*-Nitrotoluene is reduced to *o*-toluidine by iron powder and hydrochloric acid. Alkaline reduction with iron or zinc leads in a stepwise fashion to azoxy, azo, and hydrazo compounds, depending upon the reaction conditions. Nitration of *o*-nitrotoluene gives 2,5-dinitrotoluene [121-14-2] and 2,6-dinitrotoluene [606-20-2]. Chlorination of *o*-nitrotoluene in the absence of iron yields *o*-nitrobenzyl chloride [612-23-7], *o*-chlorotoluene, or *o*-chlorobenzyl chloride, depending on the reaction conditions. In the presence of iron, chlorination results in the formation of 2-nitro-6-chlorotoluene [83-42-1] and 2-nitro-4-chlorotoluene [89-59-8].

m-Nitrotoluene. *m*-Nitrotoluene [99-08-1] is a clear yellow liquid that freezes at 16.1°C. It is readily soluble in ethanol, benzene, and diethyl ether. It is soluble in most organic solvents and is only sparingly soluble in water, 0.05 g/100 g of water at 30°C. The physical properties of *m*-nitrotoluene are given in Table 7.

m-Nitrotoluene does not have an active methyl group as do the ortho and para

Table 6. Physical Properties of *o*-Nitrotoluene

Property	Value	Reference
melting point, °C		
α	-9.55	3
β (stable form)	-3.85	3
boiling point, °C		
at 101 kPa (1 atm)	221.7	5
at 0.13 kPa[a]	50.0	7
density, g/cm^3		
d_{15}^{19}	1.1622	3
d_4^{20}	1.163	4
d_4^{60}	1.124	5
refractive index, n_D^{20}	1.5474	3
surface tension (at 15°C), mN/m (= dyn/cm)	42.3	5
viscosity (at 15°C), mPa·s (= cP)	0.0262	5
heat of combustion (at constant volume), MJ/mol[b]	3.75	6
(at constant volume)		
vapor density (air = 1)	4.72	7
flash point (closed cup), °C	106	7

[a] To convert kPa to mm Hg, multiply by 7.5.

[b] To convert J to cal, divide by 4.184.

Table 7. Physical Properties of *m*-Nitrotoluene

Property	Value	Reference
melting point, °C	16.1	6
boiling point, °C		
at 101 kPa (1 atm)	231.9	7
at 98 kPa[a]	227.0	3
at 0.13 kPa[a]	50.2	7
density, d_4^{20}	1.1571	3
refractive index, n_D^{21}	1.5470	3
surface tension (at 30°C), mN/m (= dyn/cm)	39.9	5
viscosity (at 30°C), mPa·s (= cP)	0.0178	5
heat of combustion (at constant volume), MJ/mol[b]	3.732	22
vapor density (air = 1)	4.72	7
flash point (closed cup), °C	106	7

[a] To convert kPa to mm Hg, multiply by 7.5.
[b] To convert J to cal, divide by 4.184.

isomers. It is oxidized readily to *m*-nitrobenzoic acid [121-92-6] by chromic acid and more slowly by potassium hexacyanoferrate(III) in alkaline solution. *m*-Nitrobenzaldehyde [99-61-6] is the chief product of the electrolytic oxidation of *m*-nitrotoluene. Acid, neutral, or catalytic reduction of *m*-nitrotoluene yields *m*-toluidine. Nitration of *m*-nitrotoluene yields primarily 3,4-dinitrotoluene [610-39-9] and small amounts of 2,3-dinitrotoluene [602-01-7] and 2,5-dinitrotoluene [619-15-8].

p-Nitrotoluene. *p*-Nitrotoluene [99-99-0] crystallizes in colorless rhombic crystals. It is only slightly soluble in water, 0.044 g/100 g of water at 30°C; moderately soluble in methanol and ethanol; and readily soluble in acetone, diethyl ether, and benzene. The physical properties of *p*-nitrotoluene are listed in Table 8.

The methyl group of *p*-nitrotoluene is activated by the para nitro group. *p*-Nitrotoluene is oxidized to *p*-nitrobenzoic acid [62-23-7] by potassium hexacyanofer-

Table 8. Physical Properties of *p*-Nitrotoluene

Property	Value	Reference
melting point, °C	51.7	6
boiling point, °C		
at 101 kPa (1 atm)	238.5	6
at 1.2 kPa[a]	104.5	3
at 0.13 kPa[a]	53.7	7
density, g/cm^3		
d_4^{20}	1.286	4
d_4^{55}	1.123	6
d_4^{73}	1.1038	3
refractive index, $n_D^{62.5}$	1.5346	4
surface tension (at 60°C), mN/m (= dyn/cm)	36.8	5
viscosity (at 60°C), mPa·s (= cP)	0.01204	5
heat of combustion (at constant volume), MJ/mol[b]	3.718	6
vapor density (air = 1)	4.72	7
flash point (closed cup), °C	106	

[a] To convert kPa mm Hg, multiply by 7.5.
[b] To convert J to cal, divide by 4.184.

rate(III) in alkaline solution, potassium permanganate, or potassium dichromate. p-Nitrotoluene is converted to p-nitrobenzaldehyde [555-16-8] by electrolytic oxidation in an acetic acid/sulfuric acid mixture or by treatment with lead(IV) oxide in concentrated sulfuric acid. p-Nitrotoluene is reduced by iron and hydrochloric acid to p-toluidine. Alkaline reduction with iron leads to the formation of a mixture of azoxy, azo, and hydrazo compounds, depending upon the reaction conditions. Nitration of p-nitrotoluene gives 2,4-dinitrotoluene. Chlorination can occur on either the aromatic ring or the methyl group, and the resulting product depends on the catalyst and reaction conditions used. Under free-radical reaction conditions, p-nitrobenzyl chloride [100-14-1] is formed and, in the presence of iron or antimony(III) chloride, 4-nitro-2-chlorotoluene [121-86-8] is obtained. p-Nitrotoluene undergoes sulfonation, yielding 2-methyl-5-nitrobenzenesulfonic acid [32784-87-5]. Heating p-nitrotoluene with an alcoholic potassium hydroxide solution results in the formation of 4,4'-dinitrostilbene [2501-02-2].

Manufacture and Processing. Mononitrotoluenes are produced by the nitration of toluene in a manner similar to that described for nitrobenzene. The presence of the methyl group on the aromatic ring facilitates the nitration of toluene, as compared to that of benzene, and increases the ease of oxidation which results in undesirable by-products. Thus, the nitration of toluene generally is carried out at lower temperatures than the nitration of benzene to minimize oxidative side reactions. Toluene is less soluble than benzene in the acid phase; thus, vigorous agitation of the reaction mixture is necessary to maximize the interfacial area of the two phases and the mass transfer of the reactants. The rate of a typical industrial nitration can be modeled in terms of a fast reaction taking place in a zone in the aqueous phase adjacent to the interface where the reaction is diffusion controlled.

Mononitrotoluenes can be produced by either a batch or continuous process. With a typical batch process, the toluene is fed into the nitrator and cooled to ca 25°C. The nitrating acid (52–56 wt % H_2SO_4, 28–32 wt % HNO_3, and 12–20 wt % H_2O) is added slowly below the surface of the toluene, and the temperature of the reaction mixture is maintained at 25°C by adjusting the feed rate of the nitrating acid and the amount of cooling. After all of the acid is added, the temperature is raised slowly to 35–40°C. After completion of the reaction, the reaction mixture is put into a separator where the spent acid is withdrawn from the bottom and is reconcentrated. The crude product is washed in several steps with dilute caustic and then water. The product is steam distilled to remove excess toluene and then dried by distilling the remaining traces of water. The resulting product contains 55–60 wt % o-nitrotoluene, 3–4 wt % m-nitrotoluene, and 35–40 wt % p-nitrotoluene. The yield of mononitrotoluenes is ca 96%. Based on a yield of 1000 kg of mononitrotoluenes, the material requirements for this process are as follows (20):

Material	Quantity, kg
toluene	690
sulfuric acid	810
nitric acid	450
water	240
10 wt % NaOH solution	22

The separation of the isomers is carried out by a combination of fractional distillation and crystallization. In a fractional vacuum-distillation step, the distillate, which is obtained at a head temperature of 96–97°C at 1.6 kPa (12 mm Hg), is fairly pure o-nitrotoluene and can be purified further by crystallization. The meta isomer is distilled from a mixture of m- and p-nitrotoluene and can be purified further by additional distillation and crystallization steps. The bottoms product from the distillation steps is cooled in a crystallizer to obtain p-nitrotoluene.

Effort has focused on increasing the amount of the para isomer that is formed in the mononitration of toluene, since it generally is in the greatest demand of the three isomers. In a typical nitration with mixed nitric and sulfuric acids, the ratio of p-nitrotoluene/o-nitrotoluene (p/o) usually is 0.6. Nitration of toluene with nitric acid in the presence of phosphoric acid leads to an increase in formation of the para isomer with a p/o ratio of 1.11 (8,23–24). Nitration with nitric acid in the presence of various aromatic sulfonic acids either in solution or on a support, eg, diatomaceous earth, results in increased selectivity toward para substitution with p/o ratios of 0.8–1.5 (24–26). With this technique, normally a large excess of toluene and a relatively large amount of catalyst must be used with highly concentrated nitric acid. Another approach to increase the selectivity for para nitration has been the addition of an amount of anhydrous calcium sulfate, which is based on the mol of water formed in the nitration reaction; p/o ratios of 1.20–1.35 are obtained (27). A gas-phase nitration of toluene with nitric acid in the presence of a carrier substance, which is based on SiO_2 and/or Al_2O_3 and which is impregnated with a high boiling inorganic acid, has been described (28). This technique is carried out with a large excess of toluene at 0.7–6.7 kPa (5.3–50.3 mm Hg) and 100–140°C; p/o ratios of 1.2–2.0 are obtained. The effect of the different catalysts on the isomer distribution of the mononitrotoluenes is summarized in Table 9.

If pure isomers are required, the ortho and meta compounds can be prepared by indirect methods. o-Nitrotoluene can be obtained by treating 2,4-dinitrotoluene with ammonium sulfide followed by diazotization and boiling with ethanol. m-Nitrotoluene can be prepared from p-toluidine by acetylation, nitration deacetylation, diazotization, and boiling with ethanol. A fairly pure p-nitrotoluene, which has been isolated from the isomeric mixture, can be purified further by repeated crystallizations.

Economic Aspects. Annual U.S. production of the mononitrotoluenes is 20,000 metric tons, of which ca 13,000 t is the ortho isomer and 7,000 is the para isomer. The prices of o-, m-, and p-nitrotoluene in bulk, fob plant site, are $0.93/kg, $2.20/kg, and $0.88/kg, respectively (January, 1980). The mononitrotoluenes are manufactured by E. I. du Pont de Nemours & Co., Inc. and First Mississippi Corp.

Table 9. Isomer Ratio of Mononitrotoluenes with Different Nitration Catalysts

Catalyst	Para, wt %	Ortho, wt %	Meta, wt %	p/o	% Conversion[a]	Reference
normal mixed acid	38	58	4	0.6	98	
H_3PO_4	50.2	45.1	4.3	1.11	99.6	23
m-benzenedisulfonic acid	43.6	53.8	2.9	0.81	92.8	25
m-benzenedisulfonic acid on Celite 545				1.53	92	26
anhydrous $CaSO_4$	54.4	43.2	2.3	1.26	89	27
5% H_2SO_4 on Al_2O_3	62.5	34.0	3.5	1.84	61.1	28

[a] Based on nitric acid present.

Analytical and Test Methods. o-Nitrotoluene can be analyzed for purity and isomer content by ir spectroscopy with an accuracy of ca 1%. p-Nitrotoluene content can be estimated by the decomposition of the isomeric toluene diazonium chlorides since the ortho and meta isomers decompose more readily than the para isomer. A colorimetric method for determining the content of the various isomers is based upon the color which forms when the mononitrotoluenes are dissolved in sulfuric acid (29). From the absorption of the sulfuric acid solution at 436 and 305 nm, the ortho and para isomer content can be determined, and the meta isomer can be obtained by difference. However, this and other colorimetric methods are subject to possible interferences from other aromatic nitro compounds. A titrimetric method, which is based on the reduction of the nitro group with titanium(III) sulfate or chloride, can be used to determine mononitrotoluenes (19). Chromatographic methods, eg, gas chromatography or high speed liquid chromatography, are well-suited for the determination of mononitrotoluene as well as its individual isomers. Freezing points are used commonly as indicators of purity of the various isomers.

Health and Safety Factors. The toxic effects of the mononitrotoluenes are similar to but less pronounced than those described for nitrobenzene. The maximum allowable concentration for the mononitrotoluenes is 5 ppm (30 mg/m^3) (20). Mononitrotoluenes are low grade methemoglobin formers (4) and may be absorbed through the skin and respiratory tract. The toxicity of alkyl nitrobenzenes decreases with an increasing number of alkyl groups and increases with an increasing number of nitro groups. The mononitrotoluenes represent moderate fire hazards when exposed to heat or flame. The same precautions that are used in handling nitrobenzene should be used for these compounds.

Uses. o-Nitrotoluene is used in the synthesis of intermediates for azo dyes, sulfur dyes, rubber chemicals, and agricultural chemicals. Typical intermediates are o-toluidine, o-nitrobenzaldehyde, 2-nitro-4-chlorotoluene, 2-nitro-6-chlorotoluene, 2-amino-4-chlorotoluene (Fast Scarlet TR Base), and 2-amino-6-chlorotoluene (Fast Red KB Base). p-Nitrotoluene is used principally in the production of intermediates for azo and sulfur dyes. Typical intermediates are p-toluidine, p-nitrobenzaldehyde, and 4-nitro-2-chlorotoluene.

Table 10. Physical Properties of 2,4-Dinitrotoluene

Property	Value	Reference
melting point, °C	70–71	3
boiling point (at 101 kPa or 1 atm), °C	300[a]	6
density, g/cm^3		
d_4^{15}	1.521	7
d_4^{71}	1.321	6
vapor density (air = 1)	6.27	7
heat of combustion, MJ/mol[b]	3.568	6

[a] Slight decomposition.
[b] To convert J to cal, divide by 4.184.

DINITROTOLUENES

Dinitration of toluene results in the formation of a number of isomeric products and, with a typical sulfuric/nitric acid nitrating mixture, the following mixture of isomers is obtained: 78 wt % 2,4-dinitrotoluene [121-14-2], 19 wt % 2,6-dinitrotoluene [606-20-2], 2.5 wt % 3,4-dinitrotoluene [610-39-9], 1 wt % 2,3-dinitrotoluene [602-01-7], and 0.5 wt % 2,5-dinitrotoluene [619-15-8]. The dinitrotoluenes are a moderate fire and explosion hazard when exposed to heat or flame. The maximum allowable concentration in air is 1.5 mg/m^3 (0.2 ppm). Dinitrotoluenes are used as intermediates for the production of toluene diisocyanate and dyestuffs. They also are used as explosives (qv).

2,4-Dinitrotoluene. 2,4-Dinitrotoluene crystallizes in yellow needles from carbon disulfide and is soluble in a number of organic solvents. It is only slightly soluble in water, 0.03 g/100 g of water at 22°C. The physical properties of 2,4-dinitrotoluene are listed in Table 10.

2,4-Dinitrotoluene can be prepared by the nitration of p-nitrotoluene with yields of ca 96%, or it can be obtained from the direct nitration of toluene. 2,4-Dinitrotoluene is oxidized to 2,4-dinitrobenzoic acid [610-30-0] by potassium permanganate or chromic acid, and it is reduced to 2,4-diaminotoluene by iron and acetic acid. It is reduced partially by zinc chloride and hydrochloric acid to 2-amino-4-nitrotoluene [99-55-8] and by ammonium sulfide to 4-amino-2-nitrotoluene [119-32-4].

BIBLIOGRAPHY

"Nitrobenzene and Nitrotoluenes" in ECT 1st ed., Vol. 9, pp. 388–401, by H. H. Bieber and A. G. Hill, Calco Chemical Div., American Cyanamid Co.; "Nitrobenzene and Nitrotoluenes" in ECT 2nd ed., Vol. 13, pp. 834–853, by Harold J. Matsuguma, Picatinny Arsenal, Department of the Army.

1. E. Mitscherlich, *Ann. Phys. Chem.* **31,** 625 (1834).
2. A. Gero, *Textbook of Organic Chemistry*, John Wiley & Sons, Inc., New York, 1963.
3. J. R. A. Pollock and R. Stevens, eds., *Dictionary of Organic Compounds*, 4th ed., Eyre & Spottiswoode Publishers Ltd., London, Eng., 1965.
4. G. D. Clayton and F. E. Clayton, eds., *Patty's Industrial Hygiene and Toxicology*, 3rd. rev. ed., John Wiley & Sons, Inc., New York, 1981.
5. J. Timmermans, *Physico-Chemical Constants of Pure Organic Compounds*, Elsevier Publishing Co., Inc., New York, 1950.
6. J. A. Dean, *Lange's Handbook of Chemistry*, 12th ed., McGraw-Hill Book Co., New York, 1979.
7. N. R. Sax, *Dangerous Properties of Industrial Materials*, 4th ed., Reinhold Publishing Corp., New York, 1975.
8. L. F. Albright and C. Hanson, eds., *Industrial and Laboratory Nitrations*, American Chemical Society, Washington, D.C., 1976.
9. W. L. Faith, D. B. Keyes, and R. L. Clark, *Industrial Chemicals*, 3rd ed., John Wiley & Sons, Inc., New York, 1975.
10. U.S. Pat. 3,092,671 (June 4, 1963), S. B. Humphrey and D. R. Smoak (to United States Rubber Company).
11. U.S. Pat. 4,021,498 (May 3, 1977), V. Alexanderson, J. B. Trecek, and C. M. Vanderwaart (to American Cyanamid Company).
12. U.S. Pat. 3,928,975 (Dec. 23, 1975), M. W. Dassel (to E. I. du Pont de Nemours & Co., Inc.).
13. U.S. Pat. 3,981,935 (Sept. 21, 1976), R. McCall (to E. I. du Pont de Nemours & Co., Inc.).
14. U.S. Pat. 4,112,005 (Sept. 5, 1978), K. Thiem, A. Hamers, and J. Heinen (to Bayer Atkiengessellschaft).
15. U.S. Pat. 3,976,704 (Aug. 24, 1976), R. J. Vaughan (to Varen Technology).
16. G. Davis and N. Cook, *Chemtech* **7,** 626 (1977).
17. *Chemical Economics Handbook*, Stanford Research Institute, Menlo Park, Calif., Jan. 1979.

18. F. Feigl, *Spot Tests in Organic Analysis*, 5th ed., Elsevier Publishing Co., New York, 1956.
19. Y. A. Gawargious, *The Determination of Nitro and Related Functions*, Academic Press, New York, 1973.
20. *Documentation of the Threshold Limit Values*, 3rd ed., American Conference of Governmental Industrial Hygienists, 4th Printing, 1977.
21. *Fire Protection Guide on Hazardous Materials*, 5th ed., National Fire Protection Association International, Boston, Mass., 1973.
22. T. Urbanski, *Chemistry and Technology of Explosives*, Vol. 1, The MacMillan Company, New York, 1964.
23. Fr. Pat. 1,541,376 (Oct. 4, 1968), C. L. Hakansson, A. A. Arvidsson, and I. Loken (to Aktiebolag Bofors).
24. T. Kameo and O. Manabe, *Nippon Kagaku Daishi*, 1543 (1973).
25. U.S. Pat. 3,196,186 (July 20, 1965), A. W. Sogn and J. G. Natoli (to Allied Chemical Corp.).
26. T. Kameo, S. Nishimura, and O. Manabe, *Nippon Kagaku Daishi*, 122 (1974).
27. U.S. Pat. 3,957,889 (May 18, 1976), B. Milligan and D. G. Miller (to Air Products and Chemicals, Inc.).
28. U.S. Pat. 4,112,006 (Sept. 5, 1978), H. Schubert and F. Wunder (to Hoescht Atkiengesellschaft).
29. F. D. Snell and C. T. Snell, *Colorimetric Methods of Analysis*, 3rd ed., Vols. 3 and 4, D. van Nostrand Co., Inc., Princeton, N.J., 1954.

K. L. DUNLAP
Mobay Chemical Corporation

NITROFURANS.   See Antibacterial agents, synthetic; Furan derivatives.

NITROGEN

The discovery of nitrogen [7727-37-9] usually is attributed to Rutherford in 1772, although Scheele, Cavendish, and Priestly obtained it at about the same time or earlier. Nitrogen, N, at no. 7, is a nonmetallic element and exists as a colorless, odorless, diatomic gas N_2. There are two stable isotopes ^{14}N and ^{15}N that occur, respectively, at 99.62% and 0.38% which gives an average atomic weight of 14.008. Nitrogen is the lightest element in group VA of the periodic table. Although nitrogen in compounds can assume a number of valences, its chief valences are +5 and −3. The electronic configuration of the nitrogen atom in its ground state is $1s^2 2s^2 2p^2$. Nitrogen comprises almost 80 wt % of the air and occurs in the protein matter of all living things, in many organic compounds, in ammonia and ammonium salts, and in nitrate mineral deposits in Chile and Bolivia. Nitrogen in organic compounds may be determined by the Kjeldahl or Dumas methods.

Physical Properties

Gaseous nitrogen condenses to a colorless liquid at $-195.8°C$ and to a white solid at $-209.9°C$. The solid exists in the α or cubic form, which is stable below $-237.5°C$, and in the β or hexagonal form, which is stable from $-237.5°C$ to its melting point. Other properties of nitrogen are given in Table 1 (1–2).

The deviation from the perfect gas law is not great at ordinary pressures but becomes considerable at higher pressures; eg, at $0°C$, the pressure–volume (PV) product for nitrogen at 101 MPa (1000 atm) is more than twice the PV product at atmospheric pressure. Nitrogen is soluble in water to the extent of 23 pph (by vol) at $0°C$ under a partial pressure of nitrogen of 101 kPa (1 atm).

Chemical Properties

Nitrogen undergoes a variety of reactions at high temperatures (see Ammonia; Cyanamides; Nitric acid). Nitrogen reacts with ozone (qv) in a hot tube which gives nitrogen dioxide and some nitrous oxide (3–4).

$$3\,N_2 + 4\,O_3 \rightarrow 6\,NO_2$$

A mixture of the nitrogen sulfides is formed from the reaction of nitrogen with elementary sulfur in an electric discharge at $100°C$ and 102.9 kPa (772 mm Hg) (see Sulfur compounds) (5). Nitrogen reacts with a mixture of oxygen and chlorine gas at $400°C$ resulting in the formation of nitrosyl chloride (6).

Nitrogen reacts with a number of the metals, eg, the alkali metals (7), the alkaline

Table 1. Properties of Nitrogen[a]

Property	Value
mp, K	63.2
bp, K	77.35
density, kg/m^3	
gas at 273.15 K and 101.3 kPa[b]	1.2504
d_T[c] liquid	$d_T = 1160.4 - 4.55\ T$
liquid at bp	808.6
solid at 63.2 K	1028
vapor pressure[c]	
liquid[b]	$\log_{10}P_{kPa} = -339.8/T - 0.0056286\ T + 6.83537$
β form of solid[b]	$\log_{10}P_{kPa} = -381.6/T - 0.0062372\ T + 7.53585$
critical temperature, K	126.2
critical pressure, kPa[b]	3,390
specific heat between 300 and 2500 K[c,d]	$C_p = 6.76 + 0.000606\ T + 0.00000013\ T^2$
heat of fusion at mp, J/g[d]	25.6
heat of vaporization at bp, J/g[d]	199

[a] Refs. 1–2.
[b] To convert kPa to mm Hg, multiply by 7.5; To convert to atm, divide by 101.3.
[c] T = absolute temperature.
[d] To convert J/g to Btu/lb, multiply by 0.429.

earth metals (8), manganese and chromium (9), tantalum (10), tungsten (wolfram) (11), and titanium, forming nitrides (qv). Cerium carbide and uranium carbide react with nitrogen at elevated temperatures to form the corresponding metallic nitrides (12–13).

$$2\,CeC_2 + N_2 \xrightarrow{\;1250°C\;} 2\,CeN + 4\,C$$

$$3\,UC_2 + 2\,N_2 \xrightarrow{\;1940°C\;} U_3N_4 + 6\,C$$

Nitrogen dissolves in iron and steel to a very limited extent by the following reaction:

$$N_2\,(gas) \rightleftarrows 2\,N$$

where N is atomic nitrogen dissolved in iron. The solubility of nitrogen in pure liquid iron at 1600°C and under 101.3 kPa (1 atm) partial pressure is ca 0.045 wt % (14). The solubility decreases with a decrease in temperature and is affected by the presence of other elements in commercial iron and steel. It generally is accepted that the presence of oxygen and sulfur in steel reduces the rate of nitrogen transfer (15). Nitrogen increases the strength and hardness of steel but decreases its ductility and toughness (16). Significant amounts of nitrogen are present in a hardened surface layer of a nitriding grade of steel. Nitrogen content affects the grain size of aluminum-killed or deoxidized and microalloyed steels because of the formation of nitride particles resulting in increased strength and toughness. Nitrogen also is used as an austenite stabilizer and strengthener in stainless steels.

At 1500°C, nitrogen reacts with acetylene to form hydrogen cyanide (17) (see Cyanides).

$$N_2 + C_2H_2 \rightarrow 2\,HCN$$

Nitrogen reacts with calcium silicide, $CaSi_2$, at 1000°C. In addition to the silicon analogue of calcium cyanamide, approximately equimolar quantities of the silicon analogue of calcium cyanide are formed (18) (see Cyanamides).

$$2\,N_2 + 2\,CaSi_2 \rightarrow CaSiN_2 + Ca(SiN)_2 + Si$$

With a mixture of graphite and sodium carbonate at 900°C, nitrogen yields a mixture of carbon monoxide and sodium cyanide (19).

$$N_2 + 4\,C + Na_2CO_3 \rightarrow 3\,CO + 2\,NaCN$$

A similar reaction has been reported with graphite and barium carbonate (20).

Nitrogen reacts with titanium(IV) chloride in a silent electric discharge, forming a derivative of titanium chloronitride (21).

$$N_2 + 4\,TiCl_4 \rightarrow 2\,TiNCl.TiCl_4 + Cl_2$$

The action of a glow discharge on nitrogen at low pressure gives active nitrogen. A greenish-yellow glow is produced which persists after the discharge, and the gas is remarkably active chemically; it combines in the cold with mercury, sulfur, and phosphorus. This activity is believed to result from the presence of excited molecules at various energy levels (22).

Manufacture and Processing

Nitrogen is produced commercially by separating it from air. Three important methods of separation are cryogenic distillation, combustion of natural gas or propane and air, and pressure-swing adsorption (PSA). The choice of method depends primarily on the desired production capacity and nitrogen purity requirements. Cryogenic distillation is the economic choice for large-scale production and high purity requirements and is the most extensively used of the three methods. A substantial amount of lower purity inert gas still is produced from combustion of hydrocarbons, but the rapid increase in natural gas and propane prices lessens the economic attractiveness of this method. Use of the PSA process is increasing in Europe for small production requirements of 10–1000 m^3/h (350–35,300 ft^3/h).

Cryogenic Air Separation.   Most nitrogen is produced in large-tonnage plants as a coproduct with oxygen. The plants typically produce 100–4000 metric tons per day of oxygen and nitrogen gas. If all of the oxygen content of the air is separated for sale, the salable nitrogen production from the plant can be up to three times the oxygen production. The nitrogen and oxygen gas can be distributed by pipeline, or some or all of either gas can be liquefied for storage or shipment in vacuum-insulated vessels. In 1978, United States nitrogen and oxygen production was 12.61×10^6 metric tons and 16.165×10^6 t, respectively (23); ca 25 wt % of the nitrogen was liquefied for bulk shipment.

Where there is a fairly constant requirement for more than 300 m^3/h (ca 10,000 ft^3/h) of gaseous nitrogen, an on-site plant that is designed only to produce nitrogen should be considered. The basic equipment and process flow for a typical nitrogen plant is shown in Figure 1. Air that has been compressed to ca 700 kPa (ca 100 psi) is

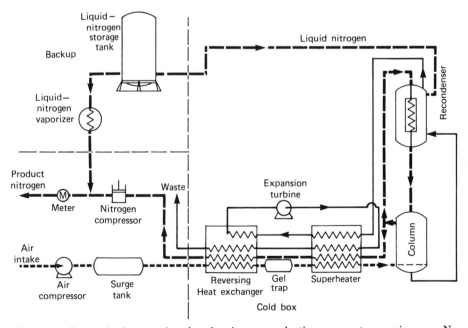

Figure 1.   Cryogenic air separation plant for nitrogen production. ——, waste; - - -, air; – – –, N_2.

supplied to a thermally insulated cold box containing reversing heat exchangers, a gel trap, a distillation column, and a superheater. The compressed air is cooled in the reversing heat exchangers against outgoing nitrogen-gas product and waste gas. All the water and most of the carbon dioxide and hydrocarbons entering with the air are removed by freezing in the reversing heat exchangers. Air and waste-gas flows are reversed at regular intervals so the accumulated water, carbon dioxide, and other components that are frozen out of the air are removed in the waste-gas stream. After passing through the reversing heat exchanger, the air is fed through a cold-end gel trap where hazardous hydrocarbons and the remaining small quantities of carbon dioxide are removed.

The clean air is cooled further in the superheater and is fed into the distillation column where it is liquefied and separated into high purity nitrogen product and waste gas containing ca 38 wt % oxygen. Both streams are warmed to ambient temperature as they are returned through the superheater and reversing heat exchanger. The high purity product is supplied to the pipeline. Refrigeration requirements for the plant are met by expanding a portion of the waste gas across an expansion turbine or by addition of liquid nitrogen from a storage tank. The product nitrogen is supplied at 350–520 kPa (50–75 psi), depending on the plant capacity. Pipeline pressures can be increased with the addition of a product compressor. Typical product nitrogen contains <30 ppm total impurities with a maximum of 5 ppm oxygen and a dew point of −60°C.

The capacities of the nitrogen plants described above are 300–7000 m^3/h (10,000–250,000 ft^3/h). In 1980, a 300-m^3/h plant had a capital cost of ca $640,000 and a 7000-m^3/h plant had a capital cost of $1,960,000. The power requirements were 0.25–0.32 (kW·h)/m^3 (0.71–0.91 kW·h/100 ft^3) depending on supply pressure. Total nitrogen costs ranged from $.08/m^3 ($2.25/1000 ft^3) for a small plant to $.025/m^3 ($0.71/1000 ft^3) for a large plant, assuming full utilization of plant capacity (see Cryogenics).

Inert-Gas Generators. Nitrogen inert-gas generators are available in capacities of 30–300 m^3/h (1,000–10,000 ft^3/h) and higher. Natural gas or propane is burned with air and the products of combustion are removed leaving purified nitrogen. A typical inert-gas system includes an air pump, air and natural-gas flow controls, a burner and combustion chamber, a refrigerant dryer, and molecular-sieve adsorbent beds (see Molecular sieves; Drying agents). Filtered air is drawn through a pump. Air and natural gas are controlled to provide a specific air–gas ratio to the burner. The gas burns in the combustion chamber in an exothermic reaction and essentially complete combustion is achieved. The burned gas contains nitrogen, carbon dioxide, water vapor, and small amounts of carbon monoxide and hydrogen.

Gases leaving the combustion chamber are cooled in a surface condenser and pass through a vapor separator where condensed water is removed. The gases then flow to a refrigerant dryer where the dew point is reduced to 4°C. Carbon dioxide and more water vapor are removed in a molecular-sieve bed; two beds are used alternately; gases are purified in one bed while the other bed is reactivated by vacuum pumping. The nitrogen product contains ca 500 ppm carbon dioxide and 1000 ppm carbon monoxide and hydrogen with a dew point of −60°C and a supply pressure slightly over atmospheric.

In 1980, a 115-m^3/h (4000 ft^3/h) nitrogen exothermic-gas generator costs $100,000, $150,000 installed. The generator uses ca 0.15 m^3 (5.3 ft^3) of natural gas, 0.14 kW·h electric power and 0.05 m^3 water/m^3 nitrogen. Total nitrogen costs average ca \$.06/m^3 (\$1.70/1000 ft^3), assuming full utilization of generator output.

Pressure-Swing Adsorption. The use of PSA systems for producing nitrogen is recent; 14 systems were reported in operation in Europe in 1978 (24). A typical system involves the use of two adsorbent beds containing a molecular-sieve coke produced from hard coal. A compressor loads one bed with air from which oxygen, water vapor, and carbon dioxide are adsorbed while nitrogen passes through to the product pipeline. At the same time, the other bed is depressurized and oxygen, water vapor, and carbon dioxide are desorbed from the molecular-sieve coke. The pressurization and depressurization of the beds are alternately reversed using control valves. Optimum loading and desorption periods are selected according to the nitrogen purity desired.

Nitrogen purity of 95–99.9 wt % can be achieved with the PSA system. Capital costs for the PSA plants in 1978 were \$40,000 for 10 m^3/h (353 ft^3/h) to \$350,000 for 1000 m^3/h (35,300 ft^3/h). Energy requirements were 0.6 (kW·h)/m^3 (1.6 kW·h/100 ft^3) of nitrogen containing 1000 ppm O_2 and a dew point below $-40°C$ (see Adsorptive separation).

Shipment

The method of shipping nitrogen from a cryogenic air-separation plant depends on the quantity and whether the use requires liquid or gaseous nitrogen. Large-volume gaseous requirements are delivered in pipelines (qv). The largest nitrogen pipeline systems in the United States are located in the Gulf Coast area of Texas and Louisiana. Three hundred forty kilometers of pipelines serve over 60 customers in the petrochemical and refining industries and transport ca 15,000 t/d of nitrogen. The pipelines are up to 0.3 m in diameter and carry nitrogen at 1750–4200 kPa (250–600 psi).

Small quantities of gaseous nitrogen are shipped in various sized steel cylinders containing up to ca 9 kg of nitrogen that is compressed to 15,200 kPa (2200 psi). Small quantities of liquid nitrogen also are shipped in various sized, vacuum-insulated containers. One popular portable container has a capacity of 120 kg and a maximum working pressure of 1620 kPa (235 psi) (25). These portable containers are delivered by truck from local distribution points. In the United States, there are over 3500 local distribution plants and supply stores selling nitrogen in liquid or gas cylinders.

Most nitrogen is shipped by bulk delivery of liquid nitrogen to storage tanks at the customer site. Liquid nitrogen is shipped in double-walled, vacuum-insulated tank trailers with capacities of up to 23 t. Capacities of typical customer storage tanks are 0.7–40 t of liquid nitrogen, which is equivalent to 570–34,000 m^3 (ca 20,000–1,200,000 ft^3) of gaseous nitrogen. The storage tanks also are vacuum-insulated with inner tanks of aluminum, 9 wt % nickel alloy steel, or austenitic stainless steel. Tank working pressures commonly are up to 1730 kPa (250 psi) and the heat leak into the tank evaporates ca 0.25–0.5 wt % of the full contents of the tank each day. Nitrogen can be withdrawn from the tanks as a liquid or it can be passed through vaporizers if it is to be supplied in a gaseous state. Ambient air, steam, or electric power is used to vaporize liquid for gas supply. 1980 prices for liquid nitrogen delivered into a supplier-owned storage tank located within 83 km of a producing plant are \$0.18–0.11/kg for volumes of 11,500–115,000 kg/mo (\$0.21–0.12/m^3 for 10,000–100,000 m^3/mo) (\$0.59–0.34/100 ft^3 for 350,000–3,500,000 ft^3/mo).

Economic Aspects

Nitrogen usually is shipped no more than a few hundred kilometers. Energy (usually electric power) and capital investment are the primary cost factors. The principal world producers with production and distribution facilities in many countries are L'Air Liquide S.A. (Fr.), Air Products and Chemicals, Inc. (U.S.), BOC International (UK) and its subsidiary Airco, Inc. (U.S.), and Union Carbide Corporation (U.S.). There are many additional producers in most countries. Although the following statistics pertain to the United States, the pattern is very similar in other industrialized countries.

Elemental nitrogen is the second-ranking industrial gas behind oxygen and is the sixth volume chemical on a tonnage basis. In 1978, 286 plants reported producing nitrogen by cryogenic air separation (23). Total United States shipments in 1978 were 7.9×10^6 metric tons with a value of 125×10^6 by pipeline and 3.1×10^6 t with a value of 206×10^6 by bulk liquid delivery.

Pipeline nitrogen production grew as it competed with inert-gas generators for large steel, chemical, and petroleum-refining applications. In the early 1960s, the United States space program created a large new demand for liquid nitrogen and cryogenic technology. When this demand declined, suppliers promoted many new industrial applications for bulk liquid nitrogen. As a result of these factors, total United States nitrogen production has grown from 44,000 t in 1950 to 610,000 t in 1960 to 5,000,000 t in 1970. Since 1970, high growth has continued at an average annual rate of 12%.

Health and Safety Factors

Safe handling of nitrogen gas or liquid requires knowledge of its properties and following safe practices; safety bulletins are available from suppliers (26). The potential hazards result from the same characteristics for which nitrogen is used. Nitrogen is nontoxic, but its use in an enclosed, inadequately ventilated area without proper venting could result in oxygen depletion and asphyxiation of people working in the area. Nitrogen vessels must be purged with air before being entered for maintenance.

The extremely cold temperatures of liquid nitrogen can quickly freeze any unprotected area of skin or eyes which may come in contact with the nitrogen or with materials cooled by it. Proper clothing, gloves, and a face shield or safety goggles should be worn when working with liquid nitrogen. Low temperatures of liquid nitrogen or cold vapor can embrittle many materials including carbon steels, most alloy steels, martensitic stainless steels, and rubber and plastics. The embrittled materials can easily fracture in use causing possible injury. Copper, brass, bronze, Monel, aluminum, and 300 series austenitic stainless steels remain ductile and are satisfactory for cryogenic applications. Equipment design must be such that liquid nitrogen is not trapped in a confined enclosure without relief valves. Pressures to 280,000 kPa (40,600 psi) can develop from heating liquid nitrogen to ambient temperature in a confined space. Conversely, vacuum conditions can occur in equipment that is cooled indirectly by liquid nitrogen.

Uses

Most uses of gaseous nitrogen depend on inert characteristics. It is used extensively in the metallurgical, chemical, and food industries as a blanket or purge to preclude oxidation during processing, storage, and packaging. Dissolved oxygen is stripped from liquids by sparging nitrogen through the liquids. Nitrogen pressure also is used for pneumatic instruments, hydraulic accumulators, stirring, and pressure transfer.

In liquid form, nitrogen is used primarily as an expendable refrigerant. Liquid nitrogen usually is delivered as a saturated liquid at ca $-195°C$. Refrigeration (qv) applications utilize its low boiling point and high heat capacity. The primary refrigeration uses are food freezing, cryopulverizing of plastics and spices, biological preservation of blood and semen, and embrittling molded-rubber parts for deflashing.

The use pattern for nitrogen in 1977 was ca 10% for low temperature uses and 90% for gas uses including 35% for chemical processing, 15% for metal refining and heat treating, 10% for electronics, and 30% for miscellaneous applications.

Chemical Industry. Nitrogen is used throughout the chemical process industry, and usually is treated as a utility. The major process utility uses are inert-gas blanketing of reactors and vessels to purge oxygen; controlling oxygen content and temperature in oxidation reactions; column and vessel pressure control; drying; catalyst and adsorbent regeneration; agitation and removal of unreacted reagents, especially monomers from reactors; and improving or maintaining color and viscosity of products.

Polymer processing is a large consumer of utility nitrogen (27–33). In the formation of acetal resins (qv) from formaldehyde, a nitrogen purge first is used to clear the reactor of oxygen (34). In the production of the resin from trioxane, the initial reaction is continued on a moving belt; a heated dry-nitrogen atmosphere is used to remove solvent, catalyst, and unreacted trioxane (35). Nitrogen also is used as a diluent to control reaction rates (36).

Petroleum Industry. In petroleum refining, nitrogen is used in the regeneration of spent reforming catalyst. The catalyst loses activity because of carbon deposition. The carbon is burned off with oxygen, and nitrogen serves as the carrier for the oxygen to control the rate of combustion (37–38). Nitrogen also is used to cool catalytic reactors in reforming and hydrocracking prior to catalyst removal and maintenance (39) (see Petroleum). Significant quantities also are consumed in blanketing storage tanks and product loading and unloading. Pipelines and caverns also are purged with nitrogen prior to initial filling or product changeovers. Large quantities of nitrogen have been used to provide pressure for enhanced oil-recovery tests (see Petroleum, enhanced oil recovery) (40).

Heat Treatment. Inert-gas generators traditionally provided atmospheres for a variety of metal heat-treating applications including annealing, hardening, carburizing, and powder metal sintering. Natural gas and air react to produce atmospheres with compositions ranging from mostly nitrogen to 40% nitrogen, 40% hydrogen, and 20% carbon monoxide by volume. A growing number of plants produce these atmospheres from nitrogen and other pure gases, eg, hydrogen, or from nitrogen and methanol which are injected directly into the furnace. The nitrogen serves to exclude air and control the concentration of active gases, eg, hydrogen which reduces metal oxides and carbon monoxide which helps control the carbon level of iron and steel parts (41–42).

Food Industry. Nitrogen is used widely in the food industry to prevent oxidative deterioration, mold growth, and insect infestation. An atmosphere with less than ca 2% oxygen kills any insects present, and maintenance of such a level for a few days destroys dormant eggs (see Food processing).

The major use of nitrogen as a refrigerant is in the freezing of food. Frozen products include meat, seafood, and baked goods. Food is frozen on stainless steel mesh belts traveling through an insulated tunnel. Freezing is accomplished in minutes; this rapid freezing is well suited for processing portion-control foods in a continuous processing line including forming, freezing, and packaging. Faster freezing results in better quality retention because of reduced dehydration, reduced bacterial growth, and less cell damage. An estimated 300,000 t of cryogens, predominantly nitrogen, was used to freeze hamburger patties in 1977 (43). Large quantities also were used to freeze pizza, cooked chicken, and shrimp.

Cryopulverizing. The use of liquid nitrogen to cool materials before and/or during size reduction (qv) has been called cryopulverizing, cryogrinding, and cryocomminution. A typical system includes a feed hopper, a vibrating or screw feeder, a liquid-nitrogen precooler, a mill, and a liquid-nitrogen flow controller (44). Approximately 50,000 t of nitrogen are used annually in the United States to cryopulverize thermoplastic and thermoset resins, scrap rubber and plastic, color concentrates, spices, gums, waxes, and reactive materials. Resins are cryopulverized to achieve small particle sizes, ie, 100–150 μm, for electrostatic spray coating. Spices, eg, mustard, cinnamon, and nutmeg, can be ground resulting in a finer product requiring no screening and exhibiting no temperature degradation (see Flavors and spices).

Cryoforming of Metals. Deformation of austenitic stainless steels at cryogenic temperatures can increase the tensile strength significantly, but the material becomes more brittle and less ductile. A process has been developed in which annealed austenitic stainless steel is deformed at a strain of 10% at ambient temperature and then is deformed at a strain of 10% at a cryogenic temperature. Tensile strength is increased more than in the earlier cryodeformation process and higher toughness is achieved (45). The process is being practiced to produce a high strength AISI 302 stainless steel wire (46). The wire can be used to make springs which are 30% stronger or 40% lighter than those made from conventionally drawn AISI 302 wire (47).

BIBLIOGRAPHY

"Nitrogen" in *ECT* 1st ed., Vol. 9, pp. 404–406, by E. S. Gould, Polytechnic Institute of Brooklyn; "Nitrogen" in *ECT* 2nd ed., Vol. 13, pp. 857–863, by J. W. Hall, Union Carbide Corporation, Linde Division.

1. D. Mann, ed., *LNG Materials and Fluids*, National Bureau of Standards, Boulder, Colorado, 1977, pp. 1–2.
2. *Cryog. Ind. Gases*, 39 (July 1976).
3. D. Barbier and D. Chalonge, *Compt. Rend.* **213**, 1010 (1941).
4. S. C. Lind and D. C. Bardwell, *J. Am. Chem. Soc.* **51**, 2751 (1929).
5. W. Moldenhauer and A. Zimmermann, *Ber.* **B62**, 2390 (1929).
6. W. Krauss and M. Saracini, *Z. Phys. Chem. (Leipzig)* **A178**, 245 (1937).
7. W. Moldenhauer and H. Mottig, *Ber.* **B62**, 1954 (1929).
8. A. von Antropoff and K. H. Kruger, *Z. Phys. Chem. (Leipzig)* **A167**, 49 (1933).
9. B. Neumann, C. Kroger, and H. Haebler, *Z. Anorg. Allgem. Chem.* **196**, 65 (1931).
10. M. R. Andrews, *J. Am. Chem. Soc.* **54**, 1845 (1932).
11. C. J. Smithhells and H. P. Rooksby, *J. Chem. Soc.*, 1882 (1927).
12. F. Fichter and C. Scholly, *Helv. Chim. Acta* **3**, 164 (1920).

13. O. Heuser, *Z. Anorg. Allgem. Chem.* **154,** 353 (1926).
14. R. Pehlke and J. F. Elliott, *Trans. Met. Soc. AIME* **218,** 1088 (1960).
15. R. Pehlke and J. F. Elliott, *Trans. Met. Soc. AIME* **227,** 844 (1963).
16. *Metals Handbook*, Vol. I, American Society for Metals, Metals Park, Ohio, 1978.
17. W. E. Garner and S. W. Saunders, *J. Chem. Soc.* **125,** 1634 (1924).
18. L. Wohler and O. Bock, *Z. Anorg. Allgem. Chem.* **134,** 221 (1924).
19. C. K. Ingold and D. Wilson, *J. Chem. Soc.* **121,** 2278 (1922).
20. O. Kuhling and O. Berkhold, *Ber.* **41,** 28 (1908).
21. L. Hock and W. Knauff, *Z. Anorg. Allgem. Chem.* **228,** 193 (1936).
22. *Thorpe's Dictionary of Applied Chemistry*, 4th ed., Vol. 8, Longmans, Green and Co., London, Eng., 1947, pp. 509–511.
23. *Current Industrial Reports, Industrial Gases 1978*, U.S. Department of Commerce, Washington, D.C., Dec. 1979, pp. 3, 9.
24. K. Knoblauch, *Chem. Eng.*, 87 (Nov. 6, 1978).
25. *N2 (F-4218)*, Union Carbide Corporation, New York, Feb. 1978, p. 5.
26. *Precautions and Safe Practices for Handling Liquefied Atmospheric Gases (F-9888)*, Union Carbide Corporation, New York.
27. U.S. Pat. 2,987,506 (June 6, 1961), F. G. Lum (to California Research Corp.).
28. U.S. Pat. 2,719,776 (Oct. 4, 1955), P. Kummel (to Inventa A.G.).
29. U.S. Pat. 2,687,552 (Aug. 31, 1954), R. Gabler (to Inventa A.G.).
30. U.S. Pat. 3,103,408 (Sept. 10, 1963), M. C. Chen and M. E. Cox (to Simoniz Co.).
31. U.S. Pat. 2,318,742 (May 11, 1943), E. C. Britton and W. J. LeFebre (to The Dow Chemical Co.).
32. U.S. Pat. 3,654,782 (Sept. 18, 1962), R. Saxon (to American Cyanamid Co.).
33. U.S. Pat. 2,654,731 (Oct. 6, 1953), H. T. Patterson (to E. I. du Pont de Nemours & Co., Inc.).
34. U.S. Pat. 3,828,286 (March 25, 1958), R. N. MacDonald (to E. I. du Pont de Nemours & Co., Inc.).
35. U.S. Pat. 2,982,758 (May 2, 1961), C. L. Michaud (to Celanese Corp. of America).
36. U.S. Pat. 2,657,200 (Oct. 27, 1953), F. C. McGrew and P. S. Pinkney (to E. I. du Pont de Nemours & Co., Inc.).
37. U.S. Pat. 2,980,631 (April 18, 1961), R. G. Craig and D. H. Stevenson (to Houdry Process Corp.).
38. U.S. Pat. 3,069,352 (Dec. 18, 1962), M. A. Moseman (to Esso Research and Engineering Co.).
39. G. Holcomb and S. Laschober, *Hydrocarbon Process.*, 163 (Sept. 1979).
40. S. Fletcher, *The Houston Post*, 22C (Sept. 3, 1978).
41. U.S. Pat. 4,139,375 (Feb. 13, 1979), J. Solomon and T. F. Kinneman (to Union Carbide Corporation).
42. U.S. Pat. 4,145,232 (Mar. 20, 1979), J. Solomon (to Union Carbide Corporation).
43. R. E. Rust, *Cryogenic Freezing of Ground Beef, Proceedings Meat Science Seminar*, Cooperative Extension Service, Iowa State University, Ames, Iowa, May 24, 1977.
44. R. B. Davis and J. Wary, *Chem. Technol.* **6,** 200 (March 1976).
45. U.S. Pat. 4,042,421 (Aug. 16, 1977), J. Van den Sype, W. A. Kilinskas, R. B. Mazzarella, and J. B. Lightstone (to Union Carbide Corporation).
46. U.S. Pat. 4,161,415 (July 17, 1979), J. Van den Sype and L. Stambaugh (to Union Carbide Corporation).
47. J. Van den Sype and W. A. Kilinskas, *Springs* **16**(2), 20 (Oct. 1977).

RONALD W. SCHROEDER
Union Carbide Corporation

NITROGEN FIXATION

To appreciate the importance of nitrogen fixation (dinitrogen fixation) fully, it is necessary to understand both the global distribution of nitrogen and its movement within the nitrogen cycle. Global inventories show that more than 99.9% of the nitrogen on earth is present as the dinitrogen molecule (N_2), of which somewhat more than 97% is trapped in primary rocks (2×10^{17} metric tons) and sedimentary rocks (4×10^{14} t) and ca 2% (4×10^{15} t) is free in the atmosphere (1–2). In comparison, all plants and animals together contain only 1×10^{10} t of nitrogen. In addition, about 1.2×10^{12} t is distributed in the form of dead organic matter about equally between land and sea. The latter contains 6×10^{11} t of soluble inorganic forms of nitrogen (excluding N_2) (1–5). Thus, only a very small proportion of the nitrogen present on earth is involved at any one time in the cycle between its usable fixed form and its inert molecular form. Various transformations, which form the basis of the nitrogen cycle, allow nitrogen to move between the atmospheric (inert) pool and the fixed (usable) terrestrial pools (Fig. 1). Nitrogen fixation is involved with the atmosphere-to-terrestrial (land or sea) direction. Nitrification and denitrification convert ammonia to nitrate and then to dinitrogen (via nitrogen oxides) which is lost to the atmosphere; leaching and erosion of soils result in the movement of fixed nitrogen between land and sea. The biological world stays ahead of a nitrogen deficiency because the fixation rate is just above the denitrification rate, but the margin is estimated to be slim (6). The effect of the above processes in the soil is shown in Figure 2. Only about one third of the available nitrate is assimilated by plants, one third is leached away, and one third is denitrified and lost to the atmosphere (7).

Molecular nitrogen is fixed (1,8–9) either by natural nonbiological and biological processes or by industrial ammonia (qv) production. The biological contribution estimates range from 100×10^6 to 175×10^6 t/yr with the latest estimate (9) at 122×10^6 t/yr. Industrial fixation contributes about 50×10^6 t/yr for fertilizer uses (10), up from ca 40×10^6 t in 1975 (11). Other processes, like lightning and combustion, are estimated to fix about 30×10^6 t/yr (11). Thus, the biological process represents the biggest share (ca 60%) of the total annual fixation rate, about two and one half times more than the industrial production of fertilizer (see Fertilizers).

The increasingly intensive cultivation of the arable land of the earth, which has been fueled by nonrenewable resources like petroleum and natural gas, is a short-range

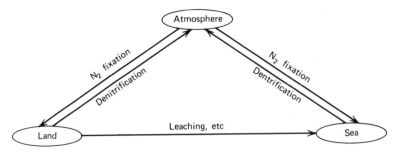

Figure 1.   Schematic representation of the mobilization of nitrogen between land, sea, and atmosphere.

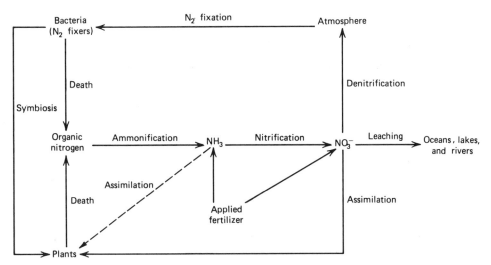

Figure 2.   Biological pathways and processes involved in the nitrogen cycle.

solution to the growing problem of global food supply, which has been exacerbated by a continuing population increase. As the availability of fertilizer nitrogen is almost always the limiting factor in crop productivity, the various processes for the fixation of atmospheric dinitrogen hold the key to the long-term solution of global food supply. In certain instances, like the legumes (peas, beans, alfalfa, etc), nature has provided a mechanism for biological interaction between the plant and a nitrogen-fixing bacterium. This symbiotic association allows the plant to receive fixed nitrogen directly from the bacterium harbored in its root nodules in exchange for a carbohydrate supply for energy. Although this approach works well, it is limited and important food crops, like the cereal grains (rice, wheat, and corn), and root and tuber crops, do not harbor symbiotic partners. Hence, for crop productivity to reach commercially acceptable levels, extensive augmentation by industrially fixed nitrogen is necessary. Such problems have encouraged present-day research into all areas of nitrogen fixation.

In the industrial arena, the search for technological advances to produce cheap hydrogen from a variety of feedstocks and to improve the overall efficiency of the process is underway. Modern ammonia plants, almost without exception, rely on natural gas or liquid hydrocarbons, such as naphtha, as the source of hydrogen for synthesis. Since hydrogen is the principal determinant in the price of nitrogenous fertilizer and the supply of natural gas is decreasing, other feedstocks for cheap hydrogen, such as coal, are considered (see Feedstocks; Hydrogen).

Considerable progress in biological nitrogen fixation has been made owing to the combined efforts of biochemists and chemists, particularly since 1970. In the early 1960s, only crude cell-free preparations were available for investigating the nitrogen-fixation phenomenon. Today, highly purified samples of the enzyme nitrogenase, which is responsible for catalyzing this process, are available. Insight into the requirements and chemistry of the process itself has increased greatly.

More recently, purely chemical processes have been devised that bind the dinitrogen molecule and, in some cases, activate it sufficiently so that reduced nitrogen compounds, ammonia and/or hydrazine, are produced on protonation. Such an approach, which is concerned only with determining the requirements for binding and reducing dinitrogen, is complemented by efforts to simulate the composition, reactivity, and spectral properties of the enzyme itself. Both approaches utilize transition-metal ions because these entities have the necessary properties for interaction with dinitrogen and are strongly indicated as key constituents of the enzyme. Until recently, the biological information necessary for effective modeling was not available. Studies with chemical systems were undertaken in order to give clues as to the requirements for dinitrogen reduction. Although even now there is no well-defined mechanism that describes biological nitrogen reduction accurately and completely, a considerable body of information is available concerning the various prosthetic groups within nitrogenase and their interrelationships. Thus, actual model-building is now not only valid but required for continued progress.

Industrial Processes

Until the early 19th century, the fixed usable nitrogen, stockpiled over millions of years by various natural processes, was enough to sustain the needs of the earth's population. Then, with the dramatic growth of large cities and populations, the demand for augmenting this supply, particularly in industrialized areas, led to the beginnings of the nitrogenous fertilizer industry. Guano, which is hardened bird droppings, was imported into Europe from Peru, as was saltpeter (sodium nitrate) from Chile. These fertilizer forms were supplemented in the industralized nations by the ammoniacal by-products from coal gas. Further increasing demand led to the invention of several processes, some of which were exploited commercially.

The first such process was the Birkeland-Eyde process for dinitrogen oxidation which was implemented in 1905 (12). In this process, air is passed through an electric arc at temperatures above 3000°C to generate nitric oxide (NO). On cooling the air stream, further oxidation gives nitrogen dioxide (NO_2) which on absorption into water gives a mixture of nitric (HNO_3) and nitrous (HNO_2) acids (see also Nitric acid).

$$N_2 + O_2 \rightleftarrows 2\,NO$$
$$2\,NO + O_2 \rightarrow 2\,NO_2$$
$$2\,NO_2 + H_2O \rightarrow HNO_3 + HNO_2$$

Only ca 2% conversion to NO occurs. This low yield and the tendency to revert to N_2 and O_2 if the product stream is not quenched rapidly are the main problems associated with this process. However, the consumption of large amounts of electricity, ca 60,000 kW·h/t N_2 fixed, and the concomitant expense to sustain the arc led to its demise. It was only economical in countries like Norway, where it was developed, and where substantial amounts of cheap hydroelectric power are available.

At about the same time, the Frank-Caro cyanamide process was commercialized (13–14). In this process, limestone is heated to produce lime which then reacts with carbon in a highly energy-demanding reaction to give calcium carbide. Reaction with dinitrogen gives calcium cyanamide, which hydrolyzes to ammonia and calcium carbonate (see Cyanamides).

$$CaCO_3 \longrightarrow CaO + CO_2$$

$$2\,NH_3 \xleftarrow{+3\,H_2O} CaCN_2 + C \xleftarrow{+N_2} CaC_2 + CO \quad \downarrow{+3\,C}$$

Even though the overall energy requirement of this process is only ca 20–25% of the arc process, the Haber-Bosch process, which was developed at about this same time, proved to be more economical.

A third process, the Serpak process, for the catalytic nitriding of aluminum, was never exploited to any significant degree, mainly because of its large energy requirements (12). A mixture of alumina, coke, and dinitrogen was heated at 1800°C to produce aluminum nitride, which on hydrolysis gave ammonia.

$$Al_2O_3 + 3\,C + N_2 \rightarrow 2\,AlN + 3\,CO$$

$$2\,AlN + 3\,H_2O \rightarrow 2\,NH_3 + Al_2O_3$$

Currently, the enormous synthetic ammonia industry employs only the Haber-Bosch process (10,12–15). It was developed in Germany in the years just before World War I. Its development was aided by the concomitant development of a simple catalyzed process for the oxidation of ammonia to nitrate which was needed at that time for the explosives industry. Dinitrogen (N_2) and dihydrogen (H_2) are combined directly and equilibrium is reached under the appropriate operating conditions. The resultant gas stream contains ca 20% ammonia.

$$N_2 + 3\,H_2 \rightleftarrows 2\,NH_3$$

When the reaction was first discovered, it required a considerably higher temperature (ca 1300°C) than used today. Thus, until Haber discovered the appropriate catalyst, it was not attractive commercially. Today, it suffers from the requirement of significant quantities of nonrenewable fossil fuels to operate.

Although ammonia itself is commonly used as a fertilizer in the United States, elsewhere it is often converted into solid or liquid fertilizers, such as urea (qv), ammonium nitrate or sulfate, and various solutions (see Ammonium compounds).

The Haber-Bosch Process.   A modern ammonia plant performs two distinct functions. The more energy-demanding and complex function is the preparation and purification of the synthesis gas, containing N_2 and H_2 in a 1:3 ratio, from a variety of feedstocks. The second function is the catalytic conversion of synthesis gas to ammonia (see Fig. 3). In the years since its commercial introduction in 1913, many process changes have been made, particularly with respect to synthesis-gas production, to lower costs and give greater efficiencies.

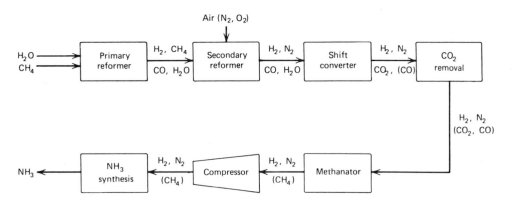

Figure 3.   The Haber-Bosch process. Gases in parentheses are minor constituents of the mixture.

Synthesis Gas Production.   Through World War II, coal was the primary raw material for ammonia synthesis, either directly as in the original water–gas plant or indirectly via coke-oven gas. Since that time, petroleum-based products have taken over and now represent ca 88% of the feedstock for all ammonia produced. Feedstock availability and price are, in these energy-conscious times, the main problems facing the industry. Hence, interest in returning to coal and its gasification as a feedstock has been renewed (see Coal, coal conversion processes). Dihydrogen for synthesis gas is produced either by steam reforming of natural gas and other lighter hydrocarbons, such as naphtha, or by the partial oxidation of heavy oils and coal. Other processes include coal carbonization, oil refining, and the electrolysis of water (10,13,16–18).

In the catalytic steam reforming of natural gas (see Fig. 3), the hydrocarbon stream, consisting principally of methane, is desulfurized first in order to avoid catalyst poisoning (see Gas, natural). Together with superheated steam, it then contacts a nickel catalyst in the primary reformer at ca 3.04 MPa (30 atm) pressure and 800°C to convert methane to dihydrogen.

$$CH_4 + H_2O \rightarrow 3\,H_2 + CO$$

Reforming is completed in a secondary reformer where air is added to elevate the temperature by partial combustion of the gas stream. The amount of air added is so adjusted that a 3:1 H_2 to N_2 ratio results downstream of the shift converter as required for ammonia synthesis. The water-gas shift converter then produces more H_2 from carbon monoxide and water. The resulting carbon dioxide is removed by various techniques (10) at the next stage.

$$CO + H_2O \rightarrow H_2 + CO_2$$

A low temperature shift process, using a zinc–chromium–copper oxide catalyst, has replaced the earlier iron oxide-catalyzed high temperature system. Any remaining CO and/or CO_2 at this stage is converted to methane over a nickel catalyst.

The partial-oxidation process differs only in the initial stages before the water-gas shift converter. Since it is a noncatalyzed process, desulfurization can be carried out further downstream. The proportions of a mixture of heavy oil, coal, etc, O_2, and steam, at a very high temperature, are so adjusted that the exit gases contain a substantial proportion of H_2 and carbon monoxide.

$$3\,C + H_2O + O_2 \rightarrow H_2 + 3\,CO$$

These gases are then fed to the water-gas converter as in the steam-reforming process, after which they are compressed to ca 20.3 MPa (ca 200 atm) for processing in the catalytic ammonia converter.

A breakthrough in this area was the development of centrifugal turbine compressors. Significant energy savings occurred compared with the reciprocating compressors, but only at ammonia outputs of greater than 600 t/d, ie, with the larger plants. Thus, all newer plants fall in the 1000–2000 t/d class. These compressors also can be run directly from steam turbines. Similarly, energy can also be saved by increasing the operating pressure for the reformers in synthesis-gas production since this lowers compression costs at the catalytic ammonia converter and improves heat recovery from the unreacted steam. This last point is in line with the general trend for ammonia plants to become independent of external electricity sources. Many recovery systems are employed in such a way that waste heat may be recaptured in a reusable form.

Catalytic Conversion to Ammonia. A large number of catalysts have been suggested for the ammonia synthesis reaction; most are transition metals representing Groups VIB, VIIB, and VIII. Haber originally proposed osmium as catalyst and achieved an equilibrium yield of ammonia of 8% at 550°C and 15.2 MPa (150 atm). However, on an industrial scale only iron, cobalt, molybdenum, and tungsten have been found to be practical. The addition of certain promoter salts and oxides favors ammonia formation. The best and most economical catalyst was found to be metallic iron, with alumina and potassium oxide as promoters. This type of catalyst is still used. The promoters prevent catalyst sintering, ie, they increase its heat stability and aid in desorption of ammonia from the surface. Of the three known iron oxides, only magnetite (Fe_3O_4) yields an efficient catalyst after reduction with H_2 to spongy iron (13) (see Catalysis).

Since the ammonia-synthesis reaction is an equilibrium, the quantity of ammonia depends on temperature, pressure, and the H_2-to-N_2 ratio. At 500°C and 20.3 MPa (200 atm), the equilibrium mixture contains 17.6% ammonia. To obtain complete conversion, the unreacted gases must be returned to the catalyst. Recirculation is preferred to using several converters in series. The compressed synthesis gas is passed into the ammonia synthesis reactor at the appropriate temperature. The ammonia formed (15–25% yield) is removed from the exit gases by condensation at about −20°C, and the gases are recirculated with fresh synthesis gas into the synthesis reactor. The ammonia must be removed continually as its presence decreases both the equilibrium yield and the reaction rate by reducing the partial pressure of the N_2–H_2 mixture. The mechanism of the synthesis reaction has never been clarified fully. However, decomposition of N_2 into nitrogen atoms on the catalyst surface is likely with the rate-determining step being either chemisorption of N_2 or its hydrogenation.

Operational Constraints and Problems. Synthetic ammonia manufacture is a mature technology and all fundamental technical problems have been solved. However, extensive know-how in the construction and operation of the facilities is required, which, although apparently simple in concept, are very complex in practice. Some of the myriad of operational parameters change rapidly, such as feedstock source or quality, and the plant operator has to adjust accordingly. Most modern facilities rely on computers to monitor and optimize performance on a continual basis. This situation can produce problems in countries lacking industrial expertise and can lead to improper design, suboptimal performance, and increased equipment down-time. These factors may well become increasingly important as more and more ammonia-producing facilities are located in hydrocarbon-rich countries which have limited experience with these large, complex industrial projects.

Biological Systems

Biological nitrogen fixation is confined to microorganisms. Only prokaryotes, ie, living things without an organized nucleus (bacteria, blue–green algae, and actinomycetes), can reduce dinitrogen to ammonia. Such bacteria can be either free-livers, such as *Azotobacter* and *Clostridium*, or can form symbiotic associations with higher plants, like the *Rhizobium*–legume system. The latter group is much more important agriculturally. In exchange for the fixed nitrogen supplied by the bacterium, the legume supplies energy in the form of carbohydrate obtained by photosynthesis. Thus, renewable solar energy powers this fertilizer production system in contrast to the nonrenewable energy sources used commercially. Ammonia from the Haber process involves an additional energy cost in transportation to the user. Furthermore, since commercial operation must be year-round to be economical whereas fertilizer application is seasonal, storage costs are incurred. Thus, as food demands increase and fossil fuel must be conserved, the exploitation of biological nitrogen fixation may well be the answer to the problem. This exploitation may encompass a whole spectrum of approaches. For example, molecular genetic techniques could be employed to engineer corn, wheat, etc, to fix enough nitrogen for its own requirements (see Genetic engineering). Another, more immediate and pragmatic approach would be the more widespread use in agriculture of known, symbiotic N_2-fixing systems and looser, more informal associations. At the other end of the spectrum is the development of new catalysts, based on nitrogenase, for alternative (or complementary) dinitrogen-reducing processes. These approaches are under intensive investigation and the chances are good that some, at least, will make substantial contributions in the future.

Plant–Bacterial Associations. The relationship of N_2-fixing bacteria with the roots of higher plants was recognized in the late 19th century (19–21). Since then, many additional associations have been found, some of which are important contributors to the nitrogen cycle (4,20–23).

Rhizobium–Legume Associations. The legumes (family Leguminoseae) include temperate and tropical flowering plants, ranging from small plants, like clover, to bushes and large trees, such as acacia. Most plants can be infected by bacteria of the genus *Rhizobium* which colonize their roots within nodules. The best known associations occur with important crops, such as the pulses (peas and beans, including soybeans), clovers, and alfalfa (1,24–25). These associations show some specificity with certain bacteria only infecting certain plants. These bacteria (termed *Rhizobia*) are divided into cross-inoculation groups based on their infecting properties (26); however, this classification is not rigid (27–28). The most promiscuous group of *Rhizobia* are the cowpea type, which infect not only the cowpea, peanut, and acacia, but also (uniquely for *Rhizobia*) a nonleguminous tropical plant (29). This recognition process and the initial stages of infection are still under active investigation (24,30–31), but it is known that the *Rhizobia* usually enter close to or via root hairs. Within the infected plant cells, the bacteria cease growing and enlarge and often take on unusual shapes to become bacteroids (24–25,32). Infection occurs as early as the appearance of the first leaves and fixation continues usually until pod filling. Certain strains of *Rhizobia* are relatively ineffective as N_2 fixers, but good agricultural practice demands the inoculation of plants by effective strains. Thus, seeds are coated with inoculum (a dried rhizobial culture in peat) before planting to ensure good fixation. This practice is important, because once a plant is infected by one strain, it affects the ability of other strains to invade the plant. Infection is also inhibited if the soil contains significant amounts of fixed nitrogen (33–34).

Rhizobia were once thought to fix dinitrogen only after becoming bacteroids within the plant nodule, but it has now been shown that certain free-living species of *Rhizobium* (typically the cowpea type) can fix N_2 in the absence of plant material (35–39). This property is manifested when the oxygen content of cultures is less than 0.5%. This low oxygen requirement mimics the role of leghemoglobin in nodules, which is to supply a constant and correct amount of oxygen to the bacteroids for metabolism without causing damage to the oxygen-sensitive nitrogenase (40–41). Although leghemoglobin, the protein that gives cut nodules their red color, is plant-produced

(42–43), nitrogenase is definitely of bacterial origin. It has very similar properties to the enzyme obtained from the free-living organisms as shown by analysis (44), cross-reactions with the *Azotobacter* proteins (45), and the products of catalysis, including H_2 evolution concomitant with substrate reduction. In some examples, H_2 evolution directly from the nodule can be detected (46). Those associations in which such evolution does not occur contain an uptake hydrogenase that recaptures the hydrogen and recyles the energy. However, a maximum of only about one half of the energy can be recovered in this way. This apparently energy-wasteful process of H_2 evolution and its recapture have been suggested as both an index of efficiency and a criterion for the selection of *Rhizobia* for agricultural use (47).

Nodulated Nonleguminous Angiosperms (Actinorhizal Associations). The best known example is *Alnus* (alder) but other woody trees and shrubs also have root nodules (more specifically called *Actinorhizae*) harboring, not *Rhizobia* but another class of microorganisms, *Actinomycetes* (48). Such plants have a wide geographical distribution and often are the first plant types to colonize poor or devastated soils (49). They have, therefore, an important ecological role and some may possibly be of great significance in biomass production. Alder is increasing in importance in timber production in the northwestern United States because the growth of Douglas fir is apparently stimulated by intercropping with alder. The nodules have a different organization to those on legumes (50). They do not contain a hemoglobinlike protein but must have some system for controlling oxygen input to the nodule.

Progress in investigating these systems has been slowed by problems with isolating the microbial symbiont. Only recently have pure cultures been obtained and used for reinfection (51). Until this time, inoculation of seedlings was effected with a suspension of crushed nodules. The *Actinomycete* symbiont is now called *Frankia* and some species have been described based on cross-inoculation groupings; however, the classification is not generally accepted. The organism has hyphae-like filaments that penetrate the plant tissue and may end in club-shaped vesicles. These vesicles may contain the nitrogenase (50). Cell-free, partially purified preparations of nitrogenase from these vesicles have been obtained, which appear to have properties similar to those of the purified enzyme from the free-living bacteria (52).

Algal Associations. Not only are blue–green algae (more properly called cyanobacteria) important free-livers, they also form associations with a variety of plants. These associations range from the lichens, involving a fungus, through liverworts and ferns to gymnosperms and an angiosperm (53–56). The water fern, *Azolla*, in association with its blue–green algal symbiont, *Anabaena azollae*, is an important contributor to rice culture. It is being used both as a green manure and in dual culture with rice. Estimates indicate that it can supply fixed nitrogen to rice in amounts comparable to those supplied by the rhizobial symbiont to the legume. It is being examined for various aquacultures (including rice) as a method of rapidly upgrading tropical agriculture (55,57–58). The fern is usually 1–2 cm wide (although *Azolla nilotica* can grow to 15–20 cm in width) and the symbiont lives in a cavity in the dorsal leaf lobe. Both the fern and alga are photosynthetic; the alga has a higher-than-usual percentage of heterocystous cells (containing nitrogenase). In nature, the fern is always associated with the alga, but it can be freed and, when provided with a fixed-nitrogen source, grown alone. The alga from this association is difficult to grow alone and there is no evidence for reinfection of an algal-free plant (59).

Only the Cycads among the gymnosperms form N_2-fixing associations. These take the form of root nodules containing the blue–green alga, *Nostoc* (53). In the angiosperms, only one plant, *Gunnera*, forms nodules with a cyanobacterium (*Nostoc*) and these are above ground at the base of the leaves (56).

Associative Symbioses. This group is constituted by some recently discovered, rather informal associations in which some interdependence exists between various grasses (family Gramineae) and certain bacteria. The best-characterized examples are the association between the tropical grass *Paspalum* and the bacterium *Azotobacter paspali*, and that between the grass *Digitaria* and *Azospirillum brasilense* (60–61). In *Paspalum*, a mucilaginous sheath forms around the root within which the bacteria live and fix dinitrogen. This association is thus not formalized in the sense that the bacteria do not invade the plant tissue. In the *Digitaria–Azospirillum* example, however, the roots are invaded but no nodule develops. The extent to which the plants benefit from the association is uncertain. *A. lipoferum*, which occurs in temperate zones, associates with certain corn and sorghum cultivars, but the effect, thus far, on the plants appears to be small (62–63). However, these associations have been discovered only recently and may become of agronomic and economic importance in the future as information relevant to the manipulation of the systems becomes available.

Other Associations. An association, involving the leaves of an angiosperm, *Psychotria*, and *Klebsiella* bacteria, exists. Here, the bacteria occupy nodules on the leaf surface, but even though known to fix dinitrogen as a free-liver, the *Klebsiellae* cease to do so in the nodule. Although some benefit is likely to be derived by each of the partners, it is not from nitrogen fixation (64). *Mycorrhizae* are symbiotic fungi which were once thought to fix dinitrogen in association with the roots of the gymnosperm, *Podocarpus*. Now, however, it appears that they only create an environment conducive to fixation by soil bacteria (23,65). Finally, *Citrobacter* infest the intestinal tract of termites and can, if necessary, fix dinitrogen there. Again, the benefit gained by the insect is likely to be very small (66).

Free-Living Microorganisms. Except for the cyanobacteria, the free-living bacteria are generally not agriculturally important. They contribute to the soil only about 0.1% of the fixed nitrogen of a leguminous association. The free-living cyanobacteria, by comparison, contribute about 2–5% as much as a leguminous association. The difference probably lies in the cyanobacteria's property of photosynthesis which relieves their dependence on the often limiting carbon (and, therefore, energy-yielding) substrates in the soil. Free-living bacteria are used for most of the research on the isolation and characterization of the enzyme nitrogenase, responsible for N_2 fixation (1,4,17,22,67). These N_2 fixers can be classified into aerobes, anaerobes, facultative anaerobes, and photosynthetic bacteria, including cyanobacteria.

Aerobes. The best known examples of this class are the *Azotobacter*. All are obligate aerobes and efficiently fix nitrogen only in air. Many other genera fix N_2 but they are usually more sensitive to oxygen, eg, *Corynebacterium* and *Azospirillum*, which require microaerophilic conditions for fixation. All have systems, some more efficient than others, that protect nitrogenase from damage by O_2 while growing aerobically. Similarly, *Rhizobium* fixes only when the microaerophilic conditions existing in the nodule are simulated in a plant-free culture (35–39). Many other genera oxidize gases such as hydrogen or methane, to derive the energy necessary for growth and fixation.

Anaerobes. These bacteria, typified by *Clostridium pasteurianum*, are found in soil and water and, as obligate anaerobes, cannot use oxygen. The clostridia metabolize glucose (and related compounds) to butyrate, carbon dioxide, and dihydrogen and many, but not all, species fix N_2. A second group of N_2-fixing anaerobes are the sulfur bacteria, such as *Desulfovibrio*. Again, not all fix N_2, but they grow by reducing sulfate (or other oxidized sulfur compounds) to sulfide which is, in part, responsible for the smell of polluted environments. *Desulfovibrio* is the only nonphotosynthetic N_2 fixer to occur naturally in the sea and is important ecologically in the formation of nitrogen-containing marine sediments.

Facultative Anaerobes. These bacteria can grow with or without dioxygen, but they fix N_2 only under anaerobic conditions. *Klebsiella* occurs in soil, water, and animal intestines and contains a number of species that fix N_2. They are related to *Escherichia coli* which, although not a naturally occurring N_2 fixer itself, can be genetically modified to fix N_2. Other genera, including *Citrobacter* and *Enterobacter* are also of this type. A second group of such N_2 fixers is *Bacillus*. This genus, like the clostridia, also has the ability to form spores to survive unfavorable conditions.

Photosynthetic Bacteria and Cyanobacteria. This group includes examples that fix N_2 under aerobic, anaerobic, and microaerophilic conditions. In contrast to the above three groups, they can use CO_2 via photosynthesis as their sole carbon source for growth. The photosynthetic anaerobes include both sulfur and nonsulfur bacteria. Red *Chromatium* oxidizes sulfur (or sulfides) to sulfate during photosynthesis to produce the reductant (and possibly energy) necessary for N_2 fixation. *Rhodospirillum* is a purple, nonsulfur, N_2-fixing genus. The cyanobacteria differ from the above types because they grow aerobically and also produce dioxygen (like plants) from photosynthesis. They protect nitrogenase in a variety of ways. Some, like *Anabaena*, have specialized cells, called heterocysts, which are incapable of evolving O_2 and where nitrogenase is located. Others, like the filamentous *Plectonema*, do not have heterocysts and fix N_2 only under lower oxygen pressures and low light intensity. However, the unicellular *Gleothece* can fix in air. The cyanobacteria, as stated above, are easily the most important contributors of fixed nitrogen to both the sea and land among these free-living organisms and, in addition, have a symbiotic contribution (55,67).

Nitrogenase. The enzyme system responsible for N_2 fixation is called nitrogenase. It was first prepared as a cell-free extract from the anaerobe *Clostridium pasteurianum* in 1960 (68). Since then, more than 20 species have yielded partially purified nitrogenase; five have been extensively purified. These extracts are highly oxygen sensitive. Nitrogenase needs both a source of reducing potential and energy (in the form of adenosine triphosphate, ATP) to operate (69–70). In general, the enzymes from all N_2-fixing bacteria are very similar (45).

Nitrogenase has been isolated in two distinct forms. One is the nitrogenase complex, which so far has been well characterized only for *Azotobacter vinelandii*. It is more stable to oxygen than the component proteins discussed below and has a molecular weight of ca 300,000. The complex contains one molybdenum–iron protein, one iron protein, and one ferredoxin (or iron–sulfur protein) which is not essential for activity (71–72). The second form results from the separation of nitrogenase into its two component proteins, designated the molybdenum–iron protein [Mo–Fe] and the iron protein [Fe] (73–74). These proteins have molecular weights of ca 230,000 and 60,000, respectively. The former contains ca 2 Mo, 30 Fe, and 30 sulfides per mole (1,75), apparently divided into four Fe_4S_4 clusters (76) and one or two FeMo cofactor centers (see below), all contained within two sets of nonidentical subunits (77). The iron protein contains four Fe atoms and four sulfides per mole in the form of a single Fe_4S_4 cluster and consists of two identical subunits (1,75,78). Neither protein shows any enzymatic activity alone (73), but when recombined, they display the same range of reactivity as the complex, crude cell-free extracts or whole cells. Maximum specific activity of [Mo–Fe] occurs at a 10:1 ratio of [Fe] to [Mo–Fe]. However, it appears that a 1:1 or 2:1 ratio of [Fe] to [Mo–Fe] is the more likely stoichiometry of nitrogenase within the cell (79–80).

Nitrogenase catalyzes the ATP-dependent reaction of a reductant with N_2 to form NH_3 plus ADP (adenosine diphosphate) and phosphate (81). ADP is an inhibitor of nitrogenase catalysis (82) by competing for the ATP-binding sites on [Fe] (see below). This inhibition is overcome by using the creatine phosphate–creatine phosphokinase system in assays to reconvert ADP to ATP (81). Reductants capable of supporting nitrogenase turnover have redox potentials more negative than -300 mV and include naturally occurring ferredoxins and flavodoxins *in vivo* (81,83–84) and sodium dithionite ($Na_2S_2O_4$), *in vitro* (70). Nitrogenase reduces a number of small-molecule organic substrates (85) in addition to N_2 (Fig. 4). In the absence (and to a lesser extent in the presence) of these substrates, H_2 is evolved (74). The optimal stoichiometry for N_2 reduction is probably as shown below; approximately four moles ATP are hydrolyzed for each pair of electrons used, a quantity independent of the substrate reduced (86).

$$N_2 + 4\ S_2O_4^{2-} + 16\ MgATP^{2-} + 24\ H_2O \xrightarrow{\text{nitrogenase}} 2\ NH_{3(aq)} + H_2$$

$$+ 8\ SO_3^{2-} + 16\ MgADP^- + 16\ HPO_4^{2-} + 24\ H^+$$

The requirement for ATP makes biological N_2 fixation an energy-consuming process and organisms preferentially use fixed nitrogen when it is available. Despite this energy requirement, biological nitrogen fixation is an attractive alternative to the Haber process, particularly if the apparently wasteful H_2 evolution can be eliminated.

Mechanistic studies, although not yet providing detailed molecular data, have

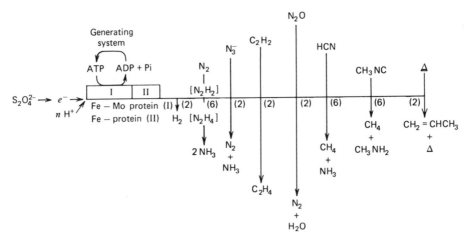

Figure 4.   Requirements for and the substrates and reduction products of nitrogenase catalysis. Pi = inorganic phosphate.

nonetheless established certain key features of nitrogenase turnover. The chemical, spectroscopic, and redox character of [Fe] is changed by the binding of 2 mol ATP (87–89). Each [Fe] then delivers a single low potential electron to [Mo–Fe] (90). The latter also may bind ATP (91) and is capable of receiving substantial numbers of low potential electrons (87,92). Extensive studies using epr, polarography, and electronic absorption spectroscopy suggest that [Fe], with two bound ATP moieties, transfers its electron to [Mo–Fe] at a potential that is sufficient to effect substrate reduction with concomitant ATP hydrolysis (88,90,92–96). Thus, [Mo–Fe] is believed to contain the N_2 binding and reducing site whereas the other components may be viewed as supplying the required reducing equivalents and energy to that site. More recently, a relatively small, iron- and molybdenum-containing entity, called the iron–molybdenum cofactor (FeMo–co), has been isolated from [Mo–Fe]. It accounts for all of the molybdenum and about half of the iron of [Mo–Fe] and may contain the substrate-reducing site of nitrogenase (97).

Iron–Molybdenum Cofactor.   FeMo–co has been isolated from a number of bacterial sources (97). It can reactivate crude extracts from cells of the mutants, *Azotobacter vinelandii UW45* or *Klebsiella pneumoniae nif B*, which without added FeMo–co are incapable of N_2 reduction. This cofactor contains molybdenum, iron, and sulfide in a 1:6–8:4–6 atomic ratio (97–99), but no proteinaceous material (99–100). It is stable in certain organic solvents but highly susceptible to inactivation by dioxygen (97). The discovery of nonheme iron associated with this cofactor differentiates it from Mo–co, the molybdenum cofactor from fungal nitrate reductase and liver sulfite oxidase (97,101). Data obtained by a combination of Mössbauer and epr spectroscopy (102–103) and molybdenum x-ray absorption spectroscopy (104) strongly suggest that the molybdenum and certain of the iron environments in [Mo–Fe] from *A. vinelandii* and *C. pasteurianum* and in FeMo–co are identical. FeMo–co is also the magnetic center responsible for the typical epr signal ($g = 4.3, 3.7, 2.0$) of [Mo–Fe] in the presence of sodium dithionite (103). Thus, FeMo–co is not destroyed on removal from [Mo–Fe] and still possesses at least some of the chemical (and possibly catalytic) properties of the enzyme. However, no nitrogen-reducing activity has, so far, been demonstrated for FeMo–co outside of the protein.

Genetics and Regulation.   Although mutants of the N_2-fixing system were reported in the 1960s, nitrogen fixation genetics was not investigated until 1971 when it was shown that the nitrogen fixation (*nif*) genes were clustered closely together on the single circular chromosome of *Klebsiella pneumoniae* (105–106). The genetic techniques of transduction and conjugation have both been used since then to transfer the *nif* genes to initially nonfixing bacteria. In transduction, the N_2-fixing cells are infected with a temperate phage which can be thought of as a nonlethal bacterial virus. The phage cleaves a portion of its first host's chromosome (the donor) and transfers it to a second host (the recipient) on subsequent infection. These originally nonfixing recipients, which gain the *nif* genes from the donor, become N_2 fixers. Conjugation involves sexual activity during which the bacteria are joined and genetic material is passed from one to the other. The transferred genetic material carries additional genetic information which confers advantageous properties on the bacterial cell. Both transduction (105) and conjugation (106–107) have successfully produced N_2-fixing bacteria from nonfixers. Thus, *Escherichia coli*, a common enteric bacterium, and *Salmonella typhimurium*, both of which do not naturally fix N_2, were modified to become N_2 fixers (when grown anaerobically). *Agrobacterium tumefaciens*, however, although it could be shown to have received the *nif* genes, still did not fix N_2, indicating that it lacked other information necessary for the *nif* genes to be expressed (108).

Plasmids are extrachromosomal genetic elements that are not essential for cell growth and replication but confer environmental advantages (eg, antibiotic production or resistance or sexual combination) when present. They occur naturally and can also be constructed artificially. Plasmids in permissive hosts can replicate. They are circular pieces of DNA and can vary in size up to 5% (or larger) of the size of the chromosome. They can be incorporated into bacterial cells by both transduction and conjugation techniques and even by simply adding isolated plasmids to a cell culture. These genetic elements offer the opportunity of transferring genes at much higher frequency than occurs by the usual molecular genetic techniques. At present, evidence is accumulating for the location of the *nif* genes on plasmids in *Rhizobium*. Certainly by inserting them into a plasmid (109), they can be successfully transferred to initially nonfixing species (108–111).

Genetic techniques have also been used to determine the number and location of the *nif* genes on the *K. pneumoniae* chromosome, ie, to map the *nif* gene cluster. Surprisingly, at least 15 genes are found to be associated in some way with nitrogen fixation (112–117). Similar studies with *Rhizobium* are underway but are still at a preliminary stage (118–119). This large number of genes, compared to the usual several genes for a biological process, indicates a great complexity, that in turn makes genetic engineering employing nitrogen fixation a difficult and complex undertaking (117). However, encouragement can be drawn from the successful use of transduction techniques to overcome the specificity barrier. For example, plasmids containing the nodulation genes for peas have been transferred from *R. leguminosarum* to certain species of *Rhizobium*, that do not normally infect peas, allowing them to nodulate successfully the roots of pea plants (119) (see Genetic engineering).

Synthesis of nitrogenase by a cell is under very tight genetic control that allows a nitrogen-fixing cell to do so only when necessary. Thus, if fixed nitrogen is present in the environment, the bacteria utilize it until it is totally depleted rather than synthesize nitrogenase to fix their own (120). The mechanism of this control aspect is not yet understood. Although glutamine synthetase is thought to play a vital role (121–123), this situation is considerably more complicated than first thought (124). Similarly, in cells that require anaerobic conditions for N_2 fixation, a control system is present that prevents the oxygen-sensitive nitrogenase from being synthesized in air (125). This system prevents the energy wastage involved in producing an enzyme that would be inactivated rapidly. Other controls are concerned with the effects of molybdenum (126) and N_2 itself (127) on nitrogenase synthesis (117,123). All these studies on free-living bacteria become even more complicated when extended to *Rhizobium* (128–129) especially when the role of the leguminous plant in symbiosis is considered (130).

Chemical Approaches

Dinitrogen-Reducing Systems. Dinitrogen, N_2, a very stable molecule, has a dissociation energy of 941 kJ/mol (225 kcal/mol) and an ionization potential of 15.6 eV, indicating that it is very difficult to cleave or to oxidize. For reduction, electrons must be added to its lowest unoccupied molecular orbital at -7 eV, which can occur only with highly electropositive metals such as lithium. However, lithium also reacts with water, suggesting that these interactions are unlikely in the aqueous environment of the natural enzymic system. Even so, such systems have achieved some success in N_2 reduction in nonaqueous solvents.

Aprotic Systems. The first active N_2-reducing system was reported in 1964 (131). Titanium tetrachloride or bis(η^5-cyclopentadienyl)titanium dichloride, $(C_5H_5)_2TiCl_2$, with ethylmagnesium bromide or lithium naphthalenide as reductant in ethyl ether (131–132) were the basic constituents of the reaction. Although a nitrogen pressure of 10.1–15.2 MPa (100–150 atm) was used, the pressure for the half-maximum reaction rate is about 50 kPa (0.5 atm). The system is partially inhibited by strongly solvating solvents, carbon monoxide, acetylene, olefins, and dihydrogen, but not by small amounts of dioxygen (133). Despite early speculation to the contrary, dinitrogen complexes are undoubtedly formed as intermediates with titanium(II) postulated as the active species (see also Coordination compounds). Although the mechanism of these reactions is still unclear, the conversion of dinitrogen into nitride is likely (134). These systems are not catalytic because the solvolysis needed to liberate ammonia destroys the active species. However, about five cycles of reduction, dinitrogen absorption, and solvolysis can be obtained with the bis(isopropoxy)titanium dichloride–diglyme–sodium system using 2-propanol for solvolysis (133). A truly catalytic effect of titanium was demonstrated with a mixture of titanium tetrachloride, metallic aluminum, and aluminum tribromide at 50°C when 200 mol ammonia/mol titanium was obtained via the catalytic nitriding of aluminum (135). These nitriding systems also can produce nitrogen-containing organic compounds. Thus, bis(η^5-cyclopentadienyl)titanium dichloride and excess phenyllithium in ether solution under dinitrogen at room temperature give, after hydrolysis, 0.15 mol aniline/mol titanium, as well as ammonia (136). When bis(η^5-cyclopentadienyl)titanium dichloride is reduced with magnesium metal in tetrahydrofuran under dinitrogen and the resultant mixture reacts with ketones, secondary amines form upon subsequent hydrolysis (137).

By careful adjustment of conditions, metal–N_2 compounds have been isolated from similar reaction mixtures. However, a number of reaction products have been isolated from the $(\eta^5\text{-}C_5H_5)_2TiCl_n$ ($n = 1, 2$)/N_2/reductant system, all of which assume an intense blue color in solution (λ_{max} ca 600 nm). The relationship among these products is unclear (138–139). The lability of the cyclopentadienyl ring is an important complicating factor which has been overcome to a great extent by pentamethyl substitution. Thus, when $[\eta^5\text{-}C_5(CH_3)_5]_2Ti$ produced from $[\eta^5\text{-}C_5(CH_3)_5]_2TiCl_2$ is used, two distinct interconvertible N_2 complexes are formed (140).

X-ray crystallography shows the mono-N_2 complex to have a linear TiN=NTi bridge with an N—N bond length of 0.116 nm (141). The tri-N_2 complex of Ti is unstable above −80°C, but its zirconium analogue, $\{[\eta^5\text{-}C_5(CH_3)_5]_2Zr(N_2)\}_2(N_2)$, is thermally stable and easily synthesized by Na–Hg reduction of $[\eta^5\text{-}C_5(CH_3)_5]_2ZrCl_2$ under N_2. It has a similar linear (ZrN=NZr) bridge (N—N bond = 0.118 nm) plus one end-on terminal N_2 ligand (N—N = 0.1115 nm) on each zirconium atom (see Fig. 5) (140,142–144).

The mono-N_2 complex, $[(\eta^5\text{-}C_5R_5)_2Ti]_2(N_2)$ (R = CH_3,H), apparently cannot produce hydrazine or ammonia directly. But, at −80°C in the presence of excess reductant, it gives ammonia (142,145). In contrast, the tri-N_2 complex, $\{[\eta^5\text{-}C_5(CH_3)_5]_2Zr(N_2)\}_2(N_2)$, reacts directly with hydrogen chloride at −80°C in toluene to liberate two N_2 molecules and produce hydrazine (0.9 mol/mol) (142,144) as does its titanium analogue (141). This dinuclear zirconium complex shows no significant structural differences either between the terminal and bridging N_2 ligands or when compared with the structure of other metal–N_2 complexes, most of which do not produce hydrazine or ammonia on protonation. Thus, the structural result itself gives no indication of the requirements for N_2 reduction. Isotopic labeling with $^{15}N_2$ has, however, given an insight into the reduction process. Only the terminal N_2 ligands exchange with $^{15}N_2$ (or CO) to give $\{[\eta^5\text{-}C_5(CH_3)_5]_2Zr(^{15}N_2)\}_2(N_2)$. However, since the hydrazine produced on protonation contains some of the nitrogen-15 label, apparently the terminal and bridging N_2 ligands are scrambled to form the product, even though a total of only one N_2 per complex is actually reduced. This scrambling can be explained by the symmetrical intermediate, $[\eta^5\text{-}C_5(CH_3)_5]_2Zr(N_2H)_2$, as shown below.

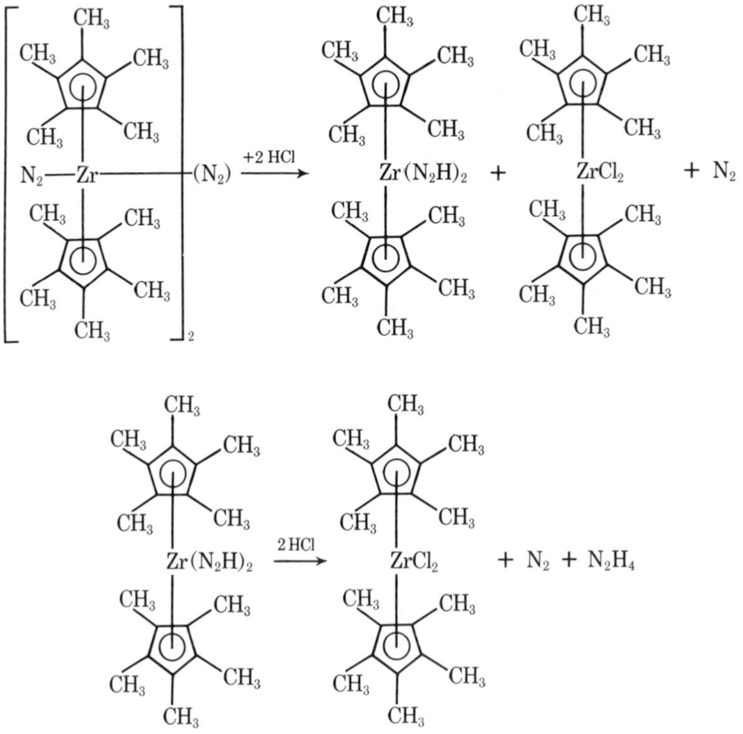

This concept gains support from the observation that no hydrazine is formed with $\{[\eta^5\text{-}C_5(CH_3)_5]_2Zr(CO)\}_2(N_2)$ (140,142–144).

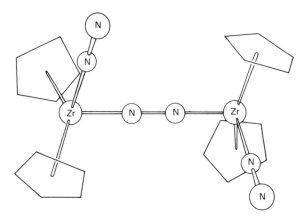

Figure 5. Molecular structure of $\{[\eta^5\text{-}C_5(CH_3)_5]_2Zr(N_2)\}_2(N_2)$ (143). Courtesy of the American Chemical Society.

When $(i\text{-}C_3H_7)MgCl$ is used with $(\eta^5\text{-}C_5H_5)_2TiCl_n$ $(n = 1, 2)$ under N_2 at $-80°C$, $[(\eta^5\text{-}C_5H_5)_2Ti(i\text{-}C_3H_7)]_2(N_2)$ is formed. This mono-N_2 compound needs to react further with $(i\text{-}C_3H_7)MgCl$ to give species that produce hydrazine or ammonia on hydrolysis (146). Similar dark-blue complexes, $[(\eta^5\text{-}C_5H_5)_2Ti(R)]_2(N_2)$, with similar properties are also formed directly from the titanium(III) species, $(\eta^5\text{-}C_5H_5)_2Ti(R)$ $(R = C_6F_5$, substituted phenyl and benzyl), and N_2 in toluene at $-20°C$ (147). A related mono-N_2 zirconium(III) compound, $\{(\eta^5\text{-}C_5H_5)_2ZrCH[Si(CH_3)_3]_2\}_2(N_2)$, in contrast, gives 0.2 mol hydrazine per mole directly on solvolysis (148).

At one time, it appeared that there was no direct connection between the highly reducing systems that produce ammonia or hydrazine from N_2 and the well-defined metal–dinitrogen complexes. However, just as metal–dinitrogen compounds have been isolated from the reducing systems, so too have a number of metal–dinitrogen compounds been degraded to ammonia and/or hydrazine. The mononuclear, tertiary phosphine complexes of molybdenum and tungsten are the best-studied examples. The early reactions of $trans\text{-}M(N_2)_2(dppe)_2$ [M = Mo, W; dppe = 1,2-bis(diphenyl-phosphino)ethane] (149) with excess acid did not appear to proceed past the hydraz-ido(2−) (N—NH$_2$) stage (150).

$$trans\text{-}Mo(N_2)_2(dppe)_2 + 2\,HBr \rightarrow Mo(N_2H_2)Br_2(dppe)_2 + N_2$$

However, a mixture of $Mo(N_2)_2(dppe)_2$ and $Mo(N_2H_2)Br_2(dppe)_2$, when heated in N-methylpyridone with aqueous HBr, formed some ammonia (151). In separate experiments, using $C_6H_5P(CH_3)_2$ or $(C_6H_5)_2PCH_3$ to form cis- and $trans\text{-}M(N_2)_2(PR_3)_4$, respectively (150), followed by treatment with acid, much higher ammonia yields are obtained. For M = W, about 2 mol ammonia per mole of complex are produced.

$$cis\text{-}W(N_2)_2[P(CH_3)_2C_6H_5]_4 \xrightarrow{\text{H}_2\text{SO}_4\text{--CH}_3\text{OH}} N_2 + 2\,NH_3 + W(VI)\text{oxides} + 4\,P(CH_3)_2C_6H_5$$

With M = Mo, only 0.7 mol ammonia is produced (152). Similar treatment of $M(N_2H_2)X_2[P(CH_3)_2C_6H_5]_3$ with H_2SO_4–CH_3OH provides similar yields of ammonia. These and related data suggest a possible stepwise sequence for the reduction and protonation of N_2 on a single molybdenum atom in nitrogenase (152). The disadvantages of this system are that the metal-containing product has not been characterized nor has the system been cycled. A more recent reaction system has overcome these problems by taking advantage of both the stabilizing effect of the chelating phosphines (dppe) and the lability of the simple phosphines. Thus, $Mo(N_2)_2[P(C_6H_5)_3][triphos]$, where triphos = $(C_6H_5)P[CH_2CH_2P(C_6H_5)_2]_2$, reacts with anhydrous HBr as follows (153):

$$2 \, Mo(N_2)_2[P(C_6H_5)_3][\text{triphos}] + 8 \, HBr \rightarrow 2 \, MoBr_3[\text{triphos}] + 2 \, NH_4Br + 3 \, N_2 + 2 \, P(C_6H_5)_3$$

The product, $MoBr_3(\text{triphos})$, is the starting material for the preparation of the metal–N_2 complex and the system can be cycled.

Production of ammonia in the Mo and W systems has certain features in common with that by Ti and Zr systems. Both have two N_2 ligands in the metal's coordination sphere but only one is reduced and in both, N_2 reduction is initiated by addition of acid. Protonation of N_2 occurs as the phosphine or N_2 ligands, which stabilize the lower oxidation states of the metals, are successively replaced by acid counterions that favor the higher oxidation states. For $M(N_2)_2(PR_3)_4$ and related compounds, this exchange encourages the transfer of more metal electron density to N_2 and thus promotes further protonation. Because the chelating diphosphines, eg, dppe are much less easily replaced, $M(N_2)_2(\text{dppe})_2$ does not produce ammonia readily. With the more easily displaced, simple phosphines, all six metal electrons are used to produce ammonia. With the mixed-phosphine system, ammonia is formed but only three electrons per atom of molybdenum are used. In the binuclear zirconium system, only four electrons are available and hydrazine is formed. Here also, electron flow from metal to N_2 could be encouraged by loss of N_2 and coordination of the appropriate acid anion.

These ammonia- and hydrazine-forming reactions appear to be conducted under mild conditions. However, the reducing power of, eg, magnesium metal ($E° = -2.4$ V) is built into these systems already during the preparation of the metal–N_2 complexes. The complete degradation of the simple phosphine complexes of Mo and W during ammonia formation does not favor catalysis. An advantage is offered by the mixed-phosphine Mo system and the zirconocene system, where the product of acid degradation is the starting material for preparation of the metal–N_2 complex. It remains to be seen if significantly milder reductants can effect the synthesis of these or similarly reactive N_2-containing species or if they can be operated cyclically or catalytically (see also Organometallics).

The $trans$-$M(N_2)_2(\text{dppe})_2$ complexes (M = Mo, W) also undergo carbon—nitrogen bond formation on reaction with alkyl or acyl halides as does the $Mo(NNH_2)$ unit with ketones (154–156).

$$trans\text{-}M(N_2)_2(\text{dppe})_2 + RX \rightarrow M(N_2R)X(\text{dppe})_2 + N_2$$

$$trans\text{-}[Mo(NNH_2)F(\text{dppe})_2]BF_4 + R_2C{=}O \rightarrow [Mo(NNCR_2)F(\text{dppe})_2]BF_4 + H_2O$$

Treatment of $Mo[N_2(C_4H_9)]Br(\text{dppe})_2$ in benzene–methanol for 10 h at 100°C under N_2 with sodium methoxide gives ammonia (0.3 mol/mol) and 0.3 mol of n-butylamine, whereas $[W\{N_2(CH_3)_2\}Br(\text{dppe})_2]Br$ gives only dimethylamine with base (157–158). Another approach to carbon—nitrogen bond formation is shown below, where the dimethylhydrazine ligand in the second equation can be displaced by N_2 to form the basis of a cyclic process (159–160).

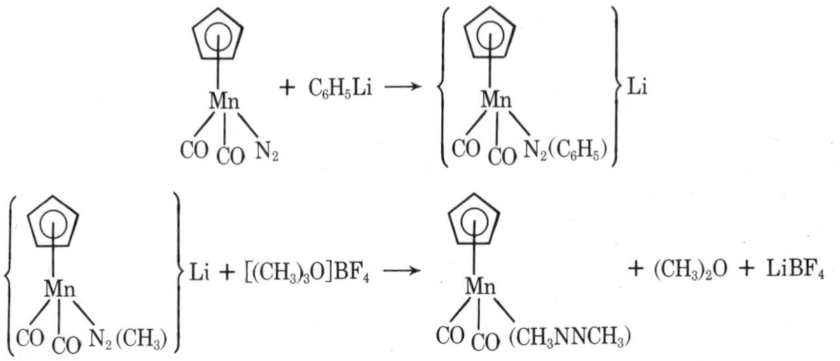

Such reaction systems may have significant industrial importance in the future for producing organic nitrogen compounds using N_2 instead of ammonia as a feedstock.

Many other examples of chemical N_2 reduction in less well-defined systems have been reported. Systems based on V, Cr, and Fe apparently produce measurable, but less than stoichiometric, quantities of hydrazine and ammonia (161). However, the system based on $MoCl_4(dppe)$–Mg–alkyl bromide is reported to absorb 106 mol N_2/Mo atom in some undefined way and to liberate diazene ($HN{=}NH$) on hydrolysis. Possibly, Mo acts as a catalyst for producing a magnesium–nitride complex (162). A similar effect may occur with $MoCl_5$–NaHg–Mg^{2+} in methanol where 3.6 mol hydrazine/Mo atom can be produced (163).

Aqueous Systems. Many strong reducing agents in the presence of derivatives of transition metals have been reported to produce minute amounts of ammonia from N_2 in aqueous solution. However, spurious results are obtained easily because low metal concentrations are used and contaminating species may occur, the Nessler test for ammonia is not specific, the system may scavenge traces of ammonia or nitrogen oxides (which are subsequently reduced) from the N_2 gas, and nitrogen-containing substances added to the reaction mixture may be degraded to ammonia. Only those systems that have been substantiated through the use of $^{15}N_2$ are discussed here.

Aqueous systems are known that reduce dinitrogen to either hydrazine or ammonia as the main product. The essential catalyst to produce hydrazine is a reduced molybdenum or vanadium salt in the presence of a reductant and a substantial amount of Mg^{2+}. Aqueous or aqueous–alcoholic solutions of sodium molybdate(VI) or oxo-trichloromolybdenum(V), when mixed with titanium trichloride as reductant and Mg^{2+} at pH 10–14, produce some hydrazine at 25°C and 0.1 MPa (1 atm) N_2. However, at 50–100°C and 5.1–15.2 MPa (50–150 atm) N_2, yields of hydrazine reach 100 mol/mol Mo. At the higher temperature, some ammonia is produced also. Vanadium(II) or chromium(II) are equally effective as the reductant. The reaction mixture is heterogeneous and it is suggested that hydroxide-bonded polynuclear entities furnish the reducing capacity from Ti(III) through Mo(III) to the N_2. This system's efficiency is about 1% of the natural system (164). Further experiments show that vanadium(II) can replace both molybdenum and titanium in this system. At alkaline pH, lower temperatures and 10.1 MPa (100 atm) N_2 pressure, quantitative reduction (0.22 mol hydrazine/mol V) occurs within minutes (164). A four-electron reaction is proposed, with a tetramer of vanadium(II) ions undergoing oxidation to V^{3+} (165).

$$(V^{2+})_4 + N_2 + 4\,H_2O \rightarrow 4\,V^{3+} + N_2H_4 + 4\,OH^-$$

At room temperature or higher, ammonia and hydrogen are the products owing to the reaction of the hydrazine with more vanadium(II) (164). The rate of N_2 reduction is faster with vanadium than with molybdenum. Both systems reduce acetylene to ethylene plus ethane and are poisoned by carbon monoxide (166). An independent reappraisal of this system confirms N_2 reduction but suggests the alternative reaction mechanism (with the same overall stoichiometry) based on a monovanadium entity as shown below (167).

$$V^{2+} + N_2 + H_2O \rightarrow VO^{2+} + N_2H_2$$

$$VO^{2+} + V^{2+} + H_2O \rightarrow 2\,V^{3+} + 2\,OH^-$$

$$\underline{2\,N_2H_2 \rightarrow N_2H_4 + N_2}$$

$$4\,V^{2+} + N_2 + 4\,H_2O \rightarrow 4\,V^{3+} + N_2H_4 + 4\,OH^-$$

A related homogeneous aqueous-alcoholic system, composed of vanadium(II) complexes of catechol and its derivatives, reduces N_2 to ammonia with concomitant H_2 evolution. Only catecholates are active in this system, which is very sensitive to pH. This system has been likened to nitrogenase. It has been suggested that both use a sequence of two four-electron reductions to evolve one H_2 for every N_2 reduced. In the first step, a tetramer of V^{2+} produces hydrazine (as above), followed by a similar four-electron step to produce H_2 and two ammonia molecules. This system resembles nitrogenase in reducing acetylene in a specific cis manner to ethylene and is inhibited by carbon monoxide (168–169).

A third, aqueous, N_2-reducing system was developed based on the knowledge that nitrogenase contains iron, molybdenum, sulfide, and thiol groups. It was reported that 3–5 μmol of NH_3 are produced from about 5 mmol Na_2MoO_4, 2.5 mmol thioglycerol, 0.1 mmol $FeSO_4 \cdot 5H_2O$, and 0.25 g $NaBH_4$ in 50 mL of borate buffer (pH 9.6) under 13.7 MPa (135 atm) N_2 (170). In the absence of molybdenum, no ammonia is obtained. Yields up to ca 0.04 mol NH_3/Mo atom are obtained with a molybdenum–cysteine complex under 0.1 MPa (1 atm) N_2. Specific stimulation of activity by ATP is reported but acids may produce the same effect (171–172). These so-called molybdothiol systems are suggested to react as follows, even though no intermediates have been isolated (173).

$$Mo(IV) + N_2 + H_2O \rightarrow OMo(VI) + N_2H_2$$

$$3\, N_2H_2 \rightarrow 2\, N_2 + H_2 + N_2H_4$$

$$Mo(IV) + H_2O + N_2H_4 \rightarrow OMo(VI) + 2\, NH_3$$

Rather more successful is the $[MoO(CN)_4(H_2O)]^{2-}$–$NaBH_4$ system which gives up to 0.3 mol ammonia per mol complex (174). Treatment of $[MoO(CN)_4(H_2O)]^{2-}$ directly with mild acid under N_2 produces 0.0003 mol NH_3/mol Mo after 48 h at 75°C. The most successful system of this type uses sodium borohydride with MoO_4^{2-}/insulin (6:1) mixtures. It produces 65 mol ammonia/mol Mo in 30 min at 23°C under 0.1 MPa (1 atm) N_2 (175).

The currently available data for all the above chemical N_2-reducing systems indicate that much further development is required in order for them to become important N_2 reduction methods. The aqueous systems are extremely difficult to characterize in detail and the various mechanisms, often contradictory from different research groups, do not have the soundest of bases. However, they do show unequivocally that N_2 can be reduced in aqueous medium. They also provide insight into the binding and activation of N_2 toward reduction and into the induction of internal redox reactions. The first step in enzymic N_2 reduction undoubtedly involves the binding of N_2 to a transition-metal site. There is nothing in the known enzymology which precludes either changes in the metal coordination sphere or protonation of bound N_2 (or both) as initiators of the redox process.

All systems now available have serious inherent disadvantages which limit or negate their utility for the reduction of N_2 under truly mild conditions. Thus, although protonation of $W(N_2)_2(PR_3)_4$ and $\{[\eta^5\text{-}C_5(CH_3)_5]_2Zr(N_2)\}_2(N_2)$ to yield NH_3 and N_2H_4, respectively, appear to be mild N_2 reductions, the metal–N_2 complexes were formed in a reaction employing extremely powerful reductants. Similarly, the aprotic systems need powerful reductants whose use would likely be precluded in a useful commercial process. The V(II) systems produce ammonia in protic media but their use is limited because V(II) cannot be regenerated under the basic conditions necessary for efficient N_2 reduction. It would seem, therefore, that these inorganic systems for N_2 reduction in their present form almost certainly do not provide a useful means of ammonia production for the near future. Nitrogenase, however, is already known to produce ammonia from N_2 under ambient conditions. This enzyme, therefore, offers a working basis upon which a practical nitrogen-fixation process could be devised and developed, perhaps concomitantly with purely chemical systems.

Nitrogenase Models. A number of proposals have been made over the years as to the composition of the active site and the mechanism of substrate reduction by nitrogenase (170,176–177) even though definitive, physicochemical information about the enzyme was lacking. For example, its uv-visible absorption spectrum is essentially featureless, its epr spectrum, although definitive, could not, until very recently (102), be interpreted, and the two molybdenum atoms per molecule could not be detected by any technique except elemental analysis. Most suggested mechanisms were based on an analysis of the products of substrate reductions. The situation was further complicated by the original Mo–cofactor concept, which suggested that the molybdenum-containing site of all molybdenum-containing enzymes known at that time was very similar, if not identical. Furthermore, it was suggested that this site could be transferred essentially intact from one enzyme to another (178–179). This latter process was achieved by acid treatment, and thus inactivation, of one molybdenum enzyme, eg, xanthine oxidase. The denatured mixture was added to an extract of an acceptor protein, like the molybdenum-deficient (and, therefore, inactive) nitrate reductase from a mutant fungus. The once molybdenum-deficient protein was now activated by the reconstitution of its molybdenum-containing site and nitrate reduction could occur. This interchange process still holds true for all molybdoenzymes (180) except nitrogenase, which has its own unique cofactor, the iron–molybdenum cofactor (see above) (97,101).

Research in this area has taken two directions: the study of FeMo–co itself and the chemical synthesis of mixed molybdenum–iron–sulfur cluster compounds. The inspiration for the latter was a molybdenum x-ray absorption spectroscopic (XAS) study, particularly the analysis of the extended x-ray absorption fine structure (EXAFS) region of the nitrogenase [Mo–Fe] proteins (104,181–183). These studies have proven invaluable by giving the first insight into the environment of molybdenum in nitrogenase. Although XAS gives only a radially averaged view around the absorber atom (in this case, molybdenum), it can accurately determine the number, distance, and type of neighboring atoms (181–182). Analysis of the EXAFS data shows 3–4 bound sulfur atoms at ca 0.236 nm and 2–3 iron atoms at a distance of ca 0.272 nm, with a somewhat more tentative assignment of 1–2 sulfur atoms at a distance of ca 0.25 nm from molybdenum (104,181). These, then, are the structural features with which the synthetic models must be compatible.

Several types of Mo–Fe–S clusters have been synthesized. The first two examples took advantage of the techniques developed for the successful synthesis of chemical models for the electron-transfer proteins called ferredoxins, which contain one (or more) Fe_4S_4 cubes with each iron in the cube also bound by a thiolate ligand (184). With MoS_4^{2-} and $Fe(SR)_3$, a more complex product results (183,185–188), consisting of two $[MoS_4(FeSR)_3]$ cubelike structures bridged via the two molybdenum atoms by sulfide and/or thiolate groups, eg, $\{[(C_2H_5SFe)_3S_4Mo]_2S(SC_2H_5)_2\}^{3-}$ and

$\{[(C_6H_5SFe)_3S_4Mo]_2(SC_6H_5)_3\}^{3-}$ (see Fig. 6). Both complexes have Mo—Fe and Mo—S distances of ca 0.273 and 0.237 nm, respectively, within each cube. A distance of 0.255 nm from Mo to the bridging thiolates was found in the latter structure. These parameters are very close to those determined for the molybdenum–iron proteins (and, therefore, inferred for FeMo–co) by XAS, with the exception that the longer distance to the bridging thiolate sulfur is not observed in the proteins. Thus, the $[(RSFe)_3S_4Mo]$ cluster might exist in nitrogenase, but, if so, cannot constitute the FeMo–co site completely because the atomic ratio of molybdenum to iron is not correct nor is the epr spectrum duplicated (183). A certain amount of ligand rearrangement is also required if the substrate is to contact the coordinatively saturated molybdenum in such a cubelike arrangement.

More recently, other Mo–Fe–S clusters have been synthesized consisting of linear arrangements of iron, molybdenum, and sulfur; examples are $[(RS)_2FeS_2MoS_2]^{2-}$, $[Cl_2FeS_2MoS_2FeCl_2]^{2-}$, and $[S_2MoS_2FeS_2MoS_2]^{3-}$ (189–192). These complexes are produced by reaction of an appropriate iron compound with MoS_4^{2-} and, where determined, have Mo—S and Mo—Fe bond distances compatible with those determined for the molybdenum site of [Mo–Fe] by XAS. The last example also has an epr signal similar to FeMo–co (192). It is not clear at this time whether these linear clusters approach FeMo–co more closely than the cubelike compounds. They suffer from similar drawbacks to the cube-types but do have a coordinatively unsaturated molybdenum atom that could allow interaction with N_2.

Apparently, the molybdenum-containing prosthetic group of nitrogenase is sufficiently small to apply current inorganic chemical techniques to structure and reactivity studies. This information, together with the knowledge that the early transition metals can activate and reduce dinitrogen, indicates clearly that the chemical synthesis of a Mo–Fe–S cluster capable of N_2 reduction is a distinct possibility. As fossil fuels become more and more scarce, such a possibility offers hope for future generations.

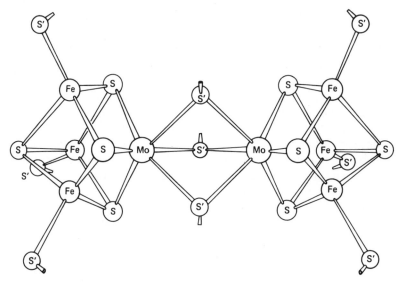

Figure 6.   Molecular structure of $[\{(C_6H_5SFe)_3MoS_4\}_2(C_6H_5S)_3]^{3-}$ (186). Courtesy of the Royal Society of Chemistry (London). C_6H_5 groups are not shown.

Outlook

The mature Haber-Bosch technology is unlikely to change substantially in the future, although coal as a feedstock will demand increasing attention. It may be, however, that the centers for commercial ammonia production will relocate, possibly to Saudi Arabia and Venezuela, where large quantities of natural gas are flared from the production of crude oil. This relocation though would not offset the main problems of high transportation and storage costs involved in ammonia production and distribution. The development of improved catalysts that work at lower temperatures and pressures than those currently used commercially is one possible prospect, although ammonia manufacturers have been searching for such a catalyst for a long time without much success. The discovery of chemical N_2 fixation under ambient conditions is more compatible with a simple, low temperature and low pressure system, possibly operated at an electrode and driven by a renewable resource, such as solar, wind, or water power (or other off-peak electrical power), located near or in irrigation streams. These systems might produce and apply ammonia continuously (in the stream) or store it as an increasingly concentrated ammoniacal solution for later application. In fact, the Birkeland-Eyde process of dinitrogen oxidation in an electric arc is being reconsidered in just such a context (193), particularly for areas where fertilizer production capacity of a few tons per year can make a significant impact on agricultural production. Thus, simple, inexpensive, small-scale systems may have a place in the future in areas where cheap hydro or solar power is available and where the high capital investment requirements for a large-scale manufacturing system cannot be justified. The developing countries, ie, remote areas, which have very limited fossil fuel supplies or fertilizer demands, are the initial targets of these systems.

Other important contributions could be made by improving the utilization of applied nitrogen fertilizer. Only about 50% is actually assimilated by plants. To this end, slow-release fertilizers can make an impact by providing a consistent, gradual fertilizer supply to the crop throughout the growing season and by ensuring that large quantities of ammonia are not available in the soil for nitrifying bacteria to convert to nitrate, which is lost by both leaching to ground water and by denitrification. Similarly, the development of nitrification and denitrification inhibitors would also prevent these ammonia losses to the atmosphere and ground water. However, the effects of such inhibitors on the nitrogen cycle is unclear, but without doubt some response elsewhere in the cycle would compensate for this change.

The exploitation of the benefits of biological fixation would demand, in the shorter term, an increased use of legumes and other symbiotic systems in agriculture. The most effective rhizobial strains and the appropriate cultivars should be matched. If these associations could be manipulated in order to start fixation earlier or to continue it later into the plant's growth, a substantial benefit would accrue. A significant effect would also be forthcoming if the ability to fix N_2 in the presence of fixed-nitrogen sources were conferred on bacteria. Cyanobacteria are remarkably self-sufficient and their presence in rice paddies, eg, should be exploited, together with the *Azolla–Anabaena* association, to enhance the vitally important rice production. The more recent discovery of the root associative (or less formalized) symbioses, also called rhizocoenoses, indicates that a readily available supply of energy-yielding compounds in the soil (near roots, in this instance) could enhance N_2 fixation by various soil bacteria. It also suggests that other, as yet undiscovered, associations might exist which could be of agricultural benefit. Studies of these systems and the more formal symbioses may demonstrate how to engineer new associations, possibly with principal food crops, which could have dramatic effects on both fertilizer usage and food production.

Finally, the genetic manipulation of nitrogen fixation appears to be the ultimate solution for both reducing fossil fuel energy inputs to fertilizer production and increasing food supplies. Genetic manipulation of the recognition and infection processes might result in new or enhanced symbiotic associations. The success of transferring the *nif* genes from one bacterial genus to another has opened up the possibility of their transfer to a crop plant. However, genetic transfer is not enough; if it were, it probably would happen naturally and N_2-fixing plants would exist. In addition, no plant containing bacterial genes has yet been grown to maturity. Perhaps the best solution would be the relocation of nitrogenase to the chloroplasts of leaves where, if properly protected from the oxygen evolved by photosynthesis, it could take advantage of directly available reducing equivalents produced from light.

Reservations have been expressed regarding the outcomes of such genetic transfers, particularly if done indiscriminantly. For example, an extremely resiliant, competitive, N_2-fixing weed might be produced which could take over agricultural lands or block waterways. However, if the nitrogen limitation were removed, it is likely that some other nutrient would limit the growth of any such genetically modified plant. The limitation would probably be phosphate or potassium or, at the extreme in very fertile soils, the carbon dioxide necessary for energy production via photosynthesis. But the benefits from new N_2-fixing crops would far outweigh the possible risks. They would offset the manufacturing, transportation, and storage costs associated with commercial fertilizer and also circumvent pollution caused by leaching, while simultaneously increasing the world food supply.

BIBLIOGRAPHY

"Nitrogen Fixation" in *ECT* 2nd ed., Suppl. Vol., pp. 604–623, by R. H. Stanley, Titanium Intermediates Ltd.

1. R. C. Burns and R. W. F. Hardy, *Nitrogen Fixation in Bacteria and Higher Plants*, Springer-Verlag, Berlin, 1975, p. 43.
2. C. C. Delwiche, *Ambio* **6,** 106 (1977).
3. R. M. Garrels, F. T. Mackenzie, and C. Hunt, *Chemical Cycles and the Global Environment: Assessing Human Influences*, Kaufman, Inc., Los Altos, Calif., 1975.
4. W. E. Newton and W. H. Orme-Johnson, eds., *Nitrogen Fixation*, University Park Press, Baltimore, Md., 1980.
5. R. H. Burris in ref. 4, Vol. 1, p. 7.
6. W. J. Payne, J. J. Rowe, and B. F. Sherr in ref. 4, Vol. 1, p. 29.
7. S. H. Wittwer in A. Hollaender, ed., *Genetic Engineering for Nitrogen Fixation*, Plenum Press, New York, 1977, p. 515.
8. C. C. Delwiche, *Sci. Am.* **223,** 136 (1970).
9. E. W. Paul, *Ecol. Bull.* (*Stockholm*) **26,** (1978).
10. *Fertilizer Manual*, International Fertilizer Development Center, Muscle Shoals, Ala., 1978.
11. K. J. Skinner, *Chem. Eng. News*, 22 (Oct. 4, 1976).
12. F. A. Ernst, *Fixation of Atmospheric Nitrogen*, van Nostrand Co., New York, 1928.
13. C. A. Vancini, *Synthesis of Ammonia*, CRC Press, Cleveland, Ohio, 1971.
14. M. Appl, *Nitrogen* **100,** 47 (1976).
15. F. Haber, *Z. Angew. Chem.* **27,** 473 (1914).
16. D. W. Nichols, P. C. Williamson, and D. R. Waggoner in ref. 4, Vol. 1, p. 43.
17. W. E. Newton and C. J. Nyman, eds., *Proc. 1st. Int. Symp. on Nitrogen Fixation*, Washington State University Press, Pullman, Wash., 1976.

18. G. C. Sweeney in ref. 17, p. 648.

19. H. Hellriegel and H. Wilfarth, *Z. Ver. Rübenzucker-Industrie Deutschen Reichs*, (1888).

20. E. B. Fred, I. L. Baldwin, and E. McCoy, *Root Nodule Bacteria and Leguminous Plants, Studies in Science No. 5*, University of Wisconsin, Madison, Wisc., 1932.

21. P. W. Wilson in J. R. Postgate, ed., *Chemistry and Biochemistry of Nitrogen Fixation*, Plenum Press, London, Eng., 1971, p. 1.

22. W. E. Newton, J. R. Postgate, and C. Rodriguez-Barrueco, eds., *Recent Developments in Nitrogen Fixation*, Academic Press, London, Eng., 1977.

23. G. Bond, *Ann. Rev. Plant Physiol.* **18,** 107 (1967).

24. J. M. Vincent in ref. 4, Vol. II, p. 103.

25. A. H. Gibson in ref. 17, p. 400.

26. D. O. Norris, *Empire J. Exp. Agric.* **24,** 247 (1956).

27. J. K. Wilson, *Soil Sci.* **58,** 61 (1944).

28. R. O. D. Dixon, *Ann. Rev. Microbiol.* **23,** 137 (1969).

29. M. J. Trinick, *Nature (London)* **244,** 459 (1973).

30. F. B. Dazzo in ref. 4, Vol. II, p. 165.

31. C. Napoli, R. Sanders, R. Carlson, and P. Albersheim in ref. 4, Vol. II, p. 189.

32. W. Newcomb in ref. 4, Vol. II, p. 87.

33. G. E. Ham in ref. 4, Vol. II, p. 131.

34. J. C. Burton in ref. 17, p. 429.

35. D. L. Keister, *J. Bact.* **123,** 1265 (1975).

36. W. G. W. Kurz and T. A. LaRue, *Nature (London)* **256,** 407 (1975).

37. J. A. McComb, J. Elliott, and M. J. Dilworth, *Nature (London)* **256,** 409 (1975).

38. J. D. Pagan, J. J. Child, W. R. Scowcroft, and A. H. Gibson, *Nature (London)* **256,** 406 (1975).

39. J. D. Tjepkema and H. J. Evans, *Biochem. Biophys. Res. Comm.* **65,** 625 (1975).

40. C. A. Appleby, F. J. Bergerson, P. K. MacNicol, G. L. Turner, B. A. Wittenberg, and J. B. Wittenberg in ref. 17, Vol. 1, p. 274.

41. N. Ellfolk, *Endeavour* **31,** 139 (1972).

42. J. D. Smith, *Biochem. J.* **44,** 585 (1949).

43. M. J. Dilworth, *Biochim. Biophys. Acta* **184,** 432 (1969).

44. D. W. Israel, R. L. Howard, H. J. Evans, and S. A. Russell, *J. Biol. Chem.* **249,** 500 (1974).

45. D. W. Emerich and R. H. Burris, *J. Bact.* **134,** 936 (1978).

46. K. R. Schubert and H. J. Evans, *Proc. Nat. Acad. Sci. USA* **73,** 1201 (1976).

47. H. J. Evans, D. W. Emerich, T. Ruiz-Argueso, R. J. Maier, and S. L. Albrecht in ref. 4, Vol. II, p. 69.

48. G. Bond in P. S. Nutman, ed., *Symbiotic Nitrogen Fixation in Plants*, Cambridge University Press, London, Eng., 1976, p. 443.

49. G. Bond in ref. 22, p. 531.

50. J. G. Torrey, *Bioscience* **28,** 586 (1978).

51. D. Callaham, P. Del Tredici, and J. G. Torrey, *Science* **199,** 899 (1978).

52. D. R. Benson, D. J. Arp, and R. H. Burris, *Science* **205,** 688 (1979).

53. J. W. Millbank in A. Quispel, ed., *The Biology of Nitrogen Fixation*, Elsevier, New York, 1974, p. 238.

54. J. W. Millbank in R. W. F. Hardy and W. S. Silver, eds., *A Treatise on Nitrogen Fixation*, John Wiley & Sons, Inc., New York, 1977, Section III, p. 125.

55. G. A. Peters, *Bioscience* **28,** 580 (1978).

56. W. B. Silvester in ref. 48, p. 521.

57. S. N. Talley and D. W. Rains in ref. 4, Vol. II, p. 311.

58. A. W. Moore, *Bot. Rev.* **35,** 17 (1969).

59. G. A. Peters, T. B. Ray, B. C. Mayne, and R. E. Toia, Jr. in ref. 4, Vol. II, p. 293.

60. J. Döbereiner and J. M. Day in ref. 17, p. 518.

61. J. Döbereiner, J. M. Day, and P. J. Dart, *J. Gen. Microbiol.* **71,** 103 (1972).

62. R. H. Burris in ref. 22, p. 487.

63. J. F. W. von Bülow and J. Döbereiner, *Proc. Nat. Acad. Sci. USA* **72,** 2389 (1974).

64. W. S. Silver, Y. M. Centifanto, and D. J. D. Nicholas, *Nature (London)* **199,** 396 (1963).

65. B. N. Richards and G. K. Voight, *Nature (London)* **201,** 310 (1964).

66. J. R. Benemann, *Science* **181,** 164 (1973).

67. W. D. P. Stewart, ed., *Nitrogen Fixation by Free-Living Organisms*, Cambridge University Press, London, Eng., 1975.

68. J. E. Carnahan, L. E. Mortenson, H. F. Mower, and J. E. Castle, *Biochim. Biophys. Acta* **44,** 520 (1960).

69. J. E. Carnahan and J. E. Castle, *Ann. Rev. Plant Physiol* **14,** 125 (1963).

70. W. A. Bulen, R. C. Burns, and J. R. LeComte, *Proc. Nat. Acad. Sci. USA* **53,** 532 (1965).

71. W. A. Bulen and J. R. LeComte, *Methods Enzymol.* **24,** 456 (1972).

72. H. Haaker and C. Veeger, *Eur. J. Biochem.* **77,** 1 (1977).

73. L. E. Mortenson, *Biochim. Biophys. Acta* **127,** 18 (1966).

74. W. A. Bulen and J. R. LeComte, *Proc. Nat. Acad. Sci. USA* **56,** 979 (1966).

75. W. H. Orme-Johnson, L. C. Davis, M. T. Henzl, B. A. Averill, N. R. Orme-Johnson, E. Münck, and R. Zimmerman in ref. 22, p. 131.

76. D. M. Kurtz, Jr., R. S. McMillan, B. K. Burgess, L. E. Mortenson, and R. H. Holm, *Proc. Nat. Acad. Sci. USA* **76,** 4986 (1979).

77. C. Kennedy, R. R. Eady, E. Kondorosi, and D. K. Rekosh, *Biochem. J.* **155,** 383 (1976).

78. W. O. Gillum, L. E. Mortenson, J.-S. Chen, and R. H. Holm, *J. Am. Chem. Soc.* **99,** 584 (1977).

79. L. C. Davis, V. K. Shah, and W. J. Brill, *Biochim. Biophys. Acta* **384,** 353 (1975).

80. V. K. Shah and W. J. Brill, *Biochim. Biophys. Acta* **305,** 445 (1973).

81. L. E. Mortenson, *Proc. Nat. Acad. Sci. USA* **52,** 272 (1964).

82. W. A. Bulen, J. R. LeComte, R. C. Burns, and J. Hinkson in A. San Pietro, ed., *Non-Heme Iron Proteins: Role in Energy Conversion*, Antioch Press, Yellow Springs, Ohio, 1965, p. 261.

83. E. Knight, Jr., and R. W. F. Hardy, *J. Biol. Chem.* **242,** 1370 (1967).

84. J. R. Benemann, D. C. Yoch, R. C. Valentine, and D. I. Arnon, *Proc. Nat. Acad. Sci. USA* **64,** 1079 (1969).

85. R. W. F. Hardy, R. C. Burns, and G. W. Parshall, *Adv. Chem. Ser.* **100,** 219 (1971).

86. G. D. Watt, W. A. Bulen, A. Burns, and K. L. Hadfield, *Biochemistry*, **14,** 4266 (1975).

87. G. D. Watt and W. A. Bulen in ref. 17, p. 248.

88. W. G. Zumft, G. Palmer, and L. E. Mortenson, *Biochim. Biophys. Acta* **292,** 413 (1973).

89. M.-Y. Tso and R. H. Burris, *Biochim. Biophys. Acta* **309,** 263 (1973).

90. T. Ljones and R. H. Burris, *Biochem. Biophys. Res. Comm.* **80,** 22 (1978).

91. R. W. Miller, R. L. Robson, M. G. Yates, and R. R. Eady, *Can. J. Biochem.* **58,** 542 (1980).

92. G. D. Watt, A. Burns, and S. Lough in ref. 4, Vol. I, p. 159.

93. G. D. Watt and A. Burns, *Biochemistry* **16,** 264 (1977).

94. R. V. Hagemen and R. H. Burris, *Biochemistry* **17,** 4117 (1978).

95. W. H. Orme-Johnson, W. D. Hamilton, T. Ljones, M.-Y. Tso, R. H. Burris, V. K. Shah, and W. J. Brill, *Proc. Nat. Acad. Sci. USA* **69,** 3142 (1972).

96. B. E. Smith, D. J. Lowe, and R. C. Bray, *Biochem. J.* **135,** 331 (1973).

97. V. K. Shah and W. J. Brill, *Proc. Nat. Acad. Sci. USA* **74,** 3249 (1977).

98. B. K. Burgess, E. I. Stiefel, and W. E. Newton, *J. Biol. Chem.* **255,** 353 (1980).

99. B. E. Smith in W. E. Newton and S. Otsuka, eds., *Molybdenum Chemistry of Biological Significance*, Plenum Press, New York, 1980, p. 179.

100. W. E. Newton, B. K. Burgess, and E. I. Stiefel in ref. 99, p. 191.

101. P. T. Pienkos, V. K. Shah, and W. J. Brill, *Proc. Natl. Acad. Sci. USA* **74,** 5468 (1977).

102. B. H. Huynh, E. Münck, and W. H. Orme-Johnson, *Biochim. Biophys. Acta* **576,** 192 (1979).

103. J. Rawlings, V. K. Shah, J. R. Chisnell, W. J. Brill, R. Zimmerman, E. Münck, and W. H. Orme-Johnson, *J. Biol. Chem.* **253,** 1001 (1978).

104. S. P. Cramer, W. O. Gillum, K. O. Hodgson, L. E. Mortenson, E. I. Stiefel, J. R. Chisnell, W. J. Brill, and V. K. Shah, *J. Am. Chem. Soc.* **100,** 3814 (1978).

105. S. Streicher, E. Gurney, and R. C. Valentine, *Proc. Nat. Acad. Sci. USA* **68,** 1174 (1971).

106. R. A. Dixon and J. R. Postgate, *Nature* (*London*) **234,** 47 (1971).

107. *Ibid.*, **237,** 102 (1972).

108. R. A. Dixon, F. C. Cannon, and A. Kondorosi, *Nature* (*London*) **260,** 268 (1976).

109. K. A. Janssen, G. E. Reidel, F. M. Ausubel, and F. C. Cannon in ref. 4, Vol. I, p. 85.

110. F. C. Cannon, R. A. Dixon, J. R. Postgate, and S. B. Primrose, *J. Gen. Microbiol.* **80,** 227, 241 (1974).

111. F. C. Cannon, G. E. Reidel, and F. M. Ausubel, *Proc. Nat. Acad. Sci. USA* **74,** 2963 (1977).

112. T. MacNeil, G. P. Roberts, D. MacNeil, M. A. Supiano, and W. J. Brill in ref. 4, Vol. I, p. 63.

113. R. Dixon, M. Merrick, M. Filser, C. Kennedy, and J. R. Postgate in ref. 4, Vol. I, p. 71.

114. T. MacNeil, D. MacNeil, G. P. Roberts, M. A. Supiano, and W. J. Brill, *J. Bact.* **136,** 253 (1978).

115. C. Kennedy, *Mol. Gen. Genet.* **157,** 199 (1977).

116. R. A. Dixon, C. Kennedy, A. Kondorosi, V. Krishnapillai, and M. Merrick, *Mol. Gen. Genet.* **157,** 189 (1977).
117. D. MacNeil, T. MacNeil, and W. J. Brill, *Bioscience* **28,** 576 (1978).
118. J. E. Beringer and D. A. Hopwood, *Nature (London)* **264,** 291 (1976).
119. A. V. Buchanon-Wollaston, J. E. Beringer, N. J. Brewin, P. R. Hirsch, and A. W. B. Johnson, *Mol. Gen. Genet.* **178,** 185 (1980).
120. R. M. Pengra and P. W. Wilson, *J. Bact.* **75,** 251 (1958).
121. S. L. Streicher, K. T. Shanmugan, F. Ausubel, C. Morandi, and R. B. Goldberg, *J. Bact.* **120,** 815 (1974); R. S. Tubb, *Nature (London)* **251,** 481 (1974).
122. R. A. Ludwig, *Proc. Nat. Acad. Sci. USA* **77,** 5817 (1980).
123. W. J. Brill, *Ann. Rev. Microbiol.* **29,** 109 (1975).
124. S. Kustu, D. Burton, E. Garcia, L. McCarter, and N. McFarland, *Proc. Nat. Acad. Sci. USA* **76,** 4576 (1979); F. M. Ausubel, S. C. Bird, K. J. Durbin, K. A. Janssen, R. F. Margolskee, and A. P. Peskin, *J. Bact.* **140,** 597 (1979).
125. R. T. St. John, V. K. Shah, and W. J. Brill, *J. Bact.* **119,** 266 (1974).
126. W. J. Brill, A. L. Steiner, and V. K. Shah, *J. Bact.* **118,** 986 (1974).
127. G. Daesch and L. E. Mortenson, *J. Bact.* **110,** 103 (1972).
128. J. M. Vincent in ref. 4, Vol. II, p. 103.
129. J. E. Beringer in ref. 17, p. 358.
130. F. B. Holl and T. A. LaRue in ref. 17, p. 391.
131. M. E. Vol'pin and V. B. Shur, *Dokl. Akad. Nauk SSSR* **156,** 1102 (1964); M. E. Vol'pin, V. B. Shur, and M. A. Ilatovskaya, *Izvest. Akad. Nauk SSSR Ser. Khim.* **19,** 1728 (1964).
132. G. Henrici-Olivé and S. Olivé, *Angew. Chem. Int. Ed.* **6,** 873 (1967).
133. E. E. van Tamelen, G. Boche, and R. Greeley, *J. Am. Chem. Soc.* **90,** 1677 (1968).
134. G. J. Leigh in ref. 22, p. 1.
135. M. E. Vol'pin, M. A. Ilatovskaya, L. V. Kosyakova, and V. B. Shur, *J. Chem. Soc. Chem. Comm.,* 1074 (1968).
136. M. E. Vol'pin, V. B. Shur, R. V. Kudryavtsev, and L. A. Prodayko, *J. Chem. Soc. Chem. Comm.,* 1038 (1968).
137. E. E. van Tamelen and H. Rudler, *J. Am. Chem. Soc.* **92,** 5253 (1970).
138. Yu. G. Borod'ko, I. N. Ivleva, L. M. Kachapina, S. I. Salienko, A. K. Shilova, and A. E. Shilov, *J. Chem. Soc. Chem. Comm.,* 1178 (1972).
139. R. H. Marvich and H. H. Brintzinger, *J. Am. Chem. Soc.* **93,** 2046 (1971); G. P. Pez and S. C. Kwan, *J. Am. Chem. Soc.* **98,** 8079 (1976).
140. J. M. Manriquez, D. R. McAllister, E. Rosenberg, A. M. Shiller, K. L. Williamson, S. L. Chan, and J. E. Bercaw, *J. Am. Chem. Soc.* **100,** 3078 (1978).
141. R. D. Sanner, D. M. Duggan, T. C. McKenzie, R. E. Marsh, and J. E. Bercaw, *J. Am. Chem. Soc.* **98,** 8358 (1976).
142. J. E. Bercaw in ref. 22, p. 25.
143. R. D. Sanner, J. M. Manriquez, R. E. Marsh, and J. E. Bercaw, *J. Am. Chem. Soc.* **98,** 8351 (1976).
144. J. M. Manriquez, R. D. Sanner, R. E. Marsh, and J. E. Bercaw, *J. Am. Chem. Soc.* **98,** 3042 (1976).
145. J. E. Bercaw, R. H. Marvich, L. G. Bell, and H. H. Brintzinger, *J. Am. Chem. Soc.* **94,** 1219 (1972).
146. Yu. G. Borod'ko, I. N. Ivleva, L. M. Kachapina, E. F. Kvashina, A. K. Shilova, and A. E. Shilov, *J. Chem. Soc. Chem. Comm.,* 169 (1973).
147. J. H. Teuben, *J. Organometal. Chem.* **57,** 159 (1973).
148. M. J. S. Gynane, J. Jeffrey, and M. F. Lappert, *J. Chem. Soc. Chem. Comm.,* 34 (1978).
149. M. Hidai, K. Tominari, and Y. Uchida, *J. Am. Chem. Soc.* **94,** 1010 (1972).
150. J. Chatt, G. A. Heath, and R. L. Richards, *J. Chem. Soc. Dalton Trans.,* 2074 (1974).
151. C. R. Brulet and E. E. van Tamelen, *J. Am. Chem. Soc.* **97,** 911 (1975).
152. J. Chatt, A. J. Pearman, and R. L. Richards, *J. Chem. Soc. Dalton Trans.,* 1852 (1977); J. Chatt, *Chemtech.* **11**(3), 162 (1981).
153. J. A. Baumann and T. A. George, *J. Am. Chem. Soc.* **102,** 6153 (1980).
154. V. W. Day, T. A. George, and S. D. A. Iske, *J. Am. Chem. Soc.* **97,** 4127 (1975).
155. J. Chatt, A. A. Diamantis, G. A. Heath, N. E. Hooper, and G. J. Leigh, *J. Chem. Soc. Dalton Trans.,* 688 (1977).
156. M. Hidai, Y. Mizobe, and Y. Uchida, *J. Am. Chem. Soc.* **98,** 7824 (1976).
157. D. C. Busby and T. A. George, *Inorg. Chim. Acta* **29,** L273 (1978).
158. P. C. Bevan, J. Chatt, G. J. Leigh, and E. G. Leelamanii, *J. Organometal. Chem.* **139,** C59 (1977).

159. D. Sellman and W. Weiss, *Angew. Chem. Int. Ed.* **16**, 880 (1977).

160. *Ibid.*, **17**, 269 (1978).

161. J. Chatt, J. R. Dilworth, and R. L. Richards, *Chem. Rev.* **78**, 589 (1978).

162. P. Sobota and B. Jezowska-Trzebiatowska, *Coord. Chem. Rev.* **26**, 71 (1978).

163. L. P. Didenko, A. G. Ovcharenko, A. E. Shilov, and A. G. Shilova, *Kinet. Katal.* **18**, 1078 (1977).

164. A. Shilov, N. Denisov, O. Efimov, N. Shuvalov, N. Shuvalova, and A. Shilova, *Nature (London)* **231**, 460 (1971).

165. N. T. Denisov, N. I. Shuvalova, and A. E. Shilov, *Kinet. Katal.* **14**, 1325 (1973).

166. L. A. Nikonova, O. N. Efimov, A. G. Ovcharenko, and A. E. Shilov, *Kinet. Katal.* **13**, 249 (1972).

167. S. I. Zones, M. R. Palmer, J. G. Palmer, J. M. Doemeny, and G. N. Schrauzer, *J. Am. Chem. Soc.* **100**, 2113 (1978).

168. L. A. Nikonova and A. E. Shilov in ref. 22, p. 41.

169. L. A. Nikonova, S. A. Isaeva, N. I. Pershikova, and A. E. Shilov, *J. Mol. Catal.* **1**, 367 (1975–1976).

170. G. N. Schrauzer, *Angew. Chem. Int. Ed.* **14**, 514 (1975).

171. G. N. Schrauzer, G. W. Kiefer, K. Tano, and P. A. Doemeny, *J. Am. Chem. Soc.* **96**, 641 (1974).

172. A. P. Khrushch, A. E. Shilov, and T. A. Vorontsova, *J. Am. Chem. Soc.* **96**, 4987 (1974).

173. G. N. Schrauzer in ref. 22, p. 109.

174. E. L. Moorehead, P. R. Robinson, T. M. Vickrey, and G. N. Schrauzer, *J. Am. Chem. Soc.* **98**, 6555 (1976).

175. B. J. Weathers, J. H. Grate, N. A. Strampach, and G. N. Schrauzer, *J. Am. Chem. Soc.* **101**, 925 (1979).

176. E. I. Stiefel in ref. 22, p. 69.

177. E. I. Stiefel, W. E. Newton, G. D. Watt, K. L. Hadfield, and W. A. Bulen, *Adv. Chem. Ser.* **162**, 353 (1977).

178. A. Nason, K.-Y. Lee, S.-S. Pan, P. A. Ketchum, A. Lamberti, and J. DeVries, *Proc. Nat. Acad. Sci. USA* **68**, 3242 (1971).

179. K.-Y. Lee, S.-S. Pan, R. H. Erickson, and A. Nason, *J. Biol. Chem.* **249**, 3941 (1974).

180. J. L. Johnson, B. Hainline, H. P. Jones, and K. V. Rajagopalan in ref. 4, Vol. I, p. 249.

181. K. O. Hodgson in ref. 4, Vol. I, p. 261.

182. T. D. Tullius, S. D. Conradson, J. M. Berg, and K. O. Hodgson in ref. 99, p. 139.

183. T. E. Wolff, J. M. Berg, C. Warrick, K. O. Hodgson, R. H. Holm, and R. B. Frankel, *J. Am. Chem. Soc.* **100**, 4630 (1978).

184. B. A. Averill, T. Herskovitz, R. H. Holm, and J. A. Ibers, *J. Am. Chem. Soc.* **95**, 3523 (1973).

185. C. D. Garner and co-workers in ref. 99, p. 203.

186. G. Christou, C. D. Garner, F. E. Mabbs, and T. J. King, *J. Chem. Soc. Chem. Comm.*, 740 (1978).

187. T. E. Wolff, J. M. Berg, K. O. Hodgson, R. B. Frankel, and R. H. Holm, *J. Am. Chem. Soc.* **101**, 4140 (1979).

188. T. E. Wolff, J. M. Berg, P. P. Power, K. O. Hodgson, and R. H. Holm, *J. Am. Chem. Soc.* **101**, 5454 (1979).

189. D. Coucouvanis, E. D. Simhon, D. Swenson, and N. C. Baenziger, *J. Chem. Soc. Chem. Comm.*, 361 (1979).

190. D. Coucouvanis, N. C. Baenziger, E. D. Simhon, P. Stremple, D. Swenson, A. Simopoulos, A. Kostikas, V. Petrouleas, and V. Papaefthymiou, *J. Am. Chem. Soc.* **102**, 1732 (1980).

191. R. M. Tieckelmann, H. C. Silvis, T. A. Kent, B. H. Huynh, J. V. Waszczak, B.-K. Teo, and B. A. Averill, *J. Am. Chem. Soc.* **102**, 5550 (1980).

192. J. W. McDonald, G. D. Friesen, and W. E. Newton, *Inorg. Chim. Acta* **46**, L79 (1980).

193. R. W. Treharne, D. R. Moles, M. R. Bruce, and C. K. McKibben, *Proc. Second Review Meeting of I.N.P.U.T.S. Project, East–West Center, Honolulu, Hawaii, 1978*, p. 279.

WILLIAM E. NEWTON
Charles F. Kettering Research Laboratory

NITROGEN SULFIDE, N_4S_4.   See Explosives.

NITROGEN TRICHLORIDE, NCl_3.   See Chloramines.

NITROPARAFFINS

Nitroparaffins (or nitroalkanes) are derivatives of the alkanes in which one hydrogen or more is replaced by the electronegative nitro group ($-NO_2$), which is attached to carbon through nitrogen. The nitroparaffins are isomeric with alkyl nitrites, RONO, which are esters of nitrous acid. The nitro group in a nitroparaffin has been shown to be symmetrical about the R—N axis, and may be represented as a resonance hybrid of the following structures:

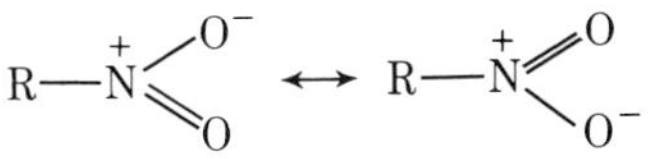

$$R-\overset{+}{N}\underset{O}{\overset{O^-}{\diagup}} \longleftrightarrow R-\overset{+}{N}\underset{O^-}{\overset{O}{\diagup}}$$

Nitroparaffins are classed as primary, RCH_2NO_2, secondary, R_2CHNO_2, and tertiary, R_3CNO_2, by the same convention used for alcohols. Primary and secondary nitroparaffins exist in tautomeric equilibrium with the enolic or aci forms.

$$RCH_2NO_2 \rightleftharpoons RCH{=}NO_2H, \text{ and } RR'CHNO_2 \rightleftharpoons RR'C{=}NO_2H$$

The nitroparaffins are named as derivatives of the corresponding hydrocarbons by using the prefix nitro to designate the $-NO_2$ group (1); eg, CH_3NO_2, nitromethane; $CH_3CH(NO_2)CH_3$, 2-nitropropane; and $CH_3CH(NO_2)_2$, 1,1-dinitroethane.

Table 1 shows the tautomeric forms of some nitroparaffins and also some trivial names.

The salts obtained from nitroparaffins and the so-called nitronic acids are identical and may be named as derivatives of either, eg, sodium salt of *aci*-nitromethane, or sodium methanenitronate [25854-38-0].

Nitromethane, nitroethane, 1-nitropropane, and 2-nitropropane are produced by a vapor-phase process developed in the 1930s (2). Some 1980 (July) prices are listed below.

Table 1. Tautomeric Isomers of Some Nitroparaffins and Some Trivial Names

Nitroparaffin	CAS Registry number	Formula	Other names
isonitroethane	[4202-72-6]	$CH_3CH{=}NO_2H$	*aci*-nitroethane ethanenitronic acid[a]
2-isonitropropane	[79-46-9]	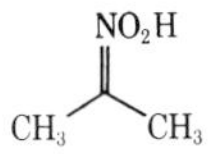 $\underset{CH_3 \quad CH_3}{\overset{NO_2H}{\diagdown\diagup}}$	*aci*-2-nitropropane 2-propanenitronic acid[a]
1-isonitro-1-nitroethane	[5923-46-6]	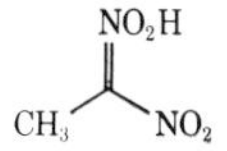 $\underset{CH_3 \quad NO_2}{\overset{NO_2H}{\diagdown\diagup}}$	*aci*-1,1-dinitroethane 1-nitro-1-ethanenitronic acid[a]
trinitromethane	[517-25-9]	$HC(NO_2)_3$	nitroform
2-nitroacetaldehyde oxime	[5653-21-4]	$O_2NCH_2CH{=}NOH$	methazonic acid

[a] Nitronic acid names conflict with IUPAC organic nomenclature rules (Section D-5.5).

	Price, $/kg	Quantity
nitropropane (mixture)	0.77	bulk
nitroethane	3.50	bulk
nitromethane	4.00	208-L (55-gal) drums, truckloads

Physical Properties

The physical constants of the lower mononitroparaffins and of a number of polynitroparaffins are listed in Tables 2–4. Most polynitroparaffins are colorless crystalline or waxlike solids at or near room temperature. They are insoluble in water and alkanes but soluble in most other organic solvents. The lower mononitroparaffins are colorless dense liquids with mild odors. The boiling points of the mononitro-paraffins are much higher than those of the isomeric nitrites; eg, the normal boiling point of nitromethane is 101.2°C whereas that of methyl nitrite is −12°C. This phe-nomenon may be attributed in large part to intermolecular hydrogen bonding. Accurate vapor-pressure determinations (3) for the lower nitroparaffins have been made and adapted to an Antoine equation (see Tables 2–3). A nomograph was constructed from these data (4). The properties of azeotropes of nitroparaffins with water or with organic liquids are given in ref. 5. Critical solution temperature data of nitroparaffins in binary and ternary systems are given in refs. 6 and 7, respectively.

The molecular configuration of the C—NO_2 group in nitromethane has been determined by electron-diffraction methods (8). These data indicate that the nitrogen atom and the atoms attached to it lie in the same plane, and that the O—N—O bond angle (127 ± 3°) is greater than the C—N—O bond angles (116 ± 3°). The spreading of the O—N—O bond angle beyond 120° is attributed to the repulsion of the negatively charged oxygen atoms of the highly polar system. For ir and mass spectra, see ref. 9 and 10, respectively.

Most organic compounds, including aromatic hydrocarbons, alcohols, esters, ketones, ethers, and carboxylic acids are miscible with nitroparaffins, whereas alkanes and cycloalkanes have limited solubility. The lower nitroparaffins are excellent solvents for coating materials, waxes, resins, gums, and dyes.

The thermal characteristics of higher nitroparaffins are quite different from those of nitromethane. The nitropropanes provide nearly twice as much heat as does ni-tromethane when burned in air or oxygen. When the only source of oxygen is that contained within the molecule, nitropropanes yield only 20% as much energy as ni-tromethane on burning.

Chemical Properties

The chemical reactions of the nitroparaffins have been discussed in depth in a number of review articles and monographs (11–17), including their utility for the synthesis of heterocyclic and other compounds (18–19).

Tautomerism. Primary and secondary mononitroparaffins are acidic substances which exist in tautomeric equilibria with their nitronic acids. The nitro isomer is weakly acidic; the nitronic acid isomer (aci form) is much more acidic. A comparison of the ionization constants of the two forms in water at 25°C is given in Table 5.

An equilibrium mixture of the isomers usually contains a much higher proportion of the true nitro compound. The equilibrium for each isomeric system is influenced by the dielectric strength and the hydrogen-acceptor characteristics of the solvent

Table 2. Physical Properties of the Lower Mononitroparaffins

Property	Nitromethane [75-52-5]	Nitroethane [79-24-3]	1-Nitropropane [108-03-2]	2-Nitropropane [79-46-9]
molecular weight[a]	61.041	75.068	89.095	89.095
boiling point at 101.3 kPa[b], °C	101.20	114.07	131.18	120.25
vapor pressure, kPa[b]				
at 20°C	3.64	2.11	1.01	1.73
at 25°C	4.89	2.79	1.36	2.40
freezing point, °C	−28.55	−89.52	−103.99	−91.32
density, g/cm^3				
at 20°C	1.138	1.051	1.001	0.988
at 30°C	1.124	1.039	0.991	0.977
coefficient of expansion per °C	0.00122	0.00112	0.00101	0.00104
refractive index, n_D^{20}	1.38188	1.39193	1.40160	1.39439
n_D^{30}	1.37738	1.38754	1.39755	1.39028
surface tension at 20°C, mN/m (= dyn/cm)	37.48	32.66	30.64	29.87
viscosity, mPa·s (= cP)				
at 10°C	0.731	0.769	0.972	0.883
at 20°C	0.647	0.677	0.844	0.770
at 25°C	0.610	0.638	0.790	0.721
at 30°C	0.576	0.602	0.740	0.677
heat of combustion (liq) at 25°C, kJ/mol[c]	−708.4	−1362	−2016	−2000
heat of vaporization (liq) at 25°C, kJ/mol[c]	38.27	41.6	43.39	41.34
at bp, kJ/mol[c]	34.4	38.0	38.5	36.8
heat of formation (liq) at 25°C, kJ/mol[c]	−113.1	−141.8	−168.0	−180.7
specific heat				
at 25°C, J/(mol·°C)[c]	106.0	138.5	175.6	175.2
at 25°C, J/(g·°C)[c]	1.74	1.85	1.97	1.97
dielectric constant at 30°C	35.87	28.06	23.24	25.52
dipole moment, C·m[d]				
gas	11.68 × 10^{-30}	11.94 × 10^{-30}	12.41 × 10^{-30}	12.44 × 10^{-30}
liquid	10.58 × 10^{-30}	10.64 × 10^{-30}		
aqueous azeotrope,				
bp, °C	83.59	87.22	91.63	88.55
wt % nitroparaffin	76.4	71.0	63.5	70.6
pH of 0.01 M aqueous solution at 25°C	6.4	6.0	6.0	6.2
solubility in water, wt %				
at 20°C	10.5	4.6	1.5	1.7
at 25°C	11.1	4.7	1.5	1.7
at 70°C	19.3	6.6	2.2	2.3
solubility of water in nitroparaffin, wt %				
at 20°C	1.8	0.9	0.6	0.5

Table 2 (*continued*)

Property	Nitromethane [75-52-5]	Nitroethane [79-24-3]	1-Nitropropane [108-03-2]	2-Nitropropane [79-46-9]
at 25°C	2.1	1.1	0.6	0.5
at 70°C	7.6	3.0	1.7	1.6
Antoine's constant[b], $\log p_{kPa} = A - B/(t + C)$				
A	6.399073	6.300057	6.252442	6.208143
B	1441.610	1435.402	1474.299	1422.898
C	226.939	220.184	215.986	218.341
critical temperature, °C	315	388	402	344
critical pressure, kPa[b]	6.309			
electrical conductivity, $10^{-7}/(\Omega \cdot cm)$	6.56	5		5

[a] Calculated.
[b] To convert kPa to mm Hg, multiply by 7.5.
[c] To convert J to cal, divide by 4.184.
[d] To convert C·m to debye (μ), divide by 3.336×10^{-30}.

Table 3. Physical Constants of C$_4$ and Higher Mononitroparaffins

Property	1-Nitrobutane [627-05-4]	2-Nitrobutane [600-24-8]	1-Nitro-2-methyl-propane [625-74-1]	2-Nitro-2-methyl-propane [594-70-7]	Nitro-cyclohexane [1122-60-7]
freezing point, °C	−81.33	glass	−76.85	26.23	−34
boiling point, °C	152.77	139.50	141.72	127.16	205.5–206
Antoine's constant[a], $\log p_{kPa} = A - B/(t + C)$					
A	6.220403	6.202795	6.199044	6.112625	
B	1523.797	1494.318	1483.643	1396.948	
C	208.778	216.542	212.905	212.989	
vapor pressure at 20°C, kPa[a]	0.36	0.77	0.64	solid	
density at 25°C, g/cm^3	0.96848	0.96036	0.95848	solid	1.0680_4^{19}
refractive index, n_D^{20}	1.41019	1.40407	1.40642	1.39175^{30}	1.4608

[a] To convert kPa to mm Hg, multiply by 7.5.

medium. The aci form is dissolved and neutralized rapidly by strong bases, and gives characteristic color reactions with ferric chloride.

Polynitroparaffins are stronger acids than the corresponding mononitroparaffins. Thus, 1,1-dinitroethane has an ionization constant of 5.6×10^{-6} in water at 20°C; trinitromethane is a typical strong acid with an ionization constant in the range of 10^{-2}–10^{-3}. Neutralization of these substances occurs rapidly, and they may be titrated readily.

Table 4. Physical Constants of Polynitro Compounds

Compound	CAS Registry No.	mp, °C	Boiling point °C	at kPa[a]	Sp gr	Refractive index, n_D^t	Solubility H₂O	Ethanol	Ethyl ether
dinitro-methane	[625-76-3]		39–40	0.266	1.524	1.4480^{20}		sol	sol
trinitro-methane	[517-25-9]	14.3 (dec)	45–47	2.933	1.5967_4^{24}	1.445511_{He}^{24}	sol	sol	sol
tetranitro-methane	[509-14-8]	13.8	125.7	101.3	1.6377_4^{21}	1.43416^{21}	insol	sol	sol
1,1-dinitro-ethane	[600-40-8]		185–186	101.3	1.3503_{23}^{23}	1.4346^{20}	sl sol	sol	sol
1,2-dinitro-ethane	[7570-26-5]	39–40	135	0.800	1.4597_4^{20}	1.4488^{20}	sl sol	sol	sol
1,1,1-tri-nitro-ethane	[595-86-8]	57	68	2.266	1.4223_4^{77}	1.4171_α^{77}	insol	sol	sol
2,2-dinitro-propane	[595-49-3]	54	185	101.3			insol	sol	sol
1,1-dinitro-cyclohex-ane	[4028-15-3]	36	142–143	4.666	1.2452_4^{21}	1.4732^{21}	insol	sol	sol

[a] To convert kPa to mm Hg, multiply by 7.5.

Table 5. Ionization Constants of the Lower Mononitroparaffins

Compound	K_{nitro}	K_{aci}
nitromethane	6.1×10^{-11}	5.6×10^{-4}
nitroethane	3.5×10^{-9}	3.9×10^{-5}
2-nitropropane	2.1×10^{-8}	7.7×10^{-6}
1-nitropropane		2.0×10^{-5}

In addition to neutralization, prolonged action of alkaline reagents can effect oxidation–reduction and extensive decomposition. 1,1-Dinitroparaffins and trini-tromethane are more stable than mononitro compounds during neutralization and subsequent regeneration, and therefore, less stringent experimental conditions are permissible.

Acidification of mononitroparaffin salts immediately gives the nitronic acid. Many nitronic acids have been isolated and stored as crystalline solids. In solution, a nitronic acid either isomerizes slowly into its more stable nitro form, or undergoes some irreversible transformation. The isomerization occurs sufficiently slowly that it can be measured by conductometric or halometric methods.

Salts. Nitroparaffin salts dissociate to form ambidentate anions, which are capable of alkylation at either the carbon or oxygen atom (20).

$$\left[\begin{array}{c} R' \\ | \\ RCNO_2 \end{array} \right]^- \rightleftharpoons \begin{array}{c} R' \\ | \\ RC \end{array} = \overset{+}{N} \begin{array}{c} O^- \\ \diagdown \\ O^- \end{array}$$

Reactions of alkyl, allylic, and benzylic halides with these salts usually give carbonyl compounds, presumably through the nitronic ester (*O*-alkylation) as an intermediate. With certain benzyl halides substituted in the para or ortho position with nitro groups, the reaction gives almost exclusively *C*-alkylation. For example, sodium 2-propanenitronate [*12384-98-4*] reacts with *p*-nitrobenzyl chloride (21) or with *p*-nitrobenzyltrimethylammonium iodide (22) to form 2-methyl-2-nitro-1-(*p*-nitrophenyl)propane [*5440-67-5*]. This reaction depends on the nature of the leaving group (20) and is inhibited by powerful electron acceptors (23). It has been considered a radical-anion (23) or chain process (24).

Alkali salts of primary nitroparaffins (but not of secondary nitroparaffins) react with acyl cyanides to yield α-nitroketones by *C*-acylation (25).

$$RCH{=}NO_2M + R'COCN \rightarrow R'COCH(NO_2)R + MCN$$

Most other acylating agents act on salts of either primary or secondary nitroparaffins by *O*-acylation, giving first the nitronic anhydrides which rearrange to give, eg, nitrosoacyloxy compounds (26).

Alkaline solutions of mononitroparaffins undergo many different reactions when stored for long periods or acidified or heated. Acidification of solutions of mononitro salts is best effected slowly at 0°C or lower with weak acids or buffered acidic mixtures, such as acetic acid–urea, carbon dioxide, or hydroxylammonium chloride. If mineral acids are used under mild conditions (eg, dilute HCl at 0°C), decomposition yields a carbonyl compound and nitrous oxide (Nef Reaction).

$$\underset{\displaystyle CH_3\overset{\displaystyle \|}{C}H}{\overset{\displaystyle NO_2Na}{}} \xrightarrow[0°C]{HCl} \underset{\displaystyle CH_3\overset{\displaystyle \|}{C}H}{\overset{\displaystyle O}{}} + N_2O$$

Reaction with nitrous acid can be used to differentiate primary, secondary, and tertiary mononitroparaffins. Primary nitroparaffins give nitrolic acids, which dissolve in alkali to form bright red salts.

$$RCH{=}NO_2H + HNO_2 \rightarrow [RCH(NO)NO_2 + H_2O] \rightarrow RC({=}NOH)NO_2 + H_2O$$

$$RC({=}NOH)NO_2 + NaOH \rightarrow RC(NO){=}NO_2Na + H_2O$$

Secondary nitroparaffins give alkali-insoluble nitroso derivatives known as pseudonitroles. As monomers in the liquid state, pseudonitroles have a characteristic blue color; as solids they exist as white crystalline dimers.

$$R_2C{=}NO_2H + HNO_2 \rightarrow R_2C(NO)NO_2 + H_2O$$

Tertiary nitroparaffins do not react with nitrous acid and no color develops.

With sodium azide, salts of secondary nitroparaffins rearrange to *N*-substituted amides (27). With SO_2, primary or secondary nitroparaffins give imidodisulfonic acid salts (28). Potassium nitroform reacts quantitatively with nitryl chloride in ether to form tetranitromethane (29).

Sodium methanenitronate reacts with phenyl isocyanate in benzene to give the readily separable sodium salts of nitroacetanilide and nitromalondianilide. Except as the salt, nitromethane is unreactive with phenyl isocyanate at temperatures up to 100°C; the higher homologues do not give condensation products that can be isolated.

Acid Hydrolysis. With hot concentrated mineral acids, primary nitroparaffins yield a fatty acid and a hydroxylamine salt.

$$CH_3CH_2CH_2NO_2 + H_2O + H_2SO_4 \rightarrow C_2H_5COOH + H_3NOH \cdot HSO_4$$

If anhydrous acid and lower temperatures are used, the intermediate hydroxamic acid can be recovered.

$$CH_3CH_2NO_2 \xrightarrow[\text{HCl}]{\text{anhydrous}} CH_3C(=NOH)OH$$

Halogenation. In the presence of alkali, chlorine replaces the hydrogen atoms on the carbon atom holding the nitro group. If more than one hydrogen atom is present, the hydrogen atoms can be replaced in stages; exhaustive chlorination of nitromethane yields chloropicrin [76-06-2] (trichloronitromethane). The chlorination can be stopped at intermediate stages. Bromination or iodination takes a similar course but bromopicrin [464-10-8] and iodopicrin [39247-25-1] tend to be less stable.

Halonitroparaffins can be prepared in which the halogen and nitro groups are not on the same carbon atom. The direct chlorination of nitroparaffins to give non-geminal substitution is promoted by irradiation in anhydrous media (30–31). For example, nitroethane yields 2-chloro-1-nitroethane [625-47-8], and 1-nitropropane yields both 2- [503-76-4] and 3-chloro-1-nitropropane [16694-52-3] by this method. With sodium iodide in acetone, 3-chloro-1-nitropropane gives the 3-iodo analogue, which, in turn, yields nitrocyclopropane [13021-02-8] on treatment with alkalies (32). Replacement of the hydroxy group in nitro alcohols (qv) with halogen yields vicinal halonitroparaffins. Action of phosphorus tribromide on nitro alcohols in dimethyl-formamide gives *vic*-bromonitroparaffins (33).

The acid chloride of *aci*-nitromethane, $CH_2=N(Cl)O$ (mp −43°C, bp 2–3°C), is formed by fusion of nitromethane and picrylpyridinium chloride (34). It is hydrolyzed to nitrosomethane, reduces potassium permanganate strongly, and exhibits no reactions characteristic of hydroxamic acids.

Condensation with Carbonyl Compounds. Primary and secondary nitroparaffins undergo aldol-type condensation with a variety of aldehydes and ketones to give nitro alcohols (11). Those derived from the lower nitroparaffins and formaldehyde are available commercially (see Nitro alcohols). Nitro alcohols can be reduced to the corresponding amino alcohols (see Alkanolamines).

These reversible condensations are catalyzed by bases or acids, such as zinc chloride and aluminum isopropoxide, or by anion-exchange resins. Ultrasonic vibrations improve the reaction rate and yield (see Ultrasonics). Condensation of aromatic aldehydes or ketones with nitroparaffins yields either the nitro alcohol or the nitro olefin, depending on the catalyst. Conjugated unsaturated aldehydes or ketones and nitroparaffins (Michael condensation) yield nitro-substituted carbonyl compounds rather than nitro alcohols. Condensations with keto aldehydes or keto esters give the substituted nitro alcohols (35); keto aldehydes condense preferentially at the aldehyde function.

Most nitroparaffins do not react with ketones but in the presence of alkoxide catalysts, nitromethane and lower aliphatic ketones give nitro alcohols; in the presence of amine catalysts dinitro compounds are obtained.

$$2\ CH_3NO_2\ +\ CH_3\overset{\overset{\textstyle O}{\|}}{C}CH_3 \xrightarrow{R_2NH} O_2NCH_2\underset{\underset{\textstyle CH_3}{|}}{\overset{\overset{\textstyle CH_3}{|}}{C}}CH_2NO_2\ +\ H_2O$$

[762-98-1]

Nitro olefins can be made in some cases by dehydration of the aromatic nitrohydroxy derivatives. Subsequent reduction yields the aromatic amine. The following three-step reaction yielding 2-amino-1-phenylbutane illustrates the synthesis of this class of valuable pharmaceutical compounds.

$$CH_3CH_2CH_2NO_2 \ + \ C_6H_5CHO \ \longrightarrow \ C_6H_5CHOHCHCH_2CH_3 \ \underset{}{\overset{-H_2O}{\longrightarrow}} \ C_6H_5CH{=}C\underset{NO_2}{\overset{CH_2CH_3}{<}} \ \overset{+H_2}{\longrightarrow}$$

with the intermediate bearing NO_2 below, and the final product:

$$C_6H_5CH_2CHCH_2CH_3$$
$$\underset{NH_2}{|}$$

In the presence of amine salts of weak acids, the nitro olefin is formed directly.

$$C_6H_5CHO \ + \ CH_3CH_2NO_2 \ \underset{HOOCCH_3}{\overset{RNH_2}{\longrightarrow}} \ C_6H_5CH{=}C\underset{NO_2}{\overset{CH_3}{<}} \ + \ H_2O$$

Mannich-Type Reactions. Secondary nitroparaffins, formaldehyde, and primary or secondary amines can react in one step to yield Mannich bases.

$$R_2CHNO_2 \ + \ CH_2O \ + \ R'_2NH \ \longrightarrow \ R_2CCH_2NR'_2 \ + \ H_2O$$
$$\underset{NO_2}{|}$$

Primary nitroparaffins react with two moles of formaldehyde and two moles of amine to yield 2-nitro-1,3-propanediamines. With excess formaldehyde, Mannich bases from primary nitroparaffins and primary amines can react further to give nitro-substituted cyclic derivatives, such as tetrahydro-1,3-oxazines or hexahydropyrimidines (36–37). Pyrolysis of salts of Mannich bases, particularly of the boron trifluoride complex (38), yields nitro olefins by loss of the amine moiety. Closely related to the Mannich reaction is the formation of sodium 2-nitrobutane-1-sulfonate [76794-27-9] by warming 1-nitropropane with formaldehyde and sodium sulfite (39).

Reduction. The lower nitroparaffins are reduced readily to the corresponding primary amines with a number of reducing agents. Partial reduction yields aldoximes, ketoximes, or N-substituted hydroxylamines. Suitable reduction methods range from iron and hydrochloric acid to high pressure hydrogenation over Raney nickel or noble-metal catalysts. Reaction conditions have been developed also for the reduction of olefinic or carbonyl groups with or without reduction of the nitro group present and for the reduction of the nitro group leaving the other groups untouched (40). Some of the products obtained are useful as pharmaceuticals.

Oxidation. Nitroparaffins are resistant to oxidation. At ordinary temperatures, they are attacked only very slowly by strong oxidizing agents such as potassium permanganate, manganese dioxide, or lead peroxide. Nitronate salts, however, are oxidized more easily. The salt of 2-nitropropane is converted to 2,3-dimethyl-2,3-dinitrobutane [3964-18-9], acetone, and nitrite ion by persulfates or electrolytic oxidation. With potassium permanganate, only acetone is recovered.

$$3 \ \underset{CH_3}{\overset{CH_3}{>}}C{=}\overset{+}{N}O_2^{-}Na \ \overset{[O]}{\longrightarrow} \ CH_3\underset{NO_2}{\overset{CH_3}{\underset{|}{C}}}{-}\underset{NO_2}{\overset{CH_3}{\underset{|}{C}}}CH_3 \ + \ CH_3\overset{O}{\overset{||}{C}}CH_3 \ + \ NaNO_2$$

[3964-18-9]

α-Nitroacetaldehyde [5007-21-6] is formed in low yield by the oxidation of nitroethane with selenium dioxide. This product is easily oxidized in air to nitroacetic acid [625-75-2], which spontaneously decarboxylates to nitromethane.

Addition to Multiple Bonds. Mono- or polynitroparaffins with a hydrogen on the carbon atom carrying the nitro group add to activated double bonds under the influence of basic catalysts (41–43). Thus, nitromethane forms tris(β-cyanoethyl)nitromethane [1466-48-4] with acrylonitrile and 2-nitropropane yields 4-methyl-4-nitrovaleronitrile [16507-00-9]. These Michael-type condensations with acrylic compounds take place in liquid ammonia without catalyst (44). In the presence of dehydrating agents, such as phenyl isocyanate or phosphorus oxychloride, nitroparaffins add to activated olefins, such as methyl acrylate, to give oxazines (45).

Nitroparaffins add 1,4 to conjugated systems; methyl vinyl ketones, for example, yield the corresponding γ-nitro ketone, which can be reduced to a γ-nitro alcohol (46). More than one vinyl group may react with primary nitroparaffins (47).

Conjugated nitro olefins can function as the acceptor to which nitroparaffins may add to form polynitro derivatives (48–49). Nitro olefins also undergo Diels-Alder condensations with compounds such as maleic anhydride or anthracene (50).

Nitroparaffins and HBF_4 react with alkynes through conjugate addition of a proton and the nitronate ion at the triple bond as illustrated by nitromethane and 5-decyne (51).

$$BuC\equiv CBu \ + \ CH_3NO_2 \xrightarrow{HBF_4} \underset{\underset{N=CH_2}{\overset{\displaystyle |}{\underset{O}{\diagdown}}}}{\overset{\overset{\displaystyle \bar{O}}{\parallel}}{BuCCHBu}}$$

Other Reactions. α-Nitroalkanoic acids or their esters can be prepared (52–54) by treating nitroparaffins with magnesium methyl carbonate, or with triisopropylaluminum and carbon dioxide. These products are reduced readily to α-amino acids.

$$CH_3NO_2 \ + \ (CH_3OCOO)_2Mg \ \longrightarrow \ \underset{O}{\overset{O}{\diagup}} \ + \ 2\ CH_3OH \ + \ CO_2$$

1,1,1-Trinitroparaffins can be prepared from 1,1-dinitroparaffins by electrolytic nitration (electrolysis in aqueous caustic sodium nitrite solution) (55). Secondary nitroparaffins dimerize on electrolytic oxidation (56); eg, 2-nitropropane yields 2,3-dimethyl-2,3-dinitrobutane, as well as some 2,2-dinitropropane. Addition of sodium nitrite to the anolyte favors formation of the former. The oxidation of salts of aci-2-nitropropane with either cationic or anionic oxidants generally gives both 2,2-dinitropropane and acetone (57); with ammonium peroxysulfate, eg, these products are formed in 53 and 14% yields, respectively. Ozone oxidation of nitroso groups gives nitro compounds; eg, 2-nitroso-2-nitropropane [5275-46-7] (propylpseudonitrole) yields 2,2-dinitropropane (58).

O-Acylation of 2-nitropropane occurs on reaction with either ketene or acetic anhydride (59) in the presence of dry sodium acetate at 70–80°C. Ketovinylation of 2-nitropropane at the 1-position occurs on treatment of sodium 2-propanenitronate with a chlorovinyl ketone (60).

Furoxans are formed by the dehydration of two moles of a nitroparaffin (61).

$$2 \ RCH_2NO_2 \xrightarrow[\text{in CHCl}_3]{POCl_3, \ (C_2H_5)_3N} \left[2 \ RC\equiv N \rightarrow O \right] \longrightarrow$$

Preparation and Manufacture

Synthetic methods suitable for preparation of a wide variety of nitroparaffins are reviewed in refs. 62–64.

A general one-step method for preparation of primary and secondary nitroparaffins from amines by oxidation with *m*-chloroperbenzoic acid in 1,2-dichloroethane has been reported (65). This method is particularly useful for laboratory quantities of a wide variety of nitroparaffins because a large number of amines are available readily from ketones by oxime reduction and because the reaction is highly specific for nitroparaffins.

Higher nitroalkanes are prepared from lower primary nitroalkanes by a one-pot synthesis (66). Successive condensations with aldehydes and acylating agents are followed by reduction with sodium borohydride. Overall conversions in the 75–80% range are reported.

The only method currently utilized commercially is vapor-phase nitration of propane, although ethane and butane also can be nitrated quite readily. The data in Table 6 show the typical distribution of nitroparaffins obtained from the nitration of propane with nitric acid at different temperatures (67). Nitrogen dioxide can be used for nitration, but its low boiling point (21°C) limits its effectiveness except at increased pressure. Nitrogen pentoxide is a powerful nitrating agent for *n*-alkanes; however, it is expensive and often gives polynitrated products.

Nitromethane, nitroethane, 1-nitropropane, and 2-nitropropane are made on a large scale at the Sterlington, Louisiana, plant of International Minerals & Chemical Corporation. In the manufacturing process (11,68–70), nitric acid reacts with excess

Table 6.   Effect of Temperature on the Nitration of Propane with Nitric Acid

Product	Nitration temperature		Type of substitution
	505–510°C	790–795°C	
nitromethane, mol %	22.0	32.3	alkyl cleavage
nitroethane, mol %	16.6	24.2	alkyl cleavage
1-nitropropane, mol %	13.2	24.2	primary
2-nitropropane, mol %	48.2	19.3	secondary
ratio of cleavage products to substitution products	0.628	1.30	

propane at 370–450°C and 0.81–1.2 MPa (8–12 atm). Stainless steel is the preferred material of construction. The reaction products are cooled and the nitroparaffins and associated by-products, such as aldehydes and ketones, condense. Propane and nitric oxide remain in the gas stream. The propane is separated and recycled to the reactor. The nitric oxide is converted to nitric acid, which, mixed with fresh acid, is fed to the reactor. The crude nitroparaffins are washed to remove oxygenated impurities, and then are fractionated to commercial-grade nitroparaffins (see Table 7).

The only other nitroparaffin manufactured on a large scale was nitrocyclohexane [1122-60-7], made by liquid-phase nitration of cyclohexane. Nitrocyclohexane was the starting material for ε-caprolactam via reduction to cyclohexanone oxime. This process has given way to other more efficient processes (see Polyamides). At present nitrocyclohexane is not being produced in large quantities for either captive use or sale.

The preparation of polynitroparaffins is reviewed in ref. 71. 2,2-Dinitropropane has been produced in pilot-plant quantities by liquid-phase nitration starting from either propane or 2-nitropropane (72–73) (see Nitration).

gem-Dinitroparaffins are made conveniently from primary or secondary mononitroparaffins by the oxidative nitration of mononitroparaffins in alkaline solution using silver ion as the oxidizing agent (Shechter-Kaplan reaction) (74–76). This reaction has been used to prepare 1,1-dinitroethane on a tonnage scale.

$$R_2C{=}NO_2^- + NO_2^- + 2\,Ag^+ \rightarrow R_2C(NO_2)_2 + 2\,Ag$$

Table 7. Typical Properties of Commercial-Grade Nitroparaffins

Property	Nitromethane	Nitroethane	1-Nitropropane[a]	2-Nitropropane[a]
distillation range at 101.3 kPa[b] (90% min), °C	100–103	112–116	129–133	119–122
vapor density (air = 1)	2.11	2.58	3.06	3.06
change of density with temperature, 0–50°C, g/(cm³·°C)	0.0014	0.0012	0.0011	0.0011
weight per L at 20°C, kg	1.13	1.05	1.00	0.9
flash point, °C (°F)				
Tag open cup	44.4 (112)	41.1 (106)	48.9 (120)	37.8 (100)
Tag closed cup	35.6 (96)	30.6 (87)	35.6 (96)	27.8 (82)
lower limit of flammability (at °C), vol %	7.3(33)	3.4(30)	2.2(34)	2.5(27)
ignition temperature, °C	418	414	420	428
evaporation rate[c]	139	121	88	110
evaporation number[d]	9	11	16	10
hydrogen bonding parameter, γ	2.5	2.5	2.5	2.5
solubility parameter, δ	12.7	11.1	10.7	10.7

[a] A mixture of 1- and 2-nitropropane is marketed as NiPar S-30.

[b] To convert kPa to mm Hg, multiply by 7.5.

[c] *n*-Butyl acetate = 100.

[d] Diethyl ether = 1.

Shipment and Storage; Specifications

The four commercial nitroparaffins are available in drums, but only nitroethane and the nitropropanes are sold in tank cars or trucks. With the exception of nitromethane, there are no restrictions on shipment or storage of the lower mononitroproparaffins.

Safety factors, however, have been of prime consideration in the development of recommendations (77–78) for the storage and safe handling of nitromethane during recovery operations and transfer in piping systems. Bulk shipments, except specified mixtures containing nitromethane and a diluent, are prohibited by DOT. Nitromethane preferably should be stored in the 208-L (55-gal) drums in which it is shipped. These containers are of lightweight construction and there is little possibility that they might develop sufficiently high internal pressure either to ignite the nitromethane or to allow it to burn as a monopropellant. Bulk-storage tanks should be isolated, buried, or barricaded to protect them from projectile impacts should an explosion occur in nearby equipment or facilities. Despite the fact that nitromethane may be detonated under certain conditions, it is not classified as an explosive for shipping purposes.

Commercial-grade nitroparaffins are shipped and stored in ordinary carbon steel. However, wet nitroparaffins containing more than 0.1–0.2% water may become discolored when stored in steel for long periods, even though corrosion is not excessive. Aluminum and stainless steel are completely resistant to corrosion by wet nitroparaffins. Storage in contact with lead, copper, or alloys containing these metals should be avoided. Polymeric materials for gaskets, hoses, and connections should be tested for their suitability before exposure to nitroparaffins.

Because of their flash points, commercial nitroparaffins are classified as flammable liquids under DOT regulations, and a red label is required.

The specifications of the four commercial nitroparaffins are given in Table 8.

Analytical Methods

The nitroparaffins have been determined by procedures such as fractionation, titration, colorimetry, infrared spectroscopy, mass spectrometry, and gas chromatography. The early analytical methods and uses of polynitroparaffins as analytical reagents are reviewed in ref. 11. More recent qualitative and quantitative methods are reviewed in ref. 79.

A titration method for primary and secondary nitroparaffins using hypochlorite gives good accuracy (±0.1%) (80). It is based on the following equation:

$$CH_3CH_2NO_2 + 2\ NaClO \rightarrow CH_3CCl_2NO_2 + 2\ NaOH$$

[594-72-9]

Table 8.   Specifications of Commercially Available Nitroparaffins

Assay	Nitromethane	Nitroethane	1-Nitropropane	2-Nitropropane
purity, min wt %[a]	95	92.5	94	94
total nitroparaffins, min wt %[a]	99	99	99	99
specific gravity at 25/25°C	1.124–1.129	1.042–1.047	0.997–0.999	0.984–0.988
acidity as acetic acid, max wt %	0.1	0.1	0.2	0.1
water, max wt %	0.1	0.2	0.1	0.1
color, max APHA	20	20	20	20

[a] Determined by gas chromatograph or mass spectrometer.

A number of colorimetric methods are available (79); however, spectroscopic methods generally are preferred.

Data on infrared curves for many nitroparaffins and their sodium salts have been reported (10,81–85). References 83 and 86–87 give uv spectra. Accurate analysis and positive identification of the components of a mixture of several nitroparaffins can be obtained by mass spectroscopy (88).

Gas chromatography is probably the most versatile method for analyzing nitroparaffins (89–91), eg, in the presence of nitric acid esters (92).

High performance liquid chromatography may be used to determine nitroparaffins by utilizing the standard uv detector at 254 nm. This method is particularly applicable to small amounts of nitroparaffins present in, eg, nitro alcohols which cannot be analyzed easily by gas chromatography.

Health and Safety Factors

Commercial nitroparaffins can, in general, be handled similarly to other common solvents, provided certain precautionary measures are observed (77,93–95).

Toxicology. Nitroparaffins are moderately toxic, as determined by oral administration to laboratory animals.

The lower nitroparaffins appear to have minimal effects when applied to skin. Low grade skin irritation experienced by plant operators from frequent exposure is probably the result of the drying effect common to many organic solvents, but no allergies or other adverse physiological effects have been reported from skin exposure. This is in direct contrast to the highly toxic nature of aromatic nitro compounds by skin absorption.

Inhalation is the chief route of worker exposure. Comparative data from acute or subchronic inhalation exposures with rats (96) indicate that nitromethane and nitroethane are the least toxic by this route and do not induce methemoglobin formation. The nitropropanes are less well tolerated; 2-nitropropane is more toxic than 1-nitropropane and is more likely to cause methemoglobinemia.

The 1978 TLVs recommended by the ACGIH (97–98) and adopted as a time-weighted average value by OSHA (99) are given in Table 9.

A comprehensive study of the tolerance of laboratory animals to vapors of 2-nitropropane was reported in 1952 (100). In 1976, NIOSH sponsored a 6-mo inhalation study using rats and rabbits (101). All animals survived exposure to nitromethane at 750 ppm and 100 ppm with no unexpected findings. 2-Nitropropane was tested at 200 ppm and 27 ppm in rabbits and male rats. No compound-related changes were found in tissues of rabbits at either dose, or in rats at 27 ppm. Liver damage was extensive in male rats exposed to 207 ppm for 6 mo, and hepatocellular carcinomas were observed.

Table 9. TLVs of Nitroparaffins

Nitroparaffin	ppm	mg/m^3
nitromethane	100	250
nitroethane	100	310
1-nitropropane	25	90
2-nitropropane	25	90

Inhalation testing with nitropropanes was extended to longer periods. No malignancies nor significant pathological changes were observed in any of the male or female rats exposed to 25 ppm of 2-nitropropane vapors for 7 h/d for 5 d/wk for 22 mo (102). Male rats exposed to 100 ppm 2-nitropropane vapors for 18 mo did show liver damage with neoplasia. No liver damage or neoplasia has been observed from inhalation of 100 ppm of 1-nitropropane vapors over a period of 18 mo.

Safe Handling. Any work area should be ventilated adequately to maintain the concentration of nitroparaffins below the accepted TLVs. Fresh-air masks should be supplied to workers entering confined areas (eg, tank cars) containing a high concentration of nitroparaffin vapors. Nitroparaffins have very high heats of adsorption on respirator canisters containing Hopcalite (a mixture of oxides of copper, cobalt, manganese, and silver present in some canisters for converting carbon monoxide to carbon dioxide); the use of respirator masks containing these substances may lead to fire in the presence of high concentrations of the nitroparaffins.

The ignition temperatures of the lower homologues are relatively high for organic solvents. When ignited, nitromethane burns with a lazy flame that often dies spontaneously, and in any case is extinguished readily with water which floats on the heavier nitromethane. Nitropropanes burn more vigorously but less so than gasoline.

Some dry-chemical fire extinguishers contain sodium or potassium bicarbonate; these should not be used on nitromethane or nitroethane fires. Dry-chemical extinguishers can be used on nitropropane fires (see also Plant safety).

Nitromethane can explode if subjected to a severe shock while under confinement in heavy-walled pressure containers. Such shock may be initiated by sudden pressurization, penetration by a projectile, or heating approximately to the critical temperature of nitromethane (315°C). These conditions, combining high pressure with high temperature, are the same as those under which nitromethane burns as a monopropellant. Certain compounds, eg, amines or strong oxidizing agents, in admixture with nitromethane, can sensitize it to decomposition by strong shock. The addition of such sensitizers should be avoided unless the nitromethane is intended as an explosive (see under Uses).

The insensitivity of nitromethane to detonation by shock under normal conditions of handling has been demonstrated by a number of full-scale tests. Sensitivity to shock is increased at elevated temperature; at about 60°C nitromethane can be detonated by a no. 8 detonating cap. Nitroethane or the nitropropanes in unconfined conditions have not been exploded by heat or shock applied under extreme test conditions.

Uses

Derivatives. The nitroparaffins are useful intermediates for the synthesis of a variety of chemical compounds. The production of nitro alcohols (qv), alkanolamines (qv), hydroxylamine, and chloropicrin consumes several metric tons of nitroparaffins annually. Nitroacetic acid, eg, is an important synthesis tool (19).

Solvents. Large volumes of nitroparaffins are consumed for use as solvents in coatings and inks, for extractions, for crystallizations, and as a reaction medium (see Solvents, industrial). The nitroparaffins are excellent solvents for cellulose esters. Resins such as vinyls, epoxies, and polyamides, and acrylic polymers generally are soluble in nitroparaffins; nitromethane, eg, is an excellent solvent for α-cyanoacrylate (see Acrylic ester polymers). Nitroparaffins can be formulated into solvent blends, eg, with alcohols or aromatic hydrocarbons. The use of nitropropanes in coatings enhances the overall performance of the coating formulation (103) (see Coatings).

In Separation Processes. Because of their compatability with a wide range of chemical compounds, nitroparaffins are employed as processing solvents. They can be used to separate closely related materials present in natural products or reaction mixtures. For example, oleic acid is difficult to separate from polyunsaturated fatty acids, but can be crystallized at −25 to −40°C from 2-nitropropane (104). Similarly, cetyl and oleyl alcohols can be separated with nitropropane at room temperature.

Nitromethane is a suitable solvent for the separation of aromatic from aliphatic compounds by solvent-extraction techniques (105) (see Extraction).

Other extraction processes utilizing nitroparaffins include the separation of lactic acid from fermentation beers, nitrocellulose from nitrating solutions, and plutonium(IV) from aqueous solutions. Nitropropanes extract rosin from pinewood at elevated temperatures. The solvent is recovered by cooling the resulting solution and allowing the rosin to precipitate. Another process employs nitromethane to remove CO_2 and H_2S from hydrocarbon gas streams (106).

Separation processes are based also on azeotropic or extractive distillations. Thus, toluene can be separated from paraffins boiling at similar temperatures by azeotropic distillation with nitromethane or 2-nitropropane. Ethylbenzene can be removed from styrene by distillation of an azeotrope with 1-nitropropane (see Azeotropic and extractive distillation). Extractive distillation with a nitroparaffin has been used in separating 1,3-butadiene from other C_4 alkenes and alkanes.

Reaction Medium. Nitromethane forms crystalline complexes with salts, such as aluminum chloride, that can produce cation free radicals (107). The lower nitroparaffins are excellent solvents for aluminum chloride in Friedel-Crafts reactions (qv) and Ziegler-type polymerizations (see Ziegler-Natta catalysts; Olefin polymers). Numerous advantages of such a system include uniform catalyst activity, homogeneous-phase reaction, convenient temperature, and specificity of product (108).

Chemical Stabilizers. The nitroparaffins, particularly nitromethane, are employed extensively to prevent the decomposition of various chlorinated and fluorinated solvents used as metal degreasers and aerosol propellants. The addition of a small amount of a mixture of nitromethane and an oxygenated solvent (eg, dioxane, butylene oxide) to 1,1,1-trichloroethane inhibits decomposition and prevents corrosive attack.

Nitroparaffins are effective antigassing agents in aluminum-paste formulations for inks, paints, and related coatings which may react with water present to release hydrogen gas which swells sealed containers (109).

Explosives. Mixtures of ammonium nitrate with nitromethane (110) or 2-nitropropane (111) are explosive; a sensitizer, such as 2-amino-2-methyl-1-propanol, may be added (112). Such explosives are employed in mining, oil wells, and seismic exploration. When 4–6% of an amine, such as diethylamine or diethylenetriamine, is added to nitromethane, a powerful cap-sensitive explosive is produced (113–114) (see Explosives and propellants).

Propellants. Various nitroparaffins are considered for use as monopropellants and bipropellants in rocket motors. Nitromethane, nitroethane, 2-nitropropane, and 2,2-dinitropropane have been tested as fuels in bipropellant systems. 2-Nitropropane gives a desirable combination of physical properties, ease of handling, and availability. The theoretical performance of 2-nitropropane with several oxidizers has been calculated (115) as follows:

Fuel	Oxidizer	Specific impulse, s	Density impulse, $(g \cdot s)/cm^3$
2-nitropropane	97% HNO_3	225	287
2-nitropropane	oxygen	256	271
JP-4 hydrocarbon	oxygen	256	263

Reciprocating Engines. Nitroparaffins have been investigated as fuels in Otto cycle engines (as contrasted with diesel cycle engines) (115). When nitroparaffins are used instead of hydrocarbons, the power output of an engine is increased because more fuel can be burned per unit of air, and the number of moles of gaseous products formed in the combustion chamber is increased (see Gasoline and other motor fuels).

Nitroparaffins have been used in hobby (model) engines and in racing cars for short distances. A typical fuel for such a use contains 25 vol % castor oil, 10 vol % methanol, 45 vol % nitromethane, and 20 vol % 2-nitropropane. This mixture is homogeneous above 24.2°C. Lowering the content of nitromethane or castor oil or both gives a mixture that is homogeneous at lower temperatures.

Unwise use of nitroparaffins in engines not capable of handling the increased power output results in burned pistons and broken crankshafts. The tendency for preignition increases with increasing concentration of nitroparaffins in nitroparaffin–methanol fuel blends. This phenomenon can be controlled by proper cooling and by employing very smooth surfaces in the combustion chambers.

The effect of nitroparaffins in diesel engine fuels has been known for many years (115–116). The addition of as little as 0.1–0.2 wt % 2,2-dinitropropane results in an increase in the cetane number of 5 to 10 units. Nitroparaffins, particularly in situations where extra power may be required for short periods of time, increase the power output of the engine and reduce the exhaust smoke. This has important potential for air-pollution abatement. Mononitroparaffins can be incorporated directly in diesel fuel up to the limit of their solubility, ie, nitromethane up to 2 wt % and nitroethane up to 18 wt %; nitropropanes are completely soluble. The lower nitroparaffins are good smoke depressants; eg, the use of 50 wt % 2-nitropropane increases peak power by 6% with an attendant drop of 50% in engine smoke. Small concentrations (0.1%) of nitropropane improve diesel fuel combustion.

Pesticides. Chloropicrin, one of the oldest industrial chemicals containing a nitro group, is a powerful lacrimator and nauseant agent (see Chemicals in war). It also has extensive use as an insecticide, soil sterilizer, and fumigant. 1,1-Dichloro-1-nitroethane has similar fumigant activity. 1-Chloro-2-nitropropane is effective as a soil fungicide (117) (see Fungicides; Insect control technology).

BIBLIOGRAPHY

"Nitroparaffins" in *ECT* 1st ed., Vol. 9, pp. 428–455, by H. Shechter, Ohio State University, and R. B. Kaplan, E. I. du Pont de Nemours & Co., Inc.; "Nitroparaffins" in *ECT* 2nd ed., Vol. 13, pp. 864–888, by Jerome L. Martin and Philip J. Baker, Jr., Commercial Solvents Corporation.

1. *IUPAC Nomenclature of Organic Chemistry*, 1979 ed., Pergamon Press, New York, p. 275.
2. U.S. Pat. 1,967,667 (July 24, 1934), H. B. Hass, E. B. Hodge, and B. M. Vanderbilt (to Purdue Research Foundation); *Ind. Eng. Chem.* **28,** 339 (1936).
3. E. E. Toops, Jr., *J. Phys. Chem.* **60,** 304 (1956).
4. B. Fader, *Chem. Process.* (*Chicago*) **19**(8), 174 (1956).

5. L. H. Horsley, *Adv. Chem. Ser.* **6**, (1952); L. H. Horsley and W. S. Tamplin, *Adv. Chem. Ser.* **35**, (1962); L. H. Horsley, *Adv. Chem. Ser.* **116**, (1973).

6. A. W. Francis, *Adv. Chem. Ser.* **31**, (1961).

7. A. W. Francis, *Liquid–Liquid Equilibriums*, John Wiley & Sons, Inc., New York, 1963, pp. 174, 214; A. W. Francis, *J. Chem. Eng. Data* **11**, 234 (1966).

8. L. O. Brockway, J. Y. Beach, and L. Pauling, *J. Am. Chem. Soc.* **57**, 2693 (1935).

9. G. Geiseler and H. Kessler, "Physikalische Eigenschaften Homologer Primärer und Stellungsisomerer Geradkettiger Nitroalkane" in T. Urbanski, ed., *Nitro Compounds: Proc. Int. Symp. Warsaw, Poland, 1963*, Pergamon Press Ltd., Oxford, Eng., 1964, pp. 187.–194.

10. R. T. Aplin, M. Fischer, D. Becher, H. Budzikiewicz, and C. Djerassi, *J. Am. Chem. Soc.* **87**, 4888 (1965); N. M. M. Nibbering, Th. J. de Boer, and H. J. Hofman, *Rec. Trav. Chim.* **84**, 481 (1965); A. V. Iogansen and G. D. Litovchenko, *Zh. Prikl. Spektroskopii, Akad. Nauk Belorussk. SSR* **21**(3), 243 (1965); *ibid.* **3**(6), 538 (1965); *Chem. Abstr.* **63**, 10858 (1965); **64**, 16837 (1966).

11. H. B. Hass and E. F. Riley, *Chem. Rev.* **32**, 373 (1943).

12. N. Levy and J. D. Rose, *Q. Rev. (London)* **1**, 358 (1947).

13. P. A. S. Smith, *The Chemistry of Open-Chain Organic Nitrogen Compounds*, Vol. II, W. A. Benjamin, Inc., New York, 1966, pp. 391–454.

14. H. H. Baer and L. Urbas, "Activating and Directing Effects of the Nitro Group in Aliphatic Systems" in H. Feuer, ed., *The Chemistry of the Nitro and Nitroso Groups*, Part 2, Interscience Publishers, a division of John Wiley & Sons, Inc., New York, 1970, pp. 75–200.

15. D. Seebach, E. W. Colvin, F. Lehr, and T. Weller, *Chimia* **33**, 1 (1979).

16. *Tetrahedron* **19**(Suppl. 1), (1963).

17. T. Urbanski, ed., *Nitro Compounds: Proc. Int. Symp. Warsaw, Poland, 1963*, Pergamon Press Ltd., Oxford, Eng., 1964.

18. T. Urbanski, *Synthesis*, 613 (1974).

19. M. T. Shipchandler, *Synthesis*, 666 (1979).

20. N. Kornblum and P. Pink in ref. 16, pp. 17–22.

21. H. B. Hass, E. J. Berry, and M. L. Bender, *J. Am. Chem. Soc.* **71**, 2290 (1949).

22. H. Shechter and R. B. Kaplan, *J. Am. Chem. Soc.* **73**, 1883 (1951).

23. R. C. Kerber, G. W. Urry, and N. Kornblum, *J. Am. Chem. Soc.* **87**, 4520 (1965).

24. G. A. Russell and W. C. Danen, *J. Am. Chem. Soc.* **88**, 5663 (1966).

25. G. B. Bachman and T. Hokama, *J. Am. Chem. Soc.* **81**, 4882 (1959).

26. E. H. White and W. J. Considine, *J. Am. Chem. Soc.* **80**, 626 (1958).

27. L. G. Donaruma and M. L. Huber, *J. Org. Chem.* **21**, 965 (1956).

28. H. L. Wehrmeister, *J. Org. Chem.* **25**, 2132 (1960).

29. T. Urbanski, Z. Novak, and E. Morag, *Bull. Acad. Polon. Sci., Ser. Sci. Chim.* **11**(2), 77 (1963).

30. U.S. Pat. 2,337,912 (Dec. 28, 1943), E. T. McBee and E. F. Riley (to Purdue Research Foundation).

31. U.S. Pat. 3,099,612 (July 30, 1963), L. A. Wilson (to Commercial Solvents Corp.).

32. U.S. Pat. 3,100,806 (Aug. 13, 1963), P. Bay (to Abbott Laboratories).

33. U.S. Pat. 3,054,829 (Sept. 18, 1962), G. B. Bachman and R. O. Downs (to Purdue Research Foundation).

34. K. Okon and G. Aluchna, *Bull. Acad. Polon. Sci., Ser. Sci. Chim. Geol. Geograph.* **7**, 83 (1959).

35. N. J. Leonard and A. B. Simon, *J. Org. Chem.* **17**, 1262 (1952).

36. M. Senkus, *J. Am. Chem. Soc.* **68**, 1611 (1946); **72**, 2967 (1950).

37. T. Urbanski and co-workers in ref. 17, pp. 195–218.

38. W. D. Emmons, W. N. Cannon, J. W. Dawson, and R. M. Ross, *J. Am. Chem. Soc.* **75**, 1993 (1953).

39. U.S. Pat. 2,477,870 (Aug. 2, 1949), M. H. Gold and L. J. Draker (to The Visking Corp.).

40. S. L. Ioffe, V. A. Tartakovskii, and S. S. Novikov, *Russ. Chem. Rev.* **35**, 19 (1966).

41. S. S. Novikov, I. S. Korsakova, and K. K. Babievskii, *Usp. Khim.* **26**, 1109 (1957); *Chem. Abstr.* **52**, 6154 (1958).

42. E. E. Hamel in ref. 16, pp. 85–95.

43. M. B. Frankel in ref. 16, pp. 213–217.

44. S. Wakamatsu and K. Shimo, *J. Org. Chem.* **27**, 1609 (1962).

45. E. Profft, *Chem. Tech. (Berlin)* **8**, 705 (1956).

46. H. Shechter, D. L. Ley, and L. Zeldin, *J. Am. Chem. Soc.* **74**, 3664 (1952).

47. H. Feuer and R. Harmetz, *J. Org. Chem.* **26**, 1061 (1961).

48. V. V. Perekalin, "Chemistry and Structure of Unsaturated Nitro Compounds" (in Russian) in ref. 17, pp. 135–157.

49. V. V. Perekalin, *Unsaturated Nitro Compounds*, Gosudarst. Nauch.-Tekh. Izdatel. Khim. Lit.,

Leningrad, USSR, 1961; *Chem. Abstr.* (to Russ. ed.) **55,** 17470 (1961); in Israel Program for Scientific Translation, Jerusalem, Israel, 1963; *Chem. Abstr.* (to Engl. ed.) **61,** 11862 (1964).

50. M. H. Gold and K. Klager in ref. 16, pp. 77–84.
51. G. V. Roitburd, W. A. Smit, A. V. Semenovsky, A. A. Shchegolev, V. F. Kucherov, O. S. Chizhov, and V. I. Kadentsev, *Tetrahedron Lett.* **48,** 4935 (1972).
52. U.S. Pat. 3,055,936 (Sept. 25, 1962), M. Stiles and H. L. Finkbeiner (to Research Corp.).
53. H. L. Finkbeiner and M. Stiles, *J. Am. Chem. Soc.* **85,** 616 (1963).
54. H. L. Finkbeiner and G. W. Wagner, *J. Org. Chem.* **28,** 215 (1963).
55. A. P. Hardt, F. G. Borgardt, W. L. Reed, and P. Noble, Jr., *Electrochem. Technol.* **1,** 375 (1963).
56. C. T. Bahner, *Ind. Eng. Chem.* **44,** 317 (1952).
57. H. Shechter and R. B. Kaplan, *J. Am. Chem. Soc.* **75,** 3980 (1953).
58. U.S. Pat. 3,267,158 (Aug. 16, 1966), A. J. Havlik (to Aerojet-General Corp.).
59. T. Urbanski and W. Gurzynska, *Roczniki Chem.* **25,** 213 (1951); *Chem. Abstr.* **46,** 7994 (1952).
60. V. F. Belyaev and R. I. Shamanovskaya, *Zh. Organ. Khim.* **1,** 1388 (1965).
61. T. Mukaiyama and T. Hoshino, *J. Am. Chem. Soc.* **82,** 5339 (1960).
62. A. V. Topchiev, *Nitration of Hydrocarbons and Other Organic Compounds*, Pergamon Press Ltd., Oxford, Eng., 1959.
63. N. Kornblum, "The Synthesis of Aliphatic and Alicyclic Nitrocompounds" in A. C. Cope, ed., *Organic Reactions*, Vol. 12, John Wiley & Sons, Inc., New York, 1962, pp. 101–156.
64. H. O. Larson, "Methods of Formation of the Nitro Group in Aliphatic and Alicyclic Systems" in ref. 14, Part 1, 1969, pp. 301–348.
65. K. E. Gilbert and W. T. Borden, *J. Org. Chem.* **44,** 659 (1979).
66. G. B. Bachman and R. J. Maleski, *J. Org. Chem.* **37,** 2810 (1972).
67. H. B. Hass and H. Shechter, *Ind. Eng. Chem.* **39,** 817 (1947).
68. F. A. Lowenheim and M. K. Moran, eds., *Faith, Keyes, and Clark's Industrial Chemicals*, 4th ed., John Wiley & Sons, Inc., New York, 1975.
69. R. N. Shreve, *The Chemical Process Industries*, McGraw-Hill Book Co., Inc., New York, 1956, p. 933.
70. J. C. Reidel, *Oil Gas J.* **54**(36), 110 (1956).
71. P. Noble, Jr., F. G. Borgardt, and W. L. Reed, *Chem. Rev.* **64,** 7 (1964).
72. U.S. Pat. 2,489,320 (Nov. 29, 1949), E. M. Nygaard and W. I. Denton (to Socony-Vacuum Oil Co.).
73. W. I. Denton, R. B. Bishop, E. M. Nygaard, and T. T. Noland, *Ind. Eng. Chem.* **40,** 381 (1948).
74. R. B. Kaplan and H. Shechter, *J. Am. Chem. Soc.* **83,** 3535 (1961).
75. U.S. Pat. 2,997,504 (Aug. 22, 1961), H. Shechter and R. B. Kaplan (to The Ohio State University Research Foundation).
76. U.S. Pat. 3,000,966 (Sept. 19, 1961), K. Klager (to Aerojet-General Corp.).
77. *Nitromethane: Storage and Handling*, NP Tech. Data Sheet No. 2, 3rd ed., International Minerals & Chemical Corp., Des Plaines, Ill., 1976.
78. *Nitroparaffins and Their Hazards*, Research Report No. 12, Committee on Fire Prevention and Engineering Standards, National Board of Fire Underwriters, New York, 1959.
79. C. J. Wassink and J. T. Allen, "Nitro Compounds, Organic" in F. D. Snell and L. S. Ettre, eds., *Encyclopedia of Industrial Chemical Analysis*, Vol. 16, John Wiley & Sons, Inc., 1972, pp. 412–448.
80. L. R. Jones and J. A. Riddick, *Anal. Chem.* **28,** 1137 (1956).
81. C. Frejacques and M. Leclercq, *Mem. Poudres* **39,** 57 (1957).
82. J. R. Nielsen and D. C. Smith, *Ind. Eng. Chem., Anal. Ed.* **15,** 609 (1943).
83. R. N. Haszeldine, *J. Chem. Soc.*, 2525 (1953).
84. N. Kornblum, H. E. Ungnade, and R. A. Smiley, *J. Org. Chem.* **21,** 377 (1956).
85. Z. Buczkowski and T. Urbanski, *Spectrochim. Acta* **18,** 1187 (1962).
86. R. L. Foley, W. M. Lee, and B. Musulin, *Anal. Chem.* **36,** 1100 (1964).
87. M. J. Kamlet and D. J. Glover, *J. Org. Chem.* **27,** 537 (1962).
88. J. C. Neerman and O. S. Knight, *Chem. Eng.* **56**(11), 125 (1949).
89. R. M. Bethea and T. D. Wheelock, *Anal. Chem.* **31,** 1834 (1959).
90. W. Biernacki and T. Urbanski, *Bull. Acad. Polon. Sci., Ser. Sci. Chem.* **10,** 601 (1962).
91. R. M. Bethea and F. S. Adams, *Anal. Chem.* **33,** 832 (1961); *J. Chromatog.* **8,** 532 (1962).
92. E. Camera, D. Pravisani, and W. Ohman, *Explosivstoffe* **13**(9), 237 (1965).
93. "Nitropropane," "Nitroethane," and "Nitromethane," *Hygienic Guide Series*, American Industrial Hygiene Association, Akron, Ohio, 1978.

94. *The Storage and Handling of Nitropropane Solvents*, *NP Tech. Data Sheet No. 20*, International Minerals & Chemical Corp., Des Plaines, Ill., 1976.

95. *Nitromethane*, Industrial Data Sheet, National Safety Council, Chicago, Ill., 1981.

96. J. Dequidt, P. Vasseur, and J. Potencier, *Bull. Soc. Pharm. Lille* 83, 131, 137 (1972); 29 (1973).

97. *Threshold Limit Values for 1978*, American Conference of Governmental Industrial Hygienists, Cincinnati, Ohio, 1978.

98. *Documentation of Threshold Limit Values*, rev. ed., Committee on Threshold Limit Values, American Conference of Governmental Industrial Hygienists, Cincinnati, Ohio, 1966, pp. 139–143.

99. *29 CFR 1910.1000*, Occupational Safety and Health Administration, U.S. Department of Labor, U.S. Government Printing Office, Washington, D.C., 1980.

100. J. F. Treon and F. R. Dutra, *A.M.A. Arch. Ind. Hyg. Occup. Med.* **5**, 52 (1952).

101. T. R. Lewis, C. E. Ulrich, and W. M. Busey, *J. Environ. Pathol. Toxicol.* **2**, 233 (1979).

102. T. B. Griffin, F. Coulston, and A. A. Stein, *Ecotoxicol. Environ. Saf.* **4**, 267 (1980).

103. *Benefit by Using the Nitropropane Solvents*, *NP Tech. Bulletin No. 43*, International Minerals & Chemical Corp., Des Plaines, Ill., 1976.

104. U.S. Pat. 3,345,389 (Oct. 3, 1967), K. T. Zilch (to Emery Industries, Inc.).

105. U.S. Pat. 3,244,762 (April 5, 1966), A. C. McKinnis (to Union Oil Co. of California).

106. U.S. Pat. 3,255,572 (June 14, 1966), L. N. Miller, O. C. Holbrook, and B. B. Woertz (to Union Oil Co. of California).

107. W. F. Forbes and P. D. Sullivan, *J. Am. Chem. Soc.* **88**, 2862 (1966).

108. L. Schmerling, *Ind. Eng. Chem.* **40**, 2072 (1948).

109. U.S. Pat. 2,848,344 (Aug. 19, 1958), M. H. Brown (to Aluminum Co. of America).

110. U.S. Pat. 2,325,064 (July 27, 1943), R. W. Lawrence (to Hercules Powder Co.).

111. Belg. Pat. 624,797 (May 14, 1963), T. W. Royer and J. S. Brower (to Aerojet-General Corp.).

112. Brit. Pat. 824,533 (Dec. 2, 1959), R. S. Egly (to Commercial Solvents Corp.).

113. Belg. Pat. 627,768 (July 30, 1963), (to Aerojet-General Corp.).

114. U.S. Pat. 3,239,395 (March 8, 1966), E. A. Laurence (to Aerojet-General Corp.).

115. R. S. Egly and E. S. Starkman, "Nitroparaffin Fuels" in J. J. McKetta, Jr., ed., *Advances in Petroleum Chemistry and Refining*, Vol. 10, Interscience Publishers, a division of John Wiley & Sons, Inc., New York, 1965, pp. 408–454.

116. R. E. Albright, F. L. Nelson, and L. Raymond, *Ind. Eng. Chem.* **41**, 929 (1949).

117. U.S. Pat. 3,078,209 (Feb. 19, 1963), J. R. Willard and E. G. Maitlen (to FMC Corp.).

PHILIP J. BAKER, JR.
ALLEN F. BOLLMEIER, JR.
International Minerals & Chemical Corporation

N-NITROSAMINES

N-Nitrosodialkylamines (*N*-nitrosamines) have been known for over 100 years (1–2). They are typical organic compounds which have been useful as synthetic intermediates or as solvents in addition to possessing interesting structural and spectroscopic properties.

N-Nitrosodimethylamine is highly hepatotoxic and carcinogenic (3–6). Most *N*-nitrosamines are carcinogenic, and no animal species which has been tested is resistant to nitrosamine-induced cancer (7). Several *N*-nitrosamines have been listed as suspected human carcinogens (8). In the mid-1960s, a case of liver damage in sheep was caused by *N*-nitrosodimethylamine (NDMA) in the animals' food, which had been preserved with nitrite (see Pet and livestock feeds) (9). It was hypothesized that the nitrite reacts with dimethylamine in the food to form the NDMA.

Properties

Many of the chemical, physical, and biological properties of more than 20 selected *N*-nitrosamines are summarized in ref. 10. *N*-Nitrosamines represent a wide range of structural types and molecular weights. The single feature common to the *N*-nitrosamines is the NNO functionality. *N*-Nitrosamines are extremely numerous since there are few restrictions on the groups that can be attached to the remaining two valences on the amine nitrogen, eg, (**1**):

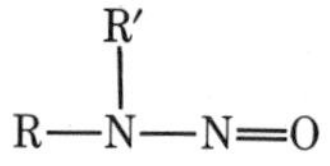

$$R{-}\underset{\underset{(1)}{|}}{\overset{\overset{R'}{|}}{N}}{-}N{=}O$$

where R,R' can be alkyl, aryl, or both; for R or R' = XCH$_2$, X can be H, alkyl, aryl, halogen, alkoxy, etc; for R or R' = aryl, various substituents may be on the ring(s); and when R or R' = H, or when X = OH, the resulting primary *N*-nitrosamines or α-hydroxy-*N*-nitrosamines generally are unstable (11–13). Examples of most other types, including derivatives of cyclic amines, eg, *N*-nitrosomorpholine (**2**) or *N*-nitrosopyrrolidine (**3**), have been characterized (6–7,10,14–21).

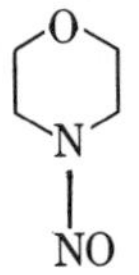

(2)    (3)

N-Nitrosamines typically are volatile solids or oils and are yellow because of absorption of visible light by the NNO group. The electron delocalization in the functionality confers sufficient double-bond character on the N—N bond, so that the *E* and *Z* isomers which result from unsymmetrical substitution, eg, (**4**) and (**5**), often can be separated (22).

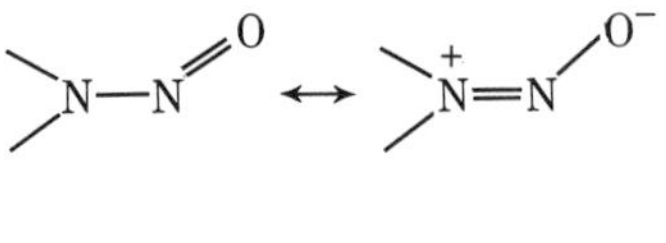

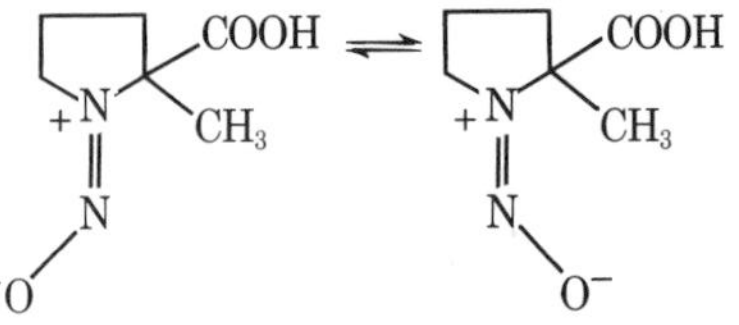

(4)   [*76530-17-1*]   (5)

The analytical properties of the *N*-nitrosamines, especially the nmr and mass spectra, vary widely depending on the substituents on the amine nitrogen (23–26). The nmr spectra are affected by the $E \leftrightarrows Z$ isomerism around the N—N partial double bond and by the axial–equatorial geometry resulting from conformational isomerism in the heterocycles (23–24). Some general spectral characteristics which might be expected for typical dialkylnitrosamines or simple heterocyclic nitrosamines are given in Table 1.

Synthesis. The classic *N*-nitrosamine synthesis is the reaction of a secondary amine with nitrite ion at ca pH 3, eg, equation 1 (28).

$$\text{>}NH + NO_2^- \xrightarrow{\text{H}^+} \text{>}N{-}N{=}O \tag{1}$$

However, primary and tertiary amines also form *N*-nitrosamines, although the yields generally are much lower (20–21). These reactions can be catalyzed by various substances, eg, sodium thiocyanate or formaldehyde (29–30). Nitrosations of higher amines, eg, dihexylamine, are enhanced by micelle formation (31). Other nitrosating agents, eg, various nitrogen oxides, nitrosate secondary amines at higher pH (32–33). *N*-Nitrosamines can be formed under some conditions by transfer of the NO group from one amine to another (eq. 2) (34).

$$\tag{2}$$

Efficient nitrosations of amines with inorganic nitrosyl compounds also have been reported (eg, eq. 3) (35). *N*-Nitrosamine formation can be inhibited in polar, ie,

$$\text{>}NH + Fe(CN)_5NO^{2-} \longrightarrow \text{>}NNO \tag{3}$$

Table 1. Summary of Characteristic Spectral Properties of *N*-Nitrosamines[a]

Property	Functionality involved	Value
uv–visible absorption, nm	NNO	λ_{max} 230–235
		330–375
ir absorption, cm^{-1}	N—N stretch	1040–1160
	N—O stretch	1430–1500
nmr absorption, τ	αCH (E)	(CCl$_4$) 5.8–5.9
		(C$_6$H$_6$) 6.1–6.32
	αCH (Z)	(CCl$_4$) 6.3–6.6
		(C$_6$H$_6$) 6.6–6.8
mass spectral fragmentation	entire molecule	M/e = M, (M-17), M-30, M-31

[a] Refs. 23–27.

aqueous, environments by ascorbic acid (vitamin C) and in nonpolar environments by α-tocopherol (vitamin E) (36–37).

   Reactions.   The chemistry of the *N*-nitrosamines is extensive (7,14,21). Most of the reactions of the nitrosamines, with respect to their biological or environmental behavior, involve one of two main reactive centers, ie, the nitroso group and the C—H bonds adjacent (α) to the amine nitrogen. The nitroso group can be removed readily in a reaction which is essentially the reverse of the nitrosation reaction (eq. 4), or it can be oxidized (eq. 5), or reduced (eq. 6) (38–40).

$$>N-N=O \longrightarrow >N-H \tag{4}$$

$$>N-N=O + CF_3CO_3H \longrightarrow >N-NO_2 \tag{5}$$

$$>N-N=O + Zn/CH_3CO_2H \longrightarrow >N-NH_2 \tag{6}$$

   Absorption of uv radiation by the NNO group produces a set of photochemical reactions. Under neutral conditions and at moderate to high concentrations, these compounds often are chemically stable; although the $E \leftrightarrows Z$ equilibrium, with respect to rotation around the N—N bond, can be affected (39). At low pH, a variety of photoproducts are formed, including compounds attributed to photoelimination, photoreduction, and photo-oxidation (40). Low concentrations of most nitrosamines, even at neutral pH, can be eliminated by prolonged irradiation at 366 nm; this technique is used in the identification of *N*-nitrosamines that are present in low concentrations in complex mixtures (41).

   Reactions at the α carbon have been of considerable interest because it is at that position that the enzymatic reaction, which is believed to initiate the events leading to a carcinogenic metabolite, occurs (5–7,42). The α hydrogens exchange readily, which apparently is the result of stabilization of an anionic intermediate by way of electron delocalization (eq. 7) (43–44):

$$H-C \Big\backslash N-N=O \underset{H_2O,\ HO^-}{\overset{D_2O,\ DO^-}{\rightleftharpoons}} D-C \Big\backslash N-N=O \tag{7}$$

This property has been exploited extensively in syntheses of *N*-nitrosamine derivatives by the reaction of electrophiles (E) with α-lithiated intermediates. These intermediates are prepared by hydrogen–lithium exchange using lithium diisopropylamide (LDA) (eq. 8) (45–46):

$$H-C \Big\backslash N-N=O \xrightarrow{LDA} LiC \Big\backslash N-N=O \xrightarrow{E} E-C \Big\backslash N-N=O \tag{8}$$

Analytical and Test Methods

The possible presence of N-nitrosamines in food and their possible *in vivo* formation has created a continuing need for the detection and confirmation of low levels of N-nitrosamines in fairly complex mixtures of organic chemicals. Many of the N-nitrosamines that have been recognized as environmentally significant are sufficiently volatile and stable for analysis by gas chromatography (gc). High performance liquid chromatographic (hplc) systems are being developed for the less volatile N-nitrosamines (14). A variety of detection techniques, eg, polarography, spectrophotometric cleavage of the NNO bond followed by detection of the resulting nitrite, alkali flame detection, electrolyte conductivity (Coulson detector), and mass spectrometry, have been utilized (47). The most significant contribution to N-nitrosamine analysis has been the Thermal Energy Analyzer which can be used for both gc and hplc (48–49). The nitrosamine is pyrolyzed to generate an NO radical which reacts with ozone to yield molecular oxygen and NO_2 in an excited state (NO_2^*). The NO_2^* returns to the ground state by release of a photon in the ir region, and the photon is detected and amplified by a photomultiplier. There are several classes of compounds, eg, alkyl nitrites and C-nitroso compounds, which give false-positive responses; however, various screening methods are being developed to distinguish these types of molecules from nitrosamines (39,50–51).

Confirmation of the identities of nitrosamine generally is accomplished by mass spectrometry (ms) (52). High resolution gc–ms, as well as gc–ms in various single-ion modes, can be used as a specific detector, especially when screening for particular nitrosamines (52) (see Analytical methods Trace and residue analysis).

Health and Safety Factors (Toxicology)

Carcinogenicity. Some of the toxicological properties of a selected group of nitrosamines are listed in Table 2. The number of nitrosamines that have been tested for carcinogenicity probably exceeds 100 (5–7,10,14,17,53). Most are carcinogenic, although the potency varies by over three orders of magnitude within the series (6,42). The mean tumorogenic dose for N-nitrosodiethylamine (NDEA), for example, is only ca 0.0006 mol/kg body weight, whereas that for N-nitrosodiethanolamine (NDELA) is nearly 1 mol/kg body weight (6). Carcinogenic effects have been observed both with single, relatively large doses and with long-term chronic exposure to lower doses (6,54). The N-nitrosamines generally are organ selective. Symmetrical dialkyl-N-nitrosamines, for example, typically are liver carcinogens and unsymmetrical dialkyl-N-nitrosamines tend to attack the esophagus. Other target organs include the nose, bladder, pancreas, lungs, and kidneys (6). Both the potency and the organ selectivity of the N-nitrosamines appear to be systematically related to their molecular structure as reflected by their solubility properties and the reactivity of the α-C—H bonds (55–56). In some cases, the carcinogenic potency can be affected by diet. The incidence of N-nitrosomethylbenzylamine-induced esophageal tumors, for example, can be increased, and the induction period decreased, by a zinc-deficient diet (57).

There are species-related and sex-related differences in both potency and organ selectivity (6–7,58). The parent nitrosamines are not carcinogenic, but require metabolic activation in order to exert their carcinogenic effect. The mechanisms involved in nitrosamine carcinogenesis are not completely understood, although the sequence

Table 2. Toxicological Properties of Representative N-Nitrosamines in the BD Rat [a]

Compound	CAS Registry Number	mg/kg		Principal target organ
		LD_{50} [a]	$\log(1/D_{50})$ [b]	
N-nitrosodimethylamine (NDMA or DMN)	[62-75-9]	40	2.3	liver
N-nitrosodiethylamine (NDEA or DEN)	[55-18-5]	280	3.2	liver, esophagus
N-nitrosodiethanolamine (NDELA)	[1116-54-7]	7500	0.05	liver
N-nitrosodipropylamine	[621-64-7]	480	2.1	liver, esophagus
N-nitrosodiisopropylamine	[601-77-4]	850	1.0	liver
N-nitrosopyrrolidine (NPYR)	[930-55-2]	900	1.4	liver
N-nitrosomorpholine (NMOR)	[59-89-2]	320	1.9	liver
N-nitrosodicyclohexylamine	[947-92-2]		c	
N-nitrosoproline	[7519-36-0]		c	
N-nitrosomethyl(benzyl)amine	[937-40-6]	18	3.1	esophagus
N-nitrosopiperidine	[100-75-4]	200	1.9	liver, esophagus
N-nitrosonornicotine	[16543-55-8]		d	

[a] Ref. 6.
[b] Ref. 42.
[c] Not carcinogenic to the BD rat.
[d] Not investigated in the BD rat.

outlined below is postulated for the metabolic activation of N-nitrosamines (5–7,12–13,59–63).

$$(6)$$ $$(7)$$

$$[(7) \rightleftharpoons R'CH_2N=NOH] \xrightarrow{H^+} [R'CH_2N=\overset{+}{N}OH_2] \xrightarrow{?} [R'CH_2\overset{+}{N}_2] + H_2O$$

$$(8) \qquad (9) \qquad (10)$$

$$(10) \xrightarrow{?} [R'CH_2{}^+]$$

$$(11)$$

There is substantial evidence for the initial enzymatic oxidation and for the alkylation of nucleic acids by the resulting electrophiles (5–6,12–13,62–65). The intervening steps, ie, intermediates (7–11), however, are hypothesized largely by analogy with the known behavior of primary nitrosamines (eg, (7)) (11,66). There is some evidence that microsomal oxidation of nitrosamines ultimately may lead to intermediates (10) and/or (11) *in vitro*, but it appears unlikely that either of these is significantly involved in nucleic acid alkylation *in vitro* (60–61). Tumors are thought to arise from mispairing during subsequent replications of the alkylated DNA molecules. Most alkylation appears to occur at the N-7 position of DNA guanine (12), but alkylation at the O-6

(12) guanine

position of guanine is more closely correlated with tumor formation than is alkylation at *N*-7 (65).

Mutagenicity. The *N*-nitrosamines are, in general, mutagenic toward standard bacterial-tester strains (67). As with carcinogenicity, enzymatic activation, with liver microsomal preparations in this case, is required. Certain substituted *N*-nitrosamines, most notably the α-acetoxy derivatives, eg, (**13**), are mutagenic without microsomal activation (12). This supports the hypothesis that enzymatic oxidation leads to the formation of unstable α-hydroxy derivatives, eg, (**14**). However, for simple *N*-nitro-

samines, no systematic relationship has been found between carcinogenicity and mutagenicity (67–70).

Environmental Aspects. Direct exposure to preformed *N*-nitrosamines in the environment is rare. *N*-nitrosamines are not widely distributed in water or in the atmosphere, although *N*-nitrosamines have been detected in certain localities (71–72). *N*-Nitrosamines are present in some pesticide preparations and, thus, may be found in areas exposed to these pesticides (73). Cigarette smoke contains *N*-nitrosamines, including *N*-nitrosodimethylamine and *N*-nitrosonornicotine (74–75). High levels of *N*-nitrosodiethanolamine often are found in cutting fluids; lower levels have been detected in cosmetics (qv) (76–77). *N*-nitrosodicyclohexylamine occurs in corrosion inhibitors that are formulated from dicyclohexylammonium nitrite (78).

The potential human hazards arising from such *N*-nitrosamine exposure, as well as those discussed below, are impossible to assess on an absolute basis and difficult to assess on a relative basis. *N*-Nitrosodimethylamine is a highly potent animal carcinogen, *N*-nitrosodiethanolamine is a weak animal carcinogen, *N*-nitrosonornicotine is intermediate, and *N*-nitrosodicyclohexylamine is noncarcinogenic in animal experiments (6). A relative risk factor, based on concentration as well as relative potency, has been proposed (79), but it is not clear how to compare relative potencies in an animal species to relative potencies in humans.

Low levels of *N*-nitrosamines occur in several types of food, including cheese, processed meats, beer, and cooked bacon (19). *N*-Nitrosopyrrolidine occurs at the highest levels, ie, 5–50 ppb, in cooked bacon, but *N*-nitrosodimethylamine and *N*-nitrosodiethylamine also are often present, although at somewhat lower levels, ie, 2–25 ppb (19). The origins of these substances are not well understood, although it is clear in the case of *N*-nitrosopyrrolidine that nitrite is involved (80). The nitrosamine levels in bacon have been reduced substantially following the development of methods for inhibiting nitrosation reactions (81).

A principal area of current concern is the possibility of *in vivo* formation of *N*-nitrosamines from amines and nitrite, which may be present in food or drinking water or which may be formed endogenously (82). Animals that have been fed amines and nitrite develop tumors that are identical with those expected from the corresponding

N-nitrosamine, and *in vivo* formation of N-nitrosamines from amines and nitrite has been demonstrated (83–85).

Nitrite formed in the body may be more significant than nitrite ingested with nitrite-preserved foods. Saliva, for example, contains a constant low (ca 7 µg/mL) level of nitrite, which can increase rapidly following ingestion of nitrate because of the presence of efficient nitrate-reducing bacteria (86–87). N-Nitrosamine formation from amines that are added to whole human saliva, without added nitrite, has been observed (88). In addition to salivary nitrite, there is recent evidence to suggest that nitrite is formed in the intestine at levels much higher than would be ingested normally (89). Other experiments have shown that N-nitrosamines are present in human feces, but it is not known whether these observations are related to one another or if either phenomenon is associated with human intestinal cancer (90).

There is no direct evidence that firmly links human cancer to N-nitrosamines, but it is unlikely that humans should be uniquely resistant to these carcinogens (91). Exposure of the average person to performed N-nitrosamines may be of minor significance with respect to those formed endogenously. Smokers probably have higher exposures than nonsmokers, but it is not known to what extent this may contribute to the elevated incidence of lung cancer among smokers. Endogenous formation of N-nitrosamines may become epidemiologically important, ie, may lead to cancer incidences above normal background levels, for populations with above-average nitrite, nitrate, or amine intakes.

BIBLIOGRAPHY

1. A. Geuther, *Lieb. Ann.* **128,** 151 (1863).
2. A. Geuther and E. Schiele, *J. Prakt. Chem.* **4,** 485 (1871).
3. H. A. Fruend, *Ann. Intern. Med.* **10,** 1144 (1937).
4. J. M. Barnes and P. N. Magee, *Brit. J. Ind. Med.* **11,** 167 (1954).
5. P. N. Magee and J. M. Barnes, *Adv. Cancer Res.* **10,** 163 (1967).
6. H. Druckrey, R. Preussmann, S. Ivankovic, and D. Schmahl, *Z. Krebsforsch.* **69,** 103 (1967).
7. P. N. Magee, R. Montesano, and R. Preussmann in C. Searles, ed., *Chemical Carcinogens*, ACS *Monograph 173*, American Chemical Society, Washington, D.C., 1976, pp. 491–625.
8. *Chem. Eng. News*, 20 (July 31, 1978).
9. J. Sakshaug, E. Sognen, M. A. Hansen, and N. Koppang, *Nature* **206,** 1261 (1965).
10. IARC Working Group on the Evaluation of Carcinogenic Risk of Chemicals To Humans, *Some N-Nitroso Compounds*, IARC Monograph No. 17, International Agency for Research on Cancer, Lyon, France, 1978.
11. J. March, *Advanced Organic Chemistry: Reactions, Mechanisms, and Structure*, McGraw-Hill, New York, 1968, pp. 783–790.
12. J. E. Baldwin, S. E. Branz, R. F. Gomez, P. L. Kraft, A. J. Sinskey, S. R. Tannenbaum, *Tetrahedron Lett.*, 333 (1976).
13. B. Gold and W. B. Linder, *J. Am. Chem. Soc.* **101,** 6772 (1979).
14. E. A. Walker, M. Castegnaro, L. Griciute, and R. E. Lyle, eds., *Environmental Aspects of N-Nitroso Compounds*, IARC Scientific Publication No. 19, International Agency for Research on Cancer, Lyon, France, 1978.
15. J.-P. Anselme, ed., *N-Nitrosamines*, ACS Symposium Series *101*, American Chemical Society, Washington, D.C., 1979.
16. H. H. Hiatt, J. D. Watson, and J. A. Winstein, eds., *Origins of Human Cancer*, Cold Spring Harbor Conference on Cell Proliferation, Cold Spring Harbor Laboratory, N.Y., 1977.
17. D. Schmähl, *Oncology* **37,** (1980).
18. E. A. Walker, P. Bogovski, and L. Griciute, eds., *Environmental N-Nitroso Compounds Analysis and Formation*, IARC Scientific Publication No. 14, International Agency for Research on Cancer, Lyon, France, 1976.

19. R. A. Scanlan, *CRC Crit. Rev. Food Technol.* **5**, 357 (1975).

20. A. L. Fridman, F. M. Mukhametshin, and S. S. Novikov, *Russ. Chem. Rev.* **40**, 34 (1971).

21. M. L. Douglass, B. L. Kabacoff, G. A. Anderson, and M. C. Cheng, *J. Soc. Cosmet. Chem.* **29**, 581 (1978).

22. W. T. Iwaoka, T. Hansen, S.-T. Hsieh, and M. C. Archer, *J. Chromatogr.* **103**, 349 (1975).

23. Y. L. Chow and C. J. Colon, *Can. J. Chem.* **46**, 2827 (1968).

24. R. Fraser and L. K. Ng, *J. Amer. Chem. Soc.* **98**, 5895 (1976).

25. W. T. Rainey, W. H. Christie, and W. Lijinsky, *Biomed. Mass Spectrom.* **5**, 395 (1978).

26. J. W. Pensabene, W. Fiddler, C. J. Dooley, R. C. Doerr, and A. E. Wasserman, *J. Agric. Food Chem.* **204**, 274 (1972).

27. R. K. Smalley and B. J. Wakefield in F. Scheinmann, ed., *An Introduction to Spectroscopic Methods for the Identification of Organic Compounds*, Vol. 1, Pergamon Press, Oxford, 1970.

28. S. S. Mirvish, *Toxicol. Appl. Pharm.* **31**, 325 (1975).

29. T. Y. Fan and S. R. Tannenbaum, *J. Agric. Food Chem.* **21**, 237 (1973).

30. L. K. Keefer and P. Roller, *Science* **181**, 1245 (1973).

31. J. D. Okun and M. C. Archer, *Oncology* **37**, 147 (1980).

32. B. C. Challis, A. Edwards, R. R. Hunma, S. A. Kyrtopoulos, and J. R. Outram in Ref. 14, p. 127.

33. B. C. Challis and S. A. Kyrtopoulos, *J. Chem. Soc. (Perkin I)*, 299 (1979).

34. S. S. Singer, *J. Org. Chem.* **43**, 4612 (1978).

35. H. Maltz, M. A. Grant, and M. C. Navaroli, *J. Org. Chem.* **36**, 363 (1971).

36. S. S. Mirvish, L. Walcave, M. Eagen, and P. Shibik, *Science* **177**, 65 (1972).

37. W. J. Mergens, J. J. Kamm, H. L. Newmark, W. Fiddler, and J. Pensabene in Ref. 14, p. 199.

38. G. Eisenbrand and R. Preussmann, *Arzneim. Forsch.* **20**, 1513 (1970).

39. C. J. Michejda, N. E. Davidson, and L. K. Keefer, *Chem. Comm.*, 633 (1976).

40. Y. L. Chow, *Acc. Chem. Res.* **6**, 354 (1973).

41. W. Fiddler, R. C. Doerr, and E. G. Piotrowski in Ref. 14, p. 33.

42. J. S. Wishnok, M. C. Archer, A. S. Edelman, and W. M. Rand, *Chem. Biol. Interact.* **20**, 43 (1978).

43. L. K. Keefer and C. H. Fodor, *J. Am. Chem. Soc.* **92**, 5747 (1970).

44. D. Seebach and D. Enders, *Angew. Chem. Int. Ed.* **14**, 15 (1975).

45. B. Renger, H.-O. Kalinowski, and D. Seebach, *Chem. Ber.* **110**, 1866 (1977).

46. D. Seebach, D. Enders, and B. Renger, *Chem. Ber.* **110**, 1852 (1977).

47. A. E. Wasserman in P. Bogovski, R. Preussmann, and E. A. Walker, eds., *N-Nitroso Compounds, Analysis and Formation*, IARC Scientific Publication No. 3, International Agency for Research on Cancer, Lyon, France, 1972.

48. D. H. Fine, H. Rufeh, and B. Gunther, *Anal. Lett.* **6**, 731 (1973).

49. I. S. Krull and M. H. Wolf, *Am. Labor.*, 85 (May 1980).

50. T. Y. Fan, I. S. Krull, R. D. Ross, M. H. Wolf, and D. H. Fine in Ref. 14, p. 3.

51. T. J. Hansen, M. C. Archer, and S. R. Tannenbaum, *Anal. Chem.* **51**, 1526 (1979).

52. T. Gough, *Analyst* **103**, 785 (1978).

53. S. S. Mirvish, *J. Toxicology Env. Health* **2**, 1267 (1977).

54. P. N. Magee and J. M. Barnes, *J. Pathol. Bacteriol.* **84**, 19 (1962).

55. J. S. Wishnok in Ref. 15, p. 153.

56. A. S. Edelman, P. L. Kraft, W. M. Rand, and J. S. Wishnok, *Chem. Biol. Interact.*, in press.

57. L. L. Y. Fong and P. M. Newberne in Ref. 14, p. 503.

58. W. Lijinsky in Ref. 15, p. 165.

59. P. D. Lawley in W. Nakahara, S. Takayama, T. Sugimura, and S. Odashima, eds., *Topics in Chemical Carcinogenesis*, University of Tokyo Press, Tokyo, Jpn., 1972, p. 272.

60. K. K. Park, J. S. Wishnok, and M. C. Archer, *Chem. Biol. Int.* **18**, 349 (1977).

61. K. K. Park, M. C. Archer, and J. S. Wishnok, *Chem. Biol. Int.* **29**, 139 (1980).

62. D. F. Heath, *Biochem. J.* **85**, 72 (1962).

63. H. Bartsch, C. Malaveille, and R. Montesano, *Cancer Res.* **35**, 644 (1975).

64. O. G. Fahmy, M. J. Fahmy, and M. Weissler, *Biochem. Pharmacol.* **24**, 2009 (1975).

65. A. E. Pegg, *Adv. Cancer Res.* **25**, 195 (1977).

66. C. J. Collins, *Acc. Chem. Res.* **4**, 315 (1971).

67. J. McCann, E. Choi, E. Yamasaki, and B. N. Ames, *Proc. Natl. Acad. Sci. USA* **72**, 5135 (1975).

68. L. B. Kier and L. M. Hall, *J. Pharm. Sci.* **65**, 1806 (1976).

69. T. K. Rao, J. A. Young, W. Lijinsky, and J. L. Eppler, *Mut. Res.* **58**, 66 (1979).

70. W. J. Dunn III and S. Wald, *J. Chem. Inf. Comput. Sci.* **21**, 8 (1981).

71. D. Shapley, *Science* **191,** 268 (1976).
72. D. H. Fine, D. P. Rounbehler, N. M. Belcher, and S. S. Epstein, *Science* **192,** 1328 (1976).
73. S. Z. Cohen, G. Zweig, M. Law, D. Wright, and W. R. Bontoyan in Ref. 10, p. 333.
74. K. D. Brunneman and D. Hoffmann in Ref. 10, p. 343.
75. K. D. Brunneman, L. Yu, and D. Hoffmann, *Cancer Res.* **37,** 3218 (1977).
76. T. Y. Fan, J. Morrison, D. P. Rounbehler, R. Ross, D. H. Fine, W. Miles, and N. P. Sen, *Science* **196,** 70 (1977).
77. T. Y. Fan, U. Goff, L. Song, D. H. Fine, G. P. Arsenault, and K. Biemann, *Food Cosmet. Toxicol.* **15,** 423 (1977).
78. M. C. Archer and J. S. Wishnok, *J. Environ. Sci. Health* **AII,** 587 (1976).
79. M. C. Archer and J. S. Wishnok, *Food Cosmet. Toxicol.* **15,** 233 (1977).
80. N. P. Sen, B. Donaldson, S. Seaman, J. R. Iyengar, and W. F. Miles, *J. Agric. Food Chem.* **24,** 397 (1976).
81. T. Fazio, D. C. Havery, and J. W. Howard in E. A. Walker, L. Griciute, M. Castegnaro, and M. Börzsönyi, eds., *N-Nitroso Compounds: Analysis, Formation, and Occurrence,* IARC Scientific Publication No. 31, International Agency for Research on Cancer, Lyon, France, 1980, p. 419.
82. J. S.Wishnok, *J. Chem. Educ.* **54,** 440 (1977).
83. J. Sander and G. Burkle, *Z. Krebsforsch.* **73,** 54 (1969).
84. M. Greenblatt, S. S. Mirvish, and B. T. So, *J. Natl. Cancer Inst.* **46,** 1029 (1971).
85. T. S. Mysliwy, E. L. Wick, M. C. Archer, R. C. Shank, and P. M. Newberne, *Brit. J. Cancer* **30,** 279 (1974).
86. S. R. Tannenbaum, A. J. Sinskey, M. Weisman, and W. W. Bishop, *J. Natl. Cancer Inst.* **53,** 79 (1974).
87. S. R. Tannenbaum, M. Weisman, and D. Fett, *Food Cosmet. Toxicol.* **14,** 549 (1976).
88. S. R. Tannenbaum, M. C. Archer, J. S. Wishnok, and W. W. Bishop, *J. Natl. Cancer Inst.* **60,** 251 (1978).
89. S. R. Tannenbaum, D. Fett, V. R. Young, P. C. Land, and W. R. Bruce, *Science* **200,** 1487 (1978).
90. A. J. Varghese, P. C. Land, R. Furrer, and W. R. Bruce in Ref. 10, p. 257.
91. *Chem. Eng. News* **59,** 30 (Apr. 13, 1981).

General References

Refs. 7, 9–21, and 32–33 are also general references.
L. R. Ember, "Nitrosamines: Assessing the Relative Risk," *Chem. Eng. News,* 20 (Mar. 31, 1980).

JOHN S. WISHNOK
Massachusetts Institute of Technology

NOBELIUM. See Actinides and transactinides.